Birkhäuser

Progress in Mathematics

Volume 330

More information about this series at http://www.springer.com/series/4848

Maria Gorelik • Vladimir Hinich • Anna Melnikov
Editors

Representations and Nilpotent Orbits of Lie Algebraic Systems

In Honour of the 75th Birthday of Tony Joseph

Birkhäuser

Editors
Maria Gorelik
Department of Mathematics
Weizmann Institute of Science
Rehovot, Israel

Vladimir Hinich
Department of Mathematics
University of Haifa
Haifa, Israel

Anna Melnikov
Department of Mathematics
University of Haifa
Haifa, Israel

ISSN 0743-1643 ISSN 2296-505X (electronic)
Progress in Mathematics
ISBN 978-3-030-23530-7 ISBN 978-3-030-23531-4 (eBook)
https://doi.org/10.1007/978-3-030-23531-4

Mathematics Subject Classification: 22-XX, 22Exx, 22E25, 22E27

This book is published under the imprint Birkhäuser, www.birkhauser-science.com by the registered company Springer Nature Switzerland AG.
The registered company address is: Gewerbestrasse 11, 6330 Cham, Switzerland

Preface

Lie theory, inaugurated through the fundamental work of Sophus Lie during the late nineteenth century, has proved central in many areas of mathematics and theoretical physics. Sophus Lie's formulation was originally in the language of analysis and geometry; however, by now, a vast algebraic counterpart of the theory has been developed. As in algebraic geometry, the deepest and most far-reaching results in Lie theory nearly always come about when geometric and algebraic techniques are combined.

A core part of Lie theory is the structure and representation theory of complex semisimple Lie algebras and Lie groups, which is an exemplary harmonious field in modern mathematics. It has deep ties to physics, and many areas of mathematics, such as combinatorics, category theory, and others. This field has inspired many generalizations, among them the representation theories of affine Lie algebras, vertex operator algebras, locally finite Lie algebras, Lie superalgebras, etc. This volume originates from a pair of sister conferences titled "Algebraic Modes of Representations" held in Israel in July 2017. The first conference took place at the Weizmann Institute of Science, Rehovot, July 16–18, and the second conference took place at the University of Haifa, July 19–23. Both conferences were dedicated to the 75th birthday of Anthony Joseph, who has been one of the leading figures in Lie Theory from the 1970s until today. The conferences were supported by the United States–Israel Binational Science Foundation and the Chorafas Institute for Scientific Exchange (Weizmann part) and by the Israel Science Foundation (Haifa part).

Joseph has had a fundamental influence on both classical representation theory and quantized representation theory. A detailed description of his work in both areas has been given in the articles by W. McGovern and D. Farkash–G. Letzter in the volume "Studies in Lie theory," Progress in Mathematics, vol. 243, Birkhauser. Concerning Joseph's contribution to classical representation theory, it is impossible not to mention his classification of primitive ideals of the universal enveloping algebra of $\mathfrak{sl}(n)$. The essential new ingredient here is the introduction of a partition of the Weyl group into left cells, corresponding to the Robinson map from the symmetric group to the standard Young tableaux. Joseph further extended this result to other simple Lie algebras using similar techniques, and this has since then become a powerful tool in Lie theory.

As for quantized representation theory, Joseph's monograph "Quantum Groups and Their Primitive Ideals," Ergebnisse der Mathematik und Ihrer Grenzgebiete, 3rd series, vol. 29, has had a fundamental influence over the field since its appearance in 1995.

The present volume contains 14 original papers covering a broad spectrum of current aspects of Lie theory. The areas discussed include primitive ideals, invariant theory, geometry of Lie group actions, crystals, quantum affine algebras, Yangians, categorification, and vertex algebras.

The authors of this volume are happy to dedicate their works to Anthony Joseph.

Rehovot, Israel — Maria Gorelik
Haifa, Israel — Vladimir Hinich
Haifa, Israel — Anna Melnikov

Contents

List of Publications

Anthony Joseph

1. (With C.A. Coulson), Spheroidal wave functions for the hydrogen atom, Proc. Phys. Soc., **90**, 887–893 (1967).
2. (With C.A. Coulson), A constant of the motion for the two-centre Kepler problem, Int. J. Quant. Chem., **1**, 337–347 (1967).
3. The influence of the mass on the bound states of molecular systems, Proc. Phys. Soc., **91**, 574–576 (1967).
4. The theory of local degeneracy, Int. J. Quant. Chem., **1**, 535–559 (1967).
5. On the determination of the exact number of bound states of a given potential, Int. J. Quant. Chem., **1**, 615–629 (1967).
6. The spin dipole, Proc. Phys. Soc., **92**, 847–852 (1967).
7. Self-adjoint ladder operators I, Rev. Mod. Phys., **99**, 829–837 (1967).
8. (With C.A. Coulson), Self-adjoint ladder operators II, Rev. Mod. Phys. **39**, 838–849 (1967).
9. Self-adjoint ladder operators III, Rev. Mod. Phys., **40**, 845–871 (1968).
10. The theory of conditional invariance. Proc. Roy. Soc. **A305**, 405–427 (1968).
11. The hemispherical box: An example of virtual symmetry, J. of Phys. **A2**, 719 (1969).
12. (With A.I. Solomon), Global and infinitesimal non-linear chiral transformations, J. Math. Phys., **11**, 748 (1970).
13. Derivations of Lie brackets and canonical quantisation, Commun. Math. Phys., **17**, 210–232 (1970).
14. A second anticommutant theorem for symmetric ternary algebras, Proc. Camb. Phil. Soc. **69**, 25–52 (1971).
15. (With M.A. Jacobs, S. Nussinov and A.A. Rangwala), An improved upper bound on the imaginary part of elastic scattering. Phys. Rev. **D2**, 1970–1974 (1970).
16. A classification of local current algebras, Commun. Math. Phys. **19**, 106–118 (1970).

17. (With W.G. Sullivan), A commutant property of symmetric ternary algebras, Proc. Camb. Phil. Soc., **70**, 1–4 (1971).
18. Commuting polynomials in quantum canonical operators and realizations of Lie algebras, J. Math. Phys. **13**, 351–357, (1972).
19. Magnetic spin monopole from the Yang-Mills field, Phys. Rev. **D5**, 313–320, (1972).
20. Realizations of Lie algebras from the canonical anticommutation relations, Nuovo Cimento, **8A**, 217–234 (1972).
21. On the dynamical solution to the Sugawara model, Nucl. Phys., **B42**, 418–436 (1972).
22. On the dynamical solution to the Sugawara model: II The Lie identities, Ann. of Phys., **73**, 417–438 (1972).
23. Recovery of the Lie identities of field theory from minimal conditions, Nucl. Phys., **B43**, 107–118 (1972).
24. An inequality relating cross sections to polarization from isospin conservation, Phys. Letters, **42B**, 368 (1972).
25. Gelfand-Kirillov dimension for algebras associated with the Weyl algebra, Ann. de l'Institut Henri Poincaré, **17**, 325–336 (1972).
26. Combinatorial analysis of the Lie identities in field theory, in *Statistical Mechanics and Field Theory*. Lectures given at the 1971 Haifa Summer School (Eds. R.N. Sen and C. Weil, Keter Publishing House, Jerusalem).
27. The Weyl algebra—semisimple and nilpotent elements, Am. J. Math. **97**, 597–615 (1975).
28. A characterization theorem for realizations of $s\ell(2)$, Proc. Camb. Phil. Soc., *75*, 119–131 (1974).
29. Proof of the Gelfand-Kirilov conjecture for solvable Lie algebras, Proc. Amer. Math. Soc. *45*. 1–10 (1974).
30. Symplectic structure in the enveloping algebra of a Lie algebra, Bull. Soc. Math. France, **102**, 75–83 (1974).
31. Minimal realizations and spectrum generating algebras, Commun. Math. Phys., **36**, 325–338 (1974).
32. Realizations in classical and quantum mechanics, Proc. of the Third Intern. Colloq. on Group Theoretical Methods in Physics (Eds. H. Bacry and A. Grossman), 1974.
33. Infinite dimensional Lie algebras in mathematics and physics, Proc. of the Third Intern. Colloq. on Group Theoretical Methods in Physics (Eds. H. Bacry and A. Grossman), 1974.
34. A generalization of the Gelfand-Kirillov conjecture, Amer. J. Math., **99**, 1151–1165 (1977).
35. The algebraic method in representation theory, in *Group Theoretical Methods in Physics* (Ed. A. Janner), LN 50 (Physics), 95–109, Springer-Verlag, New York, 1976.
36. Second commutant theorems in enveloping algebras, Amer. J. Math. **99**, 1167–1192 (1977).

37. The minimal orbit in a simple Lie algebra and its associated maximal ideal, Ann. Sci. Ec. Norm. Sup., **9**, 1–30 (1976).
38. A preparation theorem for the prime spectrum of a semisimple Lie algebra, J. of Algebra, **48**, 241–289 (1977).
39. A characteristic variety for the prime spectrum of a semisimple Lie algebra, in Non-commutative Harmonic Analysis (Ed. J. Carmona), LN 587, 102–118, Springer-Verlag, New York, 1977.
40. Sur les vecteurs de plus haut poids dans l'algèbre enveloppante d'une algèbre de Lie semisimple complexe, Comptes Rendus A **281**, 835–837 (1975).
41. Sur la classification des idéaux primitifs dans l'algèbre enveloppante d'une algèbre de Lie réductive, Comptes Rendus A **284**, 425–427 (1977).
42. On the annihilators of the simple subquotients of the principal series, Ann. Ec. Norm. Sup., **10**, 419–440 (1977).
43. A wild automorphism of $Us\ell(2)$, Math. Proc. Camb. Phil. Soc. **80**, 61–64 (1976).
44. A generalization of Quillen's lemma and its application to the Weyl algebras, Israel J. Math., **28**, 177–192 (1977).
45. Gelfand-Kirillov dimension for the annihilators of simple quotients of Verma modules, J. Lond. Math. Soc., **18**, 50–60 (1978).
46. (With L.W. Small), An additivity principle for Goldie rank, Israel J. Math. **31**, 105–114 (1978).
47. Sur la classification des ideaux primitifs dans l'algèbre enveloppante de $s\ell(n+1, \mathbb{C})$, Comptes Rendus Ser. A-287, 303–306 (1978).
48. W module structure in the primitive spectrum of the enveloping algebra of a semisimple Lie algebra, in Lecture Notes in Mathematics, 728, Springer-Verlag, 116–135 (1979).
49. Dixmier's problem for Verma and principal series submodules, J. Lond. Math. Soc. **20**, 193–204 (1979).
50. Towards the Jantzen conjecture, Compositio Math. **40**, 35–67 (1980).
51. Towards the Jantzen conjecture II, Compositio Math. **40**, 69–78 (1980).
52. Towards the Jantzen conjecture III, Compositio Math. **41**, 23–30 (1981).
53. Goldie rank in the enveloping algebra of a semisimple Lie algebra I, J. Algebra **65**, 269–283 (1980).
54. Goldie rank in the enveloping algebra of a semisimple Lie algebra II, J. Algebra **65**, 284–306 (1980).
55. Kostant's problem, Goldie rank and the Gelfand-Kirillov conjecture, Invent. Math., **56**, 191–213 (1980).
56. On the Gelfand-Kirillov conjecture for induced ideals in the semisimple case, Bull. Math. Soc. France **107**, 139–159 (1979).
57. (With O. Gabber), On the Bernstein-Gelfand-Gelfand resolution and the Duflo sum formula, Compos. Math. **43**, 107–131 (1981).
58. (With O. Gabber), towards the Kazhdan-Lusztig conjecture, Ann. Ec. Norm. Sup., **14**, 261–302 (1981).
59. Goldie rank in the enveloping algebra of a semisimple Lie algebra III, J. Algebra, **73**, 295–326 (1981).

60. The Enright functor in the Bernstein-Gelfand-Gelfand O category, Invent. Math., **67** 423–445 (1982).
61. Completion functors in the O category, in Lecture Notes, **1020**, Springer-Verlag, 1983, 80–106.
62. On the classification of primitive ideals in the enveloping algebra of a semisimple Lie algebra, in Lecture Notes, **1024**, Springer-Verlag, 30–76, 1983.
63. Primitive ideals in enveloping algebras, ICM Proceedings, PWM—North Holland, Warsaw 1984, 403–414.
64. On the variety of a highest weight module, J. Algebra, **88** (1984), 238–278.
65. On the associated variety of a primitive ideal, J. Algebra, **93** (1985), 509–523.
66. (With J.T. Stafford), Modules of k-finite vectors over semisimple Lie algebras, Proc. Lond. Math. Soc. **49** (1984), 361–384.
67. On the Demazure character formula, Ann. Ec. Norm. Sup. **18** (1985), 389–419.
68. Three topics in enveloping algebras, Proceedings of Durham Symposium, 1983, (unpublished).
69. On the Demazure character formula II, Compos. Math., **58** (1986), 259–278.
70. Kostant's problem and Goldie rank, in: Lecture Notes in Mathematics **880**, Springer-Verlag, Berlin, 1981, 249–266.
71. On the cyclicity of vectors associated with Duflo involutions, Lecture Notes in Mathematics, **1243**, 144–188, Springer-Verlag, Berlin, 1987.
72. A criterion for an ideal to be induced, J. Algebra, **110** (1987), 480–497.
73. On the multiplicities of the adjoint representation in simple quotients of an enveloping algebra of a semisimple Lie algebra, Trans. Amer. Math. Soc., **316**, (1989), 447–491.
74. Rings which are modules in the Bernstein-Gelfand-Gelfand O category, J. Algebra **113** (1988), 110–126.
75. A sum rule for scale factors in the Goldie rank polynomials, J. Algebra **118** (1988), 276–311.
76. A sum rule for scale factors in the Goldie rank polynomials, addendum, J. Algebra **118** (1988), 312–321.
77. A surjectivity theorem for rigid highest weight modules, Invent. Math. **92** (1988), 567–596.
78. The primitive spectrum of an enveloping algebra, Asterisque, **173–174** (1989), 13–53.
79. Characters for unipotent representations, J. Algebra, **130** (1990), 273–295.
80. On the characteristic polynomials for orbital varieties, Ann. Ec. Norm. Sup., **22** (1989), 569–603.
81. Rings of $\mathfrak{b}$-finite endomorphisms of simple highest weight modules are Goldie, IMCP, Ring theory, 1989, The Weizmann Science Press, Jerusalem 1990, 124–134.
82. The surjectivity theorem, characteristic polynomials and induced ideals, in The Orbit Method in Representation Theory, Eds. M. Duflo, N.V. Pedersen, M. Vergne, Birkhauser, Boston, 1990, 85–98.

83. (With T.J. Enright). An intrinsic analysis of unitarizable highest weight modules, Math. Ann., **288** (1990), 571–594.
84. Annihilators and associated varieties of unitary highest weight modules, Ann Ec. Norm. Sup., **25** (1992), 1–45.
85. Some ring theoretic techniques and open problems in enveloping algebras, Proceedings of Berkeley Conference, July 1989, Ed. S. Montgomery and L. Small, Springer-Verlag, Berlin 1992, p. 27–67.
86. (With G. Letzter). Local finiteness of the adjoint action for quantized enveloping algebras. Journal of Algebra, **153** (1992), 289–318.
87. (With G. Letzter). Separation of variables for quantized enveloping algebras. Amer. J. of Math. **116** (1994), 125–177.
88. (With G. Perets and P. Polo). Sur l'equivalence de catégories de Beilinson et Bernstein, Comptes Rendus, Paris **313** (1991), 705–709.
89. (With G. Letzter). Verma module annihilators for quantized enveloping algebras, Ann. Ec. Norm. Sup., **28** (1995), 493–526.
90. Faithfully flat embeddings for minimal primitive quotients of quantized enveloping algebras, Israel Math. Conf. Proc. **7** (1993), 79–106.
91. Idéaux premiers et primitifs de l'algèbre des fonctions sur un groupe quantique, Comptes Rendus **316** (1993), 1139–1142.
92. On the prime and primitive spectra of the algebra of functions on a quantum group, J. Algebra, **169** (1994), 441–511.
93. (With G. Letzter). Rosso's form and quantized Kac-Moody algebras, Math. Zeit. **222** (1996), 543–571.
94. Enveloping Algebras: Problems Old and New, 385–413 in Progress in Mathematics **123**, Birkhäuser, Boston 1994.
95. Some Remarks on the R-matrix, J. Algebra **180** (1996), 412–430.
96. Orbital varieties, Goldie rank polynomials and unitary highest weight modules, In B. Orsted and H. Schlichtkrull (eds.), Algebraic and Analytic Methods in Representation Theory, Academic Press, London 1997, 53–98.
97. Sur une conjecture de Feigin, Comptes Rendus, Paris **320** (1995), 1441–1444.
98. Sur les idéaux génériques sur l'algèbre des fonctions sur un groupe quantique, Comptes Rendus, Paris, **321** (1995), 135–140.
99. Preservation of Coxeter structure under finite group action, Acts du Colloque, Reims 1995, Colloques et Seminaires du SMF, 1997, 185–219.
100. (With A. Braverman). The minimal realization from deformation theory, J. Algebra **205** (1998), 13–36.
101. The orbital varieties of the minimal orbit, Ann. Ec. Norm. Sup., Serie 4, **31** (1998), 17–45.
102. On a Harish-Chandra homorphism, CRAS **324**(I), 1997, 759–764.
103. Sur l'annulateur d'un module de Verma, In B. Broer (ed.) Representation Theories and Algebraic geometry, Kluwer Academic Publishers, NATO Scientific Affairs, 1998, 237–300.
104. On the mock Peter-Weyl theorem and the Drinfeld double of a double, J. reine angew. Math. **507** (1999), 37–56.

105. The admissibility of bounded modules for an affine Lie algebra, Alg. Rept. Theory, **3**, (2000), 131–149.
106. A completion of the quantized enveloping algebra of a Kac-Moody algebra, J. Algebra, **214** (1999), 235–275.
107. On the Kostant-Parthasarathy-Ranga Rao-Varadarajan determinants, I. Injectivity and multiplicitie, J. Algebra **241** (2001), 27–45.
108. (with G. Letzter). On the KPRV determinants, II. Construction of the KPRV determinants, J. Algebra **241** (2001), 46–66.
109. (with G . Letzter and D. Todoric). On the KPRV determinants, III. Computation of the KPRV determinants, J. Algebra **241** (2001), 67–88.
110. (with G . Letzter and S. Zelikson). On the Brylinski-Kostant filtration, JAMS **14** (2000), 945–970.
111. (with F. Fauquant-Millet). Semi-invariants de l'algèbre enveloppante quantifiée d'une algèbre parabolique, Transf. Groups. **6** (2001), 125–142.
112. On an affine KPRV determinant at $q = 1$, Bull. Math. Sci., **125** (2001), 23–48.
113. (with J. Greenstein). A Chevalley-Kostant presentation of basic modules for $\mathfrak{sl}(2)^\wedge$ and the associated affine KPRV determinants at $q = 1$, Bull. Math. Soc., **125** (2001), 85–108.
114. (with W. Borho). Sheets and topology of primitive spectra for semisimple Lie algebras, J. Algebra **244** (2001), 76–167.
115. (with D. Todoric). On the quantum KPRV determinants for semisimple and affine Lie algebras, Alg. Rept. Theory **5** (2002), 57–99.
116. (with G. Letzter). Evaluation of the quantum affine PRV determinant, Int. Math. Research Notices (2002), no. 2–3, 307–322.
117. (with A. Melnikov). Quantization of Orbital Varieties in $\mathfrak{sl}_n$, Progr. in Math. Birkhauser, **213** (2003), 165–196.
118. A Decomposition Theorem for Demazure Crystals, J. Algebra, **265** (2003), 562–578.
119. On the graded injectivity of the Conze embedding, J. Algebra **265** (2003), 358–378.
120. (with W. Borho). Corrigendum to "Sheets and topology of primitive spectra for semisimple Lie algebras", J. Algebra **259** (2003), 310–311.
121. The Enigma of the missing invariants on the nilradical of a Borel, Bull. Sci. Math. 128 (2004), 433–446.
122. (with V. Hinich). Orbital variety closures and the convolution product in Borel-Moore homology, Selecta Mathematica II (2005), 9–36
123. (with F. Fauquant-Millet), Semi-centre de l'algèbre enveloppante d'une sous-algèbre parabolique d'une algèbre de Lie semi-simple, Ann.Ec. Norm. Sup. **38** (2005), 155–191.
124. Modules with a Demazure flag, In Studies in Lie Theory (J. Bernstein et al: eds) Progress in Math. 243, Birkhauser, Boston 2006, pp. 131–169.
125. Results and Problems in Enveloping Algebras arising from Quantum groups, in Studies on Advanced Mathematics (ed. V. Kaimanovich) AMS Transl. (2) **217** (2006), 87–100.

126. (with I.Heckenberger) On the left and right Brylinski-Kostant filtrations, Algebr. Represent. Theor. **12** (2009) 417–442.
127. (with Fauquant -Millet) La somme des faux degrés—un mystère en théorie des invariants -(French) [The sum of the false degrees—a mystery in the theory of invariants] Adv. Math. **217** (4) (2008), 1476–1520.
128. Goldie rank ratios and quaternionic extensions, J. Algebra **301** (2006), 781–790.
129. On semi-invariants and index for biparabolic (seaweed) algebras, I, J. Algebra **305** (2006), 487–515.
130. On semi-invariants and index for biparabolic (seaweed) algebras, II, J. Algebra **312** (2007), 158–193.
131. Parabolic Actions in type A and their eigenslices, Transf. Groups **12**(3), (2007), 515–547.
132. (with P. Lamprou) Maximal Poisson commutative subalgebras for truncated parabolic subalgebras of maximal index in $\mathfrak{sl}_n$, Transf. Groups **12** (2007), 549–571.
133. A slice theorem for truncated parabolics of index one and the Bezout equation, Bull . Sci. Math. **131** (2007), 276–290.
134. Slices for Biparabolic Coadjoint Actions in Type A, J. Algebra **319** (2008), 5060–5100.
135. Compatible adapted pairs and a common slice theorem for centralizers, Transf. Groups **13** (2008), 637–669.
136. A Pentagonal crystal, the Golden section, alcove packing and a periodic tilings, Transf. Groups **14** (2009), 557–612.
137. (with P. Lamprou) A Littelmann path model for crystals of generalized Kac-Moody algebras, Adv. Math. 221 (2009), no. 6, 2019–2058. (An updated corrected version appears in arXiv:0804.1485.)
138. (with D. Shafrir) Polynomiality of Invariants, Unimodularity and Adapted Pairs, Transf. Groups, 15, no. 4 (2010), 851–882.
139. An algebraic slice in the co-adjoint space of the Borel and the Coxeter element, Adv. Math. 227 (2011), 522–585.
140. (Edited by C. Hoyt) Consequences of the Littelmann path theory for the structure of the Kashiwara $B(\infty)$ crystal. Highlights in Lie algebraic methods, 25Ű64, Progr. Math., 295, Birkhäuser/Springer, New York, 2012.
141. (with F. Fauquant-Millet) Slices for Biparabolics of Index 1, Transform. Groups 16 (2011), no. 4, 1081–1113.
142. A direct proof of a generalized Harish-Chandra isomorphism, Transform. Groups, Transform. Groups 17 (2012), no. 2, 513Ű521.
143. Zhelobenko Invariants, Bernstein-Gelfand-Gelfand Operators and the Analogue Kostant Clifford algebra Conjecture, Transform. Groups, Transform. Groups 17 (2012), no. 3, 781–821.
144. Analogue Zhelobenko Invariants, Bernstein-Gelfand-Gelfand Operators and the Kostant Clifford algebra Conjecture, Transform. Groups Transform. Groups 17 (2012), no. 3, 823–833.

145. Some Remarks on Weierstrass Sections, Adapted Pairs and Polynomiality, In V. Dobrev, Ed. Springer Proceedings in Mathematics and Statistics, Vol. 36, pp. 39–60, 2013.
146. Modules for Relative Yangians and Kazhdan-Lusztig polynomials, Transformation Groups, Vol. 19, 105Ű129, 2014.
147. The Hidden Semi-invariant Generators of a Quasi-Frobenius Biparabolic, Transformation Groups, 19, no. 3 (2014), pp. 735–778.
148. The integrality of an adapted pair, Transformation Groups, 20 (2015), no. 3, 771–816.
149. (with F. Fauquant-Millet) Adapted pairs and Weierstrass sections, arXiv:1503.02523, Unpublished.
150. (with F. Fauquant-Millet) Adapted Pairs in type A and regular nilpotent elements, J. Algebra, 453 (2016), 1–67.
151. (with I. Heckenberger) Erratum to the Article, On the left and right Brylinski-Kostant filtrations, Algebr. Represent. Theor. 17 (2014), 1925–1926.
152. A Preparation theorem for the $B(\infty)$ crystal, Selecta Math 23, no. 2, (2017), 1309–1353.
153. Relative Yangians of Weyl Type, Selecta Math 22, no. 4, (2016), 2059–2098.
154. (with P. Lamprou) A new interpretation of the Catalan numbers arising in the theory of crystals. J. Algebra 504 (2018), 85–128.
155. (with P. Lamprou) A Substitution Theorem for the Borcherds-Weyl Group, C. R. Math. Acad. Sci. Paris 354 (2016), no. 9, 887–890.
156. Convexity Properties of the Canonical S-graphs, Israel J. Math. 226 (2018), no. 2, 827–849.
157. Trails, S-graphs and identities in Demazure modules, arXiv: 1702.00243.
158. Dual Kashiwara functions for the Kashiwara $B(\infty)$ crystal, Transf. groups (to appear).
159. Self-similar triangle tiling and representing weight diagrams in the plane, this volume.
160. Trails for Minuscule Modules and Dual Kashiwara Function for the $B(\infty)$ Crystal, Proceedings of QTS10/LT12 (Varna, 2017), Ed. V.K. Dobrev (to appear).

Book

Quantum Groups and their Primitive Ideals, Springer-Verlag, Berlin 1995.

Edited Volumes

(1) (With A. Connes, M. Duflo and R. Rentschler). *Operator algebras, unitary represents, equivalents and invariant theory*, Actes du colloque en l'honneur de Jacques Dixmier, Progr. in Math. Vol. 92 Birkhäuser, Boston 1990.
(2) (With S. Shnider). *Quantum deformations of algebras and their represents*, Israel Mathematical Conference Proceedings Vol. 7, Bar-Ilan University, 1993.
(3) (With F. Mignot, F. Murat, B. Prum and R. Rentschler). First European Congress of Mathematics, Vol. I, Progr. in Math. Vol. 119, Birkhäuser, Boston 1994.
(4) (With F. Mignot, F. Murat, B. Prum and R. Rentschler). First European Congress of Mathematics, Vol. II, Progr. in Math. Vol. 120, Birkhäuser, Boston 1994.
(5) (With F. Mignot, F. Murat, B. Prum and R. Rentschler). First European Congress of Mathematics, Vol. III, Progr. in Math. Vol. 121, Birkhäuser, Boston 1994.
(6) (With A. Melnikov and R. Rentschler). Studies in Memory of Issai Schur, Progr. in Math. Progr. in Math. Vol. 210, Birkhäuser, Boston 2002.
(7) (With S. Kumar and M. Vergne). B. Kostant, Collected papers 1955–1966, Vol. I, Springer, London 2009.
(8) (With A. Melnikov and I. Penkov). Highlights in Lie Algebraic Methods, Progr. In Math. Vol. 295, Birkhäuser, Boston 2011.

Singular Support of a Vertex Algebra and the Arc Space of Its Associated Scheme

Tomoyuki Arakawa and Andrew R. Linshaw

Dedicated to Professor Anthony Joseph on his seventy-fifth birthday

Abstract Attached to a vertex algebra $\mathcal{V}$ are two geometric objects. The *associated scheme* of $\mathcal{V}$ is the spectrum of Zhu's Poisson algebra $R_{\mathcal{V}}$. The *singular support* of $\mathcal{V}$ is the spectrum of the associated graded algebra $\mathrm{gr}(\mathcal{V})$ with respect to Li's canonical decreasing filtration. There is a closed embedding from the singular support to the arc space of the associated scheme, which is an isomorphism in many interesting cases. In this note we give an example of a non-quasi-lisse vertex algebra whose associated scheme is reduced, for which the isomorphism is not true as schemes but true as varieties.

MSC: 17B69, 81R10, 14E18

1 Introduction

Attached to a vertex algebra $\mathcal{V}$ are two geometric objects. The *associated scheme* $\tilde{X}_{\mathcal{V}}$ of $\mathcal{V}$ is the spectrum of commutative algebra $R_{\mathcal{V}}$, which is an affine Poisson scheme of finite type.[1] The *singular support* $\mathrm{SS}(\mathcal{V})$ of $\mathcal{V}$ is the spectrum of the

[1]Provided that $\mathcal{V}$ is finitely strongly generated.

T. Arakawa (✉)
Research Institute for Mathematical Sciences, Kyoto University, Kyoto, Japan
e-mail: arakawa@kurims.kyoto-u.ac.jp

A. R. Linshaw
University of Denver, Denver, CO, USA
e-mail: andrew.linshaw@du.edu

M. Gorelik et al. (eds.), *Representations and Nilpotent Orbits of Lie Algebraic Systems*, Progress in Mathematics 330,
https://doi.org/10.1007/978-3-030-23531-4_1

associated graded algebra $\mathrm{gr}(\mathcal{V})$ with respect to Li's canonical decreasing filtration, which is a *vertex Poisson* scheme of infinite type.[2] There is a closed embedding

$$\Phi : \mathrm{SS}(\mathcal{V}) \hookrightarrow (\tilde{X}_{\mathcal{V}})_\infty$$

from the singular support to the arc space $\tilde{X}_\infty$ of the associated scheme, which is an isomorphism in many interesting cases.

Originally Zhu [29] introduced the algebra $R_{\mathcal{V}}$ to define a certain finiteness condition on a vertex algebra. Recall that a vertex algebra $\mathcal{V}$ is called *lisse* (or C_2-cofinite) if $\dim \tilde{X}_{\mathcal{V}} = 0$. Using the map Φ one can show that this condition is equivalent to that $\dim \mathrm{SS}(\mathcal{V}) = 0$, and hence, the lisse condition is a natural finiteness condition [3]. It is known that lisse vertex (operator) algebras have many nice properties, such as modular invariance property of characters [23, 29], and this condition has been assumed in many significant theories of vertex (operator) algebras. However, recently non-lisse vertex algebras have caught a lot of attention due to the *Higgs branch conjecture* by Beem and Rastelli [10], which states that the reduced scheme $X_{\mathcal{V}}$ of $\tilde{X}_{\mathcal{V}}$ should be isomorphic to the Higgs branch of a four-dimensional $N = 2$ superconformal field theory $\mathcal{T}$ if $\mathcal{V}$ obtained from $\mathcal{T}$ by the correspondence discovered by [9], see the survey articles [4, 5] and the references therein.

It is natural to ask whether the map Φ is always an isomorphism, and if not, whether Φ defines an isomorphism as varieties. Very recently counterexamples to the first question were found by van Ekeren and Heluani [16] in the case that $\mathcal{V}$ is lisse in their study of chiral homology of elliptic curves. It was also shown recently in [8] that the map Φ defines an isomorphism as varieties if $\mathcal{V}$ is *quasi-lisse*, that is, the Poisson variety $X_{\mathcal{V}}$ has finitely many symplectic leaves. In this note we give an example of a non-quasi-lisse vertex algebra whose associated scheme is reduced, for which Φ is not an isomorphism of schemes, but still defines an isomorphism of varieties. We remark that by tensoring one of the lisse examples in [16] with any non-quasi-lisse vertex algebra, one can trivially obtain a non-quasi-lisse example. However, all such examples have the property that the associated scheme is nonreduced.

2 Vertex Algebras

We assume that the reader is familiar with vertex algebras, which have been discussed from various points of view in the literature [11, 17–19]. Given an element a in a vertex algebra $\mathcal{V}$, the field associated to a via the state-field correspondence is denoted by

---

[2]Unless $\mathcal{V}$ is finite-dimensional.

$$a(z) = \sum_{n \in \mathbb{Z}} a(n) z^{-n-1} \in \mathrm{End}(\mathcal{V})[[z, z^{-1}]].$$

Throughout this paper, we shall identify $\mathcal{V}$ with the corresponding space of fields. Given $a, b \in \mathcal{V}$, the *operators product expansion* (OPE) formula is given by

$$a(z)b(w) \sim \sum_{n \geq 0} (a_{(n)}b)(w)\,(z-w)^{-n-1}.$$

Here $(a_{(n)}b)(w) = \mathrm{Res}_z[a(z), b(w)](z-w)^n$ where

$$[a(z), b(w)] = a(z)b(w) \; - \; (-1)^{|a||b|} b(w)a(z),$$

and $\sim$ means equal modulo terms which are regular at $z = w$. The *normally ordered product* $: a(z)b(z) :$ is defined to be

$$a(z)_{-}b(z) \; + \; (-1)^{|a||b|} b(z)a(z)_{+},$$

where

$$a(z)_{-} = \sum_{n<0} a(n) z^{-n-1}, \qquad a(z)_{+} = \sum_{n \geq 0} a(n) z^{-n-1}.$$

We usually omit the formal variable z and write $: a(z)b(z) : = : ab :$, when no confusion can arise. For $a_1, \dots, a_k \in \mathcal{V}$, the iterated normally ordered product is defined inductively by

$$: a_1 a_2 \cdots a_k : = : a_1 \big(: a_2 \cdots a_k : \big). \tag{2.1}$$

A subset $S = \{a_i \mid i \in I\}$ of $\mathcal{V}$ is said to *strongly generate* $\mathcal{V}$, if $\mathcal{V}$ is spanned by the set of normally ordered monomials

$$: \partial^{k_1} a_{i_1} \cdots \partial^{k_m} a_{i_m} :, \qquad i_1, \dots, i_m \in I, \qquad k_1, \dots, k_m \geq 0.$$

If S is an ordered strong generating set $\{\alpha^1, \alpha^2, \dots\}$, we say that S *freely generates* $\mathcal{V}$, if $\mathcal{V}$ has a PBW basis consisting of

$$\begin{aligned} &: \partial^{k_1^1} \alpha^{i_1} \dots \partial^{k_{r_1}^1} \alpha^{i_1} \partial^{k_1^2} \alpha^{i_2} \dots \partial^{k_{r_2}^2} \alpha^{i_2} \dots \partial^{k_1^n} \alpha^{i_n} \dots \partial^{k_{r_n}^n} \alpha^{i_n} :, \qquad 1 \leq i_1 < \dots < i_n, \\ &k_1^1 \geq k_2^1 \geq \dots \geq k_{r_1}^1, \quad k_1^2 \geq k_2^2 \geq \dots \geq k_{r_2}^2, \quad \dots, \quad k_1^n \geq k_2^n \geq \dots \geq k_{r_n}^n, \\ &k_1^t > k_2^t > \dots > k_{r_t}^t \end{aligned} \tag{2.2}$$

whenever α^{i_t} is odd.

In particular, the monomials (2.2) are linearly independent, so there are no nontrivial normally ordered polynomial relations among the generators and their derivatives.

$\beta\gamma$-System The $\beta\gamma$-system $\mathcal{S}$ is freely generated by even fields β, γ satisfying

$$\begin{aligned} \beta(z)\gamma(w) &\sim (z-w)^{-1}, & \gamma(z)\beta(w) &\sim -(z-w)^{-1}, \\ \beta(z)\beta(w) &\sim 0, & \gamma(z)\gamma(w) &\sim 0. \end{aligned} \tag{2.3}$$

It has Virasoro element $L^{\mathcal{S}} = \frac{1}{2}\big(:\beta\partial\gamma: - :\partial\beta\gamma:\big)$ of central charge $c = -1$, under which β, γ are primary of weight $\frac{1}{2}$.

$\mathcal{W}_3$-Algebra The $\mathcal{W}_3$-algebra $\mathcal{W}_3^c$ with central charge c was introduced by Zamolodchikov [28]. It is an extension of the Virasoro algebra, and is freely generated by a Virasoro field L and an even weight 3 primary field W. In fact, $\mathcal{W}_3^c$ is isomorphic to the principal $\mathcal{W}$-algebra $\mathcal{W}^k(\mathfrak{sl}_3, f_{\text{prin}})$ where $c = 2 - \frac{24(k+2)^2}{k+3}$. For generic values of c, $\mathcal{W}_3^c$ is simple, but for certain special values it has a nontrivial ideal. In this paper, we only need the case $c = -2$, which is nongeneric. We shall denote the simple graded quotient of $\mathcal{W}_3^{-2}$ by $\mathcal{W}$ for the rest of the paper. Since $\mathcal{W}_3^{-2}$ has a nontrivial ideal, $\mathcal{W}$ is strongly but not freely generated by L, W.

There is a useful embedding $i : \mathcal{W} \to \mathcal{S}$ due to Wang [26], given by

$$\begin{aligned} &L \mapsto \frac{1}{2} :\beta\beta\gamma\gamma: + :\beta(\partial\gamma): - :(\partial\beta)\gamma:, \\ &W \mapsto \frac{1}{4\sqrt{2}}\Big(2 :\beta^3\gamma^3: + 9 :\beta^2(\partial\gamma)\gamma: + 3 :\beta\partial^2\gamma: - 9 :(\partial\beta)\beta\gamma^2: - 12\partial\beta)(\partial\gamma): + 3 :(\partial^2\beta)\gamma:\Big), \end{aligned} \tag{2.4}$$

and we shall identify $\mathcal{W}$ with its image in $\mathcal{S}$. In fact, $\mathcal{W}$ is precisely the subalgebra of $\mathcal{S}$ that commutes with the Heisenberg algebra generated by $:\beta\gamma:$. Note that W is normalized so that it satisfies

$$W(z)W(w) \sim -\frac{9}{8}(z-w)^{-6} + \frac{27}{8}L(w)(z-w)^{-4} + \frac{27}{16}\partial L(w)(z-w)^{-3}$$

$$+\left(\frac{9}{2} :LL: - \frac{27}{32}\partial^2 L\right)(w)(z-w)^{-2} + \left(\frac{9}{2} :(\partial L)L: - \frac{3}{16}\partial^3 L\right)(w)(z-w)^{-1}.$$

This normalization is nonstandard but convenient for our purposes.

Zhu's Commutative Algebra and the Associated Variety Given a vertex algebra $\mathcal{V}$, define

$$C(\mathcal{V}) = \text{Span}\{a_{(-2)}b \mid a, b \in \mathcal{V}\}, \qquad R_{\mathcal{V}} = \mathcal{V}/C(\mathcal{V}). \tag{2.5}$$

It is well known that $R_{\mathcal{V}}$ is a commutative, associative algebra with product induced by the normally ordered product [29]. Also, if $\mathcal{V}$ is graded by conformal weight, $R_{\mathcal{V}}$ inherits this grading. Define the associated scheme

$$\tilde{X}_{\mathcal{V}} = \mathrm{Spec}(R_{\mathcal{V}}), \tag{2.6}$$

and the associated variety

$$X_{\mathcal{V}} = \mathrm{Specm}(R_{\mathcal{V}}) = (\tilde{X}_{\mathcal{V}})_{\mathrm{red}}. \tag{2.7}$$

Here $(\tilde{X}_{\mathcal{V}})_{\mathrm{red}}$ denotes the reduced scheme of $\tilde{X}_{\mathcal{V}}$. If $\{\alpha_i \mid i \in I\}$ is a strong generating set for $\mathcal{V}$, the images of these fields in $R_{\mathcal{V}}$ will generate $R_{\mathcal{V}}$ as a ring. In particular, $R_{\mathcal{V}}$ is finitely generated if and only if $\mathcal{V}$ is strongly finitely generated.

Since the $\beta\gamma$-system $\mathcal{S}$ is freely generated by β, γ, $R_{\mathcal{S}} \cong \mathbb{C}[b, g]$, where b, g denote the images of β, γ in $R_{\mathcal{S}}$. On the other hand, since $\mathcal{W}$ is not freely generated by L, W, the structure of $R_{\mathcal{W}}$ is more complicated.

Lemma 2.1 *Let ℓ, w denote the images of L, W in $R_{\mathcal{W}}$. Then $R_{\mathcal{W}} \cong \mathbb{C}[\ell, w]/\langle w^2 - \ell^3\rangle$.*

Proof Since $\mathcal{W}$ is strongly generated by L, W, $R_{\mathcal{W}}$ is generated by ℓ, w, so $R_{\mathcal{W}} \cong \mathbb{C}[\ell, w]/I$ for some ideal I. By Lemma 2.1 of [27], we have the following normally ordered relation in $\mathcal{W}$ at weight 6:

$$:W^2: - :L^3: - \frac{7}{8} :(\partial^2 L)L: - \frac{19}{32} :(\partial L)^2: = 0. \tag{2.8}$$

Note that (2.8) differs slightly from the formula in [27] because our normalization of W is different. It follows that $w^2 - \ell^3 \in I$.

To see that $I \subseteq \langle w^2 - \ell^3\rangle$, let $p = p(\ell, w) \in I$. Without loss of generality, we may assume p is homogeneous of weight d. It must come from a normally ordered polynomial relation

$$P = P(L, \partial L, \dots, W, \partial W, \dots) = 0$$

of weight d in $\mathcal{W}$ among L, W and their derivatives. The monomials of p correspond to the normally ordered monomials of P which do not lie in $C(\mathcal{W})$, and have the form

$$:L^i W^j:, \qquad 2i + 3j = d. \tag{2.9}$$

Using (2.8) repeatedly, we can rewrite this relation in the form

$$P' = P'(L, \partial L, \dots, W, \partial W, \dots) = 0,$$

where all terms of the form (2.9) that appear either have $j = 0$ or $j = 1$. In fact, since P' is homogeneous of weight d, we must have $j = 0$ if d is even, and $j = 1$ if d is odd, so only one such term can appear. If this term appears with nonzero coefficient, as a normally ordered polynomial in β, γ and their derivatives, it will contribute the term $:\beta^{2i+3j}\gamma^{2i+3j}:$, which cannot be canceled. This contradicts $P' = 0$, so each monomial in P' must lie in $C(\mathcal{W})$. Equivalently, $p \in \langle w^2 - \ell^3\rangle$. □

3 Jet Schemes and Arc Spaces

We recall some basic facts about jet schemes, following the notation in [15]. Let X be an irreducible scheme over $\mathbb{C}$ of finite type. The first jet scheme X_1 is the total tangent space of X, and for $m > 1$ the jet schemes X_m are higher-order generalizations which are determined by their functor of points. Given a $\mathbb{C}$-algebra A, we have a bijection

$$\mathrm{Hom}(\mathrm{Spec}(A), X_m) \cong \mathrm{Hom}(\mathrm{Spec}(A[t]/\langle t^{m+1}\rangle), X).$$

Thus the $\mathbb{C}$-valued points of X_m correspond to the $\mathbb{C}[t]/\langle t^{m+1}\rangle$-valued points of X. For $p > m$, we have projections $\pi_{p,m} : X_p \to X_m$ and $\pi_{p,m} \circ \pi_{q,p} = \pi_{q,m}$ when $q > p > m$. The assignment $X \mapsto X_m$ is functorial, and a morphism $f : X \to Y$ induces $f_m : X_m \to Y_m$ for all $m \geq 1$. If X is nonsingular, X_m is irreducible and nonsingular for all m. If X, Y are nonsingular and $f : X \to Y$ is a smooth surjection, f_m is surjective for all m.

For an affine scheme $X = \mathrm{Spec}(R)$ where $R = \mathbb{C}[y_1, \dots, y_r]/\langle f_1, \dots, f_k\rangle$, X_m is also affine and we can give explicit equations for X_m as follows. Define variables $y_1^{(i)}, \dots y_r^{(i)}$ for $i = 0, \dots, m$, and define a derivation D by

$$D(y_j^{(i)}) = \begin{cases} y_j^{(i+1)} & 0 \leq i < m \\ 0 & i = m \end{cases}, \tag{3.1}$$

which specifies its action on all of $\mathbb{C}[y_1^{(i)}, \dots, y_r^{(i)}]$, for $0 \leq i \leq m$. In particular, $f_\ell^{(i)} = D^i(f_\ell)$ is a well-defined polynomial in $\mathbb{C}[y_1^{(i)}, \dots, y_r^{(i)}]$. Letting

$$R_m = \mathbb{C}[y_1^{(i)}, \dots, y_r^{(i)}]/\langle f_1^{(i)}, \dots, f_k^{(i)}\rangle,$$

we have $X_m \cong \mathrm{Spec}(R_m)$. By identifying y_j with $y_j^{(0)}$, we may identify R with a subalgebra of R_m. There is a $\mathbb{Z}_{\geq 0}$-grading on R_m which we call *height*, given by

$$R_m = \bigoplus_{n\geq 0} R_m[n], \qquad \mathrm{ht}(y_j^{(i)}) = i. \tag{3.2}$$

For all m, $R_m[0] = R$ and $R_m[n]$ is an R-module.

Given a scheme X, define

$$X_\infty = \lim_{\leftarrow} X_m, \tag{3.3}$$

which is known as the *arc space* of X. For a $\mathbb{C}$-algebra A, we have a bijection

$$\mathrm{Hom}(\mathrm{Spec}(A), X_\infty) \cong \mathrm{Hom}(\mathrm{Spec}(A[[t]]), X),$$

so the $\mathbb{C}$-valued points of X_∞ correspond to the $\mathbb{C}[[t]]$-valued points of X. If $X = \mathrm{Spec}(R)$ as above,

$$X_\infty \cong \mathrm{Spec}(R_\infty), \text{ where } R_\infty = \mathbb{C}[y_1^{(i)}, \dots, y_r^{(i)}]/\langle f_1^{(i)}, \dots, f_k^{(i)}\rangle.$$

Here $i \geq 0$, and $D(y_j^{(i)}) = y_j^{(i+1)}$ for all i.

By a theorem of Kolchin [21], X_∞ is irreducible if X is irreducible. However, even if X is irreducible and reduced, X_∞ need not be reduced. The following result is due to Sebag (see Example 8 of [24], as well as more general results in [25]), but we include a proof for the benefit of the reader.

Lemma 3.1 *For* $X = Spec(\mathbb{C}[\ell, w]/\langle w^2 - \ell^3\rangle) = \tilde{X}_{\mathcal{W}}$, X_∞ *is not reduced.*

Proof We have

$$X_\infty \cong \mathrm{Spec}(R_\infty), \qquad R_\infty = \mathbb{C}[\ell^{(0)}, \ell^{(1)}, \dots, w^{(0)}, w^{(1)}, \dots]/\langle f^{(0)}, f^{(1)}, \dots\rangle, \tag{3.4}$$

where $f^{(0)} = (\ell^{(0)})^3 - (w^{(0)})^2$. Consider the element

$$r_1 = 3\ell^{(1)}w^{(0)} - 2\ell^{(0)}w^{(1)} \in \mathbb{C}[\ell^{(0)}, \ell^{(1)}, \dots, w^{(0)}, w^{(1)}, \dots]. \tag{3.5}$$

First, $r_1 \notin \langle f^{(0)}, f^{(1)}, \dots\rangle$ since no element of this ideal has leading term of degree 2. However, $(r_1)^3 \in \langle f^{(0)}, f^{(1)}, \dots\rangle$; a calculation shows that

$$\begin{aligned}(r_1)^3 =& \Big(-81\ell^{(0)}\ell^{(1)}\ell^{(2)}w^{(0)} - \frac{27}{2}(\ell^{(0)})^2\ell^{(3)}w^{(0)} + 18(\ell^{(0)})^2\ell^{(2)}w^{(1)} - 4(w^{(1)})^3 + 15w^{(0)}w^{(1)}w^{(2)}\\ &+ 9(w^{(0)})^2w^{(3)}\Big)f^{(0)} + \Big(\frac{9}{2}(\ell^{(0)})^2\ell^{(2)}w^{(0)} + 12(\ell^{(0)})^2\ell^{(1)}w^{(1)} - 7w^{(0)}(w^{(1)})^2 - 3(w^{(0)})^2w^{(2)}\Big)f^{(1)}\\ &+ \Big(-\frac{9}{2}(\ell^{(0)})^2\ell^{(1)}w^{(0)} - 6(\ell^{(0)})^3w^{(1)} + 9(w^{(0)})^2w^{(1)}\Big)f^{(2)} + \Big(\frac{9}{2}(\ell^{(0)})^3w^{(0)} - \frac{9}{2}(w^{(0)})^3\Big)f^{(3)}.\end{aligned} \tag{3.6}$$

Therefore regarded as an element of R_∞, $r_1 \neq 0$ but $(r_1)^3 = 0$. □

It is well known that in characteristic zero, for any affine scheme X, the nilradical $\mathcal{N} \subseteq \mathcal{O}(X_\infty)$ is a differential ideal; in other words, $D(\mathcal{N}) \subseteq \mathcal{N}$. A natural question (see [20]) is whether $\mathcal{N}$ is finitely generated as a differential ideal, and whether an explicit generating set can be found. In general, $\mathcal{N}$ need not be finitely generated; this was shown for $X = \mathrm{Spec}(\mathbb{C}[x, y]/\langle xy\rangle)$ in [12]. In the example

$X = \mathrm{Spec}(\mathbb{C}[\ell, w]/\langle w^2 - \ell^3\rangle)$, a calculation shows that in addition to r_1,

$$r_2 = (w^{(1)})^2 - \frac{9}{4}\ell^{(0)}(\ell^{(1)})^2 \tag{3.7}$$

does not lie in $\langle f^{(0)}, f^{(1)}, \dots \rangle$, but $(r_2)^3$ does. So r_2 is another nontrivial element of $\mathcal{N}$. We expect that $\mathcal{N}$ is generated as a differential ideal by r_1 and r_2.

The following characterization of $\mathcal{N}$ in this example will also be useful to us.

Lemma 3.2 *Let $X = Spec(\mathbb{C}[\ell, w]/\langle w^2 - \ell^3\rangle)$ and let t be a coordinate function on $\mathbb{C}$. Consider the map*

$$\mathbb{C} \to X, \qquad t \mapsto (t^2, t^3), \tag{3.8}$$

and the induced homomorphism

$$\varphi : \mathcal{O}(X_\infty) \to \mathcal{O}(\mathbb{C}_\infty), \qquad \ell^{(0)} \mapsto (t^{(0)})^2, \qquad w^{(0)} \mapsto (t^{(0)})^3. \tag{3.9}$$

Then $\mathcal{N} = ker(\varphi)$.

Proof Since (3.8) is birational, the map $\mathbb{C}_\infty \to X_\infty$ on arc spaces induced by (3.8) is dominant, see Proposition 3.2 of [15]. Therefore $\ker(\varphi) \subseteq \mathcal{N}$. On the other hand, $\mathcal{N} \subseteq \ker(\varphi)$ since $\mathcal{O}(\mathbb{C}_\infty) \cong \mathbb{C}[t^{(0)}, t^{(1)}, \dots]$, which is an integral domain. □

4 Li's Filtration and Singular Support

For any vertex algebra $\mathcal{V}$, we have Li's canonical decreasing filtration

$$F^0(\mathcal{V}) \supseteq F^1(\mathcal{V}) \supseteq \cdots,$$

where $F^p(\mathcal{V})$ is spanned by elements of the form

$$: \partial^{n_1} a^1 \partial^{n_2} a^2 \cdots \partial^{n_r} a^r :,$$

where $a^1, \dots, a^r \in \mathcal{V}$, $n_i \geq 0$, and $n_1 + \cdots + n_r \geq p$ [22]. Clearly $\mathcal{V} = F^0(\mathcal{V})$ and $\partial F^i(\mathcal{V}) \subseteq F^{i+1}(\mathcal{V})$. Set

$$\mathrm{gr}(\mathcal{V}) = \bigoplus_{p \geq 0} F^p(\mathcal{V})/F^{p+1}(\mathcal{V}),$$

and for $p \geq 0$ let

$$\sigma_p : F^p(\mathcal{V}) \to F^p(\mathcal{V})/F^{p+1}(\mathcal{V}) \subseteq \mathrm{gr}(\mathcal{V})$$

be the projection. Note that $\mathrm{gr}(\mathcal{V})$ is a graded commutative algebra with product

$$\sigma_p(a)\sigma_q(b) = \sigma_{p+q}(a_{(-1)}b),$$

for $a \in F^p(\mathcal{V})$ and $b \in F^q(\mathcal{V})$. We say that the subspace $F^p(\mathcal{V})/F^{p+1}(\mathcal{V})$ has *height* p. Note that $\mathrm{gr}(\mathcal{V})$ has a differential ∂ defined by

$$\partial(\sigma_p(a)) = \sigma_{p+1}(\partial a),$$

for $a \in F^p(\mathcal{V})$. Finally, $\mathrm{gr}(\mathcal{V})$ has the structure of a Poisson vertex algebra [22]; for $n \geq 0$, we define

$$\sigma_p(a)_{(n)}\sigma_q(b) = \sigma_{p+q-n}a_{(n)}b.$$

Zhu's commutative algebra $R_{\mathcal{V}}$ is isomorphic to the subalgebra $F^0(\mathcal{V})/F^1(\mathcal{V}) \subseteq \mathrm{gr}(\mathcal{V})$, since $F^1(\mathcal{V})$ coincides with the space $C(\mathcal{V})$ defined by (2.5). Moreover, $\mathrm{gr}(\mathcal{V})$ is generated by $R_{\mathcal{V}}$ as a differential graded commutative algebra [22]. Since $\tilde{X}_{\mathcal{V}} = \mathrm{Spec}(R_{\mathcal{V}})$, there is always a surjective homomorphism of differential graded rings

$$\Phi_{\mathcal{V}} : \mathcal{O}((\tilde{X}_{\mathcal{V}})_\infty) \to \mathrm{gr}(\mathcal{V}), \tag{4.1}$$

where the grading on $\mathcal{O}((\tilde{X}_{\mathcal{V}})_\infty)$ is given by (3.2). Define the *singular support*

$$\mathrm{SS}(\mathcal{V}) = \mathrm{Spec}(\mathrm{gr}(\mathcal{V})), \tag{4.2}$$

which is then a subscheme of $(\tilde{X}_{\mathcal{V}})_\infty$. A natural question which was raised by Arakawa and Moreau [7] is whether the map (4.1) is always an isomorphism. This is true in many examples and it was recently shown in [8] to hold as varieties when $\mathcal{V}$ is quasi-lisse, that is, if $X_{\mathcal{V}}$ has finitely many symplectic leaves, see [6] for the details. We note that the vertex algebra $\mathcal{W}$ is not quasi-lisse.

5 Main Result

Theorem 5.1 *For the vertex algebra $\mathcal{W}$, the map $\Phi_{\mathcal{W}} : \mathcal{O}((\tilde{X}_{\mathcal{W}})_\infty) \to gr(\mathcal{W})$ is not injective, so $(\tilde{X}_{\mathcal{W}})_\infty$ and $SS(\mathcal{W})$ are not isomorphic as schemes.*

Proof As before, we use the notation

$$\mathcal{O}((\tilde{X}_{\mathcal{W}})_\infty) \cong R_\infty = \mathbb{C}[\ell^{(0)}, \ell^{(1)}, \dots, w^{(0)}, w^{(1)}, \dots]/\langle f^{(0)}, f^{(1)}, \dots\rangle.$$

We use the same notation $\partial^i L, \partial^i W$ to denote the images of the fields $\partial^i L, \partial^i W \in \mathcal{W}$ in the subspace $F^i(\mathcal{W})/F^{i+1}(\mathcal{W})$ of $\mathrm{gr}(\mathcal{W})$. We therefore may identify $\mathrm{gr}(\mathcal{W})$ with a quotient of the polynomial ring $\mathbb{C}[L, \partial L, \dots, W, \partial W, \dots]$. In this notation, $\Phi_{\mathcal{W}}(\ell^{(0)}) = L$ and $\Phi_{\mathcal{W}}(w^{(0)}) = W$.

We will show that the nilpotent elements r_1 and r_2 in $\mathcal{O}((\tilde{X}_{\mathcal{W}})_\infty)$ given by (3.5) and (3.7) lie in $\ker(\Phi_{\mathcal{W}})$. By Lemma 2.1 of [27], we have the following relation in $\mathcal{W}$ at weight 6:

$$3 : (\partial L)W : -2 : L(\partial W) : +\frac{1}{4}\partial^3 W = 0.$$

Therefore in $F^1(\mathcal{W})/F^2(\mathcal{W})$, we have the relation

$$3(\partial L)W - 2L\partial W = 0.$$

Since $\Phi_{\mathcal{W}}(r_1) = 3(\partial L)W - 2L\partial W$, $r_1 \in \ker(\Phi_{\mathcal{W}})$.

Similarly, in $\mathcal{W}$ we have the following relation in weight 8:

$$: (\partial W)^2 : -\frac{9}{4} : (\partial L)^2 L : -\frac{3}{16} : (\partial^4 L)L : -\frac{3}{8} : (\partial^3 L)(\partial L) : -\frac{9}{32} : (\partial^2 L)^2 : +\frac{1}{160}\partial^6 L = 0,$$

so in $F^2(\mathcal{W})/F^3(\mathcal{W})$ we have the relation $(\partial W)^2 - \frac{9}{4}(\partial L)^2 L = 0$, and $r_2 \in \ker(\Phi_{\mathcal{W}})$. □

Theorem 5.2 *Even though $(\tilde{X}_{\mathcal{W}})_\infty$ and $SS(\mathcal{W})$ differ as schemes, the map of varieties*

$$SS(\mathcal{W})_{red} \to ((\tilde{X}_{\mathcal{W}})_\infty)_{red}$$

induced by $\Phi_{\mathcal{W}}$ is an isomorphism.

Proof It suffices to show that the map $\varphi : \mathcal{O}((\tilde{X}_{\mathcal{W}})_\infty) \to \mathcal{O}(\mathbb{C}_\infty)$ given by (3.9) factors through the map $\Phi_{\mathcal{W}} : \mathcal{O}((\tilde{X}_{\mathcal{W}})_\infty) \to \mathrm{gr}(\mathcal{W})$, since $\ker(\varphi) = \mathcal{N}$. First, the embedding $i : \mathcal{W} \to \mathcal{S}$ given by (2.4) induces a map

$$\mathrm{gr}(i) : \mathrm{gr}(\mathcal{W}) \to \mathrm{gr}(\mathcal{S}).$$

Identifying $\mathrm{gr}(\mathcal{S})$ with $\mathbb{C}[\beta, \partial\beta, \dots, \gamma, \partial\gamma, \dots]$, this map is given on generators by

$$\mathrm{gr}(i)(L) = (\beta\gamma)^2, \qquad \mathrm{gr}(i)(W) = (\beta\gamma)^3.$$

We also have an injective map of differential graded algebras

$$\psi : \mathcal{O}(\mathbb{C}_\infty) \to \mathrm{gr}(\mathcal{S}),$$

defined on the generator $t^{(0)}$ of $\mathcal{O}(\mathbb{C}_\infty)$ by $\psi(t^{(0)}) = \beta\gamma$. Since

$$\Phi_{\mathcal{W}}(\ell^{(0)}) = (\beta\gamma)^2 = \mathrm{gr}(i)(L) = \psi((t^{(0)})^2), \qquad \Phi_{\mathcal{W}}(w^{(0)}) = (\beta\gamma)^3 = \mathrm{gr}(i)(W) = \psi((t^{(0)})^3),$$

and L, W generate $\mathrm{gr}(\mathcal{W})$ as a differential algebra, it is clear that

$$\mathrm{gr}(i)(\mathrm{gr}(\mathcal{W})) = \psi(A) \cong A = \varphi(\mathcal{O}((\tilde{X}_{\mathcal{W}})_\infty)),$$

where $A \subseteq \mathcal{O}(\mathbb{C}_\infty)$ is the subalgebra generated by $(t^{(0)})^2, (t^{(0)})^3$, and their derivatives. This completes the proof. □

In this example, we expect that $\mathrm{gr}(i) : \mathrm{gr}(\mathcal{W}) \to \mathrm{gr}(\mathcal{S})$ is injective, so that $\mathrm{gr}(\mathcal{W}) \cong A$, and in particular is reduced. However, we caution the reader that the associated graded functor is not left exact in general.

6 Failure of Associated Graded Functor to be Left Exact

Here we give an example of a simple vertex algebra $\mathcal{V}$ which has a free field realization $i : \mathcal{V} \to \mathcal{H}$ where $\mathcal{H}$ is the Heisenberg algebra, such that the induced map $\mathrm{gr}(i) : \mathrm{gr}(\mathcal{V}) \to \mathrm{gr}(\mathcal{H})$ is not injective.

First, $\mathcal{H}$ is generated by an even field α satisfying

$$\alpha(z)\alpha(w) \sim (z-w)^{-2},$$

and has Virasoro element $L = \frac{1}{2} :\alpha\alpha:$ of central charge $c = 1$. There is an action of $\mathbb{Z}_2$ sending $\alpha \mapsto -\alpha$ which preserves L, and we consider the orbifold

$$\mathcal{V} = \mathcal{H}^{\mathbb{Z}_2}.$$

By a result of Dong and Nagatomo [14], $\mathcal{V}$ is strongly generated by L together with a unique up to scalar weight 4 field primary field

$$W^4 = -\frac{1}{6\sqrt{6}} :\alpha^4: -\frac{1}{4\sqrt{6}} :(\partial\alpha)^2: +\frac{1}{6\sqrt{6}} :(\partial^2\alpha)\alpha:,$$

which is normalized so that it satisfies

$$W^4(z)W^4(w) \sim \frac{1}{4}(z-w)^{-8} + \cdots.$$

One can check by direct calculation that $\mathcal{V}$ is isomorphic to the simple, principal $\mathcal{W}$-algebra of $\mathfrak{sp}_4$ with central charge $c = 1$. It is convenient to replace W^4 with the field

$$W = \frac{35}{132} :(\partial^2\alpha)\alpha: = \frac{35\sqrt{2/3}}{33} W^4 + \frac{70}{297} :L^2: +\frac{35}{396}\partial^2 L,$$

which is not primary. A calculation shows that we have the following nontrivial relations in $\mathcal{V}$ at weights 8 and 10, respectively.

$$\begin{aligned} &:W^2: - :L^2W: + \frac{35}{132} :(\partial^2 L)L^2: - \frac{35}{264} :(\partial L)^2 L: + \frac{13265}{69696} :(\partial^4 L)L: \\ &+ \frac{19495}{139392} :(\partial^3 L)\partial L: - \frac{59}{88} :(\partial^2 L)W: - \frac{497}{352} :(\partial L)\partial W: - \frac{181}{528} :L\partial^2 W: \\ &- \frac{139}{2112}\partial^4 W + \frac{10955}{557568}\partial^6 L = 0, \end{aligned} \tag{6.1}$$

$$\begin{aligned} &:L^3W: + \frac{4455}{1024} :(\partial W)\partial W: - \frac{35}{132} :(\partial^2 L)L^3: + \frac{35}{264} :(\partial L)^2 L^2: + \frac{347}{256} :(\partial L)^2 W: - \\ &\frac{1069}{256} :(\partial L)L\partial W: - \frac{49}{16} :L^2\partial^2 W: + \frac{385}{576} :(\partial^4 L)L^2: + \frac{48965}{101376} :(\partial^3 L)(\partial L)L: - \\ &\frac{35}{44} :(\partial^2 L)^2 L: + \frac{35}{88} :(\partial^2 L)(\partial L)^2: - \frac{1687}{1536} :(\partial^4 L)W: - \frac{5939}{3072} :(\partial^3 L)(\partial W): - \\ &\frac{247}{256} :(\partial^2 L)(\partial^2 W): - \frac{10927}{6144} :(\partial L)(\partial^3 W): + \frac{779}{1536} :L\partial^4 W: + \frac{3899}{36864} :(\partial^6 L)L: + \\ &\frac{102851}{270336} :(\partial^5 L)\partial L: + \frac{7525}{67584} :(\partial^4 L)\partial^2 L: + \frac{659645}{4866048} :(\partial^3 L)^2: \\ &+ \frac{68311}{6488064}\partial^8 L - \frac{3187}{49152}\partial^6 W = 0. \end{aligned} \tag{6.2}$$

Lemma 6.1 *Let ℓ, w denote the images of L, W in $R_{\mathcal{V}}$. Then*

$$R_{\mathcal{V}} \cong \mathbb{C}[\ell, w]/I$$

where I is the ideal generated by $w(w-\ell^2)$ and $\ell^3 w$. In particular, $\tilde{X}_{\mathcal{V}} = Spec(R_{\mathcal{V}})$ is irreducible of dimension one, but is not reduced.

Proof Since $\mathcal{V}$ is strongly generated by L, W, it follows from (6.1) and (6.2) that $R_{\mathcal{V}} \cong \mathbb{C}[\ell, w]/I$ for some ideal I which contains $w(w-\ell^2)$ and $\ell^3 w$. The proof that I is generated by these two elements is similar to the proof of Lemma 2.1, and is omitted. Since I is contained in the ideal $\langle w \rangle$, the map $\mathbb{C}[\ell] \to R_{\mathcal{V}}$ is injective, and $R_{\mathcal{V}}$ has Krull dimension 1. Since w is a nontrivial nilpotent element of $R_{\mathcal{V}}$, $\tilde{X}_{\mathcal{V}}$ is not reduced. Finally, it is easy to see that the nilradical $\mathcal{N}$ of $R_{\mathcal{V}}$ is generated by w, so $\mathcal{N}$ is prime and $\tilde{X}_{\mathcal{V}}$ is irreducible. □

Corollary 6.2 *Let $i : \mathcal{V} \to \mathcal{H}$ be the inclusion. Since $gr(\mathcal{H})$ is the polynomial ring $\mathbb{C}[\alpha, \partial\alpha, \dots]$, the induced map $gr(i) : gr(\mathcal{V}) \to gr(\mathcal{H})$ is not injective.*

In fact, it is easy to verify that the image of gr(i) is just the differential polynomial algebra generated by $\mathrm{gr}(i)(L) = \frac{1}{2}\alpha^2$. Finally, we remark that as in our main example $\mathcal{W}$, the map $\Phi_{\mathcal{V}} : \mathcal{O}((\tilde{X}_{\mathcal{V}})_\infty) \to \mathrm{gr}(\mathcal{V})$ is not injective for $\mathcal{V} = \mathcal{H}^{\mathbb{Z}_2}$. For example,

$$r = \ell^{(0)}\ell^{(2)}w^{(0)} + (\ell^{(1)})^2 w^{(0)} - \frac{1}{2}(\ell^{(0)})^2 w^{(2)}$$

is a nontrivial element of $\ker(\Phi_{\mathcal{V}})$. In fact, r is nilpotent in $\mathcal{O}((\tilde{X}_{\mathcal{V}})_\infty)$ and satisfies $r^3 = 0$.

7 Universal Enveloping Vertex Algebras

Let $\mathcal{V}$ be a conformal vertex algebra with a strong generating set S, i.e., for $a, b \in S$, the all terms in the OPE $a(z)b(w)$ can be expressed as normally ordered polynomials in the elements of S and their derivatives. In the language of de Sole and Kac [13], the OPE algebra gives rise to a nonlinear conformal algebra satisfying skew-symmetry. There is a well-defined universal enveloping vertex algebra $\mathcal{UV}$ which is the initial object in the category of vertex algebras with the above strong generating set and OPE algebra. If for all fields $a, b, c \in S$ and integers $r, s \geq 0$, the Jacobi identities

$$a_{(r)}(b_{(s)}c) - (-1)^{|a||b|} b_{(s)}(a_{(r)}c) - \sum_{i=0}^{r} \binom{r}{i} (a_{(i)}b)_{(r+s-i)}c = 0, \tag{7.1}$$

hold as formal consequences of the OPE relations, this Lie conformal algebra is then called a *nonlinear Lie conformal algebra*. The main result (Theorem 3.9) of [13] is that in this case, $\mathcal{UV}$ is freely generated by S. This means that it has a PBW basis consisting of monomials in the elements of S and their derivatives.

In the examples $\mathcal{W}$ and $\mathcal{V}$ above, the universal enveloping vertex algebras are the universal $\mathcal{W}_3$-algebra with $c = -2$ and the universal $\mathcal{W}(\mathfrak{sp}_4, f_{\mathrm{prin}})$-algebra with $c = 1$, respectively. Both of these are freely generated, so the associated varieties are isomorphic to $\mathbb{C}^2$ and the map (4.1) is an isomorphism in both cases. It is natural to ask whether (4.1) is always an isomorphism for universal enveloping vertex algebras, and in this section we provide a counterexample.

In [1], Adamovic studied a class of simple vertex algebra called $\mathcal{W}(2, 2p-1)$-algebras, where $p \geq 2$ is a positive integer. They are strongly generated by a Virasoro field L with central charge $c = 1 - \frac{6(p-1)^2}{p}$, and a weight $2p - 1$ primary field W, and coincide with the singlet subalgebras of the $\mathcal{W}_{2,p}$-triplet algebras. The triplet algebras were the first examples of C_2-cofinite, nonrational vertex algebras to appear in the literature [2].

We consider the case $p = 3$, and we denote the $\mathcal{W}(2, 5)$-algebra by $\mathcal{A}$. It can be realized explicitly inside the Heisenberg algebra $\mathcal{H}$ with generator α as follows.

$$L = \frac{1}{2} : \alpha^2 : + \sqrt{\frac{2}{3}} \partial\alpha,$$
$$W = \frac{1}{4\sqrt{2}} : \alpha^5 : + \frac{5}{4\sqrt{3}} : (\partial\alpha)\alpha^3 : + \frac{5}{12\sqrt{2}} : (\partial^2\alpha)\alpha^2 : + \frac{5}{8\sqrt{2}} : (\partial\alpha)^2\alpha :$$
$$+ \frac{5}{48\sqrt{3}} : (\partial^3\alpha)\alpha : + \frac{5}{24\sqrt{3}} : (\partial^2\alpha)\partial\alpha : + \frac{1}{144\sqrt{2}} \partial^4\alpha. \tag{7.2}$$

The Virasoro field L has central charge -7, and the primary weight 5 field W satisfies

$$W(z)W(w) \sim \frac{175}{12}(z-w)^{-10} - \frac{125}{6}L(w)(z-w)^{-8} - \frac{125}{12}\partial L(w)(z-w)^{-7}$$
$$+ \left(\frac{125}{3} : LL : - \frac{125}{8}\partial^2 L\right)(w)(z-w)^{-6} + \left(\frac{125}{3} : (\partial L)L : - \frac{125}{36}\partial^3 L\right)(w)(z-w)^{-5}$$
$$+ \left(50 : L^3 : + \frac{25}{24} : (\partial L)^2 : -25 : (\partial^2 L)L : - \frac{175}{72}\partial^4 L\right)(w)(z-w)^{-4}$$
$$+ \left(75 : (\partial L)L^2 : - \frac{175}{8} : (\partial^2 L)\partial L : - \frac{125}{36} : (\partial^3 L)L : - \frac{35}{96}\partial^5 L\right)(w)(z-w)^{-3}$$
$$+ \left(\frac{25}{2} : L^4 : + \frac{1175}{48} : (\partial L)^2 L : + \frac{125}{12} : (\partial^2 L)L^2 : - \frac{775}{128} : (\partial^2 L)^2 :\right.$$
$$\left. - \frac{225}{64} : (\partial^3 L)\partial L : - \frac{175}{64} : (\partial^4 L)L : - \frac{1115}{13824}\partial^6 L\right)(z-w)^{-2}$$
$$+ \left(25 : (\partial L)L^3 : - \frac{25}{96} : (\partial L)^3 : - \frac{125}{48} : (\partial^2 L)(\partial L)L : + \frac{125}{24} : (\partial^3 L)L^2 :\right.$$
$$\left. - \frac{775}{288} : (\partial^3 L)\partial^2 L : - \frac{425}{288} : (\partial^4 L)\partial L : - \frac{115}{288} : (\partial^5 L)L : - \frac{365}{24192}\partial^7 L\right)(w)(z-w)^{-1} \tag{7.3}$$

We have the following normally ordered relations in weights 8 and 10, respectively.

$$2 : L\partial W : -5 : (\partial L)W : - \frac{1}{6}\partial^3 W = 0, \tag{7.4}$$

$$: W^2 : - : L^5 : - \frac{335}{24} : (\partial L)^2 L^2 : - \frac{25}{3} : (\partial L)L^3 : + \frac{283}{64} : (\partial^2 L)(\partial L)^2 :$$
$$+ \frac{309}{64} : (\partial^2 L)^2 L : - \frac{67}{36} : (\partial^3 L)(\partial L)L : + \frac{49}{216} : (\partial^3 L)^2 : - \frac{23}{32} : (\partial^4 L)L^2 :$$
$$+ \frac{49}{64} : (\partial^4 L)(\partial^2 L) : + \frac{249}{1280} : (\partial^5 L)\partial L : + \frac{223}{3840} : (\partial^6 L)L : + \frac{1}{504}\partial^8 L = 0. \tag{7.5}$$

It is straightforward to show using (7.4) and (7.5) that

$$R_{\mathcal{A}} \cong \mathbb{C}[\ell, w]/\langle w^2 - \ell^5\rangle. \tag{7.6}$$

Here ℓ, w denote the images of L, W in $R_{\mathcal{A}}$.

Next, let $\mathcal{U} = \mathcal{U}\mathcal{A}$ denote the universal enveloping vertex algebra of $\mathcal{A}$. By abuse of notation, we shall also denote the generators of $\mathcal{U}$ by L, W; they satisfy the same OPE relations as the generators of $\mathcal{A}$. We also denote by ℓ, w the images of L, W in $R_{\mathcal{U}}$.

Lemma 7.1 $R_{\mathcal{U}} \cong \mathbb{C}[\ell, w]/\langle w^2 - \ell^5\rangle \cong R_{\mathcal{A}}$.

Proof Using (7.3), we can compute the left side of the Jacobi identity (7.1) in the case $a = b = c = W$, $r = 4$ and $s = 3$. We find that it does not vanish identically as a consequence of the OPE relations, but instead is given by

$$\frac{9075}{16}(2 : L\partial W : -5 : (\partial L)W : -\frac{1}{6}\partial^3 W). \tag{7.7}$$

Since all Jacobi identities must hold in any vertex algebra, (7.7) must be a null vector, so that (7.4) holds in $\mathcal{U}$. Therefore the corresponding Lie conformal algebra is *not* a nonlinear Lie conformal algebra, and $\mathcal{U}$ is not freely generated by L and W. Applying the operator $W_{(2)}$ to the identity (7.4) yields a nonzero multiple of the identity (7.5). Therefore (7.5) also must hold in $\mathcal{U}$, which shows that the relation $w^2 - \ell^5$ holds in $R_{\mathcal{U}}$. Since $R_{\mathcal{A}}$ is a quotient of $R_{\mathcal{U}}$, the claim follows. □

Remark 7.2 We expect that $\mathcal{A} = \mathcal{U}$, but we do not prove this.

As in our previous example $\mathcal{W}$, even though the scheme $X_{\mathcal{U}} = \mathrm{Spec}(R_{\mathcal{U}})$ is reduced, the arc space $(X_{\mathcal{U}})_\infty$ is not. In particular,

$$r = 2\ell^{(0)}w^{(1)} - 5\ell^{(1)}w^{(0)}$$

is a nontrivial nilpotent element of $\mathcal{O}((\tilde{X}_{\mathcal{U}})_\infty)$ satisfying $r^3 = 0$, and $r \in \ker(\Phi_{\mathcal{U}})$. Therefore $\mathcal{U}$ is an example of a universal enveloping vertex algebra for which the map (4.1) fails to be injective.

Finally, via the embedding

$$\mathcal{H} \to \mathcal{S}, \qquad \alpha \mapsto \sqrt{-1} : \beta\gamma :,$$

$\mathcal{A}$ can be identified with a subalgebra of $\mathcal{S}$. By the same argument as Theorem 5.2, one can check that the map on varieties induced by (4.1) is an isomorphism. Therefore the same holds for $\mathcal{U}$.

Acknowledgements T. A. is supported by JSPS KAKENHI Grants #17H01086 and #17K18724. A. L. is supported by Simons Foundation Grant #318755. We thank Julien Sebag for helpful comments on an earlier draft of this paper.

References

1. D. Adamovic, *Classification of irreducible modules of certain subalgebras of free boson vertex algebra*, Journal of Algebra 270 (2003) 115–132.
2. D. Adamovic and A. Milas, *On the triplet vertex algebra $\mathcal{W}(p)$*, Adv. Math. 217 (2008), no. 6, 2664–2699.
3. T. Arakawa, *A remark on the C_2 cofiniteness condition on vertex algebras*, Math. Z. 270 (2012), no. 1–2, 559–575.
4. T. Arakawa, *Associated Varieties and Higgs Branches (A Survey), Contemp. Math.* 711 (2018), 37–44.
5. T. Arakawa, *Representation theory of W-algebras and Higgs branch conjecture, Proc. Int. Cong. of Math 2018 Rio de Janeiro*, Vol. 1 (1261–1278).
6. T. Arakawa and K. Kawasetsu, *Quasi-lisse vertex algebras and modular linear differential equations*, In: V. G. Kac, V. L. Popov (eds.), Lie Groups, Geometry, and Representation Theory, A Tribute to the Life and Work of Bertram Kostant, Progr. Math., 326, Birkhauser, 2018.
7. T. Arakawa and A. Moreau, *Lectures on $\mathcal{W}$-algebras*, preprint.
8. T. Arakawa and A. Moreau, *Arc spaces and chiral symplectic cores*, to appear in the special issue of Publ. Res. Inst. Math. in honor of Professor Masaki Kashiwara's 70th birthday.
9. C. Beem, M. Lemos, P. Liendo, W. Peelaers, L. Rastelli, and B. C. van Rees, *Infinite chiral symmetry in four dimensions*, Comm. Math. Phys., 336(3):1359–1433, 2015.
10. C. Beem and L. Rastelli, *Vertex operator algebras, Higgs branches, and modular differential equations*, J. High Energ. Phys. (2018) 2018:114. https://doi.org/10.1007/JHEP08(2018)114
11. R. Borcherds, *Vertex operator algebras, Kac-Moody algebras and the monster*, Proc. Nat. Acad. Sci. USA 83 (1986) 3068–3071.
12. D. Bourqui and J. Sebag, *The radical of the differential ideal generated by XY in the ring of two variable differential polynomials is not differentially finitely generated*, to appear Journal of Comm. Algebra (2017).
13. A. de Sole and V. Kac, *Freely generated vertex algebras and non-linear Lie conformal algebras*, Comm. Math. Phys. 254 (2005), no. 3, 659–694.
14. C. Dong and K. Nagatomo, *Classification of irreducible modules for the vertex operator algebra $M(1)^+$*, J. Algebra 216 (1999), no. 1, 384–404.
15. L. Ein and M. Mustata, *Jet schemes and singularities*, Algebraic geometry—Seattle 2005. Part 2, 505–546, Proc. Sympos. Pure Math., 80, Part 2, Amer. Math. Soc., Providence, RI, 2009.
16. J. van Ekeren and R. Heluani, *Chiral homology of elliptic curves and Zhu's algebra*, arXiv:1804.00017 [math.QA].
17. E. Frenkel and D. Ben-Zvi, *Vertex Algebras and Algebraic Curves*, Math. Surveys and Monographs, Vol. 88, American Math. Soc., 2001.
18. I. B. Frenkel, J. Lepowsky, and A. Meurman, *Vertex Operator Algebras and the Monster*, Academic Press, New York, 1988.
19. V. Kac, *Vertex Algebras for Beginners*, University Lecture Series, Vol. 10. American Math. Soc., 1998
20. K. Kpognon, J. Sebag, *Nilpotency in arc schemes of plane curves*, Comm. in Algebra, Vol. 45 no 5 (2017), 2195–2221
21. E. Kolchin, *Differential algebra and algebraic groups*, Academic Press, New York 1973.
22. H. Li, *Vertex algebras and vertex Poisson algebras*, Commun. Contemp. Math. 6 (2004) 61–110.
23. M. Miyamoto, *Modular invariance of vertex operator algebras satisfying C2-cofiniteness*. Duke Math. J. 122(1), 51–91 (2004).
24. J. Sebag, *Arcs schemes, derivations and Lipman's theorem*, J. Algebra 347 (2011) 173–183.
25. J. Sebag, *A remark on Berger's conjecture, Kolchin's theorem and arc schemes*, Archiv der Math., Vol. 108 no (2017), 145–150
26. W. Wang, *$\mathcal{W}_{1+\infty}$ algebra, $\mathcal{W}_3$ algebra, and Friedan-Martinec-Shenker bosonization*, Comm. Math. Phys. 195 (1998), no. 1, 95–111.

27. W. Wang, *Classification of irreducible modules of $\mathcal{W}_3$ algebra with $c = -2$*, Comm. Math. Phys. 195 (1998), no. 1, 113–128.
28. A.B. Zamolodchikov, *Infinite extra symmetries in two-dimensional conformal quantum field theory* (Russian), Teoret. Mat. Fiz. 65 (1985), 347–359. English translation, Theoret. and Math. Phys. 65 (1985), 1205–1213.
29. Y. Zhu, *Modular invariants of characters of vertex operators*, J. Amer. Soc. 9 (1996) 237–302.

On Cacti and Crystals

Arkady Berenstein, Jacob Greenstein, and Jian-Rong Li

To Anthony Joseph, with admiration

Abstract In the present work we study actions of various groups generated by involutions on the category $\mathcal{O}_q^{int}(\mathfrak{g})$ of integrable highest weight $U_q(\mathfrak{g})$-modules and their crystal bases for any symmetrizable Kac–Moody algebra $\mathfrak{g}$. The most notable of them are the cactus group and (yet conjectural) Weyl group action on any highest weight integrable module and its lower and upper crystal bases. Surprisingly, some generators of cactus groups are anti-involutions of the Gelfand–Kirillov model for $\mathcal{O}_q^{int}(\mathfrak{g})$ closely related to the remarkable quantum twists discovered by Kimura and Oya (Int Math Res Notices, 2019).

MSC: 17B37, 17B10 (primary); 20F36, 18D10, 20F55 (secondary)

1 Introduction

In the present work we study the action of various groups generated by involutions on the category $\mathcal{O}_q^{int}(\mathfrak{g})$ of integrable highest weight $U_q(\mathfrak{g})$-modules for any symmetrizable Kac–Moody algebra $\mathfrak{g}$ (the necessary notation is introduced in Sect. 2).

A. Berenstein (✉)
Department of Mathematics, University of Oregon, Eugene, OR, USA
e-mail: arkadiy@math.uoregon.edu

J. Greenstein
Department of Mathematics, University of California, Riverside, CA, USA
e-mail: jacobg@ucr.edu

J.-R. Li
Department of Mathematics, The Weizmann Institute of Science, Rehovot, Israel
e-mail: jianrong.li@uni-graz.at

M. Gorelik et al. (eds.), *Representations and Nilpotent Orbits of Lie Algebraic Systems*, Progress in Mathematics 330,
https://doi.org/10.1007/978-3-030-23531-4_2

Let $V \in \mathcal{O}_q^{int}(\mathfrak{g})$. We claim that for every node i of the Dynkin diagram I of $\mathfrak{g}$ there exists a unique linear operator σ_V^i on V such that

$$\sigma_V^i(E_i^{(k)}(u)) = E_i^{(l-k)}(u) \tag{1.1}$$

for all $l \geq k \geq 0$ and for all $u \in \ker F_i \cap \ker(K_i - q_i^{-l})$. Clearly, $(\sigma_V^i)^2 = \mathrm{id}_V$. Denote by $\mathsf{W}(V)$ the subgroup of $\mathrm{GL}_{\Bbbk}(V)$ generated by the σ_V^i, $i \in I$.

Theorem 1.1 *For any non-zero module $V \in \mathcal{O}_q^{int}(\mathfrak{g})$, the assignments*

$$\sigma_V^i \mapsto \begin{cases} 1, & i \in J(V) \\ s_i, & \text{otherwise} \end{cases}$$

where $J(V) = \{i \in I : F_i(V) = \{0\}\}$, define a homomorphism ψ_V from $\mathsf{W}(V)$ to the Weyl group W of $\mathfrak{g}$.

We prove Theorem 1.1 in Sect. 3.3 by showing that the image of ψ_V can be described in terms of a natural action of W on a certain set of extremal vectors in V. In particular, ψ_V is surjective if and only if $J(V) = \emptyset$. Moreover, we show that $\sigma_V^i = \mathrm{id}_V$ if and only if $i \in J(V)$. This suggests the following

Conjecture 1.2 The homomorphism ψ_V is injective for any $V \in \mathcal{O}_q^{int}(\mathfrak{g})$.

Clearly, it is equivalent to $(\sigma^i\sigma^j)^{m_{ij}} = \mathrm{id}_V$, $i \neq j \in I$ for appropriate choices of m_{ij}. We proved it for $m_{ij} = 2$ and we have ample evidence that this conjecture holds for $m_{ij} = 3$. We also verified it for all modules in which weight spaces of non-zero weight are one-dimensional (see Theorem 7.2). This class of modules includes all miniscule and quasi-miniscule ones. Conjecture 1.2 combined with Theorem 1.1 implies that W acts naturally and faithfully on objects in $\mathcal{O}_q^{int}(\mathfrak{g})$, which is quite surprising. Informally speaking, this conjecture asserts that Kashiwara's action of the Weyl group on crystal bases lifts to an action on the corresponding module (see Remark 5.7).

Remark 1.3 The definition (1.1) of σ_V^i makes sense for any integrable $U(\mathfrak{g})$-module where $\mathfrak{g}$ is a semisimple or a (not necessarily symmetrizable) Kac–Moody Lie algebra. The "classical" Theorem 1.1 holds verbatim. Moreover, Conjecture 1.2 implies its classical version for all (even not symmetrizable) Kac–Moody algebras.

It turns out that we can extend the group $\mathsf{W}(V)$ by adding involutions σ^J for any non-empty $J \subset I$ such that the subgroup $W_J = \langle s_i : i \in J\rangle$ is finite; we denote the set of all such J by $\mathscr{J}$. Note that $\{i\} \in \mathscr{J}$ for all $i \in I$ and in particular $\mathscr{J}$ is non-empty.

Proposition 1.4 (Proposition 4.14) *For any $V \in \mathcal{O}_q^{int}(\mathfrak{g})$, $J \in \mathscr{J}$ there exists a unique $\Bbbk$-linear map $\sigma^J = \sigma_V^J : V \to V$ such that*

(a) $\sigma^J(v) = v^J$ *for any* $v \in \bigcap_{i\in J} \ker E_i$ *where* v^J *is a distinguished element in* $\bigcap\limits_{i\in J} \ker F_i \cap U_q(\mathfrak{g}^J)v$ *defined in Proposition 4.14*(a);

(b) $\sigma^J(F_j(v)) = E_{j^\star}(\sigma^J(v))$, $\sigma^J(E_j(v)) = F_{j^\star}(\sigma^J(v))$ *for all* $j \in J$, $v \in V$ *where* $^\star : J \to J$ *is the involution on* J *induced by the longest element* $w_\circ^J$ *of* W_J *via* $s_{j^\star} = w_\circ^J s_j w_\circ^J$, $j \in J$ *(see Sect. 2.1).*

Moreover, for any morphism $f : V \to V'$ *in* $\mathcal{O}_q^{int}(\mathfrak{g})$ *the following diagram commutes*

$$\begin{array}{ccc} V & \xrightarrow{\sigma_V^J} & V \\ {\scriptstyle f}\downarrow & & \downarrow{\scriptstyle f} \\ V' & \xrightarrow{\sigma_{V'}^J} & V' \end{array} \qquad (1.2)$$

By definition, $\sigma^J = \sigma^i$ if $J = \{i\}$. The following is the main result of this paper.

Theorem 1.5 *Let* $V \in \mathcal{O}_q^{int}(\mathfrak{g})$. *Then for any* $J \in \mathscr{J}$ *we have in* $\mathrm{GL}_\Bbbk(V)$

(a) $\sigma^J \circ \sigma^J = 1$;

(b) *If* $J = J' \cup J''$ *where* J' *and* J'' *are orthogonal (that is,* $J' \cap J'' = \emptyset$ *and* $s_{j'}s_{j''} = s_{j''}s_{j'}$ *for all* $j' \in J'$, $j'' \in J''$*), then* $\sigma^J = \sigma^{J'} \circ \sigma^{J''}$*; in particular,* $\sigma^{J'} \circ \sigma^{J''} = \sigma^{J''} \circ \sigma^{J'}$ *if* $J', J'' \in \mathscr{J}$ *are orthogonal.*

(c) $\sigma^J \circ \sigma^K = \sigma^{K^\star} \circ \sigma^J$ *for any* $K \subset J$, *where* $^\star : J \to J$ *is as in Proposition 1.4*(b).

We prove Theorem 1.5 in Sect. 4.3 using appropriate modifications of Lusztig's symmetries (which we introduce in Sect. 4.1).

Following (and slightly generalizing) [29] (see also [10]), we denote Cact_W the group generated by the τ_J, $J \in \mathscr{J}$ subject to all relations of Theorem 1.5. Indeed, this definition coincides with that in [29, (1.1)] if W is finite because $\tau_J = \tau_{J'}\tau_{J''}$ for any J as in Theorem 1.5(b). By definition, the assignments $\tau_J \mapsto \sigma_V^J$, $J \in \mathscr{J}$ define a representation of Cact_W on V. In view of (1.2) we obtain the following immediate corollary of Theorem 1.5 (see Sect. 4 for the notation)

Corollary 1.6 *The group* Cact_W *acts on the category* $\mathcal{O}_q^{int}(\mathfrak{g})$ *via* $\tau_J \mapsto \sigma_\bullet^J$, $J \in \mathscr{J}$.

The study of cactus groups began with $\mathsf{Cact}_n := \mathsf{Cact}_{S_n}$ which appeared, to name but a few, in [13, 14, 16, 33, 35] in connection with the study of moduli spaces of rational curves with $n+1$ marked points and their applications in mathematical physics. It is easy to see that Cact_n is generated by involutions $\tau_{i,j} = \tau_{\{i,\dots,j-1\}}$, $1 \le i < j \le n$ subject to the relations

$$\tau_{i,j}\tau_{k,l} = \tau_{k,l}\tau_{i,j}, \qquad i < j < k < l$$

$$\tau_{i,l}\tau_{j,k} = \tau_{i+l-k,i+l-j}\tau_{i,l}, \qquad i \le j < k \le l.$$

Categorical actions of Cact_n on n-fold tensor products in symmetric coboundary categories (first introduced in [15]) were studied in [20, 34] and also implicitly in [9] where the braided structure on the category $\mathscr{O}_q^{int}(\mathfrak{g})$ was converted into a symmetric coboundary structure for any complex reductive Lie algebra $\mathfrak{g}$ (for non-abelian examples of coboundary categories, see the discussion after Theorem 1.8). It would be interesting to compare these actions of Cact_n with the one given by Corollary 1.6. We expect that they are connected in some cases via the celebrated Howe duality (see, e.g., the forthcoming paper [4]). In view of Corollary 1.6 it is natural to seek other categorical representations of Cact_W for all Coxeter groups W.

Conjecture 1.2 suggests that our representation of Cact_W on $\mathscr{O}_q^{int}(\mathfrak{g})$ is not faithful. Namely, in view of the discussion after the conjecture, we expect that the kernel $\mathsf{K}_\mathfrak{g}$ of this representation of Cact_W contains all elements $(\sigma^i\sigma^j)^{m_{ij}}$, $i \neq j \in I$. For example, if $\mathfrak{g} = \mathfrak{sl}_3$, then $\tau_{1,2}\tau_{1,3} = \tau_{1,3}\tau_{2,3}$ and so Cact_W is freely generated by involutions $\tau_{1,2}$ and $\tau_{1,3}$. It is easy to see that $\tau_{1,2} \notin \mathsf{K}_{\mathfrak{sl}_3}$, while $\tau_{1,3} \notin \mathsf{K}_{\mathfrak{sl}_3}$ by Remark 7.14. Thus, we expect that $\mathsf{K}_{\mathfrak{sl}_3} = \{(\tau_{1,2}\tau_{1,3})^{6n} : n \in \mathbb{Z}\}$.

Therefore, we can pose the following

Problem 1.7 Find the kernel $\mathsf{K}_\mathfrak{g}$ of the representation of Cact_W on $\mathscr{O}_q^{int}(\mathfrak{g})$.

To outline an approach to Problem 1.7, denote Φ_V, $V \in \mathscr{O}_q^{int}(\mathfrak{g})$ the subgroup of $GL_\Bbbk(V)$ generated by the σ_V^J, $J \in \mathscr{J}$. Then clearly $\mathsf{K}_\mathfrak{g}$ is the intersection of kernels of canonical homomorphisms $\mathsf{Cact}_W \to \Phi_V$ over all $V \in \mathscr{O}_q^{int}(\mathfrak{g})$. We show (Proposition 4.18) that $\Phi_V \cong \Phi_{\underline{V}}$ where $\underline{V} = \bigoplus_{\lambda\in P^+ : \mathrm{Hom}_{U_q(\mathfrak{g})}(V_\lambda, V)\neq 0} V_\lambda$. In particular, $\mathsf{Cact}_W/\mathsf{K}_\mathfrak{g}$ is isomorphic to $\Phi_{\mathcal{C}_q(\mathfrak{g})}$ where $\mathcal{C}_q(\mathfrak{g}) = \bigoplus_{\lambda\in P^+} V_\lambda$ is the Gelfand–Kirillov model for $\mathscr{O}_q^{int}(\mathfrak{g})$; in fact, it has a structure of an associative algebra (see Sect. 6). Thus, in view of the above we expect that $\mathsf{Cact}_3/\mathsf{K}_{\mathfrak{sl}_3}$ is isomorphic to the dihedral group of order 12. However, it is likely that $\Phi_{\mathcal{C}_q(\mathfrak{g})}$ is infinite for simple $\mathfrak{g}$ different from $\mathfrak{sl}_2$ and $\mathfrak{sl}_3$.

It turns out that the action of Cact_W on $\mathscr{O}_q^{int}(\mathfrak{g})$ descends to a permutation representation on any crystal basis of any object V (see Sect. 2.5 for definitions and notation). Thus, we obtain the following refinement of [19, Theorem 5.19].

Theorem 1.8 *Let $V \in \mathscr{O}_q^{int}(\mathfrak{g})$. Then for any lower or upper crystal basis (L, B) of V at $q = 0$ the group Φ_V preserves L and acts on B by permutations.*

We prove Theorem 1.8 in Sect. 5 by means of what we call **c**-crystal bases, which allow one to treat lower and upper crystal bases uniformly. Taking into account that B is graded by the weight lattice of $\mathfrak{g}$, all weights occur in a crystal basis of $\mathcal{C}_q(\mathfrak{g})$ and that W acts faithfully on the weight lattice, we obtain an immediate

Corollary 1.9 *The assignments $\sigma^J \mapsto w_\circ^J$, $J \in \mathscr{J}$ define a surjective homomorphism* $\mathsf{Cact}_W/\mathsf{K}_\mathfrak{g} \to W$ *which refines the natural epimorphism* $\mathsf{Cact}_W \to W$ *from* [29].

Analogously to the notion of the pure braid group, one calls the kernel of the natural homomorphism $\mathsf{Cact}_W \to W$ the pure cactus group (this term was used for Cact_n in e.g., [16, 33, 35]). Thus, Corollary 1.9 asserts that $\mathsf{K}_\mathfrak{g}$ is pure.

The involution σ_V^I was first defined in [8] for $\mathfrak{g} = \mathfrak{gl}_n$ and simple polynomial representations V_λ and explicitly computed on the corresponding crystal in [28]. In fact, it coincides with the famous Schützenberger involution (see Remark 4.11). Following a suggestion of the first author and [28], an action of Cact_n on the category of crystal bases was constructed in [20], thus turning it into a symmetric coboundary category.

We expect that to solve Problem 1.7 it suffices to find the kernels of permutation representations of Cact_W on all B.

Since $\mathsf{W}(V)$ is naturally a subgroup of Φ_V, its action on V induces an action on B by permutations which coincides with Kashiwara's crystal Weyl group (see Remark 5.7).

In case when $\mathfrak{g}$ is reductive we can refine Theorem 1.8 as follows.

Theorem 1.10 *Let $\mathfrak{g}$ be a reductive Lie algebra and let $V \in \mathcal{O}_q^{int}(\mathfrak{g})$. Then for any crystal basis (L, B) of V the involution σ_V^I preserves the corresponding upper global crystal basis $\mathbf{B}_V$ of V.*

An analogous result for $J \subsetneq I$ is weaker. We prove (Proposition 5.8) that the image of any element of $\mathbf{B}_V$ under σ^J, $J \in \mathscr{J}$ is a $\bar{\cdot}$-invariant element of V where $\bar{\cdot}$ is the anti-linear involution fixing $\mathbf{B}_V$. However, as explained in Remark 7.18, σ^J does not need to preserve $\mathbf{B}_V$ if $J \subsetneq I$. For example, if V is the 27-dimensional simple module $V_{2\rho}$ for $\mathfrak{g} = \mathfrak{sl}_3$, then the σ_V^i, $i = 1, 2$ do not preserve the canonical basis of V.

An analogue of Theorem 1.10 for a simple V and its lower global crystal basis was deduced from [30, Proposition 21.1.2] in [20, Theorem 5].

We prove Theorem 1.10 in Sect. 6.5. A central role in our argument is played by the following surprising property of σ^I on the aforementioned quantum Gelfand–Kirillov model $\mathcal{C}_q(\mathfrak{g})$ of $\mathcal{O}_q^{int}(\mathfrak{g})$.

Theorem 1.11 (Theorem 6.21) *For $\mathfrak{g}$ reductive finite dimensional, $\sigma^I_{\mathcal{C}_q(\mathfrak{g})}$ is an algebra anti-involution on $\mathcal{C}_q(\mathfrak{g})$.*

Our proofs of Theorems 1.10 and 1.11 rely in a crucial way on the properties of a remarkable quantum twist defined in [27]. We do not expect an analogous result for $J \subsetneq I$; for example, for $\mathfrak{g} = \mathfrak{sl}_3$, the σ^i, $i \in \{1, 2\}$ are not algebra anti-automorphisms of $\mathcal{C}_q(\mathfrak{g})$.

In view of Theorem 1.8 we can refine Conjecture 1.2 for every $V \in \mathcal{O}_q^{int}(\mathfrak{g})$ with $\underline{V} = \mathcal{C}_q(\mathfrak{g})$ as follows. We expect that in the notation of Theorem 1.8 the group Φ_V acts on B faithfully. Morally, this means that each element of Φ_V is semisimple in $GL_\Bbbk(V)$.

Similarly to Remark 1.3, our constructions, results, and conjectures make sense if one replaces $U_q(\mathfrak{g})$ by $U(\mathfrak{g})$ for any (symmetrizable or not) Kac–Moody algebra $\mathfrak{g}$. Some results (for example, Theorem 1.8) should be possible to rescue even when W

is not crystallographic (and so $\mathfrak{g}$ does not exist) with the aid of theory of continuous crystals initiated by A. Joseph in [21].

2 Preliminaries

2.1 Coxeter Groups

Let I be a finite set. Let W be a Coxeter group with Coxeter generators s_i, $i \in I$ subject to the relations $(s_i s_j)^{m_{ij}} = 1$ where $m_{ii} = 1$, $m_{ij} = m_{ji}$, and $m_{ij} \in \{0\} \cup \mathbb{Z}_{\geq 2}$ for $i \neq j \in I$. Let $\ell : W \to \mathbb{Z}_{\geq 0}$ be the Coxeter length function, that is, $\ell(w)$ is the minimal length of a presentation of w as a product of the s_i, $i \in I$. We say that $\mathbf{i} = (i_1, \ldots, i_r) \in I^r$ is reduced if $\ell(s_{i_1} \cdots s_{i_r}) = r$ and denote by $R(w)$ the set of reduced words for w, that is, $R(w) = \{(i_1, \ldots, i_{\ell(w)}) \in I^{\ell(w)} : w = s_{i_1} \cdots s_{i_{\ell(w)}}\}$.

Given $J \subset I$ we denote by W_J the subgroup of W generated by the s_i, $i \in J$. We will need the following standard fact (see [11, IV.1.8, Théorème 2]).

Lemma 2.1 *For any $J, J' \subset I$*

(a) $W_J \cap W_{J'} = W_{J \cap J'}$;
(b) $W_J \subset W_{J'}$ *if and only if* $J \subset J'$.

Let $\mathscr{J} = \{J \subset I, : |W_J| < \infty\}$. If $J \in \mathscr{J}$, we denote by $w_\circ^J$ the unique longest element of W_J; thus, $\ell(s_j w_\circ^J) < \ell(w_\circ^J)$ for all $j \in J$. If $I \in \mathscr{J}$, we abbreviate $w_\circ = w_\circ^I$. Given $J \in \mathscr{J}$ and $j \in J$, there exists a unique $j^\star \in J$ such that $s_{j^\star} = w_\circ^J s_j w_\circ^J$; the assignments $j \mapsto j^\star$ define an involution $^\star : J \to J$.

Given $J \subset I$, we set $J^\perp = \{i \in I \setminus J : m_{ij} = 2, \forall j \in J\} = \{i \in I \setminus J : s_i s_j = s_j s_i, \forall j \in J\}$. We say that $J, J' \subset I$ are *orthogonal* if $J \cap J' = \emptyset$ and $J' \subset J^\perp$ (whence $J \subset J'^\perp$).

Define a relation $\sim$ on I by $i \sim j$ if $i = j$ or $m_{ij} > 2$. Then the transitive closure of this relation is an equivalence on I which we still denote by $\sim$. In particular, if $i \sim i'$, then there exists a sequence (called *admissible*) $(i_0, \ldots, i_d) \in I^{d+1}$ with $i_0 = i$, $i_d = i'$ and $m_{i_{r-1}, i_r} > 2$, $1 \leq r \leq d$. Define $\operatorname{dist}(i, i')$ to be the minimal length of an admissible sequence beginning with i and ending with i'. Clearly, this defines a metric on I.

Define a topology on I by declaring that the fundamental neighborhood of each $i \in I$ is its equivalence class with respect to $\sim$. In particular, each open set is closed and vice versa and is a union of equivalence classes. For $J \subset I$ we denote by $\operatorname{cl}(J)$ its closure in that topology, that is, the union of equivalence classes of elements of J. Denote $\partial(J)$ the boundary of J, that is, the complement of J in $\operatorname{cl}(J)$. The following is immediate.

Lemma 2.2 *Let $J \subset I$. Then $I = J \cup J^\perp$ if and only if J is closed in the above topology.*

The following is a reformulation of a well-known fact [11, IV.1.9, Proposition 2]

Lemma 2.3 *Let $J \subset I$ be a closed subset. Then W is the internal direct product of W_J and $W_{J^\perp}$.*

Given a group G acting on a set X, denote by $\mathsf{K}_X(G)$ the kernel of the natural homomorphism of groups $G \to \mathrm{Bij}(X)$ induced by the action. By definition, the action of G on X is faithful if and only if $\mathsf{K}_X(G) = \{1\}$.

The following is the main result of Sect. 2.1 (which is probably known although we could not find it in the literature).

Theorem 2.4 *We have $\mathsf{K}_J := \mathsf{K}_{W/W_J}(W) = W_{I\setminus \mathrm{cl}(I\setminus J)}$ for any $J \subset I$. In particular, if I is connected and $J \subsetneq I$, then W acts faithfully on W/W_J.*

Proof The following is immediate.

Lemma 2.5 *Let G be a group and H be a subgroup of G. Then $\mathsf{K}_{G/H}(G) = \{k \in H : g^{-1}kg \in H, \forall\, g \in G\}$ is a subgroup of H.*

The following Lemmata are apparently well-known. We provide their proof for the reader's convenience.

Lemma 2.6 *Let $w \in W$ and let $J \subset I$ be such that $\ell(s_j w) = \ell(w) - 1$ for all $j \in J$. Then W_J is finite and $w = w_\circ^J w'$ for some $w' \in W$ with $\ell(w) = \ell(w') + \ell(w_\circ^J)$.*

Proof By [11, Ch. IV, Ex. 3], every $u \in W$ can be written uniquely as $[u]_J \cdot {}^J[u]$ where $[u]_J \in W_J$, ${}^J[u] \in {}^JW = \{x \in W : \ell(s_j x) > \ell(x), \forall\, j \in J\}$ and $\ell(u) = \ell([u]_J) + \ell({}^J[u])$. The uniqueness of such a presentation implies that $[s_j w]_J = s_j[w]_J$ and ${}^J[s_j w] = {}^J[w]$ for all $j \in J$ and so that $\ell(s_j[w]_J) < \ell([w]_J)$ for all $j \in J$. This implies that W_J is finite and $[w]_J$ is its longest element $w_\circ^J$. The assertion follows with $w' = {}^J[w]$. □

Lemma 2.7 *For $i \in I$ and $u \in W_{I\setminus\{i\}}$ the following are equivalent.*

(a) $u \in W_{\{i\}^\perp}$ *(in particular, $s_i u = u s_i$);*
(b) $s_i u s_i \in W_{I\setminus\{i\}}$.

Proof The implication (a) $\Longrightarrow$ (b) is obvious. To prove the opposite implication, note that the assumption in (b) implies that $s_i u = u' s_i$ for some $u' \in W_{I\setminus\{i\}}$. Then $\ell(s_i u) = \ell(u) + 1$ and $\ell(u' s_i) = \ell(u') + 1$ whence $\ell(u) = \ell(u')$. We prove the assertion

$$s_i u = u' s_i \implies u = u' \in W_{\{i\}^\perp} \tag{2.1}$$

by induction on $\ell(u) = \ell(u')$, the case $\ell(u) = \ell(u') = 0$ being obvious. If $\ell(u) = \ell(u') > 0$, then there exists $j \neq i \in I$ such that $\ell(s_j u') < \ell(u')$. Let $w = s_i u$. Then $\ell(s_j w) < \ell(w)$ and $\ell(s_i w) < \ell(w)$. Applying Lemma 2.6 to w and $J = \{i, j\}$ we conclude that $w = \underbrace{(s_i s_j \cdots)}_{m_{ij}} u'$ with $\ell(w) = m_{ij} + \ell(u')$ and so $u = \underbrace{(s_j s_i \cdots)}_{m_{ij}-1} u'$ with $\ell(u) = m_{ij} - 1 + \ell(u')$. Since $u \in W_{I\setminus\{i\}}$, a reduced word for u cannot contain i, yet for any $(i_1, \ldots, i_r) \in R(u')$, $(\underbrace{j, i, \ldots,}_{m_{ij}-1} i_1, \ldots, i_r) \in R(u)$. Thus, $m_{ij} = 2$ and so $j \in \{i\}^\perp$. Then $s_i(s_j u) = s_j s_i u = (s_j u')s_i$. Thus, $s_j u$, $s_j u'$ satisfy (2.1) and $\ell(s_j u) < \ell(u)$. Then the induction hypothesis implies that $s_j u = s_j u' \in W_{\{i\}^\perp}$ and hence $u = u' \in W_{\{i\}^\perp}$. □

Lemma 2.8 *Let i, i' be connected in I and let $(i = i_0, i_1, \ldots, i_d = i') \in I^{d+1}$ be an admissible sequence with $d = \operatorname{dist}(i, i')$. Suppose that $w \in W_{I\setminus\{i\}}$ and $s_{i_0} \cdots s_{i_d} w s_{i_d} \cdots s_{i_0} \in W_{I\setminus\{i\}}$. Then $w \in W_{\{i_0,\ldots,i_d\}^\perp}$.*

Proof The argument is by induction on d. The case $d = 0$ (that is, $i = i'$) is established in Lemma 2.7. Suppose that $d > 0$. Let $u = s_{i_1} \cdots s_{i_d} w s_{i_d} \cdots s_{i_1}$. By Lemma 2.7, $u \in W_{\{i\}^\perp}$. Since $m_{i,i_1} > 2$, $i_1 \notin \{i\}^\perp$. Thus, $u \in W_{I'\setminus\{i_1\}}$ where $I' = I \setminus \{i\}$ and $\operatorname{dist}(i_1, i') = d - 1$. By the induction hypothesis, $u \in W_{\{i_1,\ldots,i_d\}^\perp}$ and in particular $u = w$. But then $w \in W_{\{i\}^\perp} \cap W_{\{i_1,\ldots,i_d\}^\perp} = W_{\{i\}^\perp \cap \{i_1,\ldots,i_d\}^\perp} = W_{\{i_0,\ldots,i_d\}^\perp}$ where we used Lemma 2.1(a) and the observation that $J^\perp \cap J'^\perp = (J \cup J')^\perp$. □

By Lemma 2.5, $\mathsf{K}_J = \{w \in W : uwu^{-1} \in W_J, \forall u \in W\}$ and is a subgroup of W_J. Suppose that $w \in \mathsf{K}_J$; in particular, $w \in W_J$. Furthermore, using Lemma 2.7 with $u = w$ and $i \in I \setminus J$, we conclude that $w \in \bigcap_{i \in I\setminus J} W_{\{i\}^\perp} = W_{(I\setminus J)^\perp}$. Let $i' \in \partial(I \setminus J)$. By definition, there exists $i \in I \setminus J$ and an admissible sequence $(i_0, \ldots, i_d)$ with $d = \operatorname{dist}(i, i')$, $i_0 = i$ and $i_d = i'$. Since $uwu^{-1} \in W_J$ with $u = s_{i_0} \cdots s_{i_d}$, it follows from Lemma 2.8 that $w \in W_{\{i_0,\ldots,i_d\}^\perp} \subset W_{\{i'\}^\perp}$. Thus, $w \in W_{(I\setminus J)^\perp \cap \partial(I\setminus J)^\perp} = W_{(\operatorname{cl}(I\setminus J))^\perp} = W_{I\setminus \operatorname{cl}(I\setminus J)}$. We proved that $\mathsf{K}_J \subset W_{J_0}$ where $J_0 = I \setminus \operatorname{cl}(I \setminus J)$.

To complete the proof of Theorem 2.4 we need the following.

Lemma 2.9 *Let $J' \subset J$ which is closed in I. Then $W_{J'} \subset \mathsf{K}_J$.*

Proof Since J' is closed, $W_J = W_{J'} \times W_{J\setminus J'}$ and $W = W_{J'} \times W_{I\setminus J'}$ by Lemma 2.3. Then $W/W_J = W_{I\setminus J'}/W_{J\setminus J'}$. Since $W_{J'}$ acts by left multiplication in the first factor, this implies that $W_{J'}$ acts trivially on W/W_J. □

Applying Lemma 2.9 with $J' = J_0 = I \setminus \operatorname{cl}(I \setminus J)$ we conclude that $W_{J_0} \subset \mathsf{K}_J$. Thus, $\mathsf{K}_J = W_{J_0}$. This completes the proof of Theorem 2.4. □

2.2 Cartan Data and Weyl Group

In this section we mostly follow [23]. Let $A = (a_{ij})_{i,j\in I}$ be a symmetrizable generalized Cartan matrix, that is $a_{ii} = 2$, $i \in I$, $-a_{ij} \in \mathbb{Z}_{\geq 0}$ and $a_{ij} = 0 \implies a_{ji} = 0$, $i \neq j$ and $d_i a_{ij} = d_j a_{ji}$ for some $\mathbf{d} = (d_i)_{i\in I} \in \mathbb{Z}^I_{>0}$. We fix the following data:

- a finite dimensional complex vector space $\mathfrak{h}$;
- linearly independent subsets $\{\alpha_i\}_{i\in I}$ of $\mathfrak{h}^*$ and $\{\alpha_i^\vee\}_{i\in I}$ of $\mathfrak{h}$;

- a symmetric non-degenerate bilinear form $(\cdot,\cdot)$ on $\mathfrak{h}^*$, and
- a lattice $P \subset \mathfrak{h}^*$ of rank $\dim\mathfrak{h}^*$

such that

1° $\alpha_j(\alpha_i^\vee) = a_{ij}$, $i, j \in I$;
2° $(\alpha_i, \alpha_i) \in 2\mathbb{Z}_{>0}$;
3° $\lambda(\alpha_i^\vee) = 2(\lambda, \alpha_i)/(\alpha_i, \alpha_i)$ for all $\lambda \in \mathfrak{h}^*$;
4° $\alpha_i \in P$ for all $i \in I$;
5° $\lambda(\alpha_i^\vee) \in \mathbb{Z}$ for all $\lambda \in P$;
6° $(P, P) \subset \frac{1}{d}\mathbb{Z}$ for some $d \in \mathbb{Z}_{>0}$.

These assumptions imply, in particular, that $\dim\mathfrak{h} \geq 2|I| - \operatorname{rank} A$.

Denote by Q (respectively, Q^+) the subgroup (respectively, the submonoid) of P generated by the α_i. Let $P^+ = \{\lambda \in P : \lambda(\alpha_i^\vee) \in \mathbb{Z}_{\geq 0}, \ \forall i \in I\}$.

Define $\omega_i \in \mathfrak{h}^*$, $i \in I$, by $\omega_i(\alpha_j^\vee) = \delta_{i,j}$, $j \in J$ and $\omega_i(h) = 0$ for all $h \in \bigcap_{i\in I} \ker\alpha_i$. We will assume that $\omega_i \in P$, $i \in I$ and denote by P_{int} (respectively, P_{int}^+) the subgroup (respectively, the submonoid) of P generated by the ω_i, $i \in I$. Given any $J \subset I$, denote $\rho_J = \sum_{j\in J}\omega_j \in P$; we abbreviate $\rho_I = \rho$.

Let W be the Weyl group associated with the matrix A, that is, the Coxeter group with $m_{ij} = 2$ if $a_{ij} = 0$, $m_{ij} = 3$ if $a_{ij}a_{ji} = 1$, $m_{ij} = 4$ if $a_{ij}a_{ji} = 2$, $m_{ij} = 6$ if $a_{ij}a_{ji} = 3$ and $m_{ij} = 0$ if $a_{ij}a_{ji} > 3$. It is well-known that W is finite if and only if A is positive definite. It should be noted that in that case $\alpha_i \in P_{int}$ for all $i \in I$. The group W acts on $\mathfrak{h}$ (respectively, on $\mathfrak{h}^*$) by $s_i h = h - \alpha_i(h)\alpha_i^\vee$ (respectively, $s_i\lambda = \lambda - \lambda(\alpha_i^\vee)\alpha_i$), $h \in \mathfrak{h}$, $\lambda \in \mathfrak{h}^*$ and $i \in I$. Then we have $(w\lambda)(h) = \lambda(w^{-1}h)$ for all $w \in W$, $h \in \mathfrak{h}$ and $\lambda \in \mathfrak{h}^*$. Clearly, $W(P) = P$ and $P = P_{int} \oplus P^W$ where $P^W = \{\lambda \in P : w\lambda = \lambda, \ \forall w \in W\} = \{\lambda \in P : \lambda(\alpha_i^\vee) = 0, \ \forall i \in I\}$.

Given $J \subset I$ we define a linear map $\rho_J^\vee : \mathfrak{h}^* \to \mathbb{C}$ by $\rho_J^\vee(\alpha_i) = 1$, $i \in J$ and $\rho_J^\vee(\lambda) = 0$ if $(\lambda, \alpha_i) = 0$ for all $i \in J$. As before, we abbreviate $\rho_I^\vee = \rho^\vee$. If $J \in \mathscr{J}$, then it can be shown that $\rho_J^\vee(\lambda)$ is equal to $\frac{1}{2}\lambda(\sum_{h\in R_J^\vee} h)$ where $R_J^\vee = \{h \in \mathfrak{h} : h \in (\bigcup_{i\in J} W_J\alpha_i^\vee) \cap \sum_{i\in J}\mathbb{Z}_{\geq 0}\alpha_i^\vee\}$ is the set of positive co-roots of W_J. In particular, this implies that $\rho_J^\vee(P) \subset \frac{1}{2}\mathbb{Z}$.

If $J \in \mathscr{J}$, then for each $j \in J$ we have $w_\circ^J(\alpha_j) = -\alpha_{j^\star}$.

Given $\lambda \in P^+$, denote $J_\lambda = \{i \in I : \lambda(\alpha_i^\vee) = 0\} = \{i \in I : s_i\lambda = \lambda\}$. It is well-known that $\operatorname{Stab}_W \lambda = W_{J_\lambda}$ for $\lambda \in P^+$.

2.3 *Quantum Groups*

We associate with the datum $(A, \mathfrak{h}, \{\alpha_i\}_{i\in I}, \{\alpha_i^\vee\}_{i\in I})$ a complex Lie algebra $\mathfrak{g}$ generated by the e_i, f_i, $i \in I$ and $h \in \mathfrak{h}$ subject to the relations

$$[h, h'] = 0, \quad [h, e_i] = \alpha_i(h)e_i, \quad [h, f_i] = -\alpha_i(h)f_i, \quad [e_i, f_j] = \delta_{i,j}\alpha_i^\vee, \quad h, h' \in \mathfrak{h},\ i, j \in I$$

$$(\operatorname{ad} e_i)^{1-a_{ij}}(e_j) = 0 = (\operatorname{ad} f_i)^{1-a_{ij}}(f_j), \qquad i \neq j.$$

If A is positive definite, then $\mathfrak{g}$ is a reductive finite dimensional Lie algebra. For $J \subset I$ we denote by $\mathfrak{g}^J$ the subalgebra of $\mathfrak{g}$ generated by the e_i, f_i, $i \in J$ and $\mathfrak{h}$. It can also be regarded as the Lie algebra corresponding to the datum $(A|_{J\times J}, \mathfrak{h}, \{\alpha_i\}_{i\in J}, \{\alpha_i^\vee\}_{i\in J})$. In particular, if $J \in \mathscr{J}$, then $\mathfrak{g}^J$ is a reductive finite dimensional Lie algebra.

Let $\Bbbk$ be any field of characteristic zero containing $q^{\frac{1}{2d}}$ which is purely transcendental over $\mathbb{Q}$. Given any $v \in \Bbbk^\times$ with $v^2 \neq 1$ define

$$(n)_v = \frac{v^n - v^{-n}}{v - v^{-1}}, \qquad (n)_v! = \prod_{s=1}^{n} (s)_v, \qquad \binom{n}{k}_v = \prod_{s=1}^{k} \frac{(n-s+1)_v}{(s)_v}.$$

Let $q_i = q^{\frac{1}{2}(\alpha_i,\alpha_i)}$. Henceforth, given any associative algebra $\mathcal{A}$ over $\Bbbk$ and $X_i \in \mathcal{A}$, $i \in I$ denote $X_i^{(n)} := X_i^n/(n)_{q_i}!$ We will always use the convention that $X_i^{(n)} = 0$ if $n < 0$.

Define the Drinfeld-Jimbo quantum group $U_q(\mathfrak{g})$ corresponding to $\mathfrak{g}$ as the associative algebra over $\Bbbk$ with generators K_λ, $\lambda \in \frac{1}{2}P$ and E_i, F_i, $i \in I$ subject to the relations

$$K_\lambda E_i = q^{(\lambda,\alpha_i)} E_i K_\lambda, \; K_\lambda F_i = q^{-(\lambda,\alpha_i)} F_i K_\lambda, \; [E_i, F_j] = \delta_{ij} \frac{K_{\alpha_i} - K_{-\alpha_i}}{q_i - q_i^{-1}}, \quad \lambda \in \tfrac{1}{2}P, \; i, j \in I$$

$$\sum_{r+s=1-a_{ij}} (-1)^r E_i^{(r)} E_j E_i^{(s)} = 0 = \sum_{r+s=1-a_{ij}} (-1)^r F_i^{(r)} F_j F_i^{(s)} \quad i \neq j \in I.$$

This is a Hopf algebra with the "balanced" comultiplication

$$\Delta(E_i) = E_i \otimes K_{\frac{1}{2}\alpha_i} + K_{-\frac{1}{2}\alpha_i} \otimes E_i, \; \Delta(F_i) = F_i \otimes K_{\frac{1}{2}\alpha_i} + K_{-\frac{1}{2}\alpha_i} \otimes F_i, \; i \in I, \tag{2.2}$$

while $\Delta(K_\lambda) = K_\lambda \otimes K_\lambda$, $\lambda \in \frac{1}{2}P$. Denote by $U_q^+(\mathfrak{g})$ (respectively, $U_q^-(\mathfrak{g})$) the subalgebra of $U_q(\mathfrak{g})$ generated by the E_i (respectively, the F_i), $i \in I$. Then $U_q^\pm(\mathfrak{g})$ is graded by $\pm Q^+$ with $\deg E_i = \alpha_i = -\deg F_i$. Given $\nu \in Q^+$, denote by $U_q^\pm(\mathfrak{g})(\pm\nu)$ the subspace of homogeneous elements of $U_q^\pm(\mathfrak{g})$ of degree $\pm\nu$.

Given $J \subset I$ we denote by $U_q(\mathfrak{g}^J)$ the subalgebra of $U_q(\mathfrak{g})$ generated by the E_j, F_j, $j \in J$ and K_λ, $\lambda \in \frac{1}{2}P$ and set $U_q^\pm(\mathfrak{g}^J) = U_q^\pm(\mathfrak{g}) \cap U_q(\mathfrak{g}^J)$.

If $J \in \mathscr{J}$, then the algebra $U_q(\mathfrak{g}^J)$ admits an automorphism θ_J defined by $\theta_J(E_i) = F_{i^\star}$, $\theta_J(F_i) = E_{i^\star}$ and $\theta_J(K_\lambda) = K_{w_\circ^J\lambda}$, $\lambda \in \frac{1}{2}P$. If $I \in \mathscr{J}$, we abbreviate $\theta_I = \theta$.

2.4 Integrable Modules

We say that a $U_q(\mathfrak{g})$-module M is *integrable* if $M = \bigoplus_{\beta\in P} M(\beta)$ where $M(\beta) = \{m \in M : K_\lambda(m) = q^{(\lambda,\beta)}m, \forall\lambda \in \frac{1}{2}P\}$ and the E_i, F_i, $i \in I$ act locally nilpotently on M. Given any $m \in M$ we can write uniquely

$$m = \bigoplus_{\beta\in P} m(\beta) \tag{2.3}$$

where $m(\beta) \in M(\beta)$ and $m(\beta) = 0$ for all but finitely many $\beta \in P$. Denote $\operatorname{supp} m = \{\beta \in P : m(\beta) \neq 0\}$. By definition, if $m \in M(\beta)$ and $u_\pm \in U_q^\pm(\mathfrak{g})(\pm\nu)$, $\nu \in \pm Q^+$, then $u_\pm(m) \in M(\beta\pm\nu)$. We say that $m \in M(\beta)$, $\beta \in P$ is homogeneous of weight β and call $M(\beta)$ a weight subspace of M.

Definition 2.10 The category $\mathscr{O}_q^{int}(\mathfrak{g})$ is the full subcategory of the category of $U_q(\mathfrak{g})$-modules whose objects are integrable $U_q(\mathfrak{g})$-modules M with the following property: given $m \in M$, there exists $N(m) \geq 0$ such that $U_q^+(\mathfrak{g})(\nu)(m) = 0$ for all $\nu \in Q^+$ with $\rho^\vee(\nu) \geq N(m)$.

Given $V \in \mathscr{O}_q^{int}(\mathfrak{g})$, let $V_+ = \bigcap_{i\in I} \ker E_i$ where the E_i are regarded as linear endomorphisms of V. For any subset S of an object V in $\mathscr{O}_q^{int}(\mathfrak{g})$ we denote $S(\beta) = S \cap V(\beta)$ and $S_+ = S \cap V_+$.

It is well-known (see, e.g., [30, Theorem 6.2.2]) that $\mathscr{O}_q^{int}(\mathfrak{g})$ is semisimple and its simple objects are simple highest weight modules V_λ, $\lambda \in P^+$ with $(V_\lambda)_+ = V_\lambda(\lambda)$ one-dimensional. Furthermore, every $V \in \mathscr{O}_q^{int}(\mathfrak{g})$ is generated by V_+ as a $U_q(\mathfrak{g})$-module and $V_+(\lambda) \neq 0$ implies that $\lambda \in P^+$. Given $\lambda \in P^+$ and $V \in \mathscr{O}_q^{int}(\mathfrak{g})$ denote $\mathcal{I}_\lambda(V)$ the λ-isotypical component of V as a $U_q(\mathfrak{g})$-module. Thus, every simple submodule (and hence a direct summand) of $\mathcal{I}_\lambda(V)$ is isomorphic to V_λ and $\mathcal{I}_\lambda(V)_+ = V_+(\lambda)$. Furthermore, for any $v \in V_+$ we have the following equality of $U_q(\mathfrak{g})$-submodules of V

$$U_q(\mathfrak{g})(v) = \sum_{\lambda\in\operatorname{supp} v} U_q(\mathfrak{g})(v(\lambda)), \tag{2.4}$$

where the sum is direct and each summand is simple and isomorphic to V_λ.

It is immediate from the definition that every object $\mathscr{O}_q^{int}(\mathfrak{g})$ can be regarded as an object in $\mathscr{O}_q^{int}(\mathfrak{g}^J)$, $J \subset I$. Denote $P_J^+ = \{\mu \in P : \mu(\alpha_j^\vee) \in \mathbb{Z}_{\geq 0}, \forall j \in J\}$. Given $V \in \mathscr{O}_q^{int}(\mathfrak{g})$ and $\lambda_J \in P_J^+$ denote by $\mathcal{I}_{\lambda_J}^J(V)$ the λ_J-isotypical component of V as a $U_q(\mathfrak{g}^J)$-module. Clearly, $\mathcal{I}_{\lambda_J}^J(\mathcal{I}_\lambda(V)) = \mathcal{I}_\lambda(V) \cap \mathcal{I}_{\lambda_J}^J(V)$ for any $V \in \mathscr{O}_q^{int}(\mathfrak{g})$, $\lambda \in P^+$, $\lambda_J \in P_J^+$. We denote $V_+^J = \bigcap_{j\in J} \ker E_j \subset V_+$. Then $\mathcal{I}_{\lambda_J}^J(V)$ is generated by $V_+^J(\lambda_J)$ as a $U_q(\mathfrak{g}^J)$-module.

2.5 Crystal Operators, Lattices and Bases

Here we recall some necessary facts from Kashiwara's theory of crystal bases. To treat lower and upper crystal operators and lattices uniformly, we find it convenient to interpolate between them using **c**-*crystal operators* and lattices (for other generalizations, see, e.g., [12, 18]).

The following fact is standard (for example, see [30, Lemma 16.1.4]).

Lemma 2.11 *Let $V \in \mathscr{O}_q^{int}(\mathfrak{g})$ and fix $i \in I$. Then*

$$V = \bigoplus_{0\le n\le l} F_i^n(\ker E_i \cap \ker(K_{\alpha_i} - q_i^l)) = \bigoplus_{0\le n\le l} E_i^n(\ker F_i \cap \ker(K_{\alpha_i} - q_i^{-l}))$$

Let $\mathbb{D} = \{(l,k,s) \in \mathbb{Z}_{\ge 0} \times \mathbb{Z}_{\ge 0} \times \mathbb{Z} : k-l \le s \le k \le l\}$. Fix a map $\mathbf{c} : \mathbb{D} \to \mathbb{Q}(z)^\times$ and denote its value at (l,k,s) by $\mathbf{c}_{l,k,s}$. We use the convention that $\mathbf{c}_{l,k,s} = 0$ whenever $(l,k,s) \in \mathbb{Z}^3 \setminus \mathbb{D}$. Using Lemma 2.11 we can define *generalized Kashiwara operators* $\tilde{e}_{i,s}^{\mathbf{c}} \in \operatorname{End}_{\Bbbk} V$, $s \in \mathbb{Z}$ by

$$\tilde{e}_{i,s}^{\mathbf{c}}(F_i^k(u)) = \mathbf{c}_{l,k,s}(q_i) F_i^{k-s}(u), \tag{2.5}$$

for every $u \in \ker E_i \cap \ker(K_{\alpha_i} - q_i^l)$, $0 \le k \le l$. Note that under these assumptions on u, $\tilde{e}_{i,s}^{\mathbf{c}}(F_i^{(k)}(u)) \ne 0$ if and only if $(l,k,s) \in \mathbb{D}$. Clearly, like lower or upper Kashiwara operators, the generalized ones commute with morphisms in $\mathscr{O}_q^{int}(\mathfrak{g})$.

Lemma 2.12 *Let $u \in \ker E_i \cap \ker(K_{\alpha_i} - q_i^l)$ and $u' \in \ker F_i \cap \ker(K_{\alpha_i} - q_i^{-l})$, $0 \le k \le l$. Then*

$$\tilde{e}_{i,s}^{\mathbf{c}}(F_i^{(k)}(u)) = \underline{\mathbf{c}}_{l,k,s}(q_i) F_i^{(k-s)}(u), \quad \tilde{e}_{i,s}^{\mathbf{c}}(E_i^{(k)}(u')) = \underline{\mathbf{c}}_{l,l-k,s}(q_i) E_i^{(k+s)}(u'), \tag{2.6}$$

for all $(l,k,s) \in \mathbb{D}$, where $\underline{\mathbf{c}}_{l,k',s'} = \mathbf{c}_{l,k',s'}(k'-s')_z!/(k')_z!$.

Proof The first identity in (2.6) is immediate from (2.5). To prove the second, note that $E_i^{(l+1)}(u') = 0$ and so $u = E_i^{(l)}(u') \in \ker E_i \cap \ker(K_{\alpha_i} - q_i^l)$. It follows from [30, §3.4.2] that $E_i^{(k)}(u') = F_i^{(l-k)}(u)$. Using the first identity in (2.6) we obtain

$$\tilde{e}_{i,s}^{\mathbf{c}}(E_i^{(k)}(u')) = \tilde{e}_{i,s}^{\mathbf{c}}(F_i^{(l-k)}(u)) = \underline{\mathbf{c}}_{l,l-k,s}(q_i) F_i^{(l-k-s)}(u) = \underline{\mathbf{c}}_{l,l-k,s}(q_i) E_i^{(k+s)}(u').$$

□

The following is immediate.

Lemma 2.13 *Given $\mathbf{c} : \mathbb{D} \to \mathbb{Q}(z)^\times$ we have*

(a) $\tilde{e}_{i,0}^{\mathbf{c}} = \mathrm{id}_V$ *for all $V \in \mathscr{O}_q^{int}(\mathfrak{g})$ if and only if $\mathbf{c}_{l,k,0} = 1$ for all $0 \le k \le l$;*

(b) $\tilde{e}^{\mathbf{c}}_{i,t} \circ \tilde{e}^{\mathbf{c}}_{i,s} = \tilde{e}^{\mathbf{c}}_{i,s+t}$ *for all* $s, t \in \mathbb{Z}$ *with* $st \geq 0$ *and for all* $V \in \mathscr{O}^{int}_q(\mathfrak{g})$ *if and only if* $\mathbf{c}_{l,k,s+t} = \mathbf{c}_{l,k,s}\mathbf{c}_{l,k-s,t}$ *for all* $0 \leq k \leq l$, $s, t \in \mathbb{Z}$, $st \geq 0$.

The following is easy to deduce from [25, §3.1]

Lemma 2.14 *Define* $\mathbf{c}^{low}, \mathbf{c}^{up} : \mathbb{D} \to \mathbb{Q}(z)^\times$ *by*

$$\mathbf{c}^{low}_{l,k,s} = \frac{(k)_z!}{(k-s)_z!}, \qquad \mathbf{c}^{up}_{l,k,s} = \frac{(l-k+s)_z!}{(l-k)_z!}, \qquad (l,k,s) \in \mathbb{D}. \tag{2.7}$$

We have

$$(\tilde{e}^{low}_i)^s = \tilde{e}^{\mathbf{c}^{low}}_{i,s}, \qquad (\tilde{e}^{up}_i)^s = \tilde{e}^{\mathbf{c}^{up}}_{i,s}, \qquad i \in I,\ s \in \mathbb{Z},$$

where $\tilde{e}^{low}_i$ *(respectively,* $\tilde{e}^{up}_i$*) are lower (respectively, upper) Kashiwara's operators as defined in* [25, §3.1].

Fix $\mathbf{c} : \mathbb{D} \to \mathbb{Q}(z)^\times$ and let $V \in \mathscr{O}^{int}_q(\mathfrak{g})$. Let $\mathbb{A}$ be the local subring of $\mathbb{Q}(q) \subset \Bbbk$ consisting of rational functions regular at 0. Generalizing well-known definitions of Kashiwara, we say that an $\mathbb{A}$-submodule L of V is a $\mathbf{c}$-crystal lattice if $V = \Bbbk \otimes_{\mathbb{A}} L$, $L = \bigoplus_{\beta \in P}(L \cap V(\beta))$, and $\tilde{e}^{\mathbf{c}}_{i,s}(L) \subset L$ for all $i \in I$, $s \in \mathbb{Z}$.

We will be mostly interested in a special class of crystal lattices which we refer to as monomial. We need the following notation. Given $v \in V$ set

$$\mathsf{M}^{\mathbf{c}}_J(v) = \{v\} \cup \bigcup_{k \in \mathbb{Z}_{>0}} \{\tilde{e}^{\mathbf{c}}_{i_1,m_1} \cdots \tilde{e}^{\mathbf{c}}_{i_k,m_k}(v) \,:\, (i_1, \dots, i_k) \in J^k,\ (m_1, \dots, m_k) \in \mathbb{Z}^k\}.$$

We abbreviate $\mathsf{M}^{\mathbf{c}}(v) = \mathsf{M}^{\mathbf{c}}_I(v)$. We call an $\mathbb{A}$-submodule L of V a $(\mathbf{c}, J)$*-monomial lattice* if

$$L = \sum_{v_+} \mathsf{M}^{\mathbf{c}}_J(v_+)$$

where the sum is over all $v_+ \in L \cap V^J_+(\lambda_J)$, $\lambda_J \in P^+$. Clearly, L inherits a weight decomposition from V and $\tilde{e}^{\mathbf{c}}_{j,a}(L) \subset L$ for all $j \in J$, $a \in \mathbb{Z}$. In particular, if L is a $(\mathbf{c}, I)$-monomial lattice and $\Bbbk \otimes_{\mathbb{A}} L = V$, then L is a $\mathbf{c}$-crystal lattice.

Denote $\tilde{L}$ the $\mathbb{Q}$-vector space L/qL. Given $\tilde{v} \in \tilde{L}$, denote

$$\tilde{\mathsf{M}}^{\mathbf{c}}_J(\tilde{v}) = \{\tilde{v}\} \cup \bigcup_{k \in \mathbb{Z}_{>0}} \{\tilde{e}^{\mathbf{c}}_{i_1,m_1} \cdots \tilde{e}^{\mathbf{c}}_{i_k,m_k}(\tilde{v}) \,:\, (i_1, \dots, i_k) \in J^k,$$
$$(m_1, \dots, m_k) \in \mathbb{Z}^k\} \subset \tilde{L}.$$

As before, we abbreviate $\tilde{\mathsf{M}}^{\mathbf{c}}(v) = \tilde{\mathsf{M}}^{\mathbf{c}}_I(v)$.

By [24, Theorem 3] and [25, Theorem 3.3.1], if $\mathbf{c} = \mathbf{c}^{low}$ or $\mathbf{c} = \mathbf{c}^{up}$, then every object in $\mathscr{O}^{int}_q(\mathfrak{g})$ admits a $\mathbf{c}$-crystal lattice. Moreover, in that case for any $\lambda \in P^+$, and any $v_\lambda \in V_\lambda(\lambda)$ the smallest $\mathbb{A}$-submodule of V_λ containing v_λ and invariant with respect to the $\tilde{e}^{\mathbf{c}}_{i,s}$, $i \in I$, $s \in \mathbb{Z}_{<0}$ is a $\mathbf{c}$-crystal lattice.

Let L be a $\mathbf{c}$-crystal lattice of $V \in \mathscr{O}_q^{int}(\mathfrak{g})$. Clearly, operators $\tilde{e}_{i,s}^{\mathbf{c}}$ commute with the action of q on L and thus factor through to $\mathbb{Q}$-linear operators on $\tilde{L} = L/qL$ denoted by the same symbols. Similarly to [24, 25], we say that (L, B), where B is a weight basis of $\tilde{L}$, is a $\mathbf{c}$-crystal basis of V at $q = 0$ if $\tilde{e}_{i,s}^{\mathbf{c}}(B) \subset B \cup \{0\}$, $i \in I$, $s \in \mathbb{Z}$. By [24, 25], every object in $\mathscr{O}_q^{int}(\mathfrak{g})$ admits a $\mathbf{c}$-crystal basis provided that $\mathbf{c} \in \{\mathbf{c}^{up}, \mathbf{c}^{low}\}$.

The following is well-known (cf. [24, 25]).

Lemma 2.15 *Let $V \in \mathscr{O}_q^{int}(\mathfrak{g})$, $\mathbf{c} \in \{\mathbf{c}^{low}, \mathbf{c}^{up}\}$ and let (L, B) be a $\mathbf{c}$-crystal basis at $q = 0$. Then for any $J \subset I$*

(a) *L is a $(\mathbf{c}, J)$-monomial lattice;*
(b) *$\tilde{\mathsf{M}}_J^{\mathbf{c}}(b) \subset B \cup \{0\}$ for any $b \in B$;*
(c) *$B = \bigcup_{b_+ \in B_+^J} \tilde{\mathsf{M}}_J^{\mathbf{c}}(b_+) \setminus \{0\}$ where $B_+^J = \bigcap_{j \in J} \ker \tilde{e}_{j,1}^{\mathbf{c}} \subset B$;*

Remark 2.16 It is not hard to see that if for given $\mathbf{c} : \mathbb{D} \to \mathbb{Q}(z)^\times$ and $J \subset I$ Lemma 2.15(a)–(c) hold then (L, B) is a $\mathbf{c}$-crystal basis at $q = 0$ of V regarded as a $U_q(\mathfrak{g}^J)$-module.

3 Properties of σ^i and Proof of Theorem 1.1

3.1 Special Monomials in $U_q^{\pm}(\mathfrak{g})$

Given any reduced sequence $\mathbf{i} = (i_1, \ldots, i_m) \in I^m$ and $\lambda \in P^+$ we define $F_{\mathbf{i},\lambda} \in U_q^-(\mathfrak{g})$ and $E_{\mathbf{i},\lambda} \in U_q^+(\mathfrak{g})$, $\lambda \in P^+$ by

$$F_{\mathbf{i},\lambda} = F_{i_1}^{(a_1)} \cdots F_{i_m}^{(a_m)}, \qquad E_{\mathbf{i},\lambda} = E_{i_1}^{(a_1)} \cdots E_{i_m}^{(a_m)}, \tag{3.1}$$

where $a_k = a_k(\mathbf{i}, \lambda) = s_{i_{k+1}} \cdots s_{i_m} \lambda(\alpha_{i_k}^\vee) = \lambda(s_{i_m} \cdots s_{i_{k+1}} \alpha_{i_k}^\vee) \in \mathbb{Z}_{\geq 0}$.

Lemma 3.1 *Let $w \in W$, $\lambda \in P^+$, $i \in I$. Then*

(a) *$F_{\mathbf{i},\lambda} = F_{\mathbf{i}',\lambda}$ and $E_{\mathbf{i},\lambda} = E_{\mathbf{i}',\lambda}$ for any $\mathbf{i}, \mathbf{i}' \in R(w)$ and $\lambda \in P^+$. Thus, we can define $F_{w,\lambda} := F_{\mathbf{i},\lambda}$ and $E_{w,\lambda} := E_{\mathbf{i},\lambda}$ for some $\mathbf{i} \in R(w)$;*
(b) *If $\ell(s_i w) = \ell(w) + 1$, then $F_{s_i w,\lambda} = F_i^{(w\lambda(\alpha_i^\vee))} F_{w,\lambda}$ and $E_{s_i w,\lambda} = E_i^{(w\lambda(\alpha_i^\vee))} E_{w,\lambda}$;*
(c) *If $\ell(s_i w) = \ell(w) - 1$, then $F_{w,\lambda} = F_i^{(-w\lambda(\alpha_i^\vee))} F_{s_i w,\lambda}$ and $E_{w,\lambda} = E_i^{(-w\lambda(\alpha_i^\vee))} E_{s_i w,\lambda}$;*
(d) *If $s_i\lambda = \lambda$, then $F_{ws_i,\lambda} = F_{w,\lambda}$;*
(e) *$\deg F_{w,\lambda} = -\deg E_{w,\lambda} = w\lambda - \lambda$;*
(f) *Suppose that W is finite. Then $\theta(F_{w,\lambda}) = E_{w_\circ w w_\circ, -w_\circ \lambda}$.*

Proof It is well-known that $\mathbf{i}$ can be obtained from $\mathbf{i}'$ by a finite sequence of rank 2 braid moves of the form $\underbrace{s_is_j\cdots}_{m_{ij}} = \underbrace{s_js_i\cdots}_{m_{ij}}$ with m_{ij} finite. Thus, it suffices to prove part (a) in case when w is the longest element in the subgroup of W generated by $s_i, s_j, i \neq j \in I$. But in that case it was established in [30, Proposition 39.3.7].

Parts (b), (c), and (d) are obtained from part (a) by choosing appropriate reduced decompositions. To prove parts (e) and (f) we use induction on $\ell(w)$, the case $\ell(w) = 0$ being obvious. For the inductive step, suppose that $\ell(s_iw) = \ell(w) + 1$. Since $\mu(\alpha_i^\vee)\alpha_i = \mu - s_i\mu$, $\mu \in P$, by part (b) and the induction hypothesis we have $\deg F_{s_iw,\lambda} = -w\lambda(\alpha_i^\vee)\alpha_i + w\lambda - \lambda = s_iw\lambda - \lambda$. This proves the inductive step in part (e). Since $s_{i^\star} = w_\circ s_i w_\circ$ we have by Lemma 3.1(b)

$$\theta(F_{s_iw,\lambda}) = \theta(F_i^{(w\lambda(\alpha_i^\vee))}F_{w,\lambda}) = E_{i^\star}^{(w\lambda(\alpha_i^\vee))}E_{w_\circ ww_\circ,-w_\circ\lambda} = E_{i^\star}^{(-w\lambda(w_\circ\alpha_{i^\star}^\vee))}E_{w_\circ ww_\circ,-w_\circ\lambda}$$

$$= E_{i^\star}^{((w_\circ ww_\circ(-w_\circ\lambda))(\alpha_{i^\star}^\vee))}E_{w_\circ ww_\circ,-w_\circ\lambda} = E_{s_{i^\star}w_\circ ww_\circ,-w_\circ\lambda} = E_{w_\circ s_i ww_\circ,-w_\circ\lambda}.$$

The inductive step in part (f) is proven. □

3.2 Extremal Vectors

For $V \in \mathscr{O}_q^{int}(\mathfrak{g})$ and $v \in V$, denote $J(v) = \{i \in I : F_i(v) = 0\}$ and define $J(S) = \bigcap_{v\in S} J(v)$, $S \subset V$. One can show that $J(V)$ also equals to $\{i \in I : E_i(V) = \{0\}\}$. We will need the following basic properties of these sets.

Proposition 3.2 *Let $V \in \mathscr{O}_q^{int}(\mathfrak{g})$. Then*

(a) *$J(v) = J_\lambda$ for any $v \in V_+(\lambda) \setminus \{0\}$, $\lambda \in P^+$;*
(b) *There exists $v \in V_+$ such that $J(V) = J(U_q(\mathfrak{g})(v))$.*

Proof It is well-known (see, e.g., [30, Chap. 6]) that the annihilating ideal of v in $U_q^-(\mathfrak{g})$ is generated by the $F_i^{\lambda(\alpha_i^\vee)+1}$, $i \in I$. Thus, $F_i(v) = 0$ if and only if $\lambda(\alpha_i^\vee) = 0$. This proves part (a). To prove part (b), note the following obvious fact.

Lemma 3.3 *Let R be a ring and let $M = \bigoplus_{\alpha\in A} M_\alpha$ as R-modules. Let S be a subset of R. Then $\operatorname{Ann}_S M = \bigcap_{\alpha\in A'} \operatorname{Ann}_S M_\alpha = \operatorname{Ann}_S M'$ where A' is any subset of A such that for each $\alpha \in A$ there exists $\alpha' \in A'$ such that $M_\alpha \cong M_{\alpha'}$ and $M' = \bigoplus_{\alpha\in A'} M_\alpha$. In particular, if S is finite, then $\operatorname{Ann}_S M = \bigcap_{\alpha\in A_0} \operatorname{Ann}_S M_\alpha = \operatorname{Ann}_S M_0$ where A_0 is a finite subset of A' and $M_0 = \bigoplus_{\alpha\in A_0} M_\alpha$.*

Apply this Lemma to $R = U_q(\mathfrak{g})$ and $S = \{F_i : i \in I\}$, which identifies with I, and $M = V$. Clearly, $J(V) = \{i \in I : F_i \in \operatorname{Ann}_S V\}$. Since S is finite and V is a direct sum of simple modules, it follows from Lemma 3.3 that $\operatorname{Ann}_S V = \operatorname{Ann}_S V'$

where $V' = \bigoplus_{\lambda\in\Omega} U_q(\mathfrak{g})(v_\lambda)$ for some finite $\Omega \subset \{\lambda \in P^+ : \mathcal{I}_\lambda(V) \neq 0\}$ and $v_\lambda \in V_+(\lambda) \setminus \{0\}$, $\lambda \in \Omega$. Since $V' = U_q(\mathfrak{g})(v)$ with $v = \sum_{\lambda\in\Omega} v_\lambda$, part (b) follows. □

Given $v \in V_+(\lambda) \setminus \{0\}$ and $w \in W$, define the *standard extremal vectors* $[v]_w$ of v by $[v]_w := F_{w,\lambda}(v)$. Furthermore, given $v \in V_+ \setminus \{0\}$, define

$$[v]_W := \{[v(\lambda)]_w \,:\, w \in W,\, \lambda \in \operatorname{supp} v\}. \tag{3.2}$$

Proposition 3.4 *Let $V \in \mathcal{O}_q^{int}(\mathfrak{g})$. Then for any $v \in V_+ \setminus \{0\}$*

(a) *if v is homogeneous, then $[v]_w = [v]_{w'}$ if and only if $w' \in wW_{J(v)}$. In particular, the assignments $[v]_w \mapsto wW_{J(v)}$ define a bijection $\mathsf{J}_v : [v]_W \to W/W_{J(v)}$;*

(b) *for any $v \in V_+ \setminus \{0\}$ the set $[v]_W$ is linearly independent.*

Proof To prove (a), let $v \in V_+(\lambda) \setminus \{0\}$ for some $\lambda \in P^+$ and recall that $\operatorname{Stab}_W \lambda = W_{J_\lambda}$. It follows from Lemma 3.1(d) by an obvious induction on $\ell(w'')$ that $F_{ww'',\lambda} = F_{w,\lambda}$ for all $w'' \in W_{J_\lambda}$. Since $w\lambda = w'\lambda$ implies that $w' = ww''$ for some $w'' \in W_{J_\lambda}$, it follows that $F_{w',\lambda}(v) = F_{w,\lambda}(v)$. Conversely, by Lemma 3.1(e) we have $F_{w,\lambda}(v) \in V(w\lambda)$. Thus, if $w\lambda \neq w'\lambda$, then $F_{w,\lambda}(v)$ and $F_{w',\lambda}(v)$ are in different weight subspaces of V and are linearly independent (and hence not equal).

In particular, we proved that $[v(\lambda)]_W$ is linearly independent for all $v \in V_+ \setminus \{0\}$ and $\lambda \in \operatorname{supp} v$. This, together with (2.4), proves part (b). □

Proposition 3.5 *Let $V \in \mathcal{O}_q^{int}(\mathfrak{g})$. For each $i \in I$, $v \in V_+ \setminus \{0\}$ the following are equivalent.*

(a) $(\lambda, w\alpha_i) = 0$ *for all* $\lambda \in \operatorname{supp} v$, $w \in W$;

(b) $\operatorname{cl}(\{i\}) \in J(U_q(\mathfrak{g})(v))$.

Proof Let J be the neighborhood of i. In particular, $J \cup J^\perp = I$.

(a) $\Longrightarrow$ (b) We need the following Lemma.

Lemma 3.6 *For every $i \sim j \in I$, $i \neq j$ there exists $w = w_{i,j} \in W$ such that $w_{i,j}\alpha_i \in \mathbb{Z}_{>0}\alpha_j + \sum_{k\in I\setminus\{j\}} \mathbb{Z}_{\geq 0}\alpha_k$.*

Proof Let $\mathbf{i} = (i = i_0, i_1, \ldots, i_d = j) \in I^{d+1}$ be an admissible sequence with $d = \operatorname{dist}(i, j)$. In particular, this sequence is repetition free. Denote $\beta_k = s_{i_k} \cdots s_{i_1}(\alpha_i)$. We claim that $\beta_k \in \sum_{0\leq r\leq k} \mathbb{Z}_{>0}\alpha_{i_r}$. We argue by induction on k, the case $k = 0$ being obvious. For the inductive step, note that

$$\beta_k = s_{i_k}(\beta_{k-1}) \in \sum_{0\leq r\leq k-1} \mathbb{Z}_{>0} s_{i_k}\alpha_{i_r} = \sum_{0\leq r\leq k-1} \mathbb{Z}_{>0}(\alpha_{i_r} - \alpha_{i_r}(\alpha_{i_k}^\vee)\alpha_{i_k}).$$

Since $\mathbf{i}$ is admissible, $\alpha_{i_{k-1}}(\alpha_{i_k}^\vee) < 0$ while $\alpha_{i_r}(\alpha_{i_k}^\vee) \leq 0$ for all $0 \leq r \leq k-2$. Therefore, $\beta_k \in \sum_{0\leq r\leq k} \mathbb{Z}_{>0}\alpha_{i_r}$. In particular, $w = s_{i_d} \cdots s_{i_1}$ is the desired $w_{i,j} \in W$. □

Write $\lambda \in \operatorname{supp} v$ as $\lambda = \lambda' + \sum_{k\in I} l_k\omega_k$ where $\lambda' \in P^W$ and $l_k \in \mathbb{Z}_{\geq 0}$, $k \in I$. Let $j \in J$. In the notation of Lemma 3.6, we have $w_{i,j}\alpha_i = \sum_{k\in I} n_k\alpha_k$ with $n_j \in \mathbb{Z}_{>0}$ and $n_k \in \mathbb{Z}_{\geq 0}$, $k \in I \setminus \{j\}$, for some $w_{i,j} \in W$. Then $0 = 2(\lambda, w_{i,j}\alpha_i) = (\alpha_j, \alpha_j)l_j n_j + \sum_{k\in I\setminus\{j\}}(\alpha_k, \alpha_k)l_k n_k$. Since $n_j > 0$ and $n_k \geq 0$ for all $k \neq j$ this forces $l_j = 0$.

In particular, $J \subset J(v)$. Since J is closed in I, $[U_q^-(\mathfrak{g}^{J^\perp}), F_j] = 0$. Let $V' = U_q(\mathfrak{g})(v)$. Then $V' = U_q^-(\mathfrak{g})(v) = U_q^-(\mathfrak{g}^{J^\perp})(v)$ and so $J \subset J(V')$.

(b) $\Longrightarrow$ (a) Since $J \subset J(U_q(\mathfrak{g})(v))$ it follows that $(\lambda, \alpha_j) = 0$ for all $j \in J$. Since $W\alpha_j \in \sum_{j'\in J} \mathbb{Z}\alpha_{j'}$ for any $j \in J$, the assertion follows. □

Lemma 3.7 *Let $V \in \mathcal{O}_q^{int}(\mathfrak{g})$ such that $V = U_q(\mathfrak{g})(v)$ for some $v \in V_+$. The following are equivalent for $i \in I$.*

(a) $i \in J(V)$.
(b) $[v(\lambda)]_{s_i w} = [v(\lambda)]_w$ *for all* $\lambda \in \operatorname{supp} v$ *and for all* $w \in W$;

Proof The condition in part (b) implies that $s_i w\lambda = w\lambda$ for all $\lambda \in \operatorname{supp} v$ and $w \in W$. Since $s_i w\lambda = w\lambda - (w\lambda)(\alpha_i^\vee)\alpha_i$, it follows that $(\lambda, w\alpha_i) = 0$ for all $\lambda \in \operatorname{supp} v$ and $w \in W$ and so $i \in J(V)$ by Proposition 3.5.

Conversely, if $i \in J(V)$, then $F_i([v]_w) = 0$ for all $w \in W$. In particular, if $\ell(s_i w) = \ell(w) + 1$ then, since $[v]_{s_i w} = F_i^{(w\lambda(\alpha_i^\vee))}[v]_w \neq 0$ it follows that $(w\lambda, \alpha_i) = 0$ and thus $[v]_{s_i w} = [v]_w$. Similarly, if $\ell(s_i w) = \ell(w) - 1$ applying the previous argument to $w' = s_i w$, we obtain the same equality. □

3.3 *Proof of Theorem 1.1*

We will now express the action of the σ^i, $i \in I$ on extremal vectors in terms of the natural action of W on W/W_J.

Proposition 3.8 *Let $V \in \mathcal{O}_q^{int}(\mathfrak{g})$ and $v \in V_+ \setminus \{0\}$. Then*

(a) *The set $[v]_W$ is $\mathsf{W}(V)$-invariant. More precisely, $\sigma^i([v(\lambda)]_w) = [v(\lambda)]_{s_i w}$ for all $i \in I$, $w \in W$ and $\lambda \in \operatorname{supp} v$;*
(b) *The canonical image of $\mathsf{W}(V)$ in* $\operatorname{Bij}([v]_W)$ *is isomorphic to W_{J_0} where $J_0 = I \setminus \operatorname{cl}(J(U_q(\mathfrak{g})(v)))$.*

Proof To prove part (a), let $\lambda \in P^+$, $w \in W$, $i \in I$ and suppose first that $\ell(s_i w) = \ell(w) + 1$. Then $w\lambda(\alpha_i^\vee) > 0$ and $[v(\lambda)]_w \in \ker E_i$, whence $\sigma^i([v(\lambda)]_w) = F_i^{(w\lambda(\alpha_i^\vee))}([v(\lambda)]_w) = F_{s_i w,\lambda}(v(\lambda)) = [v(\lambda)]_{s_i w}$ where we used Lemma 3.1(b). If $\ell(s_i w) = \ell(w) - 1$, then $w = s_i w'$. By the above, $[v(\lambda)]_w = [v(\lambda)]_{s_i w'} = \sigma^i([v(\lambda)]_{w'}) = \sigma^i([v(\lambda)]_{s_i w})$. Since σ^i is an involution, it follows that $\sigma^i([v(\lambda)]_w) = [v(\lambda)]_{s_i w}$.

To prove (b), the bijections from Proposition 3.4(a) allow one to identify $[v]_W$ with $\bigsqcup_{\lambda\in\operatorname{supp} v} W/W_{J_\lambda}$. In particular, this induces an action of W on $[v]_W$ via $w\cdot[v(\lambda)]_{w'}=[v(\lambda)]_{ww'}$, $\lambda\in\operatorname{supp} v$, $w,w'\in W$. By part (a) the canonical images of $\mathsf{W}(V)$ and W in $\operatorname{Bij}([v]_W)$ coincide. We need the following general fact.

Lemma 3.9 *Let G be a group acting on $X=\bigsqcup_{\alpha\in A} X_\alpha$. Then the canonical image of G in* Bij(X) *is isomorphic to G/K where $K=\bigcap_{\alpha\in A} K_\alpha$ and $K_\alpha=\{g\in G : gx=x,\ \forall x\in X_\alpha\}$ is the kernel of the action of G on X_α.*

Applying this Lemma to $X=[v]_W$, $X_\lambda=[v(\lambda)]_W$ and $G=W$ and using the fact that $K_\lambda=W_{J_\lambda}$ by Theorem 2.4 we conclude that $K=\bigcap_{\lambda\in\operatorname{supp} v} W_{J_\lambda}$. By Lemma 2.1, $K=W_{J'}$ where J' is the set of $i\in I$ such that s_i fixes $[v]_W$ elementwise. By Lemma 3.7, $J'=J(V')$ where $V'=U_q(\mathfrak{g})(v)$. Since J' is closed, being an intersection of closed sets, $W/W_{J'}\cong W_{I\setminus J'}$. □

Proof of Theorem 1.1 It follows from Proposition 3.4(a) that the assignments $[v]_w\mapsto wW_J$ define a bijection $\mathsf{J} : [v]_W\to W/W_J$. This induces a group homomorphism $\xi_V : \mathsf{W}(V)\to\operatorname{Bij}(W/W_J)$ via $\xi_V(\sigma^i)(wW_J)=\mathsf{J}(\sigma^i([v]_w))=\mathsf{J}([v]_{s_iw})=s_iwW_J$. It follows that $\xi_V(\mathsf{W}(V))$ coincides with the image of W in $\operatorname{Bij}(W/W_J)$ given by the natural action. By Lemma 2.3, the latter is canonically isomorphic to $W_{I\setminus J_0}$ where $J_0=\operatorname{cl}(I\setminus J)$. □

4 Modified Lusztig Symmetries and Involutions σ^J

Let $\mathscr{C}$ be a $\Bbbk$-linear category whose objects are $\Bbbk$-vector spaces and let G be a group. An action of G on $\mathscr{C}$ is an assignment $g\mapsto g_\bullet=\{g_V : V\in\mathscr{C}\}$, $g\in G$, where $g_V\in GL_\Bbbk(V)$ such that $(gg')_V=g_V\circ g'_V$ for all $g,g'\in G$ and $V\in\mathscr{C}$ and $g_{V'}\circ f=f\circ g_V$ for any $g\in G$ and any morphism $f:V\to V'$ in $\mathscr{C}$.

Recall that the braid group Br_W associated with a Coxeter group W is generated by the T_i, $i\in I$ subject to the relations $\underbrace{T_iT_j\cdots}_{m_{ij}}=\underbrace{T_jT_i\cdots}_{m_{ij}}$ for all $i\neq j\in I$ in the notation of Sect. 2.1. In this section we discuss modified Lusztig symmetries which provide an action of Br_W associated with the Weyl group W of $\mathfrak{g}$ on the category $\mathscr{O}_q^{int}(\mathfrak{g})$ and use them to construct an action of Cact_W on the same category.

4.1 Modified Lusztig Symmetries

Given $i\in I$ and $V\in\mathscr{O}_q^{int}(\mathfrak{g})$ define $T_i^\pm\in\operatorname{End}_\Bbbk V$ by

$$T_i^+=T'_{i,1}K_{\frac{1}{2}\alpha_i},\qquad T_i^-=T''_{i,1}K_{-\frac{1}{2}\alpha_i}$$

where $T'_{i,1}, T''_{i,1} \in \operatorname{End}_{\Bbbk} V$ are Lusztig symmetries (see [30, §5.2]). We refer to these operators as *modified Lusztig symmetries*. By definition, $T_i^{\pm}(V(\beta)) = V(s_i\beta)$ and so $T_i^{\pm} K_\lambda = K_{s_i\lambda} T_i^{\pm}$, $\lambda \in \frac{1}{2}P$.

Lemma 4.1 *The assignments $T_i \mapsto T_i^+$ (respectively, $T_i \mapsto T_i^-$) define an action of Br_W on $\mathscr{O}_q^{int}(\mathfrak{g})$.*

Proof It can be deduced from [30, Proposition 39.4.3] along the lines of [2, Lemma 5.2] that the T_i^+ (and the T_i^-), $i \in I$, satisfy the defining relations of Br_W as endomorphisms of V for each $V \in \mathscr{O}_q^{int}(\mathfrak{g})$. To prove that this action of Br_W commutes with morphisms, write, using [32, §3.1], for $V \in \mathscr{O}_q^{int}(\mathfrak{g})$ and $v \in V$

$$
\begin{aligned}
T_i^+(v) &= \sum_{(a,b,c)\in\mathbb{Z}^3_{\geq 0}} (-1)^b q_i^{b-ac} K_{\frac{1}{2}(a-c-1)\alpha_i} F_i^{(a)} E_i^{(b)} F_i^{(c)} K_{\frac{1}{2}(a-c)\alpha_i}(v), \\
T_i^-(v) &= \sum_{(a,b,c)\in\mathbb{Z}^3_{\geq 0}} (-1)^b q_i^{b-ac} K_{\frac{1}{2}(c-a+1)\alpha_i} E_i^{(a)} F_i^{(b)} E_i^{(c)} K_{\frac{1}{2}(c-a)\alpha_i}(v)
\end{aligned}
\tag{4.1}
$$

where the sum is finite since V is integrable. It is now obvious that the $T_i^{\pm}$, $i \in I$ commute with homomorphisms of $U_q(\mathfrak{g})$-modules. □

It is well-known (see, e.g., [30, §39.4.7]) that the element $T_{i_1} \cdots T_{i_r}$ with $\mathbf{i} = (i_1, \ldots, i_r) \in I^r$ reduced depends only on $w = s_{i_1} \cdots s_{i_r}$ and not on $\mathbf{i}$. This allows to define the canonical section of the natural group homomorphism $\mathsf{Br}_W \to W$, $T_i \mapsto s_i$, $i \in I$, by $w \mapsto T_w := T_{i_1} \cdots T_{i_r}$ where $(i_1, \ldots, i_r) \in R(w)$. Denote $T_w^{\pm}$ the linear endomorphisms of any $V \in \mathscr{O}_q^{int}(\mathfrak{g})$ arising from Lemma 4.1 which correspond to the canonical element T_w of Br_W. The elements T_w, $T_w^{\pm}$ are characterized by the following well-known property.

Lemma 4.2 ([30, §39.4.7]) *Let $w, w' \in W$ be such that $\ell(w) + \ell(w') = \ell(ww')$. Then $T_{ww'} = T_w T_{w'}$ and $T_{ww'}^{\pm} = T_w^{\pm} \circ T_{w'}^{\pm}$ as linear endomorphisms of $V \in \mathscr{O}_q^{int}(\mathfrak{g})$.*

Proposition 4.3 *Let $V \in \mathscr{O}_q^{int}(\mathfrak{g})$. Then for any $w, w' \in W$, $\lambda \in P^+$ and $v \in V_+(\lambda)$ we have:*

(a) *if $\ell(ww') = \ell(w) + \ell(w')$, then*

$$T_w^+(F_{w',\lambda}(v)) = q^{\frac{1}{2}(w'\lambda, \rho - w^{-1}\rho)} F_{ww',\lambda}(v),$$

$$T_w^-(F_{w',\lambda}(v)) = (-1)^{\rho^\vee(w'\lambda - ww'\lambda)} q^{\frac{1}{2}(w'\lambda, \rho - w^{-1}\rho)} F_{ww',\lambda}(v).$$

(b) *if $\ell(ww') = \ell(w') - \ell(w)$, then*

$$T_w^+(F_{w',\lambda}(v)) = (-1)^{\rho^\vee(w'\lambda - ww'\lambda)} q^{-\frac{1}{2}(w'\lambda, \rho - w^{-1}\rho)} F_{ww',\lambda}(v),$$

$$T_w^-(F_{w',\lambda}(v)) = q^{-\frac{1}{2}(w'\lambda, \rho - w^{-1}\rho)} F_{ww',\lambda}(v).$$

(c) *if* $\ell(ww') = \ell(w) - \ell(w')$, *then*

$$T_w^+(F_{w',\lambda}(v)) = (-1)^{\rho^\vee(w'\lambda-\lambda)} q^{\frac{1}{2}(\lambda,2\rho-w'^{-1}(\rho+w^{-1}\rho))} F_{ww',\lambda}(v)$$

$$T_w^-(F_{w',\lambda}(v)) = (-1)^{\rho^\vee(\lambda-ww'\lambda)} q^{\frac{1}{2}(\lambda,2\rho-w'^{-1}(\rho+w^{-1}\rho))} F_{ww',\lambda}(v).$$

Proof To prove (a) we argue by induction on $\ell(w)$, the case $\ell(w) = 0$ being vacuously true. The following Lemma is the main ingredient in the proof of the inductive steps in Proposition 4.3.

Lemma 4.4 *In the notation of Proposition 4.3 we have, for all* $i \in I$

$$T_i^+(F_{w',\lambda}(v)) = \begin{cases} q^{\frac{1}{2}(w'\lambda,\alpha_i)} F_{s_i w',\lambda}(v), & (w'\lambda,\alpha_i) \geq 0, \\ (-1)^{w'\lambda(\alpha_i^\vee)} q^{-\frac{1}{2}(w'\lambda,\alpha_i)} F_{s_i w',\lambda}(v), & (w'\lambda,\alpha_i) \leq 0, \end{cases}$$
$$T_i^-(F_{w',\lambda}(v)) = \begin{cases} (-1)^{w'\lambda(\alpha_i^\vee)} q^{\frac{1}{2}(w'\lambda,\alpha_i)} F_{s_i w',\lambda}(v), & (w'\lambda,\alpha_i) \geq 0, \\ q^{-\frac{1}{2}(w'\lambda,\alpha_i)} F_{s_i w',\lambda}(v), & (w'\lambda,\alpha_i) \leq 0. \end{cases} \tag{4.2}$$

Proof Clearly, $F_{w',\lambda}(v)$ is either a highest or a lowest weight vector in the ith simple quantum $\mathfrak{sl}_2$-submodule V_m it generates where $m = |(w'\lambda, \alpha_i^\vee)|$. Then (4.2) follows from [30, Propositions 5.2.2, 5.2.3]. Namely, let V_m be the standard simple $U_{\mathbf{v}}(\mathfrak{sl}_2)$-module with the standard basis $\{z_k\}_{0\leq k\leq m}$ such that $K(z_k) = \mathbf{v}^{m-2k} z_k$ and $z_k = F^{(k)}(z_0) = E^{(m-k)}(z_m)$, $0 \leq k \leq m$. Recall that $T^+ = T_1' K^{\frac{1}{2}}$ and $T^- = T_1'' K^{-\frac{1}{2}}$. Then by [30, Propositions 5.2.2, 5.2.3] we have

$$T^+(z_k) = (-1)^k \mathbf{v}^{k(m-k)+\frac{1}{2}m} z_{m-k},\ T^-(z_k) = (-1)^{m-k} \mathbf{v}^{k(m-k)+\frac{1}{2}m} z_{m-k},\ 0 \leq k \leq m. \tag{4.3}$$

Thus, (4.2) is obtained by applying (4.3) with $k = 0$ if $w'\lambda(\alpha_i^\vee) \geq 0$ and $k = m$ if $w'\lambda(\alpha_i^\vee) \leq 0$. □

To prove the inductive steps in part (a) of Proposition 4.3, suppose that $w, w' \in W$ and $i \in I$ satisfy $\ell(s_i ww') = \ell(w) + \ell(w') + 1$. In particular, $\ell(s_i w) = \ell(w) + 1$ and $\ell(ww') = \ell(w) + \ell(w')$. Then we have, by the induction hypothesis and (4.2)

$$\begin{aligned} T_{s_i w}^+(F_{w',\lambda}(v)) &= T_i^+ T_w^+(F_{w',\lambda}(v)) = q^{\frac{1}{2}(w'\lambda,\rho-w^{-1}\rho)} T_i^+(F_{ww',\lambda}(v)) \\ &= q^{\frac{1}{2}(w'\lambda,\rho-w^{-1}\rho)+\frac{1}{2}(ww'\lambda,\alpha_i)} F_{s_i ww',\lambda}(v) = q^{\frac{1}{2}(w'\lambda,\rho-w^{-1}\rho+w^{-1}\alpha_i)} F_{s_i ww',\lambda}(v) \\ &= q^{\frac{1}{2}(w'\lambda,\rho-(s_i w)^{-1}\rho)} F_{s_i ww',\lambda}(v), \end{aligned}$$

where we used the W-invariance of $(\cdot,\cdot)$ and the obvious observation that $\alpha_i = \rho - s_i\rho$. The second identity in part (a) is proved similarly using the observation that $\mu(\alpha_i^\vee) = \rho^\vee(\mu - s_i\mu)$.

The proof of part (b) is identical, the only difference being that we assume $\ell(s_i ww') = \ell(w') - \ell(w) - 1$ which implies that $\ell(s_i w) = \ell(w) + 1$ and $\ell(ww') = \ell(w) - \ell(w')$.

To prove part (c), denote $w_1 = ww'$. Then $\ell(w) = \ell(w_1) + \ell(w'^{-1})$, $w = w_1 w'^{-1}$ and so $T_w^+ = T_{w_1}^+ \circ T_{w'^{-1}}^+$ by Lemma 4.2. Applying part (b) with $w = w'^{-1}$ and then part (a) with $w' = 1$ and $w = w_1$ we obtain

$$\begin{aligned}
T_w^+(F_{w',\lambda}(v)) &= T_{w_1}^+(T_{w'^{-1}}^+(F_{w',\lambda}(v))) \\
&= (-1)^{\rho^\vee(w'\lambda-\lambda)} q^{-\frac{1}{2}(w'\lambda,\rho-w'\rho)} T_{w_1}^+(v) \\
&= (-1)^{\rho^\vee(w'\lambda-\lambda)} q^{-\frac{1}{2}(\lambda,w'^{-1}\rho+w_1^{-1}\rho)} F_{w_1,\lambda}(v) \\
&= (-1)^{\rho^\vee(w'\lambda-\lambda)} q^{(\lambda,\rho)-\frac{1}{2}(\lambda,w'^{-1}\rho)+\frac{1}{2}(\lambda,\rho-w'^{-1}w^{-1}\rho)} F_{w_1,\lambda}(v) \\
&= (-1)^{\rho^\vee(w'\lambda-\lambda)} q^{\frac{1}{2}(\lambda,2\rho-w'^{-1}(\rho+w^{-1}\rho))} F_{ww',\lambda}(v).
\end{aligned}$$

The identity for T_w^- is proved similarly. □

Recall from [30, Chapter 37] that Br_W also acts on $U_q(\mathfrak{g})$ via Lusztig symmetries and let $T_i^{\pm}$ be the automorphisms of $U_q(\mathfrak{g})$ defined as $T_i^+ = T'_{i,1}\,\mathrm{ad}\,K_{\frac{1}{2}\alpha_i}$ and $T_i^- = T''_{i,1}\,\mathrm{ad}\,K_{-\frac{1}{2}\alpha_i}$, $i \in I$ where $\mathrm{ad}\,K_\lambda(u) = K_\lambda u K_{-\lambda}$, $\lambda \in \frac{1}{2}P$, $u \in U_q(\mathfrak{g})$.

Remark 4.5 The operators $T_i^{\pm}$, viewed as automorphisms of $U_q(\mathfrak{g})$, were already used in [2] for studying double canonical bases of $U_q(\mathfrak{g})$.

Lemma 4.6 *On $U_q(\mathfrak{g})$ we have*

(a) $T_i^{\pm} \circ \mathrm{ad}\,K_\lambda = \mathrm{ad}\,K_{s_i\lambda} \circ T_i^{\pm}$ *for all* $\lambda \in \frac{1}{2}P$, $i \in I$;
(b) $T_w^+ = T'_{w,1} \circ \mathrm{ad}\,K_{\frac{1}{2}(\rho-w^{-1}\rho)}$ *and* $T_w^- = T''_{w,1} \circ \mathrm{ad}\,K_{\frac{1}{2}(w^{-1}\rho-\rho)}$ *for all* $w \in W$.

Proof Part (a) is immediate, while part (b) follows from (a) by induction similar to that in Proposition 4.3. □

Lemma 4.7 *Suppose that W is finite. Then for all $i \in I$ we have*

$$T_{w_\circ}^{\pm}(E_i) = -q^{-\frac{1}{2}(\alpha_i,\alpha_i)} F_{i^\star} K_{\alpha_{i^\star}}, \qquad T_{w_\circ}^{\pm}(F_i) = -q^{-\frac{1}{2}(\alpha_i,\alpha_i)} E_{i^\star} K_{-\alpha_{i^\star}}.$$

Proof By [1, Lemma 2.8] and Lemma 4.6(b) we have $T_w^{\pm}(E_i) = q^{\pm\frac{1}{2}(\rho-w^{-1}\rho,\alpha_i)} E_j$ provided that $w\alpha_i = \alpha_j$. Since $(w^{-1}\rho, \alpha_i) = (\rho, w\alpha_i) = (\rho, \alpha_j)$ and $(\alpha_i, \alpha_i) = (\alpha_j, \alpha_j)$ it follows that $T_w^{\pm}(E_i) = E_j$. In particular, since $w_\circ s_i \alpha_i = \alpha_{i^\star}$ we have $T_{w_\circ s_i}^{\pm}(E_i) = E_{i^\star}$.

On the other hand, $T_{w_\circ s_i}^{\pm}(E_i) = T_{s_{i^\star} w_\circ}^{\pm}(E_i)$ and $\ell(w_\circ) = \ell(s_{i^\star} w_\circ) + 1$ whence $T_{w_\circ}^{\pm}(E_i) = T_{i^\star}^{\pm}(E_{i^\star}) = -q^{-\frac{1}{2}(\alpha_i,\alpha_i)} F_{i^\star} K_{\alpha_{i^\star}}$. The argument for $T_{w_\circ}^{\pm}(F_i)$ is similar. □

The following Lemmata are immediate from [30, Proposition 37.1.2].

Lemma 4.8 *Let $V \in \mathcal{O}_q^{int}(\mathfrak{g})$. Then $T_i^{\pm}(x(v)) = T_i^{\pm}(x)(T_i^{\pm}(v))$ for all $i \in I$, $v \in V$, $x \in U_q(\mathfrak{g})$.*

Lemma 4.9 *We have $T_w^{\pm}(\mathcal{I}_\lambda(V)) = \mathcal{I}_\lambda(V)$ and $T_w^{\pm}(V(\beta)) = V(w\beta)$ for any $w \in W$, $V \in \mathcal{O}_q^{int}(V)$, $\lambda \in P^+$ and $\beta \in P$.*

4.2 σ via Modified Lusztig Symmetries

Let $\mathfrak{g}$ be finite dimensional reductive and define

$$\sigma^{\pm}(v) = (-1)^{\rho^\vee(\lambda\mp\beta)} q^{\frac{1}{2}((\beta,\beta)-(\lambda,\lambda))-(\lambda,\rho)} T_{w_\circ}^{\pm}(v), \quad v \in V(\beta) \cap \mathcal{I}_\lambda(V). \tag{4.4}$$

The main ingredient in our proof of Theorem 1.5 is the following result which generalizes [31, Proposition 5.5] to all reductive algebras including those whose semisimple part is not necessarily simply laced.

Theorem 4.10 *Let $V \in \mathcal{O}_q^{int}(\mathfrak{g})$. Then*

(a) *$\sigma^+ = \sigma^-$ and is an involution which thus will be denoted by σ;*
(b) *$\sigma(x(v)) = \theta(x)(\sigma(v))$ for any $x \in U_q(\mathfrak{g})$, $v \in V$;*
(c) *σ commutes with morphisms in $\mathcal{O}_q^{int}(\mathfrak{g})$.*

Remark 4.11 For $\mathfrak{g} = \mathfrak{sl}_n$ the involution σ coincides with the famous Schützenberger involution on Young tableaux which was established for the first time in [8]. Thus, we can regard σ as the generalized Schützenberger involution.

Proof We need the following properties of $\sigma^{\pm}$.

Lemma 4.12 *For any $V \in \mathcal{O}_q^{int}(\mathfrak{g})$ we have*

(i) *$\sigma^{\pm}(F_{w,\lambda}(v_\lambda)) = F_{w_\circ w,\lambda}(v_\lambda)$ for any $v_\lambda \in \mathcal{I}_\lambda(V)(\lambda)$, $\lambda \in P^+$;*
(ii) *$\sigma^{\pm}(x(v)) = \theta(x)(\sigma^{\pm}(v))$ for any $x \in U_q(\mathfrak{g})$, $v \in V$;*
(iii) *$\sigma^+ = \sigma^-$ and $(\sigma^{\pm})^2 = \mathrm{id}_V$.*

Proof Let $v_\lambda \in \mathcal{I}_\lambda(V)(\lambda)$, $\lambda \in P^+$. Since $\rho + w_\circ\rho = 0$, $\rho^\vee(w_\circ\mu) = -\rho^\vee(\mu)$ for any $\mu \in P$ and $\ell(w_\circ u) = \ell(w_\circ) - \ell(u)$ for any $u \in W$, Proposition 4.3(c) with $w = w_\circ$ and $w' = u$ yields

$$T_{w_\circ}^{\pm}(F_{u,\lambda}(v_\lambda)) = (-1)^{\rho^\vee(\pm u\lambda-\lambda)} q^{(\lambda,\rho)} F_{w_\circ u,\lambda}(v_\lambda),$$

Since $F_{u,\lambda}(v_\lambda) \in V(u\lambda)$, (4.4) yields

$$\sigma^{\pm}(F_{u,\lambda}(v_\lambda)) = (-1)^{\rho^\vee(\lambda\mp u\lambda)} q^{-(\lambda,\rho)} T_{w_\circ}^{\pm}(F_{u,\lambda}(v_\lambda)) = F_{w_\circ u,\lambda}(v_\lambda).$$

This proves part (i). To prove part (ii), let $v \in V(\beta)$. Then $E_i(v) \in V(\beta + \alpha_i)$, and we obtain, by (4.4) and Lemmata 4.6, 4.7

$$\begin{aligned}\sigma^{\pm}(E_i(v)) &= (-1)^{\rho^\vee(\lambda\mp(\beta+\alpha_i))}q^{\frac{1}{2}((\beta+\alpha_i,\beta+\alpha_i)-(\lambda,\lambda))-(\lambda,\rho)}T^{\pm}_{w_\circ}(E_i(v))\\ &= -(-1)^{\rho^\vee(\lambda\mp\beta)}q^{(\beta,\alpha_i)+\frac{1}{2}((\alpha_i,\alpha_i)+(\beta,\beta)-(\lambda,\lambda))-(\lambda,\rho)}T^{\pm}_{w_\circ}(E_i)(T^{\pm}_{w_\circ}(v))\\ &= (-1)^{\rho^\vee(\lambda\mp\beta)}q^{(\beta,\alpha_i)+\frac{1}{2}((\beta,\beta)-(\lambda,\lambda))-(\lambda,\rho)}F_{i^\star}K_{i^\star}(T^{\pm}_{w_\circ}(v))\\ &= \theta(E_i)(q^{(\beta,\alpha_i+w_\circ\alpha_{i^\star})}\sigma^{\pm}(v)) = \theta(E_i)\sigma^{\pm}(v).\end{aligned}$$

The identity $\sigma^{\pm}(F_i(v)) = \theta(F_i)(\sigma^{\pm}(v))$ is proved similarly. Finally, for any $\mu \in \frac{1}{2}P$ we have $\sigma^{\pm}(K_\mu(v)) = q^{(\beta,\mu)}\sigma^{\pm}(v) = q^{(w_\circ\beta,w_\circ\mu)}\sigma^{\pm}(v) = \theta(K_\mu)(\sigma^{\pm}(v))$.

Let $\epsilon, \epsilon' \in \{+,-\}$. It follows from part (ii) that

$$\sigma^{\epsilon}\circ\sigma^{\epsilon'}(x(v)) = \sigma^{\epsilon}(\theta(x)(v)) = \theta^2(x)(v) = x(v)$$

for any $x \in U_q(\mathfrak{g})$ and $v \in V$ since θ is an involution. Thus, $\sigma^{\epsilon}\circ\sigma^{\epsilon'}$ is an endomorphism of V as a $U_q(\mathfrak{g})$-module. By part (i) we have

$$\sigma^{\epsilon}\circ\sigma^{\epsilon'}(F_{w,\lambda}(v_\lambda)) = \sigma^{\epsilon}(F_{w_\circ w,\lambda}(v_\lambda)) = F_{w,\lambda}(v_\lambda)$$

for any $w \in W$, $\lambda \in P^+$ and $v_\lambda \in \mathcal{I}_\lambda(V)(\lambda)$. In particular, $\sigma^{\epsilon}\circ\sigma^{\epsilon'}(v_\lambda) = v_\lambda$ and so $\sigma^{\epsilon}\circ\sigma^{\epsilon'}$ is the identity map on the (simple) $U_q(\mathfrak{g})$-submodule of V generated by v_λ. Since V is generated by $\bigoplus_{\lambda\in P^+}\mathcal{I}_\lambda(V)(\lambda)$ as a $U_q(\mathfrak{g})$-module, $\sigma^{\epsilon}\circ\sigma^{\epsilon'} = \mathrm{id}_V$. This proves part (iii). □

Parts (a) and (b) of Theorem 4.10 were established in Lemma 4.12. To prove part (b), note the following obvious fact.

Lemma 4.13 *Let $\xi_\bullet = \{\xi_V \in \mathrm{End}_{U_q(\mathfrak{g})} V : V \in \mathscr{O}^{int}_q(\mathfrak{g})\}$ and suppose that $\xi_\bullet$ commute with morphisms in $\mathscr{O}^{int}_q(\mathfrak{g})$, that is $\xi_{V'}\circ f = f\circ\xi_V$ for any morphism $f : V \to V'$ in $\mathscr{O}^{int}_q(\mathfrak{g})$. Let $\chi : P^+\times P \to \Bbbk$ and define $\xi^\chi_\bullet$ by $\xi^\chi_V(v) = \sum_{\beta\in\operatorname{supp} v}\chi(\lambda,\beta)\xi_V(v(\beta))$, $v \in \mathcal{I}_\lambda(V)$, $V \in \mathscr{O}^{int}_q(\mathfrak{g})$. Then $\xi^\chi_\bullet$ also commutes with morphisms in $\mathscr{O}^{int}_q(\mathfrak{g})$.*

It is immediate from the definition (4.4) of σ that $\sigma_\bullet = (T^{\pm}_\bullet)^{\chi_\pm}$ where

$$\chi_\pm(\lambda,\beta) = (-1)^{\rho^\vee(\lambda\mp\beta)}q^{\frac{1}{2}((\beta,\beta)-(\lambda,\lambda))-(\lambda,\rho)}, \quad (\lambda,\beta)\in P^+\times P.$$

Then part (c) follows from Lemma 4.13. □

4.3 Parabolic Involutions and the Proof of Theorem 1.5

In view of Theorem 4.10, given $V \in \mathscr{O}^{int}_q(\mathfrak{g})$ and $J \in \mathscr{J}$, let $\sigma^J : V \to V$ be defined by (4.4) with $U_q(\mathfrak{g})$ replaced by $U_q(\mathfrak{g}^J)$. Thus,

$$\sigma^J(v) = (-1)^{\rho_J^\vee(\lambda_J \mp \beta)} q^{-\frac{1}{2}((\lambda_J,\lambda_J)-(\beta,\beta))-(\lambda_J,\rho_J)} T^{\pm}_{w_\circ^J}(v), \tag{4.5}$$

for any $\lambda_J \in P_J^+$, $\beta \in P$ and $v \in \mathcal{I}^J_{\lambda_J}(V)(\beta)$. Note that $\mathcal{I}^J_{\lambda_J}(V)(\beta) = 0$ unless $\lambda_J - \beta \in \sum_{j\in J} \mathbb{Z}_{\geq 0}\alpha_j$. We need the following properties of σ^J.

Proposition 4.14 *Let $V \in \mathscr{O}_q^{int}(\mathfrak{g})$. Then*

(a) $\sigma^J(F_{w,\lambda_J}(v)) = F_{w_\circ^J w,\lambda}(v)$ *for any $v \in \mathcal{I}^J_{\lambda_J}(V)(\lambda_J)$, $\lambda_J \in P^+$. In particular, $\sigma^J(v) = v^J := F_{w_\circ^J,\lambda}(v)$.*
(b) *σ^J is an involution;*
(c) *σ^J commutes with morphisms in $\mathscr{O}_q^{int}(\mathfrak{g}^J)$ and satisfies $\sigma^J(x(v)) = \theta_J(x)(\sigma^J(v))$, $x \in U_q(\mathfrak{g}^J)$, $v \in V$;*
(d) *$\sigma^J(V(\beta)) = V(w_\circ^J\beta)$, $\beta \in P$;*
(e) *for any $\lambda_J \in P_J^+$, $\beta \in P$ and $v \in \mathcal{I}^J_{\lambda_J}(V)(\beta)$ we have*

$$\sigma^J(v) = (-1)^{\rho_J^\vee(-\lambda_J \mp \beta)} q^{\frac{1}{2}((\lambda_J,\lambda_J)-(\beta,\beta))+(\lambda_J,\rho_J)} (T^{\pm}_{w_\circ^J})^{-1}(v).$$

Proof Replacing $\mathfrak{g}$ by $\mathfrak{g}^J$ we obtain part (a) from Lemma 4.12(i) and parts (b), (c) from Theorem 4.10. Part (d) is immediate from (4.5) and Lemma 4.9. To prove part (e), let $v' = \sigma^J(v)$. Then $v = \sigma^J(v')$ and $v' \in \mathcal{I}^J_{\lambda_J}(V)(\beta')$ where $\beta' = w_\circ^J\beta$. Applying $(T^{\pm}_{w_\circ})^{-1}$ to (4.5) with v replaced by v' we obtain

$$(T^{\pm}_{w_\circ})^{-1}(\sigma^J(v')) = (-1)^{\rho_J^\vee(\lambda_J \mp \beta')} q^{-\frac{1}{2}((\lambda_J,\lambda_J)-(\beta',\beta'))-(\lambda_J,\rho_J)} v'.$$

Since $\rho_J^\vee(\beta') = \rho_J^\vee(w_\circ^J\beta) = -\rho_J^\vee(\beta)$ and $(\cdot,\cdot)$ is W-invariant, it follows that

$$(T^{\pm}_{w_\circ})^{-1}(v) = (-1)^{\rho_J^\vee(\lambda_J \pm \beta)} q^{-\frac{1}{2}((\lambda_J,\lambda_J)-(\beta,\beta))-(\lambda_J,\rho_J)} \sigma^J(v).$$

The assertion is now immediate. □

We need the following results.

Proposition 4.15 *For any $J \subset J' \in \mathscr{J}$, $\sigma^{J'} \circ \sigma^J = \sigma^{J^\star} \circ \sigma^{J'}$ where $^\star : J' \to J'$ is the unique involution satisfying $\alpha_{j^\star} = -w_\circ^{J'}\alpha_j$, $j \in J'$.*

Proof We may assume, without loss of generality, that $J' = I$ (and so $\mathfrak{g}$ is reductive finite dimensional). Let $w_J = w_\circ w_\circ^J$. Note that $w_\circ = w_J w_\circ^J = w_\circ^{J^\star} w_J$ and $\ell(w_\circ) = \ell(w_J) + \ell(w_\circ^J) = \ell(w_J) + \ell(w_\circ^{J^\star})$. Then by Lemma 4.2

$$T^{\pm}_{w_J} = T^{\pm}_{w_\circ} \circ (T^{\pm}_{w_\circ^J})^{-1} = (T^{\pm}_{w_\circ^{J^\star}})^{-1} \circ T^{\pm}_{w_\circ}. \tag{4.6}$$

Let $v \in V(\beta) \cap \mathcal{I}_\lambda(V) \cap \mathcal{I}^J_{\lambda_J}(V)$, $\lambda \in P^+$, $\lambda_J \in P_J^+$, $\beta \in P$. Using Lemma 4.14(e), (4.4), Lemma 4.9 and (4.6) we obtain

$$\sigma\circ\sigma^J(v)=(-1)^{\rho_J^\vee(-\lambda_J\mp\beta)}q^{\frac{1}{2}((\lambda_J,\lambda_J)-(\beta,\beta))+(\lambda_J,\rho_J)}\sigma((T^{\pm}_{w_\circ^J})^{-1}(v))$$
$$=(-1)^{\rho^\vee(\lambda\mp w_\circ^J\beta)+\rho_J^\vee(-\lambda_J\mp\beta)}q^{\frac{1}{2}((\lambda_J,\lambda_J)-(\lambda,\lambda))+(\lambda_J,\rho_J)-(\lambda,\rho)}T^{\pm}_{w_\circ}(T^{\pm}_{w_\circ^J})^{-1}(v)$$
$$=(-1)^{\rho^\vee(\lambda\mp w_\circ^J\beta)+\rho_J^\vee(-\lambda_J\mp\beta)}q^{\frac{1}{2}((\lambda_J,\lambda_J)-(\lambda,\lambda))+(\lambda_J,\rho_J)-(\lambda,\rho)}T^{\pm}_{w_J}(v).$$

Similarly,

$$\sigma^{J^\star}\circ\sigma(v)=(-1)^{\rho^\vee(\lambda\mp\beta)}q^{-\frac{1}{2}((\lambda,\lambda)-(\beta,\beta))-(\lambda,\rho)}\sigma^{J^\star}(T^{\pm}_{w_\circ}(v))$$
$$=(-1)^{\rho^\vee(\lambda\mp\beta)+\rho^\vee_{J^\star}(w_\circ\lambda_J\mp w_\circ\beta)}q^{\frac{1}{2}((\lambda_J,\lambda_J)-(\lambda,\lambda))-(w_\circ\lambda_J,\rho_{J^\star})-(\lambda,\rho)}(T^{\pm}_{w_\circ^{J^\star}})^{-1}(T^{\pm}_{w_\circ}(v))$$
$$=(-1)^{\rho^\vee(\lambda\mp\beta)+\rho_J^\vee(-\lambda_J\pm\beta)}q^{\frac{1}{2}((\lambda_J,\lambda_J)-(\lambda,\lambda))+(\lambda_J,\rho_J)-(\lambda,\rho)}T^{\pm}_{w_J}(v)$$

since $\rho^\vee_{J^\star}(w_\circ\mu)=-\rho^\vee_J(\mu)$, $\mu\in P$ and $w_\circ\rho_{J^\star}=-\rho_J$. Since $-\rho_J^\vee(\beta)=\rho_J^\vee(w_\circ^J\beta)$ and $\rho^\vee(\beta-w\beta)=\rho_J^\vee(\beta-w\beta)$ for any $w\in W_J$, it follows that $\sigma^{J^\star}\circ\sigma=\sigma\circ\sigma^J$. In particular, since $\rho_J^\vee(\lambda_J-\beta)=\rho^\vee(\lambda_J-\beta)$ we have

$$\sigma\circ\sigma^J=\sigma^{J^\star}\circ\sigma=(-1)^{\rho^\vee(\lambda\pm\lambda_J)+\rho_J^\vee(\pm\lambda_J-\lambda_J)}q^{\frac{1}{2}((\lambda_J,\lambda_J)-(\lambda,\lambda))+(\lambda_J,\rho_J)-(\lambda,\rho)}T^{\pm}_{w_J}(v),\tag{4.7}$$

that is, the right-hand side does not explicitly depend on β. □

Proposition 4.16 *Let $J,J'\in\mathscr{J}$ be orthogonal. Then $\sigma^{J\sqcup J'}=\sigma^J\circ\sigma^{J'}$.*

Proof As before we may assume, without loss of generality, that $I=J\sqcup J'$. Then $w_\circ=w_\circ^Jw_\circ^{J'}=w_\circ^{J'}w_\circ^J$ and hence $T^{\pm}_{w_\circ}=T^{\pm}_{w_\circ^J}\circ T^{\pm}_{w_\circ^{J'}}=T^{\pm}_{w_\circ^{J'}}\circ T^{\pm}_{w_\circ^J}$ by Lemma 4.2. Let $\lambda\in P^+$, $\lambda_J\in P_J^+$, $\lambda_{J'}\in P^+_{J'}$, $\beta\in P$ and $v\in\mathcal{I}_\lambda(V)\cap\mathcal{I}^J_{\lambda_J}(V)\cap\mathcal{I}^J_{\lambda_{J'}}(V)$. Then $\gamma_J=\lambda_J-\beta\in\sum_{j\in J}\mathbb{Z}_{\ge0}\alpha_j$, $\gamma_{J'}=\lambda_{J'}-\beta\in\sum_{j\in J'}\mathbb{Z}_{\ge0}\alpha_{j'}$, and $\gamma=\lambda-\beta=\gamma_J+\gamma_{J'}$. Then we can rewrite (4.5) and (4.4) as

$$\sigma^J(v)=(-1)^{\rho_J^\vee(\gamma_J)}q^{-\frac{1}{2}(\gamma_J,\gamma_J)+(\lambda_J,\gamma_J)-(\lambda_J,\rho_J)}T^+_{w_\circ^J}(v)$$
$$\sigma^{J'}(v)=(-1)^{\rho^\vee_{J'}(\gamma_{J'})}q^{-\frac{1}{2}(\gamma_{J'},\gamma_{J'})+(\lambda_{J'},\gamma_{J'})-(\lambda_{J'},\rho_{J'})}T^+_{w_\circ^{J'}}(v)$$
$$\sigma(v)=(-1)^{\rho^\vee(\gamma)}q^{-\frac{1}{2}(\gamma,\gamma)+(\lambda,\gamma)-(\lambda,\rho)}T^+_{w_\circ}(v).$$

Since $w_\circ^J(\gamma_{J'})=\gamma_{J'}$ we have

$$\sigma^J(\sigma^{J'}(v))$$
$$=(-1)^{\rho_J^\vee(\gamma_J)+\rho^\vee_{J'}(\gamma_{J'})}q^{-\frac{1}{2}((\gamma_J,\gamma_J)+(\gamma_{J'},\gamma_{J'}))-(\lambda_J,\rho_J)-(\lambda_{J'},\rho_{J'})+(\lambda_J,\gamma_J)+(\lambda_{J'},\gamma_{J'})}T^+_{w_\circ}(v)$$
$$=(-1)^{\rho^\vee(\gamma)}q^{-\frac{1}{2}(\gamma,\gamma)-(\lambda,\rho)+(\lambda,\gamma)}T^+_{w_\circ}(v)=\sigma(v),$$

since $\rho^\vee(\gamma) = \rho_J^\vee(\gamma_J) + \rho_{J'}^\vee(\gamma_{J'})$, $(\gamma,\gamma) = (\gamma_J,\gamma_J) + (\gamma_{J'},\gamma_{J'})$, $(\lambda_J,\zeta) + (\lambda_{J'},\zeta') = (\lambda_J + \lambda_{J'}, \zeta + \zeta') = (\lambda, \zeta + \zeta')$ for any $\zeta \in \sum_{j\in J} \mathbb{Q}\alpha_j$, $\zeta' \in \sum_{j\in J'} \mathbb{Q}\alpha_{j'}$. □

Proof of Theorem 1.5 Parts (a) (respectively, (b), (c)) of Theorem 1.5 were established in Lemma 4.14(b) (respectively, Proposition 4.16, Proposition 4.15). □

4.4 Kernels of Actions of Cactus Groups

For any $V \in \mathscr{O}_q^{int}(\mathfrak{g})$, denote Φ_V be the subgroup of $\mathrm{GL}_{\mathbb{k}}(V)$ generated by the σ_V^J, $J \in \mathscr{J}$. We need the following basic properties of Φ_V.

Lemma 4.17 *For any injective morphism $f : V' \to V$ in $\mathscr{O}_q^{int}(\mathfrak{g})$ the assignments $\sigma_V^J \mapsto \sigma_{V'}^J$, $J \in \mathscr{J}$ define a surjective homomorphism $f^* : \Phi_V \to \Phi_{V'}$. In particular, if f is an isomorphism then so is f^*.*

Proof Let $V'' = f(V')$. By Theorem 4.10(c), we have $\sigma_V^J \circ f = f \circ \sigma_{V'}^J$ for all $J \in \mathscr{J}$ and so the group Φ_V acts on V'', that is, there is a canonical homomorphism of groups $\rho : \Phi_V \to \mathrm{GL}_{\mathbb{k}}(V'')$. Clearly, the assignments $g \mapsto f^{-1} \circ g \circ f$, $g \in \mathrm{GL}_{\mathbb{k}}(V'')$ define an isomorphism $\rho_f : \mathrm{GL}_{\mathbb{k}}(V'') \to \mathrm{GL}_{\mathbb{k}}(V')$. Let $f^* = \rho_f \circ \rho :$ $\Phi_V \to \mathrm{GL}_{\mathbb{k}}(V')$. We claim that $f^*(\Phi_V) = \Phi_{V'}$. Indeed, $f^*(\sigma_V^J) = \sigma_{V'}^J$ for all $J \in \mathscr{J}$. Since $\Phi_{V'}$ is generated by the $\sigma_{V'}^J$, Φ_V is generated by the σ_V^J, $J \in \mathscr{J}$, and f^* is a homomorphism of groups, the assertion follows. □

Proposition 4.18 *Let $V \in \mathscr{O}_q^{int}(\mathfrak{g})$. Then $\Phi_V \cong \Phi_{\underline{V}}$ where $\underline{V} = \bigoplus_{\lambda\in P^+ : \mathrm{Hom}_{U_q(\mathfrak{g})}(V_\lambda, V)\neq 0} V_\lambda$. In particular, for any $V \in \mathscr{O}_q^{int}(\mathfrak{g})$ the group Φ_V is a quotient of $\Phi_{C_q(\mathfrak{g})}$ where $C_q(\mathfrak{g}) = \bigoplus_{\lambda\in P^+} V_\lambda$.*

Proof Fix $f_\lambda \in \mathrm{Hom}_{U_q(\mathfrak{g})}(V_\lambda, V) \setminus \{0\}$ for all $\lambda \in P^+$ with $\mathrm{Hom}_{U_q(\mathfrak{g})}(V_\lambda, V) \neq 0$ and let $f : \underline{V} \to V$ be the direct sum of these f_λ. Then f is injective. Applying Lemma 4.17 with $V' = \underline{V}$ we obtain a surjective group homomorphism $f^* : \Phi_V \to \Phi_{\underline{V}}$. It remains to prove that its kernel is trivial. We apply Lemma 3.3 with $R = \mathbb{k}[\Phi_V]$ and $S = \{g - 1 : g \in \Phi_V\} \subset R$. Since Φ_V is a subgroup of $\mathrm{GL}_{\mathbb{k}}(V)$, $\mathrm{Ann}_S V = \{0\}$. By our choice of $\underline{V}$, $M = V$ and $M' = f(\underline{V})$ satisfy the assumptions of Lemma 3.3 and so $\mathrm{Ann}_S f(\underline{V}) = \{0\}$. Since $\ker f^* = \{g \in \Phi_V : g \circ f = \mathrm{id}_{\underline{V}}\}$, it follows that $\ker f^*$ is trivial. The second assertion is immediate from the first one and Lemma 4.17. □

5 An Action of Cact_W on c-Crystal Bases and Proof of Theorem 1.8

Retain the notation of Sect. 2.5 and observe that the assignment $(l,k,s)\mapsto(l,l-k,-s)$, $(l,k,s)\in\mathbb{D}$, defines an involution on $\mathbb{D}$. The following is the main result of this section.

Theorem 5.1 *Let $\mathfrak{g}$ be reductive. Suppose that $\mathbf{c}:\mathbb{D}\to\mathbb{Q}(z)^\times$ satisfies*

$$\underline{\mathbf{c}}_{l,k,s}=\underline{\mathbf{c}}_{l,l-k,-s},\qquad \underline{\mathbf{c}}_{l,0,-l}=1,\qquad (l,k,s)\in\mathbb{D} \tag{5.1}$$

in the notation of Lemma 2.12. Then for any $V\in\mathscr{O}_q^{int}(\mathfrak{g})$

(a) *$\sigma^I(L)=L$ for any $(\mathbf{c},I)$-monomial lattice L in V;*
(b) *If (L,B) is a $\mathbf{c}$-crystal basis such that $B_+=\{b\in B:\tilde{e}^{\mathbf{c}}_{i,1}(b)=0,\ i\in I\}$ is a basis of L_+/qL_+ where $L_+=L\cap V_+$, then the induced $\mathbb{Q}$-linear map $\tilde{\sigma}^I:L/qL\to L/qL$ preserves B.*

Proof We abbreviate $\sigma=\sigma^I$, $V_+^I=V_+$ and $\mathsf{M}^{\mathbf{c}}(v_+)=\mathsf{M}^{\mathbf{c}}_I(v_+)$ for any homogeneous $v_+\in V_+$. The key ingredient of our argument is the following

Proposition 5.2 *Let $\mathfrak{g}$ be reductive, let $V\in\mathscr{O}_q^{int}(\mathfrak{g})$ and let $\mathbf{c}:\mathbb{D}\to\mathbb{Q}(z)^\times$. Then*

(a) *If $\mathbf{c}$ satisfies the first condition in (5.1) then $\sigma\circ\tilde{e}^{\mathbf{c}}_{i,s}=\tilde{e}^{\mathbf{c}}_{i^\star,-s}\circ\sigma$ in $\operatorname{End}_{\Bbbk}V$ for any $i\in I$, $s\in\mathbb{Z}$.*
(b) *If $\mathbf{c}$ satisfies (5.1), then*
$\sigma(\mathsf{M}^{\mathbf{c}}(v_+))=\mathsf{M}^{\mathbf{c}}(v_+)$ *for any homogeneous $v_+\in V_+$.*

Proof In view of Lemma 2.11 and (2.6), to prove (a) it suffices to verify the identity for all $v\in V$ of the form $v=F_i^{(k)}(u)$, $u\in\ker E_i\cap\ker(K_{\alpha_i}-q_i^l)$, $0\le k\le l$. We have

$$\sigma\circ\tilde{e}^{\mathbf{c}}_{i,s}(v)=\underline{\mathbf{c}}_{l,k,s}(q_i)\sigma(F_i^{(k-s)}(u))=\underline{\mathbf{c}}_{l,k,s}(q_i)E_{i^\star}^{(k-s)}(\sigma(u)). \tag{5.2}$$

We need the following

Lemma 5.3 $\sigma(u)\in\ker F_{i^\star}\cap\ker(K_{\alpha_{i^\star}}-q_{i^\star}^{-l})$.

Proof Indeed, $F_{i^\star}(\sigma(u))=\sigma(E_i(u))=0$ and $K_{\alpha_i^\star}(\sigma(u))=\sigma(K_{-\alpha_i}(u))=q_i^{-l}\sigma(u)=q_{i^\star}^{-l}\sigma(u)$. □

Using Lemmata 5.3 and 2.12, we obtain

$$\tilde{e}^{\mathbf{c}}_{i^\star,-s}(\sigma(v))=\tilde{e}^{\mathbf{c}}_{i^\star,-s}(E_{i^\star}^{(k)}(\sigma(u)))=\underline{\mathbf{c}}_{l,l-k,-s}(q_{i^\star})E_{i^\star}^{(k-s)}(\sigma(u)). \tag{5.3}$$

Since $q_{i^\star}=q_i$, by the assumptions of Proposition 5.2 we have $\underline{\mathbf{c}}_{l,l-k,-s}(q_{i^\star})=\underline{\mathbf{c}}_{l,k,s}(q_i)$. Then (5.2) and (5.3) imply that $\tilde{e}^{\mathbf{c}}_{i^\star,-s}(\sigma(v))=\sigma(\tilde{e}^{\mathbf{c}}_{i,s}(v))$.

To prove part (b), we need the following

Lemma 5.4 *Suppose that $\mathbf{c}_{l,0,-l} = 1/(l)_z!$ for all $l \in \mathbb{Z}_{\geq 0}$ (that is, $\underline{\mathbf{c}}_{l,0,-l} = 1$ in the notation of Lemma 2.12). Then for any $\lambda \in P^+$, $v_+ \in V_+(\lambda)$ and $\mathbf{i} = (i_1, \ldots, i_m) \in I^m$ reduced $F_{\mathbf{i},\lambda}(v_+) = \tilde{e}^{\mathbf{c}}_{i_1,-a_1(\mathbf{i},\lambda)} \cdots \tilde{e}^{\mathbf{c}}_{i_m,-a_m(\mathbf{i},\lambda)}(v_+)$ in the notation of (3.1). In particular,*

$$\sigma(v_+) = \tilde{e}^{\mathbf{c}}_{i_1,-a_1(\mathbf{i},\lambda)} \cdots \tilde{e}^{\mathbf{c}}_{i_N,-a_N(\mathbf{i},\lambda)}(v_+)$$

where $\mathbf{i} = (i_1, \ldots, i_N) \in R(w_\circ)$.

Proof We use induction on m, the case $m = 0$ being trivial. For $\mathbf{i}$ and $\lambda \in P^+$ fixed we abbreviate $a_k = a_k(\mathbf{i}, \lambda)$. For the inductive step, note that $F_{\mathbf{i},\lambda} = F_{i_1}^{(a_1)} F_{\mathbf{i}',\lambda}$ where $\mathbf{i}' = (i_2, \ldots, i_m)$ and so $F_{\mathbf{i},\lambda}(v_+) = F_{i_1}^{(a_1)}(v')$ where $v' = F_{\mathbf{i}',\lambda}(v_+) = \tilde{e}^{\mathbf{c}}_{i_2,-a_2} \cdots \tilde{e}^{\mathbf{c}}_{i_m,-a_m}(v_+)$ by the induction hypothesis. Since $v' \in \ker E_{i_1}$, it follows from assumptions of the lemma and the first identity in (2.6) with $i = i_1$, $k = 0$ and $l = a_1 = -s$ that $F_{i_1}^{(a_1)}(v') = \tilde{e}^{\mathbf{c}}_{i_1,-a_1}(v') = \tilde{e}^{\mathbf{c}}_{i_1,-a_1} \cdots \tilde{e}^{\mathbf{c}}_{i_m,-a_m}(v_+)$. Since $\sigma(v_+) = F_{w_\circ,\lambda}(v_+)$ by Lemma 4.12(i) with $w = 1$, the second assertion follows from the first and Lemma 3.1(a). □

Suppose now that $v \in \mathsf{M}^{\mathbf{c}}(v_+)$ that is $v = \tilde{e}^{\mathbf{c}}_{j_1,m_1} \cdots \tilde{e}^{\mathbf{c}}_{j_r,m_r}(v_+) \in \mathsf{M}^{\mathbf{c}}(v_+)$, for some $(j_1, \ldots, j_r) \in I^r$ and $(m_1, \ldots, m_r) \in \mathbb{Z}$. Using Lemma 5.4 and Proposition 5.2(a), we obtain

$$\begin{aligned}\sigma(v) &= \tilde{e}^{\mathbf{c}}_{j_1^\star,-m_1} \cdots \tilde{e}^{\mathbf{c}}_{j_r^\star,-m_r}(\sigma(v_+)) \\ &= \tilde{e}^{\mathbf{c}}_{j_1^\star,-m_1} \cdots \tilde{e}^{\mathbf{c}}_{j_r^\star,-m_r} \tilde{e}^{\mathbf{c}}_{i_1,-a_1} \cdots \tilde{e}^{\mathbf{c}}_{i_N,-a_N}(v_+) \in \mathsf{M}^{\mathbf{c}}(v_+),\end{aligned}$$

where $\mathbf{i} = (i_1, \ldots, i_N) \in R(w_\circ)$ and $a_k = a_k(\mathbf{i}, \lambda)$, $1 \leq k \leq N$. Thus, $\sigma(\mathsf{M}^{\mathbf{c}}(v_+)) \subset \mathsf{M}^{\mathbf{c}}(v_+)$. Since σ is an involution, it follows that $\sigma(\mathsf{M}^{\mathbf{c}}(v_+)) = \mathsf{M}^{\mathbf{c}}(v_+)$. □

Part (a) of Theorem 5.1 is immediate from Proposition 5.2(b). In particular, for each $(\mathbf{c}, I)$-monomial lattice L in V the involution σ_V induces an involution $\tilde{\sigma}$ on the $\mathbb{Q}$-vector space $\tilde{L} = L/qL$ satisfying

$$\tilde{\sigma} \circ \tilde{e}^{\mathbf{c}}_{i,s} = \tilde{e}^{\mathbf{c}}_{i^\star,-s} \circ \tilde{\sigma}. \tag{5.4}$$

The following is immediate from Proposition 5.2(b).

Corollary 5.5 *Let L be a $(\mathbf{c}, I)$-monomial lattice in V. Then $\tilde{\sigma}(\tilde{\mathsf{M}}^{\mathbf{c}}(\tilde{v}_+)) = \tilde{\mathsf{M}}^{\mathbf{c}}(\tilde{v}_+)$ for any $\tilde{v}_+ \in L_+/qL_+$.*

Using the assumptions of part (b) of Theorem 5.1 we conclude that $\bigcup_{b_+ \in B^+} \tilde{\mathsf{M}}^{\mathbf{c}}(b_+) = B \cup \{0\}$. Then it follows from Corollary 5.5 that $\tilde{\sigma}$ preserves $B \cup \{0\}$. Since $\tilde{\sigma}$ is an involution, it follows that $\tilde{\sigma}(B) = B$. This completes the proof of Theorem 5.1(b). □

Note that (5.4) implies that for any upper crystal basis (L, B) of $V \in \mathcal{O}_q^{int}(\mathfrak{g})$ the operator $\tilde{\sigma}_V^I$ satisfies

$$\tilde{\sigma} \circ (\tilde{e}_i^{up})^s = (\tilde{e}_{i^\star}^{up})^{-s} \circ \tilde{\sigma}_V^I. \tag{5.5}$$

In particular, we obtain the following

Corollary 5.6 *Let $\lambda \in P^+$ and (L_λ, B_λ) be an upper crystal basis of V_λ. If f is any non-zero map $B_\lambda \cup \{0\} \to B_\lambda \cup \{0\}$ satisfying (5.5), then $f = \tilde{\sigma}_{V_\lambda}^I|_{B_\lambda \cup \{0\}}$.*

Proof of Theorem 1.8 Note that

$$\underline{\mathbf{c}}_{l,k,s}^{low} = 1, \qquad \underline{\mathbf{c}}_{l,k,s}^{up} = \frac{(l-k+s)_z!(k-s)_z!}{(l-k)_z!(k)_z!}, \qquad (l,k,s) \in \mathbb{D}. \tag{5.6}$$

It is now immediate that (5.1) holds for $\mathbf{c} \in \{\mathbf{c}^{up}, \mathbf{c}^{low}\}$.

Furthermore, by Lemma 2.15 and Remark 2.16, Theorem 5.1 applies to every $\mathbf{c}$-crystal basis at $q = 0$ for any $J \in \mathcal{J}$ with $\mathfrak{g}$ replaced by $\mathfrak{g}^J$ and $\mathbf{c} \in \{\mathbf{c}^{up}, \mathbf{c}^{low}\}$. Thus, σ^J preserves a lower or upper crystal lattice L and $\tilde{\sigma}^J$ preserves B.

In particular, Φ_V acts on L and this action factors through to an action on L/qL and induces an action on B by permutations. □

Remark 5.7 Let L be an upper crystal lattice for $V \in \mathcal{O}_q^{int}(\mathfrak{g})$. It follows from the definition of σ_V^i that for any $v \in L(\beta)$, $\beta \in P$ we have $\sigma_V^i(v) = \tilde{e}_i^{-\beta(\alpha_i^\vee)}(v)$. In particular, the action of $\tilde{\sigma}_V^i$ on an upper crystal basis (L, B) of V coincides with Kashiwara's crystal Weyl group action (see [26]).

We conclude this section with a discussion of the action of Cact_W on upper global crystal bases. Let $\bar{\cdot}$ be any field involution of $\mathbb{k}$ such that $\overline{q^{\frac{1}{2d}}} = q^{-\frac{1}{2d}}$.

Proposition 5.8 *Let (L, B) be an upper crystal basis of $V \in \mathcal{O}_q^{int}(\mathfrak{g})$ and let $G^{up}(B)$ be the corresponding upper global crystal basis. Denote by $\bar{\cdot}$ the (unique) additive map $V \to V$ satisfying $\overline{f \cdot b} = \overline{f} \cdot b$ for all $f \in \mathbb{k}$, $b \in G^{up}(B)$. Then $\overline{\sigma^J(b)} = \sigma^J(b)$ for any $J \in \mathcal{J}$.*

Proof Denote by $\bar{\cdot}$ the ring automorphism of $U_q(\mathfrak{g})$ satisfying $\overline{E_i} = E_i$, $\overline{F_i} = F_i$, $i \in I$, $\overline{K_\lambda} = K_{-\lambda}$, $\lambda \in \frac{1}{2}P$ and $\overline{fu} = \overline{f} \cdot \overline{u}$ for all $f \in \mathbb{k}$, $u \in U_q(\mathfrak{g})$. The following is immediate from the properties of the upper global crystal basis [25].

Lemma 5.9 *The map $\bar{\cdot} : V \to V$ defined in Proposition 5.8 satisfies*

$$\overline{u(v)} = \overline{u}(\overline{v}), \qquad v \in V,\ u \in U_q(\mathfrak{g}). \tag{5.7}$$

The following is immediate.

Lemma 5.10 *Let $\eta : U_q(\mathfrak{g}) \to U_q(\mathfrak{g})$ be any algebra automorphism commuting with $\bar{\cdot} : U_q(\mathfrak{g}) \to U_q(\mathfrak{g})$ and let $V \in \mathcal{O}_q^{int}(\mathfrak{g})$ with a fixed set $V_0 \subset V$ generating V*

as a $U_q(\mathfrak{g})$*-module. Let* $\bar{\cdot}: V \to V$ *be any map satisfying* (5.7) *and let* $\sigma \in \mathrm{End}_{\Bbbk} V$ *be such that:*

(i) $\sigma(u(v)) = \eta(u)(\sigma(v))$, $u \in U_q(\mathfrak{g})$, $v \in V$;
(ii) $\sigma(\overline{v}) = \overline{\sigma(v)}$ *for any* $v \in V_0$.

Then $\sigma(\overline{v}) = \overline{\sigma(v)}$ *for all* $v \in V$.

The set $V_0 = G^{up}(B) \cap V_+^J$ generates V as a $U_q(\mathfrak{g}^J)$-module. Clearly, θ_J commutes with $\bar{\cdot}$-involution on $U_q(\mathfrak{g}^J)$. The condition (i) of Lemma 5.10 holds with $\eta = \theta_J$ by Proposition 4.14(d). By Proposition 4.14(a), $\sigma^J(b) = F_{w_\circ^J,\lambda}(b)$ for any $b \in V_0(\lambda)$, $\lambda \in P_J^+$. Since $\overline{F_{w_\circ^J,\lambda}} = F_{w_\circ^J,\lambda}$ and $\overline{b} = b$, the condition (ii) of Lemma 5.10 is also satisfied. The assertion follows by Lemma 5.10. □

6 σ^I and the Canonical Basis

6.1 Automorphisms and Skew Derivations of Localizations

Let R be a unital $\Bbbk$-algebra. Given a monoid Γ written multiplicatively and acting on R by algebra automorphisms, define the semidirect product of R with the monoidal algebra $\Bbbk[\Gamma]$ of Γ as $R \otimes \Bbbk[\Gamma]$ with the multiplication defined by

$$(r \otimes \gamma) \cdot (r' \otimes \gamma') = r(\gamma \triangleright r') \otimes \gamma\gamma', \qquad r, r' \in R,\ \gamma, \gamma' \in \Gamma,$$

where $\triangleright$ denotes the action of Γ on R. Since $(r \otimes 1)(1 \otimes \gamma) = r \otimes \gamma$, we will henceforth omit the symbol $\otimes$ when writing elements of $R \rtimes \Bbbk[\Gamma]$. In other words, $R \rtimes \Bbbk[\Gamma]$ is generated by R as a subalgebra and Γ subject to the relations

$$\gamma \cdot r = (\gamma \triangleright r) \cdot \gamma, \qquad r \in R,\ \gamma \in \Gamma.$$

The following characterization of cross products is immediate.

Lemma 6.1 *Let* $f: R \to \widehat{R}$ *be a homomorphism of* $\Bbbk$*-algebras and let* $g: \Gamma \to \widehat{R}$ *be a homomorphism of multiplicative monoids. Suppose that* R *is a* $\Bbbk[\Gamma]$*-module algebra. Then assignments* $r \cdot \gamma \mapsto f(r) \cdot g(\gamma)$, $r \in R$, $\gamma \in \Gamma$ *define a homomorphism of* $\Bbbk$*-algebras if and only if*

$$f(\gamma \triangleright r)g(\gamma) = g(\gamma)f(r), \qquad r \in R,\ \gamma \in \Gamma. \tag{6.1}$$

Let S be a submonoid of $R \setminus \{0\}$. Denote S^{op} the opposite monoid of S and denote its elements by $[s]$, $s \in S$. Suppose that R is a $\Bbbk[S^{op}]$-module algebra with $[s] \triangleright r = \Sigma_s(r)$, $s \in S$ where Σ_s is an algebra automorphism of R and assume that

$$\Sigma_s(s) = s, \qquad s \in S. \tag{6.2}$$

Denote $R[S^{-1}] := (R \rtimes \Bbbk[S^{op}])/\langle s[s] - 1 \,:\, s \in S\rangle$. We say that S as above is an Ore submonoid if

$$rs = s\Sigma_s(r), \qquad r \in R,\, s \in S. \tag{6.3}$$

We use the convention that $\Sigma_{\lambda s} = \Sigma_s$ for all $\lambda \in \Bbbk^\times$. This notation is justified by the following

Lemma 6.2 *Suppose that* (6.2) *holds. Then the following are equivalent:*

(i) *the natural homomorphism* $\mathbf{1}_{R,S} : R \to R[S^{-1}]$ *is injective;*
(ii) *S is an Ore submonoid of R, and the assignments $r \cdot [s] \mapsto rs^{-1}$, $r \in R$, $s \in S$ define an isomorphism $R[S^{-1}] \to \underline{R[S^{-1}]}$ where $\underline{R[S^{-1}]}$ is the Ore localization of R by S;*

Proof In $R \rtimes \Bbbk[S^{op}]$ we have

$$[s] \cdot r = \Sigma_s(r) \cdot [s], \qquad s \in S,\, r \in R. \tag{6.4}$$

In particular, $[s] \cdot s = s \cdot [s]$, $s \in S$. Multiplying both sides of (6.4) by s on the left and on the right we conclude that $rs = s\Sigma_s(r)$ in $R[S^{-1}]$ for all $r \in R$, $s \in S$. This identity clearly holds in $\mathbf{1}_{R,S}(R)$. Since $\mathbf{1}_{R,S}$ is injective, this implies that the corresponding identity holds in R and so S satisfies the two-sided Ore condition and so R admits the Ore localization $\underline{R[S^{-1}]}$. The assignments $r[s] \mapsto rs^{-1}$, $r \in R$, $s \in S$ define a surjective homomorphism from $R[S^{-1}] \to \underline{R[S^{-1}]}$ which is easily seen to be injective. Thus, (a) implies (b).

Conversely, the natural homomorphism $R \to \underline{R[S^{-1}]}$, $r \mapsto r \cdot 1$, $r \in R$ is injective. Since it equals the composition of $\mathbf{1}_{R,S}$ and the isomorphism $R[S^{-1}] \to \underline{R[S^{-1}]}$, it follows that $\mathbf{1}_{R,S}$ is injective. □

The following Lemma is immediate.

Lemma 6.3 *Let R be a $\Bbbk$-algebra and $S \subset R \setminus \{0\}$ be an Ore submonoid. Let R' be a $\Bbbk$-subalgebra of R and suppose that $S' \subset R' \cap S$ is an Ore submonoid of R'. Then $R'[S'^{-1}]$ is isomorphic to the subalgebra of $R[S^{-1}]$ generated by R' and $\{s'^{-1} \,:\, s' \in S'\}$.*

Lemma 6.4 *Suppose that* (6.2) *and the assumptions of Lemma 6.2(b) hold. Let $\varphi : R \to R'$ be any $\Bbbk$-algebra homomorphism, S be an Ore submonoid of R and S' be an Ore submonoid of R' such that $\varphi(S) \subset S'$. Suppose that $\Sigma'_{\varphi(s)} \circ \varphi = \varphi \circ \Sigma_s$ for all $s \in S$. Then there exists a unique homomorphism $\widehat{\varphi} : R[S^{-1}] \to R'[S'^{-1}]$ such that $\widehat{\varphi}|_R = \varphi$.*

Proof We apply Lemma 6.1 with $\Gamma = S^{op}$, $\widehat{R} = R' \rtimes \Bbbk[S'^{op}]$ and $g : S^{op} \to \widehat{R}$ defined by $g([s]) = [\varphi(s)]$. Then

$$[\varphi(s)]\varphi(r) = \Sigma'_{\varphi(s)}(\varphi(r))[\varphi(s)] = \varphi(\Sigma_s(r))[\varphi(s)] = \varphi([s] \triangleright r)[\varphi(s)],$$

and so (6.1) holds. Therefore, the assignments $r[s] \mapsto \varphi(r)[\varphi(s)]$, $r \in R$, $s \in S$, define a homomorphism $\widehat{\widehat{\varphi}} : R \rtimes \Bbbk[S^{op}] \to R' \rtimes \Bbbk[S'^{op}]$. Since $\widehat{\widehat{\varphi}}(s[s]) = \varphi(s)[\varphi(s)]$ it follows that the image of the defining ideal of $R[S^{-1}]$ under $\widehat{\widehat{\varphi}}$ is contained in the defining ideal of $R'[S'^{-1}]$. Thus, $\widehat{\widehat{\varphi}}$ factors through to the desired homomorphism $\widehat{\varphi} : R[S^{-1}] \to R'[S'^{-1}]$. □

Let $L_\pm : R \to R'$ be $\Bbbk$-algebra homomorphisms and $E : R \to R'$ be a $\Bbbk$-linear map. We say that E is an (L_+, L_-)-derivation from R to R' if $E(rr') = E(r)L_+(r') + L_-(r)E(r')$ for all $r, r' \in R$. Denote $\mathrm{Der}_{L_+,L_-}(R, R')$ the $\Bbbk$-subspace of $\mathrm{Hom}_\Bbbk(R, R')$ of (L_+, L_-)-derivations from R to R'. We refer to an (L, L^{-1})-derivation as an L-derivation and abbreviate $\mathrm{Der}_{L_+,L_-} R = \mathrm{Der}_{L_+,L_-}(R, R)$. The following is immediate.

Lemma 6.5 *Let R_0 be a generating subset of R and let $D, D' \in \mathrm{Der}_{L_+,L_-}(R, R')$. Then $D|_{R_0} = D'|_{R_0}$ implies that $D = D'$.*

Given $r' \in R'$, denote by $D^\pm_{r'}$ the linear maps $R \to R'$

$$D^-_{r'}(x) = r'L_+(x) - L_-(x)r', \qquad D^+_{r'}(x) = L_-(x)r' - r'L_+(x), \qquad x \in R. \tag{6.5}$$

Lemma 6.6 *Let $L_\pm : R \to R'$ be $\Bbbk$-algebra homomorphisms. The assignments $r' \mapsto D^+_{r'}$ (respectively, $r' \mapsto D^-_{r'}$), $r' \in R'$ define $\Bbbk$-linear maps $R' \to \mathrm{Der}_{L_+,L_-}(R, R')$.*

Proof For any $x, x' \in R$ we have

$$\begin{aligned} D^-_{r'}(xx') &= r'L_+(xx') - L_-(xx')r' \\ &= (r'L_+(x) - L_-(x)r')L_+(x') + L_-(x)(r'L_+(x') - L_-(x')r') \\ &= D^-_{r'}(x)L_+(x') + L_-(x)D^-_{r'}(x'). \end{aligned}$$

Thus, $D^-_{r'} \in \mathrm{Der}_{L_+,L_-}(R, R')$. The computation for $D^+_{r'}$ is similar and is omitted. The linearity of both maps in r' is obvious. □

6.2 The Gelfand–Kirillov Model for the Category $\mathcal{O}^{int}_q(\mathfrak{g})$

Throughout this section we mostly follow the notation from [6, Section 6]. Let Γ be the monoid P^+ written multiplicatively, with its elements denoted by v_λ, $\lambda \in P^+$. Let $\mathcal{A}_q(\mathfrak{g})$ be an isomorphic copy of $U_q^-(\mathfrak{g})$ whose generators are denoted by x_i, $i \in I$. We denote the degree of a homogeneous element $x \in \mathcal{A}_q(\mathfrak{g})$ with respect to its Q-grading by $|x| \in -Q^+$. Define an action of Γ on $\mathcal{A}_q(\mathfrak{g})$ by $v_\lambda \triangleright x = q^{(\lambda,|x|)}x$ for $x \in \mathcal{A}_q(\mathfrak{g})$ homogeneous. Let $\mathcal{B}_q(\mathfrak{g}) = \mathcal{A}_q(\mathfrak{g}) \rtimes \Bbbk[\Gamma]$. In particular, we have

$$v_\lambda x = q^{(\lambda,|x|)} x v_\lambda \tag{6.6}$$

for all $\lambda \in P^+$ and for all $x \in \mathcal{A}_q(\mathfrak{g})$ homogeneous. We extend the Q-grading on $\mathcal{A}_q(\mathfrak{g})$ to a P-grading on $\mathcal{B}_q(\mathfrak{g})$ via $|v_\lambda| = \lambda$ for $\lambda \in P^+$. Let $\mathcal{O}_q(\mathfrak{g})$ be the category of $U_q(\mathfrak{g})$-modules whose objects satisfy all assumptions on objects of $\mathcal{O}_q^{int}(\mathfrak{g})$ except that we do not assume that the $E_i, F_i, i \in I$, act locally nilpotently while assuming that all weight subspaces are finite dimensional. The following essentially coincides with [6, Lemma 6.1].

Lemma 6.7 *The algebra $\mathcal{B}_q(\mathfrak{g})$ is a module algebra in the category $\mathcal{O}_q(\mathfrak{g})$ with respect to the action given by the following formulae for all $\lambda \in \frac{1}{2}P$, $i \in I$*

- $K_\lambda(y) = q^{(\lambda,|y|)}y$ *for all homogeneous elements $y \in \mathcal{B}_q(\mathfrak{g})$ and $\lambda \in \frac{1}{2}P$;*
- $F_i(y) = \dfrac{x_i K_{\frac{1}{2}\alpha_i}(y) - K_{-\frac{1}{2}\alpha_i}(y)x_i}{q_i - q_i^{-1}}$ *for all $y \in \mathcal{B}_q(\mathfrak{g})$ and thus is a $K_{\frac{1}{2}\alpha_i}$-derivation of $\mathcal{B}_q(\mathfrak{g})$;*
- E_i *is the unique $K_{\frac{1}{2}\alpha_i}$-derivation of $\mathcal{B}_q(\mathfrak{g})$ such that $E_i(x_j) = \delta_{i,j}$ and $E_i(v_\mu) = 0$ for all $i, j \in I$, $\mu \in P^+$.*

Thus for all $x, y \in \mathcal{B}_q(\mathfrak{g})$, $i \in I$ we have

$$X_i(xy) = X_i(x)K_{\frac{1}{2}\alpha_i}(y) + K_{-\frac{1}{2}\alpha_i}(x)X_i(y) \tag{6.7}$$

and more generally, for all $n \geq 0$

$$X_i^{(n)}(xy) = \sum_{r+s=n} X_i^{(r)}K_{-\frac{s}{2}\alpha_i}(x)X_i^{(s)}K_{\frac{r}{2}\alpha_i}(y) \tag{6.8}$$

where X_i is either E_i or F_i. The following is immediate from the definition of $\mathcal{B}_q(\mathfrak{g})$ and its $U_q(\mathfrak{g})$-module structure.

Corollary 6.8 $\mathcal{B}_q(\mathfrak{g}) = \sum_{\lambda \in P^+} \mathcal{A}_q(\mathfrak{g})v_\lambda$ *where $\mathcal{A}_q(\mathfrak{g})v_\lambda$ is a $U_q(\mathfrak{g})$-submodule of $\mathcal{B}_q(\mathfrak{g})$ for each $\lambda \in P^+$ and the sum is direct.*

In the sequel we will also use E_i^* which is defined as the unique $K_{-\frac{1}{2}\alpha_i}$-derivation of $\mathcal{B}_q(\mathfrak{g})$ satisfying $E_i^*(x_j) = \delta_{i,j}$, $E_i^*(v_\lambda) = 0$ for all $\lambda \in P^+$, $j \in I$. It is easy to check that $E_i^*(x) = (E_i(x^*))^*$, $x \in \mathcal{A}_q(\mathfrak{g})$, where $^* : \mathcal{A}_q(\mathfrak{g}) \to \mathcal{A}_q(\mathfrak{g})$ is the unique anti-involution preserving the x_i, $i \in I$.

In the spirit of [22], using the decomposition from Corollary 6.8 we can define a linear map $\mathbf{j} : \mathcal{B}_q(\mathfrak{g}) \to \mathcal{A}_q(\mathfrak{g})$ by

$$\mathbf{j}(x \cdot v_\lambda) = q^{-\frac{1}{2}(\lambda,|x|)}x \tag{6.9}$$

for all $\lambda \in P^+$ and $x \in \mathcal{A}_q(\mathfrak{g})$ homogeneous. Clearly, $\mathbf{j}|_{\mathcal{A}_q(\mathfrak{g})v_\lambda}$ is a bijection onto $\mathcal{A}_q(\mathfrak{g})$.

Lemma 6.9 *For any symmetrizable Kac–Moody* $\mathfrak{g}$ *we have:*

(a) $\mathbf{j}$ *is a surjective homomorphism of* $U_q^+(\mathfrak{g})$*-modules, with respect to the action defined in Lemma 6.7.*
(b) $\mathbf{j}(x \cdot y) = q^{\frac{1}{2}(\lambda, |\mathbf{j}(y)|) - \frac{1}{2}(\mu, \mathbf{j}(|x|))}\mathbf{j}(x) \cdot \mathbf{j}(y)$ *for all* $x \in \mathcal{A}_q(\mathfrak{g})v_\lambda$, $y \in \mathcal{A}_q(\mathfrak{g})v_\mu$ *homogeneous.*

Proof Part (a) is easily checked using Corollary 6.8. To prove part (b) note that $x = q^{\frac{1}{2}(\lambda, |\mathbf{j}(x)|)}\mathbf{j}(x)v_\lambda$ for all $x \in \mathcal{A}_q(\mathfrak{g})v_\lambda$ homogeneous and so we can write

$$x \cdot y = q^{\frac{1}{2}(\lambda+\mu, |\mathbf{j}(x)|+|\mathbf{j}(y)|)}\mathbf{j}(x \cdot y)v_{\lambda+\mu} = q^{\frac{1}{2}(\lambda, |\mathbf{j}(x)|)+\frac{1}{2}(\mu, |\mathbf{j}(y)|)}\mathbf{j}(x)v_\lambda \mathbf{j}(y)v_\mu$$
$$= q^{\frac{1}{2}(\lambda, |\mathbf{j}(x)|)+\frac{1}{2}(\mu, |\mathbf{j}(y)|)+(\lambda, |\mathbf{j}(y)|)}\mathbf{j}(x) \cdot \mathbf{j}(y)v_{\lambda+\mu}.$$

The assertion is now immediate. □

Given a $U_q(\mathfrak{g})$-module M, denote by M^{int} the set of all $m \in M$ such that $U_q(\mathfrak{g})(m) \in \mathcal{O}_q^{int}(\mathfrak{g})$. The following is well-known and in fact is easy to check.

Lemma 6.10 *The assignment* $M \mapsto M^{int}$ *for every* $U_q(\mathfrak{g})$*-module* M *and* $f \mapsto f$ *for any morphism of* $U_q(\mathfrak{g})$*-modules defines an additive submonoidal functor from the tensor category of* $U_q(\mathfrak{g})$*-modules to* $\mathcal{O}_q^{int}(\mathfrak{g})$*, that is* $M^{int} \otimes N^{int} \subset (M \otimes N)^{int}$ *for any* $U_q(\mathfrak{g})$*-modules* M, N*. In particular, if* M *is an algebra object in the category of* $U_q(\mathfrak{g})$*-modules, then* M^{int} *is its* $U_q(\mathfrak{g})$*-module subalgebra and an algebra object in the category* $\mathcal{O}_q^{int}(\mathfrak{g})$.

Proposition 6.11 *For any* $\lambda \in P^+$ *the* $U_q(\mathfrak{g})$*-submodule of* $\mathcal{A}_q(\mathfrak{g})v_\lambda$ *generated by* v_λ *is naturally isomorphic to* V_λ *and coincides with* $(\mathcal{A}_q(\mathfrak{g})v_\lambda)^{int}$.

Proof Given $M \in \mathcal{O}_q(\mathfrak{g})$, define $M^\vee = \bigoplus_{\beta \in P} M^\vee(\beta)$ where $M^\vee(\beta) = \mathrm{Hom}_{\Bbbk}(M(\beta), \Bbbk)$. Endow $M^\vee$ with a $U_q(\mathfrak{g})$-module structure via $(u \cdot f)(m) = f(u^T(m))$, $u \in U_q(\mathfrak{g})$, $f \in M^\vee$, $m \in M$, where $u \mapsto u^T$, $u \in U_q(\mathfrak{g})$ is the unique anti-involution of $U_q(\mathfrak{g})$ such that $E_i{}^T = F_i$ and $K_\mu{}^T = K_\mu$, $\mu \in \frac{1}{2}P$. The following is well-known

Lemma 6.12 *For any symmetrizable Kac–Moody* $\mathfrak{g}$*, we have:*

(a) *The assignments* $M \to M^\vee$, $M \in \mathcal{O}_q(\mathfrak{g})$ *define an involutive contravariant functor on* $\mathcal{O}_q(\mathfrak{g})$.
(b) *For any* $M \in \mathcal{O}_q(\mathfrak{g})$, $(M^{int})^\vee = (M^\vee)^{int}$.
(c) *For any* $V \in \mathcal{O}_q^{int}(\mathfrak{g})$, $V^\vee$ *is naturally isomorphic to* V.

We need the following well-known fact which essentially coincides with [3, Lemma 2.10].

Lemma 6.13 *There exists a unique non-degenerate pairing* $(\!|\cdot, \cdot|\!) : U_q^-(\mathfrak{g}) \otimes \mathcal{A}_q(\mathfrak{g}) \to \Bbbk$ *such that* $(\!|F_i, x_j|\!) = \delta_{i,j}$, $i, j \in I$, $(\!|uF_i, x|\!) = (\!|u, E_i^*(x)|\!)$, $(\!|F_i u, x|\!) = (\!|u, E_i(x)|\!)$, $u \in U_q^-(\mathfrak{g})$, $x \in \mathcal{A}_q(\mathfrak{g})$ *and* $(\!|u, x|\!) = 0$ *for* $u \in U_q^-(\mathfrak{g})$, $x \in \mathcal{A}_q(\mathfrak{g})$ *homogeneous unless* $\deg u = |x|$.

Denote M_λ, $\lambda \in P$, the Verma module with highest weight λ (see, e.g., [30, §3.4.5]). For every $\lambda \in P^+$ we fix $m_\lambda \in M_\lambda(\lambda) \setminus \{0\}$. Since M_λ is free as a $U_q^-(\mathfrak{g})$-module, every element of M_λ can be written, uniquely, as um_λ for some $u \in U_q^-(\mathfrak{g})$. Let $\mathcal{M}_q(\mathfrak{g}) = \bigoplus_{\lambda\in P^+} M_\lambda$. Define $\langle\!\langle \cdot,\cdot \rangle\!\rangle : \mathcal{M}_q(\mathfrak{g}) \otimes \mathcal{B}_q(\mathfrak{g}) \to \Bbbk$ by $\langle\!\langle u(m_\lambda), xv_\mu \rangle\!\rangle = \delta_{\lambda,\mu} (\!| u, \mathbf{j}(x) |\!)$, $u \in U_q^-(\mathfrak{g})$, $x \in \mathcal{A}_q(\mathfrak{g})$, $\lambda, \mu \in P^+$. It is immediate from the definition that $\langle\!\langle \mathcal{M}_q(\mathfrak{g})(\beta), \mathcal{B}_q(\mathfrak{g})(\beta') \rangle\!\rangle = 0$, $\beta, \beta' \in P$, unless $\beta = \beta'$.

The following Lemma seems to be well-known. We provide a proof for reader's convenience.

Lemma 6.14 *The pairing $\langle\!\langle \cdot,\cdot \rangle\!\rangle$ is non-degenerate and contragredient, that is*

$$\langle\!\langle u'(m), b \rangle\!\rangle = \langle\!\langle m, u'^T(b) \rangle\!\rangle, \qquad u' \in U_q(\mathfrak{g}),\ m \in \mathcal{M}_q(\mathfrak{g}),\ b \in \mathcal{B}_q(\mathfrak{g}). \tag{6.10}$$

In particular, $\mathcal{A}_q(\mathfrak{g})v_\lambda$, $\lambda \in P^+$ naturally identifies with $M_\lambda^\vee$.

Proof The pairing $\langle\!\langle \cdot,\cdot \rangle\!\rangle$ is non-degenerate as a direct sum of non-degenerate (in view of Lemma 6.13) pairings $M_\lambda \otimes \mathcal{A}_q(\mathfrak{g})v_\lambda \to \Bbbk$. To prove that it is contragredient, it suffices to prove (6.10) for $u' \in \{K_\mu, E_i, F_i\}$, $\mu \in \frac{1}{2}P$, $i \in I$. for all $i \in I$. Moreover, we may assume without loss of generality that $m = u(m_\lambda)$ and $b = xv_\lambda$ with $u \in U_q^-(\mathfrak{g})$, $x \in \mathcal{A}_q(\mathfrak{g})$ homogeneous. We have

$$\begin{aligned}\langle\!\langle K_\mu(u(m_\lambda)), xv_\lambda \rangle\!\rangle &= q^{(\mu,\lambda+\deg u)}\delta_{\deg u,|x|}\langle\!\langle u(m_\lambda), xv_\lambda \rangle\!\rangle \\ &= \langle\!\langle u(m_\lambda), K_\mu(xv_\lambda) \rangle\!\rangle = \langle\!\langle u(m_\lambda), K_\mu^T(xv_\lambda) \rangle\!\rangle.\end{aligned}$$

Furthermore, by Lemmata 6.9(a) and 6.13 we obtain

$$\begin{aligned}\langle\!\langle F_i(u(m_\lambda)), xv_\lambda \rangle\!\rangle &= \langle\!\langle (F_iu)(m_\lambda), xv_\lambda \rangle\!\rangle = (\!| F_iu, \mathbf{j}(x) |\!) = (\!| u, E_i(\mathbf{j}(x)) |\!) \\ &= (\!| u, \mathbf{j}(E_i(x)) |\!) = \langle\!\langle u(m_\lambda), E_i(xv_\lambda) \rangle\!\rangle = \langle\!\langle u(m_\lambda), F_i^T(xv_\lambda) \rangle\!\rangle.\end{aligned}$$

In particular, we proved (6.10) for $u' \in \{K_\mu, F_i\}$, $\mu \in \frac{1}{2}P$, $i \in I$ for all $m \in \mathcal{M}_q(\mathfrak{g})$ and $b \in \mathcal{B}_q(\mathfrak{g})$.

It remains to prove that

$$\langle\!\langle E_i(m), b \rangle\!\rangle = \langle\!\langle m, F_i(b) \rangle\!\rangle,$$

for all $m \in M_\lambda$ and for all $b \in \mathcal{A}_q(\mathfrak{g})v_\lambda$ homogeneous. We argue by induction on $\rho^\vee(\lambda - \beta)$ where $m \in M_\lambda(\beta)$. If $\beta = \lambda$, then $E_i(m) = 0$ while $F_i(b) = b'$ with $|b'| = |b| - \alpha_i$. Since $|b| \in \lambda - Q^+$, $|b'| \neq \beta$ and so $\langle\!\langle m, b' \rangle\!\rangle = 0$. For the inductive step, it suffices to assume that $m = F_j(m')$ where m' is homogeneous. We have

$$\begin{aligned}&\langle\!\langle E_i(F_j(m')), b \rangle\!\rangle = \langle\!\langle [E_i, F_j](m'), b \rangle\!\rangle + \langle\!\langle (F_jE_i)(m'), b \rangle\!\rangle \\ &= \langle\!\langle m', [E_j, F_i](b) \rangle\!\rangle + \langle\!\langle m', F_iE_j(b) \rangle\!\rangle = \langle\!\langle m', E_j(F_i(b)) \rangle\!\rangle = \langle\!\langle F_j(m'), F_i(b) \rangle\!\rangle,\end{aligned}$$

where we used (6.10) for $u' = F_j$ and $u' = [E_i, F_j] = \delta_{i,j}(q_i - q_i^{-1})^{-1}(K_{\alpha_i} - K_{-\alpha_i}) = [E_j, F_i]$.

The second assertion of the Lemma is immediate from the first. □

By [30, Proposition 3.5.6], M_λ has a unique integrable quotient isomorphic to V_λ. Applying ${}^{int\vee}$ to the surjection $M_\lambda \to V_\lambda$ we obtain, by Lemmata 6.12 and 6.14, the desired isomorphism $V_\lambda \cong (M_\lambda^\vee)^{int} = (\mathcal{A}_q(\mathfrak{g})v_\lambda)^{int}$ such that $v_\lambda \mapsto v_\lambda$. □

In view of Proposition 6.11, from now on we identify V_λ, $\lambda \in P^+$, with the $U_q(\mathfrak{g})$-submodule of $\mathcal{A}_q(\mathfrak{g})v_\lambda$ generated by v_λ.

Lemma 6.15 *For any $\lambda, \mu \in P^+$ we have $V_\lambda \cdot V_\mu = V_{\lambda+\mu}$ in $\mathcal{B}_q(\mathfrak{g})$.*

Proof It is immediate from the definition of $\mathcal{B}_q(\mathfrak{g})$ and Corollary 6.8 that $V_\lambda \cdot V_\mu \subset \mathcal{A}_q(\mathfrak{g})v_{\lambda+\mu}$. Furthermore, since $V_\lambda \cdot V_\mu$ is the image of $V_\lambda \otimes V_\mu$ which is integrable, by Proposition 6.11 we have $V_\lambda \cdot V_\mu \subset (\mathcal{A}_q(\mathfrak{g})v_{\lambda+\mu})^{int} = V_{\lambda+\mu}$. As the latter is a simple $U_q(\mathfrak{g})$-module, $V_\lambda \cdot V_\mu = V_{\lambda+\mu}$. □

Denote $\mathcal{C}_q(\mathfrak{g}) = \mathcal{B}_q(\mathfrak{g})^{int}$. The following is an immediate corollary of Proposition 6.11 and Lemma 6.15.

Corollary 6.16 *$\mathcal{C}_q(\mathfrak{g})$ decomposes as $\mathcal{C}_q(\mathfrak{g}) = \sum_{\lambda \in P^+} V_\lambda$ as a $U_q(\mathfrak{g})$-module algebra.*

Proposition 6.17 *For any symmetrizable Kac–Moody $\mathfrak{g}$ and $\lambda \in P^+$ we have:*

$$\mathbf{j}(V_\lambda) = \bigcap_{i\in I} \ker E_i^{*\lambda(\alpha_i^\vee)+1}. \tag{6.11}$$

Proof Note that Lemma 6.13 yields an isomorphism of $\Bbbk$-vector spaces $\xi : \mathcal{A}_q(\mathfrak{g}) \to U_q^-(\mathfrak{g})^\vee := \bigoplus_{\gamma\in Q^+} \operatorname{Hom}_\Bbbk(U_q^-(\mathfrak{g})(-\gamma), \Bbbk)$ defined by $\xi(x)(u) = (\!| u, x |\!)$, $x \in \mathcal{A}_q(\mathfrak{g})$, $u \in U_q^-(\mathfrak{g})$. Define $\phi_\lambda^\vee : M_\lambda^\vee \to U_q^-(\mathfrak{g})^\vee$ by $\phi_\lambda^\vee(f)(u) = f(u(m_\lambda))$ for all $f \in M_\lambda^\vee$, $u \in U_q^-(\mathfrak{g})$.

Define an action of $U_q^+(\mathfrak{g})$ on $U_q^-(\mathfrak{g})^\vee$ by $(u_+ \cdot f)(u_-) := f(u_+^T u_-)$, $u_\pm \in U_q^\pm(\mathfrak{g})$. The following is an immediate consequence of Lemmata 6.13 and 6.14.

Lemma 6.18 *For any $\lambda \in P^+$, $\phi_\lambda^\vee$ is an isomorphism of $U_q^+(\mathfrak{g})$-modules. Moreover, the following diagram in the category of $U_q^+(\mathfrak{g})$-modules commutes*

$$\begin{array}{ccc} \mathcal{B}_q(\mathfrak{g}) & \xrightarrow{\ \mathbf{j}\ } & \mathcal{A}_q(\mathfrak{g}) \\ \uparrow & & \downarrow{\scriptstyle \xi} \\ M_\lambda^\vee & \xrightarrow{\ \phi_\lambda^\vee\ } & U_q^-(\mathfrak{g})^\vee \end{array} \tag{6.12}$$

where the left vertical arrow is obtained by the identification $M_\lambda^\vee \cong \mathcal{A}_q(\mathfrak{g})v_\lambda$ from Proposition 6.11.

Let $\mathcal{J}_\lambda$, $\lambda \in P^+$, be the kernel of the canonical projection of M_λ on V_λ. It is well-known (see, e.g., [30, Proposition 3.5.6]) that $\mathcal{J}_\lambda = \sum_{i\in I} U_q^-(\mathfrak{g}) F_i^{\lambda(\alpha_i^\vee)+1}(m_\lambda)$. Applying $^\vee$ to the projection $M_\lambda \to V_\lambda$ and using that $V_\lambda \cong V_\lambda^\vee$ we obtain an embedding $V_\lambda \to M_\lambda^\vee$. Note that

$$\phi_\lambda^\vee(V_\lambda) = \{f \in U_q^-(\mathfrak{g})^\vee : f\Big(\sum_{i\in I} U_q^-(\mathfrak{g}) F_i^{\lambda(\alpha_i^\vee)+1}\Big) = \{0\}\}.$$

Therefore, $\phi_\lambda^\vee(V_\lambda) = \bigcap_{i\in I} \mathcal{K}_i$ where $\mathcal{K}_i = \{f \in U_q^-(\mathfrak{g})^\vee : f(U_q^-(\mathfrak{g}) F_i^{\lambda(\alpha_i^\vee)+1}) = 0\}$. By Lemma 6.13, $\xi^{-1}(\mathcal{K}_i) = \ker E_i^{*\lambda(\alpha_i^\vee)+1}$. Using (6.12) we obtain $\mathbf{j}(V_\lambda) = \bigcap_{i\in I} \xi^{-1}(\mathcal{K}_i) = \bigcap_{i\in I} \ker E_i^{*\lambda(\alpha_i^\vee)+1}$. □

6.3 *Realization of σ^I via Quantum Twist*

Let $v_{w\lambda} = F_{w,\lambda}(v_\lambda)$, $\lambda \in P^+$ where we use the notation from Sect. 3.1 (see also Sect. 3.2). This notation agrees with that in [6, (6.3)]. We need the following

Lemma 6.19 *Let $\mathfrak{g}$ be a symmetrizable Kac–Moody algebra. Then for any $w, w' \in W$ and $\lambda, \mu \in P^+$ we have*

(a) $v_{w\lambda} \cdot v_{w\mu} = v_{w(\lambda+\mu)}$. *In particular, for any $w \in W$, the assignments $v_\lambda \mapsto v_{w\lambda}$, $\lambda \in P^+$, define a homomorphism of monoids $g_w : \Gamma \to \mathcal{B}_q(\mathfrak{g})$;*
(b) *if $\ell(w'w) = \ell(w) + \ell(w')$, then we have*

$$v_{w'\mu} \cdot v_{w'w\lambda} = q^{(w\lambda-\lambda,\mu)} v_{w'w\lambda} \cdot v_{w'\mu};$$

(c) *if $\ell(s_i w) = \ell(w) - 1$, $i \in I$, then $v_{w\lambda} x_i = q^{(w\lambda,\alpha_i)} x_i v_{w\lambda}$ for all $\lambda \in P^+$;*
(d) *if $\mathfrak{g}$ is finite dimensional, then $v_{w_\circ\lambda} x = q^{-(w_\circ\lambda,|x|)} x v_{w_\circ\lambda}$ for all $x \in \mathcal{A}_q(\mathfrak{g})$ homogeneous.*

Proof To prove (a) we use induction on $\ell(w)$, the induction base being trivial. For the inductive step, suppose that $\ell(s_i w) = \ell(w) + 1$. Then by Lemma 3.1(b) and the induction hypothesis

$$v_{s_i w(\lambda+\mu)} = F_{s_i w,\lambda+\mu}(v_{\lambda+\mu}) = F_i^{(w(\lambda+\mu)(\alpha_i^\vee))}(v_{w(\lambda+\mu)}) = F_i^{(w(\lambda+\mu)(\alpha_i^\vee))}(v_{w\lambda} \cdot v_{w\mu}).$$

Using (6.8) and observing that $F_i^{(r)}(v_{w\lambda}) F_i^{(s)}(v_{w\mu}) = 0$ if $r > w\lambda(\alpha_i^\vee)$ or $s > w\mu(\alpha_i^\vee)$ we obtain by Lemma 3.1(b)

$$\begin{aligned} v_{s_i w(\lambda+\mu)} &= \sum_{r+t=w(\lambda+\mu)(\alpha_i^\vee)} q^{\frac{1}{2}(r w\mu - t w\lambda,\alpha_i)} F_i^{(r)}(v_{w\lambda}) \cdot F_i^{(t)}(v_{w\mu}) \\ &= F_i^{(w\lambda(\alpha_i^\vee))}(v_{w\lambda}) \cdot F_i^{(w\mu(\alpha_i^\vee))}(v_{w\mu}) = v_{s_i w\lambda} \cdot v_{s_i w\mu}. \end{aligned}$$

Part (b) was established in [6, Lemma 6.4]. To prove part (c), note that if $\ell(s_i w) = \ell(w) - 1$ then $F_i(v_{w\lambda}) = 0$. Then $x_i K_{\frac{1}{2}\alpha_i}(v_{w\lambda}) - K_{-\frac{1}{2}\alpha_i}(v_{w\lambda})x_i = 0$, whence $x_i v_{w\lambda} = q^{-(\alpha_i, w\lambda)} v_{w\lambda} x_i = q^{(w\lambda, |x_i|)} v_{w\lambda} x_i$. In particular, applying part (c) with $w = w_\circ$ we obtain, using an obvious induction on $-\rho^\vee(|x|)$,

$$v_{w_\circ\lambda} x = q^{-(w_\circ\lambda, |x|)} x v_{w_\circ\lambda},$$

which yields part (d). □

Following [6, §6.1], define *generalized quantum minors* $\Delta_{w\lambda} \in \mathcal{A}_q(\mathfrak{g})$, $w \in W$, $\lambda \in P^+$ by $\Delta_{w\lambda} := \mathbf{j}(v_{w\lambda})$. In particular,

$$v_{w\lambda} = q^{\frac{1}{2}(w\lambda - \lambda, \lambda)} \Delta_{w\lambda} v_\lambda. \tag{6.13}$$

We list some properties of generalized quantum minors which will be used in the sequel.

Lemma 6.20 *Let $w, w' \in W$, $\lambda, \mu \in P^+$. Then*

(a) $\Delta_{w\lambda} \cdot \Delta_{w\mu} = q^{\frac{1}{2}(w\mu - w^{-1}\mu, \lambda)} \Delta_{w(\lambda+\mu)}$;
(b) $\Delta_{w\mu} \cdot \Delta_{ww'\lambda} = q^{(w\mu - \mu, ww'\lambda + \lambda)} \Delta_{ww'\lambda} \cdot \Delta_{w\mu}$;
(c) *If $\mathfrak{g}$ is finite-dimensional reductive, then $\Delta_{w_\circ\lambda} \cdot \Delta_{w_\circ\mu} = \Delta_{w_\circ(\lambda+\mu)}$ and $\Delta_{w_\circ\lambda} x = q^{-(w_\circ\lambda + \lambda, |x|)} x \Delta_{w_\circ\lambda}$ for any $x \in \mathcal{A}_q(\mathfrak{g})$ homogeneous.*

Proof Parts (a) and (b) follow immediately from Lemma 6.19(a) and (b), respectively, by applying Lemma 6.9(b). The first assertion of part (c) is a special case of (a). Finally, using (6.6), (6.13) and Lemma 6.19(d) we can write

$$q^{(\lambda, |x|)} \Delta_{w_\circ\lambda} x v_\lambda = \Delta_{w_\circ\lambda} v_\lambda x = q^{-(w_\circ\lambda, |x|)} x \Delta_{w_\circ\lambda} v_\lambda.$$

It remains to apply $\mathbf{j}$ and use the fact that $\mathbf{j}|_{\mathcal{A}_q(\mathfrak{g})v_\lambda}$ is injective. □

Let $\mathcal{S}_w = \{\Delta_{w\lambda} : \lambda \in P^+\}$. It follows from Lemma 6.20(c) that $\mathcal{S}_{w_\circ}$ is an abelian submonoid of $\mathcal{A}_q(\mathfrak{g})$ and in fact is an Ore submonoid with $\Sigma_{\Delta_{w_\circ\lambda}}(x_i) = q^{(\lambda, \alpha_{i^\star} - \alpha_i)} x_i$ for $\lambda \in P^+$, $i \in I$.

Define $\widehat{\mathcal{B}}_q(\mathfrak{g}) := \mathcal{B}_q(\mathfrak{g})[\mathcal{S}_{w_\circ}^{-1}]$ and let $\widehat{\mathcal{A}}_q(\mathfrak{g})$ be the subalgebra of $\widehat{\mathcal{B}}_q(\mathfrak{g})$ generated by $\mathcal{A}_q(\mathfrak{g})$, as a subalgebra, and the $\Delta_{w_\circ\lambda}^{-1}$, $\lambda \in P^+$. Clearly, $\widehat{\mathcal{A}}_q(\mathfrak{g})$ is isomorphic to $\mathcal{A}_q(\mathfrak{g})[\mathcal{S}_{w_\circ}^{-1}]$. The following is the main result of Sect. 6.

Theorem 6.21 *Let $\mathfrak{g}$ be finite dimensional. Then*

(a) *the assignments $x_i \mapsto q_i^{\frac{1}{2}(\delta_{i,i^\star} - 1)} E_{i^\star}(\Delta_{w_\circ\omega_i}) \Delta_{w_\circ\omega_i}^{-1}$, $v_\lambda \mapsto v_{w_\circ\lambda}$, $\lambda \in P^+$, define an injective algebra homomorphism $\widehat{\sigma} : \mathcal{B}_q(\mathfrak{g}) \to \widehat{\mathcal{B}}_q(\mathfrak{g})^{op}$;*
(b) *$\widehat{\sigma}(V_\lambda) = V_\lambda$ and $\widehat{\sigma}|_{V_\lambda} = \sigma_{V_\lambda}^I$. In particular, the restriction of $\widehat{\sigma}$ to $\mathcal{C}_q(\mathfrak{g})$ is an anti-involution on $\mathcal{C}_q(\mathfrak{g})$.*

Proof The first step is to construct a homomorphism of algebras $\sigma_0 : \mathcal{A}_q(\mathfrak{g}) \to \widehat{\mathcal{A}}_q(\mathfrak{g})^{op}$.

Proposition 6.22 *The assignments*

$$x_i \mapsto q_i^{-\frac{1}{2}(1-\delta_{i,i^\star})} E_{i^\star}(\Delta_{w_\circ\omega_i})\Delta^{-1}_{w_\circ\omega_i} = q_i^{\frac{1}{2}(1-\delta_{i,i^\star})} \Delta^{-1}_{w_\circ\omega_i} E_{i^\star}(\Delta_{w_\circ\omega_i}), \quad i \in I,$$

define a homomorphism $\sigma_0 : \mathcal{A}_q(\mathfrak{g}) \to \widehat{\mathcal{A}}_q(\mathfrak{g})^{op}$ *such that* $\sigma_0(\mathcal{A}_q(\mathfrak{g})(-\gamma)) \subset \widehat{\mathcal{A}}_q(\mathfrak{g})(-w_\circ\gamma)$, $\gamma \in Q^+$.

Proof Let δ be the unique involution of $\mathcal{A}_q(\mathfrak{g})$ defined by $\delta(x_i) = x_{i^\star}$, $i \in I$. Then $\kappa : \mathcal{A}_q(\mathfrak{g}) \to \mathcal{A}_q(\mathfrak{g})$ defined by $\kappa(x) = \delta(x^*) = (\delta(x))^*$, $x \in \mathcal{A}_q(\mathfrak{g})$ is an anti-involution. We need the following

Lemma 6.23 *For any* $\lambda \in P^+$, $\kappa(\Delta_{w_\circ\lambda}) = \epsilon_\lambda \Delta_{w_\circ\lambda}$ *where* $\epsilon_\lambda \in \{\pm 1\}$.

Remark 6.24 Later we will show that $\epsilon_\lambda = 1$. However, for the purposes of proving Proposition 6.22 this is irrelevant.

Proof Define

$$\mathcal{A}_q(\mathfrak{g})^\lambda := \{x \in \mathcal{A}_q(\mathfrak{g})_{w_\circ\lambda-\lambda} : (E_i^*)^{\lambda(\alpha_i^\vee)+1}(x) = 0, \forall i \in I\}.$$

It follows from the definition and Lemma 6.9 that

$$\mathcal{A}_q(\mathfrak{g})^\lambda = \mathbf{j}(V_\lambda(w_\circ\lambda)) = \Bbbk\Delta_{w_\circ\lambda}.$$

In particular, $E_i^{1-w_\circ\lambda(\alpha_i^\vee)}(\mathcal{A}_q(\mathfrak{g})^\lambda) = 0$ for all $i \in I$. Since $\kappa(E_i(x)) = E^*_{i^\star}(\kappa(x))$, it follows that $\kappa(\Delta_{w_\circ\lambda}) \in \mathcal{A}_q(\mathfrak{g})^\lambda$ and so is a multiple of $\Delta_{w_\circ\lambda}$. Since κ is an involution, the assertion follows. □

It follows from Lemmata 6.4 and 6.23 that κ lifts to an anti-involution $\widehat{\kappa}$ on $\widehat{\mathcal{A}}_q(\mathfrak{g})$. By [27, Theorem 5.4], for any $\mathbf{c} = (c_i)_{i\in I} \in (\Bbbk^\times)^I$ the assignments

$$x_i \mapsto c_i E_i^*(\Delta_{w_\circ\omega_i})\Delta^{-1}_{w_\circ\omega_i} = c_i q_i^{\delta_{i,i^\star}-1} \Delta^{-1}_{w_\circ\omega_i} E_i^*(\Delta_{w_\circ\omega_i}), \quad i \in I,$$

define a homomorphism of algebras $\zeta_{\mathbf{c}} : \mathcal{A}_q(\mathfrak{g}) \to \widehat{\mathcal{A}}_q(\mathfrak{g})^{op}$. Let $\mathbf{c}_0 = (q_i^{\frac{1}{2}(1-\delta_{i,i^\star})})_{i\in I}$ and set $\sigma_0 := \widehat{\kappa} \circ \zeta_{\mathbf{c}_0}$. Since $\widehat{\kappa}$ is an anti-involution, we have

$$\sigma_0(x_i) = q_i^{\frac{1}{2}(1-\delta_{i,i^\star})}(\kappa(\Delta_{w_\circ\omega_i}))^{-1}\kappa(E_i^*(\Delta_{w_\circ\omega_i})) = q_i^{\frac{1}{2}(1-\delta_{i,i^\star})}\Delta^{-1}_{w_\circ\omega_i} E_{i^\star}(\Delta_{w_\circ\omega_i}).$$

Thus, σ_0 is the desired homomorphism $\mathcal{A}_q(\mathfrak{g}) \to \widehat{\mathcal{A}}_q(\mathfrak{g})^{op}$. Since $|\sigma_0(x_i)| = \alpha_{i^\star} = -w_\circ\alpha_i$, it follows that $|\sigma_0(x)| = w_\circ|x|$ for all $x \in \mathcal{A}_q(\mathfrak{g})$ homogeneous. □

Now we have all the necessary ingredients to prove Theorem 6.21(a). We apply Lemma 6.1 with $R = \mathcal{A}_q(\mathfrak{g})$, $\widehat{R} = \widehat{\mathcal{B}}_q(\mathfrak{g})^{op}$, $f = \sigma_0$ and $g = g_{w_\circ}$ viewed as a homomorphism $\Gamma \to \widehat{R}$ since Γ is abelian. Take $x \in \mathcal{A}_q(\mathfrak{g})$ homogeneous. Then the following holds in $\widehat{\mathcal{B}}_q(\mathfrak{g})$

$$g_{w_\circ}(v_\lambda)\sigma_0(v_\lambda \triangleright x) = q^{(\lambda,|x|)}v_{w_\circ\lambda}\sigma_0(x) = q^{(\lambda,|x|)-(w_\circ\lambda,w_\circ|x|)}\sigma_0(x)v_{w_\circ\lambda} = \sigma_0(x)g_{w_\circ}(v_\lambda),$$

which is (6.1) in $\widehat{R}$. Then by Lemma 6.1, $\widehat{\sigma} : \mathcal{B}_q(\mathfrak{g}) \to \widehat{\mathcal{B}}_q(\mathfrak{g})^{op}$, $v_\lambda x \mapsto \sigma_0(x)v_{w_\circ\lambda}$, $x \in \mathcal{A}_q(\mathfrak{g})$, $\lambda \in P^+$, is a well-defined homomorphism of algebras. Part (a) of Theorem 6.21 is proven.

Note that the K_λ, $\lambda \in \frac{1}{2}P$, satisfy the assumptions of Lemma 6.4 and so can be lifted to automorphisms $\widehat{K}_\lambda$ of $\widehat{\mathcal{B}}_q(\mathfrak{g})$. Define

$$\widehat{F}_i(x) = \frac{x_i\widehat{K}_{\frac{1}{2}\alpha_i}(x) - \widehat{K}_{-\frac{1}{2}\alpha_i}(x)x_i}{q_i - q_i^{-1}}, \qquad \widehat{E}_i(x) = \frac{\widehat{K}_{-\frac{1}{2}\alpha_i}(x)z_i - z_i\widehat{K}_{\frac{1}{2}\alpha_i}(x)}{q_i - q_i^{-1}}, \tag{6.14}$$

where $z_i = \widehat{\sigma}(x_{i^\star}) = q_i^{\frac{1}{2}(\delta_{i,i^\star}-1)}E_i(\Delta_{w_\circ\omega_{i^\star}})\Delta_{w_\circ\omega_{i^\star}}^{-1}$.

Proposition 6.25 *We have for all $\lambda \in \frac{1}{2}P$, $i \in I$:*

(a) $\widehat{K}_\lambda|_{\mathcal{B}_q(\mathfrak{g})} = K_\lambda$, $\widehat{F}_i|_{\mathcal{B}_q(\mathfrak{g})} = F_i$ *and* $\widehat{E}_i|_{\mathcal{B}_q(\mathfrak{g})} = E_i$*;*
(b) $\widehat{K}_\lambda \circ \widehat{\sigma} = \widehat{\sigma} \circ K_{w_\circ\lambda}$, $\widehat{F}_i \circ \widehat{\sigma} = \widehat{\sigma} \circ E_{i^\star}$ *and* $\widehat{E}_i \circ \widehat{\sigma} = \widehat{\sigma} \circ F_{i^\star}$.

Proof The first and the second assertions in part (a) are obvious. Furthermore, since $\widehat{F}_i = D^-_{(q_i-q_i^{-1})^{-1}x_i}$ and $\widehat{E}_i = D^+_{(q_i-q_i^{-1})^{-1}z_i}$ with $L_\pm = \widehat{K}_{\pm\frac{1}{2}\alpha_i}$ in the notation of (6.5), we immediately obtain the following

Lemma 6.26 *$\widehat{F}_i$ and $\widehat{E}_i$, $i \in I$ are $\widehat{K}_{\frac{1}{2}\alpha_i}$-derivations of $\widehat{\mathcal{B}}_q(\mathfrak{g})$.*

Thus, by Lemma 6.5 the last assertion in part (a) is equivalent to

$$\widehat{E}_i(v_\lambda) = 0, \qquad \widehat{E}_i(x_j) = \delta_{i,j}, \qquad \lambda \in P^+,\ i, j \in I.$$

Since $|z_i| = \alpha_i$ and $z_i \in \widehat{\mathcal{A}}_q(\mathfrak{g})$, we have

$$K_{-\frac{1}{2}\alpha_i}(v_\lambda)z_i - z_iK_{\frac{1}{2}\alpha_i}(v_\lambda) = q^{-\frac{1}{2}(\alpha_i,\lambda)}v_\lambda z_i - q^{\frac{1}{2}(\lambda,\alpha_i)}z_iv_\lambda = 0.$$

Thus, $\widehat{E}_i(v_\lambda) = 0$ for all $\lambda \in P^+$. We need the following

Lemma 6.27 *The following identity holds in $\mathcal{A}_q(\mathfrak{g})$ for all $i, j \in I$*

$$q_i^{\frac{1}{2}(\delta_{i,i^\star}-1)}(q^{\frac{1}{2}(\alpha_i,\alpha_j)}x_jE_i(\Delta_{w_\circ\omega_{i^\star}}) - q^{-\frac{1}{2}(\alpha_i,\alpha_j)}q_j^{\delta_{i,j}-\delta_{i^\star,j}}E_i(\Delta_{w_\circ\omega_{i^\star}})x_j)$$
$$= \delta_{i,j}(q_i - q_i^{-1})\Delta_{w_\circ\omega_{i^\star}}.$$

Proof This results from a straightforward computation by applying E_i to the identity

$$x_j \Delta_{w_\circ \omega_{i^\star}} = q_j^{\delta_{i,j}-\delta_{i^\star,j}} \Delta_{w_\circ \omega_{i^\star}} x_j$$

which is a special case of Lemma 6.20(c). □

Since $\Delta_{w_\circ\omega_{i^\star}}^{-1} x_j \Delta_{w_\circ\omega_{i^\star}} = q^{-(w_\circ\omega_{i^\star}+\omega_{i^\star},\alpha_j)} x_j = q_j^{\delta_{i,j}-\delta_{i^\star,j}} x_j$, we can write

$$\begin{aligned}&(q_i - q_i^{-1})\widehat{E}_i(x_j)\Delta_{w_\circ\omega_{i^\star}}\\ &= q_i^{\frac{1}{2}(\delta_{i,i^\star}-1)}(q^{\frac{1}{2}(\alpha_i,\alpha_j)}x_j E_i(\Delta_{w_\circ\omega_{i^\star}}) - q^{-\frac{1}{2}(\alpha_i,\alpha_j)}q_j^{\delta_{i,j}-\delta_{i^\star,j}} E_i(\Delta_{w_\circ\omega_{i^\star}})x_j).\end{aligned}$$

Using Lemma 6.27 we conclude that $(q_i - q_i^{-1})\widehat{E}_i(x_j)\Delta_{w_\circ\omega_{i^\star}} = \delta_{i,j}(q_i - q_i^{-1})\Delta_{w_\circ\omega_{i^\star}}$ and so $\widehat{E}_i(x_j) = \delta_{i,j}$. Part (a) of Proposition 6.25 is proven.

The first assertion in Proposition 6.25(b) is immediate since $|\widehat{\sigma}(x)| = w_\circ|x|$ for $x \in \mathcal{B}_q(\mathfrak{g})$ homogeneous. Furthermore, by (6.14) we obtain for all $x \in \mathcal{B}_q(\mathfrak{g})$.

$$\begin{aligned}\widehat{\sigma}(F_i(x)) = \frac{\widehat{\sigma}(x_i \widehat{K}_{\frac{1}{2}\alpha_i}(x)) - \widehat{\sigma}(\widehat{K}_{-\frac{1}{2}\alpha_i}(x)x_i)}{q_i - q_i^{-1}} &= \frac{\widehat{K}_{-\frac{1}{2}\alpha_{i^\star}}(\widehat{\sigma}(x))z_{i^\star} - z_{i^\star}\widehat{K}_{\frac{1}{2}\alpha_{i^\star}}(\widehat{\sigma}(x))}{q_i - q_i^{-1}}\\ &= \widehat{E}_{i^\star}(\widehat{\sigma}(x)).\end{aligned}$$

It remains to prove that $\widehat{\sigma}(E_i(x)) = \widehat{F}_{i^\star}(\widehat{\sigma}(x))$ for all $x \in \mathcal{B}_q(\mathfrak{g})$. Let $D_i = \widehat{\sigma}\circ E_i - \widehat{F}_{i^\star}\circ\widehat{\sigma}$. Since $\widehat{\sigma}\circ K_{\pm\frac{1}{2}\alpha_i} = \widehat{K}_{\mp\frac{1}{2}\alpha_{i^\star}}\circ\widehat{\sigma}$ it follows that D_i is a $\widehat{K}_{\frac{1}{2}\alpha_{i^\star}}\circ\widehat{\sigma}$-derivation from $\mathcal{B}_q(\mathfrak{g})$ to $\widehat{\mathcal{B}}_q(\mathfrak{g})^{op}$. We have

$$D_i(v_\lambda) = \widehat{\sigma}(E_i(v_\lambda)) - \widehat{F}_{i^\star}(v_{w_\circ\lambda}) = 0,$$

By Proposition 6.25(a) we have

$$\delta_{i,j} = \widehat{E}_i(x_j) = \frac{q^{\frac{1}{2}(\alpha_i,\alpha_j)}x_j z_i - q^{-\frac{1}{2}(\alpha_i,\alpha_j)}z_i x_j}{q_i - q_i^{-1}} = \frac{q_j - q_j^{-1}}{q_i - q_i^{-1}}\widehat{F}_j(z_i) \qquad (6.15)$$

and so

$$D_i(x_j) = \widehat{\sigma}(E_j(x_i)) - \widehat{F}_{i^\star}(z_{j^\star}) = \delta_{i,j} - \delta_{i^\star,j^\star} = 0.$$

Thus $D_i = 0$ on generators of $\mathcal{B}_q(\mathfrak{g})$. Then $D_i = 0$ by Lemma 6.5. This completes the proof of Proposition 6.25(b). □

To prove part (b) of Theorem 6.21, we need to show that $\widehat{\sigma}(V_\lambda) \subset V_\lambda$. The following lemma results from Proposition 6.25(a) by an obvious induction.

Lemma 6.28 *For any* $b \in \mathcal{B}_q(\mathfrak{g})$, $r \geq 1$, $(i_1, \ldots, i_r) \in I^r$, $\widehat{E}_{i_1} \cdots \widehat{E}_{i_r}(b) = E_{i_1} \cdots E_{i_r}(b) \in \mathcal{B}_q(\mathfrak{g})$. *In particular, for any* $v \in V_\lambda$, $\lambda \in P^+$ *we have* $\widehat{E}_{i_1} \cdots \widehat{E}_{i_r}(v) \in V_\lambda$.

Since V_λ is spanned by the $F_{i_1} \cdots F_{i_r}(v_\lambda)$, $r \geq 0$, $(i_1, \ldots, i_r) \in I^r$, it suffices to show that $\widehat{\sigma}(F_{i_1} \cdots F_{i_r}(v_\lambda)) \in V_\lambda$. We have by Proposition 6.25(b)

$$\widehat{\sigma}(F_{i_1} \cdots F_{i_r}(v_\lambda)) = \widehat{E}_{i_1^\star} \cdots \widehat{E}_{i_r^\star}(v_{w_\circ\lambda}) \in V_\lambda$$

by Lemma 6.28 applied with $v = v_{w_\circ\lambda}$.

Consider the operator $\sigma^I_{V_\lambda} \circ \widehat{\sigma}$. Clearly, it maps v_λ to itself and commutes with the $U_q(\mathfrak{g})$-action by Proposition 6.25(b). Since V_λ is a simple $U_q(\mathfrak{g})$-module generated by v_λ, it follows that $\widehat{\sigma}|_{V_\lambda} = (\sigma^I_{V_\lambda})^{-1} = \sigma^I_{V_\lambda}$. In particular, $\widehat{\sigma}$ is an involution on each V_λ and hence an anti-involution on $\mathcal{C}_q(\lambda)$. □

Corollary 6.29 *Let* $\mathfrak{g}$ *be finite-dimensional reductive. Then* σ^I *is an anti-involution on the algebra* $\mathcal{C}_q(\mathfrak{g})$.

6.4 σ^I on Upper Global Crystal Basis

Denote $\mathbf{B}^{up}$ the dual canonical basis in $\mathcal{A}_q(\mathfrak{g})$ and denote $\mathbf{B}_\lambda$ the upper global crystal basis of V_λ. Let $\mathbf{B} = \bigsqcup_{\lambda \in P^+} \mathbf{B}_\lambda$ be the upper global crystal basis of $\mathcal{C}_q(\mathfrak{g})$.

Theorem 6.30 *For any finite dimensional reductive* $\mathfrak{g}$ *we have* $\widehat{\sigma}(\mathbf{B}) = \mathbf{B}$. *In particular,* $\sigma^I_{V_\lambda}(\mathbf{B}_\lambda) = \mathbf{B}_\lambda$.

Proof We need the following

Lemma 6.31 (See e.g. [27, Proposition 2.33]) *For any* $\lambda \in P^+$, $\mathbf{j}(\mathbf{B}_\lambda) \subset \mathbf{B}^{up}$.

In particular, since $v_{w_\circ\lambda} \in \mathbf{B}_\lambda$, it follows that $\Delta_{w_\circ\lambda} \in \mathbf{B}^{up}$.

Denote

$$\begin{aligned}\widehat{\mathbf{B}^{up}} &= \{q^{\frac{1}{2}(w_\circ\lambda+\lambda,|b|)} b\Delta^{-1}_{w_\circ\lambda} \,:\, b \in \mathbf{B}^{up},\ \lambda \in P^+\} \\ &= \{q^{-\frac{1}{2}(w_\circ\lambda+\lambda,|b|)} \Delta^{-1}_{w_\circ\lambda} b \,:\, b \in \mathbf{B}^{up},\ \lambda \in P^+\}.\end{aligned}$$

We need the following

Lemma 6.32 *In the notation of Proposition 6.22,* $\sigma_0(\mathbf{B}^{up}) \subset \widehat{\mathbf{B}^{up}}$.

Proof Note that $\kappa(\mathbf{B}^{up}) = \mathbf{B}^{up}$ since it is a composition of two involutions preserving $\mathbf{B}^{up}$. In particular, $\kappa(\Delta_{w_\circ\lambda}) = \Delta_{w_\circ\lambda}$ for all $\lambda \in P^+$.

Let $b \in \mathbf{B}^{up}$, $\lambda \in P^+$. We have

$$\begin{aligned}\widehat{\kappa}(q^{\frac{1}{2}(w_\circ\lambda+\lambda,|b|)} b\Delta^{-1}_{w_\circ\lambda}) &= q^{\frac{1}{2}(w_\circ\lambda+\lambda,|b|)} (\kappa(\Delta_{w_\circ\lambda}))^{-1} \kappa(b) \\ &= q^{-\frac{1}{2}(w_\circ\lambda+\lambda,w_\circ|\kappa(b)|} \Delta^{-1}_{w_\circ\lambda} \kappa(b) \in \widehat{\mathbf{B}^{up}}.\end{aligned}$$

By [27, Theorem 5.4] we have $\zeta_{\mathbf{c}_0}(\mathbf{B}^{up}) \subset \widehat{\mathbf{B}^{up}}$ where $\zeta_{\mathbf{c}_0} : \mathcal{A}_q(\mathfrak{g}) \to \widehat{\mathcal{A}}_q(\mathfrak{g})$ is as in the proof of Proposition 6.22. Since $\sigma_0 = \widehat{\kappa} \circ \zeta_{\mathbf{c}_0}$, the assertion follows. □

Define

$$\widetilde{\mathbf{B}} = \{q^{\frac{1}{2}(\lambda,|b|)} b v_\lambda \,:\, b \in \mathbf{B}^{up},\ \lambda \in P^+\}.$$

It is immediate from the definition that $\mathbf{j}(\widetilde{\mathbf{B}}) = \mathbf{B}^{up}$ and that $\widetilde{\mathbf{B}}$ is a basis in $\mathcal{B}_q(\mathfrak{g})$. Moreover, it follows from Lemma 6.31 that $\mathbf{B} \subset \widetilde{\mathbf{B}}$ and

$$\mathbf{B} = \mathcal{C}_q(\mathfrak{g}) \cap \widetilde{\mathbf{B}}. \tag{6.16}$$

Finally, define

$$\widehat{\mathbf{B}} = \{q^{-\frac{1}{2}(w_\circ\lambda,|b|)} b v_{w_\circ\lambda} \,:\, b \in \widehat{\mathbf{B}^{up}},\ \lambda \in P^+\} \subset \widehat{\mathcal{B}}_q(\mathfrak{g}).$$

Proposition 6.33 $\widehat{\mathbf{B}} = \{q^{\frac{1}{2}(\lambda,|b|)} b v_\lambda \,:\, b \in \widehat{\mathbf{B}^{up}},\ \lambda \in P^+\}$. *In particular,* $\widehat{\mathbf{B}}$ *is a basis of* $\widehat{\mathcal{B}}_q(\mathfrak{g})$. *Finally,* $\widetilde{\mathbf{B}} \subset \widehat{\mathbf{B}}$.

Proof We need the following

Lemma 6.34 *Let R be a $\Bbbk$-algebra and let $S \subset R \setminus \{0\}$ be a commutative Ore submonoid. Let B be a basis of R and suppose that $\widehat{B} = \{\tau_s(b)s^{-1} \,:\, b \in B,\ s \in S\}$ is a basis of $R[S^{-1}]$ where $\tau_s : R \to R$ is some family of automorphisms satisfying $\tau_{ss'} = \tau_s \circ \tau_{s'}$, $\tau_s|_S = \mathrm{id}_S$. Then $\hat{\tau}_s(\widehat{B})s^{-1} = \widehat{B}$ for any $s \in S$, where $\hat{\tau}_s$ is the unique lifting of τ_s to $R[S^{-1}]$ provided by Lemma 6.4.*

Proof Define $f_s : R[S^{-1}] \to R[S^{-1}]$ by $f_s(x) = \hat{\tau}_s(x)s^{-1}$, $x \in R[S^{-1}]$. We claim that

$$f_s \circ f_{s'} = f_{ss'}, \qquad s, s' \in S \tag{6.17}$$

and f_s is invertible with $f_s^{-1}(x) = \hat{\tau}_s^{-1}(x)s$. Indeed, for all $x \in R[S^{-1}]$ we have

$$f_s(f_{s'}(x)) = \hat{\tau}_s(\hat{\tau}_{s'}(x)s'^{-1})s^{-1} = \hat{\tau}_{ss'}(x)s'^{-1}s^{-1} = f_{ss'}(x).$$

and also $f_s(\hat{\tau}_s^{-1}(x)s) = x$ and $\hat{\tau}_{s^{-1}}(f_s(x)) = x = f_s(\hat{\tau}_{s^{-1}}(x))$.

We have $\widehat{B} = \{f_s(b) \,:\, b \in B,\ s \in S\}$ and the assertion of the lemma is equivalent to $f_s(\widehat{B}) = \widehat{B}$ for all $s \in S$. Clearly, (6.17) implies that $f_s(\widehat{B}) \subset \widehat{B}$ for all $s \in S$. To prove the opposite inclusion, let $\hat{b} \in \widehat{B}$. Write

$$f_s^{-1}(\hat{b}) = \sum_{\hat{b}' \in \widehat{B}} \lambda_{\hat{b}'} \hat{b}'.$$

Then

$$\hat{b} = \sum_{\hat{b}' \in \widehat{B}} \lambda_{\hat{b}'} f_s(\hat{b}') = \sum_{\hat{b}'' \in f_s(\widehat{B}) \subset \widehat{B}} \lambda_{f_s^{-1}(\hat{b}'')} \hat{b}'',$$

where we used that $f_s(\widehat{B}) \subset \widehat{B}$. Since $\widehat{B}$ is a basis, this implies that $\lambda_{f_s^{-1}(\hat{b}'')} = \delta_{\hat{b},\hat{b}''}$ and so $\hat{b} \in f_s(\widehat{B})$. Therefore, $\widehat{B} \subset f_s(\widehat{B})$. □

Apply this lemma to $R = \mathcal{A}_q(\mathfrak{g})$, $S = \mathcal{S}_{w_\circ}$, $B = \mathbf{B}^{up}$ and $\widehat{B} = \widehat{\mathbf{B}^{up}}$. By [27, Proposition 3.9], $\widehat{\mathbf{B}^{up}}$ is a basis of $\widehat{\mathcal{A}}_q(\mathfrak{g})$.[1] We have $\tau_{\Delta_{w_\circ\lambda}}(b) = q^{\frac{1}{2}(w_\circ\lambda+\lambda,|b|)}b$. Then all assumptions of Lemma 6.34 are satisfied. Indeed, $\tau_{\Delta_{w_\circ\lambda}}\tau_{\Delta_{w_\circ\mu}}(b) = q^{\frac{1}{2}(w_\circ(\lambda+\mu)+\lambda+\mu,|b|)}b = \tau_{\Delta_{w_\circ(\lambda+\mu)}}(b)$ and $\tau_{\Delta_{w_\circ\lambda}}(\Delta_{w_\circ\mu}) = q^{\frac{1}{2}(w_\circ\lambda+\lambda,w_\circ\mu-\mu)}\Delta_{w_\circ\mu} = \Delta_{w_\circ\mu}$, $\lambda, \mu \in P^+$. Thus by Lemma 6.34 we have, for any $\lambda \in P^+$, $\widehat{\mathbf{B}^{up}} = \hat{\tau}^{-1}_{\Delta_{w_\circ\lambda}}(\widehat{\mathbf{B}^{up}})\Delta_{w_\circ\lambda} = \{q^{-\frac{1}{2}(w_\circ\lambda+\lambda,|b|)}b\Delta_{w_\circ\lambda} : b \in \widehat{\mathbf{B}^{up}}\}$.

Since $v_{w_\circ\lambda} = q^{\frac{1}{2}(w_\circ\lambda-\lambda,\lambda)}\Delta_{w_\circ\lambda}v_\lambda$ we have

$$\begin{aligned}
\widehat{\mathbf{B}} &= \{q^{-\frac{1}{2}(w_\circ\lambda,|b|)}bv_{w_\circ\lambda} : b \in \widehat{\mathbf{B}^{up}}, \lambda \in P^+\} \\
&= \{q^{\frac{1}{2}(\lambda,|b|)}\tau^{-1}_{\Delta_{w_\circ\lambda}}(b)v_{w_\circ\lambda} : b \in \widehat{\mathbf{B}^{up}}, \lambda \in P^+\} \\
&= \{q^{\frac{1}{2}(\lambda,|b|+w_\circ\lambda-\lambda)}\tau^{-1}_{\Delta_{w_\circ\lambda}}(b)\Delta_{w_\circ\lambda}v_\lambda : b \in \widehat{\mathbf{B}^{up}}, \lambda \in P^+\} \\
&= \{q^{\frac{1}{2}(\lambda,|b'|)}b'v_\lambda : b' \in \widehat{\mathbf{B}^{up}}, \lambda \in P^+\}
\end{aligned}$$

where we denoted $b' = \tau^{-1}_{\Delta_{w_\circ\lambda}}(b)\Delta_{w_\circ\lambda}$ and observed that $|b'| = |b| + w_\circ\lambda - \lambda$. This proves the first assertion of Proposition 6.33. The second and third assertions are now immediate. □

Now we can complete the proof of Theorem 6.30. It follows from Proposition 6.33 and Lemma 6.32 that for any $b \in \mathbf{B}^{up}$, $\lambda \in P^+$ we have

$$\widehat{\sigma}(q^{\frac{1}{2}(\lambda,|b|)}bv_\lambda) = q^{\frac{1}{2}(\lambda,|b|)}v_{w_\circ\lambda}\sigma_0(b) = q^{-\frac{1}{2}(w_\circ\lambda,|\sigma_0(b)|)}\sigma_0(b)v_{w_\circ\lambda} \in \widehat{\mathbf{B}}.$$

Thus, $\widehat{\sigma}(\widetilde{\mathbf{B}}) \subset \widehat{\mathbf{B}}$. Since $\mathbf{B} \subset \widetilde{\mathbf{B}}$, it follows that $\widehat{\sigma}(\mathbf{B}) \subset \widehat{\mathbf{B}}$. Then by Theorem 6.21(b) we conclude that $\widehat{\sigma}(\mathbf{B}) \subset \widehat{\mathbf{B}} \cap \mathcal{C}_q(\mathfrak{g})$. On the other hand, since $\widehat{\mathbf{B}}$ is linearly independent by Proposition 6.33, its intersection with $\mathcal{C}_q(\mathfrak{g})$ is also linearly independent. Since $\widetilde{\mathbf{B}} \subset \widehat{\mathbf{B}}$ by Proposition 6.33, it follows from (6.16) that $\mathbf{B} = \widetilde{\mathbf{B}} \cap \mathcal{C}_q(\mathfrak{g}) \subset \widehat{\mathbf{B}} \cap \mathcal{C}_q(\mathfrak{g})$. But $\mathbf{B}$ is a basis of $\mathcal{C}_q(\mathfrak{g})$ and so $\mathbf{B} = \widehat{\mathbf{B}} \cap \mathcal{C}_q(\mathfrak{g})$. Thus, $\widehat{\sigma}(\mathbf{B}) \subset \mathbf{B}$. Since by Theorem 6.21(b) $\widehat{\sigma}$ is an involution on $\mathcal{C}_q(\mathfrak{g})$, $\widehat{\sigma}(\mathbf{B}) = \mathbf{B}$ which

[1] The difference between our notation and that of [7, 27] is in the linear automorphism of $\mathcal{A}_q(\mathfrak{g})$ defined on homogeneous elements x by $x \mapsto q^{\frac{1}{2}(|x|,|x|)-(|x|,\rho)}x$.

completes the proof of the first assertion of Theorem 6.30. The second assertion is immediate from the first and Theorem 6.21(b). □

6.5 Proof of Theorem 1.10

Proof Let V be any object in $\mathcal{O}_q^{int}(\mathfrak{g})$. Let (L^{up}, B^{up}) be an upper crystal basis of V and let $G(B^{up})$ be the corresponding upper global crystal basis (see [25]). By [25, Theorem 3.3.1] there exists a direct sum decomposition $V = \sum_j V^j$ such that $\mathbf{B}^j := G(B^{up}) \cap V^j$ is a basis of V^j and each $V^j \cong V_{\lambda_j}$, $\lambda_j \in P^+$. The latter isomorphism identifies $\mathbf{B}^j$ with $\mathbf{B}_{\lambda_j}$. Since by Theorem 4.10, σ_V^I is compatible with direct sum decompositions, the restriction of σ_V^I to V^j coincides with $\sigma_{V_j}^I$ and under the above isomorphism it identifies with $\sigma_{V_{\lambda_j}}^I$ and thus preserves $\mathbf{B}_{\lambda_j}$ by Theorem 6.30. □

7 Examples

7.1 Thin Modules

Let $\lambda \in P^+$. We say that V_λ is *quasi-miniscule* if $V_\lambda(\beta) \neq 0$ implies that $\beta \in W\lambda \cup \{0\}$. For example, V_{ω_i}, $i \in I$ for $\mathfrak{g} = \mathfrak{sl}_n$ are (quasi)-miniscule, as well as the quantum analogue of the adjoint representation of $\mathfrak{g}$.

Lemma 7.1 *Conjecture 1.2 holds for any quasi-miniscule $V = V_\lambda$.*

Proof Let $v = v(\lambda) \in V_\lambda(\lambda)$. Then in the notation of (3.2) we have $V_\lambda = \Bbbk \cdot [v]_W \oplus V_\lambda(0)$. As shown in Proposition 3.8, the action of $\mathsf{W}(V)$ on the basis $[v]_W$ of $\Bbbk \cdot [v]_W$ is given by the Weyl group action on W/W_{J_λ}. It remains to observe that $\sigma^i|_{V_\lambda(0)} = \mathrm{id}_{V_\lambda(0)}$, $i \in I$. □

This result can be extended to a larger class of modules. We say that $V \in \mathcal{O}_q^{int}(\mathfrak{g})$ is *thin* if $\dim V(\beta) \leq 1$ for all $\beta \in P \setminus \{0\}$. By definition, every quasi-miniscule module is thin. Furthermore, all modules $V_{m\omega_1}$, $V_{m\omega_n}$, $m \in \mathbb{Z}_{\geq 0}$ are thin for $\mathfrak{g} = \mathfrak{sl}_{n+1}$.

Theorem 7.2 *Conjecture 1.2 holds for thin modules.*

Proof Let (L, B) be an upper crystal basis of $V \in \mathcal{O}_q^{int}(\mathfrak{g})$ and let $G^{up}(B) \subset V$ be the corresponding global crystal basis. We say that $b \in G^{up}(B)$ of weight $\beta \in P$ is thin if either $\beta = 0$ or $V(\beta) = \Bbbk b$. Denote $G_0^{up}(B)$ the set of thin elements in $G^{up}(B)$. Clearly, V is thin if and only if $G_0^{up}(B) = G^{up}(B)$. We need the following

Proposition 7.3 *For any* $b \in G^{up}(B) \cap V(\beta)$, $\beta \in P$ *with* $\dim V(\beta) = 1$ *we have* $\Phi_V(b) \subset G^{up}(B)$.

Proof It suffices to prove that $\sigma^J(b) \in G^{up}(B)$ for all $J \in \mathscr{J}$. Since $\sigma^J(V(\beta)) = V(w_\circ^J\beta)$ and $\dim V(w_\circ^J\beta) = \dim V(\beta) = 1$, it follows that $\sigma^J(b) = cb'$ for some $b' \in G^{up}(B) \cap V(w_\circ^J\beta)$. Let $\underline{b}$ be the image of b in B under the quotient map $L \mapsto L/qL$. By Theorem 1.8, $\tilde{\sigma}^J(\underline{b}) = \underline{b}'$ and so $c \in 1 + q\mathbb{A}$. It follows from Proposition 5.8 that $\overline{c} = c$ and so $c = 1$. □

Let $g : B \to G^{up}(B)$ be Kashiwara's bijection (cf. [25]) and let $B_0 = g^{-1}(G_0^{up}(B))$. Then by Theorem 1.8 and Proposition 7.3 we have $g(\tilde{\sigma}^i(b)) = \sigma^i(g(b))$ for all $b \in B_0$. Since the action of $\tilde{\sigma}^i$ on B coincides with the action of W defined in [26], it follows from [26, Theorem 7.2.2] that $\mathsf{W}(V)$ is a homomorphic image of W.

We may assume, without loss of generality, that $V = V_\lambda$ and $J(V) = \emptyset$. In view of Proposition 3.8(b), the action of $\mathsf{W}(V)$ on the set $[v_\lambda]_W$, $v_\lambda \in V_\lambda(\lambda)$ is faithful and coincides with that of W. This implies that ψ_V from Theorem 1.1 is an isomorphism. □

7.2 Crystallizing Cactus Group Action for $\mathfrak{g} = \mathfrak{sl}_3$

We now describe combinatorial consequences of Theorem 1.8 for $\mathfrak{g} = \mathfrak{sl}_3$. It turns out that the corresponding action of Cact_{S_3} lifts to the ambient set

$$\widehat{\mathbf{M}} = \{(m_1, m_2, m_{12}, m_{21}, m_{01}, m_{02}) \in \mathbb{Z}_{\ge 0}^2 \times \mathbb{Z}^4 \ : \ m_1 m_2 = 0\}$$

where the crystal basis for $\mathcal{C}_q(\mathfrak{sl}_3)$ identifies with $\mathbf{M} = \widehat{\mathbf{M}} \cap \mathbb{Z}_{\ge 0}^6$.

We need some notation. Define $\mathrm{wt}_i : \widehat{\mathbf{M}} \to \mathbb{Z}$ by

$$\mathrm{wt}_i(\mathbf{m}) = m_{0i} - m_i + m_j - m_{ij}, \qquad \{i, j\} = \{1, 2\}.$$

Furthermore, define $e_i^r : \widehat{\mathbf{M}} \to \widehat{\mathbf{M}}$, $i \in \{1, 2\}$, $r \in \mathbb{Z}$, by

$$e_i^r(m_1, m_2, m_{12}, m_{21}, m_{01}, m_{02}) = (m_1', m_2', m_{12}', m_{21}', m_{01}', m_{02}')$$

where, for $\mathbf{m} = (m_1, m_2, m_{12}, m_{21}, m_{01}, m_{02}) \in \widehat{\mathbf{M}}$ we set

$$m_i' = [m_i - m_j - r]_+, \quad m_j' = [m_j - m_i + r]_+, \quad m_{ji}' = m_{ji}, \quad m_{0j}' = m_{0j},$$

$$m_{ij}' = m_{ij} + \min(m_i - r, m_j), \quad m_{0i}' = m_{0i} + r + \min(m_i - r, m_j), \quad \{i, j\} = \{1, 2\},$$

and $[x]_+ := \max(x, 0)$, $x \in \mathbb{Z}$. The following is well-known (cf. [5, Example 6.26]).

Lemma 7.4 *The e_i^r, $i \in \{1,2\}$, $r \in \mathbb{Z}$ satisfy*

$$e_1^r e_1^s = e_1^{r+s}, \quad e_2^r e_2^s = e_2^{r+s}, \quad e_1^r e_2^{r+s} e_1^s = e_2^s e_1^{r+s} e_2^r, \quad r,s \in \mathbb{Z}. \tag{7.1}$$

In particular, $e_i^0 = \mathrm{id}$ and $(e_i^r)^{-1} = e_i^{-r}$, $r \in \mathbb{Z}$, $i \in \{1,2\}$.

Proof Define a map $\widehat{\mathbf{k}} : \widehat{\mathbf{M}} \to \mathbb{Z}^5$ by $\widehat{\mathbf{k}}(m_1, m_2, m_{12}, m_{21}, m_{01}, m_{02}) \mapsto (a_1, a_2, a_3, l_1, l_2)$ where

$$a_1 = m_1 + m_{21}, \quad a_2 = m_2 + m_{12} + m_{21}, \quad a_3 = m_{12}, \quad l_i = m_i + m_{3-i,i} + m_{0i}$$

with $i \in \{1,2\}$. It is easy to see that $\widehat{\mathbf{k}}$ is a bijection with its inverse given by

$$(a_1, a_2, a_3, l_1, l_2) \mapsto (m_1, m_2, m_{12}, m_{21}, m_{01}, m_{02}),$$

where $m_1 = [a_1 + a_3 - a_2]_+$, $m_2 = [a_2 - a_1 - a_3]_+$, $m_{12} = a_3$, $m_{21} = \min(a_1, a_2 - a_3)$, $m_{01} = l_1 - a_1$, $m_{02} = l_2 - a_2 + \min(a_1, a_2 - a_3)$.

The action of operators e_i^r, $i \in \{1,2\}$, $r \in \mathbb{Z}$ on $\mathbb{Z}^5$ induced by this bijection coincides with the action constructed in [5, Example 6.26]

$$e_1^r(a_1, a_2, a_3, l_1, l_2) = (a_1 + [\delta - r]_+ - [\delta]_+, a_2, a_3 + [\delta]_+ - \max(\delta, r), l_1, l_2)$$
$$e_2^r(a_1, a_2, a_3, l_1, l_2) = (a_1, a_2 - r, a_3, l_1, l_2).$$

where $\delta = a_1 + a_3 - a_2$. The identities from the Lemma are now easy to obtain by using tropicalized relations for the e_i given after Definition 2.20 in [5] in the context of [5, Example 6.26]. □

Define $\underline{\sigma} = \underline{\sigma}^{\{1,2\}} : \widehat{\mathbf{M}} \to \widehat{\mathbf{M}}$ by

$$(m_1, m_2, m_{12}, m_{21}, m_{01}, m_{02}) \mapsto (m_1, m_2, m_{02}, m_{01}, m_{21}, m_{12}).$$

Clearly, $\underline{\sigma}$ is an involution and $\underline{\sigma}(\mathbf{M}) = \mathbf{M}$. Furthermore, define $\underline{\sigma}^i : \widehat{\mathbf{M}} \to \widehat{\mathbf{M}}$, $i \in \{1,2\}$ by

$$\underline{\sigma}^i(\mathbf{m}) = e_i^{-\mathrm{wt}_i(\mathbf{m})}(\mathbf{m}), \qquad \mathbf{m} \in \widehat{\mathbf{M}}.$$

Proposition 7.5 *The following identities hold in* $\mathrm{Bij}(\widehat{\mathbf{M}})$

$$\underline{\sigma}^i \circ \underline{\sigma}^i = \mathrm{id}, \qquad \underline{\sigma}^i \circ e_i^r = e_i^{-r} \circ \underline{\sigma}^i, \qquad \underline{\sigma} \circ e_i^r = e_j^{-r} \circ \underline{\sigma},$$
$$\underline{\sigma}^i \circ \underline{\sigma}^j \circ \underline{\sigma}^i = \underline{\sigma}^j \circ \underline{\sigma}^i \circ \underline{\sigma}^j, \qquad \underline{\sigma}^i \circ \underline{\sigma} = \underline{\sigma} \circ \underline{\sigma}^j,$$

where $\{i,j\} = \{1,2\}$. In particular, the assignments $\tau_{i,i+1} \mapsto \underline{\sigma}^i$, $i \in \{1,2\}$, $\tau_{13} \mapsto \underline{\sigma}$ define an action of Cact_{S_3} *on* $\widehat{\mathbf{M}}$.

Proof Since $\mathrm{wt}_i(e_i^r(\mathbf{m})) = \mathrm{wt}_i(\mathbf{m}) + 2r$ for any $\mathbf{m} \in \widehat{\mathbf{M}}$, we have

$$\underline{\sigma}^i \circ \underline{\sigma}^i(\mathbf{m}) = e_i^{-(\mathrm{wt}_i(\mathbf{m})-2\,\mathrm{wt}_i(\mathbf{m}))-\mathrm{wt}_i(\mathbf{m})}(\mathbf{m}) = \mathbf{m},$$

while

$$\underline{\sigma}^i \circ e_i^r(\mathbf{m}) = e_i^{-r-\mathrm{wt}_i(\mathbf{m})}(\mathbf{m}) = e_i^{-r} \circ \underline{\sigma}^i(\mathbf{m}).$$

To prove the third identity, note that $e_i^r(\underline{\sigma}(\mathbf{m})) = (\tilde{m}_1, \tilde{m}_2, \tilde{m}_{12}, \tilde{m}_{21}, \tilde{m}_{01}, \tilde{m}_{02})$, where

$$\tilde{m}_i = [m_i - m_j - r]_+, \quad \tilde{m}_j = [m_j - m_i + r]_+, \quad \tilde{m}_{ji} = m_{0i}, \quad \tilde{m}_{0j} = m_{ij},$$
$$\tilde{m}_{ij} = m_{0j} + \min(m_i - r, m_j), \quad \tilde{m}_{0i} = m_{ji} + \min(m_i, m_j + r), \qquad \{i, j\} = \{1, 2\},$$

which is easily seen to coincide with $\underline{\sigma}(e_j^{-r}(\mathbf{m}))$. The braid identity follows from the last relation in (7.1) (known as Verma relations) and the identity $\mathrm{wt}_j(e_i^r(\mathbf{m})) = \mathrm{wt}_j(\mathbf{m}) - r$, $\{i, j\} = \{1, 2\}$. Finally,

$$\underline{\sigma} \circ \underline{\sigma}^j(\mathbf{m}) = \underline{\sigma} \circ e_j^{-\mathrm{wt}_j(\mathbf{m})}(\mathbf{m}) = e_i^{\mathrm{wt}_j(\mathbf{m})}(\underline{\sigma}(\mathbf{m})) = e_i^{-\mathrm{wt}_i(\underline{\sigma}(\mathbf{m}))}(\underline{\sigma}(\mathbf{m})) = \underline{\sigma}^i \circ \underline{\sigma}(\mathbf{m}),$$

where we used the identity $\mathrm{wt}_i(\underline{\sigma}(\mathbf{m})) = m_{ji} - m_i + m_j - m_{0j} = -\mathrm{wt}_j(\mathbf{m})$. □

Remark 7.6 It would be interesting to define analogues of $\widehat{\mathbf{M}}$ for other $\mathfrak{g}$ and study the action of the corresponding cactus groups on $\widehat{\mathbf{M}}$. We plan to study this in a subsequent publication via the approach of [5].

Given $l_1, l_2 \in \mathbb{Z}$ define

$$\widehat{\mathbf{M}}_{l_1,l_2} = \{(m_1, m_2, m_{12}, m_{21}, m_{01}, m_{02}) \in \widehat{\mathbf{M}}$$
$$: m_{01} + m_1 + m_{21} = l_1,\ m_{02} + m_2 + m_{12} = l_2\}.$$

Clearly, $\underline{\sigma}$, $\underline{\sigma}^i$, e_i^r, $i \in \{1, 2\}$, $r \in \mathbb{Z}$ preserve $\widehat{\mathbf{M}}_{l_1,l_2}$ for any $l_1, l_2 \in \mathbb{Z}$. Set $\mathbf{M}_{l_1,l_2} = \widehat{\mathbf{M}}_{l_1,l_2} \cap \mathbf{M}$.

In view of [5, Example 6.26], $\widehat{\mathbf{k}}(\mathbf{M}_{l_1,l_2})$, where $\widehat{\mathbf{k}}$ is defined in the proof of Lemma 7.4, identifies with the upper crystal basis $B^{up}(V_{l_1\omega_1+l_2\omega_2})$ of $V_{l_1\omega_1+l_2\omega_2}$. In particular, $\widehat{\mathbf{k}}(\mathbf{M})$ identifies with the upper crystal basis $B^{up}(\mathcal{C}_2) = \bigsqcup_{\lambda \in P^+} B^{up}(V_\lambda)$ of $\mathcal{C}_2 = \mathcal{C}_q(\mathfrak{sl}_3)$. We use this identification throughout the rest of this chapter.

Proposition 7.7 *Under the above identification, the restrictions of* $\underline{\sigma}$, $\underline{\sigma}^i$, $i \in \{1, 2\}$ *to* $\mathbf{M}$ *coincide with the action of* Cact_{S_3} *on* $B^{up}(\mathcal{C}_2)$ *provided by Theorem 1.8 with* $\mathfrak{g} = \mathfrak{sl}_3$ *and* $V = \mathcal{C}_2$.

Proof It follows from Corollary 5.6 applied to $f = \underline{\sigma}$ extended to $B_\lambda \cup \{0\}$, and Proposition 7.5 that $\underline{\sigma}$ coincides with $\tilde{\sigma}^{\{1,2\}}_{V_\lambda}$ for all $\lambda \in P^+$. On the other hand, by Remark 5.7 we have $\underline{\sigma}^i = \tilde{\sigma}^{\{i\}}_{V_\lambda}$, $i \in \{1,2\}$, $\lambda \in P^+$. □

7.3 Gelfand–Kirillov Model for $\mathfrak{g} = \mathfrak{sl}_3$

Our goal here is to illustrate results and constructions from Sect. 6 for $\mathfrak{g} = \mathfrak{sl}_3$ and provide some evidence for Conjecture 1.2. We freely use the notation from Sects. 6 and 7.2. In this case the algebra $\mathcal{A}_2 = \mathcal{A}_q(\mathfrak{g})$ is generated by the x_i, $i \in \{1,2\}$ subject to the relations

$$x_i^2 x_j - (q+q^{-1}) x_i x_j x_i + x_j x_i^2 = 0, \qquad \{i,j\} = \{1,2\}. \tag{7.2}$$

Define

$$x_{ij} = \frac{q^{\frac{1}{2}} x_i x_j - q^{-\frac{1}{2}} x_j x_i}{q - q^{-1}}, \qquad \{i,j\} = \{1,2\}.$$

Then $x_i x_j = q^{\frac{1}{2}} x_{ij} + q^{-\frac{1}{2}} x_{ji}$, $\{i,j\} = \{1,2\}$ and (7.2) is equivalent to $x_i x_{ij} = q x_{ij} x_i$ or $x_i x_{ji} = q^{-1} x_{ji} x_i$, $\{i,j\} = \{1,2\}$. The following is well-known[1] (see e.g. [7]).

Lemma 7.8 *The dual canonical basis* $\mathbf{B}^{up}$ *in the algebra* $\mathcal{A}_2$ *is*

$$\begin{aligned}\mathbf{B}^{up} =& \{q^{\frac{1}{2}(m_1-m_2)(m_{21}-m_{12})} x_1^{m_1} x_2^{m_2} x_{12}^{m_{12}} x_{21}^{m_{21}} \\ & : (m_1, m_2, m_{12}, m_{21}) \in \mathbb{Z}^4_{\ge 0},\ m_1 m_2 = 0\}.\end{aligned}$$

We have $\widehat{\mathcal{A}_2} = \mathcal{A}_2[\mathcal{S}_{w_\circ}^{-1}] = \mathcal{A}_2[x_{12}^{-1}, x_{21}^{-1}]$. It follows from Lemma 7.8 that

$$\begin{aligned}\widehat{\mathbf{B}^{up}} =& \{q^{\frac{1}{2}(m_1-m_2)(m_{21}-m_{12})} x_1^{m_1} x_2^{m_2} x_{12}^{m_{12}} x_{21}^{m_{21}} \\ & : (m_1, m_2, m_{12}, m_{21}) \in \mathbb{Z}^2_{\ge 0} \times \mathbb{Z}^2,\ m_1 m_2 = 0\}.\end{aligned}$$

The following is immediate

Lemma 7.9

(a) *The algebra* $\mathcal{B}_2 := \mathcal{B}_q(\mathfrak{g})$ *is generated by* $\mathcal{A}_2$ *and* $\Bbbk[v_1, v_2]$, *where* $v_i = v_{\omega_i}$, *as subalgebras subject to the relations*

$$v_i x_j = q^{-\delta_{i,j}} x_j v_i, \qquad i,j \in \{1,2\}.$$

(b) *$\mathcal{B}_2$ is a $U_q(\mathfrak{g})$-module algebra with the $U_q(\mathfrak{g})$-action defined in Lemma 6.7.*

Abbreviate $z_i = F_i(v_i) = v_{s_i\omega_i}$ and $z_{ij} = F_iF_j(v_j) = v_{s_is_j\omega_j}$. Clearly,

$$z_i = q^{-\frac{1}{2}}x_iv_i, \qquad z_{ij} = q^{-\frac{1}{2}}x_{ij}v_j, \qquad \{i,j\} = \{1,2\}. \tag{7.3}$$

The following Lemma is an immediate consequence of Lemma 7.9

Lemma 7.10

(a) *The algebra $\mathcal{C}_2 = \mathcal{C}_q(\mathfrak{sl}_3)$ is generated by $v_1, v_2, z_1, z_2, z_{12}$ and z_{21} subject to the relations*

$$v_1v_2 = v_2v_1, \quad v_iz_j = q^{-\delta_{i,j}}z_jv_i, \quad v_iz_{12} = q^{-1}z_{12}v_i, \quad v_iz_{21} = q^{-1}z_{21}v_i,$$
$$i,j \in \{1,2\}.$$

$$z_iz_j = qv_iz_{ij} + q^{-1}z_{ji}v_j, \quad z_kz_{ij} = q^{-\delta_{j,k}}z_{ij}z_k, \quad \{i,j\} = \{1,2\},\ k \in \{1,2\}$$

and $z_{12}z_{21} = z_{21}z_{12}$.

(b) *$\mathcal{C}_2$ is a $U_q(\mathfrak{g})$-module algebra via*

$$E_k(v_i) = 0, \quad E_k(z_i) = \delta_{k,i}v_i, \quad E_k(z_{ji}) = \delta_{k,j}z_i,$$
$$F_k(v_i) = \delta_{i,k}z_i, \quad F_k(z_i) = \delta_{k,j}z_{ji}, \quad F_k(z_{ij}) = 0, \quad \{i,j\} = \{1,2\},\ k \in \{1,2\}.$$

(c) *The P-grading on $\mathcal{C}_q(\mathfrak{g})$ is given by $|v_i| = \omega_i$, $|z_i| = s_i\omega_i = \omega_i - \alpha_i$, $|z_{ij}| = s_js_i\omega_j = \omega_j - \alpha_i - \alpha_j$, $\{i,j\} = \{1,2\}$.*

The following is an immediate corollary of Theorem 6.21.

Corollary 7.11 *The assignments*

$$v_i \mapsto z_{ji}, \quad z_i \mapsto z_i, \quad z_{ij} \mapsto v_j, \qquad \{i,j\} = \{1,2\}.$$

define an anti-involution of $\mathcal{C}_2$ which coincides with $\sigma = \sigma_{\mathcal{C}_2}^{\{1,2\}}$.

Given $\mathbf{m} \in \widehat{\mathbf{M}}$, define $\mathbf{b_m} \in \mathcal{C}_2$ as

$$\mathbf{b_m} = q^{\frac{1}{2}(m_1(m_{21}-m_{01})+m_2(m_{12}-m_{02})-(m_{12}+m_{21})(m_{01}+m_{02}))}z_1^{m_1}z_2^{m_2}z_{12}^{m_{12}}z_{21}^{m_{21}}v_1^{m_{01}}v_2^{m_{02}} \tag{7.4}$$

if $\mathbf{m} \in \mathbf{M}$ and $\mathbf{b_m} = 0$ if $\mathbf{m} \in \widehat{\mathbf{M}} \setminus \mathbf{M}$. Thus, $|\mathbf{b_m}| = (m_{01} + m_1 + m_{21})\omega_1 + (m_{02} + m_2 + m_{12})\omega_2 - (m_1 + m_{12} + m_{21})\alpha_1 - (m_2 + m_{12} + m_{21})\alpha_2 = \mathrm{wt}_1(\mathbf{m})\omega_1 + \mathrm{wt}_2(\mathbf{m})\omega_2$, $\mathbf{m} \in \mathbf{M}$.

The following is a consequence of Lemma 7.10, (6.16) and Proposition 6.33

Lemma 7.12 *The upper global crystal basis $\mathbf{B}$ of $\mathcal{C}_2$ is $\mathbf{B} = \{\mathbf{b_m} : \mathbf{m} \in \mathbf{M}\}$. Furthermore, for each $\lambda = l_1\omega_1 + l_2\omega_2 \in P^+$ we have $\mathbf{B}_\lambda = \{\mathbf{b_m} : \mathbf{m} \in \mathbf{M}_{l_1,l_2}\}$.*

Moreover, under the identification of $\mathbf{M}$ *with the upper crystal basis of* $\mathcal{C}_q(\mathfrak{g})$ *(cf. Sect. 7.2), the map* $\mathbf{M} \to \mathbf{B}$ *defined by* $\mathbf{m} \mapsto \mathbf{b_m}$, $\mathbf{m} \in \mathbf{M}$ *is Kashiwara's bijection* G [25] *between an upper crystal basis of* $\mathcal{C}_2$ *and its upper global crystal basis.*

The following is an explicit form of Theorem 6.30 for $\mathfrak{g} = \mathfrak{sl}_3$ and is immediate from (7.4) and Corollary 7.11.

Lemma 7.13 *We have* $\sigma(\mathbf{b_m}) = \mathbf{b}_{\underline{\sigma}(\mathbf{m})}$ *for all* $\mathbf{m} \in \mathbf{M}$.

Remark 7.14 It is easy to check that

$$|\mathbf{B}_\lambda(0)| = |\{\mathbf{m} \in \mathbf{M}_{l_1,l_2} \,:\, \mathrm{wt}_1(\mathbf{m}) = \mathrm{wt}_2(\mathbf{m}) = 0\}| = \begin{cases} \min(l_1,l_2)+1, & l_1 \equiv l_2 \pmod 3 \\ 0, & \text{otherwise.} \end{cases}$$

It follows from the definition of $\underline{\sigma}$ that σ is trivial on $\mathbf{B}_\lambda(0)$ if and only if $\dim V_\lambda(0) = 1$ (that is, if and only if $\min(l_1,l_2) = 0$ and $\max(l_1,l_2) \in 3\mathbb{Z}_{>0}$). Thus, $\tau_{1,3} \notin \mathsf{K}_{\mathfrak{sl}_3}$ in the notation introduced after Problem 1.7. On the other hand, it is immediate from the definitions that the σ^i, $i \in \{1,2\}$ act trivially on $V_\lambda(0)$ for any $V \in \mathscr{O}_q^{int}(\mathfrak{g})$. In particular, σ is not contained in $\mathsf{W}(V_\lambda)$ if $\dim V_\lambda(0) > 1$.

In order to calculate σ^i, $i \in \{1,2\}$ we need the following result.

Lemma 7.15 *For any* $\mathbf{m} \in \mathbf{M}$, $r \geq 0$ *and* $i \in \{1,2\}$ *we have*

$$E_i^{(r)}(\mathbf{b_m}) = \binom{m_i + m_{ij}}{r}_q \mathbf{b}_{e_i^r(\mathbf{m})} + \sum_{1\le t\le r} C_t^{(r)}(m_j + m_{ij}, m_i + m_{ij})\mathbf{b}_{e_i^r(\mathbf{m})+t\mathbf{a}_i^+}$$

$$F_i^{(r)}(\mathbf{b_m}) = \binom{m_j + m_{0i}}{r}_q \mathbf{b}_{e_i^{-r}(\mathbf{m})} + \sum_{1\le t\le r} C_t^{(r)}(m_i + m_{0i}, m_j + m_{0i})\mathbf{b}_{e_i^{-r}(\mathbf{m})-t\mathbf{a}_i^+}$$

where $\mathbf{a}_1^+ = (0,0,-1,1,-1,1)$, $\mathbf{a}_2^+ = -\mathbf{a}_1^+$ *and*

$$C_t^{(r)}(c,d) = \begin{cases} \binom{c}{t}_q \binom{d-t}{r-t}_q, & d-c \geq r, \\ \binom{d-c}{t}_q \binom{d-t}{r}_q, & d-c < r, \end{cases} \tag{7.5}$$

with the convention that $\binom{k}{l}_q = 0$ *if* $k < l$.

Proof Both identities can be proven by induction on r using Lemma 7.10 and the fact that the E_i, F_i act on $\mathcal{C}_2$ by $K_{\frac{1}{2}\alpha_i}$-derivations. □

Denote

$$\mathbf{b}_{\mathbf{m}}^{(i)} = E_i^{(r_i)}(\mathbf{b}_{e_i^{-r_i}(\mathbf{m})}), \qquad r_i = m_j + m_{0i},\ i \in \{1,2\},\ \mathbf{m} \in \mathbf{M}.$$

The following is immediate from Lemma 7.15.

Lemma 7.16 *For each $i \in \{1,2\}$, $\mathbf{B}^{(i)} = \{\mathbf{b}_{\mathbf{m}}^{(i)} : \mathbf{m} \in \mathbf{M}\}$ is a basis of C_2. Moreover, for each $i \in \{1,2\}$, $\lambda = l_1\omega_1 + l_2\omega_2 \in P^+$, $\mathbf{B}_\lambda^{(i)} := \{\mathbf{b}_{\mathbf{m}}^{(i)}, : \mathbf{m} \in \mathbf{M}_{l_1,l_2}\}$ is a basis of V_λ. Finally,*

$$\sigma^i(\mathbf{b}_{\mathbf{m}}^{(i)}) = \mathbf{b}_{\underline{\sigma}^i(\mathbf{m})}^{(i)}, \qquad i \in \{1,2\},\ \mathbf{m} \in \mathbf{M}.$$

In particular, $\sigma^i(\mathbf{B}_\lambda^{(i)}) = \mathbf{B}_\lambda^{(i)}$.

Thus, σ^i, $i \in \{1,2\}$ are easy to calculate in respective bases $\mathbf{B}^{(i)}$. To attack Conjecture 1.2 we need to find the matrix of both of them in a same basis. Note the following consequence of Lemma 7.15.

Corollary 7.17 *For each $\lambda = l_1\omega_1 + l_2\omega_2 \in P^+$ we have*

$$\mathbf{b}_{\mathbf{m}}^{(i)} = \sum_{\mathbf{m}' \in \mathbf{M}_{l_1,l_2}} C_{\mathbf{m}',\mathbf{m}}^{i;\lambda} \mathbf{b}_{\mathbf{m}'}$$

where $C^{i;\lambda}$ is an $\mathbf{M}_{l_1,l_2} \times \mathbf{M}_{l_1,l_2}$-matrix given by

$$C_{\mathbf{m}',\mathbf{m}}^{i;\lambda} = \begin{cases} \binom{m_i + m_{0i} + m_j + m_{ij}}{m_i + m_{ij}}_q, & \mathbf{m}' = \mathbf{m}, \\ C_t^{(m_j+m_{0i})}(m_j + m_{ij}, m_i + m_{0i} + m_j + m_{ij}), & \mathbf{m}' - \mathbf{m} = t\mathbf{a}_i^+,\ t \in \mathbb{Z}_{>0}, \\ 0, & \textit{otherwise}. \end{cases}$$

Remark 7.18 The bases $\mathbf{B}_\lambda^{(i)}$, $i \in \{1,2\}$, $\lambda \in P^+$ are in fact Gelfand-Tsetlin bases. The matrices $C^{i;\lambda}$ appeared first in the classical limit ($q = 1$) in [17]. According to [17, Theorem 10] their entries are closely related to Clebsch-Gordan coefficients, and so one should expect that our matrices are related to quantum Clebsch-Gordan coefficients. It is easy to see that $\sigma^i(\mathbf{B}_\lambda) = \mathbf{B}_\lambda$ if and only if V_λ is thin.

Theorem 7.19 *For each $\lambda = l_1\omega_1 + l_2\omega_2 \in P^+$ and $i \in \{1,2\}$ the matrix $N^{i;\lambda}$ of σ^i with respect to the basis $\mathbf{B}_\lambda$ of V_λ is given by*

$$N^{i;\lambda} = C^{i;\lambda} P^{i;\lambda} (C^{i;\lambda})^{-1}$$

where $P^{i;\lambda} = (P_{\mathbf{m}',\mathbf{m}}^{i;\lambda})_{\mathbf{m},\mathbf{m}' \in \mathbf{M}_{l_1,l_2}}$ with

$$P_{\mathbf{m}',\mathbf{m}}^{i;\lambda} = \delta_{\mathbf{m}',\underline{\sigma}^i(\mathbf{m})}, \qquad \mathbf{m}, \mathbf{m}' \in \mathbf{M}_{l_1,l_2}.$$

Conjecture 7.20 (Conjecture 1.2 for $\mathfrak{g} = \mathfrak{sl}_3$) For each $\lambda = l_1\omega_1 + l_2\omega_2 \in P^+$ we have

$$(N^{1;\lambda}N^{2;\lambda})^3 = 1.$$

This was verified using Mathematica® for all $l_1, l_2 \in \mathbb{Z}_{\geq 0}$ such that $l_1 + l_2 \leq 14$.

Acknowledgements The first two authors are grateful to Anton Alekseev and Université de Genève, Switzerland, for their hospitality. An important part of this work was done during the second author's stay at the Weizmann Institute of Science, Israel and during the conference in honor of Anthony Joseph's 75th birthday, at the Weizmann Institute and at the University of Haifa. The authors would like to use this opportunity to thank Maria Gorelik and Anna Melnikov for organizing that wonderful event.

This work was partially supported by a BSF grant no. 2016363 (A. Berenstein), the Simons foundation collaboration grant no. 245735 (J. Greenstein) and by the Minerva foundation with funding from the Federal German Ministry for Education and Research (J.-R. Li).

References

1. A. Berenstein and J. Greenstein, *Quantum folding*, Int. Math. Res. Not. **2011** (2011), no. 21, 4821–4883, DOI 10.1093/imrn/rnq264.
2. ——, *Double canonical bases*, Adv. Math. **316** (2017), 381–468, DOI 10.1016/j.aim.2017.06.005.
3. ——, *Canonical bases of quantum Schubert cells and their symmetries*, Selecta Math. (N.S.) **23** (2017), no. 4, 2755–2799. DOI 10.1007/s00029-017-0316-8.
4. A. Berenstein, J. Greenstein, and J.-R. Li, *Monomial braidings*, (in preparation).
5. A. Berenstein and D. Kazhdan, *Geometric and unipotent crystals. II. From unipotent bicrystals to crystal bases*, Quantum groups, Contemp. Math., vol. 433, Amer. Math. Soc., Providence, RI, 2007, pp. 13–88, DOI 10.1090/conm/433/08321.
6. A. Berenstein and D. Rupel, *Quantum cluster characters of Hall algebras*, Selecta Math. (N.S.) **21** (2015), no. 4, 1121–1176, DOI 10.1007/s00029-014-0177-3.
7. A. Berenstein and A. Zelevinsky, *String bases for quantum groups of type A_r*, I. M. Gel′fand Seminar, Adv. Soviet Math., vol. 16, Amer. Math. Soc., Providence, RI, 1993, pp. 51–89.
8. ——, *Canonical bases for the quantum group of type A_r and piecewise-linear combinatorics*, Duke Math. J. **82** (1996), no. 3, 473–502, DOI 10.1215/S0012-7094-96-08221-6.
9. A. Berenstein and S. Zwicknagl, *Braided symmetric and exterior algebras*, Trans. Amer. Math. Soc. **360** (2008), no. 7, 3429–3472, DOI 10.1090/S0002-9947-08-04373-0.
10. C. Bonnafé, *Cells and cacti*, Int. Math. Res. Not. **19** (2016), 5775–5800, DOI 10.1093/imrn/rnv324.
11. N. Bourbaki, *Éléments de mathématiques. Groupes et algèbres de Lie: Chapitres 4, 5 et 6*, Masson, Paris, 1981.
12. V. Chari, D. Jakelić, and A. A. Moura, *Branched crystals and the category $\mathcal{O}$*, J. Algebra **294** (2005), no. 1, 51–72, DOI 10.1016/j.jalgebra.2005.03.008.
13. M. Davis, T. Januszkiewicz, and R. Scott, *Fundamental groups of blow-ups*, Adv. Math. **177** (2003), no. 1, 115–179, DOI 10.1016/S0001-8708(03)00075-6.
14. S. L. Devadoss, *Tessellations of moduli spaces and the mosaic operad*, Homotopy invariant algebraic structures (Baltimore, MD, 1998), Contemp. Math., vol. 239, Amer. Math. Soc., Providence, RI, 1999, pp. 91–114, DOI 10.1090/conm/239/03599.
15. V. G. Drinfel′ d, *Quasi-hopf algebras*, Algebra i Analiz **1** (1989), no. 6, 114–148.

16. R. Jiménez Rolland and J. Maya Duque, Representation stability for the pure cactus group, Contributions of Mexican mathematicians abroad in pure and applied mathematics, Contemp. Math., vol. 709, Amer. Math. Soc., Providence, RI, 2018, pp. 53–67, DOI 10.1090/conm/709/14291.
17. I. M. Gel′fand and A. V. Zelevinskiĭ, *Polyhedra in a space of diagrams and the canonical basis in irreducible representations of* $\mathfrak{gl}_3$, Funktsional. Anal. i Prilozhen. **19** (1985), no. 2, 72–75.
18. J. Greenstein and P. Lamprou, *Path model for quantum loop modules of fundamental type*, Int. Math. Res. Not. **14** (2004), 675–711, DOI 10.1155/S1073792804131917.
19. I. Halacheva, J. Kamnitzer, L. Rybnikov, and A. Weekes, *Crystals and monodromy of Bethe vectors*, available at arXiv:1708.05105.
20. A. Henriques and J. Kamnitzer, *Crystals and coboundary categories*, Duke Math. J. **132** (2006), no. 2, 191–216, DOI 10.1215/S0012-7094-06-13221-0.
21. A. Joseph, *A pentagonal crystal, the golden section, alcove packing and aperiodic tilings*, Transform. Groups **14** (2009), no. 3, 557–612, DOI 10.1007/s00031-009-9064-y.
22. A. Joseph and G. Letzter, *Local finiteness of the adjoint action for quantized enveloping algebras*, J. Algebra **153** (1992), no. 2, 289–318, DOI 10.1016/0021-8693(92)90157-H.
23. V. G. Kac, *Infinite-dimensional Lie algebras*, 2nd ed., Cambridge University Press, Cambridge, 1985.
24. M. Kashiwara, *On crystal bases of the Q-analogue of universal enveloping algebras*, Duke Math. J. **63** (1991), no. 2, 465–516, DOI 10.1215/S0012-7094-91-06321-0.
25. ——, *Global crystal bases of quantum groups*, Duke Math. J. **69** (1993), no. 2, 455–485, DOI 10.1215/S0012-7094-93-06920-7.
26. ——, *Crystal bases of modified quantized enveloping algebra*, Duke Math. J. **73** (1994), no. 2, 383–413, DOI 10.1215/S0012-7094-94-07317-1.
27. Y. Kimura and H. Oya, *Twist automorphisms on quantum unipotent cells and dual canonical bases*, available at arXiv:1701.02268.
28. A. Kirillov and A. Berenstein, *Groups generated by involutions, Gel′fand-Tsetlin patterns, and combinatorics of Young tableaux*, Algebra i Analiz **7** (1995), no. 1, 92–152.
29. I. Losev, *Cacti and cells*, available at arXiv:1506.04400.
30. G. Lusztig, *Introduction to quantum groups*, Progress in Mathematics, vol. 110, Birkhäuser Boston, Inc., Boston, MA, 1993.
31. ——, *Canonical bases arising from quantized enveloping algebras. II*, Progr. Theoret. Phys. Suppl. **102** (1990), 175–201 (1991) DOI 10.1143/PTPS.102.175. Common trends in mathematics and quantum field theories (Kyoto, 1990).
32. ——, *Problems on canonical bases*, Algebraic groups and their generalizations: quantum and infinite-dimensional methods (University Park, PA, 1991), Proc. Sympos. Pure Math., vol. 56, Amer. Math. Soc., Providence, RI, 1994, pp. 169–176.
33. L. Rybnikov, *Cactus group and monodromy of Bethe vectors*, available at arXiv:1409.0131.
34. A. Savage, *Crystals, quiver varieties, and coboundary categories for Kac–Moody algebras*, Adv. Math. **221** (2009), no. 1, 22–53, DOI 10.1016/j.aim.2008.11.016.
35. N. White, *The monodromy of real Bethe vectors for the Gaudin model*, available at arXiv:1511.04740.

Quotients for Sheets of Conjugacy Classes

Giovanna Carnovale and Francesco Esposito

To Anthony Joseph, for 50 years of significant contributions to representation theory.

Abstract We provide a description of the orbit space of a sheet S for the conjugation action of a complex simple simply connected algebraic group G. This is obtained by means of a bijection between S/G and the quotient of a shifted torus modulo the action of a subgroup of the Weyl group and it is the group analogue of a result due to Borho and Kraft. We also describe the normalisation of the categorical quotient $\overline{S}/\!/G$ for arbitrary simple G and give a necessary and sufficient condition for $\overline{S}/\!/G$ to be normal in analogy to results of Borho, Kraft and Richardson. The example of G_2 is worked out in detail.

MSC: 20G20, 20G07

1 Introduction

Sheets for the action of a connected algebraic group G on a variety X have their origin in the work of Kostant [17], who studied the union of regular orbits for the adjoint action on a semisimple Lie algebra, and in the work of Dixmier [11]. Sheets are the irreducible components of the level sets of X consisting of points whose orbits have the same dimension. In a sense they provide a natural way to collect orbits in families in order to study properties of one orbit by looking at others in its family. For the adjoint action of a complex semisimple algebraic group G on its Lie

---

G. Carnovale (✉) · F. Esposito
Dipartimento di Matematica "Tullio Levi-Civita", Torre Archimede, Padova, Italy
e-mail: carnoval@math.unipd.it; esposito@math.unipd.it

M. Gorelik et al. (eds.), *Representations and Nilpotent Orbits of Lie Algebraic Systems*, Progress in Mathematics 330,
https://doi.org/10.1007/978-3-030-23531-4_3

algebra they were deeply and systematically studied in [2, 4]. They were described as sets, their closure was well-understood, they were classified in terms of pairs consisting of a Levi subalgebra and suitable nilpotent orbit therein, and they were used to answer affirmatively to a question posed by Dixmier on the multiplicities in the module decomposition of the ring of regular functions of an adjoint orbit in $\mathfrak{sl}(n, \mathbb{C})$. If G is classical then all sheets are smooth [15, 25]. The study of sheets in positive characteristic has appeared more recently in [27].

In analogy to this construction, sheets of primitive ideals were introduced and studied by Borho and Joseph in [3], in order to describe the set of primitive ideals in a universal enveloping algebra as a countable union of maximal varieties. More recently, Losev in [19] has introduced the notion of birational sheet in a semisimple Lie algebra, he has shown that birational sheets form a partition of the Lie algebra and has applied this result in order to establish a version of the orbit method for semisimple Lie algebras. Sheets were also used in [26] in order to parametrise the set of 1-dimensional representations of finite W-algebras, with some applications also to the theory of primitive ideals. Closures of sheets appear as associated varieties of affine vertex algebras, [1].

In characterisitc zero, several results on quotients S/G and $\overline{S}/\!/G$, for a sheet S were addressed: Katsylo has shown in [16] that S/G has the structure of a quotient and is isomorphic to the quotient of an affine variety by the action of a finite group [16]; Borho has explicitly described the normalisation of $\overline{S}/\!/G$ and Richardson, Broer, Douglass-Röhrle in [6, 12, 28] have provided the list of the quotients $\overline{S}/\!/G$ that are normal.

Sheets for the conjugation action of G on itself were studied in [8] in the spirit of [4]. If G is semisimple, they are parametrised in terms of pairs consisting of a Levi subgroup of a parabolic subgroup and a suitable isolated conjugacy class therein. Here isolated means that the connected centraliser of the semisimple part of a representative is semisimple. An alternative parametrisation can be given in terms of triples consisting of a pseudo-Levi subgroup M of G, a coset in $Z(M)/Z(M)^\circ$ and a suitable unipotent class in M. Pseudo-Levi subgroups are, in good characteristic, centralisers of semisimple elements and up to conjugation they are subroot subgroups whose root system has a base in the extended Dynkin diagram of G [23]. It is also shown in [7] that sheets in G are the irreducible components of the parts in Lusztig's partition introduced in [20], whose construction is given in terms of Springer's correspondence.

Also in the group case one wants to reach a good understanding of quotients of sheets. An analogue of Katsylo's theorem was obtained for sheets containing spherical conjugacy classes and all such sheets are shown to be smooth [9]. The proof in this case relies on specific properties of the intersection of spherical conjugacy classes with Bruhat double cosets, which do not hold for general classes. Therefore, a straightforward generalisation to arbitrary sheets is not immediate. Even in absence of a Katsylo type theorem, it is of interest to understand the orbit space S/G. In this paper we address the question for G simple provided G is simply connected if the root system is of type C or D. We give a bijection between the orbit space S/G and a quotient of a shifted torus of the form $Z(M)^\circ s$ by the action of a subgroup of the Weyl group, giving a group analogue of [18, Theorem 3.6], [2, Satz 5.6]. In most cases the subgroup does not depend on the unipotent part of the triple corresponding to the given sheet although it may depend on the isogeny type of G. This is one of the difficulties when passing from the Lie algebra case to the group case. The restriction on G needed for the bijection depends on the symmetry of the extended Dynkin diagram in this case: type C and D are the only two situations in which two distinct subsets of the extended Dynkin diagram can be equivalent even if they are not of type A. We illustrate by an example in $\mathrm{HSpin}_{10}(\mathbb{C})$ that the restriction we put is necessary in order to have injectivity so our theorem is somehow optimal.

We also address some questions related to the categorical quotient $\overline{S}//G$, for a sheet in G. We obtain group analogues of the description of the normalisation of $\overline{S}//G$ from [2] and of a necessary and sufficient condition on $\overline{S}//G$ to be normal from [28]. Finally we apply our results to compute the quotients S/G of all sheets in G of type G_2 and verify which of the quotients $\overline{S}//G$ are normal. This example served as a toy example for [10] (completed while the present paper was under review) in which we list all normal quotients for G simple.

2 Basic Notions

In this paper G is a complex connected *simple* algebraic group with maximal torus T, root system Φ, weight lattice Λ, set of simple roots $\Delta = \{\alpha_1, \ldots, \alpha_\ell\}$, Weyl group $W = N(T)/T$ and corresponding Borel subgroup B. The numbering of simple roots is as in [5]. Root subgroups are denoted by X_α for $\alpha \in \Phi$ and their elements have the form $x_\alpha(\xi)$ for $\xi \in \mathbb{C}$. Let $-\alpha_0$ be the highest root and let $\tilde{\Delta} = \Delta \cup \{\alpha_0\}$. The centraliser of an element h in a closed subgroup $H \leq G$ will be denoted by H^h and the identity component of H will be indicated by H°. If $\Pi \subset \tilde{\Delta}$, we set

$$G_\Pi := \langle T, X_{\pm\alpha} \mid \alpha \in \Pi \rangle.$$

Conjugates of such groups are called pseudo-Levi subgroups. We recall from [23, §6] that if $s \in T$ then its connected centraliser $G^{s\circ}$ is conjugated to G_Π for some Π by means of an element in $N(T)$. By [14, 2.2] we have $G^s = \langle G^{s\circ}, N(T)^s \rangle$. W_Π indicates the subgroup of W generated by the simple reflections with respect to roots in Π and it is the Weyl group of G_Π.

We realise the groups $\mathrm{Sp}_{2\ell}(\mathbb{C})$, $\mathrm{SO}_{2\ell}(\mathbb{C})$ and $\mathrm{SO}_{2\ell+1}(\mathbb{C})$, respectively, as the groups of matrices of determinant 1 preserving the bilinear forms with associated matrices: $\begin{pmatrix} 0 & I_\ell \\ -I_\ell & 0 \end{pmatrix}$, $\begin{pmatrix} 0 & I_\ell \\ I_\ell & 0 \end{pmatrix}$ and $\begin{pmatrix} 1 & & \\ & & I_\ell \\ & I_\ell & \end{pmatrix}$, respectively.

If G acts on a variety X, the action of $g \in G$ on $x \in X$ will be indicated by $(g, x) \mapsto g \cdot x$. If $X = G$ with adjoint action, we thus have $g \cdot h = ghg^{-1}$. For $n \geq 0$ we shall denote by $X_{(n)}$ the union of orbits of dimension n. The nonempty sets $X_{(n)}$ are locally closed and a sheet S for the action of G on X is an irreducible component of any of these. For any $Y \subset X$ we set Y^{reg} to be the set of points of Y

whose orbit has maximal dimension. We recall the parametrisation and description of sheets for the action of G on itself by conjugation and provide the necessary background material.

A Jordan class in G is an equivalence class with respect to the equivalence relation: $g, h \in G$ with Jordan decomposition $g = su$, $h = rv$ are equivalent if up to conjugation $G^{s\circ} = G^{r\circ}$, $r \in Z(G^{s\circ})^{\circ}s$ and $G^{s\circ} \cdot u = G^{s\circ} \cdot v$. As a set, the Jordan class of $g = su$ is thus $J(su) = G \cdot ((Z(G^{s\circ})^{\circ}s)^{reg}u)$ and it is contained in some $G_{(n)}$. Jordan classes are parametrised by G-conjugacy classes of triples $(M, Z(M)^{\circ}s, M \cdot u)$ where M is a pseudo-Levi subgroup, $Z(M)^{\circ}s$ is a coset in $Z(M)/Z(M)^{\circ}$ such that $(Z(M)^{\circ}s)^{reg} \subset Z(M)^{reg}$ and $M \cdot u$ is a unipotent conjugacy class in M. They are finitely many, locally closed, G-stable, smooth, see [21, 3.1] and [8, §4] for further details.

Every sheet $S \subset G$ contains a unique dense Jordan class, hence sheets are parametrised by conjugacy classes of a subset of the triples above mentioned. More precisely, a Jordan class $J = J(su)$ is dense in a sheet if and only if it is not contained in $(\overline{J'})^{reg}$ for any Jordan class J' different from J. We recall from [8, Propositions 4.6, 4.8] that, setting $L = C_G(Z(G^{s\circ})^{\circ})$ we have

$$\overline{J(su)}^{reg} = \bigcup_{z \in Z(G^{s\circ})^{\circ}} G \cdot (sz\mathrm{Ind}_{G^{s\circ}}^{G^{zs\circ}}(G^{s\circ} \cdot u)) = \bigcup_{z \in Z(G^{s\circ})^{\circ}} \mathrm{Ind}_L^G(L \cdot (szu)), \qquad (1)$$

where $\mathrm{Ind}_{G^{s\circ}}^{G^{zs\circ}}(G^{s\circ} \cdot u)$ is Lusztig-Spaltenstein's induction of the class of u in the Levi subgroup $G^{s\circ}$ of a parabolic subgroup of $G^{zs\circ}$, see [22] and $\mathrm{Ind}_L^G(L \cdot (szu))$ is its natural generalisation to arbitrary elements. So, Jordan classes that are dense in a sheet correspond to triples where u is a rigid orbit in $G^{s\circ}$, i.e., such that its class in $G^{s\circ}$ is not induced from a conjugacy class in a proper Levi subgroup of a parabolic subgroup of $G^{s\circ}$.

A sheet consists of a single conjugacy class if and only if $\overline{S} = \overline{J(su)} = \overline{G \cdot su}$ where u is rigid in $G^{s\circ}$ and $G^{s\circ}$ is semisimple, i.e., if and only if s is isolated and u is rigid in $G^{s\circ}$. Any sheet S in G is the image through the isogeny map π of a sheet S' in the simply-connected cover G_{sc} of G, where S' is determined up to multiplication by an element in $\mathrm{Ker}(\pi)$. Also, $Z(G^{\pi(s)\circ}) = \pi(Z(G_{sc}^{s\circ}))$ and $Z(G^{\pi(s)\circ})^{\circ} = \pi(Z(G_{sc}^{s\circ})^{\circ}) = Z(G_{sc}^{s\circ})^{\circ}\mathrm{Ker}(\pi)$.

3 A Parametrisation of Orbits in a Sheet

In this section we parametrise the set S/G of conjugacy classes in a given sheet. Let $S = \overline{J(su)}^{reg}$ with $s \in T$ and $u \in U \cap G^{s\circ}$. Let $Z = Z(G^{s\circ})$ and $L = C_G(Z^{\circ})$. The latter is always a Levi subgroup of a parabolic subgroup of G, [30, Proposition 8.4.5, Theorem 13.4.2] and if Ψ_s is the root system of $G^{s\circ}$ with respect to T, then L has root system $\Psi := \mathbb{Q}\Psi_s \cap \Phi$. Assume in addition that Ψ_s has base $\Pi \subset \tilde{\Delta}$.

Let W^X denote the stabiliser of X in W for $X = Z^\circ s,\ G^{s\circ},\ Z,\ Z^\circ,\ \Pi$. We recall that $C_G(Z(G^{s\circ})^\circ s)^\circ = G^{s\circ}$. Thus, for any lift $\dot{w}$ of $w \in W^{Z^\circ s}$ we have $\dot{w} \cdot G^{s\circ} = G^{s\circ}$, so $\dot{w} \cdot Z^\circ = Z^\circ$ and therefore $\dot{w} \cdot L = L$. Thus, any $w \in W^{Z^\circ s}$ determines an automorphism of Ψ_s and Ψ. Let $\mathcal{O} = G^{s\circ} \cdot u$. We set:

$$W^{\mathcal{O}} = \{w \in W^{Z^\circ s} \mid \dot{w} \cdot \mathcal{O} = \mathcal{O}\}. \tag{2}$$

The definition is independent of the choice of the representative of each w because $T \subset G^{s\circ}$.

Lemma 1 *Let Ψ_s be the root system of $G^{s\circ}$ with respect to T and let $\Pi \subset \Delta \cup \{-\alpha_0\}$ be a base. Let W_Π be the Weyl group of $G^{s\circ}$. Then*

$$W^{Z^\circ s} = W_\Pi \rtimes (W^\Pi \cap W^{Z^\circ s}) = \{w \in W_\Pi W^\Pi \mid w \cdot (Z^\circ s) = Z^\circ s\}.$$

In particular, if $G^{s\circ}$ is a Levi subgroup of a parabolic subgroup of G, then

$$W^{Z^\circ s} = W_\Pi \rtimes W^\Pi = N_W(W_\Pi)$$

and it is independent of the isogeny class of G.

Proof We have the following chain of inclusions:

$$W^{Z^\circ s} \leq W^{G^{s\circ}} \leq W^Z \leq W^{Z^\circ}.$$

We claim that $W_{G^{s\circ}} = W_\Pi \rtimes W^\Pi$. Indeed, $W_\Pi W^\Pi \leq W^{G^{s\circ}}$ is immediate and if $w \in W^{G^{s\circ}}$ then $w\Psi_s = \Psi_s$ and $w\Pi$ is a basis for Ψ_s. Hence, there is some $\sigma \in W_\Pi$ such that $\sigma w \in W^\Pi$. By construction W^Π normalises W_Π. The elements of $W^{G^{s\circ}}$ permute the connected components of $Z = Z(G^{s\circ})$ and $W^{Z^\circ s}$ is precisely the stabiliser of $Z^\circ s$ in there. Since the elements of W_Π fix the elements of $Z(G^{s\circ})$ pointwise, they stabilise $Z^\circ s$, whence the statement. The last statement follows from the equality $W_\Pi \ltimes W^\Pi = N_W(W_\Pi)$ in [13, Corollary 3] and [23, Lemma 33] because in this case $Z^\circ s = zZ^\circ$ for some $z \in Z(G)$, so $W^{Z^\circ s} = W^{Z^\circ}$. □

Remark 1 If $G^{s\circ}$ is not a Levi subgroup of a parabolic subgroup of G, then $W^{Z^\circ s}$ might depend on the isogeny type of G. For instance, if Φ is of type C_5 and $s = \mathrm{diag}(-I_2, x, I_2, -I_2, x^{-1}, I_2) \in \mathrm{Sp}_{10}(\mathbb{C})$ for $x^2 \neq 1,\ 0$, then:

$$\Pi = \{\alpha_0, \alpha_1, \alpha_4, \alpha_5\}$$

$$Z = Z(G^{s\circ}) = \{\mathrm{diag}(\epsilon I_2, y, \eta I_2, \epsilon I_2, y^{-1}, \eta I_2),\ y \in \mathbb{C}^*, \epsilon^2 = \eta^2 = 1\},$$

$$Z^\circ s = \{\mathrm{diag}(-I_2, y, I_2, -I_2, y^{-1}, I_2),\ y \in \mathbb{C}^*\},$$

and $W^\Pi = \langle s_{\alpha_1+\alpha_2+\alpha_3+\alpha_4} s_{\alpha_2+\alpha_3}\rangle$. Since $s_{\alpha_1+\alpha_2+\alpha_3+\alpha_4} s_{\alpha_2+\alpha_3} \cdot (Z^\circ s) = -Z^\circ s$ we have $W^{Z^\circ s} = W_\Pi$. However, if $\pi : \mathrm{Sp}_{10}(\mathbb{C}) \to \mathrm{PSp}_{10}(\mathbb{C})$ is the isogeny map, then

W^{Π} preserves $\pi(Z^\circ s)$ so $W^{\pi(Z^\circ s)} = W_\Pi \rtimes W^{\Pi}$. Taking $u = 1$ gives an example in which also $W^{\mathcal{O}}$ depends on the isogeny type.

The following lemma shows that in most cases $W^{\mathcal{O}}$ can be determined without any knowledge of u.

Lemma 2 *Let $S = \overline{J(su)}^{reg}$ be a sheet. Then, $W^{Z^\circ s} = W^{\mathcal{O}}$ if and only if we are* ***not*** *in the following situation:*

G is either $\mathrm{PSp}_{2\ell}(\mathbb{C})$, $\mathrm{HSpin}_{2\ell}(\mathbb{C})$, or $\mathrm{PSO}_{2\ell}(\mathbb{C})$;
$[G^{s\circ}, G^{s\circ}]$ has two isomorphic simple factors G_1 and G_2 that are not of type A;
the components of u in G_1 and G_2 do not correspond to the same partition.

Proof The element u is rigid in $[G^{s\circ}, G^{s\circ}] \leq G^{s\circ}$ and this happens if and only if each of its components in the corresponding simple factor of $[G^{s\circ}, G^{s\circ}]$ is rigid. Rigid unipotent elements in type A are trivial [29, Proposition 5.14], therefore what matters are only the components of u in the simple factors of type different from A. In addition, rigid nilpotent classes are characteristic in simple Lie algebras [2, Lemma 3.9, Korollar 3.10], hence using Springer's equivariant isomorphism between the nilpotent cone and the unipotent variety we deduce that rigid unipotent classes are characteristic in a simple algebraic group. In particular, in classical groups rigid unipotent elements are completely determined by their partition. For all Φ different from C and D, simple factors that are not of type A are never isomorphic. Therefore $W^{Z^\circ s} = W^{\mathcal{O}}$ in all cases with a possible exception when: Φ is of type C_ℓ or D_ℓ; $[G^{s\circ}, G^{s\circ}]$ has two isomorphic factors of type different from A; and the components of u in those two factors, that are of type C_m or D_m, respectively, correspond to different partitions.

Let us assume that we are in this situation. Then, $W^{Z^\circ s} = W^{\mathcal{O}}$ if and only if the elements of $W^{Z^\circ s}$, acting as automorphisms of Ψ_s, do not interchange the two isomorphic factors in question. We have 2 isogeny classes in type C_ℓ, 3 in type D_ℓ for ℓ odd, and 4 (up to isomorphism) in type D_ℓ for ℓ even.

If Φ is of type C_ℓ and $G = \mathrm{Sp}_{2\ell}(\mathbb{C})$ up to a central factor s can be chosen to be of the form:

$$s = \mathrm{diag}(I_m, t, -I_m, I_m, t^{-1}, -I_m) \tag{3}$$

where t is a diagonal matrix in $\mathrm{GL}_{\ell-2m}(\mathbb{C})$ with eigenvalues different from ± 1. Then Π is the union of $\{\alpha_0, \ldots, \alpha_{m-1}\}$, $\{\alpha_\ell, \alpha_{\ell-1}, \ldots, \alpha_{\ell-m+1}\}$ and possibly other subsets of simple roots orthogonal to these. Then W^{Π} is the direct product of terms permuting isomorphic components of type A with the subgroup generated by $\sigma = \prod_{j=1}^{m} s_{\alpha_j+\cdots+\alpha_{\ell-j}}$. In this case the elements of $Z^\circ s$ are of the form $\mathrm{diag}(I_m, r, -I_m, I_m, r^{-1}, -I_m)$, where r has the same shape as t and $\sigma(Z^\circ s) = -Z^\circ s \neq Z^\circ s$. Thus, $W^{\Pi} \cap W^{Z^\circ s}$ does not permute the two factors of type C_m and $W^{Z^\circ s} = W^{\mathcal{O}}$.

If, instead, $G = \mathrm{PSp}_{2\ell}(\mathbb{C})$ and the sheet is $\pi(S)$, we may take $J = J(\pi(su))$ where s is as in (3). Then, σ preserves $\pi(Z^\circ s)$ and therefore $W^{\pi(Z^\circ s)} \neq W^{\pi(\mathcal{O})}$.

Let now Φ be of type D_ℓ and $G = Spin_{2\ell}(\mathbb{C})$. With notation as in [30], we may take

$$s = \left(\prod_{j=1}^{m} \alpha_j^\vee(\epsilon^j)\right)\left(\prod_{i=m+1}^{l-m-1} \alpha_i^\vee(c_i)\right)\left(\prod_{b=2}^{m} \alpha_{\ell-b}^\vee(d^2\eta^b)\right)\alpha_{\ell-1}^\vee(\eta d)\alpha_\ell(d) \quad (4)$$

with $\epsilon^2 = \eta^2 = 1$, $\epsilon \neq \eta$, and $d,\ c_i \in \mathbb{C}^*$ generic.

Here Π is the union of $\{\alpha_0, \ldots, \alpha_{m-1}\}$, $\{\alpha_\ell, \alpha_{\ell-1}, \ldots, \alpha_{\ell-m+1}\}$ and possibly other subsets of simple roots orthogonal to these. Then W^Π is the direct product of terms permuting isomorphic components of type A and $\langle\sigma\rangle$ where $\sigma = \prod_{j=1}^{m} s_{\alpha_j+\cdots+\alpha_{\ell-j+1}}$. The coset $Z^\circ s = Z_{\epsilon,\eta}$ consists of elements of the same form as (4) with constant value of ϵ and η and m, and $Z^\circ = Z_{1,1}$ consists of the elements of similar shape with $\eta = \epsilon = 1$. Then $\sigma \cdot Z_{\epsilon,\eta} = Z_{\eta,\epsilon}$, hence $\sigma \notin W^{Z^\circ s}$, so $W^{Z^\circ s}$ preserves the components of Ψ_s of type D and $W^{Z^\circ s} = W^{\mathcal{O}}$.

If $\ell = 2q$ and $G = \mathrm{HSpin}_{2\ell}(\mathbb{C})$ and $\pi : \mathrm{Spin}_{2\ell}(\mathbb{C}) \to \mathrm{HSpin}_{2\ell}(\mathbb{C})$ is the isogeny map we see from Table 1 that $\mathrm{Ker}(\pi)$ is generated by an element k such that $kZ_{\epsilon,\eta} = Z_{-\epsilon,-\eta}$, so σ as above preserves $\pi(Z^\circ s)$ whereas it does not preserve the conjugacy class of $\pi(u)$. Therefore $\sigma \in W^{\pi(Z^\circ s)} \neq W^{\pi(\mathcal{O})}$.

If $G = \mathrm{SO}_{2\ell}(\mathbb{C})$ and $\pi : \mathrm{Spin}_{2\ell}(\mathbb{C}) \to \mathrm{SO}_{2\ell}(\mathbb{C})$ is the isogeny map, then $\mathrm{Ker}(\pi)$ is generated by an element k such that $kZ_{\epsilon,\eta} = Z_{\epsilon,\eta}$. In this case σ does not preserve $\pi(Z^\circ s)$, whence $\sigma \notin W^{\pi(Z^\circ s)} = W^{\pi(\mathcal{O})}$.

If $G = \mathrm{PSO}_{2\ell}(\mathbb{C})$ and $\pi : \mathrm{Spin}_{2\ell}(\mathbb{C}) \to \mathrm{PSO}_{2\ell}(\mathbb{C})$, then by the discussion of the previous isogeny types we see that $\sigma(Z_{\epsilon,\eta}) \subset \mathrm{Ker}(\pi)Z_{\epsilon,\eta}$, so σ preserves $\pi(Z^\circ s)$ whence $\sigma \in W^{\pi(Z^\circ s)} \neq W^{\pi(\mathcal{O})}$. □

Table 1 Kernel of the isogeny map; Φ of type B_ℓ, C_ℓ or D_ℓ

Type	Parity of ℓ	Group	$\mathrm{Ker}\pi$
B_ℓ	Any	$\mathrm{SO}_{2\ell+1}(\mathbb{C})$	$\langle\alpha_\ell^\vee(-1)\rangle$
C_ℓ	Any	$\mathrm{PSp}_{2\ell}(\mathbb{C})$	$\left\langle\prod_{j \text{ odd}} \alpha_j^\vee(-1)\right\rangle = \langle -I_{2\ell}\rangle$
D_ℓ	Even	$\mathrm{PSO}_{2\ell}(\mathbb{C})$	$\left\langle\prod_{j \text{ odd}} \alpha_j^\vee(-1),\ \alpha_{\ell-1}^\vee(-1)\alpha_\ell^\vee(-1)\right\rangle$
D_ℓ	Odd	$\mathrm{PSO}_{2\ell}(\mathbb{C})$	$\left\langle\prod_{j \text{ odd}\leq\ell-2} \alpha_j^\vee(-1)\alpha_{\ell-1}^\vee(i)\alpha_\ell^\vee(i^3)\right\rangle$
D_ℓ	Any	$\mathrm{SO}_{2\ell}(\mathbb{C})$	$\langle\alpha_{\ell-1}^\vee(-1)\alpha_\ell^\vee(-1)\rangle$
D_ℓ	Even	$\mathrm{HSpin}_{2\ell}(\mathbb{C})$	$\left\langle\prod_{j \text{ odd}} \alpha_j^\vee(-1))\right\rangle$

Following [2, §5] and according to (1) we define the map

$$\begin{aligned}\theta\colon Z^\circ s &\to S/G\\ zs &\mapsto \mathrm{Ind}_L^G(L\cdot szu)\end{aligned}$$

where $L = C_G(Z(G^{s\circ})^\circ)$.

Lemma 3 *With the above notation, $\theta(zs) = \theta(w\cdot(zs))$ for every $w \in W^{\mathscr{O}}$.*

Proof Let us observe that, since $z \in Z(L)$ and $G^{s\circ} \subset L$ there holds $L^{zs\circ} = G^{s\circ}$. In particular, $G^{s\circ}$ is a Levi subgroup of a parabolic subgroup of $G^{zs\circ}$. Let U_P be the unipotent radical of a parabolic subgroup P of G with Levi factor L and let $\dot{w}$ be a representative of w in $N_G(T)$. By [8, Proposition 4.6] we have

$$\begin{aligned}\mathrm{Ind}_L^G(L\cdot(w\cdot zs)u) &= G\cdot((w\cdot zs)uU_P)^{reg}\\ &= G\cdot(zs(\dot{w}^{-1}\cdot u)U_{\dot{w}^{-1}\cdot P})^{reg}\\ &= \mathrm{Ind}_L^G(L\cdot(zs(\dot{w}^{-1}\cdot u)))\\ &= G\cdot\left(zs\,\mathrm{Ind}_{G^{s\circ}}^{G^{zs\circ}}(\dot{w}^{-1}\cdot(G^{s\circ}\cdot u))\right)\\ &= G\cdot\left(zs\,\mathrm{Ind}_{G^{s\circ}}^{G^{zs\circ}}(G^{s\circ}\cdot u)\right)\\ &= \mathrm{Ind}_L^G(L\cdot(zsu))\end{aligned}$$

where we have used that $L = \dot{w}\cdot L$ for every $w \in W^{\mathscr{O}} \le W^{Z^\circ s}$ and independence of the choice of the parabolic subgroup with Levi factor L, [8, Proposition 4.5]. □

Remark 2 The requirement that w lies in $W^{\mathscr{O}}$ rather than in $W^{Z^\circ s}$ is necessary. For instance, we consider $G = \mathrm{PSp}_{2\ell}(\mathbb{C})$ with $\ell = 2m+1$ and s the class of $\mathrm{diag}(I_m, \lambda, -I_m, I_m, \lambda^{-1}, -I_m)$ with $\lambda^4 \neq 1$ and u rigid with non-trivial component only in the subgroup $H = \langle X_{\pm\alpha_j},\ j = 0, \ldots m-1\rangle$ of $G^{s\circ}$. The element $\sigma = \prod_{j=1}^m s_{\alpha_j+\cdots+\alpha_{\ell-j}}$ lies in $W^{Z^\circ s}\setminus W^{\mathscr{O}}$. Taking $\theta(s)$ we have

$$\mathrm{Ind}_L^G(L\cdot su) = G\cdot su$$

whereas

$$\mathrm{Ind}_L^G(L\cdot(w\cdot s)u) = \mathrm{Ind}_L^G(L\cdot s(\dot{w}\cdot u)) = G\cdot(s(\dot{w}\cdot u)),$$

where $\dot{w}$ is any representative of w in $N_G(T)$. These classes would coincide only if u and $\dot{w}\cdot u$ were conjugate in G^s. They are not conjugate in $G^{s\circ}$ because they lie in different simple components. Moreover, G^s is generated by $G^{s\circ}$ and the lifts of elements in the centraliser W^s of s in W [14, 2.2], which is contained in $W^{Z^\circ s}$. Since $\lambda^4 \neq 1$ we see that the elements of W^s cannot interchange the two components of type C_m in $G^{s\circ}$. Hence,

$$\theta(s) = \mathrm{Ind}_L^G(L\cdot su) \neq \mathrm{Ind}_L^G(L\cdot(w\cdot s)u) = \theta(w\cdot s).$$

In analogy with the Lie algebra case we formulate the following theorem. The proof follows the lines of [2, Satz 5.6] but a more detailed analysis is necessary because the naive generalisation of statement [2, Lemma 5.4] from Levi subalgebras in a Levi subalgebra to Levi subgroups in a pseudo-Levi subgroup does not hold.

Theorem 1 *Assume G is simple and different from* $\mathrm{PSO}_{2\ell}(\mathbb{C})$, $\mathrm{HSpin}_{2\ell}(\mathbb{C})$ *and* $\mathrm{PSp}_{2\ell}(\mathbb{C})$, $\ell \geq 5$. *Let $S = \overline{J(su)}^{reg}$ be a sheet with $s \in T$, $Z = Z(G^{s\circ})$ and let $W^{Z^\circ s}$ be the stabiliser of $Z^\circ s$ in W. The map θ induces a bijection $\overline{\theta}$ between $Z^\circ s/W^{Z^\circ s}$ and S/G.*

Proof Recall that under our assumptions Lemma 2 gives $W^{Z^\circ s} = W^{\mathscr{O}}$. By Lemma 3, θ induces a well-defined map $\overline{\theta}\colon Z^\circ s/W^{Z^\circ s} \to S/G$. It is surjective by [8, Proposition 4.8]. We prove injectivity.

Let us assume that $\theta(zs) = \theta(z's)$ for some $z, z' \in Z^\circ$. By construction, $Z^\circ \subset T$. By [8, Proposition 4.5] we have

$$G \cdot \left(zs\left(\mathrm{Ind}_{G^{s\circ}}^{G^{zs\circ}}(G^{s\circ}\cdot u)\right)\right) = G \cdot \left(z's\left(\mathrm{Ind}_{G^{s\circ}}^{G^{z's\circ}}(G^{s\circ}\cdot u)\right)\right).$$

This implies that $z's = \sigma \cdot (zs)$ for some $\sigma \in W$. Let $\dot{\sigma} \in N(T)$ be a representative of σ. Then

$$\begin{aligned}\theta(zs) = \theta(z's) &= G \cdot \left((\sigma\cdot zs)(\mathrm{Ind}_{G^{s\circ}}^{G^{z's\circ}}(G^{s\circ}\cdot u))\right)\\ &= G \cdot \left(zs\left(\mathrm{Ind}_{\dot{\sigma}^{-1}\cdot G^{s\circ}}^{\dot{\sigma}^{-1}\cdot G^{z's\circ}}\left(\dot{\sigma}^{-1}\cdot(G^{s\circ}\cdot u)\right)\right)\right)\\ &= G \cdot \left(zs\left(\mathrm{Ind}_{\dot{\sigma}^{-1}\cdot G^{s\circ}}^{G^{zs\circ}}\left(\dot{\sigma}^{-1}\cdot(G^{s\circ}\cdot u)\right)\right)\right).\end{aligned}$$

Since the unipotent parts of $\theta(zs)$ and $\theta(z's)$ coincide, for some $x \in G^{zs}$ we have

$$x \cdot (\mathrm{Ind}_{G^{s\circ}}^{G^{zs\circ}}(G^{s\circ}\cdot u)) = \mathrm{Ind}_{\dot{\sigma}^{-1}\cdot(G^{s\circ})}^{G^{zs\circ}}\left(\dot{\sigma}^{-1}\cdot(G^{s\circ}\cdot u)\right).$$

The element x may be written as $\dot{w}g$ for some $\dot{w} \in N(T)\cap G^{zs}$ and some $g \in G^{zs\circ}$ [14, §2.2]. Hence,

$$\begin{aligned}\mathrm{Ind}_{G^{s\circ}}^{G^{zs\circ}}(G^{s\circ}\cdot u) &= \dot{w}^{-1}\cdot\left(\mathrm{Ind}_{\dot{\sigma}^{-1}\cdot(G^{s\circ})}^{G^{zs\circ}}\left(\dot{\sigma}^{-1}\cdot(G^{s\circ}\cdot u)\right)\right)\\ &= \mathrm{Ind}_{\dot{w}^{-1}\dot{\sigma}^{-1}\cdot(G^{s\circ})}^{G^{zs\circ}}\left((\dot{w}^{-1}\dot{\sigma}^{-1})\cdot(G^{s\circ}\cdot u)\right).\end{aligned}$$

Let us put

$$M := G^{zs\circ} = \langle T, X_\alpha, \alpha \in \Phi_M\rangle, \quad L_1 := G^{s\circ} = \langle T, X_\alpha, \alpha \in \Psi\rangle$$

with $\Phi_M = \bigcup_{j=1}^{l} \Phi_j$ and $\Psi = \bigcup_{i=1}^{m} \Psi_i$ the decompositions in irreducible root subsystems. We recall that L_1 and $L_2 := (\dot{w}^{-1}\dot{\sigma}^{-1}) \cdot L_1$ are Levi subgroups of some parabolic subgroups of M. We claim that if L_1 and L_2 are conjugate in M, then zs and $z's$ are $W^{Z^\circ s}$-conjugate. Indeed, under this assumption, since L_1 and L_2 contain T, there is $\dot{\tau} \in N_M(T)$ such that $L_1 = \dot{\tau} \cdot L_2 = \dot{\tau}\dot{w}^{-1}\dot{\sigma}^{-1} \cdot L_1$, so $\tau w^{-1}\sigma^{-1}(Z^\circ) = Z^\circ$. Then, $\tau w^{-1}\sigma^{-1}(z's) = zs$ and therefore

$$\tau w^{-1}\sigma^{-1} \cdot (Z^\circ s) = \tau w^{-1}\sigma^{-1} \cdot (Z^\circ z's) = Z^\circ zs = Z^\circ s.$$

Hence zs and $z's$ are $W^{Z^\circ s}$-conjugate. We show that if Φ_M has at most one component different from type A, then L_1 is always conjugate to L_2 in M. We analyse two possibilities.

Φ_j Is of Type A for Every j In this case the same holds for Ψ_i and $u = 1$. We recall that in type A induction from the trivial orbit in a Levi subgroup corresponding to a partition λ yields the unipotent class corresponding to the dual partition [29, 7.1]. Hence, equivalence of the induced orbits in each simple factor M_j of M forces $\Phi_j \cap \Psi \cong \Phi_j \cap w^{-1}\sigma^{-1}\Psi$ for every j. Invoking [2, Lemma 5.5], in each component M_i we deduce that L_1 and L_2 are M-conjugate.

There Is Exactly One Component in Φ_M Which Is Not of Type A We set it to be Φ_1. Then, there is at most one Ψ_j, say Ψ_1, which is not of type A, and $\Psi_1 \subset \Phi_1$. In this case, $w^{-1}\sigma^{-1}\Phi_1 \subset \Psi_1$. Equivalence of the induced orbits in each simple factor M_j of M forces $\Phi_j \cap \Psi \cong \Phi_j \cap w^{-1}\sigma^{-1}\Psi$ for every $j > 1$. By exclusion, the same isomorphism holds for $j = 1$. Invoking once more [2, Lemma 5.5] for each simple component, we deduce that L_1 and L_2 are M-conjugate.

Assume now that there are exactly two components of Φ_M which are not of type A. This situation can only occur if Φ is of type B_ℓ for $\ell \geq 6$, C_ℓ for $\ell \geq 4$ or D_ℓ for $\ell \geq 8$ (we recall that $D_2 = A_1 \times A_1$ and $D_3 = A_3$). By a case-by-case analysis we directly show that σ can be taken in $W^{Z^\circ s}$.

If $G = \mathrm{Sp}_{2\ell}(\mathbb{C})$ we may assume that

$$s = \mathrm{diag}(I_m, t, -I_p, I_m, t^{-1}, -I_p)$$

with $p, m \geq 2$ and t a diagonal matrix with eigenvalues different from 0 and ± 1. Then $Z^\circ s$ consists of matrices of similar form but with t invertible, so zs and $z's$ are $zs = \mathrm{diag}(I_m, h, -I_p, I_m, h^{-1}, -I_p)$ and $z's = \mathrm{diag}(I_m, g, -I_p, I_m, g^{-1}, -I_p)$, where h and g are invertible diagonal matrices. The elements zs and $z's$ are conjugate in G if and only if $\mathrm{diag}(h, h^{-1})$ and $\mathrm{diag}(g, g^{-1})$ are conjugate in $G' = \mathrm{Sp}_{2(\ell-p-m)}(\mathbb{C})$. This is the case if and only if they are conjugate in the normaliser of the torus $T' = G' \cap T$. The natural embedding $G' \to G$ given by

$$\begin{pmatrix} A & B \\ C & D \end{pmatrix} \mapsto \begin{pmatrix} I_m & & & & \\ & A & & B & \\ & & I_{p+m} & & \\ & C & & D & \\ & & & & I_p \end{pmatrix}$$

gives an embedding of $N_{G'}(T') \leq N_G(T)$ whose image lies in $W^{Z^\circ s}$. Hence, zs and $z's$ are necessarily $W^{Z^\circ s}$-conjugate. This concludes the proof of injectivity for $G = \mathrm{Sp}_{2\ell}(\mathbb{C})$.

If $G = \mathrm{Spin}_{2\ell+1}(\mathbb{C})$, then we may assume that

$$s = \left(\prod_{j=1}^{m} \alpha_j^\vee((-1)^j)\right)\left(\prod_{b=m+1}^{\ell-p-1} \alpha_b^\vee(c_b)\right)\left(\prod_{q=1}^{p} \alpha_{\ell-q}^\vee(c^2)\right)\alpha_\ell^\vee(c)$$

where $m \geq 4$, $p \geq 2$, c, $c_b \in \mathbb{C}^*$ are generic. Here $Z^\circ s$ consists of elements of the form

$$\left(\prod_{j=1}^{m} \alpha_j^\vee((-1)^j)\right)\left(\prod_{b=m+1}^{\ell-p-1} \alpha_b^\vee(d_b)\right)\left(\prod_{q=1}^{p} \alpha_{\ell-q}^\vee(d^2)\right)\alpha_\ell^\vee(d)$$

with d_b, $d \in \mathbb{C}^*$. The reflection $s_{\alpha_1+\cdots+\alpha_\ell} = s_{\varepsilon_1}$ maps any $y \in Z^\circ s$ to $y\alpha_\ell^\vee(-1) \in Z(G)Z^\circ s = Z^\circ s$.

Let us consider the natural isogeny $\pi : G \to G_{ad} = \mathrm{SO}_{2\ell+1}(\mathbb{C})$. Then

$$\pi(s) = \mathrm{diag}(1, -I_m, t, I_p, -I_m, t^{-1}, I_p)$$

where t is a diagonal matrix with eigenvalues different from 0 and ± 1. A similar calculation as in the case of $\mathrm{Sp}_{2\ell}(\mathbb{C})$ shows that $\pi(zs)$ is conjugate to $\pi(z's)$ by an element $\sigma_1 \in W^{\pi(Z^\circ s)} = W^{\pi(\mathcal{O})}$. Then, $\sigma_1 \cdot (zs) = kz's$, where $k \in Z(G)$. If $k = 1$, then we set $\sigma = \sigma_1$ whereas if $k = \alpha_\ell^\vee(-1)$ we set $\sigma = s_{\alpha_1+\cdots+\alpha_\ell}\sigma_1$. Then $\sigma \cdot (zs) = z's$ and $\sigma \cdot (Z^\circ s) = Z(G)Z^\circ s = Z^\circ s$. This concludes the proof for $\mathrm{Spin}_{2\ell+1}(\mathbb{C})$ and $\mathrm{SO}_{2\ell+1}(\mathbb{C})$.

If $G = \mathrm{Spin}_{2\ell}(\mathbb{C})$, up to multiplication by a central element we may assume that

$$s = \left(\prod_{j=m+1}^{\ell-p-1} \alpha_j^\vee(c_j)\right)\left(\prod_{q=2}^{p} \alpha_{\ell-q}^\vee((-1)^q c^2)\right)\alpha_{\ell-1}^\vee(-c)\alpha_\ell^\vee(c)$$

where m, $p \geq 4$, c, $c_j \in \mathbb{C}^*$ are generic. The elements in $Z^\circ s$ are of the form

$$\left(\prod_{j=m+1}^{\ell-p-1} \alpha_j^\vee(d_j)\right)\left(\prod_{q=2}^{p} \alpha_{\ell-q}^\vee((-1)^q d^2)\right)\alpha_{\ell-1}^\vee(-d)\alpha_\ell^\vee(d)$$

with d_j, $d \in \mathbb{C}^*$. We argue as we did for type B_ℓ, using the isogeny $\pi : G \to \mathrm{SO}_{2\ell}(\mathbb{C})$. The element $s_{\alpha_\ell} s_{\alpha_{\ell-1}}$ maps any $y \in Z^\circ s$ to $y\alpha_{\ell-1}^\vee(-1)\alpha_\ell^\vee(-1) \in \mathrm{Ker}(\pi)Z^\circ s = Z^\circ s$. The shifted torus $\pi(Z^\circ s)$ consists of elements of the form

$$\mathrm{diag}(I_m, t, -I_p, I_m, t^{-1}, -I_p)$$

where t is a diagonal matrix in $\mathrm{GL}_{2(\ell-m-p)}(\mathbb{C})$. Two elements

$$\pi(zs) = \mathrm{diag}(I_m, h, -I_p, I_m, h^{-1}, -I_p),$$
$$\pi(z's) = \mathrm{diag}(I_m, g, -I_p, I_m, g^{-1}, -I_p)$$

therein are W-conjugate if and only if $\mathrm{diag}(1, h, 1, h^{-1})$ and $(1, g, 1, g^{-1})$ are conjugate by an element σ_1 of the Weyl group W' of $G' = \mathrm{SO}_{2(\ell-m-p+1)}(\mathbb{C})$. More precisely, even if h and g may have eigenvalues equal to 1, we may choose σ_1 in the subgroup of W' that either fixes the first and the $(\ell - m - p + 2)$-th eigenvalues or interchanges them. Considering the natural embedding of G' into $\mathrm{SO}_{2\ell}(\mathbb{C})$ in a similar fashion as we did for $\mathrm{Sp}_{2\ell}(\mathbb{C})$, we show that $\sigma_1 \in W^{\pi(Z^\circ s)}$. This proves injectivity for $\mathrm{SO}_{2\ell}(\mathbb{C})$. Arguing as we did for $\mathrm{Spin}_{2\ell+1}(\mathbb{C})$ using $s_{\alpha_\ell} s_{\alpha_{\ell-1}}$ concludes the proof of injectivity for $\mathrm{Spin}_{2\ell}(\mathbb{C})$. □

The translation isomorphism $Z^\circ s \to Z^\circ$ determines a $W^{Z^\circ s}$-equivariant map where Z° is endowed with the action $w \bullet z = (w \cdot zs)s^{-1}$, which is in general not an action by automorphisms on Z°. Hence, S/G is in bijection with the quotient $Z^\circ / W^{Z^\circ s}$ of the torus Z° where the quotient is with respect to the $\bullet$ action.

Remark 3 Injectivity of $\overline{\theta}$ does not necessarily hold for the adjoint groups $G = \mathrm{PSp}_{2\ell}(\mathbb{C})$, $\mathrm{PSO}_{2\ell}(\mathbb{C})$ and for $G = \mathrm{HSpin}_{2\ell}(\mathbb{C})$. We give an example for $G = \mathrm{HSpin}_{20}(\mathbb{C})$, in which $W^{Z^\circ s} = W^{\mathcal{O}}$ and $G^{s\circ}$ is a Levi subgroup of a parabolic subgroup of G. Let $\pi : \mathrm{Spin}_{20}(\mathbb{C}) \to G$ be the isogeny with kernel K as in Table 1. Let $u = 1$ and let

$$s = \alpha_1^\vee(a)\alpha_2^\vee(a^2)\alpha_3^\vee(a^3)\alpha_4^\vee(b)\alpha_5^\vee(c)\alpha_6^\vee(d^{-2}e^2)\alpha_7^\vee(e)\alpha_8^\vee(d^2)\alpha_9^\vee(d)\alpha_{10}^\vee(-d)K$$

with a, b, c, d, $e \in \mathbb{C}^*$ sufficiently generic. Then, $G^{s\circ}$ is generated by T and the root subgroups of the subsystem with basis indexed by the black nodes in the following extended Dynkin diagram:

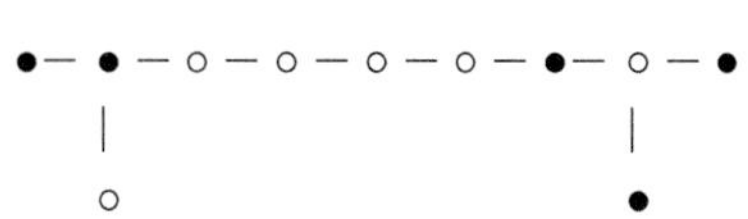

Here $Z^\circ s$ is given by elements of shape:

$$\alpha_1^\vee(a_1)\alpha_2^\vee(a_1^2)\alpha_3^\vee(a_1^3)\alpha_4^\vee(b_1)\alpha_5^\vee(c_1)\alpha_6^\vee(d_1^{-2}e_1^2)\alpha_7^\vee(e_1)\alpha_8^\vee(d_1^2)\alpha_9^\vee(d_1)\alpha_{10}^\vee(-d_1)K$$

with $a_1, b_1, c_1, d_1, e_1 \in \mathbb{C}^*$. Let

$$zs = \alpha_5^\vee(c)\alpha_6^\vee(d^2)\alpha_7^\vee(-d^2)\alpha_8^\vee(d^2)\alpha_9^\vee(d)\alpha_{10}^\vee(-d)K \in Z^\circ sK$$

obtained by setting $a_1 = b_1 = 1$, $c_1 = c$, $d_1 = d$ and $e_1 = -d^2$, and

$$z's = \alpha_5^\vee(-c)\alpha_6^\vee(d^2)\alpha_7^\vee(-d^2)\alpha_8^\vee(d^2)\alpha_9^\vee(d)\alpha_{10}^\vee(-d)K \in Z^\circ sK,$$

obtained by setting $a_1 = b_1 = 1$, $c_1 = -c$, $d_1 = d$ and $e_1 = -d^2$. The subgroup $M := G^{zs\circ} = G^{z's\circ}$ is generated by T and the root subgroups of the subsystem with basis indexed by the black nodes in the following extended Dynkin diagram:

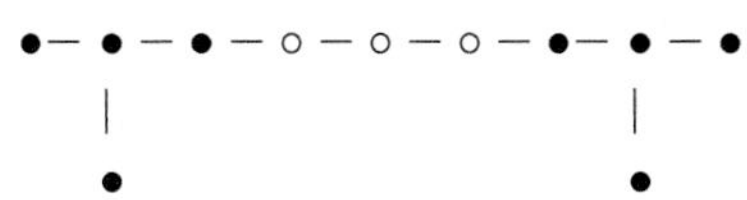

For $\sigma = \prod_{j=1}^{4} s_{\alpha_j+\cdots+\alpha_{10-j}}$ we have $\sigma \cdot zs = z's$. We claim that zs and $z's$ are not $W^{Z^\circ s}$-conjugate. Equivalently, we show that $\sigma W^{zsK} \cap W^{Z^\circ s} = \emptyset$, where W^{szK} is the stabiliser of zsK in W. Let σw be an element lying in such an intersection. We observe that if $\sigma w \in W^{Z^\circ s}$, then $\sigma w \cdot G^{s\circ} = G^{s\circ}$ hence σw cannot interchange the component of type $3A_1$ with the component of type A_2 therein. Thus, it cannot interchange the two components of type D_4 in M. However, by looking at the projection π' onto $G/Z(G) = \mathrm{PSO}_{10}(\mathbb{C})$, we see that $zsZ(G)$ is the class of the matrix

$$\mathrm{diag}(I_4, c, c^{-1}d^2, -I_4, I_4, d^{-2}c, c^{-1}, -I_4)$$

which cannot be centralised by a Weyl group element interchanging these two factors if c and d are sufficiently generic. A fortiori, this cannot happen for the class zsK. Hence, zs and $z's$ are not $W^{Z^\circ s}$-conjugate.

Let now M_1 and M_2 be the simple factors of M corresponding, respectively, to the roots $\{\alpha_j, 0 \leq j \leq 3\}$, and $\{\alpha_k, 7 \leq k \leq 10\}$, let $L_1 = M_1 \cap G^{s\circ}$ and $L_2 = M_2 \cap G^{s\circ}$. Then,

$$\theta(zs) = \mathrm{Ind}_L^G(L \cdot zs) = G \cdot \left(zs(\mathrm{Ind}_{G^{s\circ}}^M(1))\right) = G \cdot (zs(\mathrm{Ind}_{L_1}^{M_1}(1))(\mathrm{Ind}_{L_2}^{M_2}(1)))$$

and

$$\theta(z's) = \mathrm{Ind}_L^G(L \cdot z's) = G \cdot \left(z's(\mathrm{Ind}_{G^{s\circ}}^M(1)\right) = G \cdot (z's(\mathrm{Ind}_{L_1}^{M_1}(1))(\mathrm{Ind}_{L_2}^{M_2}(1))).$$

Since $\sigma(zs) = z's$ we have, for some representative $\dot{\sigma} \in N(T)$:

$$\begin{aligned}\theta(z's) &= G \cdot \left(zs(\mathrm{Ind}_{\dot{\sigma}^{-1}\cdot L_1}^{\dot{\sigma}^{-1}\cdot M_1}(1))(\mathrm{Ind}_{\dot{\sigma}^{-1}\cdot L_2}^{\dot{\sigma}^{-1}\cdot M_2}(1))\right) \\ &= G \cdot \left(zs(\mathrm{Ind}_{\dot{\sigma}^{-1}\cdot L_1}^{M_2}(1))(\mathrm{Ind}_{\dot{\sigma}^{-1}\cdot L_2}^{M_1}(1))\right).\end{aligned}$$

By [24, Example 3.1] we have $\mathrm{Ind}_{\dot{\sigma}^{-1}\cdot L_1}^{M_2}(1) = \mathrm{Ind}_{L_1}^{M_2}(1)$ and $\mathrm{Ind}_{\dot{\sigma}^{-1}\cdot L_2}^{M_1}(1) = \mathrm{Ind}_{L_1}^{M_1}(1)$ so $\theta(zs) = \theta(z's)$.

Remark 4 The parametrisation in Theorem 1 cannot be directly generalised to arbitrary Jordan classes. Indeed, if $u \in L$ is not rigid, then $L \cdot u$ is not necessarily characteristic and it may happen that for some external automorphism τ of L, the class $\tau(L \cdot u)$ differs from $L \cdot u$ even if they induce the same G-orbit. Then the map $\overline{\theta}$ is not necessarily injective.

4 The Quotient $\overline{S}//G$

In this section we discuss some properties of the categorical quotient $\overline{S}//G = \mathrm{Spec}(\mathbb{C}[\overline{S}])^G$ for G simple in any isogeny class. Since $\overline{S}//G$ parametrises only semisimple conjugacy classes it is enough to look at the so-called Dixmier sheets, i.e., the sheets containing a dense Jordan class consisting of semisimple elements. In addition, since every such Jordan class is dense in some sheet, studying the collection of $\overline{S}//G$ for S a sheet in G is the same as studying the collection of $\overline{J(s)}//G$ for $J(s)$ a semisimple Jordan class in G.

The following Theorem is a group version of [2, Satz 6.3], [18, Theorem 3.6(c)] and [28, Theorem A].

Theorem 2 *Let $S = \overline{J(s)}^{reg} \subset G$.*

1. *The normalisation of $\overline{S}//G$ is $Z(G^{s\circ})^\circ s/W^{Z^\circ s}$.*
2. *The variety $\overline{S}//G$ is normal if and only if the natural map*

$$\rho\colon \mathbb{C}[T]^W \to \mathbb{C}[Z(G^{s\circ})^\circ s]^{W^{Z^\circ s}} \tag{5}$$

induced from the restriction map $\mathbb{C}[T] \to \mathbb{C}[Z(G^{s\circ})^\circ s]$ is surjective.

Proof 1. The variety $Z(G^{s\circ})^\circ s/W^{Z^\circ s}$ is the quotient of a smooth variety (a shifted torus) by the action of a finite group, hence it is normal. Every semisimple class in $\overline{J(s)}$ meets T and $T \cap \overline{J(s)} = W \cdot (Z(G^{s\circ})^\circ s)$. Also, two elements in T are G-conjugate if and only if they are W-conjugate, hence we have an isomorphism $\overline{J(s)}//G \simeq W \cdot (Z(G^{s\circ})^\circ s)/W$ induced from the isomorphism $G//G \simeq T/W$.

We consider the morphism $\gamma\colon Z(G^{s\circ})^\circ s/W^{Z^\circ s} \to W \cdot (Z(G^{s\circ})^\circ s)/W$ induced by $zs \mapsto W \cdot (zs)$. It is surjective by construction, bijective on the dense

subset $(Z(G^{s\circ})^\circ s)^{reg}/W^{Z^\circ s}$ and finite, since the intersection of $W\cdot(zs)$ with $Z(G^{s\circ})^\circ s$ is finite. Hence γ is a normalisation morphism.
2. The variety $\overline{S}//G$ is normal if and only if the normalisation morphism is an isomorphism. This happens if and only if the composition

$$Z(G^{s\circ})^\circ s/W^{Z^\circ s}\to\overline{S}//G\subseteq G//G\simeq T/W$$

is a closed embedding, i.e., if and only if the corresponding algebra map between the rings of regular functions is surjective. □

5 An Example: Sheets and Their Quotients in Type G_2

We list here the sheets in G of type G_2 and all the conjugacy classes they contain. We shall denote by α and β, respectively, the short and the long simple roots. Since G is adjoint, by [7, Theorem 4.1] the sheets in G are in bijection with G-conjugacy classes of pairs (M,u) where M is a pseudo-Levi subgroup of G and u is a rigid unipotent element in M. The corresponding sheet is $\overline{J(su)}^{reg}$ where s is a semisimple element whose connected centraliser is M. The conjugacy classes of pseudo-Levi subgroups of G are those corresponding to the following subsets Π of the extended Dynkin diagram:

1. $\Pi=\emptyset$, so $M=T$, $u=1$, s is a regular semisimple element and S consists of all regular conjugacy classes;
2. $\Pi=\{\alpha\}$. Here $[M,M]$ is of type $\tilde{A}_1$, so $s=\alpha^\vee(\zeta)\beta^\vee(\zeta^2)=(3\alpha+2\beta)^\vee(\zeta^{-1})$ for $\zeta\neq 0,\pm1$ and $u=1$;
3. $\Pi=\{\beta\}$. Here $[M,M]$ is of type A_1 so $s=\alpha^\vee(\zeta^2)\beta^\vee(\zeta^3)=(2\alpha+\beta)^\vee(\zeta)$ for $\zeta\neq 0,\ 1\, e^{2\pi i/3},\ e^{-2\pi i/3}$ and $u=1$;
4. $\Pi=\{\alpha_0,\beta\}$. Here $[M,M]$ is of type A_2; so $u=1$ and $s=(2\alpha+\beta)^\vee(e^{2\pi i/3})$ is isolated, whence $S=G\cdot s$;
5. $\Pi=\{\alpha_0,\alpha\}$. Here $[M,M]$ is of type $\tilde{A}_1\times A_1$ so $u=1$ and $s=(3\alpha+2\beta)^\vee(-1)$ is isolated, whence $S=G\cdot s$;
6. $\Pi=\{\alpha,\ \beta\}$ so $L=G$. In this case we have three possible choices for u rigid unipotent, namely 1, $x_\alpha(1)$ or $x_\beta(1)$ (cfr. [29]). Each of these classes is a sheet on its own.

The only sheets containing more than one conjugacy classes are the regular one $S_0=G^{reg}$ corresponding to $\Pi=\emptyset$ and the two subregular ones, corresponding to $\Pi_1=\{\alpha\}$ and $\Pi_2=\{\beta\}$. For S_0 we have $Z^\circ s=T$, $W^{Z^\circ s}=W$ so S_0/G is in bijection with T/W and $\overline{S_0}//G\simeq G//G$ which is normal. For S_1 and S_2 we have:

$$S_1=\overline{J((3\alpha+2\beta)^\vee(\zeta_0))}^{reg}$$

$$=\left[\bigcup_{\zeta^2\neq 0,1} G\cdot(3\alpha+2\beta)^\vee(\zeta)\right]\cup \mathrm{Ind}_{\tilde{A}_1}^G(1)\cup G\cdot\left((3\alpha+2\beta)^\vee(-1)\mathrm{Ind}_{\tilde{A}_1}^{A_1\times\tilde{A}_1}(1)\right)$$

$$=\left[\bigcup_{\zeta^2\neq 0,1} G\cdot(3\alpha+2\beta)^\vee(\zeta)\right]\cup G\cdot\left(x_\beta(1)x_{\alpha_0}(1)\right)\cup G\cdot\left((3\alpha+2\beta)^\vee(-1)x_{\alpha_0}(1)\right)$$

for some $\zeta_0\neq 0,\pm 1$ and

$$S_2=\overline{J((2\alpha+\beta)^\vee(\xi_0))}^{reg}$$

$$=\left[\bigcup_{\xi^3\neq 0,1} G\cdot(2\alpha+\beta)^\vee(\xi)\right]\cup \mathrm{Ind}_{A_1}^G(1)\cup G\cdot\left((2\alpha+\beta)^\vee(e^{2\pi i/3})\mathrm{Ind}_{A_1}^{A_2}(1)\right)$$

$$=\left[\bigcup_{\xi^3\neq 0,1} G\cdot(2\alpha+\beta)^\vee(\xi)\right]\cup G\cdot(x_\beta(1)x_{\alpha_0}(1))\cup G\cdot\left((2\alpha+\beta)^\vee(e^{2\pi i/3})x_{\alpha_0}(1)\right)$$

for some $\xi_0\neq 0,1,e^{\pm 2\pi i/3}$.

In both cases M is a Levi subgroup of a parabolic subgroup of G. By Lemmata 1 and 2 we have $W^{Z^\circ s}=W^{\mathcal{O}}=\langle s_\alpha,s_{3\alpha+2\beta}\rangle$ for S_1 and $W^{Z^\circ s}=W^{\mathcal{O}}=\langle s_\beta,s_{2\alpha+\beta}\rangle$ for S_2. Also $Z(M)^\circ=Z(M)^\circ s$ in both cases because we are in an adjoint group and M is a Levi subgroup [23, Lemma 33], so

$$S_1/G\simeq(3\alpha+2\beta)^\vee(\mathbb{C}^\times)/\langle s_\alpha,s_{3\alpha+2\beta}\rangle\simeq(3\alpha+2\beta)^\vee(\mathbb{C}^\times)/\langle s_{3\alpha+2\beta}\rangle$$
$$S_2/G\simeq(2\alpha+\beta)^\vee(\mathbb{C}^\times)/\langle s_\beta,s_{2\alpha+\beta}\rangle\simeq(2\alpha+\beta)^\vee(\mathbb{C}^\times)/\langle s_{2\alpha+\beta}\rangle,$$

where the $\simeq$ symbols stand for the bijection $\overline{\theta}$.

Let us analyse normality of $\overline{S_1}//G$. Here, $Z(M)^\circ=(3\alpha+2\beta)^\vee(\mathbb{C}^*)\simeq\mathbb{C}^*$, so $\mathbb{C}[Z(M)^\circ]^{W^{Z^\circ s}}=\mathbb{C}[\zeta+\zeta^{-1}]$. Since G is simply connected, $\mathbb{C}[T]^W=(\mathbb{C}\Lambda)^W$ is the polynomial algebra generated by $f_1=\sum_{\gamma\in\Phi,\mathrm{short}}e^\gamma$ and $f_2=\sum_{\gamma\in\Phi,\mathrm{long}}e^\gamma$, [5, Ch.VI, §4, Théorème 1] Then,

$$\rho(f_1)(3\alpha+2\beta)^\vee(\zeta)=f_1((3\alpha+2\beta)^\vee(\zeta))=\sum_{\gamma\in\Phi,\mathrm{short}}\zeta^{(\gamma,(3\alpha+2\beta)^\vee)}=2+2\zeta+2\zeta^{-1}$$

so the restriction map is surjective and $\overline{S_1}//G$ is normal.

Let us consider normality of $\overline{S_2}//G$. Here, $Z(M)^\circ=(2\alpha+\beta)^\vee(\mathbb{C}^*)\simeq\mathbb{C}^*$, so $\mathbb{C}[Z]^\Gamma=\mathbb{C}[\zeta+\zeta^{-1}]$. Then,

$$\rho(f_1)(2\alpha+\beta)^\vee(\zeta)=f_1((2\alpha+\beta)^\vee(\zeta))$$

$$= \sum_{\gamma \in \Phi, \text{short}} \zeta^{(\gamma, (2\alpha+\beta)^\vee)} = \zeta^2 + \zeta^{-2} + 2(\zeta + \zeta^{-1})$$

whereas

$$\rho(f_2)(2\alpha + \beta)^\vee(\zeta) = f_2((2\alpha+\beta)^\vee(\zeta)) = \sum_{\gamma \in \Phi, \text{long}} \zeta^{(\gamma, (2\alpha+\beta)^\vee)} = 2 + 2\zeta^3 + 2\zeta^{-3}.$$

Let us write $y = \zeta + \zeta^{-1}$. Then, $(\zeta^2 + \zeta^{-2}) = y^2 - 2$ and $\zeta^3 + \zeta^{-3} = y^3 - 3y$ so

$$\begin{aligned} \mathrm{Im}(\rho) &= \mathbb{C}[y^2 + 2y, y^3 - 3y] = \mathbb{C}[(y+1)^2, y^3 + 3y^2 + 6y + 3 - 3y] \\ &= \mathbb{C}[(y+1)^2, (y+1)^3]. \end{aligned}$$

Hence, ρ is not surjective and $\overline{S_2}//G$ is not normal.

We observe that $\mathrm{Im}(\rho)$ is precisely the identification of the coordinate ring of $\overline{S_2}//G$ in $\mathbb{C}[T]^W$. We may thus see where this variety is not normal. We have: $\mathrm{Im}(\rho) = \mathbb{C}[(y+1)^2, (y+1)^3] \cong \mathbb{C}[Y, Z]/(Y^3 - Z^2)$ so this variety is not normal at $y + 1 = 0$, that is, for $\zeta + \zeta^{-1} + 1 = 0$. This corresponds precisely to the closed, isolated orbit $G \cdot ((2\alpha + \beta)^\vee(e^{2\pi i/3}))x_{\alpha_0}(1) = G \cdot ((2\alpha + \beta)^\vee(e^{-2\pi i/3}))x_{\alpha_0}(1)$. This example shows two phenomena: the first is that even if the sheet corresponding to the set Π_2 in Lie(G) has a normal quotient [6, Theorem 3.1], the same does not hold in the group counterpart. The second phenomenon is that the non-normality locus corresponds to an isolated class in $\overline{S_2}$. In [10] we address the general problem of normality of $\overline{S}//G$ and we prove and make use of the fact that if the categorical quotient of the closure a sheet in G is not normal, then it is certainly not normal at some isolated class.

Acknowledgements This work was partially supported by grants BIRD179758/17 "Stratifications in algebraic groups, spherical varieties, Kac Moody algebras and Kac Moody groups" and DOR1691049/16 "Representations of modular Lie algebras, algebraic groups, quantum groups and finite groups" of the University of Padova. The authors thank the referee for careful reading the first version of the manuscript and for a detailed list of suggestions for improvements of its presentation.

References

1. T. Arakawa, A. Moreau, *Sheets and associated varieties of affine vertex algebras,* Adv. Math. 320, , 157–209, (2017).
2. W. Borho, *Über Schichten halbeinfacher Lie-Algebren,* Invent. Math., 65, 283–317 (1981/82).
3. W. Borho, A. Joseph, *Sheets and topology of primitive spectra for semisimple Lie algebras,* J. Algebra 244, 76–167, (2001). Corrigendum 259, 310–311 (2003).
4. W. Borho, H. Kraft, *Über Bahnen und deren Deformationen bei linearen Aktionen reduktiver Gruppen,* Comment. Math. Helvetici, 54, 61–104 (1979).

5. N. Bourbaki, *Éléments de Mathématique. Groupes et Algèbres de Lie, Chapitres 4,5, et 6,* Masson, Paris (1981).
6. A. Broer, *Decomposition varieties in semisimple Lie algebras,* Can. J. Math. 50(5), 929–971 (1998).
7. G. Carnovale, *Lusztig's partition and sheets, with an appendix by M. Bulois,* Mathematical Research Letters, 22(3), 645–664, (2015).
8. G. Carnovale, F. Esposito, *On sheets of conjugacy classes in good characteristic,* IMRN 2012(4) , 810–828, (2012).
9. G. Carnovale, F. Esposito, *A Katsylo theorem for sheets of spherical conjugacy classes,* Representation Theory, 19, 263–280 (2015).
10. G. Carnovale; F. Esposito, *Affine hyperplane arrangements and Jordan classes,* arXiv:1807.10496, (2018).
11. J. Dixmier, *Polarisations dans les algèbres de Lie semi-simples complexes* Bull. Sci. Math. 99, 45–63, (1975).
12. J.M. Douglass, G. Röhrle, *Invariants of reflection groups, arrangements, and normality of decomposition classes in Lie algebras,* Compos. Math. 148, 921–930, (2012) .
13. R. B. Howlett, *Normalizers of parabolic subgroups of reflection groups,* J. London Math. Soc. 21, 62–80 (1980).
14. J. Humphreys, *Conjugacy Classes in Semisimple Algebraic Groups,* AMS, Providence, Rhode Island (1995).
15. A. E. Im Hof, *The sheets in a classical Lie algebra,* http://edoc.unibas.ch/257/PhDthesis, Basel (2005).
16. P.I. Katsylo, *Sections of sheets in a reductive algebraic Lie algebra,* Math. USSR Izvestyia 20(3), 449–458 (1983).
17. B. Kostant, *Lie group representations on polynomial rings,* Amer. J. Math. 85, 327–404 (1963).
18. H. Kraft, *Parametrisierung von Konjugationsklassen in sln,* Math. Ann. 234, 209–220 (1978).
19. I. Losev, *Deformation of symplectic singularities and orbit method for semisimple Lie algebras,* arXiv:1605.00592v1, (2016).
20. G. Lusztig, *On conjugacy classes in a reductive group.* In: Representations of reductive groups, 333–363, Progr. Math., 312, Birkhäuser/Springer (2015).
21. G. Lusztig, *Intersection cohomology complexes on a reductive group,* Invent. Math. 75, 205–272 (1984).
22. G. Lusztig, N. Spaltenstein, *Induced unipotent classes,* J. London Math. Soc. (2), 19, 41–52 (1979).
23. G. McNinch, E. Sommers, *Component groups of unipotent centralizers in good characteristic,* J. Algebra 270(1), 288–306 (2003).
24. A. Moreau, *Corrigendum to: On the dimension of the sheets of a reductive Lie algebra,* Journal of Lie Theory 23(4), 1075–1083 (2013).
25. D. Peterson, *Geometry of the Adjoint Representation of a Complex Semisimple Lie Algebra* PhD Thesis, Harvard University, Cambridge, Massachussetts, (1978).
26. A. Premet, L. Topley, *Derived subalgebras of centralisers and finite W-algebras,* Compos. Math. 150(9),1485–ĂŞ1548, (2014).
27. A. Premet, D. Stewart, *Rigid orbits and sheets in reductive Lie algebras over fields of prime characteristic,* Journal of the Institute of Mathematics of Jussieu, 1–31, (2016).
28. R. W. Richardson, *Normality of G-stable subvarieties of a semisimple Lie algebra,* In: Cohen *et al.*, Algebraic Groups, Utrecht 1986, Lecture Notes in Math. 1271, Springer-Verlag, New York, (1987).
29. N. Spaltenstein, *Classes Unipotentes et Sous-Groupes de Borel,* Springer-Verlag, Berlin (1982).
30. T.A. Springer, *Linear Algebraic Groups, Second Edition* Progress in Mathematics 9, Birkhäuser (1998).

About Polynomiality of the Poisson Semicentre for Parabolic Subalgebras

Florence Fauquant-Millet

I would like to dedicate this paper to Anthony Joseph for his 75th birthday, thanks to whom I discovered the world of quantum groups and then of (classical) enveloping algebras and with whom I worked a long time on this interesting subject of polynomiality of the Poisson semicentre associated to parabolic subalgebras.

Abstract Here we explain some results about the polynomiality of the Poisson semicentre for parabolic subalgebras in a complex simple Lie algebra in the particular case of maximal parabolic subalgebras in a simple Lie algebra of type B.

Mathematics Subject Classification (2000): 16W22, 17B22, 17B35

1 Introduction

Let $\mathfrak{a}$ be an algebraic complex Lie algebra and A be its adjoint group. Denote by $S(\mathfrak{a})$ the symmetric algebra of $\mathfrak{a}$ and by $S(\mathfrak{a})^A = S(\mathfrak{a})^{\mathfrak{a}} =: Y(\mathfrak{a})$ the algebra of invariants of $S(\mathfrak{a})$ under the action of A or under the adjoint action of $\mathfrak{a}$ (observe that $Y(\mathfrak{a})$ is also the Poisson centre of $S(\mathfrak{a})$ for its natural Poisson structure).

Here we are interested in the question of polynomiality of $Y(\mathfrak{a})$ especially for the particular case when $\mathfrak{a}$ is the truncated maximal parabolic subalgebra of a simple Lie algebra $\mathfrak{g}$ of type B_n.

F. Fauquant-Millet (✉)
University of Lyon, UJM Saint-Etienne, CNRS UMR 5208, Saint-Etienne, France

Institut Camille Jordan, Saint-Etienne, France
e-mail: florence.millet@univ-st-etienne.fr

M. Gorelik et al. (eds.), *Representations and Nilpotent Orbits of Lie Algebraic Systems*, Progress in Mathematics 330,
https://doi.org/10.1007/978-3-030-23531-4_4

Indeed it was not yet known whether for this case $Y(\mathfrak{a})$ is a polynomial algebra, at least for roughly half of the maximal parabolic subalgebras in this type.

To show polynomiality of $Y(\mathfrak{a})$ in this case, we construct a so-called adapted pair for $\mathfrak{a}$ (which is a generalization of a principal $\mathfrak{sl}_2$-triple for a non-reductive Lie algebra) and which was introduced in [17]. This adapted pair allows us to construct an affine and algebraic slice (which is the analogue of a Kostant slice).

The construction of an adapted pair for $\mathfrak{a}$ is based on [13, Thm. 8.6] (where a small modification is introduced). Then we need to compute an improved upper bound using [14, Lem. 6.11] to conclude that $Y(\mathfrak{a})$ is polynomial.

2 The Background

We consider a parabolic subalgebra $\mathfrak{p}$ of a complex simple Lie algebra $\mathfrak{g}$.

Set, for $\lambda \in \mathfrak{p}^*$, $S(\mathfrak{p})_\lambda := \{s \in S(\mathfrak{p}) \mid \forall x \in \mathfrak{p},\ (\operatorname{ad} x)(s) = \lambda(x)\, s\}$ where ad denotes the adjoint action of $\mathfrak{p}$ on $S(\mathfrak{p})$ which extends by derivation the adjoint action on $\mathfrak{p}$ defined by the Lie bracket.

Then the Poisson semicentre of $S(\mathfrak{p})$ is $Sy(\mathfrak{p}) = \bigoplus_{\lambda \in \mathfrak{p}^*} S(\mathfrak{p})_\lambda$, while the Poisson centre is $Y(\mathfrak{p}) = S(\mathfrak{p})_0$.

We have that $Sy(\mathfrak{p}) = S(\mathfrak{p})^{\mathfrak{p}'}$ (that is, $Sy(\mathfrak{p})$ is the algebra of invariants in $S(\mathfrak{p})$ under the adjoint action of $\mathfrak{p}'$) where $\mathfrak{p}' = [\mathfrak{p},\ \mathfrak{p}]$ and $Y(\mathfrak{p}) \subset Sy(\mathfrak{p})$.

If $\mathfrak{p} = \mathfrak{g}$, then $Y(\mathfrak{g}) = Sy(\mathfrak{g})$. Otherwise by [11, Proof of Lemma 7.9], $Y(\mathfrak{p}) = \mathbb{C} \subsetneq Sy(\mathfrak{p})$ by [3].

By [1] there exists a canonically defined truncation $\mathfrak{p}_\Lambda$ of $\mathfrak{p}$ which is the largest subalgebra of $\mathfrak{p}$ that vanishes on each weight of $Sy(\mathfrak{p})$. Then $Sy(\mathfrak{p}_\Lambda) = Y(\mathfrak{p}_\Lambda) = Sy(\mathfrak{p})$.

Fix a Cartan subalgebra $\mathfrak{h}$ of $\mathfrak{g}$, set $n = \dim \mathfrak{h} = \operatorname{rank} \mathfrak{g}$ and choose a set $\pi = \{\alpha_k\}_{1 \le k \le n}$ of simple roots. These simple roots can be expressed as linear combinations of elements ε_i, $1 \le i \le m$, of an orthonormal basis of $\mathbb{R}^m$ as in [2, Planches I–IX].

The set of roots in $\mathfrak{g}$ is denoted by $\Delta = \Delta^+ \sqcup \Delta^-$, with Δ^+, resp. Δ^-, the set of positive, resp. negative, roots. The root subspace of $\mathfrak{g}$ corresponding to $\alpha \in \Delta$ is denoted by $\mathfrak{g}_\alpha$. For $A \subset \Delta$, we denote by $\mathfrak{g}_A := \bigoplus_{\alpha \in A} \mathfrak{g}_\alpha$. For each $\alpha \in \Delta$, denote by $\alpha^\vee$ its corresponding coroot. Then we have that $\mathfrak{h} = \bigoplus_{1 \le k \le n} \mathbb{C}\alpha_k^\vee$ and $\mathfrak{g} = \mathfrak{n}^- \oplus \mathfrak{h} \oplus \mathfrak{n}$, where $\mathfrak{n}^- = \mathfrak{g}_{\Delta^-}$, resp. $\mathfrak{n} = \mathfrak{g}_{\Delta^+}$.

Fix a subset $\pi' \subsetneq \pi$ and set $\Delta^+_{\pi'} = \Delta^+ \cap \mathbb{N}\pi'$, resp. $\Delta^-_{\pi'} = \Delta^- \cap (-\mathbb{N}\pi')$. Let $\mathfrak{p} = \mathfrak{n}^- \oplus \mathfrak{h} \oplus \mathfrak{n}_{\pi'}$ be a parabolic subalgebra of $\mathfrak{g}$ where $\mathfrak{n}_{\pi'} = \mathfrak{g}_{\Delta^+_{\pi'}}$.

Then the canonical truncation $\mathfrak{p}_\Lambda$ of $\mathfrak{p}$ verifies $\mathfrak{p}_\Lambda = \mathfrak{n}^- \oplus \mathfrak{h}_\Lambda \oplus \mathfrak{n}_{\pi'}$ with $\mathfrak{h}_\Lambda \subset \mathfrak{h}$. One always has that $\mathfrak{h}_\Lambda \supset \mathfrak{h}'$, where $\mathfrak{h}' = \mathfrak{h} \cap [\mathfrak{p},\ \mathfrak{p}]$ is the Cartan subalgebra of the Levi factor $\mathfrak{g}' = \mathfrak{n}^-_{\pi'} \oplus \mathfrak{h}' \oplus \mathfrak{n}_{\pi'}$ (where $\mathfrak{n}^-_{\pi'} = \mathfrak{g}_{\Delta^-_{\pi'}}$) of $\mathfrak{p}$.

When the parabolic subalgebra $\mathfrak{p}$ is maximal, it is easily shown that $\mathfrak{h}_\Lambda = \mathfrak{h}'$.

By the Killing form K on $\mathfrak{g}$, one may identify $\mathfrak{p}^*$ with $\mathfrak{p}^- = \mathfrak{n} \oplus \mathfrak{h} \oplus \mathfrak{n}^-_{\pi'}$ and $\mathfrak{p}^*_\Lambda$ with $\mathfrak{p}^-_\Lambda = \mathfrak{n} \oplus \mathfrak{h}_\Lambda \oplus \mathfrak{n}^-_{\pi'}$.

For a $\mathfrak{h}$-module $M = \bigoplus_{\nu\in\mathfrak{h}^*} M_\nu$, resp. $M' = \bigoplus_{\nu\in\mathfrak{h}^*} M'_\nu$, with finite dimensional weight subspaces M_ν, resp. M'_ν, we define its formal character $\operatorname{ch} M = \sum_{\nu\in\mathfrak{h}^*} \dim M_\nu e^\nu$, resp. $\operatorname{ch} M' = \sum_{\nu\in\mathfrak{h}^*} \dim M'_\nu e^\nu$, and write $\operatorname{ch} M \leq \operatorname{ch} M'$ if $\dim M_\nu \leq \dim M'_\nu$ for all $\nu \in \mathfrak{h}^*$.

Theorem 1 (Upper and Lower Bounds, See [6]) *There are two polynomial algebras $\mathscr{A}$ and $\mathscr{B}$, which are also $\mathfrak{h}$-modules, such that*

$$\operatorname{ch} \mathscr{A} \leq \operatorname{ch} Sy(\mathfrak{p}) \leq \operatorname{ch} \mathscr{B}.$$

Moreover if $\operatorname{ch} \mathscr{A} = \operatorname{ch} \mathscr{B}$, then $Sy(\mathfrak{p})$ is a polynomial algebra.

For example, it occurs when $\mathfrak{g}$ is simple of type A *or* C *for any parabolic subalgebra $\mathfrak{p}$ of $\mathfrak{g}$.*

Let $E(\pi')$ be the set of $\langle ij\rangle$-orbits of π, where i and j are the involutions of π defined, for example, in [8, 2.2].

For any algebraic complex Lie algebra $\mathfrak{a}$ with A its adjoint group and any $x \in \mathfrak{a}^*$, we say that x is regular if the coadjoint orbit $A.x$ is of minimal codimension (equal to $\operatorname{ind} \mathfrak{a}$, the index of $\mathfrak{a}$). We denote by $\mathfrak{a}^*_{reg}$ the set of regular elements in $\mathfrak{a}^*$ (see [4, 1.11.6]).

Since $Sy(\mathfrak{p}_\Lambda) = Y(\mathfrak{p}_\Lambda) = Sy(\mathfrak{p})$ a theorem of Rosenlicht implies that

$$\operatorname{GKdim} Sy(\mathfrak{p}) = \operatorname{ind} \mathfrak{p}_\Lambda \tag{1}$$

and by [5, 3.2]

$$\operatorname{GKdim} Sy(\mathfrak{p}) = |E(\pi')|. \tag{2}$$

Denote by $\{\varpi_{\alpha_k}\}_{\alpha_k\in\pi}$, or simply $\{\varpi_k\}_{1\leq k\leq n}$, resp. $\{\varpi'_{\alpha_k}\}_{\alpha_k\in\pi'}$, or simply $\{\varpi'_k\}_{1\leq k\leq n|\alpha_k\in\pi'}$ the set of fundamental weights associated to π, resp. to π'. Let $\mathscr{B}_\pi$, resp. $\mathscr{B}_{\pi'}$, be the set of weights of the Poisson semicentre of $S(\mathfrak{n}\oplus\mathfrak{h})$, resp. of $S(\mathfrak{n}_{\pi'}\oplus\mathfrak{h}')$, which by [10] is a polynomial algebra in rank $\mathfrak{g}$, resp. in rank $\mathfrak{g}'$, generators whose weights are listed in [10, Tables I and II] and [6, Table] for an erratum.

For all $\Gamma \in E(\pi')$, set

$$\delta_\Gamma = -\sum_{\gamma\in\Gamma} \varpi_\gamma - \sum_{\gamma\in j(\Gamma)} \varpi_\gamma + \sum_{\gamma\in\Gamma\cap\pi'} \varpi'_\gamma + \sum_{\gamma\in i(\Gamma\cap\pi')} \varpi'_\gamma \tag{3}$$

$$\varepsilon_\Gamma = \begin{cases} 1/2 & \text{if } \Gamma = j(\Gamma),\ \sum_{\gamma\in\Gamma}\varpi_\gamma \in \mathscr{B}_\pi, \text{ and } \sum_{\gamma\in\Gamma\cap\pi'}\varpi'_\gamma \in \mathscr{B}_{\pi'}. \\ 1 & \text{otherwise.} \end{cases} \tag{4}$$

It is shown in [12, Thm. 6.7] that

$$\operatorname{ch} \mathscr{A} = \prod_{\Gamma\in E(\pi')} (1-e^{\delta_\Gamma})^{-1} \leq \operatorname{ch} Sy(\mathfrak{p}) \leq \prod_{\Gamma\in E(\pi')} (1-e^{\varepsilon_\Gamma\delta_\Gamma})^{-1} = \operatorname{ch} \mathscr{B}. \tag{5}$$

Then when $\varepsilon_\Gamma = 1$ for all $\Gamma \in E(\pi')$ we have that $\mathrm{ch}\,\mathscr{A} = \mathrm{ch}\,\mathscr{B}$, and one concludes that $Sy(\mathfrak{p}) = Y(\mathfrak{p}_\Lambda)$ is a polynomial algebra over $\mathbb{C}$. In particular, the bounds $\mathrm{ch}\,\mathscr{A}$ and $\mathrm{ch}\,\mathscr{B}$ coincide for any parabolic subalgebra of $\mathfrak{g}$ simple of type A or C.

When $\mathfrak{g}$ is simple of type B or D and $\mathfrak{p}$ is a maximal parabolic subalgebra, then the bounds coincide in roughly half of the cases.

Note that the above criterion is not a necessary condition for polynomiality of $Sy(\mathfrak{p})$.

For instance, when $\mathfrak{p} = \mathfrak{b}^- = \mathfrak{h} \oplus \mathfrak{n}^-$, then the criterion does not hold in $\mathfrak{g}$ simple of type outside A or C, but $Sy(\mathfrak{b}^-)$ is a polynomial algebra by [10].

Theorem 2 (Kostant Slice, See [18])

Assume that $\mathfrak{g}$ is a complex simple Lie algebra with $\mathfrak{h}$ a Cartan subalgebra and G the adjoint group of $\mathfrak{g}$. Identify $\mathfrak{g}$ and $\mathfrak{g}^*$ via the Killing form on $\mathfrak{g}$. Let (x, h, y) be a principal $\mathfrak{sl}_2$-triple ($h \in \mathfrak{h}$, x, y regular nilpotent in $\mathfrak{g}$, such that $[h, x] = x$ and $[h, y] = -y$). Then we have the following.

Restriction of functions gives an algebra isomorphism $Y(\mathfrak{g}) \xrightarrow{\sim} R[y+\mathfrak{g}^x]$ where $R[y+\mathfrak{g}^x]$ is the algebra of regular functions on the affine variety $y+\mathfrak{g}^x$, $\mathfrak{g}^x$ being the centralizer of the element x in $\mathfrak{g}$.

The set of degrees of the homogeneous generators of $Y(\mathfrak{g})$ is equal to the set of the eigenvalues plus one of $\mathrm{ad}\,h$ on an h-stable basis of $\mathfrak{g}^x$.

Every G-orbit in $G(y+\mathfrak{g}^x)$ meets $y+\mathfrak{g}^x$ transversally at exactly one point and $\mathfrak{g}^*$ is equal to the Zariski closure $\overline{G(y+\mathfrak{g}^x)}$ of $G(y+\mathfrak{g}^x)$: such an affine subspace $y+\mathfrak{g}^x$ is called in [15] an affine slice to the coadjoint action of $\mathfrak{g}$ - we will call it the Kostant slice. Actually the space $\mathfrak{g}^*_{reg}$ of regular elements in $\mathfrak{g}^*$ is equal to $G(y+\mathfrak{g}^x)$.

When the algebraic Lie algebra $\mathfrak{a}$ is not semisimple, there exists no principal $\mathfrak{sl}_2$-triple in general in $\mathfrak{a}$. Hence we need another notion which plays the role of a principal $\mathfrak{sl}_2$-triple in a non-reductive Lie algebra. This is the notion of an adapted pair, introduced by A. Joseph and P. Lamprou and for which we recall the following definition.

Definition 3 (Adapted Pair, See [17])

An adapted pair for $\mathfrak{a}$ is a pair $(h, y) \in \mathfrak{a} \times \mathfrak{a}^*$, with h ad-semisimple, y regular in $\mathfrak{a}^*$ and $(\mathrm{ad}\,h)(y) = -y$, where ad is the coadjoint action of $\mathfrak{a}$ on $\mathfrak{a}^*$.

Remarks 4

Adapted pairs were constructed for all truncated (bi)parabolic subalgebras in $\mathfrak{g}$ simple of type A (see [13]).

Adapted pairs need not exist, for instance they do not exist for the canonical truncation $\mathfrak{b}_\Lambda$ of the Borel subalgebra $\mathfrak{b}$ of $\mathfrak{g}$ when $\mathfrak{h}_\Lambda = \{0\}$ as it occurs when $\mathfrak{g}$ is simple of type B_n, E_7, E_8, F_4 or G_2.

Lemma 5 (Improved Upper Bound, See [14, Lem. 6.11])

Let $\mathfrak{p} = \mathfrak{n}^- \oplus \mathfrak{h} \oplus \mathfrak{n}_{\pi'}$ be a parabolic subalgebra of a simple Lie algebra $\mathfrak{g}$. Assume that there exists an adapted pair $(h, y) \in \mathfrak{h}_\Lambda \times \mathfrak{p}^*_\Lambda$ for the canonical truncation $\mathfrak{p}_\Lambda$

of $\mathfrak{p}$ *such that* $y = \sum_{\gamma \in S} x_\gamma$ *with* $x_\gamma \in \mathfrak{g}_\gamma \setminus \{0\}$, $S \subset \Delta^+ \sqcup \Delta^-_{\pi'}$ *and that* $S_{|\mathfrak{h}_\Lambda}$ *is a basis for* $\mathfrak{h}^*_\Lambda$.

Then there exists an "improved upper bound" $\mathscr{B}'$ *such that* $\mathrm{ch}\, Sy(\mathfrak{p}) \leq \mathscr{B}'$.

Assume now that equality holds.

Then restriction of functions gives an isomorphism of algebras $Sy(\mathfrak{p}) \xrightarrow{\sim} R[y + V]$ *where* V *is an* h*-stable vector space such that* $(\mathrm{ad}\, \mathfrak{p}_\Lambda)(y) \oplus V = \mathfrak{p}^*_\Lambda$ *(where* ad *is the coadjoint action). In particular* $Sy(\mathfrak{p})$ *is a polynomial algebra over* $\mathbb{C}$.

Such an affine subspace $y + V$ *is called a Weierstrass section or an algebraic slice for* $Sy(\mathfrak{p})$. *It is also an affine slice to the coadjoint action of* $\mathfrak{p}_\Lambda$ *by [7] in the sense of Theorem 2.*

With the hypotheses of the above Lemma, the improved upper bound $\mathscr{B}'$ is given by

$$\mathrm{ch}\, Sy(\mathfrak{p}) \leq \prod_{\gamma \in T} (1 - e^{-(\gamma + t(\gamma))})^{-1} = \mathscr{B}' \tag{6}$$

where $T \subset \Delta^+ \sqcup \Delta^-_{\pi'}$ is such that $\mathfrak{p}^*_\Lambda = (\mathrm{ad}\, \mathfrak{p}_\Lambda)(y) \oplus \mathfrak{g}_T$, $|T| = \mathrm{ind}\, \mathfrak{p}_\Lambda$ and for each $\gamma \in T$, $t(\gamma) \in \mathbb{Q}S$ is uniquely determined by the condition that $\gamma + t(\gamma)$ vanishes on $\mathfrak{h}_\Lambda$.

Theorem 6 (Degrees of Homogeneous Generators, See [16])

Assume that $Sy(\mathfrak{p}) = Y(\mathfrak{p}_\Lambda)$ *is a polynomial algebra in* $\ell := \mathrm{ind}\, \mathfrak{p}_\Lambda$ *generators and that* $\mathfrak{p}_\Lambda$ *admits an adapted pair* $(h,\, y) \in \mathfrak{h}_\Lambda \times \mathfrak{p}^*_\Lambda$.

- *Then* $y + V$ *is a Weierstrass section for* $Sy(\mathfrak{p})$ *(in the sense of Lemma 5) and also an affine slice to the coadjoint action of* $\mathfrak{p}_\Lambda$ *(where* V *is an* h*-stable complement of* $(\mathrm{ad}\, \mathfrak{p}_\Lambda)(y)$ *in* $\mathfrak{p}^*_\Lambda$*).*
- *If* $\{m_i\}_{i=1}^{\ell}$ *are the eigenvalues of* $\mathrm{ad}\, h$ *on a basis of* V, *then* $\{m_i + 1\}_{i=1}^{\ell}$ *are the degrees of the homogeneous generators of* $Y(\mathfrak{p}_\Lambda)$.
- *In particular,* $m_i \geq 0$.

3 Maximal Parabolic Subalgebras

From now on, we focus on a *maximal* parabolic subalgebra $\mathfrak{p}$ of $\mathfrak{g}$ simple of type B_n ($n \geq 2$) and we will explain how to construct an adapted pair for the canonical truncation $\mathfrak{p}_\Lambda$, which will provide a Weierstrass section for $Sy(\mathfrak{p}) = Y(\mathfrak{p}_\Lambda)$ by Lemma 5.

We have that $\mathfrak{p} = \mathfrak{n}^- \oplus \mathfrak{h} \oplus \mathfrak{n}_{\pi'}$ with $\pi' = \pi \setminus \{\alpha_s\}$ and the canonical truncation $\mathfrak{p}_\Lambda$ of $\mathfrak{p}$ is such that $\mathfrak{p}_\Lambda = \mathfrak{n}^- \oplus \mathfrak{h}' \oplus \mathfrak{n}_{\pi'}$ with $\mathfrak{h}' = \mathfrak{h} \cap [\mathfrak{p},\, \mathfrak{p}]$. By the Killing form on $\mathfrak{g}$, we may identify $\mathfrak{p}^*_\Lambda$ with $\mathfrak{n} \oplus \mathfrak{h}' \oplus \mathfrak{n}^-_{\pi'}$.

The bounds in Theorem 1 coincide exactly when s is odd (with Bourbaki's labeling [2]), and when $n = s = 2$ or $n = s = 4$. Then $Sy(\mathfrak{p})$ is a polynomial algebra over $\mathbb{C}$ in these cases. Moreover the weights and degrees of homogeneous generators are those of the homogeneous generators of the polynomial algebra $\mathscr{B}$, which can be computed easily : the weights are the δ_Γ, $\Gamma \in E(\pi')$, given by equality (3) and the degrees can be computed, thanks to the degrees of the homogeneous generators of the semicentre of the Borel given in [10], see [6, Cor. 5.4.2].

The case s even is more complicated since in general the bounds in Theorem 1 do not coincide. The polynomiality of $Sy(\mathfrak{p})$ will follow from the construction of an adapted pair for $\mathfrak{p}_\Lambda$, and from the fact that in this case the improved upper bound $\mathscr{B}'$ of Lemma 5 is attained. Then Theorem 6 will give the degrees of the homogeneous generators of $Sy(\mathfrak{p})$.

In [8] we already have constructed adapted pairs for truncated maximal parabolic subalgebras when the upper and lower bounds in Theorem 1 coincide. To construct such adapted pairs, we used the notion of a Heisenberg set for which we recall the definition below.

4 Construction of an Adapted Pair

Keep the above notations and hypotheses.

Definition 7 (Heisenberg Set, See [8, Def. 2]) Let $\gamma \in \Delta$. A Heisenberg set Γ_γ of centre γ is a subset of Δ such that $\gamma \in \Gamma_\gamma$ and for all $\alpha \in \Gamma_\gamma \setminus \{\gamma\}$ there exists $\alpha' \in \Gamma_\gamma$ such that $\alpha + \alpha' = \gamma$.

Example 8

Assume that $\mathfrak{g}$ is a semisimple Lie algebra with a Cartan subalgebra $\mathfrak{h}$. Denote by $(\ ,\ )$ the non-degenerate symmetric bilinear form on $\mathfrak{h}^* \times \mathfrak{h}^*$ invariant under the action of the Weyl group of $\mathfrak{g}$, induced by the Killing form on $\mathfrak{h} \times \mathfrak{h}$. Let $\Delta = \sqcup \Delta_i$ be a root system of $\mathfrak{g}$, where Δ_i are irreducible root systems with β_i the unique highest root of Δ_i.

Take $(\Delta_i)_{\beta_i} := \{\alpha \in \Delta_i \mid (\alpha, \beta_i) = 0\}$ and decompose it into irreducible root systems Δ_{ij} with highest roots β_{ij}.

Continuing we obtain a set of strongly orthogonal positive roots β_K (called the Kostant cascade for $\mathfrak{g}$) indexed by elements $K \in \mathbb{N} \cup \mathbb{N}^2 \cup \cdots$ and irreducible subsystems Δ_K of Δ.

The set $H_{\beta_K} := \{\alpha \in \Delta_K \mid (\alpha, \beta_K) > 0\}$ is a Heisenberg set of centre β_K and actually it is the maximal Heisenberg set of centre β_K which is included in Δ^+ by [10, Lem. 2.2].

We set $-H_{\beta_K} = \{\alpha \in \Delta_K \mid -\alpha \in H_{\beta_K}\}$. Let S be a subset of $\Delta^+ \sqcup \Delta^-_{\pi'}$.

We choose for all $\gamma \in S$ a Heisenberg set $\Gamma_\gamma \subset \Delta^+ \sqcup \Delta^-_{\pi'}$ of centre γ such that all Γ_γ's are disjoint.

We set $O = \bigsqcup_{\gamma \in S} \Gamma^0_\gamma$ (with $\Gamma^0_\gamma = \Gamma_\gamma \setminus \{\gamma\}$).

We set $S = S^+ \sqcup S^-$ with S^+, resp. S^-, being the subset of S containing those $\gamma \in S$ such that $\Gamma_\gamma \subset \Delta^+$, resp. $\Gamma_\gamma \subset \Delta^-_{\pi'}$.

Set $O^\pm = \bigsqcup_{\gamma \in S^\pm} \Gamma^0_\gamma$, so that $O = O^+ \sqcup O^-$.

Let $y = \sum_{\gamma \in S} x_\gamma$, with $x_\gamma \in \mathfrak{g}_\gamma \setminus \{0\}$ for all $\gamma \in S$. The set S will be said to be the support of the element y.

Choose also disjoint subsets of $\Delta^+ \sqcup \Delta^-_{\pi'}$: $T = T^+ \sqcup T^-$ and T^* ($T^+ \subset \Delta^+$ and $T^- \subset \Delta^-_{\pi'}$), also disjoint from $\Gamma = \bigsqcup_{\gamma \in S} \Gamma_\gamma$. We can now give a generalization of [13, Thm. 8.6] (see also [8, Lem. 6]).

Proposition 9 *Assume that*

(1) $S_{|\mathfrak{h}_\Lambda}$ *is a basis for* $\mathfrak{h}^*_\Lambda$.
(2) If $\alpha \in \Gamma^0_\gamma$, *with* $\gamma \in S^+$, *is such that there exists* $\beta \in O^+$, *with* $\alpha + \beta \in S$, *then* $\beta \in \Gamma^0_\gamma$ *and* $\alpha + \beta = \gamma$.
(3) If $\alpha \in \Gamma^0_\gamma$, *with* $\gamma \in S^-$, *is such that there exists* $\beta \in O^-$, *with* $\alpha + \beta \in S$, *then* $\beta \in \Gamma^0_\gamma$ *and* $\alpha + \beta = \gamma$.
(4) $\Delta^+ \sqcup \Delta^-_{\pi'} = \bigsqcup_{\gamma \in S} \Gamma_\gamma \sqcup T \sqcup T^*$.
(5) For all $\alpha \in T^*$, $\mathfrak{g}_\alpha \subset (\operatorname{ad} \mathfrak{p}_\Lambda)(y) + \mathfrak{g}_T$.
(6) $|T| = \operatorname{ind} \mathfrak{p}_\Lambda$.

Then y is regular in $\mathfrak{p}^*_\Lambda$ *and* $(\operatorname{ad} \mathfrak{p}_\Lambda)(y) \oplus \mathfrak{g}_T = \mathfrak{p}^*_\Lambda$. *Moreover if we uniquely define* $h \in \mathfrak{h}_\Lambda$ *by* $h(\gamma) = -1$ *for all* $\gamma \in S$, *then* (h, y) *is an adapted pair for* $\mathfrak{p}_\Lambda$.

Proof The proof follows that of [13, Thm. 8.6] with only a small modification. We give it for completeness. Condition (4) implies that $\mathfrak{p}_\Lambda = \mathfrak{h}_\Lambda \oplus \mathfrak{g}_{-O} \oplus \mathfrak{g}_{-S} \oplus \mathfrak{g}_{-T^*} \oplus \mathfrak{g}_{-T}$ and that $\mathfrak{p}^*_\Lambda = \mathfrak{h}_\Lambda \oplus \mathfrak{g}_O \oplus \mathfrak{g}_S \oplus \mathfrak{g}_{T^*} \oplus \mathfrak{g}_T$.

Let Φ_y denote the skew-symmetric bilinear form defined by $\Phi_y(x, x') = K(y, [x, x'])$ for all $x, x' \in \mathfrak{g}$ where recall K is the Killing form on $\mathfrak{g}$.

Conditions (2) and (3) imply by [13, Lem. 8.5] that the restriction of Φ_y to $\mathfrak{g}_{-O} \times \mathfrak{g}_{-O}$ is non-degenerate. Then $\mathfrak{g}_O \subset (\operatorname{ad} \mathfrak{g}_{-O})(y) + \mathfrak{h}_\Lambda + \mathfrak{g}_S + \mathfrak{g}_T + \mathfrak{g}_{T^*}$.

But since $O \cap S = \emptyset$ one has that for all $x \in \mathfrak{g}_O$ and $x' \in \mathfrak{g}_{-O}$, the element $x - (\operatorname{ad} x')(y)$ belongs to the orthogonal of $\mathfrak{h}_\Lambda$ for the Killing form. Then $\mathfrak{g}_O \subset (\operatorname{ad} \mathfrak{g}_{-O})(y) + \mathfrak{g}_S + \mathfrak{g}_T + \mathfrak{g}_{T^*}$.

Condition (1) implies that $\mathfrak{g}_S = (\operatorname{ad} \mathfrak{h}_\Lambda)(y)$ and that $\mathfrak{h}_\Lambda \subset (\operatorname{ad} \mathfrak{g}_{-S})(y) + \mathfrak{g}_O + \mathfrak{g}_S + \mathfrak{g}_T + \mathfrak{g}_{T^*}$. Condition (5) implies that $\mathfrak{g}_{T^*} \subset (\operatorname{ad} \mathfrak{p}_\Lambda)(y) + \mathfrak{g}_T$. Hence $\mathfrak{p}^*_\Lambda = \mathfrak{h}_\Lambda \oplus \mathfrak{g}_O \oplus \mathfrak{g}_S \oplus \mathfrak{g}_{T^*} \oplus \mathfrak{g}_T \subset (\operatorname{ad} \mathfrak{p}_\Lambda)(y) + \mathfrak{g}_T$. Finally condition (6) implies that the latter sum is direct, since $\dim \mathfrak{g}_T = \operatorname{ind} \mathfrak{p}_\Lambda \leq \operatorname{codim} (\operatorname{ad} \mathfrak{p}_\Lambda)(y)$. □

Remarks 10

(1) Proposition 9 is exactly [13, Thm. 8.6] when $T^* = \emptyset$ (then condition (5) is empty) and was used (with $T^* = \emptyset$) in the case when s is odd in [8]. Also in [8], we took all the elements of the Kostant cascade for the elements or their

opposite in S and in T, the elements of T being the simple roots or their opposite in the Kostant cascade (see [8]). Moreover for each $\gamma \in S^+$, resp. S^-, we took for the Heisenberg set Γ_γ the maximal Heisenberg set H_γ, resp. $-H_{-\gamma}$, defined in Example 8. Then the proof follows easily since by [10, Lem. 2.2] Δ^+ is the disjoint union of all the maximal Heisenberg sets H_γ's, γ being any root in the Kostant cascade, and moreover if two positive roots α, α' are such that $\alpha + \alpha' = \gamma$, with γ a root in the Kostant cascade, then α, α' both belong to the same maximal Heisenberg set H_γ.

(2) Unfortunately when s is even the strategy used for s odd does no more work. Indeed the element h of the adapted pair for $\mathfrak{p}_\Lambda$ must belong to $\mathfrak{h}_\Lambda = \mathfrak{h}'$ and then must vanish on ϖ_s. On the other hand, it must take the value -1 on each element of S. Then if the elements of the Kostant cascade $\varepsilon_{2i-1} + \varepsilon_{2i}$, for $1 \leq i \leq s/2$, would belong to S, then we would have both $h(\varepsilon_1 + \ldots + \varepsilon_s) = (-1) \times s/2$ and $h(\varepsilon_1 + \ldots + \varepsilon_s) = 0$ since $\varpi_s = \varepsilon_1 + \ldots + \varepsilon_s$.

Then we need to modify the sets S and T of the odd case and to add a set T^* which makes the verification of the regularity of y more complicated.

(3) The case $s = 2$ was already treated in [19] where polynomiality was shown, and in [14] an adapted pair for this case was also constructed. Our adapted pair is equivalent to Joseph's adapted pair in [14] in the sense of [7] since an element of the Weyl group of the Levi of $\mathfrak{p}$ sends bijectively the supports of the nilpotent elements of both adapted pairs (for more details, see [9, Remark 7.11]).

5 The Kostant Cascades

Recall that we assume that $\mathfrak{g}$ is simple of type B_n ($n \geq 2$) and that $\mathfrak{p}$ is a parabolic subalgebra of $\mathfrak{g}$ associated to $\pi' = \{\alpha_i\}_{1\leq i\leq n, i\neq s} \subset \pi = \{\alpha_i\}_{1\leq i\leq n}$. Since the case s odd was already treated in [8], we will assume that s is even.

Then the Levi factor of $\mathfrak{p}$ is semisimple of type $\mathrm{A}_{s-1} \times \mathrm{B}_{n-s}$. We will denote by β_i and $\beta_{i'}$ the elements of the Kostant cascade for $\mathfrak{g}$, by β_i' the elements of the Kostant cascade of the simple part of the Levi factor of $\mathfrak{p}$ of type A_{s-1} and by β_i'', $\beta_{i'}''$ the elements of the Kostant cascade of the simple part of the Levi factor of $\mathfrak{p}$ of type B_{n-s}.

Denote by $\{\varepsilon_i\}_{i=1}^n$ an orthonormal basis of $\mathbb{R}^n$ according to which the roots in Δ are expanded as in [2, Planche II]. Then the Kostant cascade for $\mathfrak{g}$ of type B_n is formed by the following strongly orthogonal positive roots : $\beta_i = \varepsilon_{2i-1} + \varepsilon_{2i} = \alpha_{2i-1} + 2\alpha_{2i} + \ldots + 2\alpha_n$, for all $1 \leq i \leq [n/2]$, and $\beta_{i'} = \alpha_{2i-1}$, for all $1 \leq i \leq [(n+1)/2]$.

The Kostant cascade of the simple part of the Levi factor of type A_{s-1} is formed by the following positive roots : $\beta_i' = \varepsilon_i - \varepsilon_{s+1-i} = \alpha_i + \ldots + \alpha_{s-i}$, for all $1 \leq i \leq s/2$, with $\beta_{s/2}' = \alpha_{s/2}$.

6 Examples

6.1 *First Example in Type* B_6

Here we assume that $\mathfrak{g}$ is of type B_6 and that $\pi' = \pi \setminus \{\alpha_2\}$.

We choose

$$S^+ = \{\varepsilon_2,\ \tilde{\beta}_1 = \beta_1 - \alpha_2 = \varepsilon_1 + \varepsilon_3,\ \varepsilon_4 + \varepsilon_5\}$$

$$S^- = \{-\beta_1'' = -\varepsilon_3 - \varepsilon_4,\ -\beta_2'' = -\varepsilon_5 - \varepsilon_6\}$$

$$T^+ = \{\beta_1 = \varepsilon_1 + \varepsilon_2,\ \varepsilon_1 - \varepsilon_3,\ \beta_{1'} = \varepsilon_1 - \varepsilon_2,\ \varepsilon_4 - \varepsilon_5\}$$

$$T^- = \{-\beta_{1'}'' = -\varepsilon_3 + \varepsilon_4,\ -\beta_{2'}'' = -\varepsilon_5 + \varepsilon_6\}$$

$$T^* = \{\varepsilon_6,\ \varepsilon_2 - \varepsilon_1,\ \varepsilon_2 - \varepsilon_3,\ \varepsilon_2 - \varepsilon_4,\ \varepsilon_2 - \varepsilon_5,\ \varepsilon_2 - \varepsilon_6,$$
$$\varepsilon_2 + \varepsilon_3,\ \varepsilon_2 + \varepsilon_4,\ \varepsilon_2 + \varepsilon_5,\ \varepsilon_2 + \varepsilon_6\}$$

Then the semisimple element $h \in \mathfrak{h}'$ of the adapted pair is

$$h = \varepsilon_1 - \varepsilon_2 - 2\varepsilon_3 + 3\varepsilon_4 - 4\varepsilon_5 + 5\varepsilon_6 = \alpha_1^\vee - 2\alpha_3^\vee + \alpha_4^\vee - 3\alpha_5^\vee + \alpha_6^\vee.$$

Let λ be an eigenvalue of $\operatorname{ad} h$ in $\mathfrak{p}_\Lambda^*$. Then $-\lambda$ is an eigenvalue of $\operatorname{ad} h$ in $\mathfrak{p}_\Lambda$. Denote by m_λ^* the multiplicity of λ in $\mathfrak{p}_\Lambda^*$ and by $m_{-\lambda}$ the multiplicity of $-\lambda$ in $\mathfrak{p}_\Lambda$. We have that $m_{-\lambda} = m_\lambda^*$. Set m_λ' the multiplicity of λ in $\mathfrak{h}_\Lambda \oplus \mathfrak{g}_O \oplus \mathfrak{g}_S \oplus \mathfrak{g}_{T^*} \subset \mathfrak{p}_\Lambda^*$.

Recall that the regularity of y means that $\mathfrak{p}_\Lambda^* = (\operatorname{ad} \mathfrak{p}_\Lambda)(y) \oplus \mathfrak{g}_T$, with $|T| = \operatorname{ind} \mathfrak{p}_\Lambda$.

Since $(\operatorname{ad} h)(y) = -y$, we must have that, for each eigenvalue λ of $\operatorname{ad} h$ in $\mathfrak{p}_\Lambda^*$, $m_\lambda' \le m_{\lambda+1}$.

In the table below are the multiplicities m_λ' and $m_{-\lambda}$ for each eigenvalue λ of $\operatorname{ad} h$ in $\mathfrak{p}_\Lambda^*$:

λ	-9	-8	-7	-6	-5	-4	-3	-2	-1	0
m_λ'	1	1	2	2	3	3	4	5	5	5
$m_{-\lambda}$	1	1	2	2	3	3	4	5	5	6

λ	1	2	3	4	5	6	7	8	9
m_λ'	5	4	3	3	2	2	1	1	0
$m_{-\lambda}$	5	5	4	3	3	2	2	1	1

We can see that the inequality $m_\lambda' \le m_{\lambda+1}$ is satisfied for each eigenvalue λ of $\operatorname{ad} h$ in $\mathfrak{p}_\Lambda^*$.

6.2 Second Example in Type B_6

Here we assume that $\mathfrak{g}$ is of type B_6 and that $\pi' = \pi \setminus \{\alpha_4\}$.

We choose

$$S^+ = \{\beta_1 = \varepsilon_1 + \varepsilon_2,\ \tilde{\beta}_2 = \beta_2 - \alpha_4 = \varepsilon_3 + \varepsilon_5,\ \varepsilon_4\}$$

$$S^- = \{-\tilde{\beta}_1' = -\beta_1' + \alpha_3 = \varepsilon_3 - \varepsilon_1, -\beta_1'' = -\varepsilon_5 - \varepsilon_6\}$$

$$T^+ = \{\alpha_1,\ \beta_2,\ \alpha_3,\ \alpha_3 + \alpha_4\}$$

$$T^- = \{-\alpha_5\}$$

$$T^* = \{\varepsilon_4 + \varepsilon_5,\ \varepsilon_4 + \varepsilon_6,\ \varepsilon_4 - \varepsilon_5,\ \varepsilon_4 - \varepsilon_6,\ \varepsilon_6,\ \varepsilon_4 - \varepsilon_1,\ \varepsilon_4 - \varepsilon_2,\ \varepsilon_4 - \varepsilon_3\}$$

Then the semisimple element $h \in \mathfrak{h}'$ of the adapted pair is

$$h = 3\varepsilon_1 - 4\varepsilon_2 + 2\varepsilon_3 - \varepsilon_4 - 3\varepsilon_5 + 4\varepsilon_6 = 3\alpha_1^\vee - \alpha_2^\vee + \alpha_3^\vee - 3\alpha_5^\vee + 1/2\alpha_6^\vee.$$

Adopt the same notations as in the previous example. In the table below are the multiplicities m'_λ and $m_{-\lambda}$ for each eigenvalue λ of $\operatorname{ad} h$ in $\mathfrak{p}_\Lambda^*$:

λ	-8	-7	-6	-5	-4	-3	-2	-1	0
m'_λ	1	3	1	2	4	3	2	7	7
$m_{-\lambda}$	1	3	1	2	4	3	2	7	7

λ	1	2	3	4	5	6	7
m'_λ	2	3	4	2	1	3	1
$m_{-\lambda}$	3	3	5	2	2	3	3

We can see that the inequality $m'_\lambda \leq m_{\lambda+1}$ is satisfied for each eigenvalue λ of $\operatorname{ad} h$ in $\mathfrak{p}_\Lambda^*$.

7 General Case

Consider now the general case of a maximal parabolic subalgebra $\mathfrak{p}$ associated to $\pi' = \pi \setminus \{\alpha_s\}$ in $\mathfrak{g}$ simple of type B_n $(n \geq 2)$ with s even $(2 \leq s \leq n)$.

We will show below that all conditions of Proposition 9 are satisfied, for a good choice of subsets S^+, S^-, T^+, T^-, T^* of $\Delta^+ \sqcup \Delta^-_{\pi'}$ and disjoint Heisenberg sets Γ_γ of centre γ, for all $\gamma \in S^+ \sqcup S^-$.

We choose the set $S = S^+ \sqcup S^-$ as follows :

$$S^- = \{-\tilde{\beta}'_i = -\beta'_i + \alpha_{s-i} = \varepsilon_{s-i} - \varepsilon_i,\ -\beta''_{j-s/2} = -\varepsilon_{2j-1} - \varepsilon_{2j};\ 1 \le i \le s/2 - 1,\ s/2+1 \le j \le [n/2]\}$$

If $n = s$,

$$S^+ = \{\varepsilon_s,\ \beta_i = \varepsilon_{2i-1} + \varepsilon_{2i};\ 1 \le i \le s/2 - 1\}$$

If $n > s$,

$$S^+ = \{\varepsilon_s,\ \beta_i = \varepsilon_{2i-1} + \varepsilon_{2i},\ \tilde{\beta}_{s/2} = \beta_{s/2} - \alpha_s = \varepsilon_{s-1} + \varepsilon_{s+1},\ \varepsilon_{2j} + \varepsilon_{2j+1};\ 1 \le i \le s/2 - 1,\ s/2+1 \le j \le [(n-1)/2]\}$$

One easily checks that $|S| = n - 1$. Moreover condition (1) of Proposition 9 is satisfied, thanks to the following lemma.

Lemma 11 *$S_{|\mathfrak{h}_\Lambda}$ is a basis for $\mathfrak{h}^*_\Lambda$.*

Proof It is sufficient to show that if $S = \{s_1, \dots, s_{n-1}\}$ and $\{h_1, \dots, h_{n-1}\}$ is a basis of $\mathfrak{h}_\Lambda$, then $\det(s_i(h_j))_{i,j} \neq 0$. We will prove this statement by induction on n. We choose $\{\alpha_i^\vee \mid 1 \le i \le n,\ i \neq s\}$ as a basis of $\mathfrak{h}_\Lambda$.

Add temporarily a lower subscript n to $S^\pm,\ \pi,\ \mathfrak{h}' = \mathfrak{h}_\Lambda$ to emphasize that they are defined for type B_n.

Identify an element $(x_1, x_2, \dots, x_n)$ of $\mathbb{R}^n$ with the element $(x_1, x_2, \dots, x_n, 0)$ of $\mathbb{R}^{n+1}$. Observe that $S^+_{s+1} = S^+_s \sqcup \{\varepsilon_{s-1} + \varepsilon_{s+1}\}$, whereas for n even and $n \ge s+2$ we have $S^+_{n+1} = S^+_n \sqcup \{\varepsilon_n + \varepsilon_{n+1}\}$ and for n odd we have $S^+_{n+1} = S^+_n$.

Similarly for n even, $S^-_{n+1} = S^-_n$ and for n odd, $S^-_{n+1} = S^-_n \sqcup \{-\varepsilon_n - \varepsilon_{n+1}\}$. Finally set $S_n = S^+_n \sqcup S^-_n$.

We first consider the case $n = s$.

If $n = s = 2$, then $S = \{s_1 = \varepsilon_2 = \alpha_2\}$ and $\det(s_1(\alpha_1^\vee)) = -1 \neq 0$.

Assume now that $n \ge 4$ and $n = s$. Then $S = \{\varepsilon_n,\ \varepsilon_{n-i} - \varepsilon_i,\ \beta_j \mid 1 \le i,\ j \le n/2 - 1\}$. Recall that $\{\varpi_i\}_{1 \le i \le n}$ is the set of fundamental weights of $\mathfrak{g}$. One has that for all $i \in \mathbb{N}$, $1 \le i \le n/2 - 1$, $\beta_i = \varpi_{2i} - \varpi_{2i-2}$ (where we have set $\varpi_0 = 0$) and $\varepsilon_n = -\varpi_{n-1} + 2\varpi_n$. Also, for all $i \in \mathbb{N}$, $1 \le i \le n/2 - 1$, $\varepsilon_{n-i} - \varepsilon_i = -\varpi_i + \varpi_{i-1} - \varpi_{n-1-i} + \varpi_{n-i}$.

Then, by ordering the basis of $\mathfrak{h}_\Lambda$ as $\{\alpha_2^\vee, \alpha_4^\vee, \dots, \alpha_{n-2}^\vee, \alpha_{n-1}^\vee, \alpha_1^\vee, \alpha_{n-3}^\vee, \alpha_3^\vee, \dots, \alpha_{n/2+1}^\vee, \alpha_{n/2-1}^\vee\}$ if $n/2$ is even, and as $\{\alpha_2^\vee, \alpha_4^\vee, \dots, \alpha_{n-2}^\vee, \alpha_{n-1}^\vee, \alpha_1^\vee, \alpha_{n-3}^\vee, \alpha_3^\vee, \dots, \alpha_{n/2-2}^\vee, \alpha_{n/2}^\vee\}$ if $n/2$ is odd, and by ordering elements of S as $\{\beta_1, \beta_2, \dots, \beta_{n/2-1}, \varepsilon_n, \varepsilon_{n-1} - \varepsilon_1, \varepsilon_{n-2} - \varepsilon_2, \dots, \varepsilon_{n/2+1} - \varepsilon_{n/2-1}\}$, we have that $(s_i(h_j))_{ij} = \begin{pmatrix} A & 0 \\ C & D \end{pmatrix}$ where A is an $(n/2) \times (n/2)$ lower triangular matrix with 1 everywhere on the diagonal, except the last element which is equal to -1 and D is a $(n/2-1) \times (n/2-1)$ lower triangular matrix with -1 everywhere on the diagonal. Hence $\det(s_i(h_j))_{ij} = (-1)^{n/2} \neq 0$.

For every $n \geq s$, let $\{h_1, \ldots, h_{n-1}, h_n\}$ be a basis for the truncated Cartan $\mathfrak{h}'_{n+1}$ of the truncated parabolic associated to $\pi_{n+1} \setminus \{\alpha_s\}$ in type B_{n+1}, such that $\{h_1, \ldots, h_{n-1}\}$ is a basis of the truncated Cartan $\mathfrak{h}'_n$ for the truncated parabolic associated to $\pi_n \setminus \{\alpha_s\}$ in type B_n with the identification in the beginning of the proof.

Then, using the observation in the beginning of the proof, and ordering the elements of $S_{n+1} = \{s_1, s_2, \ldots, s_n\}$ such that its first $n-1$ elements are those of S_n, we get that $\det(s_i(h_j))_{1\leq i,\, j\leq n} = (-1)^n \det(s_i(h_j))_{1\leq i,\, j\leq n-1}$, which completes the proof of the lemma. □

For the set $T = T^+ \sqcup T^-$, we choose :
If $n = s$,

$$T^+ = \{\beta_{s/2} = \varepsilon_{s-1} + \varepsilon_s,\ \beta_{i'} = \varepsilon_{2i-1} - \varepsilon_{2i} = \alpha_{2i-1};\ 1 \leq i \leq s/2\}$$
$$T^- = \emptyset$$

If $n > s$,

$$T^+ = \{\beta_{s/2} = \varepsilon_{s-1} + \varepsilon_s,\ \varepsilon_{s-1} - \varepsilon_{s+1},\ \beta_{i'} = \varepsilon_{2i-1} - \varepsilon_{2i} = \alpha_{2i-1},$$
$$\varepsilon_{s+2j} - \varepsilon_{s+2j+1} = \alpha_{s+2j};\ 1 \leq i \leq s/2,\ 1 \leq j \leq [(n-s-1)/2]\}$$
$$T^- = \{-\beta''_{i'} = -\varepsilon_{s+2i-1} + \varepsilon_{s+2i} = -\alpha_{s+2i-1};\ 1 \leq i \leq [(n-s)/2]\}$$

One easily checks that $|T| = n - s/2 + 1$. Then condition (6) of Proposition 9 is satisfied, thanks to the following lemma.

Lemma 12 *We have that* $\operatorname{ind} \mathfrak{p}_\Lambda = n - s/2 + 1$.

Proof The set of $\langle ij \rangle$-orbits in π is

$$E(\pi') = \{\{\alpha_{s/2}\},\ \{\alpha_t,\ \alpha_{s-t}\},\ \{\alpha_u\} \mid 1 \leq t \leq s/2 - 1,\ s \leq u \leq n\}.$$

Hence the lemma, by equalities (1) and (2) of Sect. 2. □

Now for the Heisenberg sets Γ_γ of centre $\gamma \in S$, we choose the following sets, where recall H_γ is the maximal Heisenberg set of centre γ given in Example 8, when γ lies in the Kostant cascade :

- for all $1 \leq i \leq s/2 - 1$, $\Gamma_{\beta_i} = H_{\beta_i}$
- $\Gamma_{\tilde{\beta}_{s/2}} = \{\tilde{\beta}_{s/2},\ \varepsilon_{s-1},\ \varepsilon_{s+1},\ \varepsilon_{s-1} \pm \varepsilon_i,\ \varepsilon_{s+1} \mp \varepsilon_i;\ s+2 \leq i \leq n\}$
- for all $s/2 + 1 \leq i \leq [(n-1)/2]$, $\Gamma_{\varepsilon_{2i}+\varepsilon_{2i+1}} = \{\varepsilon_{2i} + \varepsilon_{2i+1},\ \varepsilon_{2i},\ \varepsilon_{2i+1},\ \varepsilon_{2i} \pm \varepsilon_j,\ \varepsilon_{2i+1} \mp \varepsilon_j;\ 2i+2 \leq j \leq n\}$
- for all $1 \leq i \leq s/2 - 1$, $\Gamma_{\varepsilon_{s-i}-\varepsilon_i} = \{\varepsilon_{s-i} - \varepsilon_i,\ \varepsilon_j - \varepsilon_i,\ \varepsilon_{s-i} - \varepsilon_j;\ i+1 \leq j \leq s-i-1\}$
- for all $s/2 + 1 \leq i \leq [n/2]$, $\Gamma_{-\beta''_{i-s/2}} = -H_{\beta''_{i-s/2}}$
- $\Gamma_{\varepsilon_s} = \{\varepsilon_s\}$

By construction, all the above sets are disjoint and are Heisenberg sets.
Finally we choose :

$$\begin{cases} T^* = \{\varepsilon_n,\ \varepsilon_s - \varepsilon_i,\ \varepsilon_s + \varepsilon_j; \\ \qquad 1 \le i \le n,\ i \neq s,\ s+1 \le j \le n\} \text{ if } n \text{ even, (without } \varepsilon_n \text{ if } n = s) \\ T^* = \{-\varepsilon_n,\ \varepsilon_s - \varepsilon_i,\ \varepsilon_s + \varepsilon_j;\ 1 \le i \le n,\ i \neq s,\ s+1 \le j \le n\} \text{ if } n \text{ odd} \end{cases}$$

By construction the sets T and T^* are disjoint and also disjoint from all the Heisenberg sets given above.

Below we show that condition (4) of Proposition 9 is satisfied.

Lemma 13 *We have that* $\Delta^+ \sqcup \Delta^-_{\pi'} = \bigsqcup_{\gamma \in S} \Gamma_\gamma \sqcup T \sqcup T^*$.

Proof By [10, Lem. 2.2], we have that

$$\Delta^+ = \bigsqcup_{1 \le i \le [n/2]} H_{\beta_i} \sqcup \bigsqcup_{1 \le i \le [(n+1)/2]} H_{\beta_{i'}}.$$

Moreover for all $1 \le i \le [(n+1)/2]$, $H_{\beta_{i'}} = \{\beta_{i'}\} = \{\alpha_{2i-1}\}$.

We already have that $\bigsqcup_{1 \le i \le s/2-1} H_{\beta_i} = \bigsqcup_{1 \le i \le s/2-1} \Gamma_{\beta_i}$.

- Let $\gamma \in H_{\beta_{s/2}}$.

If $\gamma = \beta_{s/2} = \varepsilon_{s-1} + \varepsilon_s$, then $\gamma \in T$.
If $\gamma = \varepsilon_{s-1} \pm \varepsilon_i$ with $s+1 \le i \le n$, or if $\gamma = \varepsilon_{s-1}$, then $\gamma \in \Gamma_{\tilde{\beta}_{s/2}}$, unless $\gamma = \varepsilon_{s-1} - \varepsilon_{s+1}$ in which case $\gamma \in T$.
If $\gamma = \varepsilon_s \pm \varepsilon_i$ with $s+1 \le i \le n$, then $\gamma \in T^*$.
If $\gamma = \varepsilon_s$, then $\gamma \in \Gamma_{\varepsilon_s}$.

- Let $\gamma \in H_{\beta_{s/2+1}}$.

If $\gamma = \varepsilon_{s+1} \pm \varepsilon_i$, with $s+2 \le i \le n$, or if $\gamma = \varepsilon_{s+1}$, then $\gamma \in \Gamma_{\tilde{\beta}_{s/2}}$.
If $s = n-2$ and $\gamma = \varepsilon_{s+2} = \varepsilon_n$, then $\gamma \in T^*$.
If $\gamma = \varepsilon_{s+2} \pm \varepsilon_i$, with $s+3 \le i \le n$ or if $\gamma = \varepsilon_{s+2}$ (with $s \le n-3$), then $\gamma \in \Gamma_{\varepsilon_{s+2}+\varepsilon_{s+3}}$ unless $\gamma = \varepsilon_{s+2} - \varepsilon_{s+3}$ in which case $\gamma \in T$.

- Let $\gamma \in H_{\beta_i}$, with $s/2+2 \le i \le [n/2]$.

If $\gamma = \varepsilon_{2i-1} \pm \varepsilon_j$, with $2i \le j \le n$, or if $\gamma = \varepsilon_{2i-1}$, then $\gamma \in \Gamma_{\varepsilon_{2i-2}+\varepsilon_{2i-1}}$.
If $i = n/2$ (with n even), then $\gamma = \varepsilon_{2i} = \varepsilon_n \in T^*$.
If $\gamma = \varepsilon_{2i} \pm \varepsilon_j$, with $2i+1 \le j \le n$, or if $\gamma = \varepsilon_{2i}$ (with $2i \le n-1$) then $\gamma \in \Gamma_{\varepsilon_{2i}+\varepsilon_{2i+1}}$, unless $\gamma = \varepsilon_{2i} - \varepsilon_{2i+1}$ in which case $\gamma \in T$.

- Let $\gamma = \alpha_{2i-1}$, with $1 \le i \le [(n+1)/2]$.

If $1 \le i \le s/2$, then $\gamma \in T$.
If $i = s/2+1$, then $\gamma \in \Gamma_{\tilde{\beta}_{s/2}}$.
If $i \ge s/2+2$, then $\gamma \in \Gamma_{\varepsilon_{2i-2}+\varepsilon_{2i-1}}$.

Similarly one has that

$$\Delta^-_{\pi'} = \bigsqcup_{1\leq i\leq s/2} (-H_{\beta'_i}) \sqcup \bigsqcup_{1\leq i\leq [(n-s)/2]} (-H_{\beta''_i}) \sqcup \bigsqcup_{1\leq i\leq [(n-s+1)/2]} \{-\alpha_{s+2i-1}\}$$

- Let $\gamma \in (-H_{\beta'_i})$, with $1 \leq i \leq s/2$.

If $\gamma = \varepsilon_{s+1-i} - \varepsilon_j$, with $i \leq j \leq s-i$, then $\gamma \in \Gamma_{\varepsilon_{s-i+1}-\varepsilon_{i-1}}$, unless $\gamma = \varepsilon_s - \varepsilon_j$, with $1 \leq j \leq s-1$, in which case $\gamma \in T^*$.
If $\gamma = \varepsilon_j - \varepsilon_i$, with $i+1 \leq j \leq s-i$, then $\gamma \in \Gamma_{\varepsilon_{s-i}-\varepsilon_i}$.

- Let $\gamma \in (-H_{\beta''_i})$, with $1 \leq i \leq [(n-s)/2]$.

Then $\gamma \in \Gamma_{-\beta''_i}$.

- Let $\gamma = -\alpha_{s+2i-1}$, with $1 \leq i \leq [(n+1-s)/2]$.

Then $\gamma \in T^-$, unless $\gamma = -\alpha_n = -\varepsilon_n$ and n odd, in which case $\gamma \in T^*$. This completes the lemma. □

Below we check that condition (2) of Proposition 9 is satisfied.

Lemma 14 *Let $\alpha \in \Gamma^0_\gamma$, with $\gamma \in S^+$, be such that there exists $\beta \in O^+$, with $\alpha + \beta \in S$. Then $\beta \in \Gamma^0_\gamma$ and $\alpha + \beta = \gamma$.*

Proof Apply again [10, Lem. 2.2]. Let $\alpha,\ \alpha'$ be two positive roots such that $\alpha \in H_{\beta_i}$ and $\alpha' \in H_{\beta_j}$ with $1 \leq i,\ j \leq [n/2]$ and $\alpha + \alpha' \in \Delta$.

Then either $i \leq j$ and one has that $\alpha + \alpha' \in H_{\beta_i}$ or $i \geq j$ and $\alpha + \alpha' \in H_{\beta_j}$.

Moreover if $\alpha + \alpha' = \beta_i$, then $i = j$ and $\alpha,\ \alpha' \in H_{\beta_i} \setminus \{\beta_i\}$.

Assume now that $\alpha \in H_{\beta_i}$ and $\alpha' \in H_{\beta_{k'}}$ with $1 \leq i \leq [n/2]$ and $1 \leq k \leq [(n+1)/2]$ are such that $\alpha + \alpha' \in \Delta$.

Then $k = i$ or if $k = (n+1)/2$ and n odd then $k = i+1$, and $\alpha + \alpha' \in H_{\beta_i} \setminus \{\beta_i\}$.

If $\alpha \in H_{\beta_{i'}}$ and $\alpha' \in H_{\beta_{k'}}$ with $1 \leq i,\ k \leq [(n+1)/2]$, then $\alpha + \alpha' \notin \Delta$.

If $\alpha \in H_{\beta_{i'}}$ and $\alpha' \in H_{\beta_j}$ with $1 \leq j \leq [n/2]$ and $1 \leq i \leq [(n+1)/2]$ are such that $\alpha + \alpha' \in \Delta$, then $i = j$ and $\alpha + \alpha' \in H_{\beta_j} \setminus \{\beta_j\}$, or if $i = (n+1)/2$ and n odd then $i = j+1$ and $\alpha + \alpha' \in H_{\beta_j} \setminus \{\beta_j\}$.

- Let $\alpha \in \Gamma^0_{\beta_i} = H^0_{\beta_i}$, with $1 \leq i \leq s/2 - 1$, be such that there exists $\alpha' \in O^+$ with $\alpha + \alpha' \in S$. Recall the decomposition of Δ^+ given in the proof of Lemma 13.

If $\alpha' \in \Gamma^0_{\beta_j}$, with $1 \leq j \leq s/2 - 1$, then one has necessarily $i = j$. Indeed by the above if $i \leq j$ then $\alpha + \alpha' \in H_{\beta_i} \cap S$, hence $\alpha + \alpha' = \beta_i$ and then $\alpha,\ \alpha' \in H^0_{\beta_i} = \Gamma^0_{\beta_i}$. If $i > j$ then $\alpha,\ \alpha' \in \Gamma^0_{\beta_j}$, which is not possible, since the sets H_{β_i} and H_{β_j} are disjoint.

If $\alpha' \in H_{\beta_j}$ with $s/2 \leq j \leq [n/2]$, then by a similar reasoning as above, we obtain that $\alpha,\ \alpha' \in H_{\beta_i}$ which is not possible.

If $\alpha' \in H_{\beta_{j'}}$ with $1 \leq j \leq [(n+1)/2]$, then by the above $j = i$ and $\alpha + \alpha' \in H_{\beta_i} \setminus \{\beta_i\}$ which is not possible since $H_{\beta_i} \cap S = \{\beta_i\}$.

• Let $\alpha \in \Gamma^0_{\tilde{\beta}_{s/2}}$ be such that there exists $\alpha' \in O^+ \setminus \bigsqcup_{1\le i\le s/2-1}\Gamma^0_{\beta_i}$ with $\alpha+\alpha' \in S$.

Then one checks that $\alpha+\alpha' \in H_{\beta_{s/2}} \cap S$. Hence $\alpha+\alpha' = \tilde{\beta}_{s/2}$ or $\alpha+\alpha' = \varepsilon_s$. The latter case is not possible since $\alpha+\alpha' = \varepsilon_s$ implies that $\alpha = \varepsilon_{s+1}$ and $\alpha' = \varepsilon_s - \varepsilon_{s+1} \in T^*$. The former case implies that $\alpha' \in \Gamma^0_{\tilde{\beta}_{s/2}}$.

• Let $\alpha \in \Gamma^0_{\varepsilon_{2i}+\varepsilon_{2i+1}}$ with $s/2+1 \le i \le [(n-1)/2]$ be such that there exists $\alpha' \in \Gamma^0_{\varepsilon_{2j}+\varepsilon_{2j+1}}$ with $s/2+1 \le j \le [(n-1)/2]$ with $\alpha+\alpha' \in S$. Then one checks that necessarily one has $i=j$ and $\alpha+\alpha' = \varepsilon_{2i}+\varepsilon_{2i+1}$. This completes the proof.

□

We also check that condition (3) of Proposition 9 is satisfied.

Lemma 15 *Let $\alpha \in \Gamma^0_\gamma$, with $\gamma \in S^-$, be such that there exists $\beta \in O^-$, with $\alpha+\beta \in S$. Then $\beta \in \Gamma^0_\gamma$ and $\alpha+\beta = \gamma$.*

Proof Denote by π'_1, resp. π'_2, the connected component of π' corresponding to the simple roots of type A_{s-1}, resp. of type B_{n-s} and by $\Delta^-_{\pi'_1}$, resp. $\Delta^-_{\pi'_2}$ the subset of $\Delta^-_{\pi'}$ generated by π'_1, resp. by π'_2.

• Let $\alpha \in \Gamma^0_{\varepsilon_{s-i}-\varepsilon_i}$, with $1 \le i \le s/2-1$, be such that there exists $\alpha' \in O^-$ with $\alpha+\alpha' \in S$. Since $\alpha \in \Delta^-_{\pi'_1}$, there exists necessarily $1 \le j \le s/2-1$ such that $\alpha' \in \Gamma^0_{\varepsilon_{s-j}-\varepsilon_j}$. Then one easily checks that $i=j$ and that $\alpha+\alpha' = \varepsilon_{s-i}-\varepsilon_i$.

• Let $\alpha \in \Gamma^0_{-\beta''_i}$, with $1 \le i \le [(n-s)/2]$, be such that there exists $\alpha' \in O^-$ with $\alpha+\alpha' \in S$. Since $\alpha \in \Delta^-_{\pi'_2}$, there exists necessarily $1 \le j \le [(n-s)/2]$ such that $\alpha' \in \Gamma^0_{-\beta''_j}$. Then one easily checks that $i=j$ and that $\alpha+\alpha' = -\beta''_i$. This completes the proof. □

Finally we check below condition (5) : recall that we denote by x_α, $\alpha \in \Delta$, a chosen non-zero root vector in $\mathfrak{g}_\alpha$ and we will rescale the vectors x_α if necessary.

Examine first the case of $\varepsilon_s+\varepsilon_j$, with $s+1 \le j \le n$.

Lemma 16 *For all $j \in \mathbb{N}$, $s+1 \le j \le n$, we have that $x_{\varepsilon_s+\varepsilon_j} \in (\mathrm{ad}\,\mathfrak{p}_\Lambda)(y)+\mathfrak{g}_T$.*

Proof One has that $(\mathrm{ad}\,x_{\varepsilon_{s+1}})(y) = x_{\varepsilon_s+\varepsilon_{s+1}} + x_{-\varepsilon_{s+2}}$ and that $x_{-\varepsilon_{s+2}} = (\mathrm{ad}\,x_{-\varepsilon_s-\varepsilon_{s+2}})(y)$.

Hence $x_{\varepsilon_s+\varepsilon_{s+1}} = (\mathrm{ad}\,(x_{\varepsilon_{s+1}} - x_{-\varepsilon_s-\varepsilon_{s+2}}))(y) \in (\mathrm{ad}\,\mathfrak{p}_\Lambda)(y)+\mathfrak{g}_T$.

One has that $(\mathrm{ad}\,x_{\varepsilon_{s+2}})(y) = x_{\varepsilon_s+\varepsilon_{s+2}} + x_{-\varepsilon_{s+1}}$ and $x_{-\varepsilon_{s+1}} + x_{\varepsilon_{s-1}-\varepsilon_s} = (\mathrm{ad}\,x_{-\varepsilon_s-\varepsilon_{s+1}})(y)$. Hence $x_{\varepsilon_s+\varepsilon_{s+2}} = (\mathrm{ad}\,(x_{\varepsilon_{s+2}} - x_{-\varepsilon_s-\varepsilon_{s+1}}))(y) + x_{\varepsilon_{s-1}-\varepsilon_s} \in (\mathrm{ad}\,\mathfrak{p}_\Lambda)(y)+\mathfrak{g}_T$.

Let $s+3 \le j \le n-1$, j odd.

We have that $(\mathrm{ad}\,x_{\varepsilon_j})(y) = x_{\varepsilon_s+\varepsilon_j} + x_{-\varepsilon_{j+1}}$ and $x_{-\varepsilon_{j+1}} = (\mathrm{ad}\,x_{-\varepsilon_s-\varepsilon_{j+1}})(y)$. Hence $x_{\varepsilon_s+\varepsilon_j} = (\mathrm{ad}\,(x_{\varepsilon_j} - x_{-\varepsilon_s-\varepsilon_{j+1}}))(y) \in (\mathrm{ad}\,\mathfrak{p}_\Lambda)(y)+\mathfrak{g}_T$.

For $j=n$, j odd, we have that

$x_{\varepsilon_s+\varepsilon_j} = (\mathrm{ad}\,x_{\varepsilon_j})(y) \in (\mathrm{ad}\,\mathfrak{p}_\Lambda)(y)+\mathfrak{g}_T$.

Let $s+4 \le j \le n$, j even.

We have that $(\mathrm{ad}\, x_{\varepsilon_j})(y) = x_{\varepsilon_s+\varepsilon_j} + x_{-\varepsilon_{j-1}}$ and $x_{-\varepsilon_{j-1}} = (\mathrm{ad}\, x_{-\varepsilon_s-\varepsilon_{j-1}})(y)$. Hence $x_{\varepsilon_s+\varepsilon_j} = (\mathrm{ad}\,(x_{\varepsilon_j} - x_{-\varepsilon_s-\varepsilon_{j-1}}))(y) \in (\mathrm{ad}\, \mathfrak{p}_\Lambda)(y) + \mathfrak{g}_T$. □

Examine now the case of $\varepsilon_s - \varepsilon_i$, $1 \le i \le s-1$.

Lemma 17 *For all* $i \in \mathbb{N}$, $1 \le i \le s-1$, *we have that* $x_{\varepsilon_s-\varepsilon_i} \in (\mathrm{ad}\, \mathfrak{p}_\Lambda)(y) + \mathfrak{g}_T$.

Proof One has that $x_{\varepsilon_s-\varepsilon_1} = (\mathrm{ad}\ x_{\varepsilon_s-\varepsilon_{s-1}})(y) + x_{\varepsilon_s+\varepsilon_{s+1}} \in (\mathrm{ad}\ \mathfrak{p}_\Lambda)(y) + \mathfrak{g}_T$ by Lemma 16.

Moreover $(\mathrm{ad}\, x_{-\varepsilon_2})(y) = x_{\varepsilon_1} + x_{\varepsilon_s-\varepsilon_2}$ and $(\mathrm{ad}\, x_{\varepsilon_1-\varepsilon_s})(y) = x_{\varepsilon_1} + x_{\varepsilon_{s-1}-\varepsilon_s}$ hence $x_{\varepsilon_s-\varepsilon_2} = (\mathrm{ad}(x_{-\varepsilon_2} - x_{\varepsilon_1-\varepsilon_s}))(y) + x_{\varepsilon_{s-1}-\varepsilon_s} \in (\mathrm{ad}\ \mathfrak{p}_\Lambda)(y) + \mathfrak{g}_T$.

Let $4 \le i \le s-2$ with i even. One has that $(\mathrm{ad}\, x_{-\varepsilon_i})(y) = x_{\varepsilon_s-\varepsilon_i} + x_{\varepsilon_{i-1}}$. Moreover if $i \le s/2$ then $(\mathrm{ad}\, x_{\varepsilon_{i-1}-\varepsilon_s})(y) = x_{\varepsilon_{i-1}} + x_{\varepsilon_{s-i+1}-\varepsilon_s}$ and $(\mathrm{ad}\, x_{-\varepsilon_{s-i+2}-\varepsilon_s})(y) = x_{\varepsilon_{s-i+1}-\varepsilon_s}$. Hence $x_{\varepsilon_s-\varepsilon_i} = (\mathrm{ad}(x_{-\varepsilon_i} - x_{\varepsilon_{i-1}-\varepsilon_s} + x_{-\varepsilon_{s-i+2}-\varepsilon_s}))(y) \in (\mathrm{ad}\ \mathfrak{p}_\Lambda)(y) + \mathfrak{g}_T$.

If i is even and $s/2+1 \le i \le s-2$, then $x_{\varepsilon_s-\varepsilon_i} = (\mathrm{ad}(x_{-\varepsilon_i} - x_{\varepsilon_{i-1}-\varepsilon_s}))(y) \in (\mathrm{ad}\ \mathfrak{p}_\Lambda)(y) + \mathfrak{g}_T$.

Let $3 \le i \le s-3$ with i odd. One has that $(\mathrm{ad}\, x_{-\varepsilon_i})(y) = x_{\varepsilon_s-\varepsilon_i} + x_{\varepsilon_{i+1}}$ and if $i \le s/2-2$, then $(\mathrm{ad}\, x_{\varepsilon_{i+1}-\varepsilon_s})(y) = x_{\varepsilon_{i+1}} + x_{\varepsilon_{s-i-1}-\varepsilon_s}$ and $(\mathrm{ad}\, x_{-\varepsilon_{s-i-2}-\varepsilon_s})(y) = x_{\varepsilon_{s-i-1}-\varepsilon_s}$. Hence $x_{\varepsilon_s-\varepsilon_i} = (\mathrm{ad}(x_{-\varepsilon_i} - x_{\varepsilon_{i+1}-\varepsilon_s} + x_{-\varepsilon_{s-i-2}-\varepsilon_s}))(y) \in (\mathrm{ad}\ \mathfrak{p}_\Lambda)(y) + \mathfrak{g}_T$.

If $i \ge 3$, i odd and $s/2-1 \le i \le s-3$, then $x_{\varepsilon_s-\varepsilon_i} = (\mathrm{ad}(x_{-\varepsilon_i} - x_{\varepsilon_{i+1}-\varepsilon_s}))(y) \in (\mathrm{ad}\ \mathfrak{p}_\Lambda)(y) + \mathfrak{g}_T$.

Finally $x_{\varepsilon_s-\varepsilon_{s-1}} = (\mathrm{ad}(x_{-\varepsilon_{s-1}} - x_{\varepsilon_{s+1}-\varepsilon_s}))(y) \in (\mathrm{ad}\ \mathfrak{p}_\Lambda)(y) + \mathfrak{g}_T$. □

Now examine the case of $\varepsilon_s - \varepsilon_i$, with $s+1 \le i \le n$.

Lemma 18 *For all* $i \in \mathbb{N}$, $s+1 \le i \le n$, *we have that* $x_{\varepsilon_s-\varepsilon_i} \in (\mathrm{ad}\, \mathfrak{p}_\Lambda)(y) + \mathfrak{g}_T$.

Proof One has that $(\mathrm{ad}\, x_{-\varepsilon_{s+1}})(y) = x_{\varepsilon_s-\varepsilon_{s+1}} + x_{\varepsilon_{s-1}}$ and $(\mathrm{ad}\, x_{\varepsilon_{s-1}-\varepsilon_s})(y) = x_{\varepsilon_{s-1}}$. Hence $x_{\varepsilon_s-\varepsilon_{s+1}} = (\mathrm{ad}(x_{-\varepsilon_{s+1}} - x_{\varepsilon_{s-1}-\varepsilon_s}))(y) \in (\mathrm{ad}\ \mathfrak{p}_\Lambda)(y) + \mathfrak{g}_T$.

Assume that $s+2 \le i \le n-1$ and i even. One has that $(\mathrm{ad}\, x_{-\varepsilon_i})(y) = x_{\varepsilon_s-\varepsilon_i} + x_{\varepsilon_{i+1}}$ and $(\mathrm{ad}\, x_{\varepsilon_{i+1}-\varepsilon_s})(y) = x_{\varepsilon_{i+1}}$. Hence $x_{\varepsilon_s-\varepsilon_i} = (\mathrm{ad}(x_{-\varepsilon_i} - x_{\varepsilon_{i+1}-\varepsilon_s}))(y) \in (\mathrm{ad}\ \mathfrak{p}_\Lambda)(y) + \mathfrak{g}_T$.

Moreover if n is even (and $s \neq n$) then $x_{\varepsilon_s-\varepsilon_n} = (\mathrm{ad}\ x_{-\varepsilon_n})(y) \in (\mathrm{ad}\ \mathfrak{p}_\Lambda)(y) + \mathfrak{g}_T$.

Assume that $s+3 \le i \le n$ and i odd. One has that $(\mathrm{ad}\, x_{-\varepsilon_i})(y) = x_{\varepsilon_s-\varepsilon_i} + x_{\varepsilon_{i-1}}$ and $(\mathrm{ad}\, x_{\varepsilon_{i-1}-\varepsilon_s})(y) = x_{\varepsilon_{i-1}}$. Hence $x_{\varepsilon_s-\varepsilon_i} = (\mathrm{ad}(x_{-\varepsilon_i} - x_{\varepsilon_{i-1}-\varepsilon_s}))(y) \in (\mathrm{ad}\ \mathfrak{p}_\Lambda)(y) + \mathfrak{g}_T$. □

It remains to examine the case of ε_n if n is even and of $-\varepsilon_n$ if n is odd.

Lemma 19 *If n is even, then* $x_{\varepsilon_n} \in (\mathrm{ad}\, \mathfrak{p}_\Lambda)(y) + \mathfrak{g}_T$. *If n is odd, then* $x_{-\varepsilon_n} \in (\mathrm{ad}\, \mathfrak{p}_\Lambda)(y) + \mathfrak{g}_T$.

Proof Assume that n is even and $s < n$.

Then $x_{\varepsilon_n} = (\mathrm{ad}\ x_{\varepsilon_n-\varepsilon_s})(y) \in (\mathrm{ad}\ \mathfrak{p}_\Lambda)(y) + \mathfrak{g}_T$.

Now if $s = n$ then $x_{\varepsilon_n} \in \mathfrak{g}_S \subset (\mathrm{ad}\ \mathfrak{h}_\Lambda)(y)$.

Finally assume that n is odd. If $s = n-1$, then $x_{-\varepsilon_n} = (\mathrm{ad}\ x_{-\varepsilon_{n-1}-\varepsilon_n})(y) + x_{\varepsilon_{n-2}-\varepsilon_{n-1}} \in (\mathrm{ad}\ \mathfrak{p}_\Lambda)(y) + \mathfrak{g}_T$.

If $s \le n-3$, then $x_{-\varepsilon_n} = (\text{ad } x_{-\varepsilon_s-\varepsilon_n})(y) \in (\text{ad } \mathfrak{p}_\Lambda)(y) + \mathfrak{g}_T$. □

Hence condition (5) of Proposition 9 is satisfied.

By the above, all conditions of Proposition 9 are satisfied. Then one can give the following theorem.

Theorem 20 *Let $\mathfrak{g}$ be a simple Lie algebra of type B_n ($n \ge 2$). Let $\mathfrak{p}$ be a maximal parabolic subalgebra of $\mathfrak{g}$ associated to $\pi' = \pi \setminus \{\alpha_s\}$ with s even and let $\mathfrak{p}_\Lambda$ be its canonical truncation. There exists an adapted pair $(h,\ y) \in \mathfrak{h}' \times \mathfrak{p}_\Lambda^*$ for $\mathfrak{p}_\Lambda$.*

Proof It suffices to apply Proposition 9 since all its conditions are satisfied. Moreover condition (1) of Proposition 9 implies that there is a unique $h \in \mathfrak{h}_\Lambda = \mathfrak{h}'$ such that $h(\gamma) = -1$ for all $\gamma \in S$.

By a direct computation the expansion of h in terms of the elements ε_k, $1 \le k \le n$ is as follows:

$$\begin{aligned} h = &\textstyle\sum_{k=1}^{[s/4]} \big(\frac{s}{2} + 2k - 1\big)\varepsilon_{2k-1} + \sum_{k=[s/4]+1}^{s/2-1} \big(\frac{3s}{2} - 2k\big)\varepsilon_{2k-1} \\ &- \textstyle\sum_{k=1}^{[s/4]} \big(\frac{s}{2} + 2k\big)\varepsilon_{2k} - \sum_{k=[s/4]+1}^{s/2-1} \big(\frac{3s}{2} + 1 - 2k\big)\varepsilon_{2k} + \frac{s}{2}\varepsilon_{s-1} - \varepsilon_s \\ &+ \textstyle\sum_{k=1}^{[(n-s+1)/2]} \big(-2k + 1 - \frac{s}{2}\big)\varepsilon_{s+2k-1} + \sum_{k=1}^{[(n-s)/2]} \big(2k + \frac{s}{2}\big)\varepsilon_{s+2k}. \end{aligned}$$

Let us give also the expansion of h in terms of the coroots $\alpha_k^\vee$ for $1 \le k \le n$, $k \ne s$.

If n is even, one has that

$$\begin{aligned} h = &-\textstyle\sum_{k=1}^{s/2-1} k\alpha_{2k}^\vee + \sum_{k=1}^{[s/4]}(s/2+k)\alpha_{2k-1}^\vee \\ &+ \textstyle\sum_{k=[s/4]+1}^{s/2}(3s/2+1-3k)\alpha_{2k-1}^\vee - \sum_{k=s/2+1}^{n/2} k\alpha_{2k-1}^\vee \\ &+ \textstyle\sum_{k=s/2+1}^{(n-2)/2}(k-s/2)\alpha_{2k}^\vee + ((n-s)/4)\alpha_n^\vee \end{aligned}$$

If n is odd, one has that

$$\begin{aligned} h = &-\textstyle\sum_{k=1}^{s/2-1} k\alpha_{2k}^\vee + \sum_{k=1}^{[s/4]}(s/2+k)\alpha_{2k-1}^\vee \\ &+ \textstyle\sum_{k=[s/4]+1}^{s/2}(3s/2+1-3k)\alpha_{2k-1}^\vee - \sum_{k=s/2+1}^{(n-1)/2} k\alpha_{2k-1}^\vee \\ &+ \textstyle\sum_{k=s/2+1}^{(n-1)/2}(k-s/2)\alpha_{2k}^\vee - ((n+1)/4)\alpha_n^\vee \end{aligned}$$

□

8 The Polynomiality of $Sy(\mathfrak{p})$ and the Existence of a Slice

By Lemma 5 it suffices to verify that the improved upper bound $\mathscr{B}'$ mentioned in Lemma 5 is equal to the lower bound $\mathrm{ch}\,\mathscr{A}$ mentioned in Theorem 1 to conclude that $Sy(\mathfrak{p})$ is a polynomial algebra, and that we have a Weierstrass section and an affine slice too. This is the following lemma.

Lemma 21 *If $n = s$, one has that*

$$\mathrm{ch}\,\mathscr{A} = (1 - e^{-2\varpi_n})^{-2}(1 - e^{-4\varpi_n})^{-(n/2-1)} \qquad (*)$$

If $n > s$, one has that

$$\mathrm{ch}\,\mathscr{A} = (1 - e^{-\varpi_s})^{-2}(1 - e^{-2\varpi_s})^{-(n-1-s/2)} \qquad (**)$$

Proof By equality in (5) of Sect. 2, the lower bound for $\mathrm{ch}\,Sy(\mathfrak{p})$ is $\mathrm{ch}\,\mathscr{A} = \prod_{\Gamma\in E(\pi')}(1 - e^{\delta_\Gamma})^{-1}$. Recall (see proof of Lemma 12) that the set of $\langle ij\rangle$-orbits in π is $E(\pi') = \{\Gamma_{s/2} := \{\alpha_{s/2}\},\ \Gamma_t := \{\alpha_t, \alpha_{s-t}\},\ \Gamma_u := \{\alpha_u\} \mid 1 \le t \le s/2-1,\ s \le u \le n\}$. Let $\Gamma \in E(\pi')$. Since $j = \mathrm{id}_\pi$ and $i(\Gamma\cap\pi') = j(\Gamma)\cap\pi'$, one has $\delta_\Gamma = -2(\sum_{\gamma\in\Gamma}\varpi_\gamma - \sum_{\gamma\in\Gamma\cap\pi'}\varpi'_\gamma)$.

Assume first that $n = s$. Then the Levi factor of $\mathfrak{p}$ is simple of type A_{n-1} and one may check that for all $1 \le t \le n-1$, $\varpi_t - \varpi'_t = 2(t/n)\varpi_n$. Then for all $1 \le t \le n/2-1$, $\delta_{\Gamma_t} = -2(\varpi_t - \varpi'_t + \varpi_{n-t} - \varpi'_{n-t}) = -4\varpi_n$ and $\delta_{\Gamma_n} = \delta_{\Gamma_{n/2}} = -2\varpi_n$. Hence equality $(*)$ holds for $n = s$.

Assume now that $n > s$. Then the Levi factor of $\mathfrak{p}$ is the product of a simple Lie algebra of type A_{s-1} and a simple Lie algebra of type B_{n-s} (of type A_1 if $s = n-1$). For all $1 \le t \le s-1$, one checks that $\varpi_t - \varpi'_t = (t/s)\varpi_s$. Then, for all $1 \le t \le s/2-1$, one has $\delta_{\Gamma_t} = -2\varpi_s$ and $\delta_{\Gamma_{s/2}} = -\varpi_s$. On the other hand, for all $s+1 \le t \le n-1$, one has that $\varpi_t - \varpi'_t = \varpi_s$, hence $\delta_{\Gamma_t} = -2\varpi_s$. Finally $\varpi_n - \varpi'_n = (1/2)\varpi_s$ and $\delta_{\Gamma_n} = -\varpi_s$, whereas $\delta_{\Gamma_s} = -2\varpi_s$, since $\Gamma_s \cap \pi' = \emptyset$. We conclude that equality $(**)$ holds for $n > s$. □

It remains to compute the improved upper bound $\mathscr{B}'$ given by equality in (6) of Sect. 2 and to prove that this bound is equal to $(*)$ when $n = s$ or to $(**)$ when $n > s$. We verify it by a direct computation. Thus by Lemma 5 and Theorem 6 we obtain the following theorem.

Theorem 22 *Let $\mathfrak{p}$ be any maximal parabolic subalgebra of a simple Lie algebra $\mathfrak{g}$ of type B_n, $n \ge 2$ corresponding to the set of simple roots $\pi' = \pi\setminus\{\alpha_s\}$ with s even. Then $Sy(\mathfrak{p})$ is a polynomial algebra over $\mathbb{C}$ and there exists a Weierstrass section for $Sy(\mathfrak{p})$ (given by an adapted pair $(h,\ y)$ for the canonical truncation $\mathfrak{p}_\Lambda$) which is also an affine slice to the coadjoint action of $\mathfrak{p}_\Lambda$. Moreover the degrees of homogeneous generators of $Sy(\mathfrak{p})$ are the eigenvalues plus one of* $\mathrm{ad}\,h$ *on $\mathfrak{g}_T$, where $\mathfrak{g}_T$ is an h-stable complement in $\mathfrak{p}^*_\Lambda$ of* $(\mathrm{ad}\,\mathfrak{p}_\Lambda)(y)$.

Remark 23 Assume that $\mathfrak{p}$ is a maximal parabolic subalgebra of a simple Lie algebra of type D_n for which the bounds of Theorem 1 do not coincide. Then we can also prove that the Poisson semicentre $Sy(\mathfrak{p})$ is a polynomial algebra over $\mathbb{C}$, thanks to the computation of an adapted pair for the canonical truncation of $\mathfrak{p}$. This adapted pair is obtained in a similar way as in type B_n, at least when $n \ge s+2$, s even. For the extremal case, namely when n even and $n = s$ or $n-1 = s$, we can also construct an adapted pair for the canonical truncation but the construction is more complicated than in type B_n. Finally we also prove in any case that the improved upper bound $\mathscr{B}'$ is attained, hence that $Sy(\mathfrak{p})$ is a polynomial algebra over $\mathbb{C}$.

9 The Eigenvalues of ad h

Here we give the eigenvalues of ad h on $\mathfrak{g}_T$. Indeed by Theorem 22 these eigenvalues plus one are the degrees of a set of algebraically independent homogeneous generators of the Poisson semicentre $Sy(\mathfrak{p})$.

Lemma 24 *The eigenvalues of* ad h *on* $\mathfrak{g}_T$ *are :*

$s + 4i - 1 = h(\varepsilon_{2i-1} - \varepsilon_{2i})$ *for all* $i \in \mathbb{N}$, $1 \leq i \leq [s/4]$.
$3s - 4i + 1 = h(\varepsilon_{2i-1} - \varepsilon_{2i})$ *for all* $i \in \mathbb{N}$, $[s/4] + 1 \leq i \leq s/2 - 1$.
$s/2 + 1 = h(\varepsilon_{s-1} - \varepsilon_s)$.
$s/2 - 1 = h(\varepsilon_{s-1} + \varepsilon_s)$.
$s + 1 = h(\varepsilon_{s-1} - \varepsilon_{s+1})$.
$s + 4j - 1 = h(-\varepsilon_{s+2j-1} + \varepsilon_{s+2j})$, *for all* $j \in \mathbb{N}$, $1 \leq j \leq [(n - s - 1)/2]$.
$s + 4j + 1 = h(\varepsilon_{s+2j} - \varepsilon_{s+2j+1})$, *for all* $j \in \mathbb{N}$, $1 \leq j \leq [(n - s - 2)/2]$.

$$2n - s - 1 = \begin{cases} h(-\varepsilon_{n-1} + \varepsilon_n) & \text{if } n \text{ even} \\ h(\varepsilon_{n-1} - \varepsilon_n) & \text{if } n \text{ odd} \end{cases}$$

In particular we have that $s + 2k - 1$ *is an eigenvalue of* ad h *on* $\mathfrak{g}_T$, *for all* $k \in \mathbb{N}$, $1 \leq k \leq n - s$.

Proof Follows directly from the expansion of h given in the proof of Theorem 20. □

Acknowledgements I would like to mention that this work is joint with Polyxeni Lamprou. More general results about polynomiality of the Poisson semicentre associated to a maximal parabolic subalgebra (not only in type B) are established in [9]. This work is the continuation of [8] where we also constructed adapted pairs in truncated maximal parabolic subalgebras $\mathfrak{a}$, but in the case when we already knew the polynomiality of $Y(\mathfrak{a})$.

References

1. Borho, W., and P. Gabriel, and R. Rentschler, "Primideale in Einhüllenden auflösbarer Lie-Algebren (Beschreibung durch Bahnenräume)", Lecture Notes in Math. **357**, Springer-Verlag, Berlin, 1973.
2. Bourbaki, N., "Groupes et algèbres de Lie, Chapitres IV-VI", Hermann, Paris, 1968.
3. Dixmier, J., *Sur le centre de l'algèbre enveloppante d'une algèbre de Lie*, C.R. Acad. Sc. Paris **265** (1967), 408–410.
4. Dixmier, J., "Algèbres enveloppantes", Editions Jacques Gabay, les grands classiques Gauthier-Villars, Paris/Bruxelles/Montréal, 1974.
5. Fauquant-Millet, F., and A. Joseph, *Sur les semi-invariants d'une sous-algèbre parabolique d'une algèbre enveloppante quantifiée*, Transformation Groups **6** (2001), no. 2, 125–142.
6. Fauquant-Millet, F., and A. Joseph, *Semi-centre de l'algèbre enveloppante d'une sous-algèbre parabolique d'une algèbre de Lie semi-simple*, Ann. Sci. École. Norm. Sup. (4) **38** (2005), no. 2, 155–191.
7. Fauquant-Millet, F., and A. Joseph, *Adapted pairs and Weierstrass sections*, http://arxiv.org/abs/1503.02523.

8. Fauquant-Millet, F., and P. Lamprou, *Slices for maximal parabolic subalgebras of a semisimple Lie algebra*, Transformation Groups **22**, No. 4 (2017), 911–932.
9. Fauquant-Millet, F., and P. Lamprou, *Polynomiality for the Poisson Centre of Truncated Maximal Parabolic Subalgebras*, Journal of Lie Theory **28**, No. 4 (2018), 1063–1094.
10. Joseph, A., *A preparation theorem for the prime spectrum of a semisimple Lie algebra*, J. Algebra **48** (1977), 241–289.
11. Joseph, A., *On semi - invariants and index for biparabolic (seaweed) algebras I*, J. Algebra **305** (2006), no. 1, 487–515.
12. Joseph, A., *On semi - invariants and index for biparabolic (seaweed) algebras II*, J. Algebra **312** (2007), no. 1, 158–193.
13. Joseph, A., *Slices for biparabolic coadjoint actions in type A*, J. Algebra **319** (2008), no. 12, 5060–5100.
14. Joseph, A., *Compatible adapted pairs and a common slice theorem for some centralizers*, Transform. Groups **13** (2008), no. 3–4, 637–669.
15. Joseph, A., *An algebraic slice in the coadjoint space of the Borel and the Coxeter element*, Advances in Maths **227** (2011), 522–585.
16. Joseph, A., and D. Shafrir, *Polynomiality of invariants, unimodularity and adapted pairs*, Transform. Groups **15** (2010), no. 4, 851–882.
17. Joseph, A., and P. Lamprou, *Maximal Poisson commutative subalgebras for truncated parabolic subalgebras of maximal index in* $\mathfrak{sl}_n$, Transform. Groups **12** (2007), no. 3, 549–571.
18. Kostant, B., *Lie group representations on polynomial rings*, Amer. J. Math. **85** (1963), 327–404.
19. Panyushev, D., and A. Premet, and O. Yakimova, *On symmetric invariants of centralisers in reductive Lie algebras*, J. Algebra **313** (2007), no. 1, 343–391.

On Dynkin Gradings in Simple Lie Algebras

Alexander Elashvili, Mamuka Jibladze, and Victor Kac

To Tony Joseph, on his 75th birthday

MSC Codes: 17B08, 17B10, 17B20, 17B22, 17B25, 17B70, 17B80

1 Introduction

In this paper we study Dynkin gradings on simple Lie algebras arising from nilpotent elements. Specifically, we investigate Abelian subalgebras which are degree 1 homogeneous with respect to these gradings.

The study of gradings associated to nilpotent elements of simple Lie algebras is important since the finite and affine classical and quantum W-algebras are defined using these gradings. In order to study integrable systems associated to these W-algebras, it is useful to have their free field realizations. One of the ways to construct them is to use the generalized Miura map [2, 4]. This construction can be further improved by choosing an Abelian subalgebra in the term $\mathfrak{g}_1$ of the grading. That is why the description of such subalgebras, especially those having the maximal possible dimension $\frac{1}{2}\dim \mathfrak{g}_1$, is important.

We show that for each odd nilpotent orbit there always exists a canonically associated "strictly odd" nilpotent orbit, which allows us to reduce our investigations to the latter case. (Strictly odd means that all Dynkin labels are either 0 or 1.) The rest of the paper is devoted to the investigation of maximal Abelian subalgebras in $\mathfrak{g}_1$ for strictly odd nilpotent orbits in simple Lie algebras. For algebras of exceptional type we provide tables with largest possible dimensions of such subalgebras in each

A. Elashvili · M. Jibladze (✉)
Razmadze Mathematical Institute, TSU, Tbilisi, Georgia
e-mail: jib@rmi.ge

V. Kac
Department of Mathematics, MIT, Cambridge, MA, USA

M. Gorelik et al. (eds.), *Representations and Nilpotent Orbits of Lie Algebraic Systems*, Progress in Mathematics 330,
https://doi.org/10.1007/978-3-030-23531-4_5

case. For algebras of classical type, we find expressions for all possible maximal dimensions of Abelian subalgebras in $\mathfrak{g}_1$, and, based on that, characterize those nilpotent orbits for which there exists such subalgebra of half the dimension of $\mathfrak{g}_1$.

2 Recollections

Let us recall the nomenclature for nilpotent elements in a semisimple Lie algebra $\mathfrak{g}$.

Given a nilpotent element e, one chooses an $\mathfrak{sl}_2$-triple (e, h, f) for it, that is, another nilpotent element f such that $[e, f] = h$ is semisimple and the identities $[h, e] = 2e$, $[h, f] = -2f$ hold (Jacobson-Morozov theorem; see, e.g., [1]). The Dynkin grading is the eigenspace decomposition for $\operatorname{ad} h$:

$$\mathfrak{g} = \bigoplus_{j \in \mathbb{Z}} \mathfrak{g}_j.$$

Thus, a choice of a nilpotent element e defines a combinatorial object which uniquely describes the orbit of e. It is the *weighted Dynkin diagram* corresponding to e, which is the Dynkin diagram of $\mathfrak{g}$ with numbers assigned to each node. These numbers are the degrees $\alpha_i(h)$ of simple root vectors e_i with respect to the choice of a Cartan and a Borel subalgebra in such a way that h (resp. e) becomes an element of the corresponding Cartan (resp. Borel) subalgebra. The weighted Dynkin diagrams satisfy certain restrictions—for example, the weights can only be equal to 0, 1 or 2; moreover, if $\mathfrak{g}$ is simple of type A, then the weights are symmetric with respect to the center of the diagram, while for types B, C, or D there is no weight 1 occurring to the left of 2.

A nilpotent element is called *even* if there are no 1's in its weighted Dynkin diagram, *odd* if it is not even, and *strictly odd* if there are no 2's.

It is clear that for even nilpotent elements the question about Abelian subspaces in $\mathfrak{g}_1$ is trivial since $\mathfrak{g}_1$ is zero.

We will also need the following fact from [3]:

Proposition 2.1 *The degree* 1 *part* $\mathfrak{g}_1$ *of* $\mathfrak{g}$ *with respect to the grading induced by a nilpotent element* $e \in \mathfrak{g}$ *is generated as a* $\mathfrak{g}_0$*-module by those simple root vectors of* $\mathfrak{g}$ *which have weight* 1 *in the weighted Dynkin diagram corresponding to* e. □

If $\mathfrak{g}$ is a simple Lie algebra of classical type, one can assign to e another combinatorial object—a partition $\lambda_n \geqslant \lambda_{n-1} \geqslant \dots$ which records dimensions of irreducible representations of $\mathfrak{sl}_2$ into which the standard representation of $\mathfrak{g}$ decomposes as a module over its subalgebra (e, h, f). Alternatively, the partition consists of sizes of Jordan blocks in the Jordan decomposition of e as an operator acting on the standard representation of $\mathfrak{g}$. The partitions are restricted in a certain way, according to the type of $\mathfrak{g}$. For type A one may have arbitrary partitions. For types B and D, all even parts must have even multiplicity, while for type C all odd parts must have even multiplicity. These conditions are sufficient as well as necessary, that is, any partition satisfying these conditions corresponds to a nilpotent orbit in a simple Lie algebra of the respective classical type.

Let us recall how one calculates the weighted Dynkin diagram of a nilpotent element defined by a partition $\lambda = (\lambda_n \geqslant \lambda_{n-1} \geqslant \ldots)$ (cf. [6]).

Each element λ_k of the partition λ represents a copy of the λ_k-dimensional irreducible representation of $\mathfrak{sl}_2$, with eigenvalues of h equal to

$$1 - \lambda_k, 3 - \lambda_k, \ldots, \lambda_k - 3, \lambda_k - 1.$$

To obtain the weighted Dynkin diagram one collects those eigenvalues for each λ_k, arranges them in decreasing order, and takes consecutive differences.

For example, take the partition 8, 6, 3, 3, 2, 1, 1. This gives the following eigenvalues of h:

$$\begin{array}{ccccccccccc}
-7 & -5 & -3 & & -1 & & 1 & & 3 & 5 & 7 \\
 & -5 & -3 & & -1 & & 1 & & 3 & 5 & \\
 & & & -2 & & 0 & & 2 & & & \\
 & & & -2 & & 0 & & 2 & & & \\
 & & & & -1 & & 1 & & & & \\
 & & & & & 0 & & & & & \\
 & & & & & 0 & & & & &
\end{array}$$

Arranging all numbers from this table in the decreasing order gives

$$7\ 5\ 5\ 3\ 3\ 2\ 2\ 1\ 1\ 1\ 0\ 0\ 0\ 0\ -1\ -1\ -1\ -2\ -2\ -3\ -3\ -5\ -5\ -7.$$

Taking the consecutive differences then gives

$$2\ 0\ 2\ 0\ 1\ 0\ 1\ 0\ 0\ 1\ 0\ 0\ 0\ 1\ 0\ 0\ 1\ 0\ 1\ 0\ 2\ 0\ 2$$

which is already the weighted Dynkin diagram of the nilpotent in case of type A.

For the types B, C, D one has to leave only the left half of the obtained sequence (which obviously is centrally symmetric); more precisely, for an algebra of rank r, the first $r - 1$ nodes of the weighted Dynkin diagram are as stated, while the rightmost node is defined in a specific way, depending on the type. We skip this part, as it will not play any rôle for us; details can be found in, e.g., [1, Section 5.3].

For example, the same partition 8, 6, 3, 3, 2, 1, 1 also encodes a nilpotent orbit in a simple Lie algebra of type C, since all of its odd parts come with even multiplicities. Then, the weighted Dynkin diagram of this nilpotent is

$$2\ 0\ 2\ 0\ 1\ 0\ 1\ 0\ 0\ 1\ 0\ 0.$$

It is easy to see from the above procedure that the resulting weighted Dynkin diagram begins with certain sequence of 0's and 2's; if the largest part of the partition is λ_n with multiplicity m_n, and the parts of the same parity following it are λ_{n-1}

with multiplicity m_{n-1}, λ_{n-2} with multiplicity m_{n-2}, ..., λ_{n-k+1} with multiplicity m_{n-k+1}, while the next part λ_{n-k} has the opposite parity, then the first 1 appears at the $(km_n + (k-1)m_{n-1} + \ldots + 2m_{n-k+2} + m_{n-k+1})$-st place. For the type A the picture is symmetric, so one has the weights 2 and 0 at both ends of the diagram, and the weights 1 and 0 in the middle, while for the types B, C, or D the sequence of weights starts with 0 and 2 followed by a sequence of weights 0 and 1, without any further 2's.

According to the above procedure for assigning to a partition a weighted Dynkin diagram, it is easy to see the following

Proposition 2.2 *A nilpotent element in a simple Lie algebra of classical type is even iff all the parts of the corresponding partition are of the same parity; it is odd iff there are some parts with different parities, and strictly odd iff the largest part and the next largest part differ by* 1. □

3 Important Reduction

Let V and U be finite-dimensional modules over a reductive Lie algebra $\mathfrak{g}$ and let $V \otimes V \to U$ be a $\mathfrak{g}$-module homomorphism. We see this homomorphism as a $\mathfrak{g}$-equivariant algebra structure on V with values in U.

Proposition 3.1 *Suppose that there exists an Abelian subalgebra of dimension d of the algebra V. Then there exists an Abelian subalgebra of the algebra V of dimension d, spanned by weight vectors of V.*

Proof (Proposed by the Referee) It follows from Borel's fixed point theorem. Indeed, the Cartan subgroup acts on the complete variety of d-dimensional Abelian subalgebras of V, hence has a fixed point. □

Using this, in what follows we will assume throughout that for a simple Lie algebra of classical type we are given a basis in the standard representation consisting of weight vectors corresponding to the weights $\pm\varepsilon_i$, $i = 1, \ldots, n$ and moreover, for the type B, to the zero weight. In the adjoint representation, accordingly, we will have a basis corresponding to $\pm\varepsilon_i \pm \varepsilon_j$, $i \neq j$ (accounting for tensor products of basis vectors of the standard representation corresponding to $\pm\varepsilon_i$ and to $\pm\varepsilon_j$) and moreover, for the type B only, those corresponding to $\pm\varepsilon_i$ (accounting for tensor product of a basis vector corresponding to $\pm\varepsilon_i$ and that corresponding to the zero weight) and, for C only, corresponding to $\pm 2\varepsilon_i$ (accounting for the tensor product of a basis vector of the standard representation corresponding to $\pm\varepsilon_i$ with itself), $i = 1, \ldots, n$.

Proposition 3.2 *For any weighted Dynkin diagram corresponding to a nilpotent element e in a simple Lie algebra $\mathfrak{g}$, consider a subdiagram obtained as a result of erasing all nodes with weight 2. Consider the resulting subdiagram together with the remaining weights. Then all connected components of this subdiagram, except* *possibly one of them, have all weights equal to zero. Moreover, this one component (if it exists) is a weighted Dynkin diagram of some strictly odd nilpotent orbit in the diagram subalgebra $\tilde{\mathfrak{g}} \subseteq \mathfrak{g}$ of the type determined by the shape of the component.*

Proof For algebras of classical type, this is proved in Lemma 4.6 below. For an algebra of type G_2 this is clear as all nilpotent elements in it are either even or strictly odd. As for the exceptional Lie algebras of types E or F, the assertion can be seen to be true directly from looking at the Tables F4o, E6o, E7o, E8o given in the last section. □

Corollary 3.3 *For any odd nilpotent element e in a simple Lie algebra $\mathfrak{g}$ there exists a simple diagram subalgebra $\tilde{\mathfrak{g}} \subseteq \mathfrak{g}$ and a strictly odd nilpotent element $\tilde{e} \in \tilde{\mathfrak{g}}$ such that*

$$\mathfrak{g}_1(e) = \tilde{\mathfrak{g}}_1(\tilde{e}),$$

i.e., the degree 1 homogeneous parts for the grading on $\mathfrak{g}$ induced by e and for the grading on $\tilde{\mathfrak{g}}$ induced by $\tilde{e}$ coincide. In particular, these degree 1 homogeneous parts have the same Abelian subspaces.

Proof Let $\tilde{\mathfrak{g}}$ be the subalgebra corresponding to the connected component of the weighted Dynkin diagram of e as described in Proposition 3.2 above. Moreover, let $\tilde{e}$ be a representative of the orbit corresponding to the weights on this connected component—it exists by Proposition 3.2.

By construction, this subalgebra contains all simple root vectors of degree 1, and, moreover, they will be precisely the root vectors of those simple roots of $\tilde{\mathfrak{g}}$ which contribute to the degree 1 part for the grading induced by $\tilde{e}$. From Proposition 2.1 we know that $\mathfrak{g}_1(e)$ is the $\mathfrak{g}_0(e)$-module generated by these root vectors, while $\tilde{\mathfrak{g}}_1(\tilde{e})$ is the $\tilde{\mathfrak{g}}_0(\tilde{e})$-module generated by them.

Now observe that the only removed nodes which connect with an edge to some node in the remaining connected component have weight 2, so that all simple root vectors corresponding to removed nodes with weight 0 commute with every simple root vector in this component.

It follows that the $\mathfrak{g}_0(e)$-module generated by the root vectors corresponding to weight 1 nodes is no larger than the $\tilde{\mathfrak{g}}_0(\tilde{e})$-module generated by them, i. e. $\mathfrak{g}_1(e)$ coincides with $\tilde{\mathfrak{g}}_1(\tilde{e})$. □

Definition 3.4 For the orbit of an odd nilpotent element in a simple Lie algebra $\mathfrak{g}$, call its *strictly odd reduction* the nilpotent orbit in the simple Lie algebra $\tilde{\mathfrak{g}}$ obtained as in Corollary 3.3.

Given a nilpotent element $e \in \mathfrak{g}$ as in Proposition 3.2, one can explicitly construct a nilpotent element $\tilde{e} \in \tilde{\mathfrak{g}}$ from the orbit corresponding to its strictly odd reduction in the sense of Definition 3.4 as follows. The nilpotent element e clearly lies in the degree 2 subspace $\mathfrak{g}_2$ for the corresponding grading. This subspace is a $\mathfrak{g}_0$-module and it decomposes canonically into the direct sum of its submodule $[\mathfrak{g}_1, \mathfrak{g}_1]$ and the submodule $\mathfrak{g}_2(2)$ generated by the root vectors of $\mathfrak{g}$ corresponding to the simple roots of weight 2.

Proposition 3.5 *Given a nilpotent element* e*, represent it (in a unique way) as a sum* e_1+e_2 *with* $e_1 \in [\mathfrak{g}_1, \mathfrak{g}_1]$ *and* $e_2 \in \mathfrak{g}_2(2)$*. Then the weighted Dynkin diagram of* e_1 *in the subalgebra corresponding to the subdiagram described in Proposition 3.2 is given by the weights on that subdiagram.*

Proof We have a reductive group G_0 corresponding to $\mathfrak{g}_0$ acting on $\mathfrak{g}_2 = [\mathfrak{g}_1, \mathfrak{g}_1]+\mathfrak{g}_2(2)$, with the element $e = e_1 + e_2$ having an open orbit in $\mathfrak{g}_2$. This means that $[\mathfrak{g}_0, e_1 + e_2] = \mathfrak{g}_2$. But this implies that $[\mathfrak{g}_0, e_1] = [\mathfrak{g}_1, \mathfrak{g}_1]$ (and similarly for e_2). Hence $G_0 e_1$ is an open orbit in $[\mathfrak{g}_1, \mathfrak{g}_1]$.

Let us consider an intermediate subalgebra $\tilde{\mathfrak{g}} \subseteq \mathfrak{g}' \subseteq \mathfrak{g}$ corresponding to the diagram, obtained by erasing the nodes with weight 2, but leaving all other nodes together with their weights intact (this diagram can be disconnected). Proposition 3.2 easily implies that $\mathfrak{g}'$ is a direct sum of $\tilde{\mathfrak{g}}$ and of some simple algebras of type A. Hence e_1, viewed as an element of this direct sum, obviously has zero summands in all these components of type A.

On the other hand, Proposition 3.2 implies that there exists a (strictly odd) nilpotent element $\tilde{e}$ in $[\mathfrak{g}_1, \mathfrak{g}_1]$, which has the needed Dynkin diagram. Then, similarly to e_1, the element $\tilde{e}$ can also be seen as a nilpotent element in $\mathfrak{g}'$, having zero components in all remaining type A components of $\mathfrak{g}'$. It is then clear that this nilpotent element will have the weighted Dynkin diagram obtained as in Proposition 3.2. Moreover, it will have an open G_0-orbit in $[\mathfrak{g}_1, \mathfrak{g}_1]$, hence it coincides with the G_0-orbit of e_1, so $\tilde{e}$ and e_1 have the same weighted Dynkin diagram when viewed as nilpotent elements in $\mathfrak{g}'$. This implies that these elements have the same weighted Dynkin diagram with respect to $\tilde{\mathfrak{g}}$, since the latter is obtained just by throwing out type A components with zero weights only. □

Remark 3.6 It would be convenient to supplement Corollary 3.3 with an explicit construction, assigning to an $\mathfrak{sl}_2$-triple (e, f, h) corresponding to a given nilpotent orbit in $\mathfrak{g}$, an $\mathfrak{sl}_2$-triple $(\tilde{e}, \tilde{f}, \tilde{h})$ for its strictly odd reduction as in Definition 3.4. Since $\tilde{\mathfrak{g}}$ comes with a grading (determined by the weights on the corresponding subdiagram), the semisimple element $\tilde{h}$ of $\tilde{\mathfrak{g}}$ is determined by this grading, while $\tilde{f}$, which we know to exist by Corollary 3.3, is uniquely determined by $\tilde{e}$ and $\tilde{h}$. Thus having an explicit construction of $\tilde{f}$ would provide an alternative proof of Corollary 3.3 that would not require case-by-case analysis of the exceptional types. One possibility that comes to mind is to produce $\tilde{f}$ from f in the same way as we produced $\tilde{e}$ from e in Proposition 3.5—that is, take $\tilde{f} = f_1$ where $f = f_1 + f_2$ is the unique decomposition of $f \in \mathfrak{g}_{-2}$ into a sum of $f_1 \in [\mathfrak{g}_{-1}, \mathfrak{g}_{-1}]$ and $f_2 \in \mathfrak{g}_{-2}(2)$, the latter being the $\mathfrak{g}_0$-submodule of $\mathfrak{g}_{-2}$ generated by the root vectors corresponding to negatives of the simple roots with weights 2 on the initial weighted Dynkin diagram. However, as the following example shows, this does not give the correct value of $\tilde{f}$ in general.

Example 3.7 For $\mathfrak{g}$ of type D_6, consider the nilpotent orbit corresponding to the weighted Dynkin diagram 2010^{1}_{1} (and to the partition 5, 3, 2, 2). The following sum of positive root vectors

$$e := e_{1100^0_0} + e_{0111^1_0} + e_{0011^1_0} + e_{0011^0_1} + e_{0001^1_1}$$

where the subscripts denote the linear combinations of simple roots that give the corresponding positive roots, yields a representative of this orbit. The corresponding f in the $\mathfrak{sl}_2$-triple for e is the following combination of negative root vectors:

$$f := 2f_{1000^0_0} + 4f_{1100^0_0} + 2f_{0111^1_0} - 2f_{0111^0_1} + 2f_{0011^1_0} + 4f_{0011^0_1} + f_{0001^1_1},$$

where the subscripts are linear combinations of negative simple roots. Thus $h = [e, f]$ determines the grading corresponding to the above weighted Dynkin diagram. It is straightforward to check that in the degree 2 subspace $\mathfrak{g}_2$, root vectors corresponding to the combinations 1000^0_0 and 1100^0_0 of simple roots span the $\mathfrak{g}_0$-submodule $\mathfrak{g}_2(2) \subseteq \mathfrak{g}_2$ generated by the root vector of 1000^0_0, i. e. of the simple root with weight 2, while the remaining positive root vectors from $\mathfrak{g}_2$ lie in $[\mathfrak{g}_1, \mathfrak{g}_1]$. Thus, according to Proposition 3.5, a strictly odd nilpotent element $\tilde{e} = e_1$ in the diagram subalgebra $\tilde{\mathfrak{g}}$ of type D_5 corresponding to the subdiagram obtained by omitting the node with weight 2 is obtained by omitting in the sum for e the leftmost summand (the one that lies in $\mathfrak{g}_2(2)$). Thus,

$$\tilde{e} = e_{0111^1_0} + e_{0011^1_0} + e_{0011^0_1} + e_{0001^1_1}.$$

Now, if we try to choose for the companion of $\tilde{e}$ in the $\mathfrak{sl}_2$-triple the element f_1 obtained in the same way from f, i. e. by omitting in the sum for f the summands that lie in $\mathfrak{g}_{-2}(2)$, we obtain

$$f_1 = 2f_{0111^1_0} - 2f_{0111^0_1} + 2f_{0011^1_0} + 4f_{0011^0_1} + f_{0001^1_1}.$$

However, it turns out that $[e_1, f_1]$ is not the semisimple element determining the grading of $\tilde{\mathfrak{g}}$. As a matter of fact, this element is not semisimple, rather it has form

$$[e_1, f_1] = h' - e_{0100^0_0}$$

with h' in the Cartan subalgebra of $\tilde{\mathfrak{g}}$. A correct $\tilde{f}$ (the one with $[\tilde{e}, \tilde{f}] = \tilde{h}$ an element in the Cartan subalgebra of $\tilde{\mathfrak{g}}$ which gives the correct grading of $\tilde{\mathfrak{g}}$) is

$$\tilde{f} = 2f_{0111^1_0} - 2f_{0111^0_1} + 2f_{0011^0_1} + f_{0001^1_1}$$

and it is thus not obtained from f by projecting it to $[\mathfrak{g}_{-1}, \mathfrak{g}_{-1}]$ or in any other obvious way.

Let us add that there are also many examples (even for the algebras of type A) when the bracket of the projections $[e_1, f_1]$ of e and f is semisimple but does not induce the required grading on $\tilde{\mathfrak{g}}$.

4 Maximizing Abelian Subspaces

We are interested in Abelian subspaces of $\mathfrak{g}_1$. First of all, one has the following well-known fact.

Proposition 4.1 *Dimension of* $\mathfrak{g}_1$ *is even, and the largest possible dimension of an Abelian subspace in* $\mathfrak{g}_1$ *is at most* $\frac{1}{2}\dim\mathfrak{g}_1$.

Proof Let e be an element of the orbit, and choose an $\mathfrak{sl}_2$-triple (e, h, f) with $e \in \mathfrak{g}_2$, and h inducing the grading. Then one may define a bilinear form on $\mathfrak{g}_1$ via

$$(x, y)_f := \langle f, [x, y]\rangle,$$

where $\langle -, -\rangle$ is the Killing form. It is well known that the skew-symmetric form $(-, -)_f$ is nondegenerate (since $\operatorname{ad} f : \mathfrak{g}_1 \to \mathfrak{g}_{-1}$ is an isomorphism), so that dimension of $\mathfrak{g}_1$ is indeed even. Moreover any commuting elements of $\mathfrak{g}_1$ are orthogonal with respect to this form. Since such a form does not possess isotropic subspaces of more than half dimension of the space, we obtain that there are no Abelian subspaces of more than half dimension of $\mathfrak{g}_1$. □

Remark 4.2 It is known, more generally, that any homogeneous part $\mathfrak{g}_{2i-1}$ of odd degree possesses a nondegenerate skew-symmetric form—see [5, Proposition 1.2]. Thus, each $\dim\mathfrak{g}_{2i-1}$ is even, too.

We now consider the Abelian subalgebras in $\mathfrak{g}_1$, separately for the simple algebras of classical types (right now) and for the algebras of exceptional types (in Sect. 5).

We now consider the simple algebras of classical types. For the type A, it has been proved in [7] that a half-dimensional Abelian subspace in $\mathfrak{g}_1$ exists for any nilpotent orbit.

The central result of this section is the following characterization, in terms of the associated partitions, of those strictly odd nilpotent orbits in types B, C or D admitting an Abelian subspace of half the dimension in $\mathfrak{g}_1$. We will then deduce the general (not necessarily strictly odd) case, using strictly odd reduction as in Definition 3.4.

Theorem 4.3 *Given a strictly odd nilpotent element in a simple Lie algebra* $\mathfrak{g}$ *of type* B, C, *or* D, *there is an Abelian subspace of half dimension in* $\mathfrak{g}_1$ *if and only if the partition corresponding to the nilpotent element satisfies one of the following conditions:*

- *the largest part* μ *of the partition is even and there are no other even parts; moreover if* $\mathfrak{g}$ *is of type* B *then* μ *has multiplicity* 2.
- *the largest part* μ *of the partition is odd, and either there are no other odd parts, or* $\mathfrak{g}$ *is not of type* C, *and the only other parts are* $\mu - 1$ *with multiplicity* 2 *and* 1 *(with any multiplicity).*

In other words, Abelian subspaces of half dimension in $\mathfrak{g}_1$ occur precisely for those strictly odd nilpotent elements corresponding to the partitions of the following kind:

$$\begin{aligned}
&\textbf{type C}: \left[1^{2\nu_1}3^{2\nu_3}\cdots(2k-1)^{2\nu_{2k-1}}(2k)^{\nu}\right]\ (\nu_{2k-1}\nu\neq 0),\ \left[2^{\nu_2}4^{\nu_4}\cdots(2k)^{\nu_{2k}}(2k+1)^{2\nu}\right]\ (\nu_{2k}\nu\neq 0);\\
&\textbf{type B or D}: \left[2^{2\nu_2}4^{2\nu_4}\cdots(2k)^{2\nu_{2k}}(2k+1)^{\nu}\right]\quad (\nu_{2k}\nu\neq 0),\quad \left[1^{\nu_1}(2k)^2(2k+1)^{\nu}\right]\quad (\nu_{2k}\nu\neq 0);\\
&\textbf{type B}: \left[1^{\nu_1}3^{\nu_3}\cdots(2k-1)^{\nu_{2k-1}}(2k)^2\right]\quad (\nu_{2k-1}\neq 0),\\
&\textbf{type D}: \left[1^{\nu_1}3^{\nu_3}\cdots(2k-1)^{\nu_{2k-1}}(2k)^{2\nu}\right]\quad (\nu_{2k-1}\nu\neq 0).
\end{aligned}$$

Proof It will be convenient to introduce the following notations: for a partition as above, let m_k be the multiplicity of the number k in it. Moreover let S_k be the h-eigensubspace with eigenvalue k in the standard representation, and let s_k denote dimension of this subspace, i.e. multiplicity of the eigenvalue k for h.

As recalled in Sect. 1 above, the adjoint representation can be identified with the symmetric square of the standard one for type C, and with its exterior square for types B and D.

Because of this, clearly the degree 1 part of the adjoint representation is the direct sum of spaces of the form $S_k^*\otimes S_l$ with $l-k=1$, $k\geqslant 0$, and

$$\dim\mathfrak{g}_1 = s_0s_1+s_1s_2+\dots$$

Now, from the correspondence described in Sect. 2, one has

$$\begin{aligned}
s_0 &= m_1+m_3+m_5+\dots\\
s_1 &= m_2+m_4+m_6+\dots\\
s_2 &= m_3+m_5+m_7+\dots\\
s_3 &= m_4+m_6+m_8+\dots\\
&\dots\\
s_{\mu-4} &= m_{\mu-3}+m_{\mu-1}\\
s_{\mu-3} &= m_{\mu-2}+m_{\mu}\\
s_{\mu-2} &= m_{\mu-1}\\
s_{\mu-1} &= m_{\mu}
\end{aligned}\tag{1}$$

Dimension of the subspace $\mathfrak{g}_1$ of grading 1 with respect to the corresponding $\mathfrak{sl}_2$-triple is thus given by

$$\begin{aligned}
s_0s_1+s_1s_2+s_2s_3+s_3s_4+\dots &= \sum_{i,j>0} i m_i m_{i+2j-1}\\
&= m_1m_2+2m_2m_3+m_1m_4+3m_3m_4+2m_2m_5+\dots
\end{aligned}$$

Given an Abelian subspace in $\mathfrak{g}_1$, we may assume, using Proposition 3.1, that it has a basis consisting of root vectors. In particular, each of our basis vectors belongs to one of the direct summands $S_k^* \otimes S_{k+1}$.

Note that the elements of $S_{k-1}^* \otimes S_k$ commute with the elements of $S_l^* \otimes S_{l+1}$ for $l > k$; whereas, when $l = k$, we obtain a non-commuting pair as soon as our basis contains an element of the form $x \otimes y \in S_{k-1}^* \otimes S_k$ and $y' \otimes z \in S_k^* \otimes S_{k+1}$ with y and y' mutually dual basis elements. We are thus forced to choose non-intersecting subsets X_k, Y_k in the weight vector bases of S_k and include in the basis of the Abelian subspace only those $x \otimes y$ which satisfy $x \in X_{k-1}$ and $y \in Y_k$. This does not concern $k = \mu - 1$, where $\mu - 1$ is the maximal occurring eigenvalue of h (μ, as above, is the largest part of the corresponding partition): in $S_{\mu-1}$ we may choose arbitrary subset of the basis without affecting Abelianness; and since we are interested in maximal Abelian subspaces, we choose the whole basis of $S_{\mu-1}$.

Moreover, any such choice of non-intersecting subsets X_k, Y_k of bases of S_k gives indeed an Abelian subspace, and we may further assume that $X_k \cup Y_k$ is the whole basis, since otherwise our Abelian subspace would not be maximal.

The case $k = 0$ is special, and depends on the type considered.

Namely, it may happen that two basis vectors, both from $S_0^* \otimes S_1$, do not commute. Two basis elements of this subspace, being the tensor products of basis vectors corresponding to $\pm\varepsilon_i^{(0)} + \varepsilon_j^{(1)}$ and $\pm\varepsilon_k^{(0)} + \varepsilon_l^{(1)}$ respectively, will commute if and only if the sum $\pm\varepsilon_i^{(0)} + \varepsilon_j^{(1)} \pm \varepsilon_k^{(0)} + \varepsilon_l^{(1)}$ is not a root. This implies that the root vector basis of an Abelian subspace in $\mathfrak{g}_1$ cannot contain root vectors corresponding to both $\pm\varepsilon_i^{(0)} + \varepsilon_j^{(1)}$ and $\mp\varepsilon_i^{(0)} + \varepsilon_k^{(1)}$ for $j \neq k$ (since the sum of these is the root $\varepsilon_j^{(1)} + \varepsilon_k^{(1)}$).

This is the only restriction on $S_0^* \otimes S_1$ for type D. For type C, there is an additional restriction that an Abelian subspace of $\mathfrak{g}_1$ cannot contain root vectors corresponding to both $\pm\varepsilon_i^{(0)} + \varepsilon_j^{(1)}$ and $\mp\varepsilon_i^{(0)} + \varepsilon_j^{(1)}$ (since the sum of these is the root $2\varepsilon_j^{(1)}$). For type B, an additional restriction is that an Abelian subspace of $\mathfrak{g}_1$ cannot contain root vectors corresponding to both $(0+)\varepsilon_j^{(1)}$ and $(0+)\varepsilon_k^{(1)}$ for $j \neq k$ (since the sum of these is the root $\varepsilon_j^{(1)} + \varepsilon_k^{(1)}$).

It follows that to obtain a maximal Abelian subspace of $\mathfrak{g}_1$, in addition to splitting the weight vector basis of S_1 into nonintersecting subsets (X_1 and its complement Y_1), for any weights $\varepsilon_j^{(1)}$ and $\varepsilon_k^{(1)}$ corresponding to a weight basis vector in X_1 we have to pick in $S_0^* \otimes S_1$ the root basis elements corresponding either only to $\varepsilon_i^{(0)} + \varepsilon_j^{(1)}$ and $\varepsilon_i^{(0)} + \varepsilon_k^{(1)}$ or only to $-\varepsilon_i^{(0)} + \varepsilon_j^{(1)}$ and $-\varepsilon_i^{(0)} + \varepsilon_k^{(1)}$ for all possible i, but not both. Thus the maximal possible number of basis vectors from $S_0^* \otimes S_1$ which we may include in an Abelian subspace of $\mathfrak{g}_1$ is either $\left[\frac{s_0}{2}\right] x_1$ (if we choose either only $\varepsilon_i^{(0)} + \varepsilon_j^{(1)}$ or only $-\varepsilon_i^{(0)} + \varepsilon_j^{(1)}$ for all possible i and j) or s_0, provided we are not in type C and moreover X_1 consists of a single element (corresponding to some $\varepsilon_j^{(1)}$, and we choose root basis vectors corresponding to $\pm\varepsilon_i^{(0)} + \varepsilon_j^{(1)}$ for all possible i). In addition, if we are in type B, we may add one more root basis vector $v_0 \otimes v_1$ with v_0 a weight basis vector with zero weight and v_1 some weight basis vector from X_1.

Thus, we have the following possibilities for the maximal dimension of the piece of an Abelian subspace corresponding to $S_0^* \otimes S_1$:

	B	C	D
$x_1 = 0$	0	0	0
$x_1 = 1$	s_0	$\frac{s_0}{2}$	s_0
$x_1 > 1$	$\frac{s_0-1}{2}x_1 + 1$	$\frac{s_0}{2}x_1$	$\frac{s_0}{2}x_1$

This results in the following possibilities for the maximal dimension of an Abelian subspace in $\mathfrak{g}_1$:

$$\begin{aligned} &\tfrac{s_0-1}{2}x_1 + 1 + (s_1 - x_1)x_2 + (s_2 - x_2)x_3 + \ldots + (s_{\mu-3} - x_{\mu-3})x_{\mu-2} + (s_{\mu-2} - x_{\mu-2})s_{\mu-1} \\ &\qquad \text{(for type B);} \\ &\tfrac{s_0}{2}x_1 + (s_1 - x_1)x_2 + (s_2 - x_2)x_3 + \ldots + (s_{\mu-3} - x_{\mu-3})x_{\mu-2} + (s_{\mu-2} - x_{\mu-2})s_{\mu-1} \\ &\qquad \text{(for type C or D);} \\ &s_0 + (s_1 - 1)x_2 + (s_2 - x_2)x_3 + \ldots + (s_{\mu-3} - x_{\mu-3})x_{\mu-2} + (s_{\mu-2} - x_{\mu-2})s_{\mu-1} \\ &\qquad \text{(for type B or D).} \end{aligned} \tag{2}$$

where μ is the largest part of the partition.

We thus want to maximize each of these quantities for $0 \leqslant x_k \leqslant s_k$, $k = 1, \ldots, \mu - 2$. Note that each of them is linear in all of the x_k separately, hence any possible maxima are attained when every x_k is either 0 or s_k. In fact, more is true:

Lemma 4.4 *An Abelian subspace of maximal possible dimension in* $\mathfrak{g}_1$ *can be obtained either with* $x_{2j-1} = 0$, $x_{2j} = s_{2j}$ *or with* $x_{2j-1} = s_{2j-1}$, $x_{2j} = 0$ *for all* j.

Proof Looking at the subsum

$$\ldots + (s_{k-2} - x_{k-2})x_{k-1} + (s_{k-1} - x_{k-1})x_k + (s_k - x_k)x_{k+1} + \ldots$$

determining dimension of the Abelian subspace, it is easy to see that each of the following changes:

$$\begin{aligned} x_{k-1} &= 0, \qquad x_k = 0 \mapsto x_{k-1} = 0, \qquad x_k = s_k, \\ x_{k-1} &= s_{k-1}, \ x_k = s_k \mapsto x_{k-1} = s_{k-1}, \ x_k = 0 \end{aligned}$$

does not decrease the dimension of the Abelian subspace.

Indeed, these changes do not affect any other summands except those in the above subsum. The first change transforms

$$\begin{aligned} &\ldots + (s_{k-2} - x_{k-2})0 + (s_{k-1} - 0)0 + (s_k - 0)x_{k+1} + \ldots \\ \mapsto &\ldots + (s_{k-2} - x_{k-2})0 + (s_{k-1} - 0)s_k + 0x_{k+1} + \ldots, \end{aligned}$$

i.e., changes the sum by the amount equal to the change from $s_k x_{k+1}$ to $s_{k-1} s_k$. But $x_{k+1} \leqslant s_{k+1}$, and $s_{k+1} \leqslant s_{k-1}$ by (1), so that indeed the sum does not decrease.

Similarly, the second change transforms

$$\begin{aligned} &\ldots + (s_{k-2} - x_{k-2})s_{k-1} + (s_{k-1} - s_{k-1})s_k + (s_k - s_k)x_{k+1} + \ldots \\ \mapsto &\ldots + (s_{k-2} - x_{k-2})s_{k-1} + (s_{k-1} - s_{k-1})0 + (s_k - 0)x_{k+1} + \ldots, \end{aligned}$$

i.e., changes the sum by the amount equal to the change from 0 to $s_k x_{k+1}$, which is obviously a nondecreasing change.

Now using the above changes we may arrive at one of the needed choices. For simplicity, let us encode a given choice of x's by a sequence of zeroes and ones (at the kth place of the sequence stands zero if $x_k = 0$ and one if $x_k = s_k$). We are allowed to perform "local transformations" of the kind $\cdots 00 \cdots \mapsto \cdots 01 \cdots$ and $\cdots 11 \cdots \mapsto \cdots 10 \cdots$. Using one of these transformations, we can always shift the place of the leftmost occurrence of two consecutive identical symbols to the right: say, if this leftmost occurrence is $\cdots 11 \cdots$ we change it to $\cdots 10 \cdots$ and if it is $\cdots 00 \cdots$, we change it to $\cdots 01 \cdots$, and in the worst case the place of the leftmost occurrence of consecutive identical symbols still shifts to the right by at least one position. Thus, if we keep applying the appropriate transformations to the leftmost occurrence of consecutive identical symbols, we inevitably arrive either at 10101... or at 01010.... □

Applying this in (2), we obtain that the maximal possible dimension of an Abelian subspace in $\mathfrak{g}_1$ can only be equal to one of the following six expressions:

$$\begin{array}{r|rl} \frac{s_0-1}{2}s_1 + 1 + s_2s_3 + s_4s_5 + \ldots & s_1s_2 + s_3s_4 + s_5s_6 + \ldots & \text{(for type B)} \\ \frac{s_0}{2}s_1 + s_2s_3 + s_4s_5 + \ldots & s_1s_2 + s_3s_4 + s_5s_6 + \ldots & \text{(for types C, D)} \\ s_0 + s_2s_3 + s_4s_5 + \ldots & s_0 + (s_1 - 1)s_2 + s_3s_4 + s_5s_6 + \ldots & \text{(for types B, D)} \end{array}$$

To find out whether there is an Abelian subspace of half the dimension in $\mathfrak{g}_1$ is thus equivalent to finding out whether subtracting from the dimension of $\mathfrak{g}_1$, i. e. from $s_0s_1 + s_1s_2 + \ldots$, one of these sums doubled gives zero, i. e. whether one of the sums

$$\begin{array}{lr|lrl} s_0s_1 + s_1s_2 + \ldots & -2(\frac{s_0-1}{2}s_1 + 1 + s_2s_3 + s_4s_5 + \ldots) & s_0s_1 + s_1s_2 + \ldots & -2(s_1s_2 + s_3s_4 + s_5s_6 + \ldots) & \text{(B)} \\ s_0s_1 + s_1s_2 + \ldots & -2(\frac{s_0}{2}s_1 + s_2s_3 + s_4s_5 + \ldots) & s_0s_1 + s_1s_2 + \ldots & -2(s_1s_2 + s_3s_4 + s_5s_6 + \ldots) & \text{(C, D)} \\ s_0s_1 + s_1s_2 + \ldots & -2(s_0 + s_2s_3 + s_4s_5 + \ldots) & s_0s_1 + s_1s_2 + \ldots & -2(s_0 + (s_1 - 1)s_2 + s_3s_4 + s_5s_6 + \ldots) & \text{(B, D)} \end{array}$$

is zero.

Simplifying, we obtain respectively

$$\begin{array}{r|rl} s_1 - 2 + s_1s_2 - s_2s_3 + s_3s_4 - s_4s_5 + s_5s_6 - \ldots & s_0s_1 - s_1s_2 + s_2s_3 - s_3s_4 + s_4s_5 - \ldots & \text{(B)} \\ s_1s_2 - s_2s_3 + s_3s_4 - s_4s_5 + \ldots & s_0s_1 - s_1s_2 + s_2s_3 - s_3s_4 + \ldots & \text{(C, D)} \\ -2s_0 + s_0s_1 + s_1s_2 - s_2s_3 + s_3s_4 - \ldots & -2s_0 + 2s_2 + s_0s_1 - s_1s_2 + s_2s_3 - s_3s_4 + \ldots & \text{(B, D)} \end{array}$$

Rewriting this further as

$$
\begin{array}{r|l}
s_1-2+(s_1-s_3)s_2+(s_3-s_5)s_4+(s_5-s_7)s_6+\ldots & (s_0-s_2)s_1+(s_2-s_4)s_3+(s_4-s_6)s_5+\ldots\ \text{(B)}\\
(s_1-s_3)s_2+(s_3-s_5)s_4+(s_5-s_7)s_6+\ldots & (s_0-s_2)s_1+(s_2-s_4)s_3+(s_4-s_6)s_5+\ldots\ \text{(C, D)}\\
s_0(s_1-2)+(s_1-s_3)s_2+(s_3-s_5)s_4+\ldots & (s_0-s_2)(s_1-2)+(s_2-s_4)s_3+(s_4-s_6)s_5+\ldots\ \text{(B, D)}
\end{array}
$$

and taking (1) into account this can be rewritten as

$$
\begin{array}{r|r}
s_1-2+m_2s_2+m_4s_4+m_6s_6+\ldots & m_1s_1+m_3s_3+m_5s_5+\ldots\ \text{(B)}\\
m_2s_2+m_4s_4+m_6s_6+\ldots & m_1s_1+m_3s_3+m_5s_5+\ldots\ \text{(C, D)}\\
s_0(s_1-2)+m_2s_2+m_4s_4+\ldots & m_1(s_1-2)+m_3s_3+m_5s_5+\ldots\ \text{(B, D)}
\end{array}
$$

Let us now assume that our nilpotent element is strictly odd, which, in terms of the corresponding partition, means that $m_{\mu-1}>0$ (here, as before, μ is the largest nonzero part of the partition). This then implies that all multiplicities s_i are nonzero. Thus, to obtain an Abelian subspace of half the dimension of $\mathfrak{g}_1$, we have the following possibilities:

$$
\begin{array}{r|rl}
s_1=2 \text{ and } m_{2k}=0 \text{ for } 2k<\mu & m_{2k-1}=0 \text{ for } 2k-1<\mu & \text{(B)}\\
m_{2k}=0 \text{ for } 2k<\mu & m_{2k-1}=0 \text{ for } 2k-1<\mu & \text{(C, D)}\\
s_1=2 \text{ and } m_{2k}=0 \text{ for } 2k<\mu & m_1=0 \text{ or } s_1=2, \text{ and } m_{2k-1}=0 \text{ for } 1<2k-1<\mu & \text{(B, D)}
\end{array}
$$

We now make the following observations, according to the parity of μ:

- if μ is odd, then the cases in the first column are not realizable, since they require that the partition has no even parts, while, by strict oddity, both $m_{\mu-1}$ and m_μ must be nonzero;
- if μ is even, the cases in the second column are not realizable by exactly the same reason.

Taking this into account, we are left with the following cases: for μ even,

$$
\begin{array}{r|l}
m_2=m_4=\ldots=m_{\mu-2}=0,\ m_{\mu-1}>0,\ m_\mu=2 & \text{— (B)}\\
m_2=m_4=\ldots=m_{\mu-2}=0,\ m_{\mu-1}>0,\ m_\mu>0 & \text{— (C, D)}\\
m_2=m_4=\ldots=m_{\mu-2}=0,\ m_{\mu-1}>0,\ m_\mu=2 & \text{— (B, D)}
\end{array}
$$

and for μ odd,

$$
\begin{array}{r|ll}
\text{—} & m_1=m_3=\ldots=m_{\mu-2}=0,\ m_{\mu-1}>0,\ m_\mu>0 & \text{(B)}\\
\text{—} & m_1=m_3=\ldots=m_{\mu-2}=0,\ m_{\mu-1}>0,\ m_\mu>0 & \text{(C, D)}\\
\text{—} & m_3=m_5=\ldots=m_{\mu-2}=0,\ m_\mu>0 \text{ and either } m_1=0 & \text{(B, D)}\\
 & \qquad \text{or } m_2=m_4=\ldots=m_{\mu-3}=0 \text{ and } m_{\mu-1}=2 &
\end{array}
$$

Let us also observe the following:

- for μ even, the first case is subsumed by the third one;
- for μ even, the third case is subsumed by the second one for type D;
- for μ odd, the subcase $m_1 = 0$ of the third case is subsumed by the first one for type B, and by the second one for type D.

Taking all of the above into account gives the partitions as described. □

Remark 4.5 Another way to formulate the theorem is the following. In case of type C, there is exactly one parity change along the partition, while in cases B or D there might be either one or two parity changes; but if there are two parity changes, then there must be only parts equal to 1, $\mu - 1$, μ and, moreover, $\mu - 1$ must have multiplicity 2. Moreover, for the type B there is one more restriction in case there is only one parity change: namely, if the largest part is even, its multiplicity must be 2.

We now turn to the not necessarily strictly odd nilpotent orbits, using strictly odd reduction from Definition 3.4. For classical types, its reformulation in terms of partitions is as follows.

Lemma 4.6 *Let $\mathfrak{g}$ be a simple Lie algebra of classical type, and let e be a nilpotent element of $\mathfrak{g}$ corresponding to the partition $[\dots k^{m_k} \ell^{m_\ell} \dots n^{m_n}]$, with $\dots < k < \ell < \dots < n$ such that k and ℓ are of opposite parity while all the larger parts j (those with $\ell \leqslant j \leqslant n$) are of the same parity.*

Then the partition $[\dots k^{m_k}(k+1)^{m_\ell + \dots + m_n}]$ defines a strictly odd nilpotent element in a Lie algebra of the same type, and corresponds to the strictly odd reduction of e, as defined in Definition 3.4.

Proof Let us begin by noting that the modified partition is indeed suitable for the same type: if this requires that all parts of the same parity as k have even multiplicity, then we have not touched them; while if this requires that all parts of the same parity as $k+1$ are even, then ℓ and all larger parts are of the same parity as $k+1$, so each of the multiplicities $m_\ell, \dots, m_n$ was even, hence their sum is even too, and we indeed stay with the same type. Moreover, the corresponding nilpotent element is strictly odd since its largest parts are k and $k+1$.

Let us now reformulate the passage from the original partition to the modified partition in terms of weighted Dynkin diagrams. We get the following procedure: one removes all nodes (and weights) from left to right until no more 2's are left; for the types B, C, D that's all that has to be done; for the type A one has to similarly remove all 2's on the right.

This procedure precisely means leaving the connected component of the weighted Dynkin diagram that contains nonzero weights, as described in Proposition 3.2 above, so that we indeed obtain the strictly odd reduction of e. □

Corollary 4.7 *Given a nilpotent element in a simple Lie algebra $\mathfrak{g}$ of classical type* B, C, *or* D, *there is an Abelian subspace of half dimension in $\mathfrak{g}_1$ if and only if the partition corresponding to the nilpotent element satisfies the following conditions:*

type C*: there is no more than one parity change along the partition;*
types B *and* D*: there are no more than two parity changes*
and, if there is at least one parity change, then

- *if the largest part of the partition is even, then there is only one parity change, and in the* B *case moreover it must be the unique even part and must have multiplicity* 2*;*
- *if there are two parity changes, then the largest part of the partition is odd, there is a unique even part, it has multiplicity* 2*, and all smaller parts are equal to* 1.

Thus, Abelian subspaces of half dimension in $\mathfrak{g}_1$ occur precisely for nilpotent elements corresponding to partitions of one of the following kind (with $k \leqslant \ell$):

$$\begin{array}{l|l} \textbf{any type}: & [\cdots(2k-2)^{\nu_{2k-2}}(2k)^{\nu_{2k}}(2\ell+1)^{\nu_{2\ell+1}}(2\ell+3)^{\nu_{2\ell+3}}\cdots]; \\ \textbf{type C or D}: & [\cdots(2k-3)^{\nu_{2k-3}}(2k-1)^{\nu_{2k-1}}(2\ell)^{\nu_{2\ell}}(2\ell+2)^{\nu_{2\ell+2}}\cdots]; \\ \textbf{type B or D}: & \left[1^{\nu_1}(2k)^2(2\ell+1)^{\nu_{2\ell+1}}(2\ell+3)^{\nu_{2\ell+3}}\cdots\right]; \\ \textbf{type B}: & \left[\cdots(2k-3)^{\nu_{2k-3}}(2k-1)^{\nu_{2k-1}}(2\ell)^2\right], \end{array}$$

Proof This follows from Lemma 4.6. Indeed the latter shows that $\mathfrak{g}_1(e)$ for a nilpotent element e corresponding to some partition has an Abelian subspace of half dimension if and only if $\tilde{\mathfrak{g}}_1(\tilde{e})$, as described in Corollary 3.3, has such a subspace; and this happens if and only if the partition modified as in Lemma 4.6 satisfies conditions of Theorem 4.3.

It remains to note that a partition is of the indicated kind if and only if the partition obtained from it as in Lemma 4.6 satisfies conditions of Theorem 4.3. □

5 Computations

It remains to find out which of the strictly odd nilpotent orbits in simple Lie algebras of exceptional type have an Abelian subspace of half dimension in degree 1.

For that, we used the computer algebra system `GAP`. In the package `SLA` by Willem A. de Graaf included in this system one can compute with nilpotent orbits of arbitrary semisimple Lie algebras. In particular, one obtains canonical bases consisting of root vectors for the homogeneous subspaces of all degrees in the grading of the Lie algebra induced by a nilpotent element.

Using Proposition 3.1, we can determine Abelian subspaces in $\mathfrak{g}_1$ as follows. Let B be the basis of $\mathfrak{g}_1$ consisting of positive root vectors. Let us construct a graph with the set of vertices B, where two vertices e_α and e_β are connected with an edge if and only if they do not commute, that is, if and only if $\alpha + \beta$ is a root. Then, by Proposition 3.1, $\mathfrak{g}_1$ has an Abelian subspace of dimension d if and only if the basis consisting of root vectors has a subset of cardinality d consisting of pairwise commuting root vectors.

Clearly, this is equivalent to the corresponding graph having an *independent set* of cardinality d—that is, a subset consisting of d vertices such that no two of these vertices are connected by an edge. Hence, describing all possible dimensions of Abelian subspaces in $\mathfrak{g}_1$ reduces to listing all possible cardinalities of independent subsets in the corresponding graph.

There is another package, `GRAPE` by Leonard H. Soicher in `GAP`, which can be used to list all independent sets in a finite graph. Using this package, we determine independent sets of maximal possible cardinality in the graph corresponding to a nilpotent orbit.

The results are given in the tables below. A `GAP` code for computing maximal dimensions of Abelian subspaces in $\mathfrak{g}_1$ for arbitrary semisimple Lie algebras is available at [8]. In fact, the program can list all subsets of any given cardinality of pairwise commuting elements in the root vector basis.

As an illustration, we present below two cases for E_6.

Examples 5.1 The nilpotent orbit with the weighted Dynkin diagram 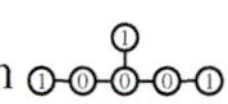 has $\mathfrak{g}_1$ of dimension 14. The corresponding graph with 14 vertices and edges connecting vertices corresponding to non-commuting root vectors in $\mathfrak{g}_1$ looks as follows:

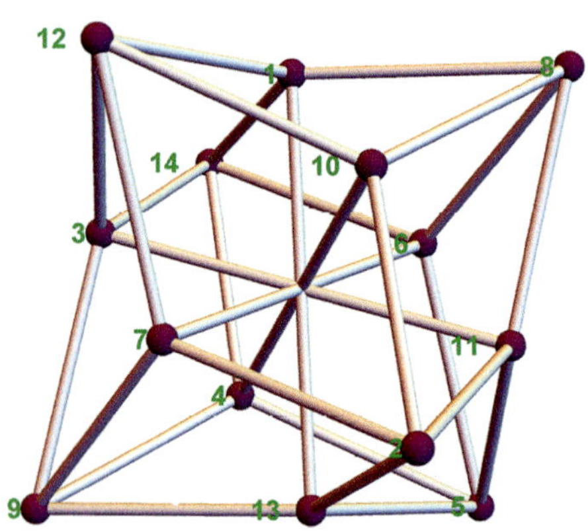

This graph has independent sets with 6 vertices, e. g. {2, 5, 8, 9, 12, 14}, but any subset on more than 6 vertices contains a pair of vertices connected with an edge, thus for this nilpotent orbit maximal dimension of an Abelian subspace is equal to 6.

Another orbit in E_6, with the diagram 0–1–0(1)–1–0, has $\mathfrak{g}_1$ of dimension 10 corresponding to the graph

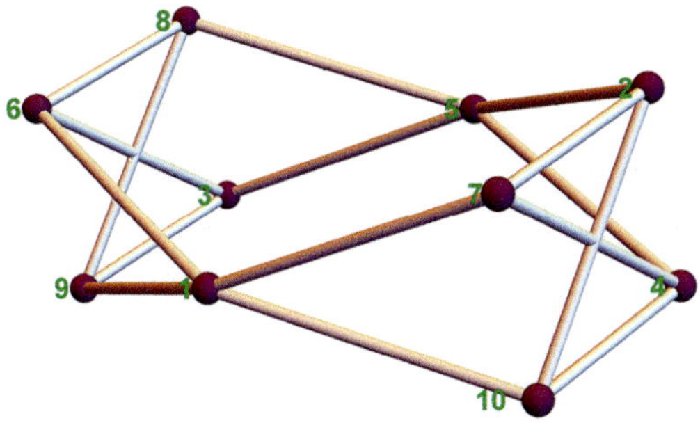

with 10 vertices. It is easy to find in this graph an independent subset with five elements – e. g. $\{1, 2, 3, 4, 8\}$.

Thus, the orbit of the first example has no Abelian subspace of half dimension in $\mathfrak{g}_1$, while that of the second example has.

Tables

Table G2s Strictly odd nilpotent orbits in G_2, all with half-Abelian $\mathfrak{g}_1$

Name	Diagram	dim $\mathfrak{g}_1$
A_1	1≡0	4
$\widetilde{A}_1$	0≡1	2

Table F4s Strictly odd nilpotent orbits in F_4

With half-Abelian $\mathfrak{g}_1$			Without half-Abelian $\mathfrak{g}_1$		
Name	Diagram	dim $\mathfrak{g}_1$	Name	Diagram	dim $\mathfrak{g}_1$ (largest dimension of an Abelian subspace)
A_1	1—0⇐0—0	14	$\widetilde{A}_1$	0—0⇐0—1	8 (2)
$A_1 + \widetilde{A}_1$	0—1⇐0—0	12	$A_2 + \widetilde{A}_1$	0—0⇐1—0	6 (2)
$C_3(a_1)$	1—0⇐1—0	6	$\widetilde{A}_2 + A_1$	0—1⇐0—1	8 (3)

Table E6s Strictly odd nilpotent orbits in E_6

With half-Abelian $\mathfrak{g}_1$			Without half-Abelian $\mathfrak{g}_1$		
Name	Diagram	dim $\mathfrak{g}_1$	Name	Diagram	dim $\mathfrak{g}_1$ (largest dimension of an Abelian subspace)
A_1	$\begin{smallmatrix}&&1&&\\0&0&0&0&0\end{smallmatrix}$	20	A_2+A_1	$\begin{smallmatrix}&&1&&\\1&0&0&0&1\end{smallmatrix}$	14 (6)
$2A_1$	$\begin{smallmatrix}&&0&&\\1&0&0&0&1\end{smallmatrix}$	16	$2A_2+A_1$	$\begin{smallmatrix}&&0&&\\1&0&1&0&1\end{smallmatrix}$	12 (5)
$3A_1$	$\begin{smallmatrix}&&0&&\\0&0&1&0&0\end{smallmatrix}$	18			
A_2+2A_1	$\begin{smallmatrix}&&0&&\\0&1&0&1&0\end{smallmatrix}$	12			
A_3+A_1	$\begin{smallmatrix}&&1&&\\0&1&0&1&0\end{smallmatrix}$	10			
A_4+A_1	$\begin{smallmatrix}&&1&&\\1&1&0&1&1\end{smallmatrix}$	8			

Table E7s Strictly odd nilpotent orbits in E_7

With half-Abelian $\mathfrak{g}_1$			Without half-Abelian $\mathfrak{g}_1$		
Name	Diagram	dim $\mathfrak{g}_1$	Name	Diagram	dim $\mathfrak{g}_1$ (largest dimension of an Abelian subspace)
A_1	$\begin{smallmatrix}&&&0&&\\0&0&0&0&0&1\end{smallmatrix}$	32	$4A_1$	$\begin{smallmatrix}&&&1&&\\1&0&0&0&0&0\end{smallmatrix}$	26 (11)
$2A_1$	$\begin{smallmatrix}&&&0&&\\0&1&0&0&0&0\end{smallmatrix}$	32	A_2+A_1	$\begin{smallmatrix}&&&0&&\\0&1&0&0&0&1\end{smallmatrix}$	24 (9)
$3A_1'$	$\begin{smallmatrix}&&&0&&\\0&0&0&0&1&0\end{smallmatrix}$	30	$2A_2+A_1$	$\begin{smallmatrix}&&&0&&\\0&1&0&0&1&0\end{smallmatrix}$	20 (8)
A_2+2A_1	$\begin{smallmatrix}&&&0&&\\0&0&0&1&0&0\end{smallmatrix}$	24	A_3+2A_1	$\begin{smallmatrix}&&&0&&\\1&0&1&0&0&1\end{smallmatrix}$	18 (7)
$(A_3+A_1)'$	$\begin{smallmatrix}&&&0&&\\0&0&0&1&0&1\end{smallmatrix}$	18	A_4+A_1	$\begin{smallmatrix}&&&0&&\\0&1&0&1&0&1\end{smallmatrix}$	14 (6)
$D_4(a_1)+A_1$	$\begin{smallmatrix}&&&1&&\\1&0&0&0&1&0\end{smallmatrix}$	16			
A_3+A_2	$\begin{smallmatrix}&&&0&&\\0&1&0&1&0&0\end{smallmatrix}$	16			

Table E8s Strictly odd nilpotent orbits in E_8

With half-Abelian $\mathfrak{g}_1$			Without half-Abelian $\mathfrak{g}_1$		
Name	Diagram	dim $\mathfrak{g}_1$	Name	Diagram	dim $\mathfrak{g}_1$ (largest dimension of an Abelian subspace)
A_1	$\begin{smallmatrix} &&&&0&& \\ 1&0&0&0&0&0&0 \end{smallmatrix}$	56	$2A_1$	$\begin{smallmatrix} &&&&0&& \\ 0&0&0&0&0&0&1 \end{smallmatrix}$	64 (22)
$3A_1$	$\begin{smallmatrix} &&&&0&& \\ 0&1&0&0&0&0&0 \end{smallmatrix}$	54	$4A_1$	$\begin{smallmatrix} &&&&1&& \\ 0&0&0&0&0&0&0 \end{smallmatrix}$	56 (21)
$A_2 + 3A_1$	$\begin{smallmatrix} &&&&0&& \\ 0&0&0&0&0&1&0 \end{smallmatrix}$	42	$A_2 + 2A_1$	$\begin{smallmatrix} &&&&0&& \\ 0&0&1&0&0&0&0 \end{smallmatrix}$	48 (16)
$A_3 + A_1$	$\begin{smallmatrix} &&&&0&& \\ 1&0&1&0&0&0&0 \end{smallmatrix}$	34	$A_2 + A_1$	$\begin{smallmatrix} &&&&0&& \\ 1&0&0&0&0&0&1 \end{smallmatrix}$	44 (17)
$A_3 + A_2 + A_1$	$\begin{smallmatrix} &&&&0&& \\ 0&0&0&0&1&0&0 \end{smallmatrix}$	30	$2A_2 + 2A_1$	$\begin{smallmatrix} &&&&0&& \\ 0&0&0&1&0&0&0 \end{smallmatrix}$	40 (16)
$A_4 + A_2 + A_1$	$\begin{smallmatrix} &&&&0&& \\ 0&0&1&0&0&1&0 \end{smallmatrix}$	24	$2A_2 + A_1$	$\begin{smallmatrix} &&&&0&& \\ 0&1&0&0&0&0&1 \end{smallmatrix}$	36 (16)
$E_7(a_5)$	$\begin{smallmatrix} &&&&0&& \\ 0&0&1&0&1&0&0 \end{smallmatrix}$	18	$A_3 + 2A_1$	$\begin{smallmatrix} &&&&0&& \\ 1&0&0&0&0&1&0 \end{smallmatrix}$	36 (15)
$A_6 + A_1$	$\begin{smallmatrix} &&&&0&& \\ 0&0&1&0&1&0&1 \end{smallmatrix}$	16	$A_3 + A_2$	$\begin{smallmatrix} &&&&0&& \\ 0&0&1&0&0&0&1 \end{smallmatrix}$	32 (13)
A_7	$\begin{smallmatrix} &&&&0&& \\ 0&1&1&0&1&0&1 \end{smallmatrix}$	14	$D_4(a_1) + A_1$	$\begin{smallmatrix} &&&&1&& \\ 0&1&0&0&0&0&0 \end{smallmatrix}$	32 (12)
			$2A_3$	$\begin{smallmatrix} &&&&0&& \\ 0&0&0&1&0&0&1 \end{smallmatrix}$	28 (13)
			$A_4 + 2A_1$	$\begin{smallmatrix} &&&&0&& \\ 1&0&0&0&1&0&0 \end{smallmatrix}$	28 (12)
			$A_4 + A_1$	$\begin{smallmatrix} &&&&0&& \\ 1&0&1&0&0&0&1 \end{smallmatrix}$	26 (10)
			$A_4 + A_3$	$\begin{smallmatrix} &&&&0&& \\ 0&1&0&0&1&0&0 \end{smallmatrix}$	24 (10)
			$A_5 + A_1$	$\begin{smallmatrix} &&&&0&& \\ 1&0&0&0&1&0&1 \end{smallmatrix}$	22 (9)
			$D_5(a_1) + A_2$	$\begin{smallmatrix} &&&&0&& \\ 1&0&1&0&0&1&0 \end{smallmatrix}$	22 (8)
			$D_6(a_2)$	$\begin{smallmatrix} &&&&1&& \\ 0&1&0&0&0&1&0 \end{smallmatrix}$	20 (9)
			$E_6(a_3) + A_1$	$\begin{smallmatrix} &&&&0&& \\ 0&1&0&1&0&0&1 \end{smallmatrix}$	20 (8)
			$D_7(a_2)$	$\begin{smallmatrix} &&&&0&& \\ 1&0&1&0&1&0&1 \end{smallmatrix}$	16 (7)

Table F4o (Non-strictly) odd nilpotent orbits in F_4, all with half-Abelian $\mathfrak{g}_1$

Name	Diagram	Strictly odd piece
B_2	$2 - \boxed{0 \Leftarrow 0 - 1}$	C_3 $(2, 1^4)$
C_3	$\boxed{1 - 0 \Leftarrow 1} - 2$	B_3 $(3, 2^2)$

Table E6o (Non-strictly) odd nilpotent orbits in E_6, all with half-Abelian $\mathfrak{g}_1$

Name	Diagram	Strictly odd piece
A_3	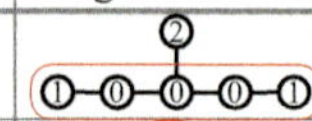	A_5
A_5	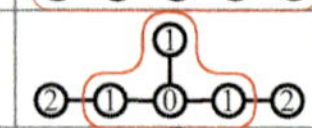	$D_4\ (3, 2^2, 1)$
$D_5(a_1)$	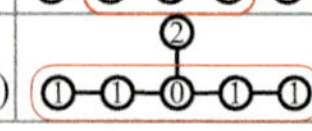	A_5

Table E7o (Non-strictly) odd nilpotent orbits in E_7

With half-Abelian $\mathfrak{g}_1$			Without half-Abelian $\mathfrak{g}_1$		
Name	Diagram	Strictly odd piece	Name	Diagram	Strictly odd piece
A_3	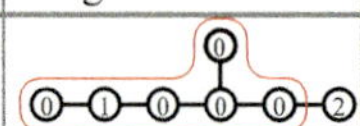	$D_6\ (2^2, 1^8)$	$D_4 + A_1$	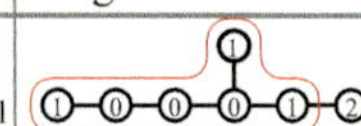	$D_6\ (3, 2^4, 1)$
$D_5(a_1)$	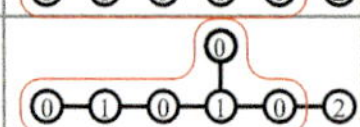	$D_6\ (3^2, 2^2, 1^2)$	$A_5 + A_1$	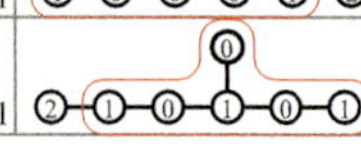	$E_6\ (2A_2 + A_1)$
A_5'	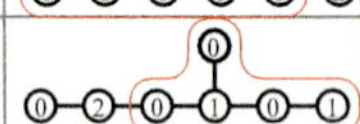	$D_5\ (3, 2^2, 1^3)$			
$D_6(a_2)$	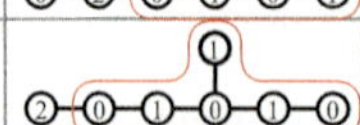	$E_6\ (A_3 + A_1)$			
$D_5 + A_1$	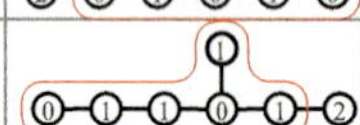	$D_6\ (4^2, 3, 1)$			
$D_6(a_1)$	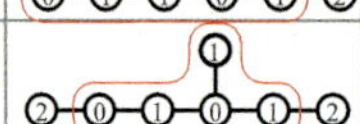	$D_5\ (3^2, 2^2)$			
D_6	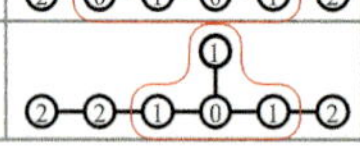	$D_4\ (3, 2^2, 1)$			

Table E8o (Non-strictly) odd nilpotent orbits in E_8

With half-Abelian $\mathfrak{g}_1$			Without half-Abelian $\mathfrak{g}_1$		
Name	Diagram	Strictly odd piece	Name	Diagram	Strictly odd piece
A_3	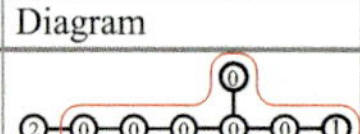	$E_7\ (A_1)$	$D_4 + A_1$		$E_7\ (4A_1)$
$D_5(a_1) + A_1$	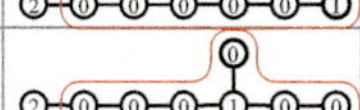	$E_7\ (A_2 + 2A_1)$	$D_5(a_1)$	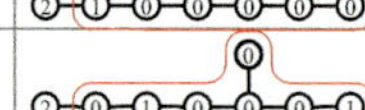	$E_7\ (A_2 + A_1)$
A_5	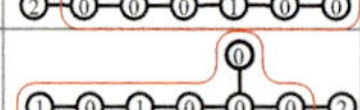	$D_7\ (3, 2^2, 1^7)$	$D_5 + A_1$	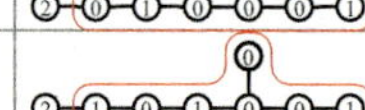	$E_7\ (A_3 + 2A_1)$
$D_6(a_1)$	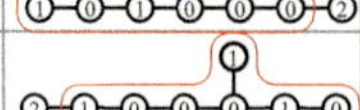	$E_7\ (D_4(a_1) + A_1)$	$E_6(a_1) + A_1$	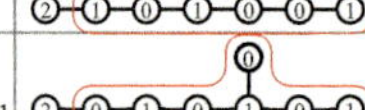	$E_7\ (A_4 + A_1)$
$E_7(a_4)$	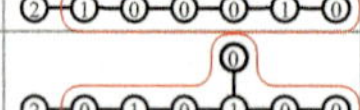	$E_7\ (A_3 + A_2)$	D_6	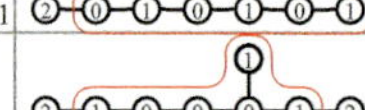	$D_6\ (3, 2^4, 1)$
$E_7(a_3)$	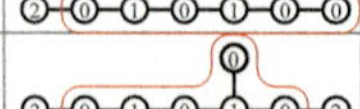	$D_6\ (3^2, 2^2, 1^2)$	$E_6 + A_1$	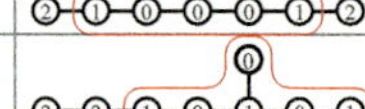	$E_6\ (2A_2 + A_1)$
D_7	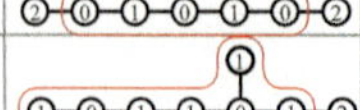	$D_7\ (5, 4^2, 1)$			
$E_7(a_2)$	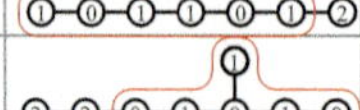	$E_6\ (A_3 + A_1)$			
$E_7(a_1)$	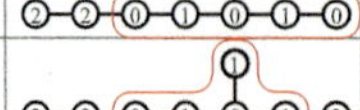	$D_5\ (3^2, 2^2)$			
E_7	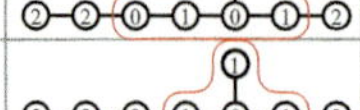	$D_4\ (3, 2^2, 1)$			

Acknowledgements The authors are grateful to the referee for highly professional work, including simplifications of proofs and improvements of exposition.

The third named author wishes to thank Daniele Valeri for discussions on generalization of Miura maps, constructed in [2].

The second named author gratefully acknowledges help of the user `marmot` from tex.stackexchange.com in producing the TEX code that was used to highlight the strictly odd pieces of weighted Dynkin diagrams in the last four tables.

References

1. D. H. Collingwood, W. M. McGovern, "Nilpotent Orbits In Semisimple Lie Algebras". Van Nostrand Reinhold, New York 1993, xi + 186 pages.
2. A. De Sole, V. G. Kac, D. Valeri, *Structure of classical (finite and affine) W-algebras*. J. Eur. Math, Soc. **18** (2016), 1873–1908.
3. A. G. Elashvili, *Centralizers of nilpotent elements in semisimple Lie algebras*. Proc. Razmadze Math. Inst. **46** (1975), 109–132 (in Russian).
4. V. G. Kac, M. Wakimoto, *Quantum reduction and representation theory of superconformal algebras*. Advances in Math. **193** (2005), 400–458.
5. D. I. Panyushev, *On spherical nilpotent orbits and beyond*. Ann. Inst. Fourier **49** (1999), 1453–1476.
6. T. A. Springer, R. Steinberg, *Conjugacy classes*. In "Seminar on Algebraic Groups and Related Finite Groups", Lecture Notes in Math. **131**, Springer Verlag, New York, 1970, 167–266.
7. T. Shoji, *Shintani descent for special linear groups*. J. Algebra **199** (1998), 175–228.
8. The `GAP` code at https://arxiv.org/src/1806.00893v1/anc/nilorg1.g

Multiplicative Slices, Relativistic Toda and Shifted Quantum Affine Algebras

Michael Finkelberg and Alexander Tsymbaliuk

To Tony Joseph on his 75th birthday, with admiration

Abstract We introduce the shifted quantum affine algebras. They map homomorphically into the quantized K-theoretic Coulomb branches of $3d$ $\mathcal{N} = 4$ SUSY quiver gauge theories. In type A, they are endowed with a coproduct, and they act on the equivariant K-theory of parabolic Laumon spaces. In type A_1, they are closely related to the type A open relativistic quantum Toda system.

Mathematics Subject Classification: 17B37, 81R10, 81T13

1 Introduction

1.1 Summary

The goal of this paper is to initiate the study of *shifted quantum affine algebras*[1] and *shifted $\mathbf{v}$-Yangians*. They arise as a tool to write down via generators and

[1] They were introduced by B. Feigin in 2010.

M. Finkelberg (✉)
National Research University Higher School of Economics, Department of Mathematics, Moscow, Russian Federation

Skolkovo Institute of Science and Technology, Moscow, Russian Federation

Institute for Information Transmission Problems, Moscow, Russian Federation

A. Tsymbaliuk
Yale University, Department of Mathematics, New Haven, CT, USA

M. Gorelik et al. (eds.), *Representations and Nilpotent Orbits of Lie Algebraic Systems*, Progress in Mathematics 330,
https://doi.org/10.1007/978-3-030-23531-4_6

relations the quantized K-theoretic Coulomb branches of $3d$ $\mathcal{N} = 4$ SUSY quiver gauge theories (see [10, Remark 3.9(2)]), similarly to the appearance of shifted Yangians in the study of the quantized Coulomb branches of $3d$ $\mathcal{N} = 4$ SUSY quiver gauge theories [10].[2] Similarly to [24], the shifted quantum affine algebras carry a coproduct, see Sect. 10 for partial results in this direction. The multiplicative analogue of the construction [4] equips the equivariant K-theory of parabolic Laumon spaces with an action of the quantized K-theoretic Coulomb branch for a type A quiver, and hence with an action of a shifted quantum affine algebra of type A. Similarly to [24], the unframed case of type A_1 quiver is closely related to the open relativistic quantum Toda system of type A.

1.2 Outline of the Paper

- In Sect. 2, we give a construction of the completed phase space of the (quasiclassical) relativistic open Toda system for arbitrary simply-connected semisimple algebraic group G via quasihamiltonian and Poisson reductions. It is a direct multiplicative analogue of the Kazhdan–Kostant construction of the (nonrelativistic) open Toda integrable system. We want to stress right away that it depends on a choice of a pair of Coxeter elements in the Weyl group W of G, via a choice of Steinberg's cross-section.[3] In the case when the two Coxeter elements coincide, the resulting completed phase space is isomorphic to the universal centralizer $\mathfrak{Z}_G^G$, see Sect. 2.3. In the case $G = SL(n)$, the universal centralizer is isomorphic to a natural n-fold cover of the moduli space of centered periodic $SU(2)$-monopoles of charge n, see Corollary 2.6.
- The conjectural quantization of the above construction of the completed phase space of the relativistic open Toda is described in Sect. 3.12. We conjecture that it is isomorphic to the corresponding spherical symmetric nil-DAHA which is realized as an equivariant K-theory of a twisted affine Grassmannian, i.e. as a sort of twisted quantized Coulomb branch (the twist is necessary in the case of non-simply-laced G). The bulk of Sect. 3 is occupied by the review of Cherednik's definition of symmetric nil-DAHA, its residue construction, and its realization as the equivariant K-theory of a twisted affine flag variety. In the simply-laced case no twist is required, and the spherical nil-DAHA in question is isomorphic to the convolution algebra $K^{G(\mathcal{O})\rtimes\mathbb{C}^\times}(\mathrm{Gr}_G)$ up to some finite extension. This convolution algebra is defined for arbitrary reductive G. In case $G = GL(n)$, this convolution algebra is likely to have a presentation via generators and relations (as a truncated shifted quantum affine algebra of type A_1), see Sect. 9. From this presentation and Proposition 11.21 we obtain a homomorphism $K^{G(\mathcal{O})\rtimes\mathbb{C}^\times}(\mathrm{Gr}_G) \to K^{L(\mathcal{O})\rtimes\mathbb{C}^\times}(\mathrm{Gr}_L)$ for any Levi subgroup

[2] We must admit right away that we were not able to prove the desired presentation of the quantized Coulomb branch for a single quiver.

[3] The appearance of Coxeter elements in the construction of relativistic Toda lattice goes back at least to [60].

$L \subset G = GL(n)$. We conjecture an existence of such a homomorphism for arbitrary Levi subgroup L in arbitrary reductive group G, but we have no clue as to a geometric construction of such a homomorphism. It would be important for a study of equivariant quantum K-theory of the flag variety $\mathcal{B}$ of G. Its analogue for the equivariant Borel-Moore homology convolution algebra $H_\bullet^{G(\mathcal{O})\rtimes\mathbb{C}^\times}(\mathrm{Gr}_G) \to H_\bullet^{L(\mathcal{O})\rtimes\mathbb{C}^\times}(\mathrm{Gr}_L)$ is constructed in [24]. However, the construction is not geometric; it uses an isomorphism with the quantum open (nonrelativistic) Toda lattice.

- Recall that for an arbitrary $3d$ $\mathcal{N} = 4$ SUSY quiver gauge theory of type ADE, the non-quantized K-theoretic Coulomb branch is identified with a multiplicative generalized slice in the corresponding affine Grassmannian [10, Remarks 3.9(2), 3.17]. These multiplicative slices are studied in detail in Sect. 4 (in the unframed case, they were studied in detail in [25]). In particular, they embed into the loop group $G(z)$, and it is likely that the image coincides with the space of scattering matrices of singular periodic monopoles [14]. The multiplication in the loop group gives rise to the multiplication of slices, which is conjecturally quantized by the coproduct of the corresponding shifted quantum affine algebras.
- In Sect. 5, we introduce the *shifted quantum affine algebras* $\mathcal{U}^{\mathrm{sc}}_{\mu^+,\mu^-}$ and $\mathcal{U}^{\mathrm{ad}}_{\mu^+,\mu^-}$ (*simply-connected* and *adjoint* versions, respectively) for any simple Lie algebra $\mathfrak{g}$ and its two coweights μ^+, μ^- (these algebras depend only on $\mu = \mu^+ + \mu^-$ up to an isomorphism). For $\mu^+ = \mu^- = 0$, they are central extensions of the standard quantum loop algebra $U_{\boldsymbol{v}}(L\mathfrak{g})$ and its adjoint version $U^{\mathrm{ad}}_{\boldsymbol{v}}(L\mathfrak{g})$. These algebras can be viewed as trigonometric versions of the shifted Yangians $\mathbf{Y}_\mu$, see [10, 24, 45].

  An alternative (but equivalent) definition of $\mathcal{U}^{\mathrm{sc}}_{\mu^+,\mu^-}$ was suggested to us by B. Feigin in Spring 2010 in an attempt to generalize the results of [7] to the K-theoretic setting (which is the subject of Sect. 12 of the present paper). In this approach, we consider an algebra with the same generators and defining relations as $U_{\boldsymbol{v}}(L\mathfrak{g})$ in the new Drinfeld realization with just one modification: the relation $[e_i(z), f_j(w)] = \frac{\delta_{ij}\delta(z/w)}{\boldsymbol{v}_i-\boldsymbol{v}_i^{-1}}\left(\psi_i^+(z) - \psi_i^-(z)\right)$ is replaced by $p_i(z)[e_i(z), f_j(w)] = \frac{\delta_{ij}\delta(z/w)}{\boldsymbol{v}_i-\boldsymbol{v}_i^{-1}}\left(\psi_i^+(z) - \psi_i^-(z)\right)$ for any collection of rational functions $\{p_i(z)\}_{i\in I}$ (here I parametrizes the set of vertices of the Dynkin diagram of $\mathfrak{g}$). For $\mathfrak{g} = \mathfrak{sl}_2$ and $\mu^+ = \mu^- \in -\mathbb{N}$, the algebra $\mathcal{U}^{\mathrm{sc}}_{\mu^+,\mu^-}$ appeared in [18, § 5.2].

  We also provide an alternative presentation of the antidominantly shifted quantum affine algebras with a finite number of generators and defining relations, see Theorem 5.5 and Appendix A for its proof. We note that this result (and its proof) also holds for any affine Lie algebra, except for type $A_1^{(1)}$. In the unshifted case, more precisely for $U_{\boldsymbol{v}}(L\mathfrak{g})$, it can be viewed as a $\boldsymbol{v}$-version of the famous Levendorskii presentation of the Yangian $Y(\mathfrak{g})$, see [47]. Motivated by Guay et al. [33], we also provide a slight modification of this presentation in Theorem A.3.
- In Sect. 6, we introduce other generators of $\mathcal{U}^{\mathrm{ad}}_{\mu^+,\mu^-}$, which can be encoded by the generating series $\{A_i^\pm(z), B_i^\pm(z), C_i^\pm(z), D_i^\pm(z)\}_{i\in I}$. We provide a complete list

of the defining relations between these generators for antidominant $\mu^+, \mu^- \in \Lambda^-$ (we use Λ^- to denote the submonoid of the coweight lattice Λ spanned by antidominant coweights), see Theorem 6.6 and Appendix B for its proof. This should be viewed as a shifted $\boldsymbol{v}$-version of the corresponding construction for Yangians of [30]. We note that while some of the relations were established (without a proof) in loc. cit., the authors did not aim at providing a complete list of the defining relations. However, a rational analogue of Theorem 6.6 provides such a list.

We would like to point out that this is one of the few places where it is essential to work with the *adjoint* version. In the simplest case, that is of $U^{\mathrm{ad}}_{\boldsymbol{v}}(L\mathfrak{sl}_2)$, these generating series coincide with the entries of the matrices $T^{\pm}(z)$ from the RTT realization of $U^{\mathrm{ad}}_{\boldsymbol{v}}(L\mathfrak{sl}_2)$, see [17] and our discussion in Sect. 11.4.

- In Sect. 7, we construct homomorphisms

$$\widetilde{\Phi}^{\underline{\lambda}}_{\mu}\colon \mathcal{U}^{\mathrm{ad}}_{0,\mu}[\mathsf{z}_1^{\pm 1},\ldots,\mathsf{z}_N^{\pm 1}] \longrightarrow \widetilde{\mathcal{A}}^{\boldsymbol{v}}_{\mathrm{frac}}[\mathsf{z}_1^{\pm 1},\ldots,\mathsf{z}_N^{\pm 1}]$$

from the adjoint version of shifted quantum affine algebras to the $\mathbb{C}(\boldsymbol{v})[\mathsf{z}_1^{\pm 1},\ldots,\mathsf{z}_N^{\pm 1}]$-algebras $\widetilde{\mathcal{A}}^{\boldsymbol{v}}_{\mathrm{frac}}[\mathsf{z}_1^{\pm 1},\ldots,\mathsf{z}_N^{\pm 1}]$ of difference operators on multidimensional tori, see Theorem 7.1 and Appendix C for its proof. Here $\underline{\lambda} = (\omega_{i_1},\ldots,\omega_{i_N})$ is a sequence of fundamental coweights, such that $\lambda - \mu$ is a sum of simple coroots with coefficients in $\mathbb{N}$, where $\lambda := \sum_{s=1}^{N} \omega_{i_s}$. This result can be viewed as a $\boldsymbol{v}$-version of the corresponding construction for shifted Yangians of [10, Theorem B.15], while the *unshifted case* of it, more precisely the case of $U_{\boldsymbol{v}}(L\mathfrak{g})$, appeared (without a proof) in [31]. For $\mathfrak{g} = \mathfrak{sl}_2$, $N = 0$ and antidominant shift, the above homomorphism made its first appearance in [18, Section 6].
- In Sect. 8, we consider the quantized K-theoretic Coulomb branch $\mathcal{A}^{\boldsymbol{v}}$ in the particular case of quiver gauge theories of ADE type (a straightforward generalization of the constructions of [9, 10], with the equivariant Borel-Moore homology replaced by the equivariant K-theory). There is a natural embedding $\mathbf{z}^*(\iota_*)^{-1}\colon \mathcal{A}^{\boldsymbol{v}} \hookrightarrow \widetilde{\mathcal{A}}^{\boldsymbol{v}}[\mathsf{z}_1^{\pm 1},\ldots,\mathsf{z}_N^{\pm 1}]$. In Theorem 8.1, we show that our homomorphism $\widetilde{\Phi}^{\underline{\lambda}}_{\mu}$ of Sect. 7 factors through the above embedding (with $\mathbb{C}[\boldsymbol{v}^{\pm 1}]$ extended to $\mathbb{C}(\boldsymbol{v})$), giving rise to a homomorphism

$$\overline{\Phi}^{\underline{\lambda}}_{\mu}\colon \mathcal{U}^{\mathrm{ad}}_{0,\mu}[\mathsf{z}_1^{\pm 1},\ldots,\mathsf{z}_N^{\pm 1}] \longrightarrow \mathcal{A}^{\boldsymbol{v}}_{\mathrm{frac}}.$$

This is a $\boldsymbol{v}$-version of the corresponding result for shifted Yangians of [10, Theorem B.18].

In Sect. 8.3, we add certain truncation relations to the relations defining $\mathcal{U}^{\mathrm{ad}}_{0,\mu}[\mathsf{z}_1^{\pm 1},\ldots,\mathsf{z}_N^{\pm 1}]$ to obtain the *truncated shifted quantum affine algebras* $\mathcal{U}^{\underline{\lambda}}_{\mu}$ such that the homomorphism $\overline{\Phi}^{\underline{\lambda}}_{\mu}$ factors through the projection and the same named homomorphism $\mathcal{U}^{\mathrm{ad}}_{0,\mu}[\mathsf{z}_1^{\pm 1},\ldots,\mathsf{z}_N^{\pm 1}] \twoheadrightarrow \mathcal{U}^{\underline{\lambda}}_{\mu} \xrightarrow{\overline{\Phi}^{\underline{\lambda}}_{\mu}} \mathcal{A}^{\boldsymbol{v}}_{\mathrm{frac}}$. We expect that $\overline{\Phi}^{\underline{\lambda}}_{\mu}\colon \mathcal{U}^{\underline{\lambda}}_{\mu} \to \mathcal{A}^{\boldsymbol{v}}_{\mathrm{frac}}$ is an isomorphism, see Conjecture 8.9.

In Sect. 8.4, we define the *shifted $\boldsymbol{v}$-Yangians* ${}^{\imath}\mathcal{Y}^{\boldsymbol{v}}_{\mu}[\mathsf{z}_1^{\pm 1}, \ldots, \mathsf{z}_N^{\pm 1}] \subset \mathcal{U}^{\mathrm{ad}}_{0,\mu}[\mathsf{z}_1^{\pm 1}, \ldots, \mathsf{z}_N^{\pm 1}]$ and their truncated quotients ${}^{\imath}\mathcal{Y}^{\lambda}_{\bar{\mu}} \subset \mathcal{U}^{\lambda}_{\bar{\mu}}$. We conjecture that $\overline{\Phi}^{\lambda}_{\mu} : {}^{\imath}\mathcal{Y}^{\lambda}_{\bar{\mu}} \to \mathcal{A}^{\boldsymbol{v}}_{\mathrm{frac}}$ is an isomorphism, see Conjecture 8.13.

One of our biggest failures is the failure to define the integral forms ${}^{\imath}\mathfrak{Y}^{\lambda}_{\bar{\mu}} \subset {}^{\imath}\mathcal{Y}^{\lambda}_{\bar{\mu}}$ and $\mathfrak{U}^{\lambda}_{\bar{\mu}} \subset \mathcal{U}^{\lambda}_{\bar{\mu}}$ over $\mathbb{C}[\boldsymbol{v}^{\pm 1}] \subset \mathbb{C}(\boldsymbol{v})$ that would (at least conjecturally) map isomorphically onto $\mathcal{A}^{\boldsymbol{v}} \subset \mathcal{A}^{\boldsymbol{v}}_{\mathrm{frac}}$. Only in the case of $\mathfrak{g} = \mathfrak{sl}_2$, making use of the $ABCD$-generators of Sect. 6, we are able to introduce the desired integral form in Sect. 9.1 (see also [29] for the integral forms for $\mathfrak{g} = \mathfrak{sl}_n$). It is worth noting that for arbitrary simply-laced $\mathfrak{g}$ and any $i \in I$, the images under $\overline{\Phi}^{\lambda}_{\mu}$ of the generators $B^{+}_{i,r}$ and $e_{i,r}$ (resp. $C^{+}_{i,r}$ and $f_{i,r}$) are the classes of dual exceptional collections of vector bundles on the corresponding minuscule Schubert varieties in the affine Grassmannian, see Remark 8.4.

The desired integral forms ${}^{\imath}\mathfrak{Y}^{\lambda}_{\bar{\mu}}$ and $\mathfrak{U}^{\lambda}_{\bar{\mu}}$ are expected to be quantizations of a certain cover ${}^{\dagger}\hat{\mathcal{W}}^{\lambda^*}_{\mu^*}$ of a multiplicative slice introduced in Sect. 4.6, see Conjecture 8.14. Here $*$ stands for the involution $\mu \mapsto -w_0\mu$ of the coweight lattice Λ.

- In Sect. 9, we prove the surjectivity of the homomorphism $\overline{\Phi}^{0}_{-n\alpha}$ in the simplest case of $\mathfrak{g} = \mathfrak{sl}_2$ and antidominant shifts, see Theorem 9.2. This identifies the slightly localized and extended quantized K-theoretic Coulomb branch $K^{\widetilde{GL}(n,\mathcal{O})\rtimes\widetilde{\mathbb{C}}^\times}_{\mathrm{loc}}(\mathrm{Gr}_{GL(n)})$ with a quotient of the localized version of the truncated shifted quantum affine algebra $\mathfrak{U}^{0}_{-n\alpha,\mathrm{loc}}$ (where $\widetilde{GL}(n)$ and $\widetilde{\mathbb{C}}^\times$ stand for the two-fold covers of $GL(n), \mathbb{C}^\times$; while the localization is obtained by inverting $1 - \boldsymbol{v}^{2m}$, $1 \le m \le n$). We reduce the proof of the isomorphism $\mathfrak{U}^{0}_{-n\alpha,\mathrm{loc}} \xrightarrow{\sim} K^{\widetilde{GL}(n,\mathcal{O})\rtimes\widetilde{\mathbb{C}}^\times}_{\mathrm{loc}}(\mathrm{Gr}_{GL(n)})$ to a verification of an identity with quantum resultants in $\mathcal{U}^{0}_{-n\alpha}$, see Remarks 9.6, and 9.12. It would be interesting to describe explicitly a basis of $\mathfrak{U}^{0}_{-n\alpha,\mathrm{loc}}$ projecting to the "canonical" basis of $K^{\widetilde{GL}(n,\mathcal{O})\rtimes\widetilde{\mathbb{C}}^\times}_{\mathrm{loc}}(\mathrm{Gr}_{GL(n)})$ formed by the classes of irreducible equivariant perverse coherent sheaves [8].
- In Sect. 10, we discuss generalizations of the classical coproducts on $U_{\boldsymbol{v}}(L\mathfrak{g})$ to the shifted setting. We start by considering the simplest case $\mathfrak{g} = \mathfrak{sl}_2$. We will denote $\mathcal{U}^{\mathrm{sc}}_{0,b\alpha/2}$ simply by $\mathcal{U}^{\mathrm{sc}}_{0,b}$ (here $b \in \mathbb{Z}$ and α is the simple positive coroot). We construct homomorphisms

$$\Delta_{b_1,b_2} : \mathcal{U}^{\mathrm{sc}}_{0,b} \longrightarrow \mathcal{U}^{\mathrm{sc}}_{0,b_1} \otimes \mathcal{U}^{\mathrm{sc}}_{0,b_2}$$

for any $b_1, b_2 \in \mathbb{Z}$, which recover the classical Drinfeld-Jimbo coproduct for $b_1 = b_2 = 0$. Our construction is parallel to the one for shifted Yangians of [24] and proceeds in two steps. First, we define such homomorphisms in the *antidominant* case $b_1, b_2 \in \mathbb{Z}_{\le 0}$, see Theorem 10.5 and Appendix D for its proof. The proof is crucially based on the aforementioned alternative presentation of the antidominantly shifted quantum affine algebras with a finite number of generators and defining relations of Theorem 5.5. Second, we use the algebra embeddings

$\iota_{n,m_1,m_2}\colon \mathcal{U}^{\mathrm{sc}}_{0,n} \hookrightarrow \mathcal{U}^{\mathrm{sc}}_{0,n+m_1+m_2}$ (here $m_1, m_2 \leq 0$) to reduce the general case to the antidominant one, see Theorem 10.10 and Appendix F for its proof. We note that our proof of injectivity of the *shift homomorphisms* ι_{n,m_1,m_2} is based on the PBW property of the shifted quantum affine algebras of $\mathfrak{sl}_2$, see Lemma 10.9 and Theorem E.2 of Appendix E.

In Sects. 10.6 and 10.7, we generalize the aforementioned case of $\mathfrak{sl}_2$ to the case of $\mathfrak{sl}_n$ $(n \geq 2)$. The idea is again to treat first the case of antidominant shifts and then deduce the general case. To achieve the former goal, it is essential to have explicit formulas for the action of the Drinfeld-Jimbo coproduct on the generators $\{e_{i,-1}, f_{i,1}, h_{i,\pm 1}\}_{i\in I}$ of $U_{\boldsymbol{v}}(L\mathfrak{sl}_n)$. This is the key technical result, stated in Theorem 10.13 and proved in Appendix G. Once this is established, it is easy to guess the formulas for the homomorphism $\Delta_{\mu_1,\mu_2}\colon \mathcal{U}^{\mathrm{sc}}_{0,\mu_1+\mu_2} \to \mathcal{U}^{\mathrm{sc}}_{0,\mu_1} \otimes \mathcal{U}^{\mathrm{sc}}_{0,\mu_2}$ in the case $\mu_1, \mu_2 \in \Lambda^-$ (antidominant), see Theorem 10.16 and its proof in Appendix H. In Theorem 10.20 we derive the construction of Δ_{μ_1,μ_2} for general $\mu_1, \mu_2 \in \Lambda$ by utilizing the algebra embeddings $\iota_{\mu,\nu_1,\nu_2}\colon \mathcal{U}^{\mathrm{sc}}_{0,\mu} \hookrightarrow \mathcal{U}^{\mathrm{sc}}_{0,\mu+\nu_1+\nu_2}$ for $\mu \in \Lambda, \nu_1, \nu_2 \in \Lambda^-$, see Theorem 10.19 and its proof in Appendix I (the latter is based on the shuffle realization of $U_{\boldsymbol{v}}(L\mathfrak{sl}_n)$ of [53, 63]).

Motivated by Finkelberg et al. [24], we expect that our construction of homomorphisms Δ_{μ_1,μ_2} can be generalized to any simply-laced $\mathfrak{g}$ and its two coweights $\mu_1, \mu_2 \in \Lambda$. However, we failed to achieve this due to a lack of explicit formulas for the Drinfeld-Jimbo coproduct of the generators $\{e_{i,-1}, f_{i,1}, h_{i,\pm 1}\}_{i\in I}$ of $U_{\boldsymbol{v}}(L\mathfrak{g})$ (even for $\mathfrak{g} = \mathfrak{sl}_n$, the formulas of Theorem 10.13 seem to be new, to our surprise).

Moreover, we expect that this coproduct extends to

$$\Delta^{\mathrm{ad}}_{\mu_1,\mu_2}\colon \mathcal{U}^{\mathrm{ad}}_{0,\mu+\mu_2}[\mathsf{z}_1^{\pm 1}, \dots, \mathsf{z}_{N_1+N_2}^{\pm 1}] \longrightarrow \mathcal{U}^{\mathrm{ad}}_{0,\mu_1}[\mathsf{z}_1^{\pm 1}, \dots, \mathsf{z}_{N_1}^{\pm 1}] \otimes \mathcal{U}^{\mathrm{ad}}_{0,\mu_2}[\mathsf{z}_{N_1+1}^{\pm 1}, \dots, \mathsf{z}_{N_1+N_2}^{\pm 1}],$$

which descends to the same named homomorphism $\Delta^{\mathrm{ad}}_{\mu_1,\mu_2}\colon \mathcal{U}^{\underline{\lambda}}_{\mu_1+\mu_2} \to \mathcal{U}^{\underline{\lambda}^{(1)}}_{\mu_1} \otimes \mathcal{U}^{\underline{\lambda}^{(2)}}_{\mu_2}$ between truncated algebras, see Conjecture 11.22. We check a particular case of this conjecture for $\mathfrak{g} = \mathfrak{sl}_2$ in Proposition 11.21, using the RTT realization of $\mathcal{U}^{\mathrm{ad}}_{0,2b}$ of Theorem 11.11.

- In Sect. 11, we discuss relativistic/trigonometric Lax matrices, the shifted RTT algebras of $\mathfrak{sl}_2$ and their relation to the shifted quantum affine algebras of $\mathfrak{sl}_2$. This yields a link between two seemingly different appearances of the RTT relations (both trigonometric and rational).

In Sect. 11.2, we recall the Kuznetsov-Tsyganov [43] local relativistic Lax matrix $L^{\boldsymbol{v},0}_i(z)$ satisfying the trigonometric RTT-relation. The complete monodromy matrix $T^{\boldsymbol{v},0}_n(z) = L^{\boldsymbol{v},0}_n(z) \cdots L^{\boldsymbol{v},0}_1(z)$ also satisfies the same relation, and its matrix coefficient $T^{\boldsymbol{v},0}_n(z)_{11}$ encodes all the hamiltonians of the q-difference quantum open Toda lattice for $GL(n)$ [19, 56].

We introduce two more local Lax matrices $L^{\boldsymbol{v},\pm 1}_i(z)$ satisfying the same trigonometric RTT-relation. They give rise to the plethora of 3^n complete monodromy matrices $T^{\boldsymbol{v}}_{\vec{k}}(z), \vec{k} \in \{-1, 0, 1\}^n$, given by the length n products of the three local Lax matrices in arbitrary order. The matrix coefficient $T_{\vec{k}}^{\mathbf{v}}(z)_{11}$ encodes the hamiltonians of the corresponding modified quantum difference Toda lattice; the quadratic hamiltonians are given by the formula (11.8). At the quasiclassical level, these integrable systems go back to [21]. We show that among these 3^n integrable systems there are no more than 3^{n-2} nonequivalent, see Lemma 11.6. It is shown in [35] that they are all obtained by the construction of [56] using arbitrary pairs of orientations of the A_{n-1} Dynkin diagram, see Remark 11.7.

In Sect. 11.4, we introduce the *shifted RTT algebras of* $\mathfrak{sl}_2$, denoted by $\mathcal{U}^{\mathrm{rtt}}_{0,-2n}$, and construct isomorphisms $\Upsilon_{0,-2n}\colon \mathcal{U}^{\mathrm{ad}}_{0,-2n} \xrightarrow{\sim} \mathcal{U}^{\mathrm{rtt}}_{0,-2n}$ for any $n \in \mathbb{N}$, see Theorem 11.8 and Theorem 11.11. For $n = 0$, this recovers the isomorphism of the *new Drinfeld* and the *RTT* realizations of the quantum loop algebra $U^{\mathrm{ad}}_{\mathbf{v}}(L\mathfrak{sl}_2)$, due to [17]. We also identify the $ABCD$ generators of $\mathcal{U}^{\mathrm{ad}}_{0,-2n}$ of Sect. 6 with the generators of $\mathcal{U}^{\mathrm{rtt}}_{0,-2n}$, see Corollary 11.10.

Viewing the Lax matrix $L_1^{\mathbf{v},-1}(z)$ as a homomorphism from $\mathcal{U}^{\mathrm{rtt}}_{0,-2}$ to the algebra of difference operators on $\mathbb{C}^\times$ and composing it with $\Upsilon_{0,-2}$, we recover the homomorphism $\widetilde{\Phi}^0_{-2}$ of Sect. 7. More generally, among all pairwise isomorphic shifted algebras $\{\mathcal{U}^{\mathrm{ad}}_{b,-2-b} | b \in \mathbb{Z}\}$ only those with $b, -2-b \leq 0$ admit an RTT realization, i.e., there are analogous isomorphisms $\Upsilon_{b,-2-b}\colon \mathcal{U}^{\mathrm{ad}}_{b,-2-b} \xrightarrow{\sim} \mathcal{U}^{\mathrm{rtt}}_{b,-2-b}$. Moreover, recasting the homomorphisms $\widetilde{\Phi}_{b,-2-b}$ (generalizations of $\widetilde{\Phi}^0_{-2}$ for $b = 0$) as the homomorphisms $\mathcal{U}^{\mathrm{rtt}}_{b,-2-b} \to \hat{\mathcal{A}}^{\mathbf{v}}_1$, we recover the other two Lax matrices $L_1^{\mathbf{v},0}(z)$ (for $b = -1$) and $L_1^{\mathbf{v},1}(z)$ (for $b = -2$).

Finally, we use the RTT presentation of $U^{\mathrm{ad}}_{\mathbf{v}}(L\mathfrak{sl}_2)$ to derive explicit formulas for the action of the Drinfeld-Jimbo coproduct on the Drinfeld *half-currents*, see Proposition 11.18 and Appendix J for its proof. We also show that the same formulas hold in the antidominantly shifted setting for the homomorphisms Δ_{b_1,b_2}, see Proposition 11.19. As a consequence of the latter, the homomorphism $\Delta^{\mathrm{ad}}_{2b_1,2b_2}$ is intertwined with the RTT coproduct $\Delta^{\mathrm{rtt}}_{2b_1,2b_2}$, see Corollary 11.20, which is used to prove the aforementioned Proposition 11.21 on the descent of $\Delta^{\mathrm{ad}}_{2b_1,2b_2}$ to the truncated versions.

- In Sect. 12, we provide yet another geometric realization of the shifted quantum affine algebras (resp. shifted Yangians) of $\mathfrak{sl}_n$ via the parabolic Laumon spaces. Roughly speaking, this arises by combining our homomorphism $\overline{\Phi}^{\lambda}_{\mu}$ of Sect. 8 (resp. $\overline{\Phi}^{\lambda}_{\mu}$ of [10, Theorem B.18]) with an action of the quantized K-theoretic (resp. cohomological) Coulomb branch $\mathcal{A}^{\mathbf{v}}_{\mathrm{frac}}$ on the localized equivariant K-theory (resp. cohomology) of parabolic Laumon spaces, constructed in [4], see Remark 12.3(c).

For any $\pi = (p_1, \ldots, p_n) \in \mathbb{Z}^n_{>0}$, we construct an action of $\mathcal{U}^{\mathrm{sc}}_{0,\mu}$, the simply-connected shifted quantum affine algebra of $\mathfrak{sl}_n$ with the shift $\mu = \sum_{j=1}^{n-1}(p_{j+1} - p_j)\omega_j$, on $M(\pi)$: the direct sum of localized equivariant K-theory of $\mathfrak{Q}_{\underline{d}}$, see Theorem 12.2. Here $\mathfrak{Q}_{\underline{d}}$ is the type π Laumon based parabolic quasiflags' space, which we recall in Sect. 12.1. In Theorem 12.6, we slightly generalize this by constructing an action of the shifted quantum affine algebra of $\mathfrak{gl}_n$ (defined in Sect. 12.7) on $M(\pi)$. In Theorem 12.4, we establish an isomorphism $M(\pi') \otimes M(\pi'') \xrightarrow{\sim} M(\pi)$ (here $\pi = \pi' + \pi''$) of $\mathcal{U}^{\mathrm{sc}}_{0,\mu}$-modules, where the action on the source arises from the formal coproduct $\widetilde{\Delta}\colon \mathcal{U}^{\mathrm{sc}}_{0,\mu} \to \mathcal{U}^{\mathrm{sc}}_{0,\mu'} \widehat{\otimes} \mathcal{U}^{\mathrm{sc}}_{0,\mu''}$, constructed in Sect. 10.1 (an analogue of the Drinfeld formal coproduct on $U_{\boldsymbol{v}}(L\mathfrak{g})$).

The *rational* counterpart of these results is established in Theorem 12.7, where we construct an action of $\mathcal{Y}^{\hbar}_{\mu}$ (the shifted Yangian of $\mathfrak{sl}_n$ with scalars extended to $\mathbb{C}(\hbar)$) on $V(\pi)$: the sum of localized equivariant cohomology of $\mathfrak{Q}_{\underline{d}}$. The dominant case ($p_1 \leq \ldots \leq p_n$) of this result was treated in [7], where the proof was deduced from the Gelfand-Tsetlin formulas of [27]. In contrast, our straightforward proof is valid for any π and, thus, gives an alternative proof of the above Gelfand-Tsetlin formulas. We also propose a $\boldsymbol{v}$-analogue of the Gelfand-Tsetlin formulas of [27], see Proposition 12.8.

Our construction can be also naturally generalized to provide the actions of the shifted quantum toroidal (resp. affine Yangian) algebras of $\mathfrak{sl}_n$ on the sum of localized equivariant K-theory (resp. cohomology) of the parabolic affine Laumon spaces, see Sect. 12.9.

In Sect. 12.10, we introduce the *Whittaker* vectors in the completions of $M(\pi)$ and $V(\pi)$:

$$\mathfrak{m} := \sum_{\underline{d}} [\mathcal{O}_{\mathfrak{Q}_{\underline{d}}}] \in M(\pi)^{\wedge} \text{ and } \mathfrak{v} := \sum_{\underline{d}} [\mathfrak{Q}_{\underline{d}}] \in V(\pi)^{\wedge}.$$

This name is motivated by their *eigenvector properties* of Proposition 12.11, Remark 12.12(c).

Motivated by the work of Brundan-Kleshchev, see [12], we expect that the truncated shifted quantum affine algebras $\mathcal{U}^{N\omega_{n-1}}_{\mu}$ of $\mathfrak{sl}_n$ should be $\boldsymbol{v}$-analogues of the finite W-algebras $W(\mathfrak{sl}_N, e_\pi)$, see [57], where $N := \sum p_i$ and $e_\pi \in \mathfrak{sl}_N$ is a nilpotent element of Jordan type π.

2 Relativistic Open Toda Lattice

2.1 *Quasihamiltonian Reduction*

Let $G \supset B \supset T$ be a reductive group with a Borel and Cartan subgroups. Let $T \subset B_- \subset G$ be the opposite Borel subgroup; let U (resp. U_-) be the unipotent radical of B (resp. B_-). We consider the *double* $D(G) = G \times G$ (see, e.g., [2, § 3.2]) equipped with an action of $G \times G$: $(u_1, u_2) \cdot (g_1, g_2) = (u_1 g_1 u_2^{-1}, u_2 g_2 u_2^{-1})$, and with a moment map $\mu = (\mu_1, \mu_2)\colon D(G) \to G \times G,\ \mu(g_1, g_2) = (g_1 g_2 g_1^{-1}, g_2^{-1})$ (see [2, Remark 3.2]). The double $D(G)$ carries a (non-closed) 2-form $\omega_D =$

$\frac{1}{2}(\mathrm{Ad}_{g_2} g_1^*\theta, g_1^*\theta) + \frac{1}{2}(g_1^*\theta, g_2^*\theta + g_2^*\overline{\theta})$ where $(\cdot,\cdot)$ is a nondegenerate invariant symmetric bilinear form on $\mathfrak{g}$, and θ (resp. $\overline{\theta}$) is the left- (resp. right-) invariant Maurer–Cartan form on G.

We choose a pair of Coxeter elements $c, c' \in W = N_G(T)/T$, and their representatives $\dot{c}, \dot{c}' \in N_G(T)$. Steinberg's cross-section $\Sigma_G^{\dot{c}} \subset G$ is defined as $Z^0(G) \cdot (U_-\dot{c} \cap \dot{c}U)$. If G is semisimple simply-connected, then the composed morphism $\Sigma_G^{\dot{c}} \hookrightarrow G \to G/\!\!/\mathrm{Ad}_G = T/W$ is an isomorphism [58, Theorem 1.4]. For arbitrary G, the composed morphism $\varrho\colon \Sigma_G^{\dot{c}} \to T/W$ is a ramified Galois cover with Galois group $\pi_1(G/Z^0(G))$. Furthermore, we consider $\Xi_G^{\dot{c}} := Z^0(G) \cdot U_-\dot{c}U_- \supset \Sigma_G^{\dot{c}}$. According to [58, § 8.9] (for a proof, see, e.g., [39]), $\Sigma_G^{\dot{c}}$ meets any U_--orbit (with respect to the conjugation action) on $\Xi_G^{\dot{c}}$ in exactly one point, and the conjugation action of U_- on $\Xi_G^{\dot{c}}$ is free, so that $\Xi_G^{\dot{c}}/\mathrm{Ad}_{U_-} \simeq \Sigma_G^{\dot{c}}$.

For example, according to [58, Example 7.4b)], for an appropriate choice of $\dot{c}$, the Steinberg cross-section $\Sigma_{SL(n)}^{\dot{c}}$ consists of the matrices with 1's just above the main diagonal, $(-1)^{n-1}$ in the bottom left corner, arbitrary entries elsewhere in the first column, and zeros everywhere else (in our conventions, B (resp. B_-) is the subgroup of upper triangular (resp. lower triangular) matrices in $SL(n)$). Hence $\Xi_{SL(n)}^{\dot{c}}$ consists of matrices with 1's just above the main diagonal, and zeros everywhere above that.

Following [26], we define the phase space of the open relativistic Toda lattice as the quasihamiltonian reduction ${}^{\dagger}\mathfrak{Z}^{c',c}(G) := \mu^{-1}(\Xi_G^{\dot{c}'} \times \mathrm{inv}(\Xi_G^{\dot{c}}))/U_- \times U_-$ where $\mathrm{inv}\colon G \to G$ is the inversion $g \mapsto g^{-1}$. The composed projection

$$\mu^{-1}(\Xi_G^{\dot{c}'} \times \mathrm{inv}(\Xi_G^{\dot{c}})) \to \mathrm{inv}(\Xi_G^{\dot{c}}) \hookrightarrow G \twoheadrightarrow G/\!\!/\mathrm{Ad}_G = T/W$$

gives rise to an integrable system $\varpi\colon {}^{\dagger}\mathfrak{Z}^{c',c}(G) \to T/W$ which factors through ${}^{\dagger}\mathfrak{Z}^{c',c}(G) \xrightarrow{\tilde{\varpi}} \Sigma_G^{\dot{c}} \xrightarrow{\varrho} T/W$.

Lemma 2.1 *If G is semisimple simply-connected, then ${}^{\dagger}\mathfrak{Z}^{c',c}(G)$ is smooth, and ω_D gives rise to a symplectic form on ${}^{\dagger}\mathfrak{Z}^{c',c}(G)$.*

Proof The morphism $\Xi_G^{\dot{c}} \to \Sigma_G^{\dot{c}} = T/W$ is smooth by [58, Theorem 1.5], so the fibered product $\Xi_G^{\dot{c}'} \times_{T/W} \Xi_G^{\dot{c}} \subset \Xi_G^{\dot{c}'} \times \Xi_G^{\dot{c}}$ is smooth. But

$$\mu\colon D(G) \supset \mu^{-1}(\Xi_G^{\dot{c}'} \times \mathrm{inv}(\Xi_G^{\dot{c}})) \to \Xi_G^{\dot{c}'} \times \mathrm{inv}(\Xi_G^{\dot{c}}) \simeq \Xi_G^{\dot{c}'} \times \Xi_G^{\dot{c}}$$

is a submersion onto $\Xi_G^{\dot{c}'} \times_{T/W} \Xi_G^{\dot{c}}$, hence $M := \mu^{-1}(\Xi_G^{\dot{c}'} \times \mathrm{inv}(\Xi_G^{\dot{c}}))$ is smooth, and its quotient modulo the free action of $U_- \times U_-$ is smooth as well.

The restriction of ω_D to M is $U_- \times U_-$-invariant, so it descends to a 2-form ω on ${}^{\dagger}\mathfrak{Z}^{c',c}(G)$. This 2-form is closed since the differential $d\omega_D = -\mu^*(\chi_1 + \chi_2)$ (see [2, Definition 2.2(B1)]) where $\chi = \frac{1}{12}(\theta, [\theta, \theta])$ is the canonical closed biinvariant 3-form on G, and χ_1 (resp. χ_2) is its pull-back from the first (resp. second) copy of G. But the restriction $\chi|_{\Xi_G^{\dot{c}}}$ vanishes identically since $(\mathfrak{b}_-, [\mathfrak{b}_-, \mathfrak{b}_-]) = 0$.

It remains to check the nondegeneracy of ω, that is given $(g_1, g_2) \in M$ to check that $\operatorname{Ker} \omega_D|_M(g_1, g_2)$ is contained in the span $v(\mathfrak{n}_- \oplus \mathfrak{n}_-)$ of tangent vectors at (g_1, g_2) arising from the action of $U_- \times U_-$. The argument in the proof of [2, Theorem 5.1] shows that $\operatorname{Ker} \omega_D|_M(g_1, g_2) \subset v(\mathfrak{g} \oplus \mathfrak{g})$. However, it is clear that $T_{(g_1,g_2)}M \cap v(\mathfrak{g} \oplus \mathfrak{g}) = v(\mathfrak{n}_- \oplus \mathfrak{n}_-)$.

The lemma is proved. □

2.2 Poisson Reduction

Note that $T \cdot \Xi_G^{\dot{c}} = \Xi_G^{\dot{c}} \cdot T = \operatorname{Ad}_T(\Xi_G^{\dot{c}}) = B_- \cdot \dot{c} \cdot B_- =: C_c$ (a Coxeter Bruhat cell). One can check that the natural morphism

$$^{\dagger}\mathfrak{Z}^{c',c}(G) = \mu^{-1}(\Xi_G^{\dot{c}'} \times \operatorname{inv}(\Xi_G^{\dot{c}}))/U_- \times U_- \to \mu^{-1}(C_{c'} \times \operatorname{inv}(C_c))/B_- \times B_-$$

is an isomorphism. Moreover, the action of $B_- \times B_-$ on $\mu^{-1}(C_{c'} \times \operatorname{inv}(C_c))$ factors through the free action of $(B_- \times B_-)/\Delta_{Z(G)}$: the quotient modulo the diagonal copy of the center of G.

The double $D(G) = G \times G$ carries the Semenov-Tian-Shansky Poisson structure [59, Section 2]. Following loc. cit., $G \times G$ with this Poisson structure is denoted by $(D_+(G), \{,\}_+)$, the *Heisenberg double*. Another Poisson structure on $G \times G$ denoted $\{,\}_-$ in *loc. cit.* is the *Drinfeld double* $D_-(G)$. The diagonal embedding $G \hookrightarrow D_-(G)$ is Poisson with respect to the standard Poisson structure on G denoted π_G in [20, § 2.1]. The dual (Semenov-Tian-Shansky) Poisson structure on G is denoted π in [20, § 2.2].

The Heisenberg double $D_+(G)$ is equipped with two commuting (left and right) dressing Poisson actions of the Drinfeld double $D_-(G)$. Restricting to the diagonal $G \hookrightarrow D_-(G)$ we obtain two commuting Poisson actions of (G, π_G) on $D_+(G)$. The multiplicative moment map of this action is nothing but $\mu\colon D_+(G) \to (G, \pi) \times (G, \pi)$ of Sect. 2.1 (a Poisson morphism). Now $C_c \subset G$ is a coisotropic subvariety [20, § 6.2] of (G, π), and $\mu^{-1}(C_{c'} \times \operatorname{inv}(C_c)) \hookrightarrow D(G)$ is a coisotropic subvariety of $(D_+(G), \{,\}_+)$. The action of $G \times G$ on $(D_+(G), \{,\}_+)$ is Poisson if $G \times G$ is equipped with the direct product of the standard Poisson-Lie structures denoted π_G in [20, § 2.1]. Note that $B_- \times B_- \subset G \times G$ is a Poisson-Lie subgroup; its Poisson structure will be denoted $\pi_{B_-} \times \pi_{B_-}$.

The characteristic distribution [20, § 6.2] of the coisotropic subvariety $\mu^{-1}(C_{c'} \times \operatorname{inv}(C_c)) \subset (D_+(G), \{,\}_+)$ coincides with the distribution defined by the tangent spaces to the $B_- \times B_-$-orbits in $\mu^{-1}(C_{c'} \times \operatorname{inv}(C_c))$. By [20, Proposition 6.7] we obtain a Poisson structure on $\mu^{-1}(C_{c'} \times \operatorname{inv}(C_c))/(B_- \times B_-) \simeq {}^{\dagger}\mathfrak{Z}^{c',c}(G)$. This Poisson structure coincides with the one arising from the symplectic form ω on ${}^{\dagger}\mathfrak{Z}^{c',c}(G)$.

2.3 *The Universal Centralizer*

Recall that the universal centralizer [49, Section 8] $\mathfrak{Z}^G_G \subset G \times \Sigma^{\dot{c}}_G$ is defined as $\mathfrak{Z}^G_G = \{(g, x) : gxg^{-1} = x\}$. In case $c = c'$ and $\dot{c} = \dot{c}'$, we have an evident embedding $\mathfrak{Z}^G_G \hookrightarrow \mu^{-1}(\Xi^{\dot{c}}_G \times \mathrm{inv}(\Xi^{\dot{c}}_G))$, and the composed morphism $\eta\colon \mathfrak{Z}^G_G \hookrightarrow \mu^{-1}(\Xi^{\dot{c}}_G \times \mathrm{inv}(\Xi^{\dot{c}}_G)) \twoheadrightarrow {}^{\dagger}\mathfrak{Z}^{c,c}(G)$. Clearly, the following diagram commutes:

$$\begin{array}{ccc} \mathfrak{Z}^G_G & \xrightarrow{\ \eta\ } & {}^{\dagger}\mathfrak{Z}^{c,c}(G) \\ \Big\downarrow{\scriptstyle \mathrm{pr}} & & {\scriptstyle \tilde{\varpi}}\Big\downarrow \\ \Sigma^{\dot{c}}_G & = & \Sigma^{\dot{c}}_G \end{array}$$

Proposition 2.2 *For semisimple simply-connected G, the morphism $\eta\colon \mathfrak{Z}^G_G \to {}^{\dagger}\mathfrak{Z}^{c,c}(G)$ is an isomorphism.*

Proof First we prove the surjectivity of η. We use the equality $U_- \times U_- = (U_- \times \{e\}) \times \Delta_{U_-}$. Given $(g_1, g_2) \in \mu^{-1}(\Xi^{\dot{c}}_G \times \mathrm{inv}(\Xi^{\dot{c}}_G))$ we first act by $(u_2, u_2) \in \Delta_{U_-}\colon (g_1, g_2) \mapsto (u_2 g_1 u_2^{-1}, u_2 g_2 u_2^{-1})$. We can find a unique u_2 such that $u_2 g_2 u_2^{-1} \in \Sigma^{\dot{c}}_G$. Let us denote the resulting $(u_2 g_1 u_2^{-1}, u_2 g_2 u_2^{-1})$ by (h_1, h_2) for brevity. Now we act by the left shift $h_1 \mapsto u_1 h_1$ which takes $h_1 h_2 h_1^{-1}$ to $u_1 h_1 h_2 h_1^{-1} u_1^{-1}$. We can find a unique u_1 such that $u_1 h_1 h_2 h_1^{-1} u_1^{-1} \in \Sigma^{\dot{c}}_G$. Now both $h_2 = u_2 g_2 u_2^{-1}$ and $u_1 h_1 h_2 h_1^{-1} u_1^{-1}$ are in $\Sigma^{\dot{c}}_G$. Being conjugate they must coincide, hence $(u_1 h_1, h_2) \in \mathfrak{Z}^G_G$.

Now if $\eta(g, x) = \eta(g', x')$, then there is $u_2 \in U_-$ such that $u_2 x u_2^{-1} = x'$, hence $x = x'$ and $u_2 = e$. Then $g' = u_1 g$ for some $u_1 \in U_-$, and both g and g' commute with x, hence $u_1 x u_1^{-1} = x$, hence $u_1 = e$, so that $g = g'$.

So η is bijective at the level of $\mathbb{C}$-points. But ${}^{\dagger}\mathfrak{Z}^{c,c}(G)$ is smooth, hence η is an isomorphism. □

Remark 2.3 For arbitrary reductive G the morphism η is an affine embedding, but it fails to be surjective already for $G = PGL(2)$ where the class of (g_1, g_2) such that $g_2 = \begin{pmatrix} a & -1 \\ 1 & 0 \end{pmatrix}$ and $g_1 g_2 g_1^{-1} = \begin{pmatrix} -a & -1 \\ 1 & 0 \end{pmatrix}$ does not lie in the image of η when $a \neq 0$. Similarly, for $G = GL(2)$, the class of (g_1, g_2) such that $g_2 = \begin{pmatrix} a & -1 \\ 1 & 0 \end{pmatrix}$ and $g_1 g_2 g_1^{-1} = \begin{pmatrix} a & 1 \\ -1 & 0 \end{pmatrix}$ does not lie in the image of η. It also follows that the natural projection ${}^{\dagger}\mathfrak{Z}^{c,c}(SL(2)) \to {}^{\dagger}\mathfrak{Z}^{c,c}(PGL(2))$ is *not* surjective.

Remark 2.4 For G semisimple simply-connected, the reduction

$$(\mathbf{D}(G), \omega_{\mathbf{D}(G)}) /\!\!/ \mathrm{diag}(G)$$

[2, Example 6.1, Remark 6.2] inherits a symplectic structure on its nonsingular locus. We have a natural morphism $\mathfrak{Z}^G_G \to (\mathbf{D}(G), \omega_{\mathbf{D}(G)}) /\!\!/ \mathrm{diag}(G)$ which is a birational isomorphism (but not an isomorphism: e.g., it contracts the centralizer of a regular unipotent element). Thus an open subvariety of $\mathfrak{Z}_G^G$ is equipped with a symplectic form pulled back from $(\mathbf{D}(G), \omega_{\mathbf{D}(G)})/\!\!/\mathrm{diag}(G)$. This form extends to a symplectic form on the entire $\mathfrak{Z}_G^G$ [8, § 2.4]. The isomorphism $\eta\colon \mathfrak{Z}_G^G \xrightarrow{\sim} {}^\dagger\mathfrak{Z}^{c,c}(G)$ is a symplectomorphism.

2.4 Comparison with the Coxeter–Toda Lattice

We compare ${}^\dagger\mathfrak{Z}^{c',c}(G)$ with the construction of [38]. Throughout this section we assume G to be semisimple simply-connected. The left action of the center $Z(G)$ on $D(G)$, $\xi \cdot (g_1, g_2) = (\xi g_1, g_2)$ gives rise to the action of $Z(G)$ on ${}^\dagger\mathfrak{Z}^{c',c}(G) = M/U_- \times U_-$ where $M = \mu^{-1}(\Xi_G^{\dot{c}'} \times \mathrm{inv}(\Xi_G^{\dot{c}})) \subset D(G) = G \times G$. We consider an open subset $M \supset \overset{\bullet}{M} := (U_- \cdot T \cdot \dot{w}_0 \cdot U_- \times G) \cap M$ given by the condition that g_1 lies in the big Bruhat cell $C_{w_0} \subset G$. Clearly, $\overset{\bullet}{M} \subset M$ is $U_- \times U_-$-invariant, and we define ${}^\dagger\overset{\bullet}{\mathfrak{Z}}{}^{c',c}(G) := \overset{\bullet}{M}/U_- \times U_-$, an open subvariety of ${}^\dagger\mathfrak{Z}^{c',c}(G)$. Let $S \subset \overset{\bullet}{M}$ be given by the condition $g_1 \in T \cdot \dot{w}_0$. Then the composed projection $S \hookrightarrow \overset{\bullet}{M} \twoheadrightarrow {}^\dagger\overset{\bullet}{\mathfrak{Z}}{}^{c',c}(G)$ is an isomorphism. Moreover, the projection $\mathrm{pr}_2\colon S \to G$ is a $Z(G)$-torsor over its image $\Xi_G^{\dot{c}} \cap \mathrm{Ad}_T(\dot{w}_0 \Xi_G^{\dot{c}'} \dot{w}_0^{-1}) = \Xi_G^{\dot{c}} \cap \mathrm{Ad}_T(U \dot{w}_0 \dot{c}' \dot{w}_0^{-1} U)$. Finally, note that the composed projection

$$\Xi_G^{\dot{c}} \cap \mathrm{Ad}_T(U \dot{w}_0 \dot{c}' \dot{w}_0^{-1} U) \hookrightarrow T \cdot U_- \cdot \dot{c} \cdot U_- \cdot T \cap T \cdot U \cdot \dot{w}_0 \dot{c}' \dot{w}_0^{-1} \cdot U \cdot T \twoheadrightarrow$$
$$\twoheadrightarrow (T \cdot U_- \cdot \dot{c} \cdot U_- \cdot T \cap T \cdot U \cdot \dot{w}_0 \dot{c}' \dot{w}_0^{-1} \cdot U \cdot T)/\mathrm{Ad}_T =: G^{\dot{c}, \dot{w}_0 \dot{c}' \dot{w}_0^{-1}}/\mathrm{Ad}_T$$

is an isomorphism. But according to [38] (see also [34]), $G^{\dot{c}, \dot{w}_0 \dot{c}' \dot{w}_0^{-1}}/\mathrm{Ad}_T$ is the phase space of the Coxeter–Toda lattice. All in all, we obtain an isomorphism (respecting the symplectic structures) $Z(G)\backslash {}^\dagger\overset{\bullet}{\mathfrak{Z}}{}^{c',c} \xrightarrow{\sim} G^{\dot{c}, \dot{w}_0 \dot{c}' \dot{w}_0^{-1}}/\mathrm{Ad}_T$.

For example, for an appropriate choice of $\dot{c}, \dot{c}' \in SL(n)$, the slice S is formed by all the tridiagonal matrices of determinant 1 with 1's just above the main diagonal, and with the invertible entries just below the main diagonal (see [34, Introduction]).

We also define an open subset $S \supset \overset{\circ}{S} := \{(g_1, g_2) \in M : g_1 \in T \cdot \dot{w}_0,\ g_2 \in U_- \cdot T \cdot U\}$. It is equipped with a projection $\mathrm{pr}_1\colon \overset{\circ}{S} \to T \cdot \dot{w}_0 \xrightarrow{\sim} T$, and with another projection $\mathrm{pr}_2\colon \overset{\circ}{S} \to U_- \cdot T \cdot U \twoheadrightarrow T$. One can check that $(\mathrm{pr}_1, \mathrm{pr}_2)\colon \overset{\circ}{S} \xrightarrow{\sim} T \times T$. We define an open subvariety ${}^\dagger\mathfrak{Z}^{c',c}(G) \supset {}^\dagger\overset{\bullet}{\mathfrak{Z}}{}^{c',c}(G) \supset {}^\dagger\overset{\circ}{\mathfrak{Z}}{}^{c',c}(G)$ as the isomorphic image of $\overset{\circ}{S}$. Thus ${}^\dagger\overset{\circ}{\mathfrak{Z}}{}^{c',c}(G) \simeq T \times T$.

2.5 Trigonometric Zastava for $SL(2)$

Recall the degree n trigonometric open zastava ${}^\dagger\overset{\circ}{Z}{}^n$ for the group $SL(2)$ (see [25]). This is the moduli space of pairs of relatively prime polynomials $(Q = z^n + q_1 z^{n-1} + \ldots + q_n,\ R = r_1 z^{n-1} + r_2 z^{n-2} + \ldots + r_n)$ such that $q_n \neq 0$. We have a morphism $\zeta\colon \mathfrak{Z}^{GL(n)}_{GL(n)} \to {}^{\dagger}\mathring{Z}^n$ taking a pair $(g, x) \in \mathfrak{Z}^{GL(n)}_{GL(n)}$ to (Q, R) where Q is the characteristic polynomial of x, and R is a unique polynomial of degree less than n such that $R(x) = g$. We denote by $\mathrm{pr}\colon {}^{\dagger}\mathring{Z}^n \to (\mathbb{C}^{\times})^{(n)}$ the morphism taking (Q, R) to the set of roots of Q.

Recall that $\Sigma^{\dot{c}}_{GL(n)} = Z^0(GL(n)) \cdot \Sigma^{\dot{c}}_{SL(n)} \simeq Z(GL(n)) \times \Sigma^{\dot{c}}_{SL(n)} = \mathbb{C}^{\times} \times \Sigma^{\dot{c}}_{SL(n)}$. We denote by $\mathsf{p}\colon \mathfrak{Z}^{GL(n)}_{GL(n)} \to \mathbb{C}^{\times}$ the composed projection $\mathfrak{Z}^{GL(n)}_{GL(n)} \to \Sigma^{\dot{c}}_{GL(n)} \to \mathbb{C}^{\times}$.

Proposition 2.5 *The following square is Cartesian:*

$$\begin{array}{ccc} \mathfrak{Z}^{GL(n)}_{GL(n)} & \xrightarrow{\ \zeta\ } & {}^{\dagger}\mathring{Z}^n \\ \big\downarrow{\scriptstyle \mathsf{p}} & & {\scriptstyle q_n}\big\downarrow \\ \mathbb{C}^{\times} & \xrightarrow{\ n\ } & \mathbb{C}^{\times} \end{array}$$

Thus $\mathfrak{Z}^{GL(n)}_{GL(n)}$ is an unramified $\mathbb{Z}/n\mathbb{Z}$-cover of ${}^{\dagger}\mathring{Z}^n$.

Proof Clear from the above discussion. □

Following [1, end of chapter 2], we consider the subvariety ${}^{\dagger}\tilde{Z}^n_1 \hookrightarrow {}^{\dagger}\mathring{Z}^n$ formed by the pairs (Q, R) such that $q_n = 1$ and the resultant of Q and R, denoted Result(Q, R), equals 1. Note that we have an evident embedding $\mathfrak{Z}^{SL(n)}_{SL(n)} \hookrightarrow \mathfrak{Z}^{GL(n)}_{GL(n)}$.

Corollary 2.6 *The restriction of the morphism ζ to $\mathfrak{Z}^{SL(n)}_{SL(n)} \subset \mathfrak{Z}^{GL(n)}_{GL(n)}$ gives rise to an isomorphism $\zeta\colon \mathfrak{Z}^{SL(n)}_{SL(n)} \xrightarrow{\sim} {}^{\dagger}\tilde{Z}^n_1$.*

Proof For $(g, x) \in \mathfrak{Z}^{GL(n)}_{GL(n)}$, the inclusion $x \in SL(n)$ is equivalent to $q_n = 1$, while we claim that the inclusion $g \in SL(n)$ is equivalent to $\mathrm{Result}(Q, R) = 1$. The latter follows by combining the equalities $gx = xg$ and $g = R(x)$ with the standard equality $\mathrm{Result}(Q, R) = \prod_{i=1}^n R(\xi_i)$, where $\{\xi_i\}_{i=1}^n$ are the roots of Q. Since $\{\xi_i\}_{i=1}^n$ are the generalized eigenvalues (taken with corresponding multiplicities) of x, it is easy to see that $\{R(\xi_i)\}_{i=1}^n$ are the generalized eigenvalues of g, hence, $\det(g) = \prod_{i=1}^n R(\xi_i)$. □

For a future use we define an unramified $\mathbb{Z}/2\mathbb{Z}$-cover ${}^{\dagger}\hat{Z}^n \to {}^{\dagger}\mathring{Z}^n$ where ${}^{\dagger}\hat{Z}^n$ is the moduli space of pairs of relatively prime polynomials $(Q = q_0z^n + q_1z^{n-1} + \ldots + q_n,\ R = r_1z^{n-1} + r_2z^{n-2} + \ldots + r_n)$ such that $q_n \cdot q_0 = (-1)^n$. The projection ${}^{\dagger}\hat{Z}^n \to {}^{\dagger}\mathring{Z}^n$ takes (Q, R) to $(q_0^{-1}Q, R)$.

Finally, there are important embeddings $\Psi\colon {}^{\dagger}\mathring{Z}^n, {}^{\dagger}\hat{Z}^n \hookrightarrow SL(2, \mathbb{C}[z])$ taking (Q, R) to a unique matrix $\begin{pmatrix} Q & \tilde{R} \\ R & \tilde{Q} \end{pmatrix}$ such that $\deg \tilde{R} \leq n > \deg \tilde{Q}$, and $\tilde{R}(0) = 0$, that is $\tilde{R} = \tilde{r}_0z^n + \tilde{r}_1z^{n-1} + \ldots + \tilde{r}_{n-1}z$. Identifying ${}^{\dagger}\hat{Z}^n$ and ${}^{\dagger}\mathring{Z}^n$ with their images inside $SL(2, \mathbb{C}[z])$, the matrix multiplication gives rise to the multiplication morphisms ${}^{\dagger}\hat{Z}^k \times {}^{\dagger}\hat{Z}^l \to {}^{\dagger}\hat{Z}^{k+l}$, ${}^{\dagger}\mathring{Z}^k \times {}^{\dagger}\mathring{Z}^l \to {}^{\dagger}\mathring{Z}^{k+l}$.

3 Quantum Relativistic Open Toda and Nil-DAHA

Throughout this section (with the exception of Sect. 3.11 dealing with $G = GL(n)$) G is an almost simple simply-connected complex algebraic group.

3.1 *Root Systems and Foldings*

Let $G^\vee$ be the Langlands dual (adjoint) group with a Cartan torus $T^\vee$. We choose a Borel subgroup $B^\vee \supset T^\vee$. It defines the set of simple positive roots $\{\alpha_i,\ i \in I\}$. Let $\mathfrak{g}^\vee$ be the Lie algebra of $G^\vee$. We realize $\mathfrak{g}^\vee$ as a *folding* of a simple simply-laced Lie algebra $\mathfrak{g}'^\vee$, i.e. as invariants of an outer automorphism σ of $\mathfrak{g}'^\vee$ preserving a Cartan subalgebra $\mathfrak{t}'^\vee \subset \mathfrak{g}'^\vee$ and acting on the root system of $(\mathfrak{g}'^\vee, \mathfrak{t}'^\vee)$. In particular, σ gives rise to the same named automorphism of the Langlands dual Lie algebras $\mathfrak{g}' \supset \mathfrak{t}'$ (note that say, if $\mathfrak{g}$ is of type B_n, then $\mathfrak{g}'$ is of type A_{2n-1}, while if $\mathfrak{g}$ is of type C_n, then $\mathfrak{g}'$ is of type D_{n+1}; in particular, $\mathfrak{g} \not\subset \mathfrak{g}'$). We choose a σ-invariant Borel subalgebra $\mathfrak{t}' \subset \mathfrak{b}' \subset \mathfrak{g}'$ such that $\mathfrak{b} = (\mathfrak{b}')^\sigma$. The corresponding set of simple roots is denoted by I'. We denote by Ξ the finite cyclic group generated by σ, and $d := |\Xi|$. Let $G' \supset T'$ denote the corresponding simply-connected Lie group and its Cartan torus. The *coinvariants* $X_*(T')_\sigma$ of σ on the coroot lattice $X_*(T')$ of $(\mathfrak{g}', \mathfrak{t}')$ coincide with the root lattice of $\mathfrak{g}^\vee$. We have an injective map $a: X_*(T')_\sigma \to X_*(T')^\sigma$ from coinvariants to invariants defined as follows: given a coinvariant α with a representative $\tilde{\alpha} \in X_*(T')$ we set $a(\alpha) := \sum_{\xi \in \Xi} \xi(\tilde{\alpha})$.

To compare with the notations of [36, § 4.4, Remark 4.5], we are in the symmetric case with $Q'_0 = Y := X^*(T^\vee) = X_*(T) = X_*(T')_\sigma$, and $Q_0 \subset X := X^*(T')_\sigma$ generated by the classes of simple roots of $T' \subset B' \subset G'$. Note that Q'_0 is generated by the classes of simple coroots of $T' \subset B' \subset G'$, and we have a canonical identification $Q_0 = Q'_0$ sending a coroot $\tilde{\alpha}$ to the corresponding root $\tilde{\alpha}^\vee$. The Weyl group W of $G \supset T$ coincides with the invariants $(W')^\sigma$ of σ on the Weyl group W' of $G' \supset T'$ (our W is denoted W_0 in [36]). The W-invariant pairing $X \times Y \to \mathbb{Q}$ defined in [36, § 4.4] is actually integer valued: $X \times Y \to \mathbb{Z}$, so that $m = 1$ (notations of loc. cit.). To compare with notations of [13, Section 1], $P := X,\ Q := Q_0$, and the natural pairing $P \times P \to \mathbb{Q}$ gives rise to the embedding $Q = Y \hookrightarrow P$. We will also need an extended lattice $Y_{\mathrm{ad}} := X_*(T_{\mathrm{ad}}) = X_*(T'_{\mathrm{ad}})_\sigma \supset Y$. Note that $\Pi := Y_{\mathrm{ad}}/Y = (X_*(T'_{\mathrm{ad}})/X_*(T'))_\sigma$. Also note that the above W-invariant identification $Q_0 = Q'_0$ extends to the W-invariant identification $Q_0 \subset X = Y_{\mathrm{ad}} \supset Q'_0$. The extended pairing $X \times Y_{\mathrm{ad}} \to \mathbb{Q}$ is no more integer valued in general, and we denote by m_{ad} the maximal denominator appearing in the values of this pairing. Finally, $R \subset X$ stands for the set of roots.

3.2 Affine Flags

We fix a primitive root of unity ζ of order $d = \mathrm{ord}(\sigma)$. We set $\mathcal{K} = \mathbb{C}((\mathsf{t})) \supset \mathcal{O} = \mathbb{C}[[\mathsf{t}]]$. The group ind-scheme $G'(\mathcal{K})$ is equipped with an automorphism ς defined as the composition of two automorphisms: a) σ on G'; b) $\mathsf{t} \mapsto \zeta\mathsf{t}$. This automorphism preserves the Iwahori subgroup $\mathbf{I}' \subset G'(\mathcal{K})$. We denote by $\mathcal{F}\ell$ the twisted affine flag space $G'(\mathcal{K})^\varsigma/(\mathbf{I}')^\varsigma$: an ind-proper ind-scheme of ind-finite type, see [55]. We denote by $\mathfrak{u} \subset \mathrm{Lie}(\mathbf{I}')^\varsigma$ its pronilpotent radical. The trivial (Tate) bundle $\underline{\mathfrak{g}'(\mathcal{K})^\varsigma}$ with the fiber $\mathfrak{g}'(\mathcal{K})^\varsigma$ over $\mathcal{F}\ell$ has a structure of an ind-scheme. It contains a profinite dimensional vector subbundle $\underline{\mathfrak{u}}$ whose fiber over a point $b \in \mathcal{F}\ell$ represented by a compact subalgebra in $\mathfrak{g}'(\mathcal{K})^\varsigma$ is the pronilpotent radical of this subalgebra. The trivial vector bundle $\underline{\mathfrak{g}'(\mathcal{K})^\varsigma}$ also contains a trivial vector subbundle $\mathfrak{u} \times \mathcal{F}\ell$. We will call $\underline{\mathfrak{u}}$ the cotangent bundle of $\mathcal{F}\ell$, and we will call the intersection $\mathbf{\Lambda} := \underline{\mathfrak{u}} \cap (\mathfrak{u} \times \mathcal{F}\ell)$ the affine Steinberg variety.

To simplify the notations we will write $\mathbf{I}$ for $(\mathbf{I}')^\varsigma$, and $\mathbf{K}$ for $G'(\mathcal{O})^\varsigma$. The convolution product on the complexified equivariant coherent K-theory $K^{\mathbb{C}^\times\times\mathbf{I}\rtimes\mathbb{C}^\times}(\mathbf{\Lambda})$ is defined as in [9, Remark 3.9(3)] (cf. [8, § 7.1] and [64, § 2.2, 2.3]). Here the first copy of $\mathbb{C}^\times$ acts by dilations in fibers of $\underline{\mathfrak{u}}$, while the second one acts by loop rotations, and $K_{\mathbb{C}^\times\times\mathbb{C}^\times}(\mathrm{pt}) = \mathbb{C}[t^{\pm 1}, q^{\pm 1}]$.

3.3 DAHA, Symmetric Case

Following [36, § 4.6], we set $\widetilde{X} := X \oplus \mathbb{Z}\delta = X^*(T')_\sigma \oplus \mathbb{Z}\delta$. This is the group of characters of $\mathbf{I} \rtimes \mathbb{C}^\times$. Note that the Picard group $\mathrm{Pic}(\mathcal{F}\ell)$ is canonically isomorphic to $X \oplus \mathbb{Z}\omega_0$. The $\mathbf{I}$-orbits on $\mathcal{F}\ell$ are parametrized by the affine Weyl group $W_a \simeq Y \rtimes W = X_*(T')_\sigma \rtimes W$. We denote by $\mathbf{\Lambda}_e \simeq \mathfrak{u}$ the closed subscheme of $\mathbf{\Lambda}$: the preimage of the one-point $\mathbf{I}$-orbit $\mathcal{F}\ell_e \subset \mathcal{F}\ell$. For $\tilde{\lambda} = (\check{\lambda}, k) \in \widetilde{X}$ we denote by $\mathcal{O}_{\mathbf{\Lambda}_e}\langle\tilde{\lambda}\rangle \in K^{\mathbb{C}^\times\times\mathbf{I}\rtimes\mathbb{C}^\times}(\mathbf{\Lambda})$ the (class of the) direct image of the structure sheaf of $\mathbf{\Lambda}_e$ twisted by the character $\tilde{\lambda}$ of $\mathbf{I} \rtimes \mathbb{C}^\times$. Let $\widetilde{I} \subset W_a$ be the set of one-dimensional $\mathbf{I}$-orbits on $\mathcal{F}\ell$. For $i \in \widetilde{I}$ we denote by $\mathcal{F}\ell_i$ the corresponding orbit, and by $\overline{\mathcal{F}\ell_i}$ its closure, isomorphic to a projective line. We denote by $\mathbf{\Lambda}_i \subset \mathbf{\Lambda}$ the closed subscheme of $\mathbf{\Lambda}$: the closure of the preimage of $\mathcal{F}\ell_i$. We denote by $\omega_{\mathbf{\Lambda}_i}$ the (class of the) direct image (wrt the closed embedding $\mathbf{\Lambda}_i \hookrightarrow \mathbf{\Lambda}$) of the inverse image (wrt the smooth projection $\mathbf{\Lambda}_i \to \overline{\mathcal{F}\ell_i}$) of the canonical line bundle on $\overline{\mathcal{F}\ell_i} \simeq \mathbb{P}^1$ equipped with the natural $\mathbb{C}^\times \times \mathbf{I} \rtimes \mathbb{C}^\times$-equivariant structure. Finally, we set $\mathsf{T}_i := -1 - t\omega_{\mathbf{\Lambda}_i} \in K^{\mathbb{C}^\times\times\mathbf{I}\rtimes\mathbb{C}^\times}(\mathbf{\Lambda})$.

Definition 3.1 (Cf. [36, Definition 5.6]) The *double affine Hecke algebra (DAHA)* $\mathcal{H}(W_a, \widetilde{X})$ is the $\mathbb{C}[q^{\pm 1}, t^{\pm 1}]$-algebra generated by $\{\mathsf{X}_{\tilde{\lambda}}, T_w | \tilde{\lambda} \in \widetilde{X}, w \in W_a\}$ with the following defining relations:

(a) T_w's satisfy the braid relations of W_a;
(b) $\mathsf{X}_{\tilde{\lambda}}\mathsf{X}_{\tilde{\mu}} = \mathsf{X}_{\tilde{\lambda}+\tilde{\mu}}$, and $\mathsf{X}_\delta = q$;

(c) $(T_i - t)(T_i + 1) = 0$ for $i \in \widetilde{I}$, where we set $T_i = T_{s_i}$;
(d) $\mathsf{X}_{\tilde{\lambda}} T_i - T_i \mathsf{X}_{\tilde{\lambda}-r\check{\alpha}_i} = (t-1)\mathsf{X}_{\tilde{\lambda}}(1 + \mathsf{X}_{-\check{\alpha}_i} + \ldots + \mathsf{X}^{r-1}_{-\check{\alpha}_i})$ where $\langle \tilde{\lambda}, \alpha_i \rangle = r \geq 0$.

Theorem 3.2 *There is a unique isomorphism* $\Phi\colon \mathcal{H}(W_a, \widetilde{X}) \xrightarrow{\sim} K^{\mathbb{C}^\times \times \mathbf{I} \rtimes \mathbb{C}^\times}(\mathbf{\Lambda})$ *such that* $\Phi(\mathsf{X}_{\tilde{\lambda}}) = \mathcal{O}_{\mathbf{\Lambda}_e}\langle \tilde{\lambda} \rangle$, *and* $\Phi(T_i) = \mathsf{T}_i$, *for any* $i \in \widetilde{I}$.

Proof Same as the one of [64, Theorem 2.5.6]. □

3.4 Nil-DAHA, Symmetric Case

The complexified equivariant K-theory $K^{\mathbf{I} \rtimes \mathbb{C}^\times}(\mathcal{F}\ell)$ forms a $\mathbb{C}[q^{\pm 1}]$-algebra with respect to the convolution. We denote by $\mathcal{O}_{\mathcal{F}\ell_e}\langle \tilde{\lambda} \rangle$ the (class of the) structure sheaf of the point orbit $\mathcal{F}\ell_e \in \mathcal{F}\ell$ twisted by a character $\tilde{\lambda} \in \widetilde{X}$. We denote by $\omega_{\overline{\mathcal{F}\ell}_i}$ the (class of the) direct image (wrt the closed embedding $\overline{\mathcal{F}\ell}_i \hookrightarrow \mathcal{F}\ell$) of the canonical line bundle on $\overline{\mathcal{F}\ell}_i$ equipped with the natural $\mathbf{I} \rtimes \mathbb{C}^\times$-equivariant structure. We set $\mathbf{T}_i := -1 - \omega_{\overline{\mathcal{F}\ell}_i} \in K^{\mathbf{I} \rtimes \mathbb{C}^\times}(\mathcal{F}\ell)$.

Definition 3.3 (Cf. [13, § 1.1]) The *nil-DAHA* $\mathcal{HH}(W_a, \widetilde{X})$ is the $\mathbb{C}[q^{\pm 1}]$-algebra generated by $\{\mathsf{X}_{\tilde{\lambda}}, \mathcal{T}_w | \tilde{\lambda} \in \widetilde{X}, w \in W_a\}$ with the following defining relations:

(a) $\mathcal{T}_w$'s satisfy the braid relations of W_a;
(b) $\mathsf{X}_{\tilde{\lambda}} \mathsf{X}_{\tilde{\mu}} = \mathsf{X}_{\tilde{\lambda}+\tilde{\mu}}$, and $\mathsf{X}_\delta = q$;
(c) $\mathcal{T}_i(\mathcal{T}_i + 1) = 0$ for $i \in \widetilde{I}$, where we set $\mathcal{T}_i = \mathcal{T}_{s_i}$;
(d) $\mathsf{X}_{\tilde{\lambda}} \mathcal{T}_i - \mathcal{T}_i \mathsf{X}_{\tilde{\lambda}-r\check{\alpha}_i} = -\mathsf{X}_{\tilde{\lambda}}(1 + \mathsf{X}_{-\check{\alpha}_i} + \ldots + \mathsf{X}^{r-1}_{-\check{\alpha}_i})$ where $\langle \tilde{\lambda}, \alpha_i \rangle = r \geq 0$.

Theorem 3.4 *There is a unique isomorphism* $\Phi\colon \mathcal{HH}(W_a, \widetilde{X}) \xrightarrow{\sim} K^{\mathbf{I} \rtimes \mathbb{C}^\times}(\mathcal{F}\ell)$ *such that* $\Phi(\mathsf{X}_{\tilde{\lambda}}) = \mathcal{O}_{\mathcal{F}\ell_e}\langle \tilde{\lambda} \rangle$, *and* $\Phi(\mathcal{T}_i) = \mathbf{T}_i$, *for any* $i \in \widetilde{I}$.

Proof Same as the one of [64, Theorem 2.5.6]. □

3.5 Extended Nil-DAHA

We consider the $2m_{\mathrm{ad}}$-fold cover $\widetilde{\mathbb{C}}^\times \to \mathbb{C}^\times$ of the loop rotation group (see the end of Sect. 3.1), and set $\widehat{\mathbf{I}} := \mathbf{I} \rtimes \widetilde{\mathbb{C}}^\times$. The group of characters of $T \times \widetilde{\mathbb{C}}^\times$ is $\widehat{X} := X \oplus \mathbb{Z}\frac{1}{2m_{\mathrm{ad}}}\delta$. The extended affine Weyl group is $W_e = Y_{\mathrm{ad}} \rtimes W = W_a \rtimes \Pi$. The extended nil-DAHA $\mathcal{HH}(W_e, \widehat{X})$ is the (extended) semidirect product $(\mathcal{HH}(W_a, \widehat{X}) \rtimes \Pi) \otimes_{\mathbb{C}[q^{\pm 1}]} \mathbb{C}[q^{\frac{\pm 1}{2m_{\mathrm{ad}}}}]$. That is, it has generators $\mathsf{X}_{\hat{\lambda}}$, $\hat{\lambda} \in \widehat{X}$, and $\mathcal{T}_i$, $i \in \widetilde{I}$, and $\pi \in \Pi$; with additional relations $\pi \mathcal{T}_i \pi^{-1} = \mathcal{T}_{\pi(i)}$, and $\pi \mathsf{X}_{\hat{\lambda}} \pi^{-1} = \mathsf{X}_{\pi(\hat{\lambda})}$.

Remark 3.5 The definition of [13, § 1.1] is equivalent to our Sect. 3.5: the generators T_i of loc. cit. correspond to $-\mathcal{T}_i - 1$; geometrically, $T_i = [\omega_{\overline{\mathcal{F}\ell}_i}]$.

3.6 Residue Construction

Let $\mathcal{A} := \mathbb{C}[q^{\frac{\pm 1}{2m_{\mathrm{ad}}}}]$, and $\mathcal{Q} := \mathbb{C}(q^{\frac{1}{2m_{\mathrm{ad}}}})$. Let $\mathcal{O}_q(T \times T)$ be an $\mathcal{A}$-algebra with generators $[\lambda, \mu]$, $\lambda, \mu \in X$, and relations $[\lambda, \mu] \cdot [\lambda', \mu'] = q^{\frac{(\mu,\lambda')-(\mu',\lambda)}{2}} [\lambda + \lambda', \mu + \mu']$. This is the subalgebra of endomorphisms of $\mathcal{A}[T]$ generated by multiplications by X_λ, $\lambda \in X$, and q-shift operators $D_q^\mu f(t) := f(q^\mu t)$ where we view q^μ as a homomorphism $\widetilde{\mathbb{C}}^\times \to T$. In other words, $D_q^\mu \mathsf{X}_\lambda = q^{(\mu,\lambda)}\mathsf{X}_\lambda$. We may and will view $\mathcal{O}_q(T \times T)$ as a subalgebra of endomorphisms of the field of rational functions $\mathcal{Q}(T)$ as well. It embeds into the subalgebra $\mathbb{C}_q(T \times T) \subset \mathrm{End}(\mathcal{Q}(T))$ generated by D_q^μ, $\mu \in X$, and multiplications by $f \in \mathcal{Q}(T)$. We consider the semidirect product $\mathbb{C}_q(T \times T) \rtimes \mathbb{C}[W]$ with respect to the diagonal action of W on $T \times T$. Inside we consider the linear subspace $\mathcal{HH}_{\mathrm{res}}(W_e, \widehat{X})$ formed by the finite sums $\sum_{w\in W}^{\mu\in X} h_{w,\mu} D_q^\mu \cdot [w]$, $h_{w,\mu} \in \mathcal{Q}(T)$, satisfying the following conditions:

(a) $h_{w,\mu}$ is regular except at the divisors $T_{\alpha,q^k} := \{t : \alpha(t) = q^k\}$, $\alpha \in R$, $k \in \mathbb{Z}$, where they are allowed to have only first order poles.
(b) $\mathrm{Res}_{T_{\alpha,q^{-k}}}(h_{w,\mu}) + \mathrm{Res}_{T_{\alpha,q^{-k}}}(h_{s_\alpha w, k\alpha + s_\alpha \mu}) = 0$ for any $\alpha \in R$.

The algebra of regular functions $\mathbb{C}[T \times \widetilde{\mathbb{C}}^\times]$ is embedded into $\mathcal{HH}_{\mathrm{res}}(W_e, \widehat{X})$ via the assignment $f \mapsto f \cdot [1]$. Furthermore, for $i \in I \subset \widetilde{I}$, we consider the *Demazure operator* [13, § 1.3] $\tau_i := \frac{1}{1-\mathsf{X}_{\alpha_i}} \cdot ([s_i] - [1]) \in \mathcal{HH}_{\mathrm{res}}(W_e, \widehat{X})$, and for $i_0 \in \widetilde{I} \setminus I$ we consider the *Demazure operator* [13, § 1.3] $\tau_{i_0} := \frac{1}{1-q\mathsf{X}_\theta^{-1}} \cdot ([s_\theta] \cdot D_q^\theta - [1]) \in \mathcal{HH}_{\mathrm{res}}(W_e, \widehat{X})$, where $\theta \in R$ is the dominant short root, $(\theta, \theta) = 2$.

Theorem 3.6

(a) *$\mathcal{HH}_{\mathrm{res}}(W_e, \widehat{X})$ is a subalgebra of $\mathbb{C}_q(T \times T) \rtimes \mathbb{C}[W]$.*
(b) *The assignment $f \mapsto f \cdot [1]$; $\mathcal{T}_i \mapsto \tau_i$, $i \in \widetilde{I}$; $\Pi \ni \pi \mapsto$ the corresponding automorphism of $\mathcal{Q}(T) = \mathbb{Q}(\widehat{X} \otimes \mathbb{C}^\times)$ (arising from the automorphism of the extended Dynkin diagram), defines an isomorphism $\varphi \colon \mathcal{HH}(W_e, \widehat{X}) \xrightarrow{\sim} \mathcal{HH}_{\mathrm{res}}(W_e, \widehat{X})$.*

Proof Same as the one of [5, Theorem 7.2]. □

Remark 3.7 Nil-DAHA $\mathcal{HH}(W_e, \widehat{X})$ is *not* isomorphic to the degeneration $\ddot{\mathsf{H}}|_{v=0}$ of [5, Section 6].

3.7 K-theory of Disconnected Flags

We define $\mathbf{I}_{\mathrm{ad}}$ as the image of $\mathbf{I}$ in $G'_{\mathrm{ad}}(\mathcal{K})^\varsigma$, and we consider the adjoint version of the affine flags $\mathcal{F}\ell_{\mathrm{ad}} := G'_{\mathrm{ad}}(\mathcal{K})^\varsigma / \mathbf{I}_{\mathrm{ad}}$. This is an ind-scheme having $|\Pi|$ connected components, each one isomorphic to $\mathcal{F}\ell$. The isomorphism of Theorem 3.4 extends to the same named isomorphism $\mathcal{HH}(W_e, \widehat{X}) \xrightarrow{\sim} K^{\mathbf{I}}(\mathcal{F}\ell_{\mathrm{ad}})$. Let us explain why the

RHS forms an algebra. We consider an algebra $K(\widehat{\mathbf{I}}\backslash G'_{\mathrm{ad}}(\mathcal{K})^{\varsigma}/\widehat{\mathbf{I}}) = K^{\widehat{\mathbf{I}}}(\mathcal{F}\ell_{\mathrm{ad}}/\Pi)$. Here we view $\Pi = Z(G'^{\sigma})$ as the center of the simply-connected group G'^{σ} acting trivially on $\mathcal{F}\ell_{\mathrm{ad}}$. Now $K^{\widehat{\mathbf{I}}}(\mathcal{F}\ell_{\mathrm{ad}}/\Pi)$ contains a subalgebra $K^{\widehat{\mathbf{I}}}(\mathcal{F}\ell_{\mathrm{ad}}/\Pi)_{\mathrm{diag}}$ formed by the classes of bi-equivariant sheaves on $\mathcal{F}\ell_{\mathrm{ad}}$ such that the Π-equivariance coincides with the $Z(G'^{\sigma})$-equivariance obtained by the restriction of $\widehat{\mathbf{I}}$-equivariance. Finally, $K^{\widehat{\mathbf{I}}}(\mathcal{F}\ell_{\mathrm{ad}}/\Pi)_{\mathrm{diag}} \simeq K^{\widehat{\mathbf{I}}}(\mathcal{F}\ell_{\mathrm{ad}})$.

3.8 Spherical Nil-DAHA

We define the new generators $\hat{\mathcal{T}}_i := -\mathcal{T}_i - 1,\ i \in \widetilde{I}$ (they correspond to the generators T_i of [13, Definition 1.1]). Geometrically, $\hat{\mathcal{T}}_i = [\omega_{\overline{\mathcal{F}\ell_i}}]$. They still satisfy the braid relations of W_a. So for any $w \in W_a$ we have a well-defined element (product of the generators) $\hat{\mathcal{T}}_w$. We also define $\hat{\mathcal{T}}'_i := \hat{\mathcal{T}}_i + 1 = -\mathcal{T}_i,\ i \in \widetilde{I}$. Geometrically, for $i \in I \subset \widetilde{I}$, we have $\hat{\mathcal{T}}'_i = \mathsf{X}_{\rho^\vee}[\mathcal{O}_{\overline{\mathcal{F}\ell_i}}]\mathsf{X}_{\rho^\vee}^{-1}$. These generators also satisfy the braid relations of W_a, so for any $w \in W_a$ we have a well-defined element (product of the generators) $\hat{\mathcal{T}}'_w$.

Given a reduced decomposition $w = s_{i_1} \cdots s_{i_l}$ we have for the class of the structure sheaf of the Schubert variety $[\mathcal{O}_{\overline{\mathcal{F}\ell_w}}] = [\mathcal{O}_{\overline{\mathcal{F}\ell_{i_1}}}] \cdots [\mathcal{O}_{\overline{\mathcal{F}\ell_{i_l}}}]$ since $\overline{\mathcal{F}\ell_w}$ has rational singularities. Hence, for $w \in W \subset W_a$, we have $[\mathcal{O}_{\overline{\mathcal{F}\ell_w}}] = \mathsf{X}_{\rho^\vee}^{-1}\hat{\mathcal{T}}'_w\mathsf{X}_{\rho^\vee}$. In particular, for the longest element $w_0 \in W$ we set $\mathbf{e} := [\mathcal{O}_{\overline{\mathcal{F}\ell_{w_0}}}] = \mathsf{X}_{\rho^\vee}^{-1}\hat{\mathcal{T}}'_{w_0}\mathsf{X}_{\rho^\vee}$, an idempotent in $\mathcal{H}\mathcal{H}(W_e, \widehat{X})$. Indeed, calculating $[\mathcal{O}_{\overline{\mathcal{F}\ell_{w_0}}}][\mathcal{O}_{\overline{\mathcal{F}\ell_{w_0}}}]$ as the pushforward of the structure sheaf from the convolution diagram $\overline{\mathcal{F}\ell_{w_0}}\widetilde{\times}\overline{\mathcal{F}\ell_{w_0}} \to \overline{\mathcal{F}\ell_{w_0}}$ we get $\mathcal{O}_{\overline{\mathcal{F}\ell_{w_0}}}$ since $R\Gamma(\overline{\mathcal{F}\ell_{w_0}}, \mathcal{O}_{\overline{\mathcal{F}\ell_{w_0}}}) = \mathbb{C}$.

We define the spherical nil-DAHA $\mathcal{H}\mathcal{H}^{\mathrm{sph}}(W_a, \widetilde{X}) := \mathbf{e}\mathcal{H}\mathcal{H}(W_a, \widetilde{X})\mathbf{e}$, and the spherical extended nil-DAHA $\mathcal{H}\mathcal{H}^{\mathrm{sph}}(W_e, \widehat{X}) := \mathbf{e}\mathcal{H}\mathcal{H}(W_e, \widehat{X})\mathbf{e}$.

3.9 Equivariant K-theory of the Affine Grassmannian

We denote by $\mathrm{Gr}_{\mathrm{ad}}$ the twisted affine Grassmannian $G'_{\mathrm{ad}}(\mathcal{K})^{\varsigma}/G'_{\mathrm{ad}}(\mathcal{O})^{\varsigma}$: an ind-proper ind-scheme of ind-finite type, see [55]. The complexified equivariant coherent K-theory $K^{\mathbf{K}\rtimes\widetilde{\mathbb{C}}^\times}(\mathrm{Gr}_{\mathrm{ad}}) = K^{G'(\mathcal{O})^{\varsigma}\rtimes\widetilde{\mathbb{C}}^\times}(\mathrm{Gr}_{\mathrm{ad}})$ forms a $\mathbb{C}[q^{\pm\frac{1}{2m_{\mathrm{ad}}}}]$-algebra with respect to the convolution (see Sect. 3.7). We have the smooth projection $p\colon \mathcal{F}\ell_{\mathrm{ad}} \to \mathrm{Gr}_{\mathrm{ad}}$, and the natural embedding $K^{\mathbf{K}\rtimes\widetilde{\mathbb{C}}^\times}(\mathrm{Gr}_{\mathrm{ad}}) \hookrightarrow K^{\mathbf{I}\rtimes\widetilde{\mathbb{C}}^\times}(\mathrm{Gr}_{\mathrm{ad}}) \overset{p^*}{\hookrightarrow} K^{\mathbf{I}\rtimes\widetilde{\mathbb{C}}^\times}(\mathcal{F}\ell_{\mathrm{ad}})$.

Corollary 3.8 *The isomorphism Φ of Sect. 3.7 takes the spherical subalgebra $\mathcal{HH}^{\mathrm{sph}}(W_e, \widehat{X}) \subset \mathcal{HH}(W_e, \widehat{X})$ isomorphically onto $K^{\mathbf{K}\rtimes\widetilde{\mathbb{C}}^\times}(\mathrm{Gr}_{\mathrm{ad}}) \subset K^{\mathbf{I}\rtimes\widetilde{\mathbb{C}}^\times}(\mathcal{F}\ell_{\mathrm{ad}})$. The right ideal $\mathbf{e}\mathcal{HH}(W_e, \widehat{X})$ corresponds to $K^{\mathbf{K}\rtimes\mathbb{C}^\times}(\mathcal{F}\ell_{\mathrm{ad}}) = (K^{\mathbf{I}\rtimes\widetilde{\mathbb{C}}^\times}(\mathcal{F}\ell_{\mathrm{ad}}))^W \subset K^{\mathbf{I}\rtimes\widetilde{\mathbb{C}}^\times}(\mathcal{F}\ell_{\mathrm{ad}})$.* □

3.10 Classical Limit

The following theorem is proved similarly to [8, Theorem 2.15]:

Theorem 3.9

(a) *The algebra $K^{\mathbf{K}}(\mathrm{Gr}_{\mathrm{ad}})$ is commutative.*
(b) *Its spectrum together with the projection onto T/W is naturally isomorphic to $\mathfrak{Z}^G_G \xrightarrow{\mathrm{pr}} T/W$.*
(c) *The Poisson structure on $K^{\mathbf{K}}(\mathrm{Gr}_{\mathrm{ad}})$ arising from the deformation $K^{\mathbf{K}\rtimes\mathbb{C}^\times}(\mathrm{Gr}_{\mathrm{ad}})$ corresponds under the above identification to the Poisson (symplectic) structure of Remark 2.4 on $\mathfrak{Z}^G_G$.* □

Corollary 3.10

(a) *The algebra $\mathcal{HH}^{\mathrm{sph}}(W_e, \widehat{X})|_{q=1}$ is commutative.*
(b) *This algebra with the subalgebra $\mathbb{C}[X]^W$ is naturally isomorphic to $\mathbb{C}[\mathfrak{Z}^G_G] \supset \mathbb{C}[T/W]$.*
(c) *The Poisson structure on $\mathcal{HH}^{\mathrm{sph}}(W_e, \widehat{X})|_{q=1}$ arising from the deformation $\mathcal{HH}^{\mathrm{sph}}(W_e, \widehat{X})$ corresponds under the above identification to the Poisson (symplectic) structure of Remark 2.4 on $\mathfrak{Z}^G_G$.* □

3.11 Nil-DAHA, General Linear Group

In case $G = GL(n) \simeq G^\vee$, the general definition of $\mathcal{HH}(W_e, \widehat{X})$ takes a particularly explicit form.

Definition 3.11 The nil-DAHA $\mathcal{HH}(GL(n))$ is the $\mathbb{C}[q^{\pm 1}]$-algebra with generators $\mathcal{T}_0, \ldots, \mathcal{T}_{n-1}, \mathsf{X}_1^{\pm 1}, \ldots, \mathsf{X}_n^{\pm 1}, \pi^{\pm 1}$, and the following relations:

(a) $\mathcal{T}_i$'s for $i \in \mathbb{Z}/n\mathbb{Z}$ satisfy the braid relations of the affine braid group of type $\tilde{A}_{n-1}$;
(b) $\mathsf{X}_i^{\pm 1}$, $i = 1, \ldots, n$, all commute;
(c) $\mathcal{T}_i(\mathcal{T}_i + 1) = 0$ for $i \in \mathbb{Z}/n\mathbb{Z}$;
(d) $\pi \mathsf{X}_i \pi^{-1} = \mathsf{X}_{i+1}$ for $i = 1, \ldots, n-1$, and $\pi \mathsf{X}_n \pi^{-1} = q\mathsf{X}_1$;
(e) $\pi \mathcal{T}_i \pi^{-1} = \mathcal{T}_{i+1}$ for $i \in \mathbb{Z}/n\mathbb{Z}$;

(f) $\mathsf{X}_{i+1}\mathcal{T}_i - \mathcal{T}_i\mathsf{X}_i = \mathsf{X}_i$, and $\mathsf{X}_i^{-1}\mathcal{T}_i - \mathcal{T}_i\mathsf{X}_{i+1}^{-1} = \mathsf{X}_{i+1}^{-1}$ for $i = 1, \ldots, n-1$;
(h) $q\mathsf{X}_1\mathcal{T}_0 - \mathcal{T}_0\mathsf{X}_n = \mathsf{X}_n$, and $q\mathsf{X}_n^{-1}\mathcal{T}_0 - \mathcal{T}_0\mathsf{X}_1^{-1} = \mathsf{X}_1^{-1}$;
(fh) $\mathsf{X}_i^{\pm 1}$ and $\mathcal{T}_j$ commute for all the pairs i, j not listed in (f,h) above.

Note that $\mathsf{X} := \mathsf{X}_1 \cdots \mathsf{X}_n$ commutes with all the $\mathcal{T}_i$'s, while $\pi\mathsf{X}\pi^{-1} = q\mathsf{X}$. For a future use we give the following

Definition 3.12 The extended nil-DAHA $\mathcal{HH}_e(GL(n))$ is the $\mathbb{C}[\boldsymbol{v}^{\pm 1}]$-algebra, $q = \boldsymbol{v}^2$, with generators $\mathcal{T}_0, \ldots, \mathcal{T}_{n-1}$, $\mathsf{X}_1^{\pm 1}, \ldots, \mathsf{X}_n^{\pm 1}$, $\pi^{\pm 1}$, $\sqrt{\mathsf{X}}^{\pm 1}$, and relations (a–fh) of Definition 3.11 plus

(i) $(\sqrt{\mathsf{X}}^{\pm 1})^2 = \mathsf{X}^{\pm 1} := \mathsf{X}_1^{\pm 1} \cdots \mathsf{X}_n^{\pm 1}$;
(j) $\sqrt{\mathsf{X}}^{\pm 1}$ commutes with all the $\mathsf{X}_i^{\pm 1}$ and all the $\mathcal{T}_i$;
(k) $\pi\sqrt{\mathsf{X}}\pi^{-1} = \boldsymbol{v}\sqrt{\mathsf{X}}$.

We interpret X_i, $i = 1, \ldots, n$, as the i-th diagonal matrix entry character of the diagonal torus $T \subset GL(n)$. It gives rise to the same named character of the Iwahori subgroup $\mathbf{I} \subset GL(n, \mathcal{K})$. We denote by $\mathcal{O}_{\mathcal{F}\ell_e}\langle \mathsf{X}_i \rangle$ the (class of the) structure sheaf of the point orbit $\mathcal{F}\ell_e \subset \mathcal{F}\ell = \mathcal{F}\ell_{GL(n)}$ (the affine flag variety of $GL(n)$) twisted by the character X_i. We denote by $\omega_{\overline{\mathcal{F}\ell_i}}$, $i = 0, \ldots, n-1$, the (class of the) direct image (wrt the closed embedding $\overline{\mathcal{F}\ell_i} \hookrightarrow \mathcal{F}\ell_{SL(n)} \hookrightarrow \mathcal{F}\ell_{GL(n)}$) of the canonical line bundle on $\overline{\mathcal{F}\ell_i}$ equipped with the natural $\mathbf{I} \rtimes \mathbb{C}^\times$-equivariant structure. We set $\mathbf{T}_i := -1 - \omega_{\overline{\mathcal{F}\ell_i}} \in K^{\mathbf{I} \rtimes \mathbb{C}^\times}(\mathcal{F}\ell)$ as in Sect. 3.4. Finally, note that the fixed point set $\mathcal{F}\ell^T$ is naturally identified with the extended affine Weyl group of $GL(n)$, that is the group of n-periodic permutations of $\mathbb{Z}$: $\sigma(k+n) = \sigma(k) + n$, and the fixed point ϖ corresponding to the shift permutation $\sigma(k) = k+1$ is a point $\mathbf{I} \rtimes \mathbb{C}^\times$-orbit $\mathcal{F}\ell_\varpi$. We denote by $\varpi \in K^{\mathbf{I} \rtimes \mathbb{C}^\times}(\mathcal{F}\ell)$ the class of the structure sheaf $\mathcal{O}_{\mathcal{F}\ell_\varpi}$.

Theorem 3.13 *There is a unique isomorphism* $\Phi\colon \mathcal{HH}(GL(n)) \xrightarrow{\sim} K^{\mathbf{I} \rtimes \mathbb{C}^\times}(\mathcal{F}\ell)$ *such that* $\Phi(\mathsf{X}_i) = \mathcal{O}_{\mathcal{F}\ell_e}\langle \mathsf{X}_i \rangle$, $i = 1, \ldots, n$, *and* $\Phi(\mathcal{T}_i) = \mathbf{T}_i$, $i = 0, \ldots, n-1$, *and* $\Phi(\pi) = \varpi$.

Proof Same as the one of [64, Theorem 2.5.6]. □

As in Sect. 3.8, we have an idempotent $\mathbf{e} = [\mathcal{O}_{\overline{\mathcal{F}\ell_{w_0}}}] \in K^{\mathbf{I} \rtimes \mathbb{C}^\times}(\mathcal{F}\ell_{SL(n)}) \subset K^{\mathbf{I} \rtimes \mathbb{C}^\times}(\mathcal{F}\ell) \simeq \mathcal{HH}(GL(n))$, and we define the spherical subalgebras $\mathcal{HH}^{\mathrm{sph}}(GL(n)) := \mathbf{e}\mathcal{HH}(GL(n))\mathbf{e}$, and $\mathcal{HH}_e^{\mathrm{sph}}(GL(n)) := \mathbf{e}\mathcal{HH}_e(GL(n))\mathbf{e}$. We also define a two-fold cover $\tilde{G} := \{(g \in GL(n),\ y \in \mathbb{C}^\times) : \det(g) = y^2\} \twoheadrightarrow G$, $\mathbf{K} := GL(n, \mathcal{O})$, $\tilde{\mathbf{K}} := \tilde{G}(\mathcal{O})$, and finally $\widetilde{\mathbb{C}}^\times$ as the two-fold cover (with coordinate $\boldsymbol{v}$) of $\mathbb{C}^\times$ (with coordinate q).

Corollary 3.14 *The isomorphism* Φ *of Theorem 3.13 takes the spherical subalgebra* $\mathcal{HH}^{\mathrm{sph}}(GL(n)) \subset \mathcal{HH}(GL(n))$ *isomorphically onto* $K^{\mathbf{K} \rtimes \mathbb{C}^\times}(\mathrm{Gr}_{GL(n)}) \subset K^{\mathbf{I} \rtimes \mathbb{C}^\times}(\mathcal{F}\ell_{GL(n)})$. *This isomorphism extends uniquely to* $\mathcal{HH}_e^{\mathrm{sph}}(GL(n)) \xrightarrow{\sim} K^{\tilde{\mathbf{K}} \rtimes \widetilde{\mathbb{C}}^\times}(\mathrm{Gr}_{GL(n)})$ *where the right-hand side is equipped with the algebra structure as in Sect. 3.7.* □

The following theorem is proved similarly to [10, Theorem 3.1, Proposition 3.18]:

Theorem 3.15

(a) *The algebras* $K^{\mathbf{K}}(\mathrm{Gr}_{GL(n)})$, $K^{\tilde{\mathbf{K}}}(\mathrm{Gr}_{GL(n)})$ *are commutative.*
(b) *The spectrum of* $K^{\mathbf{K}}(\mathrm{Gr}_{GL(n)})$ *together with the projection onto* $(\mathbb{C}^\times)^{(n)} = \mathrm{Spec}(K_{GL(n)}(\mathrm{pt}))$ *is naturally isomorphic to* ${}^\dagger\mathring{Z}^n \xrightarrow{\mathrm{pr}} (\mathbb{C}^\times)^{(n)}$ *(see Sect. 2.5).*
(c) *The spectrum of* $K^{\tilde{\mathbf{K}}}(\mathrm{Gr}_{GL(n)})$ *together with the projection onto* $\mathrm{Spec}(K^{\mathbf{K}}(\mathrm{Gr}_{GL(n)}))$ *is naturally isomorphic to* ${}^\dagger\hat{Z}^n \to {}^\dagger\mathring{Z}^n$ *(see Sect. 2.5).*
(d) *The Poisson structure on* $K^{\mathbf{K}}(\mathrm{Gr}_{GL(n)})$ *arising from the deformation* $K^{\mathbf{K}\rtimes\mathbb{C}^\times}(\mathrm{Gr}_{GL(n)})$ *corresponds under the above identification to the negative of the Poisson (symplectic) structure of [25, 34] on* ${}^\dagger\mathring{Z}^n$. *The Poisson (symplectic) structure on* $K^{\tilde{\mathbf{K}}}(\mathrm{Gr}_{GL(n)})$ *arising from the deformation* $K^{\tilde{\mathbf{K}}\rtimes\widetilde{\mathbb{C}}^\times}(\mathrm{Gr}_{GL(n)})$ *is the negative of the pull-back of the symplectic structure on* ${}^\dagger\mathring{Z}^n$. □

Corollary 3.16

(a) *The algebras* $\mathcal{HH}^{\mathrm{sph}}(GL(n))|_{q=1}$, $\mathcal{HH}^{\mathrm{sph}}_e(GL(n))|_{\mathbf{v}=1}$ *are commutative.*
(b) *The algebra* $\mathcal{HH}^{\mathrm{sph}}(GL(n))|_{q=1}$ *with the subalgebra* $\mathbb{C}[\mathsf{X}_1^{\pm 1}, \ldots, \mathsf{X}_n^{\pm 1}]^{\mathfrak{S}_n}$ *is naturally isomorphic to* $\mathbb{C}[{}^\dagger\mathring{Z}^n] \supset \mathbb{C}[(\mathbb{C}^\times)^{(n)}]$.
(c) *The Poisson structures on* $\mathcal{HH}^{\mathrm{sph}}(GL(n))|_{q=1}$, $\mathcal{HH}^{\mathrm{sph}}_e(GL(n))|_{\mathbf{v}=1}$ *arising from the deformations* $\mathcal{HH}^{\mathrm{sph}}(GL(n))$, $\mathcal{HH}^{\mathrm{sph}}_e(GL(n))$ *correspond under the above identification to the negative of the Poisson (symplectic) structures of [25, 34] on* ${}^\dagger\mathring{Z}^n$, ${}^\dagger\hat{Z}^n$. □

3.12 Quantum Poisson Reduction

Now again G is an almost simple simply-connected algebraic group. We consider Lusztig's integral form $U_q(\mathfrak{g})$ of the quantized universal enveloping algebra over $\mathbb{C}[q^{\pm 1}]$ with Cartan elements $K_\lambda,\ \lambda \in X$. It is denoted $\dot{\mathbb{U}}_{\mathcal{A}}$ in [65, § 2.2]. We extend the scalars to $\mathbb{C}[q^{\frac{\pm 1}{2m_{\mathrm{ad}}}}]$ and consider the integrable representations of $U_q(\mathfrak{g})$ with weights in X. We consider the reflection equation algebra $\mathcal{O}_q(G)$ spanned by the matrix coefficients of integrable $U_q(\mathfrak{g})$-modules (with weights in X); it is denoted $\mathbb{F}_{\mathcal{A}}$ in [65, § 2.2]. The corresponding integral form $\mathcal{D}_q(G)$ of the Heisenberg double [59, Section 3] (quantum differential operators) is denoted $\mathbb{D}_{\mathcal{A}}$ in [65, § 2.2]. The quasiclassical limit of $\mathcal{D}_q(G)$ is $D_+(G)$ with the Poisson structure $\{,\}_+$ considered in Sect. 2.2. The moment map $\mu\colon (D_+(G), \{,\}_+) \to (G, \pi) \times (G, \pi)$ is the quasiclassical limit of $\mu_q\colon U_q(\mathfrak{g}) \otimes U_q(\mathfrak{g}) \to \mathcal{D}_q(G)$ (see, e.g., [48]). The Poisson action of $(G, \pi_G) \times (G, \pi_G)$ on $D_+(G)$ is the quasiclassical limit of the comodule structure of $\mathcal{D}_q(G)$ over $\mathcal{O}_q(G) \otimes \mathcal{O}_q(G)$.

Recall the subalgebra $U_q^c(\mathfrak{n}) \subset U_q(\mathfrak{g})$ [56, § 2.2] associated to a Coxeter element c (we shall omit its dependence on $\{n_{ij}\}_{i,j\in I}$ satisfying [56, § 2.2.2]). The $U_q(\mathfrak{g})$-module $U_q(\mathfrak{g})/(U_q(\mathfrak{g})\cdot[U_q^c(\mathfrak{n}), U_q^c(\mathfrak{n})])$ is the quantization of the coisotropic subvariety $C_c \subset (G,\pi)$ of Sect. 2.2. Given a pair of Coxeter elements c, c', we consider the left ideal $\mathcal{I}_{c',c}$ of $\mathcal{D}_q(G)$ generated by $\mu_q([U_q^{c'}(\mathfrak{n}), U_q^{c'}(\mathfrak{n})] \otimes S[U_q^c(\mathfrak{n}), U_q^c(\mathfrak{n})])$ where S stands for the antipode. The invariants of $\mathcal{D}_q(G)/\mathcal{I}_{c',c}$ with respect to the coaction of $\mathcal{O}_q(B_-)\otimes\mathcal{O}_q(B_-)$ form an algebra denoted $\mathcal{O}_q({}^\dagger\mathfrak{Z}^{c',c}(G))$.

Conjecture 3.17 There is an isomorphism $\mathcal{HH}^{\mathrm{sph}}(W_e, \widehat{X}) \xrightarrow{\sim} \mathcal{O}_q({}^\dagger\mathfrak{Z}^{c,c}(G))$ equal to $\mathrm{id}_{{}^\dagger\mathfrak{Z}^{c,c}(G)}$ at $q = 1$.

4 Multiplicative Slices

4.1 Asymmetric Definition

We closely follow the exposition in [10, Section 2]. Let G be an adjoint simple complex algebraic group. We fix a Borel and a Cartan subgroup $G \supset B \supset T$. Let Λ be the coweight lattice, and let $\Lambda_+ \subset \Lambda$ be the submonoid spanned by the simple coroots $\alpha_i,\ i \in I$. The involution $\alpha \mapsto -w_0\alpha$ of Λ restricts to an involution of Λ_+ and induces an involution $\alpha_i \mapsto \alpha_{i^*}$ of the set of simple coroots. We will sometimes write $\alpha^* := -w_0\alpha$ for short. Let λ be a dominant coweight of G, and $\mu \leq \lambda$ an arbitrary coweight of G, not necessarily dominant, such that $\alpha := \lambda - \mu = \sum_{i\in I} a_i\alpha_i,\ a_i \in \mathbb{N}$. We will define the *multiplicative (trigonometric)* analogues ${}^\dagger\overline{\mathcal{W}}^\lambda_\mu$ of the generalized slices $\overline{\mathcal{W}}^\lambda_\mu$ of [10, 2(ii)].

Namely, ${}^\dagger\overline{\mathcal{W}}^\lambda_\mu$ is the moduli space of the following data:

(a) a G-bundle $\mathcal{P}$ on $\mathbb{P}^1$;

(b) a trivialization $\sigma\colon \mathcal{P}_{\mathrm{triv}}|_{\mathbb{P}^1\setminus\{1\}} \xrightarrow{\sim} \mathcal{P}|_{\mathbb{P}^1\setminus\{1\}}$ having a pole of degree $\leq \lambda$ at $1 \in \mathbb{P}^1$. This means that for an irreducible G-module $V^{\lambda^\vee}$ and the associated vector bundle $\mathcal{V}^{\lambda^\vee}_{\mathcal{P}}$ on $\mathbb{P}^1$ we have $V^{\lambda^\vee}\otimes\mathcal{O}_{\mathbb{P}^1}(-\langle\lambda,\lambda^\vee\rangle\cdot 1) \subset \mathcal{V}^{\lambda^\vee}_{\mathcal{P}} \subset V^{\lambda^\vee}\otimes\mathcal{O}_{\mathbb{P}^1}(-\langle w_0\lambda,\lambda^\vee\rangle\cdot 1)$;

(c) a reduction ϕ of $\mathcal{P}$ to a B-bundle (B-structure ϕ on $\mathcal{P}$) such that the induced T-bundle ϕ^T has degree $w_0\mu$, and the fiber of ϕ at $\infty \in \mathbb{P}^1$ is $B_- \subset G$ (with respect to the trivialization σ of $\mathcal{P}$ at $\infty \in \mathbb{P}^1$). This means in particular that for an irreducible G-module $V^{\lambda^\vee}$ and the associated vector bundle $\mathcal{V}^{\lambda^\vee}_{\mathcal{P}}$ on $\mathbb{P}^1$ we are given an invertible subsheaf $\mathcal{L}_{\lambda^\vee} \subset \mathcal{V}^{\lambda^\vee}_{\mathcal{P}}$ of degree $-\langle w_0\mu,\lambda^\vee\rangle$. We require ϕ to be transversal at $0 \in \mathbb{P}^1$ to the trivial B-structure B in $\mathcal{P}_{\mathrm{triv}}$.

4.2 Multiplicative BD Slices

Let $\underline{\lambda} = (\omega_{i_1}, \ldots, \omega_{i_N})$ be a sequence of fundamental coweights of G such that $\sum_{s=1}^N \omega_{i_s} = \lambda$. We define ${}^\dagger\overline{\mathcal{W}}^{\underline{\lambda}}_{\mu}$ as the moduli space of the following data:

(a) a collection of points $(\mathsf{z}_1, \ldots, \mathsf{z}_N) \in (\mathbb{C}^\times)^N$;
(b) a G-bundle $\mathcal{P}$ on $\mathbb{P}^1$;
(c) a trivialization (a section) σ of $\mathcal{P}$ on $\mathbb{P}^1 \setminus \{\mathsf{z}_1, \ldots, \mathsf{z}_N\}$ with a pole of degree $\leq \sum_{s=1}^N \omega_{i_s} \cdot \mathsf{z}_s$ on the complement;
(d) a reduction ϕ of $\mathcal{P}$ to a B-bundle (B-structure ϕ on $\mathcal{P}$) such that the induced T-bundle ϕ^T has degree $w_0\mu$, and the fiber of ϕ at $\infty \in \mathbb{P}^1$ is $B_- \subset G$ and transversal to B at $0 \in \mathbb{P}^1$ (with respect to the trivialization σ).

Remark 4.1 The definition of multiplicative BD slices differs from the definition of BD slices in [10, § 2(x)] only by the open condition of transversality at $0 \in \mathbb{P}^1$. Thus ${}^\dagger\overline{\mathcal{W}}^{\underline{\lambda}}_{\mu}$ is an open subvariety in $\overline{\mathcal{W}}^{\underline{\lambda}}_{\mu}$ (and similarly, ${}^\dagger\overline{\mathcal{W}}^{\lambda}_{\mu}$ is an open subvariety in $\overline{\mathcal{W}}^{\lambda}_{\mu}$). Hence, the favorable properties of the slices of [10] (e.g., the Cohen–Macaulay property) are inherited by the multiplicative slices.

4.3 A Symmetric Definition

Given arbitrary coweights μ_-, μ_+ such that $\mu_- + \mu_+ = \mu$, we consider the moduli space ${}^\dagger\overline{\mathcal{W}}^{\underline{\lambda}}_{\mu_-,\mu_+}$ of the following data:

(a) a collection of points $(\mathsf{z}_1, \ldots, \mathsf{z}_N) \in (\mathbb{C}^\times)^N$;
(b) G-bundles $\mathcal{P}_-, \mathcal{P}_+$ on $\mathbb{P}^1$;
(c) an isomorphism $\sigma : \mathcal{P}_-|_{\mathbb{P}^1\setminus\{\mathsf{z}_1,\ldots,\mathsf{z}_N\}} \xrightarrow{\sim} \mathcal{P}_+|_{\mathbb{P}^1\setminus\{\mathsf{z}_1,\ldots,\mathsf{z}_N\}}$ with a pole of degree $\leq \sum_{s=1}^N \omega_{i_s} \cdot \mathsf{z}_s$ on the complement;
(d) a trivialization of $\mathcal{P}_- = \mathcal{P}_+$ at $\infty \in \mathbb{P}^1$;
(e) a reduction ϕ_- of $\mathcal{P}_-$ to a B_--bundle (a B_--structure on $\mathcal{P}_-$) such that the induced T-bundle ϕ_-^T has degree $-w_0\mu_-$, and the fiber of ϕ_- at $\infty \in \mathbb{P}^1$ is $B \subset G$;
(f) a reduction ϕ_+ of $\mathcal{P}_+$ to a B-bundle (a B-structure on $\mathcal{P}_+$) such that the induced T-bundle ϕ_+^T has degree $w_0\mu_+$, and the fiber of ϕ_+ at $\infty \in \mathbb{P}^1$ is $B_- \subset G$. We require ϕ_- and ϕ_+ to be transversal at $0 \in \mathbb{P}^1$ (with respect to the isomorphism σ).

Note that the trivial G-bundle on $\mathbb{P}^1$ has a unique B_--reduction of degree 0 with fiber B at ∞. Conversely, a G-bundle $\mathcal{P}_-$ with a B_--structure of degree 0 is necessarily trivial, and its trivialization at ∞ uniquely extends to the whole of $\mathbb{P}^1$. Hence ${}^\dagger\overline{\mathcal{W}}^{\underline{\lambda}}_{0,\mu} = {}^\dagger\overline{\mathcal{W}}^{\underline{\lambda}}_{\mu}$.

For arbitrary ${}^\dagger\overline{\mathcal{W}}^{\underline{\lambda}}_{\mu_-,\mu_+}$, the G-bundles $\mathcal{P}_-, \mathcal{P}_+$ are identified via σ on $\mathbb{P}^1 \setminus \{\mathsf{z}_1, \ldots, \mathsf{z}_N\}$, so they are both equipped with B and B_--structures transversal around $0, \infty \in \mathbb{P}^1$, that is they are both equipped with a reduction to a T-bundle around $0, \infty \in \mathbb{P}^1$. So $\mathcal{P}_\pm = \mathcal{P}_\pm^T \times^T G$ for certain T-bundles $\mathcal{P}_\pm^T$ around $0, \infty \in \mathbb{P}^1$, trivialized at $\infty \in \mathbb{P}^1$. The modified T-bundles $'\mathcal{P}_\pm^T := \mathcal{P}_\pm^T(w_0\mu_- \cdot \infty)$ are canonically isomorphic to $\mathcal{P}_\pm^T$ off $\infty \in \mathbb{P}^1$ and trivialized at $\infty \in \mathbb{P}^1$. We define $'\mathcal{P}_\pm$ as the result of gluing $\mathcal{P}_\pm$ and $'\mathcal{P}_\pm^T \times^T G$ in the punctured neighborhood of $\infty \in \mathbb{P}^1$. Then the isomorphism $\sigma\colon {'\mathcal{P}_-}|_{\mathbb{P}^1\setminus\{\infty,\mathsf{z}_1,\ldots,\mathsf{z}_N\}} \xrightarrow{\sim} {'\mathcal{P}_+}|_{\mathbb{P}^1\setminus\{\infty,\mathsf{z}_1,\ldots,\mathsf{z}_N\}}$ extends to $\mathbb{P}^1 \setminus \{\mathsf{z}_1,\ldots,\mathsf{z}_N\}$, and $\phi_\pm$ also extends from $\mathbb{P}^1 \setminus \{\infty\}$ to a B-structure $'\phi_+$ in $'\mathcal{P}_+$ of degree $w_0\mu$ (resp. a B_--structure $'\phi_-$ on $'\mathcal{P}_-$ of degree 0).

This defines an isomorphism ${}^\dagger\overline{\mathcal{W}}{}^{\underline{\lambda}}_{\mu_-,\mu_+} \simeq {}^\dagger\overline{\mathcal{W}}{}^{\underline{\lambda}}_{\mu}$. Similarly, for the nondeformed slices we have an isomorphism ${}^\dagger\overline{\mathcal{W}}{}^{\lambda}_{\mu_-,\mu_+} \simeq {}^\dagger\overline{\mathcal{W}}{}^{\lambda}_{\mu}$.

4.4 Multiplication of Slices

Given $\lambda_1 \geq \mu_1$ and $\lambda_2 \geq \mu_2$ with λ_1, λ_2 dominant, we think of ${}^\dagger\overline{\mathcal{W}}{}^{\lambda_1}_{\mu_1}$ (resp. ${}^\dagger\overline{\mathcal{W}}{}^{\lambda_2}_{\mu_2}$) in the incarnation ${}^\dagger\overline{\mathcal{W}}{}^{\lambda_1}_{\mu_1,0}$ (resp. ${}^\dagger\overline{\mathcal{W}}{}^{\lambda_2}_{0,\mu_2}$). Note that $\mathcal{P}_-^2$ is canonically trivialized as in Sect. 4.3, and $\mathcal{P}_+^1$ is canonically trivialized for the same reason. Given $(\mathcal{P}_\pm^1, \sigma_1, \phi_\pm^1) \in {}^\dagger\overline{\mathcal{W}}{}^{\lambda_1}_{\mu_1,0}$, we change the trivialization of $\mathcal{P}_+^1$ by a (uniquely determined) element of U_- (the unipotent radical of B_-) so that the value $\phi_-^1(0)$ becomes B (while $\phi_+^1(0)$ remains equal to B_-). Now the value $\phi_-^1(\infty)$ is not B anymore; it is only transversal to B_-. In order to distinguish the data obtained by the composition with the above trivialization change, we denote them by $('\mathcal{P}_\pm^1, '\sigma_1, '\phi_\pm^1)$. Given $(\mathcal{P}_\pm^2, \sigma_2, \phi_\pm^2) \in {}^\dagger\overline{\mathcal{W}}{}^{\lambda_2}_{0,\mu_2}$, we consider $('\mathcal{P}_-^1, \mathcal{P}_+^2, \sigma_2 \circ '\sigma_1, '\phi_-^1, \phi_+^2)$ (recall that $'\mathcal{P}_+^1 = \mathcal{P}_{\mathrm{triv}} = \mathcal{P}_-^2$). These data *do not* lie in ${}^\dagger\overline{\mathcal{W}}{}^{\lambda_1+\lambda_2}_{\mu_1,\mu_2}$ since the value $'\phi_-^1(\infty)$ is not necessarily equal to B, it is only transversal to B_-. However, we change the trivialization of $'\mathcal{P}_-^1(\infty) = \mathcal{P}_+^2(\infty)$ by a (uniquely determined) element of U_-, so that the value of $'\phi_-^1(\infty)$ becomes B, and we end up in ${}^\dagger\overline{\mathcal{W}}{}^{\lambda_1+\lambda_2}_{\mu_1,\mu_2} = {}^\dagger\overline{\mathcal{W}}{}^{\lambda_1+\lambda_2}_{\mu_1+\mu_2}$.

This defines a multiplication morphism ${}^\dagger\overline{\mathcal{W}}{}^{\lambda_1}_{\mu_1} \times {}^\dagger\overline{\mathcal{W}}{}^{\lambda_2}_{\mu_2} \to {}^\dagger\overline{\mathcal{W}}{}^{\lambda_1+\lambda_2}_{\mu_1+\mu_2}$.

In particular, taking $\mu_2 = \lambda_2$ so that ${}^\dagger\overline{\mathcal{W}}{}^{\lambda_2}_{\lambda_2}$ is a point and ${}^\dagger\overline{\mathcal{W}}{}^{\lambda_1}_{\mu_1} \times {}^\dagger\overline{\mathcal{W}}{}^{\lambda_2}_{\lambda_2} = {}^\dagger\overline{\mathcal{W}}{}^{\lambda_1}_{\mu_1}$, we get a stabilization morphism ${}^\dagger\overline{\mathcal{W}}{}^{\lambda_1}_{\mu_1} \to {}^\dagger\overline{\mathcal{W}}{}^{\lambda_1+\lambda_2}_{\mu_1+\lambda_2}$.

Remark 4.2 The multiplication of slices in [10, § 2(vi)] *does not* preserve the multiplicative slices viewed as open subvarieties according to Remark 4.1 (in particular, it *does not* induce the above multiplication on multiplicative slices).

4.5 Scattering Matrix

Given a collection $(\mathsf{z}_1,\ldots,\mathsf{z}_N) \in (\mathbb{C}^\times)^N$, we define $P_{\underline{\mathsf{z}}}(z) := \prod_{s=1}^N (z - \mathsf{z}_s) \in \mathbb{C}[z]$. We also define a closed subvariety ${}^\dagger\overline{\mathcal{W}}{}^{\underline{\lambda},\underline{\mathsf{z}}}_{\mu} \subset {}^\dagger\overline{\mathcal{W}}{}^{\underline{\lambda}}_{\mu}$ as the fiber of the latter over $\underline{z} = (z_1, \ldots, z_N)$. We construct a locally closed embedding $\Psi\colon {}^\dagger\overline{\mathcal{W}}_\mu^{\lambda,\underline{z}} \hookrightarrow G[z, P^{-1}]$ into an ind-affine scheme as follows. According to Sect. 4.3, we have an isomorphism $\zeta\colon {}^\dagger\overline{\mathcal{W}}_\mu^{\lambda,\underline{z}} = {}^\dagger\overline{\mathcal{W}}_{0,\mu}^{\lambda,\underline{z}} \xrightarrow{\sim} {}^\dagger\overline{\mathcal{W}}_{\mu,0}^{\lambda,\underline{z}}$. We denote $\zeta(\mathcal{P}_\pm, \sigma, \phi_\pm)$ by $(\mathcal{P}'_\pm, \sigma', \phi'_\pm)$. Note that $\mathcal{P}_-$ and $\mathcal{P}'_+$ are trivialized, and $\mathcal{P}'_+$ is obtained from $\mathcal{P}_+$ by an application of a certain Hecke transformation at $\infty \in \mathbb{P}^1$. In particular, we obtain an isomorphism $\mathcal{P}_+|_{\mathbb{A}^1} \xrightarrow{\sim} \mathcal{P}'_+|_{\mathbb{A}^1} = \mathcal{P}_{\mathrm{triv}}|_{\mathbb{A}^1}$. As in Sect. 4.4, we change the trivialization of $\mathcal{P}'_+$ by a uniquely defined element of U_- so that the value of $\phi'_-(0)$ becomes B. Now we compose this change of trivialization with the above isomorphism $\mathcal{P}_+|_{\mathbb{A}^1} \xrightarrow{\sim} \mathcal{P}'_+|_{\mathbb{A}^1} = \mathcal{P}_{\mathrm{triv}}|_{\mathbb{A}^1}$ and with $\sigma\colon \mathcal{P}_{\mathrm{triv}}|_{\mathbb{A}^1\setminus\underline{z}} = \mathcal{P}_-|_{\mathbb{A}^1\setminus\underline{z}} \xrightarrow{\sim} \mathcal{P}_+|_{\mathbb{A}^1\setminus\underline{z}}$ to obtain an isomorphism $\mathcal{P}_{\mathrm{triv}}|_{\mathbb{A}^1\setminus\underline{z}} \xrightarrow{\sim} \mathcal{P}_{\mathrm{triv}}|_{\mathbb{A}^1\setminus\underline{z}}$, i.e. an element of $G[z, P^{-1}]$.

Here is an equivalent construction of the above embedding. Given $(\mathcal{P}_\pm, \sigma, \phi_\pm) \in {}^\dagger\overline{\mathcal{W}}_{\mu_-,\mu_+}^{\lambda,\underline{z}} = {}^\dagger\overline{\mathcal{W}}_\mu^{\lambda,\underline{z}}$, we choose a trivialization of the B-bundle $\phi_+|_{\mathbb{A}^1}$ (resp. of the B_--bundle $\phi_-|_{\mathbb{A}^1}$). This trivialization gives rise to a trivialization of the G-bundle $\mathcal{P}_+|_{\mathbb{A}^1}$ (resp. of $\mathcal{P}_-|_{\mathbb{A}^1}$), so that σ becomes an element of $G(z)$ regular at $0 \in \mathbb{P}^1$; moreover, the value of $\sigma(0)$ lies in the big Bruhat cell $B \cdot B_- \subset G$. We require that $\sigma(0) \in B \subset G$. Then σ is well-defined up to the left multiplication by an element of $B[z]$ and the right multiplication by an element of $B_{-,1}[z]$ (the kernel of evaluation at $0 \in \mathbb{P}^1\colon B_-[z] \to B_-$), i.e. σ is a well-defined element of $B[z]\backslash G(z)/B_{-,1}[z]$. Clearly, this element of $G(z)$ lies in the closure of the double coset $\overline{G[z]z^{\underline{\lambda},\underline{z}}G[z]}$ where $z^{\underline{\lambda},\underline{z}} := \prod_{s=1}^N (z - z_s)^{\omega_{i_s}}$. Moreover, it lies in $\overline{G[z]z^{\underline{\lambda},\underline{z}}G[z]} \cap \mathrm{ev}_0^{-1}(B)$. Thus we have constructed an embedding

$$\Psi'\colon {}^\dagger\overline{\mathcal{W}}_\mu^{\lambda,\underline{z}} \hookrightarrow B[z]\backslash(\overline{G[z]z^{\underline{\lambda},\underline{z}}G[z]} \cap \mathrm{ev}_0^{-1}(B))/B_{-,1}[z]$$

If we compose with an embedding $G(z) \hookrightarrow G((z^{-1}))$, then the image of Ψ' lies in $B[z]\backslash U_1[[z^{-1}]]T_1[[z^{-1}]]z^\mu U_-[[z^{-1}]]/U_{-,1}[z]$ where $U_1[[z^{-1}]] \subset U[[z^{-1}]]$ (resp. $T_1[[z^{-1}]] \subset T[[z^{-1}]]$) stands for the kernel of evaluation at $\infty \in \mathbb{P}^1$. However, the projection

$$U_1[[z^{-1}]]T_1[[z^{-1}]]z^\mu U_-[[z^{-1}]] \to B[z]\backslash U_1[[z^{-1}]]T_1[[z^{-1}]]z^\mu U_-[[z^{-1}]]/U_{-,1}[z]$$

is clearly one-to-one. Summing up, we obtain an embedding

$$\Psi\colon {}^\dagger\overline{\mathcal{W}}_\mu^{\lambda,\underline{z}} \to U_1[[z^{-1}]]T_1[[z^{-1}]]z^\mu U_-[[z^{-1}]] \cap \overline{G[z]z^{\underline{\lambda},\underline{z}}G[z]} \cap \mathrm{ev}_0^{-1}(B).$$

We claim that Ψ is an isomorphism. To see it, we construct the inverse map to ${}^\dagger\overline{\mathcal{W}}_{0,\mu}^{\lambda,\underline{z}}$: given $g(z) \in U_1[[z^{-1}]]T_1[[z^{-1}]]z^\mu U_-[[z^{-1}]] \cap \overline{G[z]z^{\underline{\lambda},\underline{z}}G[z]} \cap \mathrm{ev}_0^{-1}(B)$ we use it to glue $\mathcal{P}_+$ together with a rational isomorphism $\sigma\colon \mathcal{P}_{\mathrm{triv}} = \mathcal{P}_- \to \mathcal{P}_+$, and define ϕ_+ as the image of the standard trivial B-structure in $\mathcal{P}_{\mathrm{triv}}$ under σ.

Remark 4.3 The embedding $\overline{\mathcal{W}}^{\underline{\lambda},\underline{\mathsf{z}}}_{\mu} \hookrightarrow G(z)$ of [10, § 2(xi)] restricted to the open subvariety ${}^{\dagger}\overline{\mathcal{W}}^{\underline{\lambda},\underline{\mathsf{z}}}_{\mu} \subset \overline{\mathcal{W}}^{\underline{\lambda},\underline{\mathsf{z}}}_{\mu}$ *does not* give the above embedding ${}^{\dagger}\overline{\mathcal{W}}^{\underline{\lambda},\underline{\mathsf{z}}}_{\mu} \hookrightarrow G(z)$.

4.6 A Cover of a Slice

We define a T-torsor ${}^{\dagger}\widetilde{\mathcal{W}}^{\underline{\lambda}}_{\mu} \to {}^{\dagger}\overline{\mathcal{W}}^{\underline{\lambda}}_{\mu}$ as the moduli space of data (a–d) as in Sect. 4.2 plus

(e) a collection of nowhere vanishing sections $u_{\lambda^\vee} \in \Gamma(\mathbb{P}^1 \setminus \{\infty\}, \mathcal{L}_{\lambda^\vee})$ satisfying Plücker relations (cf. Sect. 4.1(c)).

The construction of Sect. 4.5 defines an isomorphism

$$\widetilde{\Psi}\colon {}^{\dagger}\widetilde{\mathcal{W}}^{\underline{\lambda},\underline{\mathsf{z}}}_{\mu} \xrightarrow{\sim} U_1[[z^{-1}]]T[[z^{-1}]]z^{\mu}U_-[[z^{-1}]] \cap \overline{G[z]z^{\underline{\lambda},\underline{\mathsf{z}}}G[z]} \cap \mathrm{ev}_0^{-1}(B).$$

Let $T_{[2]} \subset T$ be the subgroup of 2-torsion. For a future use we define a $T_{[2]}$-torsor ${}^{\dagger}\widetilde{\mathcal{W}}^{\underline{\lambda},\underline{\mathsf{z}}}_{\mu} \supset {}^{\dagger}\hat{\mathcal{W}}^{\underline{\lambda},\underline{\mathsf{z}}}_{\mu} \to {}^{\dagger}\overline{\mathcal{W}}^{\underline{\lambda},\underline{\mathsf{z}}}_{\mu}$ as follows. The evaluation at $0 \in \mathbb{P}^1$ gives rise to a projection $\mathrm{pr}_0\colon \overline{G[z]z^{\underline{\lambda},\underline{\mathsf{z}}}G[z]} \cap \mathrm{ev}_0^{-1}(B) \to B \to T$. The leading coefficient (at z^{μ}) gives rise to a projection $\mathrm{pr}_{\infty}\colon U_1[[z^{-1}]]T[[z^{-1}]]z^{\mu}U_-[[z^{-1}]] \to T$, and ${}^{\dagger}\hat{\mathcal{W}}^{\underline{\lambda},\underline{\mathsf{z}}}_{\mu}$ is cut out by the equation $\mathrm{pr}_0 \cdot \mathrm{pr}_{\infty} = (-1)^{\lambda-\mu} \in T_{[2]}$, where $\lambda = \sum_{s=1}^{N} \omega_{i_s}$, see Sect. 4.2. As $\underline{\mathsf{z}}$ varies, we obtain a $T_{[2]}$-torsor ${}^{\dagger}\widetilde{\mathcal{W}}^{\underline{\lambda}}_{\mu} \supset {}^{\dagger}\hat{\mathcal{W}}^{\underline{\lambda}}_{\mu} \to {}^{\dagger}\overline{\mathcal{W}}^{\underline{\lambda}}_{\mu}$.

4.7 An Example

This section is parallel to [10, § 2(xii)], but our present conventions are slightly different. Let $G = GL(2) = GL(V)$ with $V = \mathbb{C}e_1 \oplus \mathbb{C}e_2$. Let B be the stabilizer of $\mathbb{C}e_2$ (the lower triangular matrices), and let B_- be the stabilizer of $\mathbb{C}e_1$ (the upper triangular matrices). Let $N, m \in \mathbb{N}$; $\underline{\lambda}$ be an N-tuple of fundamental coweights $(0, 1)$, and $\mu = (m, N-m)$, so that $w_0\mu = (N-m, m)$. Let $\mathcal{O} := \mathcal{O}_{\mathbb{P}^1}$. We fix a collection $(\mathsf{z}_1, \ldots, \mathsf{z}_N) \in (\mathbb{C}^\times)^N$ and define $P_{\underline{\mathsf{z}}}(z) := \prod_{s=1}^{N}(z - \mathsf{z}_s) \in \mathbb{C}[z]$. Then ${}^{\dagger}\overline{\mathcal{W}}^{\underline{\lambda},\underline{\mathsf{z}}}_{\mu}$ is the moduli space of flags $(\mathcal{O} \otimes V \supset \mathcal{V} \supset \mathcal{L})$, where

(a) $\mathcal{V}$ is a 2-dimensional locally free subsheaf in $\mathcal{O} \otimes V$ coinciding with $\mathcal{O} \otimes V$ around $0, \infty \in \mathbb{P}^1$ and such that on $\mathbb{A}^1 \subset \mathbb{P}^1$ the global sections of $\det \mathcal{V}$ coincide with $P_{\underline{\mathsf{z}}}\mathbb{C}[z]e_1 \wedge e_2$ as a $\mathbb{C}[z]$-submodule of $\Gamma(\mathbb{A}^1, \det(\mathcal{O}_{\mathbb{A}^1} \otimes V)) = \mathbb{C}[z]e_1 \wedge e_2$.

(b) $\mathcal{L}$ is a line subbundle in $\mathcal{V}$ of degree $-m$, assuming the value $\mathbb{C}e_1$ at $\infty \in \mathbb{P}^1$, and such that the value of $\mathcal{L}$ at $0 \in \mathbb{P}^1$ is transversal to $\mathbb{C}e_2$. In particular, $\deg \mathcal{V}/\mathcal{L} = m - N$.

On the other hand, let us introduce a closed subvariety ${}^{\dagger}\hat{\mathcal{W}}^{\lambda,\underline{z}}_{\overline{\mu}}$ in $\mathrm{Mat}_2[z]$ formed by all the matrices $\mathsf{M} = \begin{pmatrix} A & B \\ C & D \end{pmatrix}$ such that $A(z) = a_m z^m + \ldots + a_0$, and $a_m \cdot a_0 = (-1)^m$, while $\deg C(z) < m \geq \deg B(z)$, and $B(0) = 0$; furthermore, $\det \mathsf{M} = P_{\underline{z}}(z)$.

Then we have a two-fold cover $\mho\colon {}^{\dagger}\hat{\mathcal{W}}^{\lambda,\underline{z}}_{\overline{\mu}} \to {}^{\dagger}\overline{\mathcal{W}}^{\lambda,\underline{z}}_{\overline{\mu}}$: given $\mathsf{M} \in {}^{\dagger}\hat{\mathcal{W}}^{\lambda,\underline{z}}_{\overline{\mu}}$ we view it as a transition matrix in a punctured neighborhood of $\infty \in \mathbb{P}^1$ to glue a vector bundle $\mathcal{V}$ which embeds, by construction, as a locally free subsheaf into $\mathcal{O} \otimes V$. The morphism $\mathsf{M}\mathcal{O}_{\mathbb{A}^1} e_1 \hookrightarrow \mathcal{O}_{\mathbb{A}^1} \otimes V$ naturally extends to $\infty \in \mathbb{P}^1$ with a pole of degree m, hence it extends to an embedding of $\mathcal{O}(-m \cdot \infty)$ into $\mathcal{V} \subset \mathcal{O} \otimes V$. The image of this embedding is the desired line subbundle $\mathcal{L} \subset \mathcal{V}$.

4.8 *Thick Slices*

We define *thick* multiplicative (trigonometric) slices ${}^{\dagger}\mathcal{W}_{\mu}$ as the moduli space of the following data:

(a) a G-bundle $\mathcal{P}$ on $\mathbb{P}^1$;
(b) a trivialization $\sigma\colon \mathcal{P}_{\mathrm{triv}}|_{\widehat{\mathbb{P}}^1_\infty} \xrightarrow{\sim} \mathcal{P}|_{\widehat{\mathbb{P}}^1_\infty}$ in the formal neighborhood of $\infty \in \mathbb{P}^1$;
(c) a reduction ϕ of $\mathcal{P}$ to a B-bundle (B-structure ϕ on $\mathcal{P}$) such that the induced T-bundle ϕ^T has degree $w_0\mu$, and the fiber of ϕ at $\infty \in \mathbb{P}^1$ is transversal to B (with respect to the trivialization σ of $\mathcal{P}$ at $\infty \in \mathbb{P}^1$);
(d) a collection of nowhere vanishing sections $u_{\lambda^\vee} \in \Gamma(\mathbb{P}^1 \setminus \{\infty\}, \mathcal{L}_{\lambda^\vee})$ satisfying Plücker relations (cf. Sect. 4.1(c)).

The construction of Sect. 4.6 identifies ${}^{\dagger}\mathcal{W}_{\mu}$ with the infinite type scheme (cf. [24, § 5.9])

$$ {}^{\dagger}\mathcal{W}_{\mu} \simeq U_1[[z^{-1}]]T[[z^{-1}]]z^{\mu}U_-[[z^{-1}]] \subset G((z^{-1})). \tag{4.1} $$

As the inclusion $U_1[[z^{-1}]] \hookrightarrow U((z^{-1}))$ gives rise to an isomorphism $U_1[[z^{-1}]] \simeq U[z]\backslash U((z^{-1}))$, we can identify ${}^{\dagger}\mathcal{W}_{\mu}$ with the quotient $U[z]\backslash U((z^{-1}))T[[z^{-1}]]z^{\mu}U_-((z^{-1}))/U_{-,1}[z]$, and we write π for this isomorphism. The construction of Sect. 4.5 (resp. of Sect. 4.6) defines a closed embedding ${}^{\dagger}\overline{\mathcal{W}}^{\lambda}_{\mu} \hookrightarrow {}^{\dagger}\mathcal{W}_{\mu}$ (resp. ${}^{\dagger}\hat{\mathcal{W}}^{\lambda}_{\mu} \hookrightarrow {}^{\dagger}\mathcal{W}_{\mu}$). We define the multiplication morphism $m_{\mu_1,\mu_2}\colon {}^{\dagger}\mathcal{W}_{\mu_1} \times {}^{\dagger}\mathcal{W}_{\mu_2} \to {}^{\dagger}\mathcal{W}_{\mu_1+\mu_2}$ by the formula $m_{\mu_1,\mu_2}(g_1, g_2) = \pi(g_1 g_2)$. Then the multiplication morphism $m^{\lambda_1,\lambda_2}_{\mu_1,\mu_2}\colon {}^{\dagger}\overline{\mathcal{W}}^{\lambda_1}_{\mu_1} \times {}^{\dagger}\overline{\mathcal{W}}^{\lambda_2}_{\mu_2} \to {}^{\dagger}\overline{\mathcal{W}}^{\lambda_1+\lambda_2}_{\mu_1+\mu_2}$ of Sect. 4.4 is the restriction of m_{μ_1,μ_2}. Similarly, m_{μ_1,μ_2} restricts to a multiplication ${}^{\dagger}\hat{\mathcal{W}}^{\lambda_1}_{\mu_1} \times {}^{\dagger}\hat{\mathcal{W}}^{\lambda_2}_{\mu_2} \to {}^{\dagger}\hat{\mathcal{W}}^{\lambda_1+\lambda_2}_{\mu_1+\mu_2}$.

For ν_1, ν_2 antidominant, we define the *shift maps* $\iota_{\mu,\nu_1,\nu_2}\colon {}^{\dagger}\mathcal{W}_{\mu+\nu_1+\nu_2} \to {}^{\dagger}\mathcal{W}_{\mu}$ by $g \mapsto \pi(z^{-\nu_1} g z^{-\nu_2})$.

5 Shifted Quantum Affine Algebras

Let $\mathfrak{g}$ be a simple Lie algebra, $\mathfrak{h} \subset \mathfrak{g}$ be a Cartan subalgebra of $\mathfrak{g}$, and $(\cdot,\cdot)$ be a non-degenerate invariant bilinear symmetric form on $\mathfrak{g}$ (with a square length of the shortest root equal to 2). Let $\{\alpha_i^\vee\}_{i\in I} \subset \mathfrak{h}^*$ be the simple positive roots of $\mathfrak{g}$ relative to $\mathfrak{h}$, and $c_{ij} = 2\frac{(\alpha_i^\vee,\alpha_j^\vee)}{(\alpha_i^\vee,\alpha_i^\vee)}$–the entries of the corresponding Cartan matrix. Set $d_i := \frac{(\alpha_i^\vee,\alpha_i^\vee)}{2} \in \mathbb{Z}_{>0}$ so that $d_i c_{ij} = d_j c_{ji}$ for any $i, j \in I$. Let $\nu\colon \mathfrak{h} \xrightarrow{\sim} \mathfrak{h}^*$ be the isomorphism determined by the symmetric form $(\cdot,\cdot)$ so that $\alpha_i = h_i = \nu^{-1}(\alpha_i^\vee)/d_i$ are the simple coroots of $\mathfrak{g}$.

5.1 *Algebras* $\mathcal{U}^{\mathrm{sc}}_{\mu_1,\mu_2}$ *and* $\mathcal{U}^{\mathrm{ad}}_{\mu_1,\mu_2}$

Given coweights $\mu^+,\mu^- \in \Lambda$, set $\underline{b}^\pm = \{b_i^\pm\}_{i\in I} \in \mathbb{Z}^I$ with $b_i^\pm := \alpha_i^\vee(\mu^\pm)$. Define the *simply-connected version of shifted quantum affine algebra*, denoted by $\mathcal{U}^{\mathrm{sc}}_{\mu^+,\mu^-}$ or $\mathcal{U}^{\mathrm{sc}}_{\underline{b}^+,\underline{b}^-}$, to be the associative $\mathbb{C}(\boldsymbol{v})$-algebra generated by $\{e_{i,r}, f_{i,r}, \psi^\pm_{i,\pm s_i^\pm}, (\psi^\pm_{i,\mp b_i^\pm})^{-1}\}_{i\in I}^{r\in\mathbb{Z}, s_i^\pm\geq -b_i^\pm}$ with the following defining relations (for all $i, j \in I$ and $\epsilon,\epsilon' \in \{\pm\}$):

$$[\psi_i^\epsilon(z),\psi_j^{\epsilon'}(w)] = 0,\ \psi^\pm_{i,\mp b_i^\pm}\cdot(\psi^\pm_{i,\mp b_i^\pm})^{-1} = (\psi^\pm_{i,\mp b_i^\pm})^{-1}\cdot\psi^\pm_{i,\mp b_i^\pm} = 1, \tag{U1}$$

$$(z - \boldsymbol{v}_i^{c_{ij}} w) e_i(z) e_j(w) = (\boldsymbol{v}_i^{c_{ij}} z - w) e_j(w) e_i(z), \tag{U2}$$

$$(\boldsymbol{v}_i^{c_{ij}} z - w) f_i(z) f_j(w) = (z - \boldsymbol{v}_i^{c_{ij}} w) f_j(w) f_i(z), \tag{U3}$$

$$(z - \boldsymbol{v}_i^{c_{ij}} w)\psi_i^\epsilon(z) e_j(w) = (\boldsymbol{v}_i^{c_{ij}} z - w) e_j(w)\psi_i^\epsilon(z), \tag{U4}$$

$$(\boldsymbol{v}_i^{c_{ij}} z - w)\psi_i^\epsilon(z) f_j(w) = (z - \boldsymbol{v}_i^{c_{ij}} w) f_j(w)\psi_i^\epsilon(z), \tag{U5}$$

$$[e_i(z), f_j(w)] = \frac{\delta_{ij}}{\boldsymbol{v}_i - \boldsymbol{v}_i^{-1}}\delta\left(\frac{z}{w}\right)\left(\psi_i^+(z) - \psi_i^-(z)\right), \tag{U6}$$

$$\underset{z_1,\ldots,z_{1-c_{ij}}}{\mathrm{Sym}} \sum_{r=0}^{1-c_{ij}} (-1)^r \begin{bmatrix} 1-c_{ij} \\ r \end{bmatrix}_{\boldsymbol{v}_i} e_i(z_1)\cdots e_i(z_r) e_j(w) e_i(z_{r+1})\cdots e_i(z_{1-c_{ij}}) = 0, \tag{U7}$$

$$\underset{z_1,\ldots,z_{1-c_{ij}}}{\mathrm{Sym}} \sum_{r=0}^{1-c_{ij}} (-1)^r \begin{bmatrix} 1-c_{ij} \\ r \end{bmatrix}_{\boldsymbol{v}_i} f_i(z_1)\cdots f_i(z_r) f_j(w) f_i(z_{r+1})\cdots f_i(z_{1-c_{ij}}) = 0, \tag{U8}$$

where $\boldsymbol{v}_i := \boldsymbol{v}^{d_i}$, $[a, b]_x := ab - x \cdot ba$, $[m]_{\boldsymbol{v}} := \frac{\boldsymbol{v}^m - \boldsymbol{v}^{-m}}{\boldsymbol{v} - \boldsymbol{v}^{-1}}$, $\left[\begin{smallmatrix} a \\ b \end{smallmatrix}\right]_{\boldsymbol{v}} := \frac{[a-b+1]_{\boldsymbol{v}} \cdots [a]_{\boldsymbol{v}}}{[1]_{\boldsymbol{v}} \cdots [b]_{\boldsymbol{v}}}$, $\underset{z_1,\ldots,z_s}{\mathrm{Sym}}$ stands for the symmetrization in $z_1, \ldots, z_s$, and the generating series are defined as follows:

$$e_i(z) := \sum_{r \in \mathbb{Z}} e_{i,r} z^{-r}, \ f_i(z) := \sum_{r \in \mathbb{Z}} f_{i,r} z^{-r}, \ \psi_i^{\pm}(z) := \sum_{r \geq -b_i^{\pm}} \psi_{i,\pm r}^{\pm} z^{\mp r}, \ \delta(z) := \sum_{r \in \mathbb{Z}} z^r.$$

Let us introduce another set of Cartan generators $\{h_{i,\pm r}\}_{i \in I}^{r>0}$ instead of $\{\psi_{i,\pm s_i^{\pm}}^{\pm}\}_{i \in I}^{s_i^{\pm} > -b_i^{\pm}}$ via

$$(\psi_{i,\mp b_i^{\pm}}^{\pm} z^{\pm b_i^{\pm}})^{-1} \psi_i^{\pm}(z) = \exp\left(\pm(\boldsymbol{v}_i - \boldsymbol{v}_i^{-1}) \sum_{r>0} h_{i,\pm r} z^{\mp r}\right).$$

Then, relations (U4, U5) are equivalent to the following:

$$\psi_{i,\mp b_i^{\pm}}^{\pm} e_{j,s} = \boldsymbol{v}_i^{\pm c_{ij}} e_{j,s} \psi_{i,\mp b_i^{\pm}}^{\pm}, \ [h_{i,r}, e_{j,s}] = \frac{[r c_{ij}]_{\boldsymbol{v}_i}}{r} \cdot e_{j,s+r} \ \text{for } r \neq 0, \qquad (\mathrm{U4}')$$

$$\psi_{i,\mp b_i^{\pm}}^{\pm} f_{j,s} = \boldsymbol{v}_i^{\mp c_{ij}} f_{j,s} \psi_{i,\mp b_i^{\pm}}^{\pm}, \ [h_{i,r}, f_{j,s}] = -\frac{[r c_{ij}]_{\boldsymbol{v}_i}}{r} \cdot f_{j,s+r} \ \text{for } r \neq 0. \qquad (\mathrm{U5}')$$

Let $\mathfrak{U}^{\mathrm{sc},<}_{\mu^+,\mu^-}$, $\mathfrak{U}^{\mathrm{sc},>}_{\mu^+,\mu^-}$, and $\mathfrak{U}^{\mathrm{sc},0}_{\mu^+,\mu^-}$ be the $\mathbb{C}(\boldsymbol{v})$-subalgebras of $\mathfrak{U}^{\mathrm{sc}}_{\mu^+,\mu^-}$ generated by $\{f_{i,r}\}_{i \in I}^{r \in \mathbb{Z}}$, $\{e_{i,r}\}_{i \in I}^{r \in \mathbb{Z}}$, and $\{\psi_{i,\pm s_i^{\pm}}^{\pm}, (\psi_{i,\mp b_i^{\pm}}^{\pm})^{-1}\}_{i \in I}^{s_i^{\pm} \geq -b_i^{\pm}}$, respectively. The following is proved completely analogously to [37, Theorem 2]:

Proposition 5.1

(a) (Triangular decomposition of $\mathfrak{U}^{\mathrm{sc}}_{\mu^+,\mu^-}$) The multiplication map

$$m \colon \mathfrak{U}^{\mathrm{sc},<}_{\mu^+,\mu^-} \otimes \mathfrak{U}^{\mathrm{sc},0}_{\mu^+,\mu^-} \otimes \mathfrak{U}^{\mathrm{sc},>}_{\mu^+,\mu^-} \longrightarrow \mathfrak{U}^{\mathrm{sc}}_{\mu^+,\mu^-}$$

is an isomorphism of $\mathbb{C}(\boldsymbol{v})$-vector spaces.

(b) The algebra $\mathfrak{U}^{\mathrm{sc},0}_{\mu^+,\mu^-}$ (resp. $\mathfrak{U}^{\mathrm{sc},<}_{\mu^+,\mu^-}$ and $\mathfrak{U}^{\mathrm{sc},>}_{\mu^+,\mu^-}$) is isomorphic to the $\mathbb{C}(\boldsymbol{v})$-algebra generated by $\{\psi_{i,\pm s_i^{\pm}}^{\pm}, (\psi_{i,\mp b_i^{\pm}}^{\pm})^{-1}\}_{i \in I}^{s_i^{\pm} \geq -b_i^{\pm}}$ (resp. $\{f_{i,r}\}_{i \in I}^{r \in \mathbb{Z}}$ and $\{e_{i,r}\}_{i \in I}^{r \in \mathbb{Z}}$) with the defining relations (U1) (resp. (U3, U8) and (U2, U7)). In particular, $\mathfrak{U}^{\mathrm{sc},<}_{\mu^+,\mu^-}$ and $\mathfrak{U}^{\mathrm{sc},>}_{\mu^+,\mu^-}$ are independent of $\mu^{\pm}$.

Following the terminology of [50], we also define the *adjoint version of shifted quantum affine algebra*, denoted by $\mathcal{U}^{\mathrm{ad}}_{\mu^+,\mu^-}$ or $\mathcal{U}^{\mathrm{ad}}_{\underline{b}^+,\underline{b}^-}$, by adding extra generators $\{(\phi_i^{\pm})^{\pm 1}\}_{i\in I}$ to $\mathcal{U}^{\mathrm{sc}}_{\mu^+,\mu^-}$, which satisfy the following extra relations:

$$(\psi^{\epsilon}_{i,-\epsilon b_i^{\epsilon}})^{\pm 1} = (\phi_i^{\epsilon})^{\pm 2}\cdot\prod_{j-i}(\phi_j^{\epsilon})^{\pm c_{ji}},\ (\phi_i^{\epsilon})^{\pm 1}\cdot(\phi_i^{\epsilon})^{\mp 1}=1,\ [\phi_i^{\epsilon},\phi_j^{\epsilon'}]=0, \qquad \text{(U9)}$$

$$\phi_i^{\epsilon}\psi_j^{\epsilon'}(z)=\psi_j^{\epsilon'}(z)\phi_i^{\epsilon},\ \phi_i^{\epsilon}e_j(z)=\boldsymbol{v}_i^{\epsilon\delta_{ij}}e_j(z)\phi_i^{\epsilon},\ \phi_i^{\epsilon}f_j(z)=\boldsymbol{v}_i^{-\epsilon\delta_{ij}}f_j(z)\phi_i^{\epsilon}, \qquad \text{(U10)}$$

for any $i, j\in I$ and $\epsilon,\epsilon'\in\{\pm\}$.

Both algebras $\mathcal{U}^{\mathrm{sc}}_{\mu^+,\mu^-}$ and $\mathcal{U}^{\mathrm{ad}}_{\mu^+,\mu^-}$ depend only on $\mu := \mu^+ + \mu^-$ up to an isomorphism[4]. Let $\Lambda^{\pm}\subset\Lambda$ be the submonoids spanned by $\{\pm\omega_i\}_{i\in I}$, that is, Λ^+ (resp. Λ^-) consists of dominant (resp. antidominant) coweights of Λ. We will say that the algebras $\mathcal{U}^{\mathrm{sc}}_{\mu^+,\mu^-}, \mathcal{U}^{\mathrm{ad}}_{\mu^+,\mu^-}$ are dominantly (resp. antidominantly) shifted if $\mu\in\Lambda^+$ (resp. $\mu\in\Lambda^-$). We note that $\mu\in\Lambda^+ \Leftrightarrow b_i^+ + b_i^- = \alpha_i^{\vee}(\mu)\geq 0$, $\mu\in\Lambda^- \Leftrightarrow b_i^+ + b_i^- = \alpha_i^{\vee}(\mu)\leq 0$ for all $i\in I$.

Remark 5.2 One of the key reasons to consider $\mathcal{U}^{\mathrm{ad}}_{\mu^+,\mu^-}$, not only $\mathcal{U}^{\mathrm{sc}}_{\mu^+,\mu^-}$, is to construct quantizations of the thick slices $^{\dagger}\mathcal{W}_{\mu^*}$ of Sect. 4.8 and the multiplicative slice covers $^{\dagger}\hat{\mathcal{W}}^{\underline{\lambda}^*}_{\mu^*}$ of Sect. 4.6, see our Conjecture 8.14. On the technical side, we also need an alternative set of Cartan generators, whose generating series $A_i^{\pm}(z)$ are defined via (6.1) of Sect. 6 and whose definition requires to work with $\mathcal{U}^{\mathrm{ad}}_{\mu^+,\mu^-}$ (see also Remark 6.7(b)).

Remark 5.3

(a) The elements $\{\psi^+_{i,-b_i^+}\psi^-_{i,b_i^-}\}_{i\in I}$ (resp. $\{\phi_i^+\phi_i^-\}_{i\in I}$) and their inverses are central elements of $\mathcal{U}^{\mathrm{sc}}_{\mu^+,\mu^-}$ (resp. $\mathcal{U}^{\mathrm{ad}}_{\mu^+,\mu^-}$).
(b) We have $\mathcal{U}^{\mathrm{sc}}_{0,0}/(\psi^+_{i,0}\psi^-_{i,0}-1)\simeq U_{\boldsymbol{v}}(L\mathfrak{g})$, the standard quantum loop algebra of $\mathfrak{g}$, while $\mathcal{U}^{\mathrm{ad}}_{0,0}/(\phi_i^+\phi_i^- - 1)\simeq U^{\mathrm{ad}}_{\boldsymbol{v}}(L\mathfrak{g})$, the adjoint version of $U_{\boldsymbol{v}}(L\mathfrak{g})$.
(c) We note that defining relations (U1–U8, U10) are independent of μ^+,μ^-.
(d) An equivalent definition of $\mathcal{U}^{\mathrm{sc}}_{\mu_1,\mu_2}$ was suggested to us by Boris Feigin in Spring 2010. In this definition, we take the same generators as for $U_{\boldsymbol{v}}(L\mathfrak{g})$ and just modify relation (U6) by requesting $p_i(z)[e_i(z), f_j(w)] = \frac{\delta_{ij}\delta(z/w)}{\boldsymbol{v}_i-\boldsymbol{v}_i^{-1}}\left(\psi_i^+(z)-\psi_i^-(z)\right)$ for any collection $\{p_i(z)\}_{i\in I}$ of rational functions.

[4]For example, there is an isomorphism $\mathcal{U}^{\mathrm{sc}}_{\mu^+,\mu^-}\xrightarrow{\sim}\mathcal{U}^{\mathrm{sc}}_{0,\mu^++\mu^-}$ such that $f_i(z)\mapsto f_i(z), e_i(z)\mapsto z^{b_i^+}e_i(z), \psi_i^{\pm}(z)\mapsto z^{b_i^+}\psi_i^{\pm}(z)$.

5.2 Levendorskii Type Presentation of $\mathcal{U}^{\mathrm{sc}}_{0,\mu}$ for $\mu \in \Lambda^-$

In Sect. 10, we will crucially need a presentation of the shifted quantum affine algebras via a finite number of generators and defining relations. This is the purpose of this subsection.

Fix antidominant coweights $\mu_1, \mu_2 \in \Lambda^-$ and set $\mu := \mu_1 + \mu_2$. Define $b_{1,i} := \alpha_i^\vee(\mu_1)$, $b_{2,i} := \alpha_i^\vee(\mu_2)$, $b_i := b_{1,i} + b_{2,i}$. Denote by $\hat{\mathcal{U}}_{\mu_1,\mu_2}$ the associative $\mathbb{C}(\boldsymbol{v})$-algebra generated by

$$\{e_{i,r}, f_{i,s}, (\psi^+_{i,0})^{\pm 1}, (\psi^-_{i,b_i})^{\pm 1}, h_{i,\pm 1} | i \in I, b_{2,i} - 1 \leq r \leq 0, b_{1,i} \leq s \leq 1\}$$

and with the following defining relations:

$$\begin{gathered} \{(\psi^+_{i,0})^{\pm 1}, (\psi^-_{i,b_i})^{\pm 1}, h_{i,\pm 1}\}_{i\in I} \text{ pairwise commute,} \\ (\psi^+_{i,0})^{\pm 1} \cdot (\psi^+_{i,0})^{\mp 1} = (\psi^-_{i,b_i})^{\pm 1} \cdot (\psi^-_{i,b_i})^{\mp 1} = 1, \end{gathered} \tag{$\hat{\text{U}}1$}$$

$$e_{i,r+1}e_{j,s} - \boldsymbol{v}_i^{c_{ij}} e_{i,r}e_{j,s+1} = \boldsymbol{v}_i^{c_{ij}} e_{j,s}e_{i,r+1} - e_{j,s+1}e_{i,r}, \tag{$\hat{\text{U}}2$}$$

$$\boldsymbol{v}_i^{c_{ij}} f_{i,r+1}f_{j,s} - f_{i,r}f_{j,s+1} = f_{j,s}f_{i,r+1} - \boldsymbol{v}_i^{c_{ij}} f_{j,s+1}f_{i,r}, \tag{$\hat{\text{U}}3$}$$

$$\psi^+_{i,0}e_{j,r} = \boldsymbol{v}_i^{c_{ij}} e_{j,r}\psi^+_{i,0},\ \psi^-_{i,b_i}e_{j,r} = \boldsymbol{v}_i^{-c_{ij}} e_{j,r}\psi^-_{i,b_i},\ [h_{i,\pm 1}, e_{j,r}] = [c_{ij}]_{\boldsymbol{v}_i} \cdot e_{j,r\pm 1}, \tag{$\hat{\text{U}}4$}$$

$$\psi^+_{i,0}f_{j,s} = \boldsymbol{v}_i^{-c_{ij}} f_{j,s}\psi^+_{i,0},\ \psi^-_{i,b_i}f_{j,s} = \boldsymbol{v}_i^{c_{ij}} f_{j,s}\psi^-_{i,b_i},\ [h_{i,\pm 1}, f_{j,s}] = -[c_{ij}]_{\boldsymbol{v}_i} \cdot f_{j,s\pm 1}, \tag{$\hat{\text{U}}5$}$$

$$[e_{i,r}, f_{j,s}] = 0 \text{ if } i \neq j \text{ and } [e_{i,r}, f_{i,s}] = \begin{cases} \psi^+_{i,0}h_{i,1} & \text{if } r+s=1, \\ \psi^-_{i,b_i}h_{i,-1} & \text{if } r+s=b_i-1, \\ \frac{\psi^+_{i,0} - \delta_{b_i,0}\psi^-_{i,b_i}}{\boldsymbol{v}_i - \boldsymbol{v}_i^{-1}} & \text{if } r+s=0, \\ \frac{-\psi^-_{i,b_i} + \delta_{b_i,0}\psi^+_{i,0}}{\boldsymbol{v}_i - \boldsymbol{v}_i^{-1}} & \text{if } r+s=b_i, \\ 0 & \text{if } b_i < r+s < 0, \end{cases} \tag{$\hat{\text{U}}6$}$$

$$[e_{i,0}, [e_{i,0}, \cdots, [e_{i,0}, e_{j,0}]_{\boldsymbol{v}_i^{c_{ij}}} \cdots]_{\boldsymbol{v}_i^{-c_{ij}-2}}]_{\boldsymbol{v}_i^{-c_{ij}}} = 0 \text{ for } i \neq j, \tag{$\hat{\text{U}}7$}$$

$$[f_{i,0}, [f_{i,0}, \cdots, [f_{i,0}, f_{j,0}]_{\boldsymbol{v}_i^{c_{ij}}} \cdots]_{\boldsymbol{v}_i^{-c_{ij}-2}}]_{\boldsymbol{v}_i^{-c_{ij}}} = 0 \text{ for } i \neq j, \tag{$\hat{\text{U}}8$}$$

$$[h_{i,1}, [f_{i,1}, [h_{i,1}, e_{i,0}]]] = 0,\ [h_{i,-1}, [e_{i,b_{2,i}-1}, [h_{i,-1}, f_{i,b_{1,i}}]]] = 0, \tag{$\hat{\text{U}}9$}$$

for any $i, j \in I$ and r, s such that the above relations make sense.

Remark 5.4 One can rewrite relations (Û7, Û8) in the form similar to (U7, U8) as

$$\sum_{r=0}^{1-c_{ij}} (-1)^r \begin{bmatrix} 1-c_{ij} \\ r \end{bmatrix}_{\boldsymbol{v}_i} e_{i,0}^r e_{j,0} e_{i,0}^{1-c_{ij}-r} = 0, \quad \sum_{r=0}^{1-c_{ij}} (-1)^r \begin{bmatrix} 1-c_{ij} \\ r \end{bmatrix}_{\boldsymbol{v}_i} f_{i,0}^r f_{j,0} f_{i,0}^{1-c_{ij}-r} = 0.$$

Define inductively

$$e_{i,r} := [2]_{\boldsymbol{v}_i}^{-1} \cdot \begin{cases} [h_{i,1}, e_{i,r-1}] & \text{if } r > 0, \\ [h_{i,-1}, e_{i,r+1}] & \text{if } r < b_{2,i} - 1, \end{cases}$$

$$f_{i,r} := -[2]_{\boldsymbol{v}_i}^{-1} \cdot \begin{cases} [h_{i,1}, f_{i,r-1}] & \text{if } r > 1, \\ [h_{i,-1}, f_{i,r+1}] & \text{if } r < b_{1,i}, \end{cases}$$

$$\psi_{i,r}^+ := (\boldsymbol{v}_i - \boldsymbol{v}_i^{-1}) \cdot [e_{i,r-1}, f_{i,1}] \text{ for } r > 0,$$

$$\psi_{i,r}^- := (\boldsymbol{v}_i^{-1} - \boldsymbol{v}_i) \cdot [e_{i,r-b_{1,i}}, f_{i,b_{1,i}}] \text{ for } r < b_i.$$

Theorem 5.5 *There is a unique* $\mathbb{C}(\boldsymbol{v})$*-algebra isomorphism* $\hat{\mathfrak{U}}_{\mu_1,\mu_2} \xrightarrow{\sim} \mathfrak{U}^{\mathrm{sc}}_{0,\mu}$*, such that*

$$e_{i,r} \mapsto e_{i,r},\ f_{i,r} \mapsto f_{i,r},\ \psi_{i,\pm s_i^\pm}^\pm \mapsto \psi_{i,\pm s_i^\pm}^\pm \quad \text{for } i \in I, r \in \mathbb{Z}, s_i^+ \geq 0, s_i^- \geq -b_i.$$

This provides a new presentation of $\mathfrak{U}^{\mathrm{sc}}_{0,\mu}$ via a finite number of generators and relations. The proof of this result is presented in Appendix A. Motivated by Guay et al. [33], we also provide a slight modification of this presentation of $\mathfrak{U}^{\mathrm{sc}}_{0,\mu}$ in Theorem A.3.

Remark 5.6 Theorem 5.5 can be viewed as a $\boldsymbol{v}$-version of the corresponding result for the shifted Yangians of [24, Theorem 4.3]. In the particular case $\mu_1 = \mu_2 = 0$, the latter is the standard Levendorskii presentation of the Yangian, see [47]. However, we are not aware of the reference for Theorem 5.5 even in the *unshifted case* $\mu_1 = \mu_2 = 0$.

6 *ABCD* Generators of $\mathfrak{U}^{\mathrm{ad}}_{\mu^+,\mu^-}$

In this section, we introduce an alternative set of generators of $\mathfrak{U}^{\mathrm{ad}}_{\mu^+,\mu^-}$, which will be used later in the paper (they are also of independent interest), and deduce the defining relations among them. While the definition works for any two coweights $\mu^+, \mu^- \in \Lambda$, the relations hold only for antidominant $\mu^+, \mu^- \in \Lambda^-$, which we assume from now on.

First, we define the Cartan generators $\{A^{\pm}_{i,\pm r}\}_{i\in I}^{r\geq 0}$ via

$$z^{\mp b_i^{\pm}}\psi_i^{\pm}(z)=\frac{\prod_{j-i}\prod_{p=1}^{-c_{ji}}A_j^{\pm}(\boldsymbol{v}_j^{-c_{ji}-2p}z)}{A_i^{\pm}(z)A_i^{\pm}(\boldsymbol{v}_i^{-2}z)}\ \text{with}\ A^{\pm}_{i,0}:=(\phi_i^{\pm})^{-1}, \tag{6.1}$$

where we set $A_i^{\pm}(z)=\sum_{r\geq 0}A^{\pm}_{i,\pm r}z^{\mp r}$. Using non-degeneracy of the $\boldsymbol{v}$-version of the Cartan matrix (c_{ij}) and arguing by induction in $r>0$, one can easily see that relations (6.1) for all $i\in I$ determine uniquely all $A^{\pm}_{i,\pm r}$, see Remark B.2 (cf. [30, Lemma 2.1]). An explicit formula for $A_i^{\pm}(z)$ is given by (B.2) in Appendix B.

Next, we introduce the generating series $B_i^{\pm}(z), C_i^{\pm}(z), D_i^{\pm}(z)$ via

$$B_i^{\pm}(z):=(\boldsymbol{v}_i-\boldsymbol{v}_i^{-1})A_i^{\pm}(z)e_i^{\pm}(z), \tag{6.2}$$

$$C_i^{\pm}(z):=(\boldsymbol{v}_i-\boldsymbol{v}_i^{-1})f_i^{\pm}(z)A_i^{\pm}(z), \tag{6.3}$$

$$D_i^{\pm}(z):=A_i^{\pm}(z)\psi_i^{\pm}(z)+(\boldsymbol{v}_i-\boldsymbol{v}_i^{-1})^2f_i^{\pm}(z)A_i^{\pm}(z)e_i^{\pm}(z), \tag{6.4}$$

where the *Drinfeld half-currents* are defined as follows:

$$\begin{gathered}e_i^{+}(z):=\sum_{r\geq 0}e_{i,r}z^{-r},\ e_i^{-}(z):=-\sum_{r<0}e_{i,r}z^{-r},\\ f_i^{+}(z):=\sum_{r>0}f_{i,r}z^{-r},\ f_i^{-}(z):=-\sum_{r\leq 0}f_{i,r}z^{-r}.\end{gathered} \tag{6.5}$$

It is clear that coefficients of the generating series $\{A_i^{\pm}(z), B_i^{\pm}(z), C_i^{\pm}(z), D_i^{\pm}(z)\}_{i\in I}$ together with $\{\phi_i^{\pm}\}_{i\in I}$ generate (over $\mathbb{C}(\boldsymbol{v})$) the shifted quantum affine algebra $\mathfrak{U}^{\mathrm{ad}}_{\mu^+,\mu^-}$. The following is the key result of this section.

Theorem 6.6 *Assume $\mu^+,\mu^-\in\Lambda^-$ and define $\{b_i^{\pm}\}_{i\in I}$ via $b_i^{\pm}:=\alpha_i^{\vee}(\mu^{\pm})$ as before.*

(a) The generating series $A_i^{\pm}(z), B_i^{\pm}(z), C_i^{\pm}(z), D_i^{\pm}(z)$ satisfy the following relations:

$$\begin{gathered}\phi_i^{\epsilon}A_j^{\epsilon'}(w)=A_j^{\epsilon'}(w)\phi_i^{\epsilon},\ \phi_i^{\epsilon}D_j^{\epsilon'}(w)=D_j^{\epsilon'}(w)\phi_i^{\epsilon},\\ \phi_i^{\epsilon}B_j^{\epsilon'}(w)=\boldsymbol{v}_i^{\epsilon\delta_{ij}}B_j^{\epsilon'}(w)\phi_i^{\epsilon},\ \phi_i^{\epsilon}C_j^{\epsilon'}(w)=\boldsymbol{v}_i^{-\epsilon\delta_{ij}}C_j^{\epsilon'}(w)\phi_i^{\epsilon},\end{gathered} \tag{6.6}$$

$$[A_i^{\epsilon}(z),A_j^{\epsilon'}(w)]=0, \tag{6.7}$$

$$[A_i^{\epsilon}(z),B_j^{\epsilon'}(w)]=[A_i^{\epsilon}(z),C_j^{\epsilon'}(w)]=[B_i^{\epsilon}(z),C_j^{\epsilon'}(w)]=0\ \text{for}\ i\neq j, \tag{6.8}$$

$$[B_i^\epsilon(z), B_i^{\epsilon'}(w)] = [C_i^\epsilon(z), C_i^{\epsilon'}(w)] = [D_i^\epsilon(z), D_i^{\epsilon'}(w)] = 0, \tag{6.9}$$

$$(z-w)[B_i^{\epsilon'}(w), A_i^\epsilon(z)]_{\boldsymbol{v}_i^{-1}} = (\boldsymbol{v}_i - \boldsymbol{v}_i^{-1})\left(zA_i^\epsilon(z)B_i^{\epsilon'}(w) - wA_i^{\epsilon'}(w)B_i^\epsilon(z)\right), \tag{6.10}$$

$$(z-w)[A_i^\epsilon(z), C_i^{\epsilon'}(w)]_{\boldsymbol{v}_i} = (\boldsymbol{v}_i - \boldsymbol{v}_i^{-1})\left(wC_i^{\epsilon'}(w)A_i^\epsilon(z) - zC_i^\epsilon(z)A_i^{\epsilon'}(w)\right), \tag{6.11}$$

$$(z-w)[B_i^\epsilon(z), C_i^{\epsilon'}(w)] = (\boldsymbol{v}_i - \boldsymbol{v}_i^{-1})z\left(D_i^{\epsilon'}(w)A_i^\epsilon(z) - D_i^\epsilon(z)A_i^{\epsilon'}(w)\right), \tag{6.12}$$

$$(z-w)[B_i^\epsilon(z), D_i^{\epsilon'}(w)]_{\boldsymbol{v}_i} = (\boldsymbol{v}_i - \boldsymbol{v}_i^{-1})\left(wD_i^{\epsilon'}(w)B_i^\epsilon(z) - zD_i^\epsilon(z)B_i^{\epsilon'}(w)\right), \tag{6.13}$$

$$(z-w)[D_i^{\epsilon'}(w), C_i^\epsilon(z)]_{\boldsymbol{v}_i^{-1}} = (\boldsymbol{v}_i - \boldsymbol{v}_i^{-1})\left(zC_i^\epsilon(z)D_i^{\epsilon'}(w) - wC_i^{\epsilon'}(w)D_i^\epsilon(z)\right), \tag{6.14}$$

$$(z-w)[A_i^\epsilon(z), D_i^{\epsilon'}(w)] = (\boldsymbol{v}_i - \boldsymbol{v}_i^{-1})\left(wC_i^{\epsilon'}(w)B_i^\epsilon(z) - zC_i^\epsilon(z)B_i^{\epsilon'}(w)\right), \tag{6.15}$$

$$A_i^\epsilon(z)D_i^\epsilon(\boldsymbol{v}_i^{-2}z) - \boldsymbol{v}_i^{-1}B_i^\epsilon(z)C_i^\epsilon(\boldsymbol{v}_i^{-2}z) = z^{\epsilon b_i^\epsilon}\cdot\prod_{j-i}\prod_{p=1}^{-c_{ji}} A_j^\epsilon(\boldsymbol{v}_j^{-c_{ji}-2p}z), \tag{6.16}$$

$$\begin{aligned}&(z-\boldsymbol{v}_i^{c_{ij}}w)B_i^\epsilon(z)B_j^{\epsilon'}(w) - (\boldsymbol{v}_i^{c_{ij}}z - w)B_j^{\epsilon'}(w)B_i^\epsilon(z) = \\ &zA_i^\epsilon(z)[\phi_i^+B_{i,0}^+, B_j^{\epsilon'}(w)]_{\boldsymbol{v}_i^{c_{ij}}} + wA_j^{\epsilon'}(w)[\phi_j^+B_{j,0}^+, B_i^\epsilon(z)]_{\boldsymbol{v}_i^{c_{ij}}} \text{ for } i\neq j,\end{aligned} \tag{6.17}$$

$$\begin{aligned}&(\boldsymbol{v}_i^{c_{ij}}z - w)C_i^\epsilon(z)C_j^{\epsilon'}(w) - (z-\boldsymbol{v}_i^{c_{ij}}w)C_j^{\epsilon'}(w)C_i^\epsilon(z) = \\ &-[C_i^\epsilon(z), C_{j,1}^+\phi_j^+]_{\boldsymbol{v}_i^{c_{ij}}}A_j^{\epsilon'}(w) - [C_j^{\epsilon'}(w), C_{i,1}^+\phi_i^+]_{\boldsymbol{v}_i^{c_{ij}}}A_i^\epsilon(z) \text{ for } i\neq j,\end{aligned} \tag{6.18}$$

$$\begin{aligned}&\underset{z_1,\ldots,z_{1-c_{ij}}}{\mathrm{Sym}}\left\{\prod_{a<b}(\boldsymbol{v}_iz_a - \boldsymbol{v}_i^{-1}z_b)(z_a-z_b)\cdot\right.\\ &\left.\sum_{r=0}^{1-c_{ij}}(-1)^r\begin{bmatrix}1-c_{ij}\\ r\end{bmatrix}_{\boldsymbol{v}_i} B_i^{\epsilon_1}(z_1)\cdots B_i^{\epsilon_r}(z_r)B_j^{\epsilon'}(w)B_i^{\epsilon_{r+1}}(z_{r+1})\cdots B_i^{\epsilon_{1-c_{ij}}}(z_{1-c_{ij}})\right\} = 0,\end{aligned} \tag{6.19}$$

$$\begin{aligned}&\underset{z_1,\ldots,z_{1-c_{ij}}}{\mathrm{Sym}}\left\{\prod_{a<b}(\boldsymbol{v}_iz_b - \boldsymbol{v}_i^{-1}z_a)(z_b-z_a)\cdot\right.\\ &\left.\sum_{r=0}^{1-c_{ij}}(-1)^r\begin{bmatrix}1-c_{ij}\\ r\end{bmatrix}_{\boldsymbol{v}_i} C_i^{\epsilon_1}(z_1)\cdots C_i^{\epsilon_r}(z_r)C_j^{\epsilon'}(w)C_i^{\epsilon_{r+1}}(z_{r+1})\cdots C_i^{\epsilon_{1-c_{ij}}}(z_{1-c_{ij}})\right\} = 0,\end{aligned} \tag{6.20}$$

for any $i, j \in I$ *and* $\epsilon, \epsilon', \epsilon_1, \ldots, \epsilon_{1-c_{ij}} \in \{\pm\}$.

(b) *Relations (6.6–6.20) are the defining relations. In other words, the associative* $\mathbb{C}(v)$*-algebra generated by* $\{\phi_i^\pm, A_{i,\pm r}^\pm, B_{i,r}^+, B_{i,-r-1}^-, C_{i,r+1}^+, C_{i,-r}^-, D_{i,\pm r\pm b_i^\pm}^\pm\}_{i\in I}^{r\in\mathbb{N}}$ *with the defining relations (6.6–6.20) is isomorphic to* $\mathcal{U}^{\mathrm{ad}}_{\mu^+,\mu^-}$.

We sketch the proof in Appendix B. In the unshifted case, more precisely for $U_{\boldsymbol{v}}^{\mathrm{ad}}(L\mathfrak{g})$, the above construction should be viewed as a $\boldsymbol{v}$-version of that of [30]. In *loc.cit.*, the authors introduced analogous generating series $\{A_i(u), B_i(u), C_i(u), D_i(u)\}_{i\in I}$ with coefficients in the Yangian $Y(\mathfrak{g})$ and stated (without a proof) the relations between them, similar to (6.7–6.16).[5] Meanwhile, we note that adding rational analogues of (6.17–6.20) to their list of relations, we get a complete list of the defining relations among these generating series.

Remark 6.7

(a) For $\mathfrak{g} = \mathfrak{sl}_2$, relations (6.7, 6.9–6.15) are equivalent to the RTT-relations (with the trigonometric R-matrix of (11.3)), see our proof of Theorem 11.11 below.
(b) This construction can be adapted to the setting of $\mathcal{U}^{\mathrm{sc}}_{\mu^+,\mu^-}$. First, we redefine the generating series $A_i^\pm(z) = 1 + \sum_{r>0} A_{i,\pm r}^\pm z^{\mp r}$ which have to satisfy

$$z^{\mp b_i^\pm}(\psi_{i,\mp b_i^\pm}^\pm)^{-1}\psi_i^\pm(z) = \frac{\prod_{j-i}\prod_{p=1}^{-c_{ji}} A_j^\pm(\boldsymbol{v}_j^{-c_{ji}-2p}z)}{A_i^\pm(z)A_i^\pm(\boldsymbol{v}_i^{-2}z)}. \tag{6.21}$$

Next, we define $B_i^\pm(z)$, $C_i^\pm(z)$ via formulas (6.2, 6.3). Finally, we define $D_i^\pm(z)$ via

$$D_i^\pm(z) := A_i^\pm(z)\psi_i^\pm(z) + \boldsymbol{v}_i^{\mp 1}(\boldsymbol{v}_i - \boldsymbol{v}_i^{-1})^2 f_i^\pm(z)A_i^\pm(z)e_i^\pm(z). \tag{6.22}$$

The coefficients of these generating series together with $\{(\psi_{i,-\epsilon b_i^\epsilon}^\epsilon)^{\pm 1}\}_{i\in I}^{\epsilon=\pm}$ generate $\mathcal{U}^{\mathrm{sc}}_{\mu^+,\mu^-}$. For $\mu^+, \mu^- \in \Lambda^-$ one can write a complete list of the defining relations among these generators, which look similar to (6.7–6.20).

7 Homomorphism to Difference Operators

In this section, we construct homomorphisms from the shifted quantum affine algebras to the algebras of difference operators.

[5] We note that the relation $[D_i(u), D_i(v)] = 0$ was missing in their list.

7.1 Homomorphism $\widetilde{\Phi}^{\lambda}_{\underline{\mu}}$

Let $\mathrm{Dyn}(\mathfrak{g})$ be the graph obtained from the Dynkin diagram of $\mathfrak{g}$ by replacing all multiple edges by simple ones. We fix an orientation of $\mathrm{Dyn}(\mathfrak{g})$ and we fix a dominant coweight $\lambda \in \Lambda^+$ and a coweight $\mu \in \Lambda$, such that $\lambda - \mu = \sum_{i\in I} a_i\alpha_i$ with $a_i \in \mathbb{N}$. We also fix a sequence $\underline{\lambda} = (\omega_{i_1}, \ldots, \omega_{i_N})$ of fundamental coweights, such that $\sum_{s=1}^N \omega_{i_s} = \lambda$.

Consider the associative $\mathbb{C}[\boldsymbol{v}^{\pm 1}]$-algebra $\hat{\mathcal{A}}^{\boldsymbol{v}}$ generated by $\{D_{i,r}^{\pm 1}, \mathsf{w}_{i,r}^{\pm 1/2}\}_{i\in I}^{1\le r\le a_i}$ with the defining relations (for all $i, j \in I$, $1 \le r \le a_i$, $1 \le s \le a_j$):

$$[D_{i,r}, D_{j,s}] = [\mathsf{w}_{i,r}^{1/2}, \mathsf{w}_{j,s}^{1/2}] = 0,\ D_{i,r}^{\pm 1} D_{i,r}^{\mp 1} = \mathsf{w}_{i,r}^{\pm 1/2}\mathsf{w}_{i,r}^{\mp 1/2} = 1,\ D_{i,r}\mathsf{w}_{j,s}^{1/2} = \boldsymbol{v}_i^{\delta_{ij}\delta_{rs}}\mathsf{w}_{j,s}^{1/2} D_{i,r}.$$

Let $\widetilde{\mathcal{A}}^{\boldsymbol{v}}$ be the localization of $\hat{\mathcal{A}}^{\boldsymbol{v}}$ by the multiplicative set generated by $\{\mathsf{w}_{i,r} - \boldsymbol{v}_i^m \mathsf{w}_{i,s}\}_{i\in I, m\in\mathbb{Z}}^{1\le r\ne s\le a_i} \cup \{1 - \boldsymbol{v}^m\}_{m\in\mathbb{Z}\setminus\{0\}}$ (which obviously satisfies Ore conditions). We also define their $\mathbb{C}(\boldsymbol{v})$-counterparts $\hat{\mathcal{A}}^{\boldsymbol{v}}_{\mathrm{frac}} := \hat{\mathcal{A}}^{\boldsymbol{v}} \otimes_{\mathbb{C}[\boldsymbol{v}^{\pm 1}]} \mathbb{C}(\boldsymbol{v})$ and $\widetilde{\mathcal{A}}^{\boldsymbol{v}}_{\mathrm{frac}} := \widetilde{\mathcal{A}}^{\boldsymbol{v}} \otimes_{\mathbb{C}[\boldsymbol{v}^{\pm 1}]} \mathbb{C}(\boldsymbol{v})$.

In what follows, we will work with the larger algebra $\mathcal{U}^{\mathrm{ad}}_{0,\mu}[\mathsf{z}_1^{\pm 1}, \ldots, \mathsf{z}_N^{\pm 1}]$, which is obtained from $\mathcal{U}^{\mathrm{sc}}_{0,\mu}[\mathsf{z}_1^{\pm 1}, \ldots, \mathsf{z}_N^{\pm 1}] := \mathcal{U}^{\mathrm{sc}}_{0,\mu} \otimes_{\mathbb{C}(\boldsymbol{v})} \mathbb{C}(\boldsymbol{v})[\mathsf{z}_1^{\pm 1}, \ldots, \mathsf{z}_N^{\pm 1}]$ by adding extra generators $\{(\phi_i^{\epsilon})^{\pm 1}\}_{i\in I}^{\epsilon=\pm}$ satisfying relations (U9, U10) with the only change:

$$\prod_{s:i_s=i} (-\boldsymbol{v}_i \mathsf{z}_s)^{\mp 1} \cdot (\psi^-_{i,\alpha_i^\vee(\mu)})^{\pm 1} = (\phi_i^-)^{\pm 2} \cdot \prod_{j-i} (\phi_j^-)^{\pm c_{ji}}.$$

We will also work with the larger algebras $\widetilde{\mathcal{A}}^{\boldsymbol{v}}[\mathsf{z}_1^{\pm 1}, \ldots, \mathsf{z}_N^{\pm 1}] := \widetilde{\mathcal{A}}^{\boldsymbol{v}} \otimes_{\mathbb{C}[\boldsymbol{v}^{\pm 1}]} \mathbb{C}[\boldsymbol{v}^{\pm 1}][\mathsf{z}_1^{\pm 1}, \ldots, \mathsf{z}_N^{\pm 1}]$ and $\widetilde{\mathcal{A}}^{\boldsymbol{v}}_{\mathrm{frac}}[\mathsf{z}_1^{\pm 1}, \ldots, \mathsf{z}_N^{\pm 1}] := \widetilde{\mathcal{A}}^{\boldsymbol{v}}_{\mathrm{frac}} \otimes_{\mathbb{C}(\boldsymbol{v})} \mathbb{C}(\boldsymbol{v})[\mathsf{z}_1^{\pm 1}, \ldots, \mathsf{z}_N^{\pm 1}]$.

Define

$$Z_i(z) := \prod_{1\le s\le N}^{i_s=i} \left(1 - \frac{\boldsymbol{v}_i \mathsf{z}_s}{z}\right),\ W_i(z) := \prod_{r=1}^{a_i} \left(1 - \frac{\mathsf{w}_{i,r}}{z}\right),\ W_{i,r}(z) := \prod_{1\le s\le a_i}^{s\ne r} \left(1 - \frac{\mathsf{w}_{i,s}}{z}\right),$$

$$\hat{Z}_i(z) := \prod_{1\le s\le N}^{i_s=i} \left(1 - \frac{z}{\boldsymbol{v}_i \mathsf{z}_s}\right),\ \hat{W}_i(z) := \prod_{r=1}^{a_i} \left(1 - \frac{z}{\mathsf{w}_{i,r}}\right),\ \hat{W}_{i,r}(z) := \prod_{1\le s\le a_i}^{s\ne r} \left(1 - \frac{z}{\mathsf{w}_{i,s}}\right).$$

The following is the key result of this section.

Theorem 7.1 *There exists a unique $\mathbb{C}(\boldsymbol{v})[\mathsf{z}_1^{\pm 1}, \ldots, \mathsf{z}_N^{\pm 1}]$-algebra homomorphism*

$$\widetilde{\Phi}^{\lambda}_{\underline{\mu}} : \mathcal{U}^{\mathrm{ad}}_{0,\mu}[\mathsf{z}_1^{\pm 1}, \ldots, \mathsf{z}_N^{\pm 1}] \longrightarrow \widetilde{\mathcal{A}}^{\boldsymbol{v}}_{\mathrm{frac}}[\mathsf{z}_1^{\pm 1}, \ldots, \mathsf{z}_N^{\pm 1}],$$

such that

$$e_i(z) \mapsto \frac{-\boldsymbol{v}_i}{1-\boldsymbol{v}_i^2}\prod_{t=1}^{a_i}\mathsf{w}_{i,t}\prod_{j\to i}\prod_{t=1}^{a_j}\mathsf{w}_{j,t}^{c_{ji}/2}\cdot\sum_{r=1}^{a_i}\delta\left(\frac{\mathsf{w}_{i,r}}{z}\right)\frac{Z_i(\mathsf{w}_{i,r})}{W_{i,r}(\mathsf{w}_{i,r})}\prod_{j\to i}\prod_{p=1}^{-c_{ji}}W_j(\boldsymbol{v}_j^{-c_{ji}-2p}z)D_{i,r}^{-1},$$

$$f_i(z) \mapsto \frac{1}{1-\boldsymbol{v}_i^2}\prod_{j\leftarrow i}\prod_{t=1}^{a_j}\mathsf{w}_{j,t}^{c_{ji}/2}\cdot\sum_{r=1}^{a_i}\delta\left(\frac{\boldsymbol{v}_i^2\mathsf{w}_{i,r}}{z}\right)\frac{1}{W_{i,r}(\mathsf{w}_{i,r})}\prod_{j\leftarrow i}\prod_{p=1}^{-c_{ji}}W_j(\boldsymbol{v}_j^{-c_{ji}-2p}z)D_{i,r},$$

$$\psi_i^{\pm}(z) \mapsto \prod_{t=1}^{a_i}\mathsf{w}_{i,t}\prod_{j-i}\prod_{t=1}^{a_j}\mathsf{w}_{j,t}^{c_{ji}/2}\cdot\left(\frac{Z_i(z)}{W_i(z)W_i(\boldsymbol{v}_i^{-2}z)}\prod_{j-i}\prod_{p=1}^{-c_{ji}}W_j(\boldsymbol{v}_j^{-c_{ji}-2p}z)\right)^{\pm},$$

$$(\phi_i^+)^{\pm 1} \mapsto \prod_{t=1}^{a_i}\mathsf{w}_{i,t}^{\pm 1/2},\ (\phi_i^-)^{\pm 1} \mapsto (-\boldsymbol{v}_i)^{\mp a_i}\prod_{t=1}^{a_i}\mathsf{w}_{i,t}^{\mp 1/2}.$$

We write $\gamma(z)^{\pm}$ *for the expansion of a rational function* $\gamma(z)$ *in* $z^{\mp 1}$*, respectively.*

In the *unshifted case*, more precisely for $U_{\boldsymbol{v}}(L\mathfrak{g})$, this result was stated (without a proof) in [31]. The above formulas simplify for simply-laced $\mathfrak{g}$, in which case this result can be viewed as a $\boldsymbol{v}$-version of [10, Corollary B.17]. We present the proof in Appendix C.

7.2 Homomorphism $\widetilde{\Phi}^{\lambda}_{\mu}$ in *ABC* Generators

Generalizing the construction of Sect. 6, we define new Cartan generators $\{A^{\pm}_{i,\pm r}\}_{i\in I}^{r\geq 0}$ of $\mathfrak{U}^{\mathrm{ad}}_{0,\mu}[\mathsf{z}_1^{\pm 1},\ldots,\mathsf{z}_N^{\pm 1}]$ via

$$A^{\pm}_{i,0} := (\phi_i^{\pm})^{-1},$$

$$\frac{\psi_i^+(z)}{Z_i(z)} = \frac{\prod_{j-i}\prod_{p=1}^{-c_{ji}}A_j^+(\boldsymbol{v}_j^{-c_{ji}-2p}z)}{A_i^+(z)A_i^+(\boldsymbol{v}_i^{-2}z)},$$

$$\frac{z^{\alpha_i^\vee(\mu)}\psi_i^-(z)}{\prod_{s:i_s=i}(-\boldsymbol{v}_i\mathsf{z}_s)\cdot\hat{Z}_i(z)} = \frac{\prod_{j-i}\prod_{p=1}^{-c_{ji}}A_j^-(\boldsymbol{v}_j^{-c_{ji}-2p}z)}{A_i^-(z)A_i^-(\boldsymbol{v}_i^{-2}z)},$$

where we set $A_i^{\pm}(z) := \sum_{r\geq 0}A^{\pm}_{i,\pm r}z^{\mp r}$. We also define the generating series $B_i^{\pm}(z)$, $C_i^{\pm}(z)$, and $D_i^{\pm}(z)$ via formulas (6.2), (6.3), and (6.4), respectively.

Lemma 7.2 *For antidominant* $\mu \in \Lambda^-$, *the generating series* $A_i^\pm(z), B_i^\pm(z), C_i^\pm(z), D_i^\pm(z)$ *satisfy relations (6.7–6.15).*

Proof Let c be the determinant of the Cartan matrix of $\mathfrak{g}$. Choose unique $\lambda_i^+(z) \in 1+z^{-1}\mathbb{C}(\boldsymbol{v})[\mathsf{z}_1^{\pm 1},\ldots,\mathsf{z}_N^{\pm 1}][[z^{-1}]]$, such that $Z_i(z) = \frac{\lambda_i^+(z)\lambda_i^+(\boldsymbol{v}_i^{-2}z)}{\prod_{j-i}\prod_{p=1}^{-c_{ji}}\lambda_j^+(\boldsymbol{v}_j^{-c_{ji}-2p}z)}$. Also choose $\lambda_i^-(z) \in \mathbb{C}(\boldsymbol{v}^{1/c})[\mathsf{z}_1^{\pm 1/c},\ldots,\mathsf{z}_N^{\pm 1/c}][[z]]$, such that $\hat{Z}_i(z)\cdot\prod_{s:i_s=i}(-\boldsymbol{v}_i\mathsf{z}_s) = \frac{\lambda_i^-(z)\lambda_i^-(\boldsymbol{v}_i^{-2}z)}{\prod_{j-i}\prod_{p=1}^{-c_{ji}}\lambda_j^-(\boldsymbol{v}_j^{-c_{ji}-2p}z)}$.

Then, the series $\lambda_i^\pm(z)^{-1}X_i^\pm(z)$ for $X = A, B, C, D$ are those of Sect. 6. The result follows from Theorem 6.6(a) (compare with the proof of [44, Proposition 5.5]). □

Corollary 7.3 *The following equalities hold in* $\mathcal{U}^{\mathrm{ad}}_{0,\mu}[\mathsf{z}_1^{\pm 1},\ldots,\mathsf{z}_N^{\pm 1}]$:

$$B_i^+(z) = [e_{i,0}, A_i^+(z)]_{\boldsymbol{v}_i^{-1}},\ C_i^+(z) = [z^{-1}A_i^+(z), f_{i,1}]_{\boldsymbol{v}_i^{-1}},$$

$$B_i^-(z) = [e_{i,-1}, zA_i^-(z)]_{\boldsymbol{v}_i},\ C_i^-(z) = [A_i^-(z), f_{i,0}]_{\boldsymbol{v}_i}.$$

Proof The above formula for $B_i^+(z)$ (resp. $C_i^+(z)$) follows by evaluating the terms of degree 1 (resp. 0) in w in the equality (6.10) (resp. (6.11)) with $\epsilon = \epsilon' = +$.

The formulas for $B_i^-(z), C_i^-(z)$ are proved analogously. □

The following result is straightforward.

Proposition 7.4 *The homomorphism* $\widetilde{\Phi}_\mu^\lambda$ *maps the* ABC *currents as follows:*

$$A_i^+(z) \mapsto \prod_{t=1}^{a_i}\mathsf{w}_{i,t}^{-1/2}\cdot W_i(z),\ A_i^-(z) \mapsto (-\boldsymbol{v}_i)^{a_i}\prod_{t=1}^{a_i}\mathsf{w}_{i,t}^{1/2}\cdot \hat{W}_i(z),$$

$$B_i^+(z) \mapsto \prod_{t=1}^{a_i}\mathsf{w}_{i,t}^{1/2}\prod_{j\to i}\prod_{t=1}^{a_j}\mathsf{w}_{j,t}^{c_{ji}/2}\cdot\sum_{r=1}^{a_i}\frac{W_{i,r}(z)Z_i(\mathsf{w}_{i,r})}{W_{i,r}(\mathsf{w}_{i,r})}\prod_{j\to i}\prod_{p=1}^{-c_{ji}}W_j(\boldsymbol{v}_j^{-c_{ji}-2p}\mathsf{w}_{i,r})D_{i,r}^{-1},$$

$$B_i^-(z) \mapsto -(-\boldsymbol{v}_i)^{a_i}\prod_{t=1}^{a_i}\mathsf{w}_{i,t}^{3/2}\prod_{j\to i}\prod_{t=1}^{a_j}\mathsf{w}_{j,t}^{c_{ji}/2}\cdot\sum_{r=1}^{a_i}\frac{z\hat{W}_{i,r}(z)Z_i(\mathsf{w}_{i,r})}{\mathsf{w}_{i,r}W_{i,r}(\mathsf{w}_{i,r})}\prod_{j\to i}\prod_{p=1}^{-c_{ji}}W_j(\boldsymbol{v}_j^{-c_{ji}-2p}\mathsf{w}_{i,r})D_{i,r}^{-1},$$

$$C_i^+(z) \mapsto -\prod_{t=1}^{a_i}\mathsf{w}_{i,t}^{-1/2}\prod_{j\leftarrow i}\prod_{t=1}^{a_j}\mathsf{w}_{j,t}^{c_{ji}/2}\cdot\sum_{r=1}^{a_i}\frac{\mathsf{w}_{i,r}W_{i,r}(z)}{zW_{i,r}(\mathsf{w}_{i,r})}\prod_{j\leftarrow i}\prod_{p=1}^{-c_{ji}}W_j(\boldsymbol{v}_j^{-c_{ji}-2p}\boldsymbol{v}_i^2\mathsf{w}_{i,r})D_{i,r},$$

$$C_i^-(z) \mapsto (-\boldsymbol{v}_i)^{a_i}\prod_{t=1}^{a_i}\mathsf{w}_{i,t}^{1/2}\prod_{j\leftarrow i}\prod_{t=1}^{a_j}\mathsf{w}_{j,t}^{c_{ji}/2}\cdot\sum_{r=1}^{a_i}\frac{\hat{W}_{i,r}(z)}{W_{i,r}(\mathsf{w}_{i,r})}\prod_{j\leftarrow i}\prod_{p=1}^{-c_{ji}}W_j(\boldsymbol{v}_j^{-c_{ji}-2p}\boldsymbol{v}_i^2\mathsf{w}_{i,r})D_{i,r}.$$

In particular, all these images belong to $\widetilde{\mathcal{A}}^{\boldsymbol{v}}[\mathsf{z}_1^{\pm 1},\ldots,\mathsf{z}_N^{\pm 1}] \subset \widetilde{\mathcal{A}}^{\boldsymbol{v}}_{\mathrm{frac}}[\mathsf{z}_1^{\pm 1},\ldots,\mathsf{z}_N^{\pm 1}]$.

8 K-theoretic Coulomb Branch

8.1 Quiver Gauge Theories

We follow the notations and setup of [10, Appendix A], so that $(GL(V), \mathbf{N})$ is a quiver gauge theory. As in Sect. 7, we fix a sequence $(\omega_{i_1}, \dots, \omega_{i_N})$ of fundamental coweights of G which is assumed to be simply-laced for the current discussion. We choose a basis $w_1, \dots, w_N$ in $W = \bigoplus_{i\in I} W_i$ such that $w_s \in W_{i_s}$. This defines a maximal torus $T_W \subset \prod_i GL(W_i)$, and $K_{T_W}(\mathrm{pt}) = \mathbb{C}[\mathsf{z}_1^{\pm 1}, \dots, \mathsf{z}_N^{\pm 1}]$. We consider the (quantized) K-theoretic Coulomb branch with flavor deformation $\mathcal{A}^q = K^{(GL(V)\times T_W)_{\mathcal{O}} \rtimes \mathbb{C}^\times}(\mathcal{R}_{GL(V),\mathbf{N}})$ equipped with the convolution algebra structure as in [9, Remark 3.9(3)]. It is a $K_{\mathbb{C}^\times \times T_W}(\mathrm{pt})$-algebra; we denote $K_{\mathbb{C}^\times}(\mathrm{pt}) = \mathbb{C}[q^{\pm 1}]$. We will also need $\boldsymbol{v} = q^{1/2}$, the generator of the equivariant K-theory of a point with respect to the two-fold cover $\widetilde{\mathbb{C}}^\times \to \mathbb{C}^\times$. Recall that $GL(V) = \prod_{i\in I} GL(V_i)$. We will need its 2^I-cover $\widetilde{GL}(V) = \prod_{i\in I} \widetilde{GL}(V_i)$ where $\widetilde{GL}(V_i) := \{(g \in GL(V_i),\ y \in \mathbb{C}^\times) : \det(g) = y^2\}$. We consider the extended Coulomb branch $\mathcal{A}^{\boldsymbol{v}} := K^{(\widetilde{GL}(V)\times T_W)_{\mathcal{O}} \rtimes \widetilde{\mathbb{C}}^\times}(\mathcal{R}_{GL(V),\mathbf{N}}) = \mathcal{A}^q \otimes_{K_{GL(V)\times\mathbb{C}^\times}(\mathrm{pt})} K_{\widetilde{GL}(V)\times\widetilde{\mathbb{C}}^\times}(\mathrm{pt})$. It is equipped with an algebra structure as in Sect. 3.7.

Recall from [10] that $w^*_{i,r}$ is the cocharacter of the Lie algebra of $GL(V) = \prod GL(V_i)$, which is equal to 0 except at the vertex i, and is $(0, \dots, 0, 1, 0, \dots, 0)$ at i. Here 1 is at the r-th entry ($r = 1, \dots, a_i = \dim V_i$). We denote the corresponding coordinates of T_V and $T_V^\vee$ by $\mathsf{w}_{i,r}$ and $D_{i,r}$ ($i \in I,\ 1 \le r \le a_i$). The roots are $\mathsf{w}_{i,r}\mathsf{w}_{i,s}^{-1}$ ($r \neq s$). Furthermore, $K^{(T_V\times T_W)_{\mathcal{O}} \rtimes \mathbb{C}^\times}(\mathcal{R}_{T_V,0})$ with scalars extended by $\boldsymbol{v}, \mathsf{w}_{i,r}^{\pm 1/2}$ is nothing but the algebra $\hat{\mathcal{A}}^{\boldsymbol{v}}[\mathsf{z}_1^{\pm 1}, \dots, \mathsf{z}_N^{\pm 1}] := \hat{\mathcal{A}}^{\boldsymbol{v}} \otimes_{\mathbb{C}[\boldsymbol{v}^{\pm 1}]} \mathbb{C}[\boldsymbol{v}^{\pm 1}][\mathsf{z}_1^{\pm 1}, \dots, \mathsf{z}_N^{\pm 1}]$, where $\hat{\mathcal{A}}^{\boldsymbol{v}}$ was defined in Sect. 7. We thus have an algebra embedding

$$\mathbf{z}^*(\iota_*)^{-1} \colon \mathcal{A}^{\boldsymbol{v}} \hookrightarrow \widetilde{\mathcal{A}}^{\boldsymbol{v}}[\mathsf{z}_1^{\pm 1}, \dots, \mathsf{z}_N^{\pm 1}].$$

Let $\varpi_{i,n}$ be the n-th fundamental coweight of the factor $GL(V_i)$, i.e., $w^*_{i,1} + \dots + w^*_{i,n} = (1, \dots, 1, 0, \dots, 0)$ where 1 appears n times ($1 \le n \le a_i$). Then $\mathrm{Gr}_{GL(V)}^{\varpi_{i,n}}$ is closed and isomorphic to the Grassmannian $\mathrm{Gr}(V_i, n)$ of n-dimensional quotients of V_i. Let $\mathcal{Q}_i$ be the tautological rank n quotient bundle on $\mathrm{Gr}_{GL(V)}^{\varpi_{i,n}}$. Its pull-back to $\mathcal{R}_{\varpi_{i,n}}$ is also denoted by $\mathcal{Q}_i$ for brevity. Let $\Lambda^p(\mathcal{Q}_i)$ denote the class of its p-th external power in $\mathcal{A}^{\boldsymbol{v}}$. More generally, we can consider a class $f(\mathcal{Q}_i)$ for a symmetric function f in n variables so that $\Lambda^p(\mathcal{Q}_i)$ corresponds to the p-th elementary symmetric polynomial e_p.

Similarly, we consider $\varpi^*_{i,n} = -w_0\varpi_{i,n}$, where the corresponding orbit $\mathrm{Gr}_{GL(V)}^{\varpi^*_{i,n}}$ is closed and isomorphic to the Grassmannian $\mathrm{Gr}(n, V_i)$ of n-dimensional subspaces in V_i. Let $\mathcal{S}_i$ be the tautological rank n subbundle on $\mathrm{Gr}_{GL(V)}^{\varpi^*_{i,n}}$. Its pull-back to $\mathcal{R}_{\varpi^*_{i,n}}$

is also denoted by $\mathcal{S}_i$. Now similarly to [10, (A.3), (A.5)], cf. [10, Remark A.8], we obtain

$$\mathbf{z}^*(\iota_*)^{-1}\left(f(\mathcal{Q}_i)\otimes\mathcal{O}_{\mathcal{R}_{\varpi_{i,n}}}\right)=\sum_{\substack{J\subset\{1,\dots,a_i\}\\ \#J=n}} f(\mathsf{w}_{i,J})\frac{\prod\limits_{\substack{j\leftarrow i\\ r\in J}}\prod\limits_{\substack{s=1\\ (j,s)\neq(i,r)}}^{a_j}(1-\boldsymbol{v}\mathsf{w}_{i,r}\mathsf{w}_{j,s}^{-1})}{\prod\limits_{r\in J,s\notin J}(1-\mathsf{w}_{i,s}\mathsf{w}_{i,r}^{-1})}\prod_{r\in J}D_{i,r} \tag{8.1}$$

(the appearance of $\boldsymbol{v}$ is due to the convention before [9, Remark 2.1]);

$$\mathbf{z}^*(\iota_*)^{-1}\left(f(\mathcal{S}_i)\otimes\mathcal{O}_{\mathcal{R}_{\varpi^*_{i,n}}}\right)=$$

$$\sum_{\substack{J\subset\{1,\dots,a_i\}\\ \#J=n}} f(\boldsymbol{v}^{-2}\mathsf{w}_{i,J})\prod_{\substack{r\in J\\ t:i_t=i}}(1-\boldsymbol{v}\mathsf{z}_t\mathsf{w}_{i,r}^{-1})\frac{\prod\limits_{\substack{j\to i\\ r\in J}}\prod\limits_{\substack{s=1\\ (j,s)\neq(i,r)}}^{a_j}(1-\boldsymbol{v}\mathsf{w}_{j,s}\mathsf{w}_{i,r}^{-1})}{\prod\limits_{r\in J,s\notin J}(1-\mathsf{w}_{i,r}\mathsf{w}_{i,s}^{-1})}\prod_{r\in J}D_{i,r}^{-1}, \tag{8.2}$$

where $f(\boldsymbol{v}^{-2}\mathsf{w}_{i,J})$ means that we substitute $\{\boldsymbol{v}^{-2}\mathsf{w}_{i,r}\}_{r\in J}$ to f.

Also, for the vector bundles $\Omega^p_{\varpi_{i,1}}, \Omega^p_{\varpi^*_{i,1}}$ of p-forms on $\mathrm{Gr}^{\varpi_{i,1}}_{GL(V)}, \mathrm{Gr}^{\varpi^*_{i,1}}_{GL(V)}$ we obtain

$$\mathbf{z}^*(\iota_*)^{-1}\left(\Omega^p_{\varpi_{i,1}}\otimes\mathcal{Q}_i^{\otimes p'}\otimes\mathcal{O}_{\mathcal{R}_{\varpi_{i,1}}}\right)=$$

$$\sum_{1\le r\le a_i}\mathsf{w}_{i,r}^{p'-p}\left(\sum_{\substack{J\subset\{1,\dots,a_i\}\backslash\{r\}\\ \#J=p}}\prod_{s\in J}\mathsf{w}_{i,s}\right)\frac{\prod\limits_{j\leftarrow i}\prod\limits_{\substack{s=1\\ (j,s)\neq(i,r)}}^{a_j}(1-\boldsymbol{v}\mathsf{w}_{i,r}\mathsf{w}_{j,s}^{-1})}{\prod\limits_{s\neq r}(1-\mathsf{w}_{i,s}\mathsf{w}_{i,r}^{-1})}D_{i,r}, \tag{8.3}$$

$$\mathbf{z}^*(\iota_*)^{-1}\left(\Omega^p_{\varpi^*_{i,1}}\otimes\mathcal{S}_i^{\otimes p'}\otimes\mathcal{O}_{\mathcal{R}_{\varpi^*_{i,1}}}\right)=$$

$$\sum_{1\le r\le a_i}\boldsymbol{v}^{-2p'}\mathsf{w}_{i,r}^{p'+p}\prod_{t:i_t=i}(1-\boldsymbol{v}\mathsf{z}_t\mathsf{w}_{i,r}^{-1})\left(\sum_{\substack{J\subset\{1,\dots,a_i\}\backslash\{r\}\\ \#J=p}}\prod_{s\in J}\mathsf{w}_{i,s}^{-1}\right)\frac{\prod\limits_{j\to i}\prod\limits_{\substack{s=1\\ (j,s)\neq(i,r)}}^{a_j}(1-\boldsymbol{v}\mathsf{w}_{j,s}\mathsf{w}_{i,r}^{-1})}{\prod\limits_{s\neq r}(1-\mathsf{w}_{i,r}\mathsf{w}_{i,s}^{-1})}D_{i,r}^{-1}. \tag{8.4}$$

8.2 Homomorphism $\overline{\Phi}{}^{\lambda}_{\mu}$

We set $\mathcal{A}^{\boldsymbol{v}}_{\mathrm{frac}} := \mathcal{A}^{\boldsymbol{v}} \otimes_{\mathbb{C}[\boldsymbol{v}^{\pm 1}]} \mathbb{C}(\boldsymbol{v})$. The key result of this section asserts that the homomorphism $\widetilde{\Phi}{}^{\lambda}_{\mu}$ of Theorem 7.1 factors through the above embedding $\mathbf{z}^*(\iota_*)^{-1} \colon \mathcal{A}^{\boldsymbol{v}}_{\mathrm{frac}} \hookrightarrow \widetilde{\mathcal{A}}^{\boldsymbol{v}}_{\mathrm{frac}}[\mathsf{z}_1^{\pm 1}, \ldots, \mathsf{z}_N^{\pm 1}]$, similarly to [10, Theorem B.18].

Theorem 8.1 *There exists a unique* $\mathbb{C}(\boldsymbol{v})[\mathsf{z}_1^{\pm 1}, \ldots, \mathsf{z}_N^{\pm 1}]$*-algebra homomorphism*

$$\overline{\Phi}{}^{\lambda}_{\mu} \colon \mathcal{U}^{\mathrm{ad}}_{0,\mu}[\mathsf{z}_1^{\pm 1}, \ldots, \mathsf{z}_N^{\pm 1}] \longrightarrow \mathcal{A}^{\boldsymbol{v}}_{\mathrm{frac}},$$

such that the following diagram commutes:

$$\begin{array}{ccc}
\mathcal{U}^{\mathrm{ad}}_{0,\mu}[\mathsf{z}_1^{\pm 1}, \ldots, \mathsf{z}_N^{\pm 1}] & \xrightarrow{\ \widetilde{\Phi}{}^{\lambda}_{\mu}\ } & \\
\Big\downarrow{\scriptstyle \overline{\Phi}{}^{\lambda}_{\mu}} & \searrow & \\
\mathcal{A}^{\boldsymbol{v}}_{\mathrm{frac}} & \xrightarrow[\ \mathbf{z}^*(\iota_*)^{-1}\ ]{} & \widetilde{\mathcal{A}}^{\boldsymbol{v}}_{\mathrm{frac}}[\mathsf{z}_1^{\pm 1}, \ldots, \mathsf{z}_N^{\pm 1}]
\end{array}$$

Explicitly, $\overline{\Phi}{}^{\lambda}_{\mu}$ *maps the generators as follows:*

$$e_{i,r} \mapsto \frac{(-1)^{a_i}\boldsymbol{v}}{1-\boldsymbol{v}^2} \prod_{j \to i} \prod_{t=1}^{a_j} \mathsf{w}_{j,t}^{-1/2} \cdot (\boldsymbol{v}^2 \mathcal{S}_i)^{\otimes (r+a_i)} \otimes \mathcal{O}_{\mathcal{R}_{\varpi^*_{i,1}}},$$

$$f_{i,r} \mapsto \frac{(-\boldsymbol{v})^{-\sum_{j \leftarrow i} a_j}}{1-\boldsymbol{v}^2} \prod_{j \leftarrow i} \prod_{t=1}^{a_j} \mathsf{w}_{j,t}^{1/2} \cdot \mathcal{Q}_i^{\otimes(-\sum_{j \leftarrow i} a_j)} \otimes (\boldsymbol{v}^2 \mathcal{Q}_i)^{\otimes r} \otimes \mathcal{O}_{\mathcal{R}_{\varpi_{i,1}}},$$

$$A^{+}_{i,r} \mapsto (-1)^r \prod_{t=1}^{a_i} \mathsf{w}_{i,t}^{-1/2} \cdot e_r(\{\mathsf{w}_{i,t}\}_{t=1}^{a_i}),$$

$$A^{-}_{i,-r} \mapsto (-1)^r (-\boldsymbol{v})^{a_i} \prod_{t=1}^{a_i} \mathsf{w}_{i,t}^{1/2} \cdot e_r(\{\mathsf{w}_{i,t}^{-1}\}_{t=1}^{a_i}),$$

$$\phi^{+}_{i} \mapsto \prod_{t=1}^{a_i} \mathsf{w}_{i,t}^{1/2}, \quad \phi^{-}_{i} \mapsto (-\boldsymbol{v})^{-a_i} \prod_{t=1}^{a_i} \mathsf{w}_{i,t}^{-1/2}.$$

Proof For $X \in \{e_{i,r}, f_{i,r}, A^{\pm}_{i,\pm s}, \phi^{\pm}_i \mid i \in I, r \in \mathbb{Z}, s \in \mathbb{N}\}$ consider the assignment $X \mapsto \overline{\Phi}{}^{\lambda}_{\mu}(X)$ with the right-hand side defined as above. Since $\mathbf{z}^*(\iota_*)^{-1} \colon \mathcal{A}^{\boldsymbol{v}}_{\mathrm{frac}} \hookrightarrow \widetilde{\mathcal{A}}^{\boldsymbol{v}}_{\mathrm{frac}}[\mathsf{z}_1^{\pm 1}, \ldots, \mathsf{z}_N^{\pm 1}]$ is injective and $\widetilde{\Phi}{}^{\lambda}_{\mu} \colon \mathcal{U}^{\mathrm{ad}}_{0,\mu}[\mathsf{z}_1^{\pm 1}, \ldots, \mathsf{z}_N^{\pm 1}] \to \widetilde{\mathcal{A}}^{\boldsymbol{v}}_{\mathrm{frac}}[\mathsf{z}_1^{\pm 1}, \ldots, \mathsf{z}_N^{\pm 1}]$ is an algebra homomorphism, it suffices to check that $\mathbf{z}^*(\iota_*)^{-1}\left(\overline{\Phi}_\mu^\lambda(X)\right) = \widetilde{\Phi}_\mu^\lambda(X)$ for X as above. This is a straightforward verification based on formulas (8.1) and (8.2). □

Combining Proposition 7.4 with formulas (8.3) and (8.4), we immediately find the images of the generators $\{B_{i,r}^+, C_{i,r+1}^+\}_{i\in I}^{r\geq 0}$ under $\overline{\Phi}_\mu^\lambda$.

Corollary 8.2 *For $r \in \mathbb{N}$, we have*

$$\overline{\Phi}_\mu^\lambda(B_{i,r}^+) = (-1)^{r+a_i+1}\boldsymbol{v}^{2r}\prod_{t=1}^{a_i}\mathsf{w}_{i,t}^{1/2}\prod_{j\to i}\prod_{t=1}^{a_j}\mathsf{w}_{j,t}^{-1/2}\cdot\left(\Omega_{\varpi_{i,1}^*}^{a_i-1-r}\otimes\mathcal{S}_i^{\otimes r}\otimes\mathcal{O}_{\mathcal{R}_{\varpi_{i,1}^*}}\right),$$

$$\overline{\Phi}_\mu^\lambda(C_{i,r+1}^+) = (-1)^{r+1}(-\boldsymbol{v})^{-\sum_{j\leftarrow i}a_j}\prod_{t=1}^{a_i}\mathsf{w}_{i,t}^{-1/2}\prod_{j\leftarrow i}\prod_{t=1}^{a_j}\mathsf{w}_{j,t}^{1/2}$$

$$\cdot\left(\Omega_{\varpi_{i,1}}^{r}\otimes\mathcal{Q}_i^{\otimes(r+1-\sum_{j\leftarrow i}a_j)}\otimes\mathcal{O}_{\mathcal{R}_{\varpi_{i,1}}}\right).$$

In particular, the images of $\{A_{i,r}^+, B_{i,r}^+, C_{i,r+1}^+, \phi_i^+\}_{i\in I}^{r\in\mathbb{N}}$ under $\overline{\Phi}_\mu^\lambda$ belong to $\mathcal{A}^{\boldsymbol{v}} \subset \mathcal{A}_{\mathrm{frac}}^{\boldsymbol{v}}$. In fact, the images of $\{A_{i,-r}^-, B_{i,-r-1}^-, C_{i,-r}^-, \phi_i^-\}_{i\in I}^{r\in\mathbb{N}}$ under $\overline{\Phi}_\mu^\lambda$ also belong to $\mathcal{A}^{\boldsymbol{v}}$.

Remark 8.3 (A. Weekes) In the case of shifted Yangians, the images of the generating series $B_i(z), C_i(z)$ [44, Section 5.3] in the quantized (cohomological) Coulomb branch $\mathcal{A}_\hbar$ under the homomorphism $\overline{\Phi}_\mu^\lambda$ of [10, Theorem B.18] are equal to

$$\overline{\Phi}_\mu^\lambda(B_i(z)) = (-1)^{a_i}z^{-1}\cdot c(\widetilde{\mathcal{Q}}_i, -z^{-1})\cap[\mathcal{R}_{\varpi_{i,1}^*}],$$

$$\overline{\Phi}_\mu^\lambda(C_i(z)) = (-1)^{\sum_{j\leftarrow i}a_j}z^{-1}\cdot c(\mathcal{S}_i, -z^{-1})\cap[\mathcal{R}_{\varpi_{i,1}}],$$

where $c(\mathcal{F}, z)$ denotes the Chern polynomial of a vector bundle $\mathcal{F}$. Here we view $\mathcal{Q}_i, \mathcal{S}_i$ as rank $n-1$ vector bundles on $\mathcal{R}_{\varpi_{i,1}^*}, \mathcal{R}_{\varpi_{i,1}}$, respectively, while $\widetilde{\mathcal{Q}}_i$ denotes the vector bundle $\mathcal{Q}_i$ with the equivariance structure twisted by $\hbar$.

Remark 8.4 Note that $\mathrm{Gr}_{GL(V)}^{\varpi_{i,1}} \simeq \mathbb{P}^{a_i-1} \simeq \mathrm{Gr}_{GL(V)}^{\varpi_{i,1}^*}$, and if we forget the equivariance, then up to sign, $\overline{\Phi}_\mu^\lambda(f_{i,r})$, $1\leq r\leq a_i$, is the collection of classes of pull-backs of the line bundles $\mathcal{O}_{\mathbb{P}^{a_i-1}}(1-\sum_{j\leftarrow i}a_j),\ldots,\mathcal{O}_{\mathbb{P}^{a_i-1}}(a_i-\sum_{j\leftarrow i}a_j)$, while $\overline{\Phi}_\mu^\lambda(C_{i,r}^+)$, $1\leq r\leq a_i$, is the collection of classes of pull-backs of the vector bundles $\Omega_{\mathbb{P}^{a_i-1}}^{r-1}(r-\sum_{j\leftarrow i}a_j)$. These two collections are the dual exceptional collections of vector bundles on $\mathbb{P}^{a_i-1}$ (more precisely, the former collection is left dual to the latter one). In fact, this is the historically first example of dual exceptional collections, [3]. Similarly, up to sign and forgetting equivariance, $\overline{\Phi}{}^{\lambda}_{\mu}(e_{i,r}),\ 0 \leq r < a_i$, are the classes of the exceptional collection of line bundles right dual to the exceptional collection of vector bundles whose classes are $\overline{\Phi}{}^{\lambda}_{\mu}(B^+_{i,r}),\ 0 \leq r < a_i$.

Remark 8.5 An action of the quantized K-theoretic Coulomb branch $\mathcal{A}^{\boldsymbol{v}}_{\mathrm{frac}}$ of the type A quiver gauge theory on the localized equivariant K-theory of parabolic Laumon spaces was constructed in [4]. Combining this construction with Theorem 8.1, we see that there should be a natural action of $\mathcal{U}^{\mathrm{ad}}_{0,\mu}[\mathsf{z}_1^{\pm 1},\ldots,\mathsf{z}_N^{\pm 1}]$ (with $\mathfrak{g}=\mathfrak{sl}_n$) on the aforementioned K-theory. We construct explicitly such an action of $\mathcal{U}^{\mathrm{sc}}_{0,\mu}$ in Theorem 12.2 by adapting the arguments of [61] to the current setting (the adjoint version is achieved by considering equivariant K-theory with respect to a larger torus).

8.3 Truncated Shifted Quantum Affine Algebras

We consider a 2-sided ideal $\mathcal{J}^{\lambda}_{\mu}$ of $\mathcal{U}^{\mathrm{ad}}_{0,\mu}[\mathsf{z}_1^{\pm 1},\ldots,\mathsf{z}_N^{\pm 1}]$ generated over $\mathbb{C}(\boldsymbol{v})[\mathsf{z}_1^{\pm 1},\ldots,\mathsf{z}_N^{\pm 1}]$ by the following elements:

$$A^{\pm}_{i,\pm s}\ (s > a_i),\ A^{+}_{i,0}A^{+}_{i,a_i} - (-1)^{a_i},\ A^{-}_{i,0}A^{-}_{i,-a_i} - (-1)^{a_i}\boldsymbol{v}_i^{2a_i}, \tag{8.5}$$

$$A^{-}_{i,-r} - \boldsymbol{v}_i^{a_i}A^{+}_{i,a_i-r}\ (0 \leq r \leq a_i). \tag{8.6}$$

Definition 8.6 $\mathcal{U}^{\lambda}_{\mu} := \mathcal{U}^{\mathrm{ad}}_{0,\mu}[\mathsf{z}_1^{\pm 1},\ldots,\mathsf{z}_N^{\pm 1}]/\mathcal{J}^{\lambda}_{\mu}$ is called the *truncated shifted quantum affine algebra.*

Note that the homomorphism $\widetilde{\Phi}{}^{\lambda}_{\mu}\colon \mathcal{U}^{\mathrm{ad}}_{0,\mu}[\mathsf{z}_1^{\pm 1},\ldots,\mathsf{z}_N^{\pm 1}] \to \widetilde{\mathcal{A}}^{\boldsymbol{v}}_{\mathrm{frac}}[\mathsf{z}_1^{\pm 1},\ldots,\mathsf{z}_N^{\pm 1}]$ factors through the same named homomorphism $\widetilde{\Phi}{}^{\lambda}_{\mu}\colon \mathcal{U}^{\lambda}_{\mu} \to \widetilde{\mathcal{A}}^{\boldsymbol{v}}_{\mathrm{frac}}[\mathsf{z}_1^{\pm 1},\ldots,\mathsf{z}_N^{\pm 1}]$, due to Proposition 7.4. Similarly to [10, Remark B.21], we expect this homomorphism to be injective:

Conjecture 8.7 $\widetilde{\Phi}{}^{\lambda}_{\mu}\colon \mathcal{U}^{\lambda}_{\mu} \hookrightarrow \widetilde{\mathcal{A}}^{\boldsymbol{v}}_{\mathrm{frac}}[\mathsf{z}_1^{\pm 1},\ldots,\mathsf{z}_N^{\pm 1}]$.

Remark 8.8 As a first indication of the validity of this conjecture, we note that the elements $\{B^+_{i,r}, C^+_{i,r+1}, B^-_{i,-r-1}, C^-_{i,-r}\}^{r\geq a_i}_{i\in I}$ which belong to $\mathrm{Ker}(\widetilde{\Phi}{}^{\lambda}_{\mu})$ (due to Proposition 7.4) also belong to $\mathcal{J}^{\lambda}_{\mu}$, due to Corollary 7.3 and relation (U10).

Moreover, we expect the following result:

Conjecture 8.9 $\overline{\Phi}{}^{\lambda}_{\mu}\colon \mathcal{U}^{\lambda}_{\mu} \xrightarrow{\sim} \mathcal{A}^{\boldsymbol{v}}_{\mathrm{frac}}$.

8.4 Truncated Shifted v-Yangians

Recall that $\mathfrak{g}$ is assumed to be simply-laced. Recall an explicit identification of the Drinfeld-Jimbo and the new Drinfeld realizations of the standard quantum loop algebra $U_{\boldsymbol{v}}(L\mathfrak{g})$. To this end, choose a decomposition of the highest root θ of $\mathfrak{g}$ into a sum of simple roots $\theta = \alpha^\vee_{i_1} + \alpha^\vee_{i_2} + \ldots + \alpha^\vee_{i_{h-1}}$ such that $\epsilon_k := \langle \alpha_{i_{k+1}}, \alpha^\vee_{i_1} + \ldots + \alpha^\vee_{i_k} \rangle \in \mathbb{Z}_{<0}$ for any $1 \le k \le h-2$ (here h is the Coxeter number of $\mathfrak{g}$). We encode a choice of such a decomposition by a sequence $\mathbf{i} = (i_1, \ldots, i_{h-1})$. Let $U^{\mathrm{DJ}}_{\boldsymbol{v}}(L\mathfrak{g})$ denote the Drinfeld-Jimbo quantum group of $\widehat{\mathfrak{g}}$ (affinization of $\mathfrak{g}$) with a trivial central charge, generated by $\{E_i, F_i, K^{\pm 1}_i\}_{i \in \widetilde{I}}$ (here $\widetilde{I} = I \cup \{i_0\}$ is the vertex set of the extended Dynkin diagram), see [50]. The following result is due to [16] (proved in [41]).

Theorem 8.10 *There is a $\mathbb{C}(\boldsymbol{v})$-algebra isomorphism $U^{\mathrm{DJ}}_{\boldsymbol{v}}(L\mathfrak{g}) \xrightarrow{\sim} U_{\boldsymbol{v}}(L\mathfrak{g})$, such that*

$$E_i \mapsto e_{i,0},\ F_i \mapsto f_{i,0},\ K^{\pm 1}_i \mapsto \psi^{\pm}_{i,0} \text{ for } i \in I,$$

$$E_{i_0} \mapsto [f_{i_{h-1},0}, [f_{i_{h-2},0}, \cdots, [f_{i_2,0}, f_{i_1,1}]_{\boldsymbol{v}^{\epsilon_1}} \cdots]_{\boldsymbol{v}^{\epsilon_{h-3}}}]_{\boldsymbol{v}^{\epsilon_{h-2}}} \cdot \psi^-_\theta,$$

$$F_{i_0} \mapsto (-\boldsymbol{v})^{-\epsilon} \psi^+_\theta \cdot [e_{i_{h-1},0}, [e_{i_{h-2},0}, \cdots, [e_{i_2,0}, e_{i_1,-1}]_{\boldsymbol{v}^{\epsilon_1}} \cdots]_{\boldsymbol{v}^{\epsilon_{h-3}}}]_{\boldsymbol{v}^{\epsilon_{h-2}}},$$

$$K^{\pm}_{i_0} \mapsto \psi^{\mp}_\theta,$$

where $\psi^\pm_\theta := \psi^\pm_{i_1,0} \cdots \psi^\pm_{i_{h-1},0}$, $\epsilon := \epsilon_1 + \ldots + \epsilon_{h-2}$.

In particular, the image of the negative Drinfeld-Jimbo Borel subalgebra of $U^{\mathrm{DJ}}_{\boldsymbol{v}}(L\mathfrak{g})$ generated by $\{F_i, K^{\pm 1}_i\}_{i \in \widetilde{I}}$ under the above isomorphism is the subalgebra $U^-_{\boldsymbol{v}}$ of $U_{\boldsymbol{v}}(L\mathfrak{g})$, generated by $\{f_{i,0}, (\psi^-_{i,0})^{\pm 1}, F\}_{i \in I}$ with $F := [e_{i_{h-1},0}, [e_{i_{h-2},0}, \cdots, [e_{i_2,0}, e_{i_1,-1}]_{\boldsymbol{v}^{\epsilon_1}} \cdots]_{\boldsymbol{v}^{\epsilon_{h-3}}}]_{\boldsymbol{v}^{\epsilon_{h-2}}}$. Motivated by this observation, we introduce the following definition.

Definition 8.11

(a) Fix $\mathbf{i} = (i_1, \ldots, i_{h-1})$ as above. The shifted $\boldsymbol{v}$-Yangian ${}_{\mathbf{i}}\mathcal{Y}^{\boldsymbol{v}}_\mu[\mathsf{z}^{\pm 1}_1, \ldots, \mathsf{z}^{\pm 1}_N]$ is the $\mathbb{C}(\boldsymbol{v})[\mathsf{z}^{\pm 1}_1, \ldots, \mathsf{z}^{\pm 1}_N]$-subalgebra of $\mathcal{U}^{\mathrm{ad}}_{0,\mu}[\mathsf{z}^{\pm 1}_1, \ldots, \mathsf{z}^{\pm 1}_N]$ generated by $\{f_{i,0}, (\psi^-_{i,b_i})^{\pm 1}, \hat{F}\}_{i \in I}$, where

$$\hat{F} := [e_{i_{h-1},b_{i_{h-1}}}, [e_{i_{h-2},b_{i_{h-2}}}, \cdots, [e_{i_2,b_{i_2}}, e_{i_1,b_{i_1}-1}]_{\boldsymbol{v}^{\epsilon_1}} \cdots]_{\boldsymbol{v}^{\epsilon_{h-3}}}]_{\boldsymbol{v}^{\epsilon_{h-2}}}$$

and $b_i := \alpha^\vee_i(\mu)$.

(b) The truncated shifted $\boldsymbol{v}$-Yangian ${}_{\mathbf{i}}\mathcal{Y}^\lambda_{\underline{\mu}}$ is the quotient of ${}_{\mathbf{i}}\mathcal{Y}^{\boldsymbol{v}}_\mu[\mathsf{z}^{\pm 1}_1, \ldots, \mathsf{z}^{\pm 1}_N]$ by the 2-sided ideal ${}_{\mathbf{i}}\mathcal{J}^{\lambda,+}_{\underline{\mu}} := \mathcal{J}^\lambda_{\underline{\mu}} \cap {}_{\mathbf{i}}\mathcal{Y}^{\boldsymbol{v}}_\mu[\mathsf{z}^{\pm 1}_1, \ldots, \mathsf{z}^{\pm 1}_N]$.

Remark 8.12 For $\mathfrak{g} = \mathfrak{gl}_n$ and $\mu = 0$, our definition of the $\boldsymbol{v}$-Yangian is consistent with that of the *quantum Yangian* $Y_q(\mathfrak{gl}_n)$ of [54] (in particular, independent of the choice of $\mathbf{i}$). The latter is defined via the RTT presentation, see our discussion in Appendix G, and corresponds to the subalgebra generated by the coefficients of the matrix $T^-(z)$.

Conjecture 8.13 $\overline{\Phi}^{\underline{\lambda}}_{\mu} : {}_{\mathbf{i}}\mathcal{Y}^{\underline{\lambda}}_{\mu} \xrightarrow{\sim} \mathcal{A}^{\boldsymbol{v}}_{\mathrm{frac}}$.

8.5 Integral Forms

If we believe Conjectures 8.9 and 8.13, we can transfer the integral forms $\mathcal{A}^{\boldsymbol{v}} \subset \mathcal{A}^{\boldsymbol{v}}_{\mathrm{frac}}$ to the truncated shifted quantum affine algebras and the truncated shifted $\boldsymbol{v}$-Yangians to obtain the $\mathbb{C}[\boldsymbol{v}^{\pm 1}]$-subalgebras $'\mathfrak{U}^{\underline{\lambda}}_{\mu} \subset \mathcal{U}^{\underline{\lambda}}_{\mu}$ and ${}_{\mathbf{i}}\mathfrak{Y}^{\underline{\lambda}}_{\mu} \subset {}_{\mathbf{i}}\mathcal{Y}^{\underline{\lambda}}_{\mu}$. Finally, we define the integral form $'\mathfrak{U}^{\mathrm{ad}}_{0,\mu} \subset \mathcal{U}^{\mathrm{ad}}_{0,\mu}$ as an intersection of all the preimages of $'\mathfrak{U}^{\underline{\lambda}}_{\mu}|_{\mathsf{z}_1=\ldots=\mathsf{z}_N=1}$ under projections $\mathcal{U}^{\mathrm{ad}}_{0,\mu}[\mathsf{z}_1^{\pm 1}, \ldots, \mathsf{z}_N^{\pm 1}] \twoheadrightarrow \mathcal{U}^{\underline{\lambda}}_{\mu}$ as $\underline{\lambda}$ varies, and ${}_{\mathbf{i}}\mathfrak{Y}^{\boldsymbol{v}}_{\mu} := {}'\mathfrak{U}^{\mathrm{ad}}_{0,\mu} \cap {}_{\mathbf{i}}\mathcal{Y}^{\boldsymbol{v}}_{\mu}[\mathsf{z}_1^{\pm 1}, \ldots, \mathsf{z}_N^{\pm 1}]|_{\mathsf{z}_1=\ldots=\mathsf{z}_N=1}$. Unfortunately, we cannot define these integral forms by generators and relations in general. In the case of $\mathfrak{sl}_2$ see Sect. 9.1.

Recall that $*$ stands for the involution $\mu \mapsto -w_0\mu$ of the coweight lattice Λ. Similarly to [10, Remark 3.17], one can construct an isomorphism from the non-quantized extended K-theoretic Coulomb branch Spec $K^{(\widetilde{GL(V)} \times T_W)_{\mathcal{O}}}(\mathcal{R}_{GL(V),\mathbf{N}})$ of Sect. 8.1 to the multiplicative slice cover ${}^{\dagger}\hat{\mathcal{W}}^{\underline{\lambda}^*}_{\mu^*}$ of Sect. 4.6. Its quantization is the subject of the following

Conjecture 8.14

(a) The shifted $\boldsymbol{v}$-Yangian ${}_{\mathbf{i}}\mathfrak{Y}^{\boldsymbol{v}}_{\mu}$ is a quantization of the thick multiplicative slice ${}^{\dagger}\mathcal{W}_{\mu^*}$ of Sect. 4.8, that is ${}_{\mathbf{i}}\mathfrak{Y}^{\boldsymbol{v}}_{\mu}|_{\boldsymbol{v}=1} \simeq \mathbb{C}[{}^{\dagger}\mathcal{W}_{\mu^*}]$.

(b) The truncated shifted $\boldsymbol{v}$-Yangian ${}_{\mathbf{i}}\mathfrak{Y}^{\underline{\lambda}}_{\mu}$ and the truncated shifted quantum affine algebra $'\mathfrak{U}^{\underline{\lambda}}_{\mu}$ are quantizations of the multiplicative slice cover ${}^{\dagger}\hat{\mathcal{W}}^{\underline{\lambda}^*}_{\mu^*}$ of Sect. 4.6, that is ${}_{\mathbf{i}}\mathfrak{Y}^{\underline{\lambda}}_{\mu}|_{\boldsymbol{v}=1} \simeq {}'\mathfrak{U}^{\underline{\lambda}}_{\mu}|_{\boldsymbol{v}=1} \simeq \mathbb{C}[{}^{\dagger}\hat{\mathcal{W}}^{\underline{\lambda}^*}_{\mu^*}]$.

8.6 An Example

Let $\mathfrak{g} = \mathfrak{sl}_n$, $\mu = 0$, $\underline{\lambda} = (\omega_1, \ldots, \omega_1)$ (the first fundamental coweight taken n times). Note that the symmetric group $\mathfrak{S}_n$ acts naturally on $'\mathfrak{U}^{\underline{\lambda}}_{\mu}$, permuting the parameters $\mathsf{z}_1, \ldots, \mathsf{z}_n$. This action induces the one on the quotient algebra $'\overline{\mathfrak{U}}^{\underline{\lambda}}_{\mu}$ by the relation $\mathsf{z}_1 \cdots \mathsf{z}_n = 1$. Then we expect that the evaluation homomorphism

$U_{\boldsymbol{v}}(L\mathfrak{sl}_n) \twoheadrightarrow U_{\boldsymbol{v}}(\mathfrak{sl}_n)$ [40] gives rise to an isomorphism $({}'\overline{\mathfrak{U}}^{\underline{\lambda}}_{\mu})^{\mathfrak{S}_n} \xrightarrow{\sim} {}_{\mathcal{A}}\mathbf{O}_{\mathrm{loc}}$, where ${}_{\mathcal{A}}\mathbf{O}$ is the integral form of the quantum coordinate algebra of $SL(N)$ introduced in [50, 29.5.2], and ${}_{\mathcal{A}}\mathbf{O}_{\mathrm{loc}}$ stands for its localization by inverting the quantum minors $\{c_\nu\}_{\nu\in\Lambda^+}$, see [42, 9.1.10].

9 Shifted Quantum Affine $\mathfrak{sl}_2$ and Nil-DAHA for $GL(n)$

9.1 Integral Form

In this section $\mathfrak{g} = \mathfrak{sl}_2$, whence we denote $A^\pm_{i,r}, B^\pm_{i,r}, C^\pm_{i,r}, \phi^\pm_i$ simply by $A^\pm_r, B^\pm_r, C^\pm_r, \phi^\pm$. The shift $\mu \in \Lambda = \mathbb{Z}$ is an integer. Furthermore, $\underline{\lambda} = (\omega_1, \ldots, \omega_1)$ (a collection of N copies of the fundamental coweight). The corresponding shifted quantum affine algebra is $\mathcal{U}^{\mathrm{ad}}_{0,\mu}[\mathsf{z}_1^{\pm 1}, \ldots, \mathsf{z}_N^{\pm 1}]$. We define a $\mathbb{C}[\boldsymbol{v}^{\pm 1}]$-subalgebra $\mathfrak{U}^{\mathrm{ad}}_{0,\mu}[\mathsf{z}_1^{\pm 1}, \ldots, \mathsf{z}_N^{\pm 1}] \subset \mathcal{U}^{\mathrm{ad}}_{0,\mu}[\mathsf{z}_1^{\pm 1}, \ldots, \mathsf{z}_N^{\pm 1}]$ generated by $\{A^\pm_{\pm r}, B^+_r, B^-_{-r-1}, C^+_{r+1}, C^-_{-r}, \phi^\pm\}_{r\in\mathbb{N}}$ and its quotient algebra (an integral version of the truncated shifted quantum affine algebra)

$$\mathfrak{U}^{\underline{\lambda}}_{\mu} := \mathfrak{U}^{\mathrm{ad}}_{0,\mu}[\mathsf{z}_1^{\pm 1}, \ldots, \mathsf{z}_N^{\pm 1}]/(\mathcal{I}^{\underline{\lambda}}_{\mu} \cap \mathfrak{U}^{\mathrm{ad}}_{0,\mu}[\mathsf{z}_1^{\pm 1}, \ldots, \mathsf{z}_N^{\pm 1}]).$$

Let $V = \mathbb{C}^n,\ W = \mathbb{C}^N$. According to Corollary 8.2, the homomorphism

$$\overline{\Phi}^N_{N-2n} \colon \mathcal{U}^{\mathrm{ad}}_{0,N-2n}[\mathsf{z}_1^{\pm 1}, \ldots, \mathsf{z}_N^{\pm 1}] \longrightarrow \mathcal{A}^{\boldsymbol{v}}_{\mathrm{frac}} = K^{(\widetilde{GL}(V)\times T_W)_{\mathcal{O}} \rtimes \widetilde{\mathbb{C}}^\times}(\mathcal{R}_{GL(V),\mathrm{Hom}(W,V)}) \otimes_{\mathbb{C}[\boldsymbol{v}^{\pm 1}]} \mathbb{C}(\boldsymbol{v})$$

takes $\mathfrak{U}^{\mathrm{ad}}_{0,N-2n}[\mathsf{z}_1^{\pm 1}, \ldots, \mathsf{z}_N^{\pm 1}] \subset \mathcal{U}^{\mathrm{ad}}_{0,N-2n}[\mathsf{z}_1^{\pm 1}, \ldots, \mathsf{z}_N^{\pm 1}]$ to $\mathcal{A}^{\boldsymbol{v}} \subset \mathcal{A}^{\boldsymbol{v}}_{\mathrm{frac}}$. In particular, we have $\mathfrak{U}^{\mathrm{ad}}_{0,N-2n}[\mathsf{z}_1^{\pm 1}, \ldots, \mathsf{z}_N^{\pm 1}] \subset {}'\mathfrak{U}^{\mathrm{ad}}_{0,N-2n}[\mathsf{z}_1^{\pm 1}, \ldots, \mathsf{z}_N^{\pm 1}]$ (cf. Sect. 8.5). We also define a $\mathbb{C}[\boldsymbol{v}^{\pm 1}]$-subalgebra $\mathfrak{Y}^{\boldsymbol{v}}_{N-2n}[\mathsf{z}_1^{\pm 1}, \ldots, \mathsf{z}_N^{\pm 1}] \subset \mathcal{Y}^{\boldsymbol{v}}_{N-2n}[\mathsf{z}_1^{\pm 1}, \ldots, \mathsf{z}_N^{\pm 1}]$ generated by $\{A^-_{-r}, B^-_{-r-1}, C^-_{-r}, \phi^-\}_{r\in\mathbb{N}}$. Furthermore, we define the shifted Borel $\boldsymbol{v}$-Yangian $\mathfrak{Y}^{\boldsymbol{v}}_{N-2n,-}[\mathsf{z}_1^{\pm 1}, \ldots, \mathsf{z}_N^{\pm 1}]$ as the $\mathbb{C}[\boldsymbol{v}^{\pm 1}]$-subalgebra of $\mathfrak{Y}^{\boldsymbol{v}}_{N-2n}[\mathsf{z}_1^{\pm 1}, \ldots, \mathsf{z}_N^{\pm 1}]$ generated by $\{A^-_{-r}, C^-_{-r}, \phi^-\}_{r\in\mathbb{N}}$. Finally, we have their truncated quotients $\mathfrak{Y}^{\underline{\lambda}}_{N-2n}, \mathfrak{Y}^{\underline{\lambda}}_{N-2n,-}$. We expect that

$$\mathfrak{U}^{\mathrm{ad}}_{0,N-2n}[\mathsf{z}_1^{\pm 1}, \ldots, \mathsf{z}_N^{\pm 1}] = {}'\mathfrak{U}^{\mathrm{ad}}_{0,N-2n}[\mathsf{z}_1^{\pm 1}, \ldots, \mathsf{z}_N^{\pm 1}],$$
$$\mathfrak{Y}^{\boldsymbol{v}}_{N-2n}[\mathsf{z}_1^{\pm 1}, \ldots, \mathsf{z}_N^{\pm 1}] = {}'\mathfrak{Y}^{\boldsymbol{v}}_{N-2n}[\mathsf{z}_1^{\pm 1}, \ldots, \mathsf{z}_N^{\pm 1}],$$
$$\mathfrak{U}^{\underline{\lambda}}_{N-2n} = {}'\mathfrak{U}^{\underline{\lambda}}_{N-2n},\ \mathfrak{Y}^{\underline{\lambda}}_{N-2n} = {}'\mathfrak{Y}^{\underline{\lambda}}_{N-2n}.$$

Conjecture 9.1 The natural homomorphisms induce isomorphisms

$$\mathfrak{Y}^{\underline{\lambda}}_{N-2n} \xrightarrow{\sim} \mathfrak{U}^{\underline{\lambda}}_{N-2n} \xrightarrow{\sim} \mathcal{A}^{\boldsymbol{v}}.$$

From now on, we specialize to the case $N = 0$, $\mu = -2n$. According to Corollary 3.14, the corresponding Coulomb branch $\mathcal{A}^{\boldsymbol{v}} = K^{\widetilde{GL(n,\mathcal{O})}\rtimes\widetilde{\mathbb{C}}^\times}(\mathrm{Gr}_{GL(n)})$ is nothing but the spherical extended nil-DAHA $\mathcal{H}\mathcal{H}_e^{\mathrm{sph}}(GL(n))$. We define $\mathbb{C}[\boldsymbol{v}^{\pm1}]_{\mathrm{loc}}$ inverting $(1-\boldsymbol{v}^{2m})$, $m = 1, 2, \ldots, n$. We extend the scalars to $\mathbb{C}[\boldsymbol{v}^{\pm1}]_{\mathrm{loc}}$ to obtain

$$\overline{\Phi}^0_{-2n,\mathrm{loc}} \colon \mathfrak{U}^{\mathrm{ad}}_{0,-2n,\mathrm{loc}} \longrightarrow K_{\mathrm{loc}}^{\widetilde{GL(n,\mathcal{O})}\rtimes\widetilde{\mathbb{C}}^\times}(\mathrm{Gr}_{GL(n)}).$$

The following theorem and Proposition 9.8 is a supportive evidence in favor of Conjecture 9.1.

Theorem 9.2 *$\overline{\Phi}^0_{-2n,\mathrm{loc}} \colon \mathfrak{U}^{\mathrm{ad}}_{0,-2n,\mathrm{loc}} \to K_{\mathrm{loc}}^{\widetilde{GL(n,\mathcal{O})}\rtimes\widetilde{\mathbb{C}}^\times}(\mathrm{Gr}_{GL(n)})$ is surjective.*[6]

Proof We must prove that $K_{\mathrm{loc}}^{GL(n,\mathcal{O})\rtimes\mathbb{C}^\times}(\mathrm{Gr}_{GL(n)})$ is generated by $K_{GL(n)}(\mathrm{pt}) = R(GL(n))$, and $\mathcal{O}(a)_{\varpi_1}, \mathcal{O}(a)_{\varpi_1^*}$, $a \in \mathbb{Z}$. Here $\varpi_1 = (1, 0, \ldots, 0)$ denotes the first fundamental coweight of $GL(n)$, and $\mathrm{Gr}^{\varpi_1} \simeq \mathbb{P}^{n-1}$ is the corresponding minuscule orbit, so that $\mathrm{Gr}^{\varpi_1^*} \simeq \check{\mathbb{P}}^{n-1}$. Finally, $\mathcal{Q}$ is the tautological quotient bundle on Gr^{ϖ_1}, isomorphic to the ample line bundle $\mathcal{O}(1)$ on $\mathbb{P}^{n-1}$, and $\mathcal{O}(a)_{\varpi_1}$ stands for $\mathcal{Q}^{\otimes a}$. Similarly, $\mathcal{S}$ is the tautological line subbundle on $\mathrm{Gr}^{\varpi_1^*}$ isomorphic to $\mathcal{O}(-1)$ on $\check{\mathbb{P}}^{n-1}$, and $\mathcal{O}(a)_{\varpi_1^*}$ stands for $\mathcal{S}^{\otimes -a}$. Note that $\mathcal{O}(1)_{\varpi_1}, \mathcal{O}(1)_{\varpi_1^*}$ are isomorphic to the restrictions of the determinant line bundle on $\mathrm{Gr}_{GL(n)}$.

Given an arbitrary sequence $\nu_1, \ldots, \nu_N$ with $\nu_i \in \{\varpi_1, \ldots, \varpi_n, \varpi_1^*, \ldots, \varpi_n^*\}$, the equivariant K-theory of the iterated convolution diagram

$$K^{GL(n,\mathcal{O})\rtimes\mathbb{C}^\times}(\mathrm{Gr}^{\nu_1}\widetilde{\times}\ldots\widetilde{\times}\mathrm{Gr}^{\nu_N})$$

is isomorphic to

$$K^{GL(n,\mathcal{O})\rtimes\mathbb{C}^\times}(\mathrm{Gr}^{\nu_1}) \otimes_{K_{GL(n,\mathcal{O})\rtimes\mathbb{C}^\times}(\mathrm{pt})} \cdots \otimes_{K_{GL(n,\mathcal{O})\rtimes\mathbb{C}^\times}(\mathrm{pt})} K^{GL(n,\mathcal{O})\rtimes\mathbb{C}^\times}(\mathrm{Gr}^{\nu_N}).$$

By the projection formula and rationality of singularities of $\overline{\mathrm{Gr}}^{\nu_1+\ldots+\nu_N}$, the convolution pushforward morphism

$$m_* \colon K^{GL(n,\mathcal{O})\rtimes\mathbb{C}^\times}(\mathrm{Gr}^{\nu_1}\widetilde{\times}\ldots\widetilde{\times}\mathrm{Gr}^{\nu_N}) \longrightarrow K^{GL(n,\mathcal{O})\rtimes\mathbb{C}^\times}(\overline{\mathrm{Gr}}^{\nu_1+\ldots+\nu_N})$$

is surjective. Hence in order to prove the surjectivity statement of the theorem, it suffices to express $K_{\mathrm{loc}}^{GL(n,\mathcal{O})\rtimes\mathbb{C}^\times}(\mathrm{Gr}^{\nu})$, $\nu \in \{\varpi_1, \ldots, \varpi_n, \varpi_1^*, \ldots, \varpi_n^*\}$, in terms of $\mathcal{O}(a)_{\varpi_1}, \mathcal{O}(a)_{\varpi_1^*}$, $a \in \mathbb{Z}$, and $K_{GL(n)}(\mathrm{pt})$. We will consider $\nu = \varpi_m$, $1 \le m \le n$, the case of ϖ_m^* being similar. Note that $\mathcal{O}_{\varpi_n}$ is the structure sheaf of a point $GL(n,\mathcal{O})$-orbit corresponding to the coweight $(1, \ldots, 1)$. We argue by induction in m.

[6] A stronger version of the theorem (over $\mathbb{Z}[\boldsymbol{v}^{\pm1}]$ as opposed to over $\mathbb{C}[\boldsymbol{v}^{\pm1}]_{\mathrm{loc}}$) is proved independently in [15, Corollary 2.21, Remark 2.22].

For ν as above, the Picard group of Gr^ν is $\mathbb{Z}$, and we denote the ample generator by $\mathcal{O}(1)_\nu$. It is isomorphic to the restriction of the determinant line bundle on $\mathrm{Gr}_{GL(n)}$. We start with an explicit expression for $\mathcal{O}_{\varpi_m} := \mathcal{O}_{\mathrm{Gr}^{\varpi_m}}$, $1 \le m \le n$, in terms of $\mathcal{O}(a)_{\varpi_1}$, $a \in \mathbb{Z}$. Recall that $\overline{\Phi}^0_{-2n}(f_r) = \frac{\boldsymbol{v}^{2r}}{1-\boldsymbol{v}^2}\mathcal{O}(r)_{\varpi_1}$ and $\overline{\Phi}^0_{-2n}(e_r) = \frac{(-1)^n\boldsymbol{v}^{2r+2n+1}}{1-\boldsymbol{v}^2}\mathcal{O}(-r-n)_{\varpi_1^*}$. We denote $\mathrm{ad}_x^{\boldsymbol{v}^r}\, y := [x,y]_{\boldsymbol{v}^r} = xy - \boldsymbol{v}^r yx$.

Proposition 9.3 *For any* $1 \le m \le n$, *we have*

$$\mathcal{O}_{\varpi_m} = (-1)^{\frac{m(m-1)}{2}}(1-\boldsymbol{v}^2)\overline{\Phi}^0_{-2n}(\mathrm{ad}^{\boldsymbol{v}^{2m}}_{f_{1-m}}\,\mathrm{ad}^{\boldsymbol{v}^{2(m-1)}}_{f_{3-m}}\cdots\mathrm{ad}^{\boldsymbol{v}^4}_{f_{m-3}} f_{m-1}), \tag{9.1}$$

$$\mathcal{O}_{\varpi_m^*} = (-1)^{nm+\frac{m(m+1)}{2}+1}\boldsymbol{v}^{m^2-2}(1-\boldsymbol{v}^2)\times$$
$$\overline{\Phi}^0_{-2n}(\mathrm{ad}^{\boldsymbol{v}^{-2m}}_{e_{-n+1-m}}\,\mathrm{ad}^{\boldsymbol{v}^{-2(m-1)}}_{e_{-n+3-m}}\cdots\mathrm{ad}^{\boldsymbol{v}^{-4}}_{e_{-n+m-3}} e_{-n+m-1}). \tag{9.2}$$

Proof We prove (9.1); the proof of (9.2) is similar. We will compare the images of the LHS and the RHS in $\widetilde{\mathcal{A}}^{\boldsymbol{v}}_{\mathrm{frac}}$. According to (8.1), the image of the LHS equals

$$\sum_{\#J=m}\prod_{r\in J}^{s\notin J}(1-\mathsf{w}_s\mathsf{w}_r^{-1})^{-1}\prod_{r\in J}D_r. \tag{9.3}$$

Here $J \subset \{1,\ldots,n\}$ is a subset of cardinality m. Let us denote the iterated $\boldsymbol{v}$-commutator $\mathrm{ad}^{\boldsymbol{v}^{2m}}_{f_{1-m}}\,\mathrm{ad}^{\boldsymbol{v}^{2(m-1)}}_{f_{3-m}}\cdots\mathrm{ad}^{\boldsymbol{v}^4}_{f_{m-3}} f_{m-1}$ by F_m. We want to prove

$$\widetilde{\Phi}^0_{-2n}(F_m) = (-1)^{\frac{m(m-1)}{2}}(1-\boldsymbol{v}^2)^{-1}\cdot\sum_{\#J=m}\prod_{r\in J}^{s\notin J}(1-\mathsf{w}_s\mathsf{w}_r^{-1})^{-1}\prod_{r\in J}D_r. \tag{9.4}$$

The proof proceeds by induction in m. So we assume (9.4) known for an integer $k < n$, and want to deduce (9.4) for $m = k+1$. We introduce a "shifted" $\boldsymbol{v}$-commutator $F'_k := \mathrm{ad}^{\boldsymbol{v}^{2k}}_{f_{2-k}}\,\mathrm{ad}^{\boldsymbol{v}^{2(k-1)}}_{f_{4-k}}\cdots\mathrm{ad}^{\boldsymbol{v}^4}_{f_{k-2}} f_k$. Then

$$\widetilde{\Phi}^0_{-2n}(F'_k) = (-1)^{\frac{k(k-1)}{2}}(1-\boldsymbol{v}^2)^{-1}\boldsymbol{v}^{2k}\cdot\sum_{\#J=k}\prod_{r\in J}\mathsf{w}_r\prod_{r\in J}^{s\notin J}\left(1-\frac{\mathsf{w}_s}{\mathsf{w}_r}\right)^{-1}\prod_{r\in J}D_r.$$

Now

$$\widetilde{\Phi}^0_{-2n}(F_{k+1}) = \widetilde{\Phi}^0_{-2n}([f_{-k},F'_k]_{\boldsymbol{v}^{2(k+1)}}) = [\widetilde{\Phi}^0_{-2n}(f_{-k}),\widetilde{\Phi}^0_{-2n}(F'_k)]_{\boldsymbol{v}^{2(k+1)}} =$$

$$(-1)^{\frac{k(k-1)}{2}}(1-\boldsymbol{v}^2)^{-2}\boldsymbol{v}^{2k}\cdot\left[\sum_{p=1}^{n}\frac{(\boldsymbol{v}^2\mathsf{w}_p)^{-k}}{\prod_{t\ne p}\left(1-\frac{\mathsf{w}_t}{\mathsf{w}_p}\right)}D_p,\ \sum_{\#J=k}\prod_{r\in J}\mathsf{w}_r\prod_{r\in J}^{s\notin J}\left(1-\frac{\mathsf{w}_s}{\mathsf{w}_r}\right)^{-1}\prod_{r\in J}D_r\right]_{\boldsymbol{v}^{2(k+1)}}.$$

First we check that the summands corresponding to $p \in J$ vanish. Due to the symmetry reasons, we may assume $p = 1,\ J = \{1, 2, \ldots, k\}$. Then

$$\left[\frac{(v^2 \mathsf{w}_1)^{-k}}{\prod\limits_{t>1}\left(1-\frac{\mathsf{w}_t}{\mathsf{w}_1}\right)} D_1, \prod_{r=1}^{k} \mathsf{w}_r \prod_{r\le k}^{s>k}\left(1-\frac{\mathsf{w}_s}{\mathsf{w}_r}\right)^{-1} D_1\cdots D_k\right]_{v^{2(k+1)}} =$$

$$\left[\frac{(v^2 \mathsf{w}_1)^{-k}}{\prod\limits_{t>k}\left(1-\frac{\mathsf{w}_t}{\mathsf{w}_1}\right)\prod\limits_{1<r\le k}\left(1-\frac{\mathsf{w}_r}{\mathsf{w}_1}\right)} D_1, \frac{\mathsf{w}_1\cdots \mathsf{w}_k}{\prod\limits_{s>k}\left(1-\frac{\mathsf{w}_s}{\mathsf{w}_1}\right)\prod\limits_{1<r\le k}^{s>k}\left(1-\frac{\mathsf{w}_s}{\mathsf{w}_r}\right)} D_1\cdots D_k\right]_{v^{2(k+1)}} =$$

$$\left(\frac{(v^2 \mathsf{w}_1)^{-k} v^2 \mathsf{w}_1\cdots \mathsf{w}_k}{\prod\limits_{t>k}\left(1-\frac{\mathsf{w}_t}{\mathsf{w}_1}\right)\prod\limits_{1<r\le k}\left(1-\frac{\mathsf{w}_r}{\mathsf{w}_1}\right)\prod\limits_{s>k}\left(1-v^{-2}\frac{\mathsf{w}_s}{\mathsf{w}_1}\right)\prod\limits_{1<r\le k}^{s>k}\left(1-\frac{\mathsf{w}_s}{\mathsf{w}_r}\right)} - \right.$$

$$\left. - \frac{v^{2(k+1)} \mathsf{w}_1\cdots \mathsf{w}_k (v^2\mathsf{w}_1)^{-k} v^{-2k}}{\prod\limits_{s>k}\left(1-\frac{\mathsf{w}_s}{\mathsf{w}_1}\right)\prod\limits_{1<r\le k}^{s>k}\left(1-\frac{\mathsf{w}_s}{\mathsf{w}_r}\right)\prod\limits_{t>k}\left(1-v^{-2}\frac{\mathsf{w}_t}{\mathsf{w}_1}\right)\prod\limits_{1<r\le k}\left(1-\frac{\mathsf{w}_r}{\mathsf{w}_1}\right)}\right) D_1^2 D_2\cdots D_k = 0.$$

Therefore,

$$(-1)^{\frac{k(k-1)}{2}}(1-v^2)^2\widetilde{\Phi}^0_{-2n}(F_{k+1}) =$$
$$\sum_{\#J=k}^{p\notin J}\left[\frac{(v^2\mathsf{w}_p)^{-k}}{\prod_{t\ne p}\left(1-\frac{\mathsf{w}_t}{\mathsf{w}_p}\right)} D_p, v^{2k}\prod_{r\in J}\mathsf{w}_r \prod_{r\in J}^{s\notin J}\left(1-\frac{\mathsf{w}_s}{\mathsf{w}_r}\right)^{-1}\prod_{r\in J} D_r\right]_{v^{2(k+1)}}.$$

We expand this combination of $v^{2(k+1)}$-commutators as a sum

$$\sum_{\#J=k+1}\phi_J(\mathsf{w}_1,\ldots,\mathsf{w}_n)\prod_{r\in J} D_r.$$

For the symmetry reasons, it suffices to calculate the rational function ϕ_J for a single $J = \{1, \ldots, k+1\}$. We have

$$\phi_J(\mathsf{w}_1,\ldots,\mathsf{w}_n) D_1\cdots D_{k+1} =$$

$$\sum_{r=1}^{k+1}\left[\frac{(v^2\mathsf{w}_r)^{-k}}{\prod\limits_{t\ne r}\left(1-\frac{\mathsf{w}_t}{\mathsf{w}_r}\right)} D_r, \frac{v^{2k}\mathsf{w}_1\cdots\widehat{\mathsf{w}_r}\cdots\mathsf{w}_{k+1}}{\prod\limits_{r\ne p\le k+1}^{t>k+1}\left(1-\frac{\mathsf{w}_t}{\mathsf{w}_p}\right)\prod\limits_{r\ne p\le k+1}\left(1-\frac{\mathsf{w}_r}{\mathsf{w}_p}\right)} D_1\cdots\widehat{D_r}\cdots D_{k+1}\right]_{v^{2(k+1)}} =$$

$$\sum_{r=1}^{k+1}\left(\frac{\mathsf{w}_r^{-k}\mathsf{w}_1\cdots\widehat{\mathsf{w}}_r\cdots\mathsf{w}_{k+1}}{\prod\limits_{t>k+1}\left(1-\frac{\mathsf{w}_t}{\mathsf{w}_r}\right)\prod\limits_{r\neq p\leq k+1}\left(1-\frac{\mathsf{w}_p}{\mathsf{w}_r}\right)\prod\limits_{r\neq p\leq k+1}^{t>k+1}\left(1-\frac{\mathsf{w}_t}{\mathsf{w}_p}\right)\prod\limits_{r\neq p\leq k+1}\left(1-\frac{\boldsymbol{v}^2\mathsf{w}_r}{\mathsf{w}_p}\right)}-\right.$$

$$\left.-\frac{\boldsymbol{v}^{2(k+1)}\mathsf{w}_r^{-k}\mathsf{w}_1\cdots\widehat{\mathsf{w}}_r\cdots\mathsf{w}_{k+1}}{\prod\limits_{r\neq p\leq k+1}^{t>k+1}\left(1-\frac{\mathsf{w}_t}{\mathsf{w}_p}\right)\prod\limits_{r\neq p\leq k+1}\left(1-\frac{\mathsf{w}_r}{\mathsf{w}_p}\right)\prod\limits_{t>k+1}\left(1-\frac{\mathsf{w}_t}{\mathsf{w}_r}\right)\prod\limits_{r\neq p\leq k+1}\left(1-\frac{\boldsymbol{v}^2\mathsf{w}_p}{\mathsf{w}_r}\right)}\right)D_1\cdots D_{k+1}=$$

$$-\boldsymbol{v}^{2(k+1)}\mathsf{w}_1\cdots\mathsf{w}_{k+1}\prod_{r\leq k+1}^{t>k+1}\left(1-\frac{\mathsf{w}_t}{\mathsf{w}_r}\right)^{-1}\times$$

$$\sum_{r=1}^{k+1}\left(\frac{\mathsf{w}_r^{-k-1}}{\prod\limits_{r\neq p\leq k+1}\left(1-\frac{\mathsf{w}_r}{\mathsf{w}_p}\right)\left(1-\frac{\boldsymbol{v}^2\mathsf{w}_p}{\mathsf{w}_r}\right)}-\frac{\boldsymbol{v}^{-2(k+1)}\mathsf{w}_r^{-k-1}}{\prod\limits_{r\neq p\leq k+1}\left(1-\frac{\mathsf{w}_p}{\mathsf{w}_r}\right)\left(1-\frac{\boldsymbol{v}^2\mathsf{w}_r}{\mathsf{w}_p}\right)}\right)D_1\cdots D_{k+1}.$$

This is equal to the following expression, by Lemma 9.4 below:

$$-\boldsymbol{v}^{2(k+1)}\mathsf{w}_1\cdots\mathsf{w}_{k+1}\prod_{r\leq k+1}^{t>k+1}\left(1-\frac{\mathsf{w}_t}{\mathsf{w}_r}\right)^{-1}\frac{(-1)^k(\boldsymbol{v}^2-1)}{\boldsymbol{v}^{2(k+1)}\prod\limits_{r\leq k+1}\mathsf{w}_r}D_1\cdots D_{k+1}=$$

$$(-1)^k(1-\boldsymbol{v}^2)\prod_{r\leq k+1}^{t>k+1}\left(1-\frac{\mathsf{w}_t}{\mathsf{w}_r}\right)^{-1}D_1\cdots D_{k+1}.$$

We conclude that

$$\widetilde{\Phi}^0_{-2n}(F_{k+1})=(-1)^{\frac{k(k+1)}{2}}(1-\boldsymbol{v}^2)^{-1}\cdot\sum_{\#J=k+1}\prod_{r\in J}^{s\notin J}(1-\mathsf{w}_s\mathsf{w}_r^{-1})^{-1}\prod_{r\in J}D_r,$$

and (9.4) is proved. It remains to check

Lemma 9.4 *We have*

$$\sum_{r=1}^{k+1}\left(\frac{\mathsf{w}_r^{-k-1}}{\prod\limits_{s\neq r}(1-\mathsf{w}_r/\mathsf{w}_s)(1-\boldsymbol{v}^2\mathsf{w}_s/\mathsf{w}_r)}-\frac{\boldsymbol{v}^{-2(k+1)}\mathsf{w}_r^{-k-1}}{\prod\limits_{s\neq r}(1-\mathsf{w}_s/\mathsf{w}_r)(1-\boldsymbol{v}^2\mathsf{w}_r/\mathsf{w}_s)}\right)=\frac{(-1)^k(\boldsymbol{v}^2-1)}{\boldsymbol{v}^{2(k+1)}\prod\limits_{r=1}^{k+1}\mathsf{w}_r}.$$

Proof The LHS is a degree $-k-1$ rational function of $\mathsf{w}_1,\ldots,\mathsf{w}_{k+1}$ with poles at the hyperplanes given by equations $\mathsf{w}_r-\mathsf{w}_s,\ \mathsf{w}_r-\boldsymbol{v}^2\mathsf{w}_s,\ \mathsf{w}_r\ (1\leq r\neq s\leq$

$k+1$). One can check $\operatorname{Res}_{\mathsf{w}_r-\mathsf{w}_s}\text{LHS} = \operatorname{Res}_{\mathsf{w}_r-v^2\mathsf{w}_s}\text{LHS} = 0$, so that LHS $= f\cdot\prod_{1\le r\le k+1}\mathsf{w}_r^{-1}$ for a rational function $f\in\mathbb{C}(v)$. To compute f, we specialize $\mathsf{w}_1\mapsto 0$ in the equality

$$f=\prod_{t=1}^{k+1}\mathsf{w}_t\cdot\sum_{r=1}^{k+1}\left(\frac{\prod_{s\neq r}\mathsf{w}_s}{\prod_{s\neq r}(\mathsf{w}_s-\mathsf{w}_r)(\mathsf{w}_r-v^2\mathsf{w}_s)}\cdot\frac{1}{\mathsf{w}_r}-\frac{v^{-2(k+1)}\prod_{s\neq r}\mathsf{w}_s}{\prod_{s\neq r}(\mathsf{w}_r-\mathsf{w}_s)(\mathsf{w}_s-v^2\mathsf{w}_r)}\cdot\frac{1}{\mathsf{w}_r}\right).$$

The only summands surviving under this specialization correspond to $r=1$, and so we get

$$f=\prod_{t=2}^{k+1}\mathsf{w}_t\cdot\left(\frac{\prod_{s=2}^{k+1}\mathsf{w}_s}{(-v^2)^k\cdot\prod_{s=2}^{k+1}\mathsf{w}_s^2}-\frac{v^{-2(k+1)}\cdot\prod_{s=2}^{k+1}\mathsf{w}_s}{(-1)^k\cdot\prod_{s=2}^{k+1}\mathsf{w}_s^2}\right)=(-1)^k(v^{-2k}-v^{-2(k+1)}).$$

The lemma is proved. □

The proposition is proved. □

Returning to the proof of Theorem 9.2, we need to prove that $K^{GL(n,\mathcal{O})\rtimes\mathbb{C}^\times}_{\text{loc}}(\text{Gr}^{\varpi_m})$ lies in the image $\overline{\Phi}{}^0_{-2n,\text{loc}}(\mathfrak{U}^{\text{ad}}_{0,-2n,\text{loc}})$ for $1\le m\le n$. We know that the class of the structure sheaf $\mathcal{O}_{\varpi_m}\in K^{GL(n,\mathcal{O})\rtimes\mathbb{C}^\times}_{\text{loc}}(\text{Gr}^{\varpi_m})$ lies in $\overline{\Phi}{}^0_{-2n,\text{loc}}(\mathfrak{U}^{\text{ad}}_{0,-2n,\text{loc}})$. It is also known that $K^{GL(n,\mathcal{O})\rtimes\mathbb{C}^\times}(\text{Gr}^{\varpi_m})$ as a left $K_{GL(n,\mathcal{O})\rtimes\mathbb{C}^\times}(\text{pt})$-module is generated by the classes $\Sigma^\lambda(\mathcal{Q})$ where $\mathcal{Q}$ is the tautological quotient bundle on $\text{Gr}^{\varpi_m}\simeq\text{Gr}(m,n)$, and Σ^λ is the polynomial Schur functor corresponding to a Young diagram λ with $\le m$ rows (in fact, it is enough to consider λ's with $\le n-m$ columns). Given such λ, it suffices to check that $\operatorname{Sym}\left(\mathsf{w}_1^{\lambda_1}\cdots\mathsf{w}_m^{\lambda_m}\prod_{r\le m}^{s>m}\left(1-\frac{\mathsf{w}_s}{\mathsf{w}_r}\right)^{-1}D_1\cdots D_m\right)$ lies in $\widetilde{\Phi}{}^0_{-2n,\text{loc}}(\mathfrak{U}^{\text{ad}}_{0,-2n,\text{loc}})$ (here Sym stands for the symmetrization with respect to the symmetric group $\mathfrak{S}_n$). More generally, for a Young diagram μ with $\le n$ rows we will show that $\operatorname{Sym}\left(\mathsf{w}_1^{\mu_1}\cdots\mathsf{w}_n^{\mu_n}\cdot\prod_{r\le m}^{s>m}\left(1-\frac{\mathsf{w}_s}{\mathsf{w}_r}\right)^{-1}D_1\cdots D_m\right)$ lies in $\widetilde{\Phi}{}^0_{-2n,\text{loc}}(\mathfrak{U}^{\text{ad}}_{0,-2n,\text{loc}})$. To this end, we use the *right* multiplication by $K_{GL(n,\mathcal{O})\rtimes\mathbb{C}^\times}(\text{pt})$. It suffices to check that the $K_{GL(n,\mathcal{O})\rtimes\mathbb{C}^\times}(\text{pt})_{\text{loc}}$-bimodule generated by $X_{1,m}:=\operatorname{Sym}\left(\prod_{r\le m}^{s>m}\left(1-\frac{\mathsf{w}_s}{\mathsf{w}_r}\right)^{-1}D_1\cdots D_m\right)$ contains elements $X_{F,m}:=\operatorname{Sym}\left(F\prod_{r\le m}^{s>m}\left(1-\frac{\mathsf{w}_s}{\mathsf{w}_r}\right)^{-1}D_1\cdots D_m\right)$ for any polynomial $F\in\mathbb{C}[\mathsf{w}_1,\ldots,\mathsf{w}_n]$. We can assume that $F\in\mathbb{C}[\mathsf{w}_1,\ldots,\mathsf{w}_n]^{\mathfrak{S}_m\times\mathfrak{S}_{n-m}}$, where the symmetric groups act by permuting $\{\mathsf{w}_r,\ 1\le r\le m\}$ and $\{\mathsf{w}_s,\ m+1\le s\le n\}$. Note that $\mathbb{C}[\mathsf{w}_1,\ldots,\mathsf{w}_n]^{\mathfrak{S}_m\times\mathfrak{S}_{n-m}}$ is generated by $\mathbb{C}[\mathsf{w}_1,\ldots,\mathsf{w}_m]^{\mathfrak{S}_m}$

as a left $\mathbb{C}[\mathsf{w}_1, \ldots, \mathsf{w}_n]^{\mathfrak{S}_n}$-module. Hence, it suffices to treat the case $F \in \mathbb{C}[\mathsf{w}_1, \ldots, \mathsf{w}_m]^{\mathfrak{S}_m} = \mathbb{C}[p_1, \ldots, p_m]$, where $p_k := \sum_{r=1}^m \mathsf{w}_r^k$. The latter case follows from the equality

$$\left[\sum_{r=1}^n \mathsf{w}_r^k, X_{F,m}\right] = (1 - \boldsymbol{v}^{2k}) X_{Fp_k,m}$$

for $F \in \mathbb{C}[\mathsf{w}_1, \ldots, \mathsf{w}_m]^{\mathfrak{S}_m}$.

The theorem is proved. □

Remark 9.5 The end of our proof of Theorem 9.2 is a variation of the following argument we learned from P. Etingof. We define $\mathbb{C}[\boldsymbol{v}^{\pm 1}]_{\mathrm{Loc}}$ inverting $(1 - \boldsymbol{v}^m)$, $m \in \mathbb{Z}$. We consider a $\mathbb{C}[\boldsymbol{v}^{\pm 1}]_{\mathrm{Loc}}$-algebra $\mathfrak{A}$ of finite difference operators with generators $\{\mathsf{w}_i^{\pm 1}, D_i^{\pm 1}\}_{i=1}^n$ and defining relations $D_i \mathsf{w}_j = \boldsymbol{v}^{2\delta_{ij}} \mathsf{w}_j D_i$, $[D_i, D_j] = [\mathsf{w}_i, \mathsf{w}_j] = 0$. Then the algebra of $\mathfrak{S}_n$-invariants $\mathfrak{A}^{\mathfrak{S}_n}$ is generated by its subalgebras $\mathbb{C}[\boldsymbol{v}^{\pm 1}]_{\mathrm{Loc}}[D_1^{\pm 1}, \ldots, D_n^{\pm 1}]^{\mathfrak{S}_n}$ and $\mathbb{C}[\boldsymbol{v}^{\pm 1}]_{\mathrm{Loc}}[\mathsf{w}_1^{\pm 1}, \ldots, \mathsf{w}_n^{\pm 1}]^{\mathfrak{S}_n}$.

Indeed, let $\mathfrak{B}$ be the $\mathbb{C}[\boldsymbol{v}^{\pm 1}]_{\mathrm{Loc}}$-algebra generated by $\mathsf{w}^{\pm 1}$, $D^{\pm 1}$ subject to $D\mathsf{w} = \boldsymbol{v}^2 \mathsf{w} D$. Then $\mathfrak{A} = \mathfrak{B}^{\otimes n}$ (tensor product over $\mathbb{C}[\boldsymbol{v}^{\pm 1}]_{\mathrm{Loc}}$), and $\mathfrak{A}^{\mathfrak{S}_n} = \mathrm{Sym}^n \mathfrak{B}$ (symmetric power over $\mathbb{C}[\boldsymbol{v}^{\pm 1}]_{\mathrm{Loc}}$). Now $\mathrm{Sym}^n \mathfrak{B}$ is spanned by the elements $\{b^{\otimes n}\}_{b \in \mathfrak{B}}$, and hence $\mathrm{Sym}^n \mathfrak{B}$ is generated by the elements $\{b_{(1)} + \ldots + b_{(n)}\}_{b \in \mathfrak{B}}$, where $b_{(r)} = 1 \otimes \cdots \otimes 1 \otimes b \otimes 1 \otimes \cdots \otimes 1$ (b at the r-th entry). Indeed, it suffices to verify the generation claim for an algebra $\mathbb{C}[\boldsymbol{v}^{\pm 1}]_{\mathrm{Loc}}[b]$ where it is nothing but the fundamental theorem on symmetric functions.

We conclude that $\mathrm{Sym}^n \mathfrak{B}$ is generated by the elements $\{p_{m,k} = \sum_{r=1}^n \mathsf{w}_r^m D_r^k\}_{m,k \in \mathbb{Z}}$. However, $p_{m,k} = (\boldsymbol{v}^{2mk} - 1)^{-1} [\sum_{r=1}^n D_r^k, \sum_{s=1}^n \mathsf{w}_s^m]$ for $m \neq 0 \neq k$.

Remark 9.6 Motivated by [10, Remark 3.5] we call $\mathcal{O}_{\varpi_n} \in K^{GL(n,\mathcal{O}) \rtimes \mathbb{C}^\times}(\mathrm{Gr}_{GL(n)})$ *the quantum resultant*. In fact, it is a quantization of the boundary equation for the trigonometric zastava ${}^{\dagger}\mathring{Z}^n_{SL(2)}$ which is nothing but the resultant of two polynomials. Note that, up to multiplication by an element of $\mathbb{C}[\boldsymbol{v}^{\pm 1}]$, the quantum resultant is uniquely characterized by the property

$$\mathcal{O}_{\varpi_n} \overline{\Phi}^0_{-2n}(A^{\pm}_{\pm r}) = \boldsymbol{v}^{\pm(2r-n)} \overline{\Phi}^0_{-2n}(A^{\pm}_{\pm r}) \mathcal{O}_{\varpi_n}, \ \mathcal{O}_{\varpi_n} \overline{\Phi}^0_{-2n}(f_p) = \boldsymbol{v}^{2p} \overline{\Phi}^0_{-2n}(f_p) \mathcal{O}_{\varpi_n}. \tag{9.5}$$

Remark 9.7 Here is a geometric explanation of the equality

$$\mathcal{O}(-k-1)_{\varpi_1} * \mathcal{O}_{\varpi_k} - \boldsymbol{v}^{2(k+1)} \mathcal{O}_{\varpi_k} * \mathcal{O}(-k-1)_{\varpi_1} = (-1)^k (1 - \boldsymbol{v}^2) \boldsymbol{v}^{-2(k+1)} \mathcal{O}(-1)_{\varpi_{k+1}}, \tag{9.6}$$

established as an induction step during our proof of Proposition 9.3. We have the convolution morphisms

$$\mathrm{Gr}^{\varpi_1} \widetilde{\times} \mathrm{Gr}^{\varpi_k} \xrightarrow{m} \overline{\mathrm{Gr}}^{\varpi_1 + \varpi_k} \xleftarrow{m'} \mathrm{Gr}^{\varpi_k} \widetilde{\times} \mathrm{Gr}^{\varpi_1},$$

and $\overline{\mathrm{Gr}}^{\varpi_1+\varpi_k} = \mathrm{Gr}^{\varpi_1+\varpi_k} \sqcup \mathrm{Gr}^{\varpi_{k+1}}$. Let us consider the transversal slice $\overline{\mathcal{W}}^{\varpi_1+\varpi_k}_{\varpi_{k+1}} \subset \overline{\mathrm{Gr}}^{\varpi_1+\varpi_k}$ through the point $\varpi_{k+1} = (1, \ldots, 1, 0, \ldots, 0)$ ($k+1$ 1's). It suffices to check that

$$m_*\left(\mathcal{O}(-k-1)_{\varpi_1}\widetilde{\boxtimes}\mathcal{O}_{\varpi_k}|_{m^{-1}\overline{\mathcal{W}}^{\varpi_1+\varpi_k}_{\varpi_{k+1}}}\right) - \boldsymbol{v}^{2(k+1)} m'_*\left(\mathcal{O}_{\varpi_k}\widetilde{\boxtimes}\mathcal{O}(-k-1)_{\varpi_1}|_{m'^{-1}\overline{\mathcal{W}}^{\varpi_1+\varpi_k}_{\varpi_{k+1}}}\right) =$$
$$(-1)^k(1-\boldsymbol{v}^2)\boldsymbol{v}^{-2(k+1)}\mathsf{w}_1^{-1}\cdots\mathsf{w}_{k+1}^{-1},$$

where we view $\boldsymbol{v}^{-2(k+1)}\mathsf{w}_1^{-1}\cdots\mathsf{w}_{k+1}^{-1}$ as a character of $T\times\mathbb{C}^\times$ ($T \subset GL(n)$ is the diagonal Cartan torus). According to [52, Corollary 3.4], $\overline{\mathcal{W}}^{\varpi_1+\varpi_k}_{\varpi_{k+1}}$ is naturally isomorphic to the slice $\overline{\mathcal{W}}^{\theta}_0 \subset \mathrm{Gr}_{GL(k+1)\times(\mathbb{C}^\times)^{n-k-1}}$ where $\theta = (1, 0, \ldots, 0, -1)$ is the highest coroot of $GL(k+1)$. Moreover, the preimages of $\overline{\mathcal{W}}^{\varpi_1+\varpi_k}_{\varpi_{k+1}}$ in the two convolution diagrams are isomorphic to the cotangent bundles $T^*\mathbb{P}^k$ and $T^*\check{\mathbb{P}}^k$, respectively. We will keep the following notation for the convolution morphisms restricted to the slice:

$$T^*\mathbb{P}^k \xrightarrow{m} \overline{\mathcal{W}}^{\theta}_0 \xleftarrow{m'} T^*\check{\mathbb{P}}^k.$$

Note also that $\overline{\mathcal{W}}^{\theta}_0$ is isomorphic to the minimal nilpotent orbit closure $\overline{\mathbb{O}}_{\min} \subset \mathfrak{sl}_{k+1}$. Finally, $\mathcal{O}(-k-1)_{\varpi_1}\widetilde{\boxtimes}\mathcal{O}_{\varpi_k}|_{m^{-1}\overline{\mathcal{W}}^{\varpi_1+\varpi_k}_{\varpi_{k+1}}}$ and $\mathcal{O}_{\varpi_k}\widetilde{\boxtimes}\mathcal{O}(-k-1)_{\varpi_1}|_{m'^{-1}\overline{\mathcal{W}}^{\varpi_1+\varpi_k}_{\varpi_{k+1}}}$ are isomorphic to the pull-backs of $\mathcal{O}_{\mathbb{P}^k}(-k-1)$ and $\mathcal{O}_{\check{\mathbb{P}}^k}(-k-1)$, respectively, but with *nontrivial* $\mathbb{C}^\times$-equivariant structures.

Let us explain our choice of the line bundles. According to [8, Proposition 8.2], the convolutions in question are $GL(k+1)\times\mathbb{C}^\times$-equivariant perverse coherent sheaves on $\overline{\mathbb{O}}_{\min} \subset \mathfrak{sl}_{k+1}$. Since $\dim H^k(T^*\mathbb{P}^k, \mathcal{O}_{T^*\mathbb{P}^k}(-k-1)) = 1$, while $H^k(T^*\mathbb{P}^k, \mathcal{O}_{T^*\mathbb{P}^k}(k+1)) = 0$, we have an exact sequence of perverse coherent sheaves[7] on $\overline{\mathbb{O}}_{\min} \subset \mathfrak{sl}_{k+1}$:

$$0 \to j_{!*}\mathcal{O}_{\mathbb{O}_{\min}}(-k-1)[k] \to m_*\mathcal{O}_{T^*\mathbb{P}^k}(-k-1)[k] \to \delta_0 \to 0,$$

where $j\colon \mathbb{O}_{\min} \hookrightarrow \overline{\mathbb{O}}_{\min}$ is the open embedding, and δ_0 is an irreducible skyscraper sheaf at $0 \in \overline{\mathbb{O}}_{\min}$ with certain $\mathbb{C}^\times$-equivariant structure. The same exact sequence holds for $m'_*\mathcal{O}_{T^*\check{\mathbb{P}}^k}(-k-1)[k]$, but the quotient δ_0 has a *different* $\mathbb{C}^\times$-equivariant structure.

Proposition 9.8 *The restriction of* $\widetilde{\Phi}^0_{-2n}$ *to* $\mathfrak{Y}^0_{-2n,-}$ *is injective.*

Proof Consider an ordering $A_0^- \prec A_{-1}^- \prec \ldots \prec A_{-n+1}^- \prec C_0^- \prec \ldots \prec C_{-n+1}^-$. We set $(A_0^-)^{-k} := ((-\boldsymbol{v}^2)^{-n}A_{-n}^-)^k$ for $k > 0$. For

[7] We are grateful to R. Bezrukavnikov for his explanations about perverse coherent sheaves.

$\vec{r} = (r_1, \ldots, r_{2n}) \in \mathbb{Z} \times \mathbb{N}^{2n-1}$, we define the ordered monomial $m_{\vec{r}} := (A_0^-)^{r_1}(A_{-1}^-)^{r_2} \cdots (A_{-n+1}^-)^{r_n}(C_0^-)^{r_{n+1}} \cdots (C_{-n+1}^-)^{r_{2n}}$.

Lemma 9.9 *The ordered monomials* $\{m_{\vec{r}}\}$ *span* $\mathfrak{Y}^0_{-2n,-}$.

Proof According to relations (6.7, 6.9), we have $[A_t^-, A_s^-] = [C_t^-, C_s^-] = 0$ for $s, t \leq 0$. Due to Remark 8.8, we also have $C_s^- = 0$ for $s \leq -n$. It remains to prove that all A_t^- can be taken to the left of all C_s^-. This is implied by the fact that $C_s^- A_t^-$ can be written as a linear combination of normally ordered monomials $A_{t'}^- C_{s'}^-$. The latter claim follows from relation (6.11) by induction in $\min\{-t, -s\}$. The lemma is proved. □

The following result will be proved in Sect. 9.2:

Lemma 9.10

(a) *The ordered monomials* $\{m_{\vec{r}}\}$ *form a* $K_{\mathbb{C}^\times}(\mathrm{pt})$*-basis of* $\mathfrak{Y}^0_{-2n,-}$.
(b) $\{\overline{\Phi}^0_{-2n}(m_{\vec{r}})\}$ *form a* $K_{\mathbb{C}^\times}(\mathrm{pt})$*-basis of* $\overline{\Phi}^0_{-2n}(\mathfrak{Y}^0_{-2n,-})$.

The proposition is proved. □

9.2 Positive Grassmannian

Recall the positive part of the affine Grassmannian $\mathrm{Gr}^+_{GL(n)} \subset \mathrm{Gr}_{GL(n)}$ [10, § 3(ii)] parametrizing the *sublattices* in the standard one. Recall also that $K^{GL(n,\mathcal{O})\rtimes\mathbb{C}^\times}_{\mathrm{loc}}(\mathrm{Gr}^{\varpi_1}) = K^{GL(n,\mathcal{O})\rtimes\mathbb{C}^\times}_{\mathrm{loc}}(\mathbb{P}^{n-1})$ is generated over $K_{GL(n)}(\mathrm{pt})$ by the classes of $\mathcal{O}(a)_{\varpi_1}, -n+1 \leq a \leq 0$. The proof of Theorem 9.2 shows that $\overline{\Phi}^0_{-2n,\mathrm{loc}}\colon \mathfrak{U}^{\mathrm{ad}}_{0,-2n,\mathrm{loc}} \twoheadrightarrow K^{\widetilde{GL}(n,\mathcal{O})\rtimes\widetilde{\mathbb{C}}^\times}_{\mathrm{loc}}(\mathrm{Gr}_{GL(n)})$ restricts to a surjective homomorphism $\overline{\Phi}^0_{-2n,\mathrm{loc}}\colon \mathfrak{Y}^0_{-2n,-,\mathrm{loc}} \twoheadrightarrow K^{\widetilde{GL}(n,\mathcal{O})\rtimes\widetilde{\mathbb{C}}^\times}_{\mathrm{loc}}(\mathrm{Gr}^+_{GL(n)})$.

Proposition 9.11 $\overline{\Phi}^0_{-2n,\mathrm{loc}}\colon \mathfrak{Y}^0_{-2n,-,\mathrm{loc}} \xrightarrow{\sim} K^{\widetilde{GL}(n,\mathcal{O})\rtimes\widetilde{\mathbb{C}}^\times}_{\mathrm{loc}}(\mathrm{Gr}^+_{GL(n)})$.

Proof We have to check that $\overline{\Phi}^0_{-2n,\mathrm{loc}}\colon \mathfrak{Y}^0_{-2n,-,\mathrm{loc}} \to K^{\widetilde{GL}(n,\mathcal{O})\rtimes\widetilde{\mathbb{C}}^\times}_{\mathrm{loc}}(\mathrm{Gr}^+_{GL(n)})$ is injective. To this end, note that $\mathrm{Gr}^+_{GL(n)}$ is a union of connected components numbered by nonnegative integers: $\mathrm{Gr}^+_{GL(n)} = \bigsqcup_{r\in\mathbb{N}} \mathrm{Gr}^{+,r}_{GL(n)}$, where $\mathrm{Gr}^{+,r}_{GL(n)}$ parametrizes the sublattices of codimension r in the standard one. The direct sum decomposition $K^{\widetilde{GL}(n,\mathcal{O})\rtimes\widetilde{\mathbb{C}}^\times}_{\mathrm{loc}}(\mathrm{Gr}^+_{GL(n)}) = \bigoplus_{r\in\mathbb{N}} K^{\widetilde{GL}(n,\mathcal{O})\rtimes\widetilde{\mathbb{C}}^\times}_{\mathrm{loc}}(\mathrm{Gr}^{+,r}_{GL(n)})$ is a grading of the convolution algebra. For any connected component, $K^{\widetilde{GL}(n,\mathcal{O})\rtimes\widetilde{\mathbb{C}}^\times}_{\mathrm{loc}}(\mathrm{Gr}^{+,r}_{GL(n)})$ is a free $K_{\widetilde{GL}(n,\mathcal{O})\rtimes\widetilde{\mathbb{C}}^\times}(\mathrm{pt})_{\mathrm{loc}}$-module of rank d_r, where d_r is the number of T-fixed points in $\mathrm{Gr}^{+,r}_{GL(n)}$, that is the number of weights of the irreducible $GL(n)$-module with the highest weight $(r, 0, \ldots, 0)$, isomorphic to $\mathrm{Sym}^r(\mathbb{C}^n)$. Note that all the weights of $\mathrm{Sym}^r(\mathbb{C}^n)$ have multiplicity one; in other words, $d_r = \dim \mathrm{Sym}^r(\mathbb{C}^n)$.

According to Lemma 9.9, we can introduce a grading $\mathfrak{Y}^0_{-2n,-,\mathrm{loc}} = \bigoplus_{r\in\mathbb{N}} \mathfrak{Y}^{0,r}_{-2n,-,\mathrm{loc}}$: a monomial $m_{\vec{r}}$ has degree r if $r_{n+1}+\ldots+r_{2n} = r$. It is immediate from the relations between $A^-_\bullet, C^-_\bullet$-generators that this grading is well-defined. Also, it is clear that $\overline{\Phi}^0_{-2n,\mathrm{loc}}(\mathfrak{Y}^{0,r}_{-2n,-,\mathrm{loc}}) \subset K^{\widetilde{GL}(n,\mathcal{O})\rtimes\widetilde{\mathbb{C}}^\times}_{\mathrm{loc}}(\mathrm{Gr}^{+,r}_{GL(n)})$. Meanwhile, we know from Theorem 9.2 that $\overline{\Phi}^0_{-2n,\mathrm{loc}}(\mathfrak{Y}^{0,r}_{-2n,-,\mathrm{loc}}) = K^{\widetilde{GL}(n,\mathcal{O})\rtimes\widetilde{\mathbb{C}}^\times}_{\mathrm{loc}}(\mathrm{Gr}^{+,r}_{GL(n)})$. On the other hand, we know from Lemma 9.9 that $\mathfrak{Y}^{0,r}_{-2n,-,\mathrm{loc}}$ as a left $K_{\widetilde{GL}(n,\mathcal{O})\rtimes\widetilde{\mathbb{C}}^\times}(\mathrm{pt})_{\mathrm{loc}}$-module has no more than d'_r generators, where d'_r is the number of compositions of r into n (ordered) summands. Since $d_r = d'_r$, we conclude that $\overline{\Phi}^0_{-2n,\mathrm{loc}}\colon \mathfrak{Y}^{0,r}_{-2n,-,\mathrm{loc}} \to K^{\widetilde{GL}(n,\mathcal{O})\rtimes\widetilde{\mathbb{C}}^\times}_{\mathrm{loc}}(\mathrm{Gr}^{+,r}_{GL(n)})$ must be an isomorphism, and $\mathfrak{Y}^{0,r}_{-2n,-,\mathrm{loc}}$ is a free left $K_{\widetilde{GL}(n,\mathcal{O})\rtimes\widetilde{\mathbb{C}}^\times}(\mathrm{pt})_{\mathrm{loc}}$-module of rank $d_r = d'_r$. This completes the proof of Proposition 9.11, Lemma 9.10 (and Proposition 9.8). □

Remark 9.12 One can check that the natural morphism

$$K^{\widetilde{GL}(n,\mathcal{O})\rtimes\widetilde{\mathbb{C}}^\times}(\mathrm{Gr}^+_{GL(n)})[\mathcal{O}^{-1}_{\varpi_n}] \to K^{\widetilde{GL}(n,\mathcal{O})\rtimes\widetilde{\mathbb{C}}^\times}(\mathrm{Gr}_{GL(n)})$$

is an isomorphism. Now it follows from the proof of Proposition 9.11 and Theorem 9.2 that in order to check Conjectures 8.7, 8.9 and 8.13 in our case: $\mathrm{Ker}(\widetilde{\Phi}^0_{-2n,\mathrm{loc}}) = \mathcal{J}^0_{-2n,\mathrm{loc}}$, it suffices to check the following equality in $\mathcal{U}^{\mathrm{ad}}_{0,-2n}/\mathcal{J}^0_{-2n}$:

$$-\boldsymbol{v}^{n^2-2}(1-\boldsymbol{v}^2)^2\cdot(\mathrm{ad}^{\boldsymbol{v}^{2n}}_{f_{1-n}}\,\mathrm{ad}^{\boldsymbol{v}^{2(n-1)}}_{f_{3-n}}\cdots\mathrm{ad}^{\boldsymbol{v}^4}_{f_{n-3}}\,f_{n-1})(\mathrm{ad}^{\boldsymbol{v}^{-2n}}_{e_{1-2n}}\,\mathrm{ad}^{\boldsymbol{v}^{-2(n-1)}}_{e_{3-2n}}\cdots\mathrm{ad}^{\boldsymbol{v}^{-4}}_{e_{-3}}\,e_{-1}) = 1.$$

Remark 9.13 Consider a subalgebra $\mathfrak{U}^<_{0,-2n} \subset \mathfrak{U}^{\mathrm{ad}}_{0,-2n}$ generated by $\{(\boldsymbol{v}-\boldsymbol{v}^{-1})f_s\}_{s\in\mathbb{Z}}$. Note that it is independent of n, cf. Proposition 5.1. The image $\overline{\Phi}^0_{-2n}(\mathfrak{U}^<_{0,-2n})$ in $K^{\widetilde{GL}(n,\mathcal{O})\rtimes\widetilde{\mathbb{C}}^\times}(\mathrm{Gr}_{GL(n)})$ is isomorphic to the *M-system algebra* $\mathcal{U}'_{n-1}$ of [18]. In particular, the generators $M_{m,s} \in \mathcal{U}'_{n-1}$ of [18, § 2.1] correspond to scalar multiples of the classes $\mathcal{O}(-s)_{\varpi_m} \in K^{\widetilde{GL}(n,\mathcal{O})\rtimes\widetilde{\mathbb{C}}^\times}(\mathrm{Gr}_{GL(n)})$, cf. (9.1) and [18, (2.23)].

10 Coproducts on Shifted Quantum Affine Algebras

Throughout this section, we work mainly with simply-connected shifted quantum affine algebras. However, all the results can be obviously generalized to the adjoint versions.

10.1 Drinfeld Formal Coproduct

The standard quantum loop algebra $U_{\boldsymbol{v}}(L\mathfrak{g})$ admits the Drinfeld formal coproduct

$$\widetilde{\Delta}\colon U_{\boldsymbol{v}}(L\mathfrak{g}) \longrightarrow U_{\boldsymbol{v}}(L\mathfrak{g})\widehat{\otimes}U_{\boldsymbol{v}}(L\mathfrak{g}),$$

defined in the new Drinfeld realization of $U_{\boldsymbol{v}}(L\mathfrak{g})$ via

$$\begin{aligned}\widetilde{\Delta}(e_i(z)) &:= e_i(z) \otimes 1 + \psi_i^-(z) \otimes e_i(z),\\ \widetilde{\Delta}(f_i(z)) &:= f_i(z) \otimes \psi_i^+(z) + 1 \otimes f_i(z),\\ \widetilde{\Delta}(\psi_i^\pm(z)) &:= \psi_i^\pm(z) \otimes \psi_i^\pm(z).\end{aligned} \tag{10.1}$$

Remark 10.1 Composing $\widetilde{\Delta}$ with the $\mathbb{C}^\times$-action on the first factor, D. Hernandez obtained a deformed coproduct $\Delta_\zeta : U_{\boldsymbol{v}}(L\mathfrak{g}) \to U_{\boldsymbol{v}}(L\mathfrak{g}) \otimes U_{\boldsymbol{v}}(L\mathfrak{g})((\zeta))$, where ζ is a formal variable, see [37, Section 6].

This can be obviously generalized to the shifted setting.

Lemma 10.2 *For any coweights $\mu_1^\pm, \mu_2^\pm \in \Lambda$, there is a $\mathbb{C}(\boldsymbol{v})$-algebra homomorphism*

$$\widetilde{\Delta}\colon \mathcal{U}^{\mathrm{sc}}_{\mu_1^+ + \mu_2^+, \mu_1^- + \mu_2^-} \longrightarrow \mathcal{U}^{\mathrm{sc}}_{\mu_1^+, \mu_1^-} \widehat{\otimes} \mathcal{U}^{\mathrm{sc}}_{\mu_2^+, \mu_2^-},$$

defined via (10.1).

We call this homomorphism a formal coproduct for shifted quantum affine algebras. Given two representations V_1, V_2 of $\mathcal{U}^{\mathrm{sc}}_{\mu_1^+, \mu_1^-}, \mathcal{U}^{\mathrm{sc}}_{\mu_2^+, \mu_2^-}$, respectively, we will use $V_1 \widetilde{\otimes} V_2$ to denote the representation of $\mathcal{U}^{\mathrm{sc}}_{\mu_1^+ + \mu_2^+, \mu_1^- + \mu_2^-}$ on the vector space $V_1 \otimes V_2$ induced by $\widetilde{\Delta}$, whenever the action of the infinite sums representing $\widetilde{\Delta}(e_{i,r}), \widetilde{\Delta}(f_{i,r})$ are well-defined. We will discuss a particular example of this construction in Sect. 12.6.

10.2 Drinfeld-Jimbo Coproduct

The standard quantum loop algebra $U_{\boldsymbol{v}}(L\mathfrak{g})$ also admits the Drinfeld-Jimbo coproduct

$$\Delta\colon U_{\boldsymbol{v}}(L\mathfrak{g}) \longrightarrow U_{\boldsymbol{v}}(L\mathfrak{g}) \otimes U_{\boldsymbol{v}}(L\mathfrak{g}),$$

defined in the Drinfeld-Jimbo realization of $U_{\boldsymbol{v}}(L\mathfrak{g})$ via

$$\Delta\colon E_i \mapsto E_i \otimes K_i + 1 \otimes E_i,\ F_i \mapsto F_i \otimes 1 + K_i^{-1} \otimes F_i,\ K_i^{\pm 1} \mapsto K_i^{\pm 1} \otimes K_i^{\pm 1},\ i \in \widetilde{I}.$$

Recall that $\widetilde{I} = I \cup \{i_0\}$ is the vertex set of the extended Dynkin diagram and $\{E_i, F_i, K_i^{\pm 1}\}_{i \in \widetilde{I}}$ are the standard Drinfeld-Jimbo generators of $U_{\boldsymbol{v}}^{\mathrm{DJ}}(L\mathfrak{g}) \simeq U_{\boldsymbol{v}}(L\mathfrak{g})$.

We also denote the Drinfeld-Jimbo coproduct on $U_v^{\mathrm{ad}}(L\mathfrak{g})$ by Δ^{ad}: the natural inclusion $U_v(L\mathfrak{g}) \hookrightarrow U_v^{\mathrm{ad}}(L\mathfrak{g})$ intertwines Δ and Δ^{ad}, while $\Delta^{\mathrm{ad}}(\phi_i^{\pm}) = \phi_i^{\pm} \otimes \phi_i^{\pm}$.

The goal of this section is to generalize these coproducts to the shifted setting. In other words, given $\mathfrak{g}$ and coweights $\mu_1, \mu_2 \in \Lambda$, we would like to construct homomorphisms

$$\Delta_{\mu_1,\mu_2} \colon \mathcal{U}^{\mathrm{sc}}_{0,\mu_1+\mu_2} \longrightarrow \mathcal{U}^{\mathrm{sc}}_{0,\mu_1} \otimes \mathcal{U}^{\mathrm{sc}}_{0,\mu_2},$$

which coincide with Δ in the particular case $\mu_1 = \mu_2 = 0$. We provide such a construction for the simplest case $\mathfrak{g} = \mathfrak{sl}_2$ in Sects. 10.3 ($\mu_1, \mu_2 \in \Lambda^-$) and 10.4 (general μ_1, μ_2). Using the RTT presentation of $U_v(L\mathfrak{sl}_n)$, we generalize this to obtain Δ_{μ_1,μ_2} for $\mathfrak{g} = \mathfrak{sl}_n$ in Sects. 10.6 ($\mu_1, \mu_2 \in \Lambda^-$) and 10.7 (general μ_1, μ_2).

Remark 10.3

(a) This result is nontrivial due to an absence of the Drinfeld-Jimbo type presentation of shifted quantum affine algebras.
(b) A similar coproduct for the shifted Yangians has been constructed in [24] for arbitrary simply-laced $\mathfrak{g}$.
(c) Once Δ_{μ_1,μ_2} is constructed, one should be able to immediately extend it to the homomorphism $\Delta^{\mathrm{ad}}_{\mu_1,\mu_2} \colon \mathcal{U}^{\mathrm{ad}}_{0,\mu_1+\mu_2} \to \mathcal{U}^{\mathrm{ad}}_{0,\mu_1} \otimes \mathcal{U}^{\mathrm{ad}}_{0,\mu_2}$ by setting $\Delta^{\mathrm{ad}}_{\mu_1,\mu_2}(\phi_i^{\pm}) = \phi_i^{\pm} \otimes \phi_i^{\pm}$.

10.3 Homomorphisms Δ_{b_1,b_2} for $b_1, b_2 \in \mathbb{Z}_{\leq 0}$, $\mathfrak{g} = \mathfrak{sl}_2$

We start this subsection by explicitly computing the Drinfeld-Jimbo coproduct of the Drinfeld generators $e_0, e_{-1}, f_0, f_1, \psi_0^{\pm}$ of $U_v(L\mathfrak{sl}_2)$ and $h_{\pm 1} = \pm\frac{\psi_0^{\mp}\psi_{\pm 1}^{\pm}}{v-v^{-1}}$, which generate the quantum loop algebra $U_v(L\mathfrak{sl}_2)$.

Lemma 10.4 *We have*

$$\Delta(e_0) = e_0 \otimes \psi_0^+ + 1 \otimes e_0,\ \Delta(e_{-1}) = e_{-1} \otimes \psi_0^- + 1 \otimes e_{-1},$$

$$\Delta(f_0) = f_0 \otimes 1 + \psi_0^- \otimes f_0,\ \Delta(f_1) = f_1 \otimes 1 + \psi_0^+ \otimes f_1,\ \Delta(\psi_0^{\pm}) = \psi_0^{\pm} \otimes \psi_0^{\pm},$$

$$\Delta(h_1) = h_1 \otimes 1 + 1 \otimes h_1 - (v^2 - v^{-2}) e_0 \otimes f_1,\ \Delta(h_{-1}) = h_{-1} \otimes 1 + 1 \otimes h_{-1} + (v^2 - v^{-2}) e_{-1} \otimes f_0.$$

Proof This is a straightforward computation based on the explicit identification between the Drinfeld-Jimbo and the new Drinfeld realizations of the quantum loop algebra $U_v(L\mathfrak{sl}_2)$ of Theorem 8.10: $e_0 = E_{i_1}, f_0 = F_{i_1}, \psi_0^{\pm} = K_{i_1}^{\pm 1}, e_{-1} = K_{i_1}^{-1} F_{i_0}, f_1 = E_{i_0} K_{i_1}$. □

The key result of this subsection provides analogues of Δ for antidominantly shifted quantum affine algebras of $\mathfrak{sl}_2$. For $\mu_1, \mu_2 \in \Lambda^-$, we construct homomorphisms $\Delta_{b_1,b_2}\colon \mathcal{U}^{\mathrm{sc}}_{0,b_1+b_2} \to \mathcal{U}^{\mathrm{sc}}_{0,b_1} \otimes \mathcal{U}^{\mathrm{sc}}_{0,b_2}$, where $b_1 := \alpha^\vee(\mu_1), b_2 := \alpha^\vee(\mu_2)$ (so that $b_1, b_2 \in \mathbb{Z}_{\leq 0}$).

Theorem 10.5 *For any $b_1, b_2 \in \mathbb{Z}_{\leq 0}$, there is a unique $\mathbb{C}(\boldsymbol{v})$-algebra homomorphism*

$$\Delta_{b_1,b_2}\colon \mathcal{U}^{\mathrm{sc}}_{0,b_1+b_2} \longrightarrow \mathcal{U}^{\mathrm{sc}}_{0,b_1} \otimes \mathcal{U}^{\mathrm{sc}}_{0,b_2}$$

(we will denote $\Delta = \Delta_{b_1,b_2}$ when the algebras involved are clear), such that

$$\Delta(e_r) = 1 \otimes e_r,\ \Delta(f_s) = f_s \otimes 1 \text{ for } b_2 \leq r < 0, b_1 < s \leq 0,$$

$$\Delta(e_0) = e_0 \otimes \psi_0^+ + 1 \otimes e_0,\ \Delta(e_{b_2-1}) = e_{-1} \otimes \psi_{b_2}^- + 1 \otimes e_{b_2-1},$$

$$\Delta(f_1) = f_1 \otimes 1 + \psi_0^+ \otimes f_1,\ \Delta(f_{b_1}) = f_{b_1} \otimes 1 + \psi_{b_1}^- \otimes f_0,$$

$$\Delta((\psi_0^+)^{\pm 1}) = (\psi_0^+)^{\pm 1} \otimes (\psi_0^+)^{\pm 1},\ \Delta((\psi_{b_1+b_2}^-)^{\pm 1}) = (\psi_{b_1}^-)^{\pm 1} \otimes (\psi_{b_2}^-)^{\pm 1},$$

$$\Delta(h_1) = h_1 \otimes 1 + 1 \otimes h_1 - (\boldsymbol{v}^2 - \boldsymbol{v}^{-2}) e_0 \otimes f_1,\ \Delta(h_{-1}) = h_{-1} \otimes 1 + 1 \otimes h_{-1} + (\boldsymbol{v}^2 - \boldsymbol{v}^{-2}) e_{-1} \otimes f_0.$$

These homomorphisms generalize the Drinfeld-Jimbo coproduct, since we recover the formulas of Lemma 10.4 for $b_1 = b_2 = 0$. The proof of Theorem 10.5 is presented in Appendix D and is crucially based on Theorem 5.5 which provides a presentation of the shifted quantum affine algebras via a finite number of generators and relations.

Remark 10.6 The similarity between the formulas for Δ_{b_1,b_2} of Theorem 10.5 and Drinfeld-Jimbo coproduct Δ of Lemma 10.4 can be explained as follows. Let $U_{\boldsymbol{v}}^-$ (resp. $\mathcal{U}^{\mathrm{sc},-}_{0,b_1,b_2}$) be the subalgebra of $U_{\boldsymbol{v}}(L\mathfrak{sl}_2)$ (resp. $\mathcal{U}^{\mathrm{sc}}_{0,b_1+b_2}$) generated by $\{e_{-1}, f_0, (\psi_0^-)^{\pm 1}\}$, or equivalently, by $\{e_{-r-1}, f_{-r}, (\psi_0^-)^{\pm 1}, \psi_{-r-1}^-\}_{r \in \mathbb{N}}$ (resp. by $\{e_{b_2-1}, f_{b_1}, (\psi_{b_1+b_2}^-)^{\pm 1}\}$, or equivalently, by $\{e_{b_2-r-1}, f_{b_1-r}, (\psi_{b_1+b_2}^-)^{\pm 1}, \psi_{b_1+b_2-r-1}^-\}_{r \in \mathbb{N}}$). Analogously, let $U_{\boldsymbol{v}}^+$ (resp. $\mathcal{U}^{\mathrm{sc},+}_{0,b_1,b_2}$) be the subalgebra of $U_{\boldsymbol{v}}(L\mathfrak{sl}_2)$ (resp. $\mathcal{U}^{\mathrm{sc}}_{0,b_1+b_2}$) generated by $\{e_0, f_1, (\psi_0^+)^{\pm 1}\}$ in both cases, or equivalently, by $\{e_r, f_{r+1}, (\psi_0^+)^{\pm 1}, \psi_{r+1}^+\}_{r \in \mathbb{N}}$. Then, there are unique $\mathbb{C}(\boldsymbol{v})$-algebra homomorphisms $j^{\pm}_{b_1,b_2}\colon U_{\boldsymbol{v}}^{\pm} \to \mathcal{U}^{\mathrm{sc},\pm}_{0,b_1,b_2}$, such that

$$j^+_{b_1,b_2}\colon e_0 \mapsto e_0,\ f_1 \mapsto f_1,\ (\psi_0^+)^{\pm 1} \mapsto (\psi_0^+)^{\pm 1},$$

$$j^-_{b_1,b_2}\colon e_{-1} \mapsto e_{b_2-1},\ f_0 \mapsto f_{b_1},\ (\psi_0^-)^{\pm 1} \mapsto (\psi_{b_1+b_2}^-)^{\pm 1}.$$

Moreover, the following diagram is commutative:

$$\begin{array}{ccc} U_{\boldsymbol{v}}^{\pm} & \xrightarrow{\quad\Delta\quad} & U_{\boldsymbol{v}}^{\pm} \otimes U_{\boldsymbol{v}}^{\pm} \\ \Big\downarrow {\scriptstyle \jmath^{\pm}_{b_1,b_2}} & & \Big\downarrow {\scriptstyle \jmath^{\pm}_{b_1,0} \otimes \jmath^{\pm}_{0,b_2}} \\ \mathcal{U}^{\mathrm{sc},\pm}_{0,b_1,b_2} & \xrightarrow[\quad\Delta_{b_1,b_2}\quad]{} & \mathcal{U}^{\mathrm{sc},\pm}_{0,b_1,0} \otimes \mathcal{U}^{\mathrm{sc},\pm}_{0,0,b_2} \end{array}$$

Remark 10.7 The aforementioned homomorphism Δ_{b_1,b_2} can be naturally extended to the homomorphism $\Delta^{\mathrm{ad}}_{b_1,b_2} \colon \mathcal{U}^{\mathrm{ad}}_{0,b_1+b_2} \to \mathcal{U}^{\mathrm{ad}}_{0,b_1} \otimes \mathcal{U}^{\mathrm{ad}}_{0,b_2}$ by setting $\Delta^{\mathrm{ad}}_{b_1,b_2}(\phi^{\pm}) = \phi^{\pm} \otimes \phi^{\pm}$.

10.4 Homomorphisms Δ_{b_1,b_2} for Arbitrary $b_1, b_2 \in \mathbb{Z}$, $\mathfrak{g} = \mathfrak{sl}_2$

In this subsection, we generalize the construction of Δ_{b_1,b_2} of Theorem 10.5 ($b_1, b_2 \in \mathbb{Z}_{\leq 0}$) to the general case $b_1, b_2 \in \mathbb{Z}$. We follow the corresponding construction for the shifted Yangians of [24, Theorem 4.12].

The key ingredient of our approach are the *shift homomorphisms* ι_{n,m_1,m_2} (the trigonometric analogues of the shift homomorphisms of [24]).

Proposition 10.8 *For any $n \in \mathbb{Z}$ and $m_1, m_2 \in \mathbb{Z}_{\leq 0}$, there is a unique $\mathbb{C}(\boldsymbol{v})$-algebra homomorphism $\iota_{n,m_1,m_2} \colon \mathcal{U}^{\mathrm{sc}}_{0,n} \to \mathcal{U}^{\mathrm{sc}}_{0,n+m_1+m_2}$, which maps the currents as follows:*

$$e(z) \mapsto (1-z^{-1})^{-m_1} e(z),\ f(z) \mapsto (1-z^{-1})^{-m_2} f(z),\ \psi^{\pm}(z) \mapsto (1-z^{-1})^{-m_1-m_2} \psi^{\pm}(z).$$

Proof The above assignment is obviously compatible with defining relations (U1–U8). Moreover, we have $\iota_{n,m_1,m_2} \colon \psi_0^+ \mapsto \psi_0^+,\ \psi_n^- \mapsto (-1)^{m_1+m_2} \psi^-_{n+m_1+m_2}$. □

These homomorphisms satisfy two important properties:

Lemma 10.9

(a) We have $\iota_{n+m_1+m_2,m_1',m_2'} \circ \iota_{n,m_1,m_2} = \iota_{n,m_1+m_1',m_2+m_2'}$ for any $n \in \mathbb{Z}$ and $m_1, m_2, m_1', m_2' \in \mathbb{Z}_{\leq 0}$.
(b) The homomorphism ι_{n,m_1,m_2} is injective for any $n \in \mathbb{Z}$ and $m_1, m_2 \in \mathbb{Z}_{\leq 0}$.

Part (a) is obvious, while part (b) is proved in Appendix E and follows from the PBW property for $\mathcal{U}^{\mathrm{sc}}_{0,n}$ (cf. Theorem 10.19). The following is the key result of this subsection.

Theorem 10.10 *For any $b_1, b_2 \in \mathbb{Z}$ and $b := b_1 + b_2$, there is a unique $\mathbb{C}(\boldsymbol{v})$-algebra homomorphism*

$$\Delta_{b_1,b_2} \colon \mathcal{U}^{\mathrm{sc}}_{0,b} \longrightarrow \mathcal{U}^{\mathrm{sc}}_{0,b_1} \otimes \mathcal{U}^{\mathrm{sc}}_{0,b_2},$$

such that for any $m_1, m_2 \in \mathbb{Z}_{\leq 0}$ the following diagram is commutative:

$$\begin{array}{ccc}
\mathcal{U}^{\mathrm{sc}}_{0,b} & \xrightarrow{\Delta_{b_1,b_2}} & \mathcal{U}^{\mathrm{sc}}_{0,b_1} \otimes \mathcal{U}^{\mathrm{sc}}_{0,b_2} \\
\Big\downarrow{\scriptstyle \iota_{b,m_2,m_1}} & & \Big\downarrow{\scriptstyle \iota_{b_1,0,m_1} \otimes \iota_{b_2,m_2,0}} \\
\mathcal{U}^{\mathrm{sc}}_{0,b+m_1+m_2} & \xrightarrow[\Delta_{b_1+m_1,b_2+m_2}]{} & \mathcal{U}^{\mathrm{sc}}_{0,b_1+m_1} \otimes \mathcal{U}^{\mathrm{sc}}_{0,b_2+m_2}
\end{array}$$

The proof of this theorem is presented in Appendix F and is similar to the proof of [24, Theorem 4.12].

Corollary 10.11 *For any $b_1, b_2 \in \mathbb{Z}$, we have*

$$\Delta_{b_1,b_2}(h_1) = h_1 \otimes 1 + 1 \otimes h_1 - (\boldsymbol{v}^2 - \boldsymbol{v}^{-2}) e_0 \otimes f_1,$$

$$\Delta_{b_1,b_2}(h_{-1}) = h_{-1} \otimes 1 + 1 \otimes h_{-1} + (\boldsymbol{v}^2 - \boldsymbol{v}^{-2}) e_{-1} \otimes f_0.$$

Proof In the antidominant case $b_1, b_2 \in \mathbb{Z}_{\leq 0}$, both equalities are due to our definition of Δ_{b_1,b_2} of Theorem 10.5. For general b_1, b_2, choose $m_1, m_2 \in \mathbb{Z}_{\leq 0}$ such that $b_1 + m_1, b_2 + m_2 \in \mathbb{Z}_{\leq 0}$. By the definition of ι_{b,m_2,m_1}, we have $\iota_{b,m_2,m_1}(h_{\pm 1}) = h_{\pm 1} \pm \frac{m_1+m_2}{\boldsymbol{v}-\boldsymbol{v}^{-1}}$. Meanwhile, we also have

$$\iota_{b_1,0,m_1} \otimes \iota_{b_2,m_2,0}(h_{\pm 1} \otimes 1 + 1 \otimes h_{\pm 1}) = h_{\pm 1} \otimes 1 + 1 \otimes h_{\pm 1} \pm \frac{m_1 + m_2}{\boldsymbol{v} - \boldsymbol{v}^{-1}},$$

while $\iota_{b_1,0,m_1}(e_r) = e_r$, $\iota_{b_2,m_2,0}(f_s) = f_s$ for any $r, s \in \mathbb{Z}$. The result follows by combining the formula for $\Delta_{b_1+m_1,b_2+m_2}(h_{\pm 1})$ with the commutativity of the diagram of Theorem 10.10 (we also use injectivity of the vertical arrows, due to Lemma 10.9(b)). □

The following result is analogous to [24, Proposition 4.14] and we leave its proof to the interested reader.

Lemma 10.12 *For $b = b_1 + b_2 + b_3$ with $b_1, b_3 \in \mathbb{Z}, b_2 \in \mathbb{Z}_{\leq 0}$, the following diagram is commutative:*

$$\begin{array}{ccc}
\mathcal{U}^{\mathrm{sc}}_{0,b} & \xrightarrow{\Delta_{b_1,b_2+b_3}} & \mathcal{U}^{\mathrm{sc}}_{0,b_1} \otimes \mathcal{U}^{\mathrm{sc}}_{0,b_2+b_3} \\
\Big\downarrow{\scriptstyle \Delta_{b_1+b_2,b_3}} & & \Big\downarrow{\scriptstyle \mathrm{Id} \otimes \Delta_{b_2,b_3}} \\
\mathcal{U}^{\mathrm{sc}}_{0,b_1+b_2} \otimes \mathcal{U}^{\mathrm{sc}}_{0,b_3} & \xrightarrow[\Delta_{b_1,b_2} \otimes \mathrm{Id}]{} & \mathcal{U}^{\mathrm{sc}}_{0,b_1} \otimes \mathcal{U}^{\mathrm{sc}}_{0,b_2} \otimes \mathcal{U}^{\mathrm{sc}}_{0,b_3}
\end{array}$$

10.5 *Drinfeld-Jimbo Coproduct on $U_{\boldsymbol{v}}(L\mathfrak{sl}_n)$ via Drinfeld Generators*

According to Theorem 5.5, the quantum loop algebra $U_{\boldsymbol{v}}(L\mathfrak{sl}_n)$ is generated by the elements $\{e_{i,0}, f_{i,0}, e_{i,-1}, f_{i,1}, \psi^{\pm}_{i,0}, h_{i,\pm 1}\}_{i=1}^{n-1}$. The key result of this subsection provides explicit formulas for the action of the Drinfeld-Jimbo coproduct Δ on these generators of $U_{\boldsymbol{v}}(L\mathfrak{sl}_n)$. Since $e_{i,0} = E_i, f_{i,0} = F_i, \psi^{\pm}_{i,0} = K_i^{\pm 1}$ (for $i \in I = \{1, 2, \cdots, n-1\}$), we obviously have

$$\Delta(e_{i,0}) = 1\otimes e_{i,0}+e_{i,0}\otimes\psi^{+}_{i,0},\ \Delta(f_{i,0}) = f_{i,0}\otimes 1+\psi^{-}_{i,0}\otimes f_{i,0},\ \Delta(\psi^{\pm}_{i,0}) = \psi^{\pm}_{i,0}\otimes\psi^{\pm}_{i,0}.$$

It remains to compute the coproduct of the remaining generators above.

Theorem 10.13 *Let Δ be the Drinfeld-Jimbo coproduct on $U_{\boldsymbol{v}}(L\mathfrak{sl}_n)$. Then, we have*

$$\begin{aligned}
&\Delta(h_{i,1}) = \\
&h_{i,1}\otimes 1 + 1\otimes h_{i,1} - (\boldsymbol{v}^2-\boldsymbol{v}^{-2})E^{(0)}_{i,i+1}\otimes F^{(1)}_{i+1,i} + (\boldsymbol{v}-\boldsymbol{v}^{-1})\sum_{l>i+1} E^{(0)}_{i+1,l}\otimes F^{(1)}_{l,i+1} + \\
&(\boldsymbol{v}-\boldsymbol{v}^{-1})\sum_{k<i}\boldsymbol{v}^{k+1-i}\widetilde{E}^{(0)}_{ki}\otimes F^{(1)}_{ik} + \boldsymbol{v}^{-2}(\boldsymbol{v}-\boldsymbol{v}^{-1})\sum_{l>i+1}[E^{(0)}_{i,i+1}, E^{(0)}_{i+1,l}]_{\boldsymbol{v}^3}\otimes F^{(1)}_{li} - \\
&(\boldsymbol{v}-\boldsymbol{v}^{-1})\sum_{k<i}\boldsymbol{v}^{k-i-1}[E^{(0)}_{i,i+1}, \widetilde{E}^{(0)}_{ki}]_{\boldsymbol{v}^3}\otimes F^{(1)}_{i+1,k} + \\
&(\boldsymbol{v}-\boldsymbol{v}^{-1})^2\sum_{l>i+1}^{k<i}\boldsymbol{v}^{k-i}\left(E^{(0)}_{il}\widetilde{E}^{(0)}_{ki} - E^{(0)}_{i+1,l}\widetilde{E}^{(0)}_{k,i+1}\right)\otimes F^{(1)}_{lk},
\end{aligned} \tag{10.2}$$

$$\begin{aligned}
&\Delta(h_{i,-1}) = \\
&h_{i,-1}\otimes 1 + 1\otimes h_{i,-1} + (\boldsymbol{v}^2-\boldsymbol{v}^{-2})E^{(-1)}_{i,i+1}\otimes F^{(0)}_{i+1,i} - (\boldsymbol{v}-\boldsymbol{v}^{-1})\sum_{l>i+1} E^{(-1)}_{i+1,l}\otimes F^{(0)}_{l,i+1} - \\
&(\boldsymbol{v}-\boldsymbol{v}^{-1})\sum_{k<i}\boldsymbol{v}^{i-k-1}E^{(-1)}_{ki}\otimes\widetilde{F}^{(0)}_{ik} - \boldsymbol{v}^{2}(\boldsymbol{v}-\boldsymbol{v}^{-1})\sum_{l>i+1}E^{(-1)}_{il}\otimes[F^{(0)}_{l,i+1}, F^{(0)}_{i+1,i}]_{\boldsymbol{v}^{-3}} + \\
&(\boldsymbol{v}-\boldsymbol{v}^{-1})\sum_{k<i}\boldsymbol{v}^{i+1-k}E^{(-1)}_{k,i+1}\otimes[\widetilde{F}^{(0)}_{ik}, F^{(0)}_{i+1,i}]_{\boldsymbol{v}^{-3}} - \\
&(\boldsymbol{v}-\boldsymbol{v}^{-1})^2\sum_{l>i+1}^{k<i}\boldsymbol{v}^{i-k}E^{(-1)}_{kl}\otimes\left(\widetilde{F}^{(0)}_{i+1,k}F^{(0)}_{l,i+1} - \widetilde{F}^{(0)}_{ik}F^{(0)}_{li}\right),
\end{aligned} \tag{10.3}$$

$$\Delta(e_{i,-1}) = 1 \otimes e_{i,-1} + e_{i,-1} \otimes \psi^-_{i,0} - (\boldsymbol{v}-\boldsymbol{v}^{-1}) \sum_{l>i+1} E^{(-1)}_{il} \otimes F^{(0)}_{l,i+1}\psi^-_{i,0} +$$

$$(\boldsymbol{v}-\boldsymbol{v}^{-1}) \sum_{k<i} \boldsymbol{v}^{i-k-1} E^{(-1)}_{k,i+1} \otimes \widetilde{F}^{(0)}_{ik}\psi^-_{i,0} - (\boldsymbol{v}-\boldsymbol{v}^{-1})^2 \sum_{l>i+1}^{k<i} \boldsymbol{v}^{i-k-1} E^{(-1)}_{kl} \otimes \widetilde{F}^{(0)}_{ik} F^{(0)}_{l,i+1}\psi^-_{i,0}, \tag{10.4}$$

$$\Delta(f_{i,1}) = f_{i,1} \otimes 1 + \psi^+_{i,0} \otimes f_{i,1} + \boldsymbol{v}^{-1}(\boldsymbol{v}-\boldsymbol{v}^{-1}) \sum_{l>i+1} E^{(0)}_{i+1,l}\psi^+_{i,0} \otimes F^{(1)}_{li} -$$

$$(\boldsymbol{v}-\boldsymbol{v}^{-1}) \sum_{k<i} \boldsymbol{v}^{k-i} \widetilde{E}^{(0)}_{ki}\psi^+_{i,0} \otimes F^{(1)}_{i+1,k} - (\boldsymbol{v}-\boldsymbol{v}^{-1})^2 \sum_{l>i+1}^{k<i} \boldsymbol{v}^{k-i-1} E^{(0)}_{i+1,l}\widetilde{E}^{(0)}_{ki}\psi^+_{i,0} \otimes F^{(1)}_{lk}, \tag{10.5}$$

where for $1 \le j < i \le n$ *we set*

$$\begin{aligned}
E^{(0)}_{ji} &:= [e_{i-1,0}, \cdots, [e_{j+1,0}, e_{j,0}]_{\boldsymbol{v}^{-1}} \cdots]_{\boldsymbol{v}^{-1}} = [\cdots[e_{i-1,0}, e_{i-2,0}]_{\boldsymbol{v}^{-1}}, \cdots, e_{j,0}]_{\boldsymbol{v}^{-1}},\\
F^{(0)}_{ij} &:= [f_{j,0}, \cdots, [f_{i-2,0}, f_{i-1,0}]_{\boldsymbol{v}} \cdots]_{\boldsymbol{v}} = [\cdots[f_{j,0}, f_{j+1,0}]_{\boldsymbol{v}}, \cdots, f_{i-1,0}]_{\boldsymbol{v}},\\
E^{(-1)}_{ji} &:= [e_{i-1,0}, \cdots, [e_{j+1,0}, e_{j,-1}]_{\boldsymbol{v}^{-1}} \cdots]_{\boldsymbol{v}^{-1}}\\
&= [[\cdots[e_{i-1,0}, e_{i-2,0}]_{\boldsymbol{v}^{-1}}, \cdots, e_{j+1,0}]_{\boldsymbol{v}^{-1}}, e_{j,-1}]_{\boldsymbol{v}^{-1}},\\
F^{(1)}_{ij} &:= [f_{j,1}, [f_{j+1,0}, \cdots, [f_{i-2,0}, f_{i-1,0}]_{\boldsymbol{v}} \cdots]_{\boldsymbol{v}}]_{\boldsymbol{v}} = [\cdots[f_{j,1}, f_{j+1,0}]_{\boldsymbol{v}}, \cdots, f_{i-1,0}]_{\boldsymbol{v}},\\
\widetilde{E}^{(0)}_{ji} &:= [e_{i-1,0}, \cdots, [e_{j+1,0}, e_{j,0}]_{\boldsymbol{v}} \cdots]_{\boldsymbol{v}} = [\cdots[e_{i-1,0}, e_{i-2,0}]_{\boldsymbol{v}}, \cdots, e_{j,0}]_{\boldsymbol{v}},\\
\widetilde{F}^{(0)}_{ij} &:= [f_{j,0}, \cdots, [f_{i-2,0}, f_{i-1,0}]_{\boldsymbol{v}^{-1}} \cdots]_{\boldsymbol{v}^{-1}} = [\cdots[f_{j,0}, f_{j+1,0}]_{\boldsymbol{v}^{-1}}, \cdots, f_{i-1,0}]_{\boldsymbol{v}^{-1}}.
\end{aligned} \tag{10.6}$$

The proof of this result is based on the RTT realization of $U_{\boldsymbol{v}}(L\mathfrak{sl}_n)$ and is presented in Appendix G.

Remark 10.14 The right equalities in each of the lines of (10.6) are not obvious and are established during our proof of Theorem 10.13. They play an important role in the proof of Theorem 10.16 below.

Let $U^>_{\boldsymbol{v}}(L\mathfrak{g})$ and $U^\geq_{\boldsymbol{v}}(L\mathfrak{g})$ (resp. $U^<_{\boldsymbol{v}}(L\mathfrak{g})$ and $U^\leq_{\boldsymbol{v}}(L\mathfrak{g})$) be the $\mathbb{C}(\boldsymbol{v})$-subalgebras of $U_{\boldsymbol{v}}(L\mathfrak{g})$ generated by $\{e_{i,r}\}_{i\in I}^{r\in\mathbb{Z}}$ and $\{e_{i,r}, \psi^\pm_{i,\pm s}\}_{i\in I}^{r\in\mathbb{Z}, s\in\mathbb{N}}$ (resp. $\{f_{i,r}\}_{i\in I}^{r\in\mathbb{Z}}$ and $\{f_{i,r}, \psi^\pm_{i,\pm s}\}_{i\in I}^{r\in\mathbb{Z}, s\in\mathbb{N}}$).

Corollary 10.15 *For any* $1 \le i < n$ *and* $r \in \mathbb{Z}$, *we have*

$$\Delta(h_{i,\pm1}) - h_{i,\pm1} \otimes 1 - 1 \otimes h_{i,\pm1} \in U^>_{\boldsymbol{v}}(L\mathfrak{sl}_n) \otimes U^<_{\boldsymbol{v}}(L\mathfrak{sl}_n),$$

$$\Delta(e_{i,r}) - 1 \otimes e_{i,r} \in U^>_{\boldsymbol{v}}(L\mathfrak{sl}_n) \otimes U^\leq_{\boldsymbol{v}}(L\mathfrak{sl}_n),$$

$$\Delta(f_{i,r}) - f_{i,r} \otimes 1 \in U^\geq_{\boldsymbol{v}}(L\mathfrak{sl}_n) \otimes U^<_{\boldsymbol{v}}(L\mathfrak{sl}_n).$$

Proof The claim is clear for $\Delta(h_{i,\pm 1}), \Delta(e_{i,-1}), \Delta(f_{i,1})$, due to (10.2–10.5). Applying iteratively $[\Delta(h_{i,\pm 1}), \Delta(e_{i,r})] = [2]_{\boldsymbol{v}} \cdot \Delta(e_{i,r\pm 1}), [\Delta(h_{i,\pm 1}), \Delta(f_{i,r})] = -[2]_{\boldsymbol{v}} \cdot \Delta(f_{i,r\pm 1})$, we deduce the claim for $\Delta(e_{i,r})$ and $\Delta(f_{i,r})$. □

10.6 Homomorphisms Δ_{μ_1,μ_2} for $\mu_1, \mu_2 \in \Lambda^-$, $\mathfrak{g} = \mathfrak{sl}_n$

In this subsection, we construct homomorphisms $\Delta_{\mu_1,\mu_2}\colon \mathcal{U}^{\mathrm{sc}}_{0,\mu_1+\mu_2} \to \mathcal{U}^{\mathrm{sc}}_{0,\mu_1} \otimes \mathcal{U}^{\mathrm{sc}}_{0,\mu_2}$ for $\mu_1, \mu_2 \in \Lambda^-$, which coincide with the Drinfeld-Jimbo coproduct on $U_{\boldsymbol{v}}(L\mathfrak{sl}_n)$ for $\mu_1 = \mu_2 = 0$. Set $b_{1,i} := \alpha_i^\vee(\mu_1)$ and $b_{2,i} := \alpha_i^\vee(\mu_2)$ (so that $b_{1,i}, b_{2,i} \in \mathbb{Z}_{\leq 0}$).

Theorem 10.16 *For any $\mu_1, \mu_2 \in \Lambda^-$, there is a unique $\mathbb{C}(\boldsymbol{v})$-algebra homomorphism*

$$\Delta_{\mu_1,\mu_2}\colon \mathcal{U}^{\mathrm{sc}}_{0,\mu_1+\mu_2} \longrightarrow \mathcal{U}^{\mathrm{sc}}_{0,\mu_1} \otimes \mathcal{U}^{\mathrm{sc}}_{0,\mu_2}$$

(we will denote $\Delta = \Delta_{\mu_1,\mu_2}$ when the algebras involved are clear), such that

$$\Delta(e_{i,r}) = 1 \otimes e_{i,r},\ \Delta(f_{i,s}) = f_{i,s} \otimes 1 \text{ for } b_{2,i} \leq r < 0, b_{1,i} < s \leq 0,$$

$$\Delta(e_{i,0}) = 1 \otimes e_{i,0} + e_{i,0} \otimes \psi^+_{i,0},\ \Delta(f_{i,b_{1,i}}) = f_{i,b_{1,i}} \otimes 1 + \psi^-_{i,b_{1,i}} \otimes f_{i,0},$$

$$\Delta(e_{i,b_{2,i}-1}) = 1 \otimes e_{i,b_{2,i}-1} + e_{i,-1} \otimes \psi^-_{i,b_{2,i}} - (\boldsymbol{v}-\boldsymbol{v}^{-1}) \sum_{l>i+1} E^{(-1)}_{il} \otimes F^{(0)}_{l,i+1}\psi^-_{i,b_{2,i}} +$$

$$(\boldsymbol{v}-\boldsymbol{v}^{-1}) \sum_{k<i} \boldsymbol{v}^{i-k-1} E^{(-1)}_{k,i+1} \otimes \widetilde{F}^{(0)}_{ik}\psi^-_{i,b_{2,i}} - (\boldsymbol{v}-\boldsymbol{v}^{-1})^2 \sum_{l>i+1}^{k<i} \boldsymbol{v}^{i-k-1} E^{(-1)}_{kl} \otimes \widetilde{F}^{(0)}_{ik} F^{(0)}_{l,i+1}\psi^-_{i,b_{2,i}},$$

$$\Delta(f_{i,1}) = f_{i,1} \otimes 1 + \psi^+_{i,0} \otimes f_{i,1} + \boldsymbol{v}^{-1}(\boldsymbol{v}-\boldsymbol{v}^{-1}) \sum_{l>i+1} E^{(0)}_{i+1,l}\psi^+_{i,0} \otimes F^{(1)}_{li} -$$

$$(\boldsymbol{v}-\boldsymbol{v}^{-1}) \sum_{k<i} \boldsymbol{v}^{k-i} \widetilde{E}^{(0)}_{ki}\psi^+_{i,0} \otimes F^{(1)}_{i+1,k} - (\boldsymbol{v}-\boldsymbol{v}^{-1})^2 \sum_{l>i+1}^{k<i} \boldsymbol{v}^{k-i-1} E^{(0)}_{i+1,l}\widetilde{E}^{(0)}_{ki}\psi^+_{i,0} \otimes F^{(1)}_{lk},$$

$$\Delta((\psi^+_{i,0})^{\pm 1}) = (\psi^+_{i,0})^{\pm 1} \otimes (\psi^+_{i,0})^{\pm 1},\ \Delta((\psi^-_{i,b_{1,i}+b_{2,i}})^{\pm 1}) = (\psi^-_{i,b_{1,i}})^{\pm 1} \otimes (\psi^-_{i,b_{2,i}})^{\pm 1},$$

$$\Delta(h_{i,1}) = h_{i,1} \otimes 1 + 1 \otimes h_{i,1} - (\boldsymbol{v}^2 - \boldsymbol{v}^{-2}) E^{(0)}_{i,i+1} \otimes F^{(1)}_{i+1,i} + (\boldsymbol{v} - \boldsymbol{v}^{-1}) \sum_{l>i+1} E^{(0)}_{i+1,l} \otimes F^{(1)}_{l,i+1} +$$
$$(\boldsymbol{v} - \boldsymbol{v}^{-1}) \sum_{k<i} \boldsymbol{v}^{k+1-i} \widetilde{E}^{(0)}_{ki} \otimes F^{(1)}_{ik} + \boldsymbol{v}^{-2}(\boldsymbol{v} - \boldsymbol{v}^{-1}) \sum_{l>i+1} [E^{(0)}_{i,i+1}, E^{(0)}_{i+1,l}]_{\boldsymbol{v}^3} \otimes F^{(1)}_{li} -$$
$$(\boldsymbol{v} - \boldsymbol{v}^{-1}) \sum_{k<i} \boldsymbol{v}^{k-i-1} [E^{(0)}_{i,i+1}, \widetilde{E}^{(0)}_{ki}]_{\boldsymbol{v}^3} \otimes F^{(1)}_{i+1,k} +$$
$$(\boldsymbol{v} - \boldsymbol{v}^{-1})^2 \sum_{l>i+1}^{k<i} \boldsymbol{v}^{k-i} \left(E^{(0)}_{il} \widetilde{E}^{(0)}_{ki} - E^{(0)}_{i+1,l} \widetilde{E}^{(0)}_{k,i+1} \right) \otimes F^{(1)}_{lk},$$

$$\Delta(h_{i,-1}) = h_{i,-1} \otimes 1 + 1 \otimes h_{i,-1} + (\boldsymbol{v}^2 - \boldsymbol{v}^{-2}) E^{(-1)}_{i,i+1} \otimes F^{(0)}_{i+1,i} -$$
$$(\boldsymbol{v} - \boldsymbol{v}^{-1}) \sum_{l>i+1} E^{(-1)}_{i+1,l} \otimes F^{(0)}_{l,i+1} -$$
$$(\boldsymbol{v} - \boldsymbol{v}^{-1}) \sum_{k<i} \boldsymbol{v}^{i-k-1} E^{(-1)}_{ki} \otimes \widetilde{F}^{(0)}_{ik} - \boldsymbol{v}^2 (\boldsymbol{v} - \boldsymbol{v}^{-1}) \sum_{l>i+1} E^{(-1)}_{il} \otimes [F^{(0)}_{l,i+1}, F^{(0)}_{i+1,i}]_{\boldsymbol{v}^{-3}} +$$
$$(\boldsymbol{v} - \boldsymbol{v}^{-1}) \sum_{k<i} \boldsymbol{v}^{i+1-k} E^{(-1)}_{k,i+1} \otimes [\widetilde{F}^{(0)}_{ik}, F^{(0)}_{i+1,i}]_{\boldsymbol{v}^{-3}} -$$
$$(\boldsymbol{v} - \boldsymbol{v}^{-1})^2 \sum_{l>i+1}^{k<i} \boldsymbol{v}^{i-k} E^{(-1)}_{kl} \otimes \left(\widetilde{F}^{(0)}_{i+1,k} F^{(0)}_{l,i+1} - \widetilde{F}^{(0)}_{ik} F^{(0)}_{li} \right),$$

where $E^{(0)}_{ji}, \widetilde{E}^{(0)}_{ji}, E^{(-1)}_{ji}, F^{(0)}_{ij}, \widetilde{F}^{(0)}_{ij}, F^{(1)}_{ij}$ *are defined as in (10.6).*

The proof of this result is similar to our proof of Theorem 10.5, but is much more tedious; we sketch it in Appendix H.

Remark 10.17 The similarity between the formulas for Δ_{μ_1,μ_2} of Theorem 10.16 and Δ of Theorem 10.13 can be explained via an analogue of Remark 10.6. To be more precise, let $U^{\pm}_{\boldsymbol{v}}$ be the positive/negative Borel subalgebras in the Drinfeld-Jimbo presentation of $U_{\boldsymbol{v}}(L\mathfrak{sl}_n)$, while their analogues $\mathcal{U}^{\mathrm{sc},\pm}_{0,\mu_1,\mu_2}$ (subalgebras of $\mathcal{U}^{\mathrm{sc}}_{0,\mu_1+\mu_2}$) will be introduced in Appendix H. There are natural $\mathbb{C}(\boldsymbol{v})$-algebra homomorphisms $\jmath^{\pm}_{\mu_1,\mu_2} \colon U^{\pm}_{\boldsymbol{v}} \to \mathcal{U}^{\mathrm{sc},\pm}_{0,\mu_1,\mu_2}$, see Proposition H.1. According to Proposition H.16, the following diagram is commutative:

$$\begin{array}{ccc}
U^{\pm}_{\boldsymbol{v}} & \xrightarrow{\quad \Delta \quad} & U^{\pm}_{\boldsymbol{v}} \otimes U^{\pm}_{\boldsymbol{v}} \\
\Big\downarrow {\scriptstyle \jmath^{\pm}_{\mu_1,\mu_2}} & & \Big\downarrow {\scriptstyle \jmath^{\pm}_{\mu_1,0} \otimes \jmath^{\pm}_{0,\mu_2}} \\
\mathcal{U}^{\mathrm{sc},\pm}_{0,\mu_1,\mu_2} & \xrightarrow[\quad \Delta_{\mu_1,\mu_2} \quad]{} & \mathcal{U}^{\mathrm{sc},\pm}_{0,\mu_1,0} \otimes \mathcal{U}^{\mathrm{sc},\pm}_{0,0,\mu_2}
\end{array}$$

10.7 Homomorphisms Δ_{μ_1,μ_2} for Arbitrary $\mu_1, \mu_2 \in \Lambda$, $\mathfrak{g} = \mathfrak{sl}_n$

Let us first generalize the *shift homomorphisms* of Proposition 10.8.

Lemma 10.18 *For any $\mu \in \Lambda$ and $\nu_1, \nu_2 \in \Lambda^-$, there is a unique $\mathbb{C}(\boldsymbol{v})$-algebra homomorphism $\iota_{\mu,\nu_1,\nu_2} \colon \mathcal{U}^{\mathrm{sc}}_{0,\mu} \to \mathcal{U}^{\mathrm{sc}}_{0,\mu+\nu_1+\nu_2}$, which maps the currents as follows:*

$$\iota_{\mu,\nu_1,\nu_2} \colon e_i(z) \mapsto (1 - z^{-1})^{-\alpha_i^\vee(\nu_1)} e_i(z),\ \ f_i(z) \mapsto (1 - z^{-1})^{-\alpha_i^\vee(\nu_2)} f_i(z),$$
$$\psi_i^\pm(z) \mapsto (1 - z^{-1})^{-\alpha_i^\vee(\nu_1+\nu_2)} \psi_i^\pm(z).$$

Proof The proof is analogous to that of Proposition 10.8. □

The proof of the following technical result is presented in Appendix I and is based on the shuffle realization of the quantum loop algebra $U_{\boldsymbol{v}}(L\mathfrak{sl}_n)$, see [53] (cf. [63]).

Theorem 10.19 *The homomorphism ι_{μ,ν_1,ν_2} is injective for any $\mu \in \Lambda$ and $\nu_1, \nu_2 \in \Lambda^-$.*

Combining this theorem with Corollary 10.15 and our arguments from the proof of Theorem 10.10, we get the key result of this section.

Theorem 10.20 *For any $\mu_1, \mu_2 \in \Lambda$ and $\mu := \mu_1 + \mu_2$, there is a unique $\mathbb{C}(\boldsymbol{v})$-algebra homomorphism*

$$\Delta_{\mu_1,\mu_2} \colon \mathcal{U}^{\mathrm{sc}}_{0,\mu} \longrightarrow \mathcal{U}^{\mathrm{sc}}_{0,\mu_1} \otimes \mathcal{U}^{\mathrm{sc}}_{0,\mu_2},$$

such that for any $\nu_1, \nu_2 \in \Lambda^-$ the following diagram is commutative:

$$\begin{array}{ccc}
\mathcal{U}^{\mathrm{sc}}_{0,\mu} & \xrightarrow{\ \Delta_{\mu_1,\mu_2}\ } & \mathcal{U}^{\mathrm{sc}}_{0,\mu_1} \otimes \mathcal{U}^{\mathrm{sc}}_{0,\mu_2} \\
{\scriptstyle \iota_{\mu,\nu_2,\nu_1}} \downarrow & & \downarrow {\scriptstyle \iota_{\mu_1,0,\nu_1} \otimes \iota_{\mu_2,\nu_2,0}} \\
\mathcal{U}^{\mathrm{sc}}_{0,\mu+\nu_1+\nu_2} & \xrightarrow[\ \Delta_{\mu_1+\nu_1,\mu_2+\nu_2}\ ]{} & \mathcal{U}^{\mathrm{sc}}_{0,\mu_1+\nu_1} \otimes \mathcal{U}^{\mathrm{sc}}_{0,\mu_2+\nu_2}
\end{array}$$

The following is proved analogously to Corollary 10.11:

Proposition 10.21 *For arbitrary $\mu_1, \mu_2 \in \Lambda$, the images $\Delta_{\mu_1,\mu_2}(h_{i,\pm 1})$ are given by formulas (10.2) and (10.3).*

10.8 Open Problems

Following [24], we expect that homomorphisms $\Delta_{\mu_1,\mu_2}\colon \mathfrak{U}^{\mathrm{sc}}_{0,\mu_1+\mu_2} \to \mathfrak{U}^{\mathrm{sc}}_{0,\mu_1} \otimes \mathfrak{U}^{\mathrm{sc}}_{0,\mu_2}$ (specializing to the Drinfeld-Jimbo coproduct for $\mu_1 = \mu_2 = 0$) exist for any simply-laced Lie algebra $\mathfrak{g}$ and its two coweights $\mu_1, \mu_2 \in \Lambda$. Moreover, their construction should proceed in the same way as for the aforementioned case $\mathfrak{g} = \mathfrak{sl}_n$. To be more precise, for antidominant $\mu_1, \mu_2 \in \Lambda^-$, we expect that the homomorphism Δ_{μ_1,μ_2} is characterized by the following two properties:

(a) $\Delta_{\mu_1,\mu_2}(e_{i,r}) = 1 \otimes e_{i,r},\ \Delta_{\mu_1,\mu_2}(f_{i,s}) = f_{i,s} \otimes 1$ for $\alpha_i^\vee(\mu_2) \leq r < 0, \alpha_i^\vee(\mu_1) < s \leq 0$;
(b) an analogue of the commutative diagram of Remark 10.17 holds.

For general μ_1, μ_2, we expect that the construction of Δ_{μ_1,μ_2} should be easily deduced from the antidominant case with the help of *shift homomorphisms* ι_{μ,ν_1,ν_2} ($\mu \in \Lambda, \nu_1, \nu_2 \in \Lambda^-$) as in Theorems 10.10 and 10.20.

The outlined construction of Δ_{μ_1,μ_2} for a general $\mathfrak{g}$ lacks explicit formulas for the Drinfeld-Jimbo coproduct of $\{e_{i,0}, e_{i,-1}, f_{i,0}, f_{i,1}, \psi^{\pm}_{i,0}, h_{i,\pm 1}\}_{i\in I}$–the generators of $U_{\boldsymbol{v}}(L\mathfrak{g})$, similar to those of Lemma 10.4 and Theorem 10.13.

11 Ubiquity of RTT Relations

11.1 Rational Lax Matrix

Before we proceed to the *trigonometric* setting, let us recall the classical relation between rational Lax matrices and type A quantum open Toda systems, which goes back to [28].

Let $R_{\mathrm{rat}}(z) \in \mathrm{End}(\mathbb{C}^2 \otimes \mathbb{C}^2)$ be the standard rational R-matrix:

$$R_{\mathrm{rat}}(z) = \mathrm{Id} + \frac{\hbar}{z} P, \text{ where } P \in \mathrm{End}(\mathbb{C}^2 \otimes \mathbb{C}^2) \text{ is the permutation map.}$$

Let $\hat{A}^\hbar_n$ be the associative $\mathbb{C}[\hbar]$-algebra generated by $\{\mathsf{u}_i^{\pm 1}, w_i\}_{i=1}^n$ with the defining relations $[\mathsf{u}_i, \mathsf{u}_j] = [w_i, w_j] = 0, \mathsf{u}_i^{\pm 1}\mathsf{u}_i^{\mp 1} = 1, [\mathsf{u}_i, w_j] = \delta_{ij}\hbar \mathsf{u}_i$. Define the *(local) rational Lax matrix*

$$L^\hbar_i(z) = \begin{pmatrix} z - w_i & \mathsf{u}_i^{-1} \\ -\mathsf{u}_i & 0 \end{pmatrix} \in \mathrm{Mat}(2, \hat{A}^\hbar_n[z]) \tag{11.1}$$

and introduce the *complete monodromy matrix* $T^\hbar_n(z) := L^\hbar_n(z) \cdots L^\hbar_1(z)$. Then, the monodromy matrix $T^\hbar_n(z)$ satisfies the rational RTT-relation:

$$R_{\mathrm{rat}}(z - w)(T^\hbar_n(z) \otimes 1)(1 \otimes T^\hbar_n(w)) = (1 \otimes T^\hbar_n(w))(T^\hbar_n(z) \otimes 1)R_{\mathrm{rat}}(z - w).$$

Due to this relation, the coefficients (in z) of the matrix element $T_n^\hbar(z)_{11}$ generate a commutative subalgebra of $\hat{\mathcal{A}}_n^\hbar$, known as the quantum open Toda system of $\mathfrak{gl}_n$. The coefficient of z^{n-2} equals

$$\mathsf{H}_2^{\mathrm{rat}} = \frac{1}{2}\left(\sum_{i=1}^{n} w_i\right)^2 - \frac{1}{2}\sum_{i=1}^{n} w_i^2 - \sum_{i=1}^{n-1} \mathsf{u}_i \mathsf{u}_{i+1}^{-1}. \tag{11.2}$$

We recover the standard quantum open Toda hamiltonian of $\mathfrak{sl}_n$ once we set $w_1 + \ldots + w_n = 0$.

11.2 Trigonometric/Relativistic Lax Matrices

Let $R_{\mathrm{trig}}(z) \in \mathrm{End}(\mathbb{C}^2 \otimes \mathbb{C}^2)$ be the standard trigonometric R-matrix (see [17, (3.7)]):

$$R_{\mathrm{trig}}(z) = \begin{pmatrix} 1 & 0 & 0 & 0 \\ 0 & \frac{z-1}{\boldsymbol{v}z-\boldsymbol{v}^{-1}} & \frac{z(\boldsymbol{v}-\boldsymbol{v}^{-1})}{\boldsymbol{v}z-\boldsymbol{v}^{-1}} & 0 \\ 0 & \frac{\boldsymbol{v}-\boldsymbol{v}^{-1}}{\boldsymbol{v}z-\boldsymbol{v}^{-1}} & \frac{z-1}{\boldsymbol{v}z-\boldsymbol{v}^{-1}} & 0 \\ 0 & 0 & 0 & 1 \end{pmatrix}. \tag{11.3}$$

Let $\hat{\mathcal{A}}_n^{\boldsymbol{v}}$ be the associative $\mathbb{C}(\boldsymbol{v})$-algebra generated by $\{\widetilde{\mathsf{w}}_i^{\pm 1}, D_i^{\pm 1}\}_{i=1}^n$ with the defining relations $[\widetilde{\mathsf{w}}_i, \widetilde{\mathsf{w}}_j] = [D_i, D_j] = 0$, $\widetilde{\mathsf{w}}_i^{\pm 1}\widetilde{\mathsf{w}}_i^{\mp 1} = D_i^{\pm 1}D_i^{\mp 1} = 1$, $D_i\widetilde{\mathsf{w}}_j = \boldsymbol{v}^{\delta_{ij}}\widetilde{\mathsf{w}}_j D_i$. If we set $\mathsf{w}_i^{\pm 1} = \widetilde{\mathsf{w}}_i^{\pm 2}$, we see that $\hat{\mathcal{A}}_n^{\boldsymbol{v}}$ is a particular example of the algebras $\hat{\mathcal{A}}_{\mathrm{frac}}^{\boldsymbol{v}}$ of Sect. 7. Define the *(local) relativistic Lax matrix*

$$L_i^{\boldsymbol{v},0}(z) = \begin{pmatrix} \widetilde{\mathsf{w}}_i^{-1} z^{1/2} - \widetilde{\mathsf{w}}_i z^{-1/2} & D_i^{-1} z^{1/2} \\ -D_i z^{-1/2} & 0 \end{pmatrix} \in \mathrm{Mat}(2, z^{-1/2}\hat{\mathcal{A}}_n^{\boldsymbol{v}}[z]) \tag{11.4}$$

and introduce the *complete monodromy matrix* $T_n^{\boldsymbol{v},0}(z) := L_n^{\boldsymbol{v},0}(z) \cdots L_1^{\boldsymbol{v},0}(z)$.

Lemma 11.1 *The monodromy matrix $T_n^{\boldsymbol{v},0}(z)$ satisfies the trigonometric RTT-relation:*

$$R_{\mathrm{trig}}(z/w)(T_n^{\boldsymbol{v},0}(z) \otimes 1)(1 \otimes T_n^{\boldsymbol{v},0}(w)) = (1 \otimes T_n^{\boldsymbol{v},0}(w))(T_n^{\boldsymbol{v},0}(z) \otimes 1)R_{\mathrm{trig}}(z/w).$$

Proof It suffices to check the above relation for $n = 1$. The proof in the latter case is straightforward. □

Corollary 11.2 *The coefficients (in z) of the matrix element $z^{n/2}T_n^{\boldsymbol{v},0}(z)_{11}$ generate a commutative subalgebra of $\hat{\mathcal{A}}_n^{\boldsymbol{v}}$. The coefficient of z equals*

$$\mathsf{H}_2^0 = (-1)^{n-1}\widetilde{\mathsf{w}}_1\cdots\widetilde{\mathsf{w}}_n\cdot\left(\sum_{i=1}^{n}\widetilde{\mathsf{w}}_i^{-2}+\sum_{i=1}^{n-1}\widetilde{\mathsf{w}}_i^{-1}\widetilde{\mathsf{w}}_{i+1}^{-1}D_iD_{i+1}^{-1}\right). \tag{11.5}$$

This hamiltonian is equivalent to the quadratic hamiltonian of the q-difference quantum Toda lattice of [19, (5.7)] (see also [56]) once we set $\widetilde{\mathsf{w}}_1\cdots\widetilde{\mathsf{w}}_n = 1$.

Remark 11.3 The notion of a *relativistic Lax matrix* goes back to [43]. In particular, our choice of $L_i^{\boldsymbol{v},0}(z)$ is a slight variation of their construction, which is adapted to a different choice of the trigonometric R-matrix.

Now let us consider two *(local) trigonometric Lax matrices*

$$L_i^{\boldsymbol{v},-1}(z) = \begin{pmatrix}\widetilde{\mathsf{w}}_i^{-1}-\widetilde{\mathsf{w}}_iz^{-1} & \widetilde{\mathsf{w}}_iD_i^{-1}\\ -\widetilde{\mathsf{w}}_iD_iz^{-1} & \widetilde{\mathsf{w}}_i\end{pmatrix}\in \mathrm{Mat}(2, z^{-1}\hat{\mathcal{A}}_n^{\boldsymbol{v}}[z]), \tag{11.6}$$

$$L_i^{\boldsymbol{v},1}(z) = \begin{pmatrix}\widetilde{\mathsf{w}}_i^{-1}z-\widetilde{\mathsf{w}}_i & \widetilde{\mathsf{w}}_i^{-1}D_i^{-1}z\\ -\widetilde{\mathsf{w}}_i^{-1}D_i & -\widetilde{\mathsf{w}}_i^{-1}\end{pmatrix}\in \mathrm{Mat}(2, \hat{\mathcal{A}}_n^{\boldsymbol{v}}[z]). \tag{11.7}$$

Lemma 11.4 *The Lax matrices $L_i^{\boldsymbol{v},\pm 1}(z)$ satisfy the trigonometric RTT-relation:*

$$R_{\mathrm{trig}}(z/w)(L_i^{\boldsymbol{v},\pm 1}(z)\otimes 1)(1\otimes L_i^{\boldsymbol{v},\pm 1}(w)) = (1\otimes L_i^{\boldsymbol{v},\pm 1}(w))(L_i^{\boldsymbol{v},\pm 1}(z)\otimes 1)R_{\mathrm{trig}}(z/w).$$

Proof The proof is straightforward. □

11.3 Mixed Toda Hamiltonians

Now we construct 3^n Hamiltonians generalizing H_2^0 in spirit of [21, (90)], cf. also [11, (1.1) and Section 2]. For any $\vec{k} = (k_n,\ldots,k_1)\in\{-1,0,1\}^n$, define the *mixed complete monodromy matrix*

$$T_{\vec{k}}^{\boldsymbol{v}}(z) := L_n^{\boldsymbol{v},k_n}(z)\cdots L_1^{\boldsymbol{v},k_1}(z).$$

In particular, $T_{\vec{0}}^{\boldsymbol{v}}(z) = T_n^{\boldsymbol{v},0}(z)$. Since all three matrices $L_i^{\boldsymbol{v},-1}(z), L_i^{\boldsymbol{v},0}(z), L_i^{\boldsymbol{v},1}(z)$ satisfy the RTT-relation with the same R-matrix $R_{\mathrm{trig}}(z)$, the same is true for $T_{\vec{k}}^{\boldsymbol{v}}(z)$. Hence, the coefficients (in z) of the matrix element $T_{\vec{k}}^{\boldsymbol{v}}(z)_{11}$ generate a commutative subalgebra of $\hat{\mathcal{A}}_n^{\boldsymbol{v}}$. We have

$$T_{\vec{k}}^{\boldsymbol{v}}(z)_{11} = \mathsf{H}_1^{\vec{k}}z^s+\mathsf{H}_2^{\vec{k}}z^{s+1}+\text{higher powers of } z,$$

where $s = \sum_{i=1}^{n} \frac{k_i-1}{2}$. Here $\mathsf{H}_1^{\vec{k}} = (-1)^n \widetilde{\mathsf{w}}_1 \cdots \widetilde{\mathsf{w}}_n$, while the hamiltonian $\mathsf{H}_2^{\vec{k}}$ equals

$$\mathsf{H}_2^{\vec{k}} = (-1)^{n-1}\widetilde{\mathsf{w}}_1 \cdots \widetilde{\mathsf{w}}_n \cdot \left(\sum_{i=1}^{n} \widetilde{\mathsf{w}}_i^{-2} + \sum_{i=1}^{n-1} \sigma_{i,i+1} D_i D_{i+1}^{-1} + \sum_{1 \le i < j-1 < n}^{k_{i+1}=\ldots=k_{j-1}=1} \sigma_{i,j} D_i D_j^{-1} \right), \tag{11.8}$$

where $\sigma_{i,j} := \widetilde{\mathsf{w}}_i^{-k_i-1} \widetilde{\mathsf{w}}_{i+1}^{-k_{i+1}-1} \cdots \widetilde{\mathsf{w}}_j^{-k_j-1}$.

Remark 11.5 At the classical level, the birational Bäcklund-Darboux transformations interchanging various hamiltonians $\mathsf{H}_2^{\vec{k}}$ are given in [34, Theorem 6.1].

Lemma 11.6 *For any $\vec{k}$, set $\vec{k}' = (0, k_{n-1}, \ldots, k_2, 0)$. Then, $\mathsf{H}_2^{\vec{k}}$ is equivalent to $\mathsf{H}_2^{\vec{k}'}$.*

Proof It is straightforward to see that $\mathsf{H}_2^{\vec{k}'} = \mathrm{Ad}(F(\widetilde{\mathsf{w}}_1, \ldots, \widetilde{\mathsf{w}}_n))\mathsf{H}_2^{\vec{k}}$, where $F(\widetilde{\mathsf{w}}_1, \ldots, \widetilde{\mathsf{w}}_n) = \exp(k_1 f_-(\log(\widetilde{\mathsf{w}}_1)) + k_n f_+(\log(\widetilde{\mathsf{w}}_n)))$ with $f_\pm(t) = \pm \frac{t^2}{2\log(\boldsymbol{v})} + \frac{t}{2}$. □

Remark 11.7 It follows that among the aforementioned 3^n *mixed Toda hamiltonians* $\mathsf{H}_2^{\vec{k}}$, parameterized by $\vec{k} \in \{-1, 0, 1\}^n$, there are no more than 3^{n-2} different up to equivalence. In [35] these hamiltonians are identified with the modified versions of the q-Toda hamiltonian in [19, 56], which now depend on a choice of two orientations of the Dynkin diagram of type A_{n-1} (equivalently, a choice of a pair of Coxeter elements). There are 4^{n-2} such choices, but some of them are equivalent leading to exactly 3^{n-2} inequivalent hamiltonians, which turn out to be equivalent to the aforementioned $\mathsf{H}_2^{\vec{k}}$. All the q-Toda hamiltonians of [19, 56] correspond to the pairs of coinciding orientations, i.e. to $\vec{k} = (0, \ldots, 0)$, and they share the same eigenfunction J [22, Section 3], while our mixed Toda hamiltonians do not admit the common eigenfunctions. We are grateful to P. Etingof for his suggestion to study the construction of [56] for pairs of different orientations.

11.4 Shifted RTT Algebras of $\mathfrak{sl}_2$

Fix $n \in \mathbb{N}$. Following [17] (cf. also Remark G.1), we introduce the *(trigonometric) shifted RTT algebras of* $\mathfrak{sl}_2$, denoted by $\mathcal{U}^{\mathrm{rtt}}_{0,-2n}$. These are associative $\mathbb{C}(\boldsymbol{v})$-algebras generated by

$$\{t_{11}^+[r], t_{12}^+[r], t_{21}^+[r+1], t_{22}^+[r], t_{11}^-[-m], t_{12}^-[-m-1], t_{21}^-[-m], t_{22}^-[-m-1+\delta_{n,0}]\}_{r\ge 0}^{m\ge -n} \cup$$

$$\{(t_{11}^+[0])^{-1}, (t_{11}^-[n])^{-1}\}$$

subject to the following defining relations:

$$(t_{11}^{+}[0])^{\pm 1}(t_{11}^{+}[0])^{\mp 1} = 1,\ (t_{11}^{-}[n])^{\pm 1}(t_{11}^{-}[n])^{\mp 1} = 1, \tag{R1}$$

$$R_{\mathrm{trig}}(z/w)(T^{\epsilon}(z)\otimes 1)(1\otimes T^{\epsilon'}(w)) = (1\otimes T^{\epsilon'}(w))(T^{\epsilon}(z)\otimes 1)R_{\mathrm{trig}}(z/w), \tag{R2}$$

$$\mathrm{qdet}\ T^{\pm}(z) = 1 \tag{R3}$$

for all $\epsilon, \epsilon' \in \{\pm\}$, where the two-by-two matrices $T^{\pm}(z)$ are given by

$$T^{\pm}(z) = \begin{pmatrix} T_{11}^{\pm}(z) & T_{12}^{\pm}(z) \\ T_{21}^{\pm}(z) & T_{22}^{\pm}(z) \end{pmatrix} \text{ with } T_{ij}^{\pm}(z) := \sum_r t_{ij}^{\pm}[r]z^{-r},$$

and the quantum determinant qdet is defined in a standard way as[8]

$$\mathrm{qdet}\ T^{\pm}(z) := T_{11}^{\pm}(z)T_{22}^{\pm}(\boldsymbol{v}^{-2}z) - \boldsymbol{v}^{-1}T_{12}^{\pm}(z)T_{21}^{\pm}(\boldsymbol{v}^{-2}z).$$

Note that $T^{\pm}(z)$ admits the following unique *Gauss* decomposition:

$$T^{\pm}(z) = \begin{pmatrix} 1 & 0 \\ \tilde{f}^{\pm}(z) & 1 \end{pmatrix} \begin{pmatrix} \tilde{g}_1^{\pm}(z) & 0 \\ 0 & \tilde{g}_2^{\pm}(z) \end{pmatrix} \begin{pmatrix} 1 & \tilde{e}^{\pm}(z) \\ 0 & 1 \end{pmatrix},$$

where coefficients of the *half-currents* $\tilde{e}^{\pm}(z), \tilde{f}^{\pm}(z), \tilde{g}_1^{\pm}(z), \tilde{g}_2^{\pm}(z)$ are elements of $\mathcal{U}^{\mathrm{rtt}}_{0,-2n}$.

To establish the relation between $\mathcal{U}^{\mathrm{rtt}}_{0,-2n}$ and $\mathcal{U}^{\mathrm{ad}}_{0,-2n}$ (adjoint version of the shifted quantum affine algebra of $\mathfrak{sl}_2$), recall *Drinfeld half-currents* $e^{\pm}(z), f^{\pm}(z)$ of (6.5).

Theorem 11.8

(a) The currents $\tilde{g}_1^{\pm}(z), \tilde{g}_2^{\pm}(z)$ *pairwise commute and satisfy*

$$\tilde{g}_2^{\pm}(z)\tilde{g}_1^{\pm}(\boldsymbol{v}^{-2}z) = 1.$$

(b) There exists a unique $\mathbb{C}(\boldsymbol{v})$*-algebra homomorphism* $\Upsilon_{0,-2n} \colon \mathcal{U}^{\mathrm{ad}}_{0,-2n} \to \mathcal{U}^{\mathrm{rtt}}_{0,-2n}$, *defined by*

$$e^{\pm}(z) \mapsto \tilde{e}^{\pm}(z)/(\boldsymbol{v}-\boldsymbol{v}^{-1}),\ f^{\pm}(z) \mapsto \tilde{f}^{\pm}(z)/(\boldsymbol{v}-\boldsymbol{v}^{-1}),$$

$$\psi^{\pm}(z) \mapsto \tilde{g}_2^{\pm}(z)\tilde{g}_1^{\pm}(z)^{-1},\ (\phi^{+})^{\pm 1} \mapsto (t_{11}^{+}[0])^{\mp 1},\ (\phi^{-})^{\pm 1} \mapsto \boldsymbol{v}^{\mp n}(t_{11}^{-}[n])^{\mp 1}.$$

[8] It is instructive to point out the difference with [51], where the author uses a different trigonometric R-matrix given by $R^{\mathrm{M}}_{\mathrm{trig}}(z/w) = (R_{\mathrm{trig}}(z/w)^t)^{-1}$ as well as $T^{\mathrm{M},\pm}(z) = T^{\pm}(z)^t$. For this reason, the quantum determinant $\mathrm{qdet}^{\mathrm{M}}$ of [51, Exercise 1.6] is consistent with our definition of qdet, that is, $\mathrm{qdet}^{\mathrm{M}}\ T^{\mathrm{M},\pm}(z) := T_{11}^{\mathrm{M},\pm}(z)T_{22}^{\mathrm{M},\pm}(\boldsymbol{v}^{-2}z) - \boldsymbol{v}^{-1}T_{21}^{\mathrm{M},\pm}(z)T_{12}^{\mathrm{M},\pm}(\boldsymbol{v}^{-2}z) = \mathrm{qdet}\ T^{\pm}(z)$.

(c) *For any* $b_1, b_2 \in \mathbb{Z}_{\le 0}$, *there exists a unique* $\mathbb{C}(\boldsymbol{v})$-*algebra homomorphism*

$$\Delta^{\mathrm{rtt}}_{2b_1,2b_2} : \mathcal{U}^{\mathrm{rtt}}_{0,2b_1+2b_2} \longrightarrow \mathcal{U}^{\mathrm{rtt}}_{0,2b_1} \otimes \mathcal{U}^{\mathrm{rtt}}_{0,2b_2},$$

defined by $T^{\pm}(z) \mapsto T^{\pm}(z) \otimes T^{\pm}(z)$.

Remark 11.9 The $n = 0$ case of this theorem was proved in [17], cf. Remark G.1.

Proof The verification of part (b) is analogous to the one for $n = 0$, dealt with in [17]. Once (b) is established, it is easy to see that qdet $T^{\pm}(z) = \tilde{g}_2^{\pm}(z)\tilde{g}_1^{\pm}(\boldsymbol{v}^{-2}z)$, hence (a). It is clear that $\Delta^{\mathrm{rtt}}_{2b_1,2b_2}$ is well-defined on the generators. The compatibility of $\Delta^{\mathrm{rtt}}_{2b_1,2b_2}$ with the defining relations (R1–R3) is checked analogously to the case $n = 0$. □

Recall the generating series $A^{\pm}(z), B^{\pm}(z), C^{\pm}(z), D^{\pm}(z)$ with coefficients in $\mathcal{U}^{\mathrm{ad}}_{0,-2n}$, introduced in Sect. 6.

Corollary 11.10 *The homomorphism* $\Upsilon_{0,-2n}$ *maps these generating series as follows:*

$$A^+(z) \mapsto T^+_{11}(z),\ B^+(z) \mapsto T^+_{12}(z),\ C^+(z) \mapsto T^+_{21}(z),\ D^+(z) \mapsto T^+_{22}(z),$$

$$A^-(z) \mapsto (\boldsymbol{v}z)^n T^-_{11}(z),\ B^-(z) \mapsto (\boldsymbol{v}z)^n T^-_{12}(z),\ C^-(z) \mapsto (\boldsymbol{v}z)^n T^-_{21}(z),\ D^-(z) \mapsto (\boldsymbol{v}z)^n T^-_{22}(z).$$

Proof Due to Theorem 11.8(a, b), we have

$$\Upsilon_{0,-2n}(\psi^{\pm}(z)) = 1/\tilde{g}_1^{\pm}(z)\tilde{g}_1^{\pm}(\boldsymbol{v}^{-2}z),\ \Upsilon_{0,-2n}((\phi^+)^{-1}) = t^+_{11}[0],\ \Upsilon_{0,-2n}((\phi^-)^{-1}) = \boldsymbol{v}^n t^-_{11}[n].$$

Combining this with $\psi^+(z) = \frac{1}{A^+(z)A^+(\boldsymbol{v}^{-2}z)}$, $\psi^-(z) = \frac{z^{2n}}{A^-(z)A^-(\boldsymbol{v}^{-2}z)}$, and $A_0^{\pm} = (\phi^{\pm})^{-1}$, we get $\Upsilon_{0,-2n}(A^+(z)) = \tilde{g}_1^+(z) = T^+_{11}(z)$, $\Upsilon_{0,-2n}(A^-(z)) = (\boldsymbol{v}z)^n\tilde{g}_1^-(z) = (\boldsymbol{v}z)^n T^-_{11}(z)$. The computation of the images of the remaining generating series is straightforward, e.g. $\Upsilon_{0,-2n}(B^-(z)) = (\boldsymbol{v}-\boldsymbol{v}^{-1})\Upsilon_{0,-2n}(A^-(z))\Upsilon_{0,-2n}(e^-(z)) = (\boldsymbol{v}z)^n\tilde{g}_1^-(z)\tilde{e}^-(z) = (\boldsymbol{v}z)^n T^-_{12}(z)$. □

The following is the key result of this subsection.

Theorem 11.11 *For* $n \in \mathbb{N}$, $\Upsilon_{0,-2n} : \mathcal{U}^{\mathrm{ad}}_{0,-2n} \to \mathcal{U}^{\mathrm{rtt}}_{0,-2n}$ *is an isomorphism of* $\mathbb{C}(\boldsymbol{v})$-*algebras.*

Proof Due to Theorem 11.8 and Corollary 11.10, it suffices to prove that there exists a $\mathbb{C}(\boldsymbol{v})$-algebra homomorphism $\mathcal{U}^{\mathrm{rtt}}_{0,-2n} \to \mathcal{U}^{\mathrm{ad}}_{0,-2n}$, such that

$$\begin{aligned}
&(t^+_{11}[0])^{-1} \mapsto \phi^+,\ (t^-_{11}[n])^{-1} \mapsto \boldsymbol{v}^n\phi^-,\\
&T^+_{11}(z) \mapsto A^+(z),\ T^+_{12}(z) \mapsto B^+(z),\ T^+_{21}(z) \mapsto C^+(z),\ T^+_{22}(z) \mapsto D^+(z),\\
&T^-_{11}(z) \mapsto (\boldsymbol{v}z)^{-n}A^-(z),\ T^-_{12}(z) \mapsto (\boldsymbol{v}z)^{-n}B^-(z),\\
&T^-_{21}(z) \mapsto (\boldsymbol{v}z)^{-n}C^-(z),\ T^-_{22}(z) \mapsto (\boldsymbol{v}z)^{-n}D^-(z).
\end{aligned} \tag{11.9}$$

This amounts to verifying that the assignment (11.9) preserves defining relations (R1–R3). Relation (R1) is preserved, due to $A_0^{\pm}\phi^{\pm} = \phi^{\pm}A_0^{\pm} = 1$, while (R3) is preserved, due to relation (6.16). Finally, (R2) is an equality in $\mathrm{End}(\mathbb{C}^2 \otimes \mathbb{C}^2) \otimes \mathcal{U}^{\mathrm{ad}}_{0,-2n}$ and thus can be viewed as a collection of 16 relations in $\mathcal{U}^{\mathrm{ad}}_{0,-2n}$ for each choice of $\epsilon, \epsilon' \in \{\pm\}$. It is straightforward to see that 6 of these relations follow from the rest, while the remaining 10 relations exactly match the 10 relations of (6.7, 6.9–6.15) under the assignment (11.9). □

Remark 11.12 The results of this subsection admit natural generalizations to the case of arbitrary $b_1, b_2 \in \mathbb{Z}_{\leq 0}$ such that $b_1 + b_2$ is even. In other words, one can define an analogous shifted RTT algebra of $\mathfrak{sl}_2$, denoted $\mathcal{U}^{\mathrm{rtt}}_{b_1,b_2}$, and construct a $\mathbb{C}(\boldsymbol{v})$-algebra isomorphism $\Upsilon_{b_1,b_2}\colon \mathcal{U}^{\mathrm{ad}}_{b_1,b_2} \xrightarrow{\sim} \mathcal{U}^{\mathrm{rtt}}_{b_1,b_2}$. This observation is used in Remark 11.14 below, where we provide an alternative interpretation of the Lax matrices $L_1^{\boldsymbol{v},-1}(z), L_1^{\boldsymbol{v},0}(z), L_1^{\boldsymbol{v},1}(z)$ from Sect. 11.2.

11.5 Relation Between Two Different Appearances of RTT

Recall the local trigonometric Lax matrix $L_1^{\boldsymbol{v},-1}(z)$ of (11.6). Combining the equality $\mathrm{qdet}\, L_1^{\boldsymbol{v},-1}(z) = 1$ with Lemma 11.4, we see that $L_1^{\boldsymbol{v},-1}(z)$ gives rise to an algebra homomorphism $\Phi^{\mathrm{rtt}}_{0,-2}\colon \mathcal{U}^{\mathrm{rtt}}_{0,-2} \to \hat{\mathcal{A}}_1^{\boldsymbol{v}}$ defined by $T^{\pm}(z) \mapsto L_1^{\boldsymbol{v},-1}(z)$. Recall the homomorphism $\widetilde{\Phi}^0_{-2}\colon \mathcal{U}^{\mathrm{ad}}_{0,-2} \to \hat{\mathcal{A}}_1^{\boldsymbol{v}}$ of Theorem 7.1 (where $\mathsf{w}^{1/2} = \widetilde{\mathsf{w}}$). The following is straightforward.

Lemma 11.13 *The composition $\Phi^{\mathrm{rtt}}_{0,-2} \circ \Upsilon_{0,-2}$ coincides with $\widetilde{\Phi}^0_{-2}$.*

Remark 11.14 Let us provide a similar interpretation of the other two Lax matrices $L_1^{\boldsymbol{v},0}(z)$ and $L_1^{\boldsymbol{v},1}(z)$. Recall that the algebras $\mathcal{U}^{\mathrm{ad}}_{0,-2}$ and $\mathcal{U}^{\mathrm{ad}}_{b,-2-b}$ are isomorphic for any $b \in \mathbb{Z}$. In particular, one can pull back the homomorphism $\widetilde{\Phi}^0_{-2}$ to obtain a homomorphism $\widetilde{\Phi}_{b,-2-b}\colon \mathcal{U}^{\mathrm{ad}}_{b,-2-b} \to \hat{\mathcal{A}}_1^{\boldsymbol{v}}$, explicitly given by

$$e(z) \mapsto \frac{\widetilde{\mathsf{w}}^{2+b}}{\boldsymbol{v}-\boldsymbol{v}^{-1}}\delta\left(\frac{\widetilde{\mathsf{w}}^2}{z}\right)D^{-1},\ f(z) \mapsto \frac{\widetilde{\mathsf{w}}^{b}}{1-\boldsymbol{v}^2}\delta\left(\frac{\boldsymbol{v}^2\widetilde{\mathsf{w}}^2}{z}\right)D,$$

$$\psi^{\pm}(z) \mapsto \left(\frac{\boldsymbol{v}^{-b}\widetilde{\mathsf{w}}^2 z^b}{(1-\widetilde{\mathsf{w}}^2/z)(1-\boldsymbol{v}^2\widetilde{\mathsf{w}}^2/z)}\right)^{\pm},\ (\phi^+)^{\pm 1} \mapsto \boldsymbol{v}^{\mp b/2}\widetilde{\mathsf{w}}^{\pm 1},\ (\phi^-)^{\pm 1} \mapsto -\boldsymbol{v}^{\mp(b/2+1)}\widetilde{\mathsf{w}}^{\mp 1}.$$

Due to Remark 11.12, the algebra $\mathcal{U}^{\mathrm{ad}}_{b,-2-b}$ admits an RTT realization, that is there is an isomorphism $\Upsilon_{b,-2-b}\colon \mathcal{U}^{\mathrm{ad}}_{b,-2-b} \xrightarrow{\sim} \mathcal{U}^{\mathrm{rtt}}_{b,-2-b}$, only for $b = 0, -1, -2$. Analogously to Lemma 11.13, recasting the homomorphisms $\widetilde{\Phi}_{b,-2-b}$ as the homomorphisms $\mathcal{U}^{\mathrm{rtt}}_{b,-2-b} \to \hat{\mathcal{A}}_1^{\boldsymbol{v}}$, we recover the Lax matrix $L_1^{\boldsymbol{v},0}(z)$ (for $b = -1$) and $L_1^{\boldsymbol{v},1}(z)$ (for $b = -2$). Moreover, this also explains why we had exactly three Lax matrices in Sect. 11.2.

Fix $n \geq 1$ and consider the *complete monodromy matrix* $T_n^{v,-1}(z) = L_n^{v,-1}(z)\cdots L_1^{v,-1}(z)$. Applying iteratively $\Delta^{\mathrm{rtt}}_{\bullet,\bullet}$ of Theorem 11.8(c), we get $\Delta^{\mathrm{rtt}}_n\colon \mathcal{U}^{\mathrm{rtt}}_{0,-2n} \to (\mathcal{U}^{\mathrm{rtt}}_{0,-2})^{\otimes n}$. Composing it with the homomorphism $(\Phi^{\mathrm{rtt}}_{0,-2})^{\otimes n}\colon (\mathcal{U}^{\mathrm{rtt}}_{0,-2})^{\otimes n} \to (\hat{\mathcal{A}}^v_1)^{\otimes n} \simeq \hat{\mathcal{A}}^v_n$, we obtain the homomorphism $\Phi^{\mathrm{rtt}}_{0,-2n}\colon \mathcal{U}^{\mathrm{rtt}}_{0,-2n} \to \hat{\mathcal{A}}^v_n$. The following is straightforward.

Lemma 11.15 *We have* $\Phi^{\mathrm{rtt}}_{0,-2n}(T^{\pm}(z)) = T_n^{v,-1}(z)$.

Remark 11.16 For $n > 1$, the composition $\Phi^{\mathrm{rtt}}_{0,-2n} \circ \Upsilon_{0,-2n}$ *does not* coincide with the homomorphism $\widetilde{\Phi}^0_{-2n}$ of Theorem 7.1.

Remark 11.17 The result of Lemma 11.13 admits a natural *rational* counterpart. Let $\mathbf{Y}_{-2}$ be the shifted Yangian of $\mathfrak{sl}_2$ with the shift $-\alpha$. Recall the homomorphism $\Phi^0_{-2}\colon \mathbf{Y}_{-2} \to \hat{\mathcal{A}}^\hbar_1$ of [10, Corollary B.17]. Consider a slight modification of it

$$\hat{\Phi}_{-2}\colon E(z) \mapsto (z-w)^{-1}\mathsf{u}^{-1},\ F(z) \mapsto -(z-w-\hbar)^{-1}\mathsf{u},\ H(z) \mapsto (z-w)^{-1}(z-w-\hbar)^{-1}.$$

One can also define a *(rational) shifted RTT algebra of* $\mathfrak{sl}_2$, denoted by $\mathcal{Y}^{\mathrm{rtt}}_{-2}$. This is an associative $\mathbb{C}[\hbar]$-algebra generated by $\{t_{11}[r-1], t_{12}[r], t_{21}[r], t_{22}[r+1], (t_{11}[-1])^{-1}\}_{r\geq 0}$ and with the defining relations $(t_{11}[-1])^{\pm 1}(t_{11}[-1])^{\mp 1} = 1$, $T_{11}(z)T_{22}(z-\hbar) - T_{12}(z)T_{21}(z-\hbar) = 1$, $R_{\mathrm{rat}}(z-w)(T(z)\otimes 1)(1\otimes T(w)) = (1\otimes T(w))(T(z)\otimes 1)R_{\mathrm{rat}}(z-w)$, where $T(z) = (T_{ij}(z))_{i,j=1}^2$ with $T_{ij}(z) := \sum_r t_{ij}[r]z^{-r}$. Consider the Gauss decomposition of $T(z)$:

$$T(z) = \begin{pmatrix} 1 & 0 \\ \tilde{f}(z) & 1 \end{pmatrix} \begin{pmatrix} \tilde{g}_1(z) & 0 \\ 0 & \tilde{g}_2(z) \end{pmatrix} \begin{pmatrix} 1 & \tilde{e}(z) \\ 0 & 1 \end{pmatrix}.$$

Analogously to Theorem 11.8(b), there is a $\mathbb{C}[\hbar]$-algebra homomorphism $\Upsilon^{\mathrm{rat}}_{-2}\colon \mathbf{Y}_{-2} \to \mathcal{Y}^{\mathrm{rtt}}_{-2}$, defined by $E(z) \mapsto \tilde{e}(z), F(z) \mapsto \tilde{f}(z), H(z) \mapsto \tilde{g}_2(z)\tilde{g}_1(z)^{-1}$. Composing $\Upsilon^{\mathrm{rat}}_{-2}$ with the homomorphism $\mathcal{Y}^{\mathrm{rtt}}_{-2} \to \hat{\mathcal{A}}^\hbar_1$ given by $T(z) \mapsto L_1^\hbar(z)$, we recover $\hat{\Phi}_{-2}$ from above.

11.6 Homomorphism Δ_{b_1,b_2} ($b_1, b_2 \in \mathbb{Z}_{\leq 0}$) via Drinfeld Half-Currents, $\mathfrak{g} = \mathfrak{sl}_2$

Recall the currents $e^{\pm}(z), f^{\pm}(z), \psi^{\pm}(z)$ of (6.5).

Proposition 11.18 *Let Δ be the Drinfeld-Jimbo coproduct on $U_{\boldsymbol{v}}(L\mathfrak{sl}_2)$. Then, we have*

$$\Delta(e^{\pm}(z)) = 1 \otimes e^{\pm}(z) + \sum_{r=0}^{\infty} (-\boldsymbol{v})^r (\boldsymbol{v} - \boldsymbol{v}^{-1})^{2r} \cdot e^{\pm}(z)^{r+1} \otimes f^{\pm}(\boldsymbol{v}^2 z)^r \psi^{\pm}(z), \tag{11.10}$$

$$\Delta(f^{\pm}(z)) = f^{\pm}(z) \otimes 1 + \sum_{r=0}^{\infty} (-\boldsymbol{v})^{-r} (\boldsymbol{v} - \boldsymbol{v}^{-1})^{2r} \cdot \psi^{\pm}(z) e^{\pm}(\boldsymbol{v}^2 z)^r \otimes f^{\pm}(z)^{r+1}, \tag{11.11}$$

$$\Delta(\psi^{\pm}(z)) = \sum_{r=0}^{\infty} (-1)^r [r+1]_{\boldsymbol{v}} (\boldsymbol{v} - \boldsymbol{v}^{-1})^{2r} \cdot \psi^{\pm}(z) e^{\pm}(\boldsymbol{v}^2 z)^r \otimes f^{\pm}(\boldsymbol{v}^2 z)^r \psi^{\pm}(z). \tag{11.12}$$

These formulas are analogous to those for the Yangian $Y_\hbar(\mathfrak{sl}_2)$ of [51, Exercise 3.2]. The proof of this result is based on the RTT realization of $U_{\boldsymbol{v}}(L\mathfrak{sl}_2)$ and is presented in Appendix J.

Proposition 11.19 *Let $b_1, b_2 \in \mathbb{Z}_{\leq 0}$ and $b = b_1 + b_2$. Then, the homomorphism $\Delta_{b_1,b_2} \colon \mathcal{U}^{\mathrm{sc}}_{0,b} \to \mathcal{U}^{\mathrm{sc}}_{0,b_1} \otimes \mathcal{U}^{\mathrm{sc}}_{0,b_2}$ from Theorem 10.5 also satisfies the formulas (11.10–11.12), where by abuse of notation $e^{\pm}(z)$, $f^{\pm}(z)$, $\psi^{\pm}(z)$ denote the generating series for each respective algebra.*

Proof Our proof is based on the commutative diagram of Remark 10.6:

$$\begin{array}{ccc}
U^{\pm}_{\boldsymbol{v}} & \xrightarrow{\quad\Delta\quad} & U^{\pm}_{\boldsymbol{v}} \otimes U^{\pm}_{\boldsymbol{v}} \\
\Big\downarrow {\scriptstyle \jmath^{\pm}_{b_1,b_2}} & & \Big\downarrow {\scriptstyle \jmath^{\pm}_{b_1,0} \otimes \jmath^{\pm}_{0,b_2}} \\
\mathcal{U}^{\mathrm{sc},\pm}_{0,b_1,b_2} & \xrightarrow[\quad\Delta_{b_1,b_2}\quad]{} & \mathcal{U}^{\mathrm{sc},\pm}_{0,b_1,0} \otimes \mathcal{U}^{\mathrm{sc},\pm}_{0,0,b_2}
\end{array}$$

Since $\jmath^{+}_{\bullet,\bullet} \colon e^+(z) \mapsto e^+(z)$, $f^+(z) \mapsto f^+(z)$, $\psi^+(z) \mapsto \psi^+(z)$, we immediately get the validity of (11.10–11.12) for the currents $e^+(z)$, $f^+(z)$, $\psi^+(z)$ and the homomorphism Δ_{b_1,b_2}.

Let us now treat the case of $e^-(z)$, $f^-(z)$, $\psi^-(z)$. Combining the commutativity of the above diagram (in the "−" case) with equality (11.10) yields

$$\Delta_{b_1,b_2,}(\underline{e}^-(z)) = 1 \otimes \underline{e}^-(z) + \sum_{r=0}^{\infty} (-\boldsymbol{v})^r (\boldsymbol{v} - \boldsymbol{v}^{-1})^{2r} \cdot e^-(z)^{r+1} \otimes f^-(\boldsymbol{v}^2 z)^r \psi^-(z),$$

where $\underline{e}^-(z) := e^-(z) + \sum_{r=b_2}^{-1} e_r z^{-r}$. Meanwhile, $\Delta_{b_1,b_2}(e_r) = 1 \otimes e_r$ for $b_2 \leq r \leq -1$. Hence, $\Delta_{b_1,b_2}(e^-(z))$ is given by the right-hand side of (11.10). Likewise, we get the validity of (11.11), (11.12) for the currents $f^-(z)$, $\psi^-(z)$ and the homomorphism Δ_{b_1,b_2}. □

Since our proof of (11.10–11.12) in Appendix J is based on the RTT-type coproduct $\Delta^{\mathrm{rtt}}_{0,0}$, we immediately get

Corollary 11.20 *Let $b_1, b_2 \in \mathbb{Z}_{\leq 0}$ and $b = b_1 + b_2$. The following diagram is commutative:*

$$\begin{array}{ccc} \mathcal{U}^{\mathrm{ad}}_{0,2b} & \xrightarrow{\Delta^{\mathrm{ad}}_{2b_1,2b_2}} & \mathcal{U}^{\mathrm{ad}}_{0,2b_1} \otimes \mathcal{U}^{\mathrm{ad}}_{0,2b_2} \\ {\scriptstyle \Upsilon_{0,2b}}\big\downarrow\wr & & \wr\big\downarrow{\scriptstyle \Upsilon_{0,2b_1}\otimes\Upsilon_{0,2b_2}} \\ \mathcal{U}^{\mathrm{rtt}}_{0,2b} & \xrightarrow[\Delta^{\mathrm{rtt}}_{2b_1,2b_2}]{} & \mathcal{U}^{\mathrm{rtt}}_{0,2b_1} \otimes \mathcal{U}^{\mathrm{rtt}}_{0,2b_2} \end{array}$$

11.7 Coproduct for Truncated Shifted Algebras, $\mathfrak{g} = \mathfrak{sl}_2$

For $b_1, b_2 \in \mathbb{Z}_{\leq 0}$ and $b = b_1 + b_2$, recall the homomorphism $\Delta^{\mathrm{ad}}_{2b_1,2b_2} : \mathcal{U}^{\mathrm{ad}}_{0,2b} \to \mathcal{U}^{\mathrm{ad}}_{0,2b_1} \otimes \mathcal{U}^{\mathrm{ad}}_{0,2b_2}$ of Remark 10.7. Consider the truncated versions of the algebras involved $\mathcal{U}^0_{2b}, \mathcal{U}^0_{2b_1}, \mathcal{U}^0_{2b_2}$, see Definition 8.6. The goal of this subsection is to prove the following result.

Proposition 11.21 *For $b_1, b_2 \leq 0$, the homomorphism $\Delta^{\mathrm{ad}}_{2b_1,2b_2}$ descends to the same named homomorphism $\mathcal{U}^0_{2b} \to \mathcal{U}^0_{2b_1} \otimes \mathcal{U}^0_{2b_2}$.*

Proof Define a 2-sided ideal $\mathcal{J} \subset \mathcal{U}^{\mathrm{ad}}_{0,2b_1} \otimes \mathcal{U}^{\mathrm{ad}}_{0,2b_2}$ via $\mathcal{J} := \mathcal{J}^0_{2b_1} \otimes \mathcal{U}^{\mathrm{ad}}_{0,2b_2} + \mathcal{U}^{\mathrm{ad}}_{0,2b_1} \otimes \mathcal{J}^0_{2b_2}$. It suffices to show that $\Delta^{\mathrm{ad}}_{2b_1,2b_2}(X) \in \mathcal{J}$ for every generator X of the ideal $\mathcal{J}^0_{2b}$ of (8.5–8.6). To achieve this, recall the commutative diagram of Corollary 11.20.

Case $X = A^+_s (s > -b)$ Applying the aforementioned commutative diagram to the equality $\Delta^{\mathrm{rtt}}_{2b_1,2b_2}(t^+_{11}[s]) = \sum_{s_1,s_2 \geq 0}^{s_1+s_2=s} t^+_{11}[s_1] \otimes t^+_{11}[s_2] + \sum_{s_1,s_2\geq 0}^{s_1+s_2=s} t^+_{12}[s_1] \otimes t^+_{21}[s_2]$, we get $\Delta^{\mathrm{ad}}_{2b_1,2b_2}(A^+_s) = \sum_{s_1,s_2\geq 0}^{s_1+s_2=s} A^+_{s_1} \otimes A^+_{s_2} + \sum_{s_1,s_2\geq 0}^{s_1+s_2=s} B^+_{s_1} \otimes C^+_{s_2}$. For $s_1 + s_2 = s > -b$, either $s_1 > -b_1$ or $s_2 > -b_2$. Hence, each summand in the right-hand side belongs to $\mathcal{J}$, due to Remark 8.8.

Case $X = A^+_0 A^+_{-b} - (-1)^b$ As above $\Delta^{\mathrm{ad}}_{2b_1,2b_2}(A^+_{-b}) \equiv A^+_{-b_1} \otimes A^+_{-b_2}$, where the notation $x \equiv y$ is used to denote $x - y \in \mathcal{J}$. We also have $\Delta^{\mathrm{ad}}_{2b_1,2b_2}(A^+_0) = A^+_0 \otimes A^+_0$. Thus $\Delta^{\mathrm{ad}}_{2b_1,2b_2}(A^+_0 A^+_{-b} - (-1)^b) \equiv A^+_0 A^+_{-b_1} \otimes A^+_0 A^+_{-b_2} - (-1)^b = (A^+_0 A^+_{-b_1} - (-1)^{b_1}) \otimes A^+_0 A^+_{-b_2} + (-1)^{b_1} \otimes (A^+_0 A^+_{-b_2} - (-1)^{b_2}) \equiv 0$. Hence, $\Delta^{\mathrm{ad}}_{2b_1,2b_2}(A^+_0 A^+_{-b} - (-1)^b) \in \mathcal{J}$.

Case $X = A^-_{-r} - \boldsymbol{v}^{-b} A^+_{-b-r}$ $(0 \leq r \leq -b)$ Analogously to the first case considered above, we have $\Delta^{\mathrm{ad}}_{2b_1,2b_2}(A^+_{-b-r}) \equiv \sum^{r_1+r_2=r}_{\substack{0\leq r_1\leq -b_1\\ 0\leq r_2\leq -b_2}} A^+_{-b_1-r_1} \otimes A^+_{-b_2-r_2} + \sum^{r_1+r_2=r}_{\substack{1\leq r_1\leq -b_1\\ 0\leq r_2\leq -b_2-1}} B^+_{-b_1-r_1} \otimes C^+_{-b_2-r_2}$, where the lower bounds on r_1, r_2 are due to Remark 8.8. Completely analogously, we obtain $\Delta^{\mathrm{ad}}_{2b_1,2b_2}(A^-_{-r}) \equiv$

$\sum_{\substack{0\le r_1\le -b_1\\0\le r_2\le -b_2}}^{r_1+r_2=r} A^-_{-r_1}\otimes A^-_{-r_2}+\sum_{\substack{1\le r_1\le -b_1\\0\le r_2\le -b_2-1}}^{r_1+r_2=r} B^-_{-r_1}\otimes C^-_{-r_2}$. Hence,

$$\Delta^{\mathrm{ad}}_{2b_1,2b_2}(A^-_{-r}-\boldsymbol{v}^{-b}A^+_{-b-r})\equiv \sum_{\substack{0\le r_1\le -b_1\\0\le r_2\le -b_2}}^{r_1+r_2=r}(A^-_{-r_1}\otimes A^-_{-r_2}-\boldsymbol{v}^{-b}A^+_{-b_1-r_1}\otimes A^+_{-b_2-r_2})+ \sum_{\substack{1\le r_1\le -b_1\\0\le r_2\le -b_2-1}}^{r_1+r_2=r}(B^-_{-r_1}\otimes C^-_{-r_2}-\boldsymbol{v}^{-b}B^+_{-b_1-r_1}\otimes C^+_{-b_2-r_2}). \tag{11.13}$$

The first sum of (11.13) belongs to $\mathfrak{J}$ as $A^-_{-r_1}\otimes A^-_{-r_2}-\boldsymbol{v}^{-b}A^+_{-b_1-r_1}\otimes A^+_{-b_2-r_2}=(A^-_{-r_1}-\boldsymbol{v}^{-b_1}A^+_{-b_1-r_1})\otimes A^-_{-r_2}+\boldsymbol{v}^{-b_1}A^+_{-b_1-r_1}\otimes(A^-_{-r_2}-\boldsymbol{v}^{-b_2}A^+_{-b_2-r_2})\in\mathfrak{J}$. Completely analogously, $B^-_{-r_1}\otimes C^-_{-r_2}-\boldsymbol{v}^{-b}B^+_{-b_1-r_1}\otimes C^+_{-b_2-r_2}=(B^-_{-r_1}-\boldsymbol{v}^{-b_1}B^+_{-b_1-r_1})\otimes C^-_{-r_2}+\boldsymbol{v}^{-b_1}B^+_{-b_1-r_1}\otimes(C^-_{-r_2}-\boldsymbol{v}^{-b_2}C^+_{-b_2-r_2})$. To complete the proof, it suffices to show

$$\begin{aligned}&B^-_{-r_1}-\boldsymbol{v}^{-b_1}B^+_{-b_1-r_1}\in\mathfrak{J}^0_{2b_1}\ \text{for}\ 1\le r_1\le -b_1,\\&C^-_{-r_2}-\boldsymbol{v}^{-b_2}C^+_{-b_2-r_2}\in\mathfrak{J}^0_{2b_2}\ \text{for}\ 0\le r_2\le -b_2-1.\end{aligned}\tag{11.14}$$

To prove the first inclusion of (11.14), recall that $B^+(z)=[e_0,A^+(z)]_{\boldsymbol{v}^{-1}}$, due to Corollary 7.3. Likewise (comparing the terms of degree 1 in w in the equality (6.10) with $\epsilon=-,\epsilon'=+$), we obtain $B^-(z)=[e_0,A^-(z)]_{\boldsymbol{v}^{-1}}$. Therefore,

$$B^-_{-r_1}-\boldsymbol{v}^{-b_1}B^+_{-b_1-r_1}=[e_0,A^-_{-r_1}-\boldsymbol{v}^{-b_1}A^+_{-b_1-r_1}]_{\boldsymbol{v}^{-1}}\in\mathfrak{J}^0_{2b_1}.$$

Similarly, applying the equalities $zC^\pm(z)=[A^\pm(z),f_1]_{\boldsymbol{v}^{-1}}$, we obtain

$$C^-_{-r_2}-\boldsymbol{v}^{-b_2}C^+_{-b_2-r_2}=[A^-_{-r_2-1}-\boldsymbol{v}^{-b_2}A^+_{-b_2-r_2-1},f_1]_{\boldsymbol{v}^{-1}}\in\mathfrak{J}^0_{2b_2},$$

which implies the second inclusion of (11.14). Thus, $\Delta^{\mathrm{ad}}_{2b_1,2b_2}(A^-_{-r}-\boldsymbol{v}^{-b}A^+_{-b-r})\in\mathfrak{J}$.

The cases when X is one of $A^-_{-s}(s>-b)$, $A^-_0A^-_b-(-\boldsymbol{v}^2)^{-b}$ are treated analogously to the above first two cases. This completes our proof. □

11.8 Coproduct for Truncated Shifted Algebras, General $\mathfrak{g}$

Recall the homomorphism $\Delta_{\mu_1,\mu_2}\colon\mathcal{U}^{\mathrm{sc}}_{0,\mu}\to\mathcal{U}^{\mathrm{sc}}_{0,\mu_1}\otimes\mathcal{U}^{\mathrm{sc}}_{0,\mu_2}$ of Theorem 10.20 ($\mu=\mu_1+\mu_2$, $\mathfrak{g}=\mathfrak{sl}_n$). Given $N=N_1+N_2$, this coproduct extends to

$$\Delta^{\mathrm{ad}}_{\mu_1,\mu_2}\colon\mathcal{U}^{\mathrm{ad}}_{0,\mu}[\mathsf{z}_1^{\pm1},\ldots,\mathsf{z}_N^{\pm1}]\longrightarrow\mathcal{U}^{\mathrm{ad}}_{0,\mu_1}[\mathsf{z}_1^{\pm1},\ldots,\mathsf{z}_{N_1}^{\pm1}]\otimes\mathcal{U}^{\mathrm{ad}}_{0,\mu_2}[\mathsf{z}_{N_1+1}^{\pm1},\ldots,\mathsf{z}_N^{\pm1}]$$

as in Remark 10.3(c). Given two sequences $\underline{\lambda}^{(1)} = (\omega_{i_1}, \dots, \omega_{i_{N_1}}), \underline{\lambda}^{(2)} = (\omega_{i_{N_1+1}}, \dots, \omega_{i_N})$, we concatenate them to $\underline{\lambda} = (\omega_{i_1}, \dots, \omega_{i_N})$ and consider the corresponding truncated shifted algebras $\mathcal{U}_{\overline{\mu}}^{\underline{\lambda}}, \mathcal{U}_{\overline{\mu_1}}^{\underline{\lambda}^{(1)}}, \mathcal{U}_{\overline{\mu_2}}^{\underline{\lambda}^{(2)}}$ as in Definition 8.6.

Conjecture 11.22 The aforementioned homomorphism $\Delta^{\mathrm{ad}}_{\mu_1,\mu_2}$ descends to the same named homomorphism $\Delta^{\mathrm{ad}}_{\mu_1,\mu_2} \colon \mathcal{U}_{\overline{\mu}}^{\underline{\lambda}} \to \mathcal{U}_{\overline{\mu_1}}^{\underline{\lambda}^{(1)}} \otimes \mathcal{U}_{\overline{\mu_2}}^{\underline{\lambda}^{(2)}}$.

We hope that the comultiplication $\Delta^{\mathrm{ad}}_{\mu_1,\mu_2}$ can be defined for arbitrary simply-laced $\mathfrak{g}$ (see Sect. 10.8) and descends to the truncated shifted algebras.

12 K-theory of Parabolic Laumon Spaces

12.1 Parabolic Laumon Spaces

We recall the setup of [7]. Let $\mathbf{C}$ be a smooth projective curve of genus zero. We fix a coordinate z on $\mathbf{C}$, and consider the action of $\mathbb{C}^\times$ on $\mathbf{C}$ such that $\boldsymbol{v}(z) = \boldsymbol{v}^{-2}z$. We have $\mathbf{C}^{\mathbb{C}^\times} = \{0, \infty\}$.

We consider an N-dimensional vector space W with a basis $w_1, \dots, w_N$. This defines a Cartan torus $T \subset G = GL(N) = GL(W)$. We also consider its 2^N-fold cover, the bigger torus $\widetilde{T}$, acting on W as follows: for $\widetilde{T} \ni \underline{t} = (t_1, \dots, t_N)$ we have $\underline{t}(w_i) = t_i^2 w_i$.

We fix an n-tuple of positive integers $\pi = (p_1, \dots, p_n) \in \mathbb{Z}^n_{>0}$ such that $p_1 + \ldots + p_n = N$. Let $P \subset G$ be a parabolic subgroup preserving the flag $0 \subset W_1 := \langle w_1, \dots, w_{p_1} \rangle \subset W_2 := \langle w_1, \dots, w_{p_1+p_2} \rangle \subset \cdots \subset W_{n-1} := \langle w_1, \dots, w_{p_1+\ldots+p_{n-1}} \rangle \subset W_n := W$. Let $\mathcal{B} := G/P$ be the corresponding partial flag variety.

Given an $(n-1)$-tuple of nonnegative integers $\underline{d} = (d_1, \dots, d_{n-1}) \in \mathbb{N}^{n-1}$, we consider the Laumon parabolic quasiflags' space $\mathcal{Q}_{\underline{d}}$, see [46, § 4.2]. It is the moduli space of flags of locally free subsheaves

$$0 \subset \mathcal{W}_1 \subset \cdots \subset \mathcal{W}_{n-1} \subset \mathcal{W} = W \otimes \mathcal{O}_{\mathbf{C}}$$

such that $\mathrm{rank}(\mathcal{W}_i) = p_1 + \ldots + p_i$ and $\deg(\mathcal{W}_i) = -d_i$. It is known to be a smooth connected projective variety of dimension $\dim \mathcal{B} + \sum_{i=1}^{n-1} d_i(p_i + p_{i+1})$, see [46, § 2.10].

We consider the following locally closed subvariety $\mathfrak{Q}_{\underline{d}} \subset \mathcal{Q}_{\underline{d}}$ (parabolic quasiflags based at $\infty \in \mathbf{C}$) formed by the flags

$$0 \subset \mathcal{W}_1 \subset \cdots \subset \mathcal{W}_{n-1} \subset \mathcal{W} = W \otimes \mathcal{O}_{\mathbf{C}}$$

such that $\mathcal{W}_i \subset \mathcal{W}$ is a vector subbundle in a neighborhood of $\infty \in \mathbf{C}$, and the fiber of $\mathcal{W}_i$ at ∞ equals the span $\langle w_1, \ldots, w_{p_1+\ldots+p_i} \rangle \subset W$. It is known to be a smooth connected quasiprojective variety of dimension $\sum_{i=1}^{n-1} d_i(p_i + p_{i+1})$.

12.2 Fixed Points

The group $G \times \mathbb{C}^\times$ acts naturally on $\mathcal{Q}_{\underline{d}}$, and the group $\widetilde{T} \times \mathbb{C}^\times$ acts naturally on $\mathfrak{Q}_{\underline{d}}$. The set of fixed points of $\widetilde{T} \times \mathbb{C}^\times$ on $\mathfrak{Q}_{\underline{d}}$ is finite; its description is given in [7, § 4.4].

Let $\underline{\vec{d}}$ be a collection of nonnegative integral vectors $\vec{d}_{ij} = (d_{ij}^{(1)}, \ldots, d_{ij}^{(p_j)})$, $n-1 \geq i \geq j \geq 1$, such that $d_i = \sum_{j=1}^{i} |\vec{d}_{ij}| = \sum_{j=1}^{i} \sum_{a=1}^{p_j} d_{ij}^{(a)}$, and for $i \geq k \geq j$ we have $\vec{d}_{kj} \geq \vec{d}_{ij}$, i.e., $d_{kj}^{(a)} \geq d_{ij}^{(a)}$ for any $1 \leq a \leq p_j$. Abusing notation, we denote by $\underline{\vec{d}}$ the corresponding $\widetilde{T} \times \mathbb{C}^\times$-fixed point in $\mathfrak{Q}_{\underline{d}}$:

$\mathcal{W}_1 = \mathcal{O}_{\mathbf{C}}(-d_{11}^{(1)} \cdot 0)w_1 \oplus \cdots \oplus \mathcal{O}_{\mathbf{C}}(-d_{11}^{(p_1)} \cdot 0)w_{p_1}$,

$\mathcal{W}_2 = \mathcal{O}_{\mathbf{C}}(-d_{21}^{(1)} \cdot 0)w_1 \oplus \cdots \oplus \mathcal{O}_{\mathbf{C}}(-d_{21}^{(p_1)} \cdot 0)w_{p_1} \oplus \mathcal{O}_{\mathbf{C}}(-d_{22}^{(1)} \cdot 0)w_{p_1+1} \oplus \cdots \oplus \mathcal{O}_{\mathbf{C}}(-d_{22}^{(p_2)} \cdot 0)w_{p_1+p_2}$,

$\vdots$

$\mathcal{W}_{n-1} = \mathcal{O}_{\mathbf{C}}(-d_{n-1,1}^{(1)} \cdot 0)w_1 \oplus \cdots \oplus \mathcal{O}_{\mathbf{C}}(-d_{n-1,1}^{(p_1)} \cdot 0)w_{p_1} \oplus \cdots$
$\cdots \oplus \mathcal{O}_{\mathbf{C}}(-d_{n-1,n-1}^{(1)} \cdot 0)w_{p_1+\ldots+p_{n-2}+1} \oplus \cdots \oplus \mathcal{O}_{\mathbf{C}}(-d_{n-1,n-1}^{(p_{n-1})} \cdot 0)w_{p_1+\ldots+p_{n-1}}$.

Notation Given a collection $\underline{\vec{d}}$ as above, we will denote by $\underline{\vec{d}} \pm \delta_{ij}^{(p)}$ the collection $\underline{\vec{d}}'$, such that $d_{ij}'^{(p)} = d_{ij}^{(p)} \pm 1$, while $d_{kl}'^{(a)} = d_{kl}^{(a)}$ for $(a, k, l) \neq (p, i, j)$.

12.3 Correspondences

For $i \in \{1, \ldots, n-1\}$ and $\underline{d} = (d_1, \ldots, d_{n-1})$, we set $\underline{d} + i := (d_1, \ldots, d_i + 1, \ldots, d_{n-1})$. We have a correspondence $\mathcal{E}_{\underline{d},i} \subset \mathcal{Q}_{\underline{d}} \times \mathcal{Q}_{\underline{d}+i}$ formed by the pairs $(\mathcal{W}_\bullet, \mathcal{W}'_\bullet)$ such that $\mathcal{W}'_i \subset \mathcal{W}_i$ and we have $\mathcal{W}_j = \mathcal{W}'_j$ for $j \neq i$, see [7, § 4.5]. In other words, $\mathcal{E}_{\underline{d},i}$ is the moduli space of flags of locally free sheaves

$$0 \subset \mathcal{W}_1 \subset \cdots \subset \mathcal{W}_{i-1} \subset \mathcal{W}'_i \subset \mathcal{W}_i \subset \mathcal{W}_{i+1} \subset \cdots \subset \mathcal{W}_{n-1} \subset \mathcal{W}$$

such that $\operatorname{rank}(\mathcal{W}_j) = p_1 + \ldots + p_j$ and $\deg(\mathcal{W}_j) = -d_j$, while $\operatorname{rank}(\mathcal{W}'_i) = p_1 + \ldots + p_i$ and $\deg(\mathcal{W}'_i) = -d_i - 1$. According to [46, § 2.10], $\mathcal{E}_{\underline{d},i}$ is a smooth projective algebraic variety of dimension $\dim \mathcal{B} + \sum_{i=1}^{n-1} d_i(p_i + p_{i+1}) + p_i$.

We denote by $\mathbf{p}$ (resp. $\mathbf{q}$) the natural projection $\mathcal{E}_{\underline{d},i} \to \mathcal{Q}_{\underline{d}}$ (resp. $\mathcal{E}_{\underline{d},i} \to \mathcal{Q}_{\underline{d}+i}$). We also have a map $\mathbf{s}\colon \mathcal{E}_{\underline{d},i} \to \mathbf{C}$,

$$(0 \subset \mathcal{W}_1 \subset \cdots \subset \mathcal{W}_{i-1} \subset \mathcal{W}'_i \subset \mathcal{W}_i \subset \mathcal{W}_{i+1} \subset \cdots \subset \mathcal{W}_{n-1} \subset \mathcal{W}) \mapsto \operatorname{supp}(\mathcal{W}_i/\mathcal{W}'_i).$$

The correspondence $\mathcal{E}_{\underline{d},i}$ comes equipped with a natural line bundle $\mathcal{L}_i$ whose fiber at a point

$$(0 \subset \mathcal{W}_1 \subset \cdots \subset \mathcal{W}_{i-1} \subset \mathcal{W}'_i \subset \mathcal{W}_i \subset \mathcal{W}_{i+1} \subset \cdots \subset \mathcal{W}_{n-1} \subset \mathcal{W})$$

equals $\Gamma(\mathbf{C}, \mathcal{W}_i/\mathcal{W}'_i)$. Finally, we have a transposed correspondence ${}^{\mathsf{T}}\mathcal{E}_{\underline{d},i} \subset \mathcal{Q}_{\underline{d}+i} \times \mathcal{Q}_{\underline{d}}$.

Restricting to $\mathfrak{Q}_{\underline{d}} \subset \mathcal{Q}_{\underline{d}}$, we obtain the correspondence $\mathsf{E}_{\underline{d},i} \subset \mathfrak{Q}_{\underline{d}} \times \mathfrak{Q}_{\underline{d}+i}$ together with the line bundle L_i and the natural maps $\mathbf{p}\colon \mathsf{E}_{\underline{d},i} \to \mathfrak{Q}_{\underline{d}}$, $\mathbf{q}\colon \mathsf{E}_{\underline{d},i} \to \mathfrak{Q}_{\underline{d}+i}$, $\mathbf{s}\colon \mathsf{E}_{\underline{d},i} \to \mathbf{C}\backslash\{\infty\}$. We also have a transposed correspondence ${}^{\mathsf{T}}\mathsf{E}_{\underline{d},i} \subset \mathfrak{Q}_{\underline{d}+i} \times \mathfrak{Q}_{\underline{d}}$. It is a smooth quasiprojective variety of dimension $\sum_{i=1}^{n-1} d_i(p_i + p_{i+1}) + p_i$.

12.4 Equivariant K-groups

We denote by ${}'M(\pi)$ the direct sum of equivariant (complexified) K-groups:

$${}'M(\pi) = \bigoplus_{\underline{d}} K^{\widetilde{T}\times\mathbb{C}^\times}(\mathfrak{Q}_{\underline{d}}).$$

It is a module over $K_{\widetilde{T}\times\mathbb{C}^\times}(\mathrm{pt}) = \mathbb{C}[\widetilde{T} \times \mathbb{C}^\times] = \mathbb{C}[t_1^{\pm 1}, \ldots, t_N^{\pm 1}, v^{\pm 1}]$. We define

$$M(\pi) := {}'M(\pi) \otimes_{K_{\widetilde{T}\times\mathbb{C}^\times}(\mathrm{pt})} \mathrm{Frac}(K_{\widetilde{T}\times\mathbb{C}^\times}(\mathrm{pt})).$$

It is naturally graded

$$M(\pi) = \oplus_{\underline{d}} M(\pi)_{\underline{d}}, \text{ where } M(\pi)_{\underline{d}} = K^{\widetilde{T}\times\mathbb{C}^\times}(\mathfrak{Q}_{\underline{d}}) \otimes_{K_{\widetilde{T}\times\mathbb{C}^\times}(\mathrm{pt})} \mathrm{Frac}(K_{\widetilde{T}\times\mathbb{C}^\times}(\mathrm{pt})).$$

According to the Thomason localization theorem, restriction to the $\widetilde{T} \times \mathbb{C}^\times$-fixed point set induces an isomorphism

$$K^{\widetilde{T}\times\mathbb{C}^\times}(\mathfrak{Q}_{\underline{d}}) \otimes_{K_{\widetilde{T}\times\mathbb{C}^\times}(\mathrm{pt})} \mathrm{Frac}(K_{\widetilde{T}\times\mathbb{C}^\times}(\mathrm{pt})) \xrightarrow{\sim} K^{\widetilde{T}\times\mathbb{C}^\times}(\mathfrak{Q}_{\underline{d}}^{\widetilde{T}\times\mathbb{C}^\times}) \otimes_{K_{\widetilde{T}\times\mathbb{C}^\times}(\mathrm{pt})} \mathrm{Frac}(K_{\widetilde{T}\times\mathbb{C}^\times}(\mathrm{pt})).$$

The classes of the structure sheaves $[\underline{\vec{d}}]$ of the $\widetilde{T} \times \mathbb{C}^\times$-fixed points $\underline{\vec{d}}$ (see Sect. 12.2) form a basis in $\bigoplus_{\underline{d}} K^{\widetilde{T}\times\mathbb{C}^\times}(\mathfrak{Q}_{\underline{d}}^{\widetilde{T}\times\mathbb{C}^\times}) \otimes_{K_{\widetilde{T}\times\mathbb{C}^\times}(\mathrm{pt})} \mathrm{Frac}(K_{\widetilde{T}\times\mathbb{C}^\times}(\mathrm{pt}))$. The embedding of a point $\underline{\vec{d}}$ into $\mathfrak{Q}_{\underline{d}}$ is a proper morphism, so the direct image in the equivariant K-theory is well-defined, and we will denote by $[\underline{\vec{d}}] \in M(\pi)_{\underline{d}}$ the direct image of the structure sheaf of the point $\underline{\vec{d}}$. The set $\{[\underline{\vec{d}}]\}$ forms a basis of $M(\pi)$.

12.5 Action of $\mathcal{U}_\pi^{\boldsymbol{v}}$ on $M(\pi)$

From now on, we will denote by $\mathcal{U}_\pi^{\boldsymbol{v}}$ the shifted quantum affine algebra $\mathcal{U}_{0,\mu}^{\mathrm{sc}}$ for $\mathfrak{g} = \mathfrak{sl}_n$ and $\mu = \sum_{j=1}^{n-1}(p_{j+1} - p_j)\omega_j$. We will also need the characters T_i of $\widetilde{T} \times \mathbb{C}^\times$ defined via $T_i := \prod_{j=p_1+\ldots+p_{i-1}+1}^{p_1+\ldots+p_i} t_j$. Let $\boldsymbol{v}$ stand for the character of $\widetilde{T} \times \mathbb{C}^\times : (\underline{t}, \boldsymbol{v}) \mapsto \boldsymbol{v}$.

For any $0 \leq i \leq n$, we will denote by $\underline{\mathcal{W}}_i$ the tautological $(p_1 + \ldots + p_i)$-dimensional vector bundle on $\mathfrak{Q}_{\underline{d}} \times \mathbf{C}$. Let $\varpi : \mathfrak{Q}_{\underline{d}} \times (\mathbf{C}\backslash\{\infty\}) \to \mathfrak{Q}_{\underline{d}}$ denote the standard projection. We define the generating series $\mathbf{b}_i(z)$ with coefficients in the equivariant K-theory of $\mathfrak{Q}_{\underline{d}}$ as follows:

$$\mathbf{b}_i(z) := \Lambda^{\bullet}_{-1/z}(\varpi_*(\underline{\mathcal{W}}_i|_{\mathbf{C}\backslash\{\infty\}})) = 1 + \sum_{r\geq 1} \Lambda^r(\varpi_*(\underline{\mathcal{W}}_i|_{\mathbf{C}\backslash\{\infty\}}))(-z^{-1})^r.$$

We also define the operators

$$e_{i,r} := T_{i+1}^{-1}\boldsymbol{v}^{d_{i+1}-d_i+2-i}\mathbf{p}_*((\boldsymbol{v}^i\mathsf{L}_i)^{\otimes r} \otimes \mathbf{q}^*)\colon M(\pi)_{\underline{d}} \to M(\pi)_{\underline{d}-i}, \tag{12.1}$$

$$f_{i,r} := T_i^{-1}\boldsymbol{v}^{d_i-d_{i-1}+i}\mathbf{q}_*((-\mathsf{L}_i)^{\otimes p_i} \otimes (\boldsymbol{v}^i\mathsf{L}_i)^{\otimes r} \otimes \mathbf{p}^*)\colon M(\pi)_{\underline{d}} \to M(\pi)_{\underline{d}+i}, \tag{12.2}$$

and consider the following generating series of operators on $M(\pi)$:

$$e_i(z) = \sum_{r=-\infty}^{\infty} e_{i,r}z^{-r}\colon M(\pi)_{\underline{d}} \to M(\pi)_{\underline{d}-i}[[z, z^{-1}]], \tag{12.3}$$

$$f_i(z) = \sum_{r=-\infty}^{\infty} f_{i,r}z^{-r}\colon M(\pi)_{\underline{d}} \to M(\pi)_{\underline{d}+i}[[z, z^{-1}]]. \tag{12.4}$$

We define $\psi_i^+(z)\colon M(\pi)_{\underline{d}} \to M(\pi)_{\underline{d}}[[z^{-1}]]$ and $\psi_i^-(z)\colon M(\pi)_{\underline{d}} \to z^{p_i-p_{i+1}}M(\pi)_{\underline{d}}[[z]]$ via

$$\psi_i^{\pm}(z) := T_{i+1}^{-1}T_i\boldsymbol{v}^{d_{i+1}-2d_i+d_{i-1}} \cdot \left(\frac{\mathbf{b}_{i+1}(z\boldsymbol{v}^{-i-2})\mathbf{b}_{i-1}(z\boldsymbol{v}^{-i})}{\mathbf{b}_i(z\boldsymbol{v}^{-i-2})\mathbf{b}_i(z\boldsymbol{v}^{-i})}\right)^{\pm}, \tag{12.5}$$

where as before $\gamma(z)^{\pm}$ denotes the expansion of a rational function $\gamma(z)$ in $z^{\mp 1}$, respectively.

Notation To each $\vec{\underline{d}}$, we assign a collection of $\widetilde{T} \times \mathbb{C}^\times$-weights $s_{ij}^{(a)} := t_{p_1+\ldots+p_{j-1}+a}^2\boldsymbol{v}^{-2d_{ij}^{(a)}}$.

Proposition 12.1

(a) The matrix coefficients of the operators $f_{i,r}, e_{i,r}$ in the fixed point basis $\{[\underline{\vec{d}}]\}$ of $M(\pi)$ are as follows:

$$f_{i,r[\underline{\vec{d}},\underline{\vec{d}}']} = T_i^{-1}\boldsymbol{v}^{d_i-d_{i-1}+i}(1-\boldsymbol{v}^2)^{-1}(-s_{ij}^{(a)})^{p_i}(s_{ij}^{(a)}\boldsymbol{v}^i)^r \frac{\prod_{j'\le i-1}^{a'\le p_{j'}}(1-s_{ij}^{(a)}/s_{i-1,j'}^{(a')})}{\prod_{j'\le i,a'\le p_{j'}}^{(j',a')\neq(j,a)}(1-s_{ij}^{(a)}/s_{ij'}^{(a')})}$$

if $\underline{\vec{d}}' = \underline{\vec{d}} + \delta_{ij}^{(a)}$ for certain $j \le i$, $1 \le a \le p_j$;

$$e_{i,r[\underline{\vec{d}},\underline{\vec{d}}']} = T_{i+1}^{-1}\boldsymbol{v}^{d_{i+1}-d_i+2-i}(1-\boldsymbol{v}^2)^{-1}(s_{ij}^{(a)}\boldsymbol{v}^{i+2})^r \frac{\prod_{j'\le i+1}^{a'\le p_{j'}}(1-s_{i+1,j'}^{(a')}/s_{ij}^{(a)})}{\prod_{j'\le i,a'\le p_{j'}}^{(j',a')\neq(j,a)}(1-s_{ij'}^{(a')}/s_{ij}^{(a)})}$$

if $\underline{\vec{d}}' = \underline{\vec{d}} - \delta_{ij}^{(a)}$ for certain $j \le i$, $1 \le a \le p_j$.
All the other matrix coefficients of $e_{i,r}, f_{i,r}$ vanish.

(b) The eigenvalue $\psi_i^{\pm}(z)|_{\underline{\vec{d}}}$ of $\psi_i^{\pm}(z)$ on $[\underline{\vec{d}}]$ equals

$$T_{i+1}^{-1}T_i\boldsymbol{v}^{d_{i+1}-2d_i+d_{i-1}}\left(\frac{\prod_{j\le i+1}^{a\le p_j}(1-z^{-1}\boldsymbol{v}^{i+2}s_{i+1,j}^{(a)})\prod_{j\le i-1}^{a\le p_j}(1-z^{-1}\boldsymbol{v}^i s_{i-1,j}^{(a)})}{\prod_{j\le i}^{a\le p_j}(1-z^{-1}\boldsymbol{v}^{i+2}s_{ij}^{(a)})\prod_{j\le i}^{a\le p_j}(1-z^{-1}\boldsymbol{v}^i s_{ij}^{(a)})}\right)^{\pm}.$$

The proof is straightforward and is analogous to that of [61, Proposition 2.15]. The following is the key result of this section.

Theorem 12.2 *The generating series of operators $\{\psi_i^{\pm}(z), e_i(z), f_i(z)\}_{i=1}^{n-1}$ of (12.3–12.5) acting on $M(\pi)$ satisfy the relations in $\mathfrak{U}_\pi^{\boldsymbol{v}}$, i.e., they give rise to the action of $\mathfrak{U}_\pi^{\boldsymbol{v}}$ on $M(\pi)$.*

In the particular case $\pi = 1^n$, we recover [61, Theorem 2.12].

Proof First, note that $\psi_i^+(z)$ contains only nonpositive powers of z, while $\psi_i^-(z)$ contains only powers of z bigger or equal to $p_i - p_{i+1}$ (this follows from Proposition 12.1(b)). Moreover, the coefficients of z^0 in $\psi_i^+(z)$ and of $z^{p_i-p_{i+1}}$ in $\psi_i^-(z)$ are invertible operators.

Applying Proposition 12.1, the verification of all the defining relations of $\mathfrak{U}_\pi^{\boldsymbol{v}}$, except for (U6), boils down to routine straightforward computations in the fixed point basis (compare to the proof of [61, Theorem 2.12]). The same arguments can be used to show that $[e_i(z), f_j(w)] = 0$ for $i \neq j$. It remains to prove $(\boldsymbol{v}-\boldsymbol{v}^{-1})[e_i(z), f_i(w)] = \delta\left(\frac{z}{w}\right)\left(\psi_i^+(z) - \psi_i^-(z)\right)$. Applying Proposition 12.1(a), we see that the left-hand side is diagonal in the fixed point basis and its eigenvalue on $[\underline{\vec{d}}]$ equals

$$T_{i+1}^{-1}T_i^{-1}\boldsymbol{v}^{d_{i+1}-d_{i-1}}(1-\boldsymbol{v}^2)^{-1}\cdot\delta\left(\frac{z}{w}\right)\times$$

$$\sum_{j\le i}^{a\le p_j}(-s_{ij}^{(a)})^{p_i}\left\{\boldsymbol{v}^{2p_i}\frac{\prod_{j'\le i+1}^{a'\le p_{j'}}(1-s_{i+1,j'}^{(a')}/s_{ij}^{(a)})\prod_{j'\le i-1}^{a'\le p_{j'}}(1-\boldsymbol{v}^2 s_{ij}^{(a)}/s_{i-1,j'}^{(a')})}{\prod_{j'\le i,a'\le p_{j'}}^{(j',a')\neq(j,a)}(1-s_{ij'}^{(a')}/s_{ij}^{(a)})(1-\boldsymbol{v}^2 s_{ij}^{(a)}/s_{ij'}^{(a')})}\delta\left(\frac{z}{\boldsymbol{v}^{i+2}s_{ij}^{(a)}}\right)-\right.$$

$$\left.\frac{\prod_{j'\le i+1}^{a'\le p_{j'}}(1-\boldsymbol{v}^2 s_{i+1,j'}^{(a')}/s_{ij}^{(a)})\prod_{j'\le i-1}^{a'\le p_{j'}}(1-s_{ij}^{(a)}/s_{i-1,j'}^{(a')})}{\prod_{j'\le i,a'\le p_{j'}}^{(j',a')\neq(j,a)}(1-\boldsymbol{v}^2 s_{ij'}^{(a')}/s_{ij}^{(a)})(1-s_{ij}^{(a)}/s_{ij'}^{(a')})}\delta\left(\frac{z}{\boldsymbol{v}^i s_{ij}^{(a)}}\right)\right\}.$$

To compare this expression with the eigenvalue of $\psi_i^+(z)-\psi_i^-(z)$ on $[\underline{\vec{d}}]$, it suffices to apply Lemma C.1 below to the particular case of $\gamma(z)$ chosen to be the rational function of Proposition 12.1(b).

The theorem is proved. □

Remark 12.3

(a) The above verification of (U6) by applying Lemma C.1 significantly simplifies our original indirect proof of this relation in [61].
(b) For $\pi=p^n$, this produces the action of the quantum loop algebra $U_{\boldsymbol{v}}(L\mathfrak{sl}_n)$ on $M(\pi)$.
(c) According to [4], there is an action of $\mathcal{A}^{\boldsymbol{v}}_{\mathrm{frac}}$ on $M(\pi)$. Its pull-back along the homomorphism $\overline{\Phi}{}^{\lambda}_{\mu}$ ($\underline{\lambda}=(\omega_{n-1},\ldots,\omega_{n-1})$ taken N times) yields essentially the action of $\mathfrak{U}^{\boldsymbol{v}}_{\pi}$ on $M(\pi)$ established above. In particular, the kernel $\mathrm{Ker}(\overline{\Phi}{}^{\lambda}_{\mu})=\mathrm{Ker}(\widetilde{\Phi}{}^{\lambda}_{\mu})$ acts trivially on $M(\pi)$. The first instance of that is the fact that the generators $\{A^{\pm}_{i,\pm r}: r>p_1+\ldots+p_i\}$ of $\mathfrak{U}^{\boldsymbol{v}}_{\pi}$ (see Remark 6.7(b)) act trivially on $M(\pi)$, due to the observation that the eigenvalue of $A_i^{\pm}(z)$ on $[\underline{\vec{d}}]$ equals $\prod_{j\le i}^{a\le p_j}(1-(z^{-1}\boldsymbol{v}^i s_{ij}^{(a)})^{\pm 1})$.

12.6 Tensor Products

Fix two n-tuples $\pi'=(p'_1,\ldots,p'_n)$, $\pi''=(p''_1,\ldots,p''_n)\in\mathbb{Z}^n_{>0}$ and define $\pi=(p_1,\ldots,p_n)$ via $p_i:=p'_i+p''_i\in\mathbb{Z}_{>0}$. Let $\mathfrak{U}^{\boldsymbol{v}}_{\pi'},\mathfrak{U}^{\boldsymbol{v}}_{\pi''},\mathfrak{U}^{\boldsymbol{v}}_{\pi}$ be the corresponding shifted quantum affine algebras of $\mathfrak{sl}_n$ as defined in Sect. 12.5. According to Theorem 12.2, we have natural actions of $\mathfrak{U}^{\boldsymbol{v}}_{\pi}$ on $M(\pi)$, of $\mathfrak{U}^{\boldsymbol{v}}_{\pi'}$ on $M(\pi')$, and of $\mathfrak{U}^{\boldsymbol{v}}_{\pi''}$ on $M(\pi'')$. The vector spaces $M(\pi)$ and $M(\pi')\otimes M(\pi'')$ have natural fixed point bases $\{[\underline{\vec{d}}]\}$ and $\{[\underline{\vec{d}}']\otimes[\underline{\vec{d}}'']\}$, parameterized by $\underline{\vec{d}}$ and pairs $(\underline{\vec{d}}',\underline{\vec{d}}'')$ with $\underline{\vec{d}},\underline{\vec{d}}',\underline{\vec{d}}''$ satisfying the conditions of Sect. 12.2. The assignment $(\underline{\vec{d}}',\underline{\vec{d}}'')\mapsto\underline{\vec{d}}'\cup\underline{\vec{d}}''$ defined via $(d'\cup d'')_{ij}^{(a)}=d_{ij}^{\prime(a)}$, $(d'\cup d'')_{ij}^{(p'_j+b)}=d_{ij}^{\prime\prime(b)}$ for $1\le a\le p'_j$, $1\le b\le p''_j$ provides a bijection between such pairs $(\underline{\vec{d}}',\underline{\vec{d}}'')$ and $\underline{\vec{d}}$. We also identify

$\widetilde{T}' \times \widetilde{T}'' \xrightarrow{\sim} \widetilde{T}$ via $t_{p_1+\ldots+p_{j-1}+a} = t'_{p'_1+\ldots+p'_{j-1}+a}, t_{p_1+\ldots+p_{j-1}+p'_j+b} = t''_{p''_1+\ldots+p''_{j-1}+b}$ for a, b as above. Finally, we use $\underline{\vec{0}}$ to denote the collection of zero vectors.

Recall the Drinfeld formal coproduct $\widetilde{\Delta}\colon \mathcal{U}^{\boldsymbol{v}}_{\pi} \to \mathcal{U}^{\boldsymbol{v}}_{\pi'} \widehat{\otimes} \mathcal{U}^{\boldsymbol{v}}_{\pi''}$ of Lemma 10.2.

Theorem 12.4 *There is a unique collection of* $c_{\underline{\vec{d}}',\underline{\vec{d}}''} \in \mathrm{Frac}(K_{\widetilde{T}\times\mathbb{C}^\times}(\mathrm{pt}))$ *with* $c_{\underline{\vec{0}},\underline{\vec{0}}} = 1$, *such that the map* $[\underline{\vec{d}}'] \otimes [\underline{\vec{d}}''] \mapsto c_{\underline{\vec{d}}',\underline{\vec{d}}''} \cdot [\underline{\vec{d}}' \cup \underline{\vec{d}}'']$ *induces an isomorphism* $M(\pi')\widetilde{\otimes}M(\pi'') \xrightarrow{\sim} M(\pi)$ *of* $\mathcal{U}^{\boldsymbol{v}}_{\pi}$*-representations.*

First let us make sense of the $\mathcal{U}^{\boldsymbol{v}}_{\pi}$-module $M(\pi')\widetilde{\otimes}M(\pi'')$. The action of $e_i(z)$ in the fixed point basis $\{[\underline{\vec{d}}'']\}$ of $M(\pi'')$ can be written as $e_i(z)[\underline{\vec{d}}''] = \sum_{j\le i}^{a\le p_j} a_{\underline{\vec{d}}'',\delta^{(a)}_{ij}} \delta(s^{(a)}_{ij}\boldsymbol{v}^{i+2}/z)[\underline{\vec{d}}'' - \delta^{(a)}_{ij}]$ for certain $a_{\underline{\vec{d}}'',\delta^{(a)}_{ij}} \in \mathrm{Frac}(K_{\widetilde{T}''\times\mathbb{C}^\times}(\mathrm{pt}))$. According to the comultiplication formula (10.1), we have $\widetilde{\Delta}(e_i(z))([\underline{\vec{d}}'] \otimes [\underline{\vec{d}}'']) = e_i(z)([\underline{\vec{d}}'])\otimes[\underline{\vec{d}}'']+\psi^-_i(z)([\underline{\vec{d}}'])\otimes e_i(z)([\underline{\vec{d}}''])$. The first summand is well-defined. To make sense of the second summand, we just need to apply the formula $\gamma(z)\delta(a/z) = \gamma(a)\delta(a/z)$ to the rational function $\gamma(z)$ chosen to be the eigenvalue of $\psi^-_i(z)$ on $[\underline{\vec{d}}']$. The action of $f_i(z)$ on $M(\pi')\widetilde{\otimes}M(\pi'')$ is defined analogously. Finally, the formula $\widetilde{\Delta}(\psi^\pm_i(z)) = \psi^\pm_i(z) \otimes \psi^\pm_i(z)$ provides a well-defined action of $\psi^\pm_i(z)$. These formulas endow $M(\pi') \otimes M(\pi'')$ with a well-defined action of $\mathcal{U}^{\boldsymbol{v}}_{\pi}$.

Proof According to Proposition 12.1(b), the eigenvalue of $\widetilde{\Delta}(\psi^\pm_i(z)) = \psi^\pm_i(z) \otimes \psi^\pm_i(z)$ on $[\underline{\vec{d}}'] \otimes [\underline{\vec{d}}''] \in M(\pi') \otimes M(\pi)''$ equals the eigenvalue of $\psi^\pm_i(z)$ on $[\underline{\vec{d}}' \cup \underline{\vec{d}}''] \in M(\pi)$. Hence, the map $[\underline{\vec{d}}'] \otimes [\underline{\vec{d}}''] \mapsto c_{\underline{\vec{d}}',\underline{\vec{d}}''} \cdot [\underline{\vec{d}}' \cup \underline{\vec{d}}'']$ intertwines actions of $\psi^\pm_i(z)$ for any $c_{\underline{\vec{d}}',\underline{\vec{d}}''} \in \mathrm{Frac}(K_{\widetilde{T}\times\mathbb{C}^\times}(\mathrm{pt}))$.

Consider $c_{\underline{\vec{d}}',\underline{\vec{d}}''} \in \mathrm{Frac}(K_{\widetilde{T}\times\mathbb{C}^\times}(\mathrm{pt}))$ such that $c_{\underline{\vec{0}},\underline{\vec{0}}} = 1$ and

$$\begin{aligned} \frac{c_{\underline{\vec{d}}'-\delta^{(a)}_{ij},\underline{\vec{d}}''}}{c_{\underline{\vec{d}}',\underline{\vec{d}}''}} &= (T''_{i+1})^{-1}\boldsymbol{v}^{d''_{i+1}-d''_i} \cdot \frac{\prod_{j'\le i+1}^{a'\le p''_{j'}}(1 - s''^{(a')}_{i+1,j'}/s'^{(a)}_{ij})}{\prod_{j'\le i}^{a'\le p''_{j'}}(1 - s''^{(a')}_{ij'}/s'^{(a)}_{ij})}, \\ \frac{c_{\underline{\vec{d}}',\underline{\vec{d}}''-\delta^{(a)}_{ij}}}{c_{\underline{\vec{d}}',\underline{\vec{d}}''}} &= (T'_i)^{-1}\boldsymbol{v}^{d'_i-d'_{i-1}} \cdot \frac{\prod_{j'\le i}^{a'\le p'_{j'}}(1 - \boldsymbol{v}^{-2}s'^{(a')}_{ij'}/s''^{(a)}_{ij})}{\prod_{j'\le i-1}^{a'\le p'_{j'}}(1 - \boldsymbol{v}^{-2}s'^{(a')}_{i-1,j'}/s''^{(a)}_{ij})}. \end{aligned} \tag{12.6}$$

The existence of $c_{\underline{\vec{d}}',\underline{\vec{d}}''}$ satisfying these relations as well as a verification that $[\underline{\vec{d}}'] \otimes [\underline{\vec{d}}''] \mapsto c_{\underline{\vec{d}}',\underline{\vec{d}}''} \cdot [\underline{\vec{d}}' \cup \underline{\vec{d}}'']$ intertwines actions of $e_{i,r}$ and $f_{i,r}$ are left to the interested reader. □

Remark 12.5 In the particular case $p_1 = \ldots = p_n = p$, this implies the isomorphism $M(p^n) \simeq M(1^n)^{\widetilde{\otimes} p}$ of $U_{\boldsymbol{v}}(L\mathfrak{sl}_n)$-representations. This isomorphism is reminiscent of the isomorphism between the action of the quantum toroidal algebra of $\mathfrak{gl}_1$ on the equivariant K-theory of the Gieseker moduli spaces $M(r, n)$ and the r-fold tensor product of such representation for $r = 1$, see [62, Theorem 4.6].

12.7 Shifted Quantum Affine Algebras of $\mathfrak{gl}_n$

Let $U_{\boldsymbol{v}}(\widehat{\mathfrak{gl}_n})$ be the quantum affine algebra of $\mathfrak{gl}_n$ as defined in [17, Definition 3.1], and let $U_{\boldsymbol{v}}(L\mathfrak{gl}_n)$ be the quantum loop algebra of $\mathfrak{gl}_n$, that is, $U_{\boldsymbol{v}}(L\mathfrak{gl}_n) := U_{\boldsymbol{v}}(\widehat{\mathfrak{gl}_n})/(\boldsymbol{v}^{\pm c/2}-1)$. This is an associative $\mathbb{C}(\boldsymbol{v})$-algebra generated by

$$\{X^{\pm}_{i,r}, k^{\pm}_{j,\mp s^{\pm}_j} | i=1,\ldots,n-1,\ j=1,\ldots,n,\ r\in\mathbb{Z},\ s^{\pm}_j\in\mathbb{N}\}$$

and with the defining relations as in [17, (3.3, 3.4)]. There is a natural injective $\mathbb{C}(\boldsymbol{v})$-algebra homomorphism $U_{\boldsymbol{v}}(L\mathfrak{sl}_n)\hookrightarrow U_{\boldsymbol{v}}(L\mathfrak{gl}_n)$, defined by

$$e_i(z)\mapsto \frac{X^-_i(\boldsymbol{v}^i z)}{\boldsymbol{v}-\boldsymbol{v}^{-1}},\ f_i(z)\mapsto \frac{X^+_i(\boldsymbol{v}^i z)}{\boldsymbol{v}-\boldsymbol{v}^{-1}},\ \psi^{\pm}_i(z)\mapsto (k^{\mp}_i(\boldsymbol{v}^i z))^{-1}k^{\mp}_{i+1}(\boldsymbol{v}^i z). \tag{12.7}$$

For $\pi=(p_1,\ldots,p_n)\in\mathbb{Z}^n_{>0}$, define the shifted quantum affine algebra $\mathfrak{U}^{\boldsymbol{v}}_{\pi}(\mathfrak{gl}_n)$ in the same way as $U_{\boldsymbol{v}}(L\mathfrak{gl}_n)$ except that now $s^+_j\geq -p_j$ and we formally add inverse elements $\{(k^-_{j,0})^{-1},(k^+_{j,p_j})^{-1}\}_{j=1}^n$ (as we no longer require $k^-_{j,0}k^+_{j,p_j}=1$). Note that the assignment (12.7) still gives rise to an injective[9] homomorphism $\varrho\colon \mathfrak{U}^{\boldsymbol{v}}_{\pi}\hookrightarrow \mathfrak{U}^{\boldsymbol{v}}_{\pi}(\mathfrak{gl}_n)$.

Consider the following generating series of operators on $M(\pi)$:

$$X^+_i(z) := (\boldsymbol{v}-\boldsymbol{v}^{-1})f_i(\boldsymbol{v}^{-i}z)\colon M(\pi)_{\underline{d}}\to M(\pi)_{\underline{d}+i}[[z,z^{-1}]],$$

$$X^-_i(z) := (\boldsymbol{v}-\boldsymbol{v}^{-1})e_i(\boldsymbol{v}^{-i}z)\colon M(\pi)_{\underline{d}}\to M(\pi)_{\underline{d}-i}[[z,z^{-1}]],$$

$$k^-_j(z) := T_j^{-1}\boldsymbol{v}^{d_j-d_{j-1}}\cdot(\mathbf{b}_j(z\boldsymbol{v}^{-2j})/\mathbf{b}_{j-1}(z\boldsymbol{v}^{-2j}))^+\colon M(\pi)_{\underline{d}}\to M(\pi)_{\underline{d}}[[z^{-1}]],$$

$$k^+_j(z) := T_j^{-1}\boldsymbol{v}^{d_j-d_{j-1}}\cdot(\mathbf{b}_j(z\boldsymbol{v}^{-2j})/\mathbf{b}_{j-1}(z\boldsymbol{v}^{-2j}))^-\colon M(\pi)_{\underline{d}}\to z^{-p_j}M(\pi)_{\underline{d}}[[z]]$$

with $e_i(z), f_i(z), \mathbf{b}_j(z)$ defined in Sect. 12.5.

The following is a simple generalization of Theorem 12.2.

Theorem 12.6 *The generating series of operators $X^{\pm}_i(z), k^{\pm}_j(z)$ acting on $M(\pi)$ satisfy the relations of $\mathfrak{U}^{\boldsymbol{v}}_{\pi}(\mathfrak{gl}_n)$, i.e., they give rise to the action of $\mathfrak{U}^{\boldsymbol{v}}_{\pi}(\mathfrak{gl}_n)$ on $M(\pi)$.*

The restriction of this action to the subalgebra $\mathfrak{U}^{\boldsymbol{v}}_{\pi}$ (embedded into $\mathfrak{U}^{\boldsymbol{v}}_{\pi}(\mathfrak{gl}_n)$ via ϱ) recovers the action of $\mathfrak{U}^{\boldsymbol{v}}_{\pi}$ on $M(\pi)$ of Theorem 12.2.

[9]One can prove the injectivity of ϱ by using Proposition 5.1 for both algebras. Indeed, the homomorphism ϱ is "glued" from three homomorphisms: $\varrho^{>}\colon \mathfrak{U}^{\boldsymbol{v},>}_{\pi}\to\mathfrak{U}^{\boldsymbol{v},>}_{\pi}(\mathfrak{gl}_n)$, $\varrho^{<}\colon\mathfrak{U}^{\boldsymbol{v},<}_{\pi}\to\mathfrak{U}^{\boldsymbol{v},<}_{\pi}(\mathfrak{gl}_n)$, $\varrho^0\colon\mathfrak{U}^{\boldsymbol{v},0}_{\pi}\to\mathfrak{U}^{\boldsymbol{v},0}_{\pi}(\mathfrak{gl}_n)$. The homomorphisms $\varrho^{>},\varrho^{<}$ are isomorphisms due to Proposition 5.1(b), while the injectivity of ϱ^0 is clear.

12.8 The Cohomology Case Revisited

The above results can be immediately generalized to the cohomological setting. Let $V(\pi)$ be the direct sum of localized $T \times \mathbb{C}^\times$-equivariant cohomology of type π Laumon parabolic based quasiflags' spaces:

$$V(\pi) := \bigoplus_{\underline{d}} H^\bullet_{T\times\mathbb{C}^\times}(\mathfrak{Q}_{\underline{d}}) \otimes_{H^\bullet_{T\times\mathbb{C}^\times}(\mathrm{pt})} \mathrm{Frac}(H^\bullet_{T\times\mathbb{C}^\times}(\mathrm{pt})).$$

It is a module over $\mathrm{Frac}(H^\bullet_{T\times\mathbb{C}^\times}(\mathrm{pt}))$, where $H^\bullet_{T\times\mathbb{C}^\times}(\mathrm{pt}) = \mathbb{C}[\mathrm{Lie}(T \times \mathbb{C}^\times)] = \mathbb{C}[x_1, \dots, x_N, \hbar]$.

Let $\mathcal{Y}^\hbar_\pi = \mathbf{Y}_\pi \otimes_{\mathbb{C}[\hbar]} \mathbb{C}(\hbar)$, where $\mathbf{Y}_\pi$ is the shifted Yangian of $\mathfrak{sl}_n$ in the sense of [10, Appendix B(i)]. It is the associative $\mathbb{C}(\hbar)$-algebra generated by $\{E_i^{(r+1)}, F_i^{(r+1)}, H_i^{(r+1+p_i-p_{i+1})}\}_{1\le i<n}^{r\in\mathbb{N}}$ with the same defining relations as in the standard Yangian $Y_\hbar(\mathfrak{sl}_n)$.

We define the generating series $\mathbf{a}_i(z)$with coefficients in the equivariant cohomology of $\mathfrak{Q}_{\underline{d}}$ as follows:

$$\mathbf{a}_i(z) := z^{p_1+\ldots+p_i} \cdot c(\varpi_*(\underline{\mathcal{W}}_i|_{\mathbf{C}\setminus\{\infty\}}), (-z\hbar)^{-1}),$$

where $c(\mathcal{V}, x)$ denotes the Chern polynomial (in x) of $\mathcal{V}$. We also define the operators

$$E_i^{(r+1)} := \mathbf{p}_*((c_1(\mathsf{L}_i) + i\hbar/2)^r \cdot \mathbf{q}^*) \colon V(\pi)_{\underline{d}} \to V(\pi)_{\underline{d}-i}, \tag{12.8}$$

$$F_i^{(r+1)} := (-1)^{p_i}\mathbf{q}_*((c_1(\mathsf{L}_i) + i\hbar/2)^r \cdot \mathbf{p}^*) \colon V(\pi)_{\underline{d}} \to V(\pi)_{\underline{d}+i}. \tag{12.9}$$

We define $H_i(z) = z^{p_{i+1}-p_i} + \sum_{r>p_i-p_{i+1}} H_i^{(r)}\hbar^{-r+p_i-p_{i+1}+1}z^{-r}$ via

$$H_i(z) := \left(\frac{\mathbf{a}_{i+1}(z - \frac{i+2}{2})\mathbf{a}_{i-1}(z - \frac{i}{2})}{\mathbf{a}_i(z - \frac{i+2}{2})\mathbf{a}_i(z - \frac{i}{2})}\right)^+ : V(\pi)_{\underline{d}} \to z^{p_{i+1}-p_i} V(\pi)_{\underline{d}}[[z^{-1}]]. \tag{12.10}$$

The following result is completely analogous to Theorem 12.2.

Theorem 12.7 *The operators* $\{E_i^{(r+1)}, F_i^{(r+1)}, H_i^{(r+1+p_i-p_{i+1})}\}_{1\le i<n}^{r\in\mathbb{N}}$ *of (12.8–12.10) acting on* $V(\pi)$ *satisfy the defining relations of* $\mathcal{Y}^\hbar_\pi$*, i.e., they give rise to the action of* $\mathcal{Y}^\hbar_\pi$ *on* $V(\pi)$.

A slight refinement of this theorem in the *dominant* case $p_1 \le \ldots \le p_n$ constituted the key result of [7]. In loc. cit., the authors constructed the action of the shifted Yangian of $\mathfrak{gl}_n$, denoted by $\mathcal{Y}^\hbar_\pi(\mathfrak{gl}_n)$, on $V(\pi)$. There is a natural (injective) homomorphism $\mathcal{Y}^\hbar_\pi \to \mathcal{Y}^\hbar_\pi(\mathfrak{gl}_n)$, such that $F_i^{(r+1)} \mapsto$

$\sum_{s=0}^{r}\binom{r}{s}\left(\frac{2-i}{2}\hbar\right)^{r-s}\mathsf{f}_i^{(s+1)}$, $E_i^{(r+1)}\mapsto\sum_{s=0}^{r}\binom{r}{s}\left(\frac{2-i}{2}\hbar\right)^{r-s}\mathsf{e}_i^{(s+1+p_{i+1}-p_i)}$. The pull-back of the action of [7] along this homomorphism recovers the action $\mathcal{Y}_\pi^\hbar$ on $V(\pi)$ of Theorem 12.7.

The proof of [7] was based on an explicit identification of the geometric action in the fixed point basis with the formulas of [27] for the action of $\mathcal{Y}_\pi^\hbar(\mathfrak{gl}_n)$ in the Gelfand-Tsetlin basis. The benefits of our straightforward proof of Theorem 12.7 are two-fold:

(1) we eliminate the crucial assumption $p_1\le\ldots\le p_n$ of [7],
(2) we obtain an alternative proof of the formulas of [27] (cf. Proposition 12.8 below).

Moreover, we can derive $\boldsymbol{v}$-analogues of the Gelfand-Tsetlin formulas of [27] via a certain specialization of the parameters in Proposition 12.1 as explained below. We set $t_l=\boldsymbol{v}^{\beta_l}$ for $1\le l\le N$. To a collection $\underline{\vec{d}}=(d_{ij}^{(a)})_{1\le j\le i\le n-1}^{1\le a\le p_j}$, we associate a Gelfand-Tsetlin pattern $\Lambda=\Lambda(\underline{\vec{d}})=(\lambda_{ij}^{(a)})_{1\le j\le i\le n}^{1\le a\le p_j}$ as follows: $\lambda_{nj}^{(a)}=\beta_{p_1+\ldots+p_{j-1}+a}+j-1$, $\lambda_{ij}^{(a)}=\beta_{p_1+\ldots+p_{j-1}+a}+j-1-d_{ij}^{(a)}$. Set $\lambda_j^{(a)}:=\lambda_{nj}^{(a)}$, which is independent of $\underline{\vec{d}}$. Note that the vector space $M(\pi)$ has a basis $\{[\Lambda]\}$ parametrized by $\Lambda=(\lambda_{ij}^{(a)})_{1\le j\le i\le n}^{1\le a\le p_j}$ with $\lambda_{nj}^{(a)}=\lambda_j^{(a)}$ and $\lambda_{i+1,j}^{(a)}-\lambda_{ij}^{(a)}\in\mathbb{N}$. Consider a specialization of $\{\beta_l\}_{1\le l\le N}$ such that $\lambda_j^{(a)}-\lambda_{j+1}^{(a)}\in\mathbb{N}$, while $\lambda_i^{(a)}-\lambda_j^{(b)}\notin\mathbb{Z}$ if $a\ne b$. Let S be the subset of those Λ from above such that $\lambda_{ij}^{(a)}-\lambda_{i+1,j+1}^{(a)}\in\mathbb{N}$ (note that S is finite), while $\bar{S}$ will denote the set of the remaining Gelfand-Tsetlin patterns Λ.

As before, we define

$$A_i^\pm(z):=k_1^\mp(\boldsymbol{v}^{2-i}z)k_2^\mp(\boldsymbol{v}^{4-i}z)\cdots k_i^\mp(\boldsymbol{v}^iz),$$

$$B_i^\pm(z):=(\boldsymbol{v}-\boldsymbol{v}^{-1})A_i^\pm(z)e_i^\pm(z),$$

$$C_i^\pm(z):=(\boldsymbol{v}-\boldsymbol{v}^{-1})f_i^\pm(z)A_i^\pm(z).$$

We set $\lambda_{ij}(z):=\prod_{a=1}^{p_j}(\boldsymbol{v}^{-\lambda_{ij}^{(a)}}-\boldsymbol{v}^{\lambda_{ij}^{(a)}}z^{-1})$. The next result follows from Proposition 12.1.

Proposition 12.8

(a) The vector subspace of $M(\pi)$ spanned by $\{[\Lambda]\}_{\Lambda\in\bar{S}}$ is $\mathcal{U}_\pi^{\boldsymbol{v}}(\mathfrak{gl}_n)$-invariant. We denote by $L(\pi)$ the corresponding quotient of $M(\pi)$.
(b) Let $\{\xi_\Lambda\}_{\Lambda\in S}$ be the basis of $L(\pi)$ inherited from $\{[\Lambda]\}_{\Lambda\in S}$. Then, we have:

$$A_i^\pm(\boldsymbol{v}^iz)\xi_\Lambda=\boldsymbol{v}^{m_i}\lambda_{i1}(z)\lambda_{i2}(\boldsymbol{v}^2z)\cdots\lambda_{ii}(\boldsymbol{v}^{2(i-1)}z)\xi_\Lambda,$$

$$B_i^{\pm}(\boldsymbol{v}^i \cdot \boldsymbol{v}^{2l_{ij}^{(a)}})\xi_\Lambda = -\boldsymbol{v}^{m_{i+1}-i} \cdot \lambda_{i+1,1}(\boldsymbol{v}^{2l_{ij}^{(a)}})\lambda_{i+1,2}(\boldsymbol{v}^{2(l_{ij}^{(a)}+1)}) \cdots \lambda_{i+1,i+1}(\boldsymbol{v}^{2(l_{ij}^{(a)}+i)})\xi_{\Lambda+\delta_{ij}^{(a)}},$$

$$C_i^{\pm}(\boldsymbol{v}^i \cdot \boldsymbol{v}^{2l_{ij}^{(a)}})\xi_\Lambda = \boldsymbol{v}^{m_{i-1}+i-1} \cdot \lambda_{i-1,1}(\boldsymbol{v}^{2l_{ij}^{(a)}})\lambda_{i-1,2}(\boldsymbol{v}^{2(l_{ij}^{(a)}+1)}) \cdots \lambda_{i-1,i-1}(\boldsymbol{v}^{2(l_{ij}^{(a)}+i-2)})\xi_{\Lambda-\delta_{ij}^{(a)}},$$

where $m_j := \sum_{j'=1}^{j}(j'-1)p_{j'}$ *and* $l_{ij}^{(a)} := \lambda_{ij}^{(a)} - j + 1$.

Remark 12.9

(a) In the simplest case $\pi = 1^n$, the above homomorphism $\mathcal{Y}_\pi^\hbar \to \mathcal{Y}_\pi^\hbar(\mathfrak{gl}_n)$ is the classical embedding of the Yangian of $\mathfrak{sl}_n$ into the Yangian of $\mathfrak{gl}_n$.
(b) The injectivity of the above homomorphism $\mathcal{Y}_\pi^\hbar \to \mathcal{Y}_\pi^\hbar(\mathfrak{gl}_n)$ follows from the PBW property for $\mathcal{Y}_\pi^\hbar$ (see [24, Corollary 3.15]) and its analogue for $\mathcal{Y}_\pi^\hbar(\mathfrak{gl}_n)$.
(c) We take this opportunity to correct the sign in [7, (4.2)], where the '$-$' sign should be replaced by $(-1)^{p_k}$, that is, $\mathsf{f}_k^{(r+1)} := (-1)^{p_k}\mathbf{q}_*(c_1(\mathcal{L}_k')^r \cdot \mathbf{p}^*)$.
(d) We take this opportunity to correct the typos in [23]. First, the formulas for the eigenvalues of $\mathbf{h}_i(u)$ and $\mathbf{a}_{mi}(u)$ of Theorem 3.20 and its proof should be corrected by replacing $p_{i'j'} \rightsquigarrow \hbar^{-1}p_{i'j'}$. Second, the formulas defining $\mathbf{a}_m(u)$ (Section 2.11), $\mathbf{a}_{mi}(u)$ (Section 2.13), $\mathbf{a}_{mi}(u)$ (Section 3.17) should be modified by ignoring $\mathbf{p}_*, \mathbf{q}^*$.

Remark 12.10 Let $e_\pi \in \mathfrak{gl}_N$ be a nilpotent element of Jordan type π. For $p_1 \leq \ldots \leq p_n$, Brundan-Kleshchev proved that the finite W-algebra $W(\mathfrak{gl}_N, e_\pi)$ is the quotient of $\mathcal{Y}_\pi^\hbar(\mathfrak{gl}_n)$ by the 2-sided ideal generated by $\{d_1^{(r)}\}_{r>p_1}$, see [12]. Together with Theorem 12.7 this yields a natural action of $W(\mathfrak{gl}_N, e_\pi)$ on $V(\pi)$, referred to as a finite analogue of the AGT relation in [7]. We expect that the truncated version of $\mathcal{U}_\pi^{\boldsymbol{v}}(\mathfrak{gl}_n)$ with $\lambda = N\omega_{n-1}$ should be isomorphic to the $\boldsymbol{v}$-version of the W-algebra $W(\mathfrak{gl}_N, e_\pi)$ as defined by Sevostyanov in [57].

12.9 Shifted Quantum Toroidal $\mathfrak{sl}_n$ and Parabolic Affine Laumon Spaces

The second main result of [61] provides the action of the quantum toroidal algebra $U_{\boldsymbol{v},u}(\widehat{\mathfrak{sl}}_n)$ (denoted $\ddot{\mathrm{U}}_{\boldsymbol{v}}(\widehat{\mathfrak{sl}}_n)$ in loc. cit.) on the direct sum of localized equivariant K-groups of the affine Laumon spaces $\mathcal{P}_{\underline{d}}$. The cohomological counterpart of this was established in [23], where the action of the affine Yangian $Y_{\hbar,\hbar'}(\widehat{\mathfrak{sl}}_n)$ (denoted $\widehat{Y}$ in loc. cit.) on the direct sum of localized equivariant cohomology of $\mathcal{P}_{\underline{d}}$ was constructed.

Likewise, the results of Theorems 12.2 and 12.7 can be naturally generalized to provide the actions of the shifted quantum toroidal algebra $\mathcal{U}_\pi^{\boldsymbol{v},u}$ (resp. shifted affine Yangian $\mathcal{Y}_\pi^{\hbar,\hbar'}$) on the direct sum of localized equivariant K-groups (resp. cohomology) of parabolic affine Laumon spaces. Here $\mathcal{U}_\pi^{\boldsymbol{v},u}$ is the associative

$\mathbb{C}(\boldsymbol{v}, u)$-algebra generated by $\{e_{i,r}, f_{i,r}, \psi^{\pm}_{i,\pm s_i^{\pm}} | 1 \le i \le n, r \in \mathbb{Z}, s_i^+ \ge 0, s_i^- \ge p_i - p_{i+1}\}$ and with the same defining relations as for $U_{\boldsymbol{v},u}(\widehat{\mathfrak{sl}}_n)$, while $\mathcal{Y}^{\hbar,\hbar'}_{\pi}$ is the associative $\mathbb{C}(\hbar, \hbar')$-algebra generated by $\{E_i^{(r+1)}, F_i^{(r+1)}, H_i^{(r+1+p_i-p_{i+1})} | 1 \le i \le n, r \in \mathbb{N}\}$ and with the same defining relations as for $Y_{\hbar,\hbar'}(\widehat{\mathfrak{sl}}_n)$ (here we set $p_{n+1} := p_1$). On the geometric side, the parabolic affine Laumon spaces of type π are defined similarly to the case $\pi = 1^n$. We leave details to the interested reader.

12.10 Whittaker Vector

Consider the *Whittaker vector*

$$\mathfrak{m} := \sum_{\underline{d}} [\mathcal{O}_{\mathfrak{Q}_{\underline{d}}}] \in M(\pi)^{\wedge},$$

where $M(\pi)^{\wedge} := \prod_{\underline{d}} M(\pi)_{\underline{d}}$. We also define the operators

$$e'_{i,r} := \mathbf{p}_*((\boldsymbol{v}^i \mathsf{L}_i)^{\otimes r} \otimes \mathbf{q}^*) = \boldsymbol{v}^{i-1} (k^-_{i+1,0})^{-1} e_{i,r} : M(\pi)_{\underline{d}} \to M(\pi)_{\underline{d}-i}.$$

Proposition 12.11 *For* $1 \le i \le n-1$, *we have*

$$e'_{i,0}(\mathfrak{m}) = (1-\boldsymbol{v}^2)^{-1}\mathfrak{m} \text{ and } e'_{i,1}(\mathfrak{m}) = \ldots = e'_{i,p_i-1}(\mathfrak{m}) = 0.$$

Proof According to the Bott-Lefschetz formula, we have:

(1) $\mathfrak{m} = \sum_{\vec{\underline{d}}} a_{\vec{\underline{d}}} [\vec{\underline{d}}]$, where $a_{\vec{\underline{d}}} = \prod_{w \in T_{\vec{\underline{d}}} \mathfrak{Q}_{\underline{d}}} (1-w)^{-1}$;

(2) $\frac{a_{\vec{\underline{d}}'}}{a_{\vec{\underline{d}}}} \mathbf{p}_*((\boldsymbol{v}^i \mathsf{L}_i)^{\otimes r} \otimes \mathbf{q}^*)_{[\vec{\underline{d}}', \vec{\underline{d}}]} = \mathbf{q}_*((\boldsymbol{v}^i \mathsf{L}_i)^{\otimes r} \otimes \mathbf{p}^*)_{[\vec{\underline{d}}, \vec{\underline{d}}']}$.

Set $C_{i,0} := (1-\boldsymbol{v}^2)^{-1}$ and $C_{i,r} := 0$ for $0 < r < p_i$. It suffices to prove the equality $C_{i,r} = \sum_{j \le i}^{a \le p_j} \mathbf{q}_*((\boldsymbol{v}^i \mathsf{L}_i)^{\otimes r} \otimes \mathbf{p}^*)_{[\vec{\underline{d}}, \vec{\underline{d}} + \delta_{ij}^{(a)}]}$ for any $\vec{\underline{d}}$ and any $1 \le i \le n-1, 0 \le r \le p_i - 1$. According to Proposition 12.1(a), we have

$$\mathbf{q}_*((\boldsymbol{v}^i \mathsf{L}_i)^{\otimes r} \otimes \mathbf{p}^*)_{[\vec{\underline{d}}, \vec{\underline{d}} + \delta_{ij}^{(a)}]} = (1-\boldsymbol{v}^2)^{-1} (s_{ij}^{(a)} \boldsymbol{v}^i)^r \frac{\prod_{j' \le i-1}^{a' \le p_{j'}} (1 - s_{ij}^{(a)}/s_{i-1,j'}^{(a')})}{\prod_{j' \le i, a' \le p_{j'}}^{(j',a') \ne (j,a)} (1 - s_{ij}^{(a)}/s_{ij'}^{(a')})} =$$

$$\frac{\boldsymbol{v}^i}{1-\boldsymbol{v}^2} \frac{\prod_{j' \le i}^{a' \le p_{j'}} s_{ij'}^{(a')}}{\prod_{j' \le i-1}^{a' \le p_{j'}} s_{i-1,j'}^{(a')}} \cdot (s_{ij}^{(a)} \boldsymbol{v}^i)^{r-1} \frac{\prod_{j' \le i-1}^{a' \le p_{j'}} (s_{i-1,j'}^{(a')} - s_{ij}^{(a)})}{\prod_{j' \le i, a' \le p_{j'}}^{(j',a') \ne (j,a)} (s_{ij'}^{(a')} - s_{ij}^{(a)})}.$$

For $1 \le r \le p_i - 1$, the sum

$$\sum_{j\le i}^{a\le p_j} (s_{ij}^{(a)}\mathbf{v}^i)^{r-1} \frac{\prod_{j'\le i-1}^{a'\le p_{j'}} (s_{i-1,j'}^{(a')} - s_{ij}^{(a)})}{\prod_{j'\le i, a'\le p_{j'}}^{(j',a')\neq(j,a)} (s_{ij'}^{(a')} - s_{ij}^{(a)})}$$

is a rational function in $\{s_{ij'}^{(a')}\}_{j'\le i}^{a'\le p_{j'}}$ of degree $r - p_i < 0$ and without poles. Hence, it is zero. For $r = 0$, the same arguments imply

$$\sum_{j\le i}^{a\le p_j} (s_{ij}^{(a)}\mathbf{v}^i)^{-1} \frac{\prod_{j'\le i-1}^{a'\le p_{j'}} (s_{i-1,j'}^{(a')} - s_{ij}^{(a)})}{\prod_{j'\le i, a'\le p_{j'}}^{(j',a')\neq(j,a)} (s_{ij'}^{(a')} - s_{ij}^{(a)})} = \sum_{j\le i}^{a\le p_j} (s_{ij}^{(a)}\mathbf{v}^i)^{-1} \frac{\prod_{j'\le i-1}^{a'\le p_{j'}} s_{i-1,j'}^{(a')}}{\prod_{j'\le i, a'\le p_{j'}}^{(j',a')\neq(j,a)} (s_{ij'}^{(a')} - s_{ij}^{(a)})}.$$

It remains to compute $\sum_{j\le i}^{a\le p_j} \prod_{j'\le i, a'\le p_{j'}}^{(j',a')\neq(j,a)} \frac{s_{ij'}^{(a')}}{s_{ij'}^{(a')} - s_{ij}^{(a)}}$, which is a rational function in $\{s_{ij'}^{(a')}\}_{j'\le i}^{a'\le p_{j'}}$ of degree 0 and without poles, hence, a constant. Specializing $s_{i1}^{(1)} \mapsto 0$, we see that this constant is equal to 1 (note that only one summand is nonzero under this specialization).

The proposition is proved. □

Remark 12.12

(a) For $\pi = 1^n$, this result was proved in [6, Proposition 2.31].
(b) By the same arguments, we also find $e''_{i,p_i}(\mathfrak{m}) = \frac{(-1)^{p_i-1}\mathbf{v}^{ip_i}}{1-\mathbf{v}^2}\mathfrak{m}$, where $e''_{i,r} := (k_{i,0}^{-})^2 e'_{i,r}$.
(c) Likewise, one can prove that $E_i^{(1)}(\mathfrak{v}) = \ldots = E_i^{(p_i-1)}(\mathfrak{v}) = 0$, $E_i^{(p_i)}(\mathfrak{v}) = \hbar^{-1}\mathfrak{v}$, where $\mathfrak{v} := \sum_{\underline{d}}[\mathfrak{Q}_{\underline{d}}] \in V(\pi)^\wedge$. This result was established in [7, Proposition 5.1].

Appendix A Proof of Theorem 5.5 and Its Modification

To prove Theorem 5.5, let us first note that relations ($\hat{\mathrm{U}}$1–$\hat{\mathrm{U}}$9) hold in $\mathcal{U}^{\mathrm{sc}}_{0,\mu}$. Hence, there exists an algebra homomorphism $\varepsilon\colon \hat{\mathcal{U}}_{\mu_1,\mu_2} \to \mathcal{U}^{\mathrm{sc}}_{0,\mu}$ such that $e_{i,r} \mapsto e_{i,r}, f_{i,s} \mapsto f_{i,s}, (\psi_{i,0}^{+})^{\pm 1} \mapsto (\psi_{i,0}^{+})^{\pm 1}, (\psi_{i,b_i}^{-})^{\pm 1} \mapsto (\psi_{i,b_i}^{-})^{\pm 1}, h_{i,\pm 1} \mapsto h_{i,\pm 1}$ for $i \in I, b_{2,i} - 1 \le r \le 0, b_{1,i} \le s \le 1$. Moreover, the way we defined $e_{i,r}, f_{i,r}, \psi_{i,r}^{\pm} \in \hat{\mathcal{U}}_{\mu_1,\mu_2}$ right before Theorem 5.5, it is clear that $\varepsilon\colon e_{i,r} \mapsto e_{i,r}, f_{i,r} \mapsto f_{i,r}, \psi_{i,\pm s_i^{\pm}}^{\pm} \mapsto \psi_{i,\pm s_i^{\pm}}^{\pm}$ for $i \in I, r \in \mathbb{Z}, s_i^+ \ge 0, s_i^- \ge -b_i$. In particular, ε is surjective. Injectivity of ε is equivalent to showing that relations (U1–U8) hold in $\hat{\mathcal{U}}_{\mu_1,\mu_2}$. This occupies the rest of this Appendix until A(iv), where we consider a slight modification of this presentation, see Theorem A.3 and its proof.

A(i) Derivation of Some Useful Relations in $\hat{\mathcal{U}}_{\mu_1,\mu_2}$

First, we note that (Û1, Û4, Û5) together with our definition of $e_{i,r}, f_{i,r}, \psi^+_{i,r}$ imply:

$$\psi^+_{i,0}e_{j,r} = \boldsymbol{v}_i^{c_{ij}} e_{j,r}\psi^+_{i,0},\ \psi^-_{i,b_i}e_{j,r} = \boldsymbol{v}_i^{-c_{ij}} e_{j,r}\psi^-_{i,b_i},\ [h_{i,\pm 1}, e_{j,r}] = [c_{ij}]_{\boldsymbol{v}_i}\cdot e_{j,r\pm 1}, \tag{v1}$$

$$\psi^+_{i,0}f_{j,r} = \boldsymbol{v}_i^{-c_{ij}} f_{j,r}\psi^+_{i,0},\ \psi^-_{i,b_i}f_{j,r} = \boldsymbol{v}_i^{c_{ij}} f_{j,r}\psi^-_{i,b_i},\ [h_{i,\pm 1}, f_{j,r}] = -[c_{ij}]_{\boldsymbol{v}_i}\cdot f_{j,r\pm 1}, \tag{v2}$$

$$[\psi^+_{i,0}, \psi^\pm_{j,\pm s_j^\pm}] = 0,\ [\psi^-_{i,b_i}, \psi^\pm_{j,\pm s_j^\pm}] = 0 \tag{v3}$$

for any $i, j \in I, r \in \mathbb{Z}, s_j^+ \geq 0, s_j^- \geq -b_j$.

Second, combining relations (Û1, Û4, Û5, Û6), we get

$$\begin{aligned}
&[e_{i,1}, f_{i,0}] = [e_{i,0}, f_{i,1}] = \psi^+_{i,1}/(\boldsymbol{v}_i - \boldsymbol{v}_i^{-1}),\\
&[e_{i,b_{2,i}}, f_{i,b_{1,i}-1}] = [e_{i,b_{2,i}-1}, f_{i,b_{1,i}}] = \psi^-_{i,b_i-1}/(\boldsymbol{v}_i^{-1} - \boldsymbol{v}_i).
\end{aligned} \tag{v4}$$

Note that $\psi^+_{i,1} = (\boldsymbol{v}_i - \boldsymbol{v}_i^{-1})[e_{i,0}, f_{i,1}] = (\boldsymbol{v}_i - \boldsymbol{v}_i^{-1})\psi^+_{i,0}h_{i,1}$. Hence, $[h_{i,1}, \psi^+_{i,1}] = 0$. Combining this further with (v1, v2, v4) and our definition of $\psi^+_{i,2}$, we obtain

$$[e_{i,2}, f_{i,0}] = [e_{i,1}, f_{i,1}] = [e_{i,0}, f_{i,2}] = \psi^+_{i,2}/(\boldsymbol{v}_i - \boldsymbol{v}_i^{-1}). \tag{v5}$$

Likewise, we also get

$$[e_{i,b_{2,i}}, f_{i,b_{1,i}-2}] = [e_{i,b_{2,i}-1}, f_{i,b_{1,i}-1}] = [e_{i,b_{2,i}-2}, f_{i,b_{1,i}}] = \psi^-_{i,b_i-2}/(\boldsymbol{v}_i^{-1} - \boldsymbol{v}_i). \tag{v6}$$

Third, let us point out that relation (Û9) is equivalent to

$$[h_{i,1}, \psi^+_{i,2}] = 0,\ [h_{i,-1}, \psi^-_{i,b_i-2}] = 0. \tag{v7}$$

According to the above relations, for any $i, j \in I$ we also have

$$[h_{j,-1}, \psi^+_{i,2}] = 0,\ [h_{j,1}, \psi^-_{i,b_i-2}] = 0. \tag{v8}$$

Finally, we define elements $h_{i,\pm 2} \in \hat{\mathcal{U}}_{\mu_1,\mu_2}$ as follows:

$$\begin{aligned} h_{i,2} &:= (\psi^+_{i,0})^{-1}\psi^+_{i,2}/(\boldsymbol{v}_i - \boldsymbol{v}_i^{-1}) - (\boldsymbol{v}_i - \boldsymbol{v}_i^{-1})h^2_{i,1}/2, \\ h_{i,-2} &:= (\psi^-_{i,b_i})^{-1}\psi^-_{i,b_i-2}/(\boldsymbol{v}_i^{-1} - \boldsymbol{v}_i) - (\boldsymbol{v}_i^{-1} - \boldsymbol{v}_i)h^2_{i,-1}/2. \end{aligned} \tag{A.1}$$

Due to relations ($\hat{\mathrm{U}}1$, v7, v8), for every $i, j \in I$ we have

$$[h_{i,\pm 1}, h_{i,\pm 2}] = 0, \ [h_{j,\mp 1}, h_{i,\pm 2}] = 0. \tag{v9}$$

Lemma A.1 *For any $i \in I, r \in \mathbb{Z}$, we have*

$$[h_{i,\pm 2}, e_{i,r}] = \frac{[4]_{\boldsymbol{v}_i}}{2} \cdot e_{i,r\pm 2}, \ [h_{i,\pm 2}, f_{i,r}] = -\frac{[4]_{\boldsymbol{v}_i}}{2} \cdot f_{i,r\pm 2}.$$

Proof Due to ($\hat{\mathrm{U}}2$), we have $[e_{i,0}, e_{i,-1}]_{\boldsymbol{v}_i^2} = 0$. Commuting this with $h_{i,1}$ and applying relation ($\hat{\mathrm{U}}4$), we obtain $e_{i,1}e_{i,-1} - \boldsymbol{v}_i^2 e^2_{i,0} = \boldsymbol{v}_i^2 e_{i,-1}e_{i,1} - e^2_{i,0}$. Commuting this further with $f_{i,1}$ and applying relation ($\hat{\mathrm{U}}6$), we obtain

$$\begin{aligned} &\psi^+_{i,2}e_{i,-1} - \boldsymbol{v}_i^2\psi^+_{i,1}e_{i,0} + e_{i,1}\psi^+_{i,0} - \boldsymbol{v}_i^2 e_{i,0}\psi^+_{i,1} - \delta_{b_i,0}e_{i,1}\psi^-_{i,b_i} = \\ &\boldsymbol{v}_i^2 e_{i,-1}\psi^+_{i,2} - e_{i,0}\psi^+_{i,1} + \boldsymbol{v}_i^2\psi^+_{i,0}e_{i,1} - \psi^+_{i,1}e_{i,0} - \boldsymbol{v}_i^2\delta_{b_i,0}\psi^-_{i,b_i}e_{i,1}. \end{aligned}$$

First, note that $e_{i,1}\psi^-_{i,b_i} = \boldsymbol{v}_i^2\psi^-_{i,b_i}e_{i,1}$, due to ($\hat{\mathrm{U}}4$). Second, we have

$$e_{i,1}\psi^+_{i,0} - \boldsymbol{v}_i^2 e_{i,0}\psi^+_{i,1} = \boldsymbol{v}_i^2\psi^+_{i,0}e_{i,1} - \psi^+_{i,1}e_{i,0}. \tag{v10}$$

Indeed, due to the equality $\psi^+_{i,1} = (\boldsymbol{v}_i - \boldsymbol{v}_i^{-1})\psi^+_{i,0}h_{i,1}$ and relations ($\hat{\mathrm{U}}1$, v1), we have

$$\psi^+_{i,1}e_{i,0} - \boldsymbol{v}_i^2 e_{i,0}\psi^+_{i,1} = \boldsymbol{v}_i^2(\boldsymbol{v}_i - \boldsymbol{v}_i^{-1})[2]_{\boldsymbol{v}_i} \cdot e_{i,1}\psi^+_{i,0} = (\boldsymbol{v}_i^4 - 1)e_{i,1}\psi^+_{i,0} = \boldsymbol{v}_i^2\psi^+_{i,0}e_{i,1} - e_{i,1}\psi^+_{i,0}.$$

Therefore, we get

$$\psi^+_{i,2}e_{i,-1} - \boldsymbol{v}_i^2\psi^+_{i,1}e_{i,0} = \boldsymbol{v}_i^2 e_{i,-1}\psi^+_{i,2} - e_{i,0}\psi^+_{i,1}. \tag{v11}$$

Combining the formulas $\psi^+_{i,1} = (\boldsymbol{v}_i - \boldsymbol{v}_i^{-1})\psi^+_{i,0}h_{i,1}$, $\psi^+_{i,2} = (\boldsymbol{v}_i - \boldsymbol{v}_i^{-1})\psi^+_{i,0}(h_{i,2} + \frac{\boldsymbol{v}_i - \boldsymbol{v}_i^{-1}}{2}h^2_{i,1})$ with relations ($\hat{\mathrm{U}}1$, v1, v11), we finally get $[h_{i,2}, e_{i,-1}] = \frac{[4]_{\boldsymbol{v}_i}}{2}e_{i,1}$. Commuting this relation with $h_{i,\pm 1}$ and using (v1, v9), we obtain $[h_{i,2}, e_{i,r}] = \frac{[4]_{\boldsymbol{v}_i}}{2}e_{i,r+2}$ for any $r \in \mathbb{Z}$.

Likewise, starting from the relation $[e_{i,b_{2,i}}, e_{i,b_{2,i}-1}]_{\boldsymbol{v}_i^2} = 0$ and commuting it first with $h_{i,-1}$ and then with $f_{i,b_{1,i}}$, we recover $[h_{i,-2}, e_{i,b_{2,i}}] = \frac{[4]_{\boldsymbol{v}_i}}{2} e_{i,b_{2,i}-2}$. Commuting this further with $h_{i,\pm 1}$, we get $[h_{i,-2}, e_{i,r}] = \frac{[4]_{\boldsymbol{v}_i}}{2} e_{i,r-2}$ for any $r \in \mathbb{Z}$.

The proof of $[h_{i,\pm 2}, f_{i,r}] = -\frac{[4]_{\boldsymbol{v}_i}}{2} \cdot f_{i,r\pm 2}$ is completely analogous. □

A(ii) Verification of Relations (U1–U6) with $i = j$ for $\hat{\mathcal{U}}_{\mu_1,\mu_2}$

A(ii).a Verification of (U2)

We need to prove $X^+(i; r, s) = 0$ for any $r, s \in \mathbb{Z}$, where

$$X^+(i; r, s) := [e_{i,r+1}, e_{i,s}]_{\boldsymbol{v}_i^2} + [e_{i,s+1}, e_{i,r}]_{\boldsymbol{v}_i^2}.$$

Note that $X^+(i; r, s) = X^+(i; s, r)$, and $X^+(i; -1, -1) = 0$ due to relation ($\hat{\mathrm{U}}2$).

For $a \in \{\pm 1, \pm 2\}$, we define $L_{i,a} := a/[2a]_{\boldsymbol{v}_i} \cdot \mathrm{ad}(h_{i,a}) \in \mathrm{End}(\hat{\mathcal{U}}_{\mu_1,\mu_2})$. Then, we have $L_{i,a}(X^+(i; r, s)) = X^+(i; r+a, s) + X^+(i; r, s+a)$. Set $L_i^\pm := \frac{1}{2}(L_{i,\pm 1}^2 - L_{i,\pm 2})$. Then $L_i^\pm(X^+(i; r, s)) = X^+(i; r \pm 1, s \pm 1)$. Applying iteratively L_i^+ to the equality $X^+(i; -1, -1) = 0$, we get $X^+(i; r, r) = 0$ for any $r \geq -1$. Since $2X^+(i; -1, 0) = L_{i,1}(X^+(i; -1, -1)) = 0$, we analogously get $X^+(i; r, r+1) = 0$ for $r \geq -1$. Fix $s \in \mathbb{Z}_{>0}$ and assume by induction that $X^+(i; r, r+N) = 0$ for any $r \geq -1, 0 \leq N \leq s$. Then $X^+(i; -1, s) = L_{i,1}(X^+(i; -1, s-1)) - X^+(i; 0, s-1) = 0$, due to the above assumption. Applying $(L_i^+)^{r+1}$ to the latter equality, we get $X^+(i; r, r+s+1) = 0$ for $r \geq -1$. An induction in s completes the proof of $X^+(i; r, s) = 0$ for any $r, s \geq -1$. Finally, applying iteratively L_i^-, we obtain $X^+(i; r, s) = 0$ for any $r, s \in \mathbb{Z}$.

A(ii).b Verification of (U3)

This relation is verified completely analogously to (U2).

A(ii).c Verification of (U4)

We consider the case $\epsilon = +$ (the case $\epsilon = -$ is completely analogous). We need to prove $Y^+(i; r, s) = 0$ for any $r \in \mathbb{N}, s \in \mathbb{Z}$, where

$$Y^+(i; r, s) := [\psi_{i,r+1}^+, e_{i,s}]_{\boldsymbol{v}_i^2} + [e_{i,s+1}, \psi_{i,r}^+]_{\boldsymbol{v}_i^2}.$$

The $r = s = 0$ case is due to (v10) from our proof of Lemma A.1. Moreover, the same argument also yields $Y^+(i; 0, s) = 0$ for any $s \in \mathbb{Z}$.

Note that $Y^+(i;r,s-1)+Y^+(i;s,r-1) = (\boldsymbol{v}_i-\boldsymbol{v}_i^{-1})[X^+(i;r-1,s-1),f_{i,1}] = 0$ for $r,s \geq 0$. The first equality is due to (v1) and our definition of $\psi^+_{i,r}$, while the second equality follows from $X^+(i;r-1,s-1)=0$ proved above. In particular, $Y^+(i;r,-1)+Y^+(i;0,r-1)=0$ for $r \in \mathbb{N}$.

Combining the above two observations, we find

$$Y^+(i;r,-1)=0 \text{ for any } r \in \mathbb{N}. \tag{v12}$$

Commuting iteratively the equality $Y^+(i;1,-1) = 0$ with $h_{i,\pm 1}$, we get $Y^+(i;1,s)=0$ for any $s \in \mathbb{Z}$, due to ($\hat{\mathrm{U}}$1, v1, v9).

Next, we prove the following five statements by induction in $N \in \mathbb{Z}_+$:
(A_N) $[h_{i,1},\psi^+_{i,r}]=0$ for $0 \leq r \leq N+1$;
(B_N) $[h_{i,-1},\psi^+_{i,r}]=0$ for $0 \leq r \leq N+1$;
(C_N) $[e_{i,r},f_{i,s}]=\psi^+_{i,r+s}/(\boldsymbol{v}_i-\boldsymbol{v}_i^{-1})$ for any $r,s \in \mathbb{N}$ with $1 \leq r+s \leq N+2$;
(D_N) $Y^+(i;r,s)=0$ for any $0 \leq r \leq N, s \in \mathbb{Z}$;
(E_N) $[\psi^+_{i,r},\psi^+_{i,s}]=0$ for any $r,s \geq 0$ with $r+s \leq N+2$.

Base of Induction $(N=1)$ The assertions (A_1, B_1, D_1, E_1) have been already proved above, while (C_1) follows immediately from $[h_{i,1},\psi^+_{i,2}]=0$ (cf. (v7)) and (v1, v2, v4, v5).

Induction Step Assuming $(A_N$–$E_N)$ for a given $N \in \mathbb{Z}_{>0}$, we prove $(A_{N+1}$–$E_{N+1})$.

Proof of the Induction Step Consider a polynomial algebra $B := \mathbb{C}(\boldsymbol{v})[\{x_r\}_{r=1}^\infty]$, which is $\mathbb{N}$-graded via $\deg(x_r)=r$. Define elements $\{\mathsf{h}_r\}_{r=1}^\infty$ of B via $\exp\left((\boldsymbol{v}_i-\boldsymbol{v}_i^{-1})\sum_{r=1}^\infty \mathsf{h}_r z^{-r}\right) = 1+\sum_{r=1}^\infty x_r z^{-r}$. Then, $\mathsf{h}_r = \frac{x_r}{\boldsymbol{v}_i-\boldsymbol{v}_i^{-1}} + p_r(x_1,\ldots,x_{r-1})$ with polynomials p_r satisfying $\deg(p_r(x_1,\ldots,x_{r-1}))=r$.

Using the above polynomials p_r, we define $h_{i,1},\ldots,h_{i,N+1} \in \hat{\mathcal{U}}_{\mu_1,\mu_2}$ via

$$h_{i,r} := \frac{(\psi^+_{i,0})^{-1}\psi^+_{i,r}}{\boldsymbol{v}_i-\boldsymbol{v}_i^{-1}} + p_r((\psi^+_{i,0})^{-1}\psi^+_{i,1},\ldots,(\psi^+_{i,0})^{-1}\psi^+_{i,r-1}) \text{ for } 1 \leq r \leq N+1. \tag{A.2}$$

These $h_{i,r}$ are well-defined and are independent of the choice of $N > r-1$, due to the assumption (E_N) and the aforementioned degree condition on p_r. The following is straightforward:[10]

$$[h_{i,r},e_{i,s}] = \frac{[2r]_{\boldsymbol{v}_i}}{r}\cdot e_{i,s+r} \text{ for } 1 \leq r \leq N+1, s \in \mathbb{Z}. \tag{v13}$$

[10]If we knew that $[\psi^+_{i,a},\psi^+_{i,b}]=0$ for any $0 \leq a,b \leq N+1$, then (v13) would immediately follow from (D_N) by the standard arguments. However, every monomial appearing in p_r involves only pairwise commuting $\psi^+_{i,a}$'s, due to the degree condition on p_r and the assumption (E_N). Hence, the equality (v13) follows formally from its validity in the aforementioned simpler case ($[\psi^+_{i,a},\psi^+_{i,b}]=0$ for any $0 \leq a,b \leq N+1$).

Validity of (A_{N+1}) We need to prove $[h_{i,1}, \psi^+_{i,N+2}] = 0$. According to (C_N), we have $\psi^+_{i,N+2} = (\boldsymbol{v}_i - \boldsymbol{v}_i^{-1})[e_{i,N+2-r}, f_{i,r}]$ for $0 \le r \le N+2$. Hence,

$$[h_{i,1}, \psi^+_{i,N+2}]/(\boldsymbol{v}_i^2 - \boldsymbol{v}_i^{-2}) = [e_{i,N+3-r}, f_{i,r}] - [e_{i,N+2-r}, f_{i,r+1}] \text{ for } 0 \le r \le N+2. \tag{v14}$$

Adding up these equalities for $r = 0, 1$ and using Lemma A.1 together with the assumption (C_N), we get

$$\frac{2[h_{i,1}, \psi^+_{i,N+2}]}{\boldsymbol{v}_i^2 - \boldsymbol{v}_i^{-2}} = [e_{i,N+3}, f_{i,0}] - [e_{i,N+1}, f_{i,2}]$$

$$= \frac{2}{[4]_{\boldsymbol{v}_i}} \cdot [h_{i,2}, [e_{i,N+1}, f_{i,0}]] = \frac{2[h_{i,2}, \psi^+_{i,N+1}]}{\boldsymbol{v}_i^4 - \boldsymbol{v}_i^{-4}}.$$

Likewise, adding up the equality (v14) for $r = 0, 1, \ldots, N$ and using (v13), we obtain

$$\frac{N+1}{\boldsymbol{v}_i^2 - \boldsymbol{v}_i^{-2}}[h_{i,1}, \psi^+_{i,N+2}] = \frac{N+1}{[2(N+1)]_{\boldsymbol{v}_i}} \cdot [h_{i,N+1}, [e_{i,2}, f_{i,0}]] = \frac{(N+1)[h_{i,N+1}, \psi^+_{i,2}]}{\boldsymbol{v}_i^{2(N+1)} - \boldsymbol{v}_i^{-2(N+1)}}.$$

Comparing the above two equalities, we find

$$[h_{i,1}, \psi^+_{i,N+2}] = \frac{\boldsymbol{v}_i^2 - \boldsymbol{v}_i^{-2}}{\boldsymbol{v}_i^4 - \boldsymbol{v}_i^{-4}}[h_{i,2}, \psi^+_{i,N+1}] = \frac{\boldsymbol{v}_i^2 - \boldsymbol{v}_i^{-2}}{\boldsymbol{v}_i^{2(N+1)} - \boldsymbol{v}_i^{-2(N+1)}}[h_{i,N+1}, \psi^+_{i,2}]. \tag{v15}$$

On the other hand, combining (A.2) with the assumption (E_N), we get

$$[h_{i,s}, \psi^+_{i,N+3-s}] = (\psi^+_{i,0})^{-1}[\psi^+_{i,s}, \psi^+_{i,N+3-s}]/(\boldsymbol{v}_i - \boldsymbol{v}_i^{-1}) \text{ for } 1 \le s \le N+1.$$

Hence,

$$[h_{i,1}, \psi^+_{i,N+2}] = \frac{(\psi^+_{i,0})^{-1}[\psi^+_{i,2}, \psi^+_{i,N+1}]}{(\boldsymbol{v}_i - \boldsymbol{v}_i^{-1})[2]_{\boldsymbol{v}_i^2}} = \frac{(\psi^+_{i,0})^{-1}[\psi^+_{i,2}, \psi^+_{i,N+1}]}{(\boldsymbol{v}_i - \boldsymbol{v}_i^{-1})[-N-1]_{\boldsymbol{v}_i^2}}. \tag{v16}$$

Since $[2]_{\boldsymbol{v}_i^2} \neq [-N-1]_{\boldsymbol{v}_i^2}$, the second equality of (v16) implies $[\psi^+_{i,2}, \psi^+_{i,N+1}] = 0$. Hence, $[h_{i,1}, \psi^+_{i,N+2}] = 0$, and (A_{N+1}) follows.

Validity of (B_{N+1}) We need to prove $[h_{i,-1}, \psi^+_{i,N+2}] = 0$. This follows from $[h_{i,-1}, \psi^+_{i,N+2}] = (\boldsymbol{v}_i - \boldsymbol{v}_i^{-1})[2]_{\boldsymbol{v}_i} \cdot ([e_{i,N}, f_{i,1}] - [e_{i,N+1}, f_{i,0}]) = 0$, where we used (v1, v2) in the first equality and (C_N) in the second one. Hence, (B_{N+1}) holds.

Validity of (C_{N+1}) According to (C_N), we have $\psi^+_{i,N+2} = (\boldsymbol{v}_i - \boldsymbol{v}_i^{-1})[e_{i,r}, f_{i,N+2-r}]$ for any $0 \leq r \leq N+2$. Therefore, $[h_{i,1}, \psi^+_{i,N+2}] = (\boldsymbol{v}_i^2 - \boldsymbol{v}_i^{-2})([e_{i,r+1}, f_{i,N+2-r}] - [e_{i,r}, f_{i,N+3-r}])$ due to (v1, v2). The left-hand side is zero due to (A_{N+1}) established above, hence

$$[e_{i,N+3}, f_{i,0}] = [e_{i,N+2}, f_{i,1}] = \ldots = [e_{i,1}, f_{i,N+2}] = [e_{i,0}, f_{i,N+3}].$$

Combining this with our definition $\psi^+_{i,N+3} = (\boldsymbol{v}_i - \boldsymbol{v}_i^{-1})[e_{i,N+2}, f_{i,1}]$ yields (C_{N+1}).

Validity of (D_{N+1}) Due to (A_{N+1}) and (B_{N+1}) established above, we have $[h_{i,\pm 1}, Y^+(i; N+1, s)] = [2]_{\boldsymbol{v}_i} \cdot Y^+(i; N+1, s \pm 1)$. Combining this with (v12), we see that $Y^+(i; N+1, s) = 0$ for any $s \in \mathbb{Z}$. Hence, (D_{N+1}) holds.

Validity of (E_{N+1}) We need to prove $[\psi^+_{i,r}, \psi^+_{i,N+3-r}] = 0$ for any $1 \leq r \leq N+1$. Equivalently, it suffices to prove $[h_{i,r}, \psi^+_{i,N+3-r}] = 0$ for $1 \leq r \leq N+1$. According to (C_N), we have $\psi^+_{i,N+3-r} = (\boldsymbol{v}_i - \boldsymbol{v}_i^{-1})[e_{i,N+3-r}, f_{i,0}]$. Therefore, $[h_{i,r}, \psi^+_{i,N+3-r}] = \frac{\boldsymbol{v}_i^{2r} - \boldsymbol{v}_i^{-2r}}{r} \cdot ([e_{i,N+3}, f_{i,0}] - [e_{N+3-r}, f_{i,r}]) = 0$, due to (v13) and the assertion (C_{N+1}) proved above. □

The induction step is accomplished. In particular, (D_N) completes our verification of (U4) with $i = j$.

A(ii).d Verification of (U5)

This relation is verified completely analogously to (U4).

A(ii).e Verification of (U6)

We need to prove

$$[e_{i,r}, f_{i,N-r}] = \frac{1}{\boldsymbol{v}_i - \boldsymbol{v}_i^{-1}} \cdot \begin{cases} \psi^+_{i,N} - \delta_{N,0}\delta_{b_i,0}\psi^-_{i,b_i} & \text{if } N \geq 0, \\ -\psi^-_{i,N} + \delta_{N,0}\delta_{b_i,0}\psi^+_{i,0} & \text{if } N \leq b_i, \\ 0 & \text{if } b_i < N < 0. \end{cases}$$

Note that given any value of $N \in \mathbb{Z}$, we know this equality for a certain value of $r \in \mathbb{Z}$.

Case $N > 0$ If $0 \leq r \leq N$, then $[e_{i,r}, f_{i,N-r}] = \psi^+_{i,N}/(\boldsymbol{v}_i - \boldsymbol{v}_i^{-1})$, due to (C_N). For $r < 0$, we proceed by induction in $|r|$. Due to (v1, v2), we have $[e_{i,r}, f_{i,N-r}] = [2]_{\boldsymbol{v}_i}^{-1} \cdot [[h_{i,-1}, e_{i,r+1}], f_{i,N-r}] = [2]_{\boldsymbol{v}_i}^{-1} \cdot [h_{i,-1}, [e_{i,r+1}, f_{i,N-r}]] + [e_{i,r+1}, f_{i,N-r-1}] = \psi^+_{i,N}$, where in the last equality we used the induction assumption and the equality $[h_{i,-1}, \psi^+_{i,N+1}] = 0$, due to (B_N). The case $l := N - r < 0$ is treated in the same way.

Case $N \leq 0$ We proceed by induction in $|N|$. For any $r \in \mathbb{Z}$, we have

$$[e_{i,r}, f_{i,N-r}] = [2]_{v_i}^{-1} \cdot [h_{i,-1}, [e_{i,r+1}, f_{i,N-r}]] + [e_{i,r+1}, f_{i,N-r-1}] = [e_{i,r+1}, f_{i,N-r-1}],$$

where we used the induction assumption together with (Û1, v1, v2) and $[h_{i,-1}, \psi^-_i(z)] = 0$ (the latter is proved completely analogously to (A_N)). Hence, the expression $[e_{i,r}, f_{i,N-r}]$ is independent of $r \in \mathbb{Z}$. The result follows since we know the equality holds for a certain value of r.

A(ii).f Verification of (U1)

We consider the case $\epsilon = +$ (the case $\epsilon = -$ is completely analogous). We need to prove $[\psi^+_{i,r}, \psi^+_{i,s_i^+}] = [\psi^+_{i,r}, \psi^-_{i,-s_i^-}] = 0$ for any $r, s_i^+ \geq 0, s_i^- \geq -b_i$. This is clear for $r = 0$ or $s_i^+ = 0$, or $s_i^- = -b_i$, due to (v3). Therefore, it remains to prove $[h_{i,r}, \psi^+_{i,s_i^+}] = 0$ and $[h_{i,r}, \psi^-_{i,-s_i^-}] = 0$ for $r > 0, s_i^+ > 0, s_i^- > -b_i$.

For $s_i^+ > 0$, we have $\psi^+_{i,s_i^+} = (\boldsymbol{v}_i - \boldsymbol{v}_i^{-1})[e_{i,s_i^+-1}, f_{i,1}]$, so that

$$[h_{i,r}, \psi^+_{i,s_i^+}] = \frac{[2r]_{\boldsymbol{v}_i}}{r}(\boldsymbol{v}_i - \boldsymbol{v}_i^{-1}) \cdot ([e_{i,s_i^+ + r-1}, f_{i,1}] - [e_{i,s_i^+-1}, f_{i,r+1}]) = 0,$$

where the first equality is due to (v13), while the second equality is due to relation (U6) with $i = j$ proved above.

For $s_i^- > -b_i$, we have $\psi^-_{i,-s_i^-} = (\boldsymbol{v}_i^{-1} - \boldsymbol{v}_i)[e_{i,-b_{1,i}-s_i^-}, f_{i,b_{1,i}}]$, so that

$$[h_{i,r}, \psi^-_{i,-s_i^-}] = \frac{[2r]_{\boldsymbol{v}_i}}{r}(\boldsymbol{v}_i^{-1} - \boldsymbol{v}_i) \cdot ([e_{i,r-b_{1,i}-s_i^-}, f_{i,b_{1,i}}] - [e_{i,-b_{1,i}-s_i^-}, f_{i,r+b_{1,i}}]) = 0,$$

where the first equality is due to (v13), while the second equality is due to relation (U6) with $i = j$ proved above.

This completes our verification of relations (U1–U6) with $i = j$ for $\hat{\mathcal{U}}_{\mu_1,\mu_2}$.

A(iii) Verification of Relations (U1–U8) with i ≠ j for $\hat{\mathcal{U}}_{\mu_1,\mu_2}$

A(iii).a Verification of (U2)

We need to prove $X^+(i, j; r, s) = 0$ for any $r, s \in \mathbb{Z}$, where

$$X^+(i, j; r, s) := [e_{i,r+1}, e_{j,s}]_{\boldsymbol{v}_i^{c_{ij}}} + [e_{j,s+1}, e_{i,r}]_{\boldsymbol{v}_i^{c_{ij}}}.$$

First, the equality $X^+(i, j, -1, -1) = 0$ follows from ($\hat{\text{U}}$2). Second, due to (v1) we have

$$[h_{i,1}, X^+(i, j; r, s)] = [c_{ii}]_{\boldsymbol{v}_i} \cdot X^+(i, j; r+1, s) + [c_{ij}]_{\boldsymbol{v}_i} \cdot X^+(i, j; r, s+1),$$

$$[h_{j,1}, X^+(i, j; r, s)] = [c_{ji}]_{\boldsymbol{v}_j} \cdot X^+(i, j; r+1, s) + [c_{jj}]_{\boldsymbol{v}_j} \cdot X^+(i, j; r, s+1).$$

Combining these equalities with nondegeneracy of the matrix $A_{ij} := \begin{bmatrix} [c_{ii}]_{\boldsymbol{v}_i} & [c_{ij}]_{\boldsymbol{v}_i} \\ [c_{ji}]_{\boldsymbol{v}_j} & [c_{jj}]_{\boldsymbol{v}_j} \end{bmatrix}$, we see that $X^+(i, j; r, s) = 0 \Rightarrow X^+(i, j; r+1, s) = 0, X^+(i, j; r, s+1) = 0$. Since $X^+(i, j; -1, -1) = 0$, we get $X^+(i, j; r, s) = 0$ for $r, s \geq -1$ by induction in r, s.

A similar reasoning with $h_{i,-1}, h_{j,-1}$ used instead of $h_{i,1}, h_{j,1}$ yields the implication

$$X^+(i, j; r, s) = 0 \Longrightarrow X^+(i, j; r-1, s) = 0, \ X^+(i, j; r, s-1) = 0.$$

Hence, an induction argument completes the proof of $X^+(i, j; r, s) = 0$ for any $r, s \in \mathbb{Z}$.

A(iii).b Verification of (U3)

We need to prove $X^-(i, j; r, s) = 0$ for any $r, s \in \mathbb{Z}$, where

$$X^-(i, j; r, s) := [f_{i,r+1}, f_{j,s}]_{\boldsymbol{v}_i^{-c_{ij}}} + [f_{j,s+1}, f_{i,r}]_{\boldsymbol{v}_i^{-c_{ij}}}.$$

The $r = s = 0$ case follows from ($\hat{\text{U}}$3). The general case follows from

$$X^-(i, j; r, s) = 0 \Longrightarrow X^-(i, j; r \pm 1, s) = 0, \ X^-(i, j; r, s \pm 1) = 0$$

applied iteratively to $X^-(i, j; 0, 0) = 0$, in the same vein as in the above verification of (U2).

A(iii).c Verification of (U6)

We need to prove $X(i, j; r, s) = 0$ for any $r, s \in \mathbb{Z}$, where

$$X(i, j; r, s) := [e_{i,r}, f_{j,s}].$$

First, the equality $X(i, j; 0, 0) = 0$ follows from ($\hat{\text{U}}$6). Second, due to (v1, v2) we have

$$[h_{i,\pm 1}, X(i, j; r, s)] = [c_{ii}]_{\boldsymbol{v}_i} \cdot X(i, j; r \pm 1, s) - [c_{ij}]_{\boldsymbol{v}_i} \cdot X(i, j; r, s \pm 1),$$

$$[h_{j,\pm 1}, X(i, j; r, s)] = [c_{ji}]_{\boldsymbol{v}_j} \cdot X(i, j; r \pm 1, s) - [c_{jj}]_{\boldsymbol{v}_j} \cdot X(i, j; r, s \pm 1).$$

Combining these equalities with nondegeneracy of the matrix $B_{ij} := \begin{bmatrix} [c_{ii}]_{v_i} & -[c_{ij}]_{v_i} \\ [c_{ji}]_{v_j} & -[c_{jj}]_{v_j} \end{bmatrix}$, we see that $X(i,j;r,s) = 0 \Rightarrow X(i,j;r\pm 1,s) = 0, X(i,j;r,s\pm 1) = 0$. Hence, the equality $X(i,j;r,s) = 0$ for any $r,s \in \mathbb{Z}$ follows from the $r = s = 0$ case considered above.

A(iii).d Verification of (U4)

We consider the case $\epsilon = +$ (the case $\epsilon = -$ is completely analogous). We need to prove $Y^+(i,j;r,s) = 0$ for any $r \in \mathbb{N}, s \in \mathbb{Z}$, where

$$Y^+(i,j;r,s) := [\psi^+_{i,r+1}, e_{j,s}]_{\boldsymbol{v}_i^{c_{ij}}} + [e_{j,s+1}, \psi^+_{i,r}]_{\boldsymbol{v}_i^{c_{ij}}}.$$

Due to relation (U6) (established already both for $i = j$ and $i \neq j$), we have

$$(\boldsymbol{v}_i - \boldsymbol{v}_i^{-1})[[e_{i,r+1}, e_{j,s}]_{\boldsymbol{v}_i^{c_{ij}}}, f_{i,0}] = [\psi^+_{i,r+1}, e_{j,s}]_{\boldsymbol{v}_i^{c_{ij}}},$$

$$(\boldsymbol{v}_i - \boldsymbol{v}_i^{-1})[[e_{j,s+1}, e_{i,r}]_{\boldsymbol{v}_i^{c_{ij}}}, f_{i,0}] = [e_{j,s+1}, \psi^+_{i,r} - \delta_{r,0}\delta_{b_i,0}\psi^-_{i,-b_i}]_{\boldsymbol{v}_i^{c_{ij}}} = [e_{j,s+1}, \psi^+_{i,r}]_{\boldsymbol{v}_i^{c_{ij}}}.$$

Therefore, $Y^+(i,j;r,s) = (\boldsymbol{v}_i - \boldsymbol{v}_i^{-1})[X^+(i,j;r,s), f_{i,0}] = 0$, where the last equality follows from $X^+(i,j;r,s) = 0$ proved above.

A(iii).e Verification of (U5)

We consider the case $\epsilon = +$ (the case $\epsilon = -$ is completely analogous). We need to prove $Y^-(i,j;r,s) = 0$ for any $r \in \mathbb{N}, s \in \mathbb{Z}$, where

$$Y^-(i,j;r,s) := [\psi^+_{i,r+1}, f_{j,s}]_{\boldsymbol{v}_i^{-c_{ij}}} + [f_{j,s+1}, \psi^+_{i,r}]_{\boldsymbol{v}_i^{-c_{ij}}}.$$

Analogously to our verification of (U4), we have $Y^-(i,j;r,s) = (\boldsymbol{v}_i - \boldsymbol{v}_i^{-1})[e_{i,0}, X^-(i,j;r,s)]$. Thus, the equality $Y^-(i,j;r,s) = 0$ follows from $X^-(i,j;r,s) = 0$ proved above.

A(iii).f Verification of (U1)

We consider the case $\epsilon = \epsilon' = +$ (other cases are completely analogous). Due to relation (v3), it suffices to prove $[h_{i,r}, \psi^+_{j,s}] = 0$ for $r,s \in \mathbb{Z}_{>0}$, where the elements $\{h_{i,r}\}_{r=1}^{\infty}$ were defined in (A.2).

Analogously to (v13), relations (U4, U5) imply

$$[h_{i,r}, e_{j,s}] = \frac{[rc_{ij}]_{\boldsymbol{v}_i}}{r} \cdot e_{j,s+r},\ [h_{i,r}, f_{j,s}] = -\frac{[rc_{ij}]_{\boldsymbol{v}_i}}{r} \cdot f_{j,s+r} \text{ for any } r \in \mathbb{Z}_{>0}, s \in \mathbb{Z}.$$

Hence, we have

$$[h_{i,r}, \psi^+_{j,s}] = (\boldsymbol{v}_i - \boldsymbol{v}_i^{-1})[h_{i,r}, [e_{j,s}, f_{j,0}]] = (\boldsymbol{v}_i - \boldsymbol{v}_i^{-1})\frac{[rc_{ij}]_{\boldsymbol{v}_i}}{r} \cdot ([e_{j,s+r}, f_{j,0}] - [e_{j,s}, f_{j,r}]) = 0,$$

where the first and the last equalities follow from (U6) with $i = j$ established above.

A(iii).g Verification of (U7)

In the simplest case $c_{ij} = 0$, we need to prove $[e_{i,r}, e_{j,s}] = 0$ for any $r, s \in \mathbb{Z}$. The equality $[e_{i,0}, e_{j,0}] = 0$ is due to ($\hat{\mathrm{U}}7$), while commuting it iteratively with $h_{i,\pm 1}, h_{j,\pm 1}$, we get $[e_{i,r}, e_{j,s}] = 0$, due to (v1, v2).

In general, we set $m := 1 - c_{ij}$. For any $\vec{r} = (r_1, \ldots, r_m) \in \mathbb{Z}^m$ and $s \in \mathbb{Z}$, define

$$Z^+(i, j; \vec{r}, s) := \sum_{\pi \in \mathfrak{S}_m} \sum_{t=0}^{m} (-1)^t \begin{bmatrix} m \\ t \end{bmatrix}_{\boldsymbol{v}_i} e_{i,r_{\pi(1)}} \cdots e_{i,r_{\pi(t)}} e_{j,s} e_{i,r_{\pi(t+1)}} \cdots e_{i,r_{\pi(m)}}.$$

To check (U7), we need to prove $Z^+(i, j; \vec{r}, s) = 0$ for any $\vec{r} \in \mathbb{Z}^m, s \in \mathbb{Z}$.

Let $\vec{0} = (0, \ldots, 0) \in \mathbb{Z}^m$. The equality $Z^+(i, j; \vec{0}, 0) = 0$ follows from ($\hat{\mathrm{U}}7$) (cf. Remark 5.4). Commuting $Z^+(i, j; \vec{0}, s)$ with $h_{i,\pm 1}, h_{j,\pm 1}$, and using nondegeneracy of the matrix A_{ij}, we get $Z^+(i, j; \vec{0}, s) = 0 \Rightarrow Z^+(i, j; \vec{0}, s \pm 1) = 0$. Therefore, $Z^+(i, j; \vec{0}, s) = 0$ for any $s \in \mathbb{Z}$.

Next, we prove that $Z^+(i, j; \vec{r}, s) = 0$ for any $\vec{r} = (r_1, \ldots, r_k, 0, \ldots, 0) \in \mathbb{Z}^m$, $s \in \mathbb{Z}$ by induction in $0 \leq k \leq m$. The base case $k = 0$ was just treated above. For the induction step, note that the commutator $[h_{i,r'}, Z^+(i, j; \vec{r}, s)]$ equals $\frac{(m-k)\cdot[2r']_{\boldsymbol{v}_i}}{r'} Z^+(i, j; (r_1, \ldots, r_k, r', 0, \ldots, 0), s)$ plus some other terms which are zero by the induction assumption. Hence, $Z^+(i, j; \vec{r}, s) = 0$ for any $\vec{r} \in \mathbb{Z}^m, s \in \mathbb{Z}$.

A(iii).h Verification of (U8)

Set $m := 1 - c_{ij}$. For any $\vec{r} \in \mathbb{Z}^m$, $s \in \mathbb{Z}$, define

$$Z^-(i, j; \vec{r}, s) := \sum_{\pi \in \mathfrak{S}_m} \sum_{t=0}^{m} (-1)^t \begin{bmatrix} m \\ t \end{bmatrix}_{\boldsymbol{v}_i} f_{i,r_{\pi(1)}} \cdots f_{i,r_{\pi(t)}} f_{j,s} f_{i,r_{\pi(t+1)}} \cdots f_{i,r_{\pi(m)}}.$$

Then, we need to show $Z^-(i, j; \vec{r}, s) = 0$. This is proved completely analogously to (U7).

This completes our proof of Theorem 5.5.

Remark A.2

(a) Specializing $\boldsymbol{v} \mapsto v \in \mathbb{C}^\times$ from the beginning and viewing all algebras as $\mathbb{C}$-algebras, the statement of Theorem 5.5 still holds as long as v is not a root of unity.
(b) A slightly different proof can be obtained by following the arguments in [47].
(c) We note that both Theorem 5.5 and its proof are valid also for all affine Lie algebras, except for the type $A_1^{(1)}$.

A(iv) An Alternative Presentation of $\mathcal{U}^{\mathrm{sc}}_{\mathbf{0},\mu}$ *for* $\mu \in \Lambda^-$

Inspired by the recent result [33, Theorem 2.13], we provide another realization of $\mathcal{U}^{\mathrm{sc}}_{0,\mu}$ (with $\mu \in \Lambda^-$) without the defining relation ($\hat{\mathrm{U}}$9). Following the notations of Sect. 5.2, denote by $\widetilde{\mathcal{U}}_{\mu_1,\mu_2}$ the associative $\mathbb{C}(\boldsymbol{v})$-algebra generated by

$$\{e_{i,r}, f_{i,s}, (\psi^+_{i,0})^{\pm 1}, (\psi^-_{i,b_i})^{\pm 1}, h_{i,\pm 1} | i \in I, b_{2,i} - 1 \leq r \leq 1, b_{1,i} - 1 \leq s \leq 1\}$$

with the defining relations ($\hat{\mathrm{U}}$1–$\hat{\mathrm{U}}$8). Define inductively $e_{i,r}, f_{i,r}, \psi^\pm_{i,r}$ as it was done for $\hat{\mathcal{U}}_{\mu_1,\mu_2}$ right before Theorem 5.5.

Theorem A.3 *There is a unique* $\mathbb{C}(\boldsymbol{v})$*-algebra isomorphism* $\widetilde{\mathcal{U}}_{\mu_1,\mu_2} \xrightarrow{\sim} \mathcal{U}^{\mathrm{sc}}_{0,\mu}$*, such that*

$$e_{i,r} \mapsto e_{i,r},\ f_{i,r} \mapsto f_{i,r},\ \psi^\pm_{i,\pm s_i^\pm} \mapsto \psi^\pm_{i,\pm s_i^\pm} \quad \text{for } i \in I, r \in \mathbb{Z}, s_i^+ \geq 0, s_i^- \geq -b_i.$$

Proof Due to Theorem 5.5, it suffices to show that ($\hat{\mathrm{U}}$9) can be derived from ($\hat{\mathrm{U}}$1–$\hat{\mathrm{U}}$8). We will treat only the first relation of ($\hat{\mathrm{U}}$9) (the second is completely analogous).

First, we note that relations (v1–v5) and (U2, U3, U6) with $i \neq j$ hold in $\widetilde{\mathcal{U}}_{\mu_1,\mu_2}$, since their proofs for the algebra $\hat{\mathcal{U}}_{\mu_1,\mu_2}$ were solely based on relations ($\hat{\mathrm{U}}$1–$\hat{\mathrm{U}}$6). Likewise, the equalities $Y^\pm(i, j; r, s) = 0$ from our verifications of (U4, U5) for $i \neq j$ still hold for $r \in \{0, 1\}, s \in \mathbb{Z}$.

Second, we have

$$[\psi^+_{i,2}, e_{i,0}]_{\boldsymbol{v}_i^2} + [e_{i,1}, \psi^+_{i,1}]_{\boldsymbol{v}_i^2} = 0,\ [\psi^+_{i,2}, f_{i,0}]_{\boldsymbol{v}_i^{-2}} + [f_{i,1}, \psi^+_{i,1}]_{\boldsymbol{v}_i^{-2}} = 0. \tag{v18}$$

These equalities are proved completely analogously to (v11) from our proof of Lemma A.1, but now we start from the equality $[e_{i,1}, e_{i,0}]_{\boldsymbol{v}_i^2} = 0$ rather than $[e_{i,0}, e_{i,-1}]_{\boldsymbol{v}_i^2} = 0$ (commuting it first with $h_{i,1}$ and then further with $f_{i,0}$).

Recall $h_{i,2}$ of (A.1). Analogously to Lemma A.1, we see that (v18) implies[11]

$$[h_{i,2}, e_{i,0}] = \frac{[4]_{\boldsymbol{v}_i}}{2} \cdot e_{i,2},\ [h_{i,2}, f_{i,0}] = -\frac{[4]_{\boldsymbol{v}_i}}{2} \cdot f_{i,2}. \tag{v19}$$

Likewise, the aforementioned equalities $Y^{\pm}(i, j; 1, s) = 0$ for $i \neq j, s \in \mathbb{Z}$, also imply

$$[h_{i,2}, e_{j,s}] = \frac{[2c_{ij}]_{\boldsymbol{v}_i}}{2} \cdot e_{j,s+2},\ [h_{i,2}, f_{j,s}] = -\frac{[2c_{ij}]_{\boldsymbol{v}_i}}{2} \cdot f_{j,s+2} \text{ for } i \neq j, s \in \mathbb{Z}. \tag{v20}$$

Finally, due to ($\hat{\mathrm{U}}$7, $\hat{\mathrm{U}}$8, v1, v2, v19, v20), we also get $[e_{i,r}, e_{j,s}] = [f_{i,r}, f_{j,s}] = 0$ if $c_{ij} = 0$ and $Z^{\pm}(i, j; r', 0, s) = Z^{\pm}(i, j; 1, 1, s) = 0$ if $c_{ij} = -1$ for $r, s \in \mathbb{Z}, r' \in \{0, 1, 2\}$.

In the simply-laced case, the rest of the proof follows from the next result.

Lemma A.4 *Let $i, j \in I$ be such that $c_{ij} = -1$. Then $[\psi^+_{i,1}, \psi^+_{i,2}] = 0$.*

Proof As just proved, we have $[f_{i,1}, [f_{i,1}, f_{j,0}]_{\boldsymbol{v}_i^{-1}}]_{\boldsymbol{v}_i} = 0$. Commuting this equality with $e_{j,1}$ and applying (v4) together with (U6) for $i \neq j$, we get $[f_{i,1}, [f_{i,1}, \psi^+_{j,1}]_{\boldsymbol{v}_i^{-1}}]_{\boldsymbol{v}_i} = 0$. Combining the latter equality with $\psi^+_{j,1} = (\boldsymbol{v}_j - \boldsymbol{v}_j^{-1})\psi^+_{j,0}h_{j,1} = (\boldsymbol{v}_i - \boldsymbol{v}_i^{-1})\psi^+_{j,0}h_{j,1}$ and using (v2), we find

$$[f_{i,1}, [f_{i,1}, h_{j,1}]]_{\boldsymbol{v}_i^2} = 0 \Longrightarrow [f_{i,1}, f_{i,2}]_{\boldsymbol{v}_i^2} = 0 \Longrightarrow [f_{i,2}, f_{i,1}]_{\boldsymbol{v}_i^{-2}} = 0.$$

Commuting this further with $e_{i,0}$, we obtain

$$[\psi^+_{i,2}, f_{i,1}]_{\boldsymbol{v}_i^{-2}} + [f_{i,2}, \psi^+_{i,1}]_{\boldsymbol{v}_i^{-2}} = 0.$$

Finally, we apply $[e_{i,0}, -]_{\boldsymbol{v}_i^{-2}}$ to the latter equality. In the left-hand side we get two summands computed below.

(1) We have $[e_{i,0}, [f_{i,2}, \psi^+_{i,1}]_{\boldsymbol{v}_i^{-2}}]_{\boldsymbol{v}_i^{-2}} = [[e_{i,0}, f_{i,2}], \psi^+_{i,1}]_{\boldsymbol{v}_i^{-4}} + [f_{i,2}, [e_{i,0}, \psi^+_{i,1}]_{\boldsymbol{v}_i^{-2}}]_{\boldsymbol{v}_i^{-2}}$. Due to ($\hat{\mathrm{U}}$4), $[e_{i,0}, \psi^+_{i,1}]_{\boldsymbol{v}_i^{-2}} = (\boldsymbol{v}_i^{-2} - \boldsymbol{v}_i^2)e_{i,1}\psi^+_{i,0} \Rightarrow [f_{i,2}, [e_{i,0}, \psi^+_{i,1}]_{\boldsymbol{v}_i^{-2}}]_{\boldsymbol{v}_i^{-2}} = (\boldsymbol{v}_i^{-2} - \boldsymbol{v}_i^2)[f_{i,2}, e_{i,1}]_{\boldsymbol{v}_i^{-4}}\psi^+_{i,0}$. Combining this with (v5), we thus get

$$[e_{i,0}, [f_{i,2}, \psi^+_{i,1}]_{\boldsymbol{v}_i^{-2}}]_{\boldsymbol{v}_i^{-2}} = [\psi^+_{i,2}, \psi^+_{i,1}]_{\boldsymbol{v}_i^{-4}}/(\boldsymbol{v}_i - \boldsymbol{v}_i^{-1}) + (\boldsymbol{v}_i^{-2} - \boldsymbol{v}_i^2)[f_{i,2}, e_{i,1}]_{\boldsymbol{v}_i^{-4}}\psi^+_{i,0}. \tag{v21}$$

[11] Note that we cannot deduce the statement of Lemma A.1 due to the absence of ($\hat{\mathrm{U}}$9).

(2) We have $[e_{i,0},[\psi^+_{i,2},f_{i,1}]_{\boldsymbol{v}_i^{-2}}]_{\boldsymbol{v}_i^{-2}} = [[e_{i,0},\psi^+_{i,2}]_{\boldsymbol{v}_i^{-2}},f_{i,1}]_{\boldsymbol{v}_i^{-2}} + \boldsymbol{v}_i^{-2}[\psi^+_{i,2},[e_{i,0},f_{i,1}]]$. By (v18): $[e_{i,0},\psi^+_{i,2}]_{\boldsymbol{v}_i^{-2}} = -\boldsymbol{v}_i^{-2}[\psi^+_{i,2},e_{i,0}]_{\boldsymbol{v}_i^2} = \boldsymbol{v}_i^{-2}[e_{i,1},\psi^+_{i,1}]_{\boldsymbol{v}_i^2} = \boldsymbol{v}_i^{-2}(\boldsymbol{v}_i-\boldsymbol{v}_i^{-1})[e_{i,1},h_{i,1}]_{\boldsymbol{v}_i^4}\psi^+_{i,0}$. Hence,

$$[[e_{i,0},\psi^+_{i,2}]_{\boldsymbol{v}_i^{-2}},f_{i,1}]_{\boldsymbol{v}_i^{-2}} = \boldsymbol{v}_i^{-4}(\boldsymbol{v}_i-\boldsymbol{v}_i^{-1})[e_{i,1}h_{i,1}-\boldsymbol{v}_i^4 h_{i,1}e_{i,1},f_{i,1}]\psi^+_{i,0} =$$

$$\boldsymbol{v}_i^{-4}(\boldsymbol{v}_i-\boldsymbol{v}_i^{-1})([\psi^+_{i,2},h_{i,1}]_{\boldsymbol{v}_i^4}/(\boldsymbol{v}_i-\boldsymbol{v}_i^{-1})-(\boldsymbol{v}_i+\boldsymbol{v}_i^{-1})[e_{i,1},f_{i,2}]_{\boldsymbol{v}_i^4})\psi^+_{i,0}.$$

Therefore,

$$[e_{i,0},[\psi^+_{i,2},f_{i,1}]_{\boldsymbol{v}_i^{-2}}]_{\boldsymbol{v}_i^{-2}} = \frac{[\psi^+_{i,2},\psi^+_{i,1}]}{\boldsymbol{v}_i^2(\boldsymbol{v}_i-\boldsymbol{v}_i^{-1})} + (\boldsymbol{v}_i^2-\boldsymbol{v}_i^{-2})[f_{i,2},e_{i,1}]_{\boldsymbol{v}_i^{-4}}\psi^+_{i,0} + \frac{[\psi^+_{i,2},\psi^+_{i,1}]_{\boldsymbol{v}_i^4}}{\boldsymbol{v}_i^4(\boldsymbol{v}_i-\boldsymbol{v}_i^{-1})}. \tag{v22}$$

Substituting (v22) and (v21) into $[e_{i,0},[\psi^+_{i,2},f_{i,1}]_{\boldsymbol{v}_i^{-2}} + [f_{i,2},\psi^+_{i,1}]_{\boldsymbol{v}_i^{-2}}]_{\boldsymbol{v}_i^{-2}} = 0$, we find

$$[\psi^+_{i,2},\psi^+_{i,1}]_{\boldsymbol{v}_i^{-4}} + \boldsymbol{v}_i^{-2}[\psi^+_{i,2},\psi^+_{i,1}] + \boldsymbol{v}_i^{-4}[\psi^+_{i,2},\psi^+_{i,1}]_{\boldsymbol{v}_i^4} = 0.$$

The left-hand side of this equality equals $\frac{1-\boldsymbol{v}_i^{-6}}{1-\boldsymbol{v}_i^{-2}} \cdot [\psi^+_{i,2},\psi^+_{i,1}]$. Hence, $[\psi^+_{i,1},\psi^+_{i,2}]=0$.

□

Our next result completes the proof for non-simply-laced $\mathfrak{g}$.

Lemma A.5 *If* $c_{ij}\neq 0$ *and* $[\psi^+_{i,1},\psi^+_{i,2}]=0$, *then* $[\psi^+_{j,1},\psi^+_{j,2}]=0$.

Proof Due to (v1, v2): $[h_{i,1},e_{i,r}] = \frac{[2]_{\boldsymbol{v}_i}}{[c_{ji}]_{\boldsymbol{v}_j}}\cdot[h_{j,1},e_{i,r}]$, $[h_{i,1},f_{i,r}] = \frac{[2]_{\boldsymbol{v}_i}}{[c_{ji}]_{\boldsymbol{v}_j}}\cdot[h_{j,1},f_{i,r}]$. Hence $[h_{i,1},\psi^+_{i,2}] = (\boldsymbol{v}_i-\boldsymbol{v}_i^{-1})([[h_{i,1},e_{i,1}],f_{i,1}]+[e_{i,1},[h_{i,1},f_{i,1}]]) = [2]_{\boldsymbol{v}_i}/[c_{ji}]_{\boldsymbol{v}_j}\cdot[h_{j,1},\psi^+_{i,2}]$. Therefore, $[\psi^+_{i,1},\psi^+_{i,2}]=0 \Rightarrow [h_{j,1},\psi^+_{i,2}]=0 \Rightarrow [h_{j,1},h_{i,2}]=0$ with the second implication due to (Û1). Commuting the latter equality with $f_{j,0}$, we get

$$0=[f_{j,0},[h_{j,1},h_{i,2}]] = [c_{jj}]_{\boldsymbol{v}_j}\cdot[f_{j,1},h_{i,2}] + \frac{[2c_{ij}]_{\boldsymbol{v}_i}}{2}\cdot[h_{j,1},f_{j,2}].$$

Commuting this further with $e_{j,0}$, we obtain

$$[c_{jj}]_{\boldsymbol{v}_j}\cdot[e_{j,0},[f_{j,1},h_{i,2}]] + \frac{[2c_{ij}]_{\boldsymbol{v}_i}}{2}\cdot[e_{j,0},[h_{j,1},f_{j,2}]] = 0. \tag{v23}$$

Note that

$$[e_{j,0},[f_{j,1},h_{i,2}]]=[\psi^+_{j,1},h_{i,2}]/(\boldsymbol{v}_j-\boldsymbol{v}_j^{-1})-\frac{[2c_{ij}]_{\boldsymbol{v}_i}}{2}\cdot[f_{j,1},e_{j,2}]=-\frac{[2c_{ij}]_{\boldsymbol{v}_i}}{2}\cdot[f_{j,1},e_{j,2}],$$

$$[e_{j,0},[h_{j,1},f_{j,2}]]=-[c_{jj}]_{\boldsymbol{v}_j}\cdot[e_{j,1},f_{j,2}]+[h_{j,1},\psi^+_{j,2}]/(\boldsymbol{v}_j-\boldsymbol{v}_j^{-1}),$$

$$[e_{j,2},f_{j,1}]-[e_{j,1},f_{j,2}]=[c_{jj}]^{-1}_{\boldsymbol{v}_j}\cdot[h_{j,1},[e_{j,1},f_{j,1}]]=[c_{jj}]^{-1}_{\boldsymbol{v}_j}\cdot[h_{j,1},\psi^+_{j,2}]/(\boldsymbol{v}_j-\boldsymbol{v}_j^{-1}).$$

Substituting the last three equalities into (v23), we get $\frac{[2c_{ij}]_{\boldsymbol{v}_i}}{\boldsymbol{v}_j-\boldsymbol{v}_j^{-1}}\cdot[h_{j,1},\psi^+_{j,2}]=0$.

Thus, $[h_{j,1},\psi^+_{j,2}]=0\Rightarrow[\psi^+_{j,1},\psi^+_{j,2}]=0$. □

This completes our proof of Theorem A.3. □

Appendix B Proof of Theorem 6.6

The proof of part (a) proceeds in two steps. First, we consider the simplest case $\mathfrak{g}=\mathfrak{sl}_2$. Then, we show how a general case can be easily reduced to the case of $\mathfrak{sl}_2$.

B(i) Proof of Theorem 6.6(a) for $\mathfrak{g}=\mathfrak{sl}_2$

First, let us derive an explicit formula for $A^\pm(z)$. Recall the elements $\{h_{\pm r}\}_{r=1}^\infty$ of Sect. 5, such that $z^{\mp b^\pm}(\psi^\pm_{\mp b^\pm})^{-1}\psi^\pm(z)=\exp\left(\pm(\boldsymbol{v}-\boldsymbol{v}^{-1})\sum_{r>0}h_{\pm r}z^{\mp r}\right)$. For $r\neq 0$, define $t_r:=-h_r/(1+\boldsymbol{v}^{2r})$, and set

$$A^\pm(z):=(\phi^\pm)^{-1}\cdot\exp\left(\pm(\boldsymbol{v}-\boldsymbol{v}^{-1})\sum_{r>0}t_{\pm r}z^{\mp r}\right). \tag{B.1}$$

Then, $z^{\mp b^\pm}\psi^\pm(z)=\frac{1}{A^\pm(z)A^\pm(\boldsymbol{v}^{-2}z)}$ and $A^\pm(z)$ is the unique solution with $A^\pm_0:=(\phi^\pm)^{-1}$.

Relations (6.6) and (6.7) follow immediately from (U10) and (U1), respectively, while the verification of (6.9–6.16) is based on the following result.

Lemma B.1 *For any* $\epsilon,\epsilon'\in\{\pm\}$, *we have:*

(a1) $(\boldsymbol{v}z-\boldsymbol{v}^{-1}w)A^\epsilon(z)e(w)=(z-w)e(w)A^\epsilon(z)$.
(a2) $(\boldsymbol{v}z-\boldsymbol{v}^{-1}w)A^\epsilon(z)e^{\epsilon'}(w)-(z-w)e^{\epsilon'}(w)A^\epsilon(z)=(\boldsymbol{v}-\boldsymbol{v}^{-1})wA^\epsilon(z)e^\epsilon(z)$.
(a3) $(\boldsymbol{v}z-\boldsymbol{v}^{-1}w)A^\epsilon(z)e^{\epsilon'}(w)-(z-w)e^{\epsilon'}(w)A^\epsilon(z)=(1-\boldsymbol{v}^{-2})we^\epsilon(\boldsymbol{v}^2z)A^\epsilon(z)$.
(b1) $(z-w)A^\epsilon(z)f(w)=(\boldsymbol{v}z-\boldsymbol{v}^{-1}w)f(w)A^\epsilon(z)$.
(b2) $(z-w)A^\epsilon(z)f^{\epsilon'}(w)-(\boldsymbol{v}z-\boldsymbol{v}^{-1}w)f^{\epsilon'}(w)A^\epsilon(z)=(\boldsymbol{v}^{-1}-\boldsymbol{v})zf^\epsilon(z)A^\epsilon(z)$.
(b3) $(z-w)A^\epsilon(z)f^{\epsilon'}(w)-(\boldsymbol{v}z-\boldsymbol{v}^{-1}w)f^{\epsilon'}(w)A^\epsilon(z)=(1-\boldsymbol{v}^2)zA^\epsilon(z)f^\epsilon(\boldsymbol{v}^2z)$.

(c) $(z-w)[e^{\epsilon}(z), f^{\epsilon'}(w)] = z(\psi^{\epsilon'}(w) - \psi^{\epsilon}(z))/(\boldsymbol{v}-\boldsymbol{v}^{-1})$.

(d1) $(z - \boldsymbol{v}^2 w)e^{\epsilon}(z)e^{\epsilon'}(w) - (\boldsymbol{v}^2 z - w)e^{\epsilon'}(w)e^{\epsilon}(z) = z[e_0, e^{\epsilon'}(w)]_{\boldsymbol{v}^2} + w[e_0, e^{\epsilon}(z)]_{\boldsymbol{v}^2}$.

(d2) $(z-\boldsymbol{v}^2 w)e^{\epsilon}(z)e^{\epsilon'}(w) - (\boldsymbol{v}^2 z - w)e^{\epsilon'}(w)e^{\epsilon}(z) = (1-\boldsymbol{v}^2)(we^{\epsilon}(z)^2 + ze^{\epsilon'}(w)^2)$.

(e1) $(\boldsymbol{v}^2 z - w)f^{\epsilon}(z)f^{\epsilon'}(w) - (z - \boldsymbol{v}^2 w)f^{\epsilon'}(w)f^{\epsilon}(z) = \boldsymbol{v}^2[f_1, f^{\epsilon'}(w)]_{\boldsymbol{v}^{-2}} + \boldsymbol{v}^2[f_1, f^{\epsilon}(z)]_{\boldsymbol{v}^{-2}}$.

(e2) $(\boldsymbol{v}^2 z - w)f^{\epsilon}(z)f^{\epsilon'}(w) - (z - \boldsymbol{v}^2 w)f^{\epsilon'}(w)f^{\epsilon}(z) = (\boldsymbol{v}^2 - 1)(zf^{\epsilon}(z)^2 + wf^{\epsilon'}(w)^2)$.

(f1) $(z-\boldsymbol{v}^2 w)\psi^{\epsilon}(z)e^{\epsilon'}(w) - (\boldsymbol{v}^2 z - w)e^{\epsilon'}(w)\psi^{\epsilon}(z) = (\boldsymbol{v}^{-2} - \boldsymbol{v}^2)w\psi^{\epsilon}(z)e^{\epsilon}(\boldsymbol{v}^2 z)$.

(f2) $(z-\boldsymbol{v}^2 w)\psi^{\epsilon}(z)e^{\epsilon'}(w) - (\boldsymbol{v}^2 z - w)e^{\epsilon'}(w)\psi^{\epsilon}(z) = (1-\boldsymbol{v}^4)we^{\epsilon}(\boldsymbol{v}^{-2}z)\psi^{\epsilon}(z)$.

(g1) $(\boldsymbol{v}^2 z - w)\psi^{\epsilon}(z)f^{\epsilon'}(w) - (z - \boldsymbol{v}^2 w)f^{\epsilon'}(w)\psi^{\epsilon}(z) = (\boldsymbol{v}^2 - \boldsymbol{v}^{-2})z\psi^{\epsilon}(z)f^{\epsilon}(\boldsymbol{v}^{-2}z)$.

(g2) $(\boldsymbol{v}^2 z - w)\psi^{\epsilon}(z)f^{\epsilon'}(w) - (z - \boldsymbol{v}^2 w)f^{\epsilon'}(w)\psi^{\epsilon}(z) = (\boldsymbol{v}^4 - 1)zf^{\epsilon}(\boldsymbol{v}^2 z)\psi^{\epsilon}(z)$.

Proof

(a1) According to (U4′), we have $[t_r, e_s] = \frac{\boldsymbol{v}^{-2r}-1}{r(\boldsymbol{v}-\boldsymbol{v}^{-1})}e_{s+r}$ for $r \neq 0$, $s \in \mathbb{Z}$. Combining this with (B.1), we find $A^{\pm}(z)e(w) = e(w)A^{\pm}(z)\boldsymbol{v}^{\mp 1}\exp\left(\sum_{r>0}\frac{\boldsymbol{v}^{\mp 2r}-1}{r}(w/z)^{\pm r}\right)$. The latter exponent equals $\frac{z-w}{z-\boldsymbol{v}^{-2}w}$ (in the "+" case) or $\frac{z-w}{\boldsymbol{v}^2 z - w}$ (in the "−" case), hence, (a1).

(a2, a3) First, we consider the case $\epsilon = \epsilon' = +$. Due to (a1), we have $\boldsymbol{v}A^{+}_{r+1}e_s - \boldsymbol{v}^{-1}A^{+}_r e_{s+1} = e_s A^{+}_{r+1} - e_{s+1}A^{+}_r$ for any $r \in \mathbb{N}, s \in \mathbb{Z}$. Multiplying this equality by $z^{-r}w^{-s-1}$ and summing over all $r, s \in \mathbb{N}$, we find $w^{-1}((\boldsymbol{v}z - \boldsymbol{v}^{-1}w)A^{+}(z)e^{+}(w) - (z-w)e^{+}(w)A^{+}(z)) = [e_0, A^{+}(z)]_{\boldsymbol{v}^{-1}}$. Note that the right-hand side is independent of w. Substituting either $w = z$ or $w = \boldsymbol{v}^2 z$ into the left-hand side, we get the equalities (a2) and (a3) for $\epsilon = \epsilon' = +$, respectively.

Next, we consider the case $\epsilon = \epsilon' = -$. Due to (a1), we have $\boldsymbol{v}A^{-}_{-r+1}e_{-s} - \boldsymbol{v}^{-1}A^{-}_r e_{-s+1} = e_{-s}A^{-}_{-r+1} - e_{-s+1}A^{-}_{-r}$ for any $r \in \mathbb{N}, s \in \mathbb{Z}$, where we set $A^{-}_1 := 0$. Multiplying this equality by $-z^r w^{s-1}$ and summing over all $r \in \mathbb{N}, s \in \mathbb{Z}_{>0}$, we find $w^{-1}((\boldsymbol{v}z - \boldsymbol{v}^{-1}w)A^{-}(z)e^{-}(w) - (z-w)e^{-}(w)A^{-}(z)) = [e_0, A^{-}(z)]_{\boldsymbol{v}^{-1}}$. Note that the right-hand side is independent of w. Substituting either $w = z$ or $w = \boldsymbol{v}^2 z$ into the left-hand side, we get the equalities (a2) and (a3) for $\epsilon = \epsilon' = -$, respectively.

The case $\epsilon' \neq \epsilon$ follows by combining the formula $e^{\epsilon'}(w) = e^{\epsilon}(w) + \epsilon' e(w)$ with part (a1) and the cases $\epsilon = \epsilon'$ of parts (a2, a3), established above.

(b1–b3) Parts (b1, b2, b3) are proved completely analogously to (a1, a2, a3), respectively.

(c) First, we consider the case $\epsilon = \epsilon'$. According to (U6), we have $[e_r, f_s] = \frac{\psi^{+}_{r+s}}{\boldsymbol{v}-\boldsymbol{v}^{-1}}$ for $r \geq 0, s > 0$. For $N > 0$, we have $(z-w)\sum_{s=1}^{N} w^{-s}z^{s-N} = z(w^{-N} - z^{-N})$. Hence, $(z-w)[e^{+}(z), f^{+}(w)] =$

$\sum_{N>0} z(w^{-N} - z^{-N})\frac{\psi^+_N}{v-v^{-1}} = z\frac{\psi^+(w)-\psi^+(z)}{v-v^{-1}}$. Likewise, we have $[e_{-r}, f_{-s}] = -\frac{\psi^-_{-r-s}}{v-v^{-1}}$ for $r > 0, s \geq 0$. For $N > 0$, we have $(z-w)\sum_{s=1}^{N} z^s w^{N-s} = z(z^N - w^N)$. Hence, $(z-w)[e^-(z), f^-(w)] = -\sum_{N>0} z(z^N - w^N)\frac{\psi^-_{-N}}{v-v^{-1}} = z\frac{\psi^-(w)-\psi^-(z)}{v-v^{-1}}$.

Next, we consider the case $\epsilon \neq \epsilon'$. According to (U6), we have $[e(z), f(w)] = \frac{\delta(z/w)}{v-v^{-1}}(\psi^+(z) - \psi^-(z)) = \frac{\delta(z/w)}{v-v^{-1}}(\psi^+(w) - \psi^-(w))$. Taking the terms with negative powers of w, we find $[e(z), f^+(w)] = \frac{z/w}{1-z/w}\frac{\psi^+(z)-\psi^-(z)}{v-v^{-1}} \Rightarrow (z-w)[e(z), f^+(w)] = z\frac{\psi^-(z)-\psi^+(z)}{v-v^{-1}}$, while taking the terms with nonpositive powers of z, we find $[e^+(z), f(w)] = \frac{1}{1-w/z}\frac{\psi^+(w)-\psi^-(w)}{v-v^{-1}} \Rightarrow (z-w)[e^+(z), f(w)] = z\frac{\psi^+(w)-\psi^-(w)}{v-v^{-1}}$. Combining these equalities with $(z-w)[e^+(z), f^+(w)] = z\frac{\psi^+(w)-\psi^+(z)}{v-v^{-1}}$ from above and $e^-(z) = e^+(z) - e(z)$, $f^-(z) = f^+(z) - f(z)$, we obtain the $\epsilon \neq \epsilon'$ cases of part (c).

(d1) Comparing the coefficients of $z^{-\epsilon r}w^{-\epsilon' s}$ in both sides of relation (U2), we find $e_{\epsilon r+1}e_{\epsilon' s} - v^2 e_{\epsilon r}e_{\epsilon' s+1} = v^2 e_{\epsilon' s}e_{\epsilon r+1} - e_{\epsilon' s+1}e_{\epsilon r}$ for any $r, s \in \mathbb{Z}$. Multiplying this equality by $\epsilon\epsilon' \cdot z^{-\epsilon r}w^{-\epsilon' s}$ and summing over $r \geq \delta_{\epsilon,-}$, $s \geq \delta_{\epsilon',-}$, we get (d1).

(d2) Substituting $w = z$ into the $\epsilon = \epsilon'$ case of (d1), we find $[e_0, e^{\pm}(z)]_{v^2} = (1 - v^2)e^{\pm}(z)^2$. Replacing accordingly the right-hand side of (d1), we obtain (d2).

(e1, e2) Parts (e1, e2) are proved completely analogously to (d1, d2), respectively.

(f1, f2) Parts (f1, f2) are deduced from relation (U4) in the same way as we deduced parts (a2, a3) from (a1).

(g1, g2) Parts (g1, g2) are proved completely analogously to (f1, f2), respectively.

□

Now let us verify relations (6.9–6.16) using Lemma B.1. The idea is first to use parts (a3, b2) of Lemma B.1 (resp. parts (a2, b3)) to move all the series $A^\bullet(\cdot)$ to the right (resp. to the left), and then to use Lemma B.1(c–g2) to simplify the remaining part. Since $\mathfrak{g} = \mathfrak{sl}_2$ we will drop the index i from our notation.

B(i).a Verification of the First Relation in (6.9)

We need to prove $[B^\epsilon(z), B^{\epsilon'}(w)] = 0$, or equivalently, $(z-w)[B^\epsilon(z), B^{\epsilon'}(w)] = 0$. By definition, $B^\epsilon(z)B^{\epsilon'}(w) = (v - v^{-1})^2 A^\epsilon(z)e^\epsilon(z)A^{\epsilon'}(w)e^{\epsilon'}(w)$. Applying Lemma B.1(a2), we see that

$$(z-w)B^\epsilon(z)B^{\epsilon'}(w) = (v-v^{-1})^2 A^\epsilon(z)A^{\epsilon'}(w)((v^{-1}z - vw)e^\epsilon(z)e^{\epsilon'}(w) + (v-v^{-1})ze^{\epsilon'}(w)^2).$$

Hence, the equality $(z-w)[B^{\epsilon}(z), B^{\epsilon'}(w)] = 0$ boils down to the vanishing of

$$(\boldsymbol{v}^{-1}z - \boldsymbol{v}w)e^{\epsilon}(z)e^{\epsilon'}(w) + (\boldsymbol{v}-\boldsymbol{v}^{-1})ze^{\epsilon'}(w)^2 + (\boldsymbol{v}^{-1}w - \boldsymbol{v}z)e^{\epsilon'}(w)e^{\epsilon}(z) + (\boldsymbol{v}-\boldsymbol{v}^{-1})we^{\epsilon}(z)^2,$$

which is exactly the statement of Lemma B.1(d2).

B(i).b Verification of the Second Relation in (6.9)

We need to prove $[C^{\epsilon}(z), C^{\epsilon'}(w)] = 0$, or equivalently, $(z-w)[C^{\epsilon}(z), C^{\epsilon'}(w)] = 0$. By definition, $C^{\epsilon}(z)C^{\epsilon'}(w) = (\boldsymbol{v}-\boldsymbol{v}^{-1})^2 f^{\epsilon}(z)A^{\epsilon}(z)f^{\epsilon'}(w)A^{\epsilon'}(w)$. Applying Lemma B.1(b2), we see that

$$(z-w)C^{\epsilon}(z)C^{\epsilon'}(w) = (\boldsymbol{v}-\boldsymbol{v}^{-1})^2((\boldsymbol{v}z-\boldsymbol{v}^{-1}w)f^{\epsilon}(z)f^{\epsilon'}(w) + (\boldsymbol{v}^{-1}-\boldsymbol{v})zf^{\epsilon}(z)^2)A^{\epsilon}(z)A^{\epsilon'}(w).$$

Hence, the equality $(z-w)[C^{\epsilon}(z), C^{\epsilon'}(w)] = 0$ boils down to the vanishing of

$$(\boldsymbol{v}z-\boldsymbol{v}^{-1}w)f^{\epsilon}(z)f^{\epsilon'}(w) + (\boldsymbol{v}^{-1}-\boldsymbol{v})zf^{\epsilon}(z)^2 + (\boldsymbol{v}w-\boldsymbol{v}^{-1}z)f^{\epsilon'}(w)f^{\epsilon}(z) + (\boldsymbol{v}^{-1}-\boldsymbol{v})wf^{\epsilon'}(w)^2,$$

which is exactly the statement of Lemma B.1(e2).

B(i).c Verification of the Third Relation in (6.9)

The verification of the equality $[D^{\epsilon}(z), D^{\epsilon'}(w)] = 0$ is much more cumbersome and is left to the interested reader.

B(i).d Verification of (6.10)

We need to prove $(z-w)[B^{\epsilon'}(w), A^{\epsilon}(z)]_{\boldsymbol{v}^{-1}} = (\boldsymbol{v}-\boldsymbol{v}^{-1})(zA^{\epsilon}(z)B^{\epsilon'}(w) - wA^{\epsilon'}(w)B^{\epsilon}(z))$. By definition and (6.7), the RHS equals $(\boldsymbol{v}-\boldsymbol{v}^{-1})^2A^{\epsilon}(z)A^{\epsilon'}(w)(ze^{\epsilon'}(w) - we^{\epsilon}(z))$. Meanwhile, the LHS equals $(\boldsymbol{v}-\boldsymbol{v}^{-1})(z-w)(A^{\epsilon'}(w)e^{\epsilon'}(w)A^{\epsilon}(z) - \boldsymbol{v}^{-1}A^{\epsilon}(z)A^{\epsilon'}(w)e^{\epsilon'}(w))$. We use Lemma B.1(a2) to replace the first term, so that the LHS equals

$$(\boldsymbol{v}-\boldsymbol{v}^{-1})A^{\epsilon}(z)A^{\epsilon'}(w)\left((\boldsymbol{v}z-\boldsymbol{v}^{-1}w)e^{\epsilon'}(w) - (\boldsymbol{v}-\boldsymbol{v}^{-1})we^{\epsilon}(z) - \boldsymbol{v}^{-1}(z-w)e^{\epsilon'}(w)\right),$$

which exactly coincides with the above formula for the RHS.

B(i).e Verification of (6.11)

We need to prove $(z-w)[A^{\epsilon}(z),C^{\epsilon'}(w)]_{\boldsymbol{v}} = (\boldsymbol{v}-\boldsymbol{v}^{-1})(wC^{\epsilon'}(w)A^{\epsilon}(z) - zC^{\epsilon}(z)A^{\epsilon'}(w))$. By definition and (6.7), the RHS equals $(\boldsymbol{v}-\boldsymbol{v}^{-1})^2(wf^{\epsilon'}(w) - zf^{\epsilon}(z))A^{\epsilon}(z)A^{\epsilon'}(w)$. Meanwhile, the LHS equals $(\boldsymbol{v}-\boldsymbol{v}^{-1})(z-w)(A^{\epsilon}(z)f^{\epsilon'}(w)A^{\epsilon'}(w) - \boldsymbol{v}f^{\epsilon'}(w)A^{\epsilon'}(w)A^{\epsilon}(z))$. We use Lemma B.1(b2) to replace the first term, so that the LHS equals

$$(\boldsymbol{v}-\boldsymbol{v}^{-1})\left((\boldsymbol{v}z-\boldsymbol{v}^{-1}w)f^{\epsilon'}(w)+(\boldsymbol{v}^{-1}-\boldsymbol{v})zf^{\epsilon}(z)-\boldsymbol{v}(z-w)f^{\epsilon'}(w)\right)A^{\epsilon}(z)A^{\epsilon'}(w),$$

which exactly coincides with the above formula for the RHS.

B(i).f Verification of (6.12)

We need to prove $(z-w)[B^{\epsilon}(z),C^{\epsilon'}(w)] = (\boldsymbol{v}-\boldsymbol{v}^{-1})z(D^{\epsilon'}(w)A^{\epsilon}(z) - D^{\epsilon}(z)A^{\epsilon'}(w))$. Applying the equality $A^{\epsilon}(z)e^{\epsilon}(z) = \boldsymbol{v}^{-1}e^{\epsilon}(\boldsymbol{v}^2z)A^{\epsilon}(z)$, which follows from Lemma B.1(a2), we see that the LHS equals

$$\boldsymbol{v}^{-1}(\boldsymbol{v}-\boldsymbol{v}^{-1})^2(z-w)\left(e^{\epsilon}(\boldsymbol{v}^2z)A^{\epsilon}(z)f^{\epsilon'}(w)A^{\epsilon'}(w)-f^{\epsilon'}(w)A^{\epsilon'}(w)e^{\epsilon}(\boldsymbol{v}^2z)A^{\epsilon}(z)\right).$$

Applying Lemma B.1(a3, b2) to move both $A^{\epsilon}(z)$, $A^{\epsilon'}(w)$ to the right and simplifying the resulting expression, we find that the LHS equals

$$\boldsymbol{v}^{-1}(\boldsymbol{v}-\boldsymbol{v}^{-1})^2\Big((\boldsymbol{v}z-\boldsymbol{v}^{-1}w)[e^{\epsilon}(\boldsymbol{v}^2z),f^{\epsilon'}(w)]+$$
$$(\boldsymbol{v}-\boldsymbol{v}^{-1})(zf^{\epsilon'}(w)e^{\epsilon'}(\boldsymbol{v}^2w)-(\boldsymbol{v}-\boldsymbol{v}^{-1})ze^{\epsilon}(\boldsymbol{v}^2z)f^{\epsilon}(z))\Big)A^{\epsilon}(z)A^{\epsilon'}(w).$$

Meanwhile, $D^{\epsilon}(z)=\psi^{\epsilon}(z)A^{\epsilon}(z)+\boldsymbol{v}^{-1}(\boldsymbol{v}-\boldsymbol{v}^{-1})^2f^{\epsilon}(z)e^{\epsilon}(\boldsymbol{v}^2z)A^{\epsilon}(z)$, so that the RHS equals

$$(\boldsymbol{v}-\boldsymbol{v}^{-1})\left(z(\psi^{\epsilon'}(w)-\psi^{\epsilon}(z))+\boldsymbol{v}^{-1}(\boldsymbol{v}-\boldsymbol{v}^{-1})^2z(f^{\epsilon'}(w)e^{\epsilon'}(\boldsymbol{v}^2w)-f^{\epsilon}(z)e^{\epsilon}(\boldsymbol{v}^2z))\right)A^{\epsilon}(z)A^{\epsilon'}(w).$$

Thus, the equality LHS = RHS boils down to proving

$$\boldsymbol{v}^{-2}(\boldsymbol{v}^2z-w)[e^{\epsilon}(\boldsymbol{v}^2z),f^{\epsilon'}(w)]-(1-\boldsymbol{v}^{-2})z[e^{\epsilon}(\boldsymbol{v}^2z),f^{\epsilon}(z)]=\frac{z}{\boldsymbol{v}-\boldsymbol{v}^{-1}}(\psi^{\epsilon'}(w)-\psi^{\epsilon}(z)),$$

which immediately follows by applying Lemma B.1(c) to both terms on the left.

B(i).g Verification of (6.13)

We need to prove $(z-w)[B^\epsilon(z), D^{\epsilon'}(w)]_{\boldsymbol{v}} = (\boldsymbol{v}-\boldsymbol{v}^{-1})(wD^{\epsilon'}(w)B^\epsilon(z) - zD^\epsilon(z)B^{\epsilon'}(w))$. Combining the aforementioned equality $A^\epsilon(z)e^\epsilon(z) = \boldsymbol{v}^{-1}e^\epsilon(\boldsymbol{v}^2z)A^\epsilon(z)$ with Lemma B.1(a3), we find that $(w-z)\cdot$ RHS equals

$$\Big(\boldsymbol{v}^{-1}(\boldsymbol{v}-\boldsymbol{v}^{-1})^2(w(\boldsymbol{v}^{-1}w-\boldsymbol{v}z)\psi^{\epsilon'}(w)e^\epsilon(\boldsymbol{v}^2z)+z(\boldsymbol{v}^{-1}z-\boldsymbol{v}w)\psi^\epsilon(z)e^{\epsilon'}(\boldsymbol{v}^2w))+$$

$$\boldsymbol{v}^{-1}(\boldsymbol{v}-\boldsymbol{v}^{-1})^3zw(\psi^{\epsilon'}(w)e^{\epsilon'}(\boldsymbol{v}^2w)+\psi^\epsilon(z)e^\epsilon(\boldsymbol{v}^2z))+$$

$$\boldsymbol{v}^{-2}(\boldsymbol{v}-\boldsymbol{v}^{-1})^4(w(\boldsymbol{v}^{-1}w-\boldsymbol{v}z)f^{\epsilon'}(w)e^{\epsilon'}(\boldsymbol{v}^2w)e^\epsilon(\boldsymbol{v}^2z)+z(\boldsymbol{v}^{-1}z-\boldsymbol{v}w)f^\epsilon(z)e^\epsilon(\boldsymbol{v}^2z)e^{\epsilon'}(\boldsymbol{v}^2w))+$$

$$\boldsymbol{v}^{-2}(\boldsymbol{v}-\boldsymbol{v}^{-1})^5zw(f^{\epsilon'}(w)e^{\epsilon'}(\boldsymbol{v}^2w)^2+f^\epsilon(z)e^\epsilon(\boldsymbol{v}^2z)^2)\Big)\,A^\epsilon(z)A^{\epsilon'}(w).$$

Meanwhile, using Lemma B.1(a3, b2) to move $A^\epsilon(z)$ to the right of $f^{\epsilon'}(w)e^{\epsilon'}(\boldsymbol{v}^2w)$, we find that $(w-z)\cdot$ LHS equals

$$\boldsymbol{v}^{-1}(\boldsymbol{v}-\boldsymbol{v}^{-1})(w-z)\cdot\Big((w-\boldsymbol{v}^2z)\psi^{\epsilon'}(w)e^\epsilon(\boldsymbol{v}^2z)+(\boldsymbol{v}^2-1)z\psi^{\epsilon'}(w)e^{\epsilon'}(\boldsymbol{v}^2w)+$$

$$(z-w)e^\epsilon(\boldsymbol{v}^2z)\psi^{\epsilon'}(w))\Big)\,A^\epsilon(z)A^{\epsilon'}(w)+$$

$$\boldsymbol{v}^{-2}(\boldsymbol{v}-\boldsymbol{v}^{-1})^3\cdot\Big((w-\boldsymbol{v}^2z)(w-z)f^{\epsilon'}(w)e^{\epsilon'}(\boldsymbol{v}^2w)e^\epsilon(\boldsymbol{v}^2z)+(\boldsymbol{v}^2-1)z(w-z)f^{\epsilon'}(w)e^{\epsilon'}(\boldsymbol{v}^2w)^2-$$

$$(\boldsymbol{v}z-\boldsymbol{v}^{-1}w)(\boldsymbol{v}^{-1}z-\boldsymbol{v}w)e^\epsilon(\boldsymbol{v}^2z)f^{\epsilon'}(w)e^{\epsilon'}(\boldsymbol{v}^2w)-(\boldsymbol{v}^{-1}-\boldsymbol{v})z(\boldsymbol{v}^{-1}z-\boldsymbol{v}w)e^\epsilon(\boldsymbol{v}^2z)f^\epsilon(z)e^{\epsilon'}(\boldsymbol{v}^2w)-$$

$$(\boldsymbol{v}z-\boldsymbol{v}^{-1}w)(\boldsymbol{v}-\boldsymbol{v}^{-1})we^\epsilon(\boldsymbol{v}^2z)f^{\epsilon'}(w)e^\epsilon(\boldsymbol{v}^2z)-(\boldsymbol{v}^{-1}-\boldsymbol{v})(\boldsymbol{v}-\boldsymbol{v}^{-1})zwe^\epsilon(\boldsymbol{v}^2z)f^\epsilon(z)e^\epsilon(\boldsymbol{v}^2z)\Big)\times$$

$$A^\epsilon(z)A^{\epsilon'}(w).$$

To check that the above two big expressions coincide, we first reorder some of the terms. We use Lemma B.1(f1) to move $\psi^{\epsilon'}(w)$ to the left of $e^\epsilon(\boldsymbol{v}^2z)$ via

$$(w-z)e^\epsilon(\boldsymbol{v}^2z)\psi^{\epsilon'}(w)=\psi^{\epsilon'}(w)\Big((\boldsymbol{v}^{-2}w-\boldsymbol{v}^2z)e^\epsilon(\boldsymbol{v}^2z)-(\boldsymbol{v}^{-2}-\boldsymbol{v}^2)ze^{\epsilon'}(\boldsymbol{v}^2w)\Big).$$

We also use Lemma B.1(c) to move $f^\bullet(\cdot)$ to the left of $e^\bullet(\cdot)$. After obvious cancelations, everything boils down to proving

$$(\boldsymbol{v}^{-1}z-\boldsymbol{v}w)e^\epsilon(\boldsymbol{v}^2z)e^{\epsilon'}(\boldsymbol{v}^2w)-(\boldsymbol{v}z-\boldsymbol{v}^{-1}w)e^{\epsilon'}(\boldsymbol{v}^2w)e^\epsilon(\boldsymbol{v}^2z)=(\boldsymbol{v}^{-1}-\boldsymbol{v})(ze^{\epsilon'}(\boldsymbol{v}^2w)^2+we^\epsilon(\boldsymbol{v}^2z)^2),$$

which is exactly the statement of Lemma B.1(d2).

B(i).h Verification of (6.14)

This verification is completely analogous to the above verification of (6.13) and is left to the interested reader.

B(i).i Verification of (6.15)

We need to prove $(z-w)[A^\epsilon(z), D^{\epsilon'}(w)] = (\boldsymbol{v}-\boldsymbol{v}^{-1})(wC^{\epsilon'}(w)B^\epsilon(z) - zC^\epsilon(z)B^{\epsilon'}(w))$. The LHS equals $(\boldsymbol{v}-\boldsymbol{v}^{-1})^2(z-w)(A^\epsilon(z)f^{\epsilon'}(w)A^{\epsilon'}(w)e^{\epsilon'}(w) - f^{\epsilon'}(w)A^{\epsilon'}(w)e^{\epsilon'}(w)A^\epsilon(z))$. Applying Lemma B.1(b2) to the first summand and Lemma B.1(a2) to the second summand, we see that the LHS equals

$$(\boldsymbol{v}-\boldsymbol{v}^{-1})^2((\boldsymbol{v}z-\boldsymbol{v}^{-1}w)f^{\epsilon'}(w)+(\boldsymbol{v}^{-1}-v)zf^\epsilon(z))A^\epsilon(z)A^{\epsilon'}(w)e^{\epsilon'}(w)-$$

$$(\boldsymbol{v}-\boldsymbol{v}^{-1})^2 f^{\epsilon'}(w)A^\epsilon(z)A^{\epsilon'}(w)((\boldsymbol{v}z-\boldsymbol{v}^{-1}w)e^{\epsilon'}(w)-(\boldsymbol{v}-\boldsymbol{v}^{-1})we^\epsilon(z)) =$$

$$(\boldsymbol{v}-\boldsymbol{v}^{-1})^3(wf^{\epsilon'}(w)A^{\epsilon'}(w)A^\epsilon(z)e^\epsilon(z) - zf^\epsilon(z)A^\epsilon(z)A^{\epsilon'}(w)e^{\epsilon'}(w)),$$

which obviously coincides with the RHS.

B(i).j Verification of (6.16)

We need to prove $A^\epsilon(z)D^\epsilon(\boldsymbol{v}^{-2}z) - \boldsymbol{v}^{-1}B^\epsilon(z)C^\epsilon(\boldsymbol{v}^{-2}z) = z^{\epsilon b^\epsilon}$. Due to Lemma B.1(b3), we have $f^\epsilon(\boldsymbol{v}^{-2}z)A^\epsilon(\boldsymbol{v}^{-2}z) = \boldsymbol{v}A^\epsilon(\boldsymbol{v}^{-2}z)f^\epsilon(z)$. Thus,

$$A^\epsilon(z)D^\epsilon(\boldsymbol{v}^{-2}z) = A^\epsilon(z)A^\epsilon(\boldsymbol{v}^{-2}z)(\psi^\epsilon(\boldsymbol{v}^{-2}z)+\boldsymbol{v}(\boldsymbol{v}-\boldsymbol{v}^{-1})^2 f^\epsilon(z)e^\epsilon(\boldsymbol{v}^{-2}z)),$$

$$B^\epsilon(z)C^\epsilon(\boldsymbol{v}^{-2}z) = \boldsymbol{v}(\boldsymbol{v}-\boldsymbol{v}^{-1})^2 A^\epsilon(z)e^\epsilon(z)A^\epsilon(\boldsymbol{v}^{-2}z)f^\epsilon(z).$$

According to Lemma B.1(a2), we have $e^\epsilon(z)A^\epsilon(\boldsymbol{v}^{-2}z) = \boldsymbol{v}A^\epsilon(\boldsymbol{v}^{-2}z)e^\epsilon(\boldsymbol{v}^{-2}z)$. Hence,

$$B^\epsilon(z)C^\epsilon(\boldsymbol{v}^{-2}z) = \boldsymbol{v}^2(\boldsymbol{v}-\boldsymbol{v}^{-1})^2 A^\epsilon(z)A^\epsilon(\boldsymbol{v}^{-2}z)e^\epsilon(\boldsymbol{v}^{-2}z)f^\epsilon(z).$$

Due to Lemma B.1(c), we have $-\boldsymbol{v}(\boldsymbol{v}-\boldsymbol{v}^{-1})^2[e^\epsilon(\boldsymbol{v}^{-2}z), f^\epsilon(z)] = \psi^\epsilon(z) - \psi^\epsilon(\boldsymbol{v}^{-2}z)$. Therefore, we finally get

$$A^\epsilon(z)D^\epsilon(\boldsymbol{v}^{-2}z) - \boldsymbol{v}^{-1}B^\epsilon(z)C^\epsilon(\boldsymbol{v}^{-2}z) = A^\epsilon(z)A^\epsilon(\boldsymbol{v}^{-2}z)\psi^\epsilon(\boldsymbol{v}^{-2}z)-$$

$$\boldsymbol{v}(\boldsymbol{v}-\boldsymbol{v}^{-1})^2 A^\epsilon(z)A^\epsilon(\boldsymbol{v}^{-2}z)[e^\epsilon(\boldsymbol{v}^{-2}z), f^\epsilon(z)] = A^\epsilon(z)A^\epsilon(\boldsymbol{v}^{-2}z)\psi^\epsilon(z) = z^{\epsilon b^\epsilon},$$

which completes our verification of (6.16).

B(ii) Proof of Theorem 6.6(a) for a General $\mathfrak{g}$

First, let us derive an explicit formula for $A_i^\pm(z)$. Recall the elements $\{h_{i,\pm r}\}_{i\in I}^{r>0}$ of Sect. 5, such that $z^{\mp b_i^\pm}(\psi^\pm_{i,\mp b_i^\pm})^{-1}\psi_i^\pm(z) = \exp\left(\pm(\boldsymbol{v}_i-\boldsymbol{v}_i^{-1})\sum_{r>0} h_{i,\pm r}z^{\mp r}\right)$.

For $r \neq 0$, consider the following $I \times I$ matrix $C_{\boldsymbol{v}}(r)$:

$$C_{\boldsymbol{v}}(r)_{ij} = \begin{cases} 0 & \text{if } c_{ij} = 0, \\ -1 - \boldsymbol{v}_i^{2r} & \text{if } j = i, \\ \frac{\boldsymbol{v}_j - \boldsymbol{v}_j^{-1}}{\boldsymbol{v}_i - \boldsymbol{v}_i^{-1}} \sum_{p=1}^{-c_{ji}} \boldsymbol{v}_j^{r(c_{ji}+2p)} & \text{if } j - i. \end{cases}$$

Set $t_{i,r} := \sum_{j \in I} (C_{\boldsymbol{v}}(r)^{-1})_{ij} h_{j,r}$ (matrix $C_{\boldsymbol{v}}(r)$ is invertible, due to Lemma B.3 below). Define

$$A_i^{\pm}(z) := (\phi_i^{\pm})^{-1} \cdot \exp\left(\pm(\boldsymbol{v}_i - \boldsymbol{v}_i^{-1}) \sum_{r>0} t_{i,\pm r} z^{\mp r} \right). \tag{B.2}$$

These $A_i^{\pm}(z)$ satisfy $z^{\mp b_i^{\pm}} \psi_i^{\pm}(z) = \frac{\prod_{j-i} \prod_{p=1}^{-c_{ji}} A_j^{\pm}(\boldsymbol{v}_j^{-c_{ji}-2p} z)}{A_i^{\pm}(z) A_i^{\pm}(\boldsymbol{v}_i^{-2} z)}$ as well as $A_{i,0}^{\pm} = (\phi_i^{\pm})^{-1}$. This provides an explicit formula for $A_i^{\pm}(z)$, which we referred to in Sect. 6.

Remark B.2 Comparing the coefficients of $z^{\mp r}$ ($r > 0$) in the system of equations (6.1) for all i, we see that $A_{i,\pm r}$ are recovered uniquely modulo the values of $A_{i,\pm s} (0 \leq s < r)$, due to invertibility of $C_{\boldsymbol{v}}(r)$. Therefore, an induction in r implies that the system of equations (6.1) has a unique solution $\{A_i^{\pm}(z)\}_{i \in I}$, hence, given by (B.2).

Define auxiliary $I \times I$ matrices $B_{\boldsymbol{v}}(r), D_{\boldsymbol{v}}(r)$ via $B_{\boldsymbol{v}}(r)_{ij} = \frac{[rc_{ij}]_{\boldsymbol{v}_i}}{r}$, $D_{\boldsymbol{v}}(r)_{ij} = \delta_{ij} \frac{\boldsymbol{v}_j^{-2r} - 1}{r(\boldsymbol{v}_j - \boldsymbol{v}_j^{-1})}$. The matrix $B_{\boldsymbol{v}}(r)$ is a $\boldsymbol{v}$-version of the Cartan matrix of $\mathfrak{g}$ and it is known to be invertible for any $r \neq 0$. The following is straightforward.

Lemma B.3 *For $r \neq 0$, we have $B_{\boldsymbol{v}}(r) = C_{\boldsymbol{v}}(r) D_{\boldsymbol{v}}(r)$. In particular, $C_{\boldsymbol{v}}(r)$ is invertible.*

The following result is an immediate corollary of Lemma B.3 and relations (U4′, U5′).

Lemma B.4 *For $\epsilon \in \{\pm\}$, we have:*

(a) $(\boldsymbol{v}_i z - \boldsymbol{v}_i^{-1} w) A_i^{\epsilon}(z) e_i(w) = (z - w) e_i(w) A_i^{\epsilon}(z)$, *while* $A_i^{\epsilon}(z) e_j(w) = e_j(w) A_i^{\epsilon}(z)$ *for* $j \neq i$.

(b) $(z - w) A_i^{\epsilon}(z) f_i(w) = (\boldsymbol{v}_i z - \boldsymbol{v}_i^{-1} w) f_i(w) A_i^{\epsilon}(z)$, *while* $A_i^{\epsilon}(z) f_j(w) = f_j(w) A_i^{\epsilon}(z)$ *for* $j \neq i$.

Now we are ready to sketch the proof of Theorem 6.6(a) for a general $\mathfrak{g}$.

B(ii).a Verification of (6.7) and (6.8)

Relations (6.7, 6.8) follow from Lemma B.4 and relations (U1, U6).

B(ii).b Verification of (6.9–6.16)

Let us introduce the series $\bar{A}_i^{\pm}(z)$ via $z^{\mp b_i^{\pm}}\psi_i^{\pm}(z)=\frac{1}{\bar{A}_i^{\pm}(z)\bar{A}_i^{\pm}(\boldsymbol{v}_i^{-2}z)}$, and define the generating series $\bar{B}_i^{\pm}(z),\bar{C}_i^{\pm}(z),\bar{D}_i^{\pm}(z)$ by using formulas (6.2–6.4) but with $\bar{A}_i^{\pm}(z)$ instead of $A_i^{\pm}(z)$. For a fixed i, these series satisfy the corresponding relations (6.9–6.16) of the $\mathfrak{sl}_2$ case. However, $A_i^{\pm}(z)\bar{A}_i^{\pm}(z)^{-1}$ is expressed through $\{A_j^{\pm}(z)\}_{j\neq i}$, hence, commutes with $e_i^{\epsilon}(z)$, $f_i^{\epsilon}(z)$, $A_i^{\epsilon}(z)$, due to Lemma B.4. Relations (6.9–6.16) follow (this also explains the RHS of (6.16)).

B(ii).c Verification of (6.17)

Analogously to Lemma B.1(d1), relation (U2) implies the following equality:

$$(z-\boldsymbol{v}_i^{c_{ij}}w)e_i^{\epsilon}(z)e_j^{\epsilon'}(w)-(\boldsymbol{v}_i^{c_{ij}}z-w)e_j^{\epsilon'}(w)e_i^{\epsilon}(z)=z[e_{i,0},e_j^{\epsilon'}(w)]_{\boldsymbol{v}_i^{c_{ij}}}+w[e_{j,0},e_i^{\epsilon}(z)]_{\boldsymbol{v}_i^{c_{ij}}}$$

for any $\epsilon,\epsilon'\in\{\pm\}$ (we also note that these equalities for all possible ϵ,ϵ' imply (U2)). Multiplying the above equality by $(\boldsymbol{v}_i-\boldsymbol{v}_i^{-1})(\boldsymbol{v}_j-\boldsymbol{v}_j^{-1})A_i^{\epsilon}(z)A_j^{\epsilon'}(w)$ on the left and using Lemma B.4(a), relation (6.7), and an equality $(\boldsymbol{v}_i-\boldsymbol{v}_i^{-1})e_{i,0}=\phi_i^{+}B_{i,0}^{+}$, we obtain (6.17).

B(ii).d Verification of (6.18)

Analogously to Lemma B.1(e1), relation (U3) implies the following equality:

$$(\boldsymbol{v}_i^{c_{ij}}z-w)f_i^{\epsilon}(z)f_j^{\epsilon'}(w)-(z-\boldsymbol{v}_i^{c_{ij}}w)f_j^{\epsilon'}(w)f_i^{\epsilon}(z)=-[f_j^{\epsilon'}(w),f_{i,1}]_{\boldsymbol{v}_i^{c_{ij}}}-[f_i^{\epsilon}(z),f_{j,1}]_{\boldsymbol{v}_i^{c_{ij}}}$$

for any $\epsilon,\epsilon'\in\{\pm\}$ (we also note that these equalities for all possible ϵ,ϵ' imply (U3)). Multiplying the above equality by $(\boldsymbol{v}_i-\boldsymbol{v}_i^{-1})(\boldsymbol{v}_j-\boldsymbol{v}_j^{-1})A_i^{\epsilon}(z)A_j^{\epsilon'}(w)$ on the right and using Lemma B.4(b), relation (6.7), and an equality $(\boldsymbol{v}_i-\boldsymbol{v}_i^{-1})f_{i,1}=C_{i,1}^{+}\phi_i^{+}$, we obtain (6.18).

B(ii).e Verification of (6.19)

Case $c_{ij}=0$ The equality $[B_i^{\epsilon}(z),B_j^{\epsilon'}(w)]=0$ follows immediately from Lemma B.4(a) and $[e_i^{\epsilon}(z),e_j^{\epsilon'}(w)]=0$, which is a consequence of the corresponding Serre relation (U7).

Case $c_{ij} = -1$ The corresponding Serre relation (U7) is equivalent to

$$\{e_i^{\epsilon_1}(z_1)e_i^{\epsilon_2}(z_2)e_j^{\epsilon'}(w)-(\boldsymbol{v}_i+\boldsymbol{v}_i^{-1})e_i^{\epsilon_1}(z_1)e_j^{\epsilon'}(w)e_i^{\epsilon_2}(z_2)+e_j^{\epsilon'}(w)e_i^{\epsilon_1}(z_1)e_i^{\epsilon_2}(z_2)\}+\{z_1\leftrightarrow z_2\}=0$$

for any $\epsilon_1, \epsilon_2, \epsilon' \in \{\pm\}$. Let us denote the first $\{\cdots\}$ in the LHS by $J^{\epsilon_1,\epsilon_2,\epsilon'}(z_1, z_2, w)$. Set

$$M := (\boldsymbol{v}_i-\boldsymbol{v}_i^{-1})^2(\boldsymbol{v}_j-\boldsymbol{v}_j^{-1})(\boldsymbol{v}_iz_1-\boldsymbol{v}_i^{-1}z_2)(\boldsymbol{v}_iz_2-\boldsymbol{v}_i^{-1}z_1)A_i^{\epsilon_1}(z_1)A_i^{\epsilon_2}(z_2)A_j^{\epsilon'}(w).$$

Combining the equality

$$(\boldsymbol{v}_iz_2-\boldsymbol{v}_i^{-1}z_1)A_i^{\epsilon_2}(z_2)e_i^{\epsilon_1}(z_1)=(z_2-z_1)e_i^{\epsilon_1}(z_1)A_i^{\epsilon_2}(z_2)+(\boldsymbol{v}_i-\boldsymbol{v}_i^{-1})z_1A_i^{\epsilon_2}(z_2)e_i^{\epsilon_2}(z_2)$$

(see Lemma B.1(a2)) with Lemma B.4(a), we find

$$M\cdot J^{\epsilon_1,\epsilon_2,\epsilon'}(z_1,z_2,w)=\frac{(\boldsymbol{v}_i-\boldsymbol{v}_i^{-1})z_1}{\boldsymbol{v}_iz_2-\boldsymbol{v}_i^{-1}z_1}M\cdot J^{\epsilon_2,\epsilon_2,\epsilon'}(z_2,z_2,w)+(z_2-z_1)(\boldsymbol{v}_iz_1-\boldsymbol{v}_i^{-1}z_2)\times$$
$$\{B_i^{\epsilon_1}(z_1)B_i^{\epsilon_2}(z_2)B_j^{\epsilon'}(w)-(\boldsymbol{v}_i+\boldsymbol{v}_i^{-1})B_i^{\epsilon_1}(z_1)B_j^{\epsilon'}(w)B_i^{\epsilon_2}(z_2)+B_j^{\epsilon'}(w)B_i^{\epsilon_1}(z_1)B_i^{\epsilon_2}(z_2)\}.$$

The first summand in the RHS is zero as $J^{\epsilon_2,\epsilon_2,\epsilon'}(z_2,z_2,w)=0$. Therefore, multiplying $J^{\epsilon_1,\epsilon_2,\epsilon'}(z_1,z_2,w)+J^{\epsilon_2,\epsilon_1,\epsilon'}(z_2,z_1,w)=0$ by M on the left, we obtain (6.19).

Case $c_{ij} = -2, -3$ These cases are treated similarly to $c_{ij} = -1$, but the corresponding computations become more cumbersome. We verified these cases using **MATLAB**.

B(ii).f Verification of (6.20)

This verification is analogous to that of (6.19) and is left to the interested reader.

B(iii) Proof of Theorem 6.6(b)

Part (b) of Theorem 6.6 can be obtained by reversing the above arguments. In other words, starting from the algebra generated by $(A_{i,0}^{\pm})^{-1}$ and the coefficients of the currents $A_i^{\pm}(z)$, $B_i^{\pm}(z)$, $C_i^{\pm}(z)$, $D_i^{\pm}(z)$ with the defining relations (6.6–6.20), we need to show that the elements $\phi_i^{\pm}$ and currents $e_i(z)$, $f_i(z)$, $\psi_i^{\pm}(z)$, defined via (6.1–6.4), satisfy relations (U1–U10).

This completes our proof of Theorem 6.6.

Appendix C Proof of Theorem 7.1

We denote the images of $e_i(z)$, $f_i(z)$, $\psi_i^{\pm}(z)$ under $\widetilde{\Phi}_{\mu}^{\lambda}$ by $E_i(z)$, $F_i(z)$, $\Psi_i(z)^{\pm}$. It suffices to prove that they satisfy relations (U1–U8), since relations (U9, U10) are obviously preserved by $\widetilde{\Phi}_{\mu}^{\lambda}$. While checking these relations, we will use LHS and RHS when referring to their left-hand and right-hand sides. Set $\rho_i^+ := \frac{-\boldsymbol{v}_i}{1-\boldsymbol{v}_i^2}$, $\rho_i^- := \frac{1}{1-\boldsymbol{v}_i^2}$, $W_{i,rs}(z) := \prod_{1\leq t\leq a_i}^{r\neq t\neq s} (1 - \frac{\mathsf{w}_{i,t}}{z})$.

C(i) Compatibility with (U1)

First, we check that the range of powers of z in $\psi_i^{\pm}(z)$ and $\Psi_i(z)^{\pm}$ agree. Note that

$$(1-\nu/z)^+ = 1-\nu\cdot z^{-1} \in \mathbb{C}[[z^{-1}]],\ (1/(1-\nu/z))^+ = 1+\nu z^{-1}+\nu^2 z^{-2}+\ldots \in \mathbb{C}[[z^{-1}]],$$

$$(1-\nu/z)^- = -\nu\cdot z^{-1}(1-z/\nu) \in z^{-1}\mathbb{C}[[z]],\ (1/(1-\nu/z))^- = -z/\nu - z^2/\nu^2 - \ldots \in z\mathbb{C}[[z]].$$

Therefore, $\Psi_i(z)^+$ contains only nonpositive powers of z, while $\Psi_i(z)^-$ contains only powers of z bigger or equal to

$$-\#\{s : i_s = i\} + 2a_i - \sum_{j-i} a_j(-c_{ji}) = -\alpha_i^\vee(\lambda) + \alpha_i^\vee(\lambda-\mu) = -\alpha_i^\vee(\mu) = -\alpha_i^\vee(\mu^-) = -b_i^-.$$

Moreover, the coefficients of z^0 in $\Psi_i(z)^+$ and of $z^{-b_i^-}$ in $\Psi_i(z)^-$ are invertible.

The equality $[\Psi_i(z)^{\epsilon}, \Psi_j(w)^{\epsilon'}] = 0$ follows from the commutativity of $\{\mathsf{w}_{i,r}^{\pm 1/2}\}_{i\in I}^{1\leq r\leq a_i}$.

C(ii) Compatibility with (U2)

Case $c_{ij} = 0$ The equality $[E_i(z), E_j(w)] = 0$ is obvious in this case, since $D_{i,r}^{-1}$ commute with $\mathsf{w}_{k,s}^{\pm 1/2}$ for $k = j$ or $k \to j$, while $D_{j,s}^{-1}$ commute with $\mathsf{w}_{k,r}^{\pm 1/2}$ for $k = i$ or $k \to i$.

Case $c_{ij} = 2$ We may assume $\mathfrak{g} = \mathfrak{sl}_2$ and we will drop the index i from our notation. We need to prove $(z-\boldsymbol{v}^2 w)E(z)E(w)/(\rho^+)^2 = -(w-\boldsymbol{v}^2 z)E(w)E(z)/(\rho^+)^2$.

The LHS equals

$$\boldsymbol{v}^{-2}\prod_{t=1}^{a}\mathsf{w}_t^2\cdot(z-\boldsymbol{v}^2w)\cdot\sum_{r=1}^{a}\delta\left(\frac{\mathsf{w}_r}{z}\right)\delta\left(\frac{\boldsymbol{v}^{-2}\mathsf{w}_r}{w}\right)\frac{Z(\mathsf{w}_r)Z(\boldsymbol{v}^{-2}\mathsf{w}_r)}{W_r(\mathsf{w}_r)W_r(\boldsymbol{v}^{-2}\mathsf{w}_r)}D_r^{-2}+$$

$$\boldsymbol{v}^{-2}\prod_{t=1}^{a}\mathsf{w}_t^2\cdot(z-\boldsymbol{v}^2w)\cdot\sum_{1\le r\ne s\le a}\delta\left(\frac{\mathsf{w}_r}{z}\right)\delta\left(\frac{\mathsf{w}_s}{w}\right)\frac{Z(\mathsf{w}_r)Z(\mathsf{w}_s)}{W_r(\mathsf{w}_r)W_{rs}(\mathsf{w}_s)(1-\boldsymbol{v}^{-2}\mathsf{w}_r/\mathsf{w}_s)}D_r^{-1}D_s^{-1}.$$

Using the equality

$$G(z,w)\delta\left(\frac{\nu_1}{z}\right)\delta\left(\frac{\nu_2}{w}\right)=G(\nu_1,\nu_2)\delta\left(\frac{\nu_1}{z}\right)\delta\left(\frac{\nu_2}{w}\right),\tag{C.1}$$

we see that the first sum is zero, while the second sum equals

$$\prod_{t=1}^{a}\mathsf{w}_t^2\cdot\sum_{1\le r\ne s\le a}\delta\left(\frac{\mathsf{w}_r}{z}\right)\delta\left(\frac{\mathsf{w}_s}{w}\right)\frac{Z(\mathsf{w}_r)Z(\mathsf{w}_s)}{W_{rs}(\mathsf{w}_r)W_{rs}(\mathsf{w}_s)}\frac{\boldsymbol{v}^{-2}(\mathsf{w}_r-\boldsymbol{v}^2\mathsf{w}_s)}{(1-\mathsf{w}_s/\mathsf{w}_r)(1-\boldsymbol{v}^{-2}\mathsf{w}_r/\mathsf{w}_s)}D_r^{-1}D_s^{-1}=$$

$$\prod_{t=1}^{a}\mathsf{w}_t^2\cdot\sum_{1\le r\ne s\le a}\delta\left(\frac{\mathsf{w}_r}{z}\right)\delta\left(\frac{\mathsf{w}_s}{w}\right)\frac{Z(\mathsf{w}_r)Z(\mathsf{w}_s)}{W_{rs}(\mathsf{w}_r)W_{rs}(\mathsf{w}_s)}\frac{\mathsf{w}_r\mathsf{w}_s}{\mathsf{w}_s-\mathsf{w}_r}D_r^{-1}D_s^{-1}.$$

Swapping z and w, we see that $-(w-\boldsymbol{v}^2z)E(w)E(z)/(\rho^+)^2$ equals

$$-\prod_{t=1}^{a}\mathsf{w}_t^2\cdot\sum_{1\le r\ne s\le a}\delta\left(\frac{\mathsf{w}_r}{w}\right)\delta\left(\frac{\mathsf{w}_s}{z}\right)\frac{Z(\mathsf{w}_r)Z(\mathsf{w}_s)}{W_{rs}(\mathsf{w}_r)W_{rs}(\mathsf{w}_s)}\frac{\mathsf{w}_r\mathsf{w}_s}{\mathsf{w}_s-\mathsf{w}_r}D_r^{-1}D_s^{-1}.$$

Swapping r and s in the latter sum, we get exactly the same expression as for the LHS.

Case $c_{ij}<0$ In this case, we can assume $I=\{i,j\}$ and $i\to j$. We need to prove $(z-\boldsymbol{v}_i^{c_{ij}}w)E_i(z)E_j(w)/(\rho_i^+\rho_j^+)=(\boldsymbol{v}_i^{c_{ij}}z-w)E_j(w)E_i(z)/(\rho_i^+\rho_j^+)$. The LHS equals

$$\boldsymbol{v}_i^{-c_{ij}}\prod_{t=1}^{a_i}\mathsf{w}_{i,t}^{1+c_{ij}/2}\prod_{t=1}^{a_j}\mathsf{w}_{j,t}\cdot(z-\boldsymbol{v}_i^{c_{ij}}w)\times$$

$$\sum_{1\le r\le a_i}^{1\le s\le a_j}\delta\left(\frac{\mathsf{w}_{i,r}}{z}\right)\delta\left(\frac{\mathsf{w}_{j,s}}{w}\right)\frac{Z_i(\mathsf{w}_{i,r})}{W_{i,r}(\mathsf{w}_{i,r})}D_{i,r}^{-1}\frac{Z_j(\mathsf{w}_{j,s})}{W_{j,s}(\mathsf{w}_{j,s})}\prod_{p=1}^{-c_{ij}}W_i(\boldsymbol{v}_i^{-c_{ij}-2p}w)D_{j,s}^{-1}=$$

$$\prod_{t=1}^{a_i}\mathsf{w}_{i,t}^{1+c_{ij}/2}\prod_{t=1}^{a_j}\mathsf{w}_{j,t}\cdot A(z,w)\times$$

$$\sum_{1\le r\le a_i}^{1\le s\le a_j}\delta\left(\frac{\mathsf{w}_{i,r}}{z}\right)\delta\left(\frac{\mathsf{w}_{j,s}}{w}\right)\frac{Z_i(\mathsf{w}_{i,r})Z_j(\mathsf{w}_{j,s})\prod_{p=1}^{-c_{ij}}W_{i,r}(\boldsymbol{v}_i^{-c_{ij}-2p}\mathsf{w}_{j,s})}{W_{i,r}(\mathsf{w}_{i,r})W_{j,s}(\mathsf{w}_{j,s})}D_{i,r}^{-1}D_{j,s}^{-1},$$

where $A(z,w)=\boldsymbol{v}_i^{-c_{ij}}(z-\boldsymbol{v}_i^{c_{ij}}w)\prod_{p=1}^{c_{ij}}\left(1-\frac{\boldsymbol{v}_i^{-2}z}{\boldsymbol{v}_i^{-c_{ij}-2p}w}\right)$, due to (C.1). Likewise, the RHS equals

$$\prod_{t=1}^{a_i}\mathsf{w}_{i,t}^{1+c_{ij}/2}\prod_{t=1}^{a_j}\mathsf{w}_{j,t}\cdot(\boldsymbol{v}_i^{c_{ij}}z-w)\times$$

$$\sum_{\substack{1\leq s\leq a_j\\1\leq r\leq a_i}}\delta\left(\frac{\mathsf{w}_{j,s}}{w}\right)\delta\left(\frac{\mathsf{w}_{i,r}}{z}\right)\frac{Z_j(\mathsf{w}_{j,s})}{W_{j,s}(\mathsf{w}_{j,s})}\prod_{p=1}^{-c_{ij}}W_i(\boldsymbol{v}_i^{-c_{ij}-2p}w)D_{j,s}^{-1}\frac{Z_i(\mathsf{w}_{i,r})}{W_{i,r}(\mathsf{w}_{i,r})}D_{i,r}^{-1}=$$

$$\prod_{t=1}^{a_i}\mathsf{w}_{i,t}^{1+c_{ij}/2}\prod_{t=1}^{a_j}\mathsf{w}_{j,t}\cdot B(z,w)\times$$

$$\sum_{\substack{1\leq s\leq a_j\\1\leq r\leq a_i}}\delta\left(\frac{\mathsf{w}_{i,r}}{z}\right)\delta\left(\frac{\mathsf{w}_{j,s}}{w}\right)\frac{Z_i(\mathsf{w}_{i,r})Z_j(\mathsf{w}_{j,s})\prod_{p=1}^{-c_{ij}}W_{i,r}(\boldsymbol{v}_i^{-c_{ij}-2p}\mathsf{w}_{j,s})}{W_{i,r}(\mathsf{w}_{i,r})W_{j,s}(\mathsf{w}_{j,s})}D_{i,r}^{-1}D_{j,s}^{-1},$$

where $B(z,w)=(\boldsymbol{v}_i^{c_{ij}}z-w)\prod_{p=1}^{-c_{ij}}\left(1-\frac{z}{\boldsymbol{v}_i^{-c_{ij}-2p}w}\right)$, due to (C.1).

The equality LHS = RHS follows from $A(z,w)=B(z,w)$.

C(iii) Compatibility with (U3)

Case $c_{ij}=0$ The equality $[F_i(z),F_j(w)]=0$ is obvious in this case, since $D_{i,r}$ commute with $\mathsf{w}_{k,s}^{\pm 1/2}$ for $k=j$ or $k\leftarrow j$, while $D_{j,s}$ commute with $\mathsf{w}_{k,r}^{\pm 1/2}$ for $k=i$ or $k\leftarrow i$.

Case $c_{ij}=2$ We may assume $\mathfrak{g}=\mathfrak{sl}_2$ and we will drop the index i from our notation. We need to prove $(\boldsymbol{v}^2z-w)F(z)F(w)/(\rho^-)^2=-(\boldsymbol{v}^2w-z)F(w)F(z)/(\rho^-)^2$. The LHS equals

$$(\boldsymbol{v}^2z-w)\cdot\sum_{r=1}^{a}\delta\left(\frac{\boldsymbol{v}^2\mathsf{w}_r}{z}\right)\delta\left(\frac{\boldsymbol{v}^4\mathsf{w}_r}{w}\right)\frac{1}{W_r(\mathsf{w}_r)W_r(\boldsymbol{v}^2\mathsf{w}_r)}D_r^2+$$

$$(\boldsymbol{v}^2z-w)\cdot\sum_{1\leq r\neq s\leq a}\delta\left(\frac{\boldsymbol{v}^2\mathsf{w}_r}{z}\right)\delta\left(\frac{\boldsymbol{v}^2\mathsf{w}_s}{w}\right)\frac{1}{W_r(\mathsf{w}_r)W_{rs}(\mathsf{w}_s)(1-\boldsymbol{v}^2\mathsf{w}_r/\mathsf{w}_s)}D_rD_s.$$

Using equality (C.1), we see that the first sum is zero, while the second sum equals

$$\sum_{1\le r\ne s\le a} \delta\left(\frac{\boldsymbol{v}^2 \mathsf{w}_r}{z}\right)\delta\left(\frac{\boldsymbol{v}^2 \mathsf{w}_s}{w}\right)\frac{1}{W_{rs}(\mathsf{w}_r)W_{rs}(\mathsf{w}_s)}\frac{\boldsymbol{v}^4\mathsf{w}_r-\boldsymbol{v}^2\mathsf{w}_s}{(1-\mathsf{w}_s/\mathsf{w}_r)(1-\boldsymbol{v}^2\mathsf{w}_r/\mathsf{w}_s)}D_rD_s=$$

$$\sum_{1\le r\ne s\le a} \delta\left(\frac{\boldsymbol{v}^2 \mathsf{w}_r}{z}\right)\delta\left(\frac{\boldsymbol{v}^2 \mathsf{w}_s}{w}\right)\frac{1}{W_{rs}(\mathsf{w}_r)W_{rs}(\mathsf{w}_s)}\frac{\boldsymbol{v}^2\mathsf{w}_r\mathsf{w}_s}{\mathsf{w}_s-\mathsf{w}_r}D_rD_s.$$

Swapping z and w, we see that $-(\boldsymbol{v}^2w-z)F(w)F(z)/(\rho^-)^2$ equals

$$-\sum_{1\le r\ne s\le a} \delta\left(\frac{\boldsymbol{v}^2 \mathsf{w}_r}{w}\right)\delta\left(\frac{\boldsymbol{v}^2 \mathsf{w}_s}{z}\right)\frac{1}{W_{rs}(\mathsf{w}_r)W_{rs}(\mathsf{w}_s)}\frac{\boldsymbol{v}^2\mathsf{w}_r\mathsf{w}_s}{\mathsf{w}_s-\mathsf{w}_r}D_rD_s.$$

Swapping r and s in this sum, we get exactly the same expression as for the LHS.

Case $c_{ij}<0$ In this case, we can assume $I=\{i,j\}$ and $i\to j$. Recall that $\boldsymbol{v}_i^{c_{ij}}=\boldsymbol{v}_j^{c_{ji}}$. We need to prove $(\boldsymbol{v}_j^{c_{ji}}z-w)F_i(z)F_j(w)/(\rho_i^-\rho_j^-)=(z-\boldsymbol{v}_j^{c_{ji}}w)F_j(w)F_i(z)/(\rho_i^-\rho_j^-)$. The LHS equals

$$\prod_{t=1}^{a_j}\mathsf{w}_{j,t}^{c_{ji}/2}\cdot(\boldsymbol{v}_j^{c_{ji}}z-w)\times$$

$$\sum_{1\le r\le a_i}^{1\le s\le a_j}\delta\left(\frac{\boldsymbol{v}_i^2\mathsf{w}_{i,r}}{z}\right)\delta\left(\frac{\boldsymbol{v}_j^2\mathsf{w}_{j,s}}{w}\right)\frac{1}{W_{i,r}(\mathsf{w}_{i,r})}\prod_{p=1}^{-c_{ji}}W_j(\boldsymbol{v}_j^{-c_{ji}-2p}z)D_{i,r}\frac{1}{W_{j,s}(\mathsf{w}_{j,s})}D_{j,s}=$$

$$\prod_{t=1}^{a_j}\mathsf{w}_{j,t}^{c_{ji}/2}\cdot A(z,w)\cdot\sum_{1\le r\le a_i}^{1\le s\le a_j}\delta\left(\frac{\boldsymbol{v}_i^2\mathsf{w}_{i,r}}{z}\right)\delta\left(\frac{\boldsymbol{v}_j^2\mathsf{w}_{j,s}}{w}\right)\frac{\prod_{p=1}^{-c_{ji}}W_{j,s}(\boldsymbol{v}_j^{-c_{ji}-2p}z)}{W_{i,r}(\mathsf{w}_{i,r})W_{j,s}(\mathsf{w}_{j,s})}D_{i,r}D_{j,s},$$

where $A(z,w)=(\boldsymbol{v}_j^{c_{ji}}z-w)\prod_{p=1}^{-c_{ji}}\left(1-\frac{\boldsymbol{v}_j^{-2}w}{\boldsymbol{v}_j^{-c_{ji}-2p}z}\right)$, due to (C.1). Likewise, the RHS equals

$$\boldsymbol{v}_j^{c_{ji}}\prod_{t=1}^{a_j}\mathsf{w}_{j,t}^{c_{ji}/2}\cdot(z-\boldsymbol{v}_j^{c_{ji}}w)\times$$

$$\sum_{1\le r\le a_i}^{1\le s\le a_j}\delta\left(\frac{\boldsymbol{v}_j^2\mathsf{w}_{j,s}}{w}\right)\delta\left(\frac{\boldsymbol{v}_i^2\mathsf{w}_{i,r}}{z}\right)\frac{1}{W_{j,s}(\mathsf{w}_{j,s})}D_{j,s}\frac{1}{W_{i,r}(\mathsf{w}_{i,r})}\prod_{p=1}^{-c_{ji}}W_j(\boldsymbol{v}_j^{-c_{ji}-2p}z)D_{i,r}=$$

$$\prod_{t=1}^{a_j}\mathsf{w}_{j,t}^{c_{ji}/2}\cdot B(z,w)\cdot\sum_{1\le r\le a_i}^{1\le s\le a_j}\delta\left(\frac{\boldsymbol{v}_i^2\mathsf{w}_{i,r}}{z}\right)\delta\left(\frac{\boldsymbol{v}_j^2\mathsf{w}_{j,s}}{w}\right)\frac{\prod_{p=1}^{-c_{ji}}W_{j,s}(\boldsymbol{v}_j^{-c_{ji}-2p}z)}{W_{i,r}(\mathsf{w}_{i,r})W_{j,s}(\mathsf{w}_{j,s})}D_{i,r}D_{j,s},$$

where $B(z,w)=\boldsymbol{v}_j^{c_{ji}}(z-\boldsymbol{v}_j^{c_{ji}}w)\prod_{p=1}^{-c_{ji}}\left(1-\frac{w}{\boldsymbol{v}_j^{-c_{ji}-2p}z}\right)$, due to (C.1).

The equality LHS = RHS follows from $A(z,w)=B(z,w)$.

C(iv) Compatibility with (U4)

Case $c_{ij} = 0$ The equality $[\Psi_i(z), E_j(w)] = 0$ is obvious in this case, since $D_{j,s}^{-1}$ commute with $\mathsf{w}_{k,r}^{\pm 1/2}$ for $k = i$ or $k - i$.

Case $c_{ij} = 2$ We may assume $\mathfrak{g} = \mathfrak{sl}_2$ and we will drop the index i from our notation. We need to prove $(z - \boldsymbol{v}^2 w)\Psi(z)E(w)/\rho^+ = (\boldsymbol{v}^2 z - w)E(w)\Psi(z)/\rho^+$. The LHS equals

$$\prod_{t=1}^{a} \mathsf{w}_t^2 \cdot (z - \boldsymbol{v}^2 w) \cdot \frac{Z(z)}{W(z)W(\boldsymbol{v}^{-2}z)} \sum_{r=1}^{a} \delta\left(\frac{\mathsf{w}_r}{w}\right) \frac{Z(\mathsf{w}_r)}{W_r(\mathsf{w}_r)} D_r^{-1} =$$

$$\prod_{t=1}^{a} \mathsf{w}_t^2 \cdot \sum_{r=1}^{a} \delta\left(\frac{\mathsf{w}_r}{w}\right) \frac{Z(z)Z(\mathsf{w}_r)}{W_r(\mathsf{w}_r)W_r(z)W_r(\boldsymbol{v}^{-2}z)} \frac{z - \boldsymbol{v}^2 w}{(1 - w/z)(1 - w/\boldsymbol{v}^{-2}z)} D_r^{-1},$$

due to (C.1). Likewise, the RHS equals

$$\boldsymbol{v}^{-2} \prod_{t=1}^{a} \mathsf{w}_t^2 \cdot (\boldsymbol{v}^2 z - w) \cdot \sum_{r=1}^{a} \delta\left(\frac{\mathsf{w}_r}{w}\right) \frac{Z(\mathsf{w}_r)}{W_r(\mathsf{w}_r)} D_r^{-1} \frac{Z(z)}{W(z)W(\boldsymbol{v}^{-2}z)} =$$

$$\prod_{t=1}^{a} \mathsf{w}_t^2 \cdot \sum_{r=1}^{a} \delta\left(\frac{\mathsf{w}_r}{w}\right) \frac{Z(z)Z(\mathsf{w}_r)}{W_r(\mathsf{w}_r)W_r(z)W_r(\boldsymbol{v}^{-2}z)} \frac{\boldsymbol{v}^{-2}(\boldsymbol{v}^2 z - w)}{(1 - \boldsymbol{v}^{-2}w/z)(1 - \boldsymbol{v}^{-2}w/\boldsymbol{v}^{-2}z)} D_r^{-1}.$$

The equality LHS = RHS follows.

Case $c_{ij} < 0$ In this case, we can assume $I = \{i, j\}$. There are two situations to consider: $i \to j$ and $i \leftarrow j$. Let us first treat the former case. Since $\boldsymbol{v}_i^{c_{ij}} = \boldsymbol{v}_j^{c_{ji}}$, we need to prove $(z - \boldsymbol{v}_j^{c_{ji}} w)\Psi_i(z)E_j(w)/\rho_j^+ = (\boldsymbol{v}_j^{c_{ji}} z - w)E_j(w)\Psi_i(z)/\rho_j^+$. The LHS equals

$$\prod_{t=1}^{a_i} \mathsf{w}_{i,t}^{1+c_{ij}/2} \prod_{t=1}^{a_j} \mathsf{w}_{j,t}^{1+c_{ji}/2} \cdot (z - \boldsymbol{v}_j^{c_{ji}} w) \times$$

$$\frac{Z_i(z)}{W_i(z)W_i(\boldsymbol{v}_i^{-2}z)} \prod_{p=1}^{-c_{ji}} W_j(\boldsymbol{v}_j^{-c_{ji}-2p} z) \sum_{s=1}^{a_j} \delta\left(\frac{\mathsf{w}_{j,s}}{w}\right) \frac{Z_j(\mathsf{w}_{j,s})}{W_{j,s}(\mathsf{w}_{j,s})} \prod_{p'=1}^{-c_{ij}} W_i(\boldsymbol{v}_i^{-c_{ij}-2p'} w) D_{j,s}^{-1} =$$

$$\prod_{t=1}^{a_i} \mathsf{w}_{i,t}^{1+c_{ij}/2} \prod_{t=1}^{a_j} \mathsf{w}_{j,t}^{1+c_{ji}/2} \cdot A(z, w) \times$$

$$\sum_{s=1}^{a_j} \delta\left(\frac{\mathsf{w}_{j,s}}{w}\right) \frac{Z_i(z)Z_j(\mathsf{w}_{j,s}) \prod_{p'=1}^{-c_{ij}} W_i(\boldsymbol{v}_i^{-c_{ij}-2p'} w) \prod_{p=1}^{-c_{ji}} W_{j,s}(\boldsymbol{v}_j^{-c_{ji}-2p} z)}{W_i(z)W_i(\boldsymbol{v}_i^{-2}z)W_{j,s}(\mathsf{w}_{j,s})} D_{j,s}^{-1},$$

where $A(z,w) = (z - \boldsymbol{v}_j^{c_{ji}} w)\prod_{p=1}^{-c_{ji}}\left(1 - \frac{w}{\boldsymbol{v}_j^{-c_{ji}-2p} z}\right)$. Likewise, the RHS equals

$$\boldsymbol{v}_j^{-c_{ji}} \prod_{t=1}^{a_i} \mathsf{w}_{i,t}^{1+c_{ij}/2} \prod_{t=1}^{a_j} \mathsf{w}_{j,t}^{1+c_{ji}/2} \cdot (\boldsymbol{v}_j^{c_{ji}} z - w)\times$$

$$\sum_{s=1}^{a_j} \delta\left(\frac{\mathsf{w}_{j,s}}{w}\right) \frac{Z_j(\mathsf{w}_{j,s})}{W_{j,s}(\mathsf{w}_{j,s})} \prod_{p'=1}^{-c_{ij}} W_i(\boldsymbol{v}_i^{-c_{ij}-2p'} w) D_{j,s}^{-1} \frac{Z_i(z)}{W_i(z)W_i(\boldsymbol{v}_i^{-2}z)} \prod_{p=1}^{-c_{ji}} W_j(\boldsymbol{v}_j^{-c_{ji}-2p} z) =$$

$$\prod_{t=1}^{a_i} \mathsf{w}_{i,t}^{1+c_{ij}/2} \prod_{t=1}^{a_j} \mathsf{w}_{j,t}^{1+c_{ji}/2} \cdot B(z,w)\times$$

$$\sum_{s=1}^{a_j} \delta\left(\frac{\mathsf{w}_{j,s}}{w}\right) \frac{Z_i(z)Z_j(\mathsf{w}_{j,s})\prod_{p'=1}^{-c_{ij}} W_i(\boldsymbol{v}_i^{-c_{ij}-2p'} w)\prod_{p=1}^{-c_{ji}} W_{j,s}(\boldsymbol{v}_j^{-c_{ji}-2p} z)}{W_i(z)W_i(\boldsymbol{v}_i^{-2}z)W_{j,s}(\mathsf{w}_{j,s})} D_{j,s}^{-1},$$

where $B(z,w) = \boldsymbol{v}_j^{-c_{ji}} (\boldsymbol{v}_j^{c_{ji}} z - w)\prod_{p=1}^{-c_{ji}}\left(1 - \frac{\boldsymbol{v}_j^{-2} w}{\boldsymbol{v}_j^{-c_{ji}-2p} z}\right)$.

The equality LHS = RHS follows from $A(z,w) = B(z,w)$.

The case $i \leftarrow j$ is analogous: $\Psi_i(z)$ is given by the same formula, while $E_j(w)$ differs by an absence of the factor $\prod_{t=1}^{a_i} \mathsf{w}_{i,t}^{c_{ij}/2} \cdot \prod_{p'=1}^{-c_{ij}} W_i(\boldsymbol{v}_i^{-c_{ij}-2p'} w)$. Tracing back the above calculations, it is clear that the equality still holds when this factor is dropped out.

C(v) Compatibility with (U5)

Case $c_{ij} = 0$ The equality $[\Psi_i(z), F_j(w)] = 0$ is obvious in this case, since $D_{j,s}$ commute with $\mathsf{w}_{k,r}^{\pm 1/2}$ for $k = i$ or $k - i$.

Case $c_{ij} = 2$ We may assume $\mathfrak{g} = \mathfrak{sl}_2$ and we will drop the index i from our notation. We need to prove $(\boldsymbol{v}^2 z - w)\Psi(z)F(w)/\rho^- = (z - \boldsymbol{v}^2 w)F(w)\Psi(z)/\rho^-$. The LHS equals

$$\prod_{t=1}^{a} \mathsf{w}_t \cdot (\boldsymbol{v}^2 z - w) \cdot \frac{Z(z)}{W(z)W(\boldsymbol{v}^{-2}z)} \sum_{r=1}^{a} \delta\left(\frac{\boldsymbol{v}^2 \mathsf{w}_r}{w}\right) \frac{1}{W_r(\mathsf{w}_r)} D_r =$$

$$\prod_{t=1}^{a} \mathsf{w}_t \cdot \sum_{r=1}^{a} \delta\left(\frac{\boldsymbol{v}^2 \mathsf{w}_r}{w}\right) \frac{Z(z)}{W_r(\mathsf{w}_r)W_r(z)W_r(\boldsymbol{v}^{-2}z)} \frac{\boldsymbol{v}^2 z - w}{(1 - \boldsymbol{v}^{-2}w/z)(1 - \boldsymbol{v}^{-2}w/\boldsymbol{v}^{-2}z)} D_r,$$

due to (C.1). Likewise, the RHS equals

$$\boldsymbol{v}^2 \prod_{t=1}^{a} \mathsf{w}_t \cdot (z - \boldsymbol{v}^2 w) \cdot \sum_{r=1}^{a} \delta\left(\frac{\boldsymbol{v}^2 \mathsf{w}_r}{w}\right) \frac{1}{W_r(\mathsf{w}_r)} D_r \frac{Z(z)}{W(z)W(\boldsymbol{v}^{-2}z)} =$$

$$\prod_{t=1}^{a} \mathsf{w}_t \cdot \sum_{r=1}^{a} \delta\left(\frac{\boldsymbol{v}^2 \mathsf{w}_r}{w}\right) \frac{Z(z)}{W_r(\mathsf{w}_r)W_r(z)W_r(\boldsymbol{v}^{-2}z)} \frac{\boldsymbol{v}^2(z - \boldsymbol{v}^2 w)}{(1 - w/z)(1 - w/\boldsymbol{v}^{-2}z)} D_r.$$

The equality LHS = RHS follows.

Case $c_{ij} < 0$ In this case, we can assume $I = \{i, j\}$. There are two situations to consider: $i \to j$ and $i \leftarrow j$. Let us first treat the former case. Since $\boldsymbol{v}_i^{c_{ij}} = \boldsymbol{v}_j^{c_{ji}}$, we need to prove $(\boldsymbol{v}_j^{c_{ji}} z - w)\Psi_i(z)F_j(w)/\rho_j^- = (z - \boldsymbol{v}_j^{c_{ji}} w)F_j(w)\Psi_i(z)/\rho_j^-$. The LHS equals

$$\prod_{t=1}^{a_i} \mathsf{w}_{i,t} \prod_{t=1}^{a_j} \mathsf{w}_{j,t}^{c_{ji}/2} \cdot (\boldsymbol{v}_j^{c_{ji}} z - w) \times$$

$$\frac{Z_i(z)}{W_i(z)W_i(\boldsymbol{v}_i^{-2}z)} \prod_{p=1}^{-c_{ji}} W_j(\boldsymbol{v}_j^{-c_{ji}-2p} z) \sum_{s=1}^{a_j} \delta\left(\frac{\boldsymbol{v}_j^2 \mathsf{w}_{j,s}}{w}\right) \frac{1}{W_{j,s}(\mathsf{w}_{j,s})} D_{j,s} =$$

$$\prod_{t=1}^{a_i} \mathsf{w}_{i,t} \prod_{t=1}^{a_j} \mathsf{w}_{j,t}^{c_{ji}/2} \cdot A(z, w) \cdot \sum_{s=1}^{a_j} \delta\left(\frac{\boldsymbol{v}_j^2 \mathsf{w}_{j,s}}{w}\right) \frac{Z_i(z) \prod_{p=1}^{-c_{ji}} W_{j,s}(\boldsymbol{v}_j^{-c_{ji}-2p} z)}{W_i(z)W_i(\boldsymbol{v}_i^{-2}z)W_{j,s}(\mathsf{w}_{j,s})} D_{j,s},$$

where $A(z, w) = (\boldsymbol{v}_j^{c_{ji}} z - w) \prod_{p=1}^{-c_{ji}} \left(1 - \frac{\boldsymbol{v}_j^{-2} w}{\boldsymbol{v}_j^{-c_{ji}-2p} z}\right)$. Likewise, the RHS equals

$$\boldsymbol{v}_j^{c_{ji}} \prod_{t=1}^{a_i} \mathsf{w}_{i,t} \prod_{t=1}^{a_j} \mathsf{w}_{j,t}^{c_{ji}/2} \cdot (z - \boldsymbol{v}_j^{c_{ji}} w) \times$$

$$\sum_{s=1}^{a_j} \delta\left(\frac{\boldsymbol{v}_j^2 \mathsf{w}_{j,s}}{w}\right) \frac{1}{W_{j,s}(\mathsf{w}_{j,s})} D_{j,s} \frac{Z_i(z)}{W_i(z)W_i(\boldsymbol{v}_i^{-2}z)} \prod_{p=1}^{-c_{ji}} W_j(\boldsymbol{v}_j^{-c_{ji}-2p} z) =$$

$$\prod_{t=1}^{a_i} \mathsf{w}_{i,t} \prod_{t=1}^{a_j} \mathsf{w}_{j,t}^{c_{ji}/2} \cdot B(z, w) \cdot \sum_{s=1}^{a_j} \delta\left(\frac{\boldsymbol{v}_j^2 \mathsf{w}_{j,s}}{w}\right) \frac{Z_i(z) \prod_{p=1}^{-c_{ji}} W_{j,s}(\boldsymbol{v}_j^{-c_{ji}-2p} z)}{W_i(z)W_i(\boldsymbol{v}_i^{-2}z)W_{j,s}(\mathsf{w}_{j,s})} D_{j,s},$$

where $B(z, w) = \boldsymbol{v}_j^{c_{ji}} (z - \boldsymbol{v}_j^{c_{ji}} w) \prod_{p=1}^{-c_{ji}} \left(1 - \frac{w}{\boldsymbol{v}_j^{-c_{ji}-2p} z}\right)$.

The equality LHS = RHS follows from $A(z, w) = B(z, w)$.

The case $i \leftarrow j$ is analogous: $\Psi_i(z)$ is given by the same formula, while $F_j(w)$ has an extra factor $\prod_{t=1}^{a_i} \mathsf{w}_{i,t}^{c_{ij}/2} \cdot \prod_{p'=1}^{-c_{ij}} W_i(\boldsymbol{v}_i^{-c_{ij}-2p'} w)$. The contributions of this factor into the LHS and the RHS are the same, hence, the equality still holds.

C(vi) Compatibility with (U6)

Case $c_{ij} = 0$ The equality $[E_i(z), F_j(w)] = 0$ is obvious in this case, since $D_{i,r}^{-1}$ commute with $\mathsf{w}_{k,s}^{\pm 1/2}$ for $k = i$ or $k \leftarrow j$, while $D_{j,s}$ commute with $\mathsf{w}_{k,r}^{\pm 1/2}$ for $k = i$ or $k \rightarrow i$.

Case $c_{ij} = 2$ We may assume $\mathfrak{g} = \mathfrak{sl}_2$, and we will drop the index i from our notation. We need to prove $[E(z), F(w)] = \frac{1}{\boldsymbol{v}-\boldsymbol{v}^{-1}}\delta\left(\frac{z}{w}\right)\left(\Psi(z)^+ - \Psi(z)^-\right)$. The LHS equals

$$\rho^+\rho^-\left[\prod_{t=1}^{a} \mathsf{w}_t \cdot \sum_{r=1}^{a} \delta\left(\frac{\mathsf{w}_r}{z}\right)\frac{Z(\mathsf{w}_r)}{W_r(\mathsf{w}_r)} D_r^{-1}, \sum_{s=1}^{a} \delta\left(\frac{\boldsymbol{v}^2\mathsf{w}_s}{w}\right)\frac{1}{W_s(\mathsf{w}_s)} D_s\right] = \frac{-\boldsymbol{v}}{(1-\boldsymbol{v}^2)^2}\prod_{t=1}^{a}\mathsf{w}_t \times$$

$$\left\{\sum_{r=1}^{a}\left(\delta\left(\frac{\mathsf{w}_r}{z}\right)\delta\left(\frac{\mathsf{w}_r}{w}\right)\frac{Z(\mathsf{w}_r)}{W_r(\mathsf{w}_r)W_r(\boldsymbol{v}^{-2}\mathsf{w}_r)} - \boldsymbol{v}^2\delta\left(\frac{\boldsymbol{v}^2\mathsf{w}_r}{z}\right)\delta\left(\frac{\boldsymbol{v}^2\mathsf{w}_r}{w}\right)\frac{Z(\boldsymbol{v}^2\mathsf{w}_r)}{W_r(\mathsf{w}_r)W_r(\boldsymbol{v}^2\mathsf{w}_r)}\right) + \right.$$

$$\left.\sum_{1\le r\ne s\le a}\delta\left(\frac{\mathsf{w}_r}{z}\right)\delta\left(\frac{\boldsymbol{v}^2\mathsf{w}_s}{w}\right)\frac{Z(\mathsf{w}_r)}{W_{rs}(\mathsf{w}_r)W_{rs}(\mathsf{w}_s)}\left(\frac{1}{A(z,w)} - \frac{\boldsymbol{v}^2}{B(z,w)}\right)D_r^{-1}D_s\right\},$$

where $A(z,w) = (1-\boldsymbol{v}^{-2}w/z)(1-\boldsymbol{v}^{-2}z/\boldsymbol{v}^{-2}w)$ and $B(z,w) = (1-z/\boldsymbol{v}^{-2}w)(1-w/z)$. The second sum is zero as $A(z,w) = \boldsymbol{v}^{-2}B(z,w)$.

To evaluate the RHS, we need the following standard result.

Lemma C.1 *For any rational function $\gamma(z)$ with simple poles $\{x_t\} \subset \mathbb{C}^\times$ and possibly poles of higher order at $z = 0, \infty$, the following equality holds:*

$$\gamma(z)^+ - \gamma(z)^- = \sum_t \delta\left(\frac{z}{x_t}\right)\mathrm{Res}_{z=x_t}\gamma(z)\frac{dz}{z}. \tag{C.2}$$

Proof Consider the partial fraction decomposition of $\gamma(z)$:

$$\gamma(z) = P(z) + \sum_t \frac{\nu_t}{z - x_t},$$

where $P(z)$ is a Laurent polynomial. Then $P(z)^{\pm} = P(z) \rightarrow P(z)^{+} - P(z)^{-} = 0$. Meanwhile:

$$\left(\frac{\nu_t}{z - x_t}\right)^{+} = \frac{\nu_t}{z} + \frac{\nu_t x_t}{z^2} + \frac{\nu_t x_t^2}{z^3} + \dots \text{ and } \left(\frac{\nu_t}{z - x_t}\right)^{-} = -\frac{\nu_t}{x_t} - \frac{\nu_t z}{x_t^2} - \frac{\nu_t z^2}{x_t^3} - \dots,$$

so that

$$\left(\frac{\nu_t}{z - x_t}\right)^{+} - \left(\frac{\nu_t}{z - x_t}\right)^{-} = \frac{\nu_t}{x_t}\delta\left(\frac{z}{x_t}\right) = \delta\left(\frac{z}{x_t}\right) \cdot \mathrm{Res}_{z=x_t} \frac{\nu_t}{z - x_t}\frac{dz}{z}.$$

The lemma is proved. □

Since $\Psi(z)$ is a rational function in z, which has (simple) poles only at $\{\mathsf{w}_r, \boldsymbol{v}^2\mathsf{w}_r\}_{r=1}^{a}$ and possibly poles of higher order at $z = 0, \infty$, we can apply Lemma C.1 to evaluate $\Psi(z)^{+} - \Psi(z)^{-}$:

$$\Psi(z)^{+} - \Psi(z)^{-} = \prod_{t=1}^{a} \mathsf{w}_t \cdot \sum_{r=1}^{a}\left(\delta\left(\frac{z}{\mathsf{w}_r}\right)\frac{Z(\mathsf{w}_r)}{W_r(\mathsf{w}_r)W(\boldsymbol{v}^{-2}\mathsf{w}_r)} + \delta\left(\frac{z}{\boldsymbol{v}^2\mathsf{w}_r}\right)\frac{Z(\boldsymbol{v}^2\mathsf{w}_r)}{W_r(\mathsf{w}_r)W(\boldsymbol{v}^2\mathsf{w}_r)}\right) =$$

$$\frac{1}{1-\boldsymbol{v}^2}\prod_{t=1}^{a} \mathsf{w}_t \cdot \sum_{r=1}^{a}\left(\delta\left(\frac{\mathsf{w}_r}{z}\right)\frac{Z(\mathsf{w}_r)}{W_r(\mathsf{w}_r)W_r(\boldsymbol{v}^{-2}\mathsf{w}_r)} - \boldsymbol{v}^2\delta\left(\frac{\boldsymbol{v}^2\mathsf{w}_r}{z}\right)\frac{Z(\boldsymbol{v}^2\mathsf{w}_r)}{W_r(\mathsf{w}_r)W_r(\boldsymbol{v}^2\mathsf{w}_r)}\right).$$

Hence, the RHS equals

$$\frac{1}{(\boldsymbol{v}-\boldsymbol{v}^{-1})(1-\boldsymbol{v}^2)}\prod_{t=1}^{a}\mathsf{w}_t \times$$

$$\sum_{r=1}^{a}\left(\delta\left(\frac{\mathsf{w}_r}{z}\right)\delta\left(\frac{\mathsf{w}_r}{w}\right)\frac{Z(\mathsf{w}_r)}{W_r(\mathsf{w}_r)W_r(\boldsymbol{v}^{-2}\mathsf{w}_r)} - \delta\left(\frac{\boldsymbol{v}^2\mathsf{w}_r}{z}\right)\delta\left(\frac{\boldsymbol{v}^2\mathsf{w}_r}{w}\right)\frac{\boldsymbol{v}^2 Z(\mathsf{w}_r)}{W_r(\mathsf{w}_r)W_r(\boldsymbol{v}^2\mathsf{w}_r)}\right).$$

As a result, we finally get LHS = RHS.

Case $c_{ij} < 0, i \rightarrow j$ We may assume $I = \{i, j\}$, and we need to check $[E_i(z), F_j(w)] = 0$. We have

$$\frac{[E_i(z), F_j(w)]}{\rho_i^+\rho_j^-} = \prod_{t=1}^{a_i}\mathsf{w}_{i,t} \cdot \left[\sum_{r=1}^{a_i}\delta\left(\frac{\mathsf{w}_{i,r}}{z}\right)\frac{Z_i(\mathsf{w}_{i,r})}{W_{i,r}(\mathsf{w}_{i,r})}D_{i,r}^{-1}, \sum_{s=1}^{a_j}\delta\left(\frac{\boldsymbol{v}_j^2\mathsf{w}_{j,s}}{w}\right)\frac{1}{W_{j,s}(\mathsf{w}_{j,s})}D_{j,s}\right].$$

The latter is obviously zero, since $[D_{i,r}^{-1}, \mathsf{w}_{j,s}] = 0 = [D_{j,s}, \mathsf{w}_{i,r}]$.

Case $c_{ij} < 0, i \leftarrow j$ We may assume $I = \{i, j\}$, and we need to check $E_i(z)F_j(w)/(\rho_i^+\rho_j^-) = F_j(w)E_i(z)/(\rho_i^+\rho_j^-)$. The LHS equals

$$\boldsymbol{v}_i^{-c_{ij}} \prod_{t=1}^{a_i} \mathsf{w}_{i,t}^{1+c_{ij}/2} \prod_{t=1}^{a_j} \mathsf{w}_{j,t}^{c_{ji}/2} \times$$

$$\sum_{1\le r\le a_i}^{1\le s\le a_j} \delta\left(\frac{\mathsf{w}_{i,r}}{z}\right)\delta\left(\frac{\boldsymbol{v}_j^2\mathsf{w}_{j,s}}{w}\right) \frac{Z_i(\mathsf{w}_{i,r})\prod_{p=1}^{-c_{ji}} W_j(\boldsymbol{v}_j^{-c_{ji}-2p}z)}{W_{i,r}(\mathsf{w}_{i,r})} D_{i,r}^{-1} \frac{\prod_{p'=1}^{-c_{ij}} W_i(\boldsymbol{v}_i^{-c_{ij}-2p'}w)}{W_{j,s}(\mathsf{w}_{j,s})} D_{j,s} =$$

$$\prod_{t=1}^{a_i} \mathsf{w}_{i,t}^{1+c_{ij}/2} \prod_{t=1}^{a_j} \mathsf{w}_{j,t}^{c_{ji}/2} \cdot A(z,w) \times$$

$$\sum_{1\le r\le a_i}^{1\le s\le a_j} \delta\left(\frac{\mathsf{w}_{i,r}}{z}\right)\delta\left(\frac{\boldsymbol{v}_j^2\mathsf{w}_{j,s}}{w}\right) \frac{Z_i(\mathsf{w}_{i,r})\prod_{p=1}^{-c_{ji}} W_{j,s}(\boldsymbol{v}_j^{-c_{ji}-2p}z)\prod_{p'=1}^{-c_{ij}} W_{i,r}(\boldsymbol{v}_i^{-c_{ij}-2p'}w)}{W_{i,r}(\mathsf{w}_{i,r})W_{j,s}(\mathsf{w}_{j,s})} D_{i,r}^{-1} D_{j,s},$$

where $A(z,w) = \boldsymbol{v}_i^{-c_{ij}} \prod_{p=1}^{-c_{ji}} \left(1 - \frac{\boldsymbol{v}_j^{-2}w}{\boldsymbol{v}_j^{-c_{ji}-2p}z}\right) \prod_{p'=1}^{-c_{ij}} \left(1 - \frac{\boldsymbol{v}_i^{-2}z}{\boldsymbol{v}_i^{-c_{ij}-2p'}w}\right)$.

Likewise, the RHS equals

$$\boldsymbol{v}_j^{c_{ji}} \prod_{t=1}^{a_i} \mathsf{w}_{i,t}^{1+c_{ij}/2} \prod_{t=1}^{a_j} \mathsf{w}_{j,t}^{c_{ji}/2} \times$$

$$\sum_{1\le r\le a_i}^{1\le s\le a_j} \delta\left(\frac{\boldsymbol{v}_j^2\mathsf{w}_{j,s}}{w}\right)\delta\left(\frac{\mathsf{w}_{i,r}}{z}\right) \frac{\prod_{p'=1}^{-c_{ij}} W_i(\boldsymbol{v}_i^{-c_{ij}-2p'}w)}{W_{j,s}(\mathsf{w}_{j,s})} D_{j,s} \frac{Z_i(\mathsf{w}_{i,r})\prod_{p=1}^{-c_{ji}} W_j(\boldsymbol{v}_j^{-c_{ji}-2p}z)}{W_{i,r}(\mathsf{w}_{i,r})} D_{i,r}^{-1} =$$

$$\prod_{t=1}^{a_i} \mathsf{w}_{i,t}^{1+c_{ij}/2} \prod_{t=1}^{a_j} \mathsf{w}_{j,t}^{c_{ji}/2} \cdot B(z,w) \times$$

$$\sum_{1\le r\le a_i}^{1\le s\le a_j} \delta\left(\frac{\mathsf{w}_{i,r}}{z}\right)\delta\left(\frac{\boldsymbol{v}_j^2\mathsf{w}_{j,s}}{w}\right) \frac{Z_i(\mathsf{w}_{i,r})\prod_{p=1}^{-c_{ji}} W_{j,s}(\boldsymbol{v}_j^{-c_{ji}-2p}z)\prod_{p'=1}^{-c_{ij}} W_{i,r}(\boldsymbol{v}_i^{-c_{ij}-2p'}w)}{W_{i,r}(\mathsf{w}_{i,r})W_{j,s}(\mathsf{w}_{j,s})} D_{i,r}^{-1} D_{j,s},$$

where $B(z,w) = \boldsymbol{v}_j^{c_{ji}} \prod_{p=1}^{-c_{ji}} \left(1 - \frac{w}{\boldsymbol{v}_j^{-c_{ji}-2p}z}\right) \prod_{p'=1}^{-c_{ij}} \left(1 - \frac{z}{\boldsymbol{v}_i^{-c_{ij}-2p'}w}\right)$.

The equality LHS = RHS follows from $A(z,w) = B(z,w)$.

C(vii) Compatibility with (U7)

Case $c_{ij} = 0$ In this case, $[E_i(z), E_j(w)] = 0$, due to our verification of (U2).

Case $c_{ij} < 0$ To simplify our calculations, we introduce

$$\chi_{i',r} := \prod_{t=1}^{a_{i'}} \mathsf{w}_{i',t} \cdot \prod_{j'\to i'} \prod_{t=1}^{a_{j'}} \mathsf{w}_{j',t}^{c_{j'i'}/2} \cdot \frac{Z_{i'}(\mathsf{w}_{i',r})}{W_{i',r}(\mathsf{w}_{i',r})} \prod_{j'\to i'} \prod_{p=1}^{-c_{j'i'}} W_{j'}(\boldsymbol{v}_{j'}^{-c_{j'i'}-2p} \mathsf{w}_{i',r}) D_{i',r}^{-1},$$

so that $E_{i'}(z) = \rho_{i'}^{+} \sum_{r=1}^{a_{i'}} \delta\left(\frac{\mathsf{w}_{i',r}}{z}\right) \chi_{i',r}$.

The verification of (U7) is based on the following result.

Lemma C.2 *The following relations hold:*

$$\chi_{i,r}\mathsf{w}_{j,s} = \boldsymbol{v}_i^{-2\delta_{ij}\delta_{rs}} \mathsf{w}_{j,s}\chi_{i,r} \text{ for } 1 \le r \le a_i,\ 1 \le s \le a_j,$$

$$(\mathsf{w}_{i,r_1} - \boldsymbol{v}_i^2 \mathsf{w}_{i,r_2})\chi_{i,r_1}\chi_{i,r_2} = (\boldsymbol{v}_i^2 \mathsf{w}_{i,r_1} - \mathsf{w}_{i,r_2})\chi_{i,r_2}\chi_{i,r_1} \text{ for } 1 \le r_1 \ne r_2 \le a_i,$$

$$(\mathsf{w}_{i,r} - \boldsymbol{v}_i^{c_{ij}} \mathsf{w}_{j,s})\chi_{i,r}\chi_{j,s} = (\boldsymbol{v}_i^{c_{ij}} \mathsf{w}_{i,r} - \mathsf{w}_{j,s})\chi_{j,s}\chi_{i,r} \text{ for } 1 \le r \le a_i,\ 1 \le s \le a_j.$$

Proof Follows from straightforward computations. □

With the help of this lemma, let us verify (U7) for $c_{ij} = -1$. The latter amounts to proving $[E_i(z_1), [E_i(z_2), E_j(w)]_{\boldsymbol{v}}]_{\boldsymbol{v}^{-1}}/((\rho_i^+)^2\rho_j^+) = -[E_i(z_2), [E_i(z_1), E_j(w)]_{\boldsymbol{v}}]_{\boldsymbol{v}^{-1}}/((\rho_i^+)^2\rho_j^+)$. The LHS equals

$$(1-\boldsymbol{v}^2)\left[\sum_{r_1=1}^{a_i} \delta\left(\frac{\mathsf{w}_{i,r_1}}{z_1}\right)\chi_{i,r_1}, \sum_{1\le r_2\le a_i}^{1\le s\le a_j} \delta\left(\frac{\mathsf{w}_{i,r_2}}{z_2}\right)\delta\left(\frac{\mathsf{w}_{j,s}}{w}\right)\frac{\mathsf{w}_{i,r_2}}{\mathsf{w}_{i,r_2}-\boldsymbol{v}\mathsf{w}_{j,s}}\chi_{i,r_2}\chi_{j,s}\right]_{\boldsymbol{v}^{-1}} =$$

$$\sum_{1\le r\le a_i}^{1\le s\le a_j} \delta\left(\frac{\mathsf{w}_{j,s}}{w}\right)\left\{\delta\left(\frac{\mathsf{w}_{i,r}}{z_1}\right)\delta\left(\frac{\boldsymbol{v}^{-2}\mathsf{w}_{i,r}}{z_2}\right) - \delta\left(\frac{\mathsf{w}_{i,r}}{z_2}\right)\delta\left(\frac{\boldsymbol{v}^{-2}\mathsf{w}_{i,r}}{z_1}\right)\right\}\frac{(\boldsymbol{v}^2-1)\mathsf{w}_{i,r}}{\mathsf{w}_{i,r}-\boldsymbol{v}^3\mathsf{w}_{j,s}}\chi_{i,r}^2\chi_{j,s} -$$

$$(\boldsymbol{v}^2-1)^2 \sum_{1\le r_1\ne r_2\le a_i}^{1\le s\le a_j} \delta\left(\frac{\mathsf{w}_{i,r_1}}{z_1}\right)\delta\left(\frac{\mathsf{w}_{i,r_2}}{z_2}\right)\delta\left(\frac{\mathsf{w}_{j,s}}{w}\right)\frac{A(z_1,z_2,w)}{\boldsymbol{v}^2\mathsf{w}_{i,r_1}-\mathsf{w}_{i,r_2}}\chi_{i,r_1}\chi_{i,r_2}\chi_{j,s},$$

where $A(z_1,z_2,w) = \frac{z_1z_2(z_1+z_2-(\boldsymbol{v}+\boldsymbol{v}^{-1})w)}{(z_1-\boldsymbol{v}w)(z_2-\boldsymbol{v}w)}$ and the last equality is obtained by treating separately $r_1 = r_2$ and $r_1 \ne r_2$ cases. The first sum is obviously skew-symmetric in z_1, z_2. The second sum is also skew-symmetric, due to the above relations on $\chi_{i,r}$.

The cases $c_{ij} = -2, -3$ can be treated similarly, but the corresponding computations become more cumbersome. We verified these cases using **MATLAB**.

C(viii) Compatibility with (U8)

The case $c_{ij} = 0$ is obvious. The case $c_{ij} = -1$ can be treated analogously to the above verification of (U7). The verification for the cases $c_{ij} = -2, -3$ is more cumbersome and can be performed as outlined in the verification of (U7). Our verification involved a simple computation in **MATLAB**.

This completes our proof of Theorem 7.1.

Remark C.3 Theorem 7.1 admits the following straightforward generalization. For every $i \in I$, pick two polynomials $Z_i^{(1)}(z), Z_i^{(2)}(z)$ in z^{-1} such that $Z_i(z) = Z_i^{(1)}(z)Z_i^{(2)}(z)$. There is a unique $\mathbb{C}(\boldsymbol{v})[\mathsf{z}_1^{\pm 1}, \ldots, \mathsf{z}_N^{\pm 1}]$-algebra homomorphism $\mathcal{U}^{\mathrm{ad}}_{0,\mu}[\mathsf{z}_1^{\pm 1}, \ldots, \mathsf{z}_N^{\pm 1}] \to \widetilde{\mathcal{A}}^{\boldsymbol{v}}_{\mathrm{frac}}[\mathsf{z}_1^{\pm 1}, \ldots, \mathsf{z}_N^{\pm 1}]$, such that

$$e_i(z) \mapsto \frac{-\boldsymbol{v}_i}{1-\boldsymbol{v}_i^2} \prod_{t=1}^{a_i} \mathsf{w}_{i,t} \prod_{j\to i}\prod_{t=1}^{a_j} \mathsf{w}_{j,t}^{c_{ji}/2} \cdot \sum_{r=1}^{a_i} \delta\left(\frac{\mathsf{w}_{i,r}}{z}\right) \frac{Z_i^{(1)}(\mathsf{w}_{i,r})}{W_{i,r}(\mathsf{w}_{i,r})} \prod_{j\to i}\prod_{p=1}^{-c_{ji}} W_j(\boldsymbol{v}_j^{-c_{ji}-2p}z) D_{i,r}^{-1},$$

$$f_i(z) \mapsto \frac{1}{1-\boldsymbol{v}_i^2} \prod_{j\leftarrow i}\prod_{t=1}^{a_j} \mathsf{w}_{j,t}^{c_{ji}/2} \cdot \sum_{r=1}^{a_i} \delta\left(\frac{\boldsymbol{v}_i^2\mathsf{w}_{i,r}}{z}\right) \frac{Z_i^{(2)}(\boldsymbol{v}_i^2\mathsf{w}_{i,r})}{W_{i,r}(\mathsf{w}_{i,r})} \prod_{j\leftarrow i}\prod_{p=1}^{-c_{ji}} W_j(\boldsymbol{v}_j^{-c_{ji}-2p}z) D_{i,r},$$

$$\psi_i^{\pm}(z) \mapsto \Psi_i(z)^{\pm},\ (\phi_i^+)^{\pm 1} \mapsto \prod_{t=1}^{a_i} \mathsf{w}_{i,t}^{\pm 1/2},\ (\phi_i^-)^{\pm 1} \mapsto (-\boldsymbol{v}_i)^{\mp a_i} \prod_{t=1}^{a_i} \mathsf{w}_{i,t}^{\mp 1/2}.$$

Appendix D Proof of Theorem 10.5

Due to Theorem 5.5, it suffices to check that the assignment Δ of Theorem 10.5 preserves defining relations ($\hat{\mathrm{U}}1$–$\hat{\mathrm{U}}6$, $\hat{\mathrm{U}}9$). To simplify our exposition, we will assume that $b_1, b_2 < 0$, while the case when one of them is zero is left to the interested reader (note that the case $b_1 = b_2 = 0$ has been treated in Remark 10.4). We will also work with $\bar{h}_{\pm 1} := [2]_{\boldsymbol{v}}^{-1} h_{\pm 1}$ instead of $h_{\pm 1}$, so that $[\bar{h}_{\pm 1}, e_r] = e_{r\pm 1}, [\bar{h}_{\pm 1}, f_r] = -f_{r\pm 1}$.

D(i) Compatibility with ($\hat{U}1$)

The equalities $\Delta((\psi_0^+)^{\pm 1})\Delta((\psi_0^+)^{\mp 1}) = 1$ and $\Delta((\psi_b^-)^{\pm 1})\Delta((\psi_b^-)^{\mp 1}) = 1$ follow immediately from relation ($\hat{\mathrm{U}}1$) for both $\mathcal{U}^{\mathrm{sc}}_{0,b_1}$ and $\mathcal{U}^{\mathrm{sc}}_{0,b_2}$.

The commutativity of $\Delta((\psi_0^+)^{\pm 1}), \Delta((\psi_b^-)^{\pm 1})$ between themselves and with each of $\Delta(\bar{h}_{\pm 1})$ is due to relations ($\hat{\mathrm{U}}1$, $\hat{\mathrm{U}}4$, $\hat{\mathrm{U}}5$) for both $\mathcal{U}^{\mathrm{sc}}_{0,b_1}$ and $\mathcal{U}^{\mathrm{sc}}_{0,b_2}$.

It remains to prove $[\Delta(\bar{h}_1), \Delta(\bar{h}_{-1})] = 0$. The LHS is equal to $[\bar{h}_1 \otimes 1 + 1 \otimes \bar{h}_1 - (\boldsymbol{v} - \boldsymbol{v}^{-1})e_0 \otimes f_1, \bar{h}_{-1} \otimes 1 + 1 \otimes \bar{h}_{-1} + (\boldsymbol{v} - \boldsymbol{v}^{-1})e_{-1} \otimes f_0] = (\boldsymbol{v}-\boldsymbol{v}^{-1})(e_0 \otimes f_0 - e_{-1} \otimes f_1 + e_{-1} \otimes f_1 - e_0 \otimes f_0) - (\boldsymbol{v}-\boldsymbol{v}^{-1})^2[e_0 \otimes f_1, e_{-1} \otimes f_0] = -(\boldsymbol{v} - \boldsymbol{v}^{-1})^2(e_0e_{-1} \otimes f_1f_0 - e_{-1}e_0 \otimes f_0f_1) = 0$.

Here we used ($\hat{\mathrm{U}}1$, $\hat{\mathrm{U}}4$, $\hat{\mathrm{U}}5$) for both $\mathcal{U}^{\mathrm{sc}}_{0,b_1}, \mathcal{U}^{\mathrm{sc}}_{0,b_2}$ in the first equality, while the second equality follows from $e_0e_{-1} = \boldsymbol{v}^2e_{-1}e_0$, $f_1f_0 = \boldsymbol{v}^{-2}f_0f_1$, due to ($\hat{\mathrm{U}}2$) for $\mathcal{U}^{\mathrm{sc}}_{0,b_1}$ and ($\hat{\mathrm{U}}3$) for $\mathcal{U}^{\mathrm{sc}}_{0,b_2}$.

D(ii) Compatibility with ($\hat{\mathrm{U}}2$)

We need to prove $[\Delta(e_{r+1}), \Delta(e_s)]_{\boldsymbol{v}^2} + [\Delta(e_{s+1}), \Delta(e_r)]_{\boldsymbol{v}^2} = 0$ for $b_2 - 1 \le r, s \le -1$.

Case $b_2 - 1 < r, s < -1$ Then, $[\Delta(e_{r+1}), \Delta(e_s)]_{\boldsymbol{v}^2} + [\Delta(e_{s+1}), \Delta(e_r)]_{\boldsymbol{v}^2} = 1 \otimes ([e_{r+1}, e_s]_{\boldsymbol{v}^2} + [e_{s+1}, e_r]_{\boldsymbol{v}^2}) = 0$ as the second term is zero in $\mathcal{U}^{\mathrm{sc}}_{0,b_2}$ by ($\hat{\mathrm{U}}2$).

Case $r = s = b_2 - 1$ It suffices to show that $[\Delta(e_{b_2}), \Delta(e_{b_2-1})]_{\boldsymbol{v}^2} = 0$, which follows from $[\Delta(e_{b_2}), \Delta(e_{b_2-1})]_{\boldsymbol{v}^2} = [1 \otimes e_{b_2}, e_{-1} \otimes \psi^-_{b_2} + 1 \otimes e_{b_2-1}]_{\boldsymbol{v}^2} = e_{-1} \otimes [e_{b_2}, \psi^-_{b_2}]_{\boldsymbol{v}^2} + 1 \otimes [e_{b_2}, e_{b_2-1}]_{\boldsymbol{v}^2} = 0$. The last equality follows from $[e_{b_2}, \psi^-_{b_2}]_{\boldsymbol{v}^2} = 0$ and $[e_{b_2}, e_{b_2-1}]_{\boldsymbol{v}^2} = 0$ in $\mathcal{U}^{\mathrm{sc}}_{0,b_2}$, due to ($\hat{\mathrm{U}}2$) and ($\hat{\mathrm{U}}4$), respectively.

Case $r = b_2 - 1, b_2 - 1 < s < -1$ Then, $[\Delta(e_{b_2}), \Delta(e_s)]_{\boldsymbol{v}^2} + [\Delta(e_{s+1}), \Delta(e_{b_2-1})]_{\boldsymbol{v}^2} = 1 \otimes ([e_{b_2}, e_s]_{\boldsymbol{v}^2} + [e_{s+1}, e_{b_2-1}]_{\boldsymbol{v}^2}) + e_{-1} \otimes [e_{s+1}, \psi^-_{b_2}]_{\boldsymbol{v}^2} = 0$. The last equality follows again from ($\hat{\mathrm{U}}2$) and ($\hat{\mathrm{U}}4$) for $\mathcal{U}^{\mathrm{sc}}_{0,b_2}$.

Case $r = b_2 - 1, s = -1$ Then $[\Delta(e_{b_2}), \Delta(e_{-1})]_{\boldsymbol{v}^2} = 1 \otimes [e_{b_2}, e_{-1}]_{\boldsymbol{v}^2}$ and $[\Delta(e_0), \Delta(e_{b_2-1})]_{\boldsymbol{v}^2} = [e_0 \otimes \psi^+_0 + 1 \otimes e_0, e_{-1} \otimes \psi^-_{b_2} + 1 \otimes e_{b_2-1}]_{\boldsymbol{v}^2} = e_0 \otimes [\psi^+_0, e_{b_2-1}]_{\boldsymbol{v}^2} + [e_0, e_{-1}]_{\boldsymbol{v}^2} \otimes \psi^+_0\psi^-_{b_2} + e_{-1} \otimes [e_0, \psi^-_{b_2}]_{\boldsymbol{v}^2} + 1 \otimes [e_0, e_{b_2-1}]_{\boldsymbol{v}^2} = 1 \otimes [e_0, e_{b_2-1}]_{\boldsymbol{v}^2}$ as the first three terms are zero, due to ($\hat{\mathrm{U}}2$) for $\mathcal{U}^{\mathrm{sc}}_{0,b_1}$ and ($\hat{\mathrm{U}}4$) for $\mathcal{U}^{\mathrm{sc}}_{0,b_2}$. The result follows from ($\hat{\mathrm{U}}2$) for $\mathcal{U}^{\mathrm{sc}}_{0,b_2}$.

Case $r = s = -1$ It suffices to show that $[\Delta(e_0), \Delta(e_{-1})]_{\boldsymbol{v}^2} = 0$, which follows from $[\Delta(e_0), \Delta(e_{-1})]_{\boldsymbol{v}^2} = [e_0 \otimes \psi^+_0 + 1 \otimes e_0, 1 \otimes e_{-1}]_{\boldsymbol{v}^2} = e_0 \otimes [\psi^+_0, e_{-1}]_{\boldsymbol{v}^2} + 1 \otimes [e_0, e_{-1}]_{\boldsymbol{v}^2} = 0$. The last equality follows again from relations ($\hat{\mathrm{U}}2$, $\hat{\mathrm{U}}4$) for the algebra $\mathcal{U}^{\mathrm{sc}}_{0,b_2}$.

Case $r = -1, b_2 - 1 < s < -1$ Then, $[\Delta(e_0), \Delta(e_s)]_{\boldsymbol{v}^2} = [e_0 \otimes \psi^+_0 + 1 \otimes e_0, 1 \otimes e_s]_{\boldsymbol{v}^2} = 1 \otimes [e_0, e_s]_{\boldsymbol{v}^2}$, while $[\Delta(e_{s+1}), \Delta(e_{-1})]_{\boldsymbol{v}^2} = 1 \otimes [e_{s+1}, e_{-1}]_{\boldsymbol{v}^2}$. The sum of these two terms is zero, due to ($\hat{\mathrm{U}}2$) for $\mathcal{U}^{\mathrm{sc}}_{0,b_2}$.

D(iii) Compatibility with (Û3)

We need to prove $[\Delta(f_r), \Delta(f_{s+1})]_{\boldsymbol{v}^2} + [\Delta(f_s), \Delta(f_{r+1})]_{\boldsymbol{v}^2} = 0$ for $b_1 \leq r, s \leq 0$.

Case $b_1 < r, s < 0$ Then, $[\Delta(f_r), \Delta(f_{s+1})]_{\boldsymbol{v}^2} + [\Delta(f_s), \Delta(f_{r+1})]_{\boldsymbol{v}^2} = ([f_r, f_{s+1}]_{\boldsymbol{v}^2} + [f_s, f_{r+1}]_{\boldsymbol{v}^2}) \otimes 1 = 0$ as the first term is zero in $\mathcal{U}^{\mathrm{sc}}_{0,b_1}$ by (Û3).

Case $r = s = b_1$ It suffices to show that $[\Delta(f_{b_1}), \Delta(f_{1+b_1})]_{\boldsymbol{v}^2} = 0$, which follows from $[\Delta(f_{b_1}), \Delta(f_{1+b_1})]_{\boldsymbol{v}^2} = [f_{b_1} \otimes 1 + \psi^-_{b_1} \otimes f_0, f_{1+b_1} \otimes 1]_{\boldsymbol{v}^2} = [f_{b_1}, f_{1+b_1}]_{\boldsymbol{v}^2} \otimes 1 + [\psi^-_{b_1}, f_{1+b_1}]_{\boldsymbol{v}^2} \otimes f_0 = 0$. The last equality follows from $[f_{b_1}, f_{1+b_1}]_{\boldsymbol{v}^2} = 0 = [\psi^-_{b_1}, f_{1+b_1}]_{\boldsymbol{v}^2}$, due to (Û3, Û5) for $\mathcal{U}^{\mathrm{sc}}_{0,b_1}$.

Case $r = b_1 < s < 0$ Then, $[\Delta(f_s), \Delta(f_{1+b_1})]_{\boldsymbol{v}^2} = [f_s, f_{1+b_1}]_{\boldsymbol{v}^2} \otimes 1$ and $[\Delta(f_{b_1}), \Delta(f_{s+1})]_{\boldsymbol{v}^2} = [f_{b_1}, f_{s+1}]_{\boldsymbol{v}^2} \otimes 1$ as $[\psi^-_{b_1}, f_{s+1}]_{\boldsymbol{v}^2} = 0$ in $\mathcal{U}^{\mathrm{sc}}_{0,b_1}$ by (Û5). It remains to use (Û3) for $\mathcal{U}^{\mathrm{sc}}_{0,b_1}$.

Case $r = b_1, s = 0$ Then $[\Delta(f_{b_1}), \Delta(f_1)]_{\boldsymbol{v}^2} = [f_{b_1}, f_1]_{\boldsymbol{v}^2} \otimes 1 + [f_{b_1}, \psi^+_0]_{\boldsymbol{v}^2} \otimes f_1 + [\psi^-_{b_1}, f_1]_{\boldsymbol{v}^2} \otimes f_0 + \psi^-_{b_1}\psi^+_0 \otimes [f_0, f_1]_{\boldsymbol{v}^2}$, and $[\Delta(f_0), \Delta(f_{1+b_1})]_{\boldsymbol{v}^2} = [f_0, f_{1+b_1}]_{\boldsymbol{v}^2} \otimes 1$. It remains to use $[f_{b_1}, f_1]_{\boldsymbol{v}^2} + [f_0, f_{1+b_1}]_{\boldsymbol{v}^2} = [f_{b_1}, \psi^+_0]_{\boldsymbol{v}^2} = [\psi^-_{b_1}, f_1]_{\boldsymbol{v}^2} = 0$ in $\mathcal{U}^{\mathrm{sc}}_{0,b_1}$, due (Û3) and (Û5), and $[f_0, f_1]_{\boldsymbol{v}^2} = 0$ in $\mathcal{U}^{\mathrm{sc}}_{0,b_2}$, due to (Û3).

Case $r = s = 0$ It suffices to show that $[\Delta(f_0), \Delta(f_1)]_{\boldsymbol{v}^2} = 0$, which follows from $[\Delta(f_0), \Delta(f_1)]_{\boldsymbol{v}^2} = [f_0 \otimes 1, f_1 \otimes 1 + \psi^+_0 \otimes f_1]_{\boldsymbol{v}^2} = [f_0, f_1]_{\boldsymbol{v}^2} \otimes 1 + [f_0, \psi^+_0]_{\boldsymbol{v}^2} \otimes f_1 = 0$, due to (Û3, Û5) for $\mathcal{U}^{\mathrm{sc}}_{0,b_1}$.

Case $r = 0, b_1 < s < 0$ Then $[\Delta(f_0), \Delta(f_{s+1})]_{\boldsymbol{v}^2} = [f_0, f_{s+1}]_{\boldsymbol{v}^2} \otimes 1$, and $[\Delta(f_s), \Delta(f_1)]_{\boldsymbol{v}^2} = [f_s \otimes 1, f_1 \otimes 1 + \psi^+_0 \otimes f_1]_{\boldsymbol{v}^2} = [f_s, f_1]_{\boldsymbol{v}^2} \otimes 1 + [f_s, \psi^+_0]_{\boldsymbol{v}^2} \otimes f_1$. It remains to apply the equalities $[f_0, f_{s+1}]_{\boldsymbol{v}^2} + [f_s, f_1]_{\boldsymbol{v}^2} = 0$ and $[f_s, \psi^+_0]_{\boldsymbol{v}^2} = 0$ in $\mathcal{U}^{\mathrm{sc}}_{0,b_1}$, due to (Û3) and (Û5).

D(iv) Compatibility with (Û4)

The equalities $\Delta(\psi^+_0)\Delta(e_r) = \boldsymbol{v}^2\Delta(e_r)\Delta(\psi^+_0)$ and $\Delta(\psi^-_b)\Delta(e_r) = \boldsymbol{v}^{-2}\Delta(e_r)\Delta(\psi^-_b)$ for $b_2 - 1 \leq r \leq 0$ are obvious, due to relations (Û1) and (Û4) for $\mathcal{U}^{\mathrm{sc}}_{0,b_1}, \mathcal{U}^{\mathrm{sc}}_{0,b_2}$.

Let us now verify the equality $[\Delta(\bar{h}_1), \Delta(e_r)] = \Delta(e_{r+1})$ for $b_2 - 1 \leq r \leq -1$.

Case $b_2 \leq r \leq -2$ We have $[\Delta(\bar{h}_1), \Delta(e_r)] = [\bar{h}_1 \otimes 1 + 1 \otimes \bar{h}_1 - (\boldsymbol{v} - \boldsymbol{v}^{-1})e_0 \otimes f_1, 1 \otimes e_r] = 1 \otimes e_{r+1} - (\boldsymbol{v} - \boldsymbol{v}^{-1})e_0 \otimes [f_1, e_r] = 1 \otimes e_{r+1} = \Delta(e_{r+1})$, due to (Û4, Û6) for $\mathcal{U}^{\mathrm{sc}}_{0,b_2}$.

Case $r=-1$ As above, we get $[\Delta(\bar{h}_1), \Delta(e_{-1})] = [\bar{h}_1 \otimes 1 + 1 \otimes \bar{h}_1 - (\boldsymbol{v}-\boldsymbol{v}^{-1})e_0 \otimes f_1, 1 \otimes e_{-1}] = 1 \otimes e_0 - (\boldsymbol{v}-\boldsymbol{v}^{-1})e_0 \otimes [f_1, e_{-1}] = 1 \otimes e_0 + (\boldsymbol{v}-\boldsymbol{v}^{-1})e_0 \otimes \frac{\psi_0^+}{\boldsymbol{v}-\boldsymbol{v}^{-1}} = \Delta(e_0)$.

Case $r=b_2-1$ We have $[\Delta(\bar{h}_1), \Delta(e_{b_2-1})] = [\bar{h}_1 \otimes 1 + 1 \otimes \bar{h}_1 - (\boldsymbol{v}-\boldsymbol{v}^{-1})e_0 \otimes f_1, e_{-1} \otimes \psi^-_{b_2} + 1 \otimes e_{b_2-1}] = e_0 \otimes \psi^-_{b_2} + 1 \otimes e_{b_2} - e_0 \otimes \psi^-_{b_2} - (\boldsymbol{v}-\boldsymbol{v}^{-1})[e_0 \otimes f_1, e_{-1} \otimes \psi^-_{b_2}] = 1 \otimes e_{b_2} = \Delta(e_{b_2})$, where we used $[e_0 \otimes f_1, e_{-1} \otimes \psi^-_{b_2}] = 0$ as $e_0 e_{-1} = \boldsymbol{v}^2 e_{-1} e_0$ in $\mathcal{U}^{\mathrm{sc}}_{0,b_1}$, due to $(\hat{\mathrm{U}}2)$, and $\psi^-_{b_2} f_1 = \boldsymbol{v}^2 f_1 \psi^-_{b_2}$ in $\mathcal{U}^{\mathrm{sc}}_{0,b_2}$, due to $(\hat{\mathrm{U}}5)$.

Let us now verify the equality $[\Delta(\bar{h}_{-1}), \Delta(e_r)] = \Delta(e_{r-1})$ for $b_2 \le r \le 0$.

Case $b_2<r<0$ We have $[\Delta(\bar{h}_{-1}), \Delta(e_r)] = [\bar{h}_{-1} \otimes 1 + 1 \otimes \bar{h}_{-1} + (\boldsymbol{v}-\boldsymbol{v}^{-1})e_{-1} \otimes f_0, 1 \otimes e_r] = 1 \otimes e_{r-1} + (\boldsymbol{v}-\boldsymbol{v}^{-1})e_{-1} \otimes [f_0, e_r] = 1 \otimes e_{r-1} = \Delta(e_{r-1})$, due to $(\hat{\mathrm{U}}4, \hat{\mathrm{U}}6)$ for $\mathcal{U}^{\mathrm{sc}}_{0,b_2}$.

Case $r=0$ We have $[\Delta(\bar{h}_{-1}), \Delta(e_0)] = [\bar{h}_{-1} \otimes 1 + 1 \otimes \bar{h}_{-1} + (\boldsymbol{v}-\boldsymbol{v}^{-1})e_{-1} \otimes f_0, e_0 \otimes \psi_0^+ + 1 \otimes e_0] = e_{-1} \otimes \psi_0^+ + 1 \otimes e_{-1} + (\boldsymbol{v}-\boldsymbol{v}^{-1})e_{-1} \otimes [f_0, e_0] + (\boldsymbol{v}-\boldsymbol{v}^{-1})[e_{-1} \otimes f_0, e_0 \otimes \psi_0^+] = 1 \otimes e_{-1} = \Delta(e_{-1})$, where we used $[e_{-1} \otimes f_0, e_0 \otimes \psi_0^+] = 0$ as $e_0 e_{-1} = \boldsymbol{v}^2 e_0 e_{-1}$ in $\mathcal{U}^{\mathrm{sc}}_{0,b_1}$, due to $(\hat{\mathrm{U}}2)$, and $f_0 \psi_0^+ = \boldsymbol{v}^2 \psi_0^+ f_0$ in $\mathcal{U}^{\mathrm{sc}}_{0,b_2}$, due to $(\hat{\mathrm{U}}5)$.

Case $r=b_2$ We have $[\Delta(\bar{h}_{-1}), \Delta(e_{b_2})] = [\bar{h}_{-1} \otimes 1 + 1 \otimes \bar{h}_{-1} + (\boldsymbol{v}-\boldsymbol{v}^{-1})e_{-1} \otimes f_0, 1 \otimes e_{b_2}] = 1 \otimes e_{b_2-1} + (\boldsymbol{v}-\boldsymbol{v}^{-1})e_{-1} \otimes \frac{\psi^-_{b_2}}{\boldsymbol{v}-\boldsymbol{v}^{-1}} = \Delta(e_{b_2-1})$, due to $(\hat{\mathrm{U}}4, \hat{\mathrm{U}}6)$ for $\mathcal{U}^{\mathrm{sc}}_{0,b_2}$.

D(v) Compatibility with $(\hat{\mathrm{U}}5)$

The equalities $\Delta(\psi_0^+)\Delta(f_r) = \boldsymbol{v}^{-2}\Delta(f_r)\Delta(\psi_0^+)$ and $\Delta(\psi_b^-)\Delta(f_r) = \boldsymbol{v}^2\Delta(f_r)\Delta(\psi_b^-)$ for $b_1 \le r \le 1$ are obvious, due to relations $(\hat{\mathrm{U}}1)$ and $(\hat{\mathrm{U}}5)$ for $\mathcal{U}^{\mathrm{sc}}_{0,b_1}, \mathcal{U}^{\mathrm{sc}}_{0,b_2}$.

Let us now verify the equality $[\Delta(\bar{h}_1), \Delta(f_r)] = -\Delta(f_{r+1})$ for $b_1 \le r \le 0$.

Case $b_1 < r < 0$ We have $[\Delta(\bar{h}_1), \Delta(f_r)] = [\bar{h}_1 \otimes 1 + 1 \otimes \bar{h}_1 - (\boldsymbol{v}-\boldsymbol{v}^{-1})e_0 \otimes f_1, f_r \otimes 1] = -f_{r+1} \otimes 1 - (\boldsymbol{v}-\boldsymbol{v}^{-1})[e_0, f_r] \otimes f_1 = -f_{r+1} \otimes 1 = -\Delta(f_{r+1})$, due to $(\hat{\mathrm{U}}5, \hat{\mathrm{U}}6)$ for $\mathcal{U}^{\mathrm{sc}}_{0,b_1}$.

Case $r=0$ As above, we get $[\Delta(\bar{h}_1), \Delta(f_0)] = [\bar{h}_1 \otimes 1 + 1 \otimes \bar{h}_1 - (\boldsymbol{v}-\boldsymbol{v}^{-1})e_0 \otimes f_1, f_0 \otimes 1] = -f_1 \otimes 1 - (\boldsymbol{v}-\boldsymbol{v}^{-1})[e_0, f_0] \otimes f_1 = -f_1 \otimes 1 - \psi_0^+ \otimes f_1 = -\Delta(f_1)$.

Case $r=b_1$ We have $[\Delta(\bar{h}_1), \Delta(f_{b_1})] = [\bar{h}_1 \otimes 1 + 1 \otimes \bar{h}_1 - (\boldsymbol{v}-\boldsymbol{v}^{-1})e_0 \otimes f_1, f_{b_1} \otimes 1 + \psi^-_{b_1} \otimes f_0] = -f_{1+b_1} \otimes 1 - \psi^-_{b_1} \otimes f_1 + \psi^-_{b_1} \otimes f_1 - (\boldsymbol{v}-\boldsymbol{v}^{-1})[e_0 \otimes f_1, \psi^-_{b_1} \otimes f_0] = -f_{1+b_1} \otimes 1 = -\Delta(f_{1+b_1})$, where we used $[e_0 \otimes f_1, \psi^-_{b_1} \otimes f_0] = 0$ as $f_1 f_0 = \boldsymbol{v}^{-2} f_0 f_1$ in $\mathcal{U}^{\mathrm{sc}}_{0,b_2}$, due to $(\hat{\mathrm{U}}3)$, and $\psi^-_{b_1} e_0 = \boldsymbol{v}^{-2} e_0 \psi^-_{b_1}$ in $\mathcal{U}^{\mathrm{sc}}_{0,b_1}$, due to $(\hat{\mathrm{U}}4)$.

Let us now verify the equality $[\Delta(\bar{h}_{-1}), \Delta(f_r)] = -\Delta(f_{r-1})$ for $1+b_1 \le r \le 1$.

Case $1 + b_1 < r < 1$ We have $[\Delta(\bar{h}_{-1}), \Delta(f_r)] = [\bar{h}_{-1} \otimes 1 + 1 \otimes \bar{h}_{-1} + (\boldsymbol{v} - \boldsymbol{v}^{-1})e_{-1} \otimes f_0, f_r \otimes 1] = -f_{r-1} \otimes 1 + (\boldsymbol{v} - \boldsymbol{v}^{-1})[e_{-1}, f_r] \otimes f_0 = -f_{r-1} \otimes 1 = -\Delta(f_{r-1})$, due to (Û5, Û6) for $\mathcal{U}^{\mathrm{sc}}_{0,b_1}$.

Case $r = 1$ We have $[\Delta(\bar{h}_{-1}), \Delta(f_1)] = [\bar{h}_{-1} \otimes 1 + 1 \otimes \bar{h}_{-1} + (\boldsymbol{v} - \boldsymbol{v}^{-1})e_{-1} \otimes f_0, f_1 \otimes 1 + \psi_0^+ \otimes f_1] = -f_0 \otimes 1 - \psi_0^+ \otimes f_0 + \psi_0^+ \otimes f_0 + (\boldsymbol{v} - \boldsymbol{v}^{-1})[e_{-1} \otimes f_0, \psi_0^+ \otimes f_1] = -f_0 \otimes 1 = -\Delta(f_0)$, where we used $[e_{-1} \otimes f_0, \psi_0^+ \otimes f_1] = 0$ as $f_0 f_1 = \boldsymbol{v}^2 f_1 f_0$ in $\mathcal{U}^{\mathrm{sc}}_{0,b_2}$ and $\psi_0^+ e_{-1} = \boldsymbol{v}^2 e_{-1}\psi_0^+$ in $\mathcal{U}^{\mathrm{sc}}_{0,b_1}$.

Case $r = 1 + b_1$ We have $[\Delta(\bar{h}_{-1}), \Delta(f_{1+b_1})] = [\bar{h}_{-1} \otimes 1 + 1 \otimes \bar{h}_{-1} + (\boldsymbol{v} - \boldsymbol{v}^{-1})e_{-1} \otimes f_0, f_{1+b_1} \otimes 1] = -f_{b_1} \otimes 1 - \psi_{b_1}^- \otimes f_0 = -\Delta(f_{b_1})$, due to (Û5, Û6) for $\mathcal{U}^{\mathrm{sc}}_{0,b_1}$.

D(vi) Compatibility with (Û6)

Case $b_2 \le r < 0, b_1 < s \le 0$ The equality $[\Delta(e_r), \Delta(f_s)] = 0$ is obvious.

Case $r = s = 0$ We need to prove $[\Delta(e_0), \Delta(f_0)] = \frac{1}{\boldsymbol{v}-\boldsymbol{v}^{-1}}\Delta(\psi_0^+)$. This follows from $[\Delta(e_0), \Delta(f_0)] = [e_0 \otimes \psi_0^+ + 1 \otimes e_0, f_0 \otimes 1] = [e_0, f_0] \otimes \psi_0^+ = \frac{\psi_0^+ \otimes \psi_0^+}{\boldsymbol{v}-\boldsymbol{v}^{-1}} = \frac{\Delta(\psi_0^+)}{\boldsymbol{v}-\boldsymbol{v}^{-1}}$, due to (Û6) for $\mathcal{U}^{\mathrm{sc}}_{0,b_1}$.

Case $r = 0, s = 1$ We need to prove $[\Delta_{b_1,b_2}(e_0), \Delta_{b_1,b_2}(f_1)] = \Delta_{b_1,b_2}(\psi_0^+)\Delta_{b_1,b_2}(h_1)$. This can be easily deduced from the unshifted case $b_1 = b_2 = 0$ by applying Remark 10.6. Indeed, $[\Delta_{b_1,b_2}(e_0), \Delta_{b_1,b_2}(f_1)] = [\jmath^+_{b_1,0} \otimes \jmath^+_{0,b_2}(\Delta(e_0)), \jmath^+_{b_1,0} \otimes \jmath^+_{0,b_2}(\Delta(f_1))] = \jmath^+_{b_1,0} \otimes \jmath^+_{0,b_2}(\Delta([e_0, f_1])) = \jmath^+_{b_1,0} \otimes \jmath^+_{0,b_2}(\Delta(\psi_0^+)\Delta(h_1)) = \Delta_{b_1,b_2}(\psi_0^+)\Delta_{b_1,b_2}(h_1)$, where the subscripts in Δ_{b_1,b_2} are used this time to distinguish it from the Drinfeld-Jimbo coproduct Δ.

Case $r = 0, b_1 < s < 0$ We need to prove $[\Delta(e_0), \Delta(f_s)] = 0$. This follows from $[\Delta(e_0), \Delta(f_s)] = [e_0 \otimes \psi_0^+ + 1 \otimes e_0, f_s \otimes 1] = [e_0, f_s] \otimes \psi_0^+ = 0$ as $[e_0, f_s] = 0$ in $\mathcal{U}^{\mathrm{sc}}_{0,b_1}$ by (Û6).

Case $r = 0, s = b_1$ We need to prove $[\Delta(e_0), \Delta(f_{b_1})] = 0$. This follows from $[\Delta(e_0), \Delta(f_{b_1})] = [e_0 \otimes \psi_0^+ + 1 \otimes e_0, f_{b_1} \otimes 1 + \psi_{b_1}^- \otimes f_0] = [e_0, f_{b_1}] \otimes \psi_0^+ + \psi_{b_1}^- \otimes [e_0, f_0] = -\frac{\psi_{b_1}^- \otimes \psi_0^+}{\boldsymbol{v}-\boldsymbol{v}^{-1}} + \frac{\psi_{b_1}^- \otimes \psi_0^+}{\boldsymbol{v}-\boldsymbol{v}^{-1}} = 0$, where we used $[e_0 \otimes \psi_0^+, \psi_{b_1}^- \otimes f_0] = 0$ as $\psi_0^+ f_0 = \boldsymbol{v}^{-2} f_0 \psi_0^+$ in $\mathcal{U}^{\mathrm{sc}}_{0,b_2}$, $\psi_{b_1}^- e_0 = \boldsymbol{v}^{-2} e_0 \psi_{b_1}^-$ in $\mathcal{U}^{\mathrm{sc}}_{0,b_1}$.

Case $r = -1, s = 1$ We need to prove $[\Delta(e_{-1}), \Delta(f_1)] = \frac{1}{\boldsymbol{v}-\boldsymbol{v}^{-1}}\Delta(\psi_0^+)$. This follows from $[\Delta(e_{-1}), \Delta(f_1)] = [1 \otimes e_{-1}, f_1 \otimes 1 + \psi_0^+ \otimes f_1] = \psi_0^+ \otimes [e_{-1}, f_1] = \frac{\psi_0^+ \otimes \psi_0^+}{\boldsymbol{v}-\boldsymbol{v}^{-1}} = \frac{\Delta(\psi_0^+)}{\boldsymbol{v}-\boldsymbol{v}^{-1}}$, due to (Û6) for $\mathcal{U}^{\mathrm{sc}}_{0,b_2}$.

Case $b_2 \le r < -1, s = 1$ We need to prove $[\Delta(e_r), \Delta(f_1)] = 0$. This follows from $[\Delta(e_r), \Delta(f_1)] = [1 \otimes e_r, f_1 \otimes 1 + \psi_0^+ \otimes f_1] = \psi_0^+ \otimes [e_r, f_1] = 0$ as $[e_r, f_1] = 0$ in $\mathcal{U}^{\mathrm{sc}}_{0,b_2}$ by ($\hat{\mathrm{U}}$6).

Case $r = b_2 - 1, s = 1$ We need to prove $[\Delta(e_{b_2-1}), \Delta(f_1)] = 0$. This follows from $[\Delta(e_{b_2-1}), \Delta(f_1)] = [e_{-1} \otimes \psi^-_{b_2} + 1 \otimes e_{b_2-1}, f_1 \otimes 1 + \psi_0^+ \otimes f_1] = [e_{-1}, f_1] \otimes \psi^-_{b_2} + \psi_0^+ \otimes [e_{b_2-1}, f_1] + [e_{-1} \otimes \psi^-_{b_2}, \psi_0^+ \otimes f_1] = \frac{\psi_0^+ \otimes \psi^-_{b_2}}{\boldsymbol{v}-\boldsymbol{v}^{-1}} - \frac{\psi_0^+ \otimes \psi^-_{b_2}}{\boldsymbol{v}-\boldsymbol{v}^{-1}} = 0$. Here we used $[e_{-1} \otimes \psi^-_{b_2}, \psi_0^+ \otimes f_1] = 0$ as $\psi^-_{b_2} f_1 = \boldsymbol{v}^2 f_1 \psi^-_{b_2}$ in $\mathcal{U}^{\mathrm{sc}}_{0,b_2}$, due to ($\hat{\mathrm{U}}$5), and $\psi_0^+ e_{-1} = \boldsymbol{v}^2 e_{-1} \psi_0^+$ in $\mathcal{U}^{\mathrm{sc}}_{0,b_1}$, due to ($\hat{\mathrm{U}}$4).

Case $r = b_2 - 1, s = b_1$ The proof of $[\Delta_{b_1,b_2}(e_{b_2-1}), \Delta_{b_1,b_2}(f_{b_1})] = \Delta_{b_1,b_2}(\psi_b^-)\Delta_{b_1,b_2}(h_{-1})$ can be deduced by applying Remark 10.6 analogously to the case $r = 0, s = 1$. Indeed, $[\Delta_{b_1,b_2}(e_{b_2-1}), \Delta_{b_1,b_2}(f_{b_1})] = [J^-_{b_1,0} \otimes J^-_{0,b_2}(\Delta(e_{-1})), J^-_{b_1,0} \otimes J^-_{0,b_2}(\Delta(f_0))] = J^-_{b_1,0} \otimes J^-_{0,b_2}(\Delta([e_{-1}, f_0])) = J^-_{b_1,0} \otimes J^-_{0,b_2}(\Delta(\psi_0^-)\Delta(h_{-1})) = \Delta_{b_1,b_2}(\psi_b^-)\Delta_{b_1,b_2}(h_{-1})$.

Case $r = b_2, s = b_1$ We need to prove $[\Delta(e_{b_2}), \Delta(f_{b_1})] = -\frac{1}{\boldsymbol{v}-\boldsymbol{v}^{-1}}\Delta(\psi_b^-)$. This follows from $[\Delta(e_{b_2}), \Delta(f_{b_1})] = [1 \otimes e_{b_2}, f_{b_1} \otimes 1 + \psi^-_{b_1} \otimes f_0] = \psi^-_{b_1} \otimes [e_{b_2}, f_0] = -\frac{\psi^-_{b_1} \otimes \psi^-_{b_2}}{\boldsymbol{v}-\boldsymbol{v}^{-1}} = -\frac{\Delta(\psi_b^-)}{\boldsymbol{v}-\boldsymbol{v}^{-1}}$, due to ($\hat{\mathrm{U}}$6) for $\mathcal{U}^{\mathrm{sc}}_{0,b_2}$.

Case $b_2 < r < 0, s = b_1$ We need to prove $[\Delta(e_r), \Delta(f_{b_1})] = 0$. This follows from $[\Delta(e_r), \Delta(f_{b_1})] = [1 \otimes e_r, f_{b_1} \otimes 1 + \psi^-_{b_1} \otimes f_0] = \psi^-_{b_1} \otimes [e_r, f_0] = 0$ as $[e_r, f_0] = 0$ in $\mathcal{U}^{\mathrm{sc}}_{0,b_2}$ by ($\hat{\mathrm{U}}$6).

Case $r = b_2 - 1, 1 + b_1 < s \le 0$ We need to prove $[\Delta(e_{b_2-1}), \Delta(f_s)] = 0$. This follows from $[\Delta(e_{b_2-1}), \Delta(f_s)] = [e_{-1} \otimes \psi^-_{b_2} + 1 \otimes e_{b_2-1}, f_s \otimes 1] = [e_{-1}, f_s] \otimes \psi^-_{b_2} = 0$ as $[e_{-1}, f_s] = 0$ in $\mathcal{U}^{\mathrm{sc}}_{0,b_1}$.

Case $r = b_2 - 1, s = 1 + b_1$ We need to prove $[\Delta(e_{b_2-1}), \Delta(f_{1+b_1})] = -\frac{1}{\boldsymbol{v}-\boldsymbol{v}^{-1}}\Delta(\psi_b^-)$. This follows from $[\Delta(e_{b_2-1}), \Delta(f_{1+b_1})] = [e_{-1} \otimes \psi^-_{b_2} + 1 \otimes e_{b_2-1}, f_{1+b_1} \otimes 1] = [e_{-1}, f_{1+b_1}] \otimes \psi^-_{b_2} = -\frac{\psi^-_{b_1} \otimes \psi^-_{b_2}}{\boldsymbol{v}-\boldsymbol{v}^{-1}} = -\frac{\Delta(\psi_b^-)}{\boldsymbol{v}-\boldsymbol{v}^{-1}}$, due to ($\hat{\mathrm{U}}$6) for $\mathcal{U}^{\mathrm{sc}}_{0,b_1}$.

D(vii) Compatibility with ($\hat{\mathrm{U}}$9)

Applying Remark 10.6 as we did above, we see that the equalities

$$[\Delta(h_1), [\Delta(f_1), [\Delta(h_1), \Delta(e_0)]]] = 0 \text{ and } [\Delta(h_{-1}), [\Delta(e_{b_2-1}), [\Delta(h_{-1}), \Delta(f_{b_1})]]] = 0$$

follow from the equalities $[h_1, [f_1, [h_1, e_0]]] = [2]_v \cdot [h_1, [f_1, e_1]] = [2]_v \cdot [h_1, \frac{-\psi_2^+}{v-v^{-1}}] = 0$ in U_v^+ and $[h_{-1}, [e_{-1}, [h_{-1}, f_0]]] = -[2]_v \cdot [h_{-1}, [e_{-1}, f_{-1}]] = [2]_v \cdot [h_{-1}, \frac{\psi_{-2}^-}{v-v^{-1}}] = 0$ in U_v^-, respectively.

This completes our proof of Theorem 10.5.

Appendix E Proof of Lemma 10.9(b)

E(i) PBW Property for $\mathcal{U}^{\mathrm{sc}}_{0,n}$

For $\mathcal{U}^{\mathrm{sc}}_{0,n}$, the simply-connected shifted quantum affine algebra of $\mathfrak{sl}_2$, define the PBW variables to be $\{e_s\}_{s\in\mathbb{Z}} \cup \{f_s\}_{s\in\mathbb{Z}} \cup \{\psi_r^+\}_{r>0} \cup \{\psi_{n-r}^-\}_{r>0} \cup \{(\psi_0^+)^{\pm 1}\} \cup \{(\psi_n^-)^{\pm 1}\}$. We order the elements in each group according to the decreasing order of s, r. Any expression of the form

$$e_{s_1^+} \cdots e_{s_a^+} f_{s_1^-} \cdots f_{s_b^-} \psi_{r_1^+}^+ \cdots \psi_{r_{c^+}^+}^+ \psi_{r_1^-}^- \cdots \psi_{r_{c^-}^-}^- (\psi_0^+)^{\gamma^+} (\psi_n^-)^{\gamma^-}$$

with $s_1^+ \geq \cdots \geq s_a^+$, $s_1^- \geq \cdots \geq s_b^-$, $r_1^+ \geq \cdots \geq r_{c^+}^+ > 0$, $r_1^- \leq \cdots \leq r_{c^-}^- < n$, $\gamma^\pm \in \mathbb{Z}$, $a, b, c^\pm \in \mathbb{N}$, will be referred to as the ordered monomial in the PBW variables.

The following result is easy to check using defining relations (U1–U6).

Lemma E.1 *The algebra* $\mathcal{U}^{\mathrm{sc}}_{0,n}$ *is spanned by the ordered monomials in the PBW variables.*

The key result of this section is a refinement of the previous statement.

Theorem E.2 *For any* $n \in \mathbb{Z}$, *the algebra* $\mathcal{U}^{\mathrm{sc}}_{0,n}$ *satisfies the PBW property, that is, the set of the ordered monomials in the PBW variables forms a* $\mathbb{C}(v)$*-basis of* $\mathcal{U}^{\mathrm{sc}}_{0,n}$.

E(ii) Proof of Theorem E.2

We will prove this result in four steps.

Step 1 Reduction to $\widetilde{\mathcal{U}}^{\mathrm{sc}}_{0,n}$.

Consider the associative $\mathbb{C}(v)$-algebra $\widetilde{\mathcal{U}}^{\mathrm{sc}}_{0,n}$, defined in the same way as $\mathcal{U}^{\mathrm{sc}}_{0,n}$ but without the generators $(\psi_0^+)^{-1}, (\psi_n^-)^{-1}$. Note that $\mathcal{U}^{\mathrm{sc}}_{0,n}$ is the localization of $\widetilde{\mathcal{U}}^{\mathrm{sc}}_{0,n}$ by the multiplicative set generated by ψ_0^+, ψ_n^-. Since these generators are among the PBW variables, the PBW property for $\mathcal{U}^{\mathrm{sc}}_{0,n}$ follows from the PBW property for $\widetilde{\mathcal{U}}^{\mathrm{sc}}_{0,n}$.

Step 2 PBW property for $\widetilde{\mathcal{U}}^{\mathrm{sc}}_{0,0}$.

It is well-known that the algebra $U_{\boldsymbol{v}}(L\mathfrak{sl}_2)$ satisfies the PBW property with the PBW variables chosen as $\{e_s\}_{s\in\mathbb{Z}}\cup\{f_s\}_{s\in\mathbb{Z}}\cup\{\psi^+_r\}_{r>0}\cup\{\psi^-_{-r}\}_{r>0}\cup\{(\psi^+_0)^{\pm 1}\}$. Here the elements in each group are ordered according to the decreasing order of r, s.

Lemma E.3 *There is an embedding of algebras* $\widetilde{\mathcal{U}}^{\mathrm{sc}}_{0,0} \hookrightarrow U_{\boldsymbol{v}}(L\mathfrak{sl}_2) \otimes_{\mathbb{C}(\boldsymbol{v})} \mathbb{C}(\boldsymbol{v})[t]$, *such that*

$$e_s \mapsto e_s \otimes t,\ \ f_s \mapsto f_s \otimes 1,\ \ \psi^{\pm}_{\pm r} \mapsto \psi^{\pm}_{\pm r} \otimes t.$$

Proof The above assignment obviously preserves all the defining relations of $\widetilde{\mathcal{U}}^{\mathrm{sc}}_{0,0}$. Hence, it gives rise to a homomorphism $\widetilde{\mathcal{U}}^{\mathrm{sc}}_{0,0} \to U_{\boldsymbol{v}}(L\mathfrak{sl}_2) \otimes_{\mathbb{C}(\boldsymbol{v})} \mathbb{C}(\boldsymbol{v})[t]$.

To prove the injectivity of this homomorphism, let us first note that $\widetilde{\mathcal{U}}^{\mathrm{sc}}_{0,0}$ is spanned by the ordered monomials in the PBW variables, cf. Lemma E.1. The above homomorphism maps these monomials to a subset of the basis for $U_{\boldsymbol{v}}(L\mathfrak{sl}_2) \otimes_{\mathbb{C}(\boldsymbol{v})} \mathbb{C}(\boldsymbol{v})[t]$, where we used the PBW property for $U_{\boldsymbol{v}}(L\mathfrak{sl}_2)$. Hence, the ordered monomials in the PBW variable for $\widetilde{\mathcal{U}}^{\mathrm{sc}}_{0,0}$ are linearly independent and the above homomorphism is injective. □

Our proof of Lemma E.3 implies the PBW property for $\widetilde{\mathcal{U}}^{\mathrm{sc}}_{0,0}$.

Step 3 PBW property for $\widetilde{\mathcal{U}}^{\mathrm{sc}}_{0,n}$, $n < 0$.

For $n < 0$, the algebra $\widetilde{\mathcal{U}}^{\mathrm{sc}}_{0,n}$ is obviously a quotient of $\widetilde{\mathcal{U}}^{\mathrm{sc}}_{0,0}$ by the 2-sided ideal

$$I_n := \langle \psi^-_0, \psi^-_{-1}, \ldots, \psi^-_{1+n} \rangle_{2-\text{sided}}.$$

Let I^l_n be the left ideal generated by the same elements

$$I^l_n := \langle \psi^-_0, \psi^-_{-1}, \ldots, \psi^-_{1+n} \rangle_{\text{left}}.$$

Lemma E.4 *We have* $I^l_n = I_n$.

Proof It suffices to show that I^l_n is also a right ideal. According to (U4), we have

$$\psi^-_{-r} e_s = \boldsymbol{v}^{-2}\psi^-_{-r+1}e_{s-1} - e_{s-1}\psi^-_{-r+1} + \boldsymbol{v}^{-2}e_s\psi^-_{-r},\ \ \psi^-_0 e_s = \boldsymbol{v}^{-2}e_s\psi^-_0,$$

so that the right multiplication by e_s preserves I^l_n. Similarly for f_s (need to apply (U5)), while for ψ^+_r, ψ^-_{-r} this is obvious. These elements generate $\widetilde{\mathcal{U}}^{\mathrm{sc}}_{0,0}$, hence, the claim. □

Combining the PBW property for $\widetilde{\mathcal{U}}^{\mathrm{sc}}_{0,0}$ (established in Step 2) with Lemma E.4 and $\widetilde{\mathcal{U}}^{\mathrm{sc}}_{0,n} \simeq \widetilde{\mathcal{U}}^{\mathrm{sc}}_{0,0}/I_n$, we get the PBW property for $\widetilde{\mathcal{U}}^{\mathrm{sc}}_{0,n}$.

Step 4 PBW property for $\widetilde{\mathcal{U}}^{\mathrm{sc}}_{0,n}$, $n > 0$.

The proof proceeds by induction in n. We assume that the PBW property holds for $\widetilde{\mathcal{U}}^{\mathrm{sc}}_{0,m}$ with $m < n$ and want to deduce the PBW property for $\widetilde{\mathcal{U}}^{\mathrm{sc}}_{0,n}$. Consider the homomorphism $\widetilde{\iota}_{n,-1,0}\colon \widetilde{\mathcal{U}}^{\mathrm{sc}}_{0,n} \to \widetilde{\mathcal{U}}^{\mathrm{sc}}_{0,n-1}$ defined analogously to $\iota_{n,-1,0}$ of Proposition 10.8. Explicitly,

$$\widetilde{\iota}_{n,-1,0}\colon e_s \mapsto e_s - e_{s-1},\ f_s \mapsto f_s,\ \psi_r^+ \mapsto \psi_r^+ - \psi_{r-1}^+,\ \psi_r^- \mapsto \psi_r^- - \psi_{r-1}^-,$$

where we set $\psi_{-1}^+ := 0$, $\psi_n^- := 0$ in the right-hand sides. The image of an ordered monomial in the PBW variables for $\widetilde{\mathcal{U}}^{\mathrm{sc}}_{0,n}$ under $\widetilde{\iota}_{n,-1,0}$ is a linear combination of the same ordered monomial in the PBW variables for $\widetilde{\mathcal{U}}^{\mathrm{sc}}_{0,n-1}$ with all ψ_r^- replaced by $(-\psi_{r-1}^-)$, called the leading monomial, and several other (not necessarily ordered) monomials in the PBW variables. Based on the equality $e_s e_{s-1} = \boldsymbol{v}^2 e_{s-1} e_s$ $(s \in \mathbb{Z})$, we see that rewriting these extra monomials as linear combinations of the ordered monomials in the PBW variables, all of them are actually lexicographically smaller than the leading monomial. Hence, the PBW property for $\widetilde{\mathcal{U}}^{\mathrm{sc}}_{0,n-1}$ implies the PBW property for $\widetilde{\mathcal{U}}^{\mathrm{sc}}_{0,n}$. Moreover, we immediately get the injectivity of $\widetilde{\iota}_{n,-1,0}$.

This completes our proof of Theorem E.2.

E(iii) Proof of Lemma 10.9(b)

Now we are ready to prove Lemma 10.9(b). Due to Lemma 10.9(a), it suffices to verify the injectivity of the homomorphisms $\iota_{n,-1,0}$ and $\iota_{n,0,-1}$. The former follows from the injectivity of $\widetilde{\iota}_{n,-1,0}$ from Step 4 above, while the latter can be deduced in the same way.

Appendix F Proof of Theorem 10.10

The proof of Theorem 10.10 proceeds in three steps. First, we construct Δ_{b_1,b_2} (this construction depends on a choice of sufficiently small $m_1, m_2 \leq 0$). Then, we verify that this construction is independent of the choice made. Finally, we prove the commutativity of the diagram of Theorem 10.10 for any $m_1, m_2 \in \mathbb{Z}_{\leq 0}$.

F(i) Construction of Δ_{b_1,b_2}

Fix any $m_1, m_2 \in \mathbb{Z}_{\leq 0}$ such that $b_1 + m_1, b_2 + m_2 \in \mathbb{Z}_{\leq 0}$. Consider the diagram

$$\begin{array}{ccc} \mathcal{U}^{\mathrm{sc}}_{0,b} & & \mathcal{U}^{\mathrm{sc}}_{0,b_1} \otimes \mathcal{U}^{\mathrm{sc}}_{0,b_2} \\ \Big\downarrow{\scriptstyle \iota_{b,m_2,m_1}} & & \Big\downarrow{\scriptstyle \iota_{b_1,0,m_1} \otimes \iota_{b_2,m_2,0}} \\ \mathcal{U}^{\mathrm{sc}}_{0,b+m_1+m_2} & \xrightarrow[\Delta = \Delta_{b_1+m_1,b_2+m_2}]{} & \mathcal{U}^{\mathrm{sc}}_{0,b_1+m_1} \otimes \mathcal{U}^{\mathrm{sc}}_{0,b_2+m_2} \end{array}$$

where the bottom horizontal arrow $\Delta = \Delta_{b_1+m_1,b_2+m_2}$ is defined in Theorem 10.5. Since the homomorphisms ι_{b,m_2,m_1} and $\iota_{b_1,0,m_1} \otimes \iota_{b_2,m_2,0}$ are injective, the homomorphism $\Delta_{b_1+m_1,b_2+m_2}$ gives rise to a uniquely determined homomorphism Δ_{b_1,b_2} making the above diagram commutative as far as we can prove

$$\Delta(\iota_{b,m_2,m_1}(\mathcal{U}^{\mathrm{sc}}_{0,b})) \subset (\iota_{b_1,0,m_1} \otimes \iota_{b_2,m_2,0})(\mathcal{U}^{\mathrm{sc}}_{0,b_1} \otimes \mathcal{U}^{\mathrm{sc}}_{0,b_2}). \qquad (\diamondsuit)$$

As before, we use $\mathcal{U}^{\mathrm{sc},>}_{0,b'}, \mathcal{U}^{\mathrm{sc},\geq}_{0,b'}, \mathcal{U}^{\mathrm{sc},<}_{0,b'}, \mathcal{U}^{\mathrm{sc},\leq}_{0,b'}$ to denote the $\mathbb{C}(\boldsymbol{v})$-subalgebras of $\mathcal{U}^{\mathrm{sc}}_{0,b'}$ generated by $\{e_r\}, \{e_r, \psi^{\pm}_{\pm s^{\pm}}\}, \{f_r\}, \{f_r, \psi^{\pm}_{\pm s^{\pm}}\}$, respectively. For $r \in \mathbb{Z}$, we claim that

$$\Delta(e_r) \in 1 \otimes e_r + \mathcal{U}^{\mathrm{sc},>}_{0,b_1+m_1} \otimes \mathcal{U}^{\mathrm{sc},\leq}_{0,b_2+m_2}, \quad \Delta(f_r) \in f_r \otimes 1 + \mathcal{U}^{\mathrm{sc},\geq}_{0,b_1+m_1} \otimes \mathcal{U}^{\mathrm{sc},<}_{0,b_2+m_2}. \qquad (\diamond_1)$$

This follows by combining iteratively the formulas for $\Delta(e_{-1}), \Delta(f_0), \Delta(h_{\pm 1})$ with the relations $[h_{\pm 1}, e_r] = [2]_{\boldsymbol{v}} \cdot e_{r\pm 1}, [h_{\pm 1}, f_r] = -[2]_{\boldsymbol{v}} \cdot f_{r\pm 1}$. We also note that

$$\mathcal{U}^{\mathrm{sc},\geq}_{0,b_1} \otimes \mathcal{U}^{\mathrm{sc},\leq}_{0,b_2} \subset (\iota_{b_1,0,m_1} \otimes \iota_{b_2,m_2,0})(\mathcal{U}^{\mathrm{sc}}_{0,b_1} \otimes \mathcal{U}^{\mathrm{sc}}_{0,b_2}). \qquad (\diamond_2)$$

According to $(\diamond_1)$, we get

$$\Delta(\iota_{b,m_2,m_1}(e_r)) \in 1 \otimes \sum_{s=0}^{-m_2} (-1)^s \binom{-m_2}{s} e_{r-s} + \mathcal{U}^{\mathrm{sc},>}_{0,b_1+m_1} \otimes \mathcal{U}^{\mathrm{sc},\leq}_{0,b_2+m_2}.$$

The right-hand side is an element of $(\iota_{b_1,0,m_1} \otimes \iota_{b_2,m_2,0})(\mathcal{U}^{\mathrm{sc}}_{0,b_1} \otimes \mathcal{U}^{\mathrm{sc}}_{0,b_2})$, due to $(\diamond_2)$ and the equality $1 \otimes \sum_{s=0}^{-m_2} (-1)^s \binom{-m_2}{s} e_{r-s} = (\iota_{b_1,0,m_1} \otimes \iota_{b_2,m_2,0})(1 \otimes e_r)$. Likewise,

$$\Delta(\iota_{b,m_2,m_1}(f_r)) \in \sum_{s=0}^{-m_1} (-1)^s \binom{-m_1}{s} f_{r-s} \otimes 1 + \mathcal{U}^{\mathrm{sc},\geq}_{0,b_1+m_1} \otimes \mathcal{U}^{\mathrm{sc},<}_{0,b_2+m_2}.$$

The right-hand side is an element of $(\iota_{b_1,0,m_1} \otimes \iota_{b_2,m_2,0})(\mathcal{U}^{\mathrm{sc}}_{0,b_1} \otimes \mathcal{U}^{\mathrm{sc}}_{0,b_2})$, due to $(\diamond_2)$ and the equality $\sum_{s=0}^{-m_1} (-1)^s \binom{-m_1}{s} f_{r-s} \otimes 1 = (\iota_{b_1,0,m_1} \otimes \iota_{b_2,m_2,0})(f_r \otimes 1)$. We also have

$$\Delta(\iota_{b,m_2,m_1}((\psi_0^+)^{\pm 1})) = (\iota_{b_1,0,m_1} \otimes \iota_{b_2,m_2,0})((\psi_0)^{\pm 1} \otimes (\psi_0)^{\pm 1}),$$

$$\Delta(\iota_{b,m_2,m_1}((\psi_b^-)^{\pm 1})) = (\iota_{b_1,0,m_1} \otimes \iota_{b_2,m_2,0})((\psi_{b_1}^-)^{\pm 1} \otimes (\psi_{b_2}^-)^{\pm 1}).$$

Finally, combining the relations $\psi_r^+ = (\boldsymbol{v} - \boldsymbol{v}^{-1})[e_r, f_0], \psi_{b-r}^- = (\boldsymbol{v}^{-1} - \boldsymbol{v})[e_{b-r}, f_0]$ $(r \in \mathbb{Z}_{>0})$ in $\mathcal{U}^{\mathrm{sc}}_{0,b+m_1+m_2}$ with $(\diamond_1)$ and $(\diamond_2)$, we get

$$\Delta(\psi_r^+), \Delta(\psi_{b-r}^-) \in \mathcal{U}^{\mathrm{sc},\geq}_{0,b_1+m_1} \otimes \mathcal{U}^{\mathrm{sc},\leq}_{0,b_2+m_2} \subset (\iota_{b_1,0,m_1} \otimes \iota_{b_2,m_2,0})(\mathcal{U}^{\mathrm{sc}}_{0,b_1} \otimes \mathcal{U}^{\mathrm{sc}}_{0,b_2}).$$

This completes our proof of $(\diamondsuit)$.

Therefore, we obtain the homomorphism Δ_{b_1,b_2} for the particular choice of m_1, m_2.

F(ii) Independence of the Choice of m_1, m_2

Let us now prove that the homomorphism Δ_{b_1,b_2} constructed above does not depend on the choice of m_1, m_2. To this end, fix another pair $m_1', m_2' \in \mathbb{Z}_{\leq 0}$ such that $b_1 + m_1', b_2 + m_2' \in \mathbb{Z}_{\leq 0}$, and set $m = m_1 + m_2, m' = m_1' + m_2'$.

Consider the following diagram:

$$\begin{array}{ccc}
\mathcal{U}^{\mathrm{sc}}_{0,b} & & \mathcal{U}^{\mathrm{sc}}_{0,b_1} \otimes \mathcal{U}^{\mathrm{sc}}_{0,b_2} \\
\downarrow{\scriptstyle \iota_{b,m_2,m_1}} & & \downarrow{\scriptstyle \iota_{b_1,0,m_1} \otimes \iota_{b_2,m_2,0}} \\
\mathcal{U}^{\mathrm{sc}}_{0,b+m} & \xrightarrow{\Delta_{b_1+m_1,b_2+m_2}} & \mathcal{U}^{\mathrm{sc}}_{0,b_1+m_1} \otimes \mathcal{U}^{\mathrm{sc}}_{0,b_2+m_2} \\
\downarrow{\scriptstyle \iota_{b+m,m_2',m_1'}} & & \downarrow{\scriptstyle \iota_{b_1+m_1,0,m_1'} \otimes \iota_{b_2+m_2,m_2',0}} \\
\mathcal{U}^{\mathrm{sc}}_{0,b+m+m'} & \xrightarrow{\Delta_{b_1+m_1+m_1',b_2+m_2+m_2'}} & \mathcal{U}^{\mathrm{sc}}_{0,b_1+m_1+m_1'} \otimes \mathcal{U}^{\mathrm{sc}}_{0,b_2+m_2+m_2'}
\end{array}$$

According to Lemma 10.9(a): $\iota_{b+m,m_2',m_1'} \circ \iota_{b,m_2,m_1} = \iota_{b,m_2+m_2',m_1+m_1'}$ and $(\iota_{b_1+m_1,0,m_1'} \otimes \iota_{b_2+m_2,m_2',0}) \circ (\iota_{b_1,0,m_1} \otimes \iota_{b_2,m_2,0}) = (\iota_{b_1,0,m_1+m_1'} \otimes \iota_{b_2,m_2+m_2',0})$. On the other hand, tracing back the explicit formulas for $\Delta_{b_1+m_1,b_2+m_2}$ and $\Delta_{b_1+m_1+m_1',b_2+m_2+m_2'}$ of Theorem 10.5, it is easy to check that the lower square is commutative.

The above two observations imply that the maps Δ_{b_1,b_2} are the same for both (m_1, m_2) and $(m_1 + m_1', m_2 + m_2')$. Due to the symmetry, we also see that the maps Δ_{b_1,b_2} are the same for both (m_1', m_2') and $(m_1 + m_1', m_2 + m_2')$. Therefore, the maps Δ_{b_1,b_2} are the same for both (m_1, m_2) and (m_1', m_2'). This completes our verification.

F(iii) Commutativity of the Diagram for Any $m_1, m_2 \in \mathbb{Z}_{\leq 0}$

It remains to prove the commutativity of the diagram of Theorem 10.10. To this end, choose $m'_1, m'_2 \in \mathbb{Z}_{\leq 0}$ such that $b_1 + m_1 + m'_1, b_2 + m_2 + m'_2 \in \mathbb{Z}_{\leq 0}$. Consider a diagram analogous to the previous one:

$$
\begin{array}{ccc}
\mathcal{U}^{\mathrm{sc}}_{0,b} & \xrightarrow{\quad\Delta_{b_1,b_2}\quad} & \mathcal{U}^{\mathrm{sc}}_{0,b_1} \otimes \mathcal{U}^{\mathrm{sc}}_{0,b_2} \\
\downarrow{\scriptstyle \iota_{b,m_2,m_1}} & & \downarrow{\scriptstyle \iota_{b_1,0,m_1} \otimes \iota_{b_2,m_2,0}} \\
\mathcal{U}^{\mathrm{sc}}_{0,b+m} & \xrightarrow{\quad\Delta_{b_1+m_1,b_2+m_2}\quad} & \mathcal{U}^{\mathrm{sc}}_{0,b_1+m_1} \otimes \mathcal{U}^{\mathrm{sc}}_{0,b_2+m_2} \\
\downarrow{\scriptstyle \iota_{b+m,m'_2,m'_1}} & & \downarrow{\scriptstyle \iota_{b_1+m_1,0,m'_1} \otimes \iota_{b_2+m_2,m'_2,0}} \\
\mathcal{U}^{\mathrm{sc}}_{0,b+m+m'} & \xrightarrow{\quad\Delta_{b_1+m_1+m'_1,b_2+m_2+m'_2}\quad} & \mathcal{U}^{\mathrm{sc}}_{0,b_1+m_1+m'_1} \otimes \mathcal{U}^{\mathrm{sc}}_{0,b_2+m_2+m'_2}
\end{array}
$$

By our construction, the lower square is commutative. Applying Lemma 10.9(a) as in Sect. F(ii), we also see that the outer square is commutative. Hence, the commutativity of the top square follows from the injectivity of the homomorphism $\iota_{b_1+m_1,0,m'_1} \otimes \iota_{b_2+m_2,m'_2,0}$, due to Lemma 10.9(b).

Appendix G Proof of Theorem 10.13

The proof of Theorem 10.13 proceeds in several steps. First, we recall the RTT presentation of $U_v(L\mathfrak{sl}_n)$, and derive the equalities of the right-hand sides of (10.6). Then, we compute the RTT coproduct of certain elements $\tilde{g}_i^{(\pm 1)}$ from the RTT presentation, see Theorems G.10, G.13 (this is the most technical part). This allows us to derive formulas (10.2) and (10.3). Based on these, we deduce (10.4) and (10.5).

G(i) RTT Presentation of $U_v(L\mathfrak{sl}_n)$

Let $R_{\mathrm{trig}}(z/w) \in \mathrm{End}(\mathbb{C}^n \otimes \mathbb{C}^n)$ be the standard trigonometric R-matrix of $\mathfrak{sl}_n$-type:

$$
\begin{aligned}
R_{\mathrm{trig}}(z/w) := \sum_{i=1}^{n} E_{ii} \otimes E_{ii} + \sum_{1\leq i\neq j\leq n} \frac{z-w}{vz-v^{-1}w} E_{ii} \otimes E_{jj} + \\
\sum_{1\leq j<i\leq n} \left(\frac{(v-v^{-1})z}{vz-v^{-1}w} E_{ji} \otimes E_{ij} + \frac{(v-v^{-1})w}{vz-v^{-1}w} E_{ij} \otimes E_{ji} \right)
\end{aligned}
\tag{G.1}
$$

(for $n=2$, this definition coincides with formula (11.3)).

Define the RTT algebra of $\mathfrak{sl}_n$, denoted by $U^{\mathrm{rtt}}(\mathfrak{sl}_n)$, to be the associative $\mathbb{C}(\boldsymbol{v})$-algebra generated by $\{t^{\pm}_{ij}[\pm r]\}^{r\in\mathbb{N}}_{1\le i,j\le n}$ subject to the following defining relations:

$$t^{\pm}_{ii}[0]t^{\mp}_{ii}[0]=1 \ \text{ for } 1\le i\le n,\ t^{+}_{ij}[0]=t^{-}_{ji}[0]=0 \ \text{ for } j<i, \tag{G.2}$$

$$R_{\mathrm{trig}}(z/w)(T^{\epsilon}(z)\otimes 1)(1\otimes T^{\epsilon'}(w))=(1\otimes T^{\epsilon'}(w))(T^{\epsilon}(z)\otimes 1)R_{\mathrm{trig}}(z/w), \tag{G.3}$$

$$\mathrm{qdet}\, T^{\pm}(z)=1, \tag{G.4}$$

for all $\epsilon,\epsilon'\in\{\pm\}$, where the matrices $T^{\pm}(z)\in \mathrm{Mat}_{n\times n}(U^{\mathrm{rtt}}(\mathfrak{sl}_n))$ are given by

$$T^{\pm}(z):=\sum_{i,j=1}^{n} T^{\pm}_{ij}(z)\cdot E_{ij} \ \text{ with } \ T^{\pm}_{ij}(z):=\sum_{r\ge 0} t^{\pm}_{ij}[\pm r]z^{\mp r},$$

and the quantum determinant qdet is defined in a standard way as

$$\mathrm{qdet}\, T^{\pm}(z):=\sum_{\tau\in\mathfrak{S}_n}(-\boldsymbol{v})^{-l(\tau)}T^{\pm}_{1,\tau(1)}(z)T^{\pm}_{2,\tau(2)}(\boldsymbol{v}^{-2}z)\cdots T^{\pm}_{n,\tau(n)}(\boldsymbol{v}^{2-2n}z)$$

(cf. Sect. 11.4 and a footnote there).

Remark G.1 Let us point out right away that the RTT presentation of $U_q(\widehat{\mathfrak{gl}}_n)$ (with a nontrivial central charge), given in [17, Definition 3.2], involves only three out of four relations (G.3), namely for $(\epsilon,\epsilon')=(+,+),(-,-),(-,+)$. However, as pointed out in [32, 2.3], if the central charge is trivial, then the fourth relation for $(\epsilon,\epsilon')=(+,-)$ is equivalent to the one for $(\epsilon,\epsilon')=(-,+)$. Indeed, in our notations, this follows from the equalities $R_{\mathrm{trig}}(z/w)^{-1}=R'_{\mathrm{trig}}(z/w)$, $PR'_{\mathrm{trig}}(w/z)P^{-1}=R_{\mathrm{trig}}(z/w)$, where $R'_{\mathrm{trig}}(z/w)$ is obtained from $R_{\mathrm{trig}}(z/w)$ by replacing $\boldsymbol{v}$ with $\boldsymbol{v}^{-1}$ and $P\in\mathrm{End}(\mathbb{C}^n\otimes\mathbb{C}^n)$ denotes the permutation operator.

Note that $T^{\pm}(z)$ admits the following unique Gauss decomposition:

$$T^{\pm}(z)=\widetilde{F}^{\pm}(z)\cdot\widetilde{G}^{\pm}(z)\cdot\widetilde{E}^{\pm}(z)$$

with $\widetilde{F}^{\pm}(z),\widetilde{G}^{\pm}(z),\widetilde{E}^{\pm}(z)\in\mathrm{Mat}_{n\times n}(U^{\mathrm{rtt}}(\mathfrak{sl}_n))$ of the form

$$\widetilde{F}^{\pm}(z)=\sum_i E_{ii}+\sum_{j<i}\tilde{f}^{\pm}_{ij}(z)\cdot E_{ij},\ \widetilde{G}^{\pm}(z)=\sum_i \tilde{g}^{\pm}_i(z)\cdot E_{ii},\ \widetilde{E}^{\pm}(z)=\sum_i E_{ii}+\sum_{j<i}\tilde{e}^{\pm}_{ji}(z)\cdot E_{ji}.$$

We endow $U^{\mathrm{rtt}}(\mathfrak{sl}_n)$ with the coproduct structure (also known as the RTT coproduct) via

$$\Delta^{\mathrm{rtt}}\colon U^{\mathrm{rtt}}(\mathfrak{sl}_n)\longrightarrow U^{\mathrm{rtt}}(\mathfrak{sl}_n)\otimes U^{\mathrm{rtt}}(\mathfrak{sl}_n) \ \text{ given by } \ \Delta^{\mathrm{rtt}}(T^{\pm}(z)):=T^{\pm}(z)\otimes T^{\pm}(z).$$

Theorem G.2 ([17]) *There exists a unique $\mathbb{C}(\boldsymbol{v})$-algebra isomorphism*

$$\Upsilon\colon U^{\mathrm{ad}}_{\boldsymbol{v}}(L\mathfrak{sl}_n) \xrightarrow{\sim} U^{\mathrm{rtt}}(\mathfrak{sl}_n),$$

such that

$$e^{\pm}_j(z) \mapsto \frac{\tilde{e}^{\pm}_{j,j+1}(\boldsymbol{v}^j z)}{\boldsymbol{v}-\boldsymbol{v}^{-1}},\ f^{\pm}_j(z) \mapsto \frac{\tilde{f}^{\pm}_{j+1,j}(\boldsymbol{v}^j z)}{\boldsymbol{v}-\boldsymbol{v}^{-1}},$$

$$\psi^{\pm}_j(z) \mapsto \tilde{g}^{\pm}_{j+1}(\boldsymbol{v}^j z)(\tilde{g}^{\pm}_j(\boldsymbol{v}^j z))^{-1},\ \phi^{\pm}_j \mapsto t^{\mp}_{11}[0]t^{\mp}_{22}[0]\cdots t^{\mp}_{jj}[0] \text{ for } 1 \le j < n.$$

Moreover, this isomorphism intertwines the Drinfeld-Jimbo coproduct Δ^{ad} on $U^{\mathrm{ad}}_{\boldsymbol{v}}(L\mathfrak{sl}_n)$ with the RTT coproduct Δ^{rtt} on $U^{\mathrm{rtt}}(\mathfrak{sl}_n)$.

Remark G.3 Restricting Υ to $U_{\boldsymbol{v}}(L\mathfrak{sl}_n)$, viewed as a Hopf subalgebra of $U^{\mathrm{ad}}_{\boldsymbol{v}}(L\mathfrak{sl}_n)$, we get an embedding $U_{\boldsymbol{v}}(L\mathfrak{sl}_n) \hookrightarrow U^{\mathrm{rtt}}(\mathfrak{sl}_n)$. We will deliberately refer to $U^{\mathrm{rtt}}(\mathfrak{sl}_n)$ as an RTT presentation of both algebras $U_{\boldsymbol{v}}(L\mathfrak{sl}_n)$ and $U^{\mathrm{ad}}_{\boldsymbol{v}}(L\mathfrak{sl}_n)$.

Let us express the matrix coefficients of $\widetilde{F}^{\pm}(z), \widetilde{G}^{\pm}(z), \widetilde{E}^{\pm}(z)$ as Taylor series in $z^{\mp 1}$: $\tilde{e}^{+}_{ji}(z) = \sum_{r\ge 0} \tilde{e}^{(r)}_{ji} z^{-r}$, $\tilde{e}^{-}_{ji}(z) = \sum_{r<0} \tilde{e}^{(r)}_{ji} z^{-r}$, $\tilde{f}^{+}_{ij}(z) = \sum_{r>0} \tilde{f}^{(r)}_{ij} z^{-r}$, $\tilde{f}^{-}_{ij}(z) = \sum_{r\le 0} \tilde{f}^{(r)}_{ij} z^{-r}$, $\tilde{g}^{\pm}_i(z) = \tilde{g}^{\pm}_i + \sum_{r>0} \tilde{g}^{(\pm r)}_i z^{\mp r}$. According to Theorem G.2, we have

$$\begin{aligned} &\Upsilon^{-1}(\tilde{e}^{(0)}_{j,j+1}) = (\boldsymbol{v}-\boldsymbol{v}^{-1})e_{j,0},\ \Upsilon^{-1}(\tilde{f}^{(0)}_{j+1,j}) = -(\boldsymbol{v}-\boldsymbol{v}^{-1})f_{j,0},\\ &\Upsilon^{-1}(\tilde{e}^{(-1)}_{j,j+1}) = -\boldsymbol{v}^{-j}(\boldsymbol{v}-\boldsymbol{v}^{-1})e_{j,-1}, \Upsilon^{-1}(\tilde{f}^{(1)}_{j+1,j}) = \boldsymbol{v}^{j}(\boldsymbol{v}-\boldsymbol{v}^{-1})f_{j,1}. \end{aligned} \tag{G.5}$$

The following is the key technical result of this subsection.

Proposition G.4 *For any $1 \le j < k < i \le n$, we have:*

(a) $\tilde{e}^{(0)}_{ji} = \frac{1}{\boldsymbol{v}-\boldsymbol{v}^{-1}}[\tilde{e}^{(0)}_{ki}, \tilde{e}^{(0)}_{jk}]_{\boldsymbol{v}^{-1}}$.
(b) $\tilde{f}^{(0)}_{ij} = \frac{-1}{\boldsymbol{v}-\boldsymbol{v}^{-1}}[\tilde{f}^{(0)}_{kj}, \tilde{f}^{(0)}_{ik}]_{\boldsymbol{v}}$.
(c) $\tilde{e}^{(-1)}_{ji} = \frac{1}{\boldsymbol{v}-\boldsymbol{v}^{-1}}[\tilde{e}^{(0)}_{ki}, \tilde{e}^{(-1)}_{jk}]_{\boldsymbol{v}^{-1}}$.
(d) $\tilde{f}^{(1)}_{ij} = \frac{-1}{\boldsymbol{v}-\boldsymbol{v}^{-1}}[\tilde{f}^{(1)}_{kj}, \tilde{f}^{(0)}_{ik}]_{\boldsymbol{v}}$.

Proof

(a) Comparing the matrix coefficients $\langle v_j \otimes v_k|\cdots|v_k \otimes v_i\rangle$ of both sides of the equality $R_{\mathrm{trig}}(z/w)(T^+(z)\otimes 1)(1\otimes T^+(w)) = (1\otimes T^+(w))(T^+(z)\otimes 1) R_{\mathrm{trig}}(z/w)$, we get

$$(z-w)T^+_{jk}(z)T^+_{ki}(w)+(\boldsymbol{v}-\boldsymbol{v}^{-1})zT^+_{kk}(z)T^+_{ji}(w) = (z-w)T^+_{ki}(w)T^+_{jk}(z)+(\boldsymbol{v}-\boldsymbol{v}^{-1})wT^+_{kk}(w)T^+_{ji}(z).$$

Evaluating the coefficients of z^1w^0 in both sides of this equality, we find

$$\tilde{g}_j^+\tilde{e}_{jk}^{(0)}\tilde{g}_k^+\tilde{e}_{ki}^{(0)} + (\boldsymbol{v}-\boldsymbol{v}^{-1})\tilde{g}_k^+\tilde{g}_j^+\tilde{e}_{ji}^{(0)} = \tilde{g}_k^+\tilde{e}_{ki}^{(0)}\tilde{g}_j^+\tilde{e}_{jk}^{(0)}.$$

Combining this with Lemma G.5 below, we obtain

$$(\boldsymbol{v}-\boldsymbol{v}^{-1})\tilde{g}_k^+\tilde{g}_j^+\tilde{e}_{ji}^{(0)} = \tilde{g}_k^+\tilde{g}_j^+[\tilde{e}_{ki}^{(0)}, \tilde{e}_{jk}^{(0)}]_{\boldsymbol{v}^{-1}} \Longrightarrow \tilde{e}_{ji}^{(0)} = [\tilde{e}_{ki}^{(0)}, \tilde{e}_{jk}^{(0)}]_{\boldsymbol{v}^{-1}}/(\boldsymbol{v}-\boldsymbol{v}^{-1}).$$

(b) Comparing the matrix coefficients $\langle v_i \otimes v_k|\cdots|v_k \otimes v_j\rangle$ of both sides of the equality $R_{\mathrm{trig}}(z/w)(T^-(z)\otimes 1)(1\otimes T^-(w)) = (1\otimes T^-(w))(T^-(z)\otimes 1) R_{\mathrm{trig}}(z/w)$, we get

$$(z-w)T_{ik}^-(z)T_{kj}^-(w)+(\boldsymbol{v}-\boldsymbol{v}^{-1})wT_{kk}^-(z)T_{ij}^-(w) = (z-w)T_{kj}^-(w)T_{ik}^-(z)+(\boldsymbol{v}-\boldsymbol{v}^{-1})zT_{kk}^-(w)T_{ij}^-(z).$$

Evaluating the coefficients of z^0w^1 in both sides of this equality, we find

$$-\tilde{f}_{ik}^{(0)}\tilde{g}_k^-\tilde{f}_{kj}^{(0)}\tilde{g}_j^- + (\boldsymbol{v}-\boldsymbol{v}^{-1})\tilde{g}_k^-\tilde{f}_{ij}^{(0)}\tilde{g}_j^- = -\tilde{f}_{kj}^{(0)}\tilde{g}_j^-\tilde{f}_{ik}^{(0)}\tilde{g}_k^-.$$

Combining this with Lemma G.5 below, we obtain

$$-(\boldsymbol{v}-\boldsymbol{v}^{-1})\tilde{f}_{ij}^{(0)}\tilde{g}_k^-\tilde{g}_j^- = [\tilde{f}_{kj}^{(0)}, \tilde{f}_{ik}^{(0)}]_{\boldsymbol{v}}\cdot\tilde{g}_k^-\tilde{g}_j^- \Longrightarrow \tilde{f}_{ij}^{(0)} = -[\tilde{f}_{kj}^{(0)}, \tilde{f}_{ik}^{(0)}]_{\boldsymbol{v}}/(\boldsymbol{v}-\boldsymbol{v}^{-1}).$$

(c) Comparing the matrix coefficients $\langle v_k \otimes v_j|\cdots|v_i \otimes v_k\rangle$ of both sides of the equality $R_{\mathrm{trig}}(z/w)(T^+(z)\otimes 1)(1\otimes T^-(w)) = (1\otimes T^-(w))(T^+(z)\otimes 1) R_{\mathrm{trig}}(z/w)$, we get

$$(z-w)T_{ki}^+z)T_{jk}^-(w)+(\boldsymbol{v}-\boldsymbol{v}^{-1})wT_{ji}^+(z)T_{kk}^-(w) = (z-w)T_{jk}^-(w)T_{ki}^+(z)+(\boldsymbol{v}-\boldsymbol{v}^{-1})zT_{ji}^-(w)T_{kk}^+(z).$$

Evaluating the coefficients of z^1w^1 in both sides of this equality, we find

$$\tilde{g}_k^+\tilde{e}_{ki}^{(0)}\left(\tilde{g}_j^-\tilde{e}_{jk}^{(-1)} + \sum_{j'<j}\tilde{f}_{jj'}^{(0)}\tilde{g}_{j'}^-\tilde{e}_{j'k}^{(-1)}\right) =$$
$$\left(\tilde{g}_j^-\tilde{e}_{jk}^{(-1)} + \sum_{j'<j}\tilde{f}_{jj'}^{(0)}\tilde{g}_{j'}^-\tilde{e}_{j'k}^{(-1)}\right)\tilde{g}_k^+\tilde{e}_{ki}^{(0)} + (\boldsymbol{v}-\boldsymbol{v}^{-1})\left(\tilde{g}_j^-\tilde{e}_{ji}^{(-1)} + \sum_{j'<j}\tilde{f}_{jj'}^{(0)}\tilde{g}_{j'}^-\tilde{e}_{j'i}^{(-1)}\right)\tilde{g}_k^+. \tag{G.6}$$

This equation actually implies $\tilde{g}_k^+\tilde{e}_{ki}^{(0)}\tilde{g}_j^-\tilde{e}_{jk}^{(-1)} = \tilde{g}_j^-\tilde{e}_{jk}^{(-1)}\tilde{g}_k^+\tilde{e}_{ki}^{(0)} + (\boldsymbol{v}-\boldsymbol{v}^{-1})\tilde{g}_j^-\tilde{e}_{ji}^{(-1)}\tilde{g}_k^+$. We prove this by induction in j. For $j = 1$, this is just (G.6). In general, note that for $j' < j < k < i$, the element $\tilde{f}_{jj'}^{(0)}$ commutes with $\tilde{e}_{ki}^{(0)}$ and $\tilde{g}_k^+$. The latter follows from Lemma G.5, while the equality $[\tilde{f}_{jj'}^{(0)}, \tilde{e}_{ki}^{(0)}] = 0$ follows by combining parts (a,b) from above with $[e_{a,0}, f_{b,0}] = 0$ for $a \neq b$. Hence, (G.6) implies $A(j,k,i) + \sum_{j'<j} \tilde{f}^{(0)}_{jj'} A(j',k,i) = 0$, where we set

$$A(j,k,i) := \tilde{g}^+_k \tilde{e}^{(0)}_{ki} \tilde{g}^-_j \tilde{e}^{(-1)}_{jk} - \tilde{g}^-_j \tilde{e}^{(-1)}_{jk} \tilde{g}^+_k \tilde{e}^{(0)}_{ki} - (\boldsymbol{v} - \boldsymbol{v}^{-1}) \tilde{g}^-_j \tilde{e}^{(-1)}_{ji} \tilde{g}^+_k.$$

By the induction assumption $A(j',k,i) = 0$ for $j' < j$, hence, $A(j,k,i) = 0$.

Combining this with Lemma G.5 below, we obtain

$$(\boldsymbol{v} - \boldsymbol{v}^{-1}) \tilde{g}^-_j \tilde{g}^+_k \tilde{e}^{(-1)}_{ji} = \tilde{g}^-_j \tilde{g}^+_k [\tilde{e}^{(0)}_{ki}, \tilde{e}^{(-1)}_{jk}]_{\boldsymbol{v}^{-1}} \Longrightarrow \tilde{e}^{(-1)}_{ji} = [\tilde{e}^{(0)}_{ki}, \tilde{e}^{(-1)}_{jk}]_{\boldsymbol{v}^{-1}} / (\boldsymbol{v} - \boldsymbol{v}^{-1}).$$

(d) Comparing the matrix coefficients $\langle v_k \otimes v_i | \cdots | v_j \otimes v_k \rangle$ of both sides of the equality $R_{\mathrm{trig}}(z/w)(T^+(z) \otimes 1)(1 \otimes T^-(w)) = (1 \otimes T^-(w))(T^+(z) \otimes 1) R_{\mathrm{trig}}(z/w)$, we get

$$(z-w) T^+_{kj}(z) T^-_{ik}(w) + (\boldsymbol{v} - \boldsymbol{v}^{-1}) z T^+_{ij}(z) T^-_{kk}(w) = (z-w) T^-_{ik}(w) T^+_{kj}(z) + (\boldsymbol{v} - \boldsymbol{v}^{-1}) w T^-_{ij}(w) T^+_{kk}(z).$$

Evaluating the coefficients of $z^0 w^0$ in both sides of this equality, we find

$$\left(\tilde{f}^{(1)}_{kj} \tilde{g}^+_j + \sum_{j'<j} \tilde{f}^{(1)}_{kj'} \tilde{g}^+_{j'} \tilde{e}^{(0)}_{j'j} \right) \tilde{f}^{(0)}_{ik} \tilde{g}^-_k + (\boldsymbol{v} - \boldsymbol{v}^{-1}) \left(\tilde{f}^{(1)}_{ij} \tilde{g}^+_j + \sum_{j'<j} \tilde{f}^{(1)}_{ij'} \tilde{g}^+_{j'} \tilde{e}^{(0)}_{j'j} \right) \tilde{g}^-_k =$$
$$\tilde{f}^{(0)}_{ik} \tilde{g}^-_k \left(\tilde{f}^{(1)}_{kj} \tilde{g}^+_j + \sum_{j'<j} \tilde{f}^{(1)}_{kj'} \tilde{g}^+_{j'} \tilde{e}^{(0)}_{j'j} \right). \tag{G.7}$$

This equation actually implies $\tilde{f}^{(1)}_{kj} \tilde{g}^+_j \tilde{f}^{(0)}_{ik} \tilde{g}^-_k + (\boldsymbol{v} - \boldsymbol{v}^{-1}) \tilde{f}^{(1)}_{ij} \tilde{g}^+_j \tilde{g}^-_k = \tilde{f}^{(0)}_{ik} \tilde{g}^-_k \tilde{f}^{(1)}_{kj} \tilde{g}^+_j$. We prove this by induction in j. For $j = 1$, this is just (G.7). Analogously to part (c) above, we note that the element $\tilde{e}^{(0)}_{j'j}$ commutes with $\tilde{f}^{(0)}_{ik}$ and $\tilde{g}^-_k$ for $j' < j < k < i$. Hence, (G.7) implies $B(j,k,i) + \sum_{j'<j} B(j',k,i) \tilde{e}^{(0)}_{j'j} = 0$, where we set

$$B(j,k,i) := \tilde{f}^{(1)}_{kj} \tilde{g}^+_j \tilde{f}^{(0)}_{ik} \tilde{g}^-_k + (\boldsymbol{v} - \boldsymbol{v}^{-1}) \tilde{f}^{(1)}_{ij} \tilde{g}^+_j \tilde{g}^-_k - \tilde{f}^{(0)}_{ik} \tilde{g}^-_k \tilde{f}^{(1)}_{kj} \tilde{g}^+_j.$$

By the induction assumption $B(j',k,i) = 0$ for $j' < j$, hence, $B(j,k,i) = 0$.

Combining this with Lemma G.5 below, we obtain

$$-(\boldsymbol{v} - \boldsymbol{v}^{-1}) \tilde{f}^{(1)}_{ij} \tilde{g}^-_k \tilde{g}^+_j = [\tilde{f}^{(1)}_{kj}, \tilde{f}^{(0)}_{ik}]_{\boldsymbol{v}} \cdot \tilde{g}^-_k \tilde{g}^+_j$$

$$\Longrightarrow \tilde{f}^{(1)}_{ij} = -[\tilde{f}^{(1)}_{kj}, \tilde{f}^{(0)}_{ik}]_{\boldsymbol{v}} / (\boldsymbol{v} - \boldsymbol{v}^{-1}).$$

□

Lemma G.5 *For any* $1 \leq j < i \leq n$ *and* $1 \leq a, b \leq n$, *we have:*

(a) $\tilde{g}_a^{\epsilon}\tilde{g}_b^{\epsilon'} = \tilde{g}_b^{\epsilon'}\tilde{g}_a^{\epsilon}$ *for any* $\epsilon, \epsilon' \in \{\pm\}$.
(b) $\tilde{g}_a^{\pm}\tilde{e}_{ji}^{(0)} = \boldsymbol{v}^{\pm\delta_{ai}\mp\delta_{aj}}\tilde{e}_{ji}^{(0)}\tilde{g}_a^{\pm}$.
(c) $\tilde{g}_a^{\pm}\tilde{f}_{ij}^{(0)} = \boldsymbol{v}^{\mp\delta_{ai}\pm\delta_{aj}}\tilde{f}_{ij}^{(0)}\tilde{g}_a^{\pm}$.
(d) $\tilde{g}_a^{\pm}\tilde{e}_{ji}^{(-1)} = \boldsymbol{v}^{\pm\delta_{ai}\mp\delta_{aj}}\tilde{e}_{ji}^{(-1)}\tilde{g}_a^{\pm}$.
(e) $\tilde{g}_a^{\pm}\tilde{f}_{ij}^{(1)} = \boldsymbol{v}^{\mp\delta_{ai}\pm\delta_{aj}}\tilde{f}_{ij}^{(1)}\tilde{g}_a^{\pm}$.

Proof First, we note that $t_{ii}^{\pm}[0] = \tilde{g}_i^{\pm}$. Hence, we have $\tilde{g}_i^{\pm}\tilde{g}_i^{\mp} = 1$, due to relation (G.2).

(a) Due to the above observation, it suffices to prove $\tilde{g}_a^{+}\tilde{g}_b^{+} = \tilde{g}_b^{+}\tilde{g}_a^{+}$ for $a < b$. This follows by evaluating the coefficients of z^0w^1 in the equality of the matrix coefficients $\langle v_a \otimes v_b|\cdots|v_a \otimes v_b\rangle$ of both sides of the equality $((\boldsymbol{v}z - \boldsymbol{v}^{-1}w)R_{\mathrm{trig}}(z/w))(T^+(z)\otimes 1)(1\otimes T^+(w)) = (1\otimes T^+(w))(T^+(z)\otimes 1)$ $((\boldsymbol{v}z - \boldsymbol{v}^{-1}w)R_{\mathrm{trig}}(z/w))$.
(b) Due to the above observation, it suffices to prove $\tilde{g}_a^{+}\tilde{e}_{ji}^{(0)} = \boldsymbol{v}^{\delta_{ai}-\delta_{aj}}\tilde{e}_{ji}^{(0)}\tilde{g}_a^{+}$. This follows by evaluating the coefficients of z^0w^1 in the equality of the matrix coefficients $\langle v_a \otimes v_j|\cdots|v_a \otimes v_i\rangle$ of both sides of the equality $((\boldsymbol{v}z - \boldsymbol{v}^{-1}w)R_{\mathrm{trig}}(z/w))(T^+(z)\otimes 1)(1\otimes T^+(w)) = (1\otimes T^+(w))(T^+(z)\otimes 1)((\boldsymbol{v}z - \boldsymbol{v}^{-1}w)R_{\mathrm{trig}}(z/w))$. Note that the cases $a < j, a = j, j < a < i, a = i, a > i$ have to be treated separately.
(c) Due to the above observation, it suffices to prove $\tilde{g}_a^{-}\tilde{f}_{ij}^{(0)} = \boldsymbol{v}^{\delta_{ai}-\delta_{aj}}\tilde{f}_{ij}^{(0)}\tilde{g}_a^{-}$. This follows by evaluating the coefficients of z^0w^1 in the equality of the matrix coefficients $\langle v_i \otimes v_a|\cdots|v_j \otimes v_a\rangle$ of both sides of the equality $((\boldsymbol{v}z - \boldsymbol{v}^{-1}w)R_{\mathrm{trig}}(z/w))(T^-(z)\otimes 1)(1\otimes T^-(w)) = (1\otimes T^-(w))(T^-(z)\otimes 1)((\boldsymbol{v}z - \boldsymbol{v}^{-1}w)R_{\mathrm{trig}}(z/w))$. Note that the cases $a < j, a = j, j < a < i, a = i, a > i$ have to be treated separately.
(d) Due to the above observation, it suffices to prove $\tilde{g}_a^{+}\tilde{e}_{ji}^{(-1)} = \boldsymbol{v}^{\delta_{ai}-\delta_{aj}}\tilde{e}_{ji}^{(-1)}\tilde{g}_a^{+}$. This follows by evaluating the coefficients of z^1w^1 in the equality of the matrix coefficients $\langle v_a \otimes v_j|\cdots|v_a \otimes v_i\rangle$ of both sides of the equality $((\boldsymbol{v}z - \boldsymbol{v}^{-1}w)R_{\mathrm{trig}}(z/w))(T^+(z)\otimes 1)(1\otimes T^-(w)) = (1\otimes T^-(w))(T^+(z)\otimes 1)((\boldsymbol{v}z - \boldsymbol{v}^{-1}w)R_{\mathrm{trig}}(z/w))$. Note that the cases $a < j, a = j, j < a < i, a = i, a > i$ have to be treated separately.

    Let us emphasize that this case is less trivial than part (b), due to the fact that

$$[w^1]T_{ji}^-(w) = \tilde{g}_j^-\tilde{e}_{ji}^{(-1)} + \sum_{j'<j}\tilde{f}_{jj'}^{(0)}\tilde{g}_{j'}^-\tilde{e}_{j'i}^{(-1)}.$$

    Hence, the proof proceeds by induction in j, while we also use part (c) from above.
(e) Due to the above observation, it suffices to prove $\tilde{g}_a^{-}\tilde{f}_{ij}^{(1)} = \boldsymbol{v}^{\delta_{ai}-\delta_{aj}}\tilde{f}_{ij}^{(1)}\tilde{g}_a^{-}$. This follows by evaluating the coefficients of z^0w^0 in the equality of the

matrix coefficients $\langle v_i \otimes v_u|\cdots|v_j \otimes v_a\rangle$ of both sides of the equality $((\boldsymbol{v}z - \boldsymbol{v}^{-1}w)R_{\mathrm{trig}}(z/w))(T^+(z)\otimes 1)(1\otimes T^-(w)) = (1\otimes T^-(w))(T^+(z)\otimes 1)((\boldsymbol{v}z - \boldsymbol{v}^{-1}w)R_{\mathrm{trig}}(z/w))$. Note that the cases $a < j, a = j, j < a < i, a = i, a > i$ have to be treated separately.

Analogously to part (d), this case is less trivial than part (c), due to the fact that

$$[z^{-1}]T^+_{ij}(z) = \tilde{f}^{(1)}_{ij}\tilde{g}^+_j + \sum_{j'<j} \tilde{f}^{(1)}_{ij'}\tilde{g}^+_{j'}\tilde{e}^{(0)}_{j'j}.$$

Hence, the proof proceeds by induction in j, while we also use part (b) from above. □

The following explicit formulas follow immediately from Proposition G.4.

Corollary G.6 *For any $1 \le j < i \le n$, we have:*

$$\begin{aligned}\tilde{e}^{(0)}_{ji} &= (\boldsymbol{v}-\boldsymbol{v}^{-1})^{j-i+1}[\tilde{e}^{(0)}_{i-1,i},[\tilde{e}^{(0)}_{i-2,i-1},\cdots,[\tilde{e}^{(0)}_{j+1,j+2},\tilde{e}^{(0)}_{j,j+1}]_{\boldsymbol{v}^{-1}}\cdots]_{\boldsymbol{v}^{-1}}]_{\boldsymbol{v}^{-1}} = \\ &(\boldsymbol{v}-\boldsymbol{v}^{-1})^{j-i+1}[[\cdots[\tilde{e}^{(0)}_{i-1,i},\tilde{e}^{(0)}_{i-2,i-1}]_{\boldsymbol{v}^{-1}},\cdots,\tilde{e}^{(0)}_{j+1,j+2}]_{\boldsymbol{v}^{-1}},\tilde{e}^{(0)}_{j,j+1}]_{\boldsymbol{v}^{-1}},\end{aligned} \tag{G.8}$$

$$\begin{aligned}\tilde{f}^{(0)}_{ij} &= (\boldsymbol{v}^{-1}-\boldsymbol{v})^{j-i+1}[\tilde{f}^{(0)}_{j+1,j},[\tilde{f}^{(0)}_{j+2,j+1},\cdots,[\tilde{f}^{(0)}_{i-1,i-2},\tilde{f}^{(0)}_{i,i-1}]_{\boldsymbol{v}}\cdots]_{\boldsymbol{v}}]_{\boldsymbol{v}} = \\ &(\boldsymbol{v}^{-1}-\boldsymbol{v})^{j-i+1}[[\cdots[\tilde{f}^{(0)}_{j+1,j},\tilde{f}^{(0)}_{j+2,j+1}]_{\boldsymbol{v}},\cdots,\tilde{f}^{(0)}_{i-1,i-2}]_{\boldsymbol{v}},\tilde{f}^{(0)}_{i,i-1}]_{\boldsymbol{v}},\end{aligned} \tag{G.9}$$

$$\begin{aligned}\tilde{e}^{(-1)}_{ji} &= (\boldsymbol{v}-\boldsymbol{v}^{-1})^{j-i+1}[\tilde{e}^{(0)}_{i-1,i},[\tilde{e}^{(0)}_{i-2,i-1},\cdots,[\tilde{e}^{(0)}_{j+1,j+2},\tilde{e}^{(-1)}_{j,j+1}]_{\boldsymbol{v}^{-1}}\cdots]_{\boldsymbol{v}^{-1}}]_{\boldsymbol{v}^{-1}} = \\ &(\boldsymbol{v}-\boldsymbol{v}^{-1})^{j-i+1}[[\cdots[\tilde{e}^{(0)}_{i-1,i},\tilde{e}^{(0)}_{i-2,i-1}]_{\boldsymbol{v}^{-1}},\cdots,\tilde{e}^{(0)}_{j+1,j+2}]_{\boldsymbol{v}^{-1}},\tilde{e}^{(-1)}_{j,j+1}]_{\boldsymbol{v}^{-1}},\end{aligned} \tag{G.10}$$

$$\begin{aligned}\tilde{f}^{(1)}_{ij} &= (\boldsymbol{v}^{-1}-\boldsymbol{v})^{j-i+1}[\tilde{f}^{(1)}_{j+1,j},[\tilde{f}^{(0)}_{j+2,j+1},\cdots,[\tilde{f}^{(0)}_{i-1,i-2},\tilde{f}^{(0)}_{i,i-1}]_{\boldsymbol{v}}\cdots]_{\boldsymbol{v}}]_{\boldsymbol{v}} = \\ &(\boldsymbol{v}^{-1}-\boldsymbol{v})^{j-i+1}[[\cdots[\tilde{f}^{(1)}_{j+1,j},\tilde{f}^{(0)}_{j+2,j+1}]_{\boldsymbol{v}},\cdots,\tilde{f}^{(0)}_{i-1,i-2}]_{\boldsymbol{v}},\tilde{f}^{(0)}_{i,i-1}]_{\boldsymbol{v}}.\end{aligned} \tag{G.11}$$

Recall elements $E^{(0)}_{ji}, F^{(0)}_{ij}, E^{(-1)}_{ji}, F^{(1)}_{ij} \in U_{\boldsymbol{v}}(L\mathfrak{sl}_n)$ of (10.6). Combining Corollary G.6 with (G.5), we get the following result.

Corollary G.7

(a) We have

$$\begin{aligned}&\Upsilon^{-1}(\tilde{e}^{(0)}_{ji}) = (\boldsymbol{v}-\boldsymbol{v}^{-1})E^{(0)}_{ji},\ \Upsilon^{-1}(\tilde{f}^{(0)}_{ij}) = -(\boldsymbol{v}-\boldsymbol{v}^{-1})F^{(0)}_{ij}, \\ &\Upsilon^{-1}(\tilde{e}^{(-1)}_{ji}) = -\boldsymbol{v}^{-j}(\boldsymbol{v}-\boldsymbol{v}^{-1})E^{(-1)}_{ji},\ \Upsilon^{-1}(\tilde{f}^{(1)}_{ij}) = \boldsymbol{v}^{j}(\boldsymbol{v}-\boldsymbol{v}^{-1})F^{(1)}_{ij}.\end{aligned} \tag{G.12}$$

(b) The right equalities in each of the first four lines of (10.6) hold.

To derive the right equalities of the last two lines of (10.6), we introduce

$$A^+_{ji} := \sum_{s\geq 1}\sum_{j=j_1<\ldots<j_{s+1}=i}(-1)^{s-1}\tilde{e}^{(0)}_{j_1j_2}\cdots\tilde{e}^{(0)}_{j_sj_{s+1}},$$
$$A^-_{ij} := \sum_{s\geq 1}\sum_{j=j_1<\ldots<j_{s+1}=i}(-1)^{s-1}\tilde{f}^{(0)}_{j_{s+1}j_s}\cdots\tilde{f}^{(0)}_{j_2j_1} \qquad \text{(G.13)}$$

for $1\leq j<i\leq n$. These elements will play an important role in Sect. G(ii) below.

Lemma G.8 *For any* $1\leq j<i\leq n$, *we have*

$$A^+_{ji} = (v-v^{-1})^{j-i+1}[\tilde{e}^{(0)}_{i-1,i},[\tilde{e}^{(0)}_{i-2,i-1},\cdots,[\tilde{e}^{(0)}_{j+1,j+2},\tilde{e}^{(0)}_{j,j+1}]_v\cdots]_v]_v =$$
$$(v-v^{-1})^{j-i+1}[[\cdots[\tilde{e}^{(0)}_{i-1,i},\tilde{e}^{(0)}_{i-2,i-1}]_v,\cdots,\tilde{e}^{(0)}_{j+1,j+2}]_v,\tilde{e}^{(0)}_{j,j+1}]_v, \qquad \text{(G.14)}$$

$$A^-_{ij} = (v^{-1}-v)^{j-i+1}[\tilde{f}^{(0)}_{j+1,j},[\tilde{f}^{(0)}_{j+2,j+1},\cdots,[\tilde{f}^{(0)}_{i-1,i-2},\tilde{f}^{(0)}_{i,i-1}]_{v^{-1}}\cdots]_{v^{-1}}]_{v^{-1}} =$$
$$(v^{-1}-v)^{j-i+1}[[\cdots[\tilde{f}^{(0)}_{j+1,j},\tilde{f}^{(0)}_{j+2,j+1}]_{v^{-1}},\cdots,\tilde{f}^{(0)}_{i-1,i-2}]_{v^{-1}},\tilde{f}^{(0)}_{i,i-1}]_{v^{-1}}. \qquad \text{(G.15)}$$

Proof We prove (G.14) by induction in $i-j$. The result is obvious for $i-j=1$. To perform the induction step, note that $A^+_{ji}=\tilde{e}^{(0)}_{ji}-\sum_{j<k<i}\tilde{e}^{(0)}_{jk}\cdot A^+_{ki}$. Applying the first equality of (G.8) together with the induction assumption, we get

$$(v-v^{-1})^{i-j-1}A^+_{ji} = [\tilde{e}^{(0)}_{i-1,i},[\tilde{e}^{(0)}_{i-2,i-1},\cdots,[\tilde{e}^{(0)}_{j+1,j+2},\tilde{e}^{(0)}_{j,j+1}]_{v^{-1}}\cdots]_{v^{-1}}]_{v^{-1}}-$$
$$(v-v^{-1})\sum_{j<k<i}[\tilde{e}^{(0)}_{k-1,k},\cdots,[\tilde{e}^{(0)}_{j+1,j+2},\tilde{e}^{(0)}_{j,j+1}]_{v^{-1}}\cdots]_{v^{-1}}\cdot[\tilde{e}^{(0)}_{i-1,i},\cdots,[\tilde{e}^{(0)}_{k+1,k+2},\tilde{e}^{(0)}_{k,k+1}]_v\cdots]_v.$$

Rewriting $[\tilde{e}^{(0)}_{i-1,i},X]_{v^{\pm1}}$ as $\tilde{e}^{(0)}_{i-1,i}\cdot X-v^{\pm1}X\cdot\tilde{e}^{(0)}_{i-1,i}$ and using the equality $[\tilde{e}^{(0)}_{i-1,i},\tilde{e}^{(0)}_{l,l+1}]=0$ for any $l<i-2$ (due to the quadratic Serre relations in $U^{\mathrm{ad}}_v(L\mathfrak{sl}_n)$), we immediately find

$$(v-v^{-1})^{i-j-1}A^+_{ji} = \Bigg[\tilde{e}^{(0)}_{i-1,i},[\tilde{e}^{(0)}_{i-2,i-1},\cdots,[\tilde{e}^{(0)}_{j+1,j+2},\tilde{e}^{(0)}_{j,j+1}]_{v^{-1}}\cdots]_{v^{-1}}-(v-v^{-1})\cdot$$
$$\sum_{j<k<i-1}[\tilde{e}^{(0)}_{k-1,k},\cdots,[\tilde{e}^{(0)}_{j+1,j+2},\tilde{e}^{(0)}_{j,j+1}]_{v^{-1}}\cdots]_{v^{-1}}\cdot[\tilde{e}^{(0)}_{i-2,i-1},\cdots,[\tilde{e}^{(0)}_{k+1,k+2},\tilde{e}^{(0)}_{k,k+1}]_v\cdots]_v\Bigg]_v =$$
$$[\tilde{e}^{(0)}_{i-1,i},(v-v^{-1})^{i-j-2}A^+_{j,i-1}]_v=[\tilde{e}^{(0)}_{i-1,i},[\tilde{e}^{(0)}_{i-2,i-1},\cdots,[\tilde{e}^{(0)}_{j+1,j+2},\tilde{e}^{(0)}_{j,j+1}]_v\cdots]_v]_v.$$

Note that the last equality follows from the induction assumption applied to $A^+_{j,i-1}$.

To prove that A^+_{ji} also equals the rightmost commutator of (G.14), we apply similar arguments to the equality $A^+_{ji}=\tilde{e}^{(0)}_{ji}-\sum_{j<k<i}A^+_{jk}\cdot\tilde{e}^{(0)}_{ki}$. We evaluate the

right-hand side by applying the rightmost expression of (G.8) to the terms $\tilde{e}^{(0)}_{ji}, \tilde{e}^{(0)}_{ki}$ and the induction assumption to A^+_{jk}. Rewriting $[X, \tilde{e}^{(0)}_{j,j+1}]_{\boldsymbol{v}^{\pm 1}}$ as $X \cdot \tilde{e}^{(0)}_{j,j+1} - \boldsymbol{v}^{\pm 1}\tilde{e}^{(0)}_{j,j+1} \cdot X$ and taking $\tilde{e}^{(0)}_{j,j+1}$ to the leftmost or the rightmost sides, we get the result.

The proof of (G.15) is completely analogous and is left to the interested reader. □

The following result follows by combining Lemma G.8 with formula (G.5).

Corollary G.9

(a) We have

$$\Upsilon^{-1}(A^+_{ji}) = (\boldsymbol{v} - \boldsymbol{v}^{-1})\widetilde{E}^{(0)}_{ji},\ \Upsilon^{-1}(A^-_{ij}) = -(\boldsymbol{v} - \boldsymbol{v}^{-1})\widetilde{F}^{(0)}_{ij}. \tag{G.16}$$

(b) The right equalities in the last two lines of (10.6) hold.

G(ii) Computation of $\Delta^{\mathrm{rtt}}(\tilde{g}^{(\pm 1)}_i)$

Given a Laurent series $F(z)$, we use $[z^r]F(z)$ to denote the coefficient of z^r in $F(z)$. In this subsection, we compute explicitly $\Delta^{\mathrm{rtt}}(\tilde{g}^{(\pm 1)}_i)$, see Theorems G.10 and G.13.

Theorem G.10 *For $1 \le i \le n$, we have*

$$\begin{gathered}\Delta^{\mathrm{rtt}}(\tilde{g}^{(1)}_i) = \tilde{g}^{(1)}_i \otimes \tilde{g}^+_i + \tilde{g}^+_i \otimes \tilde{g}^{(1)}_i + \sum_{l>i} \tilde{g}^+_i \tilde{e}^{(0)}_{il} \otimes \tilde{f}^{(1)}_{li} \tilde{g}^+_i + \\ \sum_{s\geq 1} \sum_{j_1<\ldots<j_{s+1}=i} (-1)^s \tilde{g}^+_i \tilde{e}^{(0)}_{j_1 j_2} \cdots \tilde{e}^{(0)}_{j_s j_{s+1}} \otimes \tilde{f}^{(1)}_{ij_1} \tilde{g}^+_i + \\ \sum_{l>i} \sum_{s\geq 1} \sum_{j_1<\ldots<j_{s+1}=i} (-1)^s \tilde{g}^+_i \tilde{e}^{(0)}_{il} \tilde{e}^{(0)}_{j_1 j_2} \cdots \tilde{e}^{(0)}_{j_s j_{s+1}} \otimes \tilde{f}^{(1)}_{lj_1} \tilde{g}^+_i .\end{gathered} \tag{G.17}$$

Proof Our starting point is the equality

$$[z^{-1}]T^+_{ii}(z) = \tilde{g}^{(1)}_i + \sum_{j<i} \tilde{f}^{(1)}_{ij} \tilde{g}^+_j \tilde{e}^{(0)}_{ji}. \tag{G.18}$$

We also note that $[z^{-1}]T^+_{ij}(z) = \tilde{f}^{(1)}_{ij} \tilde{g}^+_j + \sum_{j'<j} \tilde{f}^{(1)}_{ij'} \tilde{g}^+_{j'} \tilde{e}^{(0)}_{j'i}$ for any $i > j$. Rewriting this as $\tilde{f}^{(1)}_{ij} \tilde{g}^+_j = [z^{-1}]T^+_{ij}(z) - \sum_{j'<j} \tilde{f}^{(1)}_{ij'} \tilde{g}^+_{j'} \tilde{e}^{(0)}_{j'i}$ and applying this formula iteratively, we finally get

$$\tilde{f}^{(1)}_{ij} \tilde{g}^+_j = \sum_{s\geq 1} \sum_{j_1<\ldots<j_s=j} (-1)^{s-1} \left([z^{-1}]T^+_{ij_1}(z)\right) \tilde{e}^{(0)}_{j_1 j_2} \cdots \tilde{e}^{(0)}_{j_{s-1} j_s}. \tag{G.19}$$

Combining formulas (G.18) and (G.19), we get

$$\tilde{g}_i^{(1)} = [z^{-1}]T_{ii}^+(z) - \sum_{j<i} \left([z^{-1}]T_{ij}^+(z)\right) \cdot A_{ji}^+, \tag{G.20}$$

where A_{ji}^+ was defined in (G.13).

Thus, it remains to compute explicitly $\Delta^{\mathrm{rtt}}([z^{-1}]T_{ii}^+(z))$, $\Delta^{\mathrm{rtt}}([z^{-1}]T_{ij}^+(z))$, $\Delta^{\mathrm{rtt}}(A_{ji}^+)$ for $i > j$. Evaluating the coefficients of z^{-1} in $\Delta^{\mathrm{rtt}}(T_{ii}^+(z)) = \sum_{a=1}^n T_{ia}^+(z) \otimes T_{ai}^+(z)$, we find

$$\begin{aligned}\Delta^{\mathrm{rtt}}([z^{-1}]T_{ii}^+(z)) = \sum_{j<i} \tilde{f}_{ij}^{(1)}\tilde{g}_j^+ \otimes \tilde{g}_j^+\tilde{e}_{ji}^{(0)} + \sum_{j'<j<i} \tilde{f}_{ij'}^{(1)}\tilde{g}_{j'}^+\tilde{e}_{j'j}^{(0)} \otimes \tilde{g}_j^+\tilde{e}_{ji}^{(0)} + \\ \tilde{g}_i^{(1)} \otimes \tilde{g}_i^+ + \tilde{g}_i^+ \otimes \tilde{g}_i^{(1)} + \sum_{j<i} \tilde{f}_{ij}^{(1)}\tilde{g}_j^+\tilde{e}_{ji}^{(0)} \otimes \tilde{g}_i^+ + \sum_{j<i} \tilde{g}_i^+ \otimes \tilde{f}_{ij}^{(1)}\tilde{g}_j^+\tilde{e}_{ji}^{(0)} + \\ \sum_{l>i} \tilde{g}_i^+\tilde{e}_{il}^{(0)} \otimes \tilde{f}_{li}^{(1)}\tilde{g}_i^+ + \sum_{l>i}^{j<i} \tilde{g}_i^+\tilde{e}_{il}^{(0)} \otimes \tilde{f}_{lj}^{(1)}\tilde{g}_j^+\tilde{e}_{ji}^{(0)},\end{aligned} \tag{G.21}$$

where the first, second, and third lines in the right-hand side correspond to the contributions arising from the cases $a < i$, $a = i$, and $a > i$, respectively.

Evaluating the coefficients of z^{-1} in $\Delta^{\mathrm{rtt}}(T_{ij}^+(z)) = \sum_{a=1}^n T_{ia}^+(z) \otimes T_{aj}^+(z)$, we find

$$\begin{aligned}\Delta^{\mathrm{rtt}}([z^{-1}]T_{ij}^+(z)) = \sum_{j'<j} \tilde{f}_{ij'}^{(1)}\tilde{g}_{j'}^+ \otimes \tilde{g}_{j'}^+\tilde{e}_{j'j}^{(0)} + \sum_{j''<j'<j} \tilde{f}_{ij''}^{(1)}\tilde{g}_{j''}^+\tilde{e}_{j''j'}^{(0)} \otimes \tilde{g}_{j'}^+\tilde{e}_{j'j}^{(0)} + \\ \tilde{f}_{ij}^{(1)}\tilde{g}_j^+ \otimes \tilde{g}_j^+ + \sum_{j'<j} \tilde{f}_{ij'}^{(1)}\tilde{g}_{j'}^+\tilde{e}_{j'j}^{(0)} \otimes \tilde{g}_j^+ + \tilde{g}_i^+ \otimes \tilde{f}_{ij}^{(1)}\tilde{g}_j^+ + \sum_{j'<j} \tilde{g}_i^+ \otimes \tilde{f}_{ij'}^{(1)}\tilde{g}_{j'}^+\tilde{e}_{j'j}^{(0)} + \\ \sum_{l>i} \tilde{g}_i^+\tilde{e}_{il}^{(0)} \otimes \tilde{f}_{lj}^{(1)}\tilde{g}_j^+ + \sum_{l>i}^{j'<j} \tilde{g}_i^+\tilde{e}_{il}^{(0)} \otimes \tilde{f}_{lj'}^{(1)}\tilde{g}_{j'}^+\tilde{e}_{j'j}^{(0)},\end{aligned} \tag{G.22}$$

where the first, second, and third lines in the right-hand side correspond to the contributions arising from $a < j$, $a = j$ or i, and $a > i$, respectively. Note that for $j < a < i$ both $T_{ia}^+(z)$, $T_{aj}^+(z)$ contain only negative powers of z and hence do not contribute above.

Finally, let us compute the coproduct of A_{ji}^+.

Lemma G.11 *We have*

$$\Delta^{\mathrm{rtt}}(A_{ji}^+) = \sum_{s\geq 1} \sum_{j=j_1<\ldots<j_{s+1}=i} \sum_{r=1}^{s+1} (-1)^{s-1}\tilde{e}_{j_r j_{r+1}}^{(0)} \cdots \tilde{e}_{j_s j_{s+1}}^{(0)} \otimes \tilde{e}_{j_1 j_2}^{(0)} \cdots \tilde{e}_{j_{r-1} j_r}^{(0)} (\tilde{g}_{j_r}^+)^{-1}\tilde{g}_i^+.$$

Proof We prove this by induction in $i - j$. The base of induction $i = j + 1$ follows from the equality $A^+_{j,j+1} = \tilde{e}^{(0)}_{j,j+1}$ and Lemma G.12 below. To perform the induction step, note that

$$A^+_{ji} = \tilde{e}^{(0)}_{ji} - \sum_{j<j'<i} \tilde{e}^{(0)}_{jj'} A^+_{j'i}. \tag{G.23}$$

Next, we compute the coproduct of $\tilde{e}^{(0)}_{ji}$.

Lemma G.12 *We have*

$$\Delta^{\mathrm{rtt}}(\tilde{e}^{(0)}_{ji}) = 1 \otimes \tilde{e}^{(0)}_{ji} + \tilde{e}^{(0)}_{ji} \otimes (\tilde{g}^+_j)^{-1}\tilde{g}^+_i + \sum_{j<a<i} \tilde{e}^{(0)}_{ja} \otimes (\tilde{g}^+_j)^{-1}\tilde{g}^+_a\tilde{e}^{(0)}_{ai}.$$

Proof First, let us note that $\tilde{g}^+_j = [z^0]T^+_{jj}(z)$. Thus,

$$\Delta^{\mathrm{rtt}}(\tilde{g}^+_j) = [z^0]\left(\sum_{a=1}^{n} T^+_{ja}(z) \otimes T^+_{aj}(z)\right) = [z^0](T^+_{jj}(z) \otimes T^+_{jj}(z)) = \tilde{g}^+_j \otimes \tilde{g}^+_j.$$

We also note that $[z^0]T^+_{ji}(z) = \tilde{g}^+_j\tilde{e}^{(0)}_{ji}$. Hence, we have

$$\Delta^{\mathrm{rtt}}(\tilde{g}^+_j\tilde{e}^{(0)}_{ji}) = [z^0]\left(T^+_{jj}(z) \otimes T^+_{ji}(z) + T^+_{ji}(z) \otimes T^+_{ii}(z) + \sum_{j<a<i} T^+_{ja}(z) \otimes T^+_{ai}(z)\right) =$$
$$\tilde{g}^+_j \otimes \tilde{g}^+_j\tilde{e}^{(0)}_{ji} + \tilde{g}^+_j\tilde{e}^{(0)}_{ji} \otimes \tilde{g}^+_i + \sum_{j<a<i} \tilde{g}^+_j\tilde{e}^{(0)}_{ja} \otimes \tilde{g}^+_a\tilde{e}^{(0)}_{ai}.$$

Note that in the first equality we used $[z^0](T^+_{ja}(z) \otimes T^+_{ai}(z)) = 0$ for $a < j$ or $a > i$.

Evaluating $\Delta^{\mathrm{rtt}}(\tilde{e}^{(0)}_{ji}) = \Delta^{\mathrm{rtt}}(\tilde{g}^+_j)^{-1}\Delta^{\mathrm{rtt}}(\tilde{g}^+_j\tilde{e}^{(0)}_{ji})$ via these formulas completes our proof. □

Combining (G.23) with Lemma G.12 and applying the induction assumption to $\Delta^{\mathrm{rtt}}(A^+_{j'i})$, we immediately get the formula for $\Delta^{\mathrm{rtt}}(A^+_{ji})$ of Lemma G.11. □

Combining (G.20–G.22) with Lemma G.11, we get (G.17) after tedious computations. □

Theorem G.13 *For $1 \le i \le n$, we have*

$$\begin{aligned}\Delta^{\mathrm{rtt}}(\tilde{g}^{(-1)}_i) &= \tilde{g}^{(-1)}_i \otimes \tilde{g}^-_i + \tilde{g}^-_i \otimes \tilde{g}^{(-1)}_i + \sum_{l>i} \tilde{g}^-_i\tilde{e}^{(-1)}_{il} \otimes \tilde{f}^{(0)}_{li}\tilde{g}^-_i + \\ &\sum_{s\ge1}\sum_{j_1<\ldots<j_{s+1}=i} (-1)^s\tilde{g}^-_i\tilde{e}^{(-1)}_{ji} \otimes \tilde{f}^{(0)}_{j_{s+1}j_s}\cdots\tilde{f}^{(0)}_{j_2j_1}\tilde{g}^-_i + \\ &\sum_{l>i}\sum_{s\ge1}\sum_{j_1<\ldots<j_{s+1}=i} (-1)^s\tilde{g}^-_i\tilde{e}^{(-1)}_{jl} \otimes \tilde{f}^{(0)}_{j_{s+1}j_s}\cdots\tilde{f}^{(0)}_{j_2j_1}\tilde{f}^{(0)}_{li}\tilde{g}^-_i.\end{aligned} \tag{G.24}$$

Proof Our starting point is the equality

$$[z]T_{ii}^{-}(z)=\tilde{g}_i^{(-1)}+\sum_{j<i}\tilde{f}_{ij}^{(0)}\tilde{g}_j^{-}\tilde{e}_{ji}^{(-1)}. \tag{G.25}$$

We also note that $[z]T_{ji}^{-}(z)=\tilde{g}_j^{-}\tilde{e}_{ji}^{(-1)}+\sum_{j'<j}\tilde{f}_{jj'}^{(0)}\tilde{g}_{j'}^{-}\tilde{e}_{j'i}^{(-1)}$ for any $i>j$. Rewriting this as $\tilde{g}_j^{-}\tilde{e}_{ji}^{(-1)}=[z]T_{ji}^{-}(z)-\sum_{j'<j}\tilde{f}_{jj'}^{(0)}\tilde{g}_{j'}^{-}\tilde{e}_{j'i}^{(-1)}$ and applying this formula iteratively, we finally get

$$\tilde{g}_j^{-}\tilde{e}_{ji}^{(-1)}=\sum_{s\geq 1}\sum_{j_1<\ldots<j_s=j}(-1)^{s-1}\tilde{f}_{j_sj_{s-1}}^{(0)}\cdots\tilde{f}_{j_2j_1}^{(0)}\cdot\left([z]T_{j_1i}^{-}(z)\right). \tag{G.26}$$

Combining formulas (G.25) and (G.26), we get

$$\tilde{g}_i^{(-1)}=[z]T_{ii}^{-}(z)-\sum_{j<i}A_{ij}^{-}\cdot\left([z]T_{ji}^{-}(z)\right), \tag{G.27}$$

where A_{ij}^{-} was defined in (G.13).

Thus, it remains to compute explicitly $\Delta^{\mathrm{rtt}}([z]T_{ii}^{-}(z))$, $\Delta^{\mathrm{rtt}}([z]T_{ji}^{-}(z))$, $\Delta^{\mathrm{rtt}}(A_{ij}^{-})$ for $i>j$. Evaluating the coefficients of z^1 in $\Delta^{\mathrm{rtt}}(T_{ii}^{-}(z))=\sum_{a=1}^{n}T_{ia}^{-}(z)\otimes T_{ai}^{-}(z)$, we find

$$\begin{gathered}\Delta^{\mathrm{rtt}}([z]T_{ii}^{-}(z))=\sum_{j<i}\tilde{f}_{ij}^{(0)}\tilde{g}_j^{-}\otimes\tilde{g}_j^{-}\tilde{e}_{ji}^{(-1)}+\sum_{j'<j<i}\tilde{f}_{ij}^{(0)}\tilde{g}_j^{-}\otimes\tilde{f}_{jj'}^{(0)}\tilde{g}_{j'}^{-}\tilde{e}_{j'i}^{(-1)}+\\ \tilde{g}_i^{-}\otimes\tilde{g}_i^{(-1)}+\tilde{g}_i^{(-1)}\otimes\tilde{g}_i^{-}+\sum_{j<i}\tilde{g}_i^{-}\otimes\tilde{f}_{ij}^{(0)}\tilde{g}_j^{-}\tilde{e}_{ji}^{(-1)}+\sum_{j<i}\tilde{f}_{ij}^{(0)}\tilde{g}_j^{-}\tilde{e}_{ji}^{(-1)}\otimes\tilde{g}_i^{-}+\\ \sum_{l>i}\tilde{g}_i^{-}\tilde{e}_{il}^{(-1)}\otimes\tilde{f}_{li}^{(0)}\tilde{g}_i^{-}+\sum_{l>i}^{j<i}\tilde{f}_{ij}^{(0)}\tilde{g}_j^{-}\tilde{e}_{jl}^{(-1)}\otimes\tilde{f}_{li}^{(0)}\tilde{g}_i^{-},\end{gathered} \tag{G.28}$$

where the first, second, and third lines in the right-hand side correspond to the contributions arising from the cases $a<i$, $a=i$, and $a>i$, respectively.

Evaluating the coefficients of z^1 in $\Delta^{\mathrm{rtt}}(T_{ji}^{-}(z))=\sum_{a=1}^{n}T_{ja}^{-}(z)\otimes T_{ai}^{-}(z)$, we find

$$\begin{gathered}\Delta^{\mathrm{rtt}}([z]T_{ji}^{-}(z))=\sum_{j'<j}\tilde{f}_{jj'}^{(0)}\tilde{g}_{j'}^{-}\otimes\tilde{g}_{j'}^{-}\tilde{e}_{j'i}^{(-1)}+\sum_{j''<j'<j}\tilde{f}_{jj'}^{(0)}\tilde{g}_{j'}^{-}\otimes\tilde{f}_{j'j''}^{(0)}\tilde{g}_{j''}^{-}\tilde{e}_{j''i}^{(-1)}+\\ \tilde{g}_j^{-}\otimes\tilde{g}_j^{-}\tilde{e}_{ji}^{(-1)}+\sum_{j'<j}\tilde{g}_j^{-}\otimes\tilde{f}_{jj'}^{(0)}\tilde{g}_{j'}^{-}\tilde{e}_{j'i}^{(-1)}+\tilde{g}_j^{-}\tilde{e}_{ji}^{(-1)}\otimes\tilde{g}_i^{-}+\sum_{j'<j}\tilde{f}_{jj'}^{(0)}\tilde{g}_{j'}^{-}\tilde{e}_{j'i}^{(-1)}\otimes\tilde{g}_i^{-}+\\ \sum_{l>i}\tilde{g}_j^{-}\tilde{e}_{jl}^{(-1)}\otimes\tilde{f}_{li}^{(0)}\tilde{g}_i^{-}+\sum_{l>i}^{j'<j}\tilde{f}_{jj'}^{(0)}\tilde{g}_{j'}^{-}\tilde{e}_{j'l}^{(-1)}\otimes\tilde{f}_{li}^{(0)}\tilde{g}_i^{-},\end{gathered} \tag{G.29}$$

where the first, second, and third lines in the right-hand side correspond to the contributions arising from $a < j$, $a = j$ or i, and $a > i$, respectively. Note that for $j < a < i$ both $T^-_{ja}(z)$, $T^-_{ai}(z)$ contain only positive powers of z and hence do not contribute above.

Finally, let us compute the coproduct of A^-_{ij}.

Lemma G.14 *We have*

$$\Delta^{\mathrm{rtt}}(A^-_{ij}) = \sum_{s\geq 1}\sum_{j=j_1<\ldots<j_{s+1}=i}\sum_{r=1}^{s+1}(-1)^{s-1}\tilde{g}^-_i(\tilde{g}^-_{j_r})^{-1}\tilde{f}^{(0)}_{j_r j_{r-1}}\cdots\tilde{f}^{(0)}_{j_2 j_1}\otimes\tilde{f}^{(0)}_{j_{s+1} j_s}\cdots\tilde{f}^{(0)}_{j_{r+1} j_r}.$$

Proof We prove this by induction in $i - j$. The base of induction $i = j + 1$ follows from the equality $A^-_{j+1,j} = \tilde{f}^{(0)}_{j+1,j}$ and Lemma G.15 below. To perform the induction step, note that

$$A^-_{ij} = \tilde{f}^{(0)}_{ij} - \sum_{j<j'<i} A^-_{ij'}\tilde{f}^{(0)}_{j'j}. \tag{G.30}$$

Next, we compute the coproduct of $\tilde{f}^{(0)}_{ij}$.

Lemma G.15 *We have*

$$\Delta^{\mathrm{rtt}}(\tilde{f}^{(0)}_{ij}) = \tilde{f}^{(0)}_{ij}\otimes 1 + \tilde{g}^-_i(\tilde{g}^-_j)^{-1}\otimes\tilde{f}^{(0)}_{ij} + \sum_{j<a<i}\tilde{f}^{(0)}_{ia}\tilde{g}^-_a(\tilde{g}^-_j)^{-1}\otimes\tilde{f}^{(0)}_{aj}.$$

Proof First, let us note that $\tilde{g}^-_j = [z^0]T^-_{jj}(z)$. Thus,

$$\Delta^{\mathrm{rtt}}(\tilde{g}^-_j) = [z^0]\left(\sum_{a=1}^{n} T^-_{ja}(z)\otimes T^-_{aj}(z)\right) = [z^0](T^-_{jj}(z)\otimes T^-_{jj}(z)) = \tilde{g}^-_j\otimes\tilde{g}^-_j.$$

We also note that $[z^0]T^-_{ij}(z) = \tilde{f}^{(0)}_{ij}\tilde{g}^-_j$. Hence, we have

$$\Delta^{\mathrm{rtt}}(\tilde{f}^{(0)}_{ij}\tilde{g}^-_j) = [z^0]\left(T^-_{ij}(z)\otimes T^-_{jj}(z) + T^-_{ii}(z)\otimes T^-_{ij}(z) + \sum_{j<a<i} T^-_{ia}(z)\otimes T^-_{aj}(z)\right) =$$
$$\tilde{f}^{(0)}_{ij}\tilde{g}^-_j\otimes\tilde{g}^-_j + \tilde{g}^-_i\otimes\tilde{f}^{(0)}_{ij}\tilde{g}^-_j + \sum_{j<a<i}\tilde{f}^{(0)}_{ia}\tilde{g}^-_a\otimes\tilde{f}^{(0)}_{aj}\tilde{g}^-_j.$$

Note that in the first equality we used $[z^0](T^-_{ia}(z)\otimes T^-_{aj}(z)) = 0$ for $a < j$ or $a > i$.

Evaluating $\Delta^{\mathrm{rtt}}(\tilde{f}^{(0)}_{ij}) = \Delta^{\mathrm{rtt}}(\tilde{f}^{(0)}_{ij}\tilde{g}^-_j)\Delta^{\mathrm{rtt}}(\tilde{g}^-_j)^{-1}$ via these formulas completes our proof. □

Combining (G.30) with Lemma G.15 and applying the induction assumption to $\Delta^{\mathrm{rtt}}(A^-_{ij'})$, we immediately get the formula for $\Delta^{\mathrm{rtt}}(A^-_{ij})$ of Lemma G.14. □

Combining (G.27–G.29) with Lemma G.14, we get (G.24) after tedious computations. □

For $1 \le i \le n$, define $H_{i,\pm 1} \in U^{\mathrm{rtt}}(\mathfrak{sl}_n)$ via $H_{i,\pm 1} := (\tilde{g}_i^{\pm})^{-1}\tilde{g}_i^{(\pm 1)}$. Recall the elements A^+_{ji} and A^-_{ij} of (G.13). Combining Theorems G.10, G.13 with Lemma G.5 and the formula $\Delta^{\mathrm{rtt}}(\tilde{g}_i^{\pm}) = \tilde{g}_i^{\pm} \otimes \tilde{g}_i^{\pm}$, we get the following expressions for $\Delta^{\mathrm{rtt}}(H_{i,\pm 1})$.

Corollary G.16 *We have*

$$\Delta^{\mathrm{rtt}}(H_{i,1}) = H_{i,1}\otimes 1+1\otimes H_{i,1}+\boldsymbol{v}^{-1}\sum_{l>i}\tilde{e}^{(0)}_{il}\otimes\tilde{f}^{(1)}_{li}-\boldsymbol{v}\sum_{j<i}A^+_{ji}\otimes\tilde{f}^{(1)}_{ij}-\sum_{l>i}^{j<i}\tilde{e}^{(0)}_{il}A^+_{ji}\otimes\tilde{f}^{(1)}_{lj}, \tag{G.31}$$

$$\Delta^{\mathrm{rtt}}(H_{i,-1}) = H_{i,-1}\otimes 1+1\otimes H_{i,-1}+\boldsymbol{v}\sum_{l>i}\tilde{e}^{(-1)}_{il}\otimes\tilde{f}^{(0)}_{li}-\boldsymbol{v}^{-1}\sum_{j<i}\tilde{e}^{(-1)}_{ji}\otimes A^-_{ij}-\sum_{l>i}^{j<i}\tilde{e}^{(-1)}_{jl}\otimes A^-_{ij}\tilde{f}^{(0)}_{li}. \tag{G.32}$$

G(iii) Proof of Formula (10.2)

Recall the Hopf algebra embedding $\Upsilon\colon U_{\boldsymbol{v}}(L\mathfrak{sl}_n) \hookrightarrow U^{\mathrm{rtt}}(\mathfrak{sl}_n)$ of Theorem G.2 (see also Remark G.3). It is easy to see that

$$\Upsilon(h_{i,1}) = \frac{H_{i+1,1}-H_{i,1}}{\boldsymbol{v}^i(\boldsymbol{v}-\boldsymbol{v}^{-1})}.$$

Combining Corollaries G.7, G.9 with formula (G.31) and the fact that Υ intertwines Δ and Δ^{rtt}, we immediately get

$$\begin{gathered}
\Delta(h_{i,1}) - h_{i,1}\otimes 1 - 1\otimes h_{i,1} = \boldsymbol{v}^{-i}(\boldsymbol{v}-\boldsymbol{v}^{-1})^{-1}\times \\
\Bigg(\boldsymbol{v}^i(\boldsymbol{v}-\boldsymbol{v}^{-1})^2\sum_{l>i+1}E^{(0)}_{i+1,l}\otimes F^{(1)}_{l,i+1} - (\boldsymbol{v}-\boldsymbol{v}^{-1})^2\sum_{k<i+1}\boldsymbol{v}^{k+1}\widetilde{E}^{(0)}_{k,i+1}\otimes F^{(1)}_{i+1,k} - \\
(\boldsymbol{v}-\boldsymbol{v}^{-1})^3\sum_{k<i+1<l}\boldsymbol{v}^k E^{(0)}_{i+1,l}\widetilde{E}^{(0)}_{k,i+1}\otimes F^{(1)}_{lk} - \boldsymbol{v}^{i-1}(\boldsymbol{v}-\boldsymbol{v}^{-1})^2\sum_{l>i}E^{(0)}_{il}\otimes F^{(1)}_{li} + \\
(\boldsymbol{v}-\boldsymbol{v}^{-1})^2\sum_{k<i}\boldsymbol{v}^{k+1}\widetilde{E}^{(0)}_{ki}\otimes F^{(1)}_{ik} + (\boldsymbol{v}-\boldsymbol{v}^{-1})^3\sum_{k<i<l}\boldsymbol{v}^k E^{(0)}_{il}\widetilde{E}^{(0)}_{ki}\otimes F^{(1)}_{lk}\Bigg).
\end{gathered} \tag{G.33}$$

This formula implies (10.2) after the following simplifications:

$$\sum_{k<i<l} v^k E^{(0)}_{il} \widetilde{E}^{(0)}_{ki} \otimes F^{(1)}_{lk} - \sum_{k<i+1<l} v^k E^{(0)}_{i+1,l} \widetilde{E}^{(0)}_{k,i+1} \otimes F^{(1)}_{lk} =$$

$$\sum_{l>i+1}^{k<i} v^k (E^{(0)}_{il} \widetilde{E}^{(0)}_{ki} - E^{(0)}_{i+1,l} \widetilde{E}^{(0)}_{k,i+1}) \otimes F^{(1)}_{lk} + \sum_{k<i} v^k E^{(0)}_{i,i+1} \widetilde{E}^{(0)}_{ki} \otimes F^{(1)}_{i+1,k} - v^i \sum_{l>i+1} E^{(0)}_{i+1,l} \widetilde{E}^{(0)}_{i,i+1} \otimes F^{(1)}_{li},$$

$$- v^{-1} \sum_{l>i} E^{(0)}_{il} \otimes F^{(1)}_{li} - (v - v^{-1}) \sum_{l>i+1} E^{(0)}_{i+1,l} E^{(0)}_{i,i+1} \otimes F^{(1)}_{li} =$$

$$- v^{-1} E^{(0)}_{i,i+1} \otimes F^{(1)}_{i+1,i} + v^{-2} \sum_{l>i+1} [E^{(0)}_{i,i+1}, E^{(0)}_{i+1,l}]_{v^3} \otimes F^{(1)}_{li},$$

$$- \sum_{k<i+1} v^{k+1-i} \widetilde{E}^{(0)}_{k,i+1} \otimes F^{(1)}_{i+1,k} + (v - v^{-1}) \sum_{k<i} v^{k-i} E^{(0)}_{i,i+1} \widetilde{E}^{(0)}_{ki} \otimes F^{(1)}_{i+1,k} =$$

$$- v E^{(0)}_{i,i+1} \otimes F^{(1)}_{i+1,i} - \sum_{k<i} v^{k-i-1} [E^{(0)}_{i,i+1}, \widetilde{E}^{(0)}_{ki}]_{v^3} \otimes F^{(1)}_{i+1,k},$$

where in the second and third equalities we used

$$E^{(0)}_{il} = [E^{(0)}_{i+1,l}, E^{(0)}_{i,i+1}]_{v^{-1}}, \ \widetilde{E}^{(0)}_{k,i+1} = [E^{(0)}_{i,i+1}, \widetilde{E}^{(0)}_{ki}]_v.$$

G(iv) Proof of Formula (10.3)

The proof of (10.3) is completely analogous and is based on the formula

$$\Upsilon(h_{i,-1}) = \frac{H_{i,-1} - H_{i+1,-1}}{v^{-i}(v - v^{-1})}.$$

Combining this with Corollaries G.7, G.9, formula (G.32) and the fact that Υ intertwines Δ and Δ^{rtt}, one derives (10.3). The computations are similar to the above proof of (10.2) and are left to the interested reader.

G(v) Proof of Formula (10.4)

Recall that $[h_{i,-1}, e_{i,0}] = [2]_v \cdot e_{i,-1}$, so that

$$\Delta(e_{i,-1}) = [2]_v^{-1} \cdot [\Delta(h_{i,-1}), \Delta(e_{i,0})] = [2]_v^{-1} \cdot [\Delta(h_{i,-1}), 1 \otimes e_{i,0} + e_{i,0} \otimes \psi^+_{i,0}].$$

Applying formula (10.3) to $\Delta(h_{i,-1})$ and using Lemma G.17 below, we recover (10.4).

Lemma G.17 *For $k < i$ and $l > i+1$, the following equalities hold:*

(a) $[F^{(0)}_{l,i+1}, e_{i,0}] = 0.$
(b) $[\widetilde{F}^{(0)}_{ik}, e_{i,0}] = 0.$
(c) $[F^{(0)}_{li}, e_{i,0}] = -F^{(0)}_{l,i+1}\psi^-_{i,0}.$
(d) $[\widetilde{F}^{(0)}_{i+1,k}, e_{i,0}] = \boldsymbol{v}^{-1}\widetilde{F}^{(0)}_{ik}\psi^-_{i,0}.$
(e) $[[F^{(0)}_{l,i+1}, F^{(0)}_{i+1,i}]_{\boldsymbol{v}^{-3}}, e_{i,0}] = \frac{1-\boldsymbol{v}^{-4}}{\boldsymbol{v}-\boldsymbol{v}^{-1}}F^{(0)}_{l,i+1}\psi^-_{i,0} - \frac{1-\boldsymbol{v}^{-2}}{\boldsymbol{v}-\boldsymbol{v}^{-1}}F^{(0)}_{l,i+1}\psi^+_{i,0}.$
(f) $[[\widetilde{F}^{(0)}_{ik}, F^{(0)}_{i+1,i}]_{\boldsymbol{v}^{-3}}, e_{i,0}] = \frac{1-\boldsymbol{v}^{-4}}{\boldsymbol{v}-\boldsymbol{v}^{-1}}\widetilde{F}^{(0)}_{ik}\psi^-_{i,0} - \frac{1-\boldsymbol{v}^{-2}}{\boldsymbol{v}-\boldsymbol{v}^{-1}}\widetilde{F}^{(0)}_{ik}\psi^+_{i,0}.$
(g) $[E^{(-1)}_{i+1,l}, e_{i,0}]_{\boldsymbol{v}} = \boldsymbol{v}E^{(-1)}_{il}.$
(h) $[E^{(-1)}_{ki}, e_{i,0}]_{\boldsymbol{v}} = -\boldsymbol{v}E^{(-1)}_{k,i+1}.$
(i) $[E^{(-1)}_{il}, e_{i,0}]_{\boldsymbol{v}^{-1}} = 0.$
(j) $[E^{(-1)}_{k,i+1}, e_{i,0}]_{\boldsymbol{v}^{-1}} = 0.$
(k) $[E^{(-1)}_{kl}, e_{i,0}] = 0.$

Proof Recall that $[f_{j,0}, e_{i,0}] = \frac{\delta_{ji}}{\boldsymbol{v}-\boldsymbol{v}^{-1}}(\psi^-_{i,0} - \psi^+_{i,0})$.

Parts (a, b) are obvious as $e_{i,0}$ commutes with $f_{i+1,0}, \ldots, f_{l-1,0}$ and $f_{k,0}, \ldots, f_{i-1,0}$. Combining (a, b) with equalities $F^{(0)}_{li} = [f_{i,0}, F^{(0)}_{l,i+1}]_{\boldsymbol{v}}$ and $\widetilde{F}^{(0)}_{i+1,k} = [\widetilde{F}^{(0)}_{ik}, f_{i,0}]_{\boldsymbol{v}^{-1}}$, we get $[F^{(0)}_{li}, e_{i,0}] = [\frac{\psi^-_{i,0}-\psi^+_{i,0}}{\boldsymbol{v}-\boldsymbol{v}^{-1}}, F^{(0)}_{l,i+1}]_{\boldsymbol{v}} = -F^{(0)}_{l,i+1}\psi^-_{i,0}$ and $[\widetilde{F}^{(0)}_{i+1,k}, e_{i,0}] = [\widetilde{F}^{(0)}_{ik}, \frac{\psi^-_{i,0}-\psi^+_{i,0}}{\boldsymbol{v}-\boldsymbol{v}^{-1}}]_{\boldsymbol{v}^{-1}} = \boldsymbol{v}^{-1}\widetilde{F}^{(0)}_{ik}\psi^-_{i,0}$, which proves parts (c, d). Parts (e, f) also follow immediately from (a, b).

(g) Due to the quadratic Serre relations $e_{i,0}$ commutes with $e_{i+2,0}, \ldots, e_{l-1,0}$, hence, also with $E^{(0)}_{i+2,l}$. Meanwhile, we have $[e_{i+1,-1}, e_{i,0}]_{\boldsymbol{v}} = \boldsymbol{v}[e_{i+1,0}, e_{i,-1}]_{\boldsymbol{v}^{-1}}$, due to (U2). Thus, $[E^{(-1)}_{i+1,l}, e_{i,0}]_{\boldsymbol{v}} = [[E^{(0)}_{i+2,l}, e_{i+1,-1}]_{\boldsymbol{v}^{-1}}, e_{i,0}]_{\boldsymbol{v}} = [E^{(0)}_{i+2,l}, \boldsymbol{v}[e_{i+1,0}, e_{i,-1}]_{\boldsymbol{v}^{-1}}]_{\boldsymbol{v}^{-1}} = \boldsymbol{v}E^{(-1)}_{il}$.

(h) We have $[E^{(-1)}_{ki}, e_{i,0}]_{\boldsymbol{v}} = -\boldsymbol{v}[e_{i,0}, E^{(-1)}_{ki}]_{\boldsymbol{v}^{-1}} = -\boldsymbol{v}E^{(-1)}_{k,i+1}$.

(i) Note that $[[e_{i+1,0}, e_{i,-1}]_{\boldsymbol{v}^{-1}}, e_{i,0}]_{\boldsymbol{v}^{-1}} = \boldsymbol{v}^{-1}[[e_{i+1,-1}, e_{i,0}]_{\boldsymbol{v}}, e_{i,0}]_{\boldsymbol{v}^{-1}} = 0$, due to (U2) and (U7). Since also $e_{i,0}$ commutes with $e_{i+2,0}, \ldots, e_{l-1,0}$, we get $[E^{(-1)}_{il}, e_{i,0}]_{\boldsymbol{v}^{-1}} = 0$.

(j) As in (i), $[E^{(-1)}_{k,i+1}, e_{i,0}]_{\boldsymbol{v}^{-1}} = 0$ follows from $[[e_{i,0}, e_{i-1,0}]_{\boldsymbol{v}^{-1}}, e_{i,0}]_{\boldsymbol{v}^{-1}} = 0$, due to (U7).

(k) Comparing the matrix coefficients $\langle v_i \otimes v_k | \cdots | v_{i+1} \otimes v_l\rangle$ of both sides of the equality $R_{\mathrm{trig}}(z/w)(T^+(z) \otimes 1)(1 \otimes T^-(w)) = (1 \otimes T^-(w))(T^+(z) \otimes 1) R_{\mathrm{trig}}(z/w)$, we get

$$(z-w)T^+_{i,i+1}(z)T^-_{kl}(w) + (\boldsymbol{v}-\boldsymbol{v}^{-1})wT^+_{k,i+1}(z)T^-_{il}(w) =$$

$$(z-w)T^-_{kl}(w)T^+_{i,i+1}(z) + (\boldsymbol{v}-\boldsymbol{v}^{-1})wT^-_{k,i+1}(w)T^+_{il}(z).$$

Evaluating the coefficients of $z^1 w^1$ in both sides of this equality, we find

$$[\tilde{g}_i^+ \tilde{e}_{i,i+1}^{(0)}, \tilde{g}_k^- \tilde{e}_{kl}^{(-1)} + \sum_{j<k} \tilde{f}_{kj}^{(0)} \tilde{g}_j^- \tilde{e}_{jl}^{(-1)}] = 0.$$

Hence, by induction in k, we find $[\tilde{e}_{i,i+1}^{(0)}, \tilde{e}_{kl}^{(-1)}] = 0$, which implies $[E_{kl}^{(-1)}, e_{i,0}] = 0$.

□

This completes our proof of (10.4).

G(vi) Proof of Formula (10.5)

Recall that $[h_{i,1}, f_{i,0}] = -[2]_{\boldsymbol{v}} \cdot f_{i,1}$, so that

$$\Delta(f_{i,1}) = -[2]_{\boldsymbol{v}}^{-1} \cdot [\Delta(h_{i,1}), \Delta(f_{i,0})] = -[2]_{\boldsymbol{v}}^{-1} \cdot [\Delta(h_{i,1}), f_{i,0} \otimes 1 + \psi_{i,0}^- \otimes f_{i,0}].$$

Applying formula (10.2) to $\Delta(h_{i,1})$ and using Lemma G.18 below, we recover (10.5).

Lemma G.18 *For $k < i$ and $l > i+1$, the following equalities hold:*

(a) $[E_{i+1,l}^{(0)}, f_{i,0}] = 0.$
(b) $[\widetilde{E}_{ki}^{(0)}, f_{i,0}] = 0.$
(c) $[E_{il}^{(0)}, f_{i,0}] = \boldsymbol{v}^{-1} E_{i+1,l}^{(0)} \psi_{i,0}^+.$
(d) $[\widetilde{E}_{k,i+1}^{(0)}, f_{i,0}] = -\widetilde{E}_{ki}^{(0)} \psi_{i,0}^+.$
(e) $[[E_{i,i+1}^{(0)}, E_{i+1,l}^{(0)}]_{\boldsymbol{v}^3}, f_{i,0}] = \frac{\boldsymbol{v}^{-1}-\boldsymbol{v}^3}{\boldsymbol{v}-\boldsymbol{v}^{-1}} E_{i+1,l}^{(0)} \psi_{i,0}^+ - \frac{\boldsymbol{v}-\boldsymbol{v}^3}{\boldsymbol{v}-\boldsymbol{v}^{-1}} E_{i+1,l}^{(0)} \psi_{i,0}^-.$
(f) $[[E_{i,i+1}^{(0)}, \widetilde{E}_{ki}^{(0)}]_{\boldsymbol{v}^3}, f_{i,0}] = \frac{\boldsymbol{v}^{-1}-\boldsymbol{v}^3}{\boldsymbol{v}-\boldsymbol{v}^{-1}} \widetilde{E}_{ki}^{(0)} \psi_{i,0}^+ - \frac{\boldsymbol{v}-\boldsymbol{v}^3}{\boldsymbol{v}-\boldsymbol{v}^{-1}} \widetilde{E}_{ki}^{(0)} \psi_{i,0}^-.$
(g) $[F_{l,i+1}^{(1)}, f_{i,0}]_{\boldsymbol{v}} = -F_{li}^{(1)}.$
(h) $[F_{ik}^{(1)}, f_{i,0}]_{\boldsymbol{v}} = F_{i+1,k}^{(1)}.$
(i) $[F_{li}^{(1)}, f_{i,0}]_{\boldsymbol{v}^{-1}} = 0.$
(j) $[F_{i+1,k}^{(1)}, f_{i,0}]_{\boldsymbol{v}^{-1}} = 0.$
(k) $[F_{lk}^{(1)}, f_{i,0}] = 0.$

This lemma is proved completely analogously to Lemma G.17. The details are left to the interested reader.

This completes our proof of Theorem 10.13.

Appendix H Proof of Theorem 10.16 and Homomorphisms $j^{\pm}_{\mu_1,\mu_2}$

Our proof of Theorem 10.16 proceeds in three steps. First, we introduce subalgebras $\mathcal{U}^{\mathrm{sc},\pm}_{0,\mu_1,\mu_2}$ of $\mathcal{U}^{\mathrm{sc}}_{0,\mu_1+\mu_2}$ and construct homomorphisms $j^{\pm}_{\mu_1,\mu_2}$ which we referred to in Remark 10.17. Then, we prove Theorem 10.16, reducing some of the verifications to the case of $U_{\boldsymbol{v}}(L\mathfrak{sl}_n)$ via the aforementioned $j^{\pm}_{\mu_1,\mu_2}$. Finally, we verify the commutativity of the diagram from Remark 10.17.

Throughout this section, we assume $\mu_1, \mu_2 \in \Lambda^-$.

H(i) Homomorphisms $j^{\pm}_{\mu_1,\mu_2}$

First, we introduce subalgebras $\mathcal{U}^{\mathrm{sc},\pm}_{0,\mu_1,\mu_2}$ of $\mathcal{U}^{\mathrm{sc}}_{0,\mu_1+\mu_2}$. To this end, recall the explicit identification of the Drinfeld-Jimbo and the new Drinfeld realizations of $U_{\boldsymbol{v}}(L\mathfrak{sl}_n)$ from Theorem 8.10:

$$E_i \mapsto e_{i,0},\ F_i \mapsto f_{i,0},\ K_i^{\pm 1} \mapsto (\psi^+_{i,0})^{\pm 1} = \psi^{\pm}_{i,0} = (\psi^-_{i,0})^{\mp 1} \text{ for } 1 \leq i \leq n-1,$$

$$(K_{i_0})^{\pm 1} \mapsto (\psi^+_{1,0} \cdots \psi^+_{n-1,0})^{\mp 1},$$

$$E_{i_0} \mapsto (-\boldsymbol{v})^{-n} \cdot (\psi^+_{1,0} \cdots \psi^+_{n-1,0})^{-1} \cdot [\cdots [f_{1,1}, f_{2,0}]_{\boldsymbol{v}}, \cdots, f_{n-1,0}]_{\boldsymbol{v}},$$

$$F_{i_0} \mapsto (-\boldsymbol{v})^{n} \cdot [e_{n-1,0}, \cdots, [e_{2,0}, e_{1,-1}]_{\boldsymbol{v}^{-1}} \cdots]_{\boldsymbol{v}^{-1}} \cdot \psi^+_{1,0} \cdots \psi^+_{n-1,0}.$$

Hence, the images $U^+_{\boldsymbol{v}}$ and $U^-_{\boldsymbol{v}}$ of the Drinfeld-Jimbo Borel subalgebras are the subalgebras of $U_{\boldsymbol{v}}(L\mathfrak{sl}_n)$ generated by $\{e_{i,0}, (\psi^+_{i,0})^{\pm 1}, F^{(1)}_{n1}\}_{i=1}^{n-1}$ and $\{f_{i,0}, (\psi^+_{i,0})^{\pm 1}, E^{(-1)}_{1n}\}_{i=1}^{n-1}$, respectively.

Likewise, let $\mathcal{U}^{\mathrm{sc},+}_{0,\mu_1,\mu_2}$ and $\mathcal{U}^{\mathrm{sc},-}_{0,\mu_1,\mu_2}$ be the $\mathbb{C}(\boldsymbol{v})$-subalgebras of $\mathcal{U}^{\mathrm{sc}}_{0,\mu_1+\mu_2}$ generated by the elements $\{e_{i,0}, (\psi^+_{i,0})^{\pm 1}, F^{(1)}_{n1}\}_{i=1}^{n-1}$ and $\{f_{i,b_{1,i}}, (\psi^-_{i,b_{1,i}+b_{2,i}})^{\pm 1}, \hat{E}^{(-1)}_{1n}\}_{i=1}^{n-1}$, respectively, where as before $b_{1,i} = \alpha_i^\vee(\mu_1), b_{2,i} = \alpha_i^\vee(\mu_2), b_i = b_{1,i} + b_{2,i}$. Here, the elements $\{\hat{E}^{(-1)}_{ji}\}_{j<i}$ are defined via $\hat{E}^{(-1)}_{ji} := [e_{i-1,b_{2,i-1}}, [e_{i-2,b_{2,i-2}}, \cdots, [e_{j+1,b_{2,j+1}}, e_{j,b_{2,j}-1}]_{\boldsymbol{v}^{-1}} \cdots]_{\boldsymbol{v}^{-1}}]_{\boldsymbol{v}^{-1}}$.

Proposition H.1

(a) There is a unique $\mathbb{C}(\boldsymbol{v})$-algebra homomorphism $j^+_{\mu_1,\mu_2}\colon U^+_{\boldsymbol{v}} \to U^{\mathrm{sc},+}_{0,\mu_1,\mu_2}$, such that $e_{i,0} \mapsto e_{i,0}, (\psi^+_{i,0})^{\pm 1} \mapsto (\psi^+_{i,0})^{\pm 1}, F^{(1)}_{n1} \mapsto F^{(1)}_{n1}$.

(b) There is a unique $\mathbb{C}(\boldsymbol{v})$-algebra homomorphism $j^-_{\mu_1,\mu_2}\colon U^-_{\boldsymbol{v}} \to U^{\mathrm{sc},-}_{0,\mu_1,\mu_2}$, such that $f_{i,0} \mapsto f_{i,b_{1,i}}, (\psi^-_{i,0})^{\pm 1} \mapsto (\psi^-_{i,b_i})^{\pm 1}, E^{(-1)}_{1n} \mapsto \hat{E}^{(-1)}_{1n}$.

Proof

(a) Converting the defining relations of the positive Drinfeld-Jimbo Borel subalgebra into the new Drinfeld realization, we see that U_v^+ is generated by $\{e_{i,0}, (\psi_{i,0}^+)^{\pm 1}, F_{n1}^{(1)}\}_{i=1}^{n-1}$ with the following defining relations:

$$(\psi_{i,0}^+)^{\pm 1}\cdot(\psi_{i,0}^+)^{\mp 1}=1,\ \psi_{i,0}^+\psi_{j,0}^+=\psi_{j,0}^+\psi_{i,0}^+, \tag{H.1}$$

$$\psi_{i,0}^+ e_{j,0}=v^{c_{ij}}e_{j,0}\psi_{i,0}^+,\ \psi_{i,0}^+F_{n1}^{(1)}=v^{-\delta_{i1}-\delta_{i,n-1}}F_{n1}^{(1)}\psi_{i,0}^+, \tag{H.2}$$

$$[e_{i,0},[e_{i,0},e_{i\pm 1,0}]_v]_{v^{-1}}=0,\ [e_{i,0},e_{j,0}]=0 \text{ if } c_{ij}=0, \tag{H.3}$$

$$[e_{i,0},F_{n1}^{(1)}]=0 \text{ for } 1<i<n-1, \tag{H.4}$$

$$[e_{1,0},[e_{1,0},F_{n1}^{(1)}]]_{v^{-2}}=0,\ [e_{n-1,0},[e_{n-1,0},F_{n1}^{(1)}]]_{v^{-2}}=0, \tag{H.5}$$

$$[F_{n1}^{(1)},[F_{n1}^{(1)},e_{1,0}]]_{v^2}=0,\ [F_{n1}^{(1)},[F_{n1}^{(1)},e_{n-1,0}]]_{v^2}=0. \tag{H.6}$$

Thus, it suffices to check that these relations are preserved under the specified assignment $e_{i,0}\mapsto e_{i,0}, (\psi_{i,0}^+)^{\pm 1}\mapsto(\psi_{i,0}^+)^{\pm 1}, F_{n1}^{(1)}\mapsto F_{n1}^{(1)}$. The validity of (H.1–H.4) is obvious.

To verify the first equality of (H.5), we note that $[\psi_{1,1}^+, f_{2,0}]_v=(v^2-1)f_{2,1}\psi_{1,0}^+$, due to (U5). Combining this with (U6), we get

$$[e_{1,0},F_{n1}^{(1)}]=(v-v^{-1})^{-1}\cdot[\cdots[\psi_{1,1}^+,f_{2,0}]_v,\cdots,f_{n-1,0}]_v=vF_{n2}^{(1)}\psi_{1,0}^+.$$

Hence, $[e_{1,0},[e_{1,0},F_{n1}^{(1)}]]_{v^{-2}}=v[e_{1,0},F_{n2}^{(1)}\psi_{1,0}^+]_{v^{-2}}=v[e_{1,0},F_{n2}^{(1)}]\psi_{1,0}^+=0$, due to (U6).

The verification of the second equality of (H.5) is similar and is based on

$$[e_{n-1,0},F_{n1}^{(1)}]=\frac{[[\cdots[f_{1,1},f_{2,0}]_v,\cdots,f_{n-2,0}]_v,\psi_{n-1,0}^+-\delta_{b_{n-1},0}\psi_{n-1,0}^-]_v}{v-v^{-1}}=-vF_{n-1,1}^{(1)}\psi_{n-1,0}^+.$$

Due to the above equality $[e_{1,0},F_{n1}^{(1)}]=vF_{n2}^{(1)}\psi_{1,0}^+$ and (U4), the verification of the first equality of (H.6) boils down to the proof of $[F_{n1}^{(1)},F_{n2}^{(1)}]_v=0$. This is an equality in $\mathcal{U}_{0,\mu_1+\mu_2}^{\mathrm{sc},<}$. However, $\mathcal{U}_{0,\mu_1+\mu_2}^{\mathrm{sc},<}\simeq U_v^<(L\mathfrak{sl}_n)$, due to Proposition 5.1(b). Hence, it suffices to check this equality in $U_v(L\mathfrak{sl}_n)$. The latter follows immediately from the validity of (H.6) for U_v^+.

Due to $[e_{n-1,0},F_{n1}^{(1)}]=-vF_{n-1,1}^{(1)}\psi_{n-1,0}^+$ from above and (U4), the verification of the second equality of (H.6) boils down to the proof of $[F_{n1}^{(1)},F_{n-1,1}^{(1)}]_v=0$. Analogously to the previous verification, the latter follows from the same equality in U_v^+.

(b) The proof of part (b) is completely analogous and is left to the interested reader.

□

This completes our construction of the homomorphisms $\jmath^{\pm}_{\mu_1,\mu_2}\colon U^{\pm}_{\boldsymbol{v}} \to \mathfrak{U}^{\mathrm{sc},\pm}_{0,\mu_1,\mu_2}$, which we referred to in Remark 10.17. The following results are needed for the next subsection.

Lemma H.2

(a) For any $1 \leq j < i \leq n$, we have $E^{(0)}_{ji}, \widetilde{E}^{(0)}_{ji}, F^{(1)}_{ij} \in \mathfrak{U}^{\mathrm{sc},+}_{0,\mu_1,\mu_2}$.
(b) For any $1 \leq j < i \leq n$, define

$$\hat{F}^{\pm,(0)}_{ij} := [\cdots[f_{j,b_{1,j}}, f_{j+1,b_{1,j+1}}]_{\boldsymbol{v}^{\pm 1}}, \cdots, f_{i-1,b_{1,i-1}}]_{\boldsymbol{v}^{\pm 1}}.$$

We have $\hat{F}^{\pm,(0)}_{ij}, \hat{E}^{(-1)}_{ji} \in \mathfrak{U}^{\mathrm{sc},-}_{0,\mu_1,\mu_2}$.

Proof

(a) Since $E^{(0)}_{ji}, \widetilde{E}^{(0)}_{ji}$ are expressed via $\boldsymbol{v}^{\pm 1}$-commutators of $e_{k,0} \in \mathfrak{U}^{\mathrm{sc},+}_{0,\mu_1,\mu_2}$, we obviously get the first two inclusions. The last inclusion is clear for $(i,j) = (n,1)$. Applying iteratively $[e_{k,0}, F^{(1)}_{k+1,1}] = -\boldsymbol{v}F^{(1)}_{k1}\psi^{+}_{k,0}$, $[e_{l,0}, F^{(1)}_{il}] = \boldsymbol{v}F^{(1)}_{i,l+1}\psi^{+}_{l,0}$, we get $F^{(1)}_{ij} \in \mathfrak{U}^{\mathrm{sc},+}_{0,\mu_1,\mu_2}$ for any $j < i$.

(b) The inclusions $\hat{F}^{\pm,(0)}_{ij} \in \mathfrak{U}^{\mathrm{sc},-}_{0,\mu_1,\mu_2}$ are obvious. It remains to prove $\hat{E}^{(-1)}_{ji} \in \mathfrak{U}^{\mathrm{sc},-}_{0,\mu_1,\mu_2}$. This is clear for $(j,i) = (1,n)$. To deduce the general case, it remains to apply the equalities $[f_{i-1,b_{1,i-1}}, \hat{E}^{(-1)}_{1i}] = \hat{E}^{(-1)}_{1,i-1}\psi^{-}_{i-1,b_{i-1}}$, $[f_{l,b_{1,l}}, \hat{E}^{(-1)}_{li}] = -\hat{E}^{(-1)}_{l+1,i}\psi^{-}_{l,b_l}$. □

The proof of the following result is straightforward.

Lemma H.3 *For any $1 \leq j < i \leq n$, we have:*

$$\jmath^{+}_{\mu_1,\mu_2}\colon E^{(0)}_{ji} \mapsto E^{(0)}_{ji},\ \widetilde{E}^{(0)}_{ji} \mapsto \widetilde{E}^{(0)}_{ji},\ F^{(1)}_{ij} \mapsto F^{(1)}_{ij},\ f_{i,1} \mapsto f_{i,1},\ h_{i,1} \mapsto h_{i,1},$$

$$\jmath^{-}_{\mu_1,\mu_2}\colon F^{(0)}_{ij} \mapsto \hat{F}^{+,(0)}_{ij},\ \widetilde{F}^{(0)}_{ij} \mapsto \hat{F}^{-,(0)}_{ij},\ E^{(-1)}_{ji} \mapsto \hat{E}^{(-1)}_{ji},\ e_{i,-1} \mapsto e_{i,b_{2,i}-1},\ h_{i,-1} \mapsto h_{i,-1}.$$

H(ii) Proof of Theorem 10.16

Due to Theorem 5.5, it suffices to check that the assignment Δ of Theorem 10.16 preserves defining relations (Û1–Û9). To simplify our exposition, we will assume that μ_1, μ_2 are strictly antidominant: $b_{1,i}, b_{2,i} < 0$ for any $1 \leq i < n$. This verification is similar to the $n = 2$ case (carried out in Appendix D) and we only indicate the key technical details, see Lemmas H.4–H.15 (their proofs are similar to that of Lemma G.17 and therefore omitted). For $1 \leq a \leq b < n$, we define $\alpha^{\vee}_{[a,b]} := \alpha^{\vee}_a + \alpha^{\vee}_{a+1} + \ldots + \alpha^{\vee}_b$.

H(ii).a Compatibility with (Û1)

- The equalities $\Delta((\psi^+_{i,0})^{\pm 1})\Delta((\psi^+_{i,0})^{\mp 1}) = 1$ and $\Delta((\psi^-_{i,b_i})^{\pm 1})\Delta((\psi^-_{i,b_i})^{\mp 1}) = 1$ follow immediately from relation (Û1) for both $\mathcal{U}^{sc}_{0,\mu_1}, \mathcal{U}^{sc}_{0,\mu_2}$.
- The commutativity of $\{\Delta(\psi^+_{i,0}), \Delta(\psi^-_{i,b_i})\}_{i=1}^{n-1}$ between themselves and with $\{\Delta(h_{j,\pm 1})\}_{j=1}^{n-1}$ is due to relations (Û1, Û4, Û5) for both $\mathcal{U}^{sc}_{0,\mu_1}, \mathcal{U}^{sc}_{0,\mu_2}$.
- Finally, we verify $[\Delta(h_{i,r}), \Delta(h_{j,s})] = 0$ for $r, s \in \{\pm 1\}$. To this end, recall the homomorphism $\iota_{0,0,\mu_1} \otimes \iota_{0,\mu_2,0} \colon \mathcal{U}^{sc}_{0,0} \otimes \mathcal{U}^{sc}_{0,0} \to \mathcal{U}^{sc}_{0,\mu_1} \otimes \mathcal{U}^{sc}_{0,\mu_2}$. The key observation is that $\iota_{0,0,\mu_1} \otimes \iota_{0,\mu_2,0}(\Delta(h_{i,r})) = \Delta(h_{i,r}) + \frac{\alpha_i^\vee(\mu_1+\mu_2)}{\boldsymbol{v}^r - \boldsymbol{v}^{-r}}$ for any $i \in I, r \in \{\pm 1\}$ (cf. proof of Corollary 10.11), where by abuse of notation we use $\Delta(h_{i,r})$ to denote elements of both $\mathcal{U}^{sc}_{0,0} \otimes \mathcal{U}^{sc}_{0,0}$ and $\mathcal{U}^{sc}_{0,\mu_1} \otimes \mathcal{U}^{sc}_{0,\mu_2}$. Hence, it suffices to prove $[\Delta(h_{i,r}), \Delta(h_{j,s})] = 0$ in $\mathcal{U}^{sc}_{0,0} \otimes \mathcal{U}^{sc}_{0,0}$. The latter follows immediately from the corresponding result for $U_{\boldsymbol{v}}(L\mathfrak{sl}_n)$, in which case the assignment Δ of Theorem 10.16 coincides with the Drinfeld-Jimbo coproduct, due to Theorem 10.13.

H(ii).b Compatibility with (Û2)

We need to prove $[\Delta(e_{i,r+1}), \Delta(e_{j,s})]_{\boldsymbol{v}^{c_{ij}}} + [\Delta(e_{j,s+1}), \Delta(e_{i,r})]_{\boldsymbol{v}^{c_{ij}}} = 0$ for $b_{2,i} - 1 \leq r \leq -1, b_{2,j} - 1 \leq s \leq -1$.

Case $b_{2,i} - 1 < r \leq -1, b_{2,j} - 1 < s \leq -1$ In this case, the above sum equals $1 \otimes ([e_{i,r+1}, e_{j,s}]_{\boldsymbol{v}^{c_{ij}}} + [e_{j,s+1}, e_{i,r}]_{\boldsymbol{v}^{c_{ij}}}) = 0$, due to relations (Û2) and (Û4) for $\mathcal{U}^{sc}_{0,\mu_2}$.

Case $r = b_{2,i} - 1, b_{2,j} - 1 < s < -1$ Note that $[e_{j,s+1}, f_{a,0}] = 0$ for any $1 \leq a < n$, due to (Û6) for $\mathcal{U}^{sc}_{0,\mu_2}$. As a result, we have $[e_{j,s+1}, F^{(0)}_{ba}] = [e_{j,s+1}, \widetilde{F}^{(0)}_{ba}] = 0$ for any $1 \leq a < b \leq n$. Combining this with (Û2) and (Û4) for $\mathcal{U}^{sc}_{0,\mu_2}$, we get $[\Delta(e_{i,b_{2,i}}), \Delta(e_{j,s})]_{\boldsymbol{v}^{c_{ij}}} + [\Delta(e_{j,s+1}), \Delta(e_{i,b_{2,i}-1})]_{\boldsymbol{v}^{c_{ij}}} = 1 \otimes ([e_{i,b_{2,i}}, e_{j,s}]_{\boldsymbol{v}^{c_{ij}}} + [e_{j,s+1}, e_{i,b_{2,i}-1}]_{\boldsymbol{v}^{c_{ij}}}) = 0$ as above.

Case $r = b_{2,i} - 1, s = b_{2,j} - 1$ Due to relation (Û4) for both $\mathcal{U}^{sc}_{0,\mu_1}, \mathcal{U}^{sc}_{0,\mu_2}$, we get

$$[\Delta(e_{j,b_{2,j}}), \Delta(e_{i,b_{2,i}-1})]_{\boldsymbol{v}^{c_{ij}}} = 1 \otimes [e_{j,b_{2,j}}, e_{i,b_{2,i}-1}]_{\boldsymbol{v}^{c_{ij}}} -$$

$$(\boldsymbol{v} - \boldsymbol{v}^{-1}) \sum_{l>i+1} E^{(-1)}_{il} \otimes [e_{j,b_{2,j}}, F^{(0)}_{l,i+1}]\psi^-_{i,b_{2,i}} + (\boldsymbol{v} - \boldsymbol{v}^{-1}) \sum_{k<i} \boldsymbol{v}^{i-k-1} E^{(-1)}_{k,i+1} \otimes [e_{j,b_{2,j}}, \widetilde{F}^{(0)}_{ik}]\psi^-_{i,b_{2,i}} -$$

$$(\boldsymbol{v} - \boldsymbol{v}^{-1})^2 \sum_{l>i+1}^{k<i} \boldsymbol{v}^{i-k-1} E^{(-1)}_{kl} \otimes [e_{j,b_{2,j}}, \widetilde{F}^{(0)}_{ik} F^{(0)}_{l,i+1}]\psi^-_{i,b_{2,i}}.$$

Using this formula and Lemma H.4 below, it is straightforward to check that again we obtain $[\Delta(e_{i,b_{2,i}}), \Delta(e_{j,b_{2,j}-1})]_{\boldsymbol{v}^{c_{ij}}} + [\Delta(e_{j,b_{2,j}}), \Delta(e_{i,b_{2,i}-1})]_{\boldsymbol{v}^{c_{ij}}} = 1 \otimes ([e_{i,b_{2,i}}, e_{j,b_{2,j}-1}]_{\boldsymbol{v}^{c_{ij}}} + [e_{j,b_{2,j}}, e_{i,b_{2,i}-1}]_{\boldsymbol{v}^{c_{ij}}}) = 0$.

Lemma H.4 *For any* $1 \le k < i,\ i+1 < l \le n,\ 1 \le j < n$, *the following holds in* $\mathcal{U}^{\mathrm{sc}}_{0,\mu_2}$:

(a) $[e_{j,b_{2,j}}, F^{(0)}_{l,i+1}] = \delta_{j,i+1} F^{(0)}_{l,i+2} \psi^-_{j,b_{2,j}}$, *where we set* $F^{(0)}_{i+2,i+2} := \frac{-1}{\boldsymbol{v}-\boldsymbol{v}^{-1}}$.
(b) $[e_{j,b_{2,j}}, \widetilde{F}^{(0)}_{ik}] = -\boldsymbol{v}^{-1} \delta_{j,i-1} \widetilde{F}^{(0)}_{i-1,k} \psi^-_{j,b_{2,j}}$, *where we set* $\widetilde{F}^{(0)}_{i-1,i-1} := \frac{\boldsymbol{v}}{\boldsymbol{v}-\boldsymbol{v}^{-1}}$.

Case $r = b_{2,i} - 1, s = -1$ Clearly, $[\Delta(e_{i,b_{2,i}}), \Delta(e_{j,-1})]_{\boldsymbol{v}^{c_{ij}}} = 1 \otimes [e_{i,b_{2,i}}, e_{j,-1}]_{\boldsymbol{v}^{c_{ij}}}$ and $[\Delta(e_{j,0}), \Delta(e_{i,b_{2,i}-1})]_{\boldsymbol{v}^{c_{ij}}} = [1 \otimes e_{j,0} + e_{j,0} \otimes \psi^+_{j,0}, \Delta(e_{i,b_{2,i}-1})]_{\boldsymbol{v}^{c_{ij}}}$. We claim that as in the previous cases, one gets $[\Delta(e_{i,b_{2,i}}), \Delta(e_{j,-1})]_{\boldsymbol{v}^{c_{ij}}} + [\Delta(e_{j,0}), \Delta(e_{i,b_{2,i}-1})]_{\boldsymbol{v}^{c_{ij}}} = 1 \otimes ([e_{i,b_{2,i}}, e_{j,-1}]_{\boldsymbol{v}^{c_{ij}}} + [e_{j,0}, e_{i,b_{2,i}-1}]_{\boldsymbol{v}^{c_{ij}}}) = 0$. To this end, we note that the computations of $[1 \otimes e_{j,0}, \Delta(e_{i,b_{2,i}-1})]_{\boldsymbol{v}^{c_{ij}}}$ and $[e_{j,0} \otimes \psi^+_{j,0}, \Delta(e_{i,b_{2,i}-1})]_{\boldsymbol{v}^{c_{ij}}}$ are straightforward and are crucially based on Lemmas H.5 and H.6 below, respectively.

Lemma H.5 *For any* $1 \le k < i,\ i+1 < l \le n,\ 1 \le j < n$, *the following holds in* $\mathcal{U}^{\mathrm{sc}}_{0,\mu_2}$:

(a) $[e_{j,0}, F^{(0)}_{l,i+1}] = -\boldsymbol{v} \delta_{j,l-1} F^{(0)}_{j,i+1} \psi^+_{j,0}$, *where we set* $F^{(0)}_{i+1,i+1} := \frac{-1}{\boldsymbol{v}(\boldsymbol{v}-\boldsymbol{v}^{-1})}$.
(b) $[e_{j,0}, \widetilde{F}^{(0)}_{ik}] = \delta_{jk} \widetilde{F}^{(0)}_{i,j+1} \psi^+_{j,0}$, *where we set* $\widetilde{F}^{(0)}_{ii} := \frac{1}{\boldsymbol{v}-\boldsymbol{v}^{-1}}$.

Lemma H.6 *For any* $1 \le k < l-1 < n,\ 1 \le j < n$, *the following holds in* $\mathcal{U}^{\mathrm{sc}}_{0,\mu_1}$:
$[e_{j,0}, E^{(-1)}_{kl}]_{\boldsymbol{v}^{(\alpha^\vee_j, \alpha^\vee_{[k,l-1]})}} = \delta_{jl} E^{(-1)}_{k,l+1} - \delta_{j,k-1} E^{(-1)}_{k-1,l}$.

H(ii).c Compatibility with (Û3)

We need to prove $[\Delta(f_{i,r+1}), \Delta(f_{j,s})]_{\boldsymbol{v}^{-c_{ij}}} + [\Delta(f_{j,s+1}), \Delta(f_{i,r})]_{\boldsymbol{v}^{-c_{ij}}} = 0$ for $b_{1,i} \le r \le 0, b_{1,j} \le s \le 0$.

Case $b_{1,i} \le r < 0, b_{1,j} \le s < 0$ In this case, the above sum equals $([f_{i,r+1}, f_{j,s}]_{\boldsymbol{v}^{-c_{ij}}} + [f_{j,s+1}, f_{i,r}]_{\boldsymbol{v}^{-c_{ij}}}) \otimes 1 = 0$, due to relations (Û3) and (Û5) for $\mathcal{U}^{\mathrm{sc}}_{0,\mu_1}$.

Case $r = 0, b_{1,j} < s < 0$ Note that $[f_{j,s}, e_{a,0}] = 0$ for any $1 \le a < n$, due to (Û6) for $\mathcal{U}^{\mathrm{sc}}_{0,\mu_1}$. As a result, we have $[f_{j,s}, E^{(0)}_{ab}] = [f_{j,s}, \widetilde{E}^{(0)}_{ab}] = 0$ for any $1 \le a < b \le n$. Combining this with (Û3) and (Û5) for $\mathcal{U}^{\mathrm{sc}}_{0,\mu_1}$, we get $[\Delta(f_{i,1}), \Delta(f_{j,s})]_{\boldsymbol{v}^{-c_{ij}}} + [\Delta(f_{j,s+1}), \Delta(f_{i,0})]_{\boldsymbol{v}^{-c_{ij}}} = ([f_{i,1}, f_{j,s}]_{\boldsymbol{v}^{-c_{ij}}} + [f_{j,s+1}, f_{i,0}]_{\boldsymbol{v}^{-c_{ij}}}) \otimes 1 = 0$ as above.

Case $r = 0, s = 0$ Due to relation ($\hat{\mathrm{U}}5$) for both $\mathcal{U}^{\mathrm{sc}}_{0,\mu_1}, \mathcal{U}^{\mathrm{sc}}_{0,\mu_2}$, we get

$$[\Delta(f_{i,1}), \Delta(f_{j,0})]_{v^{-c_{ij}}} = [f_{i,1}, f_{j,0}]_{v^{-c_{ij}}} \otimes 1 + (v - v^{-1})v^{-c_{ij}-1} \sum_{l>i+1} [E^{(0)}_{i+1,l}, f_{j,0}]\psi^+_{i,0} \otimes F^{(1)}_{li} -$$

$$(v - v^{-1})v^{-c_{ij}} \sum_{k<i} v^{k-i}[\widetilde{E}^{(0)}_{ki}, f_{j,0}]\psi^+_{i,0} \otimes F^{(1)}_{i+1,k} - (v - v^{-1})^2 v^{-c_{ij}} \sum_{l>i+1}^{k<i} v^{k-i-1}[E^{(0)}_{i+1,l}\widetilde{E}^{(0)}_{ki}, f_{j,0}]\psi^+_{i,0} \otimes F^{(1)}_{lk}.$$

Using this formula and Lemma H.7 below, it is straightforward to check that we obtain $[\Delta(f_{i,1}), \Delta(f_{j,0})]_{v^{-c_{ij}}} + [\Delta(f_{j,1}), \Delta(f_{i,0})]_{v^{-c_{ij}}} = ([f_{i,1}, f_{j,0}]_{v^{-c_{ij}}} + [f_{j,1}, f_{i,0}]_{v^{-c_{ij}}}) \otimes 1 = 0$.

Lemma H.7 *For any* $1 \leq k < i,\ i+1 < l \leq n,\ 1 \leq j < n$, *the following holds in* $\mathcal{U}^{\mathrm{sc}}_{0,\mu_1}$:

(a) $[E^{(0)}_{i+1,l}, f_{j,0}] = v^{-1}\delta_{j,i+1}E^{(0)}_{i+2,l}\psi^+_{j,0}$, *where we set* $E^{(0)}_{i+2,i+2} := \frac{v}{v-v^{-1}}$.
(b) $[\widetilde{E}^{(0)}_{ki}, f_{j,0}] = -\delta_{j,i-1}\widetilde{E}^{(0)}_{k,i-1}\psi^+_{j,0}$, *where we set* $\widetilde{E}^{(0)}_{i-1,i-1} := \frac{-1}{v-v^{-1}}$.

Case $r = 0, s = b_{1,j}$ Clearly, $[\Delta(f_{j,b_{1,j}+1}), \Delta(f_{i,0})]_{v^{-c_{ij}}} = [f_{j,b_{1,j}+1}, f_{i,0}]_{v^{-c_{ij}}} \otimes 1$ and $[\Delta(f_{i,1}), \Delta(f_{j,b_{1,j}})]_{v^{-c_{ij}}} = [\Delta(f_{i,1}), f_{j,b_{1,j}} \otimes 1 + \psi^-_{j,b_{1,j}} \otimes f_{j,0}]_{v^{-c_{ij}}}$. We claim that as in the previous cases, one gets $[\Delta(f_{i,1}), \Delta(f_{j,b_{1,j}})]_{v^{-c_{ij}}} + [\Delta(f_{j,b_{1,j}+1}), \Delta(f_{i,0})]_{v^{-c_{ij}}} = ([f_{i,1}, f_{j,b_{1,j}}]_{v^{-c_{ij}}} + [f_{j,b_{1,j}+1}, f_{i,0}]_{v^{-c_{ij}}}) \otimes 1 = 0$. To this end, we note that the computations of $[\Delta(f_{i,1}), f_{j,b_{1,j}} \otimes 1]_{v^{-c_{ij}}}$ and $[\Delta(f_{i,1}), \psi^-_{j,b_{1,j}} \otimes f_{j,0}]_{v^{-c_{ij}}}$ are straightforward and are crucially based on Lemmas H.8 and H.9 below, respectively.

Lemma H.8 *For any* $1 \leq k < i,\ i+1 < l \leq n,\ 1 \leq j < n$, *the following holds in* $\mathcal{U}^{\mathrm{sc}}_{0,\mu_1}$:

(a) $[E^{(0)}_{i+1,l}, f_{j,b_{1,j}}] = -\delta_{j,l-1}E^{(0)}_{i+1,j}\psi^-_{j,b_{1,j}}$, *where we set* $E^{(0)}_{i+1,i+1} := \frac{1}{v-v^{-1}}$.
(b) $[\widetilde{E}^{(0)}_{ki}, f_{j,b_{1,j}}] = v\delta_{jk}\widetilde{E}^{(0)}_{j+1,i}\psi^-_{j,b_{1,j}}$, *where we set* $\widetilde{E}^{(0)}_{ii} := \frac{-1}{v(v-v^{-1})}$.

Lemma H.9 *For any* $1 \leq k < l-1 < n,\ 1 \leq j < n$, *the following holds in* $\mathcal{U}^{\mathrm{sc}}_{0,\mu_2}$: $[F^{(1)}_{lk}, f_{j,0}]_{v^{-(\alpha^\vee_j, \alpha^\vee_{[k,l-1]})}} = \delta_{jl}F^{(1)}_{l+1,k} - \delta_{j,k-1}F^{(1)}_{l,k-1}$.

H(ii).d Compatibility with ($\hat{\mathrm{U}}4$)

Due to relations ($\hat{\mathrm{U}}1$, $\hat{\mathrm{U}}4$, $\hat{\mathrm{U}}5$) for both $\mathcal{U}^{\mathrm{sc}}_{0,\mu_1}, \mathcal{U}^{\mathrm{sc}}_{0,\mu_2}$, we immediately obtain the equalities $\Delta(\psi^+_{i,0})\Delta(e_{j,r}) = v^{c_{ij}}\Delta(e_{j,r})\Delta(\psi^+_{i,0})$, $\Delta(\psi^-_{i,b_i})\Delta(e_{j,r}) = v^{-c_{ij}}\Delta(e_{j,r})\Delta(\psi^-_{i,b_i})$ for $b_{2,j-1} \leq r \leq 0$.

Let us now verify $[\Delta(h_{i,1}), \Delta(e_{j,r})] = [c_{ij}]_v \cdot \Delta(e_{j,r+1})$ for $b_{2,j} - 1 \leq r \leq -1$.

Case $b_{2,j} \leq r < -1$ The verification in this case follows immediately from relation ($\hat{\mathrm{U}}4$) for $\mathcal{U}^{\mathrm{sc}}_{0,\mu_2}$ combined with Lemma H.10 below.

Lemma H.10 *For any* $1 \leq a < b \leq n$, $b_{2,j} \leq r < -1$, *we have* $[F^{(1)}_{ba}, e_{j,r}] = 0$ *in* $\mathcal{U}^{\mathrm{sc}}_{0,\mu_2}$.

Case $r = -1$ Due to relation ($\hat{\mathrm{U}}4$) for $\mathcal{U}^{\mathrm{sc}}_{0,\mu_2}$, we get

$$[\Delta(h_{i,1}), \Delta(e_{j,-1})] = [c_{ij}]_{\boldsymbol{v}} \cdot 1 \otimes e_{j,0} - (\boldsymbol{v}^2 - \boldsymbol{v}^{-2}) E^{(0)}_{i,i+1} \otimes [F^{(1)}_{i+1,i}, e_{j,-1}] +$$

$$(\boldsymbol{v} - \boldsymbol{v}^{-1}) \sum_{l>i+1} E^{(0)}_{i+1,l} \otimes [F^{(1)}_{l,i+1}, e_{j,-1}] + (\boldsymbol{v} - \boldsymbol{v}^{-1}) \sum_{k<i} \boldsymbol{v}^{k+1-i} \widetilde{E}^{(0)}_{ki} \otimes [F^{(1)}_{ik}, e_{j,-1}] +$$

$$\boldsymbol{v}^{-2}(\boldsymbol{v} - \boldsymbol{v}^{-1}) \sum_{l>i+1} [E^{(0)}_{i,i+1}, E^{(0)}_{i+1,l}]_{\boldsymbol{v}^3} \otimes [F^{(1)}_{li}, e_{j,-1}] -$$

$$(\boldsymbol{v} - \boldsymbol{v}^{-1}) \sum_{k<i} \boldsymbol{v}^{k-i-1} [E^{(0)}_{i,i+1}, \widetilde{E}^{(0)}_{ki}]_{\boldsymbol{v}^3} \otimes [F^{(1)}_{i+1,k}, e_{j,-1}] +$$

$$(\boldsymbol{v} - \boldsymbol{v}^{-1})^2 \sum_{l>i+1}^{k<i} \boldsymbol{v}^{k-i} (E^{(0)}_{il} \widetilde{E}^{(0)}_{ki} - E^{(0)}_{i+1,l} \widetilde{E}^{(0)}_{k,i+1}) \otimes [F^{(1)}_{lk}, e_{j,-1}].$$

Using this formula and Lemma H.11 below, it is straightforward to check that we obtain $[\Delta(h_{i,1}), \Delta(e_{j,-1})] = [c_{ij}]_{\boldsymbol{v}} \cdot (1 \otimes e_{j,0} + e_{j,0} \otimes \psi^+_{j,0}) = [c_{ij}]_{\boldsymbol{v}} \cdot \Delta(e_{j,0})$.

Lemma H.11 *For any* $1 \leq a < b \leq n$, *we have* $[F^{(1)}_{ba}, e_{j,-1}] = \frac{-1}{\boldsymbol{v}-\boldsymbol{v}^{-1}} \delta_{ja} \delta_{j,b-1} \psi^+_{j,0}$ *in* $\mathcal{U}^{\mathrm{sc}}_{0,\mu_2}$.

Case $r = b_{2,j} - 1$ According to the next step, we have $\Delta(e_{j,b_{2,j}-1}) = \frac{[\Delta(h_{j,-1}), \Delta(e_{j,b_{2,j}})]}{[2]_{\boldsymbol{v}}}$. Apply the Jacobi identity to get $[2]_{\boldsymbol{v}} \cdot [\Delta(h_{i,1}), \Delta(e_{j,b_{2,j}-1})] = [\Delta(h_{j,-1}), [\Delta(h_{i,1}), \Delta(e_{j,b_{2,j}})]] - [\Delta(e_{j,b_{2,j}}), [\Delta(h_{i,1}), \Delta(h_{j,-1})]]$. The second summand is zero as $[\Delta(h_{i,1}), \Delta(h_{j,-1})] = 0$ by above. Due to the $r = b_{2,j}$ case considered above, we have $[\Delta(h_{i,1}), \Delta(e_{j,b_{2,j}})] = [c_{ij}]_{\boldsymbol{v}} \cdot \Delta(e_{j,b_{2,j}+1})$. It remains to apply $[\Delta(h_{j,-1}), \Delta(e_{j,b_{2,j}+1})] = [2]_{\boldsymbol{v}} \cdot \Delta(e_{j,b_{2,j}})$ as proved below.

Let us now verify the equality $[\Delta(h_{i,-1}), \Delta(e_{j,r})] = [c_{ij}]_{\boldsymbol{v}} \cdot \Delta(e_{j,r-1})$ for $b_{2,j} \leq r \leq 0$.

Case $b_{2,j} < r < 0$ The verification in this case follows immediately from relation ($\hat{\mathrm{U}}4$) for $\mathcal{U}^{\mathrm{sc}}_{0,\mu_2}$ combined with Lemma H.12 below.

Lemma H.12 *For* $1 \leq a < b \leq n, b_{2,j} < r < 0$, *we have* $[F^{(0)}_{ba}, e_{j,r}] = [\widetilde{F}^{(0)}_{ba}, e_{j,r}] = 0$ *in* $\mathcal{U}^{\mathrm{sc}}_{0,\mu_2}$.

Case $r = b_{2,j}$ For $i = j$, the verification of $[\Delta(h_{j,-1}), \Delta(e_{j,b_{2,j}})] = [2]_{\boldsymbol{v}} \cdot \Delta(e_{j,b_{2,j}-1})$ coincides with our proof of formula (10.4) from Appendix G. To prove the claim for $i \neq j$, we can either perform similar long computations or we can rather deduce from the aforementioned case $i = j$. To achieve the latter, we apply the Jacobi identity to get $[2]_{\boldsymbol{v}} \cdot [\Delta(h_{i,-1}), \Delta(e_{j,b_{2,j}})] = [\Delta(h_{j,-1}), [\Delta(h_{i,-1}), \Delta(e_{j,b_{2,j}+1})]] - [\Delta(e_{j,b_{2,j}+1}), [\Delta(h_{i,-1}), \Delta(h_{j,-1})]]$. The second summand is zero as $[\Delta(h_{i,-1}), \Delta(h_{j,-1})] = 0$ by above. Due to the $r = b_{2,j} + 1$ case considered above, we have $[\Delta(h_{i,-1}), \Delta(e_{j,b_{2,j}+1})] = [c_{ij}]_{\boldsymbol{v}} \cdot \Delta(e_{j,b_{2,j}})$. It remains to apply the aforementioned equality $[\Delta(h_{j,-1}), \Delta(e_{j,b_{2,j}})] = [2]_{\boldsymbol{v}} \cdot \Delta(e_{j,b_{2,j}-1})$.

Case $r = 0$ The verification of $[\Delta(h_{i,-1}), \Delta(e_{j,0})] = [c_{ij}]_{\boldsymbol{v}} \cdot 1 \otimes e_{j,-1}$ is similar to our proof of formula (10.4) from Appendix G. To this end, we note that the computations of $[\Delta(h_{i,-1}), 1 \otimes e_{j,0}]$ and $[\Delta(h_{i,-1}), e_{j,0} \otimes \psi^+_{j,0}]$ are straightforward and are crucially based on the above Lemmas H.5 and H.6.

H(ii).e Compatibility with (Û5)

Due to relations (Û1, Û4, Û5) for both $\mathcal{U}^{\mathrm{sc}}_{0,\mu_1}, \mathcal{U}^{\mathrm{sc}}_{0,\mu_2}$, we immediately obtain the equalities $\Delta(\psi^+_{i,0})\Delta(f_{j,r}) = \boldsymbol{v}^{-c_{ij}}\Delta(f_{j,r})\Delta(\psi^+_{i,0})$, $\Delta(\psi^-_{i,b_i})\Delta(f_{j,r}) = \boldsymbol{v}^{c_{ij}}\Delta(f_{j,r})\Delta(\psi^-_{i,b_i})$ for $b_{1,j} \leq r \leq 1$.

Let us now verify $[\Delta(h_{i,-1}), \Delta(f_{j,r})] = -[c_{ij}]_{\boldsymbol{v}} \cdot \Delta(f_{j,r-1})$ for $b_{1,j}+1 \leq r \leq 1$.

Case $b_{1,j} + 1 < r < 1$ The verification in this case follows immediately from relation (Û5) for $\mathcal{U}^{\mathrm{sc}}_{0,\mu_1}$ combined with Lemma H.13 below.

Lemma H.13 *For any* $1 \leq a < b \leq n,\ b_{1,j} + 1 < r < 1$, *we have* $[E^{(-1)}_{ab}, f_{j,r}] = 0$ *in* $\mathcal{U}^{\mathrm{sc}}_{0,\mu_1}$.

Case $r = b_{1,j} + 1$ Due to relation (Û5) for $\mathcal{U}^{\mathrm{sc}}_{0,\mu_1}$, we have

$$[\Delta(h_{i,-1}), \Delta(f_{j,b_{1,j}+1})] = -[c_{ij}]_{\boldsymbol{v}} \cdot f_{j,b_{1,j}} \otimes 1 + (\boldsymbol{v}^2 - \boldsymbol{v}^{-2})[E^{(-1)}_{i,i+1}, f_{j,b_{1,j}+1}] \otimes F^{(0)}_{i+1,i} -$$

$$(\boldsymbol{v} - \boldsymbol{v}^{-1}) \sum_{l>i+1} [E^{(-1)}_{i+1,l}, f_{j,b_{1,j}+1}] \otimes F^{(0)}_{l,i+1} - (\boldsymbol{v} - \boldsymbol{v}^{-1}) \sum_{k<i} \boldsymbol{v}^{i-k-1}[E^{(-1)}_{ki}, f_{j,b_{1,j}+1}] \otimes \widetilde{F}^{(0)}_{ik} -$$

$$\boldsymbol{v}^2(\boldsymbol{v} - \boldsymbol{v}^{-1}) \sum_{l>i+1} [E^{(-1)}_{il}, f_{j,b_{1,j}+1}] \otimes [F^{(0)}_{l,i+1}, F^{(0)}_{i+1,i}]_{\boldsymbol{v}^{-3}} +$$

$$(\boldsymbol{v} - \boldsymbol{v}^{-1}) \sum_{k<i} \boldsymbol{v}^{i+1-k}[E^{(-1)}_{k,i+1}, f_{j,b_{1,j}+1}] \otimes [\widetilde{F}^{(0)}_{ik}, F^{(0)}_{i+1,i}]_{\boldsymbol{v}^{-3}} -$$

$$(\boldsymbol{v} - \boldsymbol{v}^{-1})^2 \sum_{l>i+1}^{k<i} \boldsymbol{v}^{i-k}[E^{(-1)}_{kl}, f_{j,b_{1,j}+1}] \otimes (\widetilde{F}^{(0)}_{i+1,k} F^{(0)}_{l,i+1} - \widetilde{F}^{(0)}_{ik} F^{(0)}_{li}).$$

Using this formula and Lemma H.14 below, it is straightforward to check that we obtain $[\Delta(h_{i,-1}), \Delta(f_{j,b_{1,j}+1})] = -[c_{ij}]_{\boldsymbol{v}} \cdot (f_{j,b_{1,j}} \otimes 1 + \psi^-_{j,b_{1,j}} \otimes f_{j,0}) = -[c_{ij}]_{\boldsymbol{v}} \cdot \Delta(f_{j,b_{1,j}})$.

Lemma H.14 *For any* $1 \le a < b \le n$, *we have* $[E^{(-1)}_{ab}, f_{j,b_{1,j}+1}] = \frac{-\delta_{ja}\delta_{j,b-1}}{\boldsymbol{v}-\boldsymbol{v}^{-1}}\psi^-_{j,b_{1,j}}$ *in* $\mathcal{U}^{\mathrm{sc}}_{0,\mu_1}$.

Case $r = 1$ According to the next step, we have $\Delta(f_{j,1}) = -[2]^{-1}_{\boldsymbol{v}} \cdot [\Delta(h_{j,1}), \Delta(f_{j,0})]$. Apply the Jacobi identity to get $[2]_{\boldsymbol{v}} \cdot [\Delta(h_{i,-1}), \Delta(f_{j,1})] = [\Delta(h_{j,1}), [\Delta(h_{i,-1}), \Delta(f_{j,0})]] - [\Delta(f_{j,0}), [\Delta(h_{i,-1}), \Delta(h_{j,1})]]$. The second summand is zero as $[\Delta(h_{i,-1}), \Delta(h_{j,1})] = 0$ by above. Due to the $r = 0$ case considered above, we have $[\Delta(h_{i,-1}), \Delta(f_{j,0})] = -[c_{ij}]_{\boldsymbol{v}} \cdot \Delta(f_{j,-1})$. It remains to apply $[\Delta(h_{j,1}), \Delta(f_{j,-1})] = -[2]_{\boldsymbol{v}} \cdot \Delta(f_{j,0})$ as proved below.

Let us now verify $[\Delta(h_{i,1}), \Delta(f_{j,r})] = -[c_{ij}]_{\boldsymbol{v}} \cdot \Delta(f_{j,r+1})$ for $b_{1,j} \le r \le 0$.

Case $b_{1,j} < r < 0$ The verification in this case follows immediately from relation ($\hat{\mathrm{U}}5$) for $\mathcal{U}^{\mathrm{sc}}_{0,\mu_1}$ combined with Lemma H.15 below.

Lemma H.15 *For* $1 \le a < b \le n, b_{1,j} < r < 0$, *we have* $[E^{(0)}_{ab}, f_{j,r}] = [\widetilde{E}^{(0)}_{ab}, f_{j,r}] = 0$ *in* $\mathcal{U}^{\mathrm{sc}}_{0,\mu_1}$.

Case $r = 0$ For $i = j$, the verification of $[\Delta(h_{j,1}), \Delta(f_{j,0})] = -[2]_{\boldsymbol{v}} \cdot \Delta(f_{j,1})$ coincides with our proof of formula (10.5), sketched in Appendix G. To prove the claim for $i \neq j$, we can either perform similar long computations or we can rather deduce from the aforementioned case $i = j$. To achieve the latter, we apply the Jacobi identity to get $-[2]_{\boldsymbol{v}} \cdot [\Delta(h_{i,1}), \Delta(f_{j,0})] = [\Delta(h_{j,1}), [\Delta(h_{i,1}), \Delta(f_{j,-1})]] - [\Delta(f_{j,-1}), [\Delta(h_{i,1}), \Delta(h_{j,1})]]$. The second summand is zero as $[\Delta(h_{i,1}), \Delta(h_{j,1})] = 0$ by above. Due to the $r = -1$ case considered above, we have $[\Delta(h_{i,1}), \Delta(f_{j,-1})] = -[c_{ij}]_{\boldsymbol{v}} \cdot \Delta(f_{j,0})$. It remains to apply the aforementioned equality $[\Delta(h_{j,1}), \Delta(f_{j,0})] = -[2]_{\boldsymbol{v}} \cdot \Delta(f_{j,1})$.

Case $r = b_{1,j}$ The verification of $[\Delta(h_{i,1}), \Delta(f_{j,b_{1,j}})] = -[c_{ij}]_{\boldsymbol{v}} \cdot f_{j,b_{1,j}+1} \otimes 1$ is similar to our proof of formula (10.5), sketched in Appendix G. To this end, we note that the computations of $[\Delta(h_{i,1}), f_{j,b_{1,j}} \otimes 1]$ and $[\Delta(h_{i,1}), \psi^-_{j,b_{1,j}} \otimes f_{j,0}]$ are straightforward and are crucially based on the above Lemmas H.8 and H.9.

H(ii).f Compatibility with ($\hat{\mathrm{U}}6$)

We need to verify

$$[\Delta(e_{i,r}), \Delta(f_{j,s})] = \delta_{ij} \cdot \begin{cases} \Delta(\psi^+_{i,0})\Delta(h_{i,1}) & \text{if } r+s=1, \\ \Delta(\psi^-_{i,b_i})\Delta(h_{i,-1}) & \text{if } r+s=b_i-1, \\ \frac{\Delta(\psi^+_{i,0})}{\boldsymbol{v}-\boldsymbol{v}^{-1}} & \text{if } r+s=0, \\ \frac{-\Delta(\psi^-_{i,b_i})}{\boldsymbol{v}-\boldsymbol{v}^{-1}} & \text{if } r+s=b_i, \\ 0 & \text{otherwise,} \end{cases}$$

for $b_{2,i}-1 \leq r \leq 0,\ b_{1,j} \leq s \leq 1$, where we set $b_i := b_{1,i} + b_{2,i}$ as before.

Cases $b_{2,i}-1 < r \leq 0, b_{1,j} \leq s < 1$ Obviously follows from ($\hat{\text{U}}4$, $\hat{\text{U}}5$, $\hat{\text{U}}6$) for both $\mathcal{U}^{\mathrm{sc}}_{0,\mu_1}, \mathcal{U}^{\mathrm{sc}}_{0,\mu_2}$.

Case $b_{2,i} \leq r < -1, s = 1$ In this case, we get $[\Delta(e_{i,r}), \Delta(f_{j,1})] = 0$, due to Lemma H.10.

Case $r = -1, s = 1$ Applying Lemma H.11 from above, it is straightforward to see that we get $[\Delta(e_{i,-1}), \Delta(f_{j,1})] = \frac{\delta_{ij}}{\boldsymbol{v}-\boldsymbol{v}^{-1}}\psi^+_{i,0} \otimes \psi^+_{i,0} = \frac{\delta_{ij}}{\boldsymbol{v}-\boldsymbol{v}^{-1}}\Delta(\psi^+_{i,0})$.

Case $r = b_{2,i}-1, s = 1$ According to relation ($\hat{\text{U}}4$) verified above, we have $\Delta(e_{i,b_{2,i}-1}) = [2]_{\boldsymbol{v}}^{-1} \cdot [\Delta(h_{i,-1}), \Delta(e_{i,b_{2,i}})]$. Applying the Jacobi identity, we get $[2]_{\boldsymbol{v}} \cdot [\Delta(e_{i,b_{2,i}-1}), \Delta(f_{j,1})] = [\Delta(h_{i,-1}), [\Delta(e_{i,b_{2,i}}), \Delta(f_{j,1})]] - [\Delta(e_{i,b_{2,i}}), [\Delta(h_{i,-1}), \Delta(f_{j,1})]]$. However, both summands in the right-hand side are zero, due to the above cases and relation ($\hat{\text{U}}5$) established above.

Case $r = b_{2,i}-1, b_{1,j}+1 < s < 1$ In this case, we get $[\Delta(e_{i,b_{2,i}-1}), \Delta(f_{j,s})] = 0$, due to Lemma H.13.

Case $r = b_{2,i}-1, s = b_{1,j}+1$ Applying Lemma H.14 from above, it is straightforward to see that we get $[\Delta(e_{i,b_{2,i}-1}), \Delta(f_{j,b_{1,j}+1})] = -\frac{\delta_{ij}}{\boldsymbol{v}-\boldsymbol{v}^{-1}}\psi^-_{i,b_{1,i}} \otimes \psi^-_{i,b_{2,i}} = -\frac{\delta_{ij}}{\boldsymbol{v}-\boldsymbol{v}^{-1}}\Delta(\psi^-_{i,b_i})$.

Case $r = 0, s = 1$ Consider the homomorphism $j^+_{\mu_1,0} \otimes j^+_{0,\mu_2} \colon U^+_{\boldsymbol{v}} \otimes U^+_{\boldsymbol{v}} \to \mathcal{U}^{\mathrm{sc},+}_{0,\mu_1,0} \otimes \mathcal{U}^{\mathrm{sc},+}_{0,0,\mu_2}$. Comparing the formulas of Theorems 10.13, 10.16 and applying Lemma H.3, we get

$$[\Delta_{\mu_1,\mu_2}(e_{i,0}), \Delta_{\mu_1,\mu_2}(f_{j,1})] = [j^+_{\mu_1,0} \otimes j^+_{0,\mu_2}(\Delta(e_{i,0})), j^+_{\mu_1,0} \otimes j^+_{0,\mu_2}(\Delta(f_{j,1}))] =$$

$$j^+_{\mu_1,0} \otimes j^+_{0,\mu_2}\left([\Delta(e_{i,0}), \Delta(f_{j,1})]\right) = j^+_{\mu_1,0} \otimes j^+_{0,\mu_2}(\delta_{ij}\Delta(\psi^+_{i,0})\Delta(h_{i,1})) = \delta_{ij}\Delta_{\mu_1,\mu_2}(\psi^+_{i,0})\Delta_{\mu_1,\mu_2}(h_{i,1}),$$

where the subscripts in Δ_{μ_1,μ_2} are used this time to distinguish it from the Drinfeld-Jimbo coproduct Δ on $U_{\boldsymbol{v}}(L\mathfrak{sl}_n)$.

Case $r = b_{2,i}-1, s = b_{1,j}$ Consider the homomorphism $j^-_{\mu_1,0} \otimes j^-_{0,\mu_2} \colon U^-_{\boldsymbol{v}} \otimes U^-_{\boldsymbol{v}} \to \mathcal{U}^{\mathrm{sc},-}_{0,\mu_1,0} \otimes \mathcal{U}^{\mathrm{sc},-}_{0,0,\mu_2}$. Comparing the formulas of Theorems 10.13, 10.16 and applying Lemma H.3, we get

$$[\Delta_{\mu_1,\mu_2}(e_{i,b_{2,i}-1}), \Delta_{\mu_1,\mu_2}(f_{j,b_{1,j}})] = [j^-_{\mu_1,0} \otimes j^-_{0,\mu_2}(\Delta(e_{i,-1})), j^-_{\mu_1,0} \otimes j^-_{0,\mu_2}(\Delta(f_{j,0}))] =$$

$$j^-_{\mu_1,0} \otimes j^-_{0,\mu_2}\left([\Delta(e_{i,-1}), \Delta(f_{j,0})]\right) = j^-_{\mu_1,0} \otimes j^-_{0,\mu_2}(\delta_{ij}\Delta(\psi^-_{i,0})\Delta(h_{i,-1})) =$$

$$\delta_{ij}\Delta_{\mu_1,\mu_2}(\psi^-_{i,b_i})\Delta_{\mu_1,\mu_2}(h_{i,-1}).$$

H(ii).g Compatibility with ($\hat{\mathrm{U}}7$)

Utilizing the homomorphism $\jmath^+_{\mu_1,0}\otimes\jmath^+_{0,\mu_2}\colon U^+_v\otimes U^+_v\to\mathcal{U}^{\mathrm{sc},+}_{0,\mu_1,0}\otimes\mathcal{U}^{\mathrm{sc},+}_{0,0,\mu_2}$ as above, we get

$$[\Delta_{\mu_1,\mu_2}(e_{i,0}),[\Delta_{\mu_1,\mu_2}(e_{i,0}),\cdots,[\Delta_{\mu_1,\mu_2}(e_{i,0}),\Delta_{\mu_1,\mu_2}(e_{j,0})]_{v^{c_{ij}}}\cdots]_{v^{-c_{ij}-2}}]_{v^{-c_{ij}}}=$$

$$\jmath^+_{\mu_1,0}\otimes\jmath^+_{0,\mu_2}([\Delta(e_{i,0}),[\Delta(e_{i,0}),\cdots,[\Delta(e_{i,0}),\Delta(e_{j,0})]_{v^{c_{ij}}}\cdots]_{v^{-c_{ij}-2}}]_{v^{-c_{ij}}})=$$

$$\jmath^+_{\mu_1,0}\otimes\jmath^+_{0,\mu_2}(\Delta([e_{i,0},[e_{i,0},\cdots,[e_{i,0},e_{j,0}]_{v^{c_{ij}}}\cdots]_{v^{-c_{ij}-2}}]_{v^{-c_{ij}}}))=0,$$

where the last equality is due to the Serre relation in U^+_v (cf. Remark 5.4).

H(ii).h Compatibility with ($\hat{\mathrm{U}}8$)

Due to relation ($\hat{\mathrm{U}}8$) for $\mathcal{U}^{\mathrm{sc}}_{0,\mu_1}$, we have

$$[\Delta_{\mu_1,\mu_2}(f_{i,0}),[\Delta_{\mu_1,\mu_2}(f_{i,0}),\cdots,[\Delta_{\mu_1,\mu_2}(f_{i,0}),\Delta_{\mu_1,\mu_2}(f_{j,0})]_{v^{c_{ij}}}\cdots]_{v^{-c_{ij}-2}}]_{v^{-c_{ij}}}=$$

$$[f_{i,0},[f_{i,0},\cdots,[f_{i,0},f_{j,0}]_{v^{c_{ij}}}\cdots]_{v^{-c_{ij}-2}}]_{v^{-c_{ij}}}\otimes 1=0.$$

H(ii).i Compatibility with ($\hat{\mathrm{U}}9$)

Applying the homomorphisms $\jmath^\pm_{\mu_1,0}\otimes\jmath^\pm_{0,\mu_2}$, we see that it suffices to prove the equalities:

$$[h_{i,1},[f_{i,1},[h_{i,1},e_{i,0}]]]=0\ \text{in}\ U^+_v,\quad [h_{i,-1},[e_{i,-1},[h_{i,-1},f_{i,0}]]]=0\ \text{in}\ U^-_v.$$

These follow from $[h_{i,\pm1},\psi^\pm_{i,\pm2}]=0$ in $U^\pm_v$.

This completes our proof of Theorem 10.16.

H(iii) Relation Between Δ and Δ_{μ_1,μ_2}

The following result completes our discussion of Remark 10.17.

Proposition H.16 *The following diagram is commutative:*

$$\begin{array}{ccc}
U^\pm_v & \xrightarrow{\ \Delta\ } & U^\pm_v\otimes U^\pm_v \\
\Big\downarrow{\scriptstyle \jmath^\pm_{\mu_1,\mu_2}} & & \Big\downarrow{\scriptstyle \jmath^\pm_{\mu_1,0}\otimes\jmath^\pm_{0,\mu_2}} \\
\mathcal{U}^{\mathrm{sc},\pm}_{0,\mu_1,\mu_2} & \xrightarrow[\ \Delta_{\mu_1,\mu_2}\ ]{} & \mathcal{U}^{\mathrm{sc},\pm}_{0,\mu_1,0}\otimes\mathcal{U}^{\mathrm{sc},\pm}_{0,0,\mu_2}
\end{array}$$

Proof To simplify our computations, we will assume that μ_1, μ_2 are strictly antidominant.

(a) To prove the commutativity of the above diagram in the '+' case, it suffices to verify that $j^+_{\mu_1,0} \otimes j^+_{0,\mu_2}(\Delta(X)) = \Delta_{\mu_1,\mu_2}(\jmath^+_{\mu_1,\mu_2}(X))$ for $X \in \{e_{i,0}, (\psi^+_{i,0})^{\pm 1}, F^{(1)}_{n1}\}_{i=1}^{n-1}$. The only non-obvious verification is the one for $X = F^{(1)}_{n1}$.

The computation of $\Delta(F^{(1)}_{n1})$ is based on the computation of $\Delta^{\mathrm{rtt}}(\tilde{f}^{(1)}_{n1})$. Comparing the coefficients of z^{-1} in the equality

$$\Delta^{\mathrm{rtt}}(T^+_{n1}(z)) = T^+_{n1}(z) \otimes T^+_{11}(z) + T^+_{nn}(z) \otimes T^+_{n1}(z) + \sum_{1<i<n} T^+_{ni}(z) \otimes T^+_{i1}(z),$$

we get $\Delta^{\mathrm{rtt}}(\tilde{f}^{(1)}_{n1}\tilde{g}^+_1) = \tilde{f}^{(1)}_{n1}\tilde{g}^+_1 \otimes \tilde{g}^+_1 + \tilde{g}^+_n \otimes \tilde{f}^{(1)}_{n1}\tilde{g}^+_1$, so that $\Delta^{\mathrm{rtt}}(\tilde{f}^{(1)}_{n1}) = \tilde{f}^{(1)}_{n1} \otimes 1 + \tilde{g}^+_n(\tilde{g}^+_1)^{-1} \otimes \tilde{f}^{(1)}_{n1}$. Applying Υ^{-1} of Theorem G.2 and formula (G.12), we finally find

$$\Delta(F^{(1)}_{n1}) = F^{(1)}_{n1} \otimes 1 + \psi^+_{1,0} \cdots \psi^+_{n-1,0} \otimes F^{(1)}_{n1}.$$

Therefore, $j^+_{\mu_1,0} \otimes j^+_{0,\mu_2}(\Delta(F^{(1)}_{n1})) = F^{(1)}_{n1} \otimes 1 + \psi^+_{1,0} \cdots \psi^+_{n-1,0} \otimes F^{(1)}_{n1}$.

On the other hand, we have $\Delta_{\mu_1,\mu_2}(\jmath^+_{\mu_1,\mu_2}(F^{(1)}_{n1})) = \Delta_{\mu_1,\mu_2}(F^{(1)}_{n1})$ and

$$\Delta_{\mu_1,\mu_2}(F^{(1)}_{n1}) = [\cdots[\Delta_{\mu_1,\mu_2}(f_{1,1}), \Delta_{\mu_1,\mu_2}(f_{2,0})]_{\boldsymbol{v}}, \cdots, \Delta_{\mu_1,\mu_2}(f_{n-1,0})]_{\boldsymbol{v}}.$$

Let us first note that $[E^{(0)}_{2l}, f_{2,0}] = \boldsymbol{v}^{-1}E^{(0)}_{3l}\psi^+_{2,0}$, where we set $E^{(0)}_{33} := \frac{\boldsymbol{v}}{\boldsymbol{v}-\boldsymbol{v}^{-1}}$. Combining this with relation (U5) and the formula

$$\Delta_{\mu_1,\mu_2}(f_{1,1}) = f_{1,1} \otimes 1 + \psi^+_{1,0} \otimes f_{1,1} + \boldsymbol{v}^{-1}(\boldsymbol{v}-\boldsymbol{v}^{-1}) \sum_{l>2} E^{(0)}_{2l}\psi^+_{1,0} \otimes F^{(1)}_{l1},$$

we find

$$[\Delta_{\mu_1,\mu_2}(f_{1,1}), \Delta_{\mu_1,\mu_2}(f_{2,0})]_{\boldsymbol{v}} = [f_{1,1}, f_{2,0}]_{\boldsymbol{v}} \otimes 1 + \boldsymbol{v}^{-1}(\boldsymbol{v}-\boldsymbol{v}^{-1}) \sum_{l>2} E^{(0)}_{3l}\psi^+_{1,0}\psi^+_{2,0} \otimes F^{(1)}_{l1}.$$

Further $\boldsymbol{v}$-commuting this with $\Delta_{\mu_1,\mu_2}(f_{3,0}), \ldots, \Delta_{\mu_1,\mu_2}(f_{n-1,0})$, we finally obtain

$$\Delta_{\mu_1,\mu_2}(F^{(1)}_{n1}) = F^{(1)}_{n1} \otimes 1 + \psi^+_{1,0} \cdots \psi^+_{n-1,0} \otimes F^{(1)}_{n1}.$$

This completes our verification of $j^+_{\mu_1,0} \otimes j^+_{0,\mu_2}(\Delta(F^{(1)}_{n1})) = \Delta_{\mu_1,\mu_2}(\jmath^+_{\mu_1,\mu_2}(F^{(1)}_{n1}))$.

(b) The proof of the commutativity in the '−' case is completely analogous. □

Appendix I Proof of Theorem 10.19

Our proof of Theorem 10.19 proceeds in three steps. First, we reduce the problem to its unshifted counterpart, see Theorem I.1. To prove this theorem, we recall the shuffle realization of $U_{\boldsymbol{v}}^{>}$, see Theorem I.3. In the last and final step, we apply a simple result Proposition I.4.

I(i) Reduction to an Unshifted Case

Given $\mu \in \Lambda$ and $\nu_1, \nu_2 \in \Lambda^-$, recall the shift homomorphisms $\iota_{\mu,\nu_1,\nu_2} \colon \mathcal{U}^{\mathrm{sc}}_{0,\mu} \to \mathcal{U}^{\mathrm{sc}}_{0,\mu+\nu_1+\nu_2}$ introduced in Lemma 10.18. Note that ι_{μ,ν_1,ν_2} gives rise to the homomorphisms (restrictions)

$$\iota^{>}_{\mu,\nu_1,\nu_2} \colon \mathcal{U}^{\mathrm{sc},>}_{0,\mu} \to \mathcal{U}^{\mathrm{sc},>}_{0,\mu+\nu_1+\nu_2},\ \iota^{<}_{\mu,\nu_1,\nu_2} \colon \mathcal{U}^{\mathrm{sc},<}_{0,\mu} \to \mathcal{U}^{\mathrm{sc},<}_{0,\mu+\nu_1+\nu_2},\ \iota^{0}_{\mu,\nu_1,\nu_2} \colon \mathcal{U}^{\mathrm{sc},0}_{0,\mu} \to \mathcal{U}^{\mathrm{sc},0}_{0,\mu+\nu_1+\nu_2}.$$

Moreover, evoking the triangular decomposition of Proposition 5.1(a) for both algebras $\mathcal{U}^{\mathrm{sc}}_{0,\mu}$ and $\mathcal{U}^{\mathrm{sc}}_{0,\mu+\nu_1+\nu_2}$, we see that ι_{μ,ν_1,ν_2} is "glued" from the aforementioned three homomorphisms $\iota^{>}_{\mu,\nu_1,\nu_2}, \iota^{<}_{\mu,\nu_1,\nu_2}, \iota^{0}_{\mu,\nu_1,\nu_2}$. Hence, Theorem 10.19 is equivalent to the injectivity of these restrictions $\iota^{>}_{\mu,\nu_1,\nu_2}, \iota^{<}_{\mu,\nu_1,\nu_2}, \iota^{0}_{\mu,\nu_1,\nu_2}$. The injectivity of $\iota^{0}_{\mu,\nu_1,\nu_2}$ is clear. On the other hand, according to Proposition 5.1(b), we have $\mathcal{U}^{\mathrm{sc},>}_{0,\mu} \simeq U^{>}_{\boldsymbol{v}} \simeq \mathcal{U}^{\mathrm{sc},>}_{0,\mu+\nu_1+\nu_2}$, $\mathcal{U}^{\mathrm{sc},<}_{0,\mu} \simeq U^{<}_{\boldsymbol{v}} \simeq \mathcal{U}^{\mathrm{sc},<}_{0,\mu+\nu_1+\nu_2}$, where $U^{>}_{\boldsymbol{v}}, U^{<}_{\boldsymbol{v}}$ denote the corresponding subalgebras of $U_{\boldsymbol{v}}(L\mathfrak{sl}_n)$. As such, the injectivity of $\iota^{>}_{\mu,\nu_1,\nu_2}$ (resp. $\iota^{<}_{\mu,\nu_1,\nu_2}$) is equivalent to the injectivity of $\iota^{>}_{\nu_1} \colon U^{>}_{\boldsymbol{v}} \to U^{>}_{\boldsymbol{v}}$ (resp. $\iota^{<}_{\nu_2} \colon U^{<}_{\boldsymbol{v}} \to U^{<}_{\boldsymbol{v}}$) given by $e_i(z) \mapsto (1-z^{-1})^{-\alpha_i^\vee(\nu_1)} e_i(z)$ (resp. $f_i(z) \mapsto (1-z^{-1})^{-\alpha_i^\vee(\nu_2)} f_i(z)$) for $i \in I$.

Thus, we have reduced Theorem 10.19 to its unshifted counterpart:

Theorem I.1

(a) The homomorphism $\iota^{>}_{\nu} \colon U^{>}_{\boldsymbol{v}} \to U^{>}_{\boldsymbol{v}}$ is injective for any $\nu \in \Lambda^-$.
(b) The homomorphism $\iota^{<}_{\nu} \colon U^{<}_{\boldsymbol{v}} \to U^{<}_{\boldsymbol{v}}$ is injective for any $\nu \in \Lambda^-$.

Our proof of part (a) is crucially based on the *shuffle realization* of $U^{>}_{\boldsymbol{v}}$, which we recall next (the proof of part (b) is completely analogous).

I(ii) Shuffle Algebra (of Type A_{n-1})

Consider an $\mathbb{N}^I$-graded $\mathbb{C}(\boldsymbol{v})$-vector space $\mathbb{S} = \bigoplus_{\underline{k}=(k_1,\ldots,k_{n-1})\in\mathbb{N}^I} \mathbb{S}_{\underline{k}}$, where $\mathbb{S}_{(k_1,\ldots,k_{n-1})}$ consists of $\prod \mathfrak{S}_{k_i}$-symmetric rational functions in the variables $\{x_{i,r}\}_{i\in I}^{1\le r\le k_i}$. We also fix an $I \times I$ matrix of rational functions $(\zeta_{i,j}(z))_{i,j\in I} \in \mathrm{Mat}_{I\times I}(\mathbb{C}(z))$ by setting $\zeta_{i,j}(z) = \frac{z-\boldsymbol{v}^{-c_{ij}}}{z-1}$, where $(c_{ij})_{i,j=1}^{n-1}$ is the Cartan matrix of $\mathfrak{sl}_n$ as before. Let us now introduce the bilinear $\star$ product on $\mathbb{S}$: given $F \in \mathbb{S}_{\underline{k}}$ and $G \in \mathbb{S}_{\underline{l}}$, define $F \star G \in \mathbb{S}_{\underline{k}+\underline{l}}$ by

$$(F \star G)(x_{1,1}, \ldots, x_{1,k_1+l_1}; \ldots; x_{n-1,1}, \ldots, x_{n-1,k_{n-1}+l_{n-1}}) := \prod_{i=1}^{n-1} k_i! \cdot l_i! \times$$

$$\mathrm{Sym}_{\prod \mathfrak{S}_{k_i+l_i}} \left(F(\{x_{i,r}\}_{i\in I}^{1\le r\le k_i}) G(\{x_{i',r'}\}_{i'\in I}^{k_{i'}<r'\le k_{i'}+l_{i'}}) \cdot \prod_{i\in I}^{i'\in I} \prod_{r\le k_i}^{r'>k_{i'}} \zeta_{i,i'}(x_{i,r}/x_{i',r'}) \right).$$

Here and afterwards, given a function $f \in \mathbb{C}(\{x_{i,1}, \ldots, x_{i,m_i}\}_{i\in I})$, we define

$$\mathrm{Sym}_{\prod \mathfrak{S}_{m_i}}(f) := \prod_{i\in I} \frac{1}{m_i!} \cdot \sum_{(\sigma_1,\ldots,\sigma_{n-1})\in \mathfrak{S}_{m_1}\times\ldots\times\mathfrak{S}_{m_{n-1}}} f(\{x_{i,\sigma_i(1)}, \ldots, x_{i,\sigma_i(m_i)}\}_{i\in I}).$$

This endows $\mathbb{S}$ with a structure of an associative unital algebra with the unit $\mathbf{1} \in \mathbb{S}_{(0,\ldots,0)}$. We will be interested only in a certain subspace of $\mathbb{S}$, defined by the *pole* and *wheel conditions*:

- We say that $F \in \mathbb{S}_{\underline{k}}$ satisfies the *pole conditions* if and only if

$$F = \frac{f(x_{1,1}, \ldots, x_{n-1,k_{n-1}})}{\prod_{i=1}^{n-2} \prod_{r\le k_i}^{r'\le k_{i+1}} (x_{i,r} - x_{i+1,r'})}, \text{ where } f \in (\mathbb{C}(\boldsymbol{v})[x_{i,r}^{\pm 1}]_{i\in I}^{1\le r\le k_i})^{\prod \mathfrak{S}_{k_i}}.$$

- We say that $F \in \mathbb{S}_{\underline{k}}$ satisfies the *wheel conditions* if and only if

$$F(\{x_{i,r}\}) = 0 \text{ once } x_{i,r_1} = \boldsymbol{v} x_{i+\epsilon,l} = \boldsymbol{v}^2 x_{i,r_2} \text{ for some } \epsilon, i, r_1, r_2, l,$$

where $\epsilon \in \{\pm 1\},\ i, i+\epsilon \in I,\ 1 \le r_1, r_2 \le k_i,\ 1 \le l \le k_{i+\epsilon}$.

Let $S_{\underline{k}} \subset \mathbb{S}_{\underline{k}}$ be the subspace of all elements F satisfying these two conditions and set $S := \bigoplus_{\underline{k}\in\mathbb{N}^I} S_{\underline{k}}$. It is straightforward to check that the subspace $S \subset \mathbb{S}$ is $\star$-closed.

Definition I.2 The algebra $(S, \star)$ is called the *shuffle algebra* (of A_{n-1}-type).

The following key result, identifying this algebra with $U_{\boldsymbol{v}}^{>}$, is due to [53][12] (see also [63]).

Theorem I.3 *There is a unique $\mathbb{C}(\boldsymbol{v})$-algebra isomorphism $\Psi : U_{\boldsymbol{v}}^{>} \xrightarrow{\sim} S$ such that $e_{i,r} \mapsto x_{i,1}^r$ for any $i \in I, r \in \mathbb{Z}$.*

[12] To be more precise, [53, Theorem 1.1] establishes such a shuffle realization for the half of the quantum toroidal algebra of $\mathfrak{sl}_n$. Since the latter naturally contains $U_{\boldsymbol{v}}^{>}$ as a subalgebra, we get the claimed result.

I(iii) Proof of Theorem I.1(a)

The following result is straightforward:

Proposition I.4

(a) For any $\nu \in \Lambda^-$, there is a unique algebra homomorphism $\iota'_\nu \colon S \to S$ such that $f(\{x_{i,r}\}_{i\in I}^{1\le r\le k_i}) \mapsto \prod_{i\in I}^{1\le r\le k_i}(1-x_{i,r}^{-1})^{-\alpha_i^\vee(\nu)}\cdot f(\{x_{i,r}\}_{i\in I}^{1\le r\le k_i})$ for any $f \in S_{(k_1,\dots,k_{n-1})}$.

(b) The homomorphisms $\iota_\nu^>$ and ι'_ν are compatible: $\iota'_\nu(\Psi(X)) = \Psi(\iota_\nu^>(X))$ for any $X \in U_{\boldsymbol{v}}^>$.

(c) ι'_ν is injective.

Combining Theorem I.3 and Proposition I.4 immediately yields Theorem I.1(a).

This completes our proof of Theorem 10.19.

Appendix J Proof of Proposition 11.18

Consider the $n = 0$ case of Sect. 11.4. Let $\tilde{e}^\pm(z), \tilde{f}^\pm(z), \tilde{g}_1^\pm(z), \tilde{g}_2^\pm(z)$ be the currents entering the Gauss decomposition of $T^\pm(z)$, and set $\tilde{\psi}^\pm(z) := \tilde{g}_2^\pm(z)(\tilde{g}_1^\pm(z))^{-1}$. According to [17] (see also Theorem G.2) there is a $\mathbb{C}(\boldsymbol{v})$-algebra isomorphism

$$\Upsilon\colon U_{\boldsymbol{v}}^{\mathrm{ad}}(L\mathfrak{sl}_2) \xrightarrow{\sim} \mathfrak{U}_{0,0}^{\mathrm{rtt}}/(t_{11}^\pm[0]t_{11}^\mp[0]-1),$$

defined by

$$e^\pm(z) \mapsto \frac{\tilde{e}^\pm(\boldsymbol{v}z)}{\boldsymbol{v}-\boldsymbol{v}^{-1}},\ f^\pm(z) \mapsto \frac{\tilde{f}^\pm(\boldsymbol{v}z)}{\boldsymbol{v}-\boldsymbol{v}^{-1}},\ \psi^\pm(z) \mapsto \tilde{\psi}^\pm(\boldsymbol{v}z),\ \phi^\pm \mapsto t_{11}^\mp[0] \tag{J.1}$$

(a slight modification of $\Upsilon_{0,0}$). The isomorphism Υ intertwines coproducts $\Delta^{\mathrm{rtt}} := \Delta_{0,0}^{\mathrm{rtt}}$ and Δ^{ad}. In particular, the restriction of the pull-back of Δ^{rtt} to the subalgebra $U_{\boldsymbol{v}}(L\mathfrak{sl}_2)$ of $U_{\boldsymbol{v}}^{\mathrm{ad}}(L\mathfrak{sl}_2)$ recovers the Drinfeld-Jimbo coproduct Δ on $U_{\boldsymbol{v}}(L\mathfrak{sl}_2)$.

J(i) Computation of $\Delta(e^\pm(z))$ and $\Delta(f^\pm(z))$

The verification of formulas (11.10) and (11.11) is based on the following result.

Lemma J.1 *We have $T_{11}^\pm(z)^{-1}T_{21}^\pm(z) = \boldsymbol{v}\tilde{f}^\pm(\boldsymbol{v}^2z)$, $T_{12}^\pm(z)T_{11}^\pm(z)^{-1} = \boldsymbol{v}^{-1}\tilde{e}^\pm(\boldsymbol{v}^2z)$.*

Proof Comparing the matrix coefficients $\langle v_1 \otimes v_2|\cdots|v_1 \otimes v_1\rangle$ of both sides of the equality $R_{\mathrm{trig}}(z/w)(T^{\pm}(z)\otimes 1)(1\otimes T^{\pm}(w)) = (1\otimes T^{\pm}(w))(T^{\pm}(z)\otimes 1)R_{\mathrm{trig}}(z/w)$, we get

$$(z-w)T^{\pm}_{11}(z)T^{\pm}_{21}(w) + (\boldsymbol{v}-\boldsymbol{v}^{-1})zT^{\pm}_{21}(z)T^{\pm}_{11}(w) = (\boldsymbol{v}z-\boldsymbol{v}^{-1}w)T^{\pm}_{21}(w)T^{\pm}_{11}(z).$$

Plugging $w = \boldsymbol{v}^2 z$ into this identity, we obtain the first equality:

$$T^{\pm}_{11}(z)^{-1}T^{\pm}_{21}(z) = \boldsymbol{v}T^{\pm}_{21}(\boldsymbol{v}^2z)T^{\pm}_{11}(\boldsymbol{v}^2z)^{-1} = \boldsymbol{v}\tilde{f}^{\pm}(\boldsymbol{v}^2z).$$

Likewise, comparing the matrix coefficients $\langle v_1\otimes v_1|\cdots|v_1\otimes v_2\rangle$, we get the second equality. □

- We have $\tilde{e}^{\pm}(z) = (T^{\pm}_{11}(z))^{-1}T^{\pm}_{12}(z)$. Hence,

$$\Delta^{\mathrm{rtt}}(\tilde{e}^{\pm}(z)) = \left(T^{\pm}_{11}(z)\otimes T^{\pm}_{11}(z) + T^{\pm}_{12}(z)\otimes T^{\pm}_{21}(z)\right)^{-1}\left(T^{\pm}_{11}(z)\otimes T^{\pm}_{12}(z) + T^{\pm}_{12}(z)\otimes T^{\pm}_{22}(z)\right) =$$

$$\left(1 + T^{\pm}_{11}(z)^{-1}T^{\pm}_{12}(z)\otimes T^{\pm}_{11}(z)^{-1}T^{\pm}_{21}(z)\right)^{-1}\left(1\otimes \tilde{e}^{\pm}(z) + \tilde{e}^{\pm}(z)\otimes T^{\pm}_{11}(z)^{-1}T^{\pm}_{22}(z)\right) =$$

$$\left(\sum_{r=0}^{\infty}(-\boldsymbol{v})^r\tilde{e}^{\pm}(z)^r\otimes \tilde{f}^{\pm}(\boldsymbol{v}^2z)^r\right)\left(1\otimes\tilde{e}^{\pm}(z) + \tilde{e}^{\pm}(z)\otimes(\boldsymbol{v}\tilde{f}^{\pm}(\boldsymbol{v}^2z)\tilde{e}^{\pm}(z) + \tilde{g}^{\pm}_1(z)^{-1}\tilde{g}^{\pm}_2(z))\right) =$$

$$1\otimes\tilde{e}^{\pm}(z) + \sum_{r=0}^{\infty}(-\boldsymbol{v})^r\cdot\tilde{e}^{\pm}(z)^{r+1}\otimes\tilde{f}^{\pm}(\boldsymbol{v}^2z)^r\tilde{\psi}^{\pm}(z),$$

  where we used Lemma J.1 twice in the third equality. Applying Υ^{-1}, we recover (11.10).
- We have $\tilde{f}^{\pm}(z) = T^{\pm}_{21}(z)(T^{\pm}_{11}(z))^{-1}$. Hence,

$$\Delta^{\mathrm{rtt}}(\tilde{f}^{\pm}(z)) = \left(T^{\pm}_{21}(z)\otimes T^{\pm}_{11}(z) + T^{\pm}_{22}(z)\otimes T^{\pm}_{21}(z)\right)\left(T^{\pm}_{11}(z)\otimes T^{\pm}_{11}(z) + T^{\pm}_{12}(z)\otimes T^{\pm}_{21}(z)\right)^{-1} =$$

$$\left(\tilde{f}^{\pm}(z)\otimes 1 + T^{\pm}_{22}(z)T^{\pm}_{11}(z)^{-1}\otimes\tilde{f}^{\pm}(z)\right)\left(1 + T^{\pm}_{12}(z)T^{\pm}_{11}(z)^{-1}\otimes\tilde{f}^{\pm}(z)\right)^{-1} =$$

$$\left(\tilde{f}^{\pm}(z)\otimes 1 + (\boldsymbol{v}^{-1}\tilde{f}^{\pm}(z)\tilde{e}^{\pm}(\boldsymbol{v}^2z) + \tilde{g}^{\pm}_2(z)\tilde{g}^{\pm}_1(z)^{-1})\otimes\tilde{f}^{\pm}(z)\right)\times$$

$$\left(\sum_{r=0}^{\infty}(-\boldsymbol{v})^{-r}\tilde{e}^{\pm}(\boldsymbol{v}^2z)^r\otimes\tilde{f}^{\pm}(z)^r\right) = \tilde{f}^{\pm}(z)\otimes 1 + \sum_{r=0}^{\infty}(-\boldsymbol{v})^{-r}\cdot\tilde{\psi}^{\pm}(z)\tilde{e}^{\pm}(\boldsymbol{v}^2z)^r\otimes\tilde{f}^{\pm}(z)^{r+1},$$

  where we used Lemma J.1 twice in the third equality. Applying Υ^{-1}, we recover (11.11).

J(ii) Computation of $\Delta(\psi^{\pm}(z))$

We have $\tilde{\psi}^{\pm}(z) = \tilde{g}^{\pm}(z)^{-1}\tilde{g}_2^{\pm}(z) = T_{11}^{\pm}(z)^{-1}T_{22}^{\pm}(z) - \boldsymbol{v}\tilde{f}^{\pm}(\boldsymbol{v}^2 z)\tilde{e}^{\pm}(z)$, due to Lemma J.1. Evaluating $\Delta^{\mathrm{rtt}}(T_{11}^{\pm}(z)^{-1}T_{22}^{\pm}(z))$ as before, we get the following formula:

$$\Delta^{\mathrm{rtt}}(\tilde{\psi}^{\pm}(z)) = \sum_{r=0}^{\infty}(-1)^{r+1}\boldsymbol{v}^{r+2}\tilde{e}^{\pm}(z)^r[\tilde{e}^{\pm}(z), \tilde{f}^{\pm}(\boldsymbol{v}^2 z)] \otimes \tilde{f}^{\pm}(\boldsymbol{v}^2 z)^{r+1}\tilde{e}^{\pm}(z)+$$
$$\sum_{r=0}^{\infty}(-1)^r(\boldsymbol{v}^{r+1}\tilde{e}^{\pm}(z)^r\tilde{\psi}^{\pm}(z) - \boldsymbol{v}^{1-r}\tilde{\psi}^{\pm}(\boldsymbol{v}^2 z)\tilde{e}^{\pm}(\boldsymbol{v}^4 z)^r) \otimes \tilde{f}^{\pm}(\boldsymbol{v}^2 z)^{r+1}\tilde{e}^{\pm}(z)+$$
$$\sum_{r=0}^{\infty}(-1)^r\boldsymbol{v}^{r+1}[\tilde{e}^{\pm}(z)^r, \tilde{f}^{\pm}(\boldsymbol{v}^2 z)]\tilde{e}^{\pm}(z) \otimes \tilde{f}^{\pm}(\boldsymbol{v}^2 z)^r\tilde{\psi}^{\pm}(z)+$$
$$\sum_{r=0}^{\infty}(-1)^r\boldsymbol{v}^r\tilde{e}^{\pm}(z)^r\tilde{\psi}^{\pm}(z) \otimes \tilde{f}^{\pm}(\boldsymbol{v}^2 z)^r\tilde{\psi}^{\pm}(z)+$$
$$\sum_{r,s=0}^{\infty}(-1)^{r+s+1}\boldsymbol{v}^{-r+s+1}\tilde{\psi}^{\pm}(\boldsymbol{v}^2 z)\tilde{e}^{\pm}(\boldsymbol{v}^4 z)^r\tilde{e}^{\pm}(z)^{s+1} \otimes \tilde{f}^{\pm}(\boldsymbol{v}^2 z)^{r+s+1}\tilde{\psi}^{\pm}(z). \tag{J.2}$$

To simplify the right-hand side of this equality, we need the following result.

Lemma J.2 *We have:*

(a) $[\tilde{e}^{\pm}(z), \tilde{f}^{\pm}(w)] = \frac{(\boldsymbol{v}-\boldsymbol{v}^{-1})z}{w-z}\cdot(\tilde{\psi}^{\pm}(z) - \tilde{\psi}^{\pm}(w))$.
(b) $[\tilde{e}^{\pm}(z), \tilde{f}^{\pm}(\boldsymbol{v}^2 z)] = \frac{\tilde{\psi}^{\pm}(z)-\tilde{\psi}^{\pm}(\boldsymbol{v}^2 z)}{\boldsymbol{v}}$.
(c) $(z - \boldsymbol{v}^2 w)\tilde{\psi}^{\pm}(z)\tilde{e}^{\pm}(w) = (\boldsymbol{v}^2 z - w)\tilde{e}^{\pm}(w)\tilde{\psi}^{\pm}(z) \pm w\cdot[\tilde{e}_0, \tilde{\psi}^{\pm}(z)]_{\boldsymbol{v}^2}$.
(d) $\tilde{\psi}^{\pm}(z)\tilde{e}^{\pm}(\boldsymbol{v}^2 z) = \boldsymbol{v}^2\tilde{e}^{\pm}(\boldsymbol{v}^{-2}z)\tilde{\psi}^{\pm}(z) = \frac{\tilde{e}^{\pm}(z)\tilde{\psi}^{\pm}(z)+\tilde{\psi}^{\pm}(z)\tilde{e}^{\pm}(z)}{1+\boldsymbol{v}^{-2}}$.
(e) $(z - \boldsymbol{v}^2 w)\tilde{e}^{\pm}(z)\tilde{e}^{\pm}(w) - z\cdot[\tilde{e}_0, \tilde{e}^{\pm}(w)]_{\boldsymbol{v}^2} = (\boldsymbol{v}^2 z - w)\tilde{e}^{\pm}(w)\tilde{e}^{\pm}(z) + w\cdot [\tilde{e}_0, \tilde{e}^{\pm}(z)]_{\boldsymbol{v}^2}$.
(f) $\tilde{e}^{\pm}(\boldsymbol{v}^2 z)^2 - (1 + \boldsymbol{v}^2)\tilde{e}^{\pm}(z)\tilde{e}^{\pm}(\boldsymbol{v}^2 z) + \boldsymbol{v}^2\tilde{e}^{\pm}(z)^2 = 0$.

Proof Parts (a, c, e) follow from the corresponding relations for $e^{\pm}(z), f^{\pm}(z), \psi^{\pm}(z)$, established in Lemma B.1(c, f1, d1), respectively.

Part (b) is obtained by specializing $w = \boldsymbol{v}^2 z$ in (a). Part (d) is obtained by comparing the specializations of (c) at $w = \boldsymbol{v}^2 z$, $w = \boldsymbol{v}^{-2}z$, and $w = z$. Part (f) is obtained by comparing the specializations of (e) at $w = \boldsymbol{v}^2 z$ and $w = z$. □

The first two sums of (J.2) add up to zero, due to Lemma J.2(b, d). Applying Lemma J.2(b) to the third sum of (J.2) and Lemma J.2(d) to the last sum of (J.2), we get

$$\Delta^{\mathrm{rtt}}(\tilde{\psi}^{\pm}(z)) = \sum_{r=0}^{\infty}(-\boldsymbol{v})^r A_r(z) \otimes \tilde{f}^{\pm}(\boldsymbol{v}^2 z)^r \tilde{\psi}^{\pm}(z) \tag{J.3}$$

with

$$A_r(z) = \tilde{e}^{\pm}(z)^r \tilde{\psi}^{\pm}(z) + \tilde{e}^{\pm}(z)^{r-1}\tilde{\psi}^{\pm}(z)\tilde{e}^{\pm}(z) + \ldots + \tilde{e}^{\pm}(z)\tilde{\psi}^{\pm}(z)\tilde{e}^{\pm}(z)^{r-1} + \tilde{\psi}^{\pm}(z)\tilde{e}^{\pm}(z)^r.$$

Finally, a simple induction argument based on Lemma J.2(d, f) yields the equality

$$A_r(z) = \tilde{\psi}^{\pm}(z)\tilde{e}^{\pm}(\boldsymbol{v}^2 z)^r(1+\boldsymbol{v}^{-2}+\boldsymbol{v}^{-4}+\ldots+\boldsymbol{v}^{-2r}) = \boldsymbol{v}^{-r}[r+1]_{\boldsymbol{v}} \cdot \tilde{\psi}^{\pm}(z)\tilde{e}^{\pm}(\boldsymbol{v}^2 z)^r.$$

Plugging this into (J.3) and applying Υ^{-1}, we recover (11.12).

This completes our proof of Proposition 11.18.

Acknowledgements We are deeply grateful to M. Bershtein, R. Bezrukavnikov, A. Braverman, A. Brochier, S. Cautis, S. Cherkis, T. Dimofte, P. Etingof, B. Feigin, S. Gautam, M. Gekhtman, V. Ginzburg, D. Hernandez, D. Jordan, J. Kamnitzer, S. Khoroshkin, A. Marshakov, A. Molev, H. Nakajima, V. Pestun, L. Rybnikov, A. Shapiro, A. Weekes, M. Yakimov, whose generous help and advice was crucial in the process of our work. We also thank the anonymous referee.

M.F. was partially funded within the framework of the HSE University Basic Research Program and the Russian Academic Excellence Project '5-100'.

A.T. gratefully acknowledges support from the Simons Center for Geometry and Physics, Stony Brook University and is extremely grateful to IHES for invitation and wonderful working conditions in the summer 2017, where most of the research for this paper was performed. A.T. also thanks Yale University and Max Planck Institute for Mathematics in Bonn for the hospitality and support in the summer 2018, where the final version of this paper was completed. A.T. was partially supported by the NSF Grants DMS–1502497, DMS–1821185.

References

1. M. Atiyah, N. Hitchin, *The geometry and dynamics of magnetic monopoles*, Princeton University Press (1988).
2. A. Alekseev, A. Malkin, E. Meinrenken, *Lie group valued moment maps*, J. Differential Geom. **48** (1998), no. 3, 445–495.
3. A. Beilinson, *Coherent sheaves on* $\mathbb{P}^n$ *and problems in linear algebra*, Funct. Anal. Appl. **12** (1978), no. 3, 214–216.
4. M. Bullimore, T. Dimofte, D. Gaiotto, J. Hilburn, H. Kim, *Vortices and Vermas*, Adv. Theor. Math. Phys. **22** (2018), no. 4, 803–917.
5. V. Baranovsky, S. Evens, V. Ginzburg, *Representations of quantum tori and double-affine Hecke algebras*, preprint, arXiv:math/0005024; an abridged version *Representations of quantum tori and G-bundles on elliptic curves* published in Progr. Math. **213** (2003), 29–48.

6. A. Braverman, M. Finkelberg, *Finite difference quantum Toda lattice via equivariant K-theory*, Transf. Groups **10** (2005), no. 3–4, 363–386.
7. A. Braverman, B. Feigin, M. Finkelberg, L. Rybnikov, *A finite analog of the AGT relation I: Finite W-algebras and quasimaps' spaces*, Comm. Math. Phys. **308** (2011), no. 2, 457–478.
8. R. Bezrukavnikov, M. Finkelberg, I. Mirković, *Equivariant homology and K-theory of affine Grassmannians and Toda lattices*, Compos. Math. **141** (2005), no. 3, 746–768.
9. A. Braverman, M. Finkelberg, H. Nakajima, *Towards a mathematical definition of Coulomb branches of 3-dimensional* $\mathcal{N} = 4$ *gauge theories, II*, Adv. Theor. Math. Phys. **22** (2018), no. 5, 1071–1147; arXiv:1601.03586.
10. A. Braverman, M. Finkelberg, H. Nakajima, *Coulomb branches of 3d* $\mathcal{N} = 4$ *quiver gauge theories and slices in the affine Grassmannian* (with appendices by Alexander Braverman, Michael Finkelberg, Joel Kamnitzer, Ryosuke Kodera, Hiraku Nakajima, Ben Webster, Alex Weekes), preprint, arXiv:1604.03625.
11. M. Bertola, M. Gekhtman, *Biorthogonal Laurent polynomials, Töplitz determinants, minimal Toda orbits and isomonodromic tau functions*, Constr. Approx. **26** (2007), no. 3, 383–430.
12. J. Brundan and A. Kleshchev, *Shifted Yangians and finite W-algebras*, Adv. Math. **200** (2006), no. 1, 136–195.
13. I. Cherednik, B. Feigin, *Rogers-Ramanujan type identities and Nil-DAHA*, Adv. Math. **248** (2013), 1050–1088.
14. S. Cherkis, A. Kapustin, *Periodic monopoles with singularities and* $\mathcal{N} = 2$ *super-QCD*, Comm. Math. Phys. **234** (2003), no. 1, 1–35.
15. S. Cautis, H. Williams, *Cluster theory of the coherent Satake category*, preprint, arXiv:1801.08111.
16. V. Drinfeld, *A New realization of Yangians and quantized affine algebras*, Sov. Math. Dokl. **36** (1988), no. 2, 212–216.
17. J. Ding, I. Frenkel, *Isomorphism of two realizations of quantum affine algebra* $U_q(\widehat{\mathfrak{gl}(n)})$, Comm. Math. Phys. **156** (1993), no. 2, 277–300.
18. P. Di Francesco, R. Kedem, *Quantum Q-systems: from cluster algebras to quantum current algebras*, Lett. Math. Phys **107** (2017), no. 2, 301–341.
19. P. Etingof, *Whittaker functions on quantum groups and q-deformed Toda operators*, Amer. Math. Soc. Transl. Ser. 2 **194** (1999), 9–25.
20. S. Evens, J.-H. Lu, *Poisson geometry of the Grothendieck resolution of a complex semisimple group*, Moscow Math. Journal **7** (2007), no. 4, 613–642.
21. L. Faybusovich, M. Gekhtman, *Elementary Toda orbits and integrable lattices*, J. Math. Phys. **41** (2000), no. 5, 2905–2921.
22. B. Feigin, E. Feigin, M. Jimbo, T. Miwa, E. Mukhin, *Fermionic formulas for eigenfunctions of the difference Toda Hamiltonian*, Lett. Math. Phys. **88** (2009), no. 1–3, 39–77; correction to "Fermionic formulas for eigenfunctions of the difference Toda Hamiltonian", Lett. Math. Phys. **108** (2018), no. 7, 1779–1781.
23. B. Feigin, M. Finkelberg, A. Negut, L. Rybnikov, *Yangians and cohomology rings of Laumon spaces*, Selecta Math. (N.S.) **17** (2011), no. 3, 573–607.
24. M. Finkelberg, J. Kamnitzer, K. Pham, L. Rybnikov, A. Weekes, *Comultiplication for shifted Yangians and quantum open Toda lattice*, Adv. Math. **327** (2018), 349–389.
25. M. Finkelberg, A. Kuznetsov, L. Rybnikov, *Towards a cluster structure on trigonometric zastava* (with appendix by Galyna Dobrovolska), Selecta Math. (N.S.) **24** (2018), no. 1, 187–225.
26. V. Fock, A. Marshakov, *A note on quantum groups and relativistic Toda theory*, Nuclear Phys. B Proc. Suppl. **56B** (1997), 208–214.
27. V. Futorny, A. Molev, S. Ovsienko, *Gelfand-Tsetlin bases for representations of finite W-algebras and shifted Yangians*, in "Lie theory and its applications in physics VII", (H. D. Doebner and V. K. Dobrev, Eds), Proceedings of the VII International Workshop, Varna, Bulgaria, June 2007. Heron Press, Sofia (2008), 352–363.

28. L. Faddeev, L. Takhtajan, *The quantum method for the inverse problem and the XYZ Heisenberg model*, Russian Math. Surveys **34** (1979), no. 5, 11–68.
29. M. Finkelberg, A. Tsymbaliuk, *Shifted quantum affine algebras: integral forms in type A* (with appendices by Alexander Tsymbaliuk and Alex Weekes), Arnold Mathematical Journal (2019), https://doi.org/10.1007/s40598-019-00118-7; arXiv:1811.12137.
30. A. Gerasimov, S. Kharchev, D. Lebedev, S. Oblezin, *On a class of representations of the Yangian and moduli space of monopoles*, Comm. Math. Phys. **260** (2005), no. 3, 511–525.
31. A. Gerasimov, S. Kharchev, D. Lebedev, S. Oblezin, *On a class of representations of quantum groups*, Noncommutative geometry and representation theory in mathematical physics, Contemp. Math. **391**, Amer. Math. Soc., Providence, RI (2005), 101–110.
32. L. Gow, A. Molev, *Representations of twisted q-Yangians*, Selecta Math. (N.S.) **16** (2010), no. 3, 439–499.
33. N. Guay, H. Nakajima, C. Wendlandt, *Coproduct for Yangians of affine Kac-Moody algebras*, Adv. in Math. **338** (2018), 865–911.
34. M. Gekhtman, M. Shapiro, A. Vainshtein, *Generalized Bäcklund-Darboux transformations for Coxeter–Toda flows from a cluster algebra perspective*, Acta Math. **206** (2011), no. 2, 245–310.
35. R. Gonin, A. Tsymbaliuk, On Sevostyanov's construction of quantum difference Toda lattices, IMRN (2019), https://doi.org/10.1093/imrn/rnz083; arXiv:1804.01063.
36. M. Haiman, *Cherednik algebras, Macdonald polynomials and combinatorics*, ICM 2006 Proceedings **3**, European Math. Soc. (2006), 843–872.
37. D. Hernandez, *Representations of quantum affinizations and fusion product*, Transform. Groups **10** (2005), no. 2, 163–200.
38. T. Hoffmann, J. Kellendonk, N. Kutz, N. Reshetikhin, *Factorization dynamics and Coxeter–Toda lattices*, Comm. Math. Phys. **212** (2000), no. 2, 297–321.
39. X. He, G. Lusztig, *A generalization of Steinberg's cross section*, J. Amer. Math. Soc. **25** (2012), no. 3, 739–757.
40. M. Jimbo, *A q-analogue of $U(\mathfrak{gl}(N+1))$, Hecke algebra, and the Yang-Baxter equation*, Lett. Math. Phys. **11** (1986), no. 3, 247–252.
41. N. Jing, *On Drinfeld realization of quantum affine algebras*, in Monster and Lie Algebras, Columbus, OH, 1996 (Ohio State University Mathematical Research Institute Publications, de Gruyter, Berlin, 1998), Vol. 7, 195–206.
42. A. Joseph, *Quantum groups and their primitive ideals*, Ergebnisse der Mathematik und ihrer Grenzgebiete (3) **29**, Springer-Verlag, Berlin (1995).
43. V. Kuznetsov, A. Tsyganov, *Quantum relativistic Toda chains*, Journal of Math. Sciences **80** (1996), no. 3, 1802–1810.
44. J. Kamnitzer, P. Tingley, B. Webster, A. Weekes, O. Yacobi, *Highest weights for truncated shifted Yangians and product monomial crystals*, preprint, arXiv:1511.09131.
45. J. Kamnitzer, B. Webster, A. Weekes, O. Yacobi, *Yangians and quantizations of slices in the affine Grassmannian*, Algebra Number Theory **8** (2014), no. 4, 857–893.
46. G. Laumon, *Un analogue global du cône nilpotent*, Duke Math. Journal **57** (1988), no. 2, 647–671.
47. S. Levendorskii, On generators and defining relations of Yangians, Journal of Geometry and Physics **12** (1993), no. 1, 1–11.
48. J.-H. Lu, *Hopf algebroids and quantum groupoids*, International J. Math. **7** (1996), no. 1, 47–70.
49. G. Lusztig, *Coxeter orbits and eigenspaces of Frobenius*, Invent. Math. **38** (1976), no. 2, 101–159.
50. G. Lusztig, *Introduction to quantum groups*, Progress in Math. **110** (1993).
51. A. Molev, *Yangians and classical Lie algebras*, Mathematical Surveys and Monographs, **143**, American Mathematical Society, Providence, RI (2007).
52. A. Malkin, V. Ostrik, M. Vybornov, *The minimal degeneration singularities in the affine Grassmannians*, Duke Math. J. **126** (2005), no. 2, 233–249.
53. A. Negut, *Quantum toroidal and shuffle algebras*, preprint, arXiv:1302.6202.

54. M. Nazarov, V. Tarasov, *Yangians and Gel'fand-Zetlin bases*, Publ. RIMS Kyoto Univ. **30** (1994), no. 3, 459–478.
55. G. Pappas, M. Rapoport, *Twisted loop groups and their affine flag varieties*, Adv. Math. **219** (2008), no. 1, 118–198.
56. A. Sevostyanov, *Quantum deformation of Whittaker modules and the Toda lattice*, Duke Math. J. **105** (2000), no. 2, 211–238.
57. A. Sevostyanov, *Conjugacy classes in Weyl groups and q-W algebras*, Adv. Math. **228** (2011), no. 3, 1315–1376.
58. R. Steinberg, *Regular elements of semisimple algebraic groups*, Inst. Hautes Études Sci. Publ. Math. **25** (1965), 49–80.
59. M. Semenov-Tian-Shansky, *Poisson Lie groups, quantum duality principle, and the quantum double*, Contemp. Math. **175** (1994), 219–248.
60. M. Semenov-Tian-Shansky, A. Sevostyanov, *Drinfeld-Sokolov reduction for difference operators and deformations of* $\mathcal{W}$*-algebras. II. The general semisimple case*, Comm. Math. Phys. **192** (1998), no. 3, 631–647.
61. A. Tsymbaliuk, *Quantum affine Gelfand–Tsetlin bases and quantum toroidal algebra via* K*-theory of affine Laumon spaces*, Selecta Math. (N.S.) **16** (2010), no. 2, 173–200; for a corrected version refer to arXiv:0903.0917v3.
62. A. Tsymbaliuk, *The affine Yangian of* $\mathfrak{gl}_1$ *revisited*, Adv. Math. **304** (2017), 583–645.
63. A. Tsymbaliuk, *PBWD bases and shuffle algebra realizations for* $U_\mathbf{v}(L\mathfrak{sl}_n)$, $U_{\mathbf{v}_1,\mathbf{v}_2}(L\mathfrak{sl}_n)$, $U_\mathbf{v}(L\mathfrak{sl}(m|n))$ and their integral forms, preprint, arXiv:1808.09536.
64. M. Varagnolo, E. Vasserot, *Double affine Hecke algebras and affine flag manifolds, I.* Affine flag manifolds and principal bundles, Trends in Mathematics, Birkhäuser/Springer, Basel (2010), 233–289.
65. M. Varagnolo, E. Vasserot, *Double affine Hecke algebras at roots of unity*, Representation Theory **14** (2010), 510–600.

Some Properties of Orbital Varieties in Extremal Nilpotent Orbits

Lucas Fresse

Dedicated to Professor Anthony Joseph on the occasion of his 75th birthday

Abstract The intersection between a nilpotent orbit of a simple Lie algebra and a Borel subalgebra is always equidimensional. Its irreducible components are called orbital varieties. Orbital varieties belonging to different nilpotent orbits may have quite different behaviours. The orbital varieties of the subregular nilpotent orbit are always smooth but they have in general infinitely many B-orbits. At the opposite, the minimal nilpotent orbit is spherical but its orbital varieties may have singularities. In this paper, we characterize the orbital varieties of the subregular nilpotent orbit which have a finite number of B-orbits and we give a smoothness criterion for the orbital varieties of the minimal nilpotent orbit.

MSC: 17B08 (primary); 17B22, 17B45 14M27 (secondary)

1 Introduction

1.1 Nilpotent Orbits

Let G be a connected simple algebraic group over $\mathbb{K}$ (an algebraically closed field of characteristic zero). By $\mathfrak{g}$ we denote the Lie algebra of G, by $(g, x) \mapsto g \cdot x$ we denote the adjoint action. Let $B \subset G$ be a Borel subgroup and let $\mathfrak{n} \subset \mathfrak{g}$ be the nilpotent radical of the Lie algebra of B.

L. Fresse (✉)
Institut Élie Cartan de Lorraine, Université de Lorraine, Vandoeuvre-lès-Nancy, France
e-mail: lucas.fresse@weizmann.ac.il; lucas.fresse@univ-lorraine.fr

M. Gorelik et al. (eds.), *Representations and Nilpotent Orbits of Lie Algebraic Systems*, Progress in Mathematics 330,
https://doi.org/10.1007/978-3-030-23531-4_7

An adjoint orbit $\mathcal{O} = G\cdot x := \{g\cdot x : g \in G\}$ is called nilpotent if the intersection $\mathcal{O} \cap \mathfrak{n}$ is nonempty. The set $\mathcal{N} := G \cdot \mathfrak{n}$ is the nilpotent cone. It consists of a finite number of nilpotent orbits. We emphasize four of them:

- The regular nilpotent elements form a single orbit $\mathcal{O}_{\mathrm{reg}}$, called the regular nilpotent orbit, which is dense in $\mathcal{N}$.
- There is a single orbit $\mathcal{O}_{\mathrm{subreg}}$, called the subregular nilpotent orbit, which is dense in $\mathcal{N} \setminus \mathcal{O}_{\mathrm{reg}}$.
- There is a single nontrivial nilpotent orbit $\mathcal{O}_{\mathrm{min}}$ of minimal dimension, called the minimal nilpotent orbit; it lies in the closure of every nontrivial nilpotent orbit.
- The only closed nilpotent orbit is the trivial orbit $\mathcal{O}_{\mathrm{triv}} = \{0\}$.

1.2 Orbital Varieties

Every nilpotent orbit $\mathcal{O} \subset \mathcal{N}$ has a structure of symplectic variety, in particular its dimension $\dim \mathcal{O}$ is even. The intersection $\mathcal{O} \cap \mathfrak{n}$ is a quasi-affine variety, which is in fact equidimensional of dimension $\frac{1}{2} \dim \mathcal{O}$ (see [10]). The irreducible components of $\mathcal{O} \cap \mathfrak{n}$ are called orbital varieties. They are B-stable, Lagrangian subvarieties of $\mathcal{O}$. Orbital varieties arise in geometric representation theory, in relation with associated varieties of simple highest weight modules. We refer to the works of A. Joseph [8, 9] and the references therein. In [9], the orbital varieties of the minimal nilpotent orbit are studied with respect to their quantization properties.

Orbital varieties may be singular and may have an infinite number of B-orbits, and they have a complicated intersection pattern. There is no general classification of orbital varieties with respect to their geometrical or topological properties. We refer to [7] for some partial classifications, mainly in type A.

In this paper we study some properties of orbital varieties for an arbitrary simple algebraic group G, but in the case of the particular nilpotent orbits mentioned above.

There is not much to say about the trivial nilpotent orbit $\mathcal{O}_{\mathrm{triv}} = \{0\}$ and its sole orbital variety $\mathcal{O}_{\mathrm{triv}} \cap \mathfrak{n} = \{0\}$. In the case of the regular nilpotent orbit, the intersection $\mathcal{O}_{\mathrm{reg}} \cap \mathfrak{n}$ is a single B-orbit, hence a single B-homogeneous (and therefore smooth) orbital variety. For the remaining two extremal nilpotent orbits $\mathcal{O}_{\mathrm{min}}$ and $\mathcal{O}_{\mathrm{subreg}}$, the situation is not so straightforward. We stress the following facts:

- The minimal nilpotent orbit $\mathcal{O}_{\mathrm{min}}$ is spherical, hence every orbital variety of $\mathcal{O}_{\mathrm{min}}$ has a finite number of B-orbits. Moreover $\mathcal{O}_{\mathrm{min}} \cap \mathfrak{n}$ contains a unique closed B-orbit, which therefore lies in every orbital variety. However the orbital varieties of $\mathcal{O}_{\mathrm{min}}$ may be singular.

  In this paper, we characterize the singular orbital varieties of $\mathcal{O}_{\mathrm{min}}$.
- At the opposite, in the subregular nilpotent orbit $\mathcal{O}_{\mathrm{subreg}}$, every orbital variety is smooth; in fact, it is open in the nilradical of some minimal parabolic subalgebra. However an orbital variety of $\mathcal{O}_{\mathrm{subreg}}$ does not always contain a dense B-orbit, and two orbital varieties of $\mathcal{O}_{\mathrm{subreg}}$ rarely intersect.

In this paper, we characterize the orbital varieties of $\mathcal{O}_{\text{subreg}}$ which have a dense B-orbit (resp. a finite number of B-orbits), and we characterize the pairs of orbital varieties of $\mathcal{O}_{\text{subreg}}$ which intersect.

In particular, the results shown in this paper illustrate how orbital varieties belonging to different nilpotent orbits may have different properties. Our main results are stated in terms of roots, simple roots and biggest root: see Sect. 2.2. The orbital varieties of $\mathcal{O}_{\text{subreg}}$ can indeed be parameterized by the simple roots whereas the orbital varieties of $\mathcal{O}_{\text{min}}$ can be parameterized by the simple long roots. These parameterizations are explained in Sect. 2.1.

2 Main Results

2.1 Parameterization of Orbital Varieties

Hereafter we fix a maximal torus $T \subset B$ and let $\mathfrak{h} \subset \mathfrak{g}$ denote the corresponding Cartan subalgebra. We then consider the root system $\Phi = \Phi(\mathfrak{g}, \mathfrak{h})$, the root space decomposition

$$\mathfrak{g} = \mathfrak{h} \oplus \bigoplus_{\alpha \in \Phi} \mathfrak{g}_\alpha,$$

and the subset of positive roots Φ^+ corresponding to the choice of B and $\mathfrak{n}$, i.e., such that $\mathfrak{n} = \bigoplus_{\alpha \in \Phi^+} \mathfrak{g}_\alpha$. Let $\Pi \subset \Phi^+$ be the set of simple roots.

Let $W = W(G, T)$ be the Weyl group. By [12] there is a surjective map from W onto the set of orbital varieties of $\mathcal{N}$: for every $w \in W$, there is a unique nilpotent orbit $\mathcal{O}_w$ which intersects the linear space $\mathfrak{n} \cap (w \cdot \mathfrak{n})$ densely; then, the set $\mathcal{V}_w := \overline{B \cdot (\mathfrak{n} \cap (w \cdot \mathfrak{n}))} \cap \mathcal{O}_w$ is an orbital variety, and every orbital variety is obtained in this way. In particular $\mathcal{O}_e = \mathcal{O}_{\text{reg}}$, $\mathcal{V}_e = \mathcal{O}_{\text{reg}} \cap \mathfrak{n}$, $\mathcal{V}_{w_0} = \mathcal{O}_{w_0} = \{0\}$, where $e, w_0 \in W$, respectively, stand for the neutral element and the longest element.

The orbital varieties contained in the nilpotent orbits $\mathcal{O}_{\text{min}}$ and $\mathcal{O}_{\text{subreg}}$ have an alternative, handy parameterization, obtained as follows.

Every simple root $\alpha \in \Pi$ gives rise to a minimal parabolic subgroup P_α and a nilradical $\mathfrak{n}_\alpha = \bigoplus_{\gamma \in \Phi^+ \setminus \{\alpha\}} \mathfrak{g}_\gamma$. By [5, §4.1], we have

$$\mathfrak{n} \setminus \mathcal{O}_{\text{reg}} = \bigcup_{\alpha \in \Pi} \mathfrak{n}_\alpha, \quad \text{hence} \quad \mathcal{O}_{\text{subreg}} \cap \mathfrak{n} = \bigcup_{\alpha \in \Pi} \mathcal{O}_{\text{subreg}} \cap \mathfrak{n}_\alpha.$$

By [5, Theorem 7.1.1], for every $\alpha \in \Pi$, the intersection $\mathcal{O}_{\text{subreg}} \cap \mathfrak{n}_\alpha$ is P_α-homogeneous (thus irreducible, smooth) and dense in $\mathfrak{n}_\alpha$. This yields:

Proposition 1

(a) *The subsets* $\mathscr{V}_{\mathrm{subreg}}(\alpha) := \mathscr{O}_{\mathrm{subreg}} \cap \mathfrak{n}_\alpha$, *for* $\alpha \in \Pi$, *are exactly the irreducible components of* $\mathscr{O}_{\mathrm{subreg}} \cap \mathfrak{n}$, *i.e., the orbital varieties of* $\mathscr{O}_{\mathrm{subreg}}$.
(b) $\mathscr{V}_{\mathrm{subreg}}(\alpha)$ *is* P_α*-homogeneous (thus smooth) and its closure is the linear space* $\mathfrak{n}_\alpha$ *(thus this closure is also smooth).*
(c) *For* $\alpha, \alpha' \in \Pi$, *the orbital varieties* $\mathscr{V}_{\mathrm{subreg}}(\alpha)$ *and* $\mathscr{V}_{\mathrm{subreg}}(\alpha')$ *intersect if and only if the roots* α, α' *are not orthogonal.*
(d) $\mathscr{V}_{\mathrm{subreg}}(\alpha) = \mathscr{V}_{s_\alpha}$, *where* $s_\alpha \in W$ *is the simple reflection attached to the root* α.

Proof Parts (a) and (b) follow from the previous discussion. Part (d) follows from the definitions of $\mathscr{V}_{\mathrm{subreg}}(\alpha)$ and $\mathscr{V}_{s_\alpha}$. Let us show part (c). If the simple roots α and α' are orthogonal, then the intersection $\mathfrak{n}_\alpha \cap \mathfrak{n}_{\alpha'}$ is the nilradical of a parabolic subalgebra. By [5, Theorem 7.1.1], the maximal dimension of a nilpotent orbit intersecting $\mathfrak{n}_\alpha \cap \mathfrak{n}_{\alpha'}$ is $2 \dim \mathfrak{n}_\alpha \cap \mathfrak{n}_{\alpha'} < 2 \dim \mathfrak{n}_\alpha = \dim \mathscr{O}_{\mathrm{subreg}}$. Hence $\mathscr{V}_{\mathrm{subreg}}(\alpha) \cap \mathscr{V}_{\mathrm{subreg}}(\alpha') = \mathscr{O}_{\mathrm{subreg}} \cap \mathfrak{n}_\alpha \cap \mathfrak{n}_{\alpha'} = \emptyset$. We have shown that two orbital varieties of $\mathscr{O}_{\mathrm{subreg}}$ have an empty intersection if they correspond to simple roots which are orthogonal, i.e., which are not connected by an edge in the Dynkin diagram. By Spaltenstein [10], the variety $\mathscr{O}_{\mathrm{subreg}} \cap \mathfrak{n}$ is connected. This implies that $\mathscr{V}_{\mathrm{subreg}}(\alpha)$ and $\mathscr{V}_{\mathrm{subreg}}(\alpha')$ must intersect if there is an edge between α and α' in the Dynkin diagram. □

For every root $\alpha \in \Phi^+$ we fix a root vector $e_\alpha \in \mathfrak{g}_\alpha \setminus \{0\}$. In the simply laced cases, we say that all the roots are long. In general, let Φ_ℓ (resp., Φ_ℓ^+) stand for the set of long roots (resp., positive long roots) and let $\Pi_\ell \subset \Pi$ be the subset of simple long roots. Let $\preceq$ be the usual partial order on the root system Φ determined by the choice of the set of positive roots Φ^+. Let $\beta_{\max} \in \Phi^+$ be the biggest root, i.e., the biggest element of Φ with respect to the order $\preceq$. It is always a long root, and the root vector $e_{\beta_{\max}}$ is a representative of the minimal nilpotent orbit $\mathscr{O}_{\min}$. Note that

$$B \cdot e_{\beta_{\max}} = \mathfrak{g}_{\beta_{\max}} \setminus \{0\}, \quad \text{hence} \quad \mathscr{O}_{\min} = G \cdot e_{\beta_{\max}} = \bigcup_{w \in W} B \cdot e_{w(\beta_{\max})},$$

where the last equality follows from the Bruhat decomposition $G = \bigsqcup_{w \in W} BwB$, whence

$$\mathscr{O}_{\min} = \bigcup_{\alpha \in \Phi_\ell} B \cdot e_\alpha \quad \text{and} \quad \mathscr{O}_{\min} \cap \mathfrak{n} = \bigcup_{\alpha \in \Phi_\ell^+} B \cdot e_\alpha \tag{1}$$

since the Weyl group W acts transitively on the set of long roots. The next statement follows from [4, 8] and [3, §6.1].

Proposition 2

(a) *For every* $\alpha \in \Phi_\ell^+$, *we have*

$$\overline{B \cdot e_\alpha} \cap \mathscr{O}_{\min} = \bigcup_{\gamma \succeq \alpha} B \cdot e_\gamma$$

where the union is taken over all long roots $\gamma \in \Phi_\ell^+$ *satisfying* $\gamma \succeq \alpha$.

(b) *Thus, the subsets* $\mathcal{V}_{\min}(\alpha) := \overline{B \cdot e_\alpha} \cap \mathcal{O}_{\min}$, *for* $\alpha \in \Pi_\ell$, *are exactly the irreducible components of* $\mathcal{O}_{\min} \cap \mathfrak{n}$, *i.e., the orbital varieties of* $\mathcal{O}_{\min}$. *Every orbital variety contains in particular the orbit* $B \cdot e_{\beta_{\max}}$.
(c) *Every orbital variety* $\mathcal{V}_{\min}(\alpha)$ *(for* $\alpha \in \Pi_\ell$*) is normal, Cohen–Macaulay, and has rational singularities.*
(d) $\mathcal{V}_{\min}(\alpha) = \mathcal{V}_{s_\alpha w_0}$.

2.2 Statement of Main Results

As in Sect. 2.1, $\beta_{\max}$ stands for the biggest root. It decomposes as a sum of simple roots

$$\beta_{\max} = \sum_{\alpha \in \Pi} n(\alpha)\alpha,$$

where the coefficients $n(\alpha)$ are positive integers.

Our first main result is a smoothness criterion for the orbital varieties of the minimal nilpotent orbit. The proof is given in Sect. 3.3.

Theorem 1 *Let* $\alpha \in \Pi_\ell$ *be a simple long root and let* $\mathcal{V}_{\min}(\alpha) = \overline{B \cdot e_\alpha} \cap \mathcal{O}_{\min}$ *be the corresponding orbital variety of the minimal nilpotent orbit* $\mathcal{O}_{\min}$. *Then:*

$$\mathcal{V}_{\min}(\alpha) \text{ is smooth if and only if } n(\alpha) = 1.$$

Our second main result is a criterion of finiteness of number of B-orbits / existence of dense B-orbit for the orbital varieties of the subregular nilpotent orbit. In the result below we say that a simple root α is extremal if it belongs to only one (possibly multiple) edge of the Dynkin diagram (i.e., there is only one simple root which is not orthogonal to α). In types E_6, E_7, E_8, we consider the numbering of the simple roots determined by the following diagram (of type E_8)

$$\begin{array}{ccccccccccccc} & & & & \alpha_2 & & & & & & & & \\ & & & & | & & & & & & & & \\ \alpha_1 & — & \alpha_3 & — & \alpha_4 & — & \alpha_5 & — & \alpha_6 & — & \alpha_7 & — & \alpha_8 \end{array}$$

and its subdiagrams $\{\alpha_1, \dots, \alpha_7\}$ (of type E_7) and $\{\alpha_1, \dots, \alpha_6\}$ (of type E_6).

Theorem 2 *Let* $\alpha \in \Pi$ *be a simple root and let* $\mathcal{V}_{\text{subreg}}(\alpha) = \mathcal{O}_{\text{subreg}} \cap \mathfrak{n}_\alpha$ *be the corresponding orbital variety of the subregular nilpotent orbit* $\mathcal{O}_{\text{subreg}}$. *The following conditions are equivalent:*

(i) $\mathcal{V}_{\text{subreg}}(\alpha)$ *has a finite number of* B*-orbits;*
(ii) $\mathcal{V}_{\text{subreg}}(\alpha)$ *has a dense* B*-orbit;*
(iii) *One of the following conditions is satisfied:*

(1) *the group G is of type A or B;*
(2) *the group G is of type C or D, and α is an extremal root of the Dynkin diagram;*
(3) *the group G is of type G_2 or F_4, and α is long and extremal;*
(4) *the group G is of type E_6 (resp., E_7) and $\alpha \in \{\alpha_1, \alpha_6\}$ (resp., $\alpha = \alpha_7$).*

The proof is given in Sect. 4.

Corollary 1

(a) *In types A and C, every orbital variety of $\mathcal{O}_{\min}$ is smooth.*
(b) *In types G_2, F_4, and E_8, every orbital variety of $\mathcal{O}_{\min}$ is singular.*
(c) *In types A and B, every orbital variety of $\mathcal{O}_{\mathrm{subreg}}$ has a finite number of B-orbits.*
(d) *In type E_8, every orbital variety of $\mathcal{O}_{\mathrm{subreg}}$ has an infinite number of B-orbits.*

Remark 1 Let $\Phi^\vee = \{\alpha^\vee : \alpha \in \Phi\}$ be the dual root system of Φ and let $(\beta_{\max})^\vee = \sum_{\alpha\in\Pi} n^\vee(\alpha)\alpha^\vee$, where $n^\vee(\alpha)$ are positive integers, be the decomposition of the coroot associated to $\beta_{\max}$ in the basis $\Pi^\vee = \{\alpha^\vee : \alpha \in \Pi\}$. Then condition (iii) of Theorem 2 is equivalent to having $n^\vee(\alpha) = 1$. In this way, the conditions involved in Theorems 1 and 2 appear to be related. Similar facts (in type A but for orbital varieties of arbitrary nilpotent orbits), which suggest a relation between smooth orbital varieties and orbital varieties admitting a dense B-orbit, are also pointed out in [7].

3 Proof of Theorem 1

3.1 Notation

Recall that the root system Φ is endowed with the partial order $\preceq$ defined by letting $\alpha \preceq \beta$ if $\beta - \alpha$ is a sum of simple roots, and $\beta_{\max}$ stands for the biggest element of Φ relatively to this order.

Given two positive roots α, β, we write:

- $\alpha \lessdot \beta$ if $\beta - \alpha \in \Phi^+$;
- $\alpha \lll \beta$ if $\beta = \alpha + k\eta$ for some $\eta \in \Phi^+$ and some positive integer k.

Moreover we consider the set

$$M_{\max} := \{\alpha \in \Phi^+ : \alpha \lessdot \beta_{\max}\}.$$

The following technical lemmas can be checked case by case. The first lemma is immediate in the simply laced case (where by convention we say that all the roots are long, i.e., there is no short root).

Lemma 1 *Let γ be a short positive root such that the set*

$$\{\beta \in \Phi^+ \text{ long} : \beta \preceq \gamma\}$$

is nonempty. Then, this set contains a biggest element β_0 relatively to the order $\preceq$, and we have $\beta_0 \lessdot\gamma$.

Moreover, let us suppose that $\gamma \in M_{\max}$, so that $\beta_{\max} - \gamma \in \Phi^+$. Then, the following alternative holds:

(i) *Either $\gamma - \beta_0 = \beta_{\max} - \gamma$,*
(ii) *Or, for all root β such that $\beta_0 \preceq \beta \prec \gamma$, we have $\beta + \beta_{\max} - \gamma \notin \Phi$.*

Lemma 2 *Let α be a simple long root such that $n(\alpha) \geq 2$. Then there is a couple (γ, γ') of positive roots such that $\alpha \preceq \gamma$, $\alpha \preceq \gamma'$, and $\gamma + \gamma' = \beta_{\max}$.*

For every root α, we fix a morphism of algebraic groups $u_\alpha : \mathbb{K} \to G$ such that $hu_\alpha(s)h^{-1} = u_\alpha(\alpha(h)s)$ for all $h \in T$ and $\operatorname{Im} du_\alpha = \mathfrak{g}_\alpha$ (see [11, Lemma 7.3.3]). Note that there is a nonzero root vector $x_\alpha \in \mathfrak{g}_\alpha \setminus \{0\}$ such that

$$\operatorname{Ad} u_\alpha(s) = \exp(s \operatorname{ad} x_\alpha) \quad \text{for all } s \in \mathbb{K} \tag{1}$$

where $\operatorname{Ad} : G \to \mathrm{GL}(\mathfrak{g})$ and $\operatorname{ad} : \mathfrak{g} \to \mathfrak{gl}(\mathfrak{g})$ stand for the adjoint representations.

3.2 *Tangent Space of the Minimal Nilpotent Orbit at the Biggest Root Vector*

Recall that for each positive root α we consider a root vector $e_\alpha \in \mathfrak{g}_\alpha \setminus \{0\}$. The biggest root vector $e_{\beta_{\max}}$ is a representative of the minimal nilpotent orbit $\mathcal{O}_{\min}$. By $T_{e_{\beta_{\max}}}\mathcal{O}_{\min}$ we denote the tangent space of $\mathcal{O}_{\min}$ at $e_{\beta_{\max}}$.

Proposition 3 $T_{e_{\beta_{\max}}}\mathcal{O}_{\min} = \mathfrak{g}_{\beta_{\max}} \oplus [\mathfrak{g}_{\beta_{\max}}, \mathfrak{g}_{-\beta_{\max}}] \oplus \bigoplus_{\gamma \in M_{\max}} \mathfrak{g}_\gamma$.

Proof We first claim that

$$\dim T_{e_{\beta_{\max}}}\mathcal{O}_{\min} = \dim \mathcal{O}_{\min} = |M_{\max}| + 2. \tag{2}$$

The first equality in (2) follows from the fact that $\mathcal{O}_{\min}$ is smooth. For showing the second equality, we compute the stabilizer $\mathfrak{z}_{\mathfrak{g}}(e_{\beta_{\max}}) := \{v \in \mathfrak{g} : [v, e_{\beta_{\max}}] = 0\}$. Let $v \in \mathfrak{g}$ and let us write $v = h + \sum_{\alpha\in\Phi} v_\alpha$ where $h \in \mathfrak{h}$ and $v_\alpha \in \mathfrak{g}_\alpha$ for all $\alpha \in \Phi$. We get

$$[v, e_{\beta_{\max}}] = \beta_{\max}(h)e_{\beta_{\max}} + [v_{-\beta_{\max}}, e_{\beta_{\max}}] + \sum_{\alpha\in\Phi\setminus\{-\beta_{\max}\}} [v_\alpha, e_{\beta_{\max}}],$$

and the equality $[v, e_{\beta_{\max}}] = 0$ holds if and only if $h \in \ker \beta_{\max}$, $v_{-\beta_{\max}} = 0$, and $v_\alpha = 0$ whenever $\alpha + \beta_{\max}$ is a root. The last fact is equivalent to having $-\alpha \in M_{\max}$. Altogether, this yields

$$\dim \mathfrak{z}_{\mathfrak{g}}(e_{\beta_{\max}}) = \dim \mathfrak{g} - 2 - |M_{\max}|.$$

Since $\dim \mathscr{O}_{\min} = \dim \mathfrak{g} - \dim \mathfrak{z}_{\mathfrak{g}}(e_{\beta_{\max}})$, the verification of (2) is complete.

In view of (2), for showing the proposition, it suffices to show the inclusion

$$\mathfrak{g}_{\beta_{\max}} \oplus [\mathfrak{g}_{\beta_{\max}}, \mathfrak{g}_{-\beta_{\max}}] \oplus \bigoplus_{\gamma \in M_{\max}} \mathfrak{g}_\gamma \subset T_{e_{\beta_{\max}}} \mathscr{O}_{\min}. \tag{3}$$

There is a cocharacter $\lambda : \mathbb{K}^* \to T$ such that $\lambda(\mathbb{K}^*) \cdot e_{\beta_{\max}} = \mathbb{K}^* e_{\beta_{\max}}$. Hence $\mathbb{K}^* e_{\beta_{\max}} \subset \mathscr{O}_{\min}$, which yields the inclusion $\mathfrak{g}_{\beta_{\max}} = T_{e_{\beta_{\max}}}(\mathbb{K}^* e_{\beta_{\max}}) \subset T_{e_{\beta_{\max}}} \mathscr{O}_{\min}$. By (1), the inclusion

$$[\mathfrak{g}_{-\beta_{\max}}, \mathfrak{g}_{\beta_{\max}}] = \mathbb{K}[x_{-\beta_{\max}}, e_{\beta_{\max}}] = T_{e_{\beta_{\max}}}\big(u_{-\beta_{\max}}(\mathbb{K}) \cdot e_{\beta_{\max}}\big) \subset T_{e_{\beta_{\max}}} \mathscr{O}_{\min}$$

holds. Similarly, for every $\gamma \in M_{\max}$, letting $\gamma' := \beta_{\max} - \gamma$ (which is a positive root), the inclusion

$$\mathfrak{g}_\gamma = \mathbb{K}[x_{-\gamma'}, e_{\beta_{\max}}] = T_{e_{\beta_{\max}}}\big(u_{-\gamma'}(\mathbb{K}) \cdot e_{\beta_{\max}}\big) \subset T_{e_{\beta_{\max}}} \mathscr{O}_{\min}$$

holds. Altogether we get (3). The proof is complete. □

Remark 2

(a) More insight on the dimension formula for $\mathscr{O}_{\min}$ (see (2)) can be found in [14].
(b) An alternative proof of Proposition 3, which relies on $\mathfrak{sl}_2$-theory, can be obtained as follows. The subalgebra $\mathfrak{s} := \mathfrak{g}_{-\beta_{\max}} \oplus [\mathfrak{g}_{-\beta_{\max}}, \mathfrak{g}_{\beta_{\max}}] \oplus \mathfrak{g}_{\beta_{\max}}$ is isomorphic to $\mathfrak{sl}_2(\mathbb{K})$. The decomposition of $\mathfrak{g}$ into simple $\mathfrak{s}$-modules comprises 1-dimensional representations, 2-dimensional representations of the form $\mathfrak{g}_{-\gamma'} \oplus \mathfrak{g}_\gamma$ corresponding to the couples of positive roots (γ, γ') such that $\gamma + \gamma' = \beta_{\max}$ (i.e., $\gamma \in M_{\max}$), and a single 3-dimensional representation, namely $\mathfrak{s}$. The tangent space $T_{e_{\beta_{\max}}} \mathscr{O}_{\min}$, which coincides with $[\mathfrak{g}, e_{\beta_{\max}}]$, is then spanned by weight vectors which are not low-weight vector in any simple $\mathfrak{s}$-submodule of $\mathfrak{g}$, whence the formula stated in Proposition 3.

3.3 *Tangent Space of Orbital Varieties of the Minimal Nilpotent Orbit at the Biggest Root Vector*

Theorem 1 is implied by Proposition 4 (b) below. We need two preparatory lemmas.

Lemma 3 *For every $x \in \mathscr{O}_{\min}$, we have $\mathbb{K}^* x \subset B \cdot x$.*

Proof By (1), there is a long root vector e_β and an element $b \in B$ such that $x = b \cdot e_\beta$. For every $s \in \mathbb{K}^*$, we can find $h \in T$ such that $h \cdot e_\beta = se_\beta$, whence $sx = (bh) \cdot e_\beta \in B \cdot e_\beta = B \cdot x$. □

Recall that for every positive long root α, the root vector e_α belongs to the minimal nilpotent orbit $\mathcal{O}_{\min}$ and the biggest long root vector $e_{\beta_{\max}}$ belongs to the closure of $B \cdot e_\alpha$ (see Proposition 2 (a)).

Lemma 4 *Let $\gamma \in M_{\max}$.*

(a) *Assume that the root γ is long. Then $\mathfrak{g}_\gamma \subset T_{e_{\beta_{\max}}} \overline{B \cdot e_\gamma}$.*
(b) *Assume that γ is short and such that the set $\{\beta \in \Phi^+ \text{ long} : \beta \preceq \gamma\}$ is nonempty, hence contains a biggest element β_0 (see Lemma 1). Then $\mathfrak{g}_\gamma \subset T_{e_{\beta_{\max}}} \overline{B \cdot e_{\beta_0}}$.*

Proof Let $\gamma' := \beta_{\max} - \gamma$, which is a positive root. First we show part (a) of the lemma. In view of (1), there is an element $x_{\gamma'} \in \mathfrak{g}_{\gamma'} \setminus \{0\}$ such that

$$e_\gamma + s^{-1}[x_{\gamma'}, e_\gamma] = u_{\gamma'}(s^{-1}) \cdot e_\gamma \in B \cdot e_\gamma$$

for all $s \in \mathbb{K}^*$. Note that $[x_{\gamma'}, e_\gamma] \in \mathfrak{g}_{\beta_{\max}} \setminus \{0\} = \mathbb{K}^* e_{\beta_{\max}}$. By Lemma 3, we get $e_{\beta_{\max}} + se_\gamma \in B \cdot e_\gamma$ for all $s \in \mathbb{K}^*$. Whence the inclusion $\mathfrak{g}_\gamma \subset T_{e_{\beta_{\max}}} \overline{B \cdot e_\gamma}$.

Next let us show part (b) of the lemma. First assume that condition (i) of Lemma 1 holds, so that $\gamma = \beta_0 + \gamma'$ and $\beta_{\max} = \beta_0 + 2\gamma'$. In view of (1), we have in this case

$$e_{\beta_0} + s^{-1}[x_{\gamma'}, e_{\beta_0}] + \frac{1}{2} s^{-2}[x_{\gamma'}, [x_{\gamma'}, e_{\beta_0}]] = u_{\gamma'}(s^{-1}) \cdot e_{\beta_0} \in B \cdot e_{\beta_0} \quad \text{for all } s \in \mathbb{K}^*$$

for some $x_{\gamma'} \in \mathfrak{g}_{\gamma'} \setminus \{0\}$. Note that $[x_{\gamma'}, e_{\beta_0}] \in \mathfrak{g}_\gamma \setminus \{0\}$ while $[x_{\gamma'}, [x_{\gamma'}, e_{\beta_0}]] \in \mathfrak{g}_{\beta_{\max}} \setminus \{0\}$, hence $[x_{\gamma'}, [x_{\gamma'}, e_{\beta_0}]] = s_0^{-1} e_{\beta_{\max}}$ for some $s_0 \in \mathbb{K}^*$. Invoking also Lemma 3, this yields

$$e_{\beta_{\max}} + 2ss_0[x_{\gamma'}, e_{\beta_0}] + 2s^2 s_0 e_{\beta_0} \in B \cdot e_{\beta_0} \quad \text{for all } s \in \mathbb{K}^*.$$

Whence the inclusion $\mathfrak{g}_\gamma = \mathbb{K}[x_{\gamma'}, e_{\beta_0}] \subset T_{e_{\beta_{\max}}} \overline{B \cdot e_{\beta_0}}$.

Finally assume that condition (ii) of Lemma 1 holds. By Lemma 1, there is a positive root η and a positive integer k such that $\gamma = \beta_0 + k\eta$. Let $r \geq k$ be the integer such that $\gamma_\ell := \beta_0 + \ell\eta$ is a root for all $\ell \in \{0, 1, \ldots, r\}$ and is not a root whenever $\ell > r$. By (1), there are root vectors $e'_{\gamma_\ell} \in \mathfrak{g}_{\gamma_\ell} \setminus \{0\}$ (for $\ell = 0, 1, \ldots, r$) such that

$$u_\eta(t) \cdot e_{\beta_0} = \sum_{\ell=0}^{r} t^\ell e'_{\gamma_\ell} \quad \text{for all } t \in \mathbb{K}.$$

By assumption, $\gamma_\ell + \gamma'$ is a root if and only if $\ell = k$, and $\gamma_k + \gamma' = \gamma + \gamma' = \beta_{\max}$. Applying again (1), we get

$$t^k s^{-1} s_0^{-1} e_{\beta_{\max}} + \sum_{\ell=0}^{r} t^\ell e'_{\gamma_\ell} = u_{\gamma'}(s^{-1}) u_\eta(t) \cdot e_{\beta_0} \text{ for all } t \in \mathbb{K}, \text{ all } s \in \mathbb{K}^*,$$

for some $s_0 \in \mathbb{K}^*$. Whence (by Lemma 3)

$$e_{\beta_{\max}} + s s_0 t^{-k} \sum_{\ell=0}^{r} t^\ell e'_{\gamma_\ell} \in B \cdot e_{\beta_0} \text{ for all } t \in \mathbb{K}, \text{ all } s \in \mathbb{K}^*.$$

We deduce that

$$\sum_{\ell=0}^{r} t^\ell e'_{\gamma_\ell} \in T_{e_{\beta_{\max}}} \overline{B \cdot e_{\beta_0}} \text{ for all } t \in \mathbb{K}$$

and therefore

$$\mathfrak{g}_\gamma \subset \bigoplus_{\ell=0}^{r} \mathfrak{g}_{\gamma_\ell} \subset T_{e_{\beta_{\max}}} \overline{B \cdot e_{\beta_0}}.$$

The proof of the lemma is complete. □

Proposition 4 *Let α be a simple long root, so that $\mathscr{V}_{\min}(\alpha) := \overline{B \cdot e_\alpha} \cap \mathscr{O}_{\min}$ is an orbital variety of $\mathscr{O}_{\min}$ and $B \cdot e_{\beta_{\max}}$ is the unique closed B-orbit of $\mathscr{V}_{\min}(\alpha)$ (see Proposition 2).*

(a) $T_{e_{\beta_{\max}}} \mathscr{V}_{\min}(\alpha) = \mathfrak{g}_{\beta_{\max}} \oplus \bigoplus_{\gamma \in M_{\max},\, \gamma \succeq \alpha} \mathfrak{g}_\gamma$.
(b) *The following conditions are equivalent:*

  (i) $\mathscr{V}_{\min}(\alpha)$ *is singular;*
  (ii) $|\{\gamma \in M_{\max} : \gamma \succeq \alpha\}| > \frac{1}{2}|M_{\max}|$;
  (iii) *there exists a pair (γ, γ') of positive roots such that $\gamma \succeq \alpha$, $\gamma' \succeq \alpha$, $\gamma + \gamma' = \beta_{\max}$;*
  (iv) $n(\alpha) \geq 2$.

Proof First we show the inclusion $\supset$ in (a). Since $\mathfrak{g}_{\beta_{\max}} \setminus \{0\} = B \cdot e_{\beta_{\max}} \subset \mathscr{V}_{\min}(\alpha)$, we have $\mathfrak{g}_{\beta_{\max}} \subset T_{e_{\beta_{\max}}} \mathscr{V}_{\min}(\alpha)$. Next let $\gamma \in M_{\max}$ such that $\gamma \succeq \alpha$. If γ is a long root, then we have $B \cdot e_\gamma \subset \mathscr{V}_{\min}(\alpha)$ by Proposition 2, hence $\mathfrak{g}_\gamma \subset T_{e_{\beta_{\max}}} \mathscr{V}_{\min}(\alpha)$ by Lemma 4 (a). Assume now that γ is a short root and let β_0 be as in Lemmas 1 and 4 (b). The maximality property of β_0 implies that $\beta_0 \succeq \alpha$, whence $B \cdot e_{\beta_0} \subset \mathscr{V}_{\min}(\alpha)$ (by Proposition 2). In view of Lemma 4 (b), this yields $\mathfrak{g}_\gamma \subset T_{e_{\beta_{\max}}} \mathscr{V}_{\min}(\alpha)$. Altogether we get the inclusion $\supset$ in (a).

In view of the inclusion $\mathscr{V}_{\min}(\alpha) \subset \overline{B \cdot e_\alpha} \subset \bigoplus_{\gamma \succeq \alpha} \mathfrak{g}_\gamma$ and of Proposition 3, we also have

$$T_{e_{\beta_{\max}}} \mathscr{V}_{\min}(\alpha) \subset \Big(\bigoplus_{\gamma \succeq \alpha} \mathfrak{g}_\gamma \Big) \cap T_{e_{\beta_{\max}}} \mathscr{O}_{\min} = \mathfrak{g}_{\beta_{\max}} \oplus \bigoplus_{\gamma \in M_{\max},\, \gamma \succeq \alpha} \mathfrak{g}_\gamma,$$

and this completes the proof of part (a).

On the one hand, part (a) and Proposition 3 yield

$$\dim T_{e_{\beta_{\max}}} \mathcal{V}_{\min}(\alpha) = 1 + |\{\gamma \in M_{\max} : \gamma \succeq \alpha\}|$$

and

$$\dim \mathcal{V}_{\min}(\alpha) = \frac{1}{2} \dim \mathcal{O}_{\min} = 1 + \frac{1}{2}|M_{\max}|$$

(recall from Sect. 1.2 that we have $\dim \mathcal{V} = \frac{1}{2} \dim \mathcal{O}$ whenever $\mathcal{V}$ is an orbital variety of a nilpotent orbit $\mathcal{O}$). On the other hand, since $B \cdot e_{\beta_{\max}}$ is the unique closed B-orbit in $\mathcal{V}_{\min}(\alpha)$, we know that $\mathcal{V}_{\min}(\alpha)$ is singular if and only if $\dim T_{e_{\beta_{\max}}} \mathcal{V}_{\min}(\alpha) > \dim \mathcal{V}_{\min}(\alpha)$. The equivalence between conditions (i) and (ii) of part (b) ensues.

Since α necessarily occurs in the decomposition of $\beta_{\max}$ as a sum of simple roots, for every $\gamma \in M_{\max}$ we must have $\gamma \succeq \alpha$ or $\beta_{\max} - \gamma \succeq \alpha$. Whence $|\{\gamma \in M_{\max} : \gamma \succeq \alpha\}| \geq \frac{1}{2}|M_{\max}|$ with strict inequality if and only if there is an element $\gamma \in M_{\max}$ such that $\gamma \succeq \alpha$ and $\beta_{\max} - \gamma \succeq \alpha$, which is equivalent to the existence of a couple (γ, γ') as in (iii). Conditions (ii) and (iii) of part (b) are therefore equivalent.

The implication (iii)⇒(iv) is immediate while the inverse implication (iv)⇒(iii) follows from Lemma 2. The proof of part (b) of the statement is now complete. □

Remark 3 For every simple root $\alpha \in \Pi$, we get a maximal parabolic subgroup $P_{\max,\alpha} = \bigsqcup_{w \in W_{\max,\alpha}} BwB$, where $W_{\max,\alpha} \subset W$ is the subgroup generated by the simple reflections s_β for $\beta \in \Pi \setminus \{\alpha\}$. Assume that α is a simple long root. It is seen from Proposition 2 (a)–(b) that the orbital variety $\mathcal{V}_{\min}(\alpha)$ is $P_{\max,\alpha}$-stable, because the group $W_{\max,\alpha}$ acts on the set of long positive roots β such that $\beta \succeq \alpha$. Moreover, in the case where $n(\alpha) = 1$, the latter action is transitive. Therefore, in that case, $\mathcal{V}_{\min}(\alpha)$ is $P_{\max,\alpha}$-homogeneous. This yields another proof of the smoothness of the orbital variety $\mathcal{V}_{\min}(\alpha)$ in the case $n(\alpha) = 1$.

4 Proof of Theorem 2

As in Sect. 2.1, for every simple root α, we denote by P_α the corresponding standard minimal parabolic subgroup and by $\mathfrak{n}_\alpha$ the nilpotent radical of its Lie algebra.

As noted in Proposition 1 (b), the orbital variety $\mathcal{V}_{\text{subreg}}(\alpha) := \mathcal{O}_{\text{subreg}} \cap \mathfrak{n}_\alpha$ of $\mathcal{O}_{\text{subreg}}$ attached to α is P_α-homogeneous. In view of well-known properties of spherical varieties [2, 13] (see also [6, Lemma 1]), this fact already guarantees the equivalence between parts (i) and (ii) of Theorem 2. The purpose of this section is to prove the equivalence between parts (ii) and (iii) of Theorem 2.

4.1 Criteria of Existence of Dense B-Orbit

For a simple root α, we also denote by $\mathfrak{p}_\alpha$ the Lie algebra of the minimal parabolic subgroup P_α and by $\mathrm{Rad}(\mathfrak{p}_\alpha)$ its radical, i.e., the intersection of the Borel subalgebras of $\mathfrak{p}_\alpha$; in other words,

$$\mathrm{Rad}(\mathfrak{p}_\alpha) = \{h \in \mathfrak{h} : \alpha(h) = 0\} \oplus \mathfrak{n}_\alpha.$$

Proposition 5 *Let α be a simple root. Let $x \in \mathcal{V}_{\mathrm{subreg}}(\alpha) = \mathcal{O}_{\mathrm{subreg}} \cap \mathfrak{n}_\alpha$. The following conditions are equivalent:*

(i) *$\mathcal{V}_{\mathrm{subreg}}(\alpha)$ has a dense B-orbit;*
(ii) $\{y \in \mathfrak{p}_\alpha : [y, x] = 0\} \not\subset \mathrm{Rad}(\mathfrak{p}_\alpha)$.

Proof The orbital variety $\mathcal{V}_{\mathrm{subreg}}(\alpha)$ (which coincides with the P_α-orbit of x) has a dense B-orbit if and only if there exists an element $g \in P_\alpha$ such that $\dim B\cdot(g\cdot x) = \dim \mathcal{V}_{\mathrm{subreg}}(\alpha)$. By $\mathfrak{b}$ we denote the Lie algebra of the Borel subgroup B. Note that

$$\begin{aligned}\dim B \cdot (g \cdot x) &= \dim B - \dim\{y \in \mathfrak{b} : [y, g \cdot x] = 0\} \\ &= \dim B - \dim\{y \in g^{-1} \cdot \mathfrak{b} : [y, x] = 0\}\end{aligned}$$

and

$$\dim \mathcal{V}_{\mathrm{subreg}}(\alpha) = \dim P_\alpha \cdot x = \dim P_\alpha - \dim\{y \in \mathfrak{p}_\alpha : [y, x] = 0\}$$

hence

$$\dim \mathcal{V}_{\mathrm{subreg}}(\alpha) - \dim B\cdot(g\cdot x) = 1 - \dim\{y \in \mathfrak{p}_\alpha : [y, x] = 0\}/\{y \in g^{-1}\cdot\mathfrak{b} : [y, x] = 0\}.$$

Therefore the existence of a dense B-orbit in $\mathcal{V}_{\mathrm{subreg}}(\alpha)$ is equivalent to the existence of a Borel subalgebra $\mathfrak{b}' = g^{-1} \cdot \mathfrak{b} \subset \mathfrak{p}_\alpha$ such that $\{y \in \mathfrak{p}_\alpha : [y, x] = 0\} \not\subset \mathfrak{b}'$. This property is equivalent to condition (ii) of the statement. The proof is complete. □

Proposition 5 is an efficient criterion of existence of dense B-orbit once we know a representative x of the orbital variety $\mathcal{V}_{\mathrm{subreg}}(\alpha)$, i.e., an element x of the intersection $\mathcal{O}_{\mathrm{subreg}} \cap \mathfrak{n}_\alpha$. Such an element x is called a Richardson element of the nilradical $\mathfrak{n}_\alpha$. In the classical cases, due to the combinatorial classification of nilpotent orbits in terms of Jordan forms, it is possible to construct Richardson elements for many parabolic subalgebras; see [1]. In the classical cases, our proof of Theorem 2 relies on Proposition 5. In particular we construct Richardson elements for nilradicals of the form $\mathfrak{n}_\alpha$ (the constructions made in [1] do not apply to all the nilradicals of this form).

In the exceptional cases, out of our knowledge, there is no construction of Richardson elements. For this reason, we cannot use Proposition 5 for proving Theorem 2 in the exceptional cases (however, as a byproduct of our proof, we provide Richardson elements for certain nilradicals $\mathfrak{n}_\alpha$; see also Remark 5 below). We rely on the construction of Chevalley bases and on the following criterion.

Proposition 6 *Let α be a simple root and let $\mathscr{V}_{\mathrm{subreg}}(\alpha) = \mathscr{O}_{\mathrm{subreg}} \cap \mathfrak{n}_\alpha$ be the corresponding orbital variety of $\mathscr{O}_{\mathrm{subreg}}$. Let $x \in \mathfrak{n}_\alpha$ and let $M_\alpha(x)$ be the matrix of the linear transformation $\mathfrak{b} \to \mathfrak{n}_\alpha$, $y \mapsto [y, x]$ (relatively to some bases of $\mathfrak{b}$ and $\mathfrak{n}_\alpha$, e.g., subbases of a Chevalley basis of $\mathfrak{g}$). Then*

(a) $\dim B \cdot x = \operatorname{rank} M_\alpha(x)$.
(b) *$\mathscr{V}_{\mathrm{subreg}}(\alpha)$ has a dense B-orbit if and only if for some $x \in \mathfrak{n}_\alpha$ the rows of $M_\alpha(x)$ are linearly independent.*

Proof Part (a) is obtained as follows:

$$\dim B \cdot x = \dim B - \dim\{y \in \mathfrak{b} : [y, x] = 0\} = \dim \mathfrak{b} - \dim \ker M_\alpha(x) = \operatorname{rank} M_\alpha(x).$$

From part (a), it follows that if the rows of $M_\alpha(x)$ are linearly independent, i.e., $\operatorname{rank} M_\alpha(x) = \dim \mathfrak{n}_\alpha$, then $B \cdot x$ is a dense, open subset of $\mathfrak{n}_\alpha$, which implies that its intersection with $\mathscr{O}_{\mathrm{subreg}} \cap \mathfrak{n}_\alpha$ is nonempty; in fact, this ensures that x belongs to $\mathscr{O}_{\mathrm{subreg}} \cap \mathfrak{n}_\alpha$ (i.e., to $\mathscr{V}_{\mathrm{subreg}}(\alpha)$) since this set is B-stable. The equivalence in part (b) immediately follows from part (a) and this observation. □

Corollary 2 *We consider a connected subdiagram of the Dynkin diagram of $\mathfrak{g}$, which corresponds to a subset of simple roots $\Pi' \subset \Pi$. Let $G' \subset G$ be the connected simple algebraic subgroup corresponding to Π', let $\mathfrak{g}' \subset \mathfrak{g}$ be its Lie algebra, and let $B' \subset G'$ be the standard Borel subgroup.*

In this way, in addition to the orbital variety $\mathscr{V}_{\mathrm{subreg}}(\alpha)$ (relative to $\mathfrak{g}$), a simple root $\alpha \in \Pi'$ determines an orbital variety $\mathscr{V}'_{\mathrm{subreg}}(\alpha)$ relative to $\mathfrak{g}'$, contained in the subregular nilpotent orbit of $\mathfrak{g}'$.

If $\mathscr{V}'_{\mathrm{subreg}}(\alpha)$ has no dense B'-orbit, then $\mathscr{V}_{\mathrm{subreg}}(\alpha)$ has no dense B-orbit.

More precisely, if $\mathscr{B} := B \cdot x$ is a dense orbit of $\mathscr{V}_{\mathrm{subreg}}(\alpha)$, denoting by x' the natural projection of x onto $\mathfrak{g}'$, we have that $\mathscr{B}' := B' \cdot x'$ is a dense orbit of $\mathscr{V}_{\mathrm{subreg}}(\alpha')$. Moreover, the map $\mathscr{B} \to \mathscr{B}'$, $x \mapsto x'$ is then surjective.

Proof Let $\Phi' \subset \Phi$ be the root system generated by Π', i.e., the subset of roots which are linear combinations of the elements of Π'. Let $\Phi'^{+} := \Phi' \cap \Phi^{+}$ be the subset of positive roots. Let $\mathfrak{b}'$ be the Lie algebra of B', let $\mathfrak{n}'_\alpha \subset \mathfrak{b}'$ be the nilradical corresponding to α, let $\mathfrak{h}'$ be the standard Cartan subalgebra of $\mathfrak{g}'$. Thus

$$\mathfrak{b}' = \mathfrak{h}' \oplus \bigoplus_{\gamma \in \Phi'^{+}} \mathfrak{g}_\gamma \quad \text{and} \quad \mathfrak{n}'_\alpha = \bigoplus_{\gamma \in \Phi'^{+} \setminus \{\alpha\}} \mathfrak{g}_\gamma .$$

Recall that for each root $\gamma \in \Phi^{+}$ we consider a root vector $e_\gamma \in \mathfrak{g}_\gamma \setminus \{0\}$. Let $\{\lambda'_\gamma : \gamma \in \Pi'\} \subset \mathfrak{h}'$ and $\{\lambda_\gamma : \gamma \in \Pi\} \subset \mathfrak{h}$ be the dual bases of $\Pi' \subset \mathfrak{h}'^{*}$ and $\Pi \subset \mathfrak{h}^{*}$, respectively. Let an element $x \in \mathfrak{n}_\alpha$ and let $x' \in \mathfrak{n}'_\alpha$ be its image by the projection relative to the decomposition $\mathfrak{n}_\alpha = \mathfrak{n}'_\alpha \oplus \bigoplus_{\gamma \in \Phi^{+} \setminus \Phi'^{+}} \mathfrak{g}_\gamma$. Let $M_\alpha(x)$ be the matrix of the linear map $\mathfrak{b} \to \mathfrak{n}_\alpha$, $y \mapsto [y, x]$ in the bases $\{\lambda_\gamma : \gamma \in \Pi\} \cup \{e_\gamma :$

$\gamma \in \Phi^+\}$ (of $\mathfrak{b}$) and $\{e_\gamma : \gamma \in \Phi^+ \setminus \{\alpha\}\}$ (of $\mathfrak{n}_\alpha$). Let $M'_\alpha(x')$ be the matrix of the map $\mathfrak{b}' \to \mathfrak{n}'_\alpha$, $y \mapsto [y, x']$ in the bases $\{\lambda'_\gamma : \gamma \in \Pi'\} \cup \{e_\gamma : \gamma \in \Phi'^+\}$ (of $\mathfrak{b}'$) and $\{e_\gamma : \gamma \in \Phi'^+ \setminus \{\alpha\}\}$ (of $\mathfrak{n}'_\alpha$). Then (up to adding columns of zeros) the matrix $M'_\alpha(x')$ coincides with the submatrix of $M_\alpha(x)$ formed by the rows corresponding to the basis vectors e_γ for $\gamma \in \Phi'^+ \setminus \{\alpha\}$. Therefore, if the rows of $M'_\alpha(x')$ are linearly dependent, then so are the rows of $M_\alpha(x)$. The corollary now follows from Proposition 6 (b) (the last claim follows from the fact that the map $x \mapsto x'$ is B'-equivariant). □

Remark 4 Let $\{h, x, y\} \subset \mathfrak{g}$ be an $\mathfrak{sl}_2$-triple (i.e., $[h, x] = 2x$, $[h, y] = -2y$, and $[x, y] = h$) and let $\mathfrak{g} = \bigoplus_{i \in \mathbb{Z}} \mathfrak{g}(i)$, where $\mathfrak{g}(i) = \{z \in \mathfrak{g} : [h, z] = iz\}$. Then $\mathfrak{g}(\geq 0) := \bigoplus_{i \geq 0} \mathfrak{g}(i)$ is a parabolic subalgebra of $\mathfrak{g}$. Up to replacing the $\mathfrak{sl}_2$-triple by a conjugate, we may assume that $\mathfrak{g}(\geq 0)$ contains the Lie algebra of the Borel subgroup B.

If x belongs to the subregular nilpotent orbit $\mathscr{O}_{\mathrm{subreg}}$, then $\mathfrak{g}(\geq 0)$ is the Lie algebra of the minimal parabolic subgroup P_{α_0}, where α_0 is the simple root which has label 0 in the weighted Dynkin diagram corresponding to $\mathscr{O}_{\mathrm{subreg}}$ (see [5]). Moreover, in this case, the grading satisfies

- $\mathfrak{g}(i) = 0$ for all odd i;
- if G is of type C, D, E_6–E_8, F_4, or G_2, then $\dim \mathfrak{g}(0) = \dim \mathfrak{g}(2)$, i.e., x is distinguished.

In the case where x is distinguished, by the properties of the representations of $\mathfrak{sl}_2(\mathbb{K}) \cong \langle h, x, y \rangle_{\mathbb{K}}$, we have $\{z \in \mathfrak{g} : [z, x] = 0\} \subset \bigoplus_{i \geq 2} \mathfrak{g}(i) = \mathfrak{n}_{\alpha_0}$. Proposition 5 implies that $\mathscr{V}_{\mathrm{subreg}}(\alpha_0)$ has no dense B-orbit in this case. A different proof of the latter fact (which relies on Proposition 6 and Corollary 2) is given below.

4.2 Proof of Theorem 2 in Classical Cases

We rely on a technical lemma:

Lemma 5 *Let α be a simple root. Assume that there is an element $x \in \mathscr{O}_{\mathrm{subreg}} \cap \mathfrak{n}_\alpha$ of the form*

$$x = \sum_{\gamma \in I} x_\gamma \quad \text{with } x_\gamma \in \mathfrak{g}_\gamma \setminus \{0\}$$

where I is a subset of $\Phi^+ \setminus \{\alpha\}$ satisfying the following conditions:

(A) *There is $\gamma \in I$ such that $\gamma - \alpha \in \Phi \setminus (I \cup I')$ where $I' = I + \Phi^+$;*
(B) *There is $\gamma \in I$ such that $\gamma + \alpha \in \Phi \setminus (I \cup \hat{I}')$ where $\hat{I}' = I + (\Phi^+ \setminus \{\alpha\})$;*
(C) *For every $\delta \in \{\gamma' - \gamma : \gamma, \gamma' \in I\} \cap \Phi^+ \setminus \{\alpha\}$, there is $\beta \in I$ such that $\beta + \delta \in \Phi \setminus I$ and $(\beta + \delta) - \beta' \notin \Phi^+ \setminus \{\alpha\}$ for all $\beta' \in I \setminus \{\beta\}$;*
(D) *Φ is contained in the linear space spanned by I.*

Then $\mathcal{V}_{\mathrm{subreg}}(\alpha)$ *has no dense* B*-orbit.*

Proof In view of Proposition 5, it suffices to show the inclusion

$$\{y \in \mathfrak{p}_\alpha : [y, x] = 0\} \subset \mathfrak{n}_\alpha. \tag{1}$$

So, let $y \in \mathfrak{p}_\alpha$ such that $[y, x] = 0$, write

$$y = h + y_\alpha + y_{-\alpha} + y' \quad \text{with } h \in \mathfrak{h},\, y_\alpha \in \mathfrak{g}_\alpha,\, y_{-\alpha} \in \mathfrak{g}_{-\alpha},\, y' \in \mathfrak{n}_\alpha,$$

and let us show that $h = y_\alpha = y_{-\alpha} = 0$. Let $\gamma_1 \in I$ and $\gamma_2 \in I$ be the elements provided by conditions (A) and (B), respectively. First we see that

$$[y_{-\alpha}, x] = [-h - y_\alpha - y', x] \in \bigoplus_{\gamma \in I \cup I'} \mathfrak{g}_\gamma.$$

The vector $[y_{-\alpha}, x_{\gamma_1}]$ is the component of $[y_{-\alpha}, x]$ in the root space $\mathfrak{g}_{\gamma_1 - \alpha}$. Since $\gamma_1 - \alpha \notin I \cup I'$, we must have $[y_{-\alpha}, x_{\gamma_1}] = 0$, hence $y_{-\alpha} = 0$. Next we see that

$$[y_\alpha, x] = [-h - y', x] \in \bigoplus_{\gamma \in I \cup \hat{I}'} \mathfrak{g}_\gamma$$

and the condition $\gamma_2 + \alpha \notin I \cup \hat{I}'$ implies that $[y_\alpha, x_{\gamma_2}] = 0$; since $\gamma_2 + \alpha$ is a root, this forces $y_\alpha = 0$. Thus the relation

$$[y', x] = [-h, x] \in \bigoplus_{\gamma \in I} \mathfrak{g}_\gamma \tag{2}$$

holds. We claim that $[y', x] = 0$. Arguing by contradiction, say $[y', x] \neq 0$. Hence there are roots $\gamma \in I$ and $\delta \in \Phi^+ \setminus \{\alpha\}$ such that $[y'_\delta, x_\gamma] \neq 0$, where y'_δ is the component of y' in the root space $\mathfrak{g}_\delta$. By (2), this yields $\gamma' := \gamma + \delta \in I$. Then, let $\beta \in I$ be as in condition (C). Condition (C) implies that $[y'_\delta, x_\beta]$ is the component of $[y', x]$ in the root space $\mathfrak{g}_{\beta+\delta}$ and that it is nonzero. Since $\beta + \delta \notin I$, this contradicts (2). Therefore $[y', x] = 0$ and in turn (again by (2)) $[h, x] = 0$. The last relation implies that $h \in \bigcap_{\gamma \in I} \ker \gamma$, which, in view of condition (D), yields $h = 0$. The proof of the lemma is complete. □

Hereafter we denote by $\{\lambda_\alpha : \alpha \in \Pi\}$ the basis of the Cartan subalgebra $\mathfrak{h}$ which is dual to the basis of $\mathfrak{h}^*$ formed by the simple roots.

By $E_{i,j}^{(n)}$ we denote the elementary $n \times n$ matrix with 1 in position (i, j) and zeros elsewhere. By ${}^t a$ we denote the transpose of a matrix a. For each classical Lie algebra $\mathfrak{g}$ considered below, we consider the root datum (Φ, Φ^+) corresponding to the Cartan subalgebra $\mathfrak{h} \subset \mathfrak{g}$ formed by diagonal matrices and the Borel subalgebra $\mathfrak{b} \subset \mathfrak{g}$ formed by upper triangular matrices.

Proof (Proof of Theorem 2 in Type A) Assume that G is a simple group of type A_{n-1}, i.e., $\mathfrak{g} = \mathfrak{sl}_n(\mathbb{K})$ is the space of $n \times n$ matrices of trace zero. The subregular orbit $\mathscr{O}_{\text{subreg}}$ consists of all nilpotent matrices $x \in \mathfrak{sl}_n(\mathbb{K})$ with Jordan form $(n-1, 1)$. Let $\Phi = \{\varepsilon_i - \varepsilon_j : 1 \leq i \neq j \leq n\}$ and $\Phi^+ = \{\varepsilon_i - \varepsilon_j : 1 \leq i < j \leq n\}$. The matrix $e_{\varepsilon_i - \varepsilon_j} := E_{i,j}^{(n)}$ is a root vector for the root $\varepsilon_i - \varepsilon_j$. Let $\alpha_i := \varepsilon_i - \varepsilon_{i+1}$ $(i = 1, \ldots, n-1)$ be the simple roots.

Let a simple root $\alpha = \alpha_i$. Up to automorphism of the Dynkin diagram, we may suppose that $i < n-1$. The matrix $x := e_{\alpha_i + \alpha_{i+1}} + \sum_{j \notin \{i,i+1\}} e_{\alpha_j}$ is an element of $\mathscr{O}_{\text{subreg}} \cap \mathfrak{n}_{\alpha_i}$ and the matrix $y := \lambda_{\alpha_i} - \lambda_{\alpha_{i+1}} \in \mathfrak{p}_{\alpha_i} \backslash \mathrm{Rad}(\mathfrak{p}_{\alpha_i})$ is such that $[y, x] = 0$. From Proposition 5, we conclude that $\mathscr{V}_{\text{subreg}}(\alpha_i)$ has a dense B-orbit. □

Proof (Proof of Theorem 2 in Type B) Assume that G is a simple group of type B_m with $m \geq 2$, i.e., $\mathfrak{g} = \mathfrak{so}_n(\mathbb{K})$ with $n = 2m + 1$, seen as the subalgebra of $\mathfrak{sl}_n(\mathbb{K})$ of matrices which are skew symmetric with respect to the skew diagonal. The roots are $\Phi = \{\pm\varepsilon_i \pm \varepsilon_j : 1 \leq i < j \leq m\} \cup \{\pm\varepsilon_i : 1 \leq i \leq m\}$ and $\Phi^+ = \{\varepsilon_i \pm \varepsilon_j : 1 \leq i < j \leq m\} \cup \{\varepsilon_i : 1 \leq i \leq m\}$ and corresponding root vectors are

$$e_{\varepsilon_i - \varepsilon_j} = E_{i,j}^{(n)} - E_{n+1-j,n+1-i}^{(n)}, \; e_{\varepsilon_i + \varepsilon_j} = E_{i,n+1-j}^{(n)} - E_{j,n+1-i}^{(n)} \text{ for } 1 \leq i < j \leq m,$$

$$e_{\varepsilon_i} = E_{i,m+1}^{(n)} - E_{m+1,n+1-i}^{(n)} \quad \text{for } 1 \leq i \leq m, \quad e_{-\alpha} = {}^t e_\alpha \quad \text{for } \alpha \in \Phi^+.$$

The simple roots are $\alpha_i := \varepsilon_i - \varepsilon_{i+1}$, for $i = 1, \ldots, m-1$, and $\alpha_m := \varepsilon_m$.

The subregular orbit $\mathscr{O}_{\text{subreg}}$ consists of nilpotent matrices $x \in \mathfrak{so}_n(\mathbb{K})$ of Jordan form $(n-2, 1, 1)$. For $i \in \{1, \ldots, m-1\}$, the matrix

$$x_i := e_{\alpha_i + \alpha_{i+1}} + \sum_{j \notin \{i,i+1\}} e_{\alpha_j}$$

belongs to $\mathscr{O}_{\text{subreg}} \cap \mathfrak{n}_{\alpha_i}$. Moreover the matrix $y_i := \lambda_{\alpha_i} - \lambda_{\alpha_{i+1}}$ is an element of $\mathfrak{p}_{\alpha_i} \setminus \mathrm{Rad}(\mathfrak{p}_{\alpha_i})$ such that $[y_i, x_i] = 0$. From Proposition 5, it follows that $\mathscr{V}_{\text{subreg}}(\alpha_i)$ has a dense B-orbit. Note that x_{m-1} also belongs to $\mathscr{O}_{\text{subreg}} \cap \mathfrak{n}_{\alpha_m}$ and y_{m-1} also belongs to $\mathfrak{p}_{\alpha_m} \setminus \mathrm{Rad}(\mathfrak{p}_{\alpha_m})$, hence the orbital variety $\mathscr{V}_{\text{subreg}}(\alpha_m)$ has also a dense B-orbit. We have shown that, in type B, all the orbital varieties of $\mathscr{O}_{\text{subreg}}$ have a dense B-orbit. □

Proof (Proof of Theorem 2 in Type C) Assume G of type C_m with $m \geq 3$. We deal with the following realization of $\mathfrak{g} = \mathfrak{sp}_n(\mathbb{K})$ with $n = 2m$:

$$\mathfrak{g} = \left\{ \begin{pmatrix} a & b \\ c & -a^* \end{pmatrix} : a, b, c \text{ are } m \times m \text{ matrices}, \; b = b^*, \; c = c^* \right\}$$

where x^* stands for the transpose of x by the skew diagonal. In this case, we have the roots $\Phi = \{\pm\varepsilon_i \pm \varepsilon_j : 1 \leq i < j \leq m\} \cup \{\pm 2\varepsilon_i : 1 \leq i \leq m\}$ and $\Phi^+ = \{\varepsilon_i \pm \varepsilon_j : 1 \leq i < j \leq m\} \cup \{2\varepsilon_i : 1 \leq i \leq m\}$, and we consider the following root vectors:

$$e_{\varepsilon_i-\varepsilon_j}=E^{(n)}_{i,j}-E^{(n)}_{n+1-j,n+1-i},\ e_{\varepsilon_i+\varepsilon_j}=E^{(n)}_{i,n+1-j}+E^{(n)}_{j,n+1-i}\ \text{for } 1\le i<j\le m,$$

$$e_{2\varepsilon_i}=E^{(n)}_{i,n+1-i}\quad \text{for } 1\le i\le m,\quad e_{-\alpha}={}^te_\alpha\quad \text{for } \alpha\in\Phi^+.$$

The simple roots are $\alpha_i:=\varepsilon_i-\varepsilon_{i+1}$, for $i=1,\dots,m-1$, and $\alpha_m:=2\varepsilon_m$.

The subregular orbit $\mathscr{O}_{\rm subreg}\subset\mathfrak{sp}_n(\mathbb{K})$ consists of all nilpotent matrices $x\in\mathfrak{sp}_n(\mathbb{K})$ of Jordan form $(n-2,2)$.

The element $x_1:=e_{2\varepsilon_1}+\sum_{j=2}^m e_{\alpha_j}$ belongs to $\mathscr{O}_{\rm subreg}\cap\mathfrak{n}_{\alpha_1}$ and the element $y_1:=e_{-\alpha_1}+e_{\varepsilon_1+\varepsilon_3}$ belongs to $\mathfrak{p}_{\alpha_1}\setminus{\rm Rad}(\mathfrak{p}_{\alpha_1})$ and satisfies $[y_1,x_1]=0$. The element $x_m:=e_{2\varepsilon_{m-1}}+\sum_{j=1}^{m-1}e_{\alpha_j}$ belongs to $\mathscr{O}_{\rm subreg}\cap\mathfrak{n}_{\alpha_m}$ and satisfies $[e_{-\alpha_m},x_m]=0$. By applying Proposition 5, we get that $\mathscr{V}_{\rm subreg}(\alpha_1)$ and $\mathscr{V}_{\rm subreg}(\alpha_m)$ contain a dense B-orbit.

Finally let us show that $\mathscr{V}_{\rm subreg}(\alpha_i)$ has no dense B-orbit whenever $i\in\{2,\dots,m-1\}$. In view of Corollary 2, arguing by induction on $m\ge 3$, we may assume that $i=2$. Let $I=\{\alpha_j : j\notin\{2,3\}\}\cup\{\alpha_2+\alpha_3,2\varepsilon_3\}$ and set $x_2=\sum_{\alpha\in I}e_\alpha$. Note that $x_2\in\mathscr{O}_{\rm subreg}\cap\mathfrak{n}_{\alpha_2}$. Moreover, it is easy to see that the set I fulfills the conditions (A)–(D) of Lemma 5. It follows that $\mathscr{V}_{\rm subreg}(\alpha_2)$ has no dense B-orbit. The proof of the theorem is complete in type C. □

Proof (Proof of Theorem 2 in Type D) Assume G of type D_m for $m\ge 4$. Hence $\mathfrak{g}=\mathfrak{so}_n(\mathbb{K})$ with $n=2m$, seen as the subalgebra of $\mathfrak{sl}_n(\mathbb{K})$ formed by matrices which are skew symmetric by the skew diagonal. The roots are $\Phi=\{\pm\varepsilon_i\pm\varepsilon_j : 1\le i<j\le m\}$ and $\Phi^+=\{\varepsilon_i\pm\varepsilon_j : 1\le i<j\le m\}$ and we consider the root vectors

$$e_{\varepsilon_i-\varepsilon_j}=E^{(n)}_{i,j}-E^{(n)}_{n+1-j,n+1-i}\quad\text{and}\quad e_{\varepsilon_i+\varepsilon_j}=E^{(n)}_{i,n+1-j}-E^{(n)}_{j,n+1-i}$$

for $1\le i<j\le m$, and $e_{-\alpha}={}^te_\alpha$ for all $\alpha\in\Phi^+$. The simple roots are $\alpha_i:=\varepsilon_i-\varepsilon_{i+1}$, for $i=1,\dots,m-1$, and $\alpha_m:=\varepsilon_{m-1}+\varepsilon_m$. The subregular orbit $\mathscr{O}_{\rm subreg}$ is the set of nilpotent elements $x\in\mathfrak{so}_n(\mathbb{K})$ of Jordan form $(n-3,3)$.

The element $x_1:=e_{\varepsilon_1+\varepsilon_m}+\sum_{j=2}^m e_{\alpha_j}$ belongs to $\mathscr{O}_{\rm subreg}\cap\mathfrak{n}_{\alpha_1}$. The matrix $y_1:=2e_{-\alpha_1}-e_{\varepsilon_1+\varepsilon_4}-e_{\varepsilon_2-\varepsilon_{m-1}}-e_{\varepsilon_3-\varepsilon_m}+e_{\varepsilon_3+\varepsilon_m}$ belongs to $\mathfrak{p}_{\alpha_1}\setminus{\rm Rad}(\mathfrak{p}_{\alpha_1})$ and commutes with x_1. By Proposition 5, we deduce that $\mathscr{V}_{\rm subreg}(\alpha_1)$ has a dense B-orbit. The element $x_{m-1}:=e_{\varepsilon_{m-2}+\varepsilon_{m-1}}+\sum_{j\ne m-1}e_{\alpha_j}$ belongs to $\mathscr{O}_{\rm subreg}\cap\mathfrak{n}_{\alpha_{m-1}}$ and it commutes with $y_{m-1}:=e_{-\alpha_{m-1}}+e_{\alpha_m}\in\mathfrak{p}_{\alpha_{m-1}}\setminus{\rm Rad}(\mathfrak{p}_{\alpha_{m-1}})$, hence $\mathscr{V}_{\rm subreg}(\alpha_{m-1})$ contains a dense B-orbit. The symmetry of the Dynkin diagram guarantees that $\mathscr{V}_{\rm subreg}(\alpha_m)$ also contains a dense B-orbit.

Next we show that $\mathscr{V}_{\rm subreg}(\alpha_{m-2})$ has no dense B-orbit. In view of Corollary 2, we may assume that $m=4$. In this case $x:=e_{\alpha_1}+e_{\alpha_1+\alpha_2}+e_{\alpha_2+\alpha_3}+e_{\alpha_4}$ is an element of $\mathscr{O}_{\rm subreg}\cap\mathfrak{n}_{\alpha_2}$. Then, Lemma 5 shows that $\mathscr{V}_{\rm subreg}(\alpha_2)$ has no dense B-orbit.

Finally assume that $m\ge 5$ and let us show that the orbital variety $\mathscr{V}_{\rm subreg}(\alpha_i)$ has no dense B-orbit for $i\in\{2,\dots,m-3\}$. Invoking again Corollary 2, we may assume that $i=2$. Letting $I=\{\alpha_j : j\notin\{2,3\}\}\cup\{\alpha_2+\alpha_3,\alpha_3+\ldots+\alpha_{m-2}+\alpha_m\}$, it is easy to check that the set I fulfills conditions (A)–(D) of Lemma 5 and that the element $x := \sum_{\alpha\in I} e_\alpha$ belongs to $\mathscr{O}_{\mathrm{subreg}} \cap \mathfrak{n}_{\alpha_2}$. Hence, by Lemma 5, $\mathscr{V}_{\mathrm{subreg}}(\alpha_2)$ contains no dense B-orbit. The proof is complete in type D. □

4.3 *Proof of Theorem 2 in Exceptional Cases*

In this section, G is a simple algebraic group of exceptional type. As in Sect. 4.2, we denote by $\{\lambda_\alpha : \alpha \in \Pi\}$ the basis of the Cartan subalgebra $\mathfrak{h}$ which is dual to the basis of $\mathfrak{h}^*$ formed by the simple roots. Moreover for each exceptional type we have determined a Chevalley basis of $\mathfrak{n} = \bigoplus_{\alpha\in\Phi^+} \mathfrak{g}_\alpha$, i.e., the subbasis of a Chevalley basis of $\mathfrak{g}$. To this end, we consider the following total ordering of the positive roots:

- In types G_2 and F_4, the simple roots are ordered according to the following Dynkin diagrams:

$$G_2 : \alpha_1 \Lleftarrow \alpha_2 \qquad F_4 : \alpha_1 — \alpha_2 \Rightarrow \alpha_3 — \alpha_4$$

  In type E_8 (and in type E_6, resp., E_7), the simple roots are ordered according to the numbering of the Dynkin diagram drawn above Theorem 2 (and its subdiagram of vertices $\alpha_1, \ldots, \alpha_6$, resp., $\alpha_1, \ldots, \alpha_7$)
- Next each positive root is identified with the tuple of its coordinates in the basis Π. We consider the partial order determined by the height (i.e., the sum of the coordinates) and we order the roots of same height by the lexicographic order of the coordinates.

For instance, here is the ordered list of the positive roots in type G_2, identified with the couples of their coordinates in the basis (α_1, α_2):

$$\alpha_1 = [1, 0],\ \alpha_2 = [0, 1],\ \alpha_3 = [1, 1],\ \alpha_4 = [2, 1],\ \alpha_5 = [3, 1],\ \alpha_6 = [3, 2].$$

In each case let r be the number of simple roots, so $\alpha_1, \ldots, \alpha_r$ are the simple roots, and let n denote the number of positive roots. For $i \in \{1, \ldots, r\}$, we set $\lambda_i := \lambda_{\alpha_i}$. Finally let $(e_1, \ldots, e_n)$ be the Chevalley basis of $\mathfrak{n}$, numbered according to the total ordering of the positive roots. We fix an element

$$x = \sum_{j=1}^{n} x_j e_j \in \mathfrak{n}.$$

We consider the linear map $\mathfrak{b} \to \mathfrak{n}$, $y \mapsto [y, x]$, and we denote by $A(x)$ the matrix of this map between the bases $(\lambda_1, \ldots, \lambda_r, e_1, \ldots, e_n)$ and $(e_1, \ldots, e_n)$ of $\mathfrak{b}$ and $\mathfrak{n}$, respectively. Note that for every simple root $\alpha = \alpha_i$ (with $i \in \{1, \ldots, r\}$), the matrix $M_\alpha(x)$ of Proposition 6 is obtained from the matrix $A(x)$ by deleting the i-th row of the matrix (i.e., the row of the matrix corresponding to e_i) and replacing all the coefficients x_i by zeros. The explicit matrices $A(x)$ corresponding to the different exceptional cases are given in the Appendix.

Proof (Proof of Theorem 2 in Type G_2) On the one hand, for $x \in \mathfrak{n}_{\alpha_1}$, it is easy to see that the matrix $M_{\alpha_1}(x)$ has rank at most 4 whereas it has five rows. By Proposition 6, $\mathscr{V}_{\text{subreg}}(\alpha_1)$ has no dense B-orbit. On the other hand, the matrix $M_{\alpha_2}(e_1 + e_6)$ has five linearly independent rows. It follows from Proposition 6 that $B \cdot (e_1 + e_6)$ is a dense B-orbit of $\mathscr{V}_{\text{subreg}}(\alpha_2)$. □

Proof (Proof of Theorem 2 in Type F_4) The roots $\alpha_2, \alpha_3, \alpha_4$ generate a root system of type C_3. Hence it follows from Corollary 2 and the proof of the theorem in type C that $\mathscr{V}_{\text{subreg}}(\alpha_3)$ has no dense B-orbit.

For every $x \in \mathfrak{n}_{\alpha_2}$, the rows of the matrix $M_{\alpha_2}(x)$ corresponding to the root vectors $e_1, e_3, e_4, \ldots, e_{10}$ are linearly dependent. By Proposition 6, it follows that $\mathscr{V}_{\text{subreg}}(\alpha_2)$ has no dense B-orbit.

For $x \in \mathfrak{n}_{\alpha_4}$, the rows of the matrix $M_{\alpha_4}(x)$ corresponding to the root vectors e_j for $j \in \{1, \ldots, 16\} \setminus \{4\}$ are linearly dependent, and this shows that $\mathscr{V}_{\text{subreg}}(\alpha_4)$ has no dense B-orbit.

Finally, it can be seen that the matrix $M_{\alpha_1}(e_2 + e_3 + e_4 + e_{12})$ has linearly independent rows, and this shows that $B \cdot (e_2+e_3+e_4+e_{12})$ is dense in $\mathscr{V}_{\text{subreg}}(\alpha_1)$. The proof is complete in type F_4. □

Proof (Proof of Theorem 2 in Types E_6, E_7, E_8) In type E_8, the roots $\alpha_1, \ldots, \alpha_5$ generate a root system of type D_5 while the roots $\alpha_2, \ldots, \alpha_8$ generate a root system of type D_7. By comparing Corollary 2 and the proof of the theorem in type D, we deduce that the orbital varieties of type E_8 corresponding to $\alpha_3, \ldots, \alpha_7$ have no dense B-orbit. Arguing in the same way shows that the orbital varieties of type E_7 (resp., E_6) attached to the roots $\alpha_3, \ldots, \alpha_6$ (resp., $\alpha_3, \ldots, \alpha_5$) have no dense B-orbit.

In type E_6, letting $x_1 := \sum_{j=2}^{6} e_{\alpha_j} + e_{\alpha_1+\alpha_3+\alpha_4+\alpha_5}$, it can be seen that the matrix $M_{\alpha_1}(x_1)$ has linearly independent rows. Therefore, Proposition 6 implies that the element x_1 belongs to a dense B-orbit of $\mathscr{V}_{\text{subreg}}(\alpha_1)$. In view of the symmetry of the Dynkin diagram of type E_6, the orbital variety $\mathscr{V}_{\text{subreg}}(\alpha_6)$ has also a dense B-orbit whose representative is $\sum_{j=1}^{5} e_{\alpha_j} + e_{\alpha_3+\alpha_4+\alpha_5+\alpha_6}$. By combining Corollary 2 and the proof of Theorem 2 in type D, we obtain that, if $\mathscr{V}_{\text{subreg}}(\alpha_2)$ has a dense B-orbit, then this orbit contains an element $x \in \mathfrak{n}_{\alpha_2}$ whose natural projection on the subalgebra $\mathfrak{g}' \subset \mathfrak{g}$ of type D_5 corresponding to the set of simple roots $\Pi' = \{\alpha_1, \ldots, \alpha_5\}$ is $x' = e_{\alpha_1} + e_{\alpha_3} + e_{\alpha_4} + e_{\alpha_5} + e_{\alpha_2+\alpha_4+\alpha_5}$. For such an element x, one can see that the matrix $M_{\alpha_2}(x)$ has linearly dependent rows (the row corresponding to the root vector $e_{\alpha_2+\alpha_4+\alpha_5+\alpha_6}$ is a linear combination of the rows above it). By Proposition 6, this implies that $\mathscr{V}_{\text{subreg}}(\alpha_2)$ has no dense B-orbit. This completes the proof of the theorem in type E_6.

In type E_7, the above proof of the theorem in type E_6 and Corollary 2 imply that $\mathscr{V}_{\text{subreg}}(\alpha_2)$ has no dense B-orbit. It can be seen that the matrix $M_{\alpha_7}(x_7)$ corresponding to the element $x_7 := \sum_{j=1}^{6} e_{\alpha_j} + e_{\alpha_6+\alpha_7} + e_{\alpha_3+\alpha_4+\alpha_5+\alpha_6+\alpha_7}$ has linearly independent rows, hence Proposition 6 implies that $B \cdot x_7$ is a dense B-orbit of $\mathscr{V}_{\mathrm{subreg}}(\alpha_7)$. Finally, invoking again Corollary 2, we obtain that a dense B-orbit of $\mathscr{V}_{\mathrm{subreg}}(\alpha_1)$ (if it exists) must contain an element $x \in \mathfrak{n}_{\alpha_1}$ whose natural projection on the Lie subalgebra $\mathfrak{g}' \subset \mathfrak{g}$ of type E_6 corresponding to the set of simple roots $\Pi' = \{\alpha_1, \ldots, \alpha_6\}$ is the element x_1 written above. However, for such an element x, one can check that the row of the matrix $M_{\alpha_1}(x)$ corresponding to the root vector $e_{\alpha_1+\alpha_3+\alpha_4+\alpha_5+\alpha_6+\alpha_7}$ is a linear combination of the rows above it. This implies that $\mathscr{V}_{\mathrm{subreg}}(\alpha_1)$ has no dense B-orbit. This completes the proof of the theorem in type E_7.

In type E_8, comparing Corollary 2 with the proof of the theorem in type E_7 given above, we already deduce that $\mathscr{V}_{\mathrm{subreg}}(\alpha_1)$ and $\mathscr{V}_{\mathrm{subreg}}(\alpha_2)$ have no dense B-orbit. Invoking again Corollary 2 and the proof of the theorem in type D, we obtain that a dense B-orbit of $\mathscr{V}_{\mathrm{subreg}}(\alpha_8)$ (if it exists) should have a representative $x \in \mathfrak{n}_{\alpha_8}$ whose natural projection x' on the subalgebra $\mathfrak{g}' \subset \mathfrak{g}$ of type D_7 corresponding to the simple roots $\Pi' = \{\alpha_2, \ldots, \alpha_8\}$ is given by $x' = \sum_{j=2}^{7} e_j + e_{43}$. However, a careful calculation shows that for any such element x, the row of $M_{\alpha_8}(x)$ corresponding to the root vector e_{68} is a linear combination of the rows above it. This implies that $\mathscr{V}_{\mathrm{subreg}}(\alpha_8)$ has no dense B-orbit. The proof of the theorem is complete. □

Remark 5 Note that, in the proofs done in this section, in each case where $\mathscr{V}_{\mathrm{subreg}}(\alpha)$ has a dense B-orbit, we provide a representative of this orbit. This element is in particular a Richardson element of the nilradical $\mathfrak{n}_\alpha$.

Appendix

In this appendix, $\mathfrak{g}$ is a simple Lie algebra of exceptional type, $(\lambda_1, \ldots, \lambda_r)$ is the basis of the Cartan subalgebra $\mathfrak{h}$ which is dual to the basis of $\mathfrak{h}^*$ formed by the simple roots $\alpha_1, \ldots, \alpha_r$, $(e_1, \ldots, e_n)$ is a Chevalley basis of the maximal nilpotent subalgebra $\mathfrak{n}$ (the numbering of the vectors corresponds to the total ordering of the positive roots determined by the height, roots with the same height being ordered according to the lexicographic order of their coordinates).

Given $x = \sum_{j=1}^{n} x_j e_j$, we denote by $A(x)$ the matrix of the linear map $\mathfrak{b} \to \mathfrak{n}$, $y \mapsto [y, x]$ with respect to the bases $(\lambda_1, \ldots, \lambda_r, e_1, \ldots, e_n)$ of $\mathfrak{b}$ and $(e_1, \ldots, e_n)$ of $\mathfrak{n}$. In this appendix, we describe the matrix $A(x)$ in the different exceptional cases.

In Figs. 1 and 2, we give the matrix $A(x)$ in types G_2 and F_4, respectively. For clarity, the zero coefficients are replaced by dots. In type F_4, we write the matrix in

Fig. 1 The matrix $A(x)$ in type G_2

$$\begin{pmatrix} x_1 & \cdot & \cdot & \cdot & \cdot & \cdot & \cdot & \cdot \\ \cdot & x_2 & \cdot & \cdot & \cdot & \cdot & \cdot & \cdot \\ x_3 & x_3 & x_2 & -x_1 & \cdot & \cdot & \cdot & \cdot \\ 2x_4 & x_4 & 2x_3 & \cdot & -2x_1 & \cdot & \cdot & \cdot \\ 3x_5 & x_5 & -3x_4 & \cdot & \cdot & 3x_1 & \cdot & \cdot \\ 3x_6 & 2x_6 & \cdot & -x_5 & -3x_4 & 3x_3 & x_2 & \cdot \end{pmatrix}$$

$$A_{1,1} = \begin{pmatrix}
x_1 & \cdot & \cdot & \cdot & \cdot & \cdot & \cdot & \cdot & \cdot & \cdot & \cdot & \cdot & \cdot & \cdot \\
\cdot & x_2 & \cdot & \cdot & \cdot & \cdot & \cdot & \cdot & \cdot & \cdot & \cdot & \cdot & \cdot & \cdot \\
\cdot & \cdot & x_3 & \cdot & \cdot & \cdot & \cdot & \cdot & \cdot & \cdot & \cdot & \cdot & \cdot & \cdot \\
\cdot & \cdot & \cdot & x_4 & \cdot & \cdot & \cdot & \cdot & \cdot & \cdot & \cdot & \cdot & \cdot & \cdot \\
x_5 & x_5 & \cdot & \cdot & x_2 & -x_1 & \cdot & \cdot & \cdot & \cdot & \cdot & \cdot & \cdot & \cdot \\
\cdot & x_6 & x_6 & \cdot & \cdot & x_3 & -x_2 & \cdot & \cdot & \cdot & \cdot & \cdot & \cdot & \cdot \\
\cdot & \cdot & x_7 & x_7 & \cdot & \cdot & x_4 & -x_3 & \cdot & \cdot & \cdot & \cdot & \cdot & \cdot \\
x_8 & x_8 & x_8 & \cdot & x_6 & \cdot & -x_5 & \cdot & x_3 & -x_1 & \cdot & \cdot & \cdot & \cdot \\
\cdot & x_9 & 2x_9 & \cdot & \cdot & \cdot & 2x_6 & \cdot & \cdot & -2x_3 & \cdot & \cdot & \cdot & \cdot \\
\cdot & x_{10} & x_{10} & x_{10} & \cdot & x_7 & \cdot & -x_6 & \cdot & x_4 & -x_2 & \cdot & \cdot & \cdot \\
x_{11} & x_{11} & 2x_{11} & \cdot & x_9 & \cdot & 2x_8 & \cdot & \cdot & \cdot & \cdot & -2x_3 & -x_1 & \cdot \\
x_{12} & x_{12} & x_{12} & x_{12} & x_{10} & \cdot & \cdot & -x_8 & x_7 & \cdot & -x_5 & x_4 & \cdot & -x_1
\end{pmatrix}$$

$$A_{2,1} = \begin{pmatrix}
\cdot & x_{13} & 2x_{13} & x_{13} & \cdot & \cdot & x_{10} & -x_9 & \cdot & -x_7 & x_6 & \cdot & x_4 & -x_3 \\
x_{14} & 2x_{14} & 2x_{14} & \cdot & \cdot & x_{11} & \cdot & \cdot & -x_9 & 2x_8 & \cdot & -2x_6 & x_5 & \cdot \\
x_{15} & x_{15} & 2x_{15} & x_{15} & x_{13} & \cdot & x_{12} & -x_{11} & \cdot & \cdot & x_8 & -x_7 & \cdot & \cdot \\
\cdot & x_{16} & 2x_{16} & 2x_{16} & \cdot & \cdot & \cdot & 2x_{13} & \cdot & \cdot & -2x_{10} & \cdot & \cdot & 2x_7 \\
x_{17} & 2x_{17} & 2x_{17} & x_{17} & \cdot & x_{15} & \cdot & -x_{14} & -x_{13} & x_{12} & \cdot & -x_{10} & \cdot & x_8 \\
x_{18} & x_{18} & 2x_{18} & 2x_{18} & x_{16} & \cdot & \cdot & 2x_{15} & \cdot & \cdot & -2x_{12} & \cdot & \cdot & \cdot \\
x_{19} & 2x_{19} & 3x_{19} & x_{19} & \cdot & \cdot & x_{17} & \cdot & \cdot & -x_{15} & -x_{14} & x_{13} & x_{12} & x_{11} \\
x_{20} & 2x_{20} & 2x_{20} & 2x_{20} & x_{18} & \cdot & \cdot & 2x_{17} & -x_{16} & \cdot & \cdot & \cdot & \cdot & -2x_{12} \\
x_{21} & 2x_{21} & 3x_{21} & 2x_{21} & \cdot & \cdot & x_{20} & x_{19} & \cdot & -x_{18} & x_{17} & x_{16} & \cdot & -x_{15} \\
x_{22} & 2x_{22} & 4x_{22} & 2x_{22} & \cdot & \cdot & 2x_{21} & \cdot & \cdot & \cdot & 2x_{19} & \cdot & -x_{18} & \cdot \\
x_{23} & 3x_{23} & 4x_{23} & 2x_{23} & \cdot & x_{22} & \cdot & \cdot & \cdot & 2x_{21} & \cdot & \cdot & -x_{20} & 2x_{19} \\
2x_{24} & 3x_{24} & 4x_{24} & 2x_{24} & x_{23} & \cdot & \cdot & \cdot & x_{22} & \cdot & \cdot & 2x_{21} & \cdot & \cdot
\end{pmatrix}$$

$$A_{2,2} = \begin{pmatrix}
\cdot & \cdot & \cdot & \cdot & \cdot & \cdot & \cdot & \cdot & \cdot & \cdot & \cdot & \cdot & \cdot & \cdot \\
-x_2 & \cdot & \cdot & \cdot & \cdot & \cdot & \cdot & \cdot & \cdot & \cdot & \cdot & \cdot & \cdot & \cdot \\
x_4 & -x_3 & -x_1 & \cdot & \cdot & \cdot & \cdot & \cdot & \cdot & \cdot & \cdot & \cdot & \cdot & \cdot \\
\cdot & \cdot & -2x_4 & \cdot & \cdot & \cdot & \cdot & \cdot & \cdot & \cdot & \cdot & \cdot & \cdot & \cdot \\
\cdot & -x_6 & x_5 & x_4 & -x_2 & \cdot & \cdot & \cdot & \cdot & \cdot & \cdot & \cdot & \cdot & \cdot \\
\cdot & 2x_7 & \cdot & \cdot & -2x_4 & -x_1 & \cdot & \cdot & \cdot & \cdot & \cdot & \cdot & \cdot & \cdot \\
-x_{10} & -x_9 & -x_8 & x_7 & x_6 & \cdot & -x_3 & \cdot & \cdot & \cdot & \cdot & \cdot & \cdot & \cdot \\
\cdot & 2x_{10} & \cdot & \cdot & \cdot & x_5 & -2x_4 & -x_2 & \cdot & \cdot & \cdot & \cdot & \cdot & \cdot \\
\cdot & x_{13} & -x_{12} & \cdot & x_{10} & -x_8 & -x_7 & x_6 & -x_4 & -x_3 & \cdot & \cdot & \cdot & \cdot \\
x_{16} & \cdot & -2x_{15} & \cdot & 2x_{13} & -x_{11} & \cdot & x_9 & -2x_7 & \cdot & -2x_3 & \cdot & \cdot & \cdot \\
\cdot & \cdot & -2x_{17} & x_{16} & \cdot & -x_{14} & 2x_{13} & \cdot & -2x_{10} & x_9 & -2x_6 & -x_2 & \cdot & \cdot \\
-x_{20} & 2x_{19} & \cdot & x_{18} & -2x_{17} & \cdot & 2x_{15} & -x_{14} & -2x_{12} & x_{11} & -2x_8 & -x_5 & -x_1 & \cdot
\end{pmatrix}$$

Fig. 2 The blocks $A_{1,1}$, $A_{2,1}$, $A_{2,2}$ of the matrix $A(x)$ in type F_4

the form $A(x) = \begin{pmatrix} A_{1,1} & 0 \\ A_{2,1} & A_{2,2} \end{pmatrix}$ where $A_{1,1}$, $A_{2,1}$, and $A_{2,2}$ are the 12×14 matrices described in Fig. 2.

In type E_8, instead of drawing the matrix $A(x)$, we give the list of the roots α_j, for $j = 1, \ldots, 120$. For each root α_j, we indicate its coordinates $[\epsilon_{j,1}, \ldots, \epsilon_{j,8}]$ in the basis $(\alpha_1, \ldots, \alpha_8)$ and the couples (a, b) such that $e_j = +[e_a, e_b] = -[e_b, e_a]$. This information is sufficient for characterizing the matrix $A(x)$: the row of the matrix corresponding to e_j contains the coefficient $x_j \epsilon_{j,i}$ in the column corresponding to λ_i for all $i \in \{1, \ldots, 8\}$, and for each one of the listed couples (a, b) it contains x_b in the column corresponding to e_a and $-x_a$ in the column corresponding to e_b, and these are all the nonzero coefficients in the j-th row of the matrix.

In type E_6 (resp. E_7) the matrix $A(x)$ is obtained from the matrix $A(x)$ of type E_8 by deleting the columns corresponding to λ_7 and λ_8 (resp., the column corresponding to λ_8) and by deleting the rows and the columns corresponding to e_j whenever $(\epsilon_{j,7}, \epsilon_{j,8}) \neq (0, 0)$ (resp., whenever $\epsilon_{j,8} \neq 0$).

$\alpha_1 = [1, 0, 0, 0, 0, 0, 0, 0]$.
$\alpha_2 = [0, 1, 0, 0, 0, 0, 0, 0]$.
$\alpha_3 = [0, 0, 1, 0, 0, 0, 0, 0]$.
$\alpha_4 = [0, 0, 0, 1, 0, 0, 0, 0]$.
$\alpha_5 = [0, 0, 0, 0, 1, 0, 0, 0]$.
$\alpha_6 = [0, 0, 0, 0, 0, 1, 0, 0]$.
$\alpha_7 = [0, 0, 0, 0, 0, 0, 1, 0]$.
$\alpha_8 = [0, 0, 0, 0, 0, 0, 0, 1]$.

$\alpha_9 = [1, 0, 1, 0, 0, 0, 0, 0]$: $(1, 3)$.
$\alpha_{10} = [0, 1, 0, 1, 0, 0, 0, 0]$: $(2, 4)$.
$\alpha_{11} = [0, 0, 1, 1, 0, 0, 0, 0]$: $(3, 4)$.
$\alpha_{12} = [0, 0, 0, 1, 1, 0, 0, 0]$: $(4, 5)$.
$\alpha_{13} = [0, 0, 0, 0, 1, 1, 0, 0]$: $(5, 6)$.
$\alpha_{14} = [0, 0, 0, 0, 0, 1, 1, 0]$: $(6, 7)$.
$\alpha_{15} = [0, 0, 0, 0, 0, 0, 1, 1]$: $(7, 8)$.

$\alpha_{16} = [1, 0, 1, 1, 0, 0, 0, 0]$: $(1, 11)$, $(9, 4)$.
$\alpha_{17} = [0, 1, 1, 1, 0, 0, 0, 0]$: $(2, 11)$, $(3, 10)$.
$\alpha_{18} = [0, 1, 0, 1, 1, 0, 0, 0]$: $(2, 12)$, $(10, 5)$.
$\alpha_{19} = [0, 0, 1, 1, 1, 0, 0, 0]$: $(3, 12)$, $(11, 5)$.
$\alpha_{20} = [0, 0, 0, 1, 1, 1, 0, 0]$: $(4, 13)$, $(12, 6)$.
$\alpha_{21} = [0, 0, 0, 0, 1, 1, 1, 0]$: $(5, 14)$, $(13, 7)$.
$\alpha_{22} = [0, 0, 0, 0, 0, 1, 1, 1]$: $(6, 15)$, $(14, 8)$.

$\alpha_{23} = [1, 1, 1, 1, 0, 0, 0, 0]$: $(1, 17)$, $(2, 16)$, $(9, 10)$.
$\alpha_{24} = [1, 0, 1, 1, 1, 0, 0, 0]$: $(1, 19)$, $(9, 12)$, $(16, 5)$.
$\alpha_{25} = [0, 1, 1, 1, 1, 0, 0, 0]$: $(2, 19)$, $(3, 18)$, $(17, 5)$.
$\alpha_{26} = [0, 1, 0, 1, 1, 1, 0, 0]$: $(2, 20)$, $(10, 13)$, $(18, 6)$.
$\alpha_{27} = [0, 0, 1, 1, 1, 1, 0, 0]$: $(3, 20)$, $(11, 13)$, $(19, 6)$.
$\alpha_{28} = [0, 0, 0, 1, 1, 1, 1, 0]$: $(4, 21)$, $(12, 14)$, $(20, 7)$.
$\alpha_{29} = [0, 0, 0, 0, 1, 1, 1, 1]$: $(5, 22)$, $(13, 15)$, $(21, 8)$.

$\alpha_{30} = [1, 1, 1, 1, 1, 0, 0, 0]$: $(1, 25)$, $(2, 24)$, $(9, 18)$, $(23, 5)$.
$\alpha_{31} = [1, 0, 1, 1, 1, 1, 0, 0]$: $(1, 27)$, $(9, 20)$, $(16, 13)$, $(24, 6)$.
$\alpha_{32} = [0, 1, 1, 2, 1, 0, 0, 0]$: $(4, 25)$, $(17, 12)$, $(18, 11)$, $(19, 10)$.
$\alpha_{33} = [0, 1, 1, 1, 1, 1, 0, 0]$: $(2, 27)$, $(3, 26)$, $(17, 13)$, $(25, 6)$.
$\alpha_{34} = [0, 1, 0, 1, 1, 1, 1, 0]$: $(2, 28)$, $(10, 21)$, $(18, 14)$, $(26, 7)$.
$\alpha_{35} = [0, 0, 1, 1, 1, 1, 1, 0]$: $(3, 28)$, $(11, 21)$, $(19, 14)$, $(27, 7)$.
$\alpha_{36} = [0, 0, 0, 1, 1, 1, 1, 1]$: $(4, 29)$, $(12, 22)$, $(20, 15)$, $(28, 8)$.

$\alpha_{37} = [1, 1, 1, 2, 1, 0, 0, 0]$: $(1, 32)$, $(4, 30)$, $(18, 16)$, $(23, 12)$, $(24, 10)$.
$\alpha_{38} = [1, 1, 1, 1, 1, 1, 0, 0]$: $(1, 33)$, $(2, 31)$, $(9, 26)$, $(23, 13)$, $(30, 6)$.
$\alpha_{39} = [1, 0, 1, 1, 1, 1, 1, 0]$: $(1, 35)$, $(9, 28)$, $(16, 21)$, $(24, 14)$, $(31, 7)$.
$\alpha_{40} = [0, 1, 1, 2, 1, 1, 0, 0]$: $(4, 33)$, $(17, 20)$, $(26, 11)$, $(27, 10)$, $(32, 6)$.
$\alpha_{41} = [0, 1, 1, 1, 1, 1, 1, 0]$: $(2, 35)$, $(3, 34)$, $(17, 21)$, $(25, 14)$, $(33, 7)$.
$\alpha_{42} = [0, 1, 0, 1, 1, 1, 1, 1]$: $(2, 36)$, $(10, 29)$, $(18, 22)$, $(26, 15)$, $(34, 8)$.
$\alpha_{43} = [0, 0, 1, 1, 1, 1, 1, 1]$: $(3, 36)$, $(11, 29)$, $(19, 22)$, $(27, 15)$, $(35, 8)$.

$\alpha_{44} = [1, 1, 2, 2, 1, 0, 0, 0]$: $(3, 37)$, $(11, 30)$, $(23, 19)$, $(24, 17)$, $(25, 16)$, $(32, 9)$.
$\alpha_{45} = [1, 1, 1, 2, 1, 1, 0, 0]$: $(1, 40)$, $(4, 38)$, $(23, 20)$, $(26, 16)$, $(31, 10)$, $(37, 6)$.
$\alpha_{46} = [1, 1, 1, 1, 1, 1, 1, 0]$: $(1, 41)$, $(2, 39)$, $(9, 34)$, $(23, 21)$, $(30, 14)$, $(38, 7)$.
$\alpha_{47} = [1, 0, 1, 1, 1, 1, 1, 1]$: $(1, 43)$, $(9, 36)$, $(16, 29)$, $(24, 22)$, $(31, 15)$, $(39, 8)$.
$\alpha_{48} = [0, 1, 1, 2, 2, 1, 0, 0]$: $(5, 40)$, $(18, 27)$, $(19, 26)$, $(20, 25)$, $(32, 13)$, $(33, 12)$.
$\alpha_{49} = [0, 1, 1, 2, 1, 1, 1, 0]$: $(4, 41)$, $(17, 28)$, $(32, 14)$, $(34, 11)$, $(35, 10)$, $(40, 7)$.
$\alpha_{50} = [0, 1, 1, 1, 1, 1, 1, 1]$: $(2, 43)$, $(3, 42)$, $(17, 29)$, $(25, 22)$, $(33, 15)$, $(41, 8)$.

$\alpha_{51} = [1, 1, 2, 2, 1, 1, 0, 0]$: $(3, 45)$, $(11, 38)$, $(23, 27)$, $(31, 17)$, $(33, 16)$, $(40, 9)$, $(44, 6)$.
$\alpha_{52} = [1, 1, 1, 2, 2, 1, 0, 0]$: $(1, 48)$, $(5, 45)$, $(18, 31)$, $(20, 30)$, $(24, 26)$, $(37, 13)$, $(38, 12)$.
$\alpha_{53} = [1, 1, 1, 2, 1, 1, 1, 0]$: $(1, 49)$, $(4, 46)$, $(23, 28)$, $(34, 16)$, $(37, 14)$, $(39, 10)$, $(45, 7)$.
$\alpha_{54} = [1, 1, 1, 1, 1, 1, 1, 1]$: $(1, 50)$, $(2, 47)$, $(9, 42)$, $(23, 29)$, $(30, 22)$, $(38, 15)$, $(46, 8)$.
$\alpha_{55} = [0, 1, 1, 2, 2, 1, 1, 0]$: $(5, 49)$, $(18, 35)$, $(19, 34)$, $(28, 25)$, $(32, 21)$, $(41, 12)$, $(48, 7)$.
$\alpha_{56} = [0, 1, 1, 2, 1, 1, 1, 1]$: $(4, 50)$, $(17, 36)$, $(32, 22)$, $(40, 15)$, $(42, 11)$, $(43, 10)$, $(49, 8)$.

$\alpha_{57} = [1, 1, 2, 2, 2, 1, 0, 0]$: $(3, 52)$, $(5, 51)$, $(24, 33)$, $(25, 31)$, $(27, 30)$, $(38, 19)$, $(44, 13)$, $(48, 9)$.
$\alpha_{58} = [1, 1, 2, 2, 1, 1, 1, 0]$: $(3, 53)$, $(11, 46)$, $(23, 35)$, $(39, 17)$, $(41, 16)$, $(44, 14)$, $(49, 9)$, $(51, 7)$.
$\alpha_{59} = [1, 1, 1, 2, 2, 1, 1, 0]$: $(1, 55)$, $(5, 53)$, $(18, 39)$, $(24, 34)$, $(28, 30)$, $(37, 21)$, $(46, 12)$, $(52, 7)$.
$\alpha_{60} = [1, 1, 1, 2, 1, 1, 1, 1]$: $(1, 56)$, $(4, 54)$, $(23, 36)$, $(37, 22)$, $(42, 16)$, $(45, 15)$, $(47, 10)$, $(53, 8)$.
$\alpha_{61} = [0, 1, 1, 2, 2, 2, 1, 0]$: $(6, 55)$, $(20, 41)$, $(21, 40)$, $(33, 28)$, $(34, 27)$, $(35, 26)$, $(48, 14)$, $(49, 13)$.
$\alpha_{62} = [0, 1, 1, 2, 2, 1, 1, 1]$: $(5, 56)$, $(18, 43)$, $(19, 42)$, $(32, 29)$, $(36, 25)$, $(48, 15)$, $(50, 12)$, $(55, 8)$.

$\alpha_{63} = [1, 1, 2, 3, 2, 1, 0, 0]$: $(4, 57)$, $(12, 51)$, $(16, 48)$, $(24, 40)$, $(27, 37)$, $(32, 31)$, $(44, 20)$, $(45, 19)$, $(52, 11)$.
$\alpha_{64} = [1, 1, 2, 2, 2, 1, 1, 0]$: $(3, 59)$, $(5, 58)$, $(24, 41)$, $(25, 39)$, $(35, 30)$, $(44, 21)$, $(46, 19)$, $(55, 9)$, $(57, 7)$.

$\alpha_{65} = [1, 1, 2, 2, 1, 1, 1, 1]$: (3, 60), (11, 54), (23, 43), (44, 22), (47, 17), (50, 16), (51, 15), (56, 9), (58, 8).
$\alpha_{66} = [1, 1, 1, 2, 2, 2, 1, 0]$: (1, 61), (6, 59), (20, 46), (21, 45), (34, 31), (38, 28), (39, 26), (52, 14), (53, 13).
$\alpha_{67} = [1, 1, 1, 2, 2, 1, 1, 1]$: (1, 62), (5, 60), (18, 47), (24, 42), (36, 30), (37, 29), (52, 15), (54, 12), (59, 8).
$\alpha_{68} = [0, 1, 1, 2, 2, 2, 1, 1]$: (6, 62), (20, 50), (29, 40), (33, 36), (42, 27), (43, 26), (48, 22), (56, 13), (61, 8).

$\alpha_{69} = [1, 2, 2, 3, 2, 1, 0, 0]$: (2, 63), (10, 57), (18, 51), (23, 48), (30, 40), (32, 38), (33, 37), (44, 26), (45, 25), (52, 17).
$\alpha_{70} = [1, 1, 2, 3, 2, 1, 1, 0]$: (4, 64), (12, 58), (16, 55), (24, 49), (32, 39), (35, 37), (44, 28), (53, 19), (59, 11), (63, 7).
$\alpha_{71} = [1, 1, 2, 2, 2, 2, 1, 0]$: (3, 66), (6, 64), (21, 51), (27, 46), (38, 35), (39, 33), (41, 31), (57, 14), (58, 13), (61, 9).
$\alpha_{72} = [1, 1, 2, 2, 2, 1, 1, 1]$: (3, 67), (5, 65), (24, 50), (25, 47), (43, 30), (44, 29), (54, 19), (57, 15), (62, 9), (64, 8).
$\alpha_{73} = [1, 1, 1, 2, 2, 2, 1, 1]$: (1, 68), (6, 67), (20, 54), (29, 45), (38, 36), (42, 31), (47, 26), (52, 22), (60, 13), (66, 8).
$\alpha_{74} = [0, 1, 1, 2, 2, 2, 2, 1]$: (7, 68), (21, 56), (22, 55), (34, 43), (35, 42), (36, 41), (49, 29), (50, 28), (61, 15), (62, 14).

$\alpha_{75} = [1, 2, 2, 3, 2, 1, 1, 0]$: (2, 70), (10, 64), (18, 58), (23, 55), (30, 49), (32, 46), (41, 37), (44, 34), (53, 25), (59, 17), (69, 7).
$\alpha_{76} = [1, 1, 2, 3, 2, 2, 1, 0]$: (4, 71), (6, 70), (16, 61), (27, 53), (28, 51), (39, 40), (45, 35), (49, 31), (58, 20), (63, 14), (66, 11).
$\alpha_{77} = [1, 1, 2, 3, 2, 1, 1, 1]$: (4, 72), (12, 65), (16, 62), (24, 56), (32, 47), (43, 37), (44, 36), (60, 19), (63, 15), (67, 11), (70, 8).
$\alpha_{78} = [1, 1, 2, 2, 2, 2, 1, 1]$: (3, 73), (6, 72), (27, 54), (29, 51), (38, 43), (47, 33), (50, 31), (57, 22), (65, 13), (68, 9), (71, 8).
$\alpha_{79} = [1, 1, 1, 2, 2, 2, 2, 1]$: (1, 74), (7, 73), (21, 60), (22, 59), (34, 47), (36, 46), (39, 42), (53, 29), (54, 28), (66, 15), (67, 14).

$\alpha_{80} = [1, 2, 2, 3, 2, 2, 1, 0]$: (2, 76), (6, 75), (10, 71), (23, 61), (33, 53), (34, 51), (45, 41), (46, 40), (49, 38), (58, 26), (66, 17), (69, 14).
$\alpha_{81} = [1, 2, 2, 3, 2, 1, 1, 1]$: (2, 77), (10, 72), (18, 65), (23, 62), (30, 56), (32, 54), (44, 42), (50, 37), (60, 25), (67, 17), (69, 15), (75, 8).
$\alpha_{82} = [1, 1, 2, 3, 3, 2, 1, 0]$: (5, 76), (13, 70), (19, 66), (27, 59), (28, 57), (39, 48), (52, 35), (55, 31), (61, 24), (63, 21), (64, 20), (71, 12).
$\alpha_{83} = [1, 1, 2, 3, 2, 2, 1, 1]$: (4, 78), (6, 77), (16, 68), (27, 60), (36, 51), (45, 43), (47, 40), (56, 31), (63, 22), (65, 20), (73, 11), (76, 8).
$\alpha_{84} = [1, 1, 2, 2, 2, 2, 2, 1]$: (3, 79), (7, 78), (21, 65), (22, 64), (39, 50), (41, 47), (43, 46), (54, 35), (58, 29), (71, 15), (72, 14), (74, 9).

$\alpha_{85} = [1, 2, 2, 3, 3, 2, 1, 0]$: (2, 82), (5, 80), (13, 75), (25, 66), (33, 59), (34, 57), (46, 48), (52, 41), (55, 38), (61, 30), (64, 26), (69, 21), (71, 18).
$\alpha_{86} = [1, 2, 2, 3, 2, 2, 1, 1]$: (2, 83), (6, 81), (10, 78), (23, 68), (33, 60), (42, 51), (45, 50), (54, 40), (56, 38), (65, 26), (69, 22), (73, 17), (80, 8).
$\alpha_{87} = [1, 1, 2, 3, 3, 2, 1, 1]$: (5, 83), (13, 77), (19, 73), (27, 67), (36, 57), (47, 48), (52, 43), (62, 31), (63, 29), (68, 24), (72, 20), (78, 12), (82, 8).
$\alpha_{88} = [1, 1, 2, 3, 2, 2, 2, 1]$: (4, 84), (7, 83), (16, 74), (22, 70), (28, 65), (39, 56), (43, 53), (49, 47), (58, 36), (60, 35), (76, 15), (77, 14), (79, 11).

$\alpha_{89} = [1, 2, 2, 4, 3, 2, 1, 0]$: (4, 85), (12, 80), (20, 75), (32, 66), (34, 63), (40, 59), (52, 49), (53, 48), (55, 45), (61, 37), (69, 28), (70, 26), (76, 18), (82, 10).
$\alpha_{90} = [1, 2, 2, 3, 3, 2, 1, 1]$: (2, 87), (5, 86), (13, 81), (25, 73), (33, 67), (42, 57), (52, 50), (54, 48), (62, 38), (68, 30), (69, 29), (72, 26), (78, 18), (85, 8).
$\alpha_{91} = [1, 2, 2, 3, 2, 2, 2, 1]$: (2, 88), (7, 86), (10, 84), (22, 75), (23, 74), (34, 65), (46, 56), (49, 54), (50, 53), (58, 42), (60, 41), (79, 17), (80, 15), (81, 14).
$\alpha_{92} = [1, 1, 2, 3, 3, 2, 2, 1]$: (5, 88), (7, 87), (19, 79), (28, 72), (29, 70), (39, 62), (43, 59), (55, 47), (64, 36), (67, 35), (74, 24), (77, 21), (82, 15), (84, 12).

$\alpha_{93} = [1, 2, 3, 4, 3, 2, 1, 0]$: (3, 89), (11, 85), (19, 80), (27, 75), (32, 71), (40, 64), (41, 63), (55, 51), (57, 49), (58, 48), (61, 44), (69, 35), (70, 33), (76, 25), (82, 17).

$\alpha_{94} = [1, 2, 2, 4, 3, 2, 1, 1]$: (4, 90), (12, 86), (20, 81), (32, 73), (40, 67), (42, 63), (52, 56), (60, 48), (62, 45), (68, 37), (69, 36), (77, 26), (83, 18), (87, 10), (89, 8).

$\alpha_{95} = [1, 2, 2, 3, 3, 2, 2, 1]$: (2, 92), (5, 91), (7, 90), (25, 79), (29, 75), (34, 72), (46, 62), (50, 59), (55, 54), (64, 42), (67, 41), (74, 30), (81, 21), (84, 18), (85, 15).

$\alpha_{96} = [1, 1, 2, 3, 3, 3, 2, 1]$: (6, 92), (14, 87), (20, 84), (28, 78), (29, 76), (31, 74), (39, 68), (43, 66), (61, 47), (71, 36), (73, 35), (79, 27), (82, 22), (83, 21), (88, 13).

$\alpha_{97} = [2, 2, 3, 4, 3, 2, 1, 0]$: (1, 93), (9, 89), (16, 85), (24, 80), (31, 75), (37, 71), (45, 64), (46, 63), (57, 53), (58, 52), (59, 51), (66, 44), (69, 39), (70, 38), (76, 30), (82, 23).

$\alpha_{98} = [1, 2, 3, 4, 3, 2, 1, 1]$: (3, 94), (11, 90), (19, 86), (27, 81), (32, 78), (40, 72), (50, 63), (57, 56), (62, 51), (65, 48), (68, 44), (69, 43), (77, 33), (83, 25), (87, 17), (93, 8).

$\alpha_{99} = [1, 2, 2, 4, 3, 2, 2, 1]$: (4, 95), (7, 94), (12, 91), (32, 79), (34, 77), (36, 75), (53, 62), (55, 60), (56, 59), (67, 49), (70, 42), (74, 37), (81, 28), (88, 18), (89, 15), (92, 10).

$\alpha_{100} = [1, 2, 2, 3, 3, 3, 2, 1]$: (2, 96), (6, 95), (14, 90), (26, 84), (29, 80), (34, 78), (38, 74), (46, 68), (50, 66), (61, 54), (71, 42), (73, 41), (79, 33), (85, 22), (86, 21), (91, 13).

$\alpha_{101} = [2, 2, 3, 4, 3, 2, 1, 1]$: (1, 98), (9, 94), (16, 90), (24, 86), (31, 81), (37, 78), (45, 72), (54, 63), (57, 60), (65, 52), (67, 51), (69, 47), (73, 44), (77, 38), (83, 30), (87, 23), (97, 8).

$\alpha_{102} = [1, 2, 3, 4, 3, 2, 2, 1]$: (3, 99), (7, 98), (11, 95), (19, 91), (32, 84), (41, 77), (43, 75), (55, 65), (56, 64), (58, 62), (70, 50), (72, 49), (74, 44), (81, 35), (88, 25), (92, 17), (93, 15).

$\alpha_{103} = [1, 2, 2, 4, 3, 3, 2, 1]$: (4, 100), (6, 99), (14, 94), (26, 88), (34, 83), (36, 80), (45, 74), (53, 68), (56, 66), (61, 60), (73, 49), (76, 42), (79, 40), (86, 28), (89, 22), (91, 20), (96, 10).

$\alpha_{104} = [2, 2, 3, 4, 3, 2, 2, 1]$: (1, 102), (7, 101), (9, 99), (16, 95), (24, 91), (37, 84), (46, 77), (47, 75), (58, 67), (59, 65), (60, 64), (70, 54), (72, 53), (79, 44), (81, 39), (88, 30), (92, 23), (97, 15).

$\alpha_{105} = [1, 2, 3, 4, 3, 3, 2, 1]$: (3, 103), (6, 102), (11, 100), (14, 98), (33, 88), (41, 83), (43, 80), (51, 74), (56, 71), (58, 68), (61, 65), (76, 50), (78, 49), (84, 40), (86, 35), (91, 27), (93, 22), (96, 17).

$\alpha_{106} = [1, 2, 2, 4, 4, 3, 2, 1]$: (5, 103), (13, 99), (18, 96), (21, 94), (26, 92), (34, 87), (36, 85), (52, 74), (59, 68), (61, 67), (62, 66), (73, 55), (79, 48), (82, 42), (89, 29), (90, 28), (95, 20), (100, 12).

$\alpha_{107} = [2, 2, 3, 4, 3, 3, 2, 1]$: (1, 105), (6, 104), (9, 103), (14, 101), (16, 100), (38, 88), (46, 83), (47, 80), (51, 79), (58, 73), (60, 71), (66, 65), (76, 54), (78, 53), (84, 45), (86, 39), (91, 31), (96, 23), (97, 22).

$\alpha_{108} = [1, 2, 3, 4, 4, 3, 2, 1]$: (3, 106), (5, 105), (13, 102), (21, 98), (25, 96), (33, 92), (41, 87), (43, 85), (57, 74), (61, 72), (62, 71), (64, 68), (78, 55), (82, 50), (84, 48), (90, 35), (93, 29), (95, 27), (100, 19).

$\alpha_{109} = [2, 2, 3, 4, 4, 3, 2, 1]$: (1, 108), (5, 107), (9, 106), (13, 104), (21, 101), (30, 96), (38, 92), (46, 87), (47, 85), (57, 79), (64, 73), (66, 72), (67, 71), (78, 59), (82, 54), (84, 52), (90, 39), (95, 31), (97, 29), (100, 24).

$\alpha_{110} = [1, 2, 3, 5, 4, 3, 2, 1]$: (4, 108), (12, 105), (20, 102), (28, 98), (32, 96), (40, 92), (43, 89), (49, 87), (61, 77), (62, 76), (63, 74), (70, 68), (82, 56), (83, 55), (88, 48), (93, 36), (94, 35), (99, 27), (103, 19), (106, 11).

$\alpha_{111} = [2, 2, 3, 5, 4, 3, 2, 1]$: (1, 110), (4, 109), (12, 107), (20, 104), (28, 101), (37, 96), (45, 92), (47, 89), (53, 87), (63, 79), (66, 77), (67, 76), (70, 73), (82, 60), (83, 59), (88, 52), (94, 39), (97, 36), (99, 31), (103, 24), (106, 16).

$\alpha_{112} = [1, 3, 3, 5, 4, 3, 2, 1]$: (2, 110), (10, 108), (18, 105), (26, 102), (32, 100), (34, 98), (40, 95), (49, 90), (50, 89), (61, 81), (62, 80), (69, 74), (75, 68), (85, 56), (86, 55), (91, 48), (93, 42), (94, 41), (99, 33), (103, 25), (106, 17).

$\alpha_{113} = [2, 3, 3, 5, 4, 3, 2, 1]$: (1, 112), (2, 111), (10, 109), (18, 107), (26, 104), (34, 101), (37, 100), (45, 95), (53, 90), (54, 89), (66, 81), (67, 80), (69, 79), (75, 73), (85, 60), (86, 59), (91, 52), (94, 46), (97, 42), (99, 38), (103, 30), (106, 23).

$\alpha_{114} = [2, 2, 4, 5, 4, 3, 2, 1]$: (3, 111), (11, 109), (19, 107), (27, 104), (35, 101), (44, 96), (47, 93), (51, 92), (58, 87), (63, 84), (70, 78), (71, 77), (72, 76), (82, 65), (83, 64), (88, 57), (97, 43), (98, 39), (102, 31), (105, 24), (108, 16), (110, 9).

$\alpha_{115} = [2, 3, 4, 5, 4, 3, 2, 1]$: (2, 114), (3, 113), (17, 109), (25, 107), (33, 104), (41, 101), (44, 100), (51, 95), (54, 93), (58, 90), (69, 84), (71, 81), (72, 80), (75, 78), (85, 65), (86, 64), (91, 57), (97, 50), (98, 46), (102, 38), (105, 30), (108, 23), (112, 9).

$\alpha_{116} = [2, 3, 4, 6, 4, 3, 2, 1]$: (4, 115), (16, 112), (17, 111), (32, 107), (40, 104), (44, 103), (49, 101), (51, 99), (58, 94), (60, 93), (69, 88), (75, 83), (76, 81), (77, 80), (86, 70), (89, 65), (91, 63), (97, 56), (98, 53), (102, 45), (105, 37), (110, 23), (113, 11), (114, 10).

$\alpha_{117} = [2, 3, 4, 6, 5, 3, 2, 1]$: (5, 116), (18, 114), (19, 113), (30, 110), (32, 109), (44, 106), (48, 104), (55, 101), (57, 99), (64, 94), (67, 93), (69, 92), (75, 87), (77, 85), (82, 81), (89, 72), (90, 70), (95, 63), (97, 62), (98, 59), (102, 52), (108, 37), (111, 25), (112, 24), (115, 12).

$\alpha_{118} = [2, 3, 4, 6, 5, 4, 2, 1]$: (6, 117), (20, 115), (31, 112), (33, 111), (45, 108), (48, 107), (57, 103), (61, 101), (69, 96), (71, 94), (73, 93), (80, 87), (82, 86), (83, 85), (89, 78), (90, 76), (97, 68), (98, 66), (100, 63), (105, 52), (106, 51), (109, 40), (110, 38), (113, 27), (114, 26), (116, 13).

$\alpha_{119} = [2, 3, 4, 6, 5, 4, 3, 1]$: (7, 118), (21, 116), (34, 114), (35, 113), (46, 110), (49, 109), (58, 106), (59, 105), (61, 104), (70, 100), (71, 99), (79, 93), (80, 92), (82, 91), (88, 85), (89, 84), (95, 76), (96, 75), (97, 74), (102, 66), (103, 64), (107, 55), (108, 53), (111, 41), (112, 39), (115, 28), (117, 14).

$\alpha_{120} = [2, 3, 4, 6, 5, 4, 3, 2]$: (8, 119), (22, 117), (36, 115), (47, 112), (50, 111), (60, 108), (62, 107), (72, 103), (73, 102), (74, 101), (81, 96), (83, 95), (84, 94), (90, 88), (91, 87), (92, 86), (98, 79), (99, 78), (100, 77), (104, 68), (105, 67), (106, 65), (109, 56), (110, 54), (113, 43), (114, 42), (116, 29), (118, 15).

Acknowledgements I am grateful to the Referee for valuable comments and suggestions. Remarks 1, 2, 3, and 4 are based on her/his remarks. This work is supported in part by the ISF Grant Nr. 797/14 and by the ANR project GeoLie ANR-15-CE40-0012.

References

1. Baur, K.: Richardson elements for classical Lie algebras. J. Algebra **297**, 168–185 (2006).
2. Brion, M.: Quelques propriétés des espaces homogènes sphériques. Manuscripta Math. **55**, 191–198 (1986).
3. Bender, M., Perrin, N.: Singularities of closures of spherical B-conjugacy classes of nilpotent orbits. Preprint 2014. arxiv:1412.5654
4. Braverman, A., Joseph, A.: The minimal realization from deformation theory. J. Algebra **205**, 13–36 (1998).
5. Collingwood, D. H., McGovern, W. M.: Nilpotent orbits in semisimple Lie algebras. Van Nostrand Reinhold, New York (1993).
6. Fresse, L.: Upper triangular parts of conjugacy classes of nilpotent matrices with finite number of B-orbits. J. Math. Soc. Japan **65**, 967–992 (2013).
7. Fresse, L., Melnikov, A.: Smooth orbital varieties and orbital varieties with a dense B-orbit. Int. Math. Res. Not. **2013**, 1122–1203 (2013).
8. Joseph, A.: On the variety of a highest weight module. J. Algebra **88**, 238–278 (1984).
9. Joseph, A.: Orbital varieties of the minimal orbit. Ann. Sci. École Norm. Sup. (4) **31**, 17–45 (1998).
10. Spaltenstein, N.: On the fixed point set of a unipotent element on the variety of Borel subgroups. Topology **16**, 203–204 (1977).
11. Springer, T.: Linear algebraic groups, 2nd ed.. Progress in Mathematics, vol. 9. Birkhäuser Boston, Inc., Boston, MA (1998).
12. Steinberg, R.: On the desingularization of the unipotent variety. Invent. Math. **36**, 209–224 (1976).
13. Vinberg, È. B.: Complexity of actions of reductive groups. Funktsional. Anal, i Prilozhen. **20**, 1–13 (1986). In Russian; English translation in Funct. Anal. Appl. **20**, 1–11 (1986).
14. Wang, W.: Dimension of a minimal nilpotent orbit. Proc. Amer. Math. Soc. **127**, 935–936 (1999).

On Involutions in the Weyl Group and B-Orbit Closures in the Orthogonal Case

Mikhail V. Ignatyev

To Anthony Joseph in the occasion of his 75th birthday

Abstract We study coadjoint B-orbits on $\mathfrak{n}^*$, where B is a Borel subgroup of a complex orthogonal group G, and $\mathfrak{n}$ is the Lie algebra of the unipotent radical of B. To each basis involution w in the Weyl group W of G one can assign the associated B-orbit Ω_w. We prove that, given basis involutions σ, τ in W, if the orbit Ω_σ is contained in the closure of the orbit Ω_τ then σ is less than or equal to τ with respect to the Bruhat order on W. For a basis involution w, we also compute the dimension of Ω_w and present a conjectural description of the closure of Ω_w.

AMS Subject Classification: 17B22, 17B08, 17B30, 20F55

1 Introduction, Definitions and the Main Result

1.1 Let G be a complex reductive algebraic group, B a Borel subgroup of G, Φ the root system of G and $W = W(\Phi)$ the Weyl group of Φ. It is well-known that the Bruhat order on W encodes the cell decomposition of the flag variety G/B (see, e.g., [2]). Denote by $\mathcal{I}(\Phi)$ the poset of involutions in W (i.e., elements of W of order 2). In [19], Richardson and Springer showed that $\mathcal{I}(A_{2n})$ encodes the incidences among the closed B-orbits on the symmetric variety $\mathrm{SL}_{2n+1}(\mathbb{C})/\mathrm{SO}_{2n+1}(\mathbb{C})$. In [1], Bagno and Chernavsky presented a geometrical interpretation of the poset $\mathcal{I}(A_n)$, considering the action of the Borel subgroup of $\mathrm{GL}_n(\mathbb{C})$ on symmetric matrices by congruence. Incitti studied the poset $\mathcal{I}(\Phi)$ from a purely combinatorial point of

M. V. Ignatyev (✉)
Samara National Research University, Samara, Russia

M. Gorelik et al. (eds.), *Representations and Nilpotent Orbits of Lie Algebraic Systems*, Progress in Mathematics 330,
https://doi.org/10.1007/978-3-030-23531-4_8

view for the case of classical root system Φ (see [10–12]). In particular, he proved that this poset is graded, calculated the rank function and described the covering relation.

In [7], we presented another geometrical interpretation of $\mathcal{I}(A_{n-1})$ in terms of *coadjoint* B-orbits. Precisely, let U be the unipotent radical of B, and Let $\mathfrak{n}$ be the Lie algebra of U. Since B acts on $\mathfrak{n}$ via the adjoint action, one can consider the dual action of B on $\mathfrak{n}^*$, which is called *coadjoint*. To each involution $\sigma \in \mathcal{I}(A_{n-1})$ one can assign the B-orbit $\Omega_\sigma \subseteq \mathfrak{n}^*$ (see Sects. 1.2, 1.3 for precise definitions). By [7, Theorem 1.1], for $G = \mathrm{GL}_n(\mathbb{C})$, Ω_σ is contained in the Zariski closure $\overline{\Omega_\tau}$ of Ω_τ if and only if σ is less or equal to τ with respect to the Bruhat order $\leq_B$. In [8], completely similar results were obtained for $G = \mathrm{Sp}_{2n}(\mathbb{C})$ (i.e., for $\mathcal{I}(C_n)$), see [8, Theorem 1.1]. In some sense, these results are "dual" to Melnikov's results [15–17].

In this paper, we establish similar results for the cases $\Phi = B_n$ and $\Phi = D_n$. Namely, let G be the orthogonal group of rank n, i.e., $G = \mathrm{O}_{2n+1}(\mathbb{C})$ or $\mathrm{O}_{2n}(\mathbb{C})$ (respectively, $\Phi = B_n$ or D_n). In general, $\Omega_\sigma \subseteq \overline{\Omega_\tau}$ is *not* equivalent to $\sigma \leq_B \tau$, see Examples 2.3 and 2.8 below. On the other hand, we believe that these conditions *are* equivalent if we restrict ourselves to the case of the so-called basis involutions. An involution σ is called *basis* if there are no i such that $\sigma(i) = -i$. (Here W is standardly identified with certain subgroup of the symmetric group $S_{\pm n}$ on the $2n$ letters $1, \ldots, n, -n, \ldots, -1$, see Sect. 1.2 for the precise definition). The main result of the paper is as follows.

Theorem 1.1 *Let σ, τ be basis involutions in the Weyl group W of type B_n or D_n. If the orbit Ω_σ is contained in the Zariski closure of the orbit Ω_τ, then σ is less or equal to τ with respect to the Bruhat order on the group W.*

The paper is organized as follows. In the rest of this section we briefly recall basic facts about classical groups and collect our previous results about $\mathcal{I}(A_{n-1})$ and $\mathcal{I}(C_n)$, see Sect. 1.2. Then, we give precise definitions for the orthogonal case and formulate some Incitti's results about involutions needed in the sequel, see Sect. 1.3.

Section 2 is devoted to the proof of Theorem 1.1. Precisely, in Sect. 2.1 we prove it for B_n, see Theorem 2.2. Next, in Sect. 2.2 we prove this theorem for D_n (this requires some additional work due to the fact that the Bruhat order in this case has more complicated description than for B_n). In Sect. 2.3 we discuss the equivalence of the conditions $\Omega_\sigma \subseteq \overline{\Omega_\tau}$ and $\sigma \leq_B \tau$ for basis involutions. Namely, using Incitti's results, we present a conjectural way how to prove that if $\sigma \leq_B \tau$ then Ω_σ is contained in the closure of Ω_τ.

Finally, in Sect. 3 we discuss some related facts and conjectures. In Sect. 3.1, we obtain a formula for the dimension of the orbit Ω (see Theorem 3.1). In Sect. 3.2, a conjectural approach to orbits associated with involutions in terms of tangent cones to Schubert subvarieties of the flag variety G/B is presented.

1.2 From now on and to the end of the paper G denotes one of the classical complex algebraic groups $\mathrm{GL}_n(\mathbb{C})$, $\mathrm{O}_{2n+1}(\mathbb{C})$, $\mathrm{Sp}_{2n}(\mathbb{C})$ or $\mathrm{O}_{2n}(\mathbb{C})$. The group $\mathrm{O}_{2n+1}(\mathbb{C})$ (respectively, $\mathrm{Sp}_{2n}(\mathbb{C})$, $\mathrm{O}_{2n}(\mathbb{C})$) is realized as the subgroup of $\mathrm{GL}_{2n+1}(\mathbb{C})$ (respectively, of $\mathrm{GL}_{2n}(\mathbb{C})$) consisting of all invertible matrices g such that

$$\beta(gu, gv) = \beta(u, v)$$

for all u, v in $\mathbb{C}^{2n+1}$ (respectively, in $\mathbb{C}^{2n}$), where β is the bilinear form on $\mathbb{C}^{2n+1}$ (respectively, on $\mathbb{C}^{2n}$) defined as follows:

$$\beta(u, v) = \begin{cases} u_0v_0 + \sum_{i=1}^{n}(u_iv_{-i} + u_{-i}v_i) & \text{for } \mathrm{O}_{2n+1}(\mathbb{C}), \\ \sum_{i=1}^{n}(u_iv_{-i} - u_{-i}v_i) & \text{for } \mathrm{Sp}_{2n}(\mathbb{C}), \\ \sum_{i=1}^{n}(u_iv_{-i} + u_{-i}v_i) & \text{for } \mathrm{O}_{2n}(\mathbb{C}). \end{cases}$$

Here for $\mathrm{O}_{2n+1}(\mathbb{C})$ (respectively, for $\mathrm{Sp}_{2n}(\mathbb{C})$ and $\mathrm{O}_{2n}(\mathbb{C})$) we denote by $e_1, \ldots, e_n, e_{-n}, \ldots, e_{-1}$ (respectively, by $e_1, \ldots, e_n, e_0, e_{-n}, \ldots, e_{-1}$ and $e_1, \ldots, e_n, e_{-n}, \ldots, e_{-1}$) the standard basis of $\mathbb{C}^{2n+1}$ (respectively, of $\mathbb{C}^{2n}$), and by x_i the coordinate of a vector x corresponding to e_i.

The set of all diagonal matrices from G is a maximal torus in G; we denote it by H. Let Φ be the root system of G with respect to H. Note that Φ is of type A_{n-1} (respectively, B_n, C_n and D_n) for $\mathrm{GL}_n(\mathbb{C})$ (respectively, for $\mathrm{O}_{2n+1}(\mathbb{C})$, $\mathrm{Sp}_{2n}(\mathbb{C})$ and $\mathrm{O}_{2n}(\mathbb{C})$). The set of all upper-triangular matrices from G is a Borel subgroup of G containing H; we denote it by B. Let Φ^+ be the set of positive roots with respect to B. As usual, we identify Φ^+ with the following subset of the n-dimensional Euclidean space $\mathbb{R}^n$ with the standard inner product (see, e.g., [4]):

$$\begin{aligned} A_{n-1}^+ &= \{\epsilon_i - \epsilon_j,\ 1 \le i < j \le n\}, \\ B_n^+ &= \{\epsilon_i - \epsilon_j,\ 1 \le i < j \le n\} \\ &\quad \cup \{\epsilon_i + \epsilon_j,\ 1 \le i < j \le n\} \cup \{\epsilon_i,\ 1 \le i \le n\}, \\ C_n^+ &= \{\epsilon_i - \epsilon_j,\ 1 \le i < j \le n\} \\ &\quad \cup \{\epsilon_i + \epsilon_j,\ 1 \le i < j \le n\} \cup \{2\epsilon_i,\ 1 \le i \le n\}, \\ D_n^+ &= \{\epsilon_i - \epsilon_j,\ 1 \le i < j \le n\} \\ &\quad \cup \{\epsilon_i + \epsilon_j,\ 1 \le i < j \le n\}. \end{aligned}$$

Here $\{\epsilon_i\}_{i=1}^n$ is the standard basis of $\mathbb{R}^n$.

Denote by U the group of all strictly upper-triangular matrices from G with 1's on the diagonal, then U is the unipotent radical of B. Let $\mathfrak{g}$, $\mathfrak{h}$, $\mathfrak{b}$, $\mathfrak{n}$ be the Lie algebras of G, H, B, U, respectively. Then $\mathfrak{n}$ has a basis consisting of root vectors e_α, $\alpha \in \Phi^+$, where

$$e_{\epsilon_i} = \sqrt{2}(e_{0,i} - e_{-i,0}),\ e_{2\epsilon_i} = e_{i,-i},$$

$$e_{\epsilon_i-\epsilon_j} = \begin{cases} e_{i,j} & \text{for } A_{n-1}, \\ e_{i,j} - e_{-j,-i} & \text{for } B_n,\ C_n \text{ and } D_n, \end{cases}$$

$$e_{\epsilon_i+\epsilon_j} = \begin{cases} e_{i,-j} - e_{j,-i} & \text{for } B_n \text{ and } D_n, \\ e_{i,-j} + e_{j,-i} & \text{for } C_n, \end{cases}$$

and $e_{i,j}$ are the usual elementary matrices. For $\mathrm{O}_{2n+1}(\mathbb{C})$ (respectively, for $\mathrm{Sp}_{2n}(\mathbb{C})$ and $\mathrm{O}_{2n}(\mathbb{C})$) we index the rows (from left to right) and the columns (from top to bottom) of matrices by the numbers $1, \ldots, n, 0, -n, \ldots, -1$ (respectively, by the numbers $1, \ldots, n, -n, \ldots, -1$). Note that

$$\mathfrak{g} = \mathfrak{h} \oplus \mathfrak{n} \oplus \mathfrak{n}_-,$$

where $\mathfrak{n}_- = \langle e_{-\alpha},\ \alpha \in \Phi^+\rangle_{\mathbb{C}}$, and, by definition, $e_{-\alpha} = e_\alpha^T$. (The superscript T always indicates matrix transposition.)

Since $\{e_\alpha,\ \alpha \in \Phi^*\}$ is a basis of $\mathfrak{n}$, one can consider the dual basis $\{e_\alpha^*,\ \alpha \in \Phi^+\}$ of the dual space $\mathfrak{n}^*$. The group B acts on $\mathfrak{n}$ by the adjoint action (actually, by conjugation), so there exists the dual (*coadjoint*) action of B on $\mathfrak{n}^*$. We will denote the result of this action by $g.\lambda$ for $g \in B$, $\lambda \in \mathfrak{n}^*$. By definition,

$$\langle g.\lambda, x\rangle = \langle \lambda, g^{-1}xg\rangle,\ g \in B,\ x \in \mathfrak{n},\ \lambda \in \mathfrak{n}^*.$$

Orbits of the coadjoint action play the crucial role in representation theory of the groups B and U, see, e.g., [13, 14].

It is very convenient to identify $\mathfrak{n}^*$ with the space $\mathfrak{n}_-$ via

$$\langle \lambda, x\rangle = \operatorname{tr} \lambda x,\ \lambda \in \mathfrak{n}_-,\ x \in \mathfrak{n}.$$

For this reason, we will denote $\mathfrak{n}_-$ by $\mathfrak{n}^*$ and interpret it as the dual space of $\mathfrak{n}$. Under this identification,

$$e_\alpha^* = \begin{cases} e_\alpha^T & \text{if } \Phi = A_{n-1}, \text{ or } \Phi = C_n \text{ and } \alpha = 2\epsilon_i, \\ e_\alpha^T/4 & \text{if } \Phi = B_n \text{ and } \alpha = \epsilon_i, \\ e_\alpha^T/2 & \text{otherwise.} \end{cases}$$

Note that if $g \in B$, $\lambda \in \mathfrak{n}^*$, then $g.\lambda = (g\lambda g^{-1})_{\mathrm{low}}$, where A_{low} denotes the strictly lower-triangular part of a matrix A, i.e.,

$$(A_{\mathrm{low}})_{i,j} = \begin{cases} A_{i,j} & \text{for } i > j, \\ 0 & \text{for } i \leq j. \end{cases}$$

For a given $\lambda \in \mathfrak{n}^*$, let Ω_λ and Θ_λ denote its B-orbit and U-orbit under the coadjoint action, respectively. A subset $D \subset \Phi^+$ is called *orthogonal* if it consists of pairwise orthogonal roots. To each orthogonal subset D and each map $\xi\colon D \to \mathbb{C}^\times$ one can assign the linear forms

$$f_D = \sum_{\alpha \in D} e_\alpha^* \in \mathfrak{n}^*, \quad f_{D,\xi} = \sum_{\alpha \in D} \xi(\alpha) e_\alpha^* \in \mathfrak{n}^*.$$

(Obviously, $f_D = f_{D,\xi_1}$, where $\xi_1(\alpha) = 1$ for all $\alpha \in D$.) Given an orthogonal subset $D \subseteq \Phi^+$, we say that the orbits $\Omega_D = \Omega_{f_D}$ and $\Theta_{D,\xi} = \Theta_{f_{D,\xi}}$ are *associated* with D. Note that U-orbits associated with orthogonal subsets and their generalizations were studied, in particular, in [5, 6, 9, 18].

Now, let W be the Weyl group of Φ. We denote by s_α the reflection in W corresponding to a root α, and say that s_α is a *simple reflection* if α is a simple root. For $\Phi = A_{n-1}$, $W \cong S_n$ is isomorphic to the symmetric group on the n letters $1, \ldots, n$ via the isomorphism $s_{\epsilon_i - \epsilon_j} \mapsto (i, j)$, where (i, j) is the transposition interchanging i and j. For other classical root systems, denote by $S_{\pm n}$ the symmetric group on the $2n$ letters $1, \ldots, n, -n, \ldots, -1$ and consider the monomorphism from W to $S_{\pm n}$ defined by the formulas

$$s_{\epsilon_i - \epsilon_j} \mapsto (i, j)(-j, -i),$$
$$s_{\epsilon_i + \epsilon_j} \mapsto (i, -j)(j, -i),$$
$$s_{\epsilon_i} \mapsto (i, -i), \ s_{2\epsilon_i} \mapsto (i, -i).$$

For B_n and C_n, the image of this monomorphism coincides with the *hyperoctahedral* group, that is, the subgroup of $S_{\pm n}$ consisting of all permutations w from $S_{\pm n}$ such that $w(-i) = -w(i)$ for each $1 \leq i \leq n$. For D_n, the image of this monomorphism coincides with the *even-signed hyperoctahedral* group, that is, the subgroup of $S_{\pm n}$ consisting of all $w \in S_{\pm n}$ such that $w(-i) = -w(i)$ for each $1 \leq i \leq n$ and the number $|\{i > 0 \mid w(i) < 0\}|$ is even. We will identify W with its image under the above monomorphism.

Remark 1.2

(i) Note that every $w \in W$ is completely determined by its restriction to the subset $\{1, \ldots, n\}$. This allows us to use the usual two-line notation: if $w(i) = w_i$ for $1 \leq i \leq n$, then we will write

$$w = \begin{pmatrix} 1 & 2 & \ldots & n \\ w_1 & w_2 & \ldots & w_n \end{pmatrix}.$$

For instance, if $\Phi = D_5$, then

$$s_{\epsilon_1+\epsilon_5} s_{\epsilon_2+\epsilon_4} s_{\epsilon_2-\epsilon_4} = \begin{pmatrix} 1 & 2 & 3 & 4 & 5 \\ -5 & -2 & 3 & -4 & -1 \end{pmatrix}.$$

(ii) Note also that the set of simple roots has the following form: $\Delta = \{\alpha_1, \ldots, \alpha_n\}$, where

$$\alpha_1 = \epsilon_1 - \epsilon_2, \ \ldots, \ \alpha_{n-1} = \epsilon_{n-1} - \epsilon_n, \text{ and}$$

$$\alpha_n = \begin{cases} \epsilon_n & \text{for } B_n, \\ 2\epsilon_n & \text{for } C_n, \\ \epsilon_{n-1} + \epsilon_{n+1} & \text{for } D_n. \end{cases}$$

Recall that a *reduced decomposition* of an element $w \in W$ is an expression of w as a product of simple reflections of minimal possible length. Given $v, w \in W$, we say that v is less or equal to w with respect to the *Bruhat order*, written $v \leq_B w$, if some reduced decomposition for v is a subword of some reduced decomposition for w. It is well-known that this order plays the crucial role in many geometric aspects of theory of algebraic groups. For instance, the Bruhat order encodes the incidences among Schubert varieties

From now on and to the end of this subsection, let $G = \mathrm{GL}_n(\mathbb{C})$ or $\mathrm{Sp}_{2n}(\mathbb{C})$, i.e., $\Phi = A_{n-1}$ or C_n, respectively. There exists a nice combinatorial description of the Bruhat order on W. First, consider the case A_{n-1}. Given $w \in W$, denote by X_w the $n \times n$ matrix defined by

$$(X_w)_{i,j} = \begin{cases} 1, & \text{if } w(j) = i, \\ 0 & \text{otherwise.} \end{cases}$$

It is called the 0–1 matrix, permutation matrix or *rook placement* for w. Define the matrix R_w by putting its (i, j)th element to be equal to the rank of the lower left $(n - i + 1) \times j$ submatrix of X_w. In other words, $(R_w)_{i,j}$ is just the number or rooks located non-strictly to the South-West from (i, j).

Example 1.3 Let $n = 6$,

$$w = s_{\epsilon_1-\epsilon_4}s_{\epsilon_3-\epsilon_5} = \begin{pmatrix} 1\,2\,3\,4\,5\,6 \\ 4\,2\,5\,1\,3\,6 \end{pmatrix}.$$

Here we draw the matrices X_w and R_w (rooks are marked by $\otimes$):

$$X_w = \begin{array}{c|c|c|c|c|c|c|} & 1 & 2 & 3 & 4 & 5 & 6 \\ \hline 1 & & & & \otimes & & \\ \hline 2 & & \otimes & & & & \\ \hline 3 & & & & & \otimes & \\ \hline 4 & \otimes & & & & & \\ \hline 5 & & & \otimes & & & \\ \hline 6 & & & & & & \otimes \\ \hline \end{array}, \quad R_w = \begin{array}{c|c|c|c|c|c|c|} & 1 & 2 & 3 & 4 & 5 & 6 \\ \hline 1 & 1 & 2 & 3 & 4 & 5 & 6 \\ \hline 2 & 1 & 2 & 2 & 3 & 4 & 5 \\ \hline 3 & 1 & 1 & 2 & 2 & 3 & 4 \\ \hline 4 & 1 & 1 & 2 & 2 & 2 & 3 \\ \hline 5 & 0 & 0 & 1 & 1 & 1 & 2 \\ \hline 6 & 0 & 0 & 0 & 0 & 0 & 1 \\ \hline \end{array}.$$

Given two arbitrary matrices X, Y of the same size with integer entries, we will write $X \leq Y$ if $X_{i,j} \leq Y_{i,j}$ for all i, j. It turns out that, for $\Phi = A_{n-1}$ and v, $w \in W$, $v \leq_B w$ if and only if $R_v \leq R_w$, see, e.g., [3, Theorem 2.1.5].

For C_n, the description of the Bruhat order is very similar. Precisely, to each $w \in W$ one can assign the matrices X_w and R_w exactly by the same rule. (Here X_w and R_w are $2n \times 2n$ matrices, which rows and columns are indexed by the numbers $1, \ldots, n, -n, \ldots, -1$.) According to [3, Theorem 8.1.1], for $\Phi = C_n$ and v, $w \in W$, $v \leq_B w$ if and only if $R_v \leq R_w$, as above. In other words, the Bruhat order on W is nothing but the restriction of the Bruhat order on $S_{\pm n}$ to W.

Let $G = \mathrm{GL}_n(\mathbb{C})$ or $\mathrm{Sp}_{2n}(\mathbb{C})$, and w be an involution in W, i.e., $w \in \mathcal{I}(\Phi)$. Then w can be expressed as a product of pairwise commuting reflections. In other words, there exists an orthogonal subset $D \subseteq \Phi^+$ such that $w = \prod_{\alpha \in D} s_\alpha$. For A_{n-1}, such an expression is clearly unique. For C_n, such an expression is unique if we require D to be strongly orthogonal, which means that, given $\alpha, \beta \in D$, neither $\alpha + \beta \in \Phi^+$ nor $\alpha - \beta \in \Phi^+$. Thus, in both cases, the subset D is uniquely determined by w. We call the subset D the *support* of the involution w and denote it by $D = \mathrm{Supp}(w)$. For instance, if $\Phi = C_6$ and

$$w = \begin{pmatrix} 1 & 2 & 3 & 4 & 5 & 6 \\ 3 & -6 & 1 & -4 & -5 & -2 \end{pmatrix} = s_{\epsilon_1-\epsilon_3}s_{\epsilon_2+\epsilon_6}s_{2\epsilon_4}s_{2\epsilon_5}$$

$$= s_{\epsilon_1-\epsilon_3}s_{\epsilon_2+\epsilon_6}s_{\epsilon_4-\epsilon_5}s_{\epsilon_4+\epsilon_5},$$

then $\mathrm{Supp}(w) = \{\epsilon_1 - \epsilon_3,\ \epsilon_2 + \epsilon_6,\ 2\epsilon_4,\ 2\epsilon_5\}$.

Definition 1.4 Let $w \in \mathcal{I}(\Phi)$, and $\xi: D \to \mathbb{C}^\times$ be a map, where $D = \mathrm{Supp}(w)$. Denote

$$f_w = f_D,\ f_{w,\xi} = f_{D,\xi},\ \Omega_w = \Omega_D,\ \Theta_{w,\xi} = \Theta_{D,\xi}.$$

We say that the orbits Ω_w and $\Theta_{w,\xi}$ are associated with the involution w. Note that, thanks to [7, Lemma 2.1] and [8, Lemma 1.8],

$$\Omega_w = \bigcup_\xi \Theta_{w,\xi}.$$

To formulate the description of the incidences among B-orbits associated with involutions, we need one more partial order on $\mathcal{I}(\Phi)$. Namely, given $w \in W$, we put $R_w^* = (R_w)_{\mathrm{low}}$ and write $v \leq^* w$ for $v,\ w \in \mathcal{I}(\Phi)$ if $R_v^* \leq R_w^*$. Then, according to [7, Theorem 1.1] and [8, Theorem 1.1], we have the following result.

Theorem 1.5 *Let $\Phi = A_{n-1}$ or C_n, and $v,\ w \in \mathcal{I}(\Phi)$. Then the following conditions are equivalent*:

(i) $\Omega_v \subseteq \overline{\Omega_w}$;

(ii) $v \leq^* w$;

(iii) $v \leq_B w$.

1.3 Suppose now that $G = O_{2n+1}(\mathbb{C})$ or $O_{2n(\mathbb{C})}$, i.e., $\Phi = B_n$ or D_n, respectively. Since the Weyl group of B_n is isomorphic to the Weyl group of C_n, the Bruhat order on W for B_n can be described completely similarly to the case $\Phi = C_n$: given $v,\ w \in W$, one has $v \leq_B w$ if and only if $R_v \leq R_w$, where R_v, R_w are the $2n \times 2n$ defined in the previous subsection.

For D_n, the description of the Bruhat order is quite more complicated. Let $w \in W$. Given numbers $a, b \in \{1, 2, \ldots, n\}$, we say that $[-a, a] \times [-b, b]$ is an *empty rectangle* for w, if

$$\{i \in [\pm n] \mid |i| \geq b \text{ and } |w(i)| \geq a\} = \varnothing.$$

Here $[\pm n] = \{1, \ldots, n, -n, \ldots, -1\}$. For instance, let $n = 4$ and $w = \begin{pmatrix} 1 & 2 & 3 & 4 \\ -2 & 4 & 1 & -3 \end{pmatrix}$, then

$$X_w = \begin{array}{c|c|c|c|c|c|c|c|c|} & 1 & 2 & 3 & 4 & -4 & -3 & -2 & -1 \\ \hline 1 & & & \otimes & & & & & \\ \hline 2 & & & & & & & & \otimes \\ \hline 3 & & & & & \otimes & & & \\ \hline 4 & & \otimes & & & & & & \\ \hline -4 & & & & & & & \otimes & \\ \hline -3 & & & & \otimes & & & & \\ \hline -2 & \otimes & & & & & & & \\ \hline -1 & & & & & & \otimes & & \\ \hline \end{array},$$

and $[-4, 4] \times [-3, 3]$, $[-4, 4] \times [-4, 4]$ are empty rectangles for w.

It turns out [3, Theorem 8.2.8] that, given $v, w \in W$, $v \leq w$ if and only if

$$\begin{aligned} &\text{(i) } R_v \leq R_w; \\ &\text{(ii) for all } a, b \in \{1, \dots, n\}, \text{ if } [-a, a] \times [-b, b] \text{ is an empty rectangle} \\ &\quad \text{for both } v \text{ and } w \text{ and } (R_v)_{-(a-1),b-1} = (R_w)_{-(a-1),b-1}, \\ &\quad \text{then } (R_v)_{-(a-1),n} \equiv (R_w)_{-(a-1),n} \pmod 2. \end{aligned} \tag{1}$$

Definition 1.6 Let $\Phi = B_n$ or D_n, and $w \in \mathcal{I}(\Phi)$. The involution w is called *basis* if $|\{i \in \{1, \ \dots, \ n\} \mid w(i) = -i\}| = 0$.

We will denote the set of all basis involutions in W by $\mathcal{B}(\Phi)$. It is clear that if w is a basis involution then there exists the unique orthogonal subset $D \subseteq \Phi^+$ such that

$$w = \prod_{\alpha \in D} s_\alpha.$$

As above, we call D *the support* of w and denote $D = \mathrm{Supp}(w)$. For example, for $\Phi = B_5$,

$$w = \begin{pmatrix} 1 & 2 & 3 & 4 & 5 \\ -5 & 2 & 4 & 3 & -1 \end{pmatrix},$$

we have $\mathrm{Supp}(w) = \{\epsilon_1 + \epsilon_5, \ \epsilon_3 - \epsilon_4\}$. Now, given $w \in \mathcal{B}(\Phi)$ and a map $\xi \colon D = \mathrm{Supp}(w) \to \mathbb{C}^\times$, we define the linear forms f_w, $f_{w,\xi}$ and the orbits Ω_w and $\Theta_{w,\xi}$ exactly as in Definition 1.4. As for A_n and C_n, we say that Ω_w and $\Theta_{w,\xi}$ are *associated* with w. The goal of the paper is to prove that, given $\sigma, \ \tau \in \mathcal{B}(\Phi)$, $\Omega_\sigma \subseteq \overline{\Omega_\tau}$ implies $\sigma \leq_B \tau$. Note that this is clearly not true for arbitrary involutions in W, as it is shown in Examples 2.3, 2.8 below.

2 Proof of the Main Theorem

2.1 In this subsection, we will check that, given $\sigma, \tau \in \mathcal{B}(B_n)$, $\Omega_\sigma \subseteq \overline{\Omega_\tau}$ implies that $\sigma \leq_B \tau$. To do this, we need the following simple observation (cf. [15, Subsection 3.3], [7, Lemma 2.1], [8, Lemma 1.8], [9, Lemma 1.3]).

Lemma 2.1 *Let* $\Phi = B_n$ *or* D_n, $w \in \mathcal{B}(\Phi)$ *and* $D = \mathrm{Supp}(w)$. *Then* $\Omega_w = \bigcup \Theta_{w,\xi}$, *where the union is taken over all maps* $\xi\colon D \to \mathbb{C}^\times$.

Proof It is well-known that the map

$$\exp\colon \mathfrak{n} \to U,\ x \mapsto \sum_{k\geq 0} \frac{x^k}{k!}$$

is well-defined and is an isomorphism of affine varieties. For given $\alpha \in \Phi^+$, $s \in \mathbb{C}^\times$, put

$$x_\alpha(s) = \exp(se_\alpha),\ x_{-\alpha}(s) = x_\alpha(s)^T,$$
$$w_\alpha(s) = x_\alpha(s)x_{-\alpha}(-s^{-1})x_\alpha(s),\ h_\alpha(s) = w_\alpha(s)w_\alpha(1)^{-1}.$$

Then $h_\alpha(s)$ is a diagonal matrix from H. Furthermore, the group H is generated by $h_\alpha(s)$, $\alpha \in \Phi^+$, $s \in \mathbb{C}$, and $B = U \rtimes H$.

Let $\xi\colon D \to \mathbb{C}^\times$ be a map. To check that $\Theta_{w,\xi} \subseteq \Omega_w$, it is enough to find $h \in H$ such that $h.f_{w,\xi} = f_w$. One can easily see that if $\alpha \in D$ then

$$h_\alpha(t).f_{w,\xi} = \sum_{\beta\in D,\ \beta\neq\alpha} \xi(\beta)e_\beta^* + t^{-2}\xi(\alpha)e_\alpha^*.$$

Thus, $h.f_{w,\xi} = f_w$, where $h = \prod_{\alpha\in D} h_\alpha(\sqrt{\xi(\alpha)})$. (Here, given $s \in \mathbb{C}$, we denote by $\sqrt{s}$ a complex number such that $(\sqrt{s})^2 = s$.)

On the other hand, let $h \in H$. We claim that $h.f_{w,\xi} = f_{w,\xi'}$ for some ξ'. Indeed, since H is generated by $h_\alpha(s)$'s, $\alpha \in \Phi^+$, $s \in \mathbb{C}^\times$, we can assume without loss of generality that $h = h_\alpha(s)$ for some α and s. But in this case the statement follows immediately from the above. Since the group B is isomorphic as an algebraic group to the semi-direct product $U \rtimes H$, for a given $g \in B$, there exist unique $u \in U$, $h \in H$ such that $g = uh$. If $\xi\colon D \to \mathbb{C}^\times$ is the map such that $h.f_w = f_{w,\xi}$, then $g.f_w = u.f_{w,\xi} \in \Theta_{w,\xi}$. This concludes the proof. □

Now, if $\Phi = B_n$ or D_n, and $x \in \mathfrak{g}$, then, given $i, j \in [\pm n]$, denote

$$\pi_{i,j}(x) = \begin{pmatrix} x_{i,1} & \dots & x_{i,j} \\ \vdots & \ddots & \vdots \\ x_{-1,1} & \dots & x_{-1,j} \end{pmatrix}.$$

It is easy to see that if $w \in \mathcal{B}(\Phi)$ then $\operatorname{rk}\pi_{i,j}(\lambda) = (R_w^*)_{i,j}$ for all $\lambda \in \Omega_w$ and all lower-triangular entries (i, j) of λ (cf. [7, Lemma 2.2], [8, Lemma 2.4], [9, Theorem 1.5]). Indeed, by definition of R_w^*, $(R_w^*)_{i,j} = \operatorname{rk}\pi_{i,j}(f_w)$. Let $\xi \colon D \to \mathbb{C}^\times$ be a map. Since

$$\operatorname{rk}\pi_{i,j}(f_{w,\xi}) = \operatorname{rk}\pi_{i,j}(f_w) = (R_w^*)_{i,j},$$

it suffices to check that $\operatorname{rk}\pi_{i,j}(\lambda) = \operatorname{rk}\pi_{i,j}(u.\lambda)$ for $u \in U$, $\lambda \in \mathfrak{n}^*$. This follows immediately from the proof of [7, Lemma 2.2], because u is an upper-triangular matrix with 1's on the diagonal and λ is a lower-triangular matrix with zeroes on the diagonal. Now we are ready to prove the main result of this subsection, cf. [7, Proposition 2.3], [8, Proposition 2.5], [9, Theorem 1.5].

Theorem 2.2 *Let $\sigma, \tau \in \mathcal{B}(B_n)$. If $\Omega_\sigma \subseteq \overline{\Omega_\tau}$, then $\sigma \leq_B \tau$.*

Proof Suppose that $\sigma \not\leq_B \tau$. According to Theorem 1.5, this means that there exist i, j such that $(R_\sigma^*)_{i,j} > (R_\tau^*)_{i,j}$. Denote

$$Z = \{f \in \mathfrak{n}^* \mid \operatorname{rk}\pi_{r,s}(f) \leq (R_\tau^*)_{r,s} \text{ for all } r, s\}.$$

Clearly, Z is closed with respect to the Zariski topology. It follows from the above that $\Omega_\tau \subseteq Z$, so $\overline{\Omega_\tau} \subseteq Z$. But $f_\sigma \notin Z$, hence $\Omega_\sigma \not\subseteq Z$, a contradiction. □

Example 2.3 If involution σ is not basis then, in general, its support (and, consequently, the associated B-orbit Ω_σ) is not well-defined, because, in general, there are several different ways to represent σ as a product of pairwise commuting reflections. For example, let $n = 4$, then

$$\sigma = \begin{pmatrix} 1 & 2 & 3 & 4 \\ 1 & -2 & -3 & 4 \end{pmatrix} = s_{\epsilon_2}s_{\epsilon_3} = s_{\epsilon_2-\epsilon_3}s_{\epsilon_2+\epsilon_3}.$$

Assume for a moment that we set

$$D_\sigma = \{\epsilon_2, \epsilon_3\}$$

to be the support of σ. Then, of course, we have to say that $D_\tau = \{\epsilon_1, \epsilon_4\}$ is the support of the involution

$$\tau = \begin{pmatrix} 1 & 2 & 3 & 4 \\ -1 & 2 & 3 & -4 \end{pmatrix} = s_{\epsilon_1}s_{\epsilon_4} = s_{\epsilon_1-\epsilon_4}s_{\epsilon_1+\epsilon_4},$$

and Theorem 2.2 fails immediately. Indeed, it follows from [5, Proposition 2.1] that $\Omega_\sigma = \Omega_{s_{\epsilon_2}}$ and $\Omega_\tau = \Omega_{s_{\epsilon_1}}$, and once can easily check that $\Omega_{s_{\epsilon_2}} \subseteq \overline{\Omega_{s_{\epsilon_1}}}$ (and so $\Omega_\sigma \subseteq \overline{\Omega_\tau}$): if

$$g = x_{\epsilon_1-\epsilon_2}(-t^{-1})h_{\epsilon_1-\epsilon_2}(t^{-1}),$$

then $g.f_{\epsilon_1} \to f_{\epsilon_2}$ as $t \to 0$ (it is well-known that the Zariski closure of a constructive set coincides with its closure in the complex topology). At the same time, Theorem 1.5 claims that $\sigma \not\leq_B \tau$.

On the other hand, we may set $\{\epsilon_2 - \epsilon_3, \ \epsilon_2 + \epsilon_3\}$ to be the support of σ. (According to [20], this choice is "more canonical" than the previous one.) If we define the support of an involution in such a way, we neither have counterexamples to Theorem 2.2 nor can prove it in general.

2.2 In this subsection, we check that if $\sigma, \ \tau \in \mathcal{B}(D_n)$, then Ω_σ is contained in the closure of Ω_τ. First, applying Lemma 2.1, one can repeat the proof of Theorem 2.2 literally to show that $(R_\sigma^*)_{i,j} \leq (R_\tau^*)_{i,j}$ for all $i, \ j$. According to Theorem 1.5, this means that $(R_\sigma)_{i,j} \leq (R_\tau)_{i,j}$ for all $i, \ j$. Hence it remains to prove that the basis involutions σ and τ satisfy the second condition in (1). To do this, we need to introduce some more notation.

Let $\Phi = D_n$ and $x \in \mathfrak{g}$. Given $r \leq n$ and two ordered r-tuples $P = \{p_1, \dots, p_r\}$ and $Q = \{q_1, \dots, q_r\}$, where $1 \leq q_i \leq n$ and $p_i \in [\pm n]$ for $1 \leq i \leq r$, we denote by $\Delta_P^Q(x) = \Delta_{p_1,\dots,p_r}^{q_1,\dots,q_r}(x)$ the minor of the matrix x with the set of rows P and the set of columns Q, i.e.,

$$\Delta_P^Q(x) = \begin{vmatrix} x_{p_1,q_1} & \dots & x_{a_1,q_r} \\ \vdots & \ddots & \vdots \\ x_{p_r,q_1} & \dots & x_{p_r,q_r} \end{vmatrix}.$$

Given ordered tuples $I = \{i_1, \dots, i_r\}$, $J = \{j_1, \dots, j_s\}$, we denote by $I \cup J$ its concatenation, i.e., $I \cup J = \{i_1, \dots, i_r, j_1, \dots, j_s\}$.

If $P = \{p_1, \dots, p_r\}$ is an r-tuple and $1 \leq i < j \leq r$, then define the r-tuples $P^+[p_i, p_j]$ and $P^-[p_i, p_j]$ by the formula

$$P^+[p_i, p_j] = \{p_1, \dots, p_{i-1}, p_{i+1}, \dots, p_{j-1}, p_{j+1}, \dots, p_r, p_i, -p_i\},$$

$$P^-[p_i, p_j] = \{p_1, \dots, p_{i-1}, p_{i+1}, \dots, p_{j-1}, p_{j+1}, \dots, p_r, p_j, -p_j\}.$$

Next, if $P = \{p_1, \dots, p_r\}$ is an r-tuple, $p_i \in [\pm n]$ for $1 \leq i \leq r$, and $P' = \{p_{i_1}, \dots, p_{i_{2s}}\}$ is a tuple of even cardinality such that all p_{i_j} are distinct and belong to P, we define the set $\mathcal{S}_{P,P'}$ of r-tuples by the following rule. If $P' = \varnothing$, then $\mathcal{S}_{P,P'} = \{P\}$. For $s \geq 1$, we put

$$\mathcal{S}_{P,P'} = \bigcup_{P_0 \in \mathcal{S}_{P,P''}} \{P_0^+[p_{i_{2s-1}}, p_{i_{2s}}], P_0^-[p_{i_{2s-1}}, p_{i_{2s}}]\},$$

where, by definition, $P'' = \{p_{i_1}, \ldots, p_{i_{2s-2}}\}$. For example, if $P = \{1, 2, -3, -4\}$, $P' = \{1, -3, 2, -4\}$, then $\mathcal{S}_{P,P'}$ consists of four tuples: $\{1, -1, 2, -2\}$, $\{1, -1, -4, 4\}$, $\{-3, 3, 2, -2\}$, and $\{-3, 3, -4, 4\}$. Finally, if P, Q are r-tuples and P' is as above, we define the polynomial

$$D^Q_{P,P'}(x) = \sum_{P_0 \in \mathcal{S}_{P,P'}} \Delta^Q_{P_0}(x).$$

The following technical (but important) proposition is the key step in the proof of the main result of this subsection. Assume that $w \in \mathcal{B}(D_n)$ and $[a, -a] \times [b, -b]$, $a \geq b$, is an empty rectangle for w. Let $P = \{p_1, \ldots, p_r\}$, $Q = \{q_1, \ldots, q_s\}$ be r-tuples, $1 \leq q_i \leq b - 1$, $p_i \in [\pm n]$ for all i, where $r = (R^*_w)_{a,b-1}$. Assume that $P = I \cup J \cup K$, where each element of I (respectively, of J and K) is from a to n (respectively, from $-n$ to $-a$ and from $-a + 1$ to -1). Suppose that $|I \cup J| < n - a + 1$ or

$$\#\{\epsilon_i - \epsilon_j \in \mathrm{Supp}(w) \mid a \leq j \leq n\} \not\equiv |I| \pmod 2,$$
$$\#\{\epsilon_i + \epsilon_j \in \mathrm{Supp}(w) \mid a \leq j \leq n\} \not\equiv |J| \pmod 2.$$

(For $|I \cup J| = n - a + 1$, these two conditions are in fact equivalent.)

Proposition 2.4 *Let $P' = \{p_{i_1}, p_{i_2}, \ldots\}$ be a tuple of even cardinality with distinct elements containing in $I \cup J$. Then $D^Q_{P,P'}(\lambda) = 0$ for all $\lambda \in \Omega_w$.*

Proof Pick $\lambda \in \Omega_w$. According to Lemma 2.1, $\lambda = u.f_{w,\xi}$ for certain $u \in U$, $\xi\colon D \to \mathbb{C}^\times$, where $D = \mathrm{Supp}(w)$. Since U is generated by $x_\alpha(s)$, $\alpha \in \Phi^+$, $s \in \mathbb{C}^\times$, there exist $\alpha_1, \ldots, \alpha_k \in \Phi^+$, $s_1, \ldots, s_k \in \mathbb{C}^\times$ such that $u = x_{\alpha_1}(s_1) \ldots x_{\alpha_k}(s_k)$. The proof is by induction on k. The base $k = 0$ is trivial. Thus, we must prove the following fact: if $f \in \mathfrak{n}^*$, $\alpha \in \Phi^+$, $s \in \mathbb{C}^\times$, and all possible $D^{Q_0}_{P_0,P_0'}(f)$ are zero, then $D^Q_{P,P'}(x_\alpha(s).f) = 0$.

Given $i, j \in [\pm n]$, $s \in \mathbb{C}$, consider the elementary transformations $r^s_{i,j}$ and $c^s_{i,j}$, where, for $f \in \mathfrak{n}^*$,

$$(r^s_{i,j}(f))_{p,q} = \begin{cases} f_{i,q} + s f_{i,j} & \text{if } p = i > q, \\ f_{p,q} & \text{otherwise,} \end{cases}$$

$$(c^s_{i,j}(f))_{p,q} = \begin{cases} f_{p,j} + s f_{q,j} & \text{if } p > q = j, \\ f_{p,q} & \text{otherwise.} \end{cases}$$

Note that

$$x_{\epsilon_i - \epsilon_j}(s).f = r^s_{i,j}(r^{-s}_{-j,-i}(c^s_{j,i}(c^{-s}_{-i,-j}(f)))),$$
$$x_{\epsilon_i + \epsilon_j}(s).f = r^s_{i,-j}(r^{-s}_{j,-i}(c^s_{-i,j}(c^{-s}_{-j,i}(f)))).$$

Hence it is enough to prove that if all possible $D^{Q_0}_{P_0,P'_0}(f)$ are zero then $D^{Q}_{P,P'}(f')=0$, where $f'=r^s_{i,j}(r^{-s}_{-j,-i}(f))$, $r^s_{i,-j}(r^{-s}_{j,-i}(f))$, $c^s_{j,i}(c^{-s}_{-i,-j}(f))$ or $c^s_{-i,j}(c^{-s}_{-j,i}(f))$ for certain $i<j$, $s\in\mathbb{C}^{\times}$. We will consider this cases subsequently.

First, let $f'=r^s_{i,j}(r^{-s}_{-j,-i}(f))$. Given a tuple $T=\{t_1,t_2,\ldots\}$ with distinct elements and numbers $t_i\in T$, t, we denote by $T[t\to t_i]$ the tuple $T[t\to t_i]=\{t_1,\ldots,t_{i-1},t,t_{i+1},\ldots\}$. If $\pm i,\pm j\notin P$ (or $-i\in K$, $\pm j\notin P$, or $j\in I$, $j\notin P'$, $\pm i\notin P$, or $-i\in J$, $-i\notin P'$, $\pm j\notin P$, or $-i\in K$, $j\in I$, $j\notin P'$), then, obviously, $D^Q_{P,P'}(f')=D^Q_{P,P'}(f)=0$. If $i\in I$, $i\notin P'$, $\pm j\notin P$, then $|I[j\to i]\cup J|=|I\cup J|<n-a+1$, hence

$$D^Q_{P,P'}(f')=D^Q_{P,P'}(f)+sD^Q_{P[j\to i],P'}(f)=0.$$

If $\pm i\notin P$, $-j\in J$, $-j\notin P'$ (respectively, $\pm i\notin P$, $-j\in K$), then

$$D^Q_{P,P'}(f')=D^Q_{P,P'}(f)-sD^Q_{P[-i\to -j],P'}(f)=0,$$

because, clearly, $|I\cup J[-i\to -j]|<n-a+1$ (respectively, because $P[-i\to j]=I\cup J\cup K[-i\to j]$).

Suppose $i\in I$, $i\in P'$, $\pm j\notin P$ (the case $-i\in J$, $-i\in P'$, $\pm j\notin P$ is similar). Given a tuple $T=\{t_1,t_2,\ldots\}$ with distinct elements and a number $t_i\in T$, we denote by $T\setminus\{t_i\}$ the tuple $T\setminus\{t_i\}=\{t_1,\ldots,t_{i-1},t_{i+1},\ldots\}$. If t_i, t_j, ... are distinct elements of T, we define the tuple $T\setminus\{t_i,t_j,\ldots\}$ similarly. If $i=p_{i_{2k-1}}\in P'$ (respectively, $i=p_{i_{2k}}\in P'$) for some k, then denote $l=p_{i_{2k}}$ (respectively, $l=p_{i_{2k-1}}$). One has

$$D^Q_{P,P'}(f')=D^Q_{P,P'}(f)\pm sD^Q_{(P[j\to l])[-i\to i],P'\setminus\{i,l\}}(f)=0,$$

because $|((I\setminus\{i\})\cup(J\cup\{-i\}))[j\to l]|=|I\cup J|<n-a+1$.

Suppose now that $-j\in J$, $-j\in P'$, $\pm i\notin P$. (The cases $j\in I$, $j\in P'$, $\pm i\notin P$ and $-i\in J$, $-i\in P'$, $\pm j\notin P$ are similar.) If $-j=p_{i_{2k-1}}\in P'$ (respectively, $-j=p_{i_{2k}}\in P'$) for some k, then denote $l=p_{i_{2k}}$ (respectively, $l=p_{i_{2k-1}}$). Then

$$D^Q_{P,P'}(f')=D^Q_{P,P'}(f)\pm sD^Q_{(P[i\to l])[j\to -j],P'\setminus\{i,l\}}(f).$$

If $i\geq a$, then we can argue as in the previous paragraph. If $i<a$, then the last summand is zero because

$$|(I\cup(J\setminus\{-j\}))\setminus\{l\}|<|I\cup J|\leq n-a+1.$$

If $i,\ j \subset I, i,\ j \notin P'$, then one has

$$D^Q_{P,P'}(f') = D^Q_{P,P'}(f) + s \sum_{P_0 \in \mathcal{S}_{P,P'}} \Delta^Q_{P_0[j\to i]}(f) = 0,$$

because each minor in the last summand contains two same rows. The cases $-i,\ -j \in J, -i,\ -j \notin P'$, and $-i \in J, -i \in P', j \in I, j \notin P'$, and $-i \in J$, $-i \notin P', j \in I, j \in P'$, and $i,\ j \in I, i \in P', j \notin P'$, and $-i,\ -j \in P', -i \notin P'$, $-j \in P'$, and $-i \in K, j \in I, j \in P'$, and $-i \in K, -j \in J, -j \in P'$, and $-i \in K, -j \in J, -j \notin P'$, and $-i,\ -j \in K$ are similar to this case.

Assume that $i,\ j \in I, i \notin P', j \in P'$. (The cases $i \in I, i \in P', -j \in J$, $-j \notin P'$, and $i \in I, -j \in J, i \notin P', -j \in P'$, and $-i,\ -j \in J, -i \in P', j \notin P'$ are similar.) We see that

$$\begin{aligned} D^Q_{P,P'}(f') = D^Q_{P,P'}(f) &+ s \sum_{P_0 \in \mathcal{S}_{P,P'}, j \in P_0} \Delta^Q_{P_0[j\to i]}(f) \\ &+ s \sum_{P_0 \in \mathcal{S}_{P,P'}, j \notin P_0} \Delta^Q_{P_0[j\to i]}(f) \\ &- s \sum_{P_0 \in \mathcal{S}_{P,P'}, j \in P_0} \Delta^Q_{P_0[-i\to -j]}(f) \\ &- s^2 \sum_{P_0 \in \mathcal{S}_{P,P'}, j \in P_0} \Delta^Q_{(P_0[j\to i])[-i\to -j]}(f). \end{aligned}$$

Each minor in the second and the last summands contains two same rows. Thus,

$$\begin{aligned} D^Q_{P,P'}(f') &= s \sum_{P_0 \in \mathcal{S}_{P,P'}, j \notin P_0} \Delta^Q_{P_0[j\to i]}(f) - s \sum_{P_0 \in \mathcal{S}_{P,P'}, j \in P_0} \Delta^Q_{P_0[-i\to -j]}(f) \\ &= \pm s D^Q_{P,P'[j\to i]}(f) = 0. \end{aligned}$$

Assume now that $i,\ j \in I, i\ , j \in P'$. (The cases $-i\ , -j \in J, -i,\ -j \in P'$ and $i \in I, -j \in J, i,\ -j \in P'$, and $-i \in J, j \in I, -i,\ j \in P'$ are similar.) If $i = p_{i_{2k-1}}$ and $j = p_{i_{2k}}$, or $i = p_{i_{2k}}$ and $j = p_{i_{2k-1}}$ for certain k, then

$$\begin{aligned} D^Q_{P,P'}(f') = D^Q_{P,P'}(f) &+ s \sum_{P_0 \in \mathcal{S}_{P,P'}, i \in P_0} \Delta^Q_{P_0[j\to i]}(f) \\ - s \sum_{P_0 \in \mathcal{S}_{P,P'}, j \in P_0} \Delta^Q_{P_0[-i\to -j]}(f)& \\ = \pm s D^Q_{P[-i\to i], P'\setminus\{i,j\}}(f) &\mp s D^Q_{P[-i\to i], P'\setminus\{i,j\}}(f) = 0. \end{aligned}$$

On the other hand, suppose that such a number k does not exist. If $i = p_{i_{2k-1}}$ (respectively, $i = p_{i_{2k}}$) for certain k, then denote $i' = p_{i_{2k}}$ (respectively, $i' = p_{i_{2k-1}}$); define the number j' in a similar way using j instead of i. Then

$$\begin{aligned}D^Q_{P,P'}(f') = D^Q_{P,P'}(f) &+ s \sum_{P_0\in\mathcal{S}_{P,P'},i,j\in P_0} \Delta^Q_{P_0[j\to i]}(f)\\
&- s \sum_{P_0\in\mathcal{S}_{P,P'},i,j\in P_0} \Delta^Q_{P_0[-i\to -j]}(f)\\
&+ s \sum_{P_0\in\mathcal{S}_{P,P'},i,j'\in P_0} \Delta^Q_{P_0[j\to i]}(f)\\
&- s \sum_{P_0\in\mathcal{S}_{P,P'},i',j\in P_0} \Delta^Q_{P_0[-i\to -j]}(f)\\
&- s^2 \sum_{P_0\in\mathcal{S}_{P,P'},i,j\in P} \Delta^Q_{P_0([j\to i])(-i\to -j)}(f).\end{aligned}$$

The second, the third and the last summands are zero, because each minor in these summands contains two same rows. Thus,

$$\begin{aligned}D^Q_{P,P'}(f') &= s \sum_{P_0\in\mathcal{S}_{P,P'},i,j'\in P_0} \Delta^Q_{P_0[j\to i]}(f) - s \sum_{P_0\in\mathcal{S}_{P,P'},i',j\in P_0} \Delta^Q_{P_0[-i\to -j]}(f)\\
&= \pm s D^Q_{\widetilde{P},\widetilde{P}'}(f) = 0,\end{aligned}$$

where $\widetilde{P} = (P[-i \to i])[-j' \to j']$, $\widetilde{P}' = P' \setminus \{i, j\}$, because $\widetilde{P}$ satisfies all required conditions.

One of the most interesting cases is $i \in I, -j \in J, i, \ -j \notin P'$. Here

$$\begin{aligned}D^Q_{P,P'}(f') = D^Q_{P,P'}(f) &+ s \sum_{P_0\in\mathcal{S}_{P,P'}} \Delta^Q_{P_0[j\to i]}(f)\\
&- s \sum_{P_0\in\mathcal{S}_{P,P'}} \Delta^Q_{P_0[-i\to -j]}(f)\\
&- s^2 \sum_{P_0\in\mathcal{S}_{P,P'}} \Delta^Q_{(P_0[j\to i])[-i\to -j]}(f).\end{aligned}$$

The last summand equals $\pm s^2 D^Q_{(P[j\to i])[-i\to -j],P'}(f) = 0$. Hence

$$\begin{aligned}D^Q_{P,P'}(f') &= s \sum_{P_0\in\mathcal{S}_{P,P'}} \Delta^Q_{P_0[j\to i]}(f) - s \sum_{P_0\in\mathcal{S}_{P,P'}} \Delta^Q_{P_0[-i\to -j]}(f)\\
&= \pm s D^Q_{P,P'\cup\{i,-j\}}(f) = 0.\end{aligned}$$

The case $i \in J, j \in I, \quad i, j \notin P'$ is completely similar.

The cases $i \in I, i \notin P', -j \in K$, and $i \in I, i \notin P', -j \in K$, and $-i \in J, -i \notin P', -j \in K$, and $-i \in I, i \in P', -j \in K$ cannot occur because of the definition of I, J, K. Thus, we have considered all possible cases for $f' = r^s_{i,j}(r^{-s}_{-j,-i}(f))$.

Second, let $f' = r^s_{i,-j}(r^{-s}_{j,-i}(f))$. The proof in this case is similar to the proof for $f' = r^s_{i,j}(r^{-s}_{-j,-i}(f))$, so we skip the details. Next, let $f' = c^s_{j,i}(c^{-s}_{-i,-j}(f))$. If $j \notin Q$ then, clearly,

$$D^Q_{P,P'}(f') = D^Q_{P,P'}(f) = 0.$$

If $j \in Q, i \notin Q$, then

$$D^Q_{P,P'}(f') = D^Q_{P,P'}(f) + s D^{Q[i \to j]}_{P,P'}(f) = 0.$$

If $i, \ j \in Q$, then

$$D^Q_{P,P'}(f') = D^Q_{P,P'}(f) + s \sum_{P_0 \in \mathcal{S}_{P,P'}} \Delta^{Q[i \to j]}_{P_0}(f) = 0,$$

because each minor in the last summand contains two same columns. Finally, let $f' = c^s_{-i,j}(c^{-s}_{-j,i}(f))$, then $D^Q_{P,P'}(f') = D^Q_{P,P'}(f) = 0$. The proof is complete. $\square$

Now we can prove the main result of this subsection.

Theorem 2.5 *Let* $\sigma, \ \tau \in \mathcal{B}(D_n)$. *If* $\Omega_\sigma \subseteq \overline{\Omega}_\tau$, *then* $\sigma \leq_B \tau$.

Proof Assume that $\sigma \not\leq_B \tau$. If there exist $i, \ j$ such that $(R^*_\sigma)_{i,j} > (R^*_\tau)_{i,j}$, then, using Lemma 2.1 and arguing as in the proof of Theorem 2.2, one can check that $\Omega_\sigma \not\subseteq \overline{\Omega}_\tau$. Hence $R^*_\sigma \leq R^*_\tau$, and, by Theorem 1.5, $R_\sigma \leq R_\tau$. Thus, according to (1), there exist a, b such that $[a, -a] \times [b, -b]$ is an empty rectangle for σ and τ, $(R_\sigma)_{-(a-1),b-1} = (R_\tau)_{-(a-1),b-1}$, but $(R_\sigma)_{-(a-1),n} \not\equiv (R_\tau)_{-(a-1),n} \pmod 2$. Since σ and τ are involutions, we may assume without loss of generality that $a \geq b$.

Recall the notion of X_w for $w \in W$ from Sect. 1.3. Given an arbitrary element $w \in W$, $-n \leq p \leq q \leq -1$ and $1 \leq r \leq s \leq n$, denote

$$w^{r,s}_{p,q} = \#\{(i, j) \mid p \leq i \leq q, \ r \leq j \leq s \text{ and } (X_w)_{i,j} = 1\}.$$

In other words, $w^{r,s}_{p,q}$ is the number of rooks in X_w which rows (respectively, columns) are between p and q (respectively, between r and s). By definition of R_w,

$$\begin{aligned}(R_w)_{-(a-1),-n} = {} & w^{1,b-1}_{-(b-1),-1} + w^{1,b-1}_{-(a-1),-b} + w^{b,a-1}_{-(b-1),-1} \\ & + w^{b,a-1}_{-(a-1),-b} + w^{a,n}_{-(b-1),-1} + w^{a,n}_{-(a-1),-b}.\end{aligned}$$

If w is an involution in the Weyl group W, then $w^{1,b-1}_{-n,-a} = w^{a,n}_{-(b-1),-1}$, while the numbers $w^{1,b-1}_{-(b-1),-1}$ and $w^{b,a-1}_{-(a-1),-b}$ are even. Furthermore, in this case,

$$\begin{aligned} w^{1,b-1}_{-(a-1),-1} &= w^{1,b-1}_{-(b-1),-1} + w^{1,b-1}_{-(a-1),-b} \\ &= w^{1,a-1}_{-(b-1),-1} = w^{1,b-1}_{-(b-1),-1} + w^{b,a-1}_{-(b-1),-1}, \end{aligned}$$

so $w^{1,b-1}_{-(a-1),-b} = w^{b,a-1}_{-(b-1),-1}$. If, in addition, $[-a, a]\times[-b, b]$ is an empty rectangle for w, then $w^{a,n}_{-(a-1),-b} = 0$, thus,

$$(R_w)_{-(a-1),b-1} \equiv w^{1,b-1}_{-n,-a} \pmod 2.$$

By our assumptions, $(R_\sigma)_{-(a-1),n} \not\equiv (R_\tau)_{-(a-1),n} \pmod 2$, hence $\sigma^{1,b-1}_{-n,-a} \not\equiv \tau^{1,b-1}_{-n,-a} \pmod 2$ (this is the key observation in the proof). Denote $P = I \cup J \cup K$, $Q = \{1,\dots,b-1\}$, where

$$\begin{aligned} I &= \{i \in \{a,\dots,n\} \mid \sigma(i) \in Q\}, \\ J &= \{j \in \{-n,\dots,-a\} \mid \sigma(j) \in Q\}, \\ K &= \{k \in \{-(a-1),\dots,-1\} \mid \sigma(k) \in Q\}. \end{aligned}$$

Note that $|K| = (R_\sigma)_{-(a-1),b-1} = (R_\tau)_{-(a-1),b-1}$, while $|I|+|J| = n-a+1$ (the latter equality follows from the fact that $[-a, a] \times [-b, b]$ is an empty rectangle for σ). Moreover, it follows from $\sigma^{1,b-1}_{-n,-a} \not\equiv \tau^{1,b-1}_{-n,-a} \pmod 2$ that $\#\{\epsilon_i - \epsilon_j \in \mathrm{Supp}(\tau) \mid a \le j \le n\} \not\equiv |I| \pmod 2$ (or, equivalently, $\#\{\epsilon_i + \epsilon_j \in \mathrm{Supp}(w) \mid a \le j \le n\} \not\equiv |J| \pmod 2$). Hence, according to Proposition 2.4, $D^Q_{P,P'}(\lambda) = 0$ for all $\lambda \in \Omega_\tau$ and all subsets $P' \subset P$ of even cardinality. In particular,

$$D^Q_{P,\varnothing}(\lambda) = \Delta^Q_P(\lambda) = 0$$

for all $\lambda \in \Omega_\tau$ (and, consequently, for all $\lambda \in \overline{\Omega}_\tau$.) But $\Delta^Q_P(f_\sigma) = \pm 1$ by definition of f_σ. Thus, $\Omega_\sigma \notin \overline{\Omega}_\tau$. This concludes the proof. □

2.3 In this subsection we present a conjectural way how to prove that if $\sigma, \tau \in \mathcal{B}(\Phi)$ and $\sigma \le_B \tau$ then $\Omega_\sigma \subseteq \overline{\Omega}_\tau$. (For simplicity, we consider only the case $\Phi = B_n$ and give some additional remarks for D_n at the end of the subsection.) To do this, we need to describe the covering relation on $\mathcal{B}(\Phi)$ with respect to the Bruhat order. The covering relation on $\mathcal{I}(\Phi)$ was described by Incitti in [12]. We will state a corollary of his description for C_n in appropriate terms. To each involution $w \in \mathcal{I}(C_n)$ we assign the number

$$d(\sigma) = |\{i \in \{1, \dots, n\} \mid 2\epsilon_i \in \mathrm{Supp}(\sigma)\}|.$$

Let $\sigma,\ \tau \in \mathcal{I}(C_n)$, and $D_\sigma = \mathrm{Supp}(\sigma)$, $D_\tau = \mathrm{Supp}(\tau)$. (In fact, we will apply this corollary to B_n, but it is more convenient to formulate it for C_n, because we will use the notion of the support of an arbitrary involution.) In the tables below we consider certain special cases of "relative positions" of D_σ and D_τ in Φ^+, which are needed to formulate this corollary (Tables 1, 2, 3, 4).

Given involutions $\sigma,\ \tau \in \mathcal{I}(C_n)$, we say that the pair (τ, σ) is *of type* (or, equivalently, *belongs to case*) $a.b$ if $D_\sigma = \mathrm{Supp}(\sigma)$ and $D_\tau = \mathrm{Supp}(\tau)$ are as in the case b in Table a (for some $i,\ k,\ j,\ l$). We also say that (τ, σ) is an *admissible pair* if it is of type $a.b$ for certain $a,\ b$. From [10, Chapter 6] (see also [12, p. 76–91]) one can immediately deduce the following

Table 1 Case $d(\sigma) = d(\tau)$, first part

	$D_\sigma \setminus D_\tau$	$D_\tau \setminus D_\sigma$		$D_\sigma \setminus D_\tau$	$D_\tau \setminus D_\sigma$
1	$\epsilon_i + \epsilon_j$, $i < k < j$	$\epsilon_i + \epsilon_k$	2	$\epsilon_k - \epsilon_j$, $\epsilon_i + \epsilon_l$, $i < k < j < l$	$\epsilon_i + \epsilon_j$, $\epsilon_k - \epsilon_l$
3	$\epsilon_i - \epsilon_j$, $\epsilon_k + \epsilon_l$, $i < k < j < l$	$\epsilon_i - \epsilon_l$, $\epsilon_k + \epsilon_j$	4	$\epsilon_i - \epsilon_j$, $i < j$	$\epsilon_i + \epsilon_j$
5	$\epsilon_i + \epsilon_j$, $\epsilon_k + \epsilon_l$, $i < k < j < l$	$\epsilon_i + \epsilon_k$, $\epsilon_j + \epsilon_l$	6	$\epsilon_i + \epsilon_j$, $\epsilon_k - \epsilon_l$, $i < k < j < l$	$\epsilon_i + \epsilon_k$, $\epsilon_j - \epsilon_l$
7	$\epsilon_i + \epsilon_l$, $\epsilon_k - \epsilon_j$, $i < k < j < l$	$\epsilon_i + \epsilon_k$	8	$\epsilon_k - \epsilon_j$, $i < k < j$	$\epsilon_i - \epsilon_j$
9	$\epsilon_k + \epsilon_j$, $i < k < j$	$\epsilon_i + \epsilon_j$	10	$\epsilon_i - \epsilon_k$, $i < k < j$	$\epsilon_i - \epsilon_j$
11	$\epsilon_i - \epsilon_k$, $\epsilon_j - \epsilon_l$, $i < k < j < l$	$\epsilon_i - \epsilon_j$, $\epsilon_k - \epsilon_l$	12	$\epsilon_i - \epsilon_k$, $\epsilon_j + \epsilon_l$, $i < k < j < l$	$\epsilon_i - \epsilon_j$, $\epsilon_k + \epsilon_l$
13	$\epsilon_i + \epsilon_l$, $\epsilon_k + \epsilon_j$, $i < k < j < l$	$\epsilon_i + \epsilon_j$, $\epsilon_k + \epsilon_l$	14	$\epsilon_i - \epsilon_j$, $\epsilon_k - \epsilon_l$, $i < k < j < l$	$\epsilon_i - \epsilon_l$, $\epsilon_k - \epsilon_j$
15	$\epsilon_i - \epsilon_l$, $\epsilon_k + \epsilon_j$, $i < k < j < l$	$\epsilon_i + \epsilon_j$, $\epsilon_k - \epsilon_l$	16	$\epsilon_i - \epsilon_j$, $\epsilon_k + \epsilon_l$, $i < k < j < l$	$\epsilon_i + \epsilon_l$, $\epsilon_k - \epsilon_j$
17	$\epsilon_i - \epsilon_k$, $\epsilon_j - \epsilon_l$, $i < k < j < l$	$\epsilon_i - \epsilon_l$	18	$\epsilon_i - \epsilon_k$, $\epsilon_j + \epsilon_l$, $i < k < j < l$	$\epsilon_i + \epsilon_l$
19	$\varnothing$, $i < j$	$\epsilon_i - \epsilon_j$			

Table 2 Case $d(\sigma) = d(\tau)$, second part

	$D_\sigma \setminus D_\tau$	$D_\tau \setminus D_\sigma$		$D_\sigma \setminus D_\tau$	$D_\tau \setminus D_\sigma$
1	$2\epsilon_j$, $i < j$	$2\epsilon_i$	2	$\epsilon_i + \epsilon_j, 2\epsilon_k$, $i < k < j$	$2\epsilon_i$ $\epsilon_k + \epsilon_j$
3	$\epsilon_i - \epsilon_j, 2\epsilon_k$, $i < k < j$	$2\epsilon_i$, $\epsilon_k - \epsilon_j$	4	$\epsilon_i - \epsilon_k, 2\epsilon_j$, $i < k < j$	$2\epsilon_i$
5	$\epsilon_i - \epsilon_k, 2\epsilon_j$, $i < k < j$	$\epsilon_i - \epsilon_j$, $2\epsilon_k$	6	$\epsilon_i + \epsilon_j, 2\epsilon_k$, $i < k < j$	$\epsilon_i + \epsilon_k$, $2\epsilon_j$

Table 3 Case $d(\sigma) < d(\tau)$

	$D_\sigma \setminus D_\tau$	$D_\tau \setminus D_\sigma$		$D_\sigma \setminus D_\tau$	$D_\tau \setminus D_\sigma$
1	$\emptyset$	$2\epsilon_i$	2	$\epsilon_i + \epsilon_j$, $i < j$	$2\epsilon_i, 2\epsilon_j$
3	$\epsilon_i - \epsilon_j$, $i < j$	$2\epsilon_i$	4	$\epsilon_i + \epsilon_j$, $\epsilon_k - \epsilon_l$	$2\epsilon_i, \epsilon_k + \epsilon_j$, $i < k < j < l$
5	$\epsilon_i + \epsilon_l$, $\epsilon_k - \epsilon_j$	$2\epsilon_i, \epsilon_k + \epsilon_l$, $i < k < j < l$	6	$\epsilon_i - \epsilon_l$, $\epsilon_k - \epsilon_j$	$2\epsilon_i, \epsilon_k - \epsilon_l$, $i < k < j < l$

Table 4 Case $d(\sigma) > d(\tau)$

	$D_\sigma \setminus D_\tau$	$D_\tau \setminus D_\sigma$		$D_\sigma \setminus D_\tau$	$D_\tau \setminus D_\sigma$
1	$\epsilon_i - \epsilon_j, 2\epsilon_k$, $i < k < j$	$\epsilon_i + \epsilon_k$	2	$2\epsilon_j$	$\epsilon_i + \epsilon_j$, $i < j$
3	$2\epsilon_k, 2\epsilon_j$	$\epsilon_i + \epsilon_k$, $i < k < j$	4	$\epsilon_i - \epsilon_k, 2\epsilon_j$, $i < k < j$	$\epsilon_i + \epsilon_j$
5	$\epsilon_i - \epsilon_k$, $2\epsilon_j, 2\epsilon_l$, $i < k < j < l$	$\epsilon_i - \epsilon_l$, $\epsilon_k + \epsilon_j$	6	$\epsilon_i + \epsilon_l$, $2\epsilon_k, 2\epsilon_j$, $i < k < j < l$	$\epsilon_i + \epsilon_k$, $\epsilon_j + \epsilon_l$
7	$\epsilon_i - \epsilon_l$, $2\epsilon_k, 2\epsilon_j$, $i < k < j < l$	$\epsilon_i + \epsilon_k$, $\epsilon_j - \epsilon_l$	8	$\epsilon_i - \epsilon_j$, $2\epsilon_k, 2\epsilon_l$, $i < k < j < l$	$\epsilon_i + \epsilon_k$
9	$\epsilon_i - \epsilon_k$, $2\epsilon_j, 2\epsilon_l$, $i < k < j < l$	$\epsilon_i + \epsilon_j$			

Corollary 2.6 *Let $\sigma, \tau \in \mathcal{I}(C_n)$ and $\sigma <_B \tau$. Then there exist $\sigma_1, \ldots, \sigma_k \in \mathcal{I}(C_n)$ such that $\sigma_1 = \tau$, $\sigma_k = \sigma$ and (σ_i, σ_{i+1}) is an admissible pair for all $1 \leq i \leq k - 1$.*

Actually, this corollary does not describe the covering relation on $\mathcal{I}(C_n)$ (there are some additional conditions on σ and τ), but we will use only this part of the Incitti's description.

Since the Weyl groups of B_n and C_n coincide, we have the notion of a basis involution in $W(C_n)$; we denote the set of all basis involutions in $W(C_n)$ by $\mathcal{B}(C_n)$. We say that a pair (τ, σ) of basis involutions from $\mathcal{I}(C_n)$ is *basis-admissible* if it is of type $1.b$ for certain b.

Conjecture 2.7 Let $\sigma,\ \tau \in \mathcal{B}(C_n)$ and $\sigma <_B \tau$. Then there exist $\sigma_1, \ldots, \sigma_k \in \mathcal{B}(C_n)$ such that $\sigma_1 = \tau$, $\sigma_k = \sigma$ and (σ_i, σ_{i+1}) is basis-admissible for all $1 \leq i \leq k-1$.

We checked that this conjecture is true for $n \leq 7$. It is easy to see that if this conjecture is true for all n that $\sigma \leq_B \tau$ implies $\Omega_\sigma \subseteq \overline{\Omega}_\tau$. Indeed, we may assume without loss of generality that (τ, σ) is a basis-admissible pair. For basis-admissible pairs, the proof is case-by-case. For example, suppose that (τ, σ) is of type 1.12, i.e.,

$$\begin{aligned}\mathrm{Supp}(\sigma) \setminus \mathrm{Supp}(\tau) &= \{\epsilon_i - \epsilon_k,\ \epsilon_j + \epsilon_l\},\\ \mathrm{Supp}(\tau) \setminus \mathrm{Supp}(\sigma) &= \{\epsilon_i - \epsilon_j,\ \epsilon_k + \epsilon_j\}\end{aligned}$$

for some $1 \leq i < k < j < l \leq n$. Put

$$g(s) = x_{\epsilon_k - \epsilon_j}(t^{-2}) h_{\epsilon_i - \epsilon_j}(t^{-1}) h_{\epsilon_k + \epsilon_l}(-It^{-1})$$

and $f = g(s).f_\tau$. (Here I is the usual imaginary unit.) One can easily check by straightforward matrix calculations that, given $\alpha \in \Phi^+$,

$$f(e_\alpha) = \begin{cases} 1, & \text{if either } \alpha = \epsilon_i - \epsilon_k \text{ or } \alpha = \epsilon_j + \epsilon_l,\\ t^2, & \text{if } \alpha = \epsilon_i - \epsilon_j,\\ -t^2, & \text{if } \alpha = \epsilon_k + \epsilon_l,\\ f_\tau(e_\alpha) & \text{otherwise.} \end{cases}$$

Thus, $f \to f_\sigma$ as $t \to 0$. All other cases can be considered similarly. (For $\Phi = D_n$, one should use [10, Chapter 7] instead of [10, Chapter 6] for the description of the covering relation on $\mathcal{I}(D_n)$.)

Example 2.8 Note that $\sigma \leq_B \tau$ does not imply $\Omega_\sigma \subseteq \overline{\Omega}_\tau$ for non-basis involutions in $\mathcal{I}(B_n)$. Indeed, let $n = 4$ and

$$\sigma = \begin{pmatrix} 1 & 2 & 3 & 4 \\ -4 & 2 & 3 & -1 \end{pmatrix} = s_{\epsilon_1 + \epsilon_4},$$

$$\tau = \begin{pmatrix} 1 & 2 & 3 & 4 \\ -1 & -3 & -2 & 4 \end{pmatrix} = s_{\epsilon_1} s_{\epsilon_2 + \epsilon_3}.$$

Clearly, there is the only possibility to define the supports of these involutions, precisely,

$$\mathrm{Supp}(\sigma) = \{\epsilon_1 + \epsilon_4\},$$
$$\mathrm{Supp}(\tau) = \{\epsilon_1,\ \epsilon_2 + \epsilon_3\}.$$

One can immediately deduce from Theorem 1.5 that $\sigma <_B \tau$.

On the other hand, by definition of the coadjoint action, if $x,\ y \in \mathfrak{n}$, $f \in \mathfrak{n}^*$, then

$$\begin{aligned}((\exp x).f)(y) &= f(y) - f([x, y]) + \frac{1}{2} f([x, [x, y]]) - \dots \\ &= f(\exp \mathrm{ad}_{-x}(y)),\end{aligned}$$

where, as usual, $\mathrm{ad}_x(y) = [x, y]$. This implies that, given $\alpha \in \Phi^+$, if there are no $\beta \in \Phi^+$ such that $\alpha \leq \beta$ with respect to the natural order on Φ (i.e., $\beta - \alpha$ is zero or a sum of positive roots), then $\lambda(e_\alpha) = 0$ for all $\lambda \in \Theta_f$. Thus,

$$\lambda(e_{\epsilon_1+\epsilon_4}) = 0$$

for all $\lambda \in \Omega_\tau$, but $f_\sigma(\epsilon_1 + \epsilon_4) \neq 0$, so $\Omega_\sigma \not\subset \overline{\Omega_\tau}$.

3 Concluding Remarks

3.1 Let w be an involution from $\mathcal{B}(\Phi)$, where $\Phi = B_n$ or D_n. Being an orbit of a connected unipotent group on an affine variety $\mathfrak{n}^*$, Ω_w is a closed subvariety of $\mathfrak{n}^*$. In this subsection, we present a formula for the dimension of Ω_w (cf. [7, Proposition 4.1], [8, Theorem 3.1]). Recall the definition of the length $l(w)$ of an element $w \in W$.

Theorem 3.1 *Let $\Phi = B_n$ or D_n, and $w \in \mathcal{B}(\Phi)$. One has*

$$\dim \Omega_w = l(w).$$

Proof Denote $D = \mathrm{Supp}(w)$. We claim that if ξ_1 and ξ_2 are two distinct maps from the set D to $\mathbb{C}^\times$, then

$$\Theta_{w,\xi_1} \neq \Theta_{w,\xi_2}.$$

Indeed, for $\Phi = B_n$ (respectively, D_n) let $\widetilde{U}$ be the group of all $(2n + 1) \times (2n + 1)$ (respectively, $2n \times 2n$) upper-triangular matrices with 1's on the diagonal. Since w is an involution in $S_{\pm n}$, [18, Theorem 1.4] implies that $\widetilde{\Theta}_{w,\xi_1} \neq \widetilde{\Theta}_{w,\xi_2}$, where $\widetilde{\Theta}_{w,\xi_1}$ and $\widetilde{\Theta}_{w,\xi_2}$ denote the respective $\widetilde{U}$-orbits of f_{w,ξ_1} and f_{w,ξ_2} under the coadjoint action of $\widetilde{U}$ on the space of all lower triangular matrices with zeroes on the diagonal (see Sect. 1.2 for the definitions). Since $U \subseteq \widetilde{U}$, one has $\Theta_{w,\xi_1} \subseteq \widetilde{\Theta}_{w,\xi_1}$ and $\Theta_{w,\xi_2} \subseteq \widetilde{\Theta}_{w,\xi_2}$, hence $\Theta_{w,\xi_1} \neq \Theta_{w,\xi_2}$, as required.

Let $Z_B = \mathrm{Stab}_B\, f_w$ be the stabilizer of f_w in B. One has

$$\dim \Omega_w = \dim B - \dim Z_B.$$

Recall that $B \cong U \rtimes H$ as algebraic groups. It was shown in the proof of Lemma 2.1 that if $h \in H$, then there exists $\xi: D \to \mathbb{C}^\times$ such that $h.f_w = f_{w,\xi}$. Hence if $g = uh \in Z_B$, then

$$f_w = (uh).f_w = u.f_{w,\xi},$$

so $f_w \in \Theta_{w,\xi}$. In follows from the first paragraph of the proof that $f_w = f_{w,\xi}$. This means that the map

$$Z_U \times Z_H \to Z_B\colon (u, h) \mapsto uh$$

is an isomorphism of algebraic varieties, where $Z_U = \mathrm{Stab}_U\, f_w$ (respectively, $Z_H = \mathrm{Stab}_H\, f_w$) is the stabilizer of f_w in U (respectively, in H). Hence

$$\dim Z_B = \dim Z_U + \dim Z_H.$$

Let Θ_w be the U-orbit of the linear form f_w. Then, according to [5, Theorem 1.2], $\dim \Theta_w = l(w) - |D|$, so

$$\dim Z_U = \dim U - \dim \Theta_w = \dim U - l(w) + |D|.$$

On the other hand, put $X = \bigcup_{\xi\colon D \to \mathbb{C}^\times} \{f_{w,\xi}\}$. It follows from Lemma 2.1 and the first paragraph of the proof that

$$X = \{h.f_w, h \in H\}$$

is the H-orbit of f_w. Consequently,

$$\dim Z_H = \dim H - \dim X = \dim H - |D|,$$

because X is isomorphic as affine variety to the product of $|D|$ copies of $\mathbb{C}^\times$. Thus,

$$\begin{aligned}\dim \Omega_w &= \dim B - \dim Z_B = (\dim U + \dim H) - (\dim Z_U + \dim Z_H)\\ &= \dim U + \dim H - (\dim U - l(w) - |D|)\\ &- (\dim H - |D|) = l(w).\end{aligned}$$

The proof is complete. □

3.2 In the remainder of the paper, we briefly discuss a conjectural geometrical approach to orbits associated with involutions in terms of tangent cones to Schubert varieties. Recall that W is isomorphic to $N_G(H)/H$, where $N_G(H)$ is the normalizer of H in G. The *flag variety* $\mathcal{F} = G/B$ can be decomposed into the union $\mathcal{F} = \bigcup_{w \in W} \mathcal{X}_w^\circ$, where $\mathcal{X}_w^\circ = B\dot{w}B/B$ is called the *Schubert cell* corresponding to w. (Here $\dot{w}$ is a representative of w in $N_G(H)$.) By definition, the *Schubert variety* $\mathcal{X}_w$ is the closure of $\mathcal{X}_w^\circ$ in $\mathcal{F}$ with respect to Zariski topology. Note that $p = \mathcal{X}_{\mathrm{id}} = B/B$ is contained in $\mathcal{X}_w$ for all $w \in W$. One has $\mathcal{X}_w \subseteq \mathcal{X}_{w'}$ if and only if $w \leq_B w'$. Let T_w be the tangent space and C_w the tangent cone to $\mathcal{X}_w$ at the point p (see [2] for detailed constructions); by definition, $C_w \subseteq T_w$, and if p is a regular point of $\mathcal{X}_w$, then $C_w = T_w$. Of course, if $w \leq_B w'$, then $C_w \subseteq C_{w'}$.

Let $T = T_p\mathcal{F}$ be the tangent space to $\mathcal{F}$ at p. It can be naturally identified with $\mathfrak{n}^*$ by the following way: since $\mathcal{F} = G/B$, T is isomorphic to the factor $\mathfrak{g}/\mathfrak{b} \cong \mathfrak{n}^*$. Next, B acts on $\mathcal{F}$ by conjugation. Since p is invariant under this action, the action on $T = \mathfrak{n}^*$ is induced. One can check that this action coincides with the action of B on $\mathfrak{n}^*$ defined above. The tangent cone $C_w \subseteq T_w \subseteq T = \mathfrak{n}^*$ is B-invariant, so it splits into a union of B-orbits. Furthermore, $\overline{\Omega_\sigma} \subseteq C_\sigma$ for all $\sigma \in \mathcal{I}(C_n)$.

It is well-known that C_w is a subvariety of T_w of dimension $\dim C_w = l(w)$ [2, Chapter 2, Section 2.6]. Let $w \in \mathcal{B}(\Phi)$ for $\Phi = B_n$ or D_n. Since Ω_w is irreducible, $\overline{\Omega_w}$ is irreducible, too. Theorem 3.1 implies $\dim \overline{\Omega_w} = \dim \Omega_w = l(\sigma)$, so $\overline{\Omega_w}$ is an irreducible component of C_w of maximal dimension.

Conjecture 3.2 Let $\Phi = B_n$ or D_n, and $w \in \mathcal{B}(\Phi)$. Then the closure of the B-orbit Ω_w coincides with the tangent cone C_w to the Schubert variety $\mathcal{X}_w$ at the point $p = B/B$.

Note that this conjecture implies that if $\sigma \leq_B \tau$, then $\Omega_\sigma \subseteq \overline{\Omega_\tau}$ for $\sigma,\ \tau \in \mathcal{B}(\Phi)$.

Acknowledgements A part of this work (Sect. 3) was done during my stay at University of Haifa. I would like to express my gratitude to Prof. Dr. Anna Melnikov for her hospitality and fruitful discussions.

The work on Sect. 2 was performed at the NRU HSE with the support from the Russian Science Foundation, grant no. 16-41-01013. The work on Sect. 3 has been supported by RFBR grant no. 16-01-00154a and by ISF grant no. 797/14.

References

1. E. Bagno, Y. Cherniavsky. Congruence B-orbits and the Bruhat poset of involutions of the symmetric group. Discrete Math. **312** (2012), no. 6, 1289–1299.
2. S. Billey, V. Lakshmibai. Singular loci of Schubert varieties. Progr. in Math. **182**, Birkhäuser, 2000.
3. A. Bjorner, F. Brenti. Combinatorics of Coxeter groups. Graduate Texts in Mathematics **231**, Springer, 2005.
4. N. Bourbaki. Lie groups and Lie algebras. Chapters 4–6. Springer, 2002.
5. M.V. Ignat'ev. Orthogonal subsets of classical root systems and coadjoint orbits of unipotent groups (in Russian). Mat. Zametki **86** (2009), no. 1, 65–80. English translation: Math. Notes **86** (2009), no. 1, 65–80; arXiv: math.RT/0904.2841.

6. M.V. Ignatev. Orthogonal subsets of root systems and the orbit method (in Russian). Algebra i Analiz **22** (2010), no. 5, 104–130. English translation: St. Petersburg Math. J. **22** (2011), no. 5, 777–794; arXiv: math.RT/1007.5220.
7. M.V. Ignatyev. Combinatorics of B-orbits and the Bruhat–Chevalley order on involutions. Transformation Groups **17** (2012), no. 3, 747–780; arXiv: math.RT/1101.2189.
8. M.V. Ignatyev. The Bruhat-Chevalley order on involutions of the hyperoctahedral group and combinatorics of B-orbit closures (in Russian). Zapiski Nauchnykh Seminarov POMI **400** (2012), 166–188. English translation: J. Math. Sci. **192** (2013), no. 2, 220–231; arXiv: math.RT/1112.2624.
9. M.V. Ignatyev, A.S. Vasyukhin. Rook placements in A_n and combinatorics of B-orbit closures. J. Lie Theory **24** (2014), no. 4, 931–956; arXiv: math.RT/1310.3164.
10. F. Incitti. Bruhat order on the involutions of classical Weyl groups. Ph.D. thesis. Dipartimento di Matematika “Guido Castelnuovo”, Università di Roma “La Sapienza”, 2003.
11. F. Incitti. The Bruhat order on the involutions of the symmetric groups. J. Alg. Combin. **20** (2004), no. 3, 243–261.
12. F. Incitti. Bruhat order on classical Weyl groups: minimal chains and covering relation. European J. Combinatorics **26** (2005), 729–753.
13. A.A. Kirillov. Unitary representations of nilpotent Lie groups. Russian Math. Surveys **17** (1962), 53–110.
14. A.A. Kirillov. Lectures on the orbit method, Grad. Studies in Math. **64**, AMS, 2004.
15. A. Melnikov. B-orbit in solution to the equation $X^2 = 0$ in triangular matrices. J. Algebra **223** (2000), 101–108.
16. A. Melnikov. Description of B-orbit closures of order 2 in upper-triangular matrices. Transformation Groups **11** (2006), no. 2, 217–247.
17. A. Melnikov. B-orbits of nilpotent order 2 and link patterns. Indagationes Mathematicae **24** (2013), no.2, 443–473.
18. A.N. Panov. Involutions in S_n and associated coadjoint orbits. J. Math. Sci. **151** (2008), no. 3, 3018–3031.
19. R.W. Richardson, T.A. Springer. The Bruhat order on symmetric varieties. Geom. Dedicata **35** (1990), no. 1–3, 389–436.
20. T.A. Springer. Some remarks on involutions in Coxeter groups. Comm. Algebra **10** (1982), no. 6, 631–636.

Proper Self-Similar Triangle Tiling and Representing Weight Diagrams in the Plane

Anthony Joseph

In memory of Bar Sagi

I just want to get it over and done
I want to see tomorrow's setting sun
I want that hateful memory
To turn into the fears of yesterday.

An excerpt from "Choice", by Bar Sagi [11, XXI].

Abstract Let $P(\pi)$ denote the weight lattice of a simple Lie algebra $\mathfrak{g}$. A weight diagram D is a finite subset of $P(\pi)$ describing the weights of a simple finite dimensional representation of $\mathfrak{g}$. In (Joseph, Adv Math 22(1), 522–585, 2011; 2.8) it was noted that there exists a $\mathbb{Z}$ linear map ψ of $P(\pi)$ into $\mathbb{R}^2$. The resulting presentation of weight diagrams in the plane leads naturally to "proper self-similar triangle tilings." In (Joseph, Transform. Groups 14(3), 557–612, 2009) self-similar triangle tilings were constructed. Here this is refined by requiring the tiling to be proper. In addition a uniqueness theorem is proved.

In (Joseph, Transform. Groups, 14(3), 557–612 (2009); 8.3,10.5) it was determined exactly when ψ is injective in simple Lie algebras of types A, B. Here it is shown that ψ is injective if and only if the exponents of $\mathfrak{g}$ are prime to its Coxeter number. The possibility of representing $\psi(D)$ as the vertices coming from a proper triangle tiling is discussed.

A. Joseph (✉)
Donald Frey Professional Chair, Department of Mathematics, The Weizmann Institute of Science, Rehovot, Israel
e-mail: anthony.joseph@weizmann.ac.il

M. Gorelik et al. (eds.), *Representations and Nilpotent Orbits of Lie Algebraic Systems*, Progress in Mathematics 330,
https://doi.org/10.1007/978-3-030-23531-4_9

AMS Classification: 17B35

1 Motivation

This work was started 11 years ago out of my frustration in trying to understand the Kashiwara $B(\infty)$ crystal. In this I hoped to make use of a simple trick to draw the weight diagram of a simple Lie algebra in the plane. The result [4] was a construction of self-similar tiling with respect to triangles obtained by joining any three vertices of a regular n-gon generalizing a well-known construction for the case $n = 5$. I hoped at that time it would at least amuse my grand-daughter Bar who was 6 years old at the time. Later I found a more substantial application [5] of the above trick to constructing Weierstrass sections for the co-adjoint action of a Borel subalgebra.

About 7 years later I realized that the tiling rule in [4] had a not so satisfying iteration. To avoid this difficulty more complex patterns are needed and these are rather pleasing. I thought at that time to submit the result to an exhibition in Zefat aimed at combining mathematics and art, but I could not find a pleasing colouring of the triangles.

I put one pattern on my home page http://www.weizmann.ac.il/math/joseph/. Regrettably a pleasing colouring was also not achieved. Nevertheless the pattern adorned the conference posters of which the present volume are the proceedings. This paper gives a general rule how such patterns can be obtained and the self-similarity which results. In the second part it describes the condition under which the weight lattice is faithfully represented in the plane by the above trick and the resulting possible presentation of a weight diagram as the vertices of triangle tiling.

It turned out that the diagram in my home page had two errors. One barely noticeable, but which made it difficult to understand how the pattern had been constructed. (It took me an entire day.) The second was that my general rule given in Sect. 3.9 was not precisely followed. One may anticipate that if this rule is not followed then difficulties can occur with subsequent iterations.

In Plates 1, 2, 3, 4, 5, 6, 7, and 8 several examples of self-similar tiling are given following rule given in Sect. 3.9. All diagrams can be two-coloured except those in Plates 3, 4, 5, 9, and 10 which can just be three-coloured.

Plate 1 gives a tiling corresponding to $p_1^2T\{2, 2, 1\}$ in the notation of Sects. 3.2.1 and 3.2.2. Plate 2 gives an example from a regular hexagon and Plate 3 an example from a regular heptagon both corresponding to a two-fold iteration. Plate 4 gives a corrected version of the diagram in my home page. Plate 5 indicates why this tiling corresponds in the notation of Sects. 3.2.1 and 3.2.2, to the triangle $p_1p_2p_1T\{4, 2, 2\}$, that is to say a three-fold iteration using red, blue, and black lines to indicate successive refinements. Plate 6 gives an example from a regular nonagon again for a two-fold iteration. Plate 7 gives an example from a regular decagon for a two-fold iteration which cannot be two-coloured. In Plate 8, it is shown in this example how one uses Sect. 3.9 to obtain a three-fold iteration, specifically

Plate 1 © Anthony Joseph

Plate 2 © Anthony Joseph

to construct $p_3 p_2 p_1 T\{3, 3, 4\}$ again using a different colour for each successive refinement. A poster exhibits this last example—see Sect. 3.10.3. We remark that the result would have been different had the p_i factors been taken in a different order. Of course we can start with any n-gon, any triangle inscribed therein and any product of side lengths taken in any order. By any measure this gives a dazzling variety of different patterns. Of course one has to realize that to reproduce such patterns even if only three iterations are undertaken needs very high precision and attempting to present graphically even more iterations would be a definitive prescription for insanity.

Plate 3 © Anthony Joseph

Plate 4 © Anthony Joseph

Plate 5 © Anthony Joseph

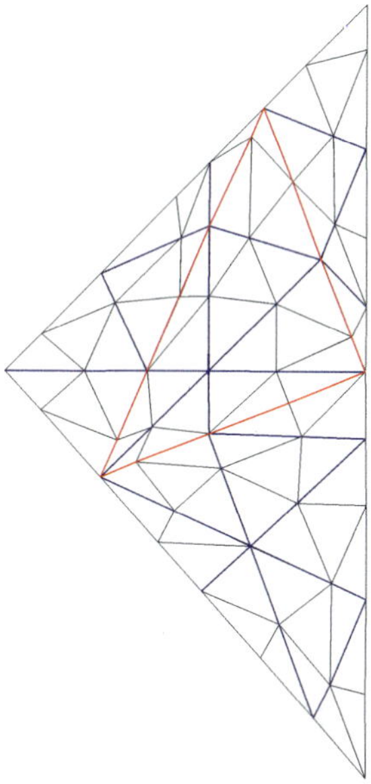

Plate 6 © Anthony Joseph

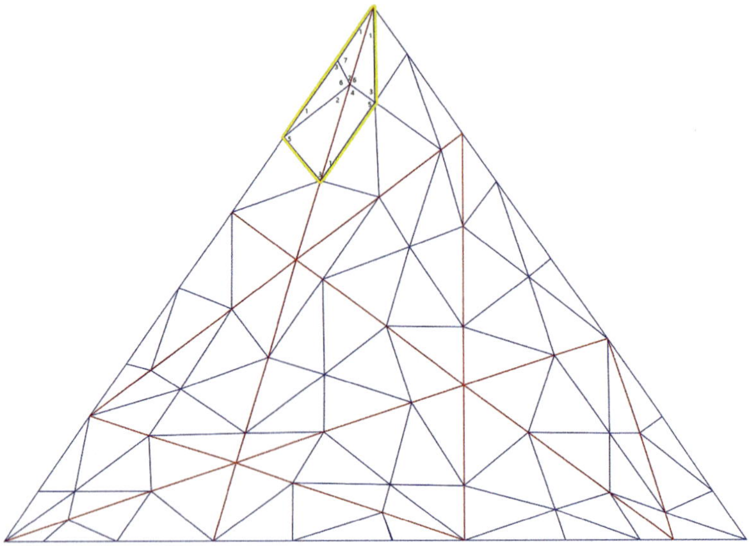

Plate 7 © Anthony Joseph

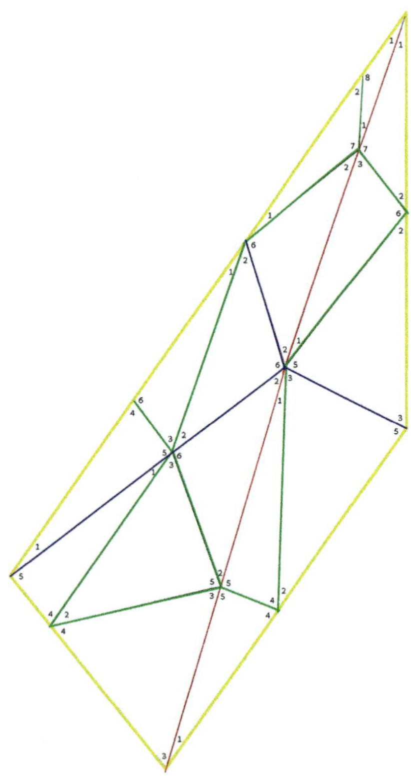

Plate 8 © Anthony Joseph

2 A Very Brief Survey of Tiling

2.1 Tilings

Tiling the plane goes back to ancient times and apart from practical purposes has been a source of great inspiration for art and design especially in cultures for which a representation of the human form is forbidden. There can be a rich mathematical theory behind such tilings though this might not have been immediately realized. For example, tiling the plane using the 19 possible crystallographic groups defined on the plane was illustrated on the walls of Alhambra palace, long before Western mathematicians had formulated the concept of a crystallographic group.

In [3] the authors describe a staggering variety of tilings and notably remark in the introduction that "We thought, naively as it turned out, that two millenia of development of plane geometry would leave little room for new ideas." Indeed this book was in time for aperiodic tiling to which a whole chapter is devoted, but too early for self-similarity which appeared, at least in its full glory [7], only 5 years after the publication of [3].

2.2 Aperiodic Tilings

The most familiar tiling is by squares, but it is also possible to use equilateral triangles. Actually the plane may be tiled by any triangle. Indeed just consider the pattern formed by three infinite sets of parallel lines.

All such tilings are periodic, the pattern being unchanged by a discrete translation subgroup with compact quotient. Actually it is easy to introduce aperiodicity. Just consider a tiling by bisected squares in which diagonals are chosen randomly. However, this tiling by right angled isosceles triangles is not essentially aperiodic in that the tiles can also be arranged by periodicity.

Essentially aperiodicity means that one or possibly a finite set S of shapes can *only* tile aperiodically. A first proposal used 3600 tiles. After much work Penrose [10] reduced this to just two. It is still an open question to know if just one suffices.

2.3 The Golden Pair

Penrose aperiodic tiling was based on the Golden Section g. It considered a pair of isosceles triangles T_1, T_2 whose side lengths were 1 and g. We call T_1, T_2 the Golden Pair. To avoid periodicity Penrose joined two copies of T_1 (resp. T_2) to form a dart (resp. kite). Adding some markings which have to be matched, achieves the required goal [10].

2.4 Self-Similarity

The periodic triangle tiling described in the first paragraph of Sect. 2.3 has the property of being self-similar, that is the basic triangle T is reproduced by a factor (called a dilation factor) which can be any positive integer t by stacking t^2 triangles. However, this is not particularly interesting.

Thurston around 1992 considered which dilation factors could be achieved using a finite set $\mathcal{S}$ of shapes. To be precise for some positive real number g and some $S \in \mathcal{S}$ one must find a tiling by the elements of $\mathcal{S}$ of gS. In this case we say that $\mathcal{S}$ admits a self-similar tiling with respect to g. Of course this process can be repeated and then the dilation factor becomes a power of g. An example given was the Golden Pair with dilation factor being the Golden Section. This last result was illustrated on the notice board at the University of Geneva where I was a guest about 12 years ago. It was a motivation for the present paper, but unfortunately I did not note down the details nor check its origins.

By a simple argument based on areas Thurston noted that the dilation factor must be an algebraic integer which is the largest real solution of its minimal polynomial (a Perron number). He further asserted that every such dilation factor could be achieved. Kenyon [7] later verified his claim. In this, some rather exotic tiles seem to be needed—see [7, Fig. 1].

2.5 Self-Similarity Using Triangles

One can ask what dilation factors are possible if all tiles are restricted to be triangles. Moreover here we can be a little more demanding. Indeed fix the minimal side length to be 1 and then demand that *every* side length appears as a dilation factor.

Let n be an integer ≥ 3 and $\mathcal{T}_n$ be the set of triangles obtained by joining any three vertices of a regular n-gon of side length 1 by line segments which do not cross. Let L_n be the set of all their side lengths and $\mathcal{L}_n$ the monoid they generate. In [4, 8.8] it was shown that $\mathcal{T}_n$ is self-similar with respect to all dilations factor from L_n. In this case we say that $\mathcal{T}_n$ admits a self-similar triangle tiling.

By repeating this process the dilation factors become not just a power of any side length but a product of any of them with any multiplicities, that is to say any element of $\mathcal{L}_n$. However, if one does this just using the construction of [4, 8.8], one finds that on the second iteration triangles may "overlap," a "bad" feature which was absent at the first iteration.

Thus let us consider a further requirement. Namely if two triangles share two common points, then they must share their entire sides defined by these two points. We call this a proper joining and we call such a pair, neighbours. Then (trivially!) the neighbour of a triangle is uniquely determined by the edge they share. In this any internal point must be the vertex of at least three triangles (or more generally elements of $\mathcal{S}$). This can fail for arbitrary triangle tiling see [4, Fig. 10]. This condition makes less sense for the tilings considered by Kenyon-Thurston [7, Fig. 1].

In Sect. 3, we show that the result noted above still holds when proper joining is imposed and we say that $\mathscr{T}_n$ admits a proper self-similar triangle tiling. However, the patterns become more complex and so more interesting. In particular the tiling cannot be two-coloured (in the usual sense that every pair of adjacent triangles is of a different colour). We do not know if it can be three-coloured, though this appears to be the case. The choice of colours to get an aesthetically pleasing arrangement is one where the mathematician must humbly yield to the artist. Thus I could only produce a rather unappealing result which became even worse on photographic reproduction. As a consequence I made no attempt to present the tiling in Zefat, back in 2011.

In the case of a triangle tiling of any shape S in the plane the vertices of the triangles which form S will be called the marked points of S.

2.6 Uniqueness

We call a proper self-similar triangle tiling, a tiling by a set $\mathscr{T}$ of triangles which are properly joined and in which the dilation factor p is any element of the monoid $\mathscr{L}$ generated by the set L of side lengths of elements of $\mathscr{T}$, normalized to contain $p_0 := 1$ as its smallest element. In this it is immediate that the set $\mathbb{N}\mathscr{L}$ is multiplicatively closed, and so the elements of L are algebraic integers. Not all algebraic integers occur but one does have infinitely many choices.

By the remark in the first paragraph of Sect. 2.3 there is a self-similar triangle tiling with dilation factor p being any positive integer. However, these are rather trivial and we exclude them by the condition that $p > 1$ and not integer. Then one may ask if necessarily there exists $n \geq 3$ such that $\mathscr{T} = \mathscr{T}_n$ and $L = L_n$ defined above. This is shown in Sect. 4, under some rather natural supplementary conditions. In this condition $(H5)$ is not too elegant and probably superfluous; but we were unable to avoid it.

2.7 Weight Diagrams

Let $\mathfrak{g}$ be a complex simple Lie algebra, $\mathfrak{h}$ a Cartan subalgebra, W the Weyl group relative to the pair $(\mathfrak{g}, \mathfrak{h})$ and $P(\pi)$ the weight lattice. In [5, 2.7] it was shown that there exists a $\mathbb{Z}C$ linear map ψ of $P(\pi)$ into $\mathbb{R}^2$. Here C is the subgroup of W generated by the pair $\{\sigma_1, \sigma_2\}$, where $\sigma := \sigma_1\sigma_2$ is a bipartite Coxeter element (see Sect. 2.8). It is isomorphic to the dihehral group D_{2c}, where c is the order of σ that is to say the Coxeter number of $\mathfrak{g}$.

Let $P^+(\pi)$ denote the set of dominant weights relative to a choice of simple roots π. Recall that to each $\lambda \in P^+(\pi)$ there exists a unique up to isomorphism simple highest weight simple $\mathfrak{g}$ module $V(\lambda)$ of highest weight λ. Let $D(\lambda)$ denote the set of weights of $V(\lambda)$. It is a finite subset of $P(\pi)$. Set $d(\lambda) = \psi(D(\lambda))$. We call $d(\lambda)$, a planar weight diagram.

Our interest in tiling was to whether a plane weight diagram can be represented as the set of marked points of a proper triangle tiling with triangles from $\mathscr{T}_c$, where we recall that c is the Coxeter number of $\mathfrak{g}$. Examples of the root diagram in A_4, A_6, B_4, B_5 for the adjoint representation as a set of marked points can be found in [4, Figs. 1,3,12,13]. In this a natural first question is whether the map ψ is injective.

Recall that after Chevalley the invariant algebra $S(\mathfrak{h})^W$ is polynomial on $\ell = \dim \mathfrak{h}$ generators which can be assumed to be homogeneous of degrees $m_j + 1 : j = 1, 2, \dots, \ell$. The $m_j : j = 1, 2, \dots, \ell$ are known as the exponents of $\mathfrak{g}$. After Coleman [2] the eigenvalues of σ on $\mathfrak{h}$ are exactly the $\exp 2\pi i m_j/c : j = 1, 2, \dots, \ell$. From the list of exponents which result, we show in Sect. 5 that ψ is injective if and only if c is prime to the exponents of $\mathfrak{g}$. No doubt the proof could be made more direct. In the case of $\mathfrak{sl}(n)$ the exponents form the set $\{1, 2, \dots, n-1\}$, and so the condition is that n must be prime.

In Sect. 6, we briefly address the question of whether a planar weight diagram can be represented by the set of marked points of a proper triangle tiling.

2.8 Weierstrass Sections

Because the Dynkin diagram of $\mathfrak{g}$ has no loops, a general result allows one to construct a "bipartite" Coxeter element $\sigma = \sigma_1\sigma_2$ where the simple reflections in the factors σ_1, σ_2 commute [1, Chap. V, Sect. 4]. If the Coxeter number c is even, then $\sigma^{c/2}$ is the unique longest element w_0 and the resulting expression is a reduced decomposition of w_0 [5, Lemma 2.8] (a result also noted in Bourbaki [1]). We call it a bipartite reduced decomposition of w_0. (As noted in [1, Chap. V, Sect. 4] one may also show that all Coxeter elements are conjugate, but unless $w_0 = -1$ this does not mean that $\sigma^{c/2} = w_0$ for *all* Coxeter elements σ. Even then the bipartite case has some useful extra properties.

In [5] we used a bipartite Coxeter element in construction of Weierstrass sections (called an algebraic slice in loc cit) for the action of a Borel subalgebra on its dual, particularly in the absence of an "adapted pair." The constructed failed only in types C, B_{2m}, F_4.

2.9 The Combinatorics of $B(\infty)$

Bipartite reduced decompositions of w_0 have been of interest in combinatorics. Recently we have shown [6, Thm. 6.3] that they are of particular interest in the description of the Kashiwara crystal $B(\infty)$, at least for $\mathfrak{g}$ classical. This brings us around in a full circle back to $B(\infty)$.

2.10 Recreation

Self-similar triangle tiling leads to many beautiful patterns which can be obtained with little effort or technical means. They do not have to satisfy proper joining (cf Sect. 2.6) even at the first iteration, for example, see [4, Fig. 10].

Proper self-similar triangle tiling is a little more subtle. In this all the triangles used can be inscribed in circles of fixed radius. Then the marked points of the tiling can be represented as intersection points of the circles and the triangles dispensed with. (We were a little too lazy to attempt to illustrate this.)

Self-similar triangle tiling can provide an amusing puzzle for children with the triangles (accurately!) cut from coloured paper backed by a magnetic sheet of the type used in placing advertising labels on the refrigerator door (usually supplied for free!). Such tiles can be easily manipulated whilst Mother (or perhaps Father!) is busy cooking. However, some of the triangles are rather thin (see Sect. 4.6.3) and could easily be swallowed by small children. Again the question of self-similarity is a little too subtle to be appreciated for those wanting to solve "finite" puzzles avoiding any venture into "general" solutions. In any case neither Bar nor her mother were particularly enamoured by the game and for me it was rather humbling to discover on the other hand that other tiling puzzles were eminently popular.

Personally I found it challenging to construct the triangle on my home page from the 82 triangles of seven different types, even when I knew the general rule described in Sect. 3.9. Lately I found it even harder to deconstruct this triangle tiling and determine which element of $p \in \mathscr{L}$ it involves. That is why I now describe in detail its construction in Plate 5.

Again by joining the two copies of the triangle on my home page, one may fill a square by the seven different triangles cut from a regular octagon using a total of 164 triangles. This would certainly make an eye-catching floor pattern for the living room if one could get a carpenter to fabricate the triangles and tiler to lay them.

Penrose tiling adorned bed covers in California (followed by a law suit). Maybe they induce sweet dreams. About 20 years ago Penrose tiles could be bought in Oxford (where Penrose was Savilian Professor of Geometry). Now a well-known global delivery firm sells a magnetic Penrose tiling pack.

One could argue that Penrose tiling formed the mathematical basis for quasi-crystals, eventually leading to D. Shechtman being awarded the Nobel prize. Although quasi-crystals live in three dimensions whilst Penrose tiling lives in two, both share an almost unnatural ubiquity of the Golden Section. Of course it was understood from early antiquity that the Golden Section appears in nature, for example, in the Vitruvian Man, exploited by Leonardo da Vinci to square the circle. My favourite consequence is the ratio of a mile to a kilometre, the former [9, mile] being 1000 paces (of a properly fed British soldier, rather than a poorly fed Roman soldier) and the latter (despite what the Paris Academy of Sciences may try to pretend) is just 1000 times the distance from the furthermost finger tips of an outstretched hand to the opposite shoulder in the same well fed soldier.

3 A Simple Construction

3.1 Integral Truncation

Fix a positive integer n. *As a subscript it will sometimes be omitted.* If k is a non-negative rational number, we let $[k]$ denote the largest integer $\leq k$.

3.1.1 The Basic Set and Its Dilations

Let S_n denote the regular n-gon of side length 1. Let L_n denote the set of lengths of lines joining the vertices of S_n and $\mathscr{L}_n$ the monoid generated by L_n.

As in Sect. 2.6, let $\mathscr{T}_n$ be the set of triangles obtained by joining any three vertices of S_n by line segments which do not cross. Observe that L_n is just the set of side lengths of elements of $\mathscr{T}_n$.

We call $\mathscr{T}_n$ the basic set.

Let S be a polygon (or "shape") and $p \geq 1$ a real number. By pS we mean the shape obtained on dilating S by p.

Our goal is to construct for all $p \in \mathscr{L}_n$ and $T \in \mathscr{T}_n$, an element of pT by properly joining elements of $\mathscr{T}_n$. Let us first spell out what this entails.

Let S be a polygon and T a triangle. Suppose that they possess a side s of common length. Then we shall denote by $S * T$ the polygon obtained by joining them along s, assuming that no overlapping of their interiors results. We remark that this notation is not entirely adequate as the pair S, T may possess several sides of common length, nor does it take account of the two possible orientations at the join. For example, if T is the smaller (resp. larger) element of the Golden Pair, then $T * T$ could be a parallelogram or a Penrose kite (resp. dart).

Of course we may wish to repeat the above procedure. In this we shall always start from elements of $\mathscr{T}_n$.

The points in the polygon so obtained which are vertices of triangles (from $\mathscr{T}$) are called its marked points. More particularly the set of marked points on its sides are called its side markings.

Then if we wish to join two such polygons to obtain a larger polygon which could have been obtained from just the basic set by proper joinings, we must not only require that the joined sides have the same length and there be no common interior but also that the markings on the joined sides match up.

Typically the above situation will arise in an inductive construction. Here one assumes for some $p \in \mathscr{L}_n$ that $\{pT | T \in \mathscr{T}_n\}$ has been constructed and view the latter as the new basic set $\mathscr{T}'$. Then for $p' \in L_n$, we construct $p'T'$ for $T' \in \mathscr{T}'$ by joining elements of $\mathscr{T}'$. Thus we must not only solve this joining problem when $\mathscr{T}'$ is our original basic set but also describe the markings on the joined sides that the elements of $\mathscr{T}'$ acquire and show that in our construction these side marking are matched. In this it will also be important to take the factors in p to be in the same order for each $T \in \mathscr{T}'$, as the resulting markings in pT will depend on this choice.

To summarize the above, let $< \mathscr{T}_n >$ denote the set of all polygons which can be obtained from $\mathscr{T}_n$ by repeated proper joinings. Given $S \in < \mathscr{T}_n >$ we may write S in the form $T_1 * T_2 * \ldots * T_n$ with $T_1, T_2, \ldots, T_n \in \mathscr{T}$.

Definition 1 A marked point of S is a vertex of some T_i. A marking of S is the set $M(S)$ of marked points of S.

3.1.2 Composite Triangles

It is clearly necessary that the distance between any two adjacent marked points on a side of $S \in < \mathscr{T}_n >$ belong to L_n. However, it is rather easy to give examples when this condition is not sufficient. Consider, for example, the case of the Golden Pair T_1, T_2. Taking the area of T_1 to be 1, then the area of T_2 must the Golden Section g—see Sect. 2.5. Consider $S = g^2 T_1$. Its long side has length $g^3 = 2g + 1$ and its short length has side $g^2 = g + 1$. This already puts some constraints on possible markings. Again $g^2 T_1$ has area $g^4 = 3g + 2$ and consequently equals all joinings of two copies of T_1 and three copies of T_2. Let v_1 denote a vertex of $g^2 T_1$ for which the sides s_2, s_3 form an angle of $\pi/5$. One cannot have a marking with points on s_2, s_3 at a distance 1 from v_1, since there is no triangle in $\mathscr{T}_5$ to fit into this corner.

In Fig. 1 we give a possible presentation of $g^2 T_1$ obtained by proper joining of elements $\mathscr{T}_5$. One may check that the remaining presentations may be obtained by flipping triangles or trapezia. One observes that all markings are side markings (which of course is exceptional). All possible side markings compatible with the above two constraints occur. A given side marking can give rise to more than one proper joining except that in this case they are related by flipping a trapezium.

Definition 2 A triangle $T \in < \mathscr{T} >$ is said to be a composed or composite triangle.

3.2 *Parametrization of the Basic Set*

3.2.1 Angle Sets

There are two natural ways to designate the elements of $\mathscr{T}_n$. The first is to label the vertices of S_n in a clockwise manner by the integers $0, 1, \ldots, n-1$, that is the natural set of representatives of $\mathbb{Z}/\mathbb{Z}n$. Let $T_{i,j,k}$ be the triangle with vertex set $0 \le i < j < k \le n-1$. Obviously we would not want to distinguish two triangles obtained by rotation, that is by adding a fixed integer to each element of the vertex set or by cyclic permutation of indices. However, it is convenient to distinguish $T_{i,j,k}$ from $T_{j,i,k}$ for scalene triangles, that is those with no equal sides. (This is required if one needs to tile the living room floor, since normally the tiles used have one shiny smooth side and one rough side.) We shall say that these two triangles are related by a *parity transformation*.

The set A_n of angles between the sides of elements of $\mathscr{T}_n$ takes the form

$$A_n = \{i\pi/n : i = 1, 2, \dots, n-2\}.$$

From now on we shall generally ignore the factor π/n. Then going in a clockwise direction the angle lying between the vertices with labels i, j is just $j-i$, if $j > i$ and $n + j - i$ if $j < i$.

We call $T_{i,j,k}$ defined above, a labelling of an element of $\mathscr{T}$ by its vertex set. We may also label this element as $T\{j-i, k-j, n+k-i\}$, that is to say by its angle set. Any two angle sets define the same triangle in $\mathscr{T}_n$ if and only if they are related by a cyclic permutation.

3.2.2 Line Sets

Let $p_i : i = 0, 1, 2, \dots, n-2$, be the length of the line segment of S_n joining the vertex 0 to the vertex $i+1$. By definition of S_n one has $p_0 = 1 = p_{n-2}$ and $L_n = \{p_i\}_{i=0}^{n-2}$. It is clear at least geometrically that the $p_i : i = 0, 1, \dots, p_{[\frac{n-2}{2}]}$ are strictly increasing. We call this the *increasing property*.

Consider the triangle bordered by the two adjacent sides of S_n (which by definition have length 1) and the third side of length p_1. In this the angle which subtends to a side of length 1 is just π/n.

Thus $g := p_1 = 2\cos\pi/n$.

By symmetry we also have

$$p_{n-2-i} = p_i, \forall i = 0, 1, \dots n-2. \tag{1}$$

This rule may be extended to all $i \in \mathbb{Z}$ by setting $p_i = 0$ for $i < 0$ and $i > n-2$. We call it *the symmetry property*. On account of the increasing and symmetry properties, it follows that the set $\{p_i\}_{i\in\mathbb{Z}}$ consists of exactly $[\frac{n}{2}]$ distinct elements, namely the elements of the set $L'_n := \{p_i\}_{i=0}^{[\frac{n-2}{2}]}$.

One may check that

$$p_1 p_{i-1} = p_i + p_{i-2}, \forall i \in \mathbb{Z}. \tag{2}$$

Of course this is well-known but we remark that a particularly simple geometric proof in the spirit of proper joinings can be found in [4, 8.7].

Up to a scaling by 2 in its argument, the ith Chebyshev polynomial P_i of the second kind may be defined through the recurrence relation

$$P_{i+1} = xP_i - P_{i-1} : \forall i \in \mathbb{N}, \tag{3}$$

together with the initial conditions $P_{-1} = 0$, $P_0 = 1$.

With this scaling $P_i : i \in \mathbb{N}$ is monic and one may check that

$$(\sin\theta)P_i(2\cos\theta) = \sin(i+1)\theta, \forall i \in \mathbb{N}.$$

Hence $p_i = P_i(g)$. This also follows from (2).

By (1) it follows that g is a root of the equation $P_{n-1} = 0$ and hence an algebraic integer. Then by (2), p_i is an algebraic integer for all $i \in \mathbb{Z}$.

The roots of $P_{n-1} = 0$ are the $2\cos t\pi/n : t = 1, 2, \ldots, n-1$, so these are also algebraic integers. Thus p_1 is the largest real root of the equation $P_{n-1} = 0$ and so by definition a Perron number. Yet the set of Perron numbers is closed under addition and multiplication [8] and hence by (2) every $p_i : i = 1, 2, \ldots, n-2$ is a Perron number and so is any $p \in \mathscr{L}_n$.

Curiously p_i^{-1} is an algebraic integer if n is odd, but this fails otherwise.

The following is well-known but we give a proof for completeness.

Lemma *The elements of L'_n are linearly independent over $\mathbb{Q}$ if and only if n is prime or a power of* 2.

Proof It follows from (3) that the elements of L'_n form a free basis of $\mathbb{Z}[x]$.

Suppose that n is odd, say $n = 2m+1$. Then $m = [\frac{n-1}{2}]$. Following say [4, 7.3], define the polynomials $Q_i(x) = P_i(x) - P_{i-1}(x)$ of degree i. (The first few are listed in [4, 7.3].) Then the relation $p_{m-1} = p_m$ becomes $Q_m(g) = 0$ and so $\mathbb{Z}[g]$ identifies with the quotient ring $Z[x]/ < Q_m >$, which is prime if and only if Q_m is irreducible. Yet by Joseph [4, Lemma 7.4], Q_m is irreducible over $\mathbb{Q}$ if and only if $n = 2m+1$ is prime.

By the first observation of the lemma, if the elements of L'_n are not linearly independent over $\mathbb{Q}$, then the $\{g^i\}_{i=0}^{m-1}$ are not linearly independent over $\mathbb{Z}$. Then g satisfies a polynomial equation $R(g) = 0$ with $\deg R < m-1$, contradicting the primeness of $Z[x]/ < Q_m >$. Conversely if Q_m is not irreducible, then there is a proper factor of R of Q_m such that $R(g) = 0$ implying that the elements of L'_n are linearly dependent over $\mathbb{Q}$.

Suppose n is even, $n = 2m$ and define polynomials $S_i(x) = P_i(x) - P_{i-2}(x)$ of degree i. The relation $p_m = p_{m-2}$ becomes $S_m(g) = 0$. By Joseph [4, 10.2(iii)], S_m is irreducible over $\mathbb{Q}$ if and only if m is a power of 2. Hence the assertion follows as in the previous case. □

Examples If $n = 4$, then $g = \sqrt{2}$ which is irrational. If $n = 5$, then g is just the Golden Section which is irrational. If $n = 6$, then $g = \sqrt{3}$ and satisfies a polynomial equation over $\mathbb{Q}$ of degree $[\frac{n-2}{2}] = 2$. We have linear independence for $n = 4, 5, 7, 8$ but not for $n = 6, 9$.

3.2.3 Determining $|\mathscr{T}_n|$

Notice that the length of the line segment in $T\{i, j, k\}$ opposite to its angle of size i is just p_{i-1} *irrespective* of the values of j, k.

In particular by the symmetry and increasing property, two elements of $\mathscr{T}$ can be properly joined along a side if and only if the angles subtending to that side are either equal or sum to π. Of course this is just a special case of a result in Euclidean geometry for triangles inscribed in a fixed circle.

One may further conclude from the above that any element of $\mathscr{T}$ is uniquely determined (up to parity for scalene triangles) by its angle set. Thus up to parity, $\mathscr{T}_n$ is in a natural bijection with the set of partitions of n into exactly three parts. Then the scalene triangles in $\mathscr{T}_n$ are in bijection with the set of partitions into three distinct parts.

Example There are just 5 partitions of 8 into three parts with two having distinct parts. Thus $|\mathscr{T}_8| = 7$. A similar computation shows that $|\mathscr{T}_{10}| = 12$.

3.3 Two Triangle Joins

The first attempt at extending the tiling by the Golden Pair is to consider joining just two triangles at one time. The following result is just [4, Lemma 8.8], when one replaces the presentation of an element $T \in \mathscr{T}$ by its angle set rather than by its vertex set.

Lemma *Fix* $0 < i < j < n$. *Then for all* t *with* $0 < t < i$ *one has*

$$p_{j-i+t-1}T\{i, j-i, n-j\} = p_{j-1}T\{t, j-i, n-(j-i)-t\} * p_{j-i-1}T\{i-t, j-(i-t), n-j\}.$$

In the above the angle of size $n-(j-i)-t$ in the first triangle is placed together with that of size $j-(i-t)$ to form a straight line. The common sides have length $p_{j-1}p_{j-i-1}$.

3.4 Insufficiency of Two Triangle Joins

Unlike the Golden Pair it is not sufficient for our basic theorem to just consider two triangle joins even when one ignores markings. Indeed let us consider what $T = T\{i, j-i, n-j\}$ can be. Since parity respects proper joinings we may identify for this purpose $T \in \mathscr{T}$ with its parity transpose.

After this identification a triangle $T \in \mathscr{T}$ is independent of the ordering of its angle set; we can assume that $0 < j-i \leq n-j \leq i < n$ and $0 < t < i$.

Lemma *Consider* $s := j-i+t-1$, *with* i, j *fixed and satisfying* $0 < j-i \leq n-j \leq i < n$. *Consider* t *taking integer values from* 1 *to* $i-1$. *Then under the above constraints,* s *takes integer values from* $j-i$ *to* $[\frac{n-2}{2}]$.

Proof Obviously s takes integer values from $j-i$ to $j-2$. Yet $2j > i+j \geq n$ and so $j > [\frac{n}{2}]$. Thus $j-2 \geq [\frac{n-2}{2}]$, as required. □

3.5 Multiple Triangle Joins

Replace $s = j - i + t - 1$ by t and then $j - i, n - j$ by j, k in Sect. 3.3. Then we conclude that $p_t T\{i, j, k\} \in \mathscr{T}_n$ as long as $[\frac{n-2}{2}] \geq t \geq \min\{i, j, k\}$. Since (1) allows us to assume that $t \leq [\frac{n-2}{2}]$, the upper bound on t causes no problem. However, the lower bound is problematic and in general one cannot express $p_t T\{i, j, k\}$ as a join of just two triangles in $\mathscr{T}$ if $t < \min\{i, j, k\}$.

To overcome this difficulty we used multiple triangle joins. Thus [4, Prop. 8.13] used the proper joining of $(t + 1)^2$ triangles in $\mathscr{T}_n$ to show that $p_t T\{i, j, k\} \in \mathscr{T}_n$ given $t < \min\{i, j, k\}$. Combined with [4, 8.8, 8.9] we were able to eliminate the above restriction on t resulting in [4, Theorem 8.14].

However, this only works for the first iteration. In the second iteration side markings are introduced on the elements $p_t T\{i, j, k\}$ and these must be matched. This is quite a non-trivial problem as already indicated by the solution. Indeed as we shall see at the first step the tiling admits a 2-colouring but this can fail at the second step.

Here we shall even modify our first iteration. Thus we use our earlier result [4, Prop. 8.13] but by noting that the solution presented there has another form obtained by reversing "orientations." Then we show that the condition $t < \min\{i, j, k\}$ in [4, Prop. 8.13] can be avoided by a process of elimination of undesirable triangles (some with non-positive angles) which we shall call ablation. One may note that this eliminates the need for Sect. 3.3 thereby making the analysis of [4] more uniform.

Ablation will also prove useful in subsequent iterations when matching of marked points must be effected.

3.6 Rigid Stacks

3.6.1 Stacks of Equilateral Triangles

Let t be an integer ≥ 0. Let T be an equilateral triangle of side length 1. Then we may write $(t + 1)T = T * T * \ldots * T$, noting that there is only one possible interpretation of the right hand side which would give this equality.

The above expresses the easy well-known fact that there is just one way to "stack" $(t+1)^2$ equilateral triangles of side length 1 to form a composite equilateral triangle T' of side length $t + 1$. In this

(i) Triangles of the stack are properly joined.
(ii) If S, T share a common side in the stack, then the angles opposite to the shared sides of S and T above are equal (and not just sum to π—cf Sect. 3.2.3).
(iii) An internal marking is at the crossing of three straight lines.

3.6.2 $\mathscr{T}$-Stacks

We call a $\mathscr{T}$-stack (of order $t+1$) an arrangement T obtained by replacing the equilateral triangles above by triangles in $\mathscr{T}$ so that (i) and (ii) above hold with of course no gaps or overlapping. A $\mathscr{T}$-stack is said to be rigid if (iii) also holds. One may check that a $\mathscr{T}$-stack of order 2 or 3 is always rigid, but rigidity may fail at order 4.

Eventually we may want to "ablate" some of the triangles in the stack if some angle value becomes equal to zero—see Sect. 3.7.

Let i, j, k be positive integers which sum to n. Let t be a positive integer $< \min\{i, j, k\}$. A rigid $\mathscr{T}$-stack T^{t+1} of order $t+1$ for $p_t T\{i, j, k\}$ was constructed in [4, 8.13], though rather abstractly. It was illustrated in [4, Figure 9], for the case $t = 3$ from which one can easily guess the general solution. A description for all $t < \min\{i, j, k\}$ is given [4, Proposition 8.13]. Notice that the upper bound on t is needed to make all angles positive. In the sequel we shall identify a $\mathscr{T}$-stack with the marked triangle it defines.

3.6.3 Angle Sums

Take t, i, j, k as Sect. 3.6.2. In particular $t < \min\{i, j, k\}$ an assumption which will be retained until the end of Sect. 3.6.

Let us describe T^{t+1} more concretely. In this we shall show that T^{t+1} is a rigid $\mathscr{T}$-stack and a realization of $p_t T\{i, j, k\}$ in $\mathscr{T}_n$.

To follow the analysis the reader may conveniently use [4, Figs. 8,9] which illustrates the cases $t = 2, 3$.

View i, j, k as being fixed and being the angle values at the corners of T^{t+1}. Then we denote by v_i the corner of T^{t+1} at which the angle has value i (in units of π/n). The directed edge going from v_i to v_j is denoted by $s_{i,j}$.

The stack of $(t+1)^2$ equilateral triangles described in Sect. 3.6.1 is formed by three sets of $t+1$ (straight) lines parallel to the external sides of T^{t+1}.

To obtain T^{t+1} from the stack of $(t+1)^2$ equilateral triangles we just modify the angles at each vertex (to the values described below—see for example [4, Fig. 9]).

This automatically ensures that (i) holds.

If (iii) holds then there are three straight lines going through each internal vertex.

These come as (three sets) of straight lines, but unlike the equilateral triangle case, the lines of a given set are not necessarily parallel to a given side of T.

Nevertheless these lines can be identified with those of the equilateral triangle stack, so we may conveniently refer them as straight lines being parallel to the respective sides of our composite triangle T^{t+1}.

The angles in each of the triangles making up T^{t+1} will take the form $\{i + u, j + v, k + w\}$ where u, v, w are integers which sum to 0 changing in the manner described below as we pass through T^{t+1}.

By continuing the lines beyond the boundary of T^{t+1} and adding a line at each external vertex, we may assume that each vertex meets six triangles whose angles are $i + u, j + v, k + w, i + u, j + v, k + w$ starting at the top and going counterclockwise. This property assures that (iii) holds. Adopt this convention.

Following the special case $t = 3$ in [4, Figure 9], we require u to decrease in units of 1 as we pass vertices along and in the direction of a line parallel to the directed edge $s_{i,j}$ or as we pass vertices along and in the direction of a line parallel to the directed edge $s_{k,i}$, whilst we require u to increase in units of 2 as we pass vertices along and in the direction of a line parallel to the directed edge $s_{j,k}$. This variation and the above assignments at each vertex ensure that on going around the three sides of composite triangle we reach the same value for a given angle—simply because $1 + 1 = 2$. It also ensures that (ii) holds.

The behaviour of v, w can be deduced from that of u by just rotating the composite triangle, that is by cyclic permutation. One checks that the result is compatible with the above assignment of angles at each vertex, again because $1 + 1 = 2$.

This completely defines the stack T^{t+1} and shows it to be a rigid $\mathscr{T}$-stack.

3.6.4 Line Sums

Recall the first sentence of Sect. 3.2.3.

Reading T^{t+1} in a clockwise manner the vertices appear in the order v_i, v_j, v_k. Moreover passing from v_i to v_j the distances between marked points form the sequence $p_{k-1-t}, p_{k-1-(t-2)}, \ldots, p_{k-1+t}$. (Notice by (1), (2) that these lengths strictly increase, rise to a maximum value which may be repeated once and then strictly decrease.) By Joseph [4, Lemma 8.12], this sum of distances from v_i to v_j is just $p_t p_{k-1}$. Through a similar result for the sides defined by the two other pairs of vertices, we conclude that T^{t+1} is a realization of $p_t T\{i, j, k\}$ in $< \mathscr{T}_n >$.

3.6.5 Orientation

We claim that the markings on the sides of T^{t+1} give it a natural clockwise orientation.

In this case when indices on the lengths p_t between adjacent marked points increase along a directed edge $s_{i,j}$ we say that it has a positive orientation, and that $s_{j,i}$ has a negative orientation. If $s_{i,j}, s_{j,k}, s_{k,i}$ all have a positive (resp. negative) orientation, we say that T^{t+1} has a clockwise (resp. anticlockwise) orientation.

In virtue of the symmetry condition (1), a side (or parts of a side) can be assigned to both orientations simultaneously. Thus as we can always take $1 \leq i < j \leq n - 2$, and if $j - i$ is even, then the indices of the $p_i, p_{i+2}, \ldots, p_j$ increase. On the other hand by the symmetry condition this is also the sequence $p_{n-i-2}, p_{n-i-4}, \ldots, p_{n-j-2}$, where the indices decrease. This phenomenon will play an important role when it comes to discussing ablation in Sect. 3.7 and in matching marked points in Sect. 3.8.

We show in the next section that there is a second rigid $\mathcal{T}$-stack S^{t+1} of order $t+1$ for $p_t T\{i, j, k\}$ with the opposite orientation. If $T\{i, j, k\}$ is scalene, this stack is *not* obtained by simply taking the mirror image because that would be a stack for $T\{i, k, j\}$. Rather it is obtained by interchanging "increase" and "decrease" in the penultimate paragraph of Sect. 3.6.3.

3.6.6 Uniqueness

One can ask if there are any other rigid $\mathcal{T}$-stacks of order $t+1$ for $p_t T\{i, j, k\}$. To answer this one must study the solutions to the equation

$$\sum_{s=0}^{t} p_{k-1+(t-2s)} = \sum_{s=0}^{t} p_{k-1+a(t-2s)}, \tag{4}$$

with $a \in \mathbb{Z}$. Obvious solutions are $a = \pm 1$. These correspond to the stacks described in Sect. 3.6.3 and the last paragraph of Sect. 3.6.5.

Let us show the case when n is prime or a power of two that the only solutions of (4) are $a = \pm 1$.

Under the above hypothesis it follows from Sect. 3.2.2 that the $p_i : i = 0, 1, \dots, [\frac{n-2}{2}]$ are linearly independent over $\mathbb{Q}$. In this case the only way that (4) can hold is for p_i on the left to cancel with p_{n-2-i} on the right via the symmetry property (1). Taking respectively $s = 0$ and $s = t$ we obtain $2(k-1) + (a-1)t = n-2 = 2(k-1) - (a-1)t$, which for $t > 0$ has only the solution $a = 1$. This is the solution we already obtained.

Let n, i, j, k be as in Sect. 3.6.2 and $t < \min\{i, j, k\}$.

Lemma *Suppose n is prime or a power of 2. Then there are exactly two rigid $\mathcal{T}$ stacks of order $t+1$.*

Proof Let T be a rigid $\mathcal{T}$ stacks of order $t+1$. Recall that T is triangulated with angle sets given by the rules (iii) and (ii) of Sect. 3.6.1.

If we only require T to just satisfy the condition on the angle sets described above and not the condition on side lengths, then for each $a \in \mathbb{Q}$, we obtain a rigid $\mathcal{T}$ stack of order $t+1$ by replacing $\{u, v, w\}$ by $\{au, av, aw\}$ in the construction of Sect. 3.6.3. Of course a must be such that all angles have positive values, but apart from this we obtain a one-parameter family of rigid $\mathcal{T}$ stacks of order $t+1$. In this we may remark that T^{t+1} (resp. S^{t+1}) is recovered by setting $a = 1$ (resp. $a = -1$).

The angle conditions (iii) and (ii) of Sect. 3.6.1 are linear relations involving the fixed values i, j, k with their number depending on t. Thus to show that there are no other rigid $\mathcal{T}$ stacks of order $t+1$ we have to show that for each value of t that there is exactly a one parameter family of solutions, equivalently fixing just one further angle value fixes all the others. Though this is an elementary exercise it is not necessarily easy because the linear relations may be over-determined.

Start from the value i of the angle at the vertex v_i of T. The vertex v opposite to v_i has angle value i by (ii). If v is an internal vertex then by (iii) it possesses a second angle of value i. Proceeding in this manner until one reaches the side $s_{i,j}$ gives $t + 1$ internal angles with value i. A similar result holds for v_j, v_k.

If $t + 1 \neq 0 \mod 3$ then there is a "central triangle" in T with angle set $\{i, j, k\}$ standing as (resp. inverted relative to) T if $t = 1 \mod 3$ (resp. $t = 2 \mod 3$). If $t = 0 \mod 3$, then there is a "central vertex" in T with angle set being two copies of $\{i, j, k\}$.

Now take $t = u + 3n > 0$ with $u = 1, 2, 3$ and $n \in \mathbb{N}$. The proof that just one additional angle value determines all others is by induction on n and is similar in the three cases.

The case $n = 0$ is a finite computation and sets the tone for the induction proof. The most convenient one we found is the following.

Recall that i is the internal angle value of a triangle in T with vertex v_i. Descending down the stack along the side $s_{i,j}$, assign $i - a$ to the top vertex of the succeeding triangle, that is to say we place $i - a$ where $i - 1$ stands on an edge vertex of [4, Figure 9]—this is case $t = u = 2$. Then either $u = 1$ or by (ii) of Sect. 3.6.1 the opposite vertex also gains the value $i - a$. Combined with our first assignment on angles, this determines a pair of angle values out of three which sum to n. In this manner all angle values are determined.

The general case proceeds by induction on n. In this we obtain a rigid $\mathscr{T}$ stack T_n of order $u + 3n + 1$, satisfying the angle relations (ii) and (iii) of Sect. 3.6.1 which we view as being embedded as the "composed central triangle" in the rigid $\mathscr{T}$ stack T_{n+1} of order $u + 3(n + 1) + 1$. Notice that by (iii) *all* six angles at the non-corner vertices of T_n are determined. Then through (ii) we obtain the values of two out of three angles of every triangle neighbour to a triangle of T_n and hence all their angle values. Repeating this procedure we obtain all the angle values on the triangles neighbour to one of these new triangles. Repeating the procedure again we obtain all angle values of T_{n+1}. This completes the induction.

Recall that the markings on the edge $s_{i,j}$ in T and in T^{t+1} (in which a is specialized to 1). These have $t + 1$ markings which subdivides its length $p_t p_{k-1}$ into t parts. Consider the angles which subtend onto these parts. In the latter case we already noted that they take the form $k + (t - 2s) : s = 0, 1, \dots, t$. Thus in the former case they must take the form $k+(t-2s)a : s = 0, 1, \dots, t$. Then $s_{i,j}$ in T^{t+1} has the correct length by [4, Lemma 8.12], as we already observed in Sect. 3.6.4. For $s_{i,j}$ in T to have the correct length we required (4) to hold. Through the hypothesis of the lemma and the remarks proceeding it, we obtain $a = \pm 1$. □

Remark Even under the hypothesis of the lemma, uniqueness can fail if t is not strictly less than $\min\{i, j, k\}$. Indeed the positively orientated $p_1 T\{2, 2, 1\}$ is given in the right hand side of Fig. 2, whilst since $T\{2, 2, 1\}$ is isosceles the negatively oriented $p_1 T\{2, 2, 1\}$ is just its mirror image. However, it also possible to present $p_1 T\{2, 2, 1\}$ as two copies of $T\{2, 2, 1\}$ with $T\{1, 3, 1\}$ sandwiched between them.

3.6.7 Orientation Revisited

In order to construct a stack with triangles oriented by their markings we need that triangles which share a common side, that is to say are neighbours, to have the opposite orientation. The simplest example is illustrated in Plate 1.

This is possible in T^{t+1} because in a stack the number of triangles which share a common internal vertex is even (namely 6). (This fact also enables a two colouring of the stack and in any case is a special case of a standard fact in algebraic topology concerning the orientation of simplexes.)

Sometimes a tiling with an incorrect orientation can be obtained by accident if not enough sides are presented to the internal faces. This would cause difficulties for further iterations.

This is what happened when I carelessly presented the tiling in my home page http://www.weizmann.ac.il/math/joseph/. The arrangement of the three triangles in the upper right hand corner differs from that given in Plate 4, which follows the recipe of Sect. 3.9. Actually this phenomenon already occurs for $n = 5$ and one can give tilings of the $p_1^2 T\{2, 2, 1\}$ other than that given in Plate 1. This is illustrated in Fig. 9.

3.7 Ablation

3.7.1 Cutting Off a Single Corner

The above analysis suffices when $t < \min\{i, j, k\}$. The general case is handled by analyzing what happens to our stack when this strict inequality fails. We retain the same rules for assigning angles in a stack given in Sect. 3.6.3. Yet this can result in some angles acquiring non-positive values. These as well as some of their neighbours will be removed by a process we call ablation.

The resulting arrangement will be called an ablated stack.

To define ablation more precisely recall the description of the stack T^{t+1} given in Sect. 3.6.

Now suppose that $t \geq i$. Then there is a marked point v_i^1 on $s_{i,j}$ for which $u = -i$, forcing the corresponding angle to be zero. Let v_i^2 denote the marked point on $s_{j,k}$ where the line parallel to $s_{k,i}$ passing through v_i^1 meets $s_{j,k}$. We call this the ith ablation line. Removal of the composite triangle whose vertices consist of v_i^1, v_j, v_i^2 is called the ith ablation (see Fig. 2), and the resulting triangle an **ablated triangle**. Such a triangle is itself a stack. It has one external vertex with angle j, one vertex with angle 0 and a third vertex with angle $n - j$. The combination of the ith, jth, kth ablations is called ablation. (All three can occur simultaneously but can be taken in any order, because they are just the cutting off corners of the stack. Ablation lines may meet on a side of the stack or cross in the interior or not at all.) One may check that ith ablation line meets (resp. crosses) jth ablation line exactly when $t - i + 1 \geq j$ (resp. $t - i + 1 > j$). Since we can assume $t \leq [\frac{n-2}{2}]$, the three ablations lines cannot all meet or cross, though some may, for example, in $p_3T\{2, 1, 5\}$.

Proposition *After the ith ablation, the following four properties hold.*

(i) *The marked point v_i^1 on $s_{i,j}$ lies on a straight line segment.*
(ii) *In all triangles, the angles $i + u$ are positive.*
(iii) *The marked point v_i^2 on $s_{j,k}$ subtends an angle of size j.*
(iv) *The new side joining v_i^1 to v_j^2 is oriented, having the same orientation as $s_{i,j}$.*

Proof Of course (i) is immediate since the angle sum of the three triangles meeting the marked point v_i^1 on $s_{i,j}$ was constructed in [4, 8.13] to be n, whilst by assumption one of these triangles has a vertex with angle $i + u = 0$. Notice that there are now just two triangles whose vertices form this marked point.

For (ii) we note that values of u increase along a line L parallel to $s_{i,k}$. Thus they take their minimal when the corresponding marked point meets $s_{i,j}$. Ablation removes those on $s_{i,j}$ where $i + u \leq 0$. Since the ith ablation removes all the triangles to the right of L (in the sense of its perpendicular), it removes all triangles in which $i + u \leq 0$. Hence the assertion.

By (i) the ith ablated stack forms a triangle with vertices v_k, v_i, v_i^2. The first two vertices have angle values k, i respectively, so the third must have angle value j, as required. Hence (iii).

We note that j is the sum of the angle values calculated as follows.

The marked point v_i^1 is exactly the ith marked point on $s_{i,j}$ counting downwards from v_i viewed as the zeroth marked point. Thus it is the $(t + 1 - i)$th marked point on $s_{i,j}$ counting upwards from v_j viewed as the zeroth marked point. Through the arrangement of triangles in a stack, it follows that v_i^2 is the $(t + 1 - i)$th marked point on $s_{j,k}$ counting leftwards from v_j viewed as the zeroth marked point. Thus the angle value in the leftmost triangle meeting v_i^2 has value $j - (t + 1 - i)$, so that in the second (rightmost) triangle meeting v_i^2, it must be $t + 1 - i$. The first can be negative, that is if $j \leq (t + 1 - i)$. In this case it will be removed by the jth ablation (in virtue of (ii) applied to the jth ablation). If equality holds in the above the ith and jth ablation lines meet on $s_{j,k}$ at v_i^2, whilst if the inequality is strict they cross at a marked point $v_{i,j}$ in the interior of the stack.

(iv) is more delicate. Indeed in terms of angle values, the orientation on the side joining v_i to the marked point v_i^1 on $s_{i,j}$ has necessarily the *opposite* orientation to its continuation from v_i^1 to v_i^2. This follows by considering the orientation of the removed composite triangle. Indeed on the common side the angle values subtended on the successive intervals defined by the marked points on the line going from v_i^2 to v_1^i must increase in units of 2 from $j - (t - i)$ to $j + t - i$. By (ii) of Sect. 3.6.1 these values are the same and change in the same way on remaining composite triangle.

Fortuitously this seemingly bad phenomenon is compensated by the symmetry property.

Indeed the values in the lengths of the line intervals (which is all that counts for matching) are the same if we replace an angle value ℓ by $n - 2 - \ell$. Then if the former decrease by 2 on going from v_i^1 to v_i^2, the latter increase on going from v_i^1 to v_i^2.

Yet we still have to check matching on the composite line obtained by joining the sides s_{v_i,v_i^1} and $s_{v_i^1,v_i^2}$. This means that we have to show that if p_r is the distance between the marked points on $s_{i,j}$ consisting of v_i^1 and its predecessor, then p_{r+2} is the distance between the marked points consisting of v_i^1 and its successor on the ith ablation line joining v_i^1 to v_i^2.

Now the first triangle above has three angle values firstly $k+w$ which equals $r+1$ by definition of p_r—see Sect. 3.2.2, secondly $j+v$ which equals $j+t+1-i$ since v_i^1 is the $(t+1-i)$th marked point on $s_{i,j}$ counting upwards from v_j and thirdly $i+u$ which equals 1 since the predecessor of v_i^1 is the $(i-1)$th marked point on $s_{i,j}$ counting downwards from v_i. Thus $r = n-(j+t+1-i)-1-1 = k+2i-t-3$.

On the other hand by the first paragraph of the proof of (iv), the second distance is $p_{j+t-i-1}$. Yet $(n-2)-(j+t-i-1) = (i+j+k-2)-(j+t-i-1) = k+2i-t-1$, so (iv) follows by the symmetry property $p_i = p_{n-2-i}$. □

3.7.2 Cutting Off Several Corners

After the jth ablation we similarly obtain a triangle with vertices v_i, v_j, v_j^2 with angle set i, j, k. When we apply both ablations in either order we obtain a triangle with vertices v_i, v_i^2, v_j^2 or $v_i, v_{i,j}, v_j^2$. By the first observation the angle subtended by the second vertex must be j.

Finally when we apply all three ablations the second of the resulting vertices is v_j^2 or $v_{i,j}$ which we already know possesses an angle of size j. The remaining two obtained by cyclic permutation of indices and so possess the angles of k, i respectively. Of course their sides are subsets of those obtained by individual ablations and therefore must be correctly oriented.

3.7.3 Positivity of Angles

From the above we obtain the

Corollary

(i) *All the angles in the triangles which make up the ablated stack are positive.*
(ii) *The ablated stack constructs the triangle $p_t T\{i, j, k\}$ with a clockwise orientation.*

3.7.4 Reversing Orientation

An ablated stack giving the required triangle $p_t T\{i, j, k\}$ with an anticlockwise orientation is similarly obtained by starting from S^{t+1}.

3.7.5 Two-Colouring at First Iteration

In an ablated stack (in contrast to a stack) some angles at the vertices may be bisected by the sides of joined triangles. Nevertheless an ablated stack can be two-coloured because it is obtained by removing corners from a stack which can be two-coloured.

From now on a "stack" will also mean an ablated stack.

On the other hand when ablated stacks are joined through the induction procedure, it may no longer be true that the number of triangles having a common vertex at an interior point is even. Thus in general a join of ablated stacks cannot be two-coloured. An example occurs in Plate 3. We do not know if they can be always three-coloured—probably. In all our examples (see plates) a three colouring can be given.

3.7.6 Failure of Two-Colouring at Subsequent Iterations

When the ablated stacks are joined through the induction procedure, property (ii) of Sect. 3.6.1 may fail. This is because we need the symmetry property to ensure correct orientation. Examples appear in Plate 3.

3.7.7 Failure of Uniqueness of Degenerate Stacks

Even when n is prime or a power of two the uniqueness property described in Sect. 3.6.6 can fail and in particular not all stacks need be obtained by ablating those described by Sect. 3.6.6. For example take $n = 5$. Then the stack for gT_2 obtained by ablation is either that given by the right hand side of Fig. 2 or (accidentally—see Sect. 3.6.5) its mirror image (which has the opposite orientation). Both have a pair neighbours of type T_2. However, one can flip a T_1, T_2 pair so that this fails. This composed triangle has neither a positive nor a negative orientation.

3.7.8 A Property of Ablated Triangles

Recall Sect. 3.7.1 and recall the notion of an ablated triangle $T \in< \mathscr{T}_n >$. It may be realized as a stack $p_t T\{i, j, k\}$ with vertices u, v, w in a clockwise labelling, starting with the angle at u being i. In the construction of Sect. 3.7.1 we ablate T if one angle of T, say at vertex u, is zero, that is to say $i = 0$.

Lemma *The distances between marked points* $u = u^1_w, u^2_w, \ldots, u^{t+1}_w = w$ *going from* u *to* w *coincide with the distances between marked points* $u = u^1_v, u^2_v, \ldots, u^{t+1}_v = v$ *those going from* u *to* v.

Proof We have only to show that for all $i \in \{1, 2, \ldots, t\}$ the subtended angle to the line with end-points u^i_w, u^{i+1}_w added to the subtended angle to the line with end-points u^i_v, u^{i+1}_v equals n. This a straightforward exercise using the variation of angles in a stack (of either orientation) given in Sect. 3.6.3 and that $j+k = n-i = n$. □

Remark This is really the same result as that given in Sect. 3.7.1(iv). Both points of view play key roles in Sect. 3.9.

3.8 Matching Marked Points

The $\mathscr{T}$ stack T^{t+1} (resp. S^{t+1}) of order $t+1$ constructing $p_t T\{i, j, k\}$ with clockwise (resp. anticlockwise) orientation constructed above is canonical in the sense that it be determined by the quadruple i, j, k, t both with respect to the triangles they contain and in their arrangement. This has the disadvantage that we are not going to obtain all possible markings on $p_t T\{i, j, k\}$, nor are we likely to obtain all possible $p_t T\{i, j, k\}$ with a given marking. However, it has the advantage that markings can be correctly aligned in the next induction step.

The formula in (2) of Sect. 3.2 iterates to give

$$p_t p_{i-1} = \sum_{j=0}^{t} p_{i-1+t-2j}, \tag{5}$$

valid for all positive integer $i > 0$, non-negative integer t with $t \leq i - 1$ and $t + i - 1 \leq n - 2$. This means that the distance between the marked points on the side subtended by angle $i\pi/n$ in $p_t T\{i, j, k\}$ (with positive orientation) gives the sequence $p_r, p_{r+2}, \ldots, p_s$ where $r = i - 1 - t, s = i - 1 + t$

Recall that $p_{n-1} = p_{-1} = 0$. Thus this sequence has a cut-off when $r = n - 1$, that is when $t = n - i$. Replacing $n - i$ by i this matches the cut-off when $s = -1$, that is when $t = i$.

In the description of $p_t T\{i, j, k\}$, these cut-offs result from ablation.

The above sequence comes in the reverse order if $p_t T\{i, j, k\}$ has an anticlockwise orientation.

In particular the sequence of indices (and hence of distances) is determined by (5) and comes in reverse order when the orientation is reversed.

We conclude that marked triangles (with markings coming from the vertices of triangles in a stack) can be joined (with a matching of markings) by a common side of length $p_{i-1} = p_{n-i-1}$ as long as they have opposite orientation and the angles opposite to this side are either equal or sum to π.

Combined with Sect. 3.6.7, this means that we may construct $p_{t'} p_t T \in < \mathcal{T}_n >$, by viewing $T' := p_t T$ as an element of $\mathcal{T}_n$ and using the construction of $p_{t'} T'$ as a rigid $\mathcal{T}$ stack of order $t' + 1$ in which neighbours (which are themselves $\mathcal{T}$ stacks of order $t + 1$) have oppositive orientation. Examples obtain from Plates 1, 2, 3, 6, 7, and 8.

3.9 The Self-Similarity Theorem

The main theorem of this section extends [4, Thm. 8.14] by taking care of the matching of marked points to ensure proper joining of triangles at each induction step.

Take $n \geq 3$ and let $< \mathcal{T}_n >$ denote the set of polygons which may be generated from $\mathcal{T}_n$ via $*$, as described in Sect. 3.1.1.

Theorem *The set* $\{pT, p \in \mathcal{L}_n, T \in \mathcal{T}_n\}$ *is contained in* $< \mathcal{T}_n >$.

Proof The proof is by induction on the number of factors in p. If $p = p_t : t = 0, 1, \dots n-2$, the assertion results from the construction in Sects. 3.5–3.7 and indeed was already obtained as one of the main results of [4]. If $p = p_{t'} p_t$, then the assertion results from the last paragraph of Sects. 3.8. The main point being that there is a reversal of sequences of indices (and hence) of lengths between marked points when orientation is reversed.

However, this procedure does *not* extend on passing from the first iteration to the second iteration, because unlike the tiling in the first iteration which may be two-coloured, the tiling in already the second iteration need not admit a two-colouring (see Plate 3). This means that neighbouring triangles need not acquire opposite orientations. Yet we claim that paradoxically one may still match marked points if orientations are properly chosen.

The proof of our claim is based on Sect. 3.7.8. Moreover it is appropriate to subdivide elements of the basic set obtained from $(k-1)$th step (Procedure 1) rather than join composed triangles obtained from $(k - 1)$th step which would necessitate matching a multitude of side markings (Procedure 2). In Procedure 1 we only need apply Sects. 3.7.1(iv) and 3.7.8 to stacks composed of triangles from the basic set which themselves replace elements of the basic step, in the proposed subdivision.

In addition to the above let us further observe that we do not need to keep track of the ablated triangles removed at each induction step but only the orientation they induce on the elements of the basic set. Moreover it is immediate that the orientation on the triangles of the basic set obtained by the kth subdivision must be given as follows.

Take the orientation induced by elements of the basic set obtained after the $(k - 1)$th subdivision. At the kth subdivision, each such triangle T is replaced by a stack. Assign to the triangle at the apex of the resulting stack the same orientation as T (regardless of whether angles become non-positive). This defines a unique orientation to each triangle in the stack.

In this each triangle which shares a common edge with the original triangle will acquire the *same* orientation as that at the apex. If there is an ablated triangle (in the sense of Sect. 3.7.1) in the stack, then by our construction this will be a composed triangle which at its apex (denoted by v_i^1 in Sect. 3.7.1) carries an element of the basic set, shares a common edge with the original triangle (on the line $v_i^1 - v_j$ in the notation of Sect. 3.7.1)and has the zero angle at the vertex v_i^1. By the above its orientation will be that of T.

It is to this ablated triangle to which Sects. 3.7.1(iv) and 3.7.8 are applied. They ensure that marked points on the ablated stack define the same orientation as that of T and that they are matched on neighbouring stacks (consisting of composed triangles) which replace the elements of the basic set obtained from the $(k-1)$th subdivision.

This extends our construction from the $(k-1)$th subdivision to the kth subdivision and concludes the proof of the theorem.

A posteriori we can further deduce that side markings must also match up in Procedure 2 and that we obtain the same result as in Procedure 1 except that the order of the dilation factors is reversed.

□

3.10 Examples

3.10.1 Tiling by Triangles Obtained from the Regular Octagon

A three-fold iteration of this process described above is given in Plates 4 and 5 representing the triangle $p_1p_2p_1T\{4, 2, 2\}$. However, since the tiling obtained from the second iteration is very nearly two-colourable, the construction of this example was not so demanding.

3.10.2 Tiling by Triangles Obtained from the Regular Decagon

For Plate 7 we used the construction of Sect. 3.9 to describe the composite triangle $p_3p_2T\{4, 3, 3\}$. Here we may use either procedure noting only that the evaluations must be carried out in the reverse order for Procedure 2 if we are to obtain the same result as in Procedure 1.

One may note that a practical advantage of Procedure 1 is that one fixes the overall size of the diagram from the start. This has the advantage of ensuring that the resulting diagram will fit on the office wall. Of course in this elements of the basic set get smaller at each iteration but then of course can be scaled up. One may note that in this there are several internal vertices which are at the meet of an odd number (in fact 5) of triangles and so the resulting diagram cannot be two-coloured.

We then considered a further p_1 dilation again carried out on the individual triangles which make up the composite triangle $p_3 p_2 T\{4, 3, 3\}$ to obtain $p_3 p_2 p_1 T\{4, 3, 3\}$.

To illustrate how Procedure 1 works, it is only necessary to do this in the neighbourhood of one internal vertex meeting 5 triangles. We chose that lying most towards the top-left hand corner of Plate 8. Now $p_3 T\{4, 3, 3\}$ is tiled by 14 triangles with the vertex in question lying on the common side of $T\{1, 4, 5\}$ given a negative orientation and $T\{3, 5, 2\}$ given a positive orientation.

In turn $p_2 T\{1, 4, 5\}$ is tiled by 4 triangles with that at its apex $T\{1, 2, 7\}$ given the orientation of $T\{1, 4, 5\}$, which is negative. On the other hand $p_2 T\{3, 5, 2\}$ is tiled by 9 triangles *counting* the one at its apex $T\{0, 6, 4\}$ given the orientation of $T\{3, 5, 2\}$ which is positive. This triangle must be ablated. Its unique neighbour $T\{1, 3, 6\}$ has hence a negative orientation and this is the one that shares a common edge with $T\{1, 2, 7\}$ obtained above and also given a negative orientation. By Sect. 3.7.8 the marked points of this common edge must match and this is indeed verified in Plate 8. In particular side markings on these neighbouring triangles with the *same* orientation are matched. This may seem surprising!

To help identify the 5 triangles being subdivided in Plate 7, we have also indicated in Plates 7 and 8, angles given in multiples of $\pi/10$ by the numbers in the corner of each triangle. The angles in the triangles in Plate 7 then appear as being subdivided in Plate 8.

3.10.3 A Poster

In a poster we give the resulting tiling of $p_3 p_2 p_1 T\{4, 3, 3\}$. It involves some roughly 300 triangles. In this the reader may not realize that this was a proper tiling from the fixed set $\mathscr{T}_{10}$ of twelve triangles (eight of which are non-congruent). Indeed it has the appearance of a random triangulation. To avoid this we use coloured lines in our poster to indicate the successive induction steps. Via Sect. 3.6.3 these are straight lines within the boundaries defined by lines of an earlier induction step.

An alternative is to assign a different colour to each of the different triangles, though in doing this some neighbours may acquire the same colour.

Achieving this with an ascetically pleasing result we leave to a reader with artistic skills. Here we remark that the poster file may be downloaded from my home page http://www.weizmann.ac.il/math/joseph/ and coloured using Paint 3D.

4 The Uniqueness of $\mathscr{T}_n$

4.1 The Goal

Assume that $\mathscr{T}$ admits a proper self-similar triangle tiling. Let $L = \{1 = p_0 < p_1 < \ldots < p_r\}$ be their side lengths, $\mathscr{L}$ the monoid generated by L and A the angle set of $\mathscr{T}$.

Given $p_i, p_j \in L$, choose $T \in \mathscr{T}$ having side length p_j. Then $p_i T$ has a side length of side $p_i p_j$, so the condition that $p_i T \in < \mathscr{T} >$ implies that there exist non-negative integers $m_{i,j}^k$ such that

$$p_i p_j = \sum_{k=1}^{r} m_{i,j}^k p_k. \tag{6}$$

In particular the $p_i : i = 1, 2, \ldots, r$ are algebraic integers. For the reasons explained in Sect. 2.7 we exclude the case where p_1 is integer.

One can ask if there exists an integer $n \geq 3$ such that $\mathscr{T} = \mathscr{T}_n$.

Two scalene triangles are similar (in the sense of Euclidean geometry) if they have the same angle set. They can be scaled so that they become congruent. However, congruence does not imply that they can be rotated into one another, except after a parity flip and this is inappropriate for tiling (the floor!). It is not clear if the condition that $\mathscr{T}$ admits a proper self-similar tiling implies that $\mathscr{T}$ is parity invariant which is of course a necessary condition for it to equal to some $\mathscr{T}_n$.

To avoid this question we assume that $\mathscr{T}$ is parity invariant from the start.

4.2 Simple Numerology

We know of no examples of proper self-similar triangle tiling besides $\mathscr{T}_n : n \geq 3$, (except for some special subsets of $\mathscr{T}_n$ and the "degenerate case" when all side lengths are integer). Indeed even in the highly structured case of $\mathscr{T}_n$ it was already quite difficult to guess how to establish self-similarity, though ultimately the proof is quite easy. Nevertheless we argue below that other quite different solutions should exist. Thus a positive answer to the uniqueness question posed in Sect. 4.1 would seem highly optimistic unless we impose some additional conditions.

Choose $p \in L$, $T \in \mathscr{T}$. By hypotheses we can write pT as a composed triangle, that is as a product with respect to $*$ of Sect. 3.1.1, by say k elements of $\mathscr{T}$. Discounting the three vertices of pT, let u (resp. v) denote the number of side markings (resp. interior marked points) of pT. Comparing the two ways to compute the total number of angles we obtain

$$1 + u + 2v = k. \tag{7}$$

The k triangles making up pT have $3k$ sides of which $u + 3$ form the sides of pT. Thus the number m of line segments joining interior points is given by

$$m = \frac{3k - u - 3}{2}. \tag{8}$$

In the k triangles, there are $2k$ angles to be determined. On the other hand from the given angles of T and the sum conditions at the $u + v$ edge markings, we obtain $u + v + 2$ linear equations for these unknowns. This number of equations is larger than $2k$ by exactly m. Of course self-similarity will impose further relations on sums of lengths of the $u+3$ line segments on the sides of pT, but this will still not make up the required number of equations. Thus we may anticipate many solutions besides those provided by the sets $\mathcal{T}_n$.

4.3 Refining the Proper Joining Condition

Let s be a common side of two properly joined triangles in $< \mathcal{T} >$. Let a, b be the angles subtending to s.

Definition The *angle condition* of proper joining at s is that either $a = b$ or $a+b = \pi$.

We may recall (cf Sect. 3.2) that this is satisfied for $\mathcal{T}_n$.

It is clear that the angle condition for proper joining introduces a further m linear equations (albeit each with two possible solutions) and one checks from (7), (8) of Sect. 4.2 that this gives a total of $2k$ linear equations for the $2k$ unknown angles.

Even if we impose the further condition that angles subtending to a common side are equal (that is excluding the alternative that they add to π), the above system of equations has considerable degeneracy particularly for a stack. A fortiori without the angle condition many more solutions will be possible.

The angle condition may be more precisely formulated in the following manner.

($H1$) There is a function $f : [0, \pi] \to \mathbb{R}$ such that for all $T \in \mathcal{T}$ the length of the side subtended by an angle in T of size a equals $f(a)$. In addition $f(a) = f(b) \Leftrightarrow a = b$, or $a + b = \pi$.

This holds for $\mathcal{T}_n$.

4.4 Completeness

A second condition which naturally complements the first is the completeness hypothesis.

($H2$) Every triangle $T \in < \mathcal{T} >$ is similar to an element of $\mathcal{T}$.

This holds for $\mathscr{T}_n$. This is because all possible angle sets appear.

If we drop $(H2)$, then certain subsets of $\mathscr{T}_n$ can give rise to a proper self-similar triangle tiling (see Sect. 4.8).

4.5 Connectedness

A pair of triangles $T, T' \in \mathscr{T}$ are said to be connected if they share a common side length. A subset $\mathscr{T}'$ of $\mathscr{T}$ is said to be connected if any two elements T', T'' lie in a chain consisting of connected pairs. Obviously $\mathscr{T}$ can be decomposed into its connected components. Again the proper joining condition implies that if T', T'' form part of some element of $< \mathscr{T} >$, then T', T'' belong to the same connected component of $\mathscr{T}$. Thus we can assume without loss of generality that

$(H3)$ $\mathscr{T}$ is connected.

4.6 Consequences of $(H1 - H3)$

4.6.1 Elementary Euclidean Geometry and Trigonometry

Recall (H1) and the notation given there. From Euclidean geometry we know that every triangle can be inscribed in a unique circle. In this a given side s of the triangle must a chord of the circle. From Trigonometry the radius of this circle must be $r = f(a)/2\sin a$. Notice that the latter also equals $f(a)/2\sin(\pi - a)$ and since sin is strictly increases in $[0, \pi/2]$ we conclude that $f(a) = f(a') \Leftrightarrow a = a'$ or $a+a' = \pi$.

Thus the angle condition of proper joining is satisfied for triangles inscribed in a fixed circle. The origin of this result is very old and the proof does not need trigonometry. For the special case $a = \pi/2$, it is attributed to Thales of Miletus who lived 300 years *before* Euclid.[1]

Conversely if two triangles are joined by a common edge then the angle condition means that they can be inscribed in the same circle. (Indeed if $a = a'$ (resp. $a+a' = \pi$) then the second triangle is laid on top of (resp. opposite to) the first. Thus we obtain.

Lemma *Assume $(H1)$ and $(H3)$ hold. Then the triangles in $\mathscr{T}$ are amongst those whose vertices lie on the circumference of a fixed circle. In particular if $T \in \mathscr{T}$ and $p \in \mathbb{R}^+ \setminus \{1\}$, then $pT \notin \mathscr{T}$.*

[1]It is believed that Thales threw a-4 day celebration (read: orgy) in honour of his discovery, though some dismiss this as myth. Certainly nowadays mathematicians have less fun.

Remark Yet it can happen (see Sect. 4.8) that a non-trivially composite triangle can lie in $\mathscr{T}$.

4.6.2 Inequalities

Recall the function f defined in (H1).

Corollary *Assume (H1) and (H3) hold. The function f is strictly increasing in $[0, \pi/2]$ and strictly decreasing in $[\pi/2, \pi]$.*

4.6.3 A Thin Triangle

Since $\mathscr{T}$ is assumed finite, so is A. We write it as $i_0 < i_1 < \ldots < i_k$. Then by definition i_0 be the unique smallest element of A. Observe that

$$\text{If } i > \pi - 2i_0, \text{ then } i \notin A. \tag{9}$$

Indeed otherwise the remaining angles of the triangle containing i would have to sum to $< 2i_0$ and at least one would have value $< i_0$.

Set $p_j = f(i_j)$. By Sect. 4.6.2 one has $p_j > p_0$ given that $i_0 < i_j < \pi - i_0$ and thus by (9) for all $i_j \in A \setminus \{i_0\}$, one has $i_0 < i_j \leq \pi - 2i_0$.

Obviously $i_0 \leq \pi/3$, the case of equality being when $\mathscr{T} = \mathscr{T}_3$.

Assume from now on that $i_0 < \pi/3$. By Sect. 4.6.2 it follows that $L \setminus \{p_0\}$ is non-empty.

We adopt the convention that $p_0 = 1$. One has $p_1 > p_0$ and we assume it is not integer. This is the case when $\mathscr{T} = \mathscr{T}_n : n \geq 4$.

Choose $T \in \mathscr{T}$ to possess the angle of smallest possible size i_0.

Lemma *Assume (H1–H3) hold. In the above notation take $p_0 = 1$ and assume p_1 non-integer. Then there exists a composed triangle $T_1 \in <\mathscr{T}>$ with angle set $\{i_0, i_0, \pi - 2i_0\}$.*

Proof Let $\{i_0, j, k\}$ be the angle set of T and consider p_1T. Let $\mathscr{S}$ denote the subset of $\mathscr{T}$ from which T may be composed.

To prove the lemma it is enough to find $T_1, T_2 \in \mathscr{S}$ with angle sets $\{i_0, j_1, j_2\}$, $\{i_0, \pi - j_1, \pi - j_2 - 2i_0\}$ respectively. Indeed by $(H1)$ the line subtended by i_0 will be the same for both triangles, which can then be joined along this line of common length to provide the required composed triangle.

Let v be a vertex in p_1T having angle value i_0. By minimality, this angle cannot be subdivided by one or more common sides of joined triangles in $\mathscr{T}$. Thus there exists a unique triangle $T_1 \in \mathscr{S}$ sharing the vertex v with angle set $\{i_0, j_1, j_2\}$. By Sect. 4.6.1 it follows that $T_1 \neq p_1T$, so T_1 is properly contained in p_1T.

Let s be the side of T_1 subtended by the angle at v. By (9), there is no triangle $T_2 \in \mathscr{T}$ with angle value $\pi - i_0$. Then (H1) implies that there is a triangle $T_2 \in \mathscr{S}$

properly joined to T_1 across s with a vertex v' angle value i_0 subtending to s. Then either T_2 is the required partner to T_1, or T_1 and T_2 form a quadrilateral with angle values $i_\ell, i_r \in]0, \pi[$ at the remaining vertices v_ℓ, v_r of T_1. Moreover $i_r + i_\ell = 2(\pi - i_0)$.

Consider the second possibility above. In this we can assume that $i_\ell \geq \pi - i_0$ without loss of generality.

Two of the sides of T_1 form part of two of the sides of $p_1 T$, whilst $p_1 T$ cannot admit i_ℓ as an angle value by (9). It follows that the side of pT_1 passing through v_ℓ must continue strictly beyond T_1. Then by the minimality of i_0 as an angle value we conclude that $i_\ell = \pi - i_0$, forcing $i_r = \pi - i_0$ also. Thus $T_1 * T_2$ is a parallelogram P with T_1, T_2 being congruent and the side of pT_1 passing through v_r must also continue strictly beyond T_1.

Now the only triangles which we can adjoin to P and can fit between the sides of $p_1 T$ are those having an angle of size i_0. Consider the triangle $T_3 \in \mathscr{S}$ properly joined to T_2 at the side s' to which its angle of size j_1 subtends. By proper joining the angle of T_3 subtending to s' equals j_1 or $\pi - j_1$. In the former case T_2, T_3 are congruent. In the latter case T_2, T_3 provide the required pair. A similar conclusion holds for the side in T_2 to which the angle of size k_1 subtends giving a fourth triangle T_4 congruent to the other three. These together form a stack of four properly joined triangles congruent to T_1 to form a composed triangle $2T_1$ contained in $p_1 T$. (see Fig. 3). Notice this forces $p_1 \geq 2$.

Since p_1 is not integer, this inclusion is strict. Then by proper joining and the minimality of p_0 we obtain a triangle T_5 (resp. T_6) properly joined to T_3 (resp. T_4) a quadrilateral with two additional angle values k_1, k_2 (resp. k_3, k_4). Let ℓ_1, ℓ_2, ℓ_3 be the angle values between the left hand side of $p_1 T$, the two quadrilaterals and the right hand side of $p_1 T$. (See Fig. 3).

One checks that $\ell_1 + \ell_2 + \ell_3 = 3i_0$. If neither quadrilateral is a triangle (and hence the required composed triangle) we obtain $\ell_1, \ell_3 > 0$.

Suppose $\ell_2 = 0$. Then T_5, T_6 form a composed triangle with one angle value equal to $2i_0$. By $(H2)$ there is a triangle in $\mathscr{T}$ with one angle value equal to $2i_0$. Then $p_1 \leq f(2i_0) = r \sin 2i_0 = 2r \sin i_0 \cos \iota_0 < 2r \sin i_0 = 2p_0 = 2$, which is a contradiction.

Suppose $\ell_2 \neq 0$, then by minimality of i_0 we must have $\ell_1 = \ell_2 = \ell_3 = i_0$. This implies that the quadrilaterals are parallelograms and that all triangles are congruent to T_1. As in the case of P, it follows that either the required pair exists or we obtain a stack of nine triangles forming the composed triangle $3T_1$.

Continuing on in this manner we obtain the required result. □

4.7 A Small Triangle

Assume that (H1)–(H3) hold.

By applying (H2) to the conclusion of Sect. 4.6.3, we conclude that there exists a triangle $T_1 \in \mathscr{T}$ with angle set $\{i_0, i_0, \pi - 2i_0\}$. By Sect. 4.6.2 the smallest element

p_0 of L is $f(i_0)$ which we have set equal to 1. Then $p := f(\pi - 2i_0) = f(2i_0)$ is an element of L and satisfies $1 < p < 2$, by the triangle inequality. More precisely as noted above $p = 2\cos i_0$.

Lemma

(i) *There exists a triangle* $T_2 \in \mathscr{T}$ *with angle set* $\{i_0, 2i_0, \pi - 3i_0\}$,
(ii) $p^2 = 1 + f(3i_0)$.
(iii) $i_0 \leq \pi/4$.
(iv) *If* $i \in A \setminus \{i_0\}$, *then* $i, \pi - i \geq 2i_0$.

Proof Consider pT_1. By Sect. 4.6.1, it cannot lie in $\mathscr{T}$ and hence must be a non-trivially composed triangle. Let $v_{\pi-2i_0}$ denote its vertex with angle value $\pi - 2i_0$ and $v_{i_0}^{\ell}$ (resp. $v_{i_0}^{r}$) its vertex with angle value i_0 lying to the left (resp. right).

Let $\mathscr{S}$ denote the subset of $\mathscr{T}$ of which pT_1 is composed.

Let s be the side of pT_1 subtended by its angle of size $\pi - 2i_0$.

By minimality, the angles of size i_0 in pT_1 cannot be subdivided by sides of triangles in $\mathscr{S}$. Since $p < 2$, the left and right sides s_ℓ, s_r of pT_1 of length p cannot admit marked points. Thus $\mathscr{T}$ admits a leftmost (resp. rightmost) marked point v^ℓ (resp. v^r) on the interior of s.

Since s_ℓ has no marked points there is a continuous piecewise linear leftmost (resp. rightmost) path s^ℓ (resp s^r) in pT_1 from v^ℓ (resp. v^r) to $v_{\pi-2i_0}$. The region bounded by s_ℓ, s, s^ℓ cannot have marked points in its interior, nor can it be subdivided by a line starting at v^ℓ. Hence it must itself be triangular. Thus s^ℓ must be a straight line with no marked points in its interior and its length must be $f(i_0) = 1$. Similarly s^r is a straight line with no marked points in its interior of length 1. Repeating this argument and using the fact that 1 is shortest length of an element of L, we conclude that $s^\ell = s^r$.

Thus pT_1 is composed of just two elements of T', T'' of $\mathscr{T}$ sharing a common side $s^\ell = s^r$ of length 1 passing from $v := v_{\pi-2i_0}$ to $v' := v^\ell = v^r$ (See Fig. 4). By the angle condition, the angle value of T' (resp. T'') at v' must be $2i_0$ (resp. $\pi - 2i_0$) or vice-versa. This gives (i).

(ii) obtains by computing the length of s in pT in the two possible ways. Indeed the side subtended by the angle value $\pi - 2i_0$ has length in T and hence has length p^2 in pT. On the other hand the contribution to the length of s coming from T' (resp. T'') is $f(\pi - 3i_0) = f(3i_0)$ (resp. 1) or vice-versa.

Since by hypothesis $i_0 < \pi/3$, we conclude from (i) and the minimality of i_0 that $\pi - 3i_0 \geq i_0$. Hence (iii).

Let $i_0' \in A$ be minimal with the property that $i_0' > i_0$.

Suppose $i_0' \leq 2i_0$. Since by (ii), $2i_0 \leq \pi/2$, we conclude from Sect. 4.6.2 that $p_1 := f(i_0') \leq p < 2$.

Repeating the argument of (i) for the composed triangle p_1T_1 we conclude that there is a triangle $T''' \in \mathscr{T}$ with angle set $\pi - i_0', i_0, i_0' - i_0$. Then by minimality of i_0 we conclude that $i_0' \geq 2i_0$. This and (9) gives (iv).

Remark 1 Note that T_2 is scalene, but the above argument is not enough to prove that both it and its parity transform appear in $\mathscr{T}$. This is why we assume that $\mathscr{T}$ is parity invariant.

Remark 2 If $i_0 = \pi/4$, then $3i_0 = \pi - i_0$, so (ii) gives $p_1 = \sqrt{2}$ which is hypotenuse of an isosceles right angled triangle with second side length 1. This proves Pythagoras' theorem for isosceles triangles by just triangle tiling. (As is well-known there is a simple proof of Pythagoras' theorem by tiling a square using four congruent right angled triangles enclosing a smaller square, but this is too easy for properly disciplining schoolchildren.) If $i_0 = \pi/5$, then $3i_0 = \pi - 2i_0$ and so $f(3i_0) = f(2i_0) = p_1$ and we obtain $p_1^2 = 1 + p_1$, which is the equation for the Golden Section.

□

4.8 An Independence Hypothesis

The proof of Sect. 4.7 not only uses the minimality of i_0 but also that the sides of length p in the composed triangle pT_1 cannot possess markings on their interior if $p < 2$. This fails in general even for L_n. For example take $n = 6$. Then $L = \{1, \sqrt{3}, 2\}$. Thus we may either write p_2T_1 as $T_2 * T_2'$, that is by joining $T_2 := T\{2, 3, 1\}$ and its parity translate $T_2' := T\{3, 2, 1\}$ along their shortest side in which case the short sides of p_2T_1 have no internal markings or as $T_1 * T_1 * T_1 * T_1$, that as a stack of four triangles $T_1 := T(1, 4, 1)$, in which they do.

To avoid the above coincidence we introduce the further hypothesis.

$(H4)$ The elements of L are linearly independent over $\mathbb{Z}$.

By Sect. 3.2.2, this is holds for L_n if and only if either n is prime or a power of 2. Thus (H4) is not altogether satisfactory, since it is not satisfied by $\mathscr{T}_n$ for all n.

One may remark that if $i_0 = \pi/6$, then already $\mathscr{T} \supset \mathscr{T}_6$ by Sects. 4.6.3 and 4.7 and the parity hypothesis, so as a consequence $(H4)$ is not needed in this case.

On the other hand $(H4)$ means that the left-hand side of (5) determines the terms appearing on its right hand side (restricted to being elements of L_n) and not just their sum. This already fails for $n = 6$, since in this case $p_2 = p_0 + p_4$.

Suppose $T \in \mathscr{T}$ is a composite of more than one triangle in $\mathscr{T}$ such that the only marked points which are not vertices, must be in the interior of T. Then for any tiling using T the condition of proper joining is not affected by replacing T as a composed triangle of elements in $\mathscr{T}$.

More specifically let $T \in \mathscr{T}$ be a composed triangle with angle set $\{a_1, a_2, a_3\}$ (reading counter-clockwise) having just four marked points, its three vertices v_1, v_2, v_3 corresponding to these angles and one marked point in its interior. Thus T is the proper join of exactly three elements $T_1, T_2, T_3 \in \mathscr{T}$ sharing a common vertex in the interior of T. (See Fig. 5.)

Applying (H1) to T_1, T_2, T_3, one deduces that for all $i = 1, 2, 3$, the angle subtending to the same edge as the angle with value a_i at v_1, must have value $\pi - a_1$ at the interior vertex of T. Such a tiling can occur with these three triangles belonging to $\mathscr{T}$ whenever a_1 can be subdivided in A into parts b_1, b_2 such that $b_3 := a_2 - b_2 = a_3 - b_1 \in A$—see Fig. 6.

An example of the above occurs in $\mathscr{T}_8$. This may be expressed by

$$T\{2,3,3\} = T\{1,5,2\} * T\{1,6,1\} * T\{1,2,5\}.$$

It follows that $\mathscr{T}_8 \setminus \{T\{2,3,3\}\}$ gives a self-similar triangle tiling satisfying (H1), (H3), (H4), (H5) of Sect. 4.10 below, but not (H2).

4.9 Computation of the Angle Set

Recall the angle set A defined in Sect. 4.5.

Proposition *Assume (H1)–(H4) hold.*

(i) *For all positive integers m with $(m+1)i_0 < \pi$ there exists a triangle $T_m \in \mathscr{T}$ with angle set $\{i_0, mi_0, \pi - (m+1)i_0\}$. Moreover*

$$p_1 f(mi_0) = f((m-1)i_0) + f((m+1)i_0). \tag{10}$$

(ii) *There exists $n \in \mathbb{N} : n \geq 3$ such that $i_0 = \pi/n$.*

(iii) $A = \{m\pi/n : m = 1, 2, \ldots, n-2\} = A_n$.

Proof T_1 is just the triangle constructed in Sect. 4.6. Moreover by Sect. 4.7 (iv), one has $i_1 = 2i_0$ and so $p_1 = f(2i_0)$ which gives (10) for $m = 1$.

Fix $p \in L \setminus \{p_0\}$ and let s be the side of pT_1 subtended by its angle of size $\pi - 2i_0$. The remaining sides of pT_1 have length p, so by $(H4)$ cannot be subdivided, and so on these sides there are no interior marked points. Thus the only side of pT_1 with interior marked points is s.

It follows that we are in the same situation described in the proof of Sect. 4.7. Thus we conclude $pT_1 = T' * T''$ for some $T', T'' \in \mathscr{T}$. Moreover these triangles both have sides of length $p_0 = 1$ and p.

We establish (i) by induction on m. The case $m = 1$ is just the triangle described in Sect. 4.6.

Assume the assertion proved for $m-1$. Then $f(\pi - mi_0) = f(mi_0) \in L$, by the conclusion of the first part of (i) and Sect. 4.6.1.

If $f(mi_0) \neq p_0$, we set $p = f(mi_0)$. It belongs to $L \setminus \{1\}$, so we may use it in the description of pT_1 as above. In view of (H1) the angle in T' (resp. T") subtending to its side of length $f(mi_0)$ has size mi_0 (resp. $\pi - mi_0$) or vise-versa. Then we can assume $T'' = T_{m-1}$ and we conclude that $T' = T_m$. This proves the first part of (i).

The second part follows by computing the length of s in the two possible ways as in the proof of Sect. 4.7(ii).

Notice that T' is a non-empty triangle lying in $\mathscr{T}$, because the interior of s has exactly one marked point. Consequently $\pi - (m+1)i_0 > 0$. Of course this cannot hold indefinitely and so we are forced to conclude that eventually $f(mi_0) = p_0$. By Sect. 4.6.1 this forces $mi_0 = \pi - i_0$. This proves (ii) with $n = m+1$.

By (i) and (ii) we conclude that $A \supset \{m\pi/n : m = 1, 2, \dots, n-2\} = A_n$. Suppose that this inclusion is strict and take $j \in A \setminus A_n$.

Repeating the above argument with $p = f(j)$, we conclude that $j + i_0 \in A \setminus A_n$. Yet this cannot continue indefinitely and the contradiction that results gives (iii). □

4.10 Orientation

Assume (H1)–(H4).

By Sect. 4.9 there exists $n \in \mathbb{N}$ with $n \geq 3$ such the angle set of $\mathscr{T}$ is A_n.

Again we can write $L = \{p_{m-1} := f(mi_0) : m = 1, 2, \dots, n-1\}$. Then (10) becomes

$$p_1 p_{m-1} = p_{m-2} + p_m, \tag{11}$$

in the conventions of Sect. 3.2. Together with the relation $p_{n-1} = p_0$, this implies that p_m is the mth Chebyshev polynomial P_m of the second kind evaluated at $2\cos \pi/n$ (in our convention for P_m). In other words $L = L_n$ also.

In view of Sect. 4.6.1 an element $T \in \mathscr{T}$ is completely determined by its angle set which takes the form $\{i\pi/n, j\pi/n, k\pi/n\}$, with (i, j, k) being a partition of n into exactly three parts. As in Sect. 3 we write $T = T\{i, j, k\}$. We have still to show that all possible choices can be made for elements of $\mathscr{T}$. That is we have shown

$$\mathscr{T} \subset \mathscr{T}_n, \text{ for some } n \in \mathbb{N}, \tag{12}$$

but we have yet to prove equality.

Fix n as in the conclusion of Sect. 4.9(ii). Then by its conclusion $T := T\{1, j, k\} \in \mathscr{T}$ for any choices of $j, k \in \mathbb{N}^+$ such that $1 + j + k = n$. We would like to show by induction on $i' \in \{1, 2, \dots, n-2\}$ that $T' := T\{i', j', k'\} \in \mathscr{T}$ for any choices of $j', k' \in \mathbb{N}^+$ such that $i' + j' + k' = n$.

In the induction hypothesis we can assume that $i' + 1 \leq n - 2$. Then we can assume that $j' + k = n$ since this only constrains j, k' to satisfy $j + k' = n - i' - 1 \geq 2$. Now we would like to join T to T' through the side subtended by the angle value j in T to the side subtended by the angle value k' in T', with the angle having value k in T to the angle having value j' in T'. This would give a triangle with angle values $1 + i', j, k'$ lying in $< \mathscr{T} >$. Then the required conclusion obtains through $(H2)$.

The only trouble with this construction is that the sides matched may not have the same lengths. Indeed that of T has length p_{j-1} which that of T' has length $p_{k'-1}$.

The above difficulty can be avoided by forming $p_{k'-1}T * p_{j-1}T'$ since now the shared side has length $p_{k'-1}p_{j-1}$.

However, in the latter construction the difficulty is that the marked points on these sides need not match up. Let us review this issue.

Take $T = T\{i, j, k\} \in \mathscr{T}$ and $p_t \in L$. By the assumed property of being a self-similar triangle tiling one has $p_t T \in < \mathscr{T} >$. Let s_i denote the side of T subtended by its angle of size $i\pi/n$. By (H4) the marked points on s_i break it into line segments whose lengths are those occurring in the right hand side of (5). However, we do not know that they occur in any particular order.

Recall the notion of orientation from Sect. 3.6.5.

$(H5)$ Given $T = T\{i, j, k\} \in \mathscr{T}$ and $p_t \in L$ then $p_t T$ with both clockwise and anticlockwise orientation belong to $< \mathscr{T} >$.

One may remark that even for the Golden Pair one can obtain a tiling of gT_2 which has neither orientation by flipping the pair T_1, T_2 in the right hand side of Fig. 2. This is even a case with n prime. This also holds for $n = 8$ as may be noted by viewing the top right hand composite triangle $p_1 T\{5, 2, 1\}$ in my home page. (This composite triangle should not have been used, a slip corrected by Plate 5, as it would cause problems if further iterations were required.)

Now by $(H5)$ we may construct $p_{k'-1}T$ with the clockwise orientation (relative to the sequence i, j, k and $p_{j-1}T'$ with the anti-clockwise orientation relative to the sequence i', j', k'. Then we can form $p_{k'-1}T * p_{j-1}T'$ as needed above since now marked points on the common edge match up.

We have thus proved the following

Theorem *Let $\mathscr{T}$ be a finite parity invariant set of triangles which admits a proper self-similar triangle tiling. Assume that $\mathscr{T}$ satisfies (H1)–(H5). Then there exists $n \in \mathbb{N}$ with $n \geq 3$ such that $\mathscr{T} = \mathscr{T}_n$.*

Remark It is plausible that $(H5)$ is not needed. It can be avoided by strengthening $(H2)$ by not requiring proper joining in the definition of $< \mathscr{T} >$.

5 Injectivity of ψ

5.1 A Coxeter Subgroup

Let us first recall the motivation for this work outlined briefly in Sect. 2.8.

Define $\mathfrak{g}$ and $\mathfrak{h}$ as in Sect. 2.8 and let $\Delta \subset \mathfrak{h}^*$ its set of non-zero roots defined relative to the pair $(\mathfrak{g}, \mathfrak{h})$ with $\pi \subset \Delta$ a choice of simple roots. We write $\pi = \{\alpha_i\}_{i=1}^{\ell}$. Let $\alpha^\vee$ be the coroot corresponding to $\alpha \in \Delta$ and s_α the simple reflection defined by $s_\alpha \lambda = \lambda - \alpha^\vee(\lambda)\alpha$, for all $\lambda \in \mathfrak{h}^*$. View the Weyl group W as the Coxeter group with generating set $\mathscr{S} = \{s_\alpha : \alpha \in \pi\}$. It is convenient to assume that card $\pi > 1$.

There is just one way to write $\mathscr{S}$ as a disjoint union of two subsets $\mathscr{S}_a, \mathscr{S}_b$ each consisting of commuting reflections. Let $\pi = \pi_a \sqcup \pi_b$ be the corresponding decomposition of the set of simple roots. Let $\mathbf{r}_a$ (resp. $\mathbf{r}_b$) denote the product of the elements of S_1 (resp. S_2). Set $\mathbf{C} =< \mathbf{r}_a, \mathbf{r}_b >$. Observe that $\mathbf{r} := \mathbf{r}_a\mathbf{r}_b$ is a Coxeter element of W, though not every Coxeter element can be put into this form. Of course as is well-known [1, Chap. V, Sect. 6, Lemma 1] all Coxeter elements are conjugate and hence conjugate to $\mathbf{r}$.

5.2 *The Basic Map*

Let $\mathbf{A}$ denote the $\ell \times \ell$ matrix obtained as the negative of the Cartan matrix *with diagonal entries set equal to zero*. By definition its off-diagonal entries are given by $\mathbf{A}_{i,j} = -\alpha_j^\vee(\alpha_i)$.

In [5, 2.6] we studied the eigenvalue equation

$$\mathbf{A}\mathbf{g} = x\mathbf{g} : \mathbf{g} = (g_1, g_2, \dots, g_\ell). \tag{13}$$

In [5, 2.6–2.8] we found a solution to (13) with $x = 2\cos\pi/c$ and the $g_i : i = 1, 2, \dots, \ell$ positive real numbers expressed as polynomials in x. These polynomials will be computed and to emphasize that the g_i are the values of polynomials at x we shall often write them as $g_i(x)$. Eventually they will satisfy a (polynomial) identity which will determine x.

The significance of this solution to (13) is described below (cf. [5, 2.7]).

First we recover the set-up described in [5, 2.5].

Let $\mathscr{A}$ be a 2×2 Cartan matrix with 2 on the diagonal and $-x$ as off-diagonal entries with $x = 2\cos\pi/n$.

Define a pair of roots (a, b) and coroots $(a^\vee, b^\vee)$ with $a^\vee(b) = b^\vee(a) = -x$. Define reflections r_a, r_b in $\mathbb{R}^2$ in the usual way by

$$r_a\mu = \mu - a^\vee(\mu)a, \quad r_b\mu = \mu - b^\vee(\mu)b.$$

Let C denote the group they generate. Recall [5, Lemma 2.7].

Lemma *There is a group homomorphism $\psi : \mathbf{C} \to C$ defined by $\psi(\mathbf{r}_a) = r_a, \psi(\mathbf{r}_b) = r_b$ and a $\mathbf{C}\mathbb{Z}$ linear map ψ of $\mathbb{Z}\pi$ into $\mathbb{Z}[x]a + \mathbb{Z}[x]b$ defined by*

$$\psi(\alpha_i) = \begin{cases} g_i(x)a : \alpha_i \in \pi_a, \\ g_i(x)b : \alpha_i \in \pi_b. \end{cases}$$

5.3 A Planar Reflection Group

Through the above lemma, one may view $\mathbf{C}$ as a reflection group on the plane with $\mathbf{r}_a, \mathbf{r}_b$ as its generating reflections. Indeed it is isomorphic to the dihedral group of order $2c$ where c is the order of $\mathbf{r}$, that is to say the Coxeter number of $\mathfrak{g}$. Define the polynomials P_n as in Sect. 3.2.2 taking $n = c$.

We may realize the pair (a, b) as unit vectors in $\mathbb{R}^2$, where b is obtained from a by counter-clockwise rotation through an angle of $\pi - \pi/c$. Set $\pi_c = \{a, b\}$ and $\Delta_c = \mathbf{C}\pi_c$.

The pair $\Delta_c, \mathbf{C}$ admits some of, though not all, the properties of a root system. As noted in [5, Lemma 2.8], one has $\mathbf{C}\pi = \Delta$ and consists of ℓ orbits of $\mathbf{r}$ each of cardinality c.

Set $r = r_a r_b$. It is an image of the Coxeter element $\mathbf{r}$ which has finite order.

The computation for the $g_i(x)$ of the above lemma (that is [5, Lemma 2.7]) is partly repeated below. In this we find a solution for an unknown variable x as a solution to a polynomial equation. It is not the only solution but it is the one we use in this lemma. This choice is necessary to make ψ a homomorphism of $\mathbf{C}$ modules.

5.4 Injectivity of ψ on Δ

Our goal is to determine when ψ extends linearly into an injection from $\mathbb{Z}\pi$ to $\mathbb{Z}[x]a+\mathbb{Z}[x]b$. Obviously this must fail if already the restriction to π is not injective. In particular we show that the latter is the case for types A_{2n+1}, B_3, D, E_6.

On the other hand it is clear that ψ is injective if and only if the $g_i(x) : \alpha_i \in \pi_a$ and the $g_i(x) : \alpha_i \in \pi_b$ are separately linearly independent over $\mathbb{Q}$.

We shall need to recall the analysis of [5, Lemma 2.6]. Take an end root α_1 of the Dynkin diagram, so that $\alpha_j : j = 1, 2, \ldots, t$ is a chain of maximal length of type A_t. Set $g_1 = 1$. Then by (13) one obtains

$$g_j(x) = P_{j-1}(x), \, j = 1, 2, \ldots, t. \tag{14}$$

5.5 The Classical Case

In type A_ℓ there is one additional relation, namely $xg_\ell(x) = g_{\ell-1}(x)$ yielding $P_\ell(x) = 0$. This has as a solution $x = 2\cos\pi/(\ell + 1)$, which is the one we take (cf. Sect. 5.3).

Notice in the above, by disjointness of the decomposition $\pi_a \sqcup \pi_b$ with respect to the above labelling, that $\psi(\alpha_i) = g_i(x)a$ for i odd and $\psi(\alpha_i) = g_i(x)b$ for i even or vice-versa. We can just consider the first case.

Suppose ℓ is odd. Then by the symmetry property (1) we obtain $\psi(\alpha_1) = g_1 a = g_\ell a = \psi(\alpha_\ell)$. This establishes the assertion of Sect. 5.4 for type $A_{2n+1} : n > 0$.

Suppose ℓ is even. One checks using the symmetry property that the set $\{g_i(x)\}_{\alpha_i \in \pi_a} = \{g_i(x)\}_{\alpha_i \in \pi_b} = \{P_{i-1}(x)\}_{i=1}^{\ell/2}$. By Sect. 3.2.2 these elements are linearly independent over $\mathbb{Q}$ if and only if $\ell + 1$ is prime.

Summarizing the above:

For A_ℓ, the map ψ is injective if and only if $\ell + 1$ is prime.

In types B_ℓ (resp. C_ℓ), we obtain $g_i(x) = P_{i-1}(x) : i < \ell$ with the additional relations $xg_{\ell-1}(x) = g_{\ell-2}(x) + 2g_\ell(x)$, $xg_\ell(x) = g_{\ell-1}(x)$ (resp. $xg_{\ell-1}(x) = g_{\ell-2}(x) + g_\ell(x)$, $xg_\ell(x)(x) = 2g_{\ell-1}(x)$). Through (3) both give $P_\ell(x) = P_{\ell-2}(x)$. Taking $n = 2\ell, i = \ell$ in [4, Lemma 2.5] gives a solution $x = 2cos\frac{\pi}{2\ell}$, which is the one we take. Then by Sect. 3.2.2 and using the symmetry property as in type A, we obtain

In types B_ℓ, C_ℓ, the map ψ is injective if and only if ℓ is a power of 2.

It might seem that types B, C are identical. However, from the above relations one finds for B_3 (resp. C_3) that $g_3 = g_1 = 1$ (resp. $g_3 = 2g_1 = 2$). Thus it is only in type B_3 that the restriction of ψ to π fails to be injective. This also obtains from the first line of [4, 10.7]. (In this $y = \cos\pi/6$ and the first line should have read $T_3(y) = y^2 - 3 = 0$.)

In type D_ℓ, we obtain three additional relations $xg_{\ell-2}(x) = g_{\ell-3}(x) + g_{\ell-1}(x) + g_\ell(x)$ and $xg_{\ell-1}(x) = g_{\ell-2}(x) = xg_\ell(x)$. Since $g_i(x) = P_{i-1}(x) : i \leq \ell - 2$, this gives by (3) that $P_{\ell-2}(x) = g_{\ell-1}(x) + g_\ell(x)$ and then that $xP_{\ell-2}(x) = P_{\ell-3}(x)$ which again by (3) gives $P_{\ell-1}(x) = P_{\ell-3}(x)$. Taking $n = 2(\ell - 1), i = \ell - 1$ in [5, Lemma 2.5] gives a solution $x = 2cos\frac{\pi}{2(\ell-1)}$, which is the one we take. (This corrects the computation in [5, Lemma 2.6] though the final result was correct.) We conclude in particular that $g_{\ell-1}(x) = g_\ell(x)$, establishing the assertion of Sect. 5.4 for type D. It particular

In types $D_\ell : \ell \geq 4$, the map ψ is not injective.

5.6 The Exceptional Case

In type E_ℓ, we can take $t = \ell - 3$ in (14). Then there are several additional relations. The first is $xg_{\ell-3}(x) = g_{\ell-4}(x) + g_{\ell-2}(x) + g_{\ell-1}(x)$, which yields $P_{\ell-3}(x) = g_{\ell-2}(x) + g_{\ell-1}(x)$. Secondly we have $xg_{\ell-2}(x) = g_{\ell-3}(x)$, $xg_{\ell-1}(x) = g_\ell(x) + g_{\ell-3}(x)$, for a suitable labelling. This gives

$$g_\ell(x) = P_{\ell-2}(x) - P_{\ell-4}(x). \tag{15}$$

Finally we have $xg_\ell(x) = g_{\ell-1}(x)$. Substitution in the previous equation gives

$$g_{\ell-1}(x) = P_{\ell-1}(x) - P_{\ell-5}(x). \tag{16}$$

Then

$$g_{\ell-2}(x) = P_{\ell-3}(x) - P_{\ell-1}(x) + P_{\ell-5}(x). \tag{17}$$

Multiplying by x, using (3) and $xg_{\ell-2}(x) = g_{\ell-3}(x)$ gives

$$g_{\ell-3}(x) = 2P_{\ell-4}(x) + P_{\ell-6}(x) - P_{\ell}(x). \tag{18}$$

Substituting from (14) results in

$$P_{\ell}(x) = P_{\ell-4}(x) + P_{\ell-6}(x). \tag{19}$$

(This makes explicit the proof of [5, 2.6(∗∗)].)

In type F_4 a similar computation (see [5, 2.6]) gives

$$g_i(x) = P_{i-1}(x) : i \leq 2, 2g_3(x) = P_2(x), 2g_4(x) = P_3(x) - P_1(x), \tag{20}$$

and then that

$$Q_4(x) := P_4(x) - P_2(x) - P_0(x) = 0. \tag{21}$$

In this Q_4 is a polynomial of degree 4. Multiplying by x^2 on both sides and using (3) gives (19) for $\ell = 6$. Multiplying the latter again by x^2 and using (5) eventually gives $P_8(x) = P_2(x)$ which by [5, Lemma 2.5] has as a solution $x = 2\cos\frac{\pi}{12}$, which is the one we take.

Notice that in type E_6 we have $g_6 = P_4(x) - P_2(x) = 1 = g_1$. This proves the last claim of first paragraph of Sect. 5.4. We conclude that

In type E_6, the map ψ is not injective.

In type E_7 we set $Q_6(x) = \frac{1}{x}Q_7(x)$ (with $Q_7(x) = P_7(x) - P_3(x) - P_1(x)$) which is a polynomial of degree 6 by say [4, 2.2]. In type E_8, we set $Q_8(x) = P_8(x) - P_4(x) - P_2(x)$, which is a polynomial of degree 8.

Recall (15) and the computation for F_4 below it. Let us just use Q to denote the appropriate polynomial Q_4, Q_6, Q_8. Then in type E_ℓ and F_4, the equation $Q(x) = 0$ has as a solution $x = 2\cos\pi/c$, where c is the Coxeter number [5, 2.7, Remark]. It is the solution we took in [5, Lemma 2.7].

If $\mathfrak{g}$ is simple with Coxeter number c, set $M(\mathfrak{g}) = \{i \in \{1, 2, \ldots, c\}$ coprime to $c\}$.

Lemma *The roots of Q are the $2\cos m/c$, where c is the Coxeter number and $m \in M(\mathfrak{g})$. In particular Q is irreducible over $\mathbb{Q}$.*

Proof This follows as in the proof of [4, Lemma 7.4(iii)]. In more detail set $z = e^{i\theta}$ with $\theta = \pi/c$ and define $\hat{Q}$ by $\hat{Q}(x) = Q(2x)$.

We already know that $\cos\theta$ which is the real part of z is a root of $\hat{Q}$. Now z is a $(2c)$th primitive root of unity and so the images z under the Galois group of $\mathbb{Q}[z]$ over $\mathbb{Q}$, form the set of $(2c)$th primitive roots of unity, which is $\{z^m\}_{m\in M}$. Taking real parts it follows that images of $\cos\theta$ under the Galois group of $\mathbb{Q}[\cos\theta]$ over $\mathbb{Q}$ form the set $R := \{\cos m\theta\}_{m\in M(\mathfrak{g})}$ of roots of $\hat{Q}$.

On the other hand taking $c = 12, 18, 30$ respectively one checks that in all cases $|M(\mathfrak{g})| = \deg \hat{Q}$. Hence R is exactly the set of roots of $\hat{Q}$, which is therefore irreducible. □

5.7 *Remaining Conclusions in the Exceptional Cases*

It remains to ascertain the possible injectivity of ψ in types E_7, E_8, F_4, G_2. Injectivity is trivial in rank 2. For injectivity in the remaining three cases, we remark that this fails if the polynomials we computed for the $g_i(x)$ are not even linearly independent over $\mathbb{Q}[x]$. (For example we already saw that this fails in type E_6.)

In type E_7, there are seven polynomials $\{g_i(x)\}_{1=1}^{7}$ which satisfy an equation of degree 6, namely $Q_6(x) = 0$. Notice this is a polynomial equation in x^2 of degree 3.

One may remark that $P_i(-x) = (-1)^i P_i(x)$. From Eqs. (14)–(18) it follows that $g_i(-x) = g_i(x)$ for $\alpha_i \in \pi_a$ whilst $g_i(-x) = -g_i(x)$ for $\alpha_i \in \pi_b$, or vice-versa. We can suppose the first holds. Then $g_i(x) : \alpha_i \in \pi_a$ are four polynomials of degree ≤ 3 in x^2. Hence their values must be linearly dependent over $\mathbb{Q}$, when x satisfies $Q(x) = 0$. Thus we obtain

In type E_7, the map ψ is not injective.

On the other hand inspection of the Eqs. (14)–(17), (20) for the g_i in types E_8, F_4 shows that the polynomials expressing them are linearly independent over $\mathbb{Q}[x]$. In view of Sect. 5.6, their values at x must be linearly independent over $\mathbb{Q}$. Hence

In types E_8, F_4, the map ψ is injective.

5.8 We may summarize the above results in the following general

Theorem *Let $\mathfrak{g}$ be a simple Lie algebra. The map ψ of Sect. 5.2 is injective if and only the exponents of $\mathfrak{g}$ are coprime to its Coxeter number c.*

Proof The exponents of $\mathfrak{g}$ are given in [1, Planches I-X]. Thus one only has to check that the exponents of $\mathfrak{g}$ are coprime to c exactly when ψ is injective as given in the above list.

Alternatively one may check that the set $M(\mathfrak{g})$ lies in the set of exponents of $\mathfrak{g}$, which is rather easy. Plausibly it has a known general proof. Then one notes that x as occurring in Sect. 5.2 is the root of a polynomial Q with rational coefficients which we can assume to be irreducible. Moreover $x = 2\cos\pi/c$. Thus as in the proof of Sect. 5.6 the remaining roots of Q take the form $2\cos m\pi/c : m \in M(\mathfrak{g})$.

Consequently $\deg Q \leq \operatorname{rank} \mathfrak{g}$ with equality if and only if $M(\mathfrak{g})$ is exactly the set of exponents of $\mathfrak{g}$.

Thus the exponents of $\mathfrak{g}$ are coprime to its Coxeter number if and only if $\deg Q = \operatorname{rank} \mathfrak{g}$.

Of course this is not quite the end of the story and a little more spadework is necessary. We still have to know that the g_i defined by (13) are given as polynomials evaluated at $x = 2\cos\pi/c$ and are "properly distributed".

The first point above was verified in [5, Lemma 2.7] being indeed part of the proof of that lemma. It was repeated in a little more detail here.

The second point is resolved if the $\{g_i(x)\}$ are linearly independent over $\mathbb{Q}[x]$ (and Q is irreducible). However, this is rather rare, though it does hold for example in types F_4, E_8. Otherwise one must check linear independence (or its failure) individually for the two sets $\{g_i\}_{\alpha_i \in \pi_a}$ and $\{g_i\}_{\alpha_i \in \pi_b}$. Of course this point was already examined in detail in Sects. 5.4–5.8 above. □

Remark Perhaps one can directly deduce that the $\psi(\alpha_i)$ are linearly independent over $\mathbb{Q}$ if and only if $\deg Q = \operatorname{rank} \mathfrak{g}$ from say the Galois group of the extension being $< \sigma >$.

6 Realizing Weight Diagrams in the Plane

6.1 The Planar Weight Diagram

Fix a simple Lie algebra $\mathfrak{g}$ and retain the notation of Sect. 5.2. For all $i = 1, 2, \ldots, \ell$, let ϖ_i be the fundamental weight correspond to α_i. If ψ is injective the lattice of all weights $P(\pi) = \oplus_{i=1}^{\ell} \mathbb{Z}\varpi_i$ can be viewed as lying in the plane. Of course this is only an image of the weight lattice if ψ is not injective.

Recall that for each $\lambda \in P^+(\pi)$ there is a finite dimensional simple highest weight $\mathfrak{g}$ module $V(\lambda)$ with highest weight λ. Let $D(\lambda)$ denote its weight diagram which we view as a finite set of points in the weight lattice, each point being the weight of a non-zero weight subspace of $V(\lambda)$. Set $d(\lambda) = \psi(D(\lambda))$, viewed as a subset of points on the plane of the plane. In this we shall generally ignore multiplicities.

If ψ is injective, then $d(\lambda)$ faithfully realizes the weight diagram of $V(\lambda)$. This can also happen "accidentally", particularly for small λ, even if ψ is not injective.

Take $\mathfrak{g}$ of type A_3, with $D(\lambda)$ its root diagram. One easily checks that $d(\lambda)$ is the root diagram of type B_2, so this is not a faithful realization of $D(\lambda)$.

By contrast if $D(\lambda)$ is the weight diagram for the fundamental module (of dimension $n+1$) in type A_n, then $d(\lambda)$ above is the regular $(n+1)$-gon and so is a faithful realization of $D(\lambda)$. For $n = 3$, it is also the weight diagram for the fundamental module of dimension 4 in type B_2.

6.2 The Role of Tiling in Planar Weight Diagrams

The problem we pose is the following.

Does there exist a proper triangle tiling from elements in $\mathscr{T}_c$ whose marked points form $d(\lambda)$?

For example in the case of the fundamental representation of $\mathfrak{sl}(n)$ one may take the zig-zag triangularization of the regular n-gon (cf [4, Figure 2]).

Further examples are obtained for the root diagrams in types A_4, A_6, B_4 in [5, Figs. 1,3,12]. These were not self-similar tilings because $d(\lambda)$ is invariant under the Coxeter group **C**. However, if one removes this symmetry by considering only fundamental chamber with respect to **C** one observes from [4, Figure 1], that a proper self-similar triangle tiling of the gT_2 results, with g is the Golden Section and T_2 the second triangle of the Golden Pair.

Answering our question is not as easy as it may first seem. Thus a natural solution might have been obtained by joining the elements of $d(\lambda)$ by line segments corresponding to the action of the roots vectors. Here one may remark that the length of any such line segment is an element of L_c, as required. However, such lines may cross (or appear to cross) at points which not images of weights. This can happen even if ψ is injective and no such lines cross in $D(\lambda)$. This is because the eye cannot distinguish between a point in $\mathbb{Q}^2[x]$ with $x = 2\cos\pi/c$ and a point in $\mathbb{R}^2$.

Another difficulty already raised in [4, Example 2] is that as λ becomes large the points of $d(\lambda)$ bunch together. This has the consequence that one has little hope of being able to describe $d(\lambda)$ as the marked points of $\mathscr{T}_c$ using just one regular c-gon. In general several c-gons of different sizes are needed. A first example occurs for the representation of $\mathfrak{sl}(5)$ with $\lambda = \alpha_1 + 2\alpha_2 + 2\alpha_3 + \alpha_4$. This is illustrated in Fig. 7. Thus writing $T_1 = T\{1, 1, 3\}$, $T_2 = T\{1, 2, 2\}$ which is the Golden pair, to describe $d(\lambda)$ we need to use $\{T_1, T_2, gT_1, gT_2, g^2T_1\}$, where g is the Golden Section. Of course we can express the last three in terms of the first two but this would introduce too many marked points. We summarize the above in the following

Conjecture Let $\mathfrak{g}$ be a simple Lie algebra and c its Coxeter number. For all $\lambda \in P^+(\pi)$ there exists a finite subset $F_\lambda \subset \mathscr{L}_c$ such that $d(\lambda)$ is the set of marked points of a proper triangle tiling using triangles from the set $\{pT : p \in F_\lambda, T \in \mathscr{T}_c\}$.

6.3 Ending on a Light Note

Take $n = 4$ and retain the notation of Sect. 6.2. In particular let g be the Golden Section.

It is interesting to compute the action of the full Weyl group on the image under ψ of the fundamental alcove. Of course this action cannot be expected to be length preserving. In [4, 9.8] we indicated how it may convert a regular pentagon into one which has been shrunk by a factor[2] of g.

Let us spell out the details as illustrated in Fig. 8.

Set $x_i = \psi(\varpi_i) : i = 1, 2, 3, 4$. They lie on two semi-infinite lines starting at the origin $\{0\}$ and forming an angle of $\pi/5$. Here x_1, x_4 can be taken to be of distance 1 from $\{0\}$ and then x_3, x_2 are at a distance g from $\{0\}$. Set $x_0 = x_5 = 0$. one checks that the distance between $x_i, x_{i+1} : i = 0, 1, 2, 3, 4$ is always 1. In particular the $\{x_i\}_{1=0}^4$ form the marked points of T_2 written as $T_1 * g^{-1}T_2$ in either of the two possible ways.

One checks that the action of $s_2s_3s_2$ which of course leaves the points $\{0, x_1, x_4\}$ fixed, simultaneously translates $\{x_2, x_3\}$ by $-(\alpha_2 + \alpha_3)$. These two new points together with those which are fixed form a regular pentagon of side length 1. On the other hand s_1s_4 leaves the points $\{0, x_2, x_3\}$ fixed and translates $\{x_1, x_4\}$ to the pair $\{x_1 - \alpha_1, x_4 - \alpha_4\}$. These two new points together with those which are fixed form a regular pentagon of side length g, sharing the origin and parts of two sides with the smaller pentagon. Through just s_1 or s_4 applied to the large pentagon we obtain a trapezium with one marked point in its interior. The latter can also be described by triangle tiling.

These operations may be described through paper folding using a cut-out of Fig. 8. Start from a large pentagon with vertices $0, s_1x_1, x_2, x_3, s_4x_4$. To apply s_1 (resp. s_4) fold the paper along the lines $0, x_2$ (resp. $0, x_3$). This gives the marked triangle. To apply $s_2s_3s_2$ fold the triangle along the line x_1, x_4. This gives the smaller pentagon (Fig. 9).

This example amused the audience of my lecture in Manchester.

It might also have amused Bar, may she be of blessed memory.

7 Figures

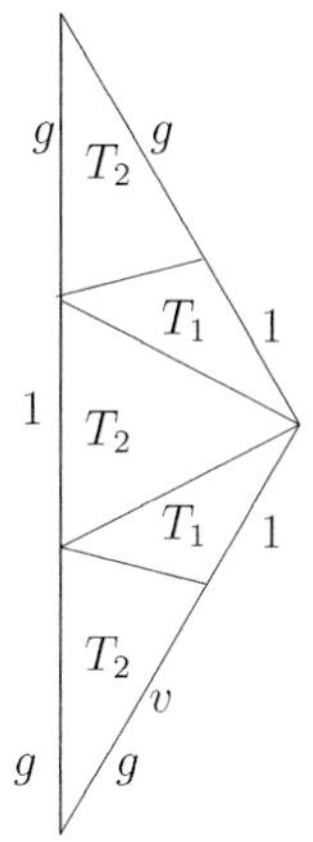

Fig. 1 A proper tiling of g^2T_1. All other proper tilings are obtained by flipping triangles and trapezia

[2]This brings to mind advice to newcomers of how to end up with a small fortune—come with a large fortune, now happily no longer true of the start-up nation.

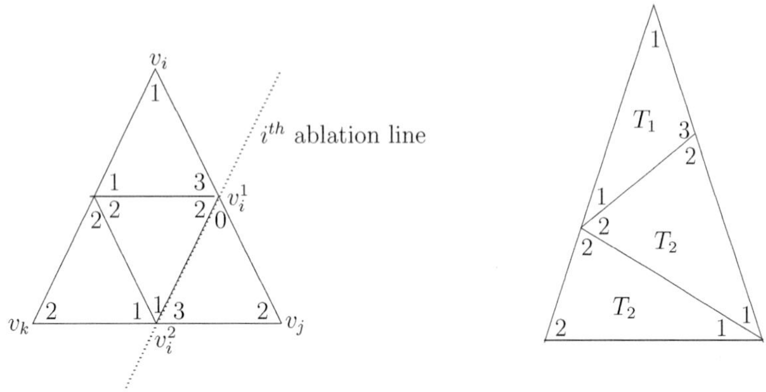

Fig. 2 An ablated stack for $t = 2, i = 1, j = k = 2$. Removing the lower right hand triangle in the left hand figure gives the right hand figure describing gT_2 with T_1, T_2 the Golden Pair and g the Golden Section

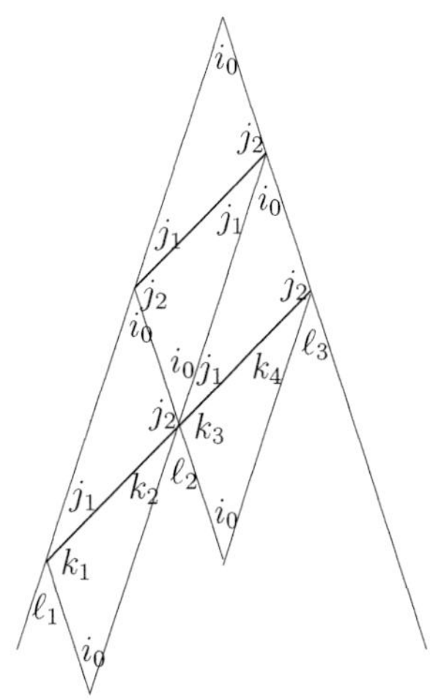

Fig. 3 From the stack of 4 copies of T_1, one computes angles to obtain $\ell_1 + \ell_2 + \ell_3 = 3i_0$

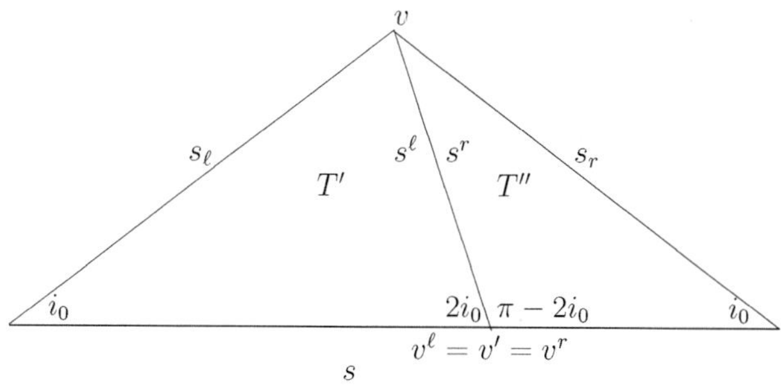

Fig. 4 The sides s^ℓ and s^r must coincide and the points v^ℓ, v^r coalesce to a single point v'. The triangles T', T'' result

Fig. 5 Applying the angle condition

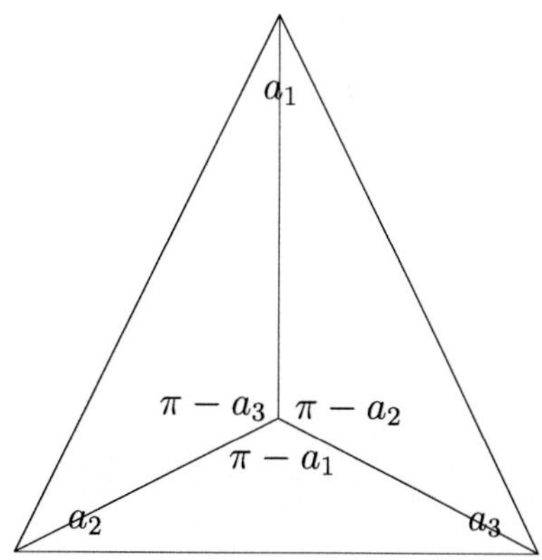

Fig. 6 In the example $b_1 = b_2 = \pi/8, b_3 = \pi/4$

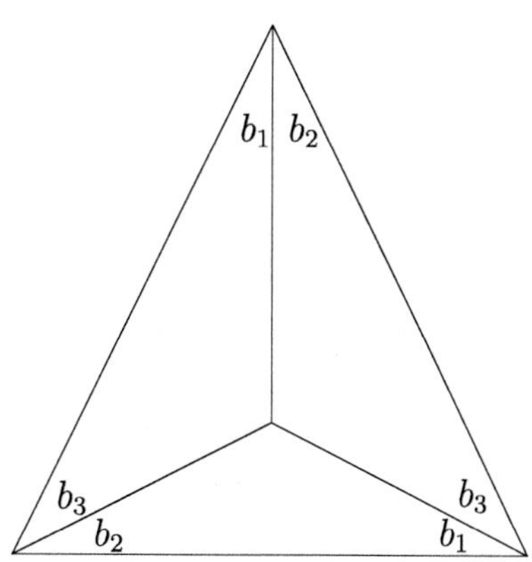

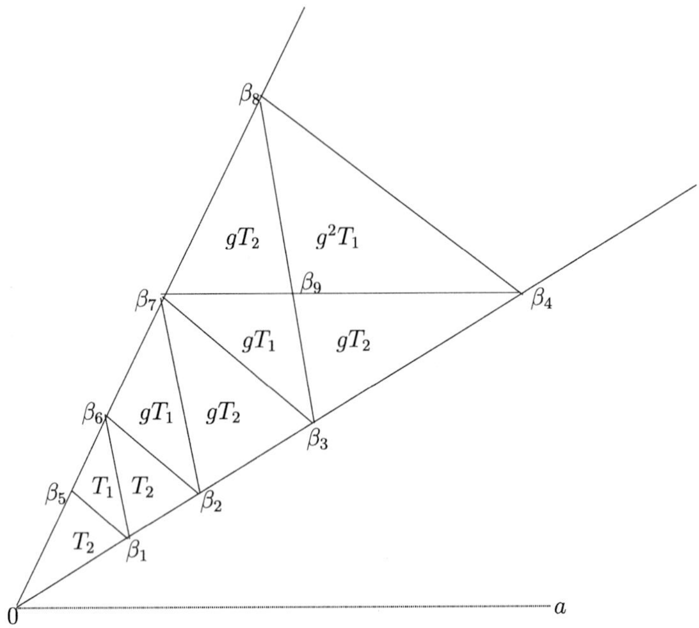

Fig. 7 The weight diagram for $V(\lambda) : \lambda = \alpha_1 + 2\alpha_2 + 2\alpha_3 + \alpha_4$ in type A_4 in the fundamental domain with respect to **C** as the marked points of a proper triangle tiling. In terms of Sects. 5.2 and 5.3, the length of a relative to the length of β_5 is $1/\sqrt{2(1-\cos\pi/6)} \approx 3.87$, whilst $\alpha_1 = a, \alpha_2 = gb, \alpha_3 = ga, \alpha_4 = b$, where b is obtained from a by counter-clockwise rotation by $\frac{5}{6}\pi$. **Legend.** $\beta_1 = \alpha_1 + \alpha_2 - \alpha_4 = a + g^{-1}b$. $\beta_2 = \alpha_3 + \alpha_4 = ga + b$. $\beta_3 = \alpha_1 + \alpha_2 + \alpha_3 = g^2a + gb$. $\beta_4 = \alpha_1 + \alpha_2 + 2\alpha_3 + \alpha_4 = g^3a + g^2b$. $\beta_5 = \alpha_1 + \alpha_4 = a + b$. $\beta_6 = \alpha_2 + \alpha_3 = g(a+b)$. $\beta_7 = \alpha_1 + \alpha_2 + \alpha_3 + \alpha_4 = g^2(a+b)$. $\beta_8 = \alpha_1 + 2\alpha_2 + 2\alpha_3 + \alpha_4 = g^3(a+b)$. $\beta_9 = \alpha_2 + 2\alpha_3 + \alpha_4 = 2ga + (1+g)b$

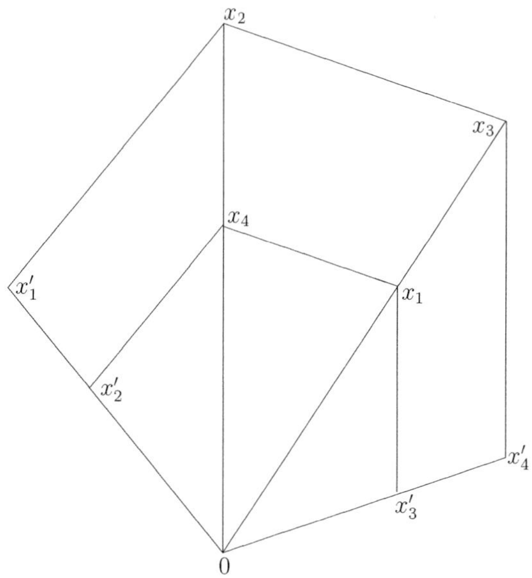

Fig. 8 Here $x_i = \psi(\varpi_i)$ with $\{\varpi_i\}_{i=1}^4$ the fundamental weights in type A_4. The fundamental alcove is the marked triangle, with marked points $0, x_1, x_2, x_3, x_4$. The action of s_1 (resp. s_4) on the fundamental alcove gives the parallelogram with an internal marking at x_4 (resp. x_1). The action of $s_1 s_4$ on the fundamental alcove gives the large pentagon and finally the action of $s_2 s_3 s_2$ on the fundamental alcove gives the small pentagon. Thus the action of $s_2 s_3 s_2 s_1 s_4$ converts the large pentagon into the small pentagon as promised in Sect. 6.3 of the text. **Legend**. $x'_1 = s_1 x_1$, $x_2 = s_2 s_3 s_2 x_2$, $x'_3 = s_2 s_3 s_2 x_3$, $x'_4 = s_4 x_4$

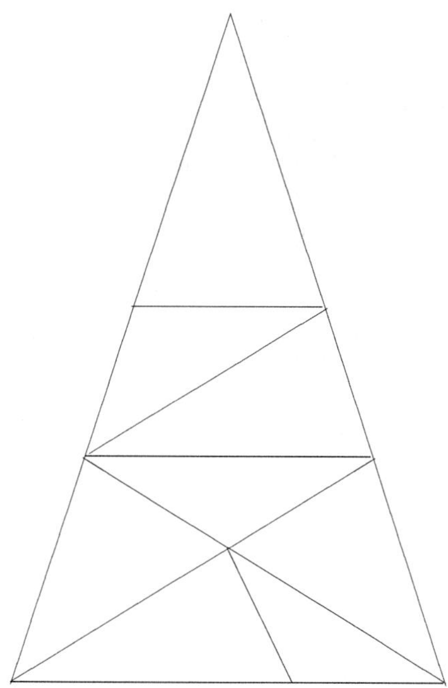

Fig. 9 A proper tiling of $p_1^2 T\{2, 2, 1\}$ *not* following the procedure of Sect. 3.9. This should be compared to Plate 1 in which this procedure is followed

8 Index of Notation

Symbols appearing frequently are given below in the paragraph they are first defined. We remark that π on its own is sometimes used as the ratio of the circumference to the diameter of a circle and very occasionally (in fact just in Sects. 5.1–5.4) as a choice of a set of simple roots for $\mathfrak{g}$.

In this n is positive integer and if fixed (that is understood) may be omitted as a subscript.

Section 2.5. $\mathscr{T}_n$.
Section 2.7. $\mathfrak{g}, \mathfrak{h}, W, P(\pi), c, P^+(\pi), D(\lambda)$.
Section 3.1. $[k]$.
Section 3.1.1. $S_n, L_n, \mathscr{L}_n, \mathscr{T}_n, T * T, M(S)$.
Section 3.2.1. $T_{i,j,k}, A_n, T\{i, j, k\}$.
Section 3.2.2. $p_i, g, L'_n.P_i$.
Section 3.6.3. $T^{t+1}, v_i, s_{i,j}$.
Section 4.3. f.
Section 5.1. $\Delta, \pi, \alpha^\vee, s_\alpha, \mathbf{r}_a, \mathbf{r}_b, \mathbf{C}, \mathbf{r}$.
Section 5.2. $\mathbf{A}, \mathbf{g}, \psi, r_a, r_b, C$.
Section 5.6. $M(\mathfrak{g})$.
Section 6.1. $\varpi_i, d(\lambda)$.

Acknowledgements I would like to thank the referee for his careful reading of the manuscript and his complimentary comments (which may not have been deserved).

I would like to thank Anna Melnikov for the encouragement she gave me in deciding to publish this paper.

I would like to thank Maria Gorelik, Vladimir Hinich and Anna Melnikov for organizing the conference for my 75th birthday and for the publication of the subsequent volume.

Efrat Levitsky made which appears in my home page. Sharon Ben Mordechai using "Illustrator" made the Plates 1, 2, 3, 4, 5, 6, 7, and 8 reproduced here as well as the poster. I would like to thank them both for their patience and dedication.

Bar Sagi died on 26th February 2017 after a terrible 4 year struggle with cancer. She was just fifteen. I name the triangle tiling considered here after her.

References

1. N. Bourbaki, Éléments de mathématiques, Groupes et algèbres de Lie, Chapitres 4–6, Masson, Paris 1981 (in French).
2. A. J. Coleman, The Betti numbers of the simple Lie groups. Canad. J. Math. 10 1958 349–356.
3. B. Grünbaum and G. C. Shephard, Tilings and patterns. W. H. Freeman and Company, New York, 1987.
4. A. Joseph, A Pentagonal crystal, the Golden section, alcove packing and aperiodic tilings. Transform. Groups 14 (2009), no. 3, 557–612.
5. A. Joseph, An algebraic slice in the coadjoint space of the Borel and the Coxeter element. Adv. Math. 227 (2011), no. 1, 522–585.
6. A. Joseph, Dual Kashiwara functions for the $B(\infty)$ crystal in the bipartite case.

7. R. Kenyon, The construction of self-similar tilings. Geom. Funct. Anal. 6 (1996), no. 3, 471–488.
8. D. A. Lind, The entropies of topological Markov shifts and a related class of algebraic integers. Ergodic Theory Dynam. Systems 4 (1984), no. 2, 283–300.
9. W. Little and collaborators, The Oxford English Dictionary, Oxford (1933).
10. R. Penrose, Pentaplexity: a class of nonperiodic tilings of the plane. Math. Intelligencer 2 (1979/80), no. 1, 32–37.
11. B. Sagi, See me soar and spread my wings, Gvanim, Tel Aviv 2018.

Closures of O_n-Orbits in the Flag Variety for GL_n

William M. McGovern

To Tony Joseph on his 75th birthday

Abstract We give a pattern avoidance criterion for the conjugates of the bottom vertex of the Bruhat graph attached to an O_n orbit $\mathcal{O}$ in the flag variety of GL_n to have degree equal to the rank of this graph as a poset. This condition is known to be necessary for rational smoothness of $\overline{\mathcal{O}}$; we conjecture that it is also sufficient, proving that this holds for n even.

Mathematics Subject Classification (1991): 22E47, 57S25

1 Introduction

Let G be a complex reductive group with Borel subgroup B and let $K = G^\theta$ be the fixed point subgroup of an involution of G. In this paper we develop the program begun in [6] and continued in [7, 8], seeking to characterize the K-orbits in G/B with rationally smooth closure via a combinatorial criterion. Here we treat the case $G = GL_n\mathbb{C}$, $K = O_n\mathbb{C}$ and give a necessary pattern avoidance criterion, conjecturing that it is also sufficient, and proving this for n even. We focus on the order ideal of orbits with closures contained in a fixed one $\bar{\mathcal{O}}$; this is an interval inside the poset of all K-orbits in G/B, ordered by containment of closures. Richardson and Springer have defined an action of the braid monoid corresponding to the Weyl group W of G on this poset [9], which we use to make it and its order ideals into graphs. Our starting point is a necessary graph-theoretic criterion for rational smoothness inspired by the work of Carrell and Peterson on

W. M. McGovern (✉)
University of Washington, Seattle, WA, USA
e-mail: mcgovern@math.washington.edu

M. Gorelik et al. (eds.), *Representations and Nilpotent Orbits of Lie Algebraic Systems*, Progress in Mathematics 330,
https://doi.org/10.1007/978-3-030-23531-4_10

Schubert varieties [3]. This criterion has been generalized by Hultman to a necessary and sufficient criterion for rational smoothness for $G = GL_{2n}\mathbb{C}$, $K = \mathrm{Sp}_{2n}\mathbb{C}$ [4]; it is *not* in general sufficient in our situation.

I would like to warmly thank Axel Hultman for suggesting the approach used in this paper.

2 Preliminaries

Set $G = GL_n\mathbb{C}$, $K = O_n\mathbb{C}$. Let B be the subgroup of upper triangular matrices in G. The quotient G/B may be identified with the variety of complete flags $V_0 \subset V_1 \subset \cdots \subset V_n$ in $\mathbb{C}^n$. The group K acts on this variety with finitely many orbits; these are parameterized by the set I_n of involutions in the symmetric group S_n[5, 9]. In more detail, let $(\cdot,\cdot)$ be the standard symmetric bilinear form on $\mathbb{C}^n$, with isometry group K. Then a flag $V_0 \subset \cdots \subset V_n$ lies in the orbit $\mathcal{O}_\pi$ corresponding to the involution π if and only if the rank r_{ij} of $(\cdot,\cdot)$ on $V_i \times V_j$ equals the cardinality $\pi_{ij} := \#\{k : 1 \le k \le i, \pi(k) \le j\}$ for all $1 \le i, j \le n$.

We use the same definition of pattern avoidance for permutations as in [8], decreeing that $\pi = \pi_1 \ldots \pi_n$ (in one-line notation) includes the pattern $\mu = \mu_1 \ldots \mu_r$ if there are indices $i_1 < i_2 < \cdots < i_r$ permuted by π such that $\pi_{i_j} > \pi_{i_k}$ if and only if $\mu_j > \mu_k$. We say that π avoids μ if it does not include μ. (The more classical definition of pattern inclusion of Billey and others [1] would not require that π permute the indices i_j. Thus by our definition the involution 65872143 does not include the pattern 2143, even though the indices 2, 1, 4, 3 occur in that order in the involution, since they are not permuted by it. We will say more about the classical definition later.)

There are well-known poset- and graph-theoretic criteria for rational smoothness of complex Schubert varieties due to Carrell and Peterson. The poset criterion does not extend to our setting but the graph one does. To state it we first recall that the partial order on I_n corresponding to inclusion of orbit closures is the reverse Bruhat order [9]. Then I_n is graded via the rank function

$$r(\pi) = \lfloor n^2/4 \rfloor - \sum_{i<\pi(i)} (\pi(i) - i - \#\{k \in \mathbb{N} : i < k < \pi(i), \pi(k) < i\})$$

where $\lfloor n^2/4 \rfloor$ denotes the greatest integer to $n^2/4$ and $r(\pi)$ equals the difference in dimension between $\mathcal{O}_\pi$ and $\mathcal{O}_c$, the unique closed orbit, corresponding to the involution $w_0 = n \ldots 1$ [9]. Let I_π be the interval consisting of all $\pi' \le \pi$ in the reverse Bruhat order. We make I_π into a graph by decreeing that the vertices μ and ν in it are adjacent if and only if either $\nu = t\mu t \ne \mu$ for some transposition t in S_n, or $\nu = t\mu$ for some transposition t in S_n with $t\mu t = \mu$ and m is even; write $\nu = t \cdot \mu$ if either of these conditions holds. We similarly make the full poset I_n into a graph as well. Then a necessary condition for $\bar{\mathcal{O}}_\pi$ to be rationally smooth is that the degree of w_0 in I_π must be $r(\pi)$; in fact, all vertices $w^{-1}w_0w$ conjugate to w_0 in W and lying in I_π must have this degree [2, 2.5]. We conjecture that this last condition is also sufficient; this has been checked for $n \leq 9$.

3 The Bad Patterns

We recast the necessary pattern avoidance condition above for rational smoothness in terms of pattern avoidance.

Theorem 1 *With notation as above, the orbit $\mathcal{O}_\pi$ has rationally singular closure whenever π contains one of the twenty-four bad patterns* 14325, 21543, 32154, 154326, 124356, 351624, 132546, 426153, 153624, 351426, 1243576, 2135467, 2137654, 4321576, 5276143, 5472163, 1657324, 4651327, 57681324, 65872143, 13247856, 34125768, 34127856, 64827153. *The same holds if π contains the pattern* 2143, *provided that there are an even number of fixed indices of π between* 21 *and* 43 *(e.g., π =* 21354687, *where the* 43 *occurs in the last two indices of π).*

Proof One checks first that the degree of w_0 in I_π is greater than $r(\pi)$ for any π in the list above, *except* for 2137654 and 4321576. The closures of the orbits corresponding to these two permutations are also rationally singular, as follows by computing that the degree of a suitable conjugate of w_0 (indexed by 7643521 in the first case and 7635421 in the second) is greater than $r(\pi)$. If π is obtained from one of the bad patterns above other than 2143 by adding one or more fixed points, then one computes that the degree of w_0 is again more than $r(\pi)$, except for 2134765, 3214576, 2137564, and 4231576, and in these cases again the degree at a suitable conjugate of w_0 is too large. Moreover, any involution π' obtained from one of the above ones by adding one fixed point has the degree of w_0 bigger than $r(\pi')$. The same holds if π is obtained from 2143 by adding two or more fixed points, with an even number of them lying between 21 and 43. If π is obtained from 2143 by adding just one fixed point not lying between 21 and 43, then the unique vertex conjugate to w_0 having this fixed point has degree larger than $r(\pi)$. Finally, if π is obtained from one of the bad patterns by adding fixed points as above and then pairs of flipped indices, then one argues as in the proof of the Lemma in [8] that some vertex in I_π has degree greater than $r(\pi)$. Hence in all cases $\bar{\mathcal{O}}_\pi$ is rationally singular. □

We specialize to the even case $n = 2m$ in our next result.

Theorem 2 *Assume that $n = 2m$ is even. The orbit $\mathcal{O}_\pi$ has rationally smooth closure if and only if the degree of w_0 in I_π is $r(\pi)$.*

Proof Set $\mathcal{O} = \mathcal{O}_\pi$. We have already noted that the degree condition is necessary, so suppose that it is satisfied. We begin by constructing a slice of $\bar{\mathcal{O}}$ to $\mathcal{O}_c$ at a particular flag, as follows. Fix a basis (e_i) of $\mathbb{C}^{2m}$ such that $(e_i, e_j) = 1$ if $i + j = 2m + 1$ and $(e_i, e_j) = 0$ otherwise, where as above $(\cdot, \cdot)$ is the symmetric form. Let (a_{ij})

be a family of complex parameters indexed by ordered pairs (i, j) satisfying either $i \leq m < j$ or $m < i < j$. We assume that $a_{ij} = a_{2m+1-j,2m+1-i}$ if $i \leq m < j$ but otherwise put no restrictions on the a_{ij}. Define a basis (b_i) of $\mathbb{C}^{2m}$ via

$$b_i = \begin{cases} e_i + \sum_{j=m+1}^{2m} a_{ij} e_j & \text{if } i \leq m \\ e_i + \sum_{j=i+1}^{2m} a_{ij} e_j & \text{otherwise} \end{cases}$$

Then the Gram matrix $G := (g_{ij} = (b_i, b_j))$ of the b_i relative to the form satisfies

$$g_{ij} = \begin{cases} 2a_{i,2m+1-j} & \text{if } i \leq j \leq m \\ g_{ji} & \text{if } j < i \leq m \\ a_{j,2m+1-i} & \text{if } i < m < j < 2m+1-i \\ 1 & \text{if } i \leq m < j = 2m+1-i \\ g_{ji} & \text{if } j \leq m < i \\ 0 & \text{otherwise} \end{cases}$$

Thus the matrix G is symmetric and has zeroes below the antidiagonal from lower left to upper right. The antidiagonal entries are all 1. Now one checks that the set $\mathcal{S}'$ of all flags $V_0 \subset \ldots \subset V_{2m}$ where (b_i) runs through all bases obtained as above from the a_{ij} and V_i is the span of $b_1, \ldots b_i$ is a slice of G/B to $\mathcal{O}_c$ at the flag f_c corresponding to the basis (e_i), which in turn corresponds to the point P where all $a_{ij} = 0$ [2, 2.1]. Intersecting $\mathcal{S}'$ with $\bar{\mathcal{O}}$ we get a slice $\mathcal{S}$ of $\bar{\mathcal{O}}$ to $\mathcal{O}_c$ at P in the sense of Brion [2, 2.1], defined by the vanishing of certain minors in the Gram matrix G. It is known ([2]) that $\mathcal{S}$ is rationally smooth (resp. smooth) at P if and only if it is rationally smooth (resp. smooth) everywhere, or if and only if $\bar{\mathcal{O}}$ is rationally smooth (resp. smooth) everywhere. (This construction works with minor modifications for odd n as well).

We now show that $\mathcal{S}$ is rationally smooth at P by verifying the conditions of [2, 1.4]. Actually we construct a rationally smooth slice $\mathcal{S}''$ for a variety $\mathcal{V}$ containing $\bar{\mathcal{O}}_\pi$, with no hypothesis on the degree of w_0 in I_π; if this hypothesis holds, then $\mathcal{V}$ coincides with $\bar{\mathcal{O}}_\pi$ and we may take $\mathcal{S}'' = \mathcal{S}$. For each conjugate $v = t \cdot w_0$ of w_0 by a transposition t with $v \not\leq \pi$, write v as $v_1 \ldots v_{2m}$ in one-line notation. Let i be the smallest index such that if $\pi_1 \ldots \pi_i$ is rearranged in ascending order as $\pi'_1 \ldots \pi'_i$ and similarly $v_1 \ldots v_i$ is rearranged as $v'_1 \ldots v'_i$, then $\pi'_j > v'_j$ for some $j \leq i$. Then there is some k such that there are more indices $\ell \leq i$ (say n_k of them) with $v_\ell \leq k$ than indices $m \leq i$ with $\pi_m \leq k$. Let j be the smallest index with $v_j \neq 2n+1-j$. Set $r := 2n+1-k+n_k-2$. If there are fewer than n_k indices less than or equal to k among $\pi_1 \ldots \pi_r$, then the minor of G consisting of those entries in rows $j, r-(n_k-2), \ldots, r-1, r$ and columns $\min(v_j+1, k-(n_k-1))$,

$k-(n_k-2), \ldots, k-1, k$ vanishes on $\bar{\mathcal{O}}_\pi$; one variable in this minor occurs to the first power and is not multiplied by any other variable. Otherwise the minor consisting of those entries in rows $j, k-(n_k-2), \ldots, k$ and columns $j, k-(n_k-2), \ldots, k$ (or just row and column k, if $n_k = 1$) vanishes on $\bar{\mathcal{O}}_\pi$ and may involve certain variables quadratically. Define the slice $\mathcal{S}''$ by the simultaneous vanishing of these minors and let $\mathcal{V}$ be the corresponding subvariety of G/B, which contains $\bar{\mathcal{O}}_\pi$.

Define an action of the m-torus $T = \mathbb{T}^m$ on the matrix G by multiplying the first m rows and columns by $t_1, \ldots, t_m$, respectively, while multiplying the last m rows and columns by $t_m^{-1}, \ldots, t_1^{-1}$, respectively; this action preserves the 1s on the antidiagonal and the vanishing of the minors that define the slice $\mathcal{S}''$. (T is just a maximal torus of K.) Then the weights of T occurring in the tangent space at P of the big slice $\mathcal{S}'$ are those of the form $2e_i$, e_i+e_j, or e_i-e_j for some $1 \le i < j \le m$ and all occur with multiplicity one. They all lie on one side of a hyperplane and P is an attractive fixed point of both $\mathcal{S}''$ and $\mathcal{S}'$. The subtori T' of T of codimension one such that the fixed point subvariety $(\mathcal{S}'')^{T'}$ of the slice $\mathcal{S}''$ under the T'-action contains more than one point correspond exactly to the conjugates $t \cdot w_0$ of w_0 not lying below π. Hence conditions [2, 1.4(ii), (iii)] are satisfied for $\mathcal{S}''$ and $\mathcal{V}$. As for [2, 1.4(i)], we find that by repeatedly slicing the slice $\mathcal{S}''$ at nonzero values of variables occurring in it, we are led to a product of varieties defined by quadratic equations in its variables, which is easily seen to be rationally smooth away from the origin. (Here is where the evenness of n is crucial; the varieties in question are not rationally smooth away from the origin if n is odd.) Thus $\mathcal{S}''$ and $\mathcal{V}$ are both rationally smooth. If the degree condition holds on w_0, then by counting dimensions we see that $\mathcal{V}$ coincides with $\bar{\mathcal{O}}_\pi$, so it too is rationally smooth. □

4 Main Result

Theorem 3 *If π avoids the bad patterns of Theorem 1, then all conjugates of w_0 in I_π have degree $r(\pi)$, so that $\bar{\mathcal{O}}_\pi$ has rationally smooth closure if n is even.*

Proof Suppose first that n is even. We first show that w_0 has degree $r(\pi)$. We have seen that the neighbors ν adjacent to any $\mu \in I_n$ (adjacent to μ) take the form either $\nu = t\mu t$ or $\nu = t\mu$, for some transposition t; accordingly we say that v is of *type 1* (resp. *type 2*) if it takes the first (resp. the second) form. Now recall that the poset I'_n of fixed-point-free involutions in S_n with the reverse Bruhat order parameterizes the poset of Sp_{2m}-orbits in G/B, ordered by inclusion of closures; moreover, the rank functions for I_n and I'_n coincide. In [8] we characterized the fixed-point-free involutions π such that the degree of w_0 in the order ideal I'_π of I'_n equals $r(\pi)$. We first claim that there is a unique minimal fixed-point-free involution $f(\pi)$ lying above π in the usual Bruhat order (so below it in the reverse one). Indeed, if π is fixed-point-free, then $f(\pi) = \pi$; otherwise, let the indices fixed by π be $i_1, \ldots, i_{2k}$ in increasing order, so that π does not fix any index between i_j and i_{j+1} for any $j \le 2k-1$. Now for every j look at the pairs of indices both lying between i_{2j-1}

and i_{2j} and flipped by π. We say that such a pair (i, ℓ) with $i < \ell$ encapsulates another one (j, k) if $i < j < k < \ell$. Let $(\ell_{j1}, \ell'_{j1}), \ldots, (\ell_{jm}, \ell'_{jm})$ enumerate all such pairs not encapsulating other ones, labelled so that $\ell_{ji} < \ell'_{ji}$ for all i and $\ell'_{ji} < \ell'_{j(i+1)}$ for $i \leq m-1$. Replace $i_{2j-1}, \ell'_{j1}, \ldots, \ell'_{jm}$ in the one-line notation of π by $\ell_{j1'}, \ldots, \ell'_{jm}, i_{2j}$, respectively, replace $\ell_{j1}, \ldots, \ell_{jm}, i_{2j}$ by $i_{2j-1}, \ell_{j1}, \ldots, \ell_{jm}$, respectively, and leave all other indices unchanged. This gives the one-line notation of $f(\pi)$. Thus, for example, if $\pi = 16754238$, then $f(\pi) = 56781234$. Now the type 1 neighbors(=adjacent vertices) of w_0 lying in I_π are the same as those lying in $I_{f(\pi)}$. We can read off the number $t(\pi)$ of type 2 neighbors of w_0 lying in I_π from the one-line notation of π, as follows. Let i_1 be the smallest index such that $\pi(i_1) \leq i_i$ and let i_2 be the smallest index such that $\pi(n+1-i_2) \geq 2n+1-i_2$. Let i be the maximum of i_1 and i_2; then $t(\pi) = m+1-i$. Now let π be an involution of even length $2m$ that either appears in the above list or is obtained from an involution in this list by adding one fixed point. One checks in all cases that either the rank difference $r(\pi) - r(f(\pi))$ is less than $t(\pi)$ or $f(\pi)$ contains a bad pattern for I'_{2m} (i.e., as a fixed-point-free involution, so that the degree of w_0 in $I'_{f(\pi)}$ is already too large). Table 1 lists the possibilities for π, making a representative choice among these possibilities if π is obtained from a bad pattern of odd length by adding one fixed point. We give first π, then its rank $r(\pi)$, then $f(\pi)$, then its rank $r(f(\pi))$ the number $t(\pi)$ of type 2 neighbors of π, and finally the difference between $r(f(\pi))$ and the degree of w_0 in I'_{2m} whenever this difference is nonzero.

Given an arbitrary involution π for which the degree d_{w_0} of w_0 in I_π is too large, either the degree of w_0 in $I'_{f(\pi)}$ must already be too large (forcing $f(\pi)$ to contain one of the seventeen bad patterns of [8]) or $t(\pi)$ must be larger than the rank difference $r(\pi) - r(f(\pi))$ (or both). The value of $t(\pi)$ is determined by the indices i_1, i_2 attached to π above. Bearing in mind the recipe for computing $f(\pi)$ from π and the list of bad patterns in [8] (none of which has length larger than eight) we see that we can replace any π for which d_{w_0} is too large by an involution it includes of length at most eight with the same property (including the indices i_1 and i_2, two fixed points of π with a pair of flipped indices lying between them, and that pair of indices). But the above patterns capture all instances where this happens for $m = 4$ (as one sees by examining all the possibilities, using, for example, the computation of the Kazhdan-Lusztig-Vogan polynomials introduced in [10] for $GL_8\mathbb{R}$ provided by the ATLAS software, available at www.liegroups.org). Thus any such π includes a pattern in the above list, as desired.

Now we show that all conjugates of w_0 in I_π also have degree $r(\pi)$. By [8, Theorem 2] the number of type 1 neighbors of any conjugate of w_0 lying below $f(\pi)$ in I_π is no greater than $r(f(\pi))$, provided this holds for w_0. One checks from the above formula for the number of type 2 neighbors of w_0 *not* lying below π that this number can only increase if w_0 is replaced by a conjugate of itself (i.e., by another fixed-point-free involution), so the degree of a conjugate of w_0 is bounded above by $r(\pi)$ whenever the degree of w_0 is, as desired.

Table 1 Bad patterns

π	$r(\pi)$	$f(\pi)$	$r(f(\pi))$	$t(\pi)$	Difference
2143	2	2143	2	1	
143256	7	341265	5	3	
215436	6	215634	5	2	
321546	6	351624	4	2	1
154326	5	456123	3	3	
124356	8	214365	6	3	
351624	4	351624	4	1	1
132546	7	351624	4	3	1
426153	4	456123	3	2	
153624	5	351624	4	1	1
351426	5	351624	4	1	1
12435768	14	21437856	11	4	
21354678	14	21563478	11	3	1
21376548	11	21678345	9	3	
43215768	11	43218765	8	2	2
52761438	8	56781234	6	3	
54721638	8	54721836	7	1	1
16573248	9	56781234	6	4	
46513278	9	46513287	8	1	1
57681324	5	57681324	5	0	1
65872143	4	65872134	4	0	1
34127856	10	34127856	10	2	1
64827153	5	64827153	5	1	1
13247856	12	34127856	10	2	1
34125768	12	34127856	10	2	1

If instead $n = 2m + 1$ is odd, then a similar but more complicated argument works. Recall first that the neighbors of $\mu \in I_\pi$ all take the form $\nu = t\mu t$ for some transposition t not commuting with μ (transpositions t commuting with μ no longer give rise to adjacent vertices in this case). We say that ν is of type 1 if the transposition t does not involve the middle index $m + 1$ and of type 2 if it does involve this index. Now given $\pi \in I_n$ there is a unique smallest $f(\pi)$ lying above π in the Bruhat order among involutions fixing the index $m + 1$ but no other. To construct $f(\pi)$, assume first that π already fixes $m + 1$; then we just apply the above recipe for f to π restricted to the other indices $1 \ldots, m, m + 2, \ldots, 2m + 1$, decreeing at the end that $f(\pi)$ also fix $m + 1$. Now assume that $\pi(m + 1) < m + 1$; if this is not the case, conjugate π by w_0 to make this hold, apply the following recipe, and then conjugate by w_0 again. Denote by i_1, i_2 the largest fixed point of π less than $m + 1$ (if there is one) and the smallest fixed point of π larger than $m + 1$ (if there is one). Enumerate the pairs of indices flipped by π not encapsulating other pairs for which the larger index is greater than or equal to $m + 1$ and less than i_2 (if it exists) as $(\ell_1, \ell'_1), \ldots, (\ell_m, \ell'_m)$ with $\ell_i < \ell'_i$ and the ℓ'_i. in increasing order, as in the previous recipe. If i_2 exists, let T consist of the set of ℓ'_i together with i_2; if i_2 does not exist, let T consist of the ℓ'_i together with ℓ_m. Likewise let S consist of the ℓ_i together with i_1 if it exists. Define a new involution π' by declaring that it fix the smallest index $m+1$ in T, flip the next smallest index of T with the smallest one in S, and so on, finally flipping the largest index in S with that in T (if i_2 does not exist), or fixing the largest index in S (if i_1 and i_2 both exist). Other indices have the same image under π' as under π. Then π' fixes $m+1$ and lies above π. Applying the above recipe to π', we get $f(\pi') = f(\pi)$. Thus, for example, if $\pi = 13245$, then S consists of the indices 1 and 2, while T consists of 3 and 4; here, $i_1 = 1, i_2 = 4$. Then π' fixes 3, flips 1 and 4, and fixes 2, whence $\pi' = 42315$; finally, $f(\pi') = f(\pi) = 45312$.

We also have an analogous formula to the one above for the number $t(\pi)$ of type 2 neighbors of w_0 in I_π. Let i_1 be the smallest index such that $\pi(i_1) \leq i_1$ and $\pi(j) \geq m+1$ for some $j \geq n+1-i_1$); similarly let i_2 be the smallest index such that $\pi(n+1-i_2) \geq n+1-i_2$ and $\pi(j) \leq m+1$ for some $j \leq i_2$. Then the number $t(\pi) = n+1-i_1-i_2$. A similar argument to the one above shows that avoiding the above list of bad patterns is sufficient to guarantee that the degree of w_0 in I_π is $r(\pi)$; similarly any conjugate of w_0 in I_π fixing $m+1$ has degree $r(\pi)$.

We now consider vertices in I_π fixing a single index other than $m+1$. For $1 \leq i \leq n$, denote by $w_0^{(i)}$ the unique involution whose one-line notation has i in the ith position and the other indices listed in decreasing order. Then either $w_0^{(i)}$ lies in I_π, or else no vertex in I_π fixes i and no other index. In the former case there is a unique smallest vertex $f^{(i)}(\pi)$ lying above π in the usual Bruhat order, constructed as above by replacing the index $m+1$ throughout by i. We construct it as above, replacing the index $m+1$ throughout by i, except that if i is not the smallest index in the set T, then $w_0^{(i)}$ does not lie above π, so that no involution fixing i alone lies above π and $f^{(i)}(\pi)$ is undefined. We argue as above that avoiding the bad patterns of Theorem 1 implies that $w_0^{(i)}$ has degree $r(\pi)$ whenever it lies in I_π and that all conjugates of it fixing only the index i have this degree as well. □

The condition that w_0 alone have degree $r(\pi)$ is *not* sufficient for rational smoothness if n is odd, as the examples $\pi = 2137654$ (cited above) and $\pi = 21435$ show. However we have

Conjecture 4 For n odd, the orbit closure $\bar{\mathcal{O}}_\pi$ is rationally smooth if and only if all conjugates of w_0 in the order ideal I_π have degree $r(\pi)$, or if and only if π avoids all bad patterns.

We also have

Conjecture 5 Avoiding all of the above patterns in the sense of this paper is equivalent to avoiding the same patterns in the classical sense

For example, we observed above that the involution 65872143 avoids the bad pattern 2143 in the sense of this paper, but it is itself another bad pattern, so does not correspond to an orbit with rationally smooth closure. Probably this conjecture can be checked by a computer without too much trouble.

Finally, we note that in our setting, unlike that of [4] and [8] smoothness and rational smoothness of orbit closures are not equivalent. The orbit closure corresponding to the involution 1324 is rationally smooth but not smooth. We hope to study smoothness of orbit closures in a future paper.

References

1. S. Billey and V. Lakshmibai, *Singular loci of Schubert varieties*, Progress in Math. **182**, Birkhäuser, Boston, 2000.
2. M. Brion, *Rational smoothness and fixed points of torus actions*, Transf. Groups **4** (1999), 127–156.
3. J. Carrell, *The Bruhat graph of a Coxeter group, a conjecture of Deodhar, ad rational smoothness of Schubert varieties*, Proc. Symp. Pure Math. **56** (1994), 53–61.
4. A. Hultman, *Criteria for rational smoothness of some symmetric orbit closures*, Adv. Math. **229** (2012), 183–200.
5. T. Matsuki and T. Oshima, *Embedding of discrete series into principal series*, in The Orbit Method in Representation Theory, Progress in Math. **82**, Birkhäuser, Boston, 1988, 147–175.
6. W. McGovern, *Closures of K-orbits in the flag variety for $U(p,q)$*, J. Alg. **322** (2009), 2709–2712.
7. W. McGovern and P. Trapa, *Pattern avoidance and smoothness of closures for orbits of a symmetric subgroup in the flag variety*, J. Alg. **322** (2009), 2713–2730.
8. W. McGovern, *Closures of K-orbits in the flag variety for $SU^*(2n)$*, Rep. Theory. **15** (2011), 568–573.
9. R. W. Richardson and T. A. Springer, *On the Bruhat order for symmetric varieties*, Geom. Dedicata **35** (1990), 389–436.
10. D. Vogan, *Irreducible characters of semisimple Lie groups III: proof of Kaxhdan-Lusztig conjecture in the integral case*. Inv. Math. **71** (1983), 381–417.

The Spin Calogero-Sutherland Model at Infinity

Maxim Nazarov

To Professor Anthony Joseph on the occasion of his 75th birthday

Abstract For $N = 1, 2, \ldots$ we consider an action of the Yangian $\mathrm{Y}(\mathfrak{gl}_n)$ on Nth symmetric power of the space of polynomials in one variable with coefficients in $\mathbb{C}^n$. This action is given by the Heckman operators (1991) via the Drinfeld functor (1986). We describe the limit of this action at $N \to \infty$. This provides another solution to the problem already considered by Khoroshkin, Matushko and Sklyanin (2017).

1 Introduction

This quantum Calogero-Sutherland model describes a system of N bosonic particles on a circle $\mathbb{R}/\pi\,\mathbb{Z}$ with the Hamiltonian [3, 19]

$$-\frac{1}{2}\sum_{j}\frac{\partial^2}{\partial q_j^2}+\sum_{i<j}\frac{\beta\,(\beta-1)}{\sin^2(q_i-q_j)} \tag{1.1}$$

where $0 \leqslant q_1, \ldots, q_N < \pi$. After conjugating by the vacuum factor

$$\Big|\prod_{i<j}\sin(q_i-q_j)\Big|^{\beta}$$

and passing to the exponential variables $x_j = \exp(2\mathrm{i}q_j)$ and the parameter $\alpha = \beta^{-1}$ the Hamiltonian (1.1) becomes

M. Nazarov (✉)
Department of Mathematics, University of York, York, UK

M. Gorelik et al. (eds.), *Representations and Nilpotent Orbits of Lie Algebraic Systems*, Progress in Mathematics 330,
https://doi.org/10.1007/978-3-030-23531-4_11

$$\frac{2}{\alpha}\,H+\frac{N^3-N}{6\alpha^2}$$

where

$$H=\alpha\sum_i(x_i\,\partial_i)^2+\sum_{i<j}\frac{x_i+x_j}{x_i-x_j}\,(x_i\,\partial_i-x_j\,\partial_j)\,. \tag{1.2}$$

Here ∂_j denotes the derivation with respect to the variable x_j. The operator H acts on the symmetric polynomials in $x_1,\dots,x_N$. It can be included into a *quantum integrable hierarchy,* that is into a ring of commuting differential operators with N generators of orders $1,\dots,N$. The joint eigenfunctions of these commuting differential operators are Jack symmetric polynomials [10].

Two different constructions of generators of this operator ring are known. The first set of generators consists of the coefficients of a certain polynomial of degree N in an auxiliary variable called the Sekiguchi-Debiard determinant [5, 16]. The second set consists of the power sums of degrees $1,\dots,N$ of the Heckman operators [9], see our Sect. 3 for their definition. These operators act on all the polynomials in $x_1,\dots,x_N$ and do not commute, yet their power sums preserve the space of symmetric polynomials. The commuting versions of the Heckman operators were found by Cherednik [4].

It is fascinating to study the limit of the Calogero-Sutherland model when the number N of particles tends to infinity. The limit of the Hamiltonian (1.2) has been known for a long time [18], but explicit description of the limit of the quantum integrable hierarchy was not available until recently. In [14] we described the limits of the generators yielded by the Sekiguchi-Debiard determinant. In [15] we described the limits of the power sums of the Heckman operators, and also identified the resulting integrable hierarchy as that of the quantum counterpart of the classical Benjamin–Ono equation. This equation describes internal waves in fluids of great depth. In [17] the same hierarchy as in [15] was obtained by another approach, namely by describing the limits of the Heckman operators themselves.

The Calogero-Sutherland model has a generalization [8] which describes N bosonic particles on a circle, each particle now having n internal degrees of freedom. Here n is any positive integer. The space of symmetric polynomials used above generalizes now to the subspace in the tensor product

$$(\mathbb{C}^n)^{\otimes N}\otimes\mathbb{C}[x_1,\dots,x_N] \tag{1.3}$$

consisting of the invariants under the simultaneous permutations of the N tensor factors $\mathbb{C}^n$ and of the variables $x_1,\dots,x_N$. Remarkably, this subspace comes [2] with an action of the Yangian $\mathrm{Y}(\mathfrak{gl}_n)$. Using either the Cherednik or the Heckman operators on $\mathbb{C}[x_1,\dots,x_N]$ this action can be obtained as a particular case of a general construction due to Drinfeld [6], see our Sect. 4. The eigenstates of this model have been studied in [20].

In the present article we consider the limit of this generalization of the Calogero-Sutherland model. This limit was already studied in [1]. Following that work, we identify the limit at $N \to \infty$ of the above-mentioned subspace of invariants in (1.3) with the bosonic Fock space $\mathcal{F}$ defined in our Sect. 2. Using the approach of [17], in Sect. 3 for any given n we describe the limits at $N \to \infty$ of the Heckman operators now acting on (1.3). This description determines the limiting action of the Yangian $\mathrm{Y}(\mathfrak{gl}_n)$ on $\mathcal{F}$, see our Sect. 4. This limiting action has been already studied in [11]. However our result has a different form, see the end of Sect. 4 for an explanation of the difference.

2 Fock Space

Fix a positive integer n. Let $\mathcal{F}$ be the commutative algebra over the complex field $\mathbb{C}$ with free generators p_{ck} where $c = 1, \ldots, n$ and $k = 0, 1, 2, \ldots$. We shall refer to $\mathcal{F}$ as to the *Fock space* with n spin degrees of freedom, see [1].

Now take the vector space $\mathbb{C}^n$ with the standard basis vectors $e_1, \ldots, e_n$. Turn $\mathbb{C}^n$ into a commutative ring by setting $e_a e_b = \delta_{ab} e_a$ for $a, b = 1, \ldots, n$ and extending this definition of multiplication on $\mathbb{C}^n$ by linearity. The element

$$e = e_1 + \ldots + e_n$$

is a unit of this ring, that is $e\, g = g$ for all $g \in \mathbb{C}^n$. The vector space $V = \mathbb{C}^n[v]$ of all polynomials in the variable v with coefficients in $\mathbb{C}^n$ then also becomes a commutative ring. The element e is a unit of the latter ring as well.

For $N = 1, 2, \ldots$ take the tensor product $V^{\otimes N}$ of N copies of the ring V. This tensor product can be naturally identified with (1.3). The symmetric group $\mathfrak{S}_N$ acts on $V^{\otimes N}$ by permuting the N tensor factors. Consider the subring $(V^{\otimes N})^{\mathfrak{S}_N} \subset V^{\otimes N}$ consisting of the elements invariant under this action. Denote by Λ_N this subring. Define a ring homomorphism

$$\mathcal{F} \to \Lambda_N \tag{2.1}$$

by mapping the identity element $1 \in \mathcal{F}$ to $e^{\otimes N}$ and also mapping the free generators $p_{ck} \in \mathcal{F}$ to the sums

$$\sum_{i=1}^{N} e^{\otimes(i-1)} \otimes e_c\, v^k \otimes e^{\otimes(N-i)} \in \Lambda_N \tag{2.2}$$

respectively. Then the sum

$$\sum_{c=1}^{n} p_{c0} \in \mathcal{F} \tag{2.3}$$

gets mapped to $N\,e^{\otimes N}$. Our homomorphism (2.1) is surjective due to the next

Proposition 2.1 *The ring Λ_N is generated by the sums* (2.2).

Proof Let $g_1, \ldots, g_N \in \mathbb{C}^n$ while $k_1, \ldots, k_N = 0, 1, 2, \ldots$. The vector space Λ_N is spanned by the sums of the tensor products

$$h_1\, v^{l_1} \otimes \ldots \otimes h_N\, v^{l_N}$$

where the summation is over all $N!$ permutations $(h_1, l_1), \ldots, (h_N, l_N)$ of a given sequence of pairs $(g_1, k_1), \ldots, (g_N, k_N)$. Let M be the number of pairs in the latter sequence which are different from $(e, 0)$. We will prove by induction on $M = 0, 1, \ldots, N$ that the sum corresponding to the $(g_1, k_1), \ldots, (g_N, k_N)$ belongs to the image of the homomorphism (2.1). Denote by S this sum. Let

$$\Lambda_N^{(M)} \subset \Lambda_N$$

be the subspace spanned by all the sums S with the given number M.

If $M = 0$ then $S = N!\,e^{\otimes N}$, that is $N!$ times the image of the identity element $1 \in \mathcal{F}$ under (2.1). Now suppose that $M > 0$. Because the sum S does not change when the sequence $(g_1, k_1), \ldots, (g_N, k_N)$ is reordered, we will assume that it is the first M pairs $(g_1, k_1), \ldots, (g_M, k_M)$ of the sequence that differ from $(e, 0)$. Then consider the product over $j = 1, \ldots, M$ of the sums

$$\sum_{i=1}^{N} e^{\otimes (i-1)} \otimes g_j\, v^{k_j} \otimes e^{\otimes (N-i)} \in \Lambda_N\,. \tag{2.4}$$

The difference between this product and $S\,/\,(N-M)!$ belongs to the subspace

$$\Lambda_N^{(0)} + \ldots + \Lambda_N^{(M-1)} \subset \Lambda_N\,.$$

Since (2.4) is a linear combination of the images (2.2) of the elements $p_{ck} \in \mathcal{F}$ with $c = 1, \ldots, n$ and $k = k_j$, we have now made the induction step. □

Proposition 2.2 *The kernels of all homomorphisms* (2.1) *with* $N = 1, 2, \ldots$ *have the zero intersection.*

Proof Consider the set of all free generators p_{ck} of the commutative ring $\mathcal{F}$. In this set of free generators we can replace p_{n0} by the sum (2.3), which will be denoted here simply by q. Take any finite linear combination of unordered monomials in the new generators of $\mathcal{F}$. Suppose that it gets mapped to zero by every homomorphism (2.1). Consider the terms in this linear combination which have the maximal total degree in all the new generators but q . Let S be the sum of these terms. Let M be their degree. If $M = 0$ then our linear combination is just a polynomial in q with complex coefficients, which for all N vanishes when mapping $q \mapsto N e^{\otimes N}$. Hence our linear combination is zero.

Suppose $M > 0$. For any $N \geqslant M$ apply to S the homomorphism (2.1). Then apply to the resulting image of S in the subspace $\Lambda_N \subset V^{\otimes N}$ the linear map $V^{\otimes N} \to V^{\otimes M}$ projecting onto the tensor product of the first M tensor factors V of $V^{\otimes N}$. Arguments similar to those of the proof of Proposition 2.1 show that the image of S in $V^{\otimes M}$ must be zero. By letting the number N vary like in the case $M = 0$ one can show that $S = 0$ then. But the equality $S = 0$ contradicts to the assumption that $M > 0$. □

We will regard the Fock space $\mathcal{F}$ as the limit at $N \to \infty$ of the ring Λ_N by using the homomorphism (2.1). The complex general linear Lie algebra $\mathfrak{gl}_n$ acts on the vector space V, and diagonally on the tensor product $V^{\otimes N}$. The latter action commutes with the action of the group $\mathfrak{S}_N$. Hence the action of $\mathfrak{gl}_n$ on $V^{\otimes N}$ preserves the subspace Λ_N . In this section we will describe the corresponding action of the Lie algebra $\mathfrak{gl}_n$ on the vector space $\mathcal{F}$. Namely, for any standard matrix unit $E_{ab} \in \mathfrak{gl}_n$ we will describe its action on $\mathcal{F}$ which makes commutative the following diagram:

$$\begin{array}{ccc} \mathcal{F} & \xrightarrow{E_{ab}} & \mathcal{F} \\ \downarrow & & \downarrow \\ \Lambda_N & \xrightarrow[E_{ab}]{} & \Lambda_N \end{array}$$

Here the vertical arrows indicate the homomorphism (2.1). It is easy to verify

Lemma 2.3 *The action of E_{ab} on V is a ring endomorphism.*

Note that the endomorphism E_{ab} does *not* preserve the element $e \in V$ unless $a = b$. We will describe the action of E_{ab} on $\mathcal{F}$ using the method of [17]. We will first consider the ring $V \otimes \mathcal{F}$. It contains $\mathcal{F}$ via the embedding

$$\iota : \mathcal{F} \to V \otimes \mathcal{F} : f \mapsto e \otimes f \tag{2.5}$$

for all $f \in \mathcal{F}$. The ring $V \otimes \mathcal{F}$ is generated by the elements $e_c\, v^k \otimes 1$ and the elements $e \otimes p_{ck}$. Let us extend (2.1) to the ring homomorphism

$$\pi_N : V \otimes \mathcal{F} \to V \otimes \Lambda_{N-1}$$

by mapping

$$e_c\, v^k \otimes 1 \mapsto e_c\, v^k \otimes e^{\otimes(N-1)}\,. \tag{2.6}$$

We have

$$\Lambda_N \subset V \otimes \Lambda_{N-1} \tag{2.7}$$

and our π_N by definition maps the element $e \otimes p_{ck} \in V \otimes \mathcal{F}$ to the sum (2.2). We will describe an operator F_{ab} on $V \otimes \mathcal{F}$ making commutative the diagram

$$\begin{array}{ccc} V \otimes \mathcal{F} & \xrightarrow{F_{ab}} & V \otimes \mathcal{F} \\ \downarrow{\scriptstyle \pi_N} & & \downarrow{\scriptstyle \pi_N} \\ V \otimes \Lambda_{N-1} & \xrightarrow[E_{ab}\otimes \mathrm{id}]{} & V \otimes \Lambda_{N-1} \end{array} \tag{2.8}$$

To this end we will introduce another ring homomorphism

$$\pi'_N : V \otimes \mathcal{F} \to V \otimes \Lambda_{N-1}$$

such that π'_N will map the element $e \otimes p_{ck} \in V \otimes \mathcal{F}$ to the sum

$$\sum_{i=2}^{N} e^{\otimes(i-1)} \otimes e_c\, v^k \otimes e^{\otimes(N-i)} \tag{2.9}$$

instead of (2.2). The homomorphism π'_N will still map (2.6) as π_N does. So

$$\begin{aligned} \pi_N(e \otimes p_{ck}) &= \pi'_N(e \otimes p_{ck}) + e_c\, v^k \otimes e^{\otimes(N-1)} \\ &= \pi'_N(e \otimes p_{ck} + e_c\, v^k \otimes 1) \end{aligned} \tag{2.10}$$

and

$$\pi'_N(e \otimes p_{ck}) = \pi_N(e \otimes p_{ck} - e_c\, v^k \otimes 1)\,. \tag{2.11}$$

By the definition of π'_N we immediately obtain commutativity of the diagram

$$\begin{array}{ccc} V \otimes \mathcal{F} & \xrightarrow{E_{ab}\otimes \mathrm{id}} & V \otimes \mathcal{F} \\ \downarrow{\scriptstyle \pi'_N} & & \downarrow{\scriptstyle \pi'_N} \\ V \otimes \Lambda_{N-1} & \xrightarrow[E_{ab}\otimes \mathrm{id}]{} & V \otimes \Lambda_{N-1} \end{array}$$

In other words, the limit at $N \to \infty$ of the operator $E_{ab} \otimes \mathrm{id}$ on $V \otimes \Lambda_{N-1}$ relative to the homomorphism π'_N is just the operator $E_{ab} \otimes \mathrm{id}$ on $V \otimes \mathcal{F}$.

Let us now turn to the homomorphism π_N. By Lemma 2.3 the operator $E_{ab} \otimes \mathrm{id}$ on $V \otimes \Lambda_{N-1}$ is a ring endomorphism. Therefore our F_{ab} will be an endomorphism of the ring $V \otimes \mathcal{F}$. Setting

$$F_{ab}(e_c\, v^k \otimes 1) = E_{ab}(e_c\, v^k) = \delta_{bc}\, e_a\, v^k \otimes 1 \tag{2.12}$$

will make the compositions $\pi_N F_{ab}$ and $(E_{ab} \otimes \mathrm{id})\, \pi_N$ coincide on the element $e_c\, v^k \otimes 1 \in V \otimes \mathcal{F}$, see (2.6) and (2.8). Again according to (2.8) we also need

$$\pi_N F_{ab}(e \otimes p_{ck}) = (E_{ab} \otimes \mathrm{id})\, \pi_N (e \otimes p_{ck})\,.$$

By (2.10) and (2.11) the right-hand side of the above displayed relations equals

$$\begin{gathered}
(E_{ab} \otimes \mathrm{id})\, \pi'_N (e \otimes p_{ck} + e_c\, v^k \otimes 1) = \\
\pi'_N (E_{ab} \otimes \mathrm{id})(e \otimes p_{ck} + e_c\, v^k \otimes 1) = \\
\pi'_N (e_a \otimes p_{ck} + \delta_{bc}\, e_a\, v^k \otimes 1) = \pi'_N ((e_a \otimes 1)(e \otimes p_{ck}) + \delta_{bc}\, e_a\, v^k \otimes 1) = \\
\pi_N ((e_a \otimes 1)(e \otimes p_{ck} - e_c\, v^k \otimes 1) + \delta_{bc}\, e_a\, v^k \otimes 1) = \\
\pi_N (e_a \otimes p_{ck} - \delta_{ac}\, e_a\, v^k \otimes 1 + \delta_{bc}\, e_a\, v^k \otimes 1)\,.
\end{gathered}$$

Hence

$$F_{ab}(e \otimes p_{ck}) = e_a \otimes p_{ck} + (\delta_{bc} - \delta_{ac})\, e_a\, v^k \otimes 1\,. \tag{2.13}$$

So the actions of F_{ab} and $E_{ab} \otimes \mathrm{id}$ on $e \otimes p_{ck}$ differ unless $\delta_{ac} = \delta_{bc}$. We get

Proposition 2.4 *The endomorphism F_{ab} of the ring $V \otimes \mathcal{F}$ defined by* (2.12) *and* (2.13) *makes commutative the diagram* (2.8).

To describe the action of E_{ab} on $\mathcal{F}$ let us now consider the linear map

$$\theta : V \otimes \mathcal{F} \to \mathcal{F} : e_c\, v^k \otimes f \mapsto p_{ck}\, f\,. \tag{2.14}$$

This is *not* a ring homomorphism, but is $\mathcal{F}$-linear relative to the embedding $\iota : \mathcal{F} \to V \otimes \mathcal{F}$ defined earlier. Moreover it makes commutative the diagram

$$\begin{array}{ccc}
V \otimes \mathcal{F} & \xrightarrow{\ \theta\ } & \mathcal{F} \\
\downarrow{\scriptstyle \pi_N} & & \downarrow \\
V \otimes \Lambda_{N-1} & \xrightarrow[\ \theta_N\ ]{} & \Lambda_N
\end{array} \tag{2.15}$$

where the rightmost vertical arrow indicates the homomorphism (2.1), while θ_N denotes the restriction of the action of the element

$$1+\sum_{i=2}^{N}(1i)\in\mathbb{C}\mathfrak{S}_N$$

to the subspace $V\otimes\Lambda_{N-1}\subset V^{\otimes N}$. Here $(1i)\in\mathfrak{S}_N$ is the transposition of 1 and i. To prove the commutativity of (2.15) observe that π_N by definition maps the subring $\mathcal{F}\subset V\otimes\mathcal{F}$ to the subring (2.7), while θ_N is Λ_N-linear. Hence it suffices to chase the element $e_c\,v^k\otimes 1\in V\otimes\mathcal{F}$ the two ways offered by the diagram (2.15). But both ways yield the same result, the sum (2.2).

Let us now place two more commutative diagrams on the left of (2.15):

$$\begin{array}{ccccccc}
\mathcal{F} & \xrightarrow{\ \iota\ } & V\otimes\mathcal{F} & \xrightarrow{\ F_{ab}\ } & V\otimes\mathcal{F} & \xrightarrow{\ \theta\ } & \mathcal{F}\\
\downarrow & & \downarrow{\scriptstyle \pi_N} & & \downarrow{\scriptstyle \pi_N} & & \downarrow\\
\Lambda_N & \longrightarrow & V\otimes\Lambda_{N-1} & \xrightarrow[E_{ab}\otimes\mathrm{id}]{} & V\otimes\Lambda_{N-1} & \xrightarrow[\theta_N]{} & \Lambda_N
\end{array}$$

Here we have the diagram (2.8) in the middle. The leftmost vertical arrow is the homomorphism (2.1), the leftmost bottom arrow is the embedding (2.7).

Theorem 2.5 *The element $E_{ab}\in\mathfrak{gl}_n$ acts on $\mathcal{F}$ as the composition $\theta\,F_{ab}\,\iota$.*

Proof The composition $\theta_N\,(E_{ab}\otimes\mathrm{id})$ acts on the subspace (2.7) as the sum

$$\sum_{i=1}^{N}\mathrm{id}^{\otimes(i-1)}\otimes E_{ab}\otimes\mathrm{id}^{\otimes(N-i)}\,.$$

Hence the theorem follows from the commutativity of the latter diagram. □

Now consider the particular case when $a=b$. By (2.13) for $c=1,\ldots,n$ and $k=0,1,2,\ldots$ we have $F_{aa}\,(e\otimes p_{ck})=e_a\otimes p_{ck}$. More generally, for any $f\in\mathcal{F}$ we have $F_{aa}\,(e\otimes f)=e_a\otimes f$ because F_{aa} is an endomorphism of the ring $V\otimes\mathcal{F}$. By Theorem 2.5 and by definition of θ we get $E_{aa}(f)=p_{a0}\,f$.

3 Heckman Operators

Let α be a complex parameter. For $i=1,\ldots,N$ consider the *Dunkl operator*

$$Y_i=\alpha\,\partial_i+\sum_{j\neq i}\frac{1}{x_i-x_j}\,(1-\sigma_{ij})$$

acting on the ring of all polynomials in the variables $x_1, \ldots, x_N$ with complex coefficients. Here ∂_i is the derivation in this ring relative to the variable x_i, while σ_{ij} is the operator on this ring exchanging the variables x_i and x_j. Note that for any permutation σ of the variables $x_1, \ldots, x_N$ we have the relation

$$\sigma^{-1} Y_i \, \sigma = Y_{\sigma(i)} \,. \tag{3.1}$$

The operators Y_i with $i = 1, \ldots, N$ pairwise commute. This fact is well known, and goes back to the work [7]. Next consider the *Heckman operator* [9]

$$Z_i = x_i \, Y_i = \alpha \, x_i \, \partial_i + \sum_{j \neq i} \frac{x_i}{x_i - x_j} (1 - \sigma_{ij}) \,.$$

The operators Z_i with $i = 1, \ldots, N$ preserve the polynomial degree, but they do not commute if $N > 1$. However, they satisfy the commutation relations

$$[\, Z_i \, , Z_j \,] = \sigma_{ij} \, (Z_i - Z_j) \,. \tag{3.2}$$

Similarly to (3.1), for any permutation σ of the N variables we have

$$\sigma^{-1} Z_i \, \sigma = Z_{\sigma(i)} \,. \tag{3.3}$$

Therefore for every $m = 1, 2, \ldots$ the operator sum

$$H_m = Z_1^m + \ldots + Z_N^m \tag{3.4}$$

commutes with σ. Hence it preserves the space of symmetric polynomials in $x_1, \ldots, x_N$. The joint eigenvectors of operators (3.4) restricted to the latter space are the *Jack polynomials* [10] corresponding to the parameter α.

Let us now regard V as the tensor product $\mathbb{C}^n \otimes \mathbb{C}[v]$ of rings. Then we can identify the ring $V^{\otimes N}$ with the tensor product of $(\mathbb{C}^n)^{\otimes N}$ by the ring of polynomials in N variables with complex coefficients. The Heckman operators act on the latter ring, and we can now extend them to $V^{\otimes N}$ so that they act on $(\mathbb{C}^n)^{\otimes N}$ trivially. More explicitly, then x_i and ∂_i in Z_i become the operators

$$e_c \, v^k \mapsto e_c \, v^{k+1} \quad \text{and} \quad e_c \, v^k \mapsto k \, e_c \, v^{k-1} \tag{3.5}$$

respectively in the ith tensor factor of $V^{\otimes N}$. Note that then σ_{ij} in Z_i acts only on the variables v in the ith and jth tensor factors of $V^{\otimes N}$. This action differs from the permutational action of the transposition $(i\,j) \in \mathfrak{S}_N$ on the tensor product $V^{\otimes N}$ unless $n = 1$.

However, when regarded as an operator on $V^{\otimes N}$, every sum (3.4) still commutes with the permutational action of the group $\mathfrak{S}_N$. So the action of this sum on $V^{\otimes N}$ preserves the subspace Λ_N. In this section we will describe the limit of the action of the sum (3.4) on Λ_N at $N \to \infty$. This limit will be an operator I_m on the vector space $\mathcal{F}$ making commutative the square diagram

$$\begin{array}{ccc} \mathcal{F} & \xrightarrow{\ I_m\ } & \mathcal{F} \\ \downarrow & & \downarrow \\ \Lambda_N & \xrightarrow[\ H_m\ ]{} & \Lambda_N \end{array} \tag{3.6}$$

Note that the operator Z_1 on $V^{\otimes N}$ preserves the subspace $V \otimes \Lambda_{N-1}$. We will first describe the limit of the action of Z_1 on this subspace. That will be an operator Z on the vector space $V \otimes \mathcal{F}$ making commutative the diagram

$$\begin{array}{ccc} V \otimes \mathcal{F} & \xrightarrow{\ Z\ } & V \otimes \mathcal{F} \\ \pi_N \downarrow & & \downarrow \pi_N \\ V \otimes \Lambda_{N-1} & \xrightarrow[\ Z_1\ ]{} & V \otimes \Lambda_{N-1} \end{array} \tag{3.7}$$

In the case $n = 1$ the operator Z was determined in [17]. We will extend this result to any n. Let D_1 and W_1 be the operators on $V^{\otimes N}$ corresponding to

$$x_1 \partial_1 \quad \text{and} \quad \sum_{j \neq 1} \frac{x_1}{x_1 - x_j} (1 - \sigma_{1j}) \tag{3.8}$$

respectively. The latter two operators act on the polynomials in the variables $x_1, \ldots, x_N$ with complex coefficients. Then $Z_1 = \alpha D_1 + W_1$ as an operator on $V^{\otimes N}$. Note that both D_1 and W_1 preserve the subspace $V \otimes \Lambda_{N-1}$.

Now introduce an operator on the vector space $V \otimes \mathcal{F}$

$$D = v \, \partial \otimes \mathrm{id} + \sum_{d=1}^{n} \sum_{l=1}^{\infty} e_d \, v^l \otimes p_{dl}^{\perp} \tag{3.9}$$

where v and ∂ are the operators (3.5) on V, respectively, while $p_{dl}^{\perp}$ denotes the product of l by the derivation in the free commutative ring $\mathcal{F}$ relative to p_{dl}. We claim that commutative is the diagram obtained by replacing Z and Z_1 in (3.7) by D and D_1, respectively. To prove this claim, observe that the operator

$\upsilon\,\partial$ is a derivation of the ring V. So it suffices to show that the compositions $\pi_N\,D$ and $D_1\,\pi_N$ coincide on any generator of the ring $V \otimes \mathcal{F}$:

$$e_c\,\upsilon^k \otimes 1 \xmapsto[D]{} k\,e_c\,\upsilon^k \otimes 1 \xmapsto[\pi_N]{} k\,e_c\,\upsilon^k \otimes e^{\otimes(N-1)}\,,$$

$$e_c\,\upsilon^k \otimes 1 \xmapsto[\pi_N]{} e_c\,\upsilon^k \otimes e^{\otimes(N-1)} \xmapsto[D_1]{} k\,e_c\,\upsilon^k \otimes e^{\otimes(N-1)}\,;$$

$$e \otimes p_{ck} \xmapsto[D]{} k\,e_c\,\upsilon^k \otimes 1 \xmapsto[\pi_N]{} k\,e_c\,\upsilon^k \otimes e^{\otimes(N-1)}\,,$$

$$e \otimes p_{ck} \xmapsto[\pi_N]{} \sum_{i=1}^{N} e^{\otimes(i-1)} \otimes e_c\,\upsilon^k \otimes e^{\otimes(N-i)} \xmapsto[D_1]{} k\,e_c\,\upsilon^k \otimes e^{\otimes(N-1)}\,.$$

Consider W_1. For $j \neq 1$ let U_j be the operator on $V^{\otimes N}$ corresponding to the summand in (3.8) with index j. Then $W_1 = U_2 + \ldots + U_N$. Observe that the restriction of the operator W_1 to the subspace $V \otimes \Lambda_{N-1}$ coincides with that of the composition $(\mathrm{id} \otimes \theta_{N-1})\,U_2$. This is because for $j = 3, \ldots, N$ the conjugation of U_2 by the action of $(2\,j) \in \mathfrak{S}_N$ on $V \otimes \Lambda_{N-1}$ yields the operator U_j, while the action of $(2\,j)$ on this subspace is trivial.

Now consider the ring $V \otimes V \otimes \mathcal{F}$. It contains $V \otimes \mathcal{F}$ as a subring via the embedding $\mathrm{id} \otimes \iota$. In particular, it contains $\mathcal{F}$ via the natural mapping $f \mapsto e \otimes e \otimes f$ for every $f \in \mathcal{F}$. Let us extend (2.1) to a homomorphism

$$\rho_N : V \otimes V \otimes \mathcal{F} \to V \otimes V \otimes \Lambda_{N-2}$$

similarly to π_N. Namely, our ρ_N maps

$$e_c\,\upsilon^k \otimes e_d\,\upsilon^l \otimes 1 \mapsto e_c\,\upsilon^k \otimes e_d\,\upsilon^l \otimes e^{\otimes(N-2)} \tag{3.10}$$

and also maps $e \otimes e \otimes p_{ck}$ to the sum (2.2). We get a commutative diagram

$$\begin{array}{ccc} V \otimes \mathcal{F} & \xrightarrow{\ \mathrm{id} \otimes \iota\ } & V \otimes V \otimes \mathcal{F} \\ \downarrow{\scriptstyle \pi_N} & & \downarrow{\scriptstyle \rho_N} \\ V \otimes \Lambda_{N-1} & \longrightarrow & V \otimes V \otimes \Lambda_{N-2} \end{array} \tag{3.11}$$

where the bottom horizontal arrow represents the natural embedding.

Further let

$$\omega : V \otimes V \otimes \mathcal{F} \to V \otimes \mathcal{F}$$

be a linear map defined by the assignment

$$e_c\, v^k \otimes e_d\, v^l \otimes f \;\mapsto\; (e_c\, v^k \otimes f)\,(e \otimes p_{dl} - e_d\, v^l \otimes 1)$$

for every $f \in \mathcal{F}$. The map ω is different from the more straightforward map

$$\mathrm{id} \otimes \theta : V \otimes V \otimes \mathcal{F} \to V \otimes \mathcal{F}\,.$$

Under the latter

$$e_c\, v^k \otimes e_d\, v^l \otimes f \;\mapsto\; e_c\, v^k \otimes p_{dl}\, f\,.$$

Later on we will also use the map $\mathrm{id} \otimes \theta$ due to the *equalizing property* below.

Lemma 3.1 *The action of ω and $\mathrm{id} \otimes \theta$ is the same on any element of the ring $V \otimes V \otimes \mathcal{F}$ divisible by $e_c \otimes e \otimes 1 - e \otimes e_c \otimes 1$ for some index c.*

Proof Any element of $V \otimes V \otimes \mathcal{F}$ is a linear combination of tensor products $e_a\, v^r \otimes e_b\, v^s \otimes f$ where $a, b = 1, \ldots, n$ and $r, s = 0, 1, 2, \ldots$ and $f \in \mathcal{F}$. Take

$$(e_c \otimes e \otimes 1 - e \otimes e_c \otimes 1)\,(e_a\, v^r \otimes e_b\, v^s \otimes f) =$$

$$\delta_{ac}\, e_a\, v^r \otimes e_b\, v^s \otimes f - e_a\, v^r \otimes \delta_{bc}\, e_b\, v^s \otimes f\,.$$

By applying the difference of maps $\mathrm{id} \otimes \theta - \omega$ to the last displayed line we get

$$\delta_{ab}\, \delta_{ac}\, v^{r+s} \otimes f - \delta_{ab}\, \delta_{bc}\, e_a\, v^{r+s} \otimes f = 0\,.$$

However, it is the map ω that makes commutative the diagram

$$\begin{array}{ccc} V \otimes V \otimes \mathcal{F} & \xrightarrow{\quad\omega\quad} & V \otimes \mathcal{F} \\ \downarrow{\scriptstyle \rho_N} & & \downarrow{\scriptstyle \pi_N} \\ V \otimes V \otimes \Lambda_{N-2} & \xrightarrow[\mathrm{id} \otimes \theta_{N-1}]{} & V \otimes \Lambda_{N-1} \end{array} \tag{3.12}$$

To prove the commutativity of (3.12) observe that π_N and ρ_N map $\mathcal{F}$, as a subring of, respectively, $V \otimes \mathcal{F}$ and $V \otimes V \otimes \mathcal{F}$, to the ring Λ_N. But the map ω is $\mathcal{F}$-linear, while the map $\mathrm{id} \otimes \theta_{N-1}$ is Λ_N-linear. The maps at all four sides of the diagram (3.12) also commute with multiplication by the elements of V in the first tensor factor of their source and target vector spaces. So it suffices to chase the element $e \otimes e_c\, v^k \otimes 1 \in V \otimes V \otimes \mathcal{F}$ the two ways offered by the diagram (3.12). Both ways yield the same result, which is the sum (2.9).

We will employ the operator U on the vector space $V \otimes V \otimes \mathcal{F}$ making commutative the diagram

$$
\begin{array}{ccc}
V \otimes V \otimes \mathcal{F} & \xrightarrow{\quad U \quad} & V \otimes V \otimes \mathcal{F} \\
\Big\downarrow{\scriptstyle \rho_N} & & \Big\downarrow{\scriptstyle \rho_N} \\
V \otimes V \otimes \Lambda_{N-2} & \xrightarrow[\quad U_2 \quad]{} & V \otimes V \otimes \Lambda_{N-2}
\end{array}
\tag{3.13}
$$

Namely, we will set

$$W = \omega\, U\, (\,\mathrm{id} \otimes \iota\,)\,. \tag{3.14}$$

Then commutative will be the diagram, obtained by replacing Z and Z_1 in (3.7) by W and W_1, respectively. To prove this claim, it suffices to place the diagrams (3.11) and (3.12), respectively on the left and on the right of (3.13). It will then follow that $Z = \alpha\, D + W$ makes commutative the diagram (3.7).

Similarly to π_N' let us introduce another ring homomorphism

$$\rho_N' : V \otimes V \otimes \mathcal{F} \to V \otimes V \otimes \Lambda_{N-2}$$

such that ρ_N' will map the element $e \otimes e \otimes p_{ck} \in V \otimes V \otimes \mathcal{F}$ to the sum

$$\sum_{i=3}^{N} e^{\otimes (i-1)} \otimes e_c\, v^k \otimes e^{\otimes (N-i)}$$

instead of (2.2). The homomorphism ρ_N' will still map (3.10) as ρ_N does. So

$$
\begin{aligned}
\rho_N\,(e \otimes e \otimes p_{ck}) &= \rho_N'\,(e \otimes e \otimes p_{ck} + e_c\, v^k \otimes e \otimes 1 + e \otimes e_c\, v^k \otimes 1\,)\,, \\
\rho_N'\,(e \otimes e \otimes p_{ck}) &= \rho_N\,(e \otimes e \otimes p_{ck} - e_c\, v^k \otimes e \otimes 1 - e \otimes e_c\, v^k \otimes 1\,)\,.
\end{aligned}
$$

For short let x and y denote the operators of multiplication by v, respectively, in the first and the second tensor factors of $V \otimes V \otimes \mathcal{F}$. Let τ be operator on $V \otimes V \otimes \mathcal{F}$ exchanging the variables v in these two tensor factors. By the definition of ρ_N' we immediately obtain commutativity of the diagram

$$
\begin{array}{ccc}
V \otimes V \otimes \mathcal{F} & \xrightarrow{\quad \frac{x}{x-y}(1-\tau) \quad} & V \otimes V \otimes \mathcal{F} \\
\Big\downarrow{\scriptstyle \rho_N'} & & \Big\downarrow{\scriptstyle \rho_N'} \\
V \otimes V \otimes \Lambda_{N-2} & \xrightarrow[\quad U_2 \quad]{} & V \otimes V \otimes \Lambda_{N-2}
\end{array}
\tag{3.15}
$$

For the purpose of determining the operator W on $V \otimes \mathcal{F}$ via (3.14) it suffices to find the action of U on the image of $\mathrm{id} \otimes \iota$, that is on the subspace

$$V \otimes e \otimes \mathcal{F} \subset V \otimes V \otimes \mathcal{F}.$$

Furthermore, the maps ρ_N and U_2 commute with multiplication by elements of the subspace $\mathbb{C}^n \subset V$ in the first tensor factors of their source and target vector spaces. Hence the operator U will have the same commuting property. Therefore it suffices to find for $l = 0, 1, 2, \ldots$ the action of U on the elements

$$e\,v^l \otimes e \otimes \prod_{(c,k)\in\mathcal{P}} p_{ck} = x^l \prod_{(c,k)\in\mathcal{P}} (e \otimes e \otimes p_{ck}) \tag{3.16}$$

where $\mathcal{P}$ is any finite collection of pairs of $c = 1, \ldots, n$ and $k = 0, 1, 2, \ldots$. This collection is unordered, but may contain same pairs with multiplicity. By the commutativity of the diagrams (3.13) and (3.15) we have

$$\rho_N\, U\Big(x^l \prod_{(c,k)\in\mathcal{P}} (e \otimes e \otimes p_{ck})\Big) = U_2\, \rho_N \Big(x^l \prod_{(c,k)\in\mathcal{P}} (e \otimes e \otimes p_{ck})\Big) =$$

$$U_2\, \rho_N' \Big(x^l \prod_{(c,k)\in\mathcal{P}} (e \otimes e \otimes p_{ck} + e_c\, v^k \otimes e \otimes 1 + e \otimes e_c\, v^k \otimes 1)\Big) =$$

$$\rho_N' \Big(\frac{x}{x-y}\Big(x^l \prod_{(c,k)\in\mathcal{P}} (e \otimes e \otimes p_{ck} + e_c\, v^k \otimes e \otimes 1 + e \otimes e_c\, v^k \otimes 1)$$

$$-\, y^l \prod_{(c,k)\in\mathcal{P}} (e \otimes e \otimes p_{ck} + e_c \otimes e\, v^k \otimes 1 + e\, v^k \otimes e_c \otimes 1)\Big)\Big) =$$

$$\rho_N \Big(\frac{x}{x-y}\Big(x^l \prod_{(c,k)\in\mathcal{P}} (e \otimes e \otimes p_{ck}) - y^l \prod_{(c,k)\in\mathcal{P}} (e \otimes e \otimes p_{ck} +$$

$$e_c \otimes e\, v^k \otimes 1 + e\, v^k \otimes e_c \otimes 1 - e_c\, v^k \otimes e \otimes 1 - e \otimes e_c\, v^k \otimes 1)\Big)\Big) =$$

$$\rho_N \Big(\frac{x}{x-y}\Big(x^l \prod_{(c,k)\in\mathcal{P}} (e \otimes e \otimes p_{ck}) - y^l \prod_{(c,k)\in\mathcal{P}} (e \otimes e \otimes p_{ck} +$$

$$(y^k - x^k)(e_c \otimes e \otimes 1 - e \otimes e_c \otimes 1))\Big)\Big).$$

This calculation shows that the operator U maps the element (3.16) to

$$\frac{x}{x-y}\Big(x^l \prod_{(c,k)\in\mathcal{P}} (e \otimes e \otimes p_{ck}) - y^l \prod_{(c,k)\in\mathcal{P}} (e \otimes e \otimes p_{ck} +$$

$$(y^k - x^k)(e_c \otimes e \otimes 1 - e \otimes e_c \otimes 1))\Big). \tag{3.17}$$

Let us apply ω to the latter element. By applying the difference $\mathrm{id} \otimes \theta - \omega$ to

$$\frac{x}{x-y}\Big(x^l \prod_{(c,k)\in\mathcal{P}} (e \otimes e \otimes p_{ck}) - y^l \prod_{(c,k)\in\mathcal{P}} (e \otimes e \otimes p_{ck}) \Big) = \frac{x\,(x^l - y^l)}{x-y} \prod_{(c,k)\in\mathcal{P}} (e \otimes e \otimes p_{ck})$$

we get the element

$$e\, v^l \otimes \prod_{(c,k)\in\mathcal{P}} p_{ck} \in V \otimes \mathcal{F} \tag{3.18}$$

multiplied by l. This multiplication by l amounts to applying to (3.18) the operator $v\,\partial \otimes \mathrm{id}$. The element (3.16) is just the image of (3.18) under $\mathrm{id} \otimes \iota$. Now by repeatedly using Lemma 3.1 we get the operator equality on $V \otimes \mathcal{F}$

$$\omega\, U\,(\mathrm{id} \otimes \iota) = (\mathrm{id} \otimes \theta)\, U\,(\mathrm{id} \otimes \iota) - v\,\partial \otimes \mathrm{id}\,. \tag{3.19}$$

Here we also used the fact that the map ω commutes with multiplication by elements of the subspace $\mathbb{C}^n \subset V$ in the first tensor factor of its source and target vector spaces, like the operator U does.

Now let $p^*_{dl} = \alpha\, p^{\perp}_{dl}$, that is the product of $\alpha\, l$ by the derivation in $\mathcal{F}$ relative to the generator p_{dl}. Then we can recall that our $Z = \alpha\, D + W$ and combine (3.9),(3.14) and (3.19) to get the following principal result.

Theorem 3.2 *The diagram* (3.7) *is made commutative by the operator*

$$Z = (\alpha - 1)\, v\,\partial \otimes \mathrm{id} + \sum_{d=1}^{n} \sum_{l=1}^{\infty} e_d\, v^l \otimes p^*_{dl} + (\mathrm{id} \otimes \theta)\, U\,(\mathrm{id} \otimes \iota)$$

where ι and θ are defined by (2.5) *and* (2.14). *The operator U on $V \otimes V \otimes \mathcal{F}$ commutes with multiplication by elements of the subspace $\mathbb{C}^n \subset V$ in the first tensor factor, and maps* (3.16) *to the element displayed in two lines* (3.17).

Corollary 3.3 *For $m = 1, 2, \ldots$ the diagram* (3.6) *is made commutative by*

$$I_m = \theta\, Z^m\, \iota\,.$$

Proof For any $i = 2, \ldots, N$ the conjugation of the operator Z_1^m by the action of $(1i) \in \mathfrak{S}_N$ on $V^{\otimes N}$ yields the operator Z_i^m. Therefore the composition $\theta_N\, Z_1^m$ acts on the subspace $\Lambda_N \subset V^{\otimes N}$ as the operator sum (3.4). Now the required statement follows from the commutativity of the composite diagram

$$
\begin{array}{ccccccc}
\mathcal{F} & \xrightarrow{\ \iota\ } & V \otimes \mathcal{F} & \xrightarrow{\ Z^m\ } & V \otimes \mathcal{F} & \xrightarrow{\ \theta\ } & \mathcal{F} \\
\downarrow & & \downarrow{\scriptstyle \pi_N} & & \downarrow{\scriptstyle \pi_N} & & \downarrow \\
\Lambda_N & \longrightarrow & V \otimes \Lambda_{N-1} & \xrightarrow[\ Z_1^m\ ]{} & V \otimes \Lambda_{N-1} & \xrightarrow[\ \theta_N\ ]{} & \Lambda_N
\end{array}
$$

Here we use the commutativity of the diagrams (2.15) and (3.7). □

4 Yangian Action

Consider the *Yangian* $\mathrm{Y}(\mathfrak{gl}_n)$. This is a complex unital associative algebra with an infinite family of generators $T_{ab}^{(1)}, T_{ab}^{(2)}, \ldots$ where $a, b = 1, \ldots, n$. Now let u be another variable. Introduce the formal power series in u^{-1}

$$T_{ab}(u) = \delta_{ab} + T_{ab}^{(1)} u^{-1} + T_{ab}^{(2)} u^{-2} + \ldots \tag{4.1}$$

with the coefficients in $\mathrm{Y}(\mathfrak{gl}_n)$. Using both the variables u and v, the defining relations in the algebra $\mathrm{Y}(\mathfrak{gl}_n)$ can be written as

$$(u - v)\,[\, T_{ab}(u), T_{cd}(v)\,] = T_{cb}(u) T_{ad}(v) - T_{cb}(v) T_{ad}(u)\,. \tag{4.2}$$

If $n = 1$ then the algebra $\mathrm{Y}(\mathfrak{gl}_n)$ is commutative by this definition. The next proposition is a particular case of a general construction due to Drinfeld [6].

Proposition 4.1 *The algebra* $\mathrm{Y}(\mathfrak{gl}_n)$ *acts on vector space* Λ_N *so that* $T_{ab}^{(m+1)}$ *with* $m = 0, 1, 2, \ldots$ *acts as the operator sum*

$$\sum_{i=1}^{N} \mathrm{id}^{\otimes (i-1)} \otimes E_{ab} \otimes \mathrm{id}^{\otimes (N-i)} \cdot (-Z_i)^m\,. \tag{4.3}$$

Note that the operator Z_i on $V^{\otimes N}$ by its definition commutes with the action of the Lie algebra $\mathfrak{gl}_n$ on any of the N tensor factors V. Further, due to the relations (3.3) the operator (4.3) commutes with the permutational action of the group $\mathfrak{S}_N$ on $V^{\otimes N}$. So the operator (4.3) preserves the subspace Λ_N. To prove Proposition 4.1

it now remains to verify that the restrictions of these operators to Λ_N satisfy the relations (4.2). To this end one employs the series

$$\delta_{ab} + \sum_{m=0}^{\infty} \sum_{i=1}^{N} \mathrm{id}^{\otimes(i-1)} \otimes E_{ab} \otimes \mathrm{id}^{\otimes(N-i)} \cdot (-Z_i)^m \, u^{-m-1} =$$

$$\delta_{ab} + \sum_{i=1}^{N} \mathrm{id}^{\otimes(i-1)} \otimes E_{ab} \otimes \mathrm{id}^{\otimes(N-i)} \cdot (u + Z_i)^{-1}$$

with operator coefficients (4.3) and applies the commutation relations (3.2). For the details of the verification of (4.2) see [12, Section 1].

Note that for any fixed $m = 0, 1, 2, \ldots$ the operator (3.4) on $V^{\otimes N}$ equals $(-1)^m$ times the sum of operators (4.3) over $a = b = 1, \ldots, n$. By using the results of the previous sections, we can now describe the limit of the action of $\mathrm{Y}(\mathfrak{gl}_n)$ on Λ_N defined in Proposition 4.1 at $N \to \infty$. This limit is an action of the algebra $\mathrm{Y}(\mathfrak{gl}_n)$ on the Fock space $\mathcal{F}$ determined by the next theorem.

Theorem 4.2 *The algebra* $\mathrm{Y}(\mathfrak{gl}_n)$ *acts on the vector space* $\mathcal{F}$ *so that* $T_{ab}^{(m+1)}$ *with* $m = 0, 1, 2, \ldots$ *acts as the composition* $\theta \, (-Z)^m \, F_{ab} \, \iota$ *.*

Proof The composition $\theta_N \, (E_{ab} \otimes \mathrm{id}) \, (-Z_1)^m$ acts on the subspace (2.7) as the operator sum (4.3). Hence the theorem follows from Proposition 4.1 by using the commutativity of the diagrams (2.8), (2.15) and (3.7). □

Other limits at $N \to \infty$ of the operators $E_{ab} \otimes \mathrm{id}$ and Z_1 on $V \otimes \Lambda_{N-1}$ were computed in [11]. Comparing our Theorems 2.5 and 3.2 with the results of [11] shows that these limits were defined by the homomorphism π_N' instead of π_N used in (3.7). This however entails changing our ι to the homomorphism

$$\iota' : \mathcal{F} \to V \otimes \mathcal{F} : \, p_{ck} \mapsto e \otimes p_{ck} + e_c \, v^k \otimes 1 \, .$$

Further, once π_N is changed to π_N' in (2.15), our linear map θ also needs to be changed, to keep the latter diagram commutative. The changed linear map

$$\theta' : \, (e_d \, v^l \otimes 1) \prod_{(c,k) \in \mathcal{P}} (e \otimes p_{ck} + e_c \, v^k \otimes 1) \mapsto \, p_{dl} \prod_{(c,k) \in \mathcal{P}} p_{ck}$$

for any pair (d, l) and for any collection $\mathcal{P}$ of pairs (c, k) as in (3.16) above.

Indeed, after receiving a preliminary version of the present article which included the above remark, Sergey Khoroshkin verified that the counterparts from [11] of our operators F_{ab} and Z on $V \otimes \mathcal{F}$ can be rewritten as

$$F_{ab}' = \varepsilon \, F_{ab} \, \varepsilon^{-1} \quad \text{and} \quad Z' = \varepsilon \, Z \, \varepsilon^{-1}$$

where ε is the ring automorphism of $V \otimes \mathcal{F}$ identical on $V \otimes 1$ such that

$$\varepsilon : e \otimes p_{ck} \mapsto e \otimes p_{ck} + e_c \, v^k \otimes 1 \, .$$

Since $\iota' = \varepsilon\,\iota$ and $\theta' = \theta\,\varepsilon^{-1}$ by the definition of the automorphism ε, then for any $m = 0, 1, 2, \ldots$ we get the equalities of operators on $\mathcal{F}$

$$\theta'(Z')^m \, \iota' = \theta \, Z^m \, \iota$$

and

$$\theta'(Z')^m \, F'_{ab} \, \iota' = \theta \, Z^m \, F_{ab} \, \iota \, .$$

By Corollary 3.3 and by Theorem 4.2, these equalities show that the limits at $N \to \infty$ of the operators H_m on Λ_N, and of the action of the algebra $\mathrm{Y}(\mathfrak{gl}_n)$ on Λ_N, are the same in [11] as in the present article. This should be the case, because the mapping (2.1) which defined the limits in [11] is the same as ours.

Acknowledgement I thank Sergey Khoroshkin and Evgeny Sklyanin for helpful conversations. The present work has been supported by the EPSRC grant N023919.

References

1. H. Awata, Y. Matsuo and T. Yamamoto, *Collective field description of spin Calogero-Sutherland models*, J. Phys. **A29** (1996), 3089–3098.
2. D. Bernard, M. Gaudin, F. D. M. Haldane and V. Pasquier, *Yang-Baxter equation in long-range interacting systems*, J. Phys. **A26** (1993), 5219–5236.
3. F. Calogero, *Ground state of a one-dimensional N-body system*, J. Math. Phys. **10** (1969), 2197–2200.
4. I. Cherednik, *A unification of Knizhnik-Zamolodchikov and Dunkl operators via affine Hecke algebras*, Invent. Math. **106** (1991), 411–431.
5. A. Debiard, *Polynômes de Tchébychev et de Jacobi dans un espace euclidien de dimension p*, C. R. Acad. Sc. Paris **I296** (1983), 529–532.
6. V. G. Drinfeld, *Degenerate affine Hecke algebras and Yangians*, Funct. Analysis Appl. **20** (1986), 56–58.
7. C. F. Dunkl, *Differential-difference operators associated to reflection groups*, Trans. Amer. Math. Soc. **311** (1989), 167–183.
8. Z. N. C. Ha and F. D. M. Haldane, *Models with inverse-square exchange*, Phys. Rev. **B46** (1992), 9359–9368.
9. G. J. Heckman, *An elementary approach to the hypergeometric shift operators of Opdam*, Invent. Math. **103** (1991), 341–350.
10. H. Jack, *A class of symmetric polynomials with a parameter*, Proc. Royal Soc. Edinburgh **A69** (1970), 1–18.
11. S. M. Khoroshkin, M. G. Matushko and E. K. Sklyanin, *On spin Calogero-Moser system at infinity*, J. Phys. **A50** (2017), 115203.
12. S. Khoroshkin and M. Nazarov, *Yangians and Mickelsson algebras I*, Transformation Groups **11** (2006), 625–658.

13. S. Khoroshkin and M. Nazarov, *On the functor of Arakawa, Suzuki and Tsuchiya*, Adv. Stud. Pure Math. **76** (2018), 275–302.
14. M. L. Nazarov and E. K. Sklyanin, *Sekiguchi-Debiard operators at infinity*, Commun. Math. Phys. **324** (2013), 831–849.
15. M. L. Nazarov and E. K. Sklyanin, *Integrable hierarchy of the quantum Benjamin-Ono equation*, Symmetry, Integrability and Geometry: Methods and Applications **9** (2013), 078.
16. J. Sekiguchi, *Zonal spherical functions on some symmetric spaces*, Publ. Res. Inst. Math. Sci. **12** (1977), 455–459.
17. A. N. Sergeev and A. P. Veselov, *Dunkl operators at infinity and Calogero-Moser systems*, Int. Math. Res. Notices (2015), 10959–10986.
18. R. Stanley, *Some combinatorial properties of Jack symmetric functions*, Adv. Math. **77** (1989), 76–115.
19. B. Sutherland, *Exact results for a quantum many-body problem in one dimension II*, Phys. Rev. **A5** (1972), 1372–1376.
20. K. Takemura and D. Uglov, *The orthogonal eigenbasis and norms of eigenvectors in the spin Calogero-Sutherland model*, J. Phys. **A30** (1997), 3685–3717.

Semi-Direct Products Involving Sp_{2n} or $Spin_n$ with Free Algebras of Symmetric Invariants

Dmitri I. Panyushev and Oksana S. Yakimova

Dedicated to A. Joseph on the occasion of his 75th birthday

Abstract This is a part of an ongoing project, the goal of which is to classify all semi-direct products $\mathfrak{s} = \mathfrak{g} \ltimes V$ such that $\mathfrak{g}$ is a simple Lie algebra, V is a $\mathfrak{g}$-module, and $\mathfrak{s}$ has a free algebra of symmetric invariants. In this paper, we obtain such a classification for the representations of the orthogonal and symplectic algebras.

MSC: 17B63, 14L30, 17B20, 22E46

1 Introduction

Let $\Bbbk$ be a field with $\operatorname{char}\Bbbk = 0$. Let S be an algebraic group defined over $\Bbbk$ with $\mathfrak{s} = \mathsf{Lie}\, S$. The invariants of S in the symmetric algebra $\mathcal{S}(\mathfrak{s}) = \Bbbk[\mathfrak{s}^*]$ of $\mathfrak{s}$ (= the symmetric invariants of $\mathfrak{s}$ or of S) are denoted by $\Bbbk[\mathfrak{s}^*]^S$ or $\mathcal{S}(\mathfrak{s})^S$. If S is connected, then we also write $\Bbbk[\mathfrak{s}^*]^{\mathfrak{s}}$ or $\mathcal{S}(\mathfrak{s})^{\mathfrak{s}}$ for them.

Let $\mathfrak{g}$ be a reductive Lie algebra. Symmetric invariants of $\mathfrak{g}$ over $\bar{\Bbbk}$ belong to the classical area of Representation Theory and Invariant Theory, where the most striking and influential results were obtained by Chevalley and Kostant in the 1950s and 1960s. Then pioneering insights of Kostant and Joseph revealed that the symmetric invariants of certain non-reductive subalgebras of $\mathfrak{g}$ can explicitly

D. I. Panyushev (✉)
Institute for Information Transmission Problems of the R.A.S, Moscow, Russia
e-mail: panyushev@iitp.ru

O. S. Yakimova
Universität zu Köln Mathematisches Institut, Köln, Deutschland
e-mail: yakimova.oksana@uni-koeln.de

M. Gorelik et al. (eds.), *Representations and Nilpotent Orbits of Lie Algebraic Systems*, Progress in Mathematics 330,
https://doi.org/10.1007/978-3-030-23531-4_12

be described and that they are very helpful for understanding representations of $\mathfrak{g}$ itself, see [5, 7, 9]. This has opened a brave new world, full of adventures and hidden treasures. Hopefully, we have found (and presented here) some of them.

Although the study of $\mathcal{S}(\mathfrak{s})^S$ is hopeless in general, there are several classes of non-reductive algebras that are still tractable. One of them is obtained via a semi-direct product construction from finite-dimensional representations of reductive groups, which is the main topic of this article, see Sect. 3 below. Another interesting class of non-reductive algebras consists of truncated biparabolic subalgebras [6], see also [3] and the references therein. Yet another class consists of the centralisers of nilpotent elements of $\mathfrak{g}$, see [12]. Remarkably, some truncated bi-parabolic subalgebras or centralisers occur also as semi-direct products.

In [22], the following problem has been proposed:

To classify the representations V of ***simple*** *algebraic groups G with* $\mathsf{Lie}\, G = \mathfrak{g}$ *such that the ring of symmetric invariants of the semi-direct product* $\mathfrak{s} = \mathfrak{g} \ltimes V$ *is polynomial.*

It is easily seen that if $\mathfrak{s}$ has this property, then $\Bbbk[V^*]^G$ is also a polynomial ring. (But not vice versa!) Therefore, the suitable representations (G, V) are contained in the list of "coregular representations" of simple algebraic groups, see [1, 17]. If a generic stabiliser for (G, V) is trivial, then $\Bbbk[\mathfrak{s}^*]^S \simeq \Bbbk[V^*]^G$. Therefore, it suffices to handle only "coregular representations" with non-trivial generic stabilisers. The latter can be determined with the help of Elashvili's tables [2]. As it should have been expected, type A is the most difficult case. The solution for just one particular item, $V = m(\mathbb{C}^n)^* \oplus k\mathbb{C}^n$ for $G = \mathrm{SL}_n$, occupies the whole paper [23]. This certainly means that obtaining classification in the SL_n-case requires considerable effort. Although the results of [23] are formulated over $\mathbb{C}$, we notice that they are actually valid over an arbitrary field of characteristic zero. The case of exceptional groups G bas been considered in [14]. The next logical step is to look at the symplectic and orthogonal groups G, which is done in this paper. To a great extent, our classification results rely on the theory developed by the second author in [22].

Let us give a brief outline of the paper. In Sect. 2, we gather some properties of the arbitrary coadjoint representations, whereas in Sect. 3, we stick to the coadjoint representations of semi-direct products and describe our classification techniques. After a brief interlude in Sect. 4 devoted to an example in type A, we dwell upon the classification of the suitable representations V of the orthogonal (Sect. 5) and symplectic (Sect. 6) groups. Our results are summarised in Theorem 3.13 and Tables 1, 2. We are taking a somewhat unusual approach towards a classification and trying to present the essential ideas for **all** pairs (G, V) under consideration. Many pairs can be handled using general theorems presented in Sect. 3, but some others require lengthy elaborated ad hoc considerations, see e.g. Theorem 5.13. It appears *a posteriori* that, for **all** representations V of $G = \mathrm{Sp}_{2n}$ with polynomial ring $\Bbbk[V^*]^{\mathrm{Sp}_{2n}}$, the algebra of symmetric invariants $\mathcal{S}(\mathfrak{s})^S$ is also polynomial. In most of the $\mathfrak{sp}_{2n}$-cases, we explicitly describe the basic invariants. There is an interesting connection with the invariants of certain centralisers. In particular, if $V = \Bbbk^{2n}$ is the standard (defining) representations of Sp_{2n}, then there is a kind of matryoshka-like structure between the invariants of the semi-direct product and the symmetric invariants of the centraliser of the minimal nilpotent orbit in $\mathfrak{sp}_{2n-2}$.

Notation Let an algebraic group Q act on an irreducible affine variety X. Then $\Bbbk[X]^Q$ stands for the algebra of Q-invariant regular functions on X and $\Bbbk(X)^Q$ is the field of Q-invariant rational functions. If $\Bbbk[X]^Q$ is finitely generated, then $X/\!\!/Q := \mathsf{Spec}\,\Bbbk[X]^Q$. Whenever $\Bbbk[X]^Q$ is a graded polynomial ring, the elements of any set of algebraically independent homogeneous generators will be referred to as *basic invariants*. If V is a Q-module and $v \in V$, then $\mathfrak{q}_v = \{\xi \in \mathfrak{q} \mid \xi{\cdot}v = 0\}$ is the *stabiliser* of v in $\mathfrak{q}$ and $Q_v = \{g \in Q \mid g{\cdot}v = v\}$ is the *isotropy group* of v in Q.

Let X be an irreducible variety (e.g. a vector space). We say that a property holds for "generic $x \in X$" if that property holds for all points of an open subset of X. An open subset is said to be *big*, if its complement does not contain divisors.

Write $\mathfrak{heis}_n$, $n \geqslant 0$, for the Heisenberg Lie algebra of dimension $2n+1$.

2 Preliminaries on the Coadjoint Representations

Let Q be a connected algebraic group and $\mathfrak{q} = \mathsf{Lie}\, Q$. The *index* of $\mathfrak{q}$ is

$$\operatorname{ind}\mathfrak{q} = \min_{\gamma\in\mathfrak{q}^*} \dim \mathfrak{q}_\gamma,$$

where $\mathfrak{q}_\gamma$ is the stabiliser of γ in $\mathfrak{q}$. In view of Rosenlicht's theorem [20, § 2.3], $\operatorname{ind}\mathfrak{q} = \operatorname{tr.deg}\Bbbk(\mathfrak{q}^*)^Q$. If $\operatorname{ind}\mathfrak{q} = 0$, then $\Bbbk[\mathfrak{q}^*]^Q = \Bbbk$. For a reductive $\mathfrak{g}$, one has $\operatorname{ind}\mathfrak{g} = \operatorname{rk}\mathfrak{g}$. In this case, $(\dim\mathfrak{g} + \operatorname{rk}\mathfrak{g})/2$ is the dimension of a Borel subalgebra of $\mathfrak{g}$. For an arbitrary $\mathfrak{q}$, set $\boldsymbol{b}(\mathfrak{q}) := (\operatorname{ind}\mathfrak{q} + \dim\mathfrak{q})/2$.

One defines the *singular set* $\mathfrak{q}^*_{\mathsf{sing}}$ of $\mathfrak{q}^*$ by

$$\mathfrak{q}^*_{\mathsf{sing}} = \{\gamma \in \mathfrak{q}^* \mid \dim\mathfrak{q}_\gamma > \operatorname{ind}\mathfrak{q}\}.$$

Set also $\mathfrak{q}^*_{\mathsf{reg}} := \mathfrak{q}^* \setminus \mathfrak{q}^*_{\mathsf{sing}}$. Further, $\mathfrak{q}$ is said to have the "codim–2" property (= to satisfy the "codim–2" condition), if $\dim\mathfrak{q}^*_{\mathsf{sing}} \leqslant \dim\mathfrak{q} - 2$. We say that $\mathfrak{q}$ satisfies *the Kostant regularity criterion* (= ***KRC***) if the following properties hold for $\mathcal{S}(\mathfrak{q})^Q$ and $\xi \in \mathfrak{g}^*$:

- $\mathcal{S}(\mathfrak{q})^Q = \Bbbk[f_1,\dots,f_l]$ is a graded polynomial ring (with basic invariants $f_1,\dots,f_l$);
- $\xi \in \mathfrak{q}^*_{\mathsf{reg}}$ if and only if $(df_1)_\xi,\dots,(df_l)_\xi$ are linearly independent.

Every reductive Lie algebra has the "codim–2" property and satisfies ***KRC***.

Observe that $(df)_\xi \in \mathfrak{q}_\xi$ for each $f \in \Bbbk[\mathfrak{q}^*]^Q$.

Theorem 2.1 (cf. [11, Theorem 1.2]) *If $\mathfrak{q}$ has the "codim–2" property, $\operatorname{tr.deg}\mathcal{S}(\mathfrak{q})^Q = \operatorname{ind}\mathfrak{q} = l$, and there are algebraically independent $f_1,\dots,f_l \in \mathcal{S}(\mathfrak{q})^Q$ such that $\sum_{i=1}^{l}\deg f_i = \boldsymbol{b}(\mathfrak{q})$, then $f_1,\dots,f_l$ freely generate $\mathcal{S}(\mathfrak{q})^Q$ and the* ***KRC*** *holds for $\mathfrak{q}$.*

Suppose that Q acts on an affine variety X. Then $f \in \Bbbk[X]$ is a *semi-invariant* of Q if $g{\cdot}f \in \Bbbk f$ for each $g \in Q$. A semi-invariant is said to be *proper* if it is not an invariant. If Q has no non-trivial characters (all 1-dimensional representations of Q are trivial), then it has no proper semi-invariants. In particular, if Q is a semi-direct product of a semisimple and a unipotent group, then all its semi-invariants are invariants. We record a well-known observation:

- if Q has no proper semi-invariants in $\mathcal{S}(\mathfrak{q})$, then $\Bbbk(\mathfrak{q}^*)^Q = \mathrm{Quot}\,(\Bbbk[\mathfrak{q}^*]^Q)$ and hence $\mathrm{tr.deg}\,\Bbbk[\mathfrak{q}^*]^Q = \mathrm{ind}\,\mathfrak{q}$.

Theorem 2.2 (cf. [8, Prop. 5.2]) *Suppose that Q has no proper semi-invariants in $\mathcal{S}(\mathfrak{q})$ and $\mathcal{S}(\mathfrak{q})^Q$ is freely generated by $f_1, \ldots, f_l$. Then the differentials $df_1, \ldots, df_l$ are linearly independent on a big open subset of $\mathfrak{q}^*$.*

For any Lie algebra $\mathfrak{q}$ defined over $\Bbbk$, set $\mathfrak{q}_{\overline{\Bbbk}} := \mathfrak{q}\otimes_{\Bbbk}\overline{\Bbbk}$. Then $\mathcal{S}(\mathfrak{q}_{\overline{\Bbbk}})^{\mathfrak{q}_{\overline{\Bbbk}}} = \mathcal{S}(\mathfrak{q})^{\mathfrak{q}}\otimes_{\Bbbk}\overline{\Bbbk}$. If we extend the field, then a set of the generating invariants over $\Bbbk$ is again a set of the generating invariants over $\overline{\Bbbk}$. In the other direction, having a minimal set $\mathcal{M}$ of homogeneous generators over $\overline{\Bbbk}$, any $\Bbbk$-basis of $\langle\mathcal{M}\rangle_{\overline{\Bbbk}} \cap \mathcal{S}(\mathfrak{q})$ is a minimal set of generators over $\Bbbk$. The properties like "being a polynomial ring" do not change under field extensions. The results in this paper are valid over fields that are not algebraically closed, but in the proofs we may safely assume that $\Bbbk = \overline{\Bbbk}$.

3 On the Coadjoint Representations of a Semi-Direct Product

For semi-direct products, there are some specific approaches to the symmetric invariants. Our convention is that G is always a connected **reductive** group and $\mathfrak{g} = \mathrm{Lie}\, G$, whereas a group Q is not necessarily reductive and $\mathfrak{q} = \mathrm{Lie}\, Q$. In this section, either $\mathfrak{s} = \mathfrak{g}\ltimes V$ or $\mathfrak{s} = \mathfrak{q}\ltimes V$, where V is a finite-dimensional G- or Q-module. Then S is a connected algebraic group with $\mathrm{Lie}\, S = \mathfrak{s}$. For instance, $S = Q \ltimes \exp(V)$.

The vector space decomposition $\mathfrak{s} = \mathfrak{q}\oplus V$ leads to $\mathfrak{s}^* = \mathfrak{q}^*\oplus V^*$. For $\mathfrak{q} = \mathfrak{g}$, we identify $\mathfrak{g}$ with $\mathfrak{g}^*$. Each element $x \in V^*$ is considered as a point of $\mathfrak{s}^*$ that is zero on $\mathfrak{q}$. We have $\exp(V){\cdot}x = \mathrm{ad}^*(V){\cdot}x + x$, where each element of $\mathrm{ad}^*(V){\cdot}x$ is zero on V. Note that $\mathrm{ad}^*(V){\cdot}x \subset \mathrm{Ann}\,(\mathfrak{q}_x) \subset \mathfrak{q}^*$ and $\dim(\mathrm{ad}^*(V){\cdot}x)$ is equal to $\dim(\mathrm{ad}^*(\mathfrak{q}){\cdot}x) = \dim\mathfrak{q} - \dim\mathfrak{q}_x$. Therefore $\mathrm{ad}^*(V){\cdot}x = \mathrm{Ann}\,(\mathfrak{q}_x)$.

There is a general formula [16] for the index of $\mathfrak{s} = \mathfrak{q}\ltimes V$:

$$\mathrm{ind}\,\mathfrak{s} = \dim V - (\dim\mathfrak{q} - \dim\mathfrak{q}_x) + \mathrm{ind}\,\mathfrak{q}_x \quad \text{with } x \in V^* \text{generic}. \tag{1}$$

The decomposition $\mathfrak{s} = \mathfrak{q}\oplus V$ defines the bi-grading on $\mathcal{S}(\mathfrak{s})$ and it appears that $\mathcal{S}(\mathfrak{s})^S$ is a bi-homogeneous subalgebra, cf. [11, Theorem 2.3(i)].

For any $x \in V^*$, the affine space $\mathfrak{q}^*{+}x$ is $\exp(V)$-stable and Q_x-stable. Further, there is the restriction homomorphism

$$\psi_x : \Bbbk[\mathfrak{s}^*]^S \to \Bbbk[\mathfrak{q}^*{+}x]^{Q_x \ltimes \exp(V)} \simeq \mathcal{S}(\mathfrak{q}_x)^{Q_x}.$$

The existence of the isomorphism $\Bbbk[\mathfrak{q}^*{+}x]^{\exp(V)} \simeq \mathcal{S}(\mathfrak{q}_x)$ is proven in [22]. If we choose x as the origin in $\mathfrak{q}^*{+}x$, then actually $\psi_x(H) \in \mathcal{S}(\mathfrak{q}_x)$ for each $H \in \Bbbk[\mathfrak{s}^*]^{\exp(V)}$, see [22, Prop. 2.7].

Suppose that $Q \lhd \tilde{Q}$ and there is an action of $\tilde{Q}$ on V that extends the Q-action. Set $\tilde{\mathfrak{s}} = \tilde{\mathfrak{q}} \ltimes V$, $\tilde{S} = \tilde{Q} \ltimes \exp(V)$.

Lemma 3.1 *We have $\mathcal{S}(\mathfrak{s})^{\tilde{S}} \subset \mathcal{S}(\mathfrak{s})^S$ and $H \in \mathcal{S}(\tilde{\mathfrak{s}})^{\tilde{S}}$ lies in $\mathcal{S}(\mathfrak{s})$ if and only if the restriction of H to $\tilde{\mathfrak{q}}^*{+}x$ lies in $\mathcal{S}(\mathfrak{q}_x)$ for a generic $x \in V^*$.*

Proof The inclusion $\mathcal{S}(\mathfrak{s})^{\tilde{S}} \subset \mathcal{S}(\mathfrak{s})^S$ is obvious. Now let $\mathfrak{m}$ be a vector space complement of $\mathfrak{q}$ in $\tilde{\mathfrak{q}}$. Then $\mathcal{S}(\tilde{\mathfrak{s}}) = \mathcal{S}(\mathfrak{q}) \otimes \mathcal{S}(\mathfrak{m}) \otimes \mathcal{S}(V)$. If H does not lie in $\mathcal{S}(\mathfrak{q}) \otimes \mathcal{S}(V)$, then $H|_{\tilde{\mathfrak{q}}^*+x}$ does not lie in $\mathcal{S}(\mathfrak{q})$ for any x from a non-empty open subset of V^*.

Finally, suppose that $H \in \mathcal{S}(\mathfrak{s})^{\exp(V)}$. Then $H|_{\tilde{\mathfrak{q}}^*+x}$ lies in $\mathcal{S}(\tilde{\mathfrak{q}}_x)$ by [22, Prop. 2.7]. Clearly, $\mathcal{S}(\mathfrak{q}) \cap \mathcal{S}(\tilde{\mathfrak{q}}_x) = \mathcal{S}(\mathfrak{q}_x)$. □

Proposition 3.2 (Prop. 3.11 in [22]) *Let Q be a connected algebraic group acting on a finite-dimensional vector space V. Set $\mathfrak{s} = \mathfrak{q} \ltimes V$. Suppose that Q has no proper semi-invariants in $\Bbbk[\mathfrak{s}^*]^{\exp(V)}$ and $\Bbbk[\mathfrak{s}^*]^S$ is a polynomial ring in $\operatorname{ind}\mathfrak{s}$ variables. For generic $x \in V^*$, we then have*

- *the restriction map $\psi : \Bbbk[\mathfrak{s}^*]^S \to \Bbbk[\mathfrak{q}^*{+}x]^{Q_x \ltimes \exp(V)} \simeq \mathcal{S}(\mathfrak{q}_x)^{Q_x}$ is onto;*
- *$\mathcal{S}(\mathfrak{q}_x)^{Q_x}$ coincides with $\mathcal{S}(\mathfrak{q}_x)^{\mathfrak{q}_x}$;*
- *$\mathcal{S}(\mathfrak{q}_x)^{Q_x}$ is a polynomial ring in $\operatorname{ind}\mathfrak{q}_x$ variables.*

Note that Q is not assumed to be reductive and Q_x is not assumed to be connected in the above proposition!

Let now V be a G-module. By a classical result of Richardson, there is a non-empty open subset $\Omega \subset V^*$ such that the stabilisers G_x are conjugate in G for all $x \in \Omega$, see, e.g., [20, Theorem 7.2]. In this situation (any representative of the conjugacy class of) G_x is called a *generic isotropy group*, denoted $\mathsf{g.i.g.}(G : V^*)$, and $\mathfrak{g}_x = \operatorname{Lie} G_x$ is a *generic stabiliser* for the G-action on V^*.

If G is semisimple and V is a reducible G-module, say $V = V_1 \oplus V_2$, then there is a trick that allows us to relate the polynomiality property for the symmetric invariants of $\mathfrak{s} = \mathfrak{g} \ltimes V$ to a smaller semi-direct product. The precise statement is as follows.

Proposition 3.3 (cf. [14, Prop. 3.5]) *With $\mathfrak{s} = \mathfrak{g} \ltimes (V_1 \oplus V_2)$ as above, let H be a generic isotropy group for $(G : V_1^*)$. If $\Bbbk[\mathfrak{s}^*]^S$ is a polynomial ring, then so is $\Bbbk[\tilde{\mathfrak{q}}^*]^{\tilde{Q}}$, where $\tilde{Q} = H \ltimes \exp(V_2)$ or $H^\circ \ltimes \exp(V_2)$.*

The above passage from $\mathfrak{s}$ to $\tilde{\mathfrak{q}}$, i.e., from $(G, V_1 \oplus V_2)$ to (H°, V_2) is called a *reduction*, and we denote it by $(G, V_1 \oplus V_2) \longrightarrow (H^\circ, V_2)$ in the diagrams below. This proposition is going to be used as a tool for proving that $\Bbbk[\mathfrak{s}^*]^S$ is not polynomial.

In what follows, the irreducible representations of simple groups are often identified with their highest weights, using the Vinberg–Onishchik numbering of the fundamental weights [19]. For instance, if $\varphi_1, \dots, \varphi_n$ are the fundamental weights of a simple algebraic group G, then $V = \varphi_i + 2\varphi_j$ stands for the direct sum of three simple G-modules, with highest weights φ_i (once) and φ_j (twice). A full notation is $V = V_{\varphi_i} + 2V_{\varphi_j}$. Note that adding a trivial 1-dimensional G-module $\Bbbk$ to V does not affect the polynomiality property for $\mathfrak{s}$.

Example 3.4 There is a diagram (tree) of reductions:

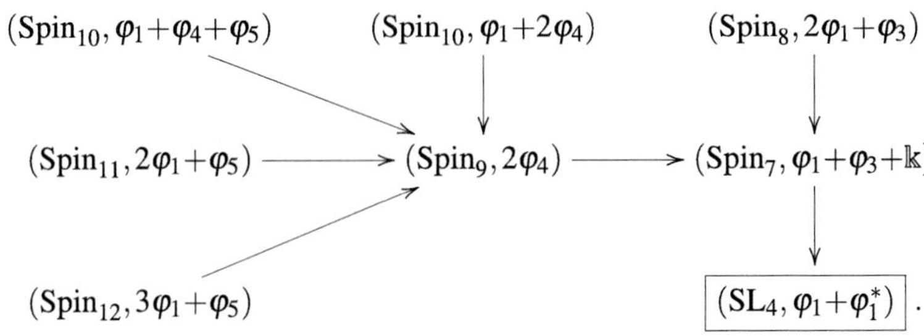

For instance, the first horizontal arrow means that for $G = \mathrm{Spin}_{11}$ and $V_1 = 2\varphi_1$, we have $\mathsf{g.i.g.}(G, V_1) = \mathrm{Spin}_9$ and the restriction of $V_2 = \varphi_5$ to $H = \mathrm{Spin}_9$ is the H-module $2\varphi_4$. The terminal item (in the box) does not have the polynomiality property by [23]. Therefore all the items here do not have the polynomiality property by Proposition 3.3.

The action $(G : V)$ is said to be *stable* if the union of closed G-orbits is dense in V. Then $\mathsf{g.i.g.}(G : V)$ is necessarily reductive.

We mention the following good situation. Suppose that G is semisimple. If a generic stabiliser for the G-action on V^* is reductive, then the action $(G : V^*)$ is stable [20, § 7]. Moreover, S has only trivial characters and no proper semi-invariants.

Example 3.5 (cf. [22, Example 3.6]) If G is semisimple, $\mathfrak{g}_x = \mathfrak{sl}_2$ for $x \in V^*$ generic, and $\Bbbk[V^*]^G$ is a polynomial ring, then $\mathcal{S}(\mathfrak{s})^{\mathfrak{s}}$ is a polynomial ring.

We say that $\dim V /\!\!/ G$ is the *rank* of the pair (G, V). For (G, V^*) of rank one, we have two general results.

Consider the following assumptions on G and V:

($\diamondsuit$) the action $(G : V^*)$ is stable, $\Bbbk[V^*]^G$ is a polynomial ring, $\Bbbk[\mathfrak{g}_\xi^*]^{G_\xi}$ is a polynomial ring for generic $\xi \in V^*$, and G has no proper semi-invariants in $\Bbbk[V^*]$.

Theorem 3.6 ([14, Theorem 2.3]) *Suppose that G and V satisfy condition* ($\diamondsuit$) *and $V^* /\!\!/ G = \mathbb{A}^1$, i.e., $\Bbbk[V^*]^G = \Bbbk[F]$ for some homogeneous F. Let L be a generic* *isotropy group for* $(G : V^*)$. *Assume further that* $D = \{x \in V^* \mid F(x) = 0\}$ *contains an open* G*-orbit, say* $G{\cdot}y$, *ind* $\mathfrak{g}_y = \text{ind}\,\mathfrak{l} =: \ell$, *and* $\mathcal{S}(\mathfrak{g}_y)^{G_y}$ *is a polynomial ring in* ℓ *variables with the same degrees of generators as* $\mathcal{S}(\mathfrak{l})^L$. *Then* $\Bbbk[\mathfrak{s}^*]^S$ *is a polynomial ring in* ind $\mathfrak{s} = \ell + 1$ *variables.*

Lemma 3.7 *Suppose that* G *is semisimple,* $\Bbbk[V^*]^G = \Bbbk[F]$ *and a generic isotropy group for* $(G : V^*)$, *say* L, *is connected and is either of type* B_2 *or* G_2. *Then* $\mathfrak{s} = \mathfrak{g} \ltimes V$ *has the polynomiality property.*

Proof Let $x \in V^*$ be generic and $G_x = L$, hence $\mathfrak{g}_x = \mathfrak{l}$. By [22, Lemma 3.5], there are irreducible bi-homogeneous S-invariants H_1 and H_2 such that their restrictions to $\mathfrak{g} + x = \mathfrak{g}^* + x$ yield the basic symmetric invariants of $\mathfrak{l}$ under the isomorphism $\Bbbk[\mathfrak{g}^*{+}x]^{G_x \ltimes \exp(V)} \simeq \mathcal{S}(\mathfrak{g}_x)^{G_x}$. Furthermore, $\Bbbk[\mathfrak{s}^*]^S = \Bbbk[F, H_1, H_2]$ if and only if H_1 and H_2 are algebraically independent over $\Bbbk[D]^G = \Bbbk$ on $\mathfrak{g} \times D$, where D is the zero set of F. W.l.o.g., we may assume that $\deg_{\mathfrak{g}} H_1 = 2$ and $\deg_{\mathfrak{g}} H_2 = 4$ (if $L = \mathsf{B}_2$) or $\deg_{\mathfrak{g}} H_2 = 6$ (if $L = \mathsf{G}_2$). We may also assume that a non-trivial relation among $H_1|_{\mathfrak{g}\times D}, H_2|_{\mathfrak{g}\times D}$ is homogeneous w.r.t. $\mathfrak{g}$ and therefore boils down to $\frac{H_1^\alpha}{H_2} \equiv a \mod (F)$ for $\alpha \in \{2, 3\}$, depending on L, and $a \in \Bbbk$. Such a relation means that H_2 is chosen wrongly and has to be replaced by a polynomial $(H_2 - aH_1^\alpha)/F^r$ with the largest possible $r \geqslant 1$. This modification decreases the total degree of H_2 and hence it cannot be performed infinitely many times. □

The following result holds for actions of arbitrary rank.

Lemma 3.8 *Suppose that* G *is semisimple,* $\Bbbk[V^*]^G$ *is a polynomial ring and a generic isotropy group for* $(G : V^*)$ *is a connected group of type* A_2. *Assume further that, for any* G*-stable divisor* $D \subset V^*$ *and a generic point* $y \in D$, *we have* $\dim \mathcal{S}^2(\mathfrak{g}_y)^{G_y} = \dim \mathcal{S}^3(\mathfrak{g}_y)^{G_y} = 1$ *and that these unique (up to a scalar) invariants are algebraically independent. Then* $\mathfrak{s} = \mathfrak{g} \ltimes V$ *has the polynomiality property.*

Proof The statement readily follows from [22, Lemma 3.5]. □

3.1 Yet Another Case of a Surjective Restriction

By Proposition 3.2, if $x \in V^*$ is generic, then the restriction homomorphism $\psi_x : \Bbbk[\mathfrak{s}^*]^S \to \Bbbk[\mathfrak{q}^*{+}x]^{Q_x \ltimes \exp(V)}$ is surjective, whenever $\Bbbk[\mathfrak{s}^*]^S$ is a polynomial ring and Q has no proper semi-invariants in $\Bbbk[\mathfrak{s}^*]^{\exp(V)}$. On the other hand, ψ_x is surjective for generic $x \in V^*$ if $Q = G$ is reductive and the G-action on V^* is stable [22, Theorem 2.8]. It is likely that the surjectivity holds for a wider class of semi-direct products.

Suppose that $\Bbbk$ is algebraically closed. Take Q and V such that $\dim(Q{\cdot}\xi) = \dim V - 1$ for generic $\xi \in V^*$. Assume that $\Bbbk[V^*]^Q \neq \Bbbk$. Then $\Bbbk[V^*]^Q = \Bbbk[F]$, where F is a homogeneous polynomial of degree $N \geqslant 1$, $\Bbbk(V^*)^Q = \Bbbk(F)$, and F separates generic Q-orbits on V^*. Hence $\Bbbk\xi \cap Q{\cdot}\xi = \{ax \mid a \in \Bbbk, a^N = 1\}$

for generic $\xi \in V^*$. Let $N_Q(\Bbbk\xi)$ be the normaliser of the line $\Bbbk\xi$. Then $N_Q(\Bbbk\xi) = C_N \times Q_\xi$, where $C_N \subset \Bbbk^\times$ is a cyclic group of order N. Let C_N act on V faithfully, then $\tilde{Q} := C_N \times Q$ acts on V and $\tilde{Q}_\xi \simeq C_N \times Q_\xi$. If $H \in \Bbbk[\mathfrak{s}^*]^Q$ is homogeneous in V, then $\psi_\xi(H)$ is an eigenvector of $C_N \subset \tilde{Q}_\xi$ and the corresponding eigenvalue depends only on $\deg_V H$.

Theorem 3.9 (Generalised Surjectivity or the "Rank-One Argument") *Let Q be a connected algebraic group with* Lie $Q = \mathfrak{q}$. *Suppose that V is a Q-module such that Q has no proper semi-invariants in $\Bbbk[V^*]$ and $\Bbbk[V^*]^Q = \Bbbk[F]$ with $F \notin \Bbbk$. Set $\mathfrak{s} = \mathfrak{q} \ltimes V$, $S = Q \ltimes \exp(V)$. Then the natural homomorphism*

$$\psi_\xi : \Bbbk[\mathfrak{s}^*]^S \to \Bbbk[\mathfrak{q}^* + \xi]^{Q_\xi \ltimes \exp(V)} \simeq \mathcal{S}(\mathfrak{q}_\xi)^{Q_\xi}$$

is onto for generic $\xi \in V^*$. Moreover, if $h \in \Bbbk[\mathfrak{q}^* + \xi]^{Q_\xi \ltimes \exp(V)}$ is a semi-invariant of $N_Q(\Bbbk\xi)$, then there is a homogeneous in V polynomial $H \in \Bbbk[\mathfrak{s}^*]^S$ with $\psi_\xi(H) = h$.

Proof Let S act on an irreducible variety X. A classical result of Rosenlicht [20, § 2.3] implies that the functions $f_1, \ldots, f_m \in \Bbbk(X)^S$ generate $\Bbbk(X)^S$ if and only if they separate generic S-orbits on X. Let $U \subset \mathfrak{s}^*$ be a non-empty open subset such that for every two different orbits $S{\cdot}u, S{\cdot}u' \subset U$, there is $\mathbf{f} \in \Bbbk(\mathfrak{s}^*)^S$ separating them, meaning that $\mathbf{f}$ takes finite values at u, u' and $\mathbf{f}(u) \neq \mathbf{f}(u')$. Then $U \cap (\mathfrak{q}^* + \xi) \neq \varnothing$ for generic $\xi \in V^*$ and hence generic $Q_\xi \ltimes \exp(V)$-orbits on $\mathfrak{q}^* + \xi$ are separated by rational S-invariants for any such ξ. In other words, for every $h \in \Bbbk(\mathfrak{q}^* + \xi)^{Q_\xi \ltimes \exp(V)}$ there is $\tilde{\boldsymbol{r}} \in \Bbbk(\mathfrak{s}^*)^S$ such that $\psi_\xi(\tilde{\boldsymbol{r}}) := \tilde{\boldsymbol{r}}|_{\mathfrak{q}^* + \xi} = h$.

The same principle applies to the group $\Bbbk^\times \times S$, where $\Bbbk^\times$ acts on V by $t{\cdot}v = tv$ for all $t \in \Bbbk^\times$, $v \in V$. A rational invariant of $(\Bbbk^\times \times Q)_\xi \ltimes \exp(V)$ on $\mathfrak{q}^* + \xi$ extends to a rational $(\Bbbk^\times \times S)$-invariant on $\mathfrak{s}^*$.

The absence of proper semi-invariants implies that $\Bbbk(V^*)^Q = \Bbbk(F)$. Hence a generic Q-orbit on V^* is of dimension $\dim V - 1$. Assume that F is homogeneous and set $N := \deg F$.

Choose a generic point $\xi \in V^*$ with $F(\xi) \neq 0$ and with $\dim(Q{\cdot}\xi) = \dim V - 1$. Then $N_Q(\Bbbk\xi) = C_N \times Q_\xi$. As above, set $\tilde{Q} := C_N \times Q$ and also $\tilde{S} := C_N \times S$. We regard $\tilde{Q}$ as a subgroup of $\Bbbk^\times \times Q$. Now $\tilde{Q}_\xi = (\Bbbk^\times \times Q)_\xi$.

The group $C_N \subset \tilde{Q}_\xi$ acts on $\Bbbk[\mathfrak{q}^* + \xi]^{Q_\xi \ltimes \exp(V)}$ and this action is diagonalisable. Suppose that $h \in \Bbbk[\mathfrak{q}^* + \xi]^{Q_\xi \ltimes \exp(V)}$ is an eigenvector of C_N. First we show that there is $r \in \Bbbk(\mathfrak{s}^*)^S$ such that $\psi_\xi(r)$ is an eigenvector of $C_N \subset \tilde{Q}_\xi$ with the same weight as h.

Recall that h extends to a rational S-invariant $\tilde{\boldsymbol{r}} \in \Bbbk(\mathfrak{s}^*)^S$. The group C_N is finite, hence $\tilde{\boldsymbol{r}}$ is contained in a finite-dimensional C_N-stable vector space and thereby $\tilde{\boldsymbol{r}}$ is a sum of rational S-invariant C_N-eigenvectors. Since a copy of C_N sitting in $\tilde{Q}$ stabilises ξ, we can replace $\tilde{\boldsymbol{r}}$ with a suitable C_N-semi-invariant component. By a standard argument, this new $\tilde{\boldsymbol{r}}$ is a ratio of two regular $\tilde{S}$-semi-invariants, say $\tilde{\boldsymbol{r}} = q/f$ now. Each bi-homogenous w.r.t. $\mathfrak{s} = \mathfrak{q} \oplus V$ component of q (or f) is again a semi-invariant of $\tilde{S}$ of the same weight as q (or f). Let us replace f (and q) with any of its non-zero bi-homogenous components. The resulting rational function r has the same weight as $\tilde{r}$. In particular, r is an S-invariant. Thus, we have found the required rational function. Since r is a semi-invariant of $\Bbbk^{\times}$, it is defined on a non-empty open subset of $\mathfrak{q}^*\times Q{\cdot}x$ for each $x \in V^*$ such that $F(x) \neq 0$ and $\dim(Q{\cdot}x) = \dim V - 1$.

Set $\bar{r} := \psi_\xi(r) \in \Bbbk(\mathfrak{q}^*+\xi)$. Then $h/\bar{r} \subset \Bbbk(\mathfrak{q}^*+\xi)^{\tilde{Q}_\xi \ltimes \exp(V)}$ and therefore extends to a rational $(\Bbbk^{\times}\times S)$-invariant on $\mathfrak{s}^*$. Multiplying the extension by r, we obtain a rational S-invariant R, which is also an eigenvector of $\Bbbk^{\times}$. Let $R = H/P$, where $H, P \in \Bbbk[\mathfrak{s}^*]$ are relatively prime. Then both H and P are homogenous in V. Note that R is defined on $\mathfrak{q}^*+\xi$, therefore also on $\mathfrak{q}^*\times Q{\cdot}\xi$ and finally on $\mathfrak{q}\times\Bbbk^{\times}(Q{\cdot}\xi)$, because $R(\eta+a\xi) = a^k R(\eta+\xi)$ for some $k \in \mathbb{Z}$ and for all $a \in \Bbbk^{\times}$, $\eta \in \mathfrak{q}^*$. Hence P is a polynomial in F, more explicitly, $P = F^d$ fore some $d \geqslant 0$. Multiplying R by $\frac{F^d}{F(\xi)^d}$ yields the required pre-image H. □

Remark 3.10 Since $\Bbbk[V^*]^Q = \Bbbk[F]$ and there are no proper Q-semi-invariants in $\Bbbk[V^*]$, $\mathfrak{q}^*\times Q{\cdot}\xi$ is a big open subset of

$$Y_\alpha = \{\mathfrak{q}^*+x \mid F(x) = F(\xi)\} = \{\gamma \in \mathfrak{s}^* \mid F(\gamma) = \alpha\},$$

where $\alpha = F(\xi)$. For a reductive group G, one knows that any regular G-invariant on a closed G-stable subset $Y \subset X$ of an affine G-variety X extends to a regular G-invariant on X. Assuming that the image of Q in $\mathrm{GL}(V^*)$ is reductive, we could present a different proof of Theorem 3.9, similar to the proof of Theorem 2.8 in [22].

3.2 Tables and Classification Tools

Our goal is to classify the pairs (G, V) such that G is either Spin_n or Sp_{2n} and the semi-direct product $\mathfrak{s} = \mathfrak{g} \ltimes V$ has a Free Algebra of symmetric invariants, (FA) for short. We also say that (G, V) is a *positive* (resp. *negative*) case, if the property (FA) **is** (resp. **is not**) satisfied for $\mathfrak{s}$.

Example 3.11 If G is arbitrary semisimple, then $\mathfrak{g} \ltimes \mathfrak{g}^{\mathsf{ab}}$, where $\mathfrak{g}^{\mathsf{ab}}$ is an Abelian ideal isomorphic to $\mathfrak{g}$ as a $\mathfrak{g}$-module, always has (FA) [18]. Therefore we exclude the adjoint representations from our further consideration.

- If $\Bbbk[\mathfrak{s}^*]^S$ is a polynomial ring, then so is $\Bbbk[V^*]^G$ [11, Section 2 (A)] (cf. [22, Section 3]). For this reason, we only have to examine all representations of G with polynomial rings of invariants.
- Since the algebras $\Bbbk[V]^G$ and $\Bbbk[V^*]^G$ (as well as $\mathcal{S}(\mathfrak{g} \ltimes V)^{G\ltimes V}$ and $\mathcal{S}(\mathfrak{g} \ltimes V^*)^{G\ltimes V^*}$) are isomorphic, it suffices to keep track of either V or V^*. The same principle applies to the two half-spin representations in type $\mathbf{D}_{2m}$.

Example 3.12 If a generic stabiliser for $(G{:}V^*)$ is trivial, then $\Bbbk[\mathfrak{s}^*]^S \simeq \Bbbk[V^*]^G$ [10, Theorem 6.4] (cf. [22, Example 3.1]). Therefore all such semi-direct products have (FA).

We are lucky that there is a classification of the representations of the simple algebraic groups with non-trivial generic stabilisers obtained by Elashvili [2]. In addition, the two independent classifications in [1, 17] provide the list of representations of simple algebraic groups with polynomial rings of invariants. Combining them, we obtain the representations in Tables 1 and 2.

Explanations to the Tables As in [1, 2, 13, 14, 22], we use the Vinberg–Onishchik numbering of fundamental weights, see [19, Table 1]. In both tables, $\mathfrak{h}$ is a generic stabiliser for $(G : V)$ and the last column indicates whether (FA) is satisfied for $\mathfrak{s}$ or not. Naturally, the positive cases are marked with '$+$'. This last column represents the main results of the article. The ring $\Bbbk[V^*]^G$ is always a polynomial ring in $\dim V/\!\!/G$ variables. If the expression for $\dim V/\!\!/G$ is bulky, then it is not included in Table 1. However, one always has $\dim V/\!\!/G = \dim V - \dim G + \dim \mathfrak{h}$. If $\mathfrak{s}$ has (FA), then $\operatorname{ind}\mathfrak{s} = \dim V/\!\!/G + \operatorname{ind}\mathfrak{h}$ is the total number of the basic invariants in $\Bbbk[\mathfrak{s}^*]^S$. The symbol $\mathbf{U}_n$ in Table 2 stands for a commutative Lie subalgebra of dimension n that consists of nilpotent elements.

Our classification is summarised in the following

Theorem 3.13 *Let G be either* Spin_n *or* Sp_{2n}, *V a finite-dimensional rational G-module, and $\mathfrak{s} = \mathfrak{g} \ltimes V$. Then $\Bbbk[\mathfrak{s}^*]^S$ is a free algebra if and only if one of the following conditions is satisfied:*

(i) $V = \mathfrak{g}$;
(ii) *V or V^* occurs in Tables 1 and 2, and the last column is marked with '$+$'. It is also possible to permute φ_5 and φ_6 for* $\mathbf{D}_6$, *and take any permutation of $\varphi_1, \varphi_3, \varphi_4$ for* $\mathbf{D}_4$.
(iii) *$\Bbbk[V]^G$ is a free algebra and* g.i.g.$(G : V)$ *is finite, i.e., (G, V) is contained in the lists of [1, 17], but is not contained in the tables of [2].*

- Generic stabilisers for the representations in the tables are taken from [2]. To verify that the generic isotropy groups are connected, we use Proposition 4.10 and Remark 4.11 in [17]. In case of reducible representations, this can be combined with the group analogue of [2, Lemma 2].
- Apart from a generic isotropy group for $(G : V^*)$, we often have to compute the isotropy group G_y, where y is a generic point of a G-stable divisor $D \subset V^*$, cf. Theorem 3.6. Mostly this is done by *ad hoc* methods. Also the following observation is very helpful. Any divisor $D \subset V_1 \oplus V_2$ projects dominantly to at least one factor V_i. Hence it contains a subset of the form $\{x_i\} \times D_{i'}$, where $x_i \in V_i$ is generic, $D_{i'} \subset V_{i'}$ is a divisor, and $\{i, i'\} = \{1, 2\}$.
- Another major ingredient in obtaining the classification is (the presence of) the "codim–2" property for $\mathfrak{s}$. Some methods for checking the "codim–2" condition are presented in [13, Sect. 4]. Similarly to the Raïs formula, see Eq. (1), we also

Table 1 The representations of the orthogonal groups with polynomial ring $\Bbbk[V]^G$ and non-trivial generic stabilisers

№	G	V	$\dim V$	$\dim V/\!\!/G$	$\mathfrak{h}$	$\operatorname{ind}\mathfrak{s}$	(FA)
1	SO_n	$m\varphi_1,\ m<n-1$	mn	$\frac{m(m+1)}{2}$	$\mathfrak{so}_{n-m}$	$\binom{m+1}{2}+[\frac{n-m}{2}]$	+
2a	$\mathbf{B}_3$	φ_3	8	1	$\mathbf{G}_2$	3	+
2b		$m\varphi_1+m'\varphi_3$ $2\leqslant m+m'\leqslant 3$ $m'>0$	$7m+8m'$		$\mathbf{A}_{4-m-m'}$		−, if (1, 1)
3a	$\mathbf{B}_4$	φ_4	16	1	$\mathbf{B}_3$	4	+
3b		$\varphi_1+\varphi_4$	25	3	$\mathbf{G}_2$	5	+
3c		$2\varphi_1+\varphi_4$	34	6	$\mathbf{A}_2$	8	+
3d		$3\varphi_1+\varphi_4$	43	10	$\mathbf{A}_1$	11	+
3e		$2\varphi_4$	32	4	$\mathbf{A}_2$	6	−
3f		$\varphi_1+2\varphi_4$	41	8	$\mathbf{A}_1$	9	+
4	$\mathbf{B}_5$	$m\varphi_1+\varphi_5,$ $0\leqslant m\leqslant 3$	$32+11m$	$1+m+m^2$	$\mathbf{A}_{4-m}$	$5+m^2$	+, if $m=0,3$ −, if $m=1,2$
5a	$\mathbf{B}_6$	φ_6	64	2	$\mathbf{A}_2+\mathbf{A}_2$	6	+
5b		$\varphi_1+\varphi_6$	77	5	$\mathbf{A}_1+\mathbf{A}_1$	7	+
6a	$\mathbf{D}_4$	$\varphi_1+\varphi_3$	16	2	$\mathbf{G}_2$	4	+
6b		$m\varphi_1+\varphi_3,\ m=2,3$	$8(m+1)$		$\mathbf{A}_{4-m}$		+, if $m=3$
6c		$m\varphi_1+\varphi_3+\varphi_4$ $m=1,2$	$8(m+2)$		$\mathbf{A}_{3-m}$		+
7a	$\mathbf{D}_5$	φ_4	16	0	$\mathfrak{so}_7\ltimes V_{\varphi_3}$	3	+
7b		$\varphi_1+\varphi_4$	26	2	$\mathbf{B}_3$	5	+
7c		$2\varphi_1+\varphi_4$	36	5	$\mathbf{G}_2$	7	+
7d		$m\varphi_1+\varphi_4,\ m=3,4$	$16+10m$		$\mathbf{A}_{5-m}$		+
7e		$2\varphi_4$	32	1	$\mathbf{G}_2$	3	+
7f		$m\varphi_1+2\varphi_4,\ m=1,2$	$32+10m$		$\mathbf{A}_{3-m}$		+, if $m=2$
7g		$3\varphi_4$ or $2\varphi_4+\varphi_5$	48	6	$\mathbf{A}_1$	7	+
7h		$m\varphi_1+\varphi_4+\varphi_5$ $0\leqslant m\leqslant 2$	$32+10m$	$2+2m+m^2$	$\mathbf{A}_{3-m}$	$5+m+m^2$	−, if $m\leqslant 1$ +, if $m=2$
8a	$\mathbf{D}_6$	$m\varphi_1+\varphi_5$ $0\leqslant m\leqslant 4$	$32+12m$	$1+m^2$	$\mathbf{A}_{5-m}$	$6-m+m^2$	+, if $m=0,4$ −, if $1\leqslant m\leqslant 3$
8b		$2\varphi_5$	64	7	$3\mathbf{A}_1$	10	+
8c		$\varphi_5+\varphi_6$	64	4	$2\mathbf{A}_1$	6	+
9a	$\mathbf{D}_7$	φ_6	64	1	$2\mathbf{G}_2$	5	+
9b		$m\varphi_1+\varphi_6,\ m=1,2$	$64+14m$		$2\mathbf{A}_{3-m}$		+

have

$$\dim\mathfrak{s}_{\gamma+y}=\dim(\mathfrak{g}_y)_{\bar\gamma}+(\dim V-\dim(G{\cdot}y)),$$

where $y\in V^*$, $\gamma\in\mathfrak{g}$, and $\bar\gamma=\gamma|_{\mathfrak{g}_y}$, cf. [22, Eq. (3·1)]. Therefore, $\mathfrak{s}$ has the "codim–2" property if and only if

Table 2 The representations of the symplectic group with polynomial ring $\Bbbk[V]^G$ and non-trivial generic stabilisers

№	G	V	$\dim V$	$\dim V /\!\!/ G$	$\mathfrak{h}$	$\operatorname{ind}\mathfrak{s}$	(FA)
1	$\mathbf{C}_n$	$m\varphi_1$, $m\leqslant 2n-1$	$2mn$	$\binom{m}{2}$	$\mathbf{C}_{n-l}$, $m=2l$; $\mathbf{C}_{n-l}\ltimes\mathfrak{heis}_{n-l}$, $m=2l-1$	$\binom{m}{2}+n-[\frac{m}{2}]$	+
2	$\mathbf{C}_n$	φ_2	$2n^2-n-1$	$n-1$	$n\mathbf{A}_1$	$2n-1$	+
3	$\mathbf{C}_n$	$\varphi_1+\varphi_2$	$2n^2+n-1$	$n-1$	$\mathbf{U}_n$	$2n-1$	+
4	$\mathbf{C}_3$	φ_3	14	1	$\mathbf{A}_2$	3	+
5		$\varphi_1+\varphi_3$	20	2	$\mathbf{A}_1$	3	+
6		$2\varphi_2$	28	8	$\mathfrak{t}_1$	9	+

(i) $\mathfrak{g}_x$ with $x \in V^*$ generic has the "codim–2" property and
(ii) for any divisor $D \subset V$, $\operatorname{ind}\mathfrak{g}_y + (\dim V - \dim(G{\cdot}y)) = \operatorname{ind}\mathfrak{s}$ holds for all points y of a non-empty open subset $U \subset D$, cf. [22, Eq. (3·2)].

- Finally, we recall an important class of semi-direct products. Let $\mathfrak{g} = \mathfrak{g}_0 \oplus \mathfrak{g}_1$ be a $\mathbb{Z}_2$-grading of $\mathfrak{g}$, i.e., $(\mathfrak{g}, \mathfrak{g}_0)$ is a symmetric pair. Then the semi-direct product $\mathfrak{s} = \mathfrak{g}_0 \ltimes \mathfrak{g}_1^{\mathsf{ab}}$, where $[\mathfrak{g}_1^{\mathsf{ab}}, \mathfrak{g}_1^{\mathsf{ab}}] = 0$, is called the $\mathbb{Z}_2$-*contraction* of $\mathfrak{g}$ related to the symmetric pair $(\mathfrak{g}, \mathfrak{g}_0)$. Set $l = \operatorname{rk}\mathfrak{g}$ and let $H_1, \ldots, H_l$ be a set of the basic symmetric invariants of $\mathfrak{g}$. Let $H_i^\bullet$ denote the bi-homogeneous component of H_i that has the highest $\mathfrak{g}_1$-degree. Then $H_i^\bullet$ is an $\mathfrak{s}$-invariant in $\mathcal{S}(\mathfrak{s})$ [11]. We say that a $\mathbb{Z}_2$-contraction $\mathfrak{s} = \mathfrak{g}_0 \ltimes \mathfrak{g}_1^{\mathsf{ab}}$ is *good* if $\mathcal{S}(\mathfrak{s})^{\mathfrak{s}}$ is freely generated by the polynomials $H_1^\bullet, \ldots, H_l^\bullet$ for some well-chosen generators $\{H_i\}$. Note that $\deg H_i = \deg H_i^\bullet$ for the usual degree.

4 An Example in Type A

The example considered in this section will be needed below in our treatment of $G = \mathrm{SO}_n$. It can also be regarded as a small step towards the classification in type A.

Suppose that $G = \mathrm{SL}_n \subset \mathrm{GL}_n = \tilde{G}$ and $V = \bigwedge^2 \Bbbk^n \oplus (\bigwedge^2 \Bbbk^n)^*$. Then $\tilde{\mathfrak{s}} = \tilde{\mathfrak{g}} \ltimes V$ is the $\mathbb{Z}_2$-contraction of $\mathfrak{so}_{2n}$ related to the symmetric pair $(\mathfrak{so}_{2n}, \mathfrak{gl}_n)$. By [21, Theorem 4.5], this $\mathbb{Z}_2$-contraction is good and satisfies ***KRC***. Our goal is to describe $\mathcal{S}(\mathfrak{s})^{\mathfrak{s}}$ using the known description for $\tilde{\mathfrak{s}}$. Let us denote the basic symmetric invariants of $\tilde{\mathfrak{s}}$ by $H_1, \ldots, H_\ell, F_1, \ldots, F_r$, where $\deg F_i = 2i$ and $\Bbbk[F_1, \ldots, F_r] = \Bbbk[V^*]^{\mathrm{GL}_n}$. Then necessary $\ell = [\frac{n+1}{2}]$, $r = [\frac{n}{2}]$.

Proposition 4.1 *If $n = 2r$ is even, then $\mathcal{S}(\mathfrak{s})^{\mathfrak{s}}$ is freely generated by* $H_1, \ldots, H_r, F_1, \ldots, F_{r-1}, F_r', F_{r+1}$ *with* $\deg F_r' = \deg F_{r+1} = r$.

Proof Since n is even, the generic isotropy group of the GL_n-action on V^* is $(\mathrm{SL}_2)^r$ and it lies in SL_n. Therefore each H_i lies in $\mathcal{S}(\mathfrak{s})$, see Lemma 3.1. The new generators F_r', F_{r+1} are the pfaffians on $\bigwedge^2 \Bbbk^n$ and $(\bigwedge^2 \Bbbk^n)^*$, respectively. We have

$$\left(\sum_{i=1}^{r} \deg H_i + \sum_{j=1}^{r-1} \deg F_j\right) + 2r = \frac{\dim \mathfrak{s} + \operatorname{ind} \mathfrak{s}}{2}.$$

The generic isotropy groups of $(G{:}\bigwedge^2 \Bbbk^n)$ and $(\tilde{G}{:}\bigwedge^2 \Bbbk^n)$ are the same and $\tilde{\mathfrak{s}}$ has the "codim-2" property by [11]. Therefore $\mathfrak{s}$ has the "codim–2" property as well. The polynomials $F_1, \dots, F_{r-1}, F_r', F_{r+1}$ freely generate $\Bbbk[V^*]^G$ [1, 17] and the other generators, $H_1, \dots, H_r$, are algebraically independent over $\Bbbk[V^*]$. Therefore Theorem 2.1 applies and provides the result. □

The case of an odd n is much more difficult, because a generic stabiliser for $(G{:}V)$ is not reductive. We conjecture that $\mathcal{S}(\mathfrak{s})^{\mathfrak{s}}$ is still a polynomial ring, but the proof would require a subtle detailed analysis of the generators $H_1, \dots, H_\ell$. Since that case is not used in this paper, we postpone the exploration. Note only that if $n = 3$, then there is an isomorphism $\bigwedge^2 \Bbbk^3 \simeq (\Bbbk^3)^*$. The pair $(\mathrm{SL}_3, \Bbbk^3 \oplus (\Bbbk^3)^*)$ was considered in [23], where it is shown that the corresponding $\mathfrak{s}$ has (FA).

5 The Classification for the Orthogonal Algebra

In this section, $G = \mathrm{Spin}_n$. We classify the finite-dimensional rational representations $(G : V)$ such that g.i.g.$(G : V)$ is infinite and the symmetric invariants of $\mathfrak{s} = \mathfrak{g} \ltimes V$ form a polynomial ring. The answer is given in Table 1.

5.1 The Negative Cases in Table 1

Most of the negative cases (i.e., those having '–' in column (FA) in Table 1) are justified by Proposition 3.3 and the reductions of Example 3.4. Another similar diagram is presented below:

$$\begin{array}{ccc} (\mathrm{Spin}_{12}, 2\varphi_1 + \varphi_5) & & \\ & \searrow & \\ (\mathrm{Spin}_{11}, \varphi_1 + \varphi_5) & \longrightarrow & \boxed{(\mathrm{Spin}_{10}, \varphi_4 + \varphi_5)}\,. \end{array} \qquad (2)$$

That is, our next step is to show that $(\mathrm{Spin}_{10}, \varphi_4 + \varphi_5)$ does not have (FA). Once this is done, we will know that all the cases in Diagram (2) are indeed negative. Afterwards, only one negative case is left, namely $(\mathrm{Spin}_{12}, \varphi_1 + \varphi_5)$.

Theorem 5.1 *The semi-direct product* $\mathfrak{s} = \mathfrak{so}_{10} \ltimes (\varphi_4 + \varphi_5)$ *does not have* (FA).

Proof Here $G = \mathrm{Spin}_{10}$ is a subgroup of $\mathrm{Spin}_{11} \subset \mathrm{GL}(V)$ and $V \simeq V^*$ as a Spin_{11}-module. A generic isotropy group in Spin_{11} is SL_5. A generic isotropy group in Spin_{10} is SL_4. There is a divisor $D \subset V$ such that G_y is connected and $\mathfrak{g}_y = \mathfrak{sl}_3 \ltimes \mathfrak{heis}_3$ for a generic point $y \in D$. The stabiliser $\mathfrak{g}_y$ is obtained as an intersection of $\mathfrak{sl}_5$ and a specially chosen $\mathfrak{so}_{10} \subset \mathfrak{so}_{11}$.

Assume that $\mathfrak{s}$ has (FA). Then $\mathcal{S}(\mathfrak{s})^{\mathfrak{s}} = \Bbbk[H_1, H_2, H_3, F_1, F_2]$, where $\Bbbk[V^*]^G = \Bbbk[F_1, F_2]$. According to Proposition 3.2, the restrictions $H_i|_{\mathfrak{g}+x}$ are generators of $\mathcal{S}(\mathfrak{sl}_4)^{\mathrm{SL}_4}$ for $x \in V^*$ generic. Therefore we may assume that $\deg_{\mathfrak{g}} H_i = i+1$. By Theorem 2.2, there is $y \in D$ with G_y as above such that the differentials $dF_1, dF_2, dH_1, dH_2, dH_3$ are linearly independent on a non-empty open subset of $\mathfrak{g} + y$ that is stable w.r.t. $G_y \ltimes \exp(V)$.

Take $\xi = \gamma + y$ with $\gamma \in \mathfrak{g}$ generic. Replacing γ by another point in $\gamma + \mathrm{ad}^*(V)y$ we may safely assume that γ is zero on $\mathrm{Ann}\,(\mathfrak{g}_y)$. Let $\bar{\gamma}$ stand for the restriction of γ to $\mathfrak{g}_y$. Then $\mathfrak{s}_\xi = (\mathfrak{g}_y)_{\bar{\gamma}} \oplus \Bbbk^2 = (\mathfrak{t}_2 \oplus \Bbbk z) \oplus \Bbbk^2$, where $\Bbbk z$ is the centre of $\mathfrak{heis}_3$, $\mathfrak{t}_2$ is a Cartan subalgebra of $\mathfrak{sl}_3$, and $\Bbbk^2 \subset V$.

We have $(dF_i)_\xi \in \mathfrak{s}_\xi \cap V = \Bbbk^2$. At the same time $(dH_i)_\xi = \eta_i + u_i$, where $u_i \in V$, $\eta_i \in \mathfrak{g}$, and η_i is the differential of $H_i|_{\mathfrak{g}+y}$ at γ. Since γ was chosen to be generic, the elements η_1, η_2, η_3 are linearly independent. Hence the restrictions $\mathbf{h}_i := H_i|_{\mathfrak{g}+y}$ are algebraically independent.

It can be easily seen that $\mathrm{ind}\,\mathfrak{g}_y = 3$ and that $\mathfrak{g}_y$ satisfies the "codim–2" condition. Since $\deg \mathbf{h}_i = i+1$, we have $\mathcal{S}(\mathfrak{g}_y)^{G_y} = \Bbbk[\mathbf{h}_1, \mathbf{h}_2, \mathbf{h}_3]$ by Theorem 2.1. But $z \in \mathcal{S}(\mathfrak{g}_y)^{G_y}$ and $\deg z = 1$. A contradiction! □

Theorem 5.2 *The semi-direct product* $\mathfrak{s} = \mathfrak{so}_{12} \ltimes (\varphi_1 + \varphi_5)$ *does not have* (FA).

Proof Here $G = \mathrm{Spin}_{12}$ and a generic isotropy group for the G-action on V_{φ_5} (resp. $V_{\varphi_1} \oplus V_{\varphi_5}$) is SL_6 (resp. SL_5). Let f be a Spin_{12}-invariant quadratic form on $V_{\varphi_1} \simeq V^*_{\varphi_1}$. Then $D = \{f = 0\} \times V^*_{\varphi_5}$ is a G-stable divisor in V^*. It can be verified that, for a generic point $y \in D$, one has $\mathfrak{g}_y = \mathfrak{sl}_4 \ltimes \mathfrak{heis}_4$ and G_y is connected. As in the proof of Theorem 5.1, $\mathcal{S}(\mathfrak{g}_y)^{G_y}$ has an element of degree 1, i.e., it is not generated by symmetric invariants of degrees 2, 3, 4, 5, but it would have been if $\mathfrak{s}$ had (FA). □

5.2 The Positive Cases in Table 1

We now proceed to the positive cases. Note first that all the instances, where $\mathfrak{h}$ is of type $\mathbf{A}_1$, are covered by Example 3.5.

Proposition 5.3 (Item 1) *The semi-direct product* $\mathfrak{s} = \mathfrak{so}_n \ltimes m\Bbbk^n$ *with* $m < n$ *has* (FA).

Proof We have $G \lhd \tilde{G}$ with $\tilde{G} = \mathrm{SO}_n \times \mathrm{SO}_m$ and $\mathfrak{s} \lhd \tilde{\mathfrak{s}}$ for $\tilde{\mathfrak{s}} = \tilde{\mathfrak{g}} \ltimes V$. The Lie algebra $\tilde{\mathfrak{s}}$ is the $\mathbb{Z}_2$-contraction of $\mathfrak{so}_{n+m}$ related to the symmetric subalgebra $\mathfrak{so}_n \oplus \mathfrak{so}_m$. Let $x \in V^*$ be generic. Then $\tilde{G}_x = G_x = \mathrm{SO}_{n-m}$. According to [11], $\Bbbk[\tilde{\mathfrak{s}}^*]^{\tilde{\mathfrak{s}}} =$

$\Bbbk[V^*]^{\tilde{G}}[H_1, \ldots, H_\ell]$ is a polynomial ring, $\ell = [\frac{n-m}{2}]$. By Lemma 3.1, $H_i \in \mathcal{S}(\mathfrak{s})$ for every i. Next, $\mathfrak{s}$ has the "codim–2" property if $m = 1$ by [11], hence $\mathfrak{s}$ always has it. The polynomials H_i are algebraically independent over $\Bbbk(V^*)$ and $\Bbbk[V^*]^G$ has $\frac{m(m+1)}{2}$ generators of degree 2. Thereby we have $\operatorname{ind}\mathfrak{s}$ algebraically independent homogeneous invariants with the total sum of degrees being equal to

$$m(m+1) + \sum_{i=1}^{\ell} \deg H_i = m(m+1) + \boldsymbol{b}(\tilde{\mathfrak{s}}) - m(m+1) = \boldsymbol{b}(\tilde{\mathfrak{s}}) = \boldsymbol{b}(\mathfrak{so}_{n+m}) = \boldsymbol{b}(\mathfrak{s}).$$

According to Theorem 2.1, $\Bbbk[\mathfrak{s}^*]^S = \Bbbk[V^*]^G[H_1, \ldots, H_\ell]$. □

Theorem 5.4 (Item 9b) *The semi-direct product $\mathfrak{s} = \mathfrak{so}_{14} \ltimes (\varphi_1 + \varphi_6)$ has (FA).*

Proof Here $G = \mathrm{Spin}_{14}$ and the pair $(\mathrm{Spin}_{14}, V_{\varphi_6}^*)$ is of rank one. Let $v \in V_{\varphi_6}^*$ be a generic point. Then $G_v = L \times L$, where L is the connected group of type $\mathbf{G}_2$. By Theorem 3.9, the restriction homomorphism

$$\psi_v : \Bbbk[\mathfrak{s}^*]^S \to \Bbbk[\mathfrak{g}^* \oplus V_{\varphi_1}^* + v]^{G_v \ltimes \exp(V)} \simeq \Bbbk[\mathfrak{g}_v^* \oplus V_{\varphi_1}^*]^{G_v \ltimes \exp(V_{\varphi_1})}$$

is surjective. Furthermore, $V_{\varphi_1} \simeq \Bbbk^{14} = \Bbbk^7 \oplus \Bbbk^7$ as an $L \times L$-module, where each $\Bbbk^7$ is a simplest irreducible $\mathbf{G}_2$-module. Hence $G_v \ltimes \exp(V_{\varphi_1}) = Q \times Q$, where $Q = L \ltimes \exp(\Bbbk^7)$. The group Q has a free algebra of symmetric invariants and $\operatorname{ind}\mathfrak{q} = 3$ [14].

There are irreducible tri-homogeneous polynomials $H_1, \ldots, H_6 \in \Bbbk[\mathfrak{s}^*]^S$ such that, for a generic point $v \in V_{\varphi_6}^*$, their images $h_i = \psi_v(H_i)$ generate $\mathcal{S}(\mathfrak{q}\times\mathfrak{q})^{Q\times Q}$. Let f be a basic G-invariant in $\Bbbk[V_{\varphi_6}^*]$.

Although the group $G \ltimes \exp(\Bbbk^{14})$ is not reductive, we can argue in the spirit of [22, Section 2] and conclude that $\Bbbk[\mathfrak{s}^*]^S[\frac{1}{f}] = \Bbbk[H_1, \ldots, H_6, f, \frac{1}{f}]$. Then the equality

$$\Bbbk[\mathfrak{s}^*]^S = \Bbbk[H_1, \ldots, H_6, f]$$

holds if and only if the restrictions of the polynomials $\{H_i\}$ to $V_{\varphi_1}^* \times D$ are algebraically independent, where $D = \{f = 0\} \subset V_{\varphi_6}^*$.

Let $G{\cdot}y \subset D$ be the dense open orbit. Then G_y is connected and $\mathfrak{g}_y = \mathfrak{l} \ltimes \mathfrak{l}^{\mathsf{ab}}$ is the Takiff Lie algebra in type $\mathbf{G}_2$, $\mathfrak{l} = \mathsf{Lie}\, L$. There is only one possible embedding of $\mathfrak{g}_y$ into $\mathfrak{so}_{14}$. Under the non-Abelian $\mathfrak{l}$ the space $\Bbbk^{14}$ decomposes as a sum of two 7-dimensional simple modules. The Abelian ideal $\mathfrak{l}^{\mathsf{ab}}$ takes one copy of $\Bbbk^7$ into another. In other words, $\mathfrak{g}_y \ltimes \Bbbk^{14} = \mathfrak{q} \ltimes \mathfrak{q}^{\mathsf{ab}}$. By [15, Example 4.1], Theorem 2.2 of the same paper [15] applies to $\mathfrak{q}$ and guarantees us that the symmetric invariants of $\mathfrak{q}\ltimes\mathfrak{q}^{\mathsf{ab}}$ form a polynomial ring in 6 generators, where the degrees of the basic invariants are the same as in the case of $\mathfrak{q} \oplus \mathfrak{q}$.

It remains to observe that the proof of [14, Theorem 2.3] can be repeated for the semi-direct product $(G \ltimes \exp(V_{\varphi_1})) \ltimes \exp(V_{\varphi_6})$ producing a suitable modification of the elements $H_1, \ldots, H_6$, cf. Theorem 3.6. □

Corollary 5.5 (Item 5a) *The reduction*

$$(\mathrm{Spin}_{14}, \varphi_1 + \varphi_6) \longrightarrow (\mathrm{Spin}_{13}, \varphi_6)$$

shows that also $(\mathrm{Spin}_{13}, \varphi_6)$ *has* (FA)*, see Proposition 3.3.*

Theorem 5.6 *The semi-direct product* $\mathfrak{s} = \mathfrak{so}_{14} \ltimes (2\varphi_1 + \varphi_6)$ *has* (FA)*.*

Proof Here $G = \mathrm{Spin}_{14}$ and the proof follows the same lines as the proof of Theorem 5.4. We split the group S as $(G \ltimes \exp(2V_{\varphi_1})) \ltimes \exp(V_{\varphi_6})$. Now $Q = L \ltimes \exp(2\Bbbk^7)$ and again $G_v \ltimes \exp(2V_{\varphi_1}) = Q \times Q$. By [14], $\mathfrak{q}$ has (FA) and the "codim–2" property. Here $\mathrm{ind}\,\mathfrak{q} = 4$ and we have eight polynomials $H_i \in \Bbbk[\mathfrak{s}^*]^S$ such that their restrictions to $\mathfrak{g} \oplus (2V^*_{\varphi_1}) + v$ generate $\mathcal{S}(\mathfrak{q} \oplus \mathfrak{q})^{Q \times Q}$. These polynomials are tri-homogeneous w.r.t. the decomposition $\mathfrak{s} = \mathfrak{g} \oplus 2V_{\varphi_1} \oplus V_{\varphi_6}$. Again $\mathfrak{g}_v \ltimes (V_{\varphi_1} \oplus V_{\varphi_1}) = \mathfrak{q} \ltimes \mathfrak{q}^{\mathsf{ab}}$, [15, Theorem 2.2] applies to $\mathfrak{q}$ and assures that the symmetric invariants of $\mathfrak{q} \ltimes \mathfrak{q}^{\mathsf{ab}}$ form a polynomial ring in 8 generators, where the degrees of the basic invariants are the same as in the case of $\mathfrak{q} \oplus \mathfrak{q}$. □

Corollary 5.7 *The reductions*

$$(\mathrm{Spin}_{14}, 2\varphi_1 + \varphi_6) \longrightarrow (\mathrm{Spin}_{13}, \varphi_1 + \varphi_6 + \Bbbk) \longrightarrow (\mathrm{Spin}_{12}, \varphi_5 + \varphi_6 + \Bbbk)$$

show that the pairs $(\mathrm{Spin}_{13}, \varphi_1 + \varphi_6)$ *and* $(\mathrm{Spin}_{12}, \varphi_5 + \varphi_6)$ *also have* (FA)*, see Proposition 3.3.*

Many representations in types $\mathbf{D}_4$, $\mathbf{B}_4$, and $\mathbf{B}_3$ are covered by reductions from $\mathbf{D}_5$. Among the type $\mathbf{D}_5$ cases, the following one is easy to handle.

Example 5.8 (Item7a) The pair $(\mathbf{D}_5, \varphi_4)$ is of rank zero and therefore the open Spin_{10}-orbit in $\Bbbk^{10}$ is big. The existence of the isomorphism $\Bbbk[\mathfrak{g} + x]^{G_x \ltimes \exp(V)} \simeq \mathcal{S}(\mathfrak{g}_x)^{G_x}$ [22] shows that $\mathcal{S}(\mathfrak{s})^{\mathfrak{s}} \simeq \mathcal{S}(\mathfrak{h})^H$, where H is the isotropy group of an element in the open orbit and $\mathfrak{h} = \mathsf{Lie}\,H$. In order to be more explicit, H is connected and $\mathfrak{h} = \mathfrak{so}_7 \ltimes \Bbbk^8$, where $\mathfrak{so}_7$ acts on $\Bbbk^8$ via the spin-representation. The algebra $\mathcal{S}(\mathfrak{h})^{\mathfrak{h}}$ is free by [22, Example 3.8]. By a coincidence, the semi-direct product encoded by $(\mathbf{D}_5, \varphi_4)$ is also a truncated maximal parabolic subalgebra $\mathfrak{p}$ of $\mathbf{E}_6$. The symmetric invariants of $\mathfrak{p}$ are studied in [3] and by a computer aided calculation it is shown there that $\mathcal{S}(\mathfrak{p})^{\mathfrak{p}}$ is a polynomial ring with three generators of degrees 6, 8, and 18.

Below we list the 'top' pairs that have to be treated individually. They are divided into two classes, in the first class $\dim V/\!\!/G = 1$ and in the second $\dim V/\!\!/G > 1$.

$$\begin{cases} \text{Rank one pairs: } (\mathbf{B}_5, \varphi_5), (\mathbf{D}_5, 2\varphi_4), (\mathbf{D}_6, \varphi_5), (\mathbf{D}_7, \varphi_6); \\ \text{higher rank pairs: } (\mathbf{D}_5, \varphi_1 + \varphi_4), (\mathbf{D}_5, 2\varphi_1 + \varphi_4), (\mathbf{D}_5, 3\varphi_1 + \varphi_4), (\mathbf{D}_6, 2\varphi_5). \end{cases} \tag{3}$$

Theorem 5.9 *The rank one pairs listed in* (3) *have* (FA).

Proof In case of $(\mathbf{D}_5, 2\varphi_4)$ a generic stabiliser is of type $\mathbf{G}_2$. This pair is covered by Lemma 3.7. For the other three pairs, many conditions of Theorem 3.6 are satisfied. For each pair, there is an open orbit $G{\cdot}y \subset D$, where D stands for the zero set of the generator $F \in \Bbbk[V^*]^G$. It remains to inspect the symmetric invariants of G_y.

A generic isotropy group for $(\mathbf{B}_5, \varphi_5)$ is SL_5, G_y is connected, and $\mathfrak{g}_y$ is a $\mathbb{Z}_2$-contraction of $\mathfrak{sl}_5$, the semi-direct product $\mathfrak{so}_5 \ltimes V_{\varphi_1^2}$, which is a good $\mathbb{Z}_2$-contraction [11].

A generic isotropy group for $(\mathbf{D}_6, \varphi_5)$ is SL_6, G_y is connected, and $\mathfrak{g}_y$ is a $\mathbb{Z}_2$-contraction of $\mathfrak{sl}_6$, the semi-direct product $\mathfrak{sp}_6 \ltimes V_{\varphi_2}$, which is a good $\mathbb{Z}_2$-contraction [21, Theorem 4.5].

A generic isotropy group for $(\mathbf{D}_7, \varphi_6)$ is $L \times L$, where L is the connected group of type $\mathbf{G}_2$, G_y is connected, and $\mathfrak{g}_y$ is the Takiff algebra $\mathfrak{l} \ltimes \mathfrak{l}^{\mathsf{ab}}$, where $\mathfrak{l} = \mathsf{Lie}\, L$. The basic symmetric invariants of $\mathfrak{g}_y$ have the same degrees as in the case of $\mathfrak{l} \oplus \mathfrak{l}$ [18].

□

Example 5.10 (Item 7d) For the pair $(\mathbf{D}_5, 3\varphi_1 + \varphi_4)$, a generic isotropy group is connected and is of type $\mathbf{A}_2$. Let $D \subset V^*$ be a G-invariant divisor. Then there are at least two copies of $\Bbbk^{10}$ in V^* such that the projection of D on each of them is surjective. For a generic $y \in D$, $G_y = (\mathrm{Spin}_8)_{\tilde{y}}$, where $\tilde{y}$ is a generic point of a Spin_8-invariant divisor $\tilde{D} \subset V_{\varphi_1} \oplus V_{\varphi_3} \oplus V_{\varphi_4}$ (here highest weights of Spin_8 are meant). Continuing the computation one obtains that $G_y = L_v$, where L is the connected group of type $\mathbf{G}_2$ and v is a highest weight vector in $\Bbbk^7$. The group L_v has a free algebra of symmetric invariants generated in degrees 2 and 3, see [14, Lemma 3.9]. Therefore Lemma 3.8 applies.

The remaining three higher rank pairs listed in (3) require elaborate arguments. For all of them, Theorem 3.9 will be the starting point. Note that the pair $(\mathrm{SO}_n, \Bbbk^n)$ is of rank one. We let $(\ ,\ )$ denote a non-degenerate SO_n-invariant scalar product on $\Bbbk^n$.

Theorem 5.11 (Item 7b) *The semi-direct product* $\mathfrak{s} = \mathfrak{so}_{10} \ltimes (\varphi_1 + \varphi_4)$ *has* (FA).

Proof Here $G = \mathrm{Spin}_{10}$ and we use the reduction

$$(\mathrm{Spin}_{10}, \varphi_1{+}\varphi_4) \to (\mathrm{Spin}_9, \varphi_4) \tag{4}$$

in the increasing direction, starting from the smaller representation and its invariants. By Theorem 3.9, the restriction homomorphism

$$\psi_v : \Bbbk[\mathfrak{s}^*]^{\mathfrak{s}} \to \Bbbk[\mathfrak{g}^* \oplus V^*_{\varphi_4} + v]^{G_v \ltimes \exp(V)} \simeq \Bbbk[\mathfrak{g}^*_v \oplus V^*_{\varphi_4}]^{G_v \ltimes \exp(V_{\varphi_4})}$$

is surjective for generic $v \in V^*_{\varphi_1}$. Here $G_v = \mathrm{Spin}_9$. The group $Q = G_v \ltimes \exp(V_{\varphi_4})$ has a free algebra of symmetric invariants [11, Theorem 4.7]. More explicitly, $\mathcal{S}(\mathfrak{q})^Q$ is generated by $(\ ,\ )$ on $\Bbbk^{16}$ and three bi-homogeneous polynomials $\mathbf{h}_1$, $\mathbf{h}_2$, $\mathbf{h}_3$ of bi-degrees $(2,4)$, $(4,4)$, $(6,6)$. Note that each generator is unique up to a non-zero scalar. Whenever $(\xi,\xi) \neq 0$ for $\xi \in V^*_{\varphi_4}$, we have $\mathbf{h}_i|_{\mathfrak{so}_9+\xi} = \Delta_{2i}$, where each Δ_{2i} is a basic symmetric invariant of $\mathfrak{so}_7 = (\mathfrak{so}_9)_\xi$. The generators Δ_{2i} are now fixed and they do not depend on the choice of ξ.

Take $H_i \in \Bbbk[\mathfrak{s}^*]^{\mathfrak{s}}$ with $\psi_v(H_i) = \mathbf{h}_i$. Without loss of generality, we may assume that H_i is homogeneous w.r.t. to $\mathfrak{g}$ and V_{φ_4}. The uniqueness of the basic symmetric $\mathfrak{q}$-invariants, allows us to take a suitable tri-homogeneous component of each H_i, see Theorem 3.9. Now assume that each H_i is irreducible. Whenever $(\xi,\xi) \neq 0$ for $\xi \in \Bbbk^{16}$ and $(\eta,\eta) \neq 0$ for $\eta \in \Bbbk^{10}$, we have $H_i|_{\mathfrak{g}+x} = a_x \Delta_{2i}$, where $x = \eta + \xi$ and $a_x \in \Bbbk^{\times}$.

According to [22, Lemma 3.5(ii)], we have $\mathcal{S}(\mathfrak{s})^{\mathfrak{s}} = \Bbbk[V^*]^G[H_1, H_2, H_3]$ if and only if the restrictions $H_i|_{\mathfrak{g}\times D}$ are algebraically independent over $\Bbbk[D]^G$ for each G-invariant divisor $D \subset V^*$.

If D contains a point $av + \xi$ with $\xi \in \Bbbk^{16}$ and $a \neq 0$, a relation among $H_i|_{\mathfrak{g}\times D}$ leads to a relation among the restrictions of $\mathbf{h}_i$ to $\mathfrak{so}_9 \times \tilde{D}$ for some Spin_9-invariant divisor $\tilde{D} \subset \Bbbk^{16}$. Moreover, this new relation is over $\Bbbk[\{v\}\times D]^{G_v} = \Bbbk$. Since the polynomials $\mathbf{h}_i$ freely generate $\mathcal{S}(\mathfrak{q})^{\mathfrak{q}}$ over $\Bbbk[V^*_{\varphi_4}]^{\mathrm{Spin}_9}$, nothing of this sort can happen. Therefore there is a unique suspicious divisor, namely, the divisor $D = \tilde{D} \times \Bbbk^{16}$, where $\tilde{D} = \{u \in \Bbbk^{10} \mid (u,u) = 0\}$.

Since each H_i is irreducible, it is non-zero on $\mathfrak{g}\times D$. Therefore there is a point $\xi \in \Bbbk^{16}$ such that $(\xi,\xi) \neq 0$ and $H_i|_{\mathfrak{g}\times\tilde{D}\times\{\xi\}} \neq 0$ for all i. Here $G_\xi = \mathrm{Spin}_7 \ltimes \exp(\Bbbk^8)$ and $\Bbbk^{10} \subset V^*$ decomposes as $\Bbbk \oplus \Bbbk^8 \oplus \Bbbk$ under G_ξ. The Abelian ideal $\Bbbk^8$ of $\mathfrak{g}_\xi$ takes $\Bbbk$ to $\Bbbk^8$ and then $\Bbbk^8$ to another copy of $\Bbbk$. Note that the vectors in each copy of $\Bbbk$ are isotropic. Take $u \neq 0$ in the first copy and $u' \neq 0$ in the second copy of $\Bbbk$. Set $\eta_t = u + tu'$, $x_t = \eta_t + \xi$ for $t \in \Bbbk$, $y = u + \xi$. Then $G_{x_t} = G_y \simeq \mathrm{Spin}_7$.

We have $(\eta_t,\eta_t) \neq 0$ for $t \neq 0$ and hence $H_i|_{\mathfrak{g}+x_t} = a_t \Delta_{2i} \neq 0$, whenever $t \neq 0$. Here $a_t \Delta_{2i} \in \mathcal{S}(\mathfrak{g}_{x_t}) = \mathcal{S}(\mathfrak{g}_y)$. Clearly $H_i|_{\mathfrak{g}\times\{y\}} = \lim\limits_{t\to 0} a_t \Delta_{2i}$ and it is either zero or a non-zero scalar multiple of Δ_{2i}. If the second possibility takes place for all i, when the restrictions of H_i to $\mathfrak{g} \times D$ are algebraically independent over $\Bbbk[D]$. Thus, it remains to prove that $H_i|_{\mathfrak{g}\times\{y\}} \neq 0$ for all i.

Assume that $H_i|_{\mathfrak{g}\times\{y\}} = 0$. Then H_i vanishes on $\mathfrak{g}\times G_\xi{\cdot}u\times\{\xi\}$ and also on $\mathfrak{g}\times G_\xi{\cdot}\Bbbk u\times\{\xi\}$, since H_i is tri-homogeneous. The subset $G_\xi{\cdot}\Bbbk u$ is dense in $\tilde{D}$ (it equals $\tilde{D}\setminus\{0\}$), hence H_i vanishes on $\mathfrak{g}\times\tilde{D}\times\{\xi\}$, too. However, this contradicts the choice of ξ. □

Theorem 5.12 (Item 7c) *If* $\mathfrak{s}$ *is given by* $(\mathbf{D}_5, 2\varphi_1 + \varphi_4)$, *then* $\Bbbk[\mathfrak{s}^*]^S = \Bbbk[V^*]^G[H_2, H_6]$ *is a polynomial ring and the multi-degrees of* H_i *are* $(2, 2, 2, 4)$, $(6, 4, 4, 8)$.

Proof For this pair, the chain of reductions is

$$(\mathrm{Spin}_{10}, 2\varphi_1{+}\varphi_4) \to (\mathrm{Spin}_9, \varphi_1{+}\varphi_4{+}\Bbbk) \to (\mathrm{Spin}_8, \varphi_3{+}\varphi_4{+}\Bbbk) \to (\mathrm{Spin}_7, \varphi_3{+}\Bbbk) \tag{5}$$

and again we are tracing the chain from the smaller groups to the larger.

By [22, Prop. 3.10], the symmetric invariants of $\mathrm{Spin}_7 \ltimes \exp(V_{\varphi_3})$ are freely generated by the following three polynomials: the scalar product $(\, . \, , . \,)$ on $V_{\varphi_3} \simeq \Bbbk^8$, $\mathbf{h}_2$, and $\mathbf{h}_6$. Here the bi-degrees of the last two are $(2, 2)$, $(6, 4)$. We are lucky that all three generators are unique (up to a scalar) and $\mathfrak{so}_7 \ltimes V_{\varphi_3}$ has the "codim–2" property. One can easily deduce that all items in (5) have the "codim–2" property. A generic isotropy group for $(\mathrm{Spin}_7 : V_{\varphi_3})$, say L, is the connected simple group of type $\mathbf{G}_2$. Take $u \in V_{\varphi_3}$ with $(u, u) \neq 0$. Then $(\mathfrak{so}_7)_u = \mathfrak{l} = \mathrm{Lie}\, L$. Let $h_2, h_6 \in \mathcal{S}((\mathfrak{so}_7)_u)$ be the restrictions of $\mathbf{h}_2$, $\mathbf{h}_6$ to $\mathfrak{so}_7{+}u$. Then h_2 and h_6 generate $\mathcal{S}(\mathfrak{l})^L$. We have $\dim \mathcal{S}^2(\mathfrak{l})^L = 1$, the generator of degree 2 is unique (up to a non-zero scalar). In the space $\mathcal{S}^6(\mathfrak{l})^L = \Bbbk h_2^3 \oplus \Bbbk h_6$, the generator h_6 is characterised by the property that it is the restriction of an invariant of $\mathrm{Spin}_7 \ltimes \exp(V_{\varphi_3})$ of bi-degree $(6, 4)$. This property does not depend on the choice of u.

Consider next $\mathfrak{s}_2 := \mathfrak{so}_8 \ltimes (V_1 \oplus V_2)$, where $V_1 = V_{\varphi_3}$, $V_2 = V_{\varphi_4}$. Choose $v \in V_1^*$ with $(v, v) \neq 0$. By Theorem 3.9, there are $\hat{h}_2, \hat{h}_6 \in \mathcal{S}(\mathfrak{s}_2)^{\mathfrak{s}_2}$ such that $\hat{h}_i|_{\mathfrak{so}^8 \oplus V_2^* + v} = \mathbf{h}_i$. One can safely replace $\hat{h}_2$ by its component of degrees 2 in $\mathfrak{so}_8$, 2 in V_2 and replace $\hat{h}_6$ by its component of degrees 6 in $\mathfrak{so}_8$, 4 in V_1. The uniqueness of generators in the case of $\mathrm{Spin}_7 \ltimes \exp(V_{\varphi_3})$ allows also to take tri-homogeneous components. Suppose now that each $\hat{h}_i$ is irreducible. Set $a_i = \deg_{V_2} \hat{h}_i$. Choose $v_2 \in V_2^*$ with $(v_2, v_2) \neq 0$. The restriction $\hat{h}_2|_{\mathfrak{so}_8 \oplus V_2 + v_2}$ is an invariant of bi-degree $(2, a_2)$ and either $a_2 = 2$ or this restriction is divisible by the invariant of bi-degree $(0, 2)$. In the last case, $\hat{h}_2$ is divisible by a generator of $\Bbbk[V_2]^{\mathrm{SO}_8}$. A contradiction. Since $\hat{h}_6|_{\mathfrak{so}_8 + v + v_2} = h_6$ and since in addition $\hat{h}_6$ is irreducible, the restriction $\hat{h}_6|_{\mathfrak{so}_8 \oplus V_1^* + v_2}$ is an invariant of bi-degree $(6, 4)$, i.e., $a_6 = 4$. Making use of Theorem 2.1, we conclude that $\Bbbk[\mathfrak{s}_2^*]^{\mathfrak{s}_2} = \Bbbk[V_1^* \oplus V_2^*]^{\mathrm{Spin}_8}[\hat{h}_2, \hat{h}_6]$.

The Spin_9-actions on $V_{\varphi_1} = \Bbbk^9$ and $V_{\varphi_4} = \Bbbk^{16}$ are of rank one. By [2], $\mathsf{g.i.g.}(\mathrm{Spin}_9 : V_{\varphi_4}) = \mathrm{Spin}_7$, and $\Bbbk^9|_{\mathrm{Spin}_7}$ is the Spin_7-module $V_{\varphi_3} \oplus \Bbbk$. The restriction homomorphism $\Bbbk[V_{\varphi_1}^* \oplus V_{\varphi_4}^*]^{\mathrm{Spin}_9} \to \Bbbk[V_{\varphi_3}^* \oplus \Bbbk]^{\mathrm{Spin}_7}$ is onto. Using Theorem 3.9 and the reductions

$$\begin{array}{ccccc} & & (\mathrm{Spin}_8, \varphi_3{+}\varphi_4) & & \\ & \nearrow & & \searrow & \\ (\mathrm{Spin}_9, \varphi_1{+}\varphi_4) & & & & (L, 0) \\ & \searrow & & \nearrow & \\ & & (\mathrm{Spin}_7, \varphi_3{+}\Bbbk) & & \end{array} \tag{6}$$

we prove that there are algebraically independent over $\Bbbk[V^*_{\varphi_1}\oplus V^*_{\varphi_4}]$ symmetric invariants of tri-degrees (2, 2, 4), (6, 4, 8) w.r.t. $\mathfrak{so}_9 \oplus \Bbbk^9 \oplus \Bbbk^{16}$. They generate the ring of symmetric invariants related to $(\mathrm{Spin}_9, \varphi_1{+}\varphi_4)$ over $\Bbbk[V^*_{\varphi_1}\oplus V^*_{\varphi_9}]^{\mathrm{Spin}_9}$ by Theorem 2.1.

One can make a reduction step from $\mathfrak{s}$ to $(\mathrm{Spin}_9, V_{\varphi_1}\oplus V_{\varphi_9})$ using either of the two copies of V_{φ_1}. This allows one to find algebraically independent over $\Bbbk[V^*]$ polynomials $H_2, H_6 \in \Bbbk[\mathfrak{s}^*]^{\mathfrak{s}}$ of multi-degrees (2, 2, 2, 4) and (6, 4, 4, 8), respectively. The basic invariants on V^* are of degrees 2, 2, 2, 3, 3. Thus, the total sum of degrees is

$$10 + 22 + 12 = 44 \quad \text{and} \quad \dim \mathfrak{s} + \operatorname{ind} \mathfrak{s} = 45 + 20 + 16 + 7 = 88.$$

Therefore, by Theorem 2.1, we have $\Bbbk[\mathfrak{s}^*]^S = \Bbbk[V^*]^G[H_2, H_6]$. □

The case of $(\mathbf{D}_6, 2\varphi_5)$ is very complicated. We begin by introducing some notation and stating a few facts related to this pair. First, $V_{\varphi_5} \simeq V^*_{\varphi_5}$ as a G-module. Second, the representation of G on V_{φ_5} is of rank one and $\Bbbk[V^*_{\varphi_5}]^G = \Bbbk[F]$, where F is a homogeneous polynomial of degree 4. It would be convenient to write $V = V_1 \oplus V_2$, where each V_i is a copy of V_{φ_5} and let F stand for the generator of $\Bbbk[V_1^*]^G$. Further, there is a natural action of SL_2 on V. We suppose that $\begin{pmatrix} 0 & 0 \\ 1 & 0 \end{pmatrix} \cdot V_2 = 0$ and that $\begin{pmatrix} 0 & 0 \\ 1 & 0 \end{pmatrix} \cdot V_1 = V_2$ for $\begin{pmatrix} 0 & 0 \\ 1 & 0 \end{pmatrix} \in \mathfrak{sl}_2$. The ring $\Bbbk[V^*]^G$ has 7 generators:

$$F = F_{(4,0)}, F_{(3,1)}, F_{(2,2)}, F_{(1,3)}, F_{(0,4)}, F_{(1,1)}, F_{(3,3)}.$$

Here $F_{(\alpha,\beta)}$ stands for a particular G-invariant in $\mathcal{S}^{\alpha}(V_1)\mathcal{S}^{\beta}(V_2)$. It is assumed that the first five polynomials build an irreducible SL_2-module and that the last two are SL_2-invariants.

We let SL_2 act on $\mathfrak{g}$ trivially and thus obtain an action of SL_2 on $\Bbbk[\mathfrak{s}^*]^S$. Note that if $H \in \Bbbk[\mathfrak{s}^*]$ and $\deg_{V_1} H > \deg_{V_2} H$, then $\begin{pmatrix} 0 & 0 \\ 1 & 0 \end{pmatrix} \cdot H \neq 0$.

Let $v \in V_1^*$ be a generic point and

$$\psi_v : \Bbbk[\mathfrak{s}^*]^S \to \Bbbk[\mathfrak{g}\oplus V_2^*{+}v]^{G_v \ltimes \exp(V)} \simeq \mathcal{S}(\mathfrak{g}_v \ltimes V_2)^{G_v \ltimes \exp(V_2)}$$

be the corresponding restriction homomorphism. Here $G_v = \mathrm{SL}_6$ and

$$V_2 = \textstyle\bigwedge^2 \Bbbk^6 \oplus (\bigwedge^2 \Bbbk^6)^* \oplus 2\Bbbk$$

as a G_v-module. Set $\mathfrak{q} = \mathfrak{g}_v \ltimes V_2$. By Proposition 4.1, $\mathcal{S}(\mathfrak{q})^{\mathfrak{q}}$ is a polynomial ring and $\Bbbk[\mathfrak{q}^*]^{\mathfrak{q}} = \Bbbk[V_2^*]^{\mathrm{SL}_6}[\mathbf{h}_1, \mathbf{h}_2, \mathbf{h}_3]$, where the generators $\mathbf{h}_i$ are of bi-degrees (2, 4), (2, 6), (2, 8).

Let $N_G(\Bbbk v)$ be the normaliser of the line $\Bbbk v$. Then $N_G(\Bbbk v) = C_4{\times}G_v$, where $C_4 = \langle\zeta\rangle$ is a cyclic group of order 4. It is not difficult to see that $\mathrm{Ad}\,(\zeta)A = -A^t$ for each $A \in \mathfrak{g}_v$ and that $\zeta{\cdot}\mathbf{h}_k = (-1)^k\mathbf{h}_k$ for each $k \in \{1, 2, 3\}$. The element ζ^2 multiplies $\psi_v(F_{(1,3)})$ and $\psi_v(F_{(1,1)})$ by -1, the product $\psi_v(F_{(1,3)})\psi_v(F_{(1,1)})$ is a C_4-invariant.

There is a cyclic group of order 4 in $N_G(\Bbbk v) \times \mathrm{GL}(V_1^*)$ that stabilisers v. This means that if $H \in \Bbbk[\mathfrak{s}^*]^S$ is homogeneous in V_1, then $\psi_v(H)$ is an eigenvector of $C_4 \subset N_G(\Bbbk v)$ and the corresponding eigenvalue depends only on $\deg_{V_1} H$.

Theorem 5.13 (Item 8b) *If* $\mathfrak{s}$ *is given by the pair* $(\mathbf{D}_6, 2\varphi_5)$, *then* $\Bbbk[\mathfrak{s}^*]^S = \Bbbk[V^*]^G[H_1, H_2, H_3]$ *is a polynomial ring and the tri-degrees of* H_i *are* $(2, 4, 4)$, $(2, 6, 6)$, $(2, 8, 8)$.

Proof According to Theorem 3.9, there are homogeneous in V_1 elements $H_1, H_2, H_3 \in \mathcal{S}(\mathfrak{s})^{\mathfrak{s}}$ such that $\psi_v(H_i) = \mathbf{h}_i$. There is no harm in assuming that these polynomials are tri-homogeneous. Suppose that $b_i = \deg_{V_1} H_i$ is the minimal possible. Set $a_i = \deg_{V_2} H_i$. Then $a_1 = 4$, $a_2 = 6$, $a_3 = 8$. The eigenvalues of ζ on $\mathbf{h}_i$ indicate that $a_i \equiv b_i \pmod 4$ for each i.

Suppose for the moment that $a_i = b_i$ for all i. It is not difficult to see that $\mathfrak{s}$ satisfies the "codim–2" condition. The elements $\mathbf{h}_1, \mathbf{h}_2, \mathbf{h}_3$ are algebraically independent over $\Bbbk(V_2)$, hence H_1, H_2, H_3 are algebraically independent over $\Bbbk(V^*)$. Thus, we have ten algebraically independent homogeneous invariants. The total sum of their degrees is

$$2 + 6 + 20 + 10 + 14 + 18 = 70 \text{ and } \dim\mathfrak{s} + \mathrm{ind}\,\mathfrak{s} = 66 + 64 + 10 = 140.$$

Thereby $\Bbbk[\mathfrak{s}^*]^S = \Bbbk[V^*]^G[H_1, H_2, H_3]$ by Theorem 2.1. It remains to show that the assumption is correct.

For a generic $v' \in V_2^*$, $\mathfrak{g}_{v'}{\ltimes}V_1 \simeq \mathfrak{q}$ and each $H_i|_{\mathfrak{g}\oplus V_1^*+v'}$ is a symmetric invariant of $\mathfrak{g}_{v'}{\ltimes}V_1$ of degree 2 in $\mathfrak{g}_{v'}$. Since the restrictions $H_i|_{\mathfrak{g}+v+v'}$ are the basic symmetric invariants of $G_{v+v'} = (\mathrm{SL}_2)^3$, the restrictions of H_i to $\mathfrak{g}\oplus V_1^*{+}v'$ are algebraically independent over $\Bbbk[V_1^*]$. Thereby $\sum b_i \geqslant 18$ and $b_i \geqslant 4$ for each i. Moreover, if $b_1 = 4$, then $b_2 \geqslant 6$.

Set $\tilde{H}_i := \begin{pmatrix} 0\,0 \\ 1\,0 \end{pmatrix}{\cdot}H_i$. If $b_i > a_i$, then $\tilde{H}_i \neq 0$. We have $\psi_v(\tilde{H}_i) \in \mathcal{S}(\mathfrak{q})^{\mathfrak{q}}$ and $\deg_{\mathfrak{g}_v} \psi_v(\tilde{H}_i) = 2$. Therefore $\psi_v(\tilde{H}_i)$ is a linear combination of $\mathbf{h}_j$ with coefficients from $\Bbbk[V_2^*]^{\mathrm{SL}_6}$. Moreover, each coefficient is an eigenvector of ζ. The first element, H_1, can be handled easily.

Assume that $\tilde{H}_1 \neq 0$. Then $\psi_v(\tilde{H}_1) = \mathbf{f}\mathbf{h}_1$ with non-zero $\mathbf{f} \in V_2^{G_v}$ and this $\mathbf{f}$ is an eigenvector of ζ. Since $\deg_{V_1} \tilde{H}_1 \equiv 3 \pmod 4$, $\mathbf{f} = \psi_v(F_{(3,1)})$ (up to a non-zero scalar). Since $\psi_{av}(\frac{\tilde{H}_1}{F_{3,1}}) = a^{b_1-4}\mathbf{h}_1$ for each $a \in \Bbbk^{\times}$ and since $F_{(3,1)}$ and F are coprime, we have $\frac{\tilde{H}_1}{F_{3,1}} \in \Bbbk[\mathfrak{s}^*]^S$. Also $\psi_v(\frac{\tilde{H}_1}{F_{3,1}}) = \mathbf{h}_1$. Clearly $\deg_{V_1} \frac{\tilde{H}_1}{F_{3,1}} = b_1{-}4 < b_1$. A contradiction with the choice of H_1. We have established that $\psi_v(\tilde{H}_1) = 0$. Hence $b_1 = 4$ and H_1 is an SL_2-invariant.

Certain further precautions are needed. It may happen that H_2 (or H_3) does not lie in a simple SL_2-module. In that case we replace H_2 (or H_3) by a suitable (and suitably normalised) component of the same tri-degree, which lies in a simple SL_2-module and which restricts to $\mathbf{h}_2 + p$ with $p \in \mathcal{S}^2(V_2)\mathbf{h}_1$ (or to $\mathbf{h}_3 + p$ with $p \in \mathcal{S}^4(V_2)\mathbf{h}_1 \oplus \mathcal{S}^2(V_2)\mathbf{h}_2$) on $\mathfrak{g}\oplus V_2^* + v$. One may say that $\mathbf{h}_2$ was (or $\mathbf{h}_2$ and $\mathbf{h}_3$ were) changed as well, so that the conditions $\psi_v(H_i) = \mathbf{h}_i$ are not violated. We also normalise F in such a way that $F(v) = 1$. Some other normalisations are done below without mentioning.

Assume that $\tilde{H}_2 \neq 0$ and that $\psi_v(\tilde{H}_2) \in \mathcal{S}^3(V_1)\mathbf{h}_1$. Then $\tilde{H}_2 \in \Bbbk(V^*)H_1$ and so does H_2, which is equal to $\begin{pmatrix} 0 & 1 \\ 0 & 0 \end{pmatrix} \cdot \tilde{H}_2$ up to a non-zero scalar. A contradiction, here $\psi_v(H_2) \neq \mathbf{h}_2$. Knowing that H_2 is an SL_2-invariant, we can use a similar argument in order to prove that $\psi_v(\tilde{H}_3) \notin \mathcal{S}^5(V_1)\mathbf{h}_1 \oplus \mathcal{S}^3(V_1)\mathbf{h}_2$ in case $\tilde{H}_3 \neq 0$.

We will see below that if $b_i > a_i$, then $\tilde{H}_i = \mathbf{H}_i + \frac{F_{(3,1)}H_i}{F}$, where $\mathbf{H}_2 \in \Bbbk(V^*)^G H_1$ and $\mathbf{H}_3 \in \Bbbk(V^*)^G H_1 \oplus \Bbbk(V^*)^G H_2$. Recall that F and $F_{(3,1)}$ are coprime. In case $\mathbf{H}_i \in \Bbbk[\mathfrak{s}^*]$, we can replace H_i with $\frac{H_i}{F} \in \Bbbk[\mathfrak{s}^*]$ decreasing $\deg_{V_1} H_i$ by 4. The main difficulties lie with non-regular $\mathbf{H}_i$.

Modification for H_2 Assume that $b_2 > 6$. Then $\psi_v(\tilde{H}_2) = \mathbf{f}_3\mathbf{h}_1 + \mathbf{f}_{(3,1)}\mathbf{h}_2$ with $\mathbf{f}_3 \in \mathcal{S}^3(V_2)^{G_v}$, $\mathbf{f}_{(3,1)} \in V_2^{G_v}$ and $\mathbf{f}_{(3,1)} \neq 0$. Both coefficients are eigenvectors of ζ. We have $\mathbf{f}_{(3,1)} = \psi_v(F_{(3,1)})$ and $\mathbf{f}_3$ is the image of

$$c_1 F_{(1,3)} + F'_{(5,3)} + c_2 F^3_{(3,1)},$$

where $c_1, c_2 \in \Bbbk$ and $F'_{(5,3)}$ is some G-invariant in $\mathcal{S}^5(V_1)\mathcal{S}^3(V_2)$. Set $\delta := \frac{b_2-6}{4}$ and

$$\mathbf{H}_2 := (c_1 F^\delta F_{(1,3)} + F^{\delta-1}F'_{(5,3)} + c_2 F^{\delta-2} F^3_{(3,1)})H_1.$$

Then $\psi_{av}(\tilde{H}_2 - \mathbf{H}_2) = a^{b_2-1}\mathbf{f}_{(3,1)}\mathbf{h}_2$ for all $a \in \Bbbk^\times$. If $\mathbf{H}_2 \notin \Bbbk[\mathfrak{s}^*]$, then $\delta = 1$ and $c_2 \neq 0$. Here

$$\tilde{H}_2 - c_1 F F_{(1,3)} H_1 - F'_{(5,3)} H_1 - c_2 \frac{F^3_{(3,1)} H_1}{F} = \frac{F_{(3,1)} H_2}{F}$$

and

$$\frac{F_{(3,1)} H_2 + c_2 F^3_{(3,1)} H_1}{F} \in \Bbbk[\mathfrak{s}^*].$$

Since F and $F_{(3,1)}$ are coprime, we have

$$\hat{H}_2 = \frac{H_2 + c_2 F^2_{(3,1)} H_1}{F} \in \Bbbk[\mathfrak{s}^*].$$

In this case we replace $\mathbf{h}_2$ with $\mathbf{h}_2+c_2\mathbf{f}^2_{(3,1)}\mathbf{h}_1$ and H_2 with $\hat{H}_2$. This does not violate the property $\zeta^2{\cdot}\mathbf{h}_2 = -\mathbf{h}_2$. Now $\deg_{V_1} H_2 = \deg_{V_2} H_2 = 6$. If $\begin{pmatrix} 0 & 1 \\ 0 & 0 \end{pmatrix} \cdot H_2 \neq 0$, then this is an invariant of tri-degree $(2, 7, 5)$ and hence lies in $\Bbbk(V^*)H_1$. But then also $H_2 \in \Bbbk(V^*)H_1$. This new contradiction shows that H_2 is an SL_2-invariant.

Modification for H_3 Now we know that $b_2 = 6$ and therefore $b_3 \geqslant 8$. Assume that $b_3 > 8$. Then

$$\psi_v(\tilde{H}_3) = \mathbf{f}'_5\mathbf{h}_1 + \mathbf{f}'_3\mathbf{h}_2 + \mathbf{f}_{(3,1)}\mathbf{h}_3$$

with $\mathbf{f}'_k \in \mathcal{S}^k(V_2)^{G_v}$, $\mathbf{f}_{(3,1)} \in V_2^{G_v}$. All three coefficients are eigenvectors of ζ. Studying the eigenvalues one concludes that $\mathbf{f}_{(3,1)} = \psi_v(F_{(3,1)})$, $\mathbf{f}'_3$ is the image of $s_1F_{(1,3)} + F'_{(5,3)} + s_2F^3_{(3,1)}$, where $F'_{(5,3)} \in \mathcal{S}^5(V_1)\mathcal{S}^3(V_2)$, $s_i \in \Bbbk$, and finally $\mathbf{f}'_5$ is the image of a rather complicated expression $\sum\limits_{j=0}^{3} F'_{(4j+3,5)}$. Set $\nu := \frac{b_3-8}{4}$ and

$$\mathbf{H}_3 := (\sum_{j=0}^{3} F'_{(4j+3,5)}F^{\nu-j})H_1 + (s_1F_{(1,3)}F^{\nu} + F'_{(5,3)}F^{\nu-1} + s_2F^3_{(3,1)}F^{\nu-2})H_2.$$

As above, $\tilde{H}_3 - \mathbf{H}_3 = \frac{F_{(3,1)}H_3}{F}$. If $\mathbf{H}_3 \notin \Bbbk[\mathfrak{s}^*]$, then $\nu = 2$ or $\nu = 1$.

Suppose that $\nu = 2$ and that $F'_{(15,5)} \neq 0$. Then $F'_{(15,5)} = F^5_{(3,1)}$ (up to a non-zero scalar) and

$$\frac{F_{(3,1)}H_3}{F} + \frac{F^5_{(3,1)}H_1}{F} \in \Bbbk[\mathfrak{s}^*] \text{ leading to } \frac{H_3 + F^4_{(3,1)}H_1}{F} \in \Bbbk[\mathfrak{s}^*].$$

Modifying $\mathbf{h}_3$ and H_3 accordingly, we obtain a new H_3 with $\deg_{V_1} H_3 \leqslant 12$.

Suppose now that $\nu = 1$. If $F'_{(15,5)} \neq 0$, then we obtain $\frac{F^5_{(3,1)}H_1}{F} \in \Bbbk[\mathfrak{s}^*]$, which cannot be the case. Thereby $F'_{(15,5)} = 0$ and

$$\frac{F_{(3,1)}H_3}{F} + \frac{F'_{(11,5)}H_1}{F} + s_2\frac{F^3_{(3,1)}H_2}{F} \in \Bbbk[\mathfrak{s}^*].$$

Since $2{\times}5 = 10 < 11$ and since $\psi_v(F_{(4,0)}) = 1$, the polynomial $F'_{(11,5)}$ is divisible by $F_{(3,1)}$, say $F'_{(11,5)} = F_{(3,1)}\mathbf{F}$. Now

$$\frac{H_3 + \mathbf{F}H_1 + s_2F^2_{(3,1)}H_2}{F} \in \Bbbk[\mathfrak{s}^*].$$

This allows us to replace H_3, modifying $\mathbf{h}_3$ at the same time, by a polynomial of tri-degree (2, 8, 8) keeping the property $\psi_v(H_3) = \mathbf{h}_3$. □

Corollary 5.14 *Suppose that* $\tilde{\mathfrak{s}} = \tilde{\mathfrak{g}} \ltimes V$ *is given by the pair* $(\mathrm{Spin}_{12} \times \mathrm{SL}_2, V_{\varphi_5} \otimes \Bbbk^2)$. *Then* $\tilde{\mathfrak{s}}$ *has* (FA) *and* $\Bbbk[\tilde{\mathfrak{s}}^*]^{\tilde{\mathfrak{s}}} = \Bbbk[V^*]^{\tilde{G}}[H_1, H_2, H_3]$, *where the bi-degrees of* H_i *are* (2, 8), (2, 12), (2, 16).

Proof Let $\mathfrak{s} = \mathfrak{g} \ltimes V$ and $\Bbbk[\mathfrak{s}^*]^S = \Bbbk[V^*]^G[H_1, H_2, H_3]$ be as in Theorem 5.13. Then $\tilde{\mathfrak{g}}_x = \mathfrak{g}_x$ for generic $x \in V^*$. Hence $\Bbbk[\tilde{\mathfrak{s}}^*]^{\tilde{S}} \subset \Bbbk[\mathfrak{s}^*]$ by Lemma 3.1. According to the proof of Theorem 5.13, H_1 and H_2 are SL_2-invariants, i.e., they are $\tilde{S}$-invariants, and also $\tilde{H}_3 \notin \mathcal{S}^8(V)H_1 \oplus \mathcal{S}^4(V)H_2$ if $\tilde{H}_3 \neq 0$. At the same time the tri-degree of $\tilde{H}_3$ is $(2, 7, 9)$ if $\tilde{H}_3 \neq 0$. Combining these two observations, we see that $\tilde{H}_3 = 0$, H_3 is an SL_2-invariant, and $\Bbbk[\tilde{\mathfrak{s}}^*]^{\tilde{\mathfrak{s}}} = \Bbbk[V^*]^{\tilde{G}}[H_1, H_2, H_3]$. Since $\Bbbk[V^*]^{\tilde{G}}$ is a polynomial ring, the result follows. □

Proposition 5.15 *All the remaining cases marked with '+' in Table 1 are indeed positive.*

Proof Making further use of Proposition 3.3, we see that all the remaining cases are covered by reductions from G of type $\mathbf{D}_5$, see Diagrams (4), (5), and also

$$\begin{array}{ccc} (\mathrm{Spin}_{10}, 3\varphi_1+\varphi_4) \longrightarrow & (\mathrm{Spin}_9, 2\varphi_1+\varphi_4+2\Bbbk) & \\ & \downarrow & \\ & (\mathrm{Spin}_8, \varphi_1+\varphi_3+\varphi_4+\Bbbk) & \longrightarrow (\mathrm{Spin}_7, 2\varphi_3), \end{array}$$

where the initial pair is positive by Example 5.10. □

6 The Classification for the Symplectic Algebra

In this section, $G = \mathrm{Sp}_{2n}$. We classify the finite-dimensional rational representations $(G : V)$ such that g.i.g.$(G : V)$ is infinite and the symmetric invariants of $\mathfrak{s} = \mathfrak{g} \ltimes V$ form a polynomial ring. The answer is given in Table 2. Surprisingly, all the possible candidates for $\mathfrak{s} = \mathfrak{g} \ltimes V$ do have (FA).

Let $e \in \mathfrak{g}$ be a nilpotent element and $\mathfrak{g}_e \subset \mathfrak{g}$ its centraliser. Then $\mathfrak{g}_e$ has (FA) by [12]. This does not seem to be relevant to our current task, but it is.

The nilpotent element e can be included into an $\mathfrak{sl}_2$-triple $\{e, h, f\} \subset \mathfrak{g}$ and this gives rise to the decomposition $\mathfrak{g} = \Bbbk f \oplus e^{\perp}$, where $e^{\perp}$ is the subspace orthogonal to e w.r.t. the Killing form of $\mathfrak{g}$. Let $\Delta_k \in \mathcal{S}(\mathfrak{sp}_{2n})$ be the sum of the principal k-minors. We write the highest f-component of Δ_k as ${}^e\Delta_k f^d$. Then $\{{}^e\Delta_k \mid k \text{ even}, 2 \leqslant k \leqslant 2n\}$ is a set of the basic symmetric invariants of $\mathfrak{g}_e$ [12, Theorem 4.4].

Let now e be a minimal nilpotent element. Then $\mathfrak{g}_e = \mathfrak{sp}_{2n-2} \ltimes \mathfrak{heis}_{n-1}$. Restricting $H \in \mathcal{S}(\mathfrak{g}_e)^{\mathfrak{g}_e}$ to the hyperplane in $\mathfrak{g}_e^*$, where $e = 0$, we obtain a symmetric invariant of $\mathfrak{s} := \mathfrak{sp}_{2n-2} \ltimes \Bbbk^{2n-2}$.

Let H_i be the restriction of ${}^e\Delta_{2i+2}$ to the hyperplane $e = 0$.

Lemma 6.1 *The algebra of symmetric invariants of* $\mathfrak{s} = \mathfrak{sp}_{2n-2}\ltimes\Bbbk^{2n-2}$ *is freely generated by the polynomials* H_i *as above with* $1 \leqslant i \leqslant n-1$.

Proof Set $n' = n-1$. The group $G' = \mathrm{Sp}_{2n'}$ acts on $V^* \simeq V = \Bbbk^{2n'}$ with an open orbit, which consists of all non-zero vectors of V^*. Therefore $\mathcal{S}(\mathfrak{s})^{\mathfrak{s}} \simeq \mathcal{S}(\mathfrak{h})^H$, where

$$H = (\mathrm{Sp}_{2n'})_v = \mathrm{Sp}_{2n'-2}\ltimes \exp(\mathfrak{heis}_{n'-1})$$

and $v \in V$ is non-zero. By a coincidence, $\mathfrak{h} = \mathfrak{g}'_{e'}$, where $e' \in \mathfrak{g}'$ is a minimal nilpotent element. We have to show that $\psi_v(H_i)$ form a set of the basic symmetric invariants of $\mathfrak{h}$ for the usual restriction $\psi_v : \Bbbk[\mathfrak{s}]^{\mathfrak{s}} \to \Bbbk[(\mathfrak{g}')^*+v]^{G'\ltimes\exp(V)} \simeq \mathcal{S}(\mathfrak{h})^{\mathfrak{h}}$.

Note that the f-degree of each Δ_k with even k is one, see [12] and the matrix description of elements of $f+\mathfrak{g}_e$ presented in Fig. 1. Further, ${}^e\Delta_{2i+2}$ is a sum $e\Delta'_{2i}+H_i$, where $\Delta'_{2i} \in \mathcal{S}(\mathfrak{g}')$. Choosing $v = (1, 0, \ldots, 0)^t$, one readily sees that $\psi_v(H_i) = {}^{e'}\Delta'_{2i}$. This concludes the proof. □

Remark 6.2 We have a nice matryoshka-like structure. Starting from $\mathfrak{g}_e$ with $\mathfrak{g} = \mathfrak{sp}_{2n+2}$ and restricting the symmetric invariants to the hyperplane $e = 0$ one obtains the symmetric invariants of the semi-direct product $\mathfrak{sp}_{2n}\ltimes\Bbbk^{2n}$. By passing to the stabiliser of a generic point $x \in V^*$ with $V = \Bbbk^{2n}$, one comes back to $(\mathfrak{sp}_{2n'})_{e'}$ with $n' = n-1$. And so on.

Suppose now that $e \in \mathfrak{g}$ is given by the partition $(2^m, 1^{2n})$, $\mathfrak{g} = \mathfrak{sp}_{2m+2n}$. Then $\mathfrak{g}_e = (\mathfrak{so}_m\oplus\mathfrak{sp}_{2n})\ltimes(\Bbbk^m\otimes\Bbbk^{2n} \oplus \mathcal{S}^2\Bbbk^m)$ and the nilpotent radical of $\mathfrak{g}_e$ is two-step nilpotent. Suppose that m is odd. Set $Y := \mathrm{Ann}\,(\mathcal{S}^2\Bbbk^m) \subset \mathfrak{g}_e^*$ and let $\tilde{H}_i$ be the restriction to Y of ${}^e\Delta_k$ with $k = 3m+2i-1$.

Lemma 6.3 *For* $1 \leqslant i \leqslant \left(n - \frac{m-1}{2}\right)$, *we have* $\tilde{H}_i \in \mathcal{S}(\mathfrak{s})^{\mathfrak{s}}$, *where* $\mathfrak{s} = \mathfrak{sp}_{2n}\ltimes m\Bbbk^{2n}$.

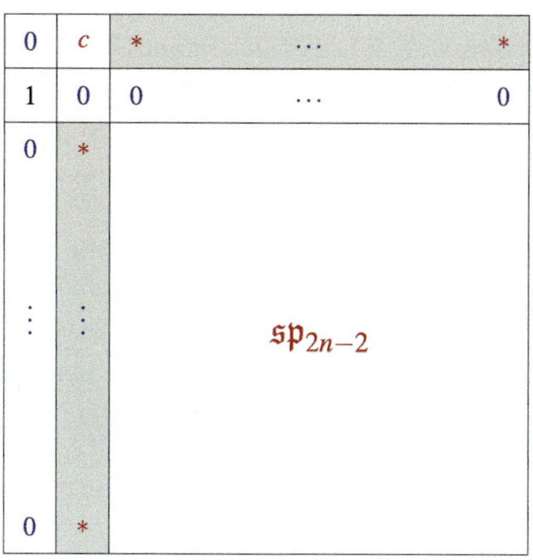

Fig. 1 Elements of $f + \mathfrak{g}_e$

Fig. 2 Elements of $f + \mathfrak{g}_e \subset \mathfrak{sp}_{2m+2n}$

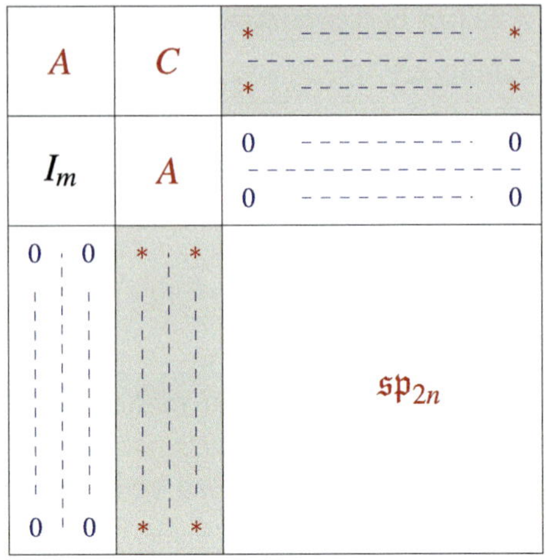

Proof By the construction, each $\tilde{H}_i$ is $\mathfrak{g}_e$-invariant. Note that $\mathfrak{g}_e$ acts on Y as the semi-direct product $(\mathfrak{so}_m \oplus \mathfrak{sp}_{2n}) \ltimes \Bbbk^m \otimes \Bbbk^{2n}$. For each even k with $k \geqslant 2m$, the f-degree of Δ_k is m [12]. For the corresponding $\tilde{H}_i$, this means that $\tilde{H}_i \in \mathcal{S}(\mathfrak{s})$, see also Fig. 2, where $C \in \mathcal{S}^2 \Bbbk^m$. □

Theorem 6.4 *All semi-direct products associated with pairs listed in Table 2 have* (FA)*.*

Proof We begin with Item 1.

Suppose that m is even. Set $\tilde{G} := \mathrm{Sp}_{2n} \times \mathrm{Sp}_m$ and $\tilde{S} := \tilde{G} \ltimes \exp(V)$. Then $G \lhd \tilde{G}$ and $S \lhd \tilde{S}$. The Lie algebra $\tilde{\mathfrak{s}} = \mathrm{Lie}\,\tilde{S}$ is the $\mathbb{Z}_2$-contraction of $\mathfrak{sp}_{2n+m}$ related to the symmetric pair $(\mathfrak{sp}_{2n+m}, \mathfrak{sp}_{2n} \oplus \mathfrak{sp}_m)$. Let $\Delta_k \in \mathcal{S}(\mathfrak{sp}_{2n+m})$ be the sum of the principal k-minors and let $\Delta_k^\bullet$ be the highest V-component of Δ_k. The elements $\Delta_k^\bullet$ with even k, $2m < k \leqslant 2n+m$, belong to a set of the algebraically independent generators of $\mathcal{S}(\tilde{\mathfrak{s}})^{\tilde{\mathfrak{s}}}$, see [21, Theorem 4.5]. For a generic point $x \in V^*$, their restrictions $\Delta_k^\bullet|_{\tilde{\mathfrak{g}}+x}$ form a generating set for the symmetric invariants of $(\mathfrak{sp}_{2n})_x = \mathfrak{sp}_{2n-m}$. Hence $\Delta_k^\bullet \in \mathcal{S}(\mathfrak{s})^S$ by Lemma 3.1. According to [22, Lemma 3.5(ii)], these elements $\Delta_k^\bullet$ (freely) generate $\mathcal{S}(\mathfrak{s})^{\mathfrak{s}}$ over $\Bbbk[V^*]^G$ if and only if their restrictions to $\mathfrak{g} \times D$ are algebraically independent over $\Bbbk[D]^G$ for any G-invariant divisor $D \subset V^*$.

In case $\mathrm{Sp}_m \cdot D$ is open in V^*, the restrictions of the elements $\Delta_k^\bullet$ to $\mathfrak{g} + y$ are algebraically independent for a generic point $y \in D$. If $\mathrm{Sp}_m \cdot D$ is not open in V^*, then D is $\tilde{G}$-invariant and the restrictions of $\Delta_k^\bullet$ to $\mathfrak{g} \times D$ are algebraically independent over $\Bbbk[D]^{\tilde{G}}$ by [22, Lemma 3.5(ii)] applied to $\tilde{\mathfrak{s}}$. If there is a non-trivial relation among these restrictions and not all the coefficients are $\tilde{G}$-invariant, then one can apply an element of $\tilde{G}$ to the relation and by taking a suitable linear combination obtain a smaller non-trivial one. Thus, a minimal non-trivial relation among the restrictions must have $\tilde{G}$ invariant coefficients. Hence the restrictions of $\Delta_k^\bullet$ to $\mathfrak{g} \times D$ are also algebraically independent over $\Bbbk[D]^G$.

Suppose now that m is odd. Consider the standard embedding $\mathfrak{sl}_{2n} \subset \mathfrak{sl}_{2n} \times \mathfrak{sl}_m \subset \mathfrak{sl}_{2n+m}$. The defining representation of Sp_{2n} on $\Bbbk^{2n}$ is self-dual. Therefore we can embed $V \simeq V^*$ into $m\Bbbk^{2n} \oplus m(\Bbbk^{2n})^*$ diagonally. This gives rise to $\mathfrak{s}^* = \mathfrak{g} \oplus V \subset \mathfrak{sl}_{2n+m}$. Let $\Delta_k \in \mathcal{S}(\mathfrak{sl}_{2n+m})$ be the sum of the principal k-minors and $\Delta_k^\bullet$ the highest V-component of the restriction $\Delta_k|_{\mathfrak{s}^*}$. Note that in case $m = 1$, we have $\Delta_k^\bullet = -H_i$, where H_i is the same as in Lemma 6.1 and $k = 2i+1$. For $m \geqslant 3$, $\Delta_k^\bullet$ is equal to $\pm\tilde{H}_i$, where $\tilde{H}_i$ is the same as in Lemma 6.3 and $k = 2m+2i-1$. Suppose that $m \geqslant 3$.

Fix a G-stable decomposition $V = V_1 \oplus V_2$ with $V_1 = \Bbbk^{2n}$. Then there is the corresponding decomposition $V^* = V_1^* \oplus V_2^*$. Choose a generic $v \in V_2^*$ and consider the usual restriction homomorphism

$$\psi_v \colon \Bbbk[\mathfrak{s}^*]^S \to \Bbbk[\mathfrak{g} \oplus V_1^* + v]^{G_v \ltimes \exp(V)} \simeq \mathcal{S}(\mathfrak{g}_v \ltimes V_1)^{G_v \ltimes \exp(V_1)}.$$

Here $G_v = \mathrm{Sp}_{2n-m+1}$. Setting $n' := n - \frac{m-1}{2}$, we obtain $\mathfrak{g}_v \ltimes V_1 = (\mathfrak{sp}_{2n'} \ltimes \Bbbk^{2n'}) \oplus \Bbbk^{m-1}$. If $k = 2m+2i-1$, then the restriction of $\Delta_k^\bullet$ to $\mathfrak{g} \oplus V_1^* + v$ is equal to cH_i, where $c \in \Bbbk^\times$ and H_i is the same symmetric invariant of $\mathfrak{sp}_{2n'} \ltimes \Bbbk^{2n'}$ as in Lemma 6.1.

The ring $\Bbbk[V^*]^G$ is freely generated by $\binom{m}{2}$ polynomials F_j of degree 2. We may (and will) assume that the first $m-1$ elements F_j lie in $V_1 \otimes V_2$ and that the remaining ones (freely) generate $\Bbbk[V_2^*]^G$. Then $\psi_v(F_j) \in \Bbbk$ for $j \geqslant m$ and $\langle \psi_v(F_j) \mid 1 \leqslant j \leqslant m-1 \rangle_\Bbbk$ is the Abelian direct summand $\Bbbk^{m-1}$ of $\mathfrak{g}_v \ltimes V_1$. We see that $F_1, \ldots, F_{m-1}, \Delta_{2m+1}^\bullet, \ldots, \Delta_{2n+m}^\bullet$ are algebraically independent over $\Bbbk[V_2^*]$. Hence

$$\left\{F_j \mid 1 \leqslant j \leqslant \binom{m}{2}\right\} \cup \{\Delta_k^\bullet \mid k \text{ odd}, 2m < k \leqslant 2n+m\}$$

is a set of algebraically independent homogeneous invariants. Our goal is to prove that this is a generating set.

There is a big open subset $U \subset V^*$ such that G_v is a generic isotropy group for $(G{:}V^*)$ for each $v \in U$. Here $G_v = (\mathrm{Sp}_{2n'})_e$ with $2n' = 2n-m+1$ and $e \in \mathfrak{sp}_{2n'}$ being a minimal nilpotent element. The algebra $\mathfrak{g}_v$ has the "codim–2" property by [12] and hence $\mathfrak{s}$ has the "codim–2" property as well.

Finally we calculate the sum of the degrees of the proposed generators. There are $\binom{m}{2}$ invariants of degree 2, the minors $\Delta_k^\bullet$ are of degrees $2m+1, 2m+3, \ldots, m+2n$. Summing up

$$2\binom{m}{2} + \frac{1}{2}\left(n - \frac{m-1}{2}\right)(2n+3m+1) = \frac{1}{2}\operatorname{ind}\mathfrak{s} + n^2 + \frac{n}{2} + nm = \frac{\operatorname{ind}\mathfrak{s} + \dim\mathfrak{s}}{2}.$$

Applying Theorem 2.1, we can conclude that $\mathcal{S}(\mathfrak{s})^{\mathfrak{s}}$ is freely generated by the polynomials F_j and $\Delta_k^\bullet$.

Item 2 is a $\mathbb{Z}_2$-contraction of SL_{2n}, and this contraction is good, see [21, Theorem 4.5].

Item 4 can be covered by Theorem 3.6 (or Lemma 3.8), this pair (G, V^*) is of rank one. There is an open orbit $G{\cdot}y \subset D$, where D stands for the zero set of the generator $F \in \Bbbk[V^*]^G$. A generic isotropy group for $(G{:}V^*)$ is SL_3, G_y is connected, and $\mathfrak{g}_y$ is equal to $\mathfrak{sl}_2 \ltimes \mathcal{S}^4\Bbbk^2$, see [4]. This $\mathfrak{g}_y$ is a good $\mathbb{Z}_2$-contraction of $\mathfrak{sl}_3$ [11].

Item 5 is covered by Example 3.5.

Item 6 is treated in [13, Appendix A], there it is shown that this pair has (FA).

The final challenge is to describe the symmetric invariants for item 3. A certain similarity with item 2 will help. Now $V = V_1 \oplus V_2$ with $V_1 = \Bbbk^{2n}$, $V_2 = V_{\varphi_2}$. Set $\mathfrak{s}_2 := \mathfrak{g} \ltimes V_2$ (this is the semi-direct product in line 2). According to [21], $\Bbbk[\mathfrak{s}_2^*]^{\mathfrak{s}_2} = \Bbbk[V_2^*]^G[\mathbf{h}_1, \ldots, \mathbf{h}_n]$, where each $\mathbf{h}_i$ is bi-homogeneous and $\deg_{\mathfrak{g}} \mathbf{h}_i = 2$. In other words, $\mathbf{h}_i \in (\mathcal{S}^2(\mathfrak{g}) \otimes \mathcal{S}(V_2))^G$. In $\mathcal{S}^2(V_1)$, there is a unique copy of $\mathfrak{g}$, which gives rise to embeddings $\iota\colon \mathcal{S}^2(\mathfrak{g}) \to \mathfrak{g} \otimes \mathcal{S}^2(V_1)$ and

$$\tilde{\iota}\colon (\mathcal{S}^2(\mathfrak{g}) \otimes \mathcal{S}(V_2))^G \to (\mathfrak{g} \otimes \mathcal{S}^2(V_1) \otimes \mathcal{S}(V_2))^G.$$

Set $H_i := \tilde{\iota}(\mathbf{h}_i)$.

Each H_i is a G-invariant by the construction. Next we check that it is also a V-invariant. Take a generic point $v \in V_2^*$. Then $\mathfrak{g}_v$ is a direct sum of n copies of $\mathfrak{sl}_2$ and under $\mathfrak{g}_v$ the space V_1 decomposes into a direct sum of n copies of $\Bbbk^2$. The restriction of $\mathbf{h}_i$ to $\mathfrak{g} + v$ is an element of $\mathcal{S}^2(\mathfrak{g}_v)^{\mathfrak{g}_v} \subset \mathcal{S}^2(\mathfrak{g}) \subset \mathfrak{g} \otimes \mathfrak{g}$. If we regard this restriction as a bi-linear function on $\mathfrak{g} \otimes \mathfrak{g}$, then its value on (A, B) for $A, B \in \mathfrak{g}$ can be calculated as follows. From each matrix we cut the $\mathfrak{sl}_2$ pieces A_j, B_j, $1 \leqslant j \leqslant n$, corresponding to the $\mathfrak{sl}_2$ summands of $\mathfrak{g}_v$ and take a linear combination $\sum \alpha_{i,j} \mathrm{tr}\,(A_j B_j)$. With a slight abuse of notation we set $\mathbf{h}_i(A, B, v) := \sum \alpha_{i,j} \mathrm{tr}\,(A_j B_j)$.

The restriction of H_i to $\mathfrak{g} \oplus V_1^* + v$ is an element of $(\mathfrak{g} \otimes \mathcal{S}^2(V_1))^{\mathfrak{g}_v}$. Take $\xi \in V_1^*$. Let $B(\xi) \in \mathfrak{g}$ be the projection of ξ^2 to $\mathfrak{g} \subset \mathcal{S}^2(V_1)$. Then

$$H_i(A + \xi + v) = \mathbf{h}_i(A, B(\xi), v).$$

Write $\xi = \xi_1 + \ldots + \xi_n$, where each ξ_j lies in its $\mathfrak{g}_v$-stable copy of $\Bbbk^2$. Then $\xi_j \otimes \xi_k$ with $j \neq k$ is orthogonal to $\mathfrak{g}_v \subset \mathfrak{g} \subset \mathcal{S}^2(V_1)$. Furthermore, $\mathrm{tr}\,(A_j B(\xi)_j) = \det(\xi_j | A_j \xi_j)$. Therefore

$$H_i(A + \xi + v) = \sum \alpha_{i,j} \det(\xi_j | A_j \xi_j).$$

We see that $H_i|_{\mathfrak{g} \oplus V_1^* + v}$ lies in $\mathcal{S}(\mathfrak{g}_v \ltimes V_1)$ and therefore is a V_2-invariant [22]. Moreover, this restriction is a V_1-invariant by [23]. Since these assertions hold for a generic vector $v \in V_2^*$, each H_i is a V-invariant. From the case of $\mathfrak{s}_2$, we know that the matrix $(\alpha_{i,j})$ is non-degenerate. Hence the invariants H_i are algebraically independent over $\Bbbk(V_2^*)$. Note that $\Bbbk[V_2^*]^G = \Bbbk[V^*]^G$. Further, $\deg H_i = \deg \mathbf{h}_i + 1$.

If we sum over all (suggested) generators, then the result is $(\dim\mathfrak{s}_2 + \mathrm{ind}\,\mathfrak{s}_2)/2 + n$ and this is exactly $(\dim\mathfrak{s} + \mathrm{ind}\,\mathfrak{s})/2$.

In order to use Theorem 2.1, it remains to prove that $\mathfrak{s}$ has the "codim–2" property. Let $D \subset V_2^*$ be a G-invariant divisor and let $y \in D$ be a generic point. If $G_y \neq (\mathrm{SL}_2)^n$, then $G_y = (\mathrm{SL}_2)^{n-2} \times (\mathrm{SL}_2 \ltimes \exp(\mathcal{S}^2\Bbbk^2))$. In particular, $\dim(G{\cdot}y) = \dim V_2 - (n-1)$. If $\mathfrak{q}$ is the Lie algebra of $Q = \mathrm{SL}_2 \ltimes \exp(\mathcal{S}^2\Bbbk^2)$, then $\mathfrak{q} = \mathfrak{sl}_2 \ltimes \mathfrak{sl}_2^{\mathsf{ab}}$. We have

$$G_y \ltimes \exp(V_1) = (\mathrm{SL}_2 \ltimes \exp(\Bbbk^2))^{n-2} \times (Q \ltimes \exp(\Bbbk^4))$$

and $\mathfrak{q}\ltimes\Bbbk^4 = \mathfrak{sl}_2 \ltimes ((\Bbbk^2 \oplus \mathcal{S}^2\Bbbk^2) \oplus \Bbbk^2)$ with the unique non-zero commutator $[\Bbbk^2, \mathcal{S}^2\Bbbk^2] = \Bbbk^2$. An easy computation shows that $\mathrm{ind}\,(\mathfrak{q}\ltimes\Bbbk^4) = 2$. Thereby $\mathrm{ind}\,(\mathfrak{g}_y \ltimes V_1) = n$ and hence $\mathfrak{g}\oplus V_1^* \times D \cap \mathfrak{s}^*_{\mathsf{reg}} \neq \varnothing$, cf. [22, Eq. (3·2)]. The Lie algebra $\mathfrak{s}$ does have the "codim–2" property. □

Acknowledgements The first author is partially supported by the RFBR grant № 16-01-00818. The second author is funded by the Deutsche Forschungsgemeinschaft (DFG, German Research Foundation)—project number 330450448.

References

1. О.М. Адамович, Е.О. Головина. Простые линейные группы Ли, имеющие свободную алгебру инвариантов, в Сб.: "*Вопросы теории групп и гомологической алгебры*", Вып. **2**, с. 3–41. Ярославль 1979 (Russian). Englich translation: O.M. ADAMOVICH and E.O. GOLOVINA. Simple linear Lie groups having a free algebra of invariants, *Selecta Math. Sov.*, **3** (1984), 183–220.
2. А.Г. Элашвили. Канонический вид и стационарные подалгебры точек общего положения для простых линейных групп Ли, *Функц. анализ и прилож.*, т. **6**, № 1 (1972), 51–62 (Russian). English translation: A.G. ELASHVILI. Canonical form and stationary subalgebras of points of general position for simple linear Lie groups, *Funct. Anal. Appl.*, **6** (1972), 44–53.
3. F. FAUQUANT-MILLET and P. LAMPROU. Polynomiality for the Poisson centre of truncated maximal parabolic subalgebras, *J. Lie Theory*, **28** (2018), 1063–1094.
4. J.-I. IGUSA. A classification of spinors up to dimension twelve. *Amer. J. Math.* **92** (1970), 997–1028.
5. A. JOSEPH. A preparation theorem for the prime spectrum of a semisimple Lie algebra, *J. Algebra*, **48** (1977), 241–289.
6. A. JOSEPH. On semi-invariants and index for biparabolic (seaweed) algebras, II, *J. Algebra*, **312** (2007), 158–193.
7. A. JOSEPH. An algebraic slice in the coadjoint space of the Borel and the Coxeter element, *Adv. Math.*, **227** (2011), no. 1, 522–585.
8. A. JOSEPH and D. SHAFRIR. Polynomiality of invariants, unimodularity and adapted pairs, *Transformation Groups*, **15** (2010), no. 4, 851–882.
9. B. KOSTANT. The cascade of orthogonal roots and the coadjoint structure of the nilradical of a Borel subgroup of a semisimple Lie group, *Mosc. Math. J.* **12** (2012), no. 3, 605–620.
10. D. PANYUSHEV. Semi-direct products of Lie algebras and their invariants, *Publ. R.I.M.S.*, **43**, no. 4 (2007), 1199–1257.

11. D. Panyushev. On the coadjoint representation of $\mathbb{Z}_2$-contractions of reductive Lie algebras, *Adv. Math.*, **213** (2007), 380–404.
12. D. Panyushev, A. Premet and O. Yakimova. On symmetric invariants of centralisers in reductive Lie algebras, *J. Algebra*, **313** (2007), 343–391.
13. D. Panyushev and O. Yakimova. Semi-direct products of Lie algebras and covariants, *J. Algebra*, **490** (2017), 283–315.
14. D. Panyushev and O. Yakimova. Symmetric invariants related to representations of exceptional simple groups, Труды Моск. Матем. Об-ва, т. **78** (2017), вып. 2, с. 195–207 (= *Trans. Mosc. Math. Soc.*, **78** (2017), 161–170).
15. D. Panyushev and O. Yakimova. Takiff algebras with polynomial rings of symmetric invariants, *Transformation Groups*, to appear (arxiv:1710.03180v1[math.RT], 17 pp.).
16. M. Raïs. L'indice des produits semi-directs $E \times_\rho \mathfrak{g}$, *C.R. Acad. Sc. Paris, Ser. A* t.**287** (1978), 195–197.
17. G. Schwarz. Representations of simple Lie groups with regular rings of invariants, *Invent. Math.*, **49** (1978), 167–191.
18. S.J. Takiff. Rings of invariant polynomials for a class of Lie algebras, *Trans. Amer. Math. Soc.* **160** (1971), 249–262.
19. Э.Б. Винберг, А.Л. Онищик. *Семинар по группам Ли и алгебраическим группам*. Москва: "Наука" 1988 (Russian). English translation: A.L. Onishchik and E.B. Vinberg. "Lie groups and algebraic groups", Berlin: Springer, 1990.
20. Э.Б. Винберг, В.Л. Попов. *"Теория Инвариантов"*, В: Современные проблемы математики. Фундаментальные направления, т. 55, стр. 137–309. Москва: ВИНИТИ 1989 (Russian). English translation: V.L. Popov and E.B. Vinberg. "Invariant theory", In: *Algebraic Geometry IV* (Encyclopaedia Math. Sci., vol. 55, pp.123–284) Berlin Heidelberg New York: Springer 1994.
21. O. Yakimova. One-parameter contractions of Lie—Poisson brackets, *J. Eur. Math. Soc.*, **16** (2014), 387–407.
22. O. Yakimova. Symmetric invariants of $\mathbb{Z}_2$-contractions and other semi-direct products, *Int. Math. Res. Notices*, (2017) 2017 (6): 1674–1716.
23. O. Yakimova. Some semi-direct products with free algebras of symmetric invariants, In: F. Callegaro et al. (eds.) *"Perspectives in Lie theory"*, Springer INdAM series, vol. **19** (2017), 266–279.

Primitive Ideals of $\mathrm{U}(\mathfrak{sl}(\infty))$ and the Robinson–Schensted Algorithm at Infinity

Ivan Penkov and Alexey Petukhov

To Anthony Joseph on the occasion of his 75th birthday

Abstract We present an algorithm which computes the annihilator in $\mathrm{U}(\mathfrak{sl}(\infty))$ of any simple highest weight $\mathfrak{sl}(\infty)$-module $L_{\mathfrak{b}}(\lambda)$. This algorithm is based on an infinite version of the Robinson–Schensted algorithm.

MSC 2010: 17B65, 05E10, 16D60

1 Background Results

The description of primitive ideals of the enveloping algebra $\mathrm{U}(\mathfrak{sl}(n))$ for $n \geq 2$ is nowadays a classical result. Duflo's Theorem [8], applied to $\mathfrak{sl}(n)$, claims that, for every fixed Borel subalgebra $\mathfrak{b} \subset \mathfrak{sl}(n)$, any primitive ideal of $\mathrm{U}(\mathfrak{sl}(n))$ is the annihilator of a simple $\mathfrak{b}$-highest weight $\mathfrak{sl}(n)$-module. Since (by a well-known generalization of Schur's Lemma) any primitive ideal intersects the centre $\mathrm{Z}(\mathfrak{sl}(n))$ of $\mathrm{U}(\mathfrak{sl}(n))$ in a maximal ideal of $\mathrm{Z}(\mathfrak{sl}(n))$, and since there are only finitely many nonisomorphic simple $\mathfrak{b}$-highest weight modules with fixed action of $\mathrm{Z}(\mathfrak{sl}(n))$, Duflo's theorem reduces the problem of classifying primitive ideals to a finite problem. Indeed, the Weyl group S_n of $\mathfrak{sl}(n)$ surjects to the set of primitive ideals I

I. Penkov (✉)
Jacobs University Bremen, Bremen, Germany
e-mail: i.penkov@jacobs-university.de

A. Petukhov
Institute for Information Transmission Problems, Moscow, Russia
e-mail: alex-2@yandex.ru

M. Gorelik et al. (eds.), *Representations and Nilpotent Orbits of Lie Algebraic Systems*, Progress in Mathematics 330,
https://doi.org/10.1007/978-3-030-23531-4_13

with fixed intersection $I \cap \mathrm{Z}(\mathfrak{g})$, and the problem of describing the primitive ideals of $\mathrm{U}(\mathfrak{sl}(n))$ is equivalent to the problem of describing the fibres of this surjection.

It was Anthony Joseph who solved this latter problem by reducing it to the Robinson–Schensted algorithm.

The purpose of our current paper is to establish a combinatorial counterpart of Joseph's result for the infinite-dimensional Lie algebra $\mathfrak{sl}(\infty)$. More precisely, we provide an algorithm for computing the primitive ideal of any simple highest weight $\mathfrak{sl}(\infty)$-module. This algorithm is our proposed "Robinson–Schensted algorithm at infinity".

We start with a brief survey of previous results on the primitive ideals of $\mathrm{U}(\mathfrak{g}_\infty)$ for direct limit Lie algebras $\mathfrak{g}_\infty$, putting in this way the current paper into context.

The Lie algebra $\mathfrak{sl}(\infty)$ is defined as the direct limit of an arbitrary chain of embeddings

$$\mathfrak{sl}(2) \hookrightarrow \mathfrak{sl}(3) \hookrightarrow \mathfrak{sl}(4) \hookrightarrow \dots .$$

More generally, one may consider an arbitrary chain of embeddings of simple Lie algebras

$$\mathfrak{g}_1 \hookrightarrow \mathfrak{g}_2 \hookrightarrow \dots \hookrightarrow \mathfrak{g}_n \hookrightarrow \mathfrak{g}_{n+1} \hookrightarrow \dots \qquad (1)$$

and its direct limit $\mathfrak{g}_\infty = \varinjlim \mathfrak{g}_n$.

An embedding $\mathfrak{g}_i \hookrightarrow \mathfrak{g}_{i+1}$ as in (1) is *diagonal* if the branching rule for the natural $\mathfrak{g}_{i+1}$-modules (the nontrivial simple $\mathfrak{g}_{i+1}$-modules of minimal dimension) involves only natural and trivial modules over $\mathfrak{g}_i$. The direct limits of chains of diagonal embeddings are known as *diagonal Lie algebras* and are classified by Baranov and Zhilinskii [1]. Furthermore, diagonal Lie algebras can be split into *nonfinitary diagonal* Lie algebras and *finitary* Lie algebras, the latter being (up to isomorphism) just three Lie algebras: $\mathfrak{sl}(\infty)$, $\mathfrak{o}(\infty)$, $\mathfrak{sp}(\infty)$. The finitary Lie algebras $\mathfrak{g}_\infty$ are defined as the direct limits of chains (1) where $\mathfrak{g}_n = \mathfrak{sl}(n+1)$, $\mathfrak{o}(n)$, $\mathfrak{sp}(2n)$, respectively.

The classification problem for nondiagonal Lie algebras $\mathfrak{g}_\infty$ appears to be wild. Nevertheless, one can make the following strong statement about primitive ideals in $\mathrm{U}(\mathfrak{g}_\infty)$:

If $\mathfrak{g}_\infty$ is nondiagonal, i.e., there are infinitely many nondiagonal embeddings in the chain (1), the only proper two-sided ideals in $\mathrm{U}(\mathfrak{g}_\infty)$ are the augmentation ideal and the zero ideal.

This statement is known as Baranov's conjecture and is proved in [14].

For nonfinitary diagonal Lie algebras $\mathfrak{g}_\infty$, a classification of two-sided ideals is obtained by Zhilinskii [23]. Here there are two-sided ideals I different from the augmentation ideal, however a characteristic feature of this case is that all quotients $\mathrm{U}(\mathfrak{g}_\infty)/I$ are locally finite dimensional. By definition, this means that the quotients $\mathrm{U}(\mathfrak{g}_n)/(\mathrm{U}(\mathfrak{g}_n) \cap I)$ are finite dimensional. A similar result has been established in the recent paper [18] also for the Witt Lie algebra (which is not a direct limit of finite-dimensional Lie algebras), and this leads us to the thought that the above results might extend to a larger class of infinite-dimensional Lie algebras. That could be a subject of future research.

None of the above results apply to the three finitary Lie algebras $\mathfrak{sl}(\infty)$, $\mathfrak{o}(\infty)$, $\mathfrak{sp}(\infty)$. The problem of classifying primitive ideals in the enveloping algebras $U(\mathfrak{sl}(\infty))$, $U(\mathfrak{o}(\infty))$ and $U(\mathfrak{sp}(\infty))$ has been open for some time, and was recently solved in [17] for $U(\mathfrak{sl}(\infty))$. Here is a brief history of the problem. It was posed by A. Zalesskii, who saw it as a problem analogous to classifying primitive (and two-sided) ideals in the group algebra of S_∞. Indeed, the latter problem admits a relatively straightforward combinatorial solution, and suggests a method for constructing primitive ideals of $U(\mathfrak{sl}(\infty))$. One considers *coherent local systems of simple* $\mathfrak{sl}(n)$*-modules*: such coherent local systems, *c.l.s.* for short, consist of nonempty sets of isomorphism classes $[L_n^\alpha]$ of simple finite-dimensional $\mathfrak{sl}(n)$-modules L_n^α for each $n \geq 2$, such that each $\mathfrak{sl}(n)$-module $L_n^{\alpha_0}$ branches over $\mathfrak{sl}(n-1)$ as a sum of $\mathfrak{sl}(n-1)$-modules among L_{n-1}^α, and every L_{n-1}^α arises from a suitable $L_n^{\alpha_0}$. The joint annihilator in $U(\mathfrak{sl}(\infty))$ of such a c.l.s. (i.e. the union over n of all joint annihilators $\cap_\alpha \mathrm{Ann}_{U(\mathfrak{sl}(n))} L_n^\alpha$ is a two-sided ideal of $U(\mathfrak{sl}(\infty))$. Furthermore, one can prove that if a c.l.s. is *irreducible*, i.e., is not a proper union of two sub-c.l.s., then its annihilator is a primitive ideal. As an important step in Zalesskii's program, A. Zhilinskii classified all c.l.s. (and, in particular, irreducible c.l.s.). Unfortunately, Zhilinskii's work is not widely available as his main paper [23] (based on [22]) is a preprint in Russian. We have given a summary of Zhilinskii's classification of c.l.s. in our survey paper [16], see also [14].

As a next step, we determined in [14] which simple c.l.s. have the same annihilator, and completed in this way the classification of primitive ideals of $U(\mathfrak{sl}(\infty))$ arising from c.l.s. We call these ideals *integrable* primitive ideals (an equivalent definition is given in [14, 16]).

The next step was made in our work [17] where we proved that any primitive ideal of $U(\mathfrak{sl}(\infty))$ is integrable, providing finally a classification of primitive ideals of $U(\mathfrak{sl}(\infty))$. The proof is based on three pillars: our study of associated pro-varieties of primitive ideals in [14], Joseph's original classification of primitive ideals in $U(\mathfrak{sl}(n))$, and certain new combinatorial facts relating "precoherent local systems" of representations of $\mathfrak{sl}(n)$ for $n \geq 1$ to coherent local systems introduced above. These latter facts use heavily the Gelfand-Tsetlin branching rule.

The final result is as follows:

Primitive ideals of $U(\mathfrak{sl}(\infty))$ *are naturally parameterized by quadruples*

$$(r, g, X, Y)$$

where r, g *are nonnegative integers and* X, Y *are Young diagrams.*

The integer r is the *rank* and represents the associated pro-variety of a primitive ideal, see [14]. The integer g is the *Grassmann number*. We call it so as it arises naturally from direct limits of exterior powers of defining $\mathfrak{sl}(n)$-modules, i.e., of direct limits of fundamental $\mathfrak{sl}(n)$-modules. More precisely, a *semiinfinite funda-*

mental $\mathfrak{sl}(\infty)$*-module* is a direct limit of fundamental $\mathfrak{sl}(n)$-modules whose degrees and codegrees both tend to infinity (there are uncountably many nonisomorphic semiinfinite modules), see [9]. The annihilators of all semiinfinite fundamental $\mathfrak{sl}(\infty)$-modules coincide, and the corresponding ideal is labeled by $(0, 1, \emptyset, \emptyset)$.

Finally, the Young diagrams X, Y also arise in a straightforward manner: the primitive ideal with coordinates $(0, 0, X, Y)$ is the annihilator of the simple tensor module $V_{X,Y}$; this module is defined as the socle of the tensor product $S_X(V) \otimes S_Y(V_*)$ where V and V_* are the two defining representations of $\mathfrak{sl}(\infty)$ (finitary column vectors and finitary row vectors) and $S_Z(\cdot)$ is the Schur functor associated to a Young diagram Z, see [6] and [19].

An essential difference with the case of $\mathfrak{sl}(n)$ is that the annihilator in $\mathrm{U}(\mathfrak{sl}(\infty))$ of most simple $\mathfrak{sl}(\infty)$-modules is equal to zero. Therefore one can think of simple $\mathfrak{sl}(\infty)$-modules with nonzero annihilators as *small*. Examples of small simple modules are the above mentioned modules $V_{X,Y}$, semiinfinite fundamental representations, and also direct limits of growing symmetric powers of defining representations of $\mathfrak{sl}(n)$ for $n \to \infty$. A small simple $\mathfrak{sl}(\infty)$-module does not need to be integrable, i.e., does not need to be a direct limit of finite-dimensional $\mathfrak{sl}(n)$-modules for $n \to \infty$. For instance, in [9] it is shown that any simple weight $\mathfrak{sl}(\infty)$-module with bounded weight multiplicities is small. However, our classification of primitive ideals implies that the annihilator of any small simple $\mathfrak{sl}(\infty)$-module is also the annihilator of a, possibly nonisomorphic, simple integrable $\mathfrak{sl}(\infty)$-module. This is a truly infinite-dimensional effect.

2 Our Goal in the Present Paper

We are now ready to explain the purpose of the paper. Despite the fact that primitive ideals of $\mathrm{U}(\mathfrak{sl}(\infty))$ are classified, the existing literature does not explain how to compute the annihilator of an arbitrary simple highest weight module $L_{\mathfrak{b}}(\lambda)$, i.e., how to find the quadruple (r, g, X, Y) corresponding to the ideal $\mathrm{Ann}_{\mathrm{U}(\mathfrak{sl}(\infty))} L_{\mathfrak{b}}(\lambda)$, for a given splitting Borel subalgebra $\mathfrak{b} \subset \mathfrak{sl}(\infty)$ and a character λ of $\mathfrak{b}$. Solving this problem is our aim in the present work. In the case of $\mathfrak{sl}(n)$, the analogous problem is solved by applying the Robinson–Schensted algorithm to the weight $\lambda + \rho$, and below we present the corresponding "infinite version" of this algorithm.

In the work [15] we have established an important preliminary result: we have found a necessary and sufficient condition on the pair $(\mathfrak{b}, \lambda)$ for the annihilator $\mathrm{Ann}_{\mathrm{U}(\mathfrak{sl}(\infty))} L_{\mathfrak{b}}(\lambda)$ to be nonzero. Recall that a splitting Borel subalgebra containing a fixed splitting Cartan subalgebra (for instance, the diagonal matrices in $\mathfrak{sl}(\infty)$) is given by an arbitrary total order $\prec$ on a countable set, see [7]. We denote this set by Θ. Theorem 3.1 in [15] asserts that $\mathrm{Ann}_{\mathrm{U}(\mathfrak{sl}(\infty))} L_{\mathfrak{b}}(\lambda) \neq 0$ if and only if Θ can be split as a finite disjoint union

$$\Theta = \Theta_1 \sqcup \ldots \sqcup \Theta_k$$

such that $i \prec j$ for any pair $i \in \Theta_s$, $j \in \Theta_t$ with $s \prec t$, and the restriction of λ to Θ_s is a constant $\lambda(s)$ for any $s < k$, satisfying $\lambda(s) - \lambda(t) \in \mathbb{Z}$ if both Θ_s, Θ_t are infinite.

The above makes it clear that in order to compute $\mathrm{Ann}_{U(\mathfrak{sl}(\infty))} L_{\mathfrak{b}}(\lambda)$ we need to provide an algorithm which transforms a given pair $(\mathfrak{b}, \lambda)$, where $\mathfrak{b}$ is a splitting Borel subalgebra and λ is a weight such that $\mathrm{Ann}_{U(\mathfrak{sl}(\infty))} L_{\mathfrak{b}}(\lambda) \neq 0$, to the quadruple corresponding to the primitive ideal (r, g, X, Y) of $\mathrm{Ann}_{U(\mathfrak{sl}(\infty))} L_{\mathfrak{b}}(\lambda)$. This is precisely what we do: we construct a version of the Robinson–Schensted algorithm which performs the above task.

3 Preliminaries

3.1 *Robinson–Schensted Algorithm and* $\mathfrak{sl}(n)$

3.1.1 Notation

We fix an algebraically closed field $\mathbb{F}$ of characteristic 0. If V is a vector space over $\mathbb{F}$, we set $V^* = \mathrm{Hom}_{\mathbb{F}}(V, \mathbb{F})$. All ideals in associative $\mathbb{F}$-algebras are assumed to be two-sided. We use the notions of Young diagrams and partitions as synonyms; when writing a Young diagram as a partition $(p_1 \geq p_2 \geq \ldots \geq p_n > 0)$, the integers p_i are the row lengths of the diagram.

We identify $\mathfrak{sl}(n)$ with the set of traceless $n \times n$-matrices. The elementary matrices

$$e_{i,j} \text{ for } 1 \leq i \neq j \leq n, \quad e_{i,i} - e_{i+1,i+1} \text{ for } 1 \leq i \leq n-1$$

form a basis of $\mathfrak{sl}(n)$. We fix the Cartan subalgebra $\mathfrak{h}_n$ of diagonal matrices and the Borel subalgebra $\mathfrak{b}_n$ of upper triangular matrices. To any linear function $\lambda \in \mathfrak{h}_n^*$ we attach the linear map

$$\lambda' : \mathfrak{b}_n \to \mathbb{F}, \quad e_{ij} \mapsto 0 \text{ for } i \neq j, \text{ and } \lambda'|_{\mathfrak{h}_n} = \lambda.$$

We denote by $\mathbb{F}_\lambda$ the one-dimensional $\mathfrak{b}_n$-module defined by λ'. Set

$$M(\lambda) := U(\mathfrak{sl}(n)) \otimes_{U(\mathfrak{b}_n)} \mathbb{F}_\lambda.$$

Let $L(\lambda)$ be the unique simple quotient of $M(\lambda)$, and

$$I(\lambda) := \mathrm{Ann} L(\lambda).$$

We identify the vector space $\mathbb{F}^n$ with the space of functions

$$f : \{1, \ldots, n\} \to \mathbb{F}.$$

For any function $f \in \mathbb{F}^n$ there exists a unique $\lambda_f \in \mathfrak{h}_n^*$ such that

$$\lambda_f(e_{ii} - e_{jj}) = f(i) - f(j).$$

Therefore to any function $f \in \mathbb{F}^n$ we can attach the primitive ideal

$$I(f) := I(\lambda_f) \subset \mathrm{U}(\mathfrak{sl}(n)).$$

The Weyl group W_n of the pair $(\mathfrak{sl}(n), \mathfrak{h}_n)$ is the symmetric group S_n, and its action on $\mathfrak{h}_n^*$ is induced by its action on $\mathbb{F}^n$ via permutations. The shifted action of S_n on $\mathbb{F}^n$, denoted by $\sigma \cdot f$, is defined as

$$\sigma \cdot f := \sigma(f + \rho_n) - \rho_n$$

where $\rho_n := (-1, -2, \ldots, -n)$.

3.1.2 Joseph's Description of Primitive Ideals

Let $\mathrm{PrimU}(\mathfrak{sl}(n))$ be the set of primitive ideals of $\mathrm{U}(\mathfrak{sl}(n))$. Duflo's Theorem implies that the map

$$\psi : \mathbb{F}^n \to \mathrm{PrimU}(\mathfrak{sl}(n)), \quad f \mapsto I(f)$$

is surjective. A description of $\mathrm{PrimU}(\mathfrak{sl}(n))$, based on the description of the fibres of ψ, is due to Joseph [10], see also [2–4, 12].

As a first step of this description, one attaches to $f \in \mathbb{F}^n$ a subgroup $W_n(f) \subset W_n$ called *the integral Weyl subgroup of* f. The subgroup $W_n(f)$ is a parabolic subgroup of S_n, and therefore is a product of permutation groups. As a second step, one defines an element $w(f) \in W(\lambda)$. In the regular case, this element $w(f)$ produces f from its dominant representative. For the singular case we refer the reader to [4]. The third step consists of applying the Robinson–Schensted algorithm to each factor of the element $w(f)$ with respect to the decomposition of $W_n(f)$ as a direct product of symmetric groups. For each factor of $w(f)$ this algorithm produces a pair of semistandard Young tableaux called *recording* tableau and *insertion* tableau.

The original result of Joseph [11] claims that $\psi(f_1) = \psi(f_2)$ if and only if

(1) f_1 and f_2 define the same character of $\mathrm{Z}(\mathfrak{sl}(n))$, i.e., there exists $k \in \mathbb{F}$ and a permutation $\sigma \in S_n$ such that $\sigma \cdot f_1 = f_2 + k$,
(2) the recording tableau of each factor of $w(f_1) \in W_n(f_1)$ coincides with the recording tableau of the corresponding factor of $w(f_2 + k)$ under σ.

For the purpose of considering the limit $n \to \infty$, it is convenient to restate Joseph's result in terms of f only without referring to $w(f)$. We do this in Theorem 3.2 below.

3.1.3 Admissible Interchanges

For $a, b \in \mathbb{F}$ we write $a >_{\mathbb{Z}} b$ whenever $a - b \in \mathbb{Z}_{>0}$. The notations $a <_{\mathbb{Z}} b, a \geq_{\mathbb{Z}} b, a \leq_{\mathbb{Z}} b$ have similar meaning.

Let $f_1, f_2 \in \mathbb{F}^n$. We say that f_1 and f_2 are *connected by the ith admissible interchange* if

$$f_1(j) = f_2(j),\ j \neq i, i+1, \quad f_1(i) = f_2(i+1),\ f_1(i+1) = f_2(i),\ 1 \leq i \leq n-1,$$

and one of the following conditions is satisfied:

(1) $f_1(i+1) - f_1(i) \notin \mathbb{Z}$,
(2) $i \leq n-2$ and $f_1(i+1) >_{\mathbb{Z}} f_1(i+2) \geq_{\mathbb{Z}} f_1(i)$,
(2′) $i \leq n-2$ and $f_1(i) >_{\mathbb{Z}} f_1(i+2) \geq_{\mathbb{Z}} f_1(i+1)$,
(3) $i \geq 2$ and $f_1(i+1) \geq_{\mathbb{Z}} f_1(i-1) >_{\mathbb{Z}} f_1(i)$,
(3′) $i \geq 2$ and $f_1(i) \geq_{\mathbb{Z}} f_1(i-1) >_{\mathbb{Z}} f_1(i+1)$.

It can be easily checked that f_1 is connected with f_2 by the ith admissible interchange if and only if f_2 is connected with f_1 by the ith admissible interchange. These admissible interchanges are known in the context of the Robinson–Schensted algorithm, see Theorem 3.2 below.

We say that f_1 and f_2 are connected by the *shifted ith admissible interchange* if the sequences $f_1 + \rho_n$ and $f_2 + \rho_n$ are connected by the ith admissible interchange.

3.1.4 A Version of Robinson–Schensted Algorithm for Finite Sequences

The Robinson–Schensted algorithm is a classical object of twentieth century mathematics and has different versions. As a reference for "the standard algorithm" we use [13]. This algorithm works with a finite sequence of nonrepeating integers, however we note that one can apply the standard algorithm to any nonrepeating finite sequence of elements of a totally ordered set $(\mathcal{S}, \prec)$. The output of this procedure consists of a Young tableau filled by elements of $\mathcal{S}$ (recording tableau) and a Young tableau of the same shape filled by positive integers (insertion tableau). The recording tableau is standard with respect to $\prec$ and the insertion tableau is standard with respect to $<$.

If a sequence consists of elements of several distinct totally ordered sets $(\mathcal{S}_i, \prec_i)$, we can split the sequence into subsequences of elements of $\mathcal{S}_i$ (one for each set) and apply the algorithm separately to such sequences. The output consists of a collection of pairs of tableaux-one pair per set $\mathcal{S}_i$.

In our case, the totally ordered sets $(\mathcal{S}_i, \prec)$ will be of the form $(a + \mathbb{Z}) \times \mathbb{Z}$, $a \in \mathbb{F}$, with the order

$$(a, i) \prec (b, j) \iff [(a >_{\mathbb{Z}} b) \text{ or } (a = b, i > j)]. \tag{1}$$

Let $f_1, \dots, f_n \in \mathbb{F}$ be a finite sequence. We attach to $f_1, \dots, f_n$ the sequence

$$(f_1, 1), \dots, (f_n, n), \tag{2}$$

and split (2) into totally ordered subsets of $(a + \mathbb{Z}) \times \mathbb{Z}$ as above. We then apply the standard Robinson–Schensted algorithm to (2). The output consists of a collection of pairs of tableaux. The recording tableau in a pair is filled by $\{(f_i, i)\}_{1 \le i \le n}$ and the insertion tableau is filled by $1, \dots, n$. As a last step we replace the pairs (f_i, i) in all recording tableaux by f_i and discard all insertion tableaux. The resulting tableaux have strictly decreasing rows and nonincreasing columns (the corner of a tableau being in the upper-left position). This is a consequence of the inequality inversion in the left and right-hand sides of formula (1).

In what follows, by *RS-algorithm*, we mean the above procedure. We denote by $RS(f_1, \dots, f_n)$ its output. We set also

$$J(f_1, \dots, f_n) := RS(f_1 - 1, \dots, f_n - n);$$

$J(f_1, \dots, f_n)$ reflects the shift of $f_1, \dots, f_n$ by "ρ".

Example 3.1 Consider the sequence $3, 4, 4, \alpha$, where $\alpha \notin \mathbb{Z}$. We have

$$J(3, 4, \alpha, 5) = RS(2, 2, \alpha - 3, 1).$$

Next, we attach to the sequence $2, 2, \alpha - 3, 1$ the sequence

$$(2, 1), (2, 2), (\alpha - 3, 3), (1, 4) \tag{3}$$

of elements of $\mathbb{F} \times \mathbb{Z}$. We have

$$(2, 2) \prec (2, 1) \prec (1, 4), \tag{4}$$

and the element $(\alpha - 3, 3)$ is incomparable with the elements of (4). We apply the RS-algorithm to the sequence (3) step-by-step from left to right:

$$(2,1) \mapsto \{(\boxed{(2,1)}, \boxed{1})\}, \quad ((2,1),(2,2)) \mapsto \{(\begin{array}{|c|}\hline (2,2)\\ \hline (2,1)\\ \hline\end{array}, \begin{array}{|c|}\hline 2\\ \hline 1\\ \hline\end{array})\},$$

$$((2,1),(2,2),(\alpha-3,3)) \mapsto \{(\begin{array}{|c|}\hline (2,2)\\ \hline (2,1)\\ \hline\end{array}, \begin{array}{|c|}\hline 2\\ \hline 1\\ \hline\end{array}), (\boxed{(\alpha-3,3)}, \boxed{3})\},$$

$$((2,1),(2,2),(\alpha-3,3),(1,4)) \mapsto \{(\begin{array}{|c|c|}\hline (2,2) & (1,4)\\ \hline (2,1) \\ \cline{1-1}\end{array}, \begin{array}{|c|c|}\hline 2 & 4\\ \hline 1 \\ \cline{1-1}\end{array}), (\boxed{(\alpha-3,3)}, \boxed{3})\}.$$

The result is

$$J(3,4,\alpha,5)=\left\{\begin{array}{|c|c|}\hline 2 & 1 \\ \hline 2 & \multicolumn{1}{c}{} \\ \cline{1-1}\end{array}\,,\ \begin{array}{|c|}\hline \alpha-3 \\ \hline\end{array}\right\}.$$

Theorem 3.2 (An Equivalent Form of Joseph's Theorem) *The following conditions are equivalent for sequences* $f_1,\dots,f_n\in\mathbb{F}$, $f'_1,\dots,f'_n\in\mathbb{F}$*:*

(1) $I(f_1,\dots,f_n)=I(f'_1,\dots,f'_n)$,
(2) $\exists k\in\mathbb{F}: J(f_1,\dots,f_n)=J(f'_1+k,\dots,f'_n+k)$,
(3) there exists $k\in\mathbb{F}$ *so that the sequences*

$$f_1,\dots,f_n \text{ and } f'_1+k,\dots,f'_n+k$$

are connected by a series of shifted admissible interchanges.

Proof This is implied by the results of [11] and [13, Exercise 4 on page 65]. □

In what follows, it will be convenient to encode Young tableaux via sequences.

Notation 3.3 To a Young tableau T with n boxes filled by elements of $a+\mathbb{Z}$ we attach the sequence $\operatorname{seq}(T)\in\mathbb{F}^n$ which consists of the rows of T ordered in the inverse lexicographical order (shorter rows come first; among rows of equal length, rows with smaller first element come first).

It is straightforward to check that

$$RS(\operatorname{seq}(T))=T.$$

This implies that T can be encoded by $\operatorname{seq}(T)$. If $T_1,\dots,T_s$ is a sequence of tableaux, we set $\operatorname{seq}(T_1,\dots,T_s)$ to be the concatenation of the sequences $\operatorname{seq}(T_1),\dots,\operatorname{seq}(T_s)$.

Example 3.4 If $T=\begin{array}{|c|c|c|}\hline 4+a & 2+a & 1+a \\ \hline 4+a & 1+a & \multicolumn{1}{c}{} \\ \cline{1-2} 4+a & 1+a & \multicolumn{1}{c}{} \\ \cline{1-2} 3+a & \multicolumn{1}{c}{} & \multicolumn{1}{c}{} \\ \cline{1-1}\end{array}$, then $\operatorname{seq}(T)=(3+a,4+a,1+a,4+a,1+a,4+a,2+a,1+a)$. If

$$T_1=\begin{array}{|c|c|}\hline 7+a & -4+a \\ \hline -8+a & \multicolumn{1}{c}{} \\ \cline{1-1}\end{array}\,,\qquad T_2=\begin{array}{|c|c|}\hline -4+b & -6+b \\ \hline -5+b & \multicolumn{1}{c}{} \\ \cline{1-1}\end{array}$$

with $a-b\notin\mathbb{Z}$, then $\operatorname{seq}(T_1,T_2)=(-8+a,7+a,-4+a,-5+b,-4+b,-6+b)$.

3.2 Coherent Local Systems and Their Annihilators

We now recall some results on c.l.s. The definition of c.l.s. is given in Sect. 1. In the current section we write $Q = \{Q_n\}$ for a c.l.s., where $Q_n = \{[L_n^\alpha]\}$ for some simple finite-dimensional modules L_n^α. Since each L_n^α is determined by its dominant $\mathfrak{b}_n$-highest weight λ_n^α, we can write $\{\lambda_n^\alpha\}$ instead. It is convenient to think of the highest weights λ_n^α as functions $f^\alpha \in \mathbb{Z}^n_{\geq 0} \subset \mathbb{F}^n$ with the normalizing conditions

$$f^\alpha(1) \geq f^\alpha(2) \geq \ldots \geq f^\alpha(n) = 0$$

or, equivalently, as partitions with at most $n-1$ parts. In this notation, $Q_n = \{f^\alpha\}$.

The *annihilator* $I(Q)$ of a c.l.s. Q is the ideal $\cup_n(\cap_\alpha \mathrm{Ann}_{\mathrm{U}(\mathfrak{sl}(n))} L_n^\alpha) \subset \mathrm{U}(\mathfrak{sl}(\infty))$.

Define functions $f_{k,n} \in \mathbb{Z}^n_{\geq 0}$ by setting

$$f_{k,n}(i) := \begin{cases} 1 & \text{if } i \leq k \\ 0 & \text{otherwise} \end{cases}.$$

The set of c.l.s. is partially ordered and forms a lattice:

$$Q \subset Q' = \{Q_n \subset Q'_n\}, \quad Q \cap Q' = \{Q_n \cap Q'_n\}.$$

In addition, Zhilinskii defines the following Cartan product on c.l.s.:

$$(Q'Q'')_n := \{f \in \mathbb{Z}^n \mid f = f' + f'' \text{ for some } f' \in (Q')_n, f'' \in (Q'')_n\},$$

see [21, Subsection 2.1]. A main result of Zhilinskii is that any irreducible c.l.s. is a Cartan product of basic c.l.s. The latter are denoted by $\mathcal{L}_i, \mathcal{R}_i, \mathcal{L}_i^\infty, \mathcal{R}_i^\infty, \mathcal{E}, \mathcal{E}^\infty$, and are defined as follows:

$\mathcal{E}^\infty$ is the c.l.s. consisting of all integral dominant weights on all levels,

$$(\mathcal{L}_i)_n := \{f_{k,n}\}_{0 \leq k \leq i}, \quad (\mathcal{L}_i^\infty)_n := \{f \in (\mathcal{E}^\infty)_n \mid f(k) = 0 \text{ for } k > i\},$$

$$(\mathcal{R}_i)_n := \{f_{k,n}\}_{n-i \leq k \leq n}, \quad (\mathcal{R}_i^\infty)_n := \{f \in (\mathcal{E}^\infty)_n \mid f(k) = f(n-i) \text{ for } k \leq n-i\},$$

$$\mathcal{E}_n := \{f_{k,n}\}_{0 \leq k < n}.$$

We can now state

Proposition 3.5 (Unique Factorization Property [21, Theorem 2.3.1]) *Any proper irreducible c.l.s., i.e., any irreducible c.l.s. non-equal $\mathcal{E}^\infty$, can be expressed uniquely as a Cartan product in the following form:*

$$\mathrm{cls}(r', r'', g, X, Y) := (\mathcal{L}_{r'}^{\infty}\mathcal{L}_{r'+1}^{l_1-l_2}\mathcal{L}_{r'+2}^{l_2-l_3}\dots\mathcal{L}_{r'+s}^{l_s-0})\,\mathcal{E}^{g}\,(\mathcal{R}_{r''}^{\infty}\mathcal{R}_{r''+1}^{r_1-r_2}\mathcal{R}_{r''+2}^{r_2-r_3}\dots\mathcal{R}_{r''+t}^{r_t-0}) \tag{5}$$

where r', r'', g are nonnegative integers, and

$$X = (l_1 \geq \dots \geq l_s > 0), \quad Y = (r_1 \geq \dots \geq r_t > 0)$$

are Young diagrams. Here, for $r' = 0$, $\mathcal{L}_{r'}^{\infty}$ is assumed to be the c.l.s. $\mathcal{T}$ consisting of the one-dimensional $\mathfrak{sl}(n)$-module at all levels, and similarly $\mathcal{R}_{r''}^{\infty}$ is assumed to equal $\mathcal{T}$ for $r'' = 0$.

As we have shown in [14], the annihilator $I(\mathrm{cls}(r', r'', g, X, Y))$ depends on the following four parameters

$$r := r' + r'', g, X, Y, \tag{6}$$

and all such annihilators are in a natural bijection with quadruples (6) where $r, g \in \mathbb{Z}_{\geq 0}$, and X, Y are arbitrary Young diagrams. We set

$$I(r, g, X, Y) := I(\mathrm{cls}(r, 0, g, X, Y)).$$

It follows from [21] that $I(r, g, X, Y)$ is a primitive ideal of $\mathrm{U}(\mathfrak{sl}(\infty))$. The main result of [17] claims that the ideals $I(r, g, X, Y)$ exhaust all nonzero proper primitive ideals of $\mathrm{U}(\mathfrak{sl}(\infty))$.

Next, following Zhilinskii, we attach to any basic c.l.s. Q a sequence $\gamma(Q; \cdot)$ of $\mathfrak{sl}(2n)$-modules by displaying the respective highest weights:

$$\gamma(\mathcal{L}_i; n) := f_{i,2n},\ 2n > i, \quad \gamma(\mathcal{L}_i^{\infty}; n) := (2i-1) f_{i,2n},\ 2n > i,$$

$$\gamma(\mathcal{R}_i; n) := f_{2n-i,2n},\ 2n > i, \quad \gamma(\mathcal{R}_i^{\infty}; n) := (2i-1) f_{2n-i,2n},\ 2n > i,$$

$$\gamma(\mathcal{E}; n) := f_{n,2n},\ n > 0.$$

Using (5) and the rule $\gamma(Q'Q''; n) := \gamma(Q'; n)+\gamma(Q''; n)$, we extend the definition of $\gamma(Q; n)$ to all proper irreducible c.l.s. Q.

To state the next lemma we need to define a *precoherent local system* (*p.l.s.* for short) Q. That consists of nonempty sets Q_n of isomorphism classes $[L_n^{\alpha}]$ of simple finite-dimensional $\mathfrak{sl}(n)$-modules L_n^{α} for each $n \geq 2$, such that each $\mathfrak{sl}(n)$-module $L_n^{\alpha_0}$ branches over $\mathfrak{sl}(n-1)$ as a sum of $\mathfrak{sl}(n-1)$-modules among L_{n-1}^{α}. For a p.l.s. we do not require that every L_{n-1}^{α} with $[L_{n-1}^{\alpha}] \in Q_{n-1}$ appear in the $\mathfrak{sl}(n-1)$-decomposition of a suitable $L_n^{\alpha_0}$ with $[L_n^{\alpha_0}] \in Q_n$.

Lemma 3.6 *Let $Q' = \{Q'_n\}$ be a p.l.s. and let Q be an irreducible c.l.s. such that $\gamma(Q; n) \in (Q')_{2n}$ for $n \gg 0$. Then $Q \subset Q'$.*

Proof The statement is implied by [21, Lemma 2.3.2]. □

3.3 Highest Weight $\mathfrak{sl}(\infty)$-Modules and Their Annihilators

In what follows we identify $\mathfrak{sl}(\infty)$ with the Lie algebra of traceless matrices $(a_{ij})_{i,j\in\Theta}$ such that each matrix has finitely many nonzero entries. We fix the splitting Cartan subalgebra $\mathfrak{h}$ of diagonal matrices (a detailed discussion of Cartan subalgebras of $\mathfrak{sl}(\infty)$ see in [5]). Any subset S of Θ defines a subalgebra $\mathfrak{sl}(S)$ spanned by

$$\{e_{ij}\}_{i\neq j\in S}, \quad \{e_{ii} - e_{jj}\}_{i,j\in S}.$$

If S is infinite then $\mathfrak{sl}(S) \cong \mathfrak{sl}(\infty)$, and $\mathfrak{sl}(S) \cong \mathfrak{sl}(|S|)$ if S is finite, where $|S|$ is the cardinality of S.

A total order $\prec$ on S defines a splitting Borel subalgebra

$$\mathfrak{b}^S(\prec) := \mathrm{span}\{e_{ii} - e_{jj}\}_{i,j\in S} + \mathrm{span}\{e_{ij}\}_{i\prec j\in S},$$

of $\mathfrak{sl}(S)$, see [15] for more details.

A function $f : S \to \mathbb{F}$ defines a character λ_f^S of $\mathfrak{b}^S(\prec)$ such that

$$\lambda_f^S(e_{ij}) = 0 \text{ for } i \prec j, \quad \lambda_f^S(e_{ii} - e_{jj}) = f(i) - f(j).$$

Let $\mathbb{F}_f^S$ be the respective one-dimensional $\mathfrak{b}^S(\prec)$-module and let

$$M_\prec^S(f) := M_\prec^S(\lambda_f) := \mathrm{U}(\mathfrak{sl}(S)) \otimes_{\mathrm{U}(\mathfrak{b}^S(\prec))} \mathbb{F}_f^S.$$

Denote by $L_\prec^S(f) := L_\prec^S(\lambda_f)$ the unique simple quotient of $M_\prec^S(\lambda_f) = M_\prec^S(f)$. Put

$$I_\prec^S(f) := \mathrm{Ann}_{\mathrm{U}(\mathfrak{sl}(S))} L_\prec^S(f).$$

If $F \subset \Theta$ is a finite subset, then $L_\prec^F(f)$ is the $\mathfrak{sl}(F)$-module $L(f|_F)$ where the totally ordered set $(F, \prec)$ is naturally identified with $(\{1, \ldots, n\}, <)$. In what follows, when given a total order $\prec$ on Θ and a function $f : \Theta \to \mathbb{F}$, we will use the above notations $M_\prec^\Theta(f), L_\prec^\Theta(f), I_\prec^\Theta(f)$ having in mind that $\prec$ defines an order on S and f defines a function on S via restriction. Whenever $S = \Theta$ we write simply $\mathfrak{b}(\prec), M_\prec(f), L_\prec(f), I_\prec(f)$. Note also that in Sect. 1 our notation $L_{\mathfrak{b}}(\lambda)$ for a simple highest weight module displayed explicitly the relevant Borel subalgebra $\mathfrak{b}$ and the highest weight λ, so $L_\prec(f)$ is another notation for $L_{\mathfrak{b}(\prec)}(\lambda_f)$.

We will be particularly interested in several special kinds of splitting Borel subalgebras.

Definition 3.7 We say that $\mathfrak{b}^S(\prec)$ is a *Dynkin* Borel subalgebra if $(S, \prec)$ is isomorphic as an ordered set to $(\mathbb{Z}_{>0}, <)$, $(\mathbb{Z}_{<0}, <)$ or $(\mathbb{Z}, <)$. This is equivalent to the condition that every root of $\mathfrak{b}^S(\prec)$ is a finite sum of simple roots, see [20].

Let $\Theta_1, \Theta_2 \subset \Theta$ be two subsets. We write $\Theta_1 \prec \Theta_2$ if $s_1 \prec s_2$ for any $s_1 \in \Theta_1$ and $s_2 \in \Theta_2$.

Definition 3.8 We say that $\mathfrak{b}^S(\prec)$ is an *ideal* Borel subalgebra if S can be partitioned into subsets

$$S_1 \prec S_2 \prec S_3$$

such that $(S_1, \prec) \cong (\mathbb{Z}_{>0}, <)$ and $(S_3, \prec) \cong (\mathbb{Z}_{<0}, <)$.

Definition 3.9 Let $S \subset \Theta$ be a subset. We say that $f \in \mathbb{F}^S$ is *$\prec$-locally constant on* S if there exists a partition $S_1 \prec \ldots \prec S_t$ of S such that $f|_{S_i}$ is constant for every S_i, $1 \le i \le t$. We say that $f \in \mathbb{F}^S$ is *almost integral on* S if there exists a finite set $F \subset S$ such that $f(i) - f(j) \in \mathbb{Z}$ for all $i, j \in S \backslash F$.

Theorem 3.10 ([15, Theorem 9]) *The following conditions are equivalent:*

(1) $I_\prec(f) \neq 0$,
(2) f *is $\prec$-locally constant and almost integral on* Θ.

The next proposition relates the computation of the annihilators of simple highest weight $\mathfrak{sl}(\infty)$-modules to the computation of the annihilators of simple highest weight $\mathfrak{sl}(n)$-modules for finite n.

Proposition 3.11 ([15, Lemma 5.7]) *Let* $F_1 \subset F_2 \subset \ldots \subset F_n \subset \ldots$ *be an infinite sequence of finite subsets of* Θ, *and let* $S := \cup_i F_i$. *Then*

$$I_\prec^S(f) = \cup_n(\cap_{i \ge n} I_\prec^{F_i}(f)).$$

Corollary 3.12 *Let* $F_1 \subset F_2 \subset \ldots \subset F_n \subset \ldots$ *be an infinite sequence of finite sets such that* $S = \cup_i F_i$. *Let* $f' \in \mathbb{F}^S$ *be a function such that one of the following holds:*

- $I_\prec^{F_i}(f) = I_\prec^{F_i}(f')$ *for all* $i \in \mathbb{Z}_{>0}$,
- $I_\prec^{F_i}(f) = I_\prec^{F_i}(f')$ *for all but finitely many* $i \in \mathbb{Z}_{>0}$.

Then $I_\prec^S(f) = I_\prec^S(f')$.

Corollary 3.13 *Let* $F_1 \subset F_2 \subset \ldots \subset F_n \subset \ldots$ *be an infinite sequence of finite sets such that* $S = \cup_i F_i$. *Let* $f' \in \mathbb{F}^S$ *be a function such that one of the following holds:*

- $f|_{F_i}$ *and* $f'|_{F_i}$ *are connected by a series of shifted admissible interchanges for all* $i \in \mathbb{Z}_{>0}$,

– *$f|_{F_i}$ and $f'|_{F_i}$ are connected by a series of shifted admissible interchanges for all but finitely many $i \in \mathbb{Z}_{>0}$.*

Then $I^S_{\prec}(f) = I^S_{\prec}(f')$.

Corollary 3.14 *Let $\Theta_1 \sqcup \Theta_2 \sqcup \ldots \sqcup \Theta_t$ be a partition of Θ and let $f \in \mathbb{F}^{\Theta}$ be a function such that $f|_{\Theta_i}$ is constant. Assume that $\prec_1, \prec_2$ are total orders on Θ such that*

$$\Theta_1 \prec_1 \Theta_2 \prec_1 \ldots \prec_1 \Theta_t \text{ and } \Theta_1 \prec_2 \Theta_2 \prec_2 \ldots \prec_2 \Theta_t. \tag{7}$$

Then $I_{\prec_1}(f) = I_{\prec_2}(f)$.

Definition 3.15 Let $\prec_1, \prec_2$ be total orders on Θ, and $f \in \mathbb{F}^{\Theta}$ be a function. We say that $\prec_1$ and $\prec_2$ are *f-equivalent* if f is locally constant with respect to a partition $\Theta_1 \sqcup \ldots \sqcup \Theta_t$ of Θ and this partition satisfies (7).

Corollary 3.14 claims that $I_{\prec_1}(f) = I_{\prec_2}(f)$ for f-equivalent total orders $\prec_1, \prec_2$.

Definition 3.16 Let $f \in \mathbb{F}^{\Theta}$ be a function, almost integral and locally constant, and let $\Theta_1 \prec \ldots \prec \Theta_t$ be some partition of Θ. We say that this partition is *f-preferred* if

$$(\Theta_1, \prec) \cong (\mathbb{Z}_{>0}, <), (\Theta_t, \prec) \cong (\mathbb{Z}_{<0}, <), (\Theta_i, \prec) \cong (\mathbb{Z}, <) \text{ for } 1 < i < t,$$

and for all i there exist $s_i^- \in \Theta_i$ and $s_{i+1}^+ \in \Theta_{i+1}$ such that $f(s) = f(s')$ for all $s \in \Theta_i, s_i^- \prec s$, and $s' \in \Theta_{i+1}, s' \prec s_{i+1}^+$. We say that a total order $\prec_f$ is *f-preferred* if there exists a partition $\Theta_1 \prec_f \ldots \prec_f \Theta_t$ which is f-preferred with respect to $\prec_f$.

Let $f \in \mathbb{F}^{\Theta}$ be an almost integral and locally finite function with respect to a partition $\Theta_1 \prec \ldots \prec \Theta_t$ of Θ. It is easy to construct an f-preferred order $\prec_f$ on Θ such that $\prec_f$ is f-equivalent to $\prec$. Indeed, let $i_1, i_2, \ldots, i_q$ be the set of indices such that $\Theta_{i_1}, \ldots, \Theta_{i_q}$ are infinite. We split each ordered set Θ_{i_k} into two infinite sets $\Theta^l_{i_k}, \Theta^r_{i_k}$ so that

$$\Theta^l_{i_1} \prec \Theta^r_{i_1} \prec \ldots \prec \Theta^l_{i_q} \prec \Theta^r_{i_q}.$$

As a result, Θ equals the disjoint union

$$(\Theta_1 \sqcup \Theta_2 \sqcup \ldots \sqcup \Theta^l_{i_1}) \sqcup (\Theta^r_{i_1} \sqcup \Theta_{i_1+1} \sqcup \ldots \sqcup \Theta^l_{i_2}) \sqcup \ldots \sqcup (\Theta^r_{i_q} \sqcup \Theta_{i_q+1} \sqcup \ldots \sqcup \Theta_t), \tag{8}$$

and we have

$$\Theta_1 \prec \Theta_2 \prec \ldots \prec \Theta^l_{i_1} \prec \Theta^r_{i_1} \prec \Theta_{i_1+1} \prec \ldots \prec \Theta^l_{i_2} \prec \ldots \prec \Theta^r_{i_q} \prec \Theta_{i_q+1} \prec \ldots \prec \Theta_t.$$

The desired order $\prec_f$ will be f-preferred with respect to the decomposition (8). To introduce $\prec_f$, we start with $(\Theta_1 \sqcup \Theta_2 \sqcup \ldots \sqcup \Theta^l_{i_1})$ and replace the given order $\prec$ by an order $\prec_f$ isomorphic to $(\mathbb{Z}_{>0}, >)$ such that $\Theta_1 \prec_f \Theta_2 \prec_f \ldots \prec_f \Theta^l_{i_1}$. Next, for $(\Theta^r_{i_1} \sqcup \Theta_2 \sqcup \ldots \sqcup \Theta^l_{i_1})$ we replace the given order $\prec$ by an order $\prec_f$ isomorphic to $(\mathbb{Z}, >)$ such that $\Theta^r_{i_1} \prec_f \Theta_{i_1+1} \prec_f \ldots \prec_f \Theta^l_{i_2}$. We repeat this last step $q-2$ times. Finally, at the right end $(\Theta^r_{i_q} \sqcup \Theta_2 \sqcup \ldots \sqcup \Theta_{i_t})$ we replace the order $\prec$ by an order $\prec_f$ isomorphic to $(\mathbb{Z}_{<0}, <)$ such that $\Theta^r_{i_q} \prec_f \Theta_{i_q+1} \prec_f \ldots \prec_f \Theta_{i_t}$. The so obtained order $\prec_f$ is f-preferred and is f-equivalent to the original order. Therefore, Corollary 3.14 implies

$$I_{\prec}(f) = I_{\prec_f}(f).$$

4 Robinson–Schensted Algorithm at Infinity

In what follows we extend the RS-algorithm to stably decreasing infinite sequences. Overall, the procedure is very similar to (and is based on) the one given in Sect. 3.1.4. We consider functions $f \in \mathbb{F}^{\mathbb{Z}_{>0}}, \mathbb{F}^{\mathbb{Z}_{<0}}, \mathbb{F}^{\mathbb{Z}}$ and identify them with the respective sequences

$$f(1), f(2), \ldots,$$

$$\ldots, f(-2), f(-1)$$

$$\ldots, f(-1), f(0), f(1), \ldots$$

Admissible interchanges for functions $f \in \mathbb{F}^{\mathbb{Z}_{>0}}, \mathbb{F}^{\mathbb{Z}_{<0}}, \mathbb{F}^{\mathbb{Z}}$ are defined as in Sect. 3.1.3.

Definition 4.1 We say that f is *stably decreasing* if $f(i) >_{\mathbb{Z}} f(i+1)$ for $|i| \gg 0$.

The formula $\rho(i) = -i$ defines three different functions $\rho_{\mathbb{Z}_{<0}} \in \mathbb{F}^{\mathbb{Z}_{<0}}, \rho_{\mathbb{Z}} \in \mathbb{F}^{\mathbb{Z}}, \rho_{\mathbb{Z}_{>0}} \in \mathbb{F}^{\mathbb{Z}_{>0}}$.

The following "insertion operation" inserts a given $f_1 \in \mathbb{F}^r$ into positions $i_1 < i_2 < \ldots < i_r$ of f_2:

$$\operatorname{ins}(i_1, \ldots, i_r; f_1, f_2)(i) := \begin{cases} f_2(i) & \text{if } i < i_1, \\ f_1(t) & \text{if } i = i_t, 1 \le t \le r, \\ f_2(i-t) & \text{if } i_t < i < i_{t+1}, 1 \le t \le r-1 \\ f_2(i-r) & \text{if } i > i_r \end{cases},$$

for $f_2 \in \mathbb{F}^{\mathbb{Z}}$, or $f_2 \in \mathbb{F}^{\mathbb{Z}_{>0}}$ and $i_1 \ge 0$,

$$\operatorname{ins}(i_1,\dots,i_r;f_1,f_2)(i) := \begin{cases} f_2(i+r) & \text{if } i < i_1 \\ f_1(t) & \text{if } i = i_t, 1 \le t \le r, \\ f_2(i+r-t) & \text{if } i_t < i < i_{t+1}, 1 \le t \le r-1 \\ f_2(i) & \text{if } i_r < i \end{cases},$$

for $f_2 \in \mathbb{F}^{\mathbb{Z}_{<0}}, i_r \le 0$.

Remark 4.2 Since the (shifted) admissible interchanges and the respective equivalence classes are defined for all sequences regardless of any stabilization conditions, it could be an interesting combinatorial problem to study the corresponding equivalence classes.

4.1 *Left-Infinite Case* ($\mathbb{Z}_{<0}$)

Consider a stably decreasing sequence $f \in \mathbb{F}^{\mathbb{Z}_{<0}}$. We now explain how to apply the infinite RS-algorithm to f. What we do is simply apply the RS-algorithm consecutively to the finite tails $f(-n),\dots,f(0)$ of f. Then, for $n \ll 0$, the RS-algorithm will keep modifying only one of the tableaux in the outputs of previous steps. This follows from the fact for $n \ll 0$ the numbers $f(n)$ are in same integrality class.

Next, note that, since f is stably decreasing, this modification will amount to adding the box $\boxed{f(-n-1)}$ to the left-hand side of the first row. In this way, the output $RS(f)$ of our algorithm consists of several (possibly none) finite Young tableaux and one tableau whose first row is infinite and all other rows are finite. Denote the infinite tableau by T_1 and the other tableaux by $T_2,\dots,T_s$. Denote the first row of T_1 by $\overline{\operatorname{seq}}(f)$, T_1 without the first row by T_1'. Set $\underline{\operatorname{seq}}(f) := \operatorname{seq}(T_1', T_2, T_3, \dots, T_s)$. Then it is straightforward to check that

$$RS(f) = RS(\operatorname{ins}(i_1,\dots,i_r;\underline{\operatorname{seq}}(f),\overline{\operatorname{seq}}(f))), \tag{1}$$

where r is the number of elements in $\underline{\operatorname{seq}}(f)$ and $i_1, i_2, \dots, i_r$, are integers such that

$$i_{k+1} > i_k + 1, \overline{\operatorname{seq}}(f)_{i_r} >_{\mathbb{Z}} \underline{\operatorname{seq}}(f)_k \text{ or } \overline{\operatorname{seq}}(f)_{i_r} - \underline{\operatorname{seq}}(f)_k \notin \mathbb{Z} \text{ for all } k \le r \tag{2}$$

(the condition on i_r is satisfied for $i_r \ll 0$). The equality (1) plays an important role in our main result below.

4.2 *Right-Infinite Case* ($\mathbb{Z}_{>0} = -\mathbb{Z}_{<0}$)

Let $f \in \mathbb{F}^{\mathbb{Z}_{>0}}$ be a stably decreasing function. It is clear that the sequence

$$f^* := (\ldots, -f(3), -f(2), -f(1))$$

is an element of $\mathbb{F}^{\mathbb{Z}_{<0}}$ and is stably decreasing. If $g \in \mathbb{F}^{\mathbb{Z}_{<0}}$, we set g^* to be the sequence

$$-g(0), -g(-1), -g(-2), \ldots$$

Then $(f^*)^* = f$ for $f \in \mathbb{F}^{\mathbb{Z}_{>0}}$ or $\mathbb{F}^{\mathbb{Z}_{<0}}$.

In this case, we have

$$RS(f^*) = RS(\mathrm{ins}(i_1, \ldots, i_s; \underline{\mathrm{seq}}(f^*)^*, \overline{\mathrm{seq}}(f^*)^*)^*),$$

where s is the number of elements in $\underline{\mathrm{seq}}(f^*)$ and $i_1, \ldots, i_r$ satisfy the mirror image of (2)

$$i_{k+1} > i_k+1, \overline{\mathrm{seq}}(f^*)_{-i_r} >_{\mathbb{Z}} \underline{\mathrm{seq}}(f^*)_k \text{ or } \overline{\mathrm{seq}}(f^*)_{-i_r} - \underline{\mathrm{seq}}(f^*)_k \notin \mathbb{Z} \text{ for all } k \leq r.$$

Remark 4.3 In the procedure presented in this subsection, we apply the RS-algorithm inductively starting from the "infinite tail" of our sequence f. It also makes sense to apply the RS-algorithm starting from the beginning of the sequence f. The result will differ by an analogue of the Schutzenberger involution, see [13].

4.3 *Two-Sided Case* ($\mathbb{Z}$)

Consider a stably decreasing almost integral sequence $f \in \mathbb{F}^{\mathbb{Z}}$. We say that f is *almost integral* if $f(n_+) - f(n_-) \in \mathbb{Z}$ for $n_- \ll 0$ and $n_+ \gg 0$.

Assume f is almost integral. To apply the infinite RS-algorithm to f, all we have to do is to apply the RS-algorithm to "middle" finite subsequences $f(n_-), \ldots, f(n_+)$ of f where $n_- \to -\infty$ and $n_+ \to +\infty$. Note that for $n_- \ll 0$ and $n_+ \gg 0$ the RS-algorithm will keep modifying only one of the tableaux in the outputs of previous steps. This follows from the fact for $n_- \ll 0$ and $n_+ \gg 0$ the numbers $f(n_-)$, $f(n_+)$ are in same integrality class.

Next, note that since f is stably decreasing this modification will amount to adding the boxes $\boxed{f(n_- - 1)}$ to the left-hand side or $\boxed{f(n_+ + 1)}$ to the right-hand side of the first row. In this way, the output $RS(f)$ of our algorithm consists of several finite Young tableaux (possibly none) and one tableau whose first row is infinite and all other rows are finite. Denote the infinite tableau by T_1 and the other tableaux by $T_2, \ldots, T_s$. Denote the first row of T_1 by $\overline{\mathrm{seq}}(f)$, T_1 without the first row by T_1'. Note that the identification of two-sided sequences with $\mathbb{F}^{\mathbb{Z}}$ is unique only up to a shift, and we fix this shift in such a way that $f(i) = \overline{\text{seq}}(f)_i$ for $i \ll 0$.

Set $\underline{\text{seq}}(f) := \text{seq}(T_1', T_2, T_3, \ldots, T_s)$. Then we point out that (1) holds also in this case where r is the number of elements in $\underline{\text{seq}}(f)$ and $i_1, i_2, \ldots, i_r$ satisfy (2).

4.4 Admissible Interchanges and Robinson–Schensted Algorithm at Infinity

The next proposition is an infinite-dimensional version of the equivalence of claims (1) and (3) in Theorem 3.2.

Proposition 4.4 *For stably decreasing functions f, f' from $\mathbb{F}^{\mathbb{Z}_{>0}}$, $\mathbb{F}^{\mathbb{Z}_{<0}}$ or $\mathbb{F}^{\mathbb{Z}}$, the following conditions are equivalent:*

(a) f and f' are connected by a series of admissible interchanges,
(b) $RS(f) = RS(f')$.

Proof Elementary and straightforward. □

5 Two Attributes of an Ideal in $\mathrm{U}(\mathfrak{sl}(\infty))$

In this section, we introduce a sequence of algebraic varieties associated with an ideal $I \subset \mathrm{U}(\mathfrak{sl}(\infty))$, as well as a c.l.s. associated with I.

Recall that $\mathrm{Z}(\mathfrak{sl}(n))$ stands for the centre of $\mathrm{U}(\mathfrak{sl}(n))$. Denote by Irr_n the set of isomorphism classes of simple finite-dimensional $\mathrm{U}(\mathfrak{sl}(n))$-modules.

Lemma 5.1 (cf. [4, Subsection 3.1]) *Let I_1, I_2 be ideals of $\mathrm{U}(\mathfrak{sl}(n))$. Then the following conditions are equivalent:*

(a) $I_1 + I_2 = \mathrm{U}(\mathfrak{sl}(n))$,
(b) $1 \in I_1 + I_2$,
(c) $(I_1 \cap \mathrm{Z}(\mathfrak{sl}(n))) + (I_2 \cap \mathrm{Z}(\mathfrak{sl}(n))) = \mathrm{Z}(\mathfrak{sl}(n))$,
(d) $1 \in (I_1 \cap \mathrm{Z}(\mathfrak{sl}(n))) + (I_2 \cap \mathrm{Z}(\mathfrak{sl}(n)))$.

Proof It is clear that (a) is equivalent to (b), and that (c) is equivalent to (d). Hence it is enough to prove that (b) is equivalent to (d).

As an $\mathfrak{sl}(n)$-module with the adjoint structure, $\mathrm{U}(\mathfrak{sl}(n))$ is locally finite, and is an infinite direct sum of $\mathfrak{sl}(n)$-isotypic components $\mathrm{U}(\mathfrak{sl}(n))_\lambda$ where λ runs over the entire set Irr_n. Hence,

$$I_1 = \oplus_{\lambda \in \mathrm{Irr}_n} (\mathrm{U}(\mathfrak{sl}(n))_\lambda \cap I_1), \quad I_2 = \oplus_{\lambda \in \mathrm{Irr}_n} (\mathrm{U}(\mathfrak{sl}(n))_\lambda \cap I_2),$$

and

$$((I_1 + I_2) \cap \mathrm{Z}(\mathfrak{sl}(n))) = (I_1 + I_2)^{\mathfrak{sl}(n)} = (I_1)^{\mathfrak{sl}(n)} + (I_2)^{\mathfrak{sl}(n)}$$
$$= (\mathrm{Z}(\mathfrak{sl}(n)) \cap I_1) + (\mathrm{Z}(\mathfrak{sl}(n)) \cap I_2),$$

where $*^{\mathfrak{sl}(n)}$ stands for $\mathfrak{g}$-invariants. This implies that (b) is equivalent to (d). □

Lemma 5.2 *Let I be an ideal of $\mathrm{U}(\mathfrak{sl}(n))$ and L be a simple finite-dimensional $\mathfrak{sl}(n)$-module. If*

$$I \cap \mathrm{Z}(\mathfrak{sl}(n)) \subset (\mathrm{Z}(\mathfrak{sl}(n)) \cap \mathrm{Ann}_{\mathrm{U}(\mathfrak{sl}(n))} L), \tag{1}$$

then $I \subset \mathrm{Ann}_{\mathrm{U}(\mathfrak{sl}(n))} L$.

Proof Note that (1) implies

$$1 \notin ((I \cap \mathrm{Z}(\mathfrak{sl}(n))) + (\mathrm{Z}(\mathfrak{sl}(n)) \cap \mathrm{Ann}_{\mathrm{U}(\mathfrak{sl}(n))} L)) = \mathrm{Z}(\mathfrak{sl}(n)) \cap \mathrm{Ann}_{\mathrm{U}(\mathfrak{sl}(n))} L.$$

Moreover, it follows from Lemma 5.1 that $1 \notin (I + \mathrm{Ann}_{\mathrm{U}(\mathfrak{sl}(n))} L)$. It is a well-known result that there exists a unique maximal ideal m of $\mathrm{U}(\mathfrak{sl}(n))$ containing $\mathrm{Z}(\mathfrak{sl}(n)) \cap \mathrm{Ann}_{\mathrm{U}(\mathfrak{sl}(n))} L$, see [4, Subsection 1.1]. Clearly, $\mathrm{Ann}_{\mathrm{U}(\mathfrak{sl}(n))} L$ is maximal, and hence

$$I + \mathrm{Ann}_{\mathrm{U}(\mathfrak{sl}(n))} L \subset m = \mathrm{Ann}_{\mathrm{U}(\mathfrak{sl}(n))} L.$$

This implies $I \subset \mathrm{Ann}_{\mathrm{U}(\mathfrak{sl}(n))} L$. □

Fix $I \subset \mathrm{U}(\mathfrak{sl}(\infty))$. For any $n \geq 2$, we set

$$Q_n(I) := \{[L] \in \mathrm{Irr}_n \mid I \cap \mathrm{U}(\mathfrak{sl}(n)) \subset \mathrm{Ann}_{\mathrm{U}(\mathfrak{sl}(n))} L\}.$$

The union of $Q_n(I)$ is a p.l.s., see Sect. 3.2. Proposition 4.8 of [17] implies that there exists a c.l.s. $Q(I)$ such that $Q(I)_n = Q_n(I)$ for $n \gg 0$. Such c.l.s. $Q(I)$ is clearly unique.

A theorem of Harish-Chandra claims that $\mathrm{Z}(\mathfrak{sl}(n))$ is isomorphic to the S_n-invariants $\mathrm{S}(\mathfrak{h}_n)^{S_n}$ in the symmetric algebra $\mathrm{S}(\mathfrak{h}_n)^{S_n}$. Therefore the radical ideals of $\mathrm{Z}(\mathfrak{sl}(n))$ are in one-to-one correspondence with the S_n-invariant subvarieties of $\mathfrak{h}_n^*$. Let $f \in \mathbb{F}^n$ be a function. Then the ideal $I(f) \cap \mathrm{Z}(\mathfrak{sl}(n))$ is maximal; it corresponds to the S_n-orbit of the weight $\lambda_f + \rho_n$ where

$$\rho_n := \lambda_{n,n-1,\ldots,1}.$$

Let I be an ideal of $\mathrm{U}(\mathfrak{sl}(\infty))$. Consider $I \cap \mathrm{Z}(\mathfrak{sl}(n))$. Clearly, $I \cap \mathrm{Z}(\mathfrak{sl}(n))$ is an ideal of $\mathrm{Z}(\mathfrak{sl}(n))$ and $\sqrt{I \cap \mathrm{Z}(\mathfrak{sl}(n))}$ is a radical ideal; it is identified with the S_n-stable subvariety $\mathrm{ZVar}_n(I)$ of $\mathfrak{h}_n^*$.

The variety $\mathrm{ZVar}_n(I)$ and the set $Q_n(I)$ are related as follows.

Proposition 5.3 *Let I be a primitive ideal of* $\mathrm{U}(\mathfrak{sl}(\infty))$. *Then*

(a) $\mathrm{ZVar}_n(I)$ *equals the Zariski closure of the set*

$$\{w(\lambda_f + \rho_n) \mid [L(f)] \in Q_n(I), w \in S_n\}.$$

(b) Let f be a dominant function such that $\lambda_f + \rho_n \in \mathrm{ZVar}_n(I)$. Then

$$[L(f)] \in Q_n(I).$$

Proof If I is primitive, then I is locally integrable, see [17, Section 4], which means that

$$I \cap \mathrm{U}(\mathfrak{sl}(n)) = \cap_{[L^\alpha] \in Q(I)_n} \mathrm{Ann}_{\mathrm{U}(\mathfrak{sl}(n))} L^\alpha.$$

This implies

$$I \cap \mathrm{Z}(\mathfrak{sl}(n)) = \cap_{[L^\alpha] \in Q(I)_n} (\mathrm{Ann}_{\mathrm{U}(\mathfrak{sl}(n))} L^\alpha \cap \mathrm{Z}(\mathfrak{sl}(n)),$$

and (a) follows.

We proceed to (b). The condition $\lambda_f + \rho_n \in \mathrm{ZVar}_n(I)$ implies

$$I \cap \mathrm{Z}(\mathfrak{sl}(n)) \subset (\mathrm{Z}(\mathfrak{sl}(n)) \cap \mathrm{Ann}_{\mathrm{U}(\mathfrak{sl}(n))} L(f)).$$

To finish the proof we use Lemma 5.2. □

Consider $f \in \mathbb{F}^\Theta$ together with an arbitrary total order $\prec$ on Θ. Put

$$\mathrm{F}_{\prec,n}(f) := \{(f(i_1) - 1, \dots, f(i_n) - n) \in \mathbb{F}^n \mid i_1 \prec \dots \prec i_n \in \Theta\}.$$

Lemma 5.4 *We have*

$$\lambda_g \in \mathrm{ZVar}_n(I_\prec(f))$$

for all $g \in \mathrm{F}_{\prec,n}(f)$.

Proof Consider a finite subset $F = \{i_1, \dots, i_n\}$ of Θ. It is clear that $L^F_\prec(f)$ is an $\mathfrak{sl}(F)$-subquotient of $L_\prec(f)$. This implies that

$$I_\prec(f) \cap \mathrm{U}(\mathfrak{sl}(n)) \subset I_\prec(f),$$

and hence that

$$I_\prec(f) \cap \mathrm{Z}(\mathfrak{sl}(n)) \subset I_\prec(f) \cap \mathrm{Z}(\mathfrak{sl}(n)).$$

The latter inclusion is equivalent to the desired statement. □

Corollary 5.5 *Assume that f and f' are connected by a series of admissible interchanges. Then*

$$\lambda_g \in \mathrm{ZVar}_n(I_{\prec}(f))$$

for all $g \in \mathrm{F}_{\prec,n}(f')$.

6 The Main Result for Dynkin Borel Subalgebras

Assume that $\mathfrak{b}(\prec)$ is a Dynkin Borel subalgebra. This means that we can identify the ordered set $(\Theta, \prec)$ with one of the three ordered sets $(\mathbb{Z}_{>0}, <)$, $(\mathbb{Z}_{<0}, <)$, $(\mathbb{Z}, <)$.

Let $f \in \mathbb{F}^{\mathbb{Z}_{<0}}, \mathbb{F}^{\mathbb{Z}_{>0}}, \mathbb{F}^{\mathbb{Z}}$ be a locally constant function. Clearly, this is equivalent to

$$\exists N \in \mathbb{Z}_{>0} : f(i) = f(i+1) \text{ for all } |i| \geq N. \tag{1}$$

We fix such an N. Put

$$h^{\pm}(f) := \lim_{n \to \pm\infty} f(n)$$

cf. (1). Note that if $(\Theta, \prec) \cong (\mathbb{Z}_{<0}, <)$ or $(\Theta, \prec) \cong (\mathbb{Z}_{>0}, <)$ then any locally constant function $f \in \mathbb{F}^{\Theta}$ is almost integral. If $f \in \mathbb{F}^{\mathbb{Z}}$ is almost integral and locally constant, then $h^+(f) - h^-(f)$ is an integer.

For a locally constant function $f \in \mathbb{F}^{\mathbb{Z}_{<0}}$, we set

$$f^+ := (\ldots, f(i) + i, \ldots, f(-2) + 2, f(-1) + 1).$$

Then f^+ is a stably decreasing function in $\mathbb{F}^{\Theta}$. It is easy to see that

$$\overline{\mathrm{seq}}(f^+)_i = h^-(f) + |\underline{\mathrm{seq}}(f^+)| - i \text{ for } i \leq -N$$

or, equivalently,

$$\overline{\mathrm{seq}}(f^+)^*_i = -h^-(f) - |\underline{\mathrm{seq}}(f^+)| - i \text{ for } i \geq N.$$

Hence the function

$$\overline{\mathrm{seq}}(f^+)^* + h^-(f) + |\underline{\mathrm{seq}}(f^+)| - \rho_{\mathbb{Z}_{>0}}, \tag{2}$$

where $h^-(f)$, $|\underline{\mathrm{seq}}(f^+)|$ are constant functions, is nonincreasing and is stably equal to zero. The nonzero values of the function (2) form a partition which we denote by $Y(f)$.

Proposition 6.1 *Let $f \in \mathbb{F}^{\mathbb{Z}_{<0}}$ be a locally constant function. Then*

$$I_<(f) = I(r(f), 0, \emptyset, Y(f))$$

where $r := r(f) := |\underline{\text{seq}}(f^+)|$.

Proof It is enough to prove

(a) $I(r(f), 0, \emptyset, Y(f)) \subset I_<(f)$,
and
(b) $I_<(f) \subset I(r(f), 0, \emptyset, Y(f))$.

Statement (a) is equivalent to

$$I(r(f), 0, \emptyset, Y(f)) \subset I_<(f') \tag{3}$$

for some f' such that f and f' are connected by a series of admissible interchanges. We pick f' as in (1) with $i_1, \ldots, i_r$ satisfying (2). Then we apply [15, Lemma 5.4] and Proposition 3.11 to the inserted variables. This shows (a).

Theorem 3.2 of [15] implies that (b) is equivalent to
(b′) $Q_n(I(r(f), 0, \emptyset, Y(f))) \subset Q_n(I_<(f))$ for $n \gg 0$.
According to [14, Lemma 7.6c)] we have

$$Q_n(I(r(f), 0, \emptyset, Y(f))) = \cup_{r'+r''=r} \text{cls}(r', r'', \emptyset, Y(f)).$$

Hence we need to prove that

$$\text{cls}(r', r'', 0, \emptyset, Y(f))_n \subset Q_n(I_<(f))$$

for any $n \gg 0$ and all nonnegative integers r', r'' such that $r' + r'' = r$.

We fix r', r'' with $r' + r'' = r$. Let the partition $Y(f)$ be $(l_1 \geq \ldots \geq l_s > 0)$. We also fix $n \geq r + s$. Then Lemma 5.4 asserts that

$$\lambda_g \in \text{ZVar}_n(I_<(f)), \tag{4}$$

for any $\lambda_g \in F_{<,n}(f)$.

We will now make use of Corollary 3.13 which allows us to replace f in the formula (4) by any f' which is connected with f by a series of shifted admissible interchanges. Let $i_1, \ldots, i_r$ be integers satisfying condition (2). Consider the subset

$$F_{i_1,\ldots,i_r} := \{i_1, \ldots, i_r, -(n-r), \ldots, -1\} \subset \Theta. \tag{5}$$

Define $f_{i_1,\ldots,i_r}$ by the requirement that $(f_{i_1,\ldots,i_r})^+$ equals the right-hand side of (1) applied to $f^+, i_1, \ldots, i_r$. The order of the elements of $F_{i_1,\ldots,i_r}$ in (5) allows us to consider $f_{i_1,\ldots,i_r}|_{F_{i_1,\ldots,i_r}}$ as a vector in $\mathbb{F}^n$. Since f and $f_{i_1,\ldots,i_r}$ are connected by a series of shifted admissible interchanges, Corollary 5.5 implies that

$$\lambda_{g'} \in \text{ZVar}_n(I_<(f)) = \text{ZVar}_n(I_<(f_{i_1,\ldots,i_r})),$$

where

$$g'(k) = \begin{cases} \underline{\operatorname{seq}}(f^+)_k + i_k - k \\ \quad \text{if } 1 \le k \le r \\ \overline{\operatorname{seq}}(f^+)_{k-1-n+s} + (k-1-n+s) - k = h^-(f) + r - k \\ \quad\quad\quad \text{if } r < k \le n-s \\ \overline{\operatorname{seq}}(f^+)_{k-1-n+s} + (k-1-n+s) - k = h^-(f) - r - k - l_{n+1-k} \\ \quad\quad\quad\quad \text{if } n-s < k \le n \end{cases}.$$

For all choices of negative integers $i_1, \ldots, i_r$ satisfying (2), the above weights $\lambda_{g'}$ form a subset of $\mathrm{ZVar}_n(I_<(f))$ whose Zariski closure contains the set $\lambda_{g''}$ for any g'' of the form

$$g''(k) = \begin{cases} i_k & \text{if } 1 \le k \le r \\ h^-(f) + r - k & \text{if } r < k \le n-s \\ h^-(f) + r - k - l_{n+1-k} & \text{if } n-s < k \le n \end{cases}$$

where now $i_1, \ldots, i_r \in \mathbb{F}$ are arbitrary. Therefore, Proposition 5.3 implies

$$[L(i_1, \ldots, i_{r'}, l_1, l_2, \ldots, l_s, 0, \ldots, 0, -j_{r''}, \ldots, -j_2, -j_1)] \in Q(I_<(f))_n \qquad (6)$$

for all positive integers $i_1 \ge i_2 \ge \ldots \ge i_{r'}$, $j_1 \ge \ldots \ge j_{r''}$ such that $i_r \ge l_1$, $j_{r''} \ge 0$. Consequently,

$$\gamma(\mathrm{cls}(r', r'', 0, \emptyset, Y(f)); n) \in Q(I_<(f))_{2n}. \qquad (7)$$

Now Lemma 3.6 implies b′), and the proof is complete. □

Proposition 6.2 *Let $f \in \mathbb{F}^{\mathbb{Z}_{>0}}$ be a locally constant function. Then*

$$I_<(f) = I(r(f^*), 0, Y(f^*), \emptyset).$$

Proof This proposition can be proved by repeating the proof of Proposition 6.1 and making some obvious changes. For a shorter proof, note that the outer automorphism

$$e_{ij} \mapsto -e_{ji}$$

of $\mathfrak{sl}(\infty)$ interchanges the simple modules $L_<(f)$ and $L_>(-f) \cong L_<(f^*)$ ($(\mathbb{Z}_{>0}, >)$ is isomorphic to $(\mathbb{Z}_{<0}, <)$ and thus $L_>(-f) \cong L_<(f^*)$), and interchanges the ideals $I(r(f^*), 0, Y(f^*), \emptyset)$ and $I(r(f^*), 0, \emptyset, Y(f^*))$. Therefore the statement also follows from Proposition 6.1. □

Consider now the case $(\Theta, \prec) = (\mathbb{Z}, <)$. It is clear that if $f \in \mathbb{F}^{\mathbb{Z}}$ is a locally constant function, then

$$f^+ := (\dots, f(-1)+1, f(0), f(1)-1, \dots, f(i)-i, \dots)$$

is an element of $\mathbb{F}^{\mathbb{Z}}$ and is stably decreasing.

Proposition 6.3 *Let $f \in \mathbb{F}^{\mathbb{Z}}$ be a locally constant function. Let*

$$r := r(f) = |\underline{\mathrm{seq}}(f^+)|.$$

Then

$$I_<(f) = I(r, h^-(f) - h^+(f) + r, \emptyset, \emptyset).$$

Proof The proof follows the same idea as the proof of Proposition 6.1. Below we highlight the necessary changes.

The inclusion

$$I(r, h^-(f) - h^+(f) + r, \emptyset, \emptyset) \subset I_<(f)$$

is equivalent to

$$I(r, h^-(f) - h^+(f) + r, \emptyset, \emptyset) \subset I_<(f') \tag{8}$$

where f' is as in (1). By applying [15, Lemma 5.4] and Proposition 3.11 to the inserted variables we establish (8).

Next, Theorem 3.2 of [15] implies that the inclusion

$$I_<(f) \subset I(r, h^-(f) - h^+(f) + r, \emptyset, \emptyset)$$

is equivalent to the inclusions

$$Q_n(I(r, h^-(f) - h^+(f) + r, \emptyset, \emptyset)) \subset Q_n(I_<(f)) \text{ for } n \gg 0.$$

As in the proof of Proposition 6.1, it suffices to show that

$$\cup_{r'+r''=r} \mathrm{cls}(r', r'', h^-(f) - h^+(f) + r, \emptyset, \emptyset) \subset Q_n(I_<(f)) \text{ for } n \gg 0.$$

We fix nonnegative integers r', r'' with $r' + r'' = r$. For $n \geq r$ Lemma 5.4 asserts that

$$\lambda_g \in \mathrm{ZVar}_n(I_<(f)), \tag{9}$$

for any $\lambda_g \in F_{<,n}(f)$. We now replace f in formula (9) by an appropriate f' with $I_<(f) = I_<(f')$. Let $i_1, \dots, i_r$ be integers satisfying (2). Consider the subset

$$F_{i_1,\dots,i_r} := \{i_1, \dots, i_r, -(n-r') - N, \dots, -1 - N, 1 + N, 2 + N, \dots, (n - r'') + N\} \subset \Theta. \tag{10}$$

Define $f_{i_1,\dots,i_r}$ by the requirement that $(f_{i_1,\dots,i_r})^+$ equals the right-hand side of (1) applied to $f^+, i_1, \dots, i_r$. The order of the elements of $F_{i_1,\dots,i_r}$ in (10) allows us to consider $f_{i_1,\dots,i_r}|F_{i_1,\dots,i_r}$ as a vector in $\mathbb{F}^{2n}$. Then $I_<(f) = I_<(f_{i_1,\dots,i_r})$. Moreover, Corollary 5.5 implies that

$$\lambda_{g'} \in \mathrm{ZVar}_{2n}(I_<(f)) = \mathrm{ZVar}_{2n}(I_<(f_{i_1,\dots,i_r})),$$

where

$$g'(k) = \begin{cases} \underline{\mathrm{seq}}(f^+)_k + i_k - k \\ \quad \text{if } 1 \le k \le r \\ \overline{\mathrm{seq}}(f^+)_{k-1-N-n-r''} + (k-1-N-n-r'') - k = h^-(f) - k \\ \quad \text{if } r < k \le n + r'' \\ \overline{\mathrm{seq}}(f^+)_{k-n-r''+N} + (k-n-r''+N) - k = h^+(f) - k \\ \quad \text{if } n + r'' < k \le 2n \end{cases}.$$

For all integers $i_1, \dots, i_r$ satisfying (2), the above weights $\lambda_{g'}$ form a subset of $\mathrm{ZVar}_{2n}(I_<(f))$ whose Zariski closure contains the set $\lambda_{g''}$ for any g'' of the form

$$g''(k) = \begin{cases} i_k & \text{if } 1 \le k \le r \\ \overline{\mathrm{seq}}(f^+)_k - k = h^-(f) - k & \text{if } r < k \le n \\ \overline{\mathrm{seq}}(f^+)_k - k = h^+(f) - k & \text{if } n < k \le 2n \end{cases}$$

where now $i_1, \dots, i_r \in \mathbb{F}$ are arbitrary. Therefore Proposition 5.3 implies

$$[L(i_1, \dots, i_{r'}, \underbrace{h^-(f), \dots, h^-(f)}_{(n-r')\text{-times}}, \underbrace{h^+(f), \dots, h^+(f)}_{(n-r'')\text{-times}}, -j_{r''}, \dots, j_1)] \in Q(I_<(f))_n.$$

for all positive integers $i_1 \ge i_2 \ge \dots \ge i_{r'}$, $j_1 \ge \dots \ge j_{r''}$ such that

$$i_{r'} \ge h^-(f), -j_{r''} \le h^+(f).$$

Consequently,

$$\gamma(\mathrm{cls}(r', r'', 0, \emptyset, \emptyset); n) \in Q(I_<(f))_{2n}.$$

We complete the proof by applying Lemma 3.6. □

7 Main Results

We are now ready to state the general result. Consider a given simple highest weight module $L_\prec(f)$ where $f \in \mathbb{F}^\Theta$ is a $\prec$-locally constant and almost integral function. Let $\prec_f$ be a total order on Θ such that $\prec_f$ is f-equivalent to $\prec$, and

$\prec_f$ is f-preferred with respect to a partition $\Theta_1 \sqcup \ldots \sqcup \Theta_t$ of Θ, see Sect. 3.3. Propositions 6.1, 6.2, 6.3 imply that

$$I_{\prec_f}^{\Theta_1}(f) = I(r_1, 0, X, \emptyset), \quad I_{\prec_f}^{\Theta_t}(f) = I(r_t, 0, \emptyset, Y),$$

$$I_{\prec_f}^{\Theta_i}(f) = I(r_i, g_i, \emptyset, \emptyset), 1 < i < t$$

for appropriate nonnegative integers $r_1, \ldots, r_t, g_1, \ldots, g_t$ and Young diagrams X, Y.

Theorem 7.1 *We have* $I_\prec(f) = I(r_1 + \ldots + r_t, g_2 + \ldots + g_{t-1}, X, Y)$.

Proof The proof follows the same lines as the proofs of Propositions 6.1, 6.2, 6.3. One first proves the inclusion

$$I(r_1 + \ldots + r_t, g_2 + \ldots + g_{t-1}, X, Y) \subset I_\prec(f)$$

by the same argument as above.

For the opposite inclusion, one considers functions $f_{i_1,\ldots,i_r}$ which, restricted to Θ_i, coincide with the corresponding functions constructed in the proofs of Propositions 6.1, 6.2, 6.3. This means that the integers $i_1, \ldots, i_r$ arise as a union of independently chosen t subsets of integers. With this modification, the argument goes through almost verbatim. □

8 Examples

8.1 Annihilator of Nonintegrable Bounded Highest Weight Modules

Assume that $(\Theta, \prec) = (\mathbb{Z}_{>0}, <)$. Fix $\alpha \in \mathbb{F}, n \in \mathbb{Z}_{\geq 1}$ and consider the $<$-locally constant function

$$f := (\underbrace{-1, \ldots, -1}_{(n-1) \text{ times}}, \alpha, 0, 0, 0, 0, \ldots). \tag{1}$$

Then

$$h(f) = 0, \quad f^+ = (-1, -2, -3, \ldots, -n, \alpha - n, -(n+1), -(n+2), \ldots),$$

$$(f^+)^* = (\ldots, (n+2), (n+1), n - \alpha, n, \ldots, 3, 2, 1),$$

$$\underline{\mathrm{seq}}((f^+)^*) = n - \alpha, \quad \overline{\mathrm{seq}}((f^+)^*) = (\ldots, (n+2), (n+1), n, \ldots, 1).$$

Hence $r = 1, Y = \emptyset$, and

$$I_<(f) = I(1, 0, \emptyset, \emptyset).$$

Assume next that $(\Theta, \prec) = (\mathbb{Z}_{<0}, <)$. Fix $\alpha \in \mathbb{F}$, $n \in \mathbb{Z}_{\geq 1}$ and consider

$$f := (\dots, -1, -1, \alpha, \underbrace{0, \dots, 0}_{(n-1)\text{ times}}), \tag{2}$$

Here

$$h(f) = -1, \quad f^+ = (\dots, n, n-1, (n-1)+\alpha, (n-2), \dots, 2, 1, 0),$$

$$\underline{\text{seq}}((f^+)^*) = n - 1 + \alpha, \quad \overline{\text{seq}}((f^+)^*) = (\dots, n, (n-1), (n-2), \dots, 0).$$

Hence $r = 1$, $X = \emptyset$, and again

$$I_<(f) = I(1, 0, \emptyset, \emptyset).$$

Finally let $(\Theta, \prec) = (\mathbb{Z}, <)$. Fix $\alpha \in \mathbb{F}$ and consider

$$f := (\dots, -1, -1, \alpha, 0, 0, \dots) \tag{3}$$

where $f(n) = \alpha$. We have

$$h(f) = -1, \quad f^+ = (\dots, 2-n, 1-n, 0-n, \alpha - n, -1-n, -2-n, \dots),$$

$$\underline{\text{seq}}((f^+)^*) = \alpha - n, \quad \overline{\text{seq}}(f^+) = (\dots, 2-n, 1-n, 0-n, -1-n, -2-n, \dots).$$

Hence $r = 1$, $g = h(f) + r = 0$, and again

$$I_<(f) = I(1, 0, \emptyset, \emptyset).$$

The above computations show that the simple highest weight modules with highest weights (1), (2), (3) share the same annihilator in $U(\mathfrak{sl}(\infty))$, namely the primitive ideal $I(1, 0, \emptyset, \emptyset)$. Moreover, in [9] it is proved that any simple nonintegrable highest weight $\mathfrak{sl}(\infty)$-module with bounded weight multiplicities is isomorphic to one of the above highest weight modules, for a suitable choice of identification of $(\Theta, \prec)$ with $(\mathbb{Z}_{>0}, <)$, $(\mathbb{Z}_{<0}, <)$, or $(\mathbb{Z}, <)$. Note that the simple modules with highest weights (1), (2), (3) are multiplicity free over $\mathfrak{h}$.

8.2 Annihilator of Semiinfinite Fundamental Representations

Let L be a direct limit of exterior powers $\Lambda^{k_n}(V_n)$ of V_n where V_n is a defining $\mathfrak{sl}(n)$-module, and k_n is a nondecreasing sequence satisfying $1 < k_n < n$ and such that $k_{n+1} = k_n$ or $k_{n+1} = k_n + 1$. Assume that

$$\lim k_n = \lim(n - k_n) = \infty.$$

Then one can show [9] that L is a highest weight $\mathfrak{sl}(\infty)$-module for an appropriately chosen Borel subalgebra $\mathfrak{b}(\prec)$, and that the highest weight $f \in \mathbb{F}^{\Theta}$ of L can be chosen to take only values 1 and 0. Moreover, the Borel subalgebra $\mathfrak{b}(\prec)$ can be chosen to be a Dynkin Borel subalgebra such that $(\Theta, \prec) = (\mathbb{Z}, <)$. Then, by Proposition 6.3,

$$I_{\prec}(f) = I(0, 1, \emptyset, \emptyset).$$

In Sect. 1 we referred to L as a semiinfinite fundamental representation of $\mathfrak{sl}(\infty)$.

8.3 Annihilators of a Class of Modules Containing all Simple Tensor Modules

Consider the case when $\mathfrak{b}$ is ideal with respect to a partition $\Theta_1 \prec \Theta_2 \prec \Theta_3$ of Θ such that Θ_2 is empty. Assume that f has finitely many nonzero coordinates. The construction at the end of Sect. 3.3 provides an f-preferred partition $\Theta'_1 \prec \Theta'_2 \prec \Theta'_3$ of Θ for which $f|_{\Theta'_2}$ equals zero. This implies that

$$I_{\prec}^{\Theta'_1}(f) = I(r_1, 0, X, \emptyset), \quad I_{\prec}^{\Theta'_2}(f) = I(0, 0, \emptyset, \emptyset), \quad I_{\prec}^{\Theta'_3}(f) = I(r_3, 0, \emptyset, Y)$$

for some $r_1, r_2 \in \mathbb{Z}_{\geq 0}$ and some Young diagrams X, Y. Therefore, $I_{\prec}(f) = I_*(r_1 + r_3, 0, X, Y)$ by Theorem 7.1. It is easy to check that the primitive ideals obtained in this way run over all ideals of the form $I(r, 0, X, Y)$ for arbitrary r, X, Y. The case when $r_1 = r_3 = 0$ corresponds to the case of the simple tensor modules $V_{X,Y}$ mentioned in Sect. 1.

Acknowledgements We thank Professors Maria Gorelik and Anna Melnikov for inviting us to participate and present the results of this paper at the sister conferences in Israel celebrating Anthony Joseph's 75th birthday. We are grateful to Professor Anthony Joseph for his kind attention to our work. Both authors have been supported in part by DFG grant PE 980/6-1, and the second author has been supported in part by RFBR grant 16-01-00818.

References

1. A. Baranov, A. Zhilinskii, *Diagonal direct limits of simple Lie algebras*, Comm. Algebra **27** (1999), 2749–2766.
2. D. Barbasch, D. Vogan, *Primitive ideals and orbital integrals in complex classical groups*, Math. Ann. **259** (1982), 153–199.
3. D. Barbasch, D. Vogan, *Primitive Ideals and Orbital Integrals in Complex Exceptional Groups*, Journ. Algebra **80** (1983), 350–382.
4. W. Borho, J.-C. Jantzen, *Über primitive Ideale in der Einhüllenden einer halbeinfacher Lie-algebra*, Inv. Math. **39** (1977), 1–53.

5. E. Dan-Cohen, I. Penkov, V. Serganova, *A Koszul category of representations of finitary Lie algebras*, Adv. Math. **289** (2016), 250–278.
6. E. Dan-Cohen, I. Penkov, S. Snyder, *Cartan subalgebras of root-reductive Lie algebras*, Journ. Algebra, **308** (2007), 583–611.
7. I. Dimitrov, I. Penkov, *Weight modules of direct limit Lie algebras*, IMRN 1999, no. 5, 223–249.
8. M. Duflo, *Sur la classification des idéaux primitifs dans l'algèbre enveloppante d'une algèbre de Lie semisimple*, Ann. of Math. **105** (1977), 107–120.
9. D. Grantcharov, I. Penkov, *Simple bounded weight modules of* $\mathfrak{sl}(\infty)$, $\mathfrak{o}(\infty)$, $\mathfrak{sp}(\infty)$, arXiv:1807.01899.
10. A. Joseph, *On the associated variety of a primitive ideal*, Journ. Algebra **93** (1985), 509–523.
11. A. Joseph, *Sur la classification des idéaux primitifs dans l'algébre enveloppante de sl(n+1, C)*, C. R. Acad. Sci. Paris Sér. A-B **287**, no 5 (1978), 303–306.
12. A. Joseph, *A characteristic variety for the primitive spectrum of a semisimple Lie algebra*, Non-commutative harmonic analysis (Actes Colloq., Marseille-Luminy, 1976), Lecture Notes in Math., Vol. 587, Springer, Berlin, 1977, 102–118.
13. D. Knuth, The art of computer programming. Volume 3. Sorting and searching, Addison-Wesley Series in Computer Science and Information Processing, Addison-Wesley, 1973.
14. I. Penkov, A. Petukhov, *On ideals in the enveloping algebra of a locally simple Lie algebra*, IMRN **13** (2015), 5196–5228.
15. I. Penkov, A. Petukhov, *Annihilators of highest weight* $\mathfrak{sl}(\infty)$*-modules*, Transformation groups **21** (2016), 821–849.
16. I. Penkov, A. Petukhov, *On ideals in* $U(\mathfrak{sl}(\infty))$, $U(\mathfrak{o}(\infty))$, $U(\mathfrak{sp}(\infty))$, Representation theory - current trends and perspectives, EMS Series of Congress Reports, European Mathematical Society (EMS), 2016, 565–602.
17. I. Penkov, A. Petukhov, *Primitive ideals of* $U(\mathfrak{sl}(\infty))$, Bulletin LMS **50** (2018), 443–448.
18. A. Petukhov, S. Sierra, *Ideals in the enveloping algebra of the positive Witt algebra*, arXiv:1710.10029.
19. I. Penkov, K. Styrkas, *Tensor representations of infinite-dimensional root-reductive Lie algebras*, in Developments and Trends in Infinite-Dimensional Lie Theory, Progress in Mathematics 288, Birkhaüser, 2011, 127–150.
20. T. Nampaisarn, *Categories O for Dynkin Borel Subalgebras of Root-Reductive Lie Algebras*, arXiv:1706.05950.
21. A. Zhilinskii, *Coherent systems of representations of inductive families of simple complex Lie algebras* (Russian), preprint of Academy of Belarussian SSR, ser. 38 (438), Minsk, 1990.
22. A. Zhilinskii, *Coherent finite-type systems of inductive families of nondiagonal inclusions* (Russian), Dokl. Acad. Nauk Belarusi **36:1**(1992), 9–13, 92.
23. A. Zhilinskii, *On the lattice of ideals in the universal enveloping algebra of a diagonal Lie algebra* (Russian), preprint, Minsk, 2011.

Prime Spectra of Abelian 2-Categories and Categorifications of Richardson Varieties

Kent Vashaw and Milen Yakimov

To Anthony Joseph on his 75th birthday, with admiration

Abstract We describe a general framework for prime, completely prime, semiprime, and primitive ideals of an abelian 2-category. This provides a non-commutative version of Balmer's prime spectrum of a tensor triangulated category. These notions are based on containment conditions in terms of thick subcategories of an abelian category and thick ideals of an abelian 2-category. We prove categorical analogs of the main properties of noncommutative prime spectra. Similar notions, starting with Serre subcategories of an abelian category and Serre ideals of an abelian 2-category, are developed. They are linked to Serre prime spectra of $\mathbb{Z}_+$-rings. As an application, we construct a categorification of the quantized coordinate rings of open Richardson varieties for symmetric Kac–Moody groups, by constructing Serre completely prime ideals of monoidal categories of modules of the KLR algebras, and by taking Serre quotients with respect to them.

Mathematics Subject Classification (2000): Primary: 18D05; Secondary: 18E10, 16P40, 17B37

1 Introduction

1.1 Noncommutative Categorical Prime Spectra

Balmer's notion of a prime spectrum of a tensor triangulated category [1, 2] is a major tool in homological algebra, representation theory, algebraic topology, and

---

K. Vashaw · M. Yakimov (✉)
Department of Mathematics, Louisiana State University, Baton Rouge, LA, USA
e-mail: kvasha1@lsu.edu; yakimov@math.lsu.edu

M. Gorelik et al. (eds.), *Representations and Nilpotent Orbits of Lie Algebraic Systems*, Progress in Mathematics 330,
https://doi.org/10.1007/978-3-030-23531-4_14

other areas. It is defined for triangulated categories with a symmetric monoidal structure. As noted in [4], Balmer's construction and results more generally apply to braided monoidal triangulated categories.

The notion of a prime spectrum of a braided monoidal triangulated category is a categorical version of the notion of a prime spectrum of a commutative ring. In the classical case of noncommutative rings, there are four different notions of primality [11]: prime, completely prime, semiprime, and primitive spectra. In this paper we develop categorical notions of all of them, and prove analogs of many of their main properties. We do this in the abelian setting. However, instead of simply considering abelian monoidal categories, we work with the more general setting of abelian 2-categories. It is necessary to consider this more general setting, because many of the monoidal categorifications of noncommutative algebras that have been constructed so far are in the setting of 2-categories, rather than monoidal categories, see [18, 29, 34]. Categorifications via 2-categories are even needed for relatively small algebras such as the idempotented version of the quantized universal enveloping algebra of $\mathfrak{sl}_2$; we refer the reader to [29] for a very informative review of this particular categorification.

1.2 Thick and Prime Ideals of Abelian 2-Categories

A *2-category* is a category enriched over the category of 1-categories. In other words, a 2-category $\mathcal{T}$ has the property that for every two objects A_1, A_2 of it, the morphisms $\mathcal{T}(A_1, A_2)$ form a 1-category and satisfy natural identity conditions. A 2-category with one object is the same thing as a strict monoidal category. An *abelian 2-category* is such a category for which the 1-categories $\mathcal{T}(A_1, A_2)$ are abelian and the composition bifunctors are biexact. We work with small 2-categories, i.e., with 2-categories $\mathcal{T}$ whose objects form a set and for which all 1-categories $\mathcal{T}(A_1, A_2)$ are small. We denote by $\mathcal{T}_1$ the isomorphism classes of 1-morphisms of $\mathcal{T}$. For two subsets $X, Y \subseteq \mathcal{T}_1$, denote by

$X \circ Y$ the set of isomorphism classes of 1-morphims of $\mathcal{T}$ having representatives of the form fg for f and g representing classes in X and Y such that fg is defined.

The different versions of prime ideals of abelian 2-categories which we develop are based on the notion of a *thick subcategory of an abelian category* and its 2-incarnation, the notion of a *thick ideal of an abelian 2-category*. Recall that a thick (sometimes called wide) subcategory of an abelian category is a nonempty full subcategory which is closed under taking kernels, cokernels, and extensions.

A *thick ideal* $\mathcal{I}$ of an abelian 2-category $\mathcal{T}$ is a collection of subcategories $\mathcal{I}(A_1, A_2)$ of $\mathcal{T}(A_1, A_2)$ for all objects A_1, A_2 of $\mathcal{T}$ such that

1. $\mathcal{I}(A_1, A_2)$ are thick subcategories of the abelian categories $\mathcal{T}(A_1, A_2)$ and
2. the composition bifunctors of $\mathcal{T}$ restrict to bifunctors

$$\mathcal{T}(A_2, A_3) \times \mathcal{I}(A_1, A_2) \to \mathcal{I}(A_1, A_3) \quad \text{and} \quad \mathcal{I}(A_2, A_3) \times \mathcal{T}(A_1, A_2) \to \mathcal{I}(A_1, A_3)$$

for all objects A_1, A_2, A_3 of $\mathcal{T}$.

We call a proper thick ideal $\mathcal{P}$ of an abelian 2-category $\mathcal{T}$

(p) *prime* if, for all thick ideals $\mathcal{I}$ and $\mathcal{J}$ of $\mathcal{T}$, $\mathcal{I}_1 \circ \mathcal{J}_1 \subseteq \mathcal{P}_1$ implies that either $\mathcal{I} \subseteq \mathcal{P}$ or $\mathcal{J} \subseteq \mathcal{P}$,
(sp) *semiprime* if it is an intersection of prime ideals,
(cp) *completely prime* if, for all $f, g \in \mathcal{T}_1$, $f \circ g \subseteq \mathcal{P}_1$ implies that either $f \in \mathcal{P}_1$ or $g \in \mathcal{P}_1$. Note that the set $f \circ g$ is either empty or is a singleton.

We obtain categorical versions of the main properties of prime, semiprime, and completely prime ideals of noncommutative rings. In Sect. 3 it is proved that the following are equivalent for a proper thick ideal $\mathcal{P}$ of an abelian 2-category $\mathcal{T}$:

(p1) $\mathcal{P}$ is a prime ideal;
(p2) If $f, g \in \mathcal{T}_1$ and $f \circ \mathcal{T}_1 \circ g \subseteq \mathcal{P}_1$, then either $f \in \mathcal{P}_1$ or $g \in \mathcal{P}_1$;
(p3) If $\mathcal{I}$ and $\mathcal{J}$ are any thick ideals properly containing $\mathcal{P}$, then $\mathcal{I}_1 \circ \mathcal{J}_1 \not\subseteq \mathcal{P}_1$;
(p4) If $\mathcal{I}$ and $\mathcal{J}$ are any left thick ideals of $\mathcal{T}$ such that $\mathcal{I}_1 \circ \mathcal{J}_1 \subseteq \mathcal{P}_1$, then either $\mathcal{I} \subseteq \mathcal{P}$ or $\mathcal{J} \subseteq \mathcal{P}$.

We call the set of such thick ideals of $\mathcal{T}$ the *prime spectrum* of $\mathcal{T}$, to be denoted by $\operatorname{Spec}(\mathcal{T})$, and define a *Zariski type topology* on it. For a *multiplicative subset* $\mathcal{M}$ of $\mathcal{T}_1$ (see Definition 3.13) and a proper thick ideal $\mathcal{I}$ of $\mathcal{T}$ such that $\mathcal{I}_1 \cap \mathcal{M} = \varnothing$, we prove that every *maximal element* of the set

$$X(\mathcal{M}, \mathcal{I}) := \{\mathcal{K} \text{ a thick ideal of } \mathcal{T} \mid \mathcal{K} \supseteq \mathcal{I} \text{ and } \mathcal{K}_1 \cap \mathcal{M} = \varnothing\}$$

is a prime ideal of $\mathcal{T}$. This implies that $\operatorname{Spec}(\mathcal{T})$ is *non-empty* for every abelian 2-category.

Categorical versions of simple, noetherian, and weakly noetherian noncommutative rings are given in Sect. 4. There we prove that for every weakly noetherian abelian 2-category $\mathcal{T}$ and a proper thick ideal $\mathcal{I}$ of $\mathcal{T}$, there exist finitely many minimal Serre prime ideals over $\mathcal{I}$ and there is a finite list of minimal prime ideals over $\mathcal{I}$ (possibly with repetition) $\mathcal{P}^{(1)}, \ldots, \mathcal{P}^{(m)}$ such that the product

$$\mathcal{P}_1^{(1)} \circ \ldots \circ \mathcal{P}_1^{(m)} \subseteq \mathcal{I}_1.$$

In Sect. 5 we prove a categorical version of the Levitzki–Nagata theorem for semiprime ideals, and furthermore show that the following are equivalent for a proper thick ideal $\mathcal{Q}$ of $\mathcal{T}$:

(sp1) $\mathcal{Q}$ is semiprime;
(sp2) If $f \in \mathcal{T}_1$ and $f \circ \mathcal{T}_1 \circ f \subseteq \mathcal{Q}_1$, then $f \in \mathcal{Q}_1$;
(sp3) If $\mathcal{I}$ is any thick ideal of $\mathcal{T}$ such that $\mathcal{I}_1 \circ \mathcal{I}_1 \subseteq \mathcal{Q}_1$, then $\mathcal{I} \subseteq \mathcal{Q}$;
(sp4) If $\mathcal{I}$ is any thick ideal properly containing $\mathcal{Q}$, then $\mathcal{I}_1 \circ \mathcal{I}_1 \nsubseteq \mathcal{Q}_1$;
(sp5) If $\mathcal{I}$ is any left thick ideal of $\mathcal{T}$ such that $\mathcal{I}_1 \circ \mathcal{I}_1 \subseteq \mathcal{Q}_1$, then $\mathcal{I} \subseteq \mathcal{Q}$.

1.3 Serre Prime Ideals of 2-Categories and Ideals of $\mathbb{Z}_+$-Rings

Serre subcategories of abelian categories are a particular type of thick subcategories. A thick ideal $\mathcal{I}$ of an abelian 2-category $\mathcal{T}$ will be called a *Serre ideal* if $\mathcal{I}(A_1, A_2)$ is a Serre subcategory of $\mathcal{T}(A_1, A_2)$ for all objects A_1, A_2 of $\mathcal{T}$. For those ideals one can consider the *Serre quotient* $\mathcal{T}/\mathcal{I}$ which is an abelian 2-category under a mild condition on $\mathcal{I}$.

We define a *Serre prime* (resp. *semiprime, completely prime*) *ideal* of an abelian 2-category $\mathcal{T}$ to be a prime (resp. semiprime, completely prime) ideal which is a Serre ideal. Section 6 treats in detail these ideals, and proves that they are characterized by similar to (p3)–(p4) and (sp3)–(sp5) properties as in Sect. 1.2, but with thick ideals replaced by Serre ideals. In other words, these kinds of ideals can be defined entirely based on the notion of Serre ideals of abelian 2-categories, just like the more general prime ideals are defined in terms of thick ideals.

The set of Serre prime ideals of $\mathcal{T}$, denoted by Serre-Spec$(\mathcal{T})$, has an induced topology from Spec$(\mathcal{T})$. This topology is shown to be intrinsically given in terms of Serre ideals of $\mathcal{T}$. If $\mathcal{C}$ is a strict abelian monoidal category, an alternative topology which more closely resembles the topology of Balmer in [1] can also be put on Serre-Spec$(\mathcal{C})$. Under this topology, Serre-Spec$(\mathcal{C})$ is a ringed space.

Denote $\mathbb{Z}_+ := \{0, 1, \ldots\}$. The Grothendieck ring $K_0(\mathcal{T})$ of an abelian 2-category $\mathcal{T}$ is a $\mathbb{Z}_+$*-ring* in the terminology of [7, Ch. 3], see Definitions 2.3 and 6.8. In Sect. 6.4 we define the notions of *Serre ideals* and *Serre prime* (*semiprime* and *completely prime*) *ideals* of a $\mathbb{Z}_+$*-ring* R. The set of Serre prime ideals of R, denoted by Serre-Spec(R), is equipped with a Zariski type topology.

It is proved in Sect. 6 that, for an abelian 2-category $\mathcal{T}$ with the property that every 1-morphism of $\mathcal{T}$ has finite length, the functor K_0 induces bijections between the sets of Serre ideals, Serre prime (semiprime and completely prime ideals) of the abelian 2-category $\mathcal{T}$ and the $\mathbb{Z}_+$-ring $K_0(\mathcal{T})$. Furthermore, the map

$$K_0 : \text{Serre-Spec}(\mathcal{T}) \to \text{Serre-Spec}(K_0(\mathcal{T}))$$

is shown to be a homeomorphism.

For a Serre prime (resp. semiprime, completely prime) ideal $\mathcal{I}$ of $\mathcal{T}$, the Serre quotient $\mathcal{T}/\mathcal{I}$ is a prime (resp. semiprime, domain) abelian 2-category. If every 1-morphism of $\mathcal{T}$ has finite length, then

$$K_0(\mathcal{T}/\mathcal{I}) \cong K_0(\mathcal{T})/K_0(\mathcal{I}).$$

The point now is that if we have a categorification of a $\mathbb{Z}_+$-ring R via an abelian 2-category $\mathcal{T}$ (i.e., $K_0(\mathcal{T}) \cong R$) and I is a Serre ideal of R, then there is a unique Serre ideal $\mathcal{I}$ of *$\mathcal{T}$ such that $K_0(\mathcal{I}) = I$. Furthermore, the Serre quotient $\mathcal{T}/\mathcal{I}$ categorifies the $\mathbb{Z}_+$-ring R/I,* i.e., *$K_0(\mathcal{T}/\mathcal{I}) \cong K_0(\mathcal{T})/K_0(\mathcal{I})$. If I is a Serre prime (resp. semiprime, completely prime) ideal of the $\mathbb{Z}_+$-ring R, then the Serre quotient $\mathcal{T}/\mathcal{I}$ is a prime (resp. semiprime, domain) abelian 2-category.* We view this construction as a general way of constructing monoidal categorifications of $\mathbb{Z}_+$-rings out of known ones by taking Serre quotients. This is illustrated in Sect. 9 in the case of the *quantized coordinate rings of open Richardson varieties* for symmetric Kac–Moody algebras.

We expect that, in addition, Serre prime ideals of abelian 2-categories and $\mathbb{Z}_+$-rings will provide a framework for finding intrinsic connections between prime ideals of noncommutative algebras and totally positive parts of algebraic varieties. In the case of the algebras of quantum matrices, such a connection was previously found by exhibiting related explicit generating sets for prime ideals of the noncommutative algebras and minors defining totally positive cells [10].

Primitive ideals of abelian 2-categories $\mathcal{T}$ are introduced in Sect. 7 as the annihilation ideals of simple exact 2-representations in the setting of [35, 36], where it is proved that all such ideals are Serre prime ideals of $\mathcal{T}$.

1.4 Prime Spectra of Additive 2-Categories

One can develop analogous (but much simpler) theory of different forms of prime ideals of an *additive 2-category* $\mathcal{T}$, which is a 2-category such that $\mathcal{T}(A_1, A_2)$ are additive categories for $A_1, A_2 \in \mathcal{T}$ and the compositions

$$\mathcal{T}(A_2, A_3) \times \mathcal{T}(A_1, A_2) \to \mathcal{T}(A_1, A_3)$$

are additive bifunctors for $A_1, A_2, A_3 \in \mathcal{T}$.

This can be done by following exactly the same route as Sects. 3–5 but based off the notion of a *thick ideal of an additive 2-category*. Call a full subcategory of an additive category *thick* if it is closed under direct sums, direct summands, and isomorphisms. A *thick ideal* $\mathcal{I}$ of an additive 2-category $\mathcal{T}$ is a collection of subcategories $\mathcal{I}(A_1, A_2)$ of $\mathcal{T}(A_1, A_2)$ for all objects A_1, A_2 of $\mathcal{T}$ such that

1. $\mathcal{I}(A_1, A_2)$ are thick subcategories of the additive categories $\mathcal{T}(A_1, A_2)$ and
2. $\mathcal{T}_1 \circ \mathcal{I}_1 \subset \mathcal{I}_1, \mathcal{I}_1 \circ \mathcal{T}_1 \subseteq \mathcal{I}_1$.

Using the same conditions on containments with respect to thick ideals and 1-morphisms as in Sects. 3–5, one defines *prime, semiprime, and completely prime ideals of additive 2-categories* and proves analogs of the results in those sections (though in a simpler way than the abelian setting). One also analogously defines a Zariski topology on the set Spec($\mathcal{T}$) of prime ideals of $\mathcal{T}$ by using containments with respect to thick subcategories of $\mathcal{T}$. There are no analogs of the Serre type ideals in this setting.

If an additive 2-category $\mathcal{T}$ has the property that each of its 1-morphisms has a unique decomposition as a finite set of indecomposables (e.g., if all additive categories $\mathcal{T}(A_1, A_2)$ are Krull–Schmidt), then the split Grothendieck ring $K_0^{sp}(\mathcal{T})$ is a $\mathbb{Z}_+$-ring, see Remark 2.4. Similarly to Sect. 6.5, for such additive 2-categories $\mathcal{T}$, one shows that the map $K_0^{sp}(-)$ gives

- a bijection between the sets of thick, prime, semiprime, and completely prime ideals of $\mathcal{T}$ and the sets of Serre ideals, Serre prime, semiprime, and completely prime ideals of the $\mathbb{Z}_+$-ring $K_0^{sp}(\mathcal{T})$, and
- a homeomorphism Spec($\mathcal{T}$) $\to$ Serre-Spec($K_0^{sp}(\mathcal{T})$).

Call the annihilation ideal of a simple 2-representation of an additive 2-category $\mathcal{T}$ (in the setting of [35, 36]) a *primitive ideal of* $\mathcal{T}$. Similarly to Sect. 7, one shows that each such ideal is a prime ideal of $\mathcal{T}$.

In a forthcoming publication we obtain analogs of the results in the paper for (noncommutative) prime spectra of triangulated 2-categories.

1.5 Categorifications of Richardson Varieties via Prime Serre Quotients

We finish with an important example of Serre completely prime ideals of abelian 2-categories that can be used to categorify the quantized coordinate rings of certain closures of open Richardson varieties. For a symmetrizable Kac–Moody group G, a pair of opposite Borel subgroups $B_\pm$ and Weyl group elements $u \leq w$, the corresponding open Richardson variety is defined as the intersections of opposite Schubert cells in the full flag variety of G,

$$R_{u,w} := (B_- u B_+)/B_+ \cap (B_+ w B_+)/B_+ \subset G/B_+.$$

They have been used in a wide range of settings in representation theory, Schubert calculus, total positivity, Poisson geometry, and mathematical physics. For symmetric Kac–Moody groups, Leclerc [30] constructed a cluster algebra inside the coordinate ring of each Richardson variety of the same dimension. In the quantum situation, Lenagan and the second named author constructed large families of toric frames for all quantized coordinate rings of Richardson varieties that generate those rings [31].

Recently, for each symmetrizable Kac–Moody algebra $\mathfrak{g}$, Kashiwara, Kim, Oh, and Park [24] constructed a monoidal categorification of the quantization of a closure of $R_{u,w}$ in terms of a monoidal subcategory of the category of graded, finite dimensional representations of the Khovanov–Lauda–Rouquier (KLR) algebras associated to $\mathfrak{g}$. Their construction uses Leclerc's interpretation of the coordinate ring of a closure of $R_{u,w}$ in terms of a double invariant subalgebra.

Denote by $\overline{R}_{u,w}$ the closure of $R_{u,w}$ in the Schubert cell $(B_+ wB_+)/B_+ \subset G/B_+$. We construct a monoidal categorification of the quantization $U_u^-[w]/I_u(w)$ of the coordinate ring of $\overline{R}_{u,w}$ used in the construction of toric frames in [31]. Here $U_q^-[w]$ are the quantum Schubert cell algebras [5, 33] and $I_u(w)$ are the homogeneous completely prime ideals of these algebras that arose in the classification of their prime spectra in [43]. This classification was based on the fundamental works of Anthony Joseph on the spectra of quantum groups [16, 17] from the early 90s. It was proved in [19, 26, 38] that certain monoidal subcategories $\mathcal{C}_w$ of the categories of graded, finite dimensional modules of the KLR algebras associated to $\mathfrak{g}$ categorify the dual integral form $U_{\mathcal{A}}^-[w]^\vee$ where $\mathcal{A} := \mathbb{Z}[q^{\pm 1}]$. We prove that for a symmetrizable Kac–Moody algebra $\mathfrak{g}$, the ideals $I_w(u) \cap U_{\mathcal{A}}^-[w]^\vee$ have bases that are subsets of the upper global/canonical basis of $U_{\mathcal{A}}^-[w]^\vee$. From this we deduce that for symmetric $\mathfrak{g}$, $I_w(u) \cap U_{\mathcal{A}}^-[w]^\vee$ are Serre completely prime ideals of the $\mathbb{Z}_+$-ring $U_{\mathcal{A}}^-[w]^\vee$. The bijection from Sect. 1.3 implies that the monoidal category $\mathcal{C}_w$ has a Serre completely prime ideal $\mathcal{I}_u(w)$ such that $K_0(\mathcal{I}_u(w)) = I_w(u)$, and thus, the Serre quotient $\mathcal{C}_w/\mathcal{I}_u(w)$ categorifies $U_{\mathcal{A}}^-[w]^\vee/(I_w(u) \cap U_{\mathcal{A}}^-[w]^\vee)$:

$$K_0(\mathcal{C}_w/\mathcal{I}_u(w)) \cong U_{\mathcal{A}}^-[w]^\vee/(I_w(u) \cap U_{\mathcal{A}}^-[w]^\vee).$$

It is an important problem to connect the categorification of Kashiwara, Kim, Oh, and Park [24] of open Richardson varieties (via subcategories of KLR modules) to ours (via Serre quotients of categories of KLR modules).

2 Abelian 2-Categories and Categorification

This section contains background material on (abelian) 2-categories and categorification of algebras.

2.1 2-Categories

A category $\mathcal{T}$ is said to be *enriched over a monoidal category* $\mathcal{M}$ if the space of morphisms between any two objects of $\mathcal{T}$ is an object in $\mathcal{M}$ and $\mathcal{T}$ satisfies natural axioms which relate composition of morphisms in $\mathcal{T}$ and the identity morphisms of objects of $\mathcal{T}$ to the monoidal structure of $\mathcal{M}$. We refer the reader to [25] for details.

A *2-category* is a category enriched over the category of 1-categories. This means that for a 2-category $\mathcal{T}$, given two objects A_1, A_2 of it, the morphisms $\mathcal{T}(A_1, A_2)$ form a 1-category. The objects of these categories are denoted by the same symbol $\mathcal{T}(A_1, A_2)$ – they are the 1-morphisms of $\mathcal{T}$. The morphisms of the categories $\mathcal{T}(A_1, A_2)$ are the 2-morphisms of $\mathcal{T}$. For a pair of 1-morphisms $f, g \in \mathcal{T}(A_1, A_2)$, we will denote by $\mathcal{T}(f, g)$ the 2-morphisms between f and g, i.e., the morphisms between the objects f and g in the category $\mathcal{T}(A_1, A_2)$.

We have 2 types of compositions of 1- and 2-morphisms. We follow the notation of [29]:

1. For a pair of objects A_1, A_2 of $\mathcal{T}$, the composition of morphisms in the category $\mathcal{T}(A_1, A_2)$ is called *vertical composition* of 2-morphisms of $\mathcal{T}$. In the *globular representation* of $\mathcal{T}$, such a composition is given by the following diagram

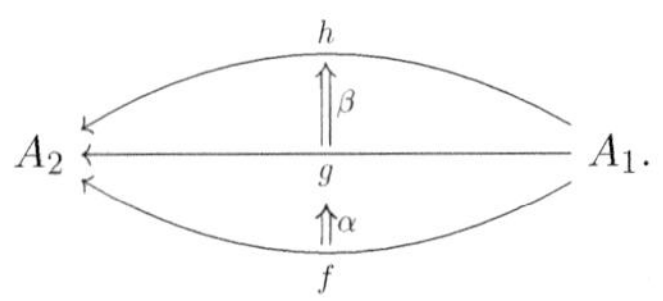

The vertical composition of the 2-morphisms $\alpha \in \mathcal{T}(f, g)$ and $\beta \in \mathcal{T}(g, h)$ will be denoted by $\beta\alpha \in \mathcal{T}(f, h)$, where f, g, h are objects of $\mathcal{T}(A_1, A_2)$.

2. For each three objects A_1, A_2, A_3 of $\mathcal{T}$, we have a bifunctor of 1-categories

$$\mathcal{T}(A_2, A_3) \times \mathcal{T}(A_1, A_2) \to \mathcal{T}(A_1, A_3). \tag{2.1}$$

The resulting composition of 1- and 2-morphisms of $\mathcal{T}$ is called *horizontal composition*. In the globular representation of $\mathcal{T}$, these compositions are given by the diagram

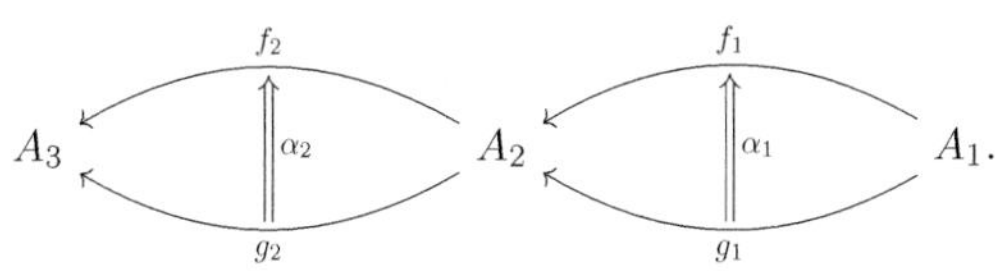

In this notation, the horizontal composition of 2-morphisms will be denoted by $\alpha_2 * \alpha_1$. The horizontal composition of 1-morphisms will be denoted by $f_2 f_1$.

A 2-category $\mathcal{T}$ has identity 1-morphisms $1_A \in \mathcal{T}(A, A)$ (for its objects $A \in \mathcal{T}$). The compositions and identity morphisms satisfy natural associativity and identity axioms [29, 34], which are equivalent to the definition of 2-categories in the language of enriched categories.

2-categories are generalizations of *monoidal categories*, in the sense that a strict monoidal category is the same thing as a 2-category with one object:

To a strict monoidal category $\mathcal{M}$, one associates a 2-category $\mathcal{T}$ with one object A by taking $\mathcal{T}(A, A) := \mathcal{M}$. The tensor product in $\mathcal{M}$ is used to define composition of 1-morphisms of $\mathcal{T}$. For $f, g \in \mathcal{M} = \mathcal{T}(A, A)$, one defines the 2-morphisms $\mathcal{T}(f, g) := \mathcal{M}(f, g)$. All 2-categories with 1 object arise in this way.

Recall that a 1-category $\mathcal{C}$ is called *small* if its objects form a set and $\mathcal{C}(A_1, A_2)$ is a set for all pairs of objects $A_1, A_2 \in \mathcal{C}$. *Throughout the paper we work with small 2-categories $\mathcal{T}$, which are 2-categories satisfying the conditions that the objects of $\mathcal{T}$ form a set and $\mathcal{T}(A_1, A_2)$ is a small 1-category for all pairs of objects A_1, A_2 of $\mathcal{T}$.*

The set of objects of such a 2-category $\mathcal{T}$ will be denoted by the same symbol $\mathcal{T}$. The set of 1-morphisms of $\mathcal{T}$ will be denoted by $\mathcal{T}_1$.

2.2 *Abelian 2-Categories and Categorification*

Definition 2.1 We will say that a 2-category $\mathcal{T}$ is an *abelian 2-category* if $\mathcal{T}(A_1, A_2)$ are abelian categories for all $A_1, A_2 \in \mathcal{T}$ and the compositions

$$\mathcal{T}(A_2, A_3) \times \mathcal{T}(A_1, A_2) \to \mathcal{T}(A_1, A_3)$$

are exact bifunctors for all $A_1, A_2, A_3 \in \mathcal{T}$.

More generally, for a ring $\mathbb{k}$, we will say that $\mathcal{T}$ is a *$\mathbb{k}$-linear abelian 2-category* if $\mathcal{T}(A_1, A_2)$ are $\mathbb{k}$-linear abelian categories for $A_1, A_2 \in \mathcal{T}$.

A *multiring category* in the terminology of [7, Definition 4.2.3] is precisely a *$\mathbb{k}$-linear abelian 2-category with one object*.

Remark 2.2 Let $\mathbb{k}$ be a field. Recall that a $\mathbb{k}$-linear abelian category $\mathcal{C}$ is called *locally finite* if it is Hom-finite (i.e., $\dim_{\mathbb{k}} \mathcal{C}(A_1, A_2) < \infty$ for all $A_1, A_2 \in \mathcal{C}$) and each object of $\mathcal{C}$ has finite length; we refer the reader to [7, §1.8] for details. Let $\mathcal{LFA}b_{ex}$ be the monoidal category of locally finite abelian categories equipped with the Deligne tensor product ([6] and [7, §1.11]) and morphisms given by exact functors.

In this terminology, a $\mathbb{k}$-linear abelian 2-category $\mathcal{T}$ with the property that the 1-categories $\mathcal{T}(A_1, A_2)$ are locally finite for all $A_1, A_2 \in \mathcal{T}$ is the same thing as a category which is enriched over the monoidal category $\mathcal{LFA}b_{ex}$. This is easy to verify, the only key step being the universality property of the Deligne tensor product with respect to exact functors [7, Proposition 1.11.2(v)].

We will denote by $K_0(\mathcal{C})$ the Grothendieck group of an abelian category $\mathcal{C}$. To each abelian 2-category $\mathcal{T}$ one associates the pre-additive category $K_0(\mathcal{T})$ whose objects are the objects of $\mathcal{T}$ and morphisms are

$$K_0(\mathcal{T})(A_1, A_2) := K_0(\mathcal{T}(A_1, A_2)) \quad \text{for } A_1, A_2 \in \mathcal{T}.$$

Given a pre-additive category $\mathcal{F}$, one says that the 2-category $\mathcal{T}$ categorifies $\mathcal{F}$ if $K_0(\mathcal{T}) \cong \mathcal{F}$ as pre-additive categories.

To a pre-additive category $\mathcal{F}$, one associates a ring with elements

$$\oplus_{A_1, A_2 \in \mathcal{F}} \mathcal{F}(A_1, A_2).$$

The product in the ring is the composition of morphisms when it makes sense and 0 otherwise. In particular, the identity morphisms 1_A are idempotents of the ring for all objects $A \in \mathcal{F}$. By abuse of notation, this ring is denoted by the same symbol $\mathcal{F}$ as the original category.

Definition 2.3 For an abelian 2-category $\mathcal{T}$, the ring $K_0(\mathcal{T})$ is called the Grothendieck ring of $\mathcal{T}$. We say that $\mathcal{T}$ categorifies an S-algebra R, for a commutative ring S, if $K_0(\mathcal{T}) \otimes_{\mathbb{Z}} S \cong R$.

Often, it is not sufficient to consider multiring categories (abelian monoidal categories) to obtain categorifications of algebras, and one needs the more general setting of 2-categories.

Remark 2.4 An additive 2-category is a 2-category $\mathcal{T}$ such that $\mathcal{T}(A_1, A_2)$ are additive categories for all $A_1, A_2 \in \mathcal{T}$ and the compositions

$$\mathcal{T}(A_2, A_3) \times \mathcal{T}(A_1, A_2) \to \mathcal{T}(A_1, A_3)$$

are additive bifunctors for all $A_1, A_2, A_3 \in \mathcal{T}$. For such a category $\mathcal{T}$, one defines the pre-additive category $K_0^{sp}(\mathcal{T})$ whose objects are the objects of $\mathcal{T}$ and morphisms are the split Grothendieck groups

$$K_0^{sp}(\mathcal{T}(A_1, A_2))$$

of the additive categories $\mathcal{T}(A_1, A_2)$ for $A_1, A_2 \in \mathcal{T}$.

We say that an additive 2-category $\mathcal{T}$ categorifies an S-algebra R if $K_0^{sp}(\mathcal{T}) \otimes_{\mathbb{Z}} S \cong R$.

3 The Prime Spectrum

In this section we define the prime spectrum of an abelian 2-category and a Zariski type topology on it. We prove two equivalent characterizations of prime ideals, extending theorems from classical ring theory. We also prove that maximal elements of the sets of ideals not intersecting multiplicative sets of 1-morphisms of 2-categories are prime ideals.

3.1 Thick Ideals of Abelian 2-Categories

Definition 3.1 A *weak subcategory* $\mathcal{I}$ of a 2-category $\mathcal{T}$ is

(1) a subcollection $\mathcal{I}$ of objects of $\mathcal{T}$ and
(2) a collection of subcategories $\mathcal{I}(A_1, A_2)$ of $\mathcal{T}(A_1, A_2)$ for $A_1, A_2 \in \mathcal{I}$,

such that the composition bifunctors (2.1) restrict to bifunctors

$$\mathcal{I}(A_2, A_3) \times \mathcal{I}(A_1, A_2) \to \mathcal{I}(A_1, A_3)$$

for $A_1, A_2, A_3 \in \mathcal{I}$.

A weak subcategory $\mathcal{I}$ of a 2-category $\mathcal{T}$ is not necessarily a 2-category on its own because it might not contain the identity morphisms 1_A for its objects $A \in \mathcal{I}$. Apart from this, a weak subcategory of a 2-category satisfies the other axioms for 2-categories. The relationship of a weak subcategory to a 2-category is the same as the relationship of a subring to a unital ring R. In the latter case, the subring does not need to contain the unit of R.

Definition 3.2

(1) A *thick subcategory* of an abelian category is a nonempty full subcategory which is closed under taking kernels, cokernels, and extensions.
(2) A *thick weak subcategory* of an abelian 2-category $\mathcal{T}$ is a weak subcategory $\mathcal{I}$ of $\mathcal{T}$ having the same set of objects and such that for any pair of objects $A_1, A_2 \in \mathcal{T}$, $\mathcal{I}(A_1, A_2)$ is a thick subcategory of the abelian category $\mathcal{T}(A_1, A_2)$.
(3) A *thick ideal* of an abelian 2-category $\mathcal{T}$ is a thick weak subcategory $\mathcal{I}$ of $\mathcal{T}$ such that, for all 1-morphisms f in $\mathcal{T}$ and g in $\mathcal{I}$, the compositions fg, gf are 1-morphisms of $\mathcal{I}$ whenever they are defined.

Sometimes the term *wide subcategory* is used instead of *thick*, see, for instance, [14].

Every thick subcategory of an abelian category is closed under isomorphisms and taking direct summands of its objects (because one can take the kernels of idempotent endomorphisms of its objects).

For a thick weak subcategory $\mathcal{I}$ of an abelian 2-category $\mathcal{T}$, $\mathcal{I}(A_1, A_2)$ is an abelian category for every pair of objects $A_1, A_2 \in \mathcal{I}$ with respect to the same kernels and cokernels as the ambient abelian category $\mathcal{T}(A_1, A_2)$.

In part (3), the compositions of 1-morphisms that are used are the horizontal compositions discussed in Sect. 2.1. More explicitly, a thick subcategory $\mathcal{I}$ of $\mathcal{T}$ is a thick ideal if for all $f_1 \in \mathcal{T}(A_1, A_2)$, $g_2 \in \mathcal{I}(A_2, A_3)$ and $f_3 \in \mathcal{T}(A_3, A_4)$, we have

$$g_2 f_1 \in \mathcal{I}(A_1, A_3) \quad \text{and} \quad f_3 g_2 \in \mathcal{I}(A_2, A_4),$$

where $A_1, A_2, A_3, A_4 \in \mathcal{T}$.

Remark 3.3 Let $\mathcal{I}$ and $\mathcal{J}$ be a pair of thick weak subcategories of $\mathcal{T}$. Then

$$\mathcal{I} \subseteq \mathcal{J} \quad \text{if and only if} \quad \mathcal{I}_1 \subseteq \mathcal{J}_1.$$

In particular,

$$\mathcal{I} = \mathcal{J} \quad \text{if and only if} \quad \mathcal{I}_1 = \mathcal{J}_1.$$

Example 3.4 There exists a unique thick ideal of every 2-category $\mathcal{T}$ whose set of 1-morphisms consists of the 0 objects of the abelian categories $\mathcal{T}(A_1, A_2)$ for all $A_1, A_2 \in \mathcal{T}$. This thick ideal will be denoted by $0_{\mathcal{T}}$. Every other thick ideal of $\mathcal{T}$ contains $0_{\mathcal{T}}$.

For two subsets $X, Y \subseteq \mathcal{T}_1$, denote by

$X \circ Y$ the set of isomorphism classes of 1-morphims of $\mathcal{T}$ having representatives of the form fg for f and g representing classes in X and Y such that fg is defined.

In general, $X \circ Y$ can be empty. For $f, g \in \mathcal{T}_1$ the composition $f \circ g$ is either empty or consists of one element.

In this notation, a thick weak subcategory $\mathcal{I}$ of $\mathcal{T}$ is a thick ideal if and only if

$$\mathcal{T}_1 \circ \mathcal{I}_1 \subseteq \mathcal{I}_1 \quad \text{and} \quad \mathcal{I}_1 \circ \mathcal{T}_1 \subseteq \mathcal{I}_1.$$

Definition 3.5 A *thick left (respectively right) ideal* of an abelian 2-category $\mathcal{T}$ is a thick weak subcategory $\mathcal{I}$ of $\mathcal{T}$ such that

$$\mathcal{T}_1 \circ \mathcal{I}_1 \subseteq \mathcal{I}_1 \quad (\text{respectively } \mathcal{I}_1 \circ \mathcal{T}_1 \subseteq \mathcal{I}_1).$$

Remark 3.6 Let A_1, A_2, B_1, B_2 be four objects of an abelian 2-category $\mathcal{T}$ such that

$$A_i \cong B_i \quad \text{for} \quad i = 1, 2.$$

Then for every thick ideal $\mathcal{I}$ of $\mathcal{T}$, we have (noncanonical) isomorphisms of abelian categories

$$\mathcal{I}(A_1, A_2) \cong \mathcal{I}(B_1, B_2). \tag{3.1}$$

Indeed, let $f_i \in \mathcal{T}(A_i, B_i)$ and $g_i \in \mathcal{T}(B_i, A_i)$ be such that

$$f_i g_i \cong 1_{B_i} \quad \text{and} \quad g_i f_i \cong 1_{A_i}$$

for $i = 1, 2$ (where the isomorphisms are in the categories $\mathcal{T}(B_i, B_i)$ and $\mathcal{T}(A_i, A_i)$). The functor giving the equivalence (3.1) is defined by

$$h \mapsto f_2 h g_1$$

on the level of objects $h \in \mathcal{I}(A_1, A_1)$ and

$$\alpha \mapsto 1_{f_2} * \alpha * 1_{g_1}$$

on the level of morphisms.

3.2 Prime Ideals of Abelian 2-Categories

A thick ideal $\mathcal{I}$ of $\mathcal{T}$ will be called *proper* if $\mathcal{I} \neq \mathcal{T}$; by Remark 3.3 this is the same as $\mathcal{I}_1 \subsetneq \mathcal{T}_1$.

Definition 3.7 We call $\mathcal{P}$ a *prime ideal* of $\mathcal{T}$ if $\mathcal{P}$ is a proper thick ideal of $\mathcal{T}$ with the property that for every pair of thick ideals $\mathcal{I}$ and $\mathcal{J}$ of $\mathcal{T}$,

$$\mathcal{I}_1 \circ \mathcal{J}_1 \subseteq \mathcal{P}_1 \quad \Rightarrow \quad \mathcal{I} \subseteq \mathcal{P} \text{ or } \mathcal{J} \subseteq \mathcal{P}.$$

The set of all prime ideals $\mathcal{P}$ of an abelian 2-category $\mathcal{T}$ will be called the *prime spectrum* of $\mathcal{T}$ and will be denoted by $\operatorname{Spec}(\mathcal{T})$.

By Remark 3.3, the property on the right side of the implication can be replaced with $\mathcal{I}_1 \subseteq \mathcal{P}_1$ or $\mathcal{J}_1 \subseteq \mathcal{P}_1$.

3.3 Two Equivalent Characterizations of Prime Ideals

The following lemma is straightforward.

Lemma 3.8 *The intersection of any family of thick ideals is a thick ideal.*

If $\mathcal{M}$ is a collection of 1-morphisms of $\mathcal{T}$ (i.e., $\mathcal{M} \subseteq \mathcal{T}_1$), let $\langle \mathcal{M} \rangle$ denote the smallest thick ideal of $\mathcal{T}$ containing $\mathcal{M}$, which exists by the previous lemma.

Lemma 3.9 *For every two collections $\mathcal{M}, \mathcal{N} \subseteq \mathcal{T}_1$ of 1-morphisms of an abelian 2-category $\mathcal{T}$, we have*

$$\langle \mathcal{M} \rangle_1 \circ \langle \mathcal{N} \rangle_1 \subseteq \langle \mathcal{M} \circ \mathcal{T}_1 \circ \mathcal{N} \rangle_1. \tag{3.2}$$

Proof We will first show that

$$\langle \mathcal{M} \rangle_1 \circ \mathcal{N} \subseteq \langle \mathcal{M} \circ \mathcal{T}_1 \circ \mathcal{N} \rangle_1. \tag{3.3}$$

The 1-morphisms of $\langle \mathcal{M} \rangle$ are obtained from the elements of $\mathcal{M}$ by successive taking of kernels and cokernels (of 2-morphisms between these elements), and extensions (between these elements), as well as compositions on the left and the right by elements in $\mathcal{T}_1$. We need to show that those operations, composed on the right with the elements of $\mathcal{N}$, yield elements of the right-hand side.

(1) Suppose that $\alpha : f \to g$ is a 2-morphism for $f, g \in \mathcal{T}_1$ with the property that

$$fn, gn \in \langle \mathcal{M} \circ \mathcal{T} \circ \mathcal{N} \rangle_1 \quad \text{for all} \quad n \in \mathcal{T}_1 \circ \mathcal{N}.$$

Note that, for example, every 1-morphism in $\mathcal{M}$ has this property. Let $\kappa : k \to f$ be the kernel of α. Since exact functors preserve kernels, $\kappa * \mathrm{id}_n : kn \to fn$ is the kernel of $\alpha * \mathrm{id}_n : fn \to gn$. The thickness property of $\langle \mathcal{M} \circ \mathcal{T}_1 \circ \mathcal{N} \rangle$ implies that $kn \in \langle \mathcal{M} \circ \mathcal{T}_1 \circ \mathcal{N} \rangle_1$ for all $n \in \mathcal{T}_1 \circ \mathcal{N}$.

Symmetrically, one shows that if $\gamma : g \to c$ is the cokernel of α, then $cn \in \langle \mathcal{M} \circ \mathcal{T}_1 \circ \mathcal{N} \rangle_1$ for all $n \in \mathcal{T}_1 \circ \mathcal{N}$.

(2) Next, assume that

$$0 \to f \to g \to h \to 0$$

is an exact sequence in one of the abelian categories $\mathcal{T}(A_1, A_2)$, where f, h have the property that $fn, hn \in \langle \mathcal{M} \circ \mathcal{T}_1 \circ \mathcal{N} \rangle_1$ for all $n \in \mathcal{T}_1 \circ \mathcal{N}$. Since horizontal composition in $\mathcal{T}$ is exact, for any $n \in \mathcal{T}_1 \circ \mathcal{N}$, we get a short exact sequence

$$0 \to fn \to gn \to hn \to 0.$$

Since the first and last terms are in $\langle \mathcal{M} \circ \mathcal{T}_1 \circ \mathcal{N} \rangle_1$, so is the middle term.

Combining (1)–(2) and the fact that $\langle \mathcal{M} \circ \mathcal{T}_1 \circ \mathcal{N} \rangle_1$ is stable under left compositions with elements of $\mathcal{T}_1$ yields (3.3). Analogously, we derive (3.2) from (3.3) by using $\langle \mathcal{M} \rangle_1$ in place of $\mathcal{M}$. □

Theorem 3.10 *A proper thick ideal $\mathcal{P}$ of an abelian 2-category $\mathcal{T}$ is prime if and only if for all $m, n \in \mathcal{T}_1$, $m \circ \mathcal{T}_1 \circ n \subseteq \mathcal{P}_1$ implies that either $m \in \mathcal{P}_1$ or $n \in \mathcal{P}_1$.*

Proof Suppose $\mathcal{P}$ is a prime ideal of $\mathcal{T}$, and that $m \circ \mathcal{T} \circ n \subseteq \mathcal{P}$ for some $m, n \in \mathcal{T}_1$. Then by the previous lemma,

$$\langle m \rangle_1 \circ \langle n \rangle_1 \subseteq \langle m \circ \mathcal{T}_1 \circ n \rangle_1 \subseteq \mathcal{P}_1,$$

and so by primeness of $\mathcal{P}$, $\langle m \rangle \subseteq \mathcal{P}_1$ or $\langle n \rangle \subseteq \mathcal{P}_1$. Therefore, m or n is in $\mathcal{P}_1$.

Now suppose $\mathcal{P}$ is a proper thick ideal of $\mathcal{T}$ with the property that for all $m, n \in \mathcal{T}_1$, $m \circ \mathcal{T}_1 \circ n \subseteq \mathcal{P}_1$ implies that either $m \in \mathcal{P}_1$ or $n \in \mathcal{P}_1$. Let $\mathcal{I}$ and $\mathcal{J}$ be a pair of thick ideals of $\mathcal{T}$ such that

$$\mathcal{I}_1 \circ \mathcal{J}_1 \subseteq \mathcal{P}_1 \quad \text{and} \quad \mathcal{J}_1 \not\subseteq \mathcal{P}_1.$$

Then there is some $j \in \mathcal{J}_1$ with $j \notin \mathcal{P}_1$. However, $i \circ \mathcal{T}_1 \circ j \subseteq \mathcal{P}_1$ for any $i \in \mathcal{I}_1$ since $i \circ \mathcal{T}_1 \subseteq \mathcal{I}_1$, $j \in \mathcal{J}_1$, and $\mathcal{I}_1 \circ \mathcal{J}_1 \subseteq \mathcal{P}_1$. The assumed property of $\mathcal{P}$ implies that $\mathcal{I}_1 \subseteq \mathcal{P}_1$. Therefore $\mathcal{I} \subseteq \mathcal{P}$ by Remark 3.3. □

It is easy to see that Theorem 3.10 implies the following:

Proposition 3.11 *A proper thick ideal $\mathcal{P}$ of an abelian 2-category $\mathcal{T}$ is prime if and only if for every pair of right thick ideals $\mathcal{I}$ and $\mathcal{J}$ of $\mathcal{T}$*

$$\mathcal{I}_1 \circ \mathcal{J}_1 \subseteq \mathcal{P}_1 \quad \Rightarrow \quad \mathcal{I} \subseteq \mathcal{P} \text{ or } \mathcal{J} \subseteq \mathcal{P}.$$

A similar characterization holds using left thick ideals.

Theorem 3.12 *A proper thick ideal $\mathcal{P}$ of an abelian 2-category $\mathcal{T}$ is prime if and only if for all thick ideals $\mathcal{I}, \mathcal{J}$ of $\mathcal{T}$ properly containing $\mathcal{P}$, we have that $\mathcal{I}_1 \circ \mathcal{J}_1 \not\subseteq \mathcal{P}_1$.*

Proof The implication ⇒ is clear. Suppose $\mathcal{P}$ is a proper thick ideal which is not prime. Then there exist some thick ideals $\mathcal{I}$ and $\mathcal{J}$ of $\mathcal{T}$ with $\mathcal{I}_1 \circ \mathcal{J}_1 \subseteq \mathcal{P}_1$ and $\mathcal{I}, \mathcal{J} \not\subseteq \mathcal{P}$. Set

$$\mathcal{M} := \mathcal{P}_1 \cup \mathcal{I}_1 \quad \text{and} \quad \mathcal{N} := \mathcal{P}_1 \cup \mathcal{J}_1.$$

By Remark 3.3, $\mathcal{P}_1$ is properly contained in both $\langle \mathcal{M} \rangle_1$ and $\langle \mathcal{N} \rangle_1$. Lemma 3.9 implies that

$$\langle \mathcal{M} \rangle \circ \langle \mathcal{N} \rangle \subseteq \langle \mathcal{M} \circ \mathcal{T}_1 \circ \mathcal{N} \rangle. \tag{3.4}$$

Observe that

$$\mathcal{M} \circ \mathcal{T}_1 \circ \mathcal{N} \subseteq \mathcal{P}_1, \tag{3.5}$$

by the following. Consider the composition itj for some $i \in \mathcal{M}, t \in \mathcal{T}_1, j \in \mathcal{N}$. So, $i \in \mathcal{I}$ or $\mathcal{P}_1$; likewise, $j \in \mathcal{I}$ or $\mathcal{P}_1$. If at least one of the two 1-morphism i, j is in $\mathcal{P}_1$, we have $itj \in \mathcal{P}_1$ since $\mathcal{P}$ is a thick ideal; if $i \in \mathcal{I}_1$ and $j \in \mathcal{J}_1$, then $i \circ t \in \mathcal{I}_1$, so $itj \in \mathcal{I}_1 \circ \mathcal{J}_1 \subseteq \mathcal{P}_1$ by assumption.

Therefore $\langle \mathcal{M} \rangle$ and $\langle \mathcal{N} \rangle$ are thick ideals properly containing $\mathcal{P}$ and $\langle \mathcal{M} \rangle_1 \circ \langle \mathcal{N} \rangle_1 \subseteq \mathcal{P}_1$ (the last inclusion follows from (3.4)–(3.5) and the minimality of the thick ideal $\langle - \rangle$). Hence, $\mathcal{P}$ does not have the stated property, which completes the proof of the theorem. □

3.4 Relation to Maximal Ideals

Definition 3.13 A nonempty set $\mathcal{M} \subseteq \mathcal{T}_1$ will be called *multiplicative* if $\mathcal{M}$ is a subset of non-zero equivalence classes of objects of $\mathcal{T}(A, A)$ for some object A of $\mathcal{T}$ and $\mathcal{M} \circ \mathcal{M} \subseteq \mathcal{M}$.

The condition that $\mathcal{M} \subseteq \mathcal{T}(A, A)$ means that all 1-morphism in $\mathcal{M}$ are composable. Let us explain the motivation for this condition. Let R be a ring and $\{e_s\}$ be a collection of orthogonal idempotents. If M is a multiplicative subset such that

$$M \subseteq \bigcup_{s,t} e_s R e_t$$

then $M \subseteq e_s R e_s$ for some s, because otherwise M will contain the 0 element of R.

Theorem 3.14 *Assume that $\mathcal{M}$ is a multiplicative subset of $\mathcal{T}_1$ for an abelian 2-category $\mathcal{T}$ and that $\mathcal{I}$ is a proper thick ideal of $\mathcal{T}$ such that $\mathcal{I}_1 \cap \mathcal{M} = \varnothing$.*

Let $\mathcal{P}$ be a maximal element of the collection of thick ideals of $\mathcal{T}$ containing $\mathcal{I}$ and intersecting $\mathcal{M}$ trivially, equipped with the inclusion relation, i.e., $\mathcal{P}$ is a maximal element of the set

$$X(\mathcal{M}, \mathcal{I}) := \{\mathcal{K} \text{ a thick ideal of } \mathcal{T} \mid \mathcal{K} \supseteq \mathcal{I} \text{ and } \mathcal{K}_1 \cap \mathcal{M} = \varnothing\}.$$

Then $\mathcal{P}$ is prime.

Proof Fix such an ideal $\mathcal{P}$. Suppose $\mathcal{Q}$ and $\mathcal{R}$ are thick ideals properly containing $\mathcal{P}$. By Theorem 3.12, it is enough to show that $\mathcal{Q} \circ \mathcal{R} \nsubseteq \mathcal{P}$. Since

$$\mathcal{I} \subseteq \mathcal{P} \subseteq \mathcal{Q} \quad \text{and} \quad \mathcal{I} \subseteq \mathcal{P} \subseteq \mathcal{R},$$

both $\mathcal{Q}_1$ and $\mathcal{R}_1$ must intersect nontrivially with $\mathcal{M}$, by the maximality assumption on $\mathcal{P}$. Let $q \in \mathcal{Q}_1 \cap \mathcal{M}$ and $r \in \mathcal{R}_1 \cap \mathcal{M}$. If $\mathcal{Q} \circ \mathcal{R} \subseteq \mathcal{P}$, then we would obtain that $qr \in \mathcal{P}$, because by the definition of multiplicative subset of $\mathcal{T}_1$, each two elements of $\mathcal{M}$ are composable. However, since $qr \in \mathcal{M}$, this contradictions with the assumption that $\mathcal{P}_1 \cap \mathcal{M} = \varnothing$. □

Remark 3.15 The set $X(\mathcal{M}, \mathcal{I})$ from Theorem 3.14 is nonempty because $\mathcal{I} \in X(\mathcal{M}, \mathcal{I})$. The union of an ascending chain of thick ideals in the set $X(\mathcal{M}, \mathcal{I})$ is obviously a thick ideal of $\mathcal{T}$. By Zorn's lemma, the set $X(\mathcal{M}, \mathcal{I})$ from Theorem 3.14 always contains at least one maximal element.

Corollary 3.16

(1) *For each proper thick ideal $\mathcal{I}$ of an abelian 2-category $\mathcal{T}$, there exists a prime ideal $\mathcal{P}$ of $\mathcal{T}$ that contains $\mathcal{I}$.*

(2) *Let $\mathcal{M}$ be a multiplicative set of an abelian 2 category $\mathcal{T}$. Every maximal element of the set of thick ideals $\mathcal{K}$ of $\mathcal{T}$ such that $\mathcal{K}_1 \cap \mathcal{M} = \varnothing$ is a prime ideal. The set of such thick ideals contains at least one maximal element.*

Proof (1) Since the thick ideal $\mathcal{I}$ is proper, there exists an object $A \in \mathcal{T}$ such that $1_A \notin \mathcal{I}_1$. Indeed, otherwise

$$\mathcal{T}(B, A) = \mathcal{T}(B, A) \circ 1_A = \mathcal{I}(B, A)$$

for all objects $A, B \in \mathcal{T}$. The statement of part (1) follows from Theorem 3.14 applied to $\mathcal{M} := \{1_A\}$ for an object $A \in \mathcal{T}$ such that $1_A \notin \mathcal{I}$.

(2) For each multiplicative subset $\mathcal{M}$ of an abelian 2-category $\mathcal{T}$, the thick ideal $0_{\mathcal{T}}$ from Example 3.4 intersects $\mathcal{M}$ trivially. This part follows from Theorem 3.14 applied to the thick ideal $\mathcal{I} := 0_{\mathcal{T}}$. □

The second part of the corollary, applied to the multiplicative subset $\mathcal{M} := \{1_A\}$ for an object $A \in \mathcal{T}$, implies the following:

Corollary 3.17 *The prime spectrum of every abelian 2-category $\mathcal{T}$ is nonempty.*

Definition 3.18

(1) An abelian 2-category $\mathcal{T}$ will be called *prime* if $0_{\mathcal{T}}$ is a prime ideal of $\mathcal{T}$.
(2) An abelian 2-category $\mathcal{T}$ will be called *simple* if the only proper thick ideal of $\mathcal{T}$ is $0_{\mathcal{T}}$.

Corollary 3.17 implies that every simple abelian 2-category $\mathcal{T}$ is prime.

3.5 The Zariski Topology

Definition 3.19 Define the family of closed sets $V(\mathcal{I}) := \{\mathcal{P} \in \mathrm{Spec}(\mathcal{T}) \mid \mathcal{P} \supseteq \mathcal{I}\}$ of $\mathrm{Spec}(\mathcal{T})$ for all thick ideals $\mathcal{I}$.

Remark 3.20 This topology is different from the one considered by Balmer [1]. The main reason for which we consider it is to ensure good behavior under the K_0 map, see Theorem 6.12(4).

Lemma 3.21 *For each abelian 2-category $\mathcal{T}$, the above family of closed sets turns* $\mathrm{Spec}(\mathcal{T})$ *into a topological space. The corresponding topology will be called the Zariski topology of* $\mathrm{Spec}(\mathcal{T})$.

It is easy to verify for that for every pair of thick ideals $\mathcal{I}, \mathcal{J}$ of $\mathcal{T}$ and for every (possibly infinite) collection $\{\mathcal{I}_s\}$ of thick ideals of $\mathcal{T}$, similarly to the classical situation, we have

$$V(\mathcal{I}) \cup V(\mathcal{J}) = V(\langle \mathcal{I}_1 \circ \mathcal{J}_1 \rangle) \quad \text{and}$$

$$\bigcap_i V(\mathcal{I}_s) = V\left(\left\langle \bigcup_i (\mathcal{I}_s)_1 \right\rangle\right).$$

Finally, we also have $V(\mathcal{T}) = \varnothing$ and $V(0_{\mathcal{T}}) = \mathrm{Spec}(\mathcal{T})$.

3.6 An Example

Let Γ be a nonempty set and $\Bbbk$ be an arbitrary field. Let ${}_{\Bbbk}\mathrm{Vect}_{\Bbbk}$ be the category of finite dimensional $\Bbbk$-vector spaces considered as $(\Bbbk, \Bbbk)$-bimodules. Let $\{\Bbbk_a \mid a \in \Gamma\}$ be a collection of fields isomorphic to $\Bbbk$ and indexed by Γ.

There is a unique $\Bbbk$-linear abelian 2-category $\mathcal{M}_\Gamma(\Bbbk)$ whose set of objects is Γ and such that

$$\mathcal{M}_\Gamma(\Bbbk)(a, b) := {}_{\Bbbk_b}\mathrm{Vect}_{\Bbbk_a} \quad \text{for} \quad a, b \in \Gamma.$$

Its composition bifunctors are given by

$$- \otimes_{\Bbbk_b} - : \mathcal{M}_\Gamma(\Bbbk)(b, c) \times \mathcal{M}_\Gamma(\Bbbk)(a, b) \to \mathcal{M}_\Gamma(\Bbbk)(a, c).$$

Its Grothendieck group is $K_0(\mathcal{M}) \cong M_\Gamma(\mathbb{Z})$ – the ring of square matrices with finitely many nonzero integer entries whose rows and columns are indexed by Γ. In the terminology of Sect. 2.2, $\mathcal{M}_\Gamma(\Bbbk)$ is a categorification of the matrix ring $M_\Gamma(\Bbbk')$ for any field $\Bbbk'$.

Analogously to the classical situation, we show:

Lemma 3.22 *The abelian 2-categories $\mathcal{M}_\Gamma(\Bbbk)$ are simple (and thus prime).*

Proof Let $\mathcal{I}$ be a thick ideal of $\mathcal{M}_\Gamma(\Bbbk)$ that properly contains the 0-ideal $0_{\mathcal{M}_\Gamma(\Bbbk)}$. Then for some $a, b \in \Gamma$,

$$\mathcal{I}(a, b) \neq 0.$$

Since $\mathcal{I}$ is thick, $\mathcal{I}(a, b)$ is a nonzero subcategory of ${}_{\Bbbk_b}\mathrm{Vect}_{\Bbbk_a}$ that is closed under taking direct summands. Hence $\mathcal{I}(a, b)$ contains the 1-dimensional vector space in ${}_{\Bbbk_b}\mathrm{Vect}_{\Bbbk_a}$, and so,

$$\mathcal{I}(a, b) = {}_{\Bbbk_b}\mathrm{Vect}_{\Bbbk_a} = \mathcal{M}_\Gamma(\Bbbk)(a, b).$$

Since all objects in $\mathcal{M}_\Gamma(\Bbbk)$ are isomorphic to each other, Remark 3.6 implies that

$$\mathcal{I}(a', b') = \mathcal{M}_\Gamma(\Bbbk)(a', b')$$

for all $a', b' \in \Gamma$. Thus $\mathcal{I} = \mathcal{M}_\Gamma(\Bbbk)$, which completes the proof. □

4 Minimal Primes in Noetherian Abelian 2-Categories

In this section we define noetherian abelian 2-categories $\mathcal{T}$, and prove that for all proper thick ideals $\mathcal{I}$ of $\mathcal{T}$, there exist finitely many minimal primes over $\mathcal{I}$ and the product of their 1-morphism sets (with repetitions) is contained in $\mathcal{I}_1$.

4.1 Noetherian Abelian 2-Categories

Definition 4.1

(1) An abelian 2-category will be called *left (resp. right) noetherian* if it satisfies the ascending chain condition on thick left (resp. right) ideals.
(2) An abelian 2-category will be called *noetherian* if it is both left and right noetherian.
(3) An abelian 2-category will be called *weakly noetherian* if it satisfies the ascending chain condition on (two-sided) thick ideals.

More concretely, an abelian 2-category is noetherian if for every chain of thick left ideals

$$\mathcal{I} \subseteq \mathcal{I}_2 \subseteq \dots$$

there exists an integer k such that $\mathcal{I}_k = \mathcal{I}_{k+1} = \dots$ and such a property is also satisfied for ascending chains of thick right ideals.

4.2 Existence of Minimal Primes

Lemma 4.2 *In any abelian 2-category $\mathcal{T}$, for every thick ideal $\mathcal{I}$ and every prime ideal $\mathcal{P}$ containing $\mathcal{I}$, there is a minimal prime $\mathcal{P}'$ such that*

$$\mathcal{I} \subseteq \mathcal{P}' \subseteq \mathcal{P}.$$

Proof Let χ denote the set of primes which contain $\mathcal{I}$ and are contained in $\mathcal{P}$. We will use Zorn's lemma to produce a minimal element of this set. We first show that any nonempty chain in χ has a lower bound in χ. Take a nonempty chain of prime ideals in χ, say

$$\mathcal{P}^{(1)} \supseteq \mathcal{P}^{(2)} \supseteq \dots$$

Then define $\mathcal{Q} = \cap_{i=1}^{\infty} \mathcal{P}^{(i)}$. Since each $\mathcal{P}^{(i)}$ contains $\mathcal{I}$ and is contained in $\mathcal{P}$, $\mathcal{Q}$ is a thick ideal which has these properties. It remains to show that $\mathcal{Q}$ is a prime ideal. Take $f, g \in \mathcal{T}_1$ such that $f \circ \mathcal{T}_1 \circ g \subset \mathcal{Q}_1$, and $f \notin \mathcal{Q}_1$. Then f is not in some $\mathcal{P}_1^{(i)}$. Therefore, $f \notin \mathcal{P}_1^{(j)}$ for $j \geq i$, and by the primeness of $\mathcal{P}^{(j)}$, $g \in \mathcal{P}_1^{(j)}$, for $j \geq i$ as well. Therefore, $g \in \mathcal{P}_1^{(k)}$ for all k, and thus, $g \in \mathcal{Q}_1$. This implies that $\mathcal{Q}$ is prime, and Zorn's lemma completes the proof. □

4.3 Finiteness and Product Properties of Minimal Primes

Theorem 4.3 *In a weakly noetherian abelian 2-category $\mathcal{T}$, for every proper thick ideal $\mathcal{I}$, there exist finitely many minimal prime ideals over $\mathcal{I}$. Furthermore, there exists a finite list of minimal prime ideals over $\mathcal{I}$ (potentially with repetition) $\mathcal{P}^{(1)}, \dots, \mathcal{P}^{(m)}$ such that the product*

$$\mathcal{P}_1^{(1)} \circ \dots \circ \mathcal{P}_1^{(m)} \subseteq \mathcal{I}_1.$$

Proof Denote the set

$$\chi := \{\mathcal{I} \text{ a proper thick ideal of } \mathcal{T} \mid \nexists \text{ prime ideals } \mathcal{P}^{(1)}, \dots, \mathcal{P}^{(m)} \supseteq \mathcal{I} \\ \text{such that } \mathcal{P}_1^{(1)} \circ \dots \circ \mathcal{P}_1^{(m)} \subseteq \mathcal{I}_1\}.$$

Suppose that χ is nonempty. By the weakly noetherian property of $\mathcal{T}$, there exists a maximal element of χ (because every ascending chain in χ eventually stabilizes). Let $\mathcal{I}$ be a maximal element of χ. The ideal $\mathcal{I}$ cannot be prime, since $\mathcal{I} \in \chi$. By Theorem 3.12 there exist proper thick ideals $\mathcal{J}$ and $\mathcal{K}$ such that

$$\mathcal{J}_1 \circ \mathcal{K}_1 \subset \mathcal{I}_1,$$

where $\mathcal{J}$ and $\mathcal{K}$ both properly contain $\mathcal{I}$. The latter property of $\mathcal{J}$ and $\mathcal{K}$ and the maximality of $\mathcal{I}$ imply that $\mathcal{J}, \mathcal{K} \notin \chi$. Hence, there exist two collection of prime ideals $\mathcal{P}^{(1)}, \dots, \mathcal{P}^{(m)} \supseteq \mathcal{J}$ and $\mathcal{Q}^{(1)}, \dots, \mathcal{Q}^{(n)} \supseteq \mathcal{K}$ such that

$$\mathcal{P}_1^{(1)} \circ \dots \circ \mathcal{P}_1^{(m)} \subseteq \mathcal{J}_1 \quad \text{and} \quad \mathcal{Q}_1^{(1)} \circ \dots \circ \mathcal{Q}_1^{(n)} \subseteq \mathcal{K}_1.$$

Then

$$\mathcal{P}_1^{(1)} \circ \ldots \circ \mathcal{P}_1^{(m)} \circ \mathcal{Q}_1^{(1)} \circ \ldots \circ \mathcal{Q}_1^{(n)} \subset \mathcal{I}_1,$$

giving a contradiction, since the ideals $\mathcal{P}^{(i)}$ and $\mathcal{Q}^{(j)}$ are prime and contain $\mathcal{I}$.

Hence, χ is empty. In other words, for every proper thick ideal $\mathcal{I}$ of $\mathcal{T}$ there exist prime ideals $\mathcal{P}^{(1)}, \ldots, \mathcal{P}^{(m)} \supseteq \mathcal{I}$ such that

$$\mathcal{P}_1^{(1)} \circ \ldots \circ \mathcal{P}_1^{(m)} \subseteq \mathcal{I}_1. \tag{4.1}$$

Applying Lemma 4.2, we obtain that for each $\mathcal{P}^{(i)}$, there exists a minimal prime $\overline{\mathcal{P}}^{(i)}$ over $\mathcal{I}$ such that $\overline{\mathcal{P}}^{(i)} \subseteq \mathcal{P}^{(i)}$. Combining this with (4.1) gives

$$\overline{\mathcal{P}}_1^{(1)} \circ \ldots \circ \overline{\mathcal{P}}_1^{(m)} \subseteq \mathcal{I}_1$$

for the minimal primes $\overline{\mathcal{P}}^{(1)}, \ldots, \overline{\mathcal{P}}^{(m)}$ of $\mathcal{I}$.

Finally, we claim that every minimal prime ideal $\mathcal{P}$ over $\mathcal{I}$ is in the list $\overline{\mathcal{P}}^{(1)}, \ldots, \overline{\mathcal{P}}^{(m)}$. This implies that there are only finitely many primes of $\mathcal{T}$ that are minimal over $\mathcal{I}$. Indeed, we have

$$\overline{\mathcal{P}}_1^{(1)} \circ \ldots \circ \overline{\mathcal{P}}_1^{(m)} \subseteq \mathcal{P}_1,$$

and by the primeness of $\mathcal{P}$, we have

$$\mathcal{I} \subseteq \mathcal{P}^{(i)} \subseteq \mathcal{P}$$

for some i. Since $\mathcal{P}$ is minimal over $\mathcal{I}$, $\mathcal{P}^{(i)} = \mathcal{P}$. □

The following corollary follows from applying Theorem 4.3 to $0_{\mathcal{T}}$.

Corollary 4.4 *A weakly noetherian abelian 2-category has finitely many minimal prime ideals.*

We also have the following corollary of Theorem 4.3:

Corollary 4.5 *For a weakly noetherian abelian 2-category $\mathcal{T}$, all closed subsets of* Spec$(\mathcal{T})$ *(with respect to the Zariski topology) are finite intersections of finitely many sets of the form $V(\mathcal{P})$ for prime ideals $\mathcal{P}$ of $\mathcal{T}$, recall Sect. 3.5.*

5 The Completely Prime Spectrum and the Semiprime Spectrum

In this section we define the notions of completely prime and semiprime ideals of abelian 2-categories, and give equivalent characterizations, one of which is an extension of the Levitzki–Nagata theorem from noncommutative ring theory.

5.1 Completely Prime Ideals

Definition 5.1 A thick ideal $\mathcal{P}$ of an abelian 2-category $\mathcal{T}$ will be called *completely prime* when it has the property that for all $f, g \in \mathcal{T}_1$:

$$f \circ g \subseteq \mathcal{P}_1 \quad \Rightarrow \quad f \in \mathcal{P}_1 \text{ or } g \in \mathcal{P}_1.$$

This is equivalent to saying that for all 1-morphisms f and g of $\mathcal{T}$, if fg is not defined or fg is a 1-morphism in $\mathcal{P}$, then f or g is a 1-morphism in $\mathcal{P}$. The stronger assumption, including the case of the condition when fg is not defined, is needed to get the correct analog of a completely prime ideal of an algebra with a set of orthogonal idempotents. Let R be a ring and $\{e_s\}$ be a collection of orthogonal idempotents. If I is a completely prime ideal of R such that

$$I \subseteq \bigoplus_{s,t} e_s R e_t,$$

then for all $s, t \neq t', s'$,

$$\text{either} \quad e_s R e_t \subseteq I \quad \text{or} \quad e_{t'} R e_{s'} \subseteq I,$$

because $(e_s R e_t)(e_{t'} R e_{s'}) = 0$.

Theorem 3.10 implies the following:

Corollary 5.2 *Every completely prime ideal of an abelian 2-category is prime.*

Proof Assume that $\mathcal{P}$ is a completely prime ideal of $\mathcal{T}$. Let $f \in \mathcal{T}(A_3, A_4)$ and $g \in \mathcal{T}(A_1, A_2)$ be such that

$$f \circ \mathcal{T}_1 \circ g \subseteq \mathcal{P}_1.$$

If $A_2 \neq A_3$, then fg is not defined and the assumption on $\mathcal{P}$ gives that either $f \in \mathcal{P}_1$ or $g \in \mathcal{P}_1$. If $A_2 = A_3$, then

$$fg = f 1_{A_2} g \in f \circ \mathcal{T}_1 \circ g \subseteq \mathcal{P}_1,$$

and, again by the assumption on $\mathcal{P}$, we have that either $f \in \mathcal{P}_1$ or $g \in \mathcal{P}_1$. □

For every abelian 2-category $\mathcal{T}$, given an object A of $\mathcal{T}$, consider the 2-subcategory $\mathcal{T}_A$ of $\mathcal{T}$ having one object A and such that $\mathcal{T}_A(A, A) := \mathcal{T}(A, A)$. It is an abelian 2-category with one object (i.e., a multiring category). The next lemma shows that the completely prime ideals of an abelian 2-category $\mathcal{T}$ are classified in terms of the completely prime ideals of these multiring categories.

Lemma 5.3 *Let $\mathcal{T}$ be an abelian 2-category.*

(1) *If $\mathcal{P}$ is a completely prime ideal of $\mathcal{T}$, then there exists an object A of $\mathcal{T}$ and a completely prime ideal $\mathcal{Q}$ of the multiring category $\mathcal{T}_A$ such that*

$$\mathcal{P}(B, C) = \begin{cases} \mathcal{Q}(A, A), & \text{if } B = C = A \\ \mathcal{T}(B, C), & \text{otherwise.} \end{cases} \tag{5.1}$$

(2) *If A is an object of $\mathcal{T}$ and $\mathcal{Q}$ is a completely prime ideal of $\mathcal{T}_A$ such that*

$$\mathcal{T}(B, A) \circ \mathcal{T}(A, B) \subseteq \mathcal{Q}(A, A) \tag{5.2}$$

for every object B of $\mathcal{T}$, then (5.1) *defines a completely prime ideal $\mathcal{P}$ of $\mathcal{T}$.*

Proof (1) Since $\mathcal{P}$ is a proper thick ideal of $\mathcal{T}$, there exists an object A of $\mathcal{T}$ such that $1_A \notin \mathcal{P}_1$. (Otherwise $\mathcal{P}_1$ will contain all 1-morphisms of $\mathcal{T}$ because

$$\mathcal{T}(B, A) \circ 1_A = \mathcal{T}(B, A). \tag{5.3}$$

This will contradict the properness of $\mathcal{P}$.) Obviously

$$\mathcal{Q}(A, A) := \mathcal{P}(A, A)$$

defines a completely prime ideal of the multiring category $\mathcal{T}_A$. It remains to show that $\mathcal{P}$ is given by (5.1) in terms of $\mathcal{Q}$.

If B is an object of $\mathcal{T}$ which is different from A, then the composition $1_A 1_B$ is not defined and $1_A \notin \mathcal{P}_1$, hence $1_B \in \mathcal{P}_1$. It follows that $\mathcal{P}$ is given by (5.1) by an argument similar to (5.3).

(2) The condition (5.2) ensures that the weak thick subcategory $\mathcal{P}$ of $\mathcal{T}$ given by (5.1) is a thick ideal of $\mathcal{T}$. Its complete primeness is easy to show. □

Definition 5.4 A multiring category $\mathcal{T}$ will be called a *domain* if its zero ideal $0_\mathcal{T}$ is completely prime, i.e., if

$$M \otimes N \cong 0 \quad \Rightarrow \quad M \cong 0 \text{ or } N \cong 0$$

for all objects M of $\mathcal{T}$.

An abelian 2-category $\mathcal{T}$ will be called *prime*, it its zero ideal $0_\mathcal{T}$ is prime.

Example 5.5 Let H be a Hopf algebra over a field $\Bbbk$. Denote by $H-$mod the category of finite dimensional H-modules. It is a $\Bbbk$-linear multiring category. This category is a domain: if $V, W \in H-$mod are such that $V \otimes W \cong 0$, then

$$\dim V \dim W = 0.$$

Therefore, either $\dim V = 0$ or $\dim W = 0$. So, either $V \cong 0$ or $W \cong 0$. □

Let $\mathcal{T}$ be an abelian 2-category. The same proof shows that if

(1) $\eta : \mathcal{T}_1 \to R$ is a map such that R is a domain and $\eta(fg) = \eta(f)\eta(g)$ for all $f, g \in \mathcal{T}$ for which the composition is defined, and
(2) $\mathcal{I}$ is a thick ideal of $\mathcal{T}$ such that $\mathcal{I}_1 = \eta^{-1}(0)$,

then $\mathcal{I}$ is a completely prime ideal of $\mathcal{T}$.

5.2 Semiprime Ideals

Definition 5.6 A thick ideal of an abelian 2-category will be called *semiprime* if it is an intersection of prime ideals. An abelian 2-category $\mathcal{T}$ will be called *semiprime*, it its zero ideal $0_{\mathcal{T}}$ is semiprime.

Theorem 4.3 implies that in a weak noetherian abelian 2-category every semiprime ideal is the intersection of the finitely many minimal primes over it.

The following theorem is a categorical version of the Levitzki–Nagata theorem.

Theorem 5.7 *A thick ideal $\mathcal{Q}$ is semiprime if and only if for all $f \in \mathcal{T}_1$,*

$$f \circ \mathcal{T}_1 \circ f \subseteq \mathcal{Q}_1 \quad \Rightarrow \quad f \in \mathcal{Q}_1. \tag{5.4}$$

Proof First, suppose $\mathcal{Q} = \bigcap_s \mathcal{P}^{(s)}$ for some collection $\{\mathcal{P}^{(s)}\}$ of primes of $\mathcal{T}$. Suppose $f \in \mathcal{T}_1$, and $f \circ \mathcal{T}_1 \circ f \subseteq \mathcal{Q}_1$. By primeness, $f \in \mathcal{P}_1^{(s)}$ for all s. Therefore, $f \in \mathcal{Q}_1$.

For the other direction, suppose that $\mathcal{Q}$ is a thick ideal of $\mathcal{T}$ having the property (5.4). Choose an element

$$g \in \mathcal{T}_1, \quad g \notin \mathcal{Q}_1,$$

and set $g_0 := g$. It follows from (5.4) that $g_0 \circ \mathcal{T}_1 \circ g_0 \not\subseteq \mathcal{Q}_1$. Choose

$$g_1 \in g_0 \circ \mathcal{T}_1 \circ g_0, \quad g_1 \notin \mathcal{Q}_1.$$

Again, since $g_1 \notin \mathcal{Q}_1$, the condition (5.4) implies that $g_1 \circ \mathcal{T}_1 \circ g_1 \not\subseteq \mathcal{Q}_1$. Proceeding inductively in this manner, we construct a sequence of 1-morphisms $g_0, g_1, \ldots$ of $\mathcal{T}$ such that

$$g_i \in g_{i-1} \circ \mathcal{T}_1 \circ g_{i-1}, \quad g_i \notin \mathcal{Q}_1. \tag{5.5}$$

Since $g_i \in g_{i-1} \circ \mathcal{T} \circ g_{i-1}$, we have $g_i \circ \mathcal{T} \circ g_i \subseteq g_{i-1} \circ \mathcal{T} \circ g_{i-1}$. Consider the set S of thick ideals $\mathcal{I}$ of $\mathcal{T}$ such that

$$\mathcal{Q} \subseteq \mathcal{I} \quad \text{and} \quad g_i \notin \mathcal{I}_1 \quad \text{for all} \quad i = 0, 1, \ldots.$$

This set is nonempty because $\mathcal{Q} \in S$. Since the union of a chain of thick ideals is a thick ideal, we can apply Zorn's lemma to get that S contains a maximal element. Denote one such element by $\mathcal{P}$. The proper thick ideal $\mathcal{P}$ is prime. Indeed, if $\mathcal{J}$ and $\mathcal{K}$ are thick ideals that properly contain $\mathcal{P}$, then by maximality of $\mathcal{P}$, there are some $g_j \in \mathcal{J}_1$ and $g_k \in \mathcal{K}_1$. If m is the max of j and k, then g_m is in both $\mathcal{J}_1$ and $\mathcal{K}_1$ by the first property in (5.5). Hence,

$$g_{m+1} \in g_m \circ \mathcal{T}_1 \circ g_m \subseteq \mathcal{J}_1 \circ \mathcal{K}_1 \quad \text{and} \quad g_{m+1} \notin \mathcal{P}_1.$$

Therefore, $\mathcal{J}_1 \circ \mathcal{K}_1 \nsubseteq \mathcal{P}_1$, and by Theorem 3.12, $\mathcal{P}$ is prime. For every element $g \in \mathcal{T}_1$ that is not in $\mathcal{Q}_1$, we have produced a prime $\mathcal{P}^{(g)}$ of $\mathcal{T}$ such that

$$\mathcal{Q} \subseteq \mathcal{P}^{(g)} \quad \text{and} \quad g \notin \mathcal{P}^{(g)}.$$

Therefore,

$$\mathcal{Q} = \bigcap_{g \in \mathcal{T}_1 \setminus \mathcal{Q}_1} \mathcal{P}^{(g)},$$

which completes the proof of the theorem. □

Theorem 5.8 *Suppose $\mathcal{Q}$ is a proper thick ideal in an abelian 2-category $\mathcal{T}$. Then the following are equivalent:*

(1) *$\mathcal{Q}$ is semiprime;*
(2) *If $\mathcal{I}$ is any thick ideal of $\mathcal{T}$ such that $\mathcal{I}_1 \circ \mathcal{I}_1 \subseteq \mathcal{Q}_1$, then $\mathcal{I} \subseteq \mathcal{Q}$;*
(3) *If $\mathcal{I}$ is any thick ideal properly containing $\mathcal{Q}$, then $\mathcal{I}_1 \circ \mathcal{I}_1 \nsubseteq \mathcal{Q}_1$;*
(4) *If $\mathcal{I}$ is any right thick ideal of $\mathcal{T}$ such that $\mathcal{I}_1 \circ \mathcal{I}_1 \subseteq \mathcal{Q}_1$, then $\mathcal{I} \subseteq \mathcal{Q}$;*
(5) *If $\mathcal{I}$ is any left thick ideal of $\mathcal{T}$ such that $\mathcal{I}_1 \circ \mathcal{I}_1 \subseteq \mathcal{Q}_1$, then $\mathcal{I} \subseteq \mathcal{Q}$.*

Proof (1) ⇒ (4): Suppose $\mathcal{Q}$ is semiprime and $\mathcal{I}$ is a right thick ideal with $\mathcal{I}_1 \circ \mathcal{I}_1 \subset \mathcal{Q}_1$. Take any $i \in \mathcal{I}_1$. Then $i \circ t \in \mathcal{I}$ for all $t \in \mathcal{T}_1$. Therefore, $i \circ \mathcal{T}_1 \circ i \in \mathcal{Q}_1$. Theorem 5.7 implies that $i \in \mathcal{Q}$. Hence, $\mathcal{I}_1 \subseteq \mathcal{Q}_1$, and thus $\mathcal{I} \subseteq \mathcal{Q}$ by Remark 3.3.

(1) ⇒ (5): This follows from a symmetric argument.

(4) ⇒ (5) and (3): This is clear, since a thick ideal is also a right thick ideal, and a left thick ideal.

(3) $\Rightarrow$ (2): Suppose (3) holds, and $\mathcal{I}$ is a thick ideal with $\mathcal{I}_1 \circ \mathcal{I}_1 \subseteq \mathcal{Q}_1$. Then $\langle \mathcal{I}_1 \cup \mathcal{Q}_1 \rangle$ is a thick ideal containing $\mathcal{Q}$. Since

$$(\mathcal{I}_1 \cup \mathcal{Q}_1) \circ (\mathcal{I}_1 \cup \mathcal{Q}_1) = (\mathcal{I}_1 \circ \mathcal{I}_1) \cup (\mathcal{Q}_1 \circ \mathcal{I}_1) \cup (\mathcal{I}_1 \circ \mathcal{Q}_1) \cup (\mathcal{Q}_1 \circ \mathcal{Q}_1) \subseteq \mathcal{Q}_1,$$

applying Lemma 3.9, we obtain

$$\langle \mathcal{I}_1 \cup \mathcal{Q}_1 \rangle_1 \circ \langle \mathcal{I}_1 \cup \mathcal{Q}_1 \rangle_1 \subseteq \langle (\mathcal{I}_1 \cup \mathcal{Q}_1) \circ \mathcal{T}_1 \circ (\mathcal{I}_1 \cup \mathcal{Q}_1) \rangle_1 = \langle (\mathcal{I}_1 \cup \mathcal{Q}_1) \circ (\mathcal{I}_1 \cup \mathcal{Q}_1) \rangle_1 \subseteq \mathcal{Q}_1.$$

From the assumption that the ideal $\mathcal{Q}$ has the property (3) and the fact that $\langle \mathcal{I}_1 \cup \mathcal{Q}_1 \rangle$ is a thick ideal containing $\mathcal{Q}$, we get that $\langle \mathcal{I}_1 \cup \mathcal{Q}_1 \rangle = \mathcal{Q}$. Therefore, $\mathcal{I}_1 \subseteq \mathcal{Q}_1$, and thus $\mathcal{I} \subseteq \mathcal{Q}$ by Remark 3.3.

(2) $\Rightarrow$ (1): Suppose (2) holds, and $f \in \mathcal{T}$ is a 1-morphism such that $f \circ \mathcal{T}_1 \circ f \subseteq \mathcal{Q}_1$. Lemma 3.9 implies that

$$\langle f \rangle_1 \circ \langle f \rangle_1 \subseteq \langle f \circ \mathcal{T}_1 \circ f \rangle_1 \subseteq \mathcal{Q}_1.$$

Therefore, by (2), $\langle f \rangle_1 \subseteq \mathcal{Q}_1$, and so, $f \in \mathcal{Q}_1$. Hence, $\mathcal{Q}_1$ is semiprime. □

We have the following corollary from the characterizations (4) and (5) of semiprime ideals in the previous theorem. For a subset $X \subseteq \mathcal{T}_1$, denote by $X^{\circ n} := X \circ \cdots \circ X$ the n-fold composition power.

Lemma 5.9 *If $\mathcal{Q}$ is a semiprime ideal of the abelian 2-category $\mathcal{T}$, and $\mathcal{I}$ is a right or left thick ideal with $(\mathcal{I}_1)^{\circ n} \subseteq \mathcal{Q}_1$, then $\mathcal{I} \subseteq \mathcal{Q}$.*

Proof We prove the statement by induction on n. For $n \geq 2$, we have

$$((\mathcal{I}_1)^{\circ(n-1)})^{\circ 2} = (\mathcal{I}_1)^{\circ n} \circ (\mathcal{I}_1)^{n-2} \subseteq \mathcal{Q}_1.$$

Theorem 5.8 implies that $(\mathcal{I}_1)^{\circ(n-1)} \subseteq \mathcal{Q}_1$, and so by the inductive assumption, $\mathcal{I} \subseteq \mathcal{Q}$. □

6 The Serre Prime Spectra of Abelian 2-Categories and $\mathbb{Z}_+$-Rings

In this section we define and investigate the notions of Serre prime, semiprime, and completely prime ideals of abelian 2-categories and $\mathbb{Z}_+$-rings. We establish that the corresponding topological spaces for abelian 2-categories and $\mathbb{Z}_+$-rings are homeomorphic. We also describe the relations of the first set of notions to the notions of prime, completely prime, and semiprime ideals of abelian 2-categories, and the second set of notions to the prime spectra of rings.

6.1 Serre Ideals of Abelian 2-Categories

Recall that a Serre subcategory of an abelian 1-category is a subcategory which is closed under subobjects, quotients, and extensions. Every Serre subcategory $\mathcal{I}$ of an abelian category $\mathcal{C}$ is thick, and in particular, is closed under isomorphisms. For such a subcategory, one forms the Serre quotient $\mathcal{C}/\mathcal{I}$ which has a canonical structure of abelian category [42, §10.3]. By [37, Theorem 5], for every Serre subcategory $\mathcal{I}$ of an abelian category $\mathcal{C}$, we have the exact sequence

$$K_0(\mathcal{I}) \to K_0(\mathcal{C}) \to K_0(\mathcal{C}/\mathcal{I}) \to 0. \tag{6.1}$$

Definition 6.1

(1) We call a thick ideal $\mathcal{I}$ of an abelian 2-category $\mathcal{T}$ a *Serre ideal* if for every two objects $A_1, A_2 \in \mathcal{T}$,

$$\mathcal{I}(A_1, A_2) \text{ is a Serre subcategory of } \mathcal{T}(A_1, A_2).$$

(2) A *Serre prime (resp. semiprime, completely prime) ideal* $\mathcal{P}$ of an abelian 2-category $\mathcal{T}$ is a prime (resp. semiprime, completely prime) ideal which is a Serre ideal.

In the terminology of Definition 3.1, a Serre ideal of an abelian 2-category $\mathcal{T}$ is a weak subcategory $\mathcal{I}$ with the same set of objects such that

(1) for any pair of objects $A_1, A_2 \in \mathcal{T}$, $\mathcal{I}(A_1, A_2)$ is a Serre subcategory of the abelian category $\mathcal{T}(A_1, A_2)$ and
(2) $\mathcal{I}_1 \circ \mathcal{T}_1 \subseteq \mathcal{I}_1$, $\mathcal{T}_1 \circ \mathcal{I}_1 \subseteq \mathcal{I}_1$.

We will say that $\mathcal{I}$ is a *left (resp. right) Serre ideal* of $\mathcal{T}$ if condition (1) is satisfied and $\mathcal{T}_1 \circ \mathcal{I}_1 \subseteq \mathcal{I}_1$ (resp. $\mathcal{I}_1 \circ \mathcal{T}_1 \subseteq \mathcal{I}_1$).

Proposition 6.2 *For every Serre ideal $\mathcal{I}$ of an abelian 2-category $\mathcal{T}$ such that $1_A \notin \mathcal{I}(A, A)$ for all objects $A \in \mathcal{T}$, one can form the Serre quotient $\mathcal{T}/\mathcal{I}$ with the same set of objects, with the morphism 1-categories*

$$(\mathcal{T}/\mathcal{I})(A_1, A_2) := \mathcal{T}(A_1, A_2)/\mathcal{I}(A_1, A_2) \quad \textit{for} \quad A_1, A_2 \in \mathcal{T},$$

and with identity 1-morphisms given by the images of 1_A. This quotient is an abelian 2-category.

The proof of the proposition is direct, using (6.1) and the following well-known fact:

If, for $i = 1, 2$, $\mathcal{C}_i$ are abelian categories, $\mathcal{I}_i$ are Serre subcategories, and $F : \mathcal{C}_1 \to \mathcal{C}_2$ is an exact functor such that $F(\mathcal{I}_1) \subseteq \mathcal{I}_2$, then the induced functor $\overline{F} : \mathcal{C}_1/\mathcal{I}_1 \to \mathcal{C}_2/\mathcal{I}_2$ is exact.

This follows from the commutativity of the square diagram consisting of the compositions of functors $\mathcal{C}_1 \to \mathcal{C}_1/\mathcal{I}_1 \overset{\overline{F}}{\to} \mathcal{C}_2/\mathcal{I}_2$ and $\mathcal{C}_1 \overset{F}{\to} \mathcal{C}_2 \to \mathcal{C}_2/\mathcal{I}_2$, the exactness of the projection functors $\mathcal{C}_i \to \mathcal{C}_i/\mathcal{I}_i$ (see [42, Exercise 10.3.2(4)]), and the fact that every exact sequence in $\mathcal{C}_1/\mathcal{I}_1$ is isomorphic to one coming from an exact sequence in $\mathcal{C}_1$, [8].

It is easy to prove that, similarly to the ring theoretic case, we have the following:

Lemma 6.3 *A proper Serre ideal $\mathcal{P}$ of a multiring category $\mathcal{T}$ is completely prime, if and only if the Serre quotient $\mathcal{T}/\mathcal{I}$ is a domain in the sense of Definition 5.4.*

Analogously to Lemmas 3.8 and 3.9 one proves the following result. We leave the details to the reader.

Lemma 6.4 *Let $\mathcal{T}$ be an abelian 2-category.*

(1) The intersection of any family of Serre ideals of $\mathcal{T}$ is a Serre ideal of $\mathcal{T}$. In particular, for any subset $\mathcal{M} \subseteq \mathcal{T}_1$, there exists a unique minimal Serre ideal of $\mathcal{T}$ containing $\mathcal{M}$; it will be denoted by $\langle \mathcal{M} \rangle^S$.

(2) For $\mathcal{M}, \mathcal{N} \subseteq \mathcal{T}_1$, we have

$$\langle \mathcal{M} \rangle_1^S \circ \langle \mathcal{N} \rangle_1^S \subseteq \langle \mathcal{M} \circ \mathcal{T}_1 \circ \mathcal{N} \rangle_1^S.$$

6.2 Serre Prime Ideals of Abelian 2-Categories

Similarly to the proofs of Theorems 3.10, 3.12, 3.14 and 5.8, using Lemma 6.4, one proves the following result:

Theorem 6.5 *Let $\mathcal{T}$ be an abelian 2-category.*

(1) *The following are equivalent for a proper Serre ideal $\mathcal{P}$ of $\mathcal{T}$:*

  (a) *$\mathcal{P}$ is a Serre prime ideal;*
  (b) *If $\mathcal{I}$ and $\mathcal{J}$ are any Serre ideals of $\mathcal{T}$ such that $\mathcal{I}_1 \circ \mathcal{J}_1 \subseteq \mathcal{P}_1$, then either $\mathcal{I} \subseteq \mathcal{P}$ or $\mathcal{J} \subseteq \mathcal{P}$;*
  (c) *If $\mathcal{I}$ and $\mathcal{J}$ are any Serre ideals properly containing $\mathcal{P}$, then $\mathcal{I}_1 \circ \mathcal{J}_1 \not\subseteq \mathcal{P}_1$;*
  (d) *If $\mathcal{I}$ and $\mathcal{J}$ are any left Serre ideals of $\mathcal{T}$ such that $\mathcal{I}_1 \circ \mathcal{J}_1 \subseteq \mathcal{P}_1$, then either $\mathcal{I} \subseteq \mathcal{P}$ or $\mathcal{J} \subseteq \mathcal{P}$.*

(2) *Let $\mathcal{M}$ be a nonempty multiplicative subset of $\mathcal{T}_1$ (cf. Definition 3.13) and $\mathcal{I}$ be a Serre ideal of $\mathcal{T}$ such that $\mathcal{I}_1 \cap \mathcal{M} = \varnothing$. Let $\mathcal{P}$ be a maximal element of the collection of Serre ideals of $\mathcal{T}$ containing $\mathcal{I}$ and intersecting $\mathcal{M}$ trivially, equipped with the inclusion relation, i.e., $\mathcal{P}$ is a maximal element of the set*

$$X(\mathcal{M}, \mathcal{I}) := \{\mathcal{K} \text{ a Serre ideal of } \mathcal{T} \mid \mathcal{K} \supseteq \mathcal{I} \text{ and } \mathcal{K}_1 \cap \mathcal{M} = \varnothing\}.$$

Then $\mathcal{P}$ is Serre prime ideal.

(3) *The following are equivalent for a proper Serre ideal $\mathcal{Q}$ of $\mathcal{T}$:*

(a) *$\mathcal{Q}$ is a Serre semiprime ideal;*
(b) *If $\mathcal{I}$ is any Serre ideal of R such that $\mathcal{I}_1 \circ \mathcal{I}_1 \subseteq \mathcal{Q}_1$, then $\mathcal{I} \subseteq \mathcal{Q}$;*
(c) *If $\mathcal{I}$ is any Serre ideal properly containing $\mathcal{Q}$, then $\mathcal{I}_1 \circ \mathcal{I}_1 \nsubseteq \mathcal{Q}_1$;*
(d) *If $\mathcal{I}$ is any left Serre ideal of $\mathcal{T}$ such that $\mathcal{I}_1 \circ \mathcal{I}_1 \subseteq \mathcal{Q}_1$, then $\mathcal{I} \subseteq \mathcal{Q}$.*

In the proof of part (1) of the theorem, the key step is to show that a proper Serre ideal $\mathcal{I}$ of $\mathcal{T}$ satisfying the property (b) is a Serre prime ideal of $\mathcal{T}$. This is proved by showing that property (b) implies that for all $m, n \in \mathcal{T}_1$,

$$m \circ \mathcal{T}_1 \circ n \subseteq \mathcal{P}_1 \quad \Rightarrow m \in \mathcal{P}_1 \ \text{ or } \ n \in \mathcal{P}_1.$$

This fact is verified by repeating the proof of Theorem 3.10, but using Lemma 6.4(2) in place of Lemma 3.9.

The set $X(\mathcal{M}, \mathcal{I})$ in part (3) of the theorem is nonempty because $\mathcal{I} \in X(\mathcal{M}, \mathcal{I})$. The union of an ascending chain of Serre ideals in the set $X(\mathcal{M}, \mathcal{I})$ is obviously a Serre ideal of $\mathcal{T}$. By Zorn's lemma, the set $X(\mathcal{M}, \mathcal{I})$ always contains at least one maximal element.

Similarly to the proof of Corollary 3.16, we obtain:

Corollary 6.6 *For every proper Serre ideal $\mathcal{I}$ of an abelian 2-category $\mathcal{T}$, there exists a Serre prime ideal $\mathcal{P}$ of $\mathcal{T}$ that contains $\mathcal{I}$.*

Analogously to the proof of Theorem 4.3 one proves the following:

Proposition 6.7 *For every abelian 2-category $\mathcal{T}$ satisfying the ACC on (2-sided) Serre ideals, given a proper Serre ideal $\mathcal{I}$ of $\mathcal{T}$, there exist finitely many minimal Serre prime ideals over $\mathcal{I}$. Furthermore, there is a finite list of minimal Serre prime ideals over $\mathcal{I}$ (possibly with repetition) $\mathcal{P}^{(1)}, \ldots, \mathcal{P}^{(m)}$ such that the product*

$$\mathcal{P}_1^{(1)} \circ \ldots \circ \mathcal{P}_1^{(m)} \subseteq \mathcal{I}_1.$$

Let Serre-Spec$(\mathcal{T})$ denote the set of Serre prime ideals of an abelian 2-category $\mathcal{T}$. Similarly to Sect. 3.5, one shows that it is a topological space with closed subsets given by

$$V^S(\mathcal{I}) = \{\mathcal{P} \in \text{Serre-Spec}(\mathcal{T}) \mid \mathcal{P} \supseteq \mathcal{I}\}$$

for the Serre ideals $\mathcal{I}$ of $\mathcal{T}$. We will refer to this as to the Zariski topology of Serre-Spec$(\mathcal{T})$. Proposition 6.7 implies that, if $\mathcal{T}$ satisfies the ACC on Serre ideals, then every closed subset of Serre-Spec$(\mathcal{T})$ is a finite intersection of subsets of the form $V^S(\mathcal{P})$ for some $\mathcal{P} \in$ Serre-Spec$(\mathcal{T})$. In particular, this property holds for all weakly noetherian abelian 2-categories $\mathcal{T}$.

The set-theoretic inclusion

$$\text{Serre-Spec}(\mathcal{T}) \hookrightarrow \text{Spec}(\mathcal{T}) \tag{6.2}$$

realizes Serre-Spec($\mathcal{T}$) as a topological subspace of Spec($\mathcal{T}$). Indeed, Lemma 6.4(1) implies that for every thick ideal $\mathcal{I}$ of $\mathcal{T}$ we have

$$V(\mathcal{I}) \cap \text{Serre-Spec}(\mathcal{T}) = V^S(\langle \mathcal{I} \rangle^S).$$

6.3 The Serre Prime Spectrum as a Ringed Space

For the following subsection, assume that $\mathcal{C}$ is an abelian monoidal category. In the case when it is strict this is the same as an abelian 2-category with one object. All constructions and results in the paper are valid for abelian monoidal categories without the strictness assumption. By Remark 3.20 and the embedding (6.2), the Zariski topology we have thus far endowed Serre-Spec($\mathcal{C}$) which is different from the topology used by Balmer in [1]. The motivation for this consists of the applications to categorification, which we develop below. However, if $\mathcal{C}$ is an abelian monoidal category, we can consider an analogue of Balmer's topology on Serre-Spec($\mathcal{C}$), where we define the closed sets of Serre-Spec($\mathcal{C}$) to be

$$V_B^S(X) = \{\mathcal{P} \in \text{Serre-Spec}(\mathcal{C}) \mid X \cap \mathcal{P} = \emptyset\}$$

for any set of objects X in $\mathcal{C}$. Analogously to Section 2 of [1], one shows that this collection defines a topological space. It may be equipped with a sheaf of commutative rings in a similar manner to [1]. Let $U = V^c$ be an open set of Serre-Spec($\mathcal{C}$), where $V = V_B^S(X)$ for some family of objects X. Let $\mathcal{C}_V = \bigcap_{\mathcal{P} \in U} \mathcal{P}$. Note that $\mathcal{C}_V$ is a Serre ideal, since it is an intersection of Serre ideals. We define a presheaf of commutative rings in the following way:

$$U \mapsto \text{End}_{\mathcal{C}/\mathcal{C}_V}(\overline{1}, \overline{1}),$$

where $\overline{1}$ is the image of 1 (the unit object of $\mathcal{C}$ with respect to the monoidal product) in the Serre quotient $\mathcal{C}/\mathcal{C}_V$. Recalling Proposition 6.2, $\mathcal{C}/\mathcal{C}_V$ has a canonical structure as an abelian monoidal category. By, e.g., Proposition 2.2.10 in [7], $\text{End}_{\mathcal{C}/\mathcal{C}_V}(\overline{1}, \overline{1})$ is a commutative ring. The sheafification of this presheaf gives Serre-Spec($\mathcal{C}$) the structure of a ringed space. The question about the construction of a ringed space structure on the spectra of abelian monoidal categories was raised by Michael Wemyss.

6.4 $\mathbb{Z}_+$-Rings and Their Serre Prime Ideals

Recall that $\mathbb{Z}_+ := \{0, 1, \ldots\}$.

We will use the following slightly weaker terminology for $\mathbb{Z}_+$-rings compared to [7, Definition 3.1.1]:

Definition 6.8 We will call a ring R a $\mathbb{Z}_+$-ring if it is a free abelian group and has a $\mathbb{Z}$-basis $\{b_\gamma \mid \gamma \in \Gamma\}$ such that for all $\alpha, \beta \in \Gamma$,

$$b_\alpha b_\beta = \sum_{\gamma \in \Gamma} n^\gamma_{\alpha,\beta} b_\gamma$$

for some $n^\gamma_{\alpha,\beta} \in \mathbb{Z}_+$.

In addition, [7, Definition 3.1.1] requires that a $\mathbb{Z}_+$-ring R be a unital ring and

$$1 = \sum_{\gamma \in \Gamma} n_\gamma b_\gamma \quad \text{for some} \quad n_\gamma \in \mathbb{Z}_+. \tag{6.3}$$

We do not require a $\mathbb{Z}_+$-ring to be unital and to have the above additional property in order to apply the notion to the Grothendieck rings of abelian 2-categories with infinitely many objects.

For a $\mathbb{Z}_+$-ring R, denote

$$R_+ := \bigoplus_{\gamma \in \Gamma} \mathbb{Z}_+ b_\gamma.$$

For $r, s \in R$, denote

$$r \leq s \quad \text{if} \quad s - r \in R_+.$$

Definition 6.9 Let R be a $\mathbb{Z}_+$-ring.

(1) A left (resp. right) ideal I of R will be called a *a left (resp. right) Serre ideal* if it has the properties that

$$I = (I \cap R_+) - (I \cap R_+) \quad \text{and} \quad s \in R_+, r \in I \cap R_+, s \leq r \Rightarrow s \in I. \tag{6.4}$$

(2) A Serre ideal of R is a 2-sided ideal I of R which satisfies (6.4).
(3) A Serre prime ideal of R is a proper Serre ideal P of R that has the property that

$$IJ \subseteq P \quad \Rightarrow \quad I \subseteq P \text{ or } J \subseteq P$$

for all Serre ideals I, J of R.
(4) A Serre semiprime ideal of R is an ideal which is the intersection of Serre prime ideals.
(5) A Serre completely prime ideal of R is a proper Serre ideal P that has the property that for all $r, s \in R_+$,

$$rs \in P \quad \Rightarrow \quad r \in P \quad \text{or} \quad s \in P.$$

For a subgroup I (under addition) of a $\mathbb{Z}_+$-ring R, the property (6.4) is equivalent to

$$I = \bigoplus_{\gamma \in \Gamma'} \mathbb{Z} b_\gamma \quad \text{for some subset} \quad \Gamma' \subseteq \Gamma. \tag{6.5}$$

In particular, the right- and 2-sided Serre ideals of R satisfy (6.5). Using this fact one easily proves the following theorem, by following the strategy of the proofs of Proposition 3.1, Theorem 3.7, and Corollary 3.8 in [11].

Theorem 6.10 *Let R be a $\mathbb{Z}_+$-ring.*

(1) *The following are equivalent for a proper Serre ideal P of R:*

(a) *P is a Serre prime ideal;*
(b) *If I and J are two Serre ideals of R properly containing P, then $IJ \not\subseteq P$;*
(c) *If I and J are two left Serre ideals of R such that $IJ \subseteq P$, then either $I \subseteq P$ or $J \subseteq P$;*
(d) *For all $\alpha, \beta \in \Gamma$,*

$$b_\alpha R b_\beta \subseteq P \quad \Rightarrow \quad b_\alpha \in P \ \text{ or } \ b_\beta \in P.$$

(2) *A proper Serre ideal P of R is a completely prime Serre ideal if and only if for all $\alpha, \beta \in \Gamma$,*

$$b_\alpha b_\beta \in P \quad \Rightarrow \quad b_\alpha \in P \ \text{ or } \ b_\beta \in P.$$

(3) *The following are equivalent for a proper Serre ideal Q of R:*

(a) *Q is a Serre semiprime ideal;*
(b) *If I is any Serre ideal of R such that $I^2 \subseteq Q$, then $I \subseteq Q$;*
(c) *If I is any Serre ideal of R properly containing Q, then $I^2 \not\subseteq Q$;*
(d) *For all $r \in R_+$,*

$$rRr \subseteq P \quad \Rightarrow \quad r \in P.$$

Denote by Serre-Spec(R) the set of Serre prime ideals of a $\mathbb{Z}_+$-ring R. Similarly to Sect. 6.2, one proves that Serre-Spec(R) is a topological space with closed subsets

$$V^S(I) = \{P \in \text{Serre-Spec}(\mathcal{T}) \mid P \supseteq I\}$$

for the Serre ideals I of R. We will call this the Zariski topology of Serre-Spec(R).

6.5 $\mathbb{Z}_+$-Rings and Abelian 2-Categories

For an abelian category $\mathcal{C}$ denote by $\mathcal{C}_s$ the equivalence classes of its simple objects.

Lemma 6.11 *Assume that $\mathcal{C}$ is an abelian category in which every object has finite length. Then the following hold:*

(1) Every two Jordan-Hölder series of an object of $\mathcal{C}$ contains the same collections of simple subquotients counted with multiplicities and, as a consequence,

$$K_0(\mathcal{C}) \cong \bigoplus_{A \in \mathcal{C}_s} \mathbb{Z}[A].$$

(2) The Serre subcategories of $\mathcal{C}$ are in bijection with the subsets of $\mathcal{C}_s$. The Serre subcategory corresponding to a subset $X \subseteq \mathcal{C}_s$ is the full subcategory $\mathcal{S}(X)$ of $\mathcal{C}$ whose objects have Jordan-Hölder series with simple subquotients isomorphic to objects in X.

(3) For every Serre subcategory $\mathcal{I}$ of $\mathcal{C}$,

$$K_0(\mathcal{C}/\mathcal{I}) \cong K_0(\mathcal{C})/K_0(\mathcal{I}).$$

Proof The first part of the lemma is [7, Theorem 1.5.4].

(2) Clearly, for every subset $X \subseteq \mathcal{C}_s$, the subcategory $\mathcal{S}(X)$ of $\mathcal{C}$ is Serre. Assume that $\mathcal{I}$ is a Serre subcategory of $\mathcal{C}$. Denote by X the isomorphism classes of simple objects of $\mathcal{C}$ which belong to $\mathcal{I}$. Since $\mathcal{I}$ is closed under taking subquotients and isomorphisms, $\mathcal{I} \subseteq \mathcal{S}(X)$. Because $\mathcal{I}$ is closed under extensions, $\mathcal{S}(X) \subseteq \mathcal{I}$. Thus, $\mathcal{I} = \mathcal{S}(X)$.

(3) It easily follows from parts (1) and (2) that the first map in (6.1) is injective. The resulting short exact sequence from (6.1) implies the third part of the lemma.

□

For an abelian 2-category $\mathcal{T}$, denote by $(\mathcal{T}_1)_s$ the isomorphism classes of simple 1-morphisms of $\mathcal{T}$. Recall Definition 3.1. For a subset $X \subseteq (\mathcal{T}_1)_s$, denote by $\mathcal{S}(X)$ the (unique) weak subcategory of $\mathcal{T}$ such that

$$\mathcal{S}(X)(A_1, A_2) := \mathcal{S}(X \cap \mathcal{T}(A_1, A_2))$$

for all $A_1, A_2 \in \mathcal{T}$.

Theorem 6.12 *Assume that $\mathcal{T}$ is an abelian 2-category with the property that every 1-morphism of $\mathcal{T}$ has finite length. (In other words, every object of $\mathcal{T}(A_1, A_2)$ has finite length for all objects $A_1, A_2 \in \mathcal{T}$.) Then the following hold:*

(1) The weak subcategories $\mathcal{I}$ of $\mathcal{T}$ with the property that $\mathcal{I}(A_1, A_2)$ is a Serre subcategory of $\mathcal{T}(A_1, A_2)$ for all $A_1, A_2 \in \mathcal{T}$ are parametrized by the subsets of $(\mathcal{T}_1)_s$. For $X \subseteq (\mathcal{T}_1)_s$, the corresponding subcategory is $\mathcal{S}(X)$.

(2) *The Grothendieck ring $K_0(\mathcal{T})$ is a $\mathbb{Z}_+$-ring and*

$$K_0(\mathcal{T}) \cong \bigoplus_{f \in (\mathcal{T}_1)_s} \mathbb{Z}[f].$$

If, in addition, $\mathcal{T}$ has finitely many objects, then $K_0(\mathcal{T})$ has the property (6.3) *and, more precisely,*

$$1 = \sum_{A \in \mathcal{T}} [1_A].$$

(3) *The map K_0 defines a bijection between the sets of left (resp. right, 2-sided) Serre ideals of $\mathcal{T}$ and of $K_0(\mathcal{T})$.*
(4) *The map K_0 defines a homeomorphism*

$$K_0 : \text{Serre-Spec}(\mathcal{T}) \xrightarrow{\cong} \text{Serre-Spec}(K_0(\mathcal{T})).$$

It is a bijection between the subsets of completely prime (resp. semiprime) ideals of $\mathcal{T}$ and $K_0(\mathcal{T})$.

Proof Part (1) follows from Lemma 6.11(2).

(2) The fact that $K_0(\mathcal{T})$ is a $\mathbb{Z}_+$-ring follows from the fact that for every abelian category $\mathcal{C}$ and $B \in \mathcal{C}$,

$$[B] \in \bigoplus_{A \in \mathcal{C}_s} \mathbb{Z}_+[A].$$

The second statement in part (2) is obvious.

(3) We consider the case of left Serre ideals, the other two cases being analogous. Let $\mathcal{I}$ be a left Serre ideal of $\mathcal{T}$. By part (1) of the theorem, $\mathcal{I} = \mathcal{S}(X)$ for some $X \subseteq (\mathcal{T}_1)_s$. Therefore, the subset

$$K_0(\mathcal{I}) = \bigoplus_{f \in X} \mathbb{Z}[f] \subseteq K_0(\mathcal{T})$$

has the property (6.4). Since $\mathcal{T}_1 \circ \mathcal{I}_1 \subseteq \mathcal{I}_1$, we have $K_0(\mathcal{T})K_0(\mathcal{I}) \subseteq K_0(\mathcal{I})$, and thus, $K_0(\mathcal{I})$ is a left Serre ideal of $K_0(\mathcal{T})$.

Next, let I be a left Serre ideal of $K_0(\mathcal{T})$. By (6.5),

$$I = \bigoplus_{f \in X} \mathbb{Z}[f]$$

for some $X \subseteq (\mathcal{T}_1)_s$. Let $\mathcal{I}$ be the weak subcategory $\mathcal{S}(X)$ of $\mathcal{T}$. Clearly, $K_0(\mathcal{I}) = I$. To show that $\mathcal{I}$ is a left Serre ideal of $\mathcal{T}$, it remains to prove that $\mathcal{T}_1 \circ \mathcal{I}_1 \subseteq \mathcal{I}_1$, i.e., that

$$g_2 f_1 \in \mathcal{I}(A_1, A_3) \quad \text{for all} \quad g_2 \in \mathcal{T}(A_2, A_3),\, f_1 \in \mathcal{I}(A_1, A_2)$$

for all objects A_1, A_2, A_3 of $\mathcal{T}$. Since I is a left Serre ideal,

$$[g_2 f_1] = [g_2][f_1] \in I = \bigoplus_{f \in X} \mathbb{Z}[f],$$

and thus, $g_2 f_1 \in \mathcal{S}(X) = \mathcal{I}$.

It is straightforward to verify that the above two maps $\mathcal{I} \mapsto K_0(\mathcal{I})$ and $I \mapsto \mathcal{I}$ are inverse bijections between the left Serre ideals of $\mathcal{T}$ and $K_0(\mathcal{T})$.

(4) Similarly to part (3) one proves that the map K_0 defines a bijection between the prime (resp. completely prime, semiprime) ideals of the abelian 2-category $\mathcal{T}$ and the $\mathbb{Z}_+$-ring $K_0(\mathcal{T})$. In the first case one uses the characterization of Serre prime ideals of an abelian 2-category in Theorem 6.5(1)(b) vs. the definition of Serre prime ideals of a $\mathbb{Z}_+$-ring. In the second case one uses the definitions of completely prime ideals in the two settings. In the third case one uses the characterizations of Serre semiprime ideals in the two settings given in Theorems 6.5(3)(b) and 6.10(3)(b).

The fact that the map

$$K_0 : \text{Serre-Spec}(\mathcal{T}) \to \text{Serre-Spec}(K_0(\mathcal{T}))$$

is a homeomorphism follows from the definitions of the collections of closed sets in the two cases in terms of Serre ideals and the bijection in part (3) of the theorem. □

We have the following immediate corollary of part (3) of the theorem and Lemma 6.11:

Corollary 6.13 *Let $\mathcal{T}$ be an abelian 2-category which is a categorification of the $\mathbb{k}$-algebra $R \otimes_{\mathbb{Z}} \mathbb{k}$ for a $\mathbb{Z}_+$-ring R. If I is a Serre ideal of R and $\mathcal{I}$ is the unique Serre ideal of $\mathcal{T}$ with $K_0(\mathcal{I}) = I$ as in Theorem 6.12(3), then $\mathcal{T}/\mathcal{I}$ is a categorification of the $\mathbb{k}$-algebra $(R/I) \otimes_{\mathbb{Z}} \mathbb{k}$.*

6.6 Serre Prime Ideals of $\mathbb{Z}_+$-Rings vs. Prime Ideals

Let R be a $\mathbb{Z}_+$-ring. In general, Serre-Spec(R) is not a subset of the prime spectrum Spec(R) of R. Similarly a Serre completely prime ideal of R is not necessarily a completely prime ideal of R (in the classical sense), and a Serre semiprime ideal of R is not necessarily a semiprime ideal of R. The point in all three cases is that the notions of Serre type are formulated in terms of inclusion properties concerning elements of R_+, while the classical notions are formulated in terms of inclusion properties concerning elements of the full ring R.

Example 6.14 Consider the commutative $\mathbb{Z}_+$-ring $R := \mathbb{Z}[x]/(x^2-1)$ with positive $\mathbb{Z}$-basis $\{1, x\}$. The 0-ideal of R is Serre prime while it is not a prime ideal of R.

Example 6.15 Consider the setting of Example 5.5 and assume that H is a finite dimensional Hopf algebra over the field $\Bbbk$. The 0 ideal of $H-$mod is Serre completely prime, and by Theorem 6.12(4), 0 is a Serre completely prime ideal of $K_0(H-\text{mod})$. However, 0 is not a completely prime ideal of the ring $K_0(H-\text{mod})$ because $K_0(H-\text{mod})\otimes_{\mathbb{Z}}\mathbb{Q}$ is a finite dimensional algebra over $\mathbb{Q}$ and thus definitely has 0 divisors (unless this algebra is isomorphic to the field $\mathbb{Q}$). Furthermore the 0 ideal of $K_0(H-\text{mod})$ is not even semiprime, except for the special case when the algebra $K_0(H-\text{mod}) \otimes_{\mathbb{Z}} \mathbb{Q}$ is semisimple (because the radical of this finite dimensional algebra is nilpotent). □

On the other hand, the following lemma provides a simple but important fact about getting Serre prime (resp. completely prime, semiprime) ideals of a $\mathbb{Z}_+$-ring R from particular types of prime (resp. completely prime, semiprime) ideals of a R in the classical sense.

Lemma 6.16 *Assume that R is a $\mathbb{Z}_+$-ring with a positive basis $\{b_\gamma \mid \gamma \in \Gamma\}$. If*

$$I = \bigoplus_{\gamma\in\Gamma'} \mathbb{Z}b_\gamma \quad \text{for some subset} \quad \Gamma' \subseteq \Gamma$$

and I is a prime (resp. completely prime, semiprime) ideal of R in the classical sense, then I is a Serre prime (resp. completely prime, semiprime) ideals of R.

Proof The first property of I is equivalent to the one in (6.4). The assumption that I is a prime (resp. completely prime, semiprime) ideal of R implies that it satisfies the condition (b) in Theorem 6.10(1) in the first case, the condition in Theorem 6.10(2) in the second case, and the condition (d) in Theorem 6.10(3) in the third case. For example, if I is a semiprime ideal of R in the classical sense, it satisfies the condition (d) in Theorem 6.10(3) for all $r \in R$. Now the lemma follows from Theorem 6.10. □

Remark 6.17 Let R be a $\mathbb{Z}_+$-ring categorified by an abelian 2-category $\mathcal{T}$. By Theorem 6.12(3) and Lemma 6.16, the prime ideals of R that are categorifiable are precisely the ones that are thick; that is the set

$$\text{Spec}(R) \cap \text{Serre-Spec}(R).$$

7 The Primitive Spectrum

In this section we describe the relationship between the annihilation ideals of simple 2-representations of abelian 2-categories and the Serre prime ideals of these categories.

7.1 2-Representations

Following Mazorchuk–Miemietz [35], define a *2-representation of a 2-category* $\mathcal{T}$ to be a strict 2-functor $\mathcal{F}$ from $\mathcal{T}$ to Cat, the 2-category of all small categories. That is, $\mathcal{F}$ sends objects of $\mathcal{T}$ to small categories, 1-morphisms of $\mathcal{T}$ to functors between categories, and 2-morphisms of $\mathcal{T}$ to natural transformations between functors.

Recall that the category of additive functors between two abelian categories has a canonical structure of an abelian category.

Definition 7.1 A 2-representation $\mathcal{F}$ of a 2-category $\mathcal{T}$ will be called *exact*, if

(1) $\mathcal{F}(A)$ is an abelian category for every object A in $\mathcal{T}$;
(2) $\mathcal{F}(f)$ is an additive functor for all 1-morphisms f in $\mathcal{T}$;
(3) For any exact sequence of 1-morphisms in $\mathcal{T}$,

$$0 \to f \to g \to h \to 0,$$

the sequence

$$0 \to \mathcal{F}(f) \to \mathcal{F}(g) \to \mathcal{F}(h) \to 0$$

is an exact sequence of 1-morphisms in Cat.

Following Mazorchuk, Miemietz, and Zhang [36, Section 3.3], we call a 2-representation $\mathcal{F}$ *simple* if the collection of categories

$$\{\mathcal{F}(A) \mid A \in \mathcal{T}\}$$

has no nonzero proper $\mathcal{T}$-invariant ideals. Such an ideal X is a subset of the disjoint union of the set of morphisms of the categories $\mathcal{F}(A)$ for $A \in \mathcal{T}$ with the following properties:

(1) ab and ba are in X for all $a \in X$ and all morphisms b in $\mathcal{F}(A)$ such that the composition is well-defined;
(2) $\mathcal{F}(f)(a) \in X$ for all $f \in \mathcal{T}_1$ and $a \in X$;
(3) There is some morphism $a \in X$ which is not a zero morphism.

7.2 Annihilation Ideals of 2-Representations

Definition 7.2 Given an exact 2-representation $\mathcal{F}$ of the abelian 2-category $\mathcal{T}$, define its *annihilation ideal* $\mathrm{Ann}(\mathcal{F})$ to be the weak subcategory of $\mathcal{T}$ having the same set of objects, set of 1-morphisms given by

$$\mathrm{Ann}(\mathcal{F})_1 := \{f \in \mathcal{T}_1 \mid \mathcal{F}(f) \text{ is a zero functor}\},$$

and set of 2 morphisms

$$\mathrm{Ann}(\mathcal{F})(f, g) := \mathcal{F}(f, g) \quad \text{for all} \quad f, g \in \mathrm{Ann}(\mathcal{F})_1.$$

Lemma 7.3 *The annihilation ideal* $\mathrm{Ann}(\mathcal{F})$ *of every exact 2-representation* $\mathcal{F}$ *of an abelian 2-category* $\mathcal{T}$ *is a Serre ideal of* $\mathcal{T}$.

Proof The proof is a direct verification of the necessary properties.

To verify the ideal property of $\mathrm{Ann}(\mathcal{F})$, choose $f \in \mathrm{Ann}(\mathcal{F})_1$ and $g \in \mathcal{T}_1$ such that the composition is defined. Then $\mathcal{F}(f)$ is a zero functor, and therefore, $\mathcal{F}(fg) = \mathcal{F}(f)\mathcal{F}(g)$ is also a zero functor. So, $fg \in \mathrm{Ann}(\mathcal{F})_1$. Likewise, $\mathcal{T}_1 \circ \mathrm{Ann}(\mathcal{F})_1 \subseteq \mathrm{Ann}(\mathcal{F})_1$.

To verify that $\mathrm{Ann}(\mathcal{F})(A_1, A_2)$ is a Serre subcategory of the abelian category $\mathcal{T}(A_1, A_2)$ for all objects A_1 and A_2 of $\mathcal{T}$, consider an exact sequence $0 \to f \to g \to h \to 0$ in $\mathcal{T}(A_1, A_2)$. By Definition 7.1(3), $0 \to \mathcal{F}(f) \to \mathcal{F}(g) \to \mathcal{F}(h) \to 0$ is an exact sequence in the abelian category of additive functors between the abelian categories $\mathcal{F}(A_1)$ and $\mathcal{F}(A_2)$.

If $f, h \in \mathrm{Ann}(\mathcal{F})$, then $\mathcal{F}(f)$ and $\mathcal{F}(h)$ are both the zero functor and $\mathcal{F}(g)$ must also be the zero functor. Hence, $g \in \mathrm{Ann}(\mathcal{F})_1$. Likewise, assuming instead that $g \in \mathrm{Ann}(\mathcal{F})_1$, we get that $f, h \in \mathrm{Ann}(\mathcal{F})_1$. Hence, $\mathrm{Ann}(\mathcal{F})$ is a Serre ideal of $\mathcal{T}$. □

Finally we have the following theorem, analogous to the relationship between prime ideals of rings and annihilators of simple representations, see e.g. [11, Proposition 3.12].

Theorem 7.4 *Suppose* $\mathcal{F}$ *is a simple exact 2-representation of an abelian 2-category* $\mathcal{T}$. *Then* $\mathrm{Ann}(\mathcal{F})$ *is a Serre prime ideal of* $\mathcal{T}$.

Proof We use the assumption of the simplicity of $\mathcal{F}$ to show that $\mathrm{Ann}(\mathcal{F})$ satisfies the condition in Theorem 3.12, form which we obtain that $\mathrm{Ann}(\mathcal{F})$ is a prime ideal of $\mathcal{T}$. The fact that $\mathrm{Ann}(\mathcal{F})$ is a Serre ideal was established in Lemma 7.3.

Suppose that $\mathcal{I}$ and $\mathcal{J}$ are thick ideals such that

$$\mathcal{I}_1 \circ \mathcal{J}_1 \subseteq \mathrm{Ann}(\mathcal{F})_1,$$

and neither $\mathcal{I}$ nor $\mathcal{J}$ is contained in $\mathrm{Ann}(\mathcal{F})$. Then we claim that the set

$$X := \{a(\mathcal{F}(j)(b))c \mid a, b, c \text{ morphisms such that the composition is defined, } j \in \mathcal{J}_1\}$$

forms a nonempty $\mathcal{T}$-invariant ideal, contradicting the simplicity of $\mathcal{F}$. It is clear that this set is an ideal, i.e., closed under composition on the left and right by any morphisms of Cat with appropriate source and target. We must show that it is invariant under $\mathcal{T}$, that it is nonzero, and that it is a proper subset of all morphisms of the categories $\mathcal{F}(A)$ for all objects A of $\mathcal{T}$.

First, assume that $g \in \mathcal{T}_1$. Then

$$\mathcal{F}(g)(a(\mathcal{F}(j)(b))c) = \mathcal{F}(g)(a)\mathcal{F}(g)(\mathcal{F}(j)(b))\mathcal{F}(g)(c) = \mathcal{F}(g)(a)\mathcal{F}(gj)(b)\mathcal{F}(g)(c),$$

which is clearly in X whenever the composition is defined, since $gj \in \mathcal{J}_1$. Hence, X is $\mathcal{T}$-invariant.

Next, we show that X is a proper subset. For all $i \in \mathcal{I}_1$ and $a(\mathcal{F}(j)(b))c \in X$, we have

$$\mathcal{F}(i)(a(\mathcal{F}(j)(b))c) = \mathcal{F}(i)(a)\mathcal{F}(ij)(b)\mathcal{F}(i)(c) = \mathcal{F}(i)(a)0\mathcal{F}(i)(c) = 0$$

whenever the composition is defined. If X equals the set of all morphisms of the collection of abelian categories $\{\mathcal{F}(A) \mid A \in \mathcal{T}\}$, then this would imply that $\mathcal{F}(i)$ is a zero functor for all $i \in \mathcal{I}_1$. Therefore, $\mathcal{I}_1 \subseteq \text{Ann}(\mathcal{F})_1$. Applying Remark 3.3 and the assumption that $\mathcal{I}$ is a thick ideal gives that $\mathcal{I}$ is contained in $\text{Ann}(\mathcal{F})$, which is a contradiction.

By a similar argument, one shows that X contains nonzero morphisms. Since $\mathcal{J}$ is not contained in the annihilator of $\mathcal{F}$ by assumption, there is some $j \in \mathcal{J}_1$ such that $\mathcal{F}(j)$ is not the zero functor, and hence there is some morphism b such that $\mathcal{F}(j)(b)$ is a nonzero morphism. Then by letting a and b be the appropriate identity morphisms, we see that $\mathcal{F}(j)(b)$ is a nonzero morphism in X.

Therefore, X is a nonzero, proper $\mathcal{T}$-invariant ideal, which contradicts our assumption that $\mathcal{F}$ is simple. This gives that $\text{Ann}(\mathcal{F})$ is a Serre prime ideal of $\mathcal{T}$. □

Definition 7.5 The *primitive spectrum* of an abelian 2-category $\mathcal{T}$, denoted $\text{Prim}(\mathcal{T})$, is the subset of $\text{Serre-Spec}(\mathcal{T})$ consisting of all primes $\mathcal{P}$ for which there exists a simple exact 2-representation $\mathcal{F}$ of $\mathcal{T}$ with $\mathcal{P} = \text{Ann}(\mathcal{F})$.

8 Quantum Schubert Cell Algebras, Canonical Bases, and Prime Ideals

This section contains background material on quantum groups and quantum Schubert cell algebras, and their canonical bases defined by Kashiwara and Lusztig. We recall facts about the homogeneous completely prime ideals of the quantum Schubert cell algebras and their relations to quantizations of Richardson varieties.

8.1 Quantum Groups, Canonical Bases, and Quantum Schubert Cell Algebras

Let $\mathfrak{g}$ be a (complex) symmetrizable Kac–Moody algebra with Cartan matrix $(a_{ij})_{i,j=1}^{r}$ and Cartan subalgebra $\mathfrak{t} \subset \mathfrak{g}$. Denote the Weyl group of $\mathfrak{g}$ by W. Let

$\{\alpha_i \mid 1 \le i \le r\} \subset \mathfrak{t}^*$ and $\{s_i \mid 1 \le i \le r\}$ be the sets of simple roots of $\mathfrak{g}$ and simple reflections of W, respectively. Denote by $\{\alpha_i^\vee \mid 1 \le i \le r\} \subset \mathfrak{t}$ and $\{\varpi_i \mid 1 \le i \le r\} \subset \mathfrak{t}^*$ the sets of simple coroots and fundamental weights of $\mathfrak{g}$. Thus, $\langle \alpha_i^\vee, \alpha_j \rangle = a_{ij}$. Let $(.,.)$ be a nondegenerate symmetric bilinear form on $\mathfrak{t}^*$ satisfying

$$\langle \alpha_i^\vee, \lambda \rangle = \frac{2(\alpha_i, \lambda)}{(\alpha_i, \alpha_i)} \text{ for } \lambda \in \mathfrak{t}^* \quad \text{and} \quad (\alpha_i, \alpha_i) = 2 \text{ for short roots } \alpha_i.$$

Then

$$d_i := \frac{(\alpha_i, \alpha_i)}{2} \in \mathbb{Z}_+.$$

Let

$$Q, \ P, \ P_+ \subset \mathfrak{t}^*$$

be the root and weight lattices of $\mathfrak{g}$, and the set of its dominant integral weights. Denote

$$P^\vee := \{h \in \mathfrak{t} \mid \langle h, P \rangle \subset \mathbb{Z}\} \subset \mathfrak{t} \quad \text{and} \quad Q_+ := \bigoplus \mathbb{Z}_+ \alpha_i \subset \mathfrak{t}^*.$$

As it is standard, we will assume that $(.,.)$ is chosen so that $(P, P) \subset \mathbb{Q}$. The induced symmetric bilinear form on $\mathfrak{t}$ will be also denoted by $(.,.)$.

Let $U_q(\mathfrak{g})$ be the quantized universal enveloping algebra of $\mathfrak{g}$ over $\mathbb{Q}(q)$ with generators e_i, f_i, q^h for $1 \le i \le r$, $h \in P^\vee$ and relations as in [20]. We will use the Hopf algebra structure of $U_q(\mathfrak{g})$ with coproduct given by

$$\Delta(e_i) = e_i \otimes 1 + q^{d_i \alpha_i^\vee} \otimes e_i, \quad \Delta(f_i) = f_i \otimes q^{-d_i \alpha_i^\vee} + 1 \otimes f_i, \quad \Delta(q^h) = q^h \otimes q^h \tag{8.1}$$

for $h \in P^\vee$, $1 \le i \le r$. Let $U_q^\pm(\mathfrak{g})$ and $U_q^0(\mathfrak{g})$ be the unital subalgebras of $U_q(\mathfrak{g})$ generated by $\{e_i \mid 1 \le i \le r\}$ (resp. $\{f_i \mid 1 \le i \le r\}$) and $\{q^h \mid h \in P^\vee\}$. Denote the (symmetric) q-integers and factorials

$$q_i := q^{d_i}, \quad [n]_i := \frac{q_i^n - q_i^{-n}}{q_i - q_i^{-1}} \quad \text{and} \quad [n]_i! := [1]_i \dots [k]_i.$$

Denote by $*$ and φ the $\mathbb{Q}(q)$-linear anti-automorphisms of $U_q(\mathfrak{g})$ defined by

$$e_i^* := e_i, \quad f_i^* := f_i, \quad (q^h)^* := q^{-h}, \quad \text{and}$$
$$\varphi(e_i) := f_i, \quad \varphi(f_i) := e_i, \quad \varphi(q^h) := q^h$$

for $1 \le i \le r, h \in P^\vee$. The composition $\varphi^* := \varphi \circ * = * \circ \varphi$, which is a $\mathbb{Q}(q)$-linear automorphism of $U_q(\mathfrak{g})$, satisfies

$$\varphi^*(e_i) = f_i, \quad \varphi^*(f_i) = e_i \quad \text{and} \quad \varphi^*(q^h) = q^{-h}.$$

This composition is denoted by $^\vee$ in [22]; we will use the above notation to avoid interference with later notation.

The Hopf algebra $U_q(\mathfrak{g})$ is graded by the root lattice Q by setting

$$\deg e_i = \alpha_i, \quad \deg f_i = -\alpha_i, \quad \deg q^h = 0. \tag{8.2}$$

The homogeneous components of a subspace Y of $U_q(\mathfrak{g})$ of degree $\gamma \in Q$ will be denoted by Y_γ. Denote by e_i'' the $\mathbb{Q}(q)$-linear skew-derivations of $U_q^-(\mathfrak{g})$ such that

$$e_i''(f_j) = \delta_{ij} \quad \text{and} \quad e_i''(xy) = e_i''(x)y + q^{-(\alpha_i,\gamma)}xe_i''(y) \quad \text{for} \quad x \in U_q^-(\mathfrak{g})_\gamma, y \in U_q^-(\mathfrak{g}).$$

Kashiwara's (nondegenerate, symmetric) bilinear form

$$(-,-)_K : U_q^-(\mathfrak{g}) \times U_q^-(\mathfrak{g}) \to \mathbb{Q}(q)$$

is defined by

$$(1,1)_K = 1 \quad \text{and} \quad (f_i x, y)_K = (x, e_i''(y))_K$$

for all $1 \le i \le r$ and $x, y \in U_q^-(\mathfrak{g})$.

Remark 8.1 This differs slightly from the conventional choice for Kashiwara's form $\langle -,- \rangle_K : U_q^-(\mathfrak{g}) \times U_q^-(\mathfrak{g}) \to \mathbb{Q}(q)$, which is defined by

$$\langle 1,1 \rangle_K = 1 \quad \text{and} \quad \langle f_i x, y \rangle_K = \langle x, e_i'(y) \rangle_K$$

for all $1 \le i \le r$ and $x, y \in U_q^-(\mathfrak{g})$ in terms of the $\mathbb{Q}(q)$-linear skew-derivations e_i' of $U_q^-(\mathfrak{g})$ given by

$$e_i'(f_j) = \delta_{ij} \quad \text{and} \quad e_i'(xy) = e_i'(x)y + q^{(\alpha_i,\gamma)}xe_i'(y) \quad \text{for} \quad x \in U_q^-(\mathfrak{g})_\gamma, y \in U_q^-(\mathfrak{g}).$$

The two forms are related by

$$(x,y)_K = \overline{\langle \overline{x}, \overline{y} \rangle}_K \quad \text{for} \quad x, y \in U_q^-(\mathfrak{g}), \tag{8.3}$$

where $x \mapsto \overline{x}$ denotes the $\mathbb{Q}$-linear automorphism of $\mathbb{Q}(q)$ given by $\bar{q} = q^{-1}$ and the bar involution of $U_q(\mathfrak{g})$ (the skew-linear automorphism of $U_q(\mathfrak{g})$ given by $\overline{f_i} = f_i$). Using (8.3), one converts dualization results with respect to one form to such results for the other.

Let $\mathcal{A} := \mathbb{Z}[q^{\pm 1}]$ and $U^{\pm}_{\mathcal{A}}(\mathfrak{g})$ be the (divided power) *integral forms* of $U^{\pm}_q(\mathfrak{g})$, which are the $\mathcal{A}$-subalgebras of $U^{\pm}_q(\mathfrak{g})$ generated by $e_i^{(k)} = e_i^k/[k]_i!$ and $f_i^{(k)} = f_i^k/[k]_i!$ for $1 \leq i \leq r$, $k \in \mathbb{Z}_+$, respectively. The *dual integral form* $U^-_{\mathcal{A}}(\mathfrak{g})^\vee$ of $U^-_q(\mathfrak{g})$ is the $\mathcal{A}$-subalgebra given by

$$U^-_{\mathcal{A}}(\mathfrak{g})^\vee = \{x \in U^-_q(\mathfrak{g}) \mid (x, U^-_{\mathcal{A}}(\mathfrak{g}))_K \subset \mathcal{A}\}.$$

Kashiwara [20] and Lusztig [32] defined the *canonical/lower global basis* of $U^{\pm}_{\mathcal{A}}(\mathfrak{g})$ and the *dual canonical/upper global basis* of $U^{\pm}_{\mathcal{A}}(\mathfrak{g})^\vee$. These bases have a number of remarkable properties; for instance, they descend to bases of integrable highest weight modules by acting on highest weight vectors. We will denote by $\mathbf{B}^{\mathrm{low}}_{\pm}$ the lower global basis of $U^{\pm}_{\mathcal{A}}(\mathfrak{g})$ and by $\mathbf{B}^{\mathrm{up}}_{-}$ the upper global basis of $U^-_{\mathcal{A}}(\mathfrak{g})$.

The lower global basis $\mathbf{B}^{\mathrm{low}}_{-}$ and the upper global basis $\mathbf{B}^{\mathrm{up}}_{-}$ form a pair of dual bases of $U^-_{\mathcal{A}}(\mathfrak{g})$ and $U^-_{\mathcal{A}}(\mathfrak{g})^\vee$ with respect to the pairing $(-,-)_K$. For $b \in \mathbf{B}^{\mathrm{low}}_{-}$, denote by $b^\vee \in \mathbf{B}^{\mathrm{up}}_{-}$ the corresponding dual element, so

$$(b, c^\vee)_K = \delta_{b,c} \quad \text{for} \quad b, c \in \mathbf{B}^{\mathrm{low}}_{-}. \tag{8.4}$$

The lower global bases $\mathbf{B}^{\mathrm{low}}_{\pm}$ satisfy the invariance properties

$$(\mathbf{B}^{\mathrm{low}}_{\pm})^* = \mathbf{B}^{\mathrm{low}}_{\pm} \quad \text{and} \quad \varphi(\mathbf{B}^{\mathrm{low}}_{\pm}) = \varphi^*(\mathbf{B}^{\mathrm{low}}_{\pm}) = \mathbf{B}^{\mathrm{low}}_{\mp}, \tag{8.5}$$

see [21, Theorem 2.1.1], [22, Theorem 4.3.2] and [19, Theorem 8.3.4].

To each Weyl group element w, one associates the quantum Schubert cell algebras $U^-_q[w] \subseteq U^-_q(\mathfrak{g})$. They can be defined in two ways. Starting from a reduced expression

$$w = s_{i_1} \dots s_{i_N}$$

of w, consider the roots

$$\beta_1 := \alpha_{i_1}, \beta_2 := s_{i_1}(\alpha_{i_2}), \dots, \beta_N := s_{i_1} \dots s_{i_{N-1}}(\alpha_{i_N})$$

and the root vectors

$$\{f_{\beta_j} := T_{i_1} \dots T_{i_{j-1}}(f_j) \mid 1 \leq j \leq N\} \tag{8.6}$$

using Lustig's braid group action [15, 33] on $U_q(\mathfrak{g})$. De Concini, Kac, and Procesi [5], and Lusztig [33, §40.2] defined the algebra $U^-_q[w]$ as the unital $\mathbb{Q}(q)$-subalgebra of $U^-_q(\mathfrak{g})$ with generating set (8.6), and proved that this is independent on the choice of reduced expression of w. Berenstein and Greenstein [3] conjectured that

$$U_q^-[w] = U_q^-(\mathfrak{g}) \cap T_w(U_q^+(\mathfrak{g})),$$

and Kimura [28] and Tanisaki [40] proved this property. It can be used as a second definition of the algebras $U_q^-[w]$. Kimura proved [27, Theorem 4.5] that

$$\mathbf{B}_-^{\mathrm{up}}[w] := \mathbf{B}_-^{\mathrm{up}} \cap U_q^-[w] \tag{8.7}$$

is an $\mathcal{A}$-basis of the $\mathcal{A}$-algebra

$$U_{\mathcal{A}}^-[w]^\vee := U_q^-[w] \cap U_{\mathcal{A}}^-(\mathfrak{g})^\vee.$$

We will refer to this algebra as to the *dual integral form* of $U_q^-[w]$. The set $\mathbf{B}_-^{\mathrm{up}}[w]$ is called the *upper global basis* of $U_{\mathcal{A}}^-[w]^\vee$.

8.2 Homogeneous Completely Prime Ideals of the Algebras $U_q^-[w]$

Denote the Hopf subalgebras $U^{\geq 0} := U_q^+(\mathfrak{g})U_q^0(\mathfrak{g})$ and $U^{\leq 0} := U_q^-(\mathfrak{g})U_q^0(\mathfrak{g})$ of $U_q(\mathfrak{g})$. The Rosso-Tanisaki form

$$(-,-)_{RT} : U^{\leq 0} \times U^{\geq 0} \to \mathbb{Q}(q^{1/d})$$

(for an appropriate $d \in \mathbb{Z}_+$) is the Hopf algebra pairing satisfying

$$(y, xx')_{RT} = (\Delta(y), x' \otimes x)_{RT}, \quad (yy', x)_{RT} = (y \otimes y', \Delta(x))_{RT}$$

for $y, y' \in U^{\leq 0}$, $x, x' \in U^{\geq 0}$, and normalized by

$$(f_i, e_j)_{RT} = \delta_{ij}, \quad (q^h, q^{h'})_{RT} = q^{-(h,h')}, \quad (f_i, q^{h'})_{RT} = (q^h, e_i)_{RT} = 0$$

for $1 \leq i, j \leq r, h, h' \in P^\vee$. We have $(U_q^-(\mathfrak{g}), U_q^+(\mathfrak{g}))_{RT} = \mathbb{Q}(q)$.

This is a slightly different normalization than the usual one [15, Eq. (6.12)(2)] needed in order to match this form to Kashiwara's one. The two normalizations for $(-,-)_{RT}$ are related to each other by a Hopf algebra automorphism of $U^{\leq 0}$ coming from the torus action associated to its Q-grading.

For $\gamma \in Q_+$ let

$$\{x_{\gamma,i}\} \quad \text{and} \quad \{y_{\gamma,i}\}$$

be a set of dual bases of $(U_q^-(\mathfrak{g}))_{-\gamma}$ and $(U_q^+(\mathfrak{g}))_\gamma$ with respect to $(-,-)_{RT}$. The *quasi-R-matrix* of $U_q(\mathfrak{g})$ is

$$\mathcal{R} := \sum_{\gamma \in Q_+} \sum_i y_{\gamma,i} \otimes x_{\gamma,i} \in U_q^+(\mathfrak{g}) \widehat{\otimes} U_q^-(\mathfrak{g})$$

where the completed tensor product is with respect to the descending filtration [33, §4.1.1].

For $\lambda \in P_+$, we will denote by $V(\lambda)$ the irreducible $U_q(\mathfrak{g})$-module of highest weight λ and by $V(\lambda)^\circ$ its restricted dual

$$V(\lambda)^\circ := \oplus_{\nu \in P} (V(\lambda)_\nu)^* \tag{8.8}$$

where

$$V(\lambda)_\nu := \{v \in V(\lambda) \mid q^h v = q^{\langle \nu, h \rangle} v, \ \forall h \in P^\vee\} \quad \text{for} \quad \nu \in P$$

are the (finite dimensional) weight spaces of $V(\lambda)$. Let v_λ be a fixed highest weight vector of $V(\lambda)$. Denote by $\mathbf{B}(\lambda)^{\mathrm{low}}$ the *lower global basis* of (the integral form $U_{\mathcal{A}}^-(\mathfrak{g}) v_\lambda$) of $V(\lambda)$. It is an $\mathcal{A}$-basis of $U_{\mathcal{A}}^-(\mathfrak{g}) v_\lambda$ and a $\mathbb{Q}(q)$-basis of $V(\lambda)$. For $w \in W$, let $v_{w\lambda}$ be the unique element of $V(\lambda)_{w\lambda}$ which belongs to $\mathbf{B}(\lambda)^{\mathrm{low}}$. Let

$$V_w^\pm(\lambda) := U_q^\pm(\mathfrak{g}) v_{w\lambda} \subseteq V(\lambda)$$

be the associated *Demazure modules*. For $v \in V(\lambda)$ and $\xi \in V(\lambda)^*$, denote the corresponding *matrix coefficient* of $V(\lambda)$ considered as a functional on $U_q(\mathfrak{g})$:

$$c_{\xi,v} \in (U_q(\mathfrak{g}))^* \quad \text{given by} \quad c_{\xi,v}(x) = \xi(x \cdot v) \quad \text{for} \ \ x \in U_q(\mathfrak{g}).$$

A subspace U of $U_q(\mathfrak{g})$ will be called *homogeneous* if

$$U = \bigoplus_{\gamma \in Q} U_\gamma \quad \text{where} \quad U_\gamma := U \cap U_q(\mathfrak{g})_\gamma.$$

Theorem 8.2 ([43, Theorem 3.1(a)], [12, Theorem 6.5(a)]) *Let $\mathfrak{g}$ be a symmetrizable Kac–Moody algebra and $w \in W$ be a Weyl group element. For all $u \in W$ such that $u \leq w$, the set*

$$I_w(u) = \{\langle c_{\xi, v_{w\lambda}} \otimes \mathrm{id}, \mathcal{R} \rangle^* \mid \xi \in V(\lambda)^\circ, \xi \perp V_u^-(\lambda), \lambda \in P_+\}$$

is a homogeneous completely prime ideal of $U_q^-[w]$.

The proof of this theorem extensively used the works of Joseph [16, 17] and Gorelik [13].

The Rosso-Tanisaki form $(-,-)_{RT}$ satisfies

$$(y^*, x^*)_{RT} = (y, x)_{RT} \quad \text{for} \quad y \in U^{\leq 0}, x \in U^{\geq 0},$$

[15, Lemma 6.16]. Therefore,

$$\mathcal{R}^{*\otimes *} = \mathcal{R}, \tag{8.9}$$

and thus, the ideals $I_w(u)$ are also given by

$$I_w(u) = \{\langle (c_{\xi, v_{w\lambda}} \circ *) \otimes \mathrm{id}, \mathcal{R}\rangle \mid \xi \in V(\lambda)^\circ, \xi \perp V_u^-(\lambda), \lambda \in P_+\}.$$

We will need the following relation between the bilinear forms $(-,-)_K$ and $(-,-)_{RT}$, and the corresponding expression for $\mathcal{R}$ in terms of global bases.

Proposition 8.3 *Let $\mathfrak{g}$ be a symmetrizable Kac–Moody algebra. For all $x_1, x_2 \in U_q^-(\mathfrak{g})$, we have*

$$(x_1, \varphi^*(x_2))_{RT} = (x_1, x_2)_K. \tag{8.10}$$

The quasi-R-matrix of $U_q(\mathfrak{g})$ is given by

$$\mathcal{R} = \sum_{b \in \mathbf{B}_-^{\mathrm{low}}} \varphi^*(b) \otimes b^\vee = \sum_{b \in \mathbf{B}_-^{\mathrm{low}}} \varphi(b) \otimes (b^\vee)^*, \tag{8.11}$$

recall (8.4).

Proof The Q-grading of $U^{\leq 0}$ specializes to a $\mathbb{Z}_+$-grading via the group homomorphism $Q \to \mathbb{Z}$ given by $\alpha_i \mapsto -1$. The corresponding graded components will be denoted by $(U^{\leq 0})_l$. Set

$$(U^{\leq 0})_{\geq l} := (U^{\leq 0})_l \oplus (U^{\leq 0})_{l+1} \oplus \dots$$

For $x := f_{i_1} \dots f_{i_k}$ and $h := d_{i_1}\alpha_{i_1}^\vee + \dots + d_{i_k}\alpha_{i_k}^\vee$, we have

$$\Delta(x) - x \otimes q^{-h} - \sum_{j=1}^{k} q^{(\alpha_{i_j}, \alpha_{i_1} + \dots + \alpha_{i_{j-1}})} f_{i_1} \dots f_{i_{j-1}} f_{i_{j+1}} \dots f_{i_k} \otimes f_{i_j} q^{-h + d_{i_j}\alpha_{i_j}^\vee} \in U^{\leq 0} \otimes (U^{\leq 0})_{\geq 2},$$

i.e.,

$$\Delta(x) - x \otimes q^{-h} - \sum_{i=1}^{r} e_i''(x) \otimes f_i q^{-h + d_i \alpha_i^\vee} \in U^{\leq 0} \otimes (U^{\leq 0})_{\geq} 2.$$

This property, and the two properties of the Rosso-Tanisaki form

$$((U^{\leq 0})_\gamma, (U^{\geq 0})_\nu)_{RT} = 0 \quad \text{for} \quad \gamma + \nu \neq 0,$$

$$(xq^h, yq^{h'})_{RT} = q^{-(h,h')}(x, y)_{RT} \quad \text{for} \quad x \in U_q^-(\mathfrak{g}), y \in U_q^+(\mathfrak{g}), h, h' \in P^\vee$$

(see [15, Eqs. 6.13(1)–(2)]) imply that the bilinear form $\langle -, - \rangle$ on U_q^- given by

$$\langle x, y \rangle := (x, \varphi^*(y))_{RT}$$

satisfies $\langle x, f_i y \rangle = \langle e_i''(x), y \rangle$ for $x, y \in U_q^-(\mathfrak{g})$ and $1 \le i \le r$. The uniqueness property of the Kashiwara form implies that this form equals $(-, -)_K$, which proves (8.10).

The invariance property (8.5), the relation (8.10) between the bilinear forms $(-, -)_{RT}$ and $(-, -)_K$, and the orthogonality property (8.4) imply that

$$\{\varphi^*(b) \mid b \in \mathbf{B}_-^{\mathrm{low}}\} \quad \text{and} \quad \{b^\vee \mid b \in \mathbf{B}_-^{\mathrm{low}}\}$$

are a pair of dual bases of $U_q^+(\mathfrak{g})$ and $U_q^-(\mathfrak{g})$ with respect to the pairing $(-, -)_{RT}$. This gives the first equality in (8.11). The second equality follows from (8.9) and the first invariance property in (8.5). □

8.3 *Quantizations of Richardson Varieties*

Let G be the Kac–Moody group (over $\mathbb{C}$) corresponding to $\mathfrak{g}$. Let $B_\pm$ be opposite Borel subgroups of G. For $u, w \in W$, the *open Richardson variety* associated to the pair (u, w) is the locally closed subset of the flag variety G/B_+ defined by

$$R_{u,w} := (B_- u B_+)/B_+ \cap (B_+ w B_+)/B_+.$$

It is nonempty if and only if $u \le w$ in which case it has dimension $\ell(w) - \ell(u)$ (in terms of the standard length function $\ell : W \to \mathbb{Z}_+$). We have the stratifications of G/B_+ into unions of Schubert cells

$$G/B_+ = \coprod_{w \in W} (B_+ w B_+)/B_+ = \coprod_{u \in W} (B_- u B_+)/B_+$$

and open Richardson varieties

$$G/B_+ = \coprod_{\substack{u \le w \\ u, w \in W}} R_{u,w}.$$

Denote the closure

$$\overline{R}_{u,w} := \mathrm{Cl}_{(B_+ w B_+)/B_+}(R_{u,w})$$

of $R_{u,w}$ in the Schubert cell $(B_+ w B_+)/B_+$.

For $w \in W$, define the dual extremal vectors

$$\xi_{w\lambda} \in V(\lambda)^*_{w\lambda} \quad \text{by} \quad \xi_{w\lambda}(v_{w\lambda}) = 1$$

(keeping in mind that $\dim V(\lambda)_{w\lambda} = 1$). Denote the image of the corresponding extremal matrix coefficient in $U_q^-[w]$:

$$\Delta_{\lambda, w\lambda} := \langle c_{\xi, v_{w\lambda}} \otimes \mathrm{id}, \mathcal{R} \rangle^* \in U_q^-[w] \quad \text{for} \quad \lambda \in P_+.$$

Proposition 8.4 *For all symmetrizable Kac–Moody algebras $\mathfrak{g}$ and $u \leq w \in W$, the factor ring $U_q^-[w]/I_w(u)$ is a quantization of the coordinate ring $\mathbb{C}[\overline{R}_{u,w}]$ of the closure of the open Richardson variety $R_{u,w}$ in the Schubert cell $(B_+ w B_+)/B_+$. The localization*

$$(U^+[w]/I_w(u))[\Delta^{-1}_{\varpi_i, w\varpi_i}, 1 \leq i \leq r]$$

of this ring is a quantization of the coordinate ring $\mathbb{C}[R_{u,w}]$.

These facts were stated in [44, pp. 274–275] for finite dimensional complex simple Lie algebras $\mathfrak{g}$, but the proofs given there carry over to the symmetrizable Kac–Moody case directly.

9 Categorifying Richardson Varieties

In this section, we prove that the ideals $I_w(u) \cap U_{\mathcal{A}}^-[w]^\vee$ of $U_{\mathcal{A}}^-[w]^\vee$ are Serre completely prime ideals for all symmetric Kac–Moody algebras $\mathfrak{g}$ and $u \leq w \in W$. We then use Theorem 6.12(4) to construct a (domain) multiring category which categorifies the quantization of the coordinate ring of the closure of the open Richardson variety $R_{u,w}$ in the Schubert cell $(B_+ w B_+)/B_+$. This category is obtained as a factor of a multiring category consisting of graded, finite dimensional representations of the corresponding KLR algebras.

9.1 *The Categorifications of $U_{\mathcal{A}}^-(\mathfrak{g})^\vee$ and $U_{\mathcal{A}}^-[w]^\vee$, and Relations to Dual Canonical bases*

For each symmetrizable Kac–Moody algebra $\mathfrak{g}$, Khovanov, Lauda [26] and Rouquier [38] defined a family of (graded) quiver Hecke algebras over a base field $\Bbbk$, which we will call KLR algebras. They proved that the category $\mathcal{C}$ which is the direct sum of the categories of finite dimensional graded modules of the KLR algebras associated to $\mathfrak{g}$ has the following properties:

Theorem 9.1 (Khovanov–Lauda [26] and Rouquier [38]) *For each symmetrizable Kac–Moody algebra* $\mathfrak{g}$ *and base field* $\Bbbk$, $\mathcal{C}$ *is a* $\Bbbk$*-linear multiring category such that*

$$K_0(\mathcal{C}) \cong U^-_{\mathcal{A}}(\mathfrak{g})^\vee. \tag{9.1}$$

The action of q *on the right-hand side and the shift of grading autoequivalence of* $\mathcal{C}$ *are related via*

$$[M(k)] = q^k[M] \quad \textit{for all objects } M \textit{ of } \mathcal{C}.$$

The theorem implies that for every symmetrizable Kac–Moody algebra $\mathfrak{g}$, $U^-_{\mathcal{A}}(\mathfrak{g})^\vee$ is a $\mathbb{Z}_+$-ring with positive $\mathbb{Z}_+$-basis $\{[M]\}$ where M runs over the isomorphism classes of the simple objects of $\mathcal{C}$. (Here we disregard the structure of $U^-_{\mathcal{A}}(\mathfrak{g})^\vee$ as an $\mathcal{A}$-algebra and view it just as a ring.) For symmetric Kac–Moody algebras $\mathfrak{g}$, the relation between this basis and the upper global basis of $U^-_{\mathcal{A}}(\mathfrak{g})^\vee$ is given by the next theorem. For it we recall that the dual of each graded finite dimensional representation of a KLR algebra has a canonical structure of a KLR module which is also graded, finite dimensional. This gives a canonical duality endofunctor of $\mathcal{C}$.

Theorem 9.2 (Varagnolo–Vasserot [41] and Rouquier [39]) *For each symmetric Kac–Moody algebra* $\mathfrak{g}$ *and base field* $\Bbbk$ *of characteristic 0, under the isomorphism* (9.1)*, the upper global basis corresponds to the set of isomorphism classes of the self-dual simple modules in the category* $\mathcal{C}$.

The theorem implies that in these cases $U^-_{\mathcal{A}}(\mathfrak{g})^\vee$ is a $\mathbb{Z}_+$-ring with positive $\mathbb{Z}_+$-basis

$$q^{\mathbb{Z}}\mathbf{B}^{\mathrm{up}}_-.$$

For each symmetric Kac–Moody algebra $\mathfrak{g}$ and $w \in W$, in [19, §11.2] Kang, Kashiwara, Kim, and Oh constructed a monoidal subcategory $\mathcal{C}_w$ of $\mathcal{C}$ as the smallest monoidal Serre subcategory closed under shifts, containing a certain set of simple self-dual modules of the KLR algebras ([19, Definition 11.2.1]) and, using [9], they proved:

Theorem 9.3 (Kang–Kashiwara–Kim–Oh [19]) *For each symmetric Kac–Moody algebra* $\mathfrak{g}$,

$$K_0(\mathcal{C}_w) \cong U^-_{\mathcal{A}}[w]^\vee. \tag{9.2}$$

Combining Theorems 9.2 and 9.3 gives that, under the isomorphism (9.2), the elements of the upper global basis $\mathbf{B}^{\mathrm{up}}_-[w]^\vee$ of $U^-_{\mathcal{A}}[w]^\vee$ (recall (8.7)) correspond to the isomorphism classes of the simple self-dual objects of $\mathcal{C}_w$. In particular, $U^-_{\mathcal{A}}[w]^\vee$ is a $\mathbb{Z}_+$-ring with a positive $\mathbb{Z}$-basis

$$q^{\mathbb{Z}}\mathbf{B}^{\mathrm{up}}_-[w].$$

9.2 Serre Completely Prime Ideals of the $\mathbb{Z}_+$-Rings $U^-_{\mathcal{A}}[w]$ and the Multiring Categories $\mathcal{C}_w$

For $u \leq w$, denote the ideals

$$I_w(u)^\vee_{\mathcal{A}} := I_w(u) \cap U^-_{\mathcal{A}}[w]^\vee$$

of $U^-_{\mathcal{A}}[w]$. Theorem 8.2 implies that for every symmetrizable Kac–Moody algebra $\mathfrak{g}$, $I_w(u)^\vee_{\mathcal{A}}$ are completely prime ideals of $U^-_{\mathcal{A}}[w]$ in the classical sense. The following is the main result of this section.

Theorem 9.4

(1) Let $\mathfrak{g}$ be a symmetrizable Kac–Moody algebra and $u \leq w \in W$. The ideal $I_w(u)^\vee_{\mathcal{A}}$ has an $\mathcal{A}$-basis given by

$$\mathbf{B}^{\mathrm{up}}_-[w] \cap I_w(u)^\vee_{\mathcal{A}}.$$

Furthermore, it is a Serre completely prime ideal of the $\mathbb{Z}_+$-ring $U^-_{\mathcal{A}}[w]^\vee$.

Denote by $X_w(u)$ the set of isomorphism classes of self-dual simple objects M of $\mathcal{C}_w$ such that $[M] \in I_w(u)^\vee_{\mathcal{A}}$. Let

$$\mathcal{I}_w(u) := \mathcal{S}(X_w(u)[k], k \in \mathbb{Z})$$

be the full subcategory of $\mathcal{C}_w$ whose objects have Jordan-Hölder series with simple subquotients isomorphic to shifts of objects in $X_w(u)$ as in Lemma 6.11(2).

(2) Let $\mathfrak{g}$ be a symmetric Kac–Moody algebra and $u \leq w \in W$. For all base fields $\Bbbk$ of characteristic 0, $\mathcal{I}_w(u)$ are Serre completely prime ideals of the $\Bbbk$-linear multiring category $\mathcal{C}_w$. For the corresponding Serre quotient $\mathcal{C}_w/\mathcal{I}_w(u)$, we have

$$K_0(\mathcal{C}_w/\mathcal{I}_w(u)) \cong U^-_{\mathcal{A}}[w]^\vee/I_w(u)^\vee_{\mathcal{A}}.$$

By the first part of Theorem 9.4(1),

$$\left(U^-_{\mathcal{A}}[w]^\vee/I_w(u)^\vee_{\mathcal{A}}\right) \otimes_{\mathcal{A}} \mathbb{Q}(q) \cong U^-[w]/I_w(u)$$

and by Proposition 8.4, $U^-[w]/I_w(u)$ is a quantization of the coordinate ring $\mathbb{C}[\overline{R}_{u,w}]$ of the closure of the Richardson variety $R_{u,w}$ in the Schubert cell $(B_+wB_+)/B_+$. This fact and Theorem 9.4(2) imply that the Serre quotient $\mathcal{C}_w/\mathcal{I}_w(u)$, which is a domain in the sense of Definition 5.4, is a monoidal categorification of the quantization of $\mathbb{C}[\overline{R}_{u,w}]$.

9.3 *Proof of Theorem 9.4*

Recall that $\mathbf{B}(\lambda)^{\mathrm{low}}$ denoted the lower global basis of the irreducible module $V(\lambda)$ for $\lambda \in P_+$. We will need two facts about the lower global bases of Demazure modules proved by Kashiwara:

Theorem 9.5 (Kashiwara [21]) *For every symmetrizable Kac-Moody algebra* $\mathfrak{g}$ *and dominant integral weight* $\lambda \in P_+$*, the intersection*

$$\mathbf{B}_w^{\pm}(\lambda)^{\mathrm{low}} := \mathbf{B}(\lambda)^{\mathrm{low}} \cap V_w^{\pm}(\lambda)$$

is a $\mathbb{Q}(q)$*-basis of the Demazure module* $V_w^{\pm}(\lambda)$.

The sets $\mathbf{B}_w^{\pm}(\lambda)^{\mathrm{low}}$ are called the lower global bases of the Demazure modules $V_w^{\pm}(\lambda)$. The plus case was proved in [21, Proposition 3.2.3(i)] and the minus in [21, Proposition 4.1]. The following theorem describes the relationship between the canonical/lower global bases $\mathbf{B}_w^{+}(\lambda)$ of the Demazure modules and the action of the canonical/lower global bases of $U_{\mathcal{A}}^{+}(\mathfrak{g})$ acting on the corresponding extremal weight vectors.

Theorem 9.6 (Kashiwara [22, 23]) *Let* $\mathfrak{g}$ *be a symmetrizable Kac-Moody algebra* $\mathfrak{g}$ *and* $\lambda \in P_+$ *be a dominant integral weight. Denote the subset*

$$\mathfrak{B}_w^{+}(\lambda)^{\mathrm{low}} := \{b \in \mathbf{B}_+^{\mathrm{low}} \mid b \cdot v_{w\lambda} \neq 0\}$$

of the lower global basis of $U_{\mathcal{A}}^{+}(\mathfrak{g})$*. Then there is a bijection between this set and the lower global basis of the Demazure module* $V_w^{+}(\lambda)$ *given by*

$$\eta_w : \mathfrak{B}_w^{+}(\lambda)^{\mathrm{low}} \xrightarrow{\cong} \mathbf{B}_w^{+}(\lambda)^{\mathrm{low}} \quad \textit{given by} \quad \eta_w(b) := b \cdot v_{w\lambda}.$$

The corresponding fact to this theorem for the negative Demazure modules (where everywhere plus is replaced by minus) was proved in [21, Proposition 4.1].

The following proposition is a stronger form of the statement of the first part of Theorem 9.4(1).

Proposition 9.7 *For all symmetrizable Kac–Moody algebras* $\mathfrak{g}$*, and* $u \leq w \in W$*, the ideal* $I_w(u)$ *of the quantum Schubert cell algebra* $U_q^-[w]$ *has a* $\mathbb{Q}(q)$*-basis given by*

$$\bigcup_{\lambda \in P_+} \left\{ b^{\vee} \mid b \in \varphi^{-1}\eta_w^{-1}\left(\mathbf{B}_w^{+}(\lambda)^{\mathrm{low}} \backslash \mathbf{B}_u^{-}(\lambda)^{\mathrm{low}}\right)\right\}.$$

Proof For $\lambda \in P_+$, consider the basis of $V(\lambda)^{\circ}$ (cf. (8.8)) which is dual to the lower global basis $B(\lambda)^{\mathrm{low}}$ of $V(\lambda)$. Given $v \in B(\lambda)^{\mathrm{low}}$, denote by $v^{\vee}$ the corresponding dual element, so

$$v_1^{\vee}(v_2) = \delta_{v_1 v_2} \quad \text{for} \quad v_1, v_2 \in B(\lambda)^{\mathrm{low}}.$$

Theorem 9.5 implies that

$$\{\xi \in V(\lambda)^\circ \mid \xi \perp V_u^-(\lambda)\} = \underset{\mathbb{Q}(q)}{\mathrm{Span}}\{\mathbf{B}_w^+(\lambda)^{\mathrm{low}} \backslash \mathbf{B}_u^-(\lambda)^{\mathrm{low}}\}$$
$$\oplus \underset{\mathbb{Q}(q)}{\mathrm{Span}}\{\mathbf{B}(\lambda)^{\mathrm{low}} \backslash (\mathbf{B}_w^+(\lambda)^{\mathrm{low}} \cup \mathbf{B}_u^-(\lambda)^{\mathrm{low}})\}.$$

For $v \in \mathbf{B}(\lambda)^{\mathrm{low}} \backslash (\mathbf{B}_w^+(\lambda)^{\mathrm{low}} \cup \mathbf{B}_u^-(\lambda)^{\mathrm{low}})$, we have $v^\vee \perp V_w^+(\lambda)$, and thus,

$$\langle c_{v^\vee, v_{w\lambda}} \otimes \mathrm{id}, \mathcal{R}\rangle = 0.$$

Therefore, the subspace

$$\{\langle c_{\xi, v_{w\lambda}} \otimes \mathrm{id}, \mathcal{R}\rangle^* \mid \xi \in V(\lambda)^\circ \mid \xi \perp V_u^-(\lambda)\} \subset I_w(u)$$

is spanned by

$$\{\langle c_{v^\vee, v_{w\lambda}} \otimes \mathrm{id}, \mathcal{R}\rangle^* \mid v \in \mathbf{B}_w^+(\lambda)^{\mathrm{low}} \backslash \mathbf{B}_u^-(\lambda)^{\mathrm{low}}\}.$$

The proposition now follows from the identity

$$\langle c_{v^\vee, v_{w\lambda}} \otimes \mathrm{id}, \mathcal{R}\rangle^* = \left(\varphi^{-1}\eta_w^{-1}(v)\right)^\vee \quad \text{for} \quad v \in \mathbf{B}_w^+(\lambda)^{\mathrm{low}}. \tag{9.3}$$

To show this, first note that Theorem 9.6 implies that for $v \in \mathbf{B}_w^+(\lambda)^{\mathrm{low}}$ and $b \in \mathbf{B}_+^{\mathrm{low}}$,

$$\langle v^\vee, b \cdot v_{w\lambda}\rangle = \begin{cases} 1, & \text{if } b = \eta_w^{-1}(v) \\ 0, & \text{otherwise.} \end{cases}$$

Using this and the second part of Proposition 8.3, for $v \in \mathbf{B}_w^+(\lambda)^{\mathrm{low}}$, we obtain

$$\langle c_{v^\vee, v_{w\lambda}} \otimes \mathrm{id}, \mathcal{R}\rangle^* = \sum_{b \in \mathbf{B}_-^{\mathrm{low}}} \langle v^\vee, \varphi(b) \cdot v_{w\lambda}\rangle b^\vee = \left(\varphi^{-1}\eta_w^{-1}(v)\right)^\vee,$$

which shows (9.3) and completes the proof of the proposition. □

Proof of Theorem 9.4 (1) The first statement in part (1) follows from Proposition 9.7. By Theorem 8.2, $I_w(u)$ is a completely prime ideal of $U_q^-[w]$, and therefore, the contraction $I_w(u) \cap U_{\mathcal{A}}^-[w]$ is a completely prime ideal of $U_{\mathcal{A}}^-[w]$. The ideal $I_w(u) \cap U_{\mathcal{A}}^-[w]$ has a $\mathbb{Z}$-basis consisting of elements that belong to $q^{\mathbb{Z}}\mathbf{B}_-^{\mathrm{up}}[w]$, which, by Theorem 9.2, is precisely the positive basis of the $\mathbb{Z}_+$-ring $U_{\mathcal{A}}^-[w]$. Now we apply Lemma 6.16, which gives that $I_w(u) \cap U_{\mathcal{A}}^-[w]$ is a Serre completely prime ideal of $U_{\mathcal{A}}^-[w]$.

Part (2) follows from part (1), Theorem 6.12(4) (applied to the category $\mathcal{C}_w$) and the isomorphism $K_0(\mathcal{C}_w) \cong U_{\mathcal{A}}^{-}[w]^{\vee}$ from Theorem 9.3. □

It is possible that Theorem 9.4(2) holds for symmetrizable Kac–Moody algebras $\mathfrak{g}$ by arguments that avoid the use of Theorem 9.2.

Acknowledgements We are grateful to Masaki Kashiwara for communicating to us his proof of Theorem 9.6 and to Ken Goodearl, Birge Huisgen-Zimmermann, Osamu Iyama, Peter Jørgensen, Bernhard Keller, Bernard Leclerc, Vanessa Miemietz, and Michael Wemyss for helpful discussions on thick subcategories of abelian and triangulated categories, canonical bases, and small 2-categories. We are thankful to the referee for the thorough reading of the paper and for making a number of valuable suggestions which helped us to improve the paper.

The research of K.V. was supported by a Board of Regents LSU fellowship and NSF grant DMS-1601862. The research of M.Y. was supported by NSF grant DMS-1601862 and Bulgarian Science Fund grant DN 02-5.

References

1. P. Balmer, *The spectrum of prime ideals in tensor triangulated categories*. J. Reine Angew. Math. **588** (2005), 149–168.
2. P. Balmer, *Spectra, spectra, spectra—tensor triangular spectra versus Zariski spectra of endomorphism rings*, Algebr. Geom. Topol. **10** (2010), 1521–1563.
3. A. Berenstein and J. Greenstein, *Double canonical bases*, Adv. Math. **316** (2017), 381–468.
4. B. D. Boe, J. R. Kujawa, and D. K. Nakano, *Tensor triangular geometry for quantum groups*, preprint arXiv:1702.01289.
5. C. De Concini, V. G. Kac, and C. Procesi, *Some quantum analogues of solvable Lie groups*, in: *Geometry and analysis (Bombay, 1992)*, pp. 41–65. Tata Inst. Fund. Res., Bombay, 1995.
6. P. Deligne, *Catégories tannakiennes*, in: *The Grothendieck Festschrift, Vol. II*, Progr. in Math. vol. 87, pp. 111–195. Birkhäuser, Boston, MA, 1990.
7. P. Etingof, S. Gelaki, D. Nikshych, and V. Ostrik, *Tensor categories*, Math. Surveys and Monographs vol. 205, Amer. Math. Soc., Providence, RI, 2015.
8. P. Gabriel, *Des catégories abéliennes*, Bull. Soc. Math. France **90** (1962), 323–448.
9. C. Geiß, B. Leclerc, and J. Schröer. *Cluster structures on quantum coordinate rings*, Selecta Math. (N.S.) **19** (2013), 337–397.
10. K. L. Goodearl, S. Launois, and T. H. Lenagan, *Torus-invariant prime ideals in quantum matrices, totally nonnegative cells and symplectic leaves*, Math. Z. **269** (2011), 29–45.
11. K. R. Goodearl and R. B. Warfield, Jr., *An introduction to noncommutative Noetherian rings*, 2nd ed., London Math. Soc. Stud. Texts vol. 61, Cambridge Univ. Press, Cambridge, 2004.
12. K. R. Goodearl and M. T. Yakimov, *Integral quantum cluster structures*, preprint.
13. M. Gorelik, *The prime and the primitive spectra of a quantum Bruhat cell translate*, J. Algebra, **227** (2000), 211–253.
14. C. Ingalls and H. Thomas, *Noncrossing partitions and representations of quivers*, Compos. Math. **145** (2009), 1533–1562.
15. J. C. Jantzen, *Lectures on quantum groups*, Grad. Stud. in Math. vol. 6, Amer. Math. Soc., Providence, RI, 1996.
16. A. Joseph, *On the prime and primitive spectra of the algebra of functions on a quantum group*, J. Algebra **169** (1994), 441–511.
17. A. Joseph, *Quantum groups and their primitive ideals*, Ergebnisse der Mathematik und ihrer Grenzgebiete (3) vol. 29, Springer-Verlag, Berlin, 1995.

18. J. Kamnitzer, *Categorification of Lie algebras (after Rouquier, Khovanov-Lauda,...)*, Astérisque **361** (2014), Exp. No. 1072, 397–419.
19. S.-J. Kang, M. Kashiwara, M. Kim, and S.-J. Oh, *Monoidal categorification of cluster algebras*, J. Amer. Math. Soc. **31** (2018), 349–426.
20. M. Kashiwara, *On crystal bases of the Q-analogue of universal enveloping algebras*, Duke Math. J. **63** (1991), 465–516.
21. M. Kashiwara, *The crystal base and Littelmann's refined Demazure character formula*, Duke Math. J. **71** (1993), 839–858.
22. M. Kashiwara, *Crystal bases of modified quantized enveloping algebra*, Duke Math. J. **73** (1994), 383–413.
23. M. Kashiwara, *private communication*.
24. M. Kashiwara, M. Kim, S.-J. Oh, and E. Park, *Monoidal categories associated with strata of flag manifolds*, Adv. Math. **328** (2018), 959–1009.
25. G. M. Kelly, *Basic concepts of enriched category theory*, London Math. Soc. Lect. Note Series vol. 64, Cambridge Univ. Press, Cambridge-New York, 1982.
26. M. Khovanov and A. D. Lauda, *A diagrammatic approach to categorification of quantum groups. I*, Represent. Theory **13** (2009), 309–347.
27. Y. Kimura, *Quantum unipotent subgroup and dual canonical basis*, Kyoto J. Math. **52** (2012), 277–331.
28. Y. Kimura, *Remarks on quantum unipotent subgroups and the dual canonical basis*, Pacific J. Math. **286** (2017), 125–151.
29. A. D. Lauda, *An introduction to diagrammatic algebra and categorified quantum* $\mathfrak{sl}_2$, Bull. Inst. Math. Acad. Sin. (N.S.) **7** (2012), 165–270.
30. B. Leclerc, *Cluster structures on strata of flag varieties*, Adv. Math. **300** (2016), 190–228.
31. T. Lenagan and M. Yakimov, *Prime factors of quantum Schubert cell algebras and clusters for quantum Richardson varieties*, J. Reine Angew. Math. DOI: 10.1515/crelle-2016-0046.
32. G. Lusztig, *Canonical bases arising from quantized enveloping algebras*, J. Amer. Math. Soc. **3** (1990), 447–498.
33. G. Lusztig, *Introduction to quantum groups*, Progr. in Math. vol. 110, Birkhäuser, Boston, MA, 1993.
34. V. Mazorchuk, *Lectures on algebraic categorification*, QGM Master Class Series, Europ. Math. Soc. (EMS), Zürich, 2012.
35. V. Mazorchuk and V. Miemietz, *Cell 2-representations of finitary 2-categories*, Compos. Math. **147** (2011), 1519–1545.
36. V. Mazorchuk, V. Miemietz, and X. Zhang, *Pyramids and 2-representations*, preprint arXiv:1705.03174.
37. D. Quillen, *Higher algebraic K-theory. I.*, Lecture Notes in Math. vol. 341, pp. 85–147, 1973.
38. R. Rouquier, *2-Kac-Moody algebras*, preprint arXiv:0812.5023.
39. R. Rouquier, *Quiver Hecke algebras and 2-Lie algebras*, Algebra Colloq. **19** (2012), 359–410.
40. T. Tanisaki, *Modules over quantized coordinate algebras and PBW-bases*, J. Math. Soc. Japan **69** (2017), 1105–1156.
41. M. Varagnolo and E. Vasserot, *Canonical bases and KLR-algebras*, J. Reine Angew. Math. **659** (2011), 67–100.
42. C. A. Weibel, *An introduction to homological algebra*, Cambridge Stud. in Adv. Math. vol. 38, Cambridge Univ. Press, Cambridge, 1994.
43. M. Yakimov, *Invariant prime ideals in quantizations of nilpotent Lie algebras*, Proc. Lond. Math. Soc. (3) **101** (2010), 454–476.
44. M. Yakimov, *Strata of prime ideals of De Concini-Kac-Procesi algebras and Poisson geometry*, in: *New trends in noncommutative algebra*, Contemp. Math. vol. 562, pp. 265–278. Amer. Math. Soc., Providence, RI, 2012.
45. M. Yakimov, *Spectra and catenarity of multi-parameter quantum Schubert cells*, Glasg. Math. J., **55(A)** (2013), 169–194.

Printed by Printforce, the Netherlands